PURINE AND PYRIMIDINE
METABOLISM IN MAN VIII

ADVANCES IN EXPERIMENTAL MEDICINE AND BIOLOGY

PURINE AND PYRIMIDINE METABOLISM IN MAN VIII

Edited by

Amrik Sahota

Indiana University School of Medicine
Indianapolis, Indiana

and

Milton W. Taylor

Indiana University
Bloomington, Indiana

Springer Science+Business Media, LLC

Library of Congress Cataloging-in-Publication Data

Purine and pyrimidine metabolism in man VIII / edited Amrik Sahota,
Milton W. Taylor.
       p.   cm. -- (Advances in experimental medicine and biology :
v. 370)
    "Proceedings of the Eighth International Symposium on Purine and
Pyrimidine Metabolism in Man, held May 22-27, 1994 in Bloomington,
Indiana"--T.p. verso.
    Includes bibliographical references and index.
    ISBN 978-1-4613-6105-3   ISBN 978-1-4615-2584-4 (eBook)
    DOI 10.1007/978-1-4615-2584-4

    1. Purines--Metabolism--Disorders--Congresses. 2. Pyrimidines-
-Metabolism--Disorders--Congresses. 3. Purines--Metabolism-
-Congresses. 4. Pyrimidines--Metabolism--Congresses.   I. Sahota,
Amrik. II. Taylor, Milton W. III. International Symposium on
Purine and Pyrimidine Metabolism in Man (8th : 1994 : Bloomington,
Ind.) IV. Title: Purine and pyrimidine metabolism in man eight.
V. Title: Purine and pyrimidine metabolism in man 8. VI. Series.
    [DNLM: 1. Purines--metabolism--congresses. 2. Pyrimidines-
-metabolism--congresses.   W1 AD559 v.370 1994 / QU 58 P9855 1944]
RC632.P87P87 1994
616.3'9--dc20   95-13913
                                                        CIP

Proceedings of the Eighth International Symposium on Purine and Pyrimidine Metabolism in Man,
held May 22–27, 1994, in Bloomington, Indiana

ISBN 978-1-4613-6105-3

© 1994 Springer Science+Business Media New York
Originally published by Plenum Press, New York in 1994
Softcover reprint of the hardcover 1st edition 1994

DEDICATION

These proceedings are dedicated to Dr. H. Anne Simmonds, whose contributions to inborn errors of purine and pyrimidine metabolism, especially those involving the renal and immune systems, have been exemplary and outstanding. She has been a highly respected, and highly productive contributor to this field. Her research has influenced many scientists the world over, including the editors of this volume. It would be a rare event indeed if one did not see a new visiting scientist in her lab every few months. Her collaborations with clinical colleagues have contributed immensely to patient care. The editors owe a personal debt of gratitude to her, since without the discovery of the clinical importance of the APRT gene (for which Anne was primarily responsible), they would not have studied this gene for so long and in such detail.

Anne was born and raised on a remote farm in North Auckland, New Zealand. She did not receive any formal primary education other than through Correspondence School. She had to go away to complete her high school education and went on to obtain a MSc in organic chemistry. She declined a scholarship for a PhD, believing that without practical experience it would be meaningless, and instead joined the Waikato Hospital as a trainee in clinical laboratory sciences. While studying for her MSc in clinical sciences, she organized the first local walking and skiing club. These recreational activities have been with her to this day. In fact, on the rare occasions Anne is not in the lab or at a conference, she can most likely be found enjoying her favorite pastime. It has been noticed more than once that after a brief time on the ski slopes, Anne approaches her research with even greater vigor and intensity.

After obtaining her degree in clinical sciences, Anne spent some time gaining postgraduate experience in various hospital laboratories in Europe and North America, trying to figure out what she really wanted to do. She returned to New Zealand, and in 1963 found herself in the renal unit of an Auckland hospital. This provided her with her first introduction to pyrimidines and purines and, as everyone in the field knows, the rest is history. To quote Anne: "It all began when I was asked to help Dr. David Becroft, head of the Princess Mary Children's Hospital, to investigate a baby boy, half-Maori and half-Irish, who was failing to thrive, had an untreatable anemia, and was passing unidentifiable crystals. My job was to identify the crystals. This child is the longest survivor with the rare pyrimidine disorder, hereditary orotic aciduria." Anne then went on to study the effects of allopurinol on patients with renal disease, and during the course of these studies she developed methods for the separation and identification of purine and pyrimidine metabolites that would be considered nothing less than "gold standards." She then moved to Britain and continued this work at Guy's Hospital. It was for this work, supported by the Wellcome Foundation, that Anne received her PhD from London University.

In 1970, she was invited to set up the Purine Research Laboratory at Guy's Hospital to study purine metabolism in gout and renal failure. The laboratory has since become an ad hoc diagnostic center for diseases of purine and pyrimidine metabolism. Many of the first cases of specific genetic defects involving purines and pyrimidines in Britain and in many other countries have been diagnosed in her laboratory. Despite these successes, financial support for the laboratory has not been easy to come by. Support for the first three years came from an anonymous donor, and we should point out that the laboratory has been run for all these years on three-year contracts. This obviously placed a great deal of stress on Anne, but it is a testimony to her scientific acumen and her courage and stamina, that she continued to find support for herself, her research, and her staff. She has worked tirelessly to improve the career structure and working conditions for non-clinical scientists in clinical departments. One very important outcome of this effort has been the formation of the Association of Researchers in Medicine and Science (ARMS), an organization whose aim is to define and improve the career structure for basic scientists and clinicians in clinical research.

In addition to her independent description of APRT deficiency, Anne was the first to identify that deoxyadenosine, rather than adenosine, was the toxic metabolite accumulating in ADA deficiency, now the first enzyme deficiency for which gene therapy is being tested in humans. She has always been convinced of the importance of metabolism and has refused to be diverted to current bandwagons. She is also a dedicated European and was a founding member of the European Society for the Study of Purine and Pyrimidine Metabolism in Man (ESSPPM) with the objective of providing low cost meetings to attract and educate young investigators in this important field of metabolism.

Anne is indeed a pioneer in purine and pyrimidine metabolism, and as so often happens to pioneers, recognition takes a long time to come. On behalf of the International Organizing Committee, the Local Planning Committee, and her friends and colleagues throughout the world, we dedicate this symposium and its proceedings to Dr. H. Anne Simmonds.

PREFACE

These volumes record the presentations made at the VIII International Symposium on Purine and Pyrimidine Metabolism in Man held at Indiana University, Bloomington, USA from May 22- May 27, 1994. This was a continuation of meetings held every three years with the idea of bringing clinicians and basic scientists together, which we hope results in cross-fertilization of ideas. Some of the papers presented in this volume represent oral contributions and others are from posters, but we emphasize that both are considered of equal merit.

As is obvious from a perusal of the titles of the papers there has been a shift in the focus of this meeting, which reflects a general shift in the area of purine and pyrimidine metabolism. The emphasis has definitely shifted to gene structure and molecular genetics, with the beginnings we hope of gene therapy as an important branch of this area of science. Although many of the inherited diseases discussed in this text can be treated with drugs, the major thrust in the future will be in gene therapy, where the gene (or cDNA) will be used to treat the patient with enzyme deficiency, particularly if the patient is young.

As can be seen from the list of authors there is a remarkable degree of international cooperation in this area across countries and continents. We thank the many participants who have attended these symposia many times, and we welcome the large group of scientists from Eastern Europe who are attending this meeting for the first time.

We would like to acknowledge the financial support from a number of companies, in particular Burroughs Wellcome, Gensia, and Amgen, which were the largest contributors. We also wish to thank Dr. M. Lowengrub, Dean of the College of Arts and Sciences for financial support, and hosting the meeting. Among other sponsors, we would like to mention the Indiana University Research and Graduate School. Other contributors included Agouran Pharmaceuticals, Berlex, Biochem Therapeutics, British Bio-tech, Gilead Sciences, Harlan-Sprague, Marion Merrell Dow, Pfizer, Sterling Winthrop and the Upjohn Company. The availability of such funds allowed us to offer 22 full or partial travel grants. We also wish to thank the Monroe County Convention and Visitors Bureau for assistance in organizing trips in the area.

Organizing such a symposium involves many people and takes a considerable amount of effort. In particular, the review of abstracts, their organization into symposium topics, and selection of travel award recipients based on the quality of abstracts were formidable tasks. We thank the Abstract Review Committee (Drs. Jeff Davidson, Mitch Turker, Carol Caporelli and Joseph Cory) for doing a wonderful job. The overall planning of the program was organized by Dr. George Weber, with the day-to-day coordination being done by Dr. Milton Taylor and Dr. Amrick Sahota. We owe a debt of thanks to Kevin Knerr from the Indiana University Conference Bureau for his handling of day-to-day problems, and the amount of attention he has given to this meeting throughout the conference and its planning stages. We would also like to recognize Ginger Calloway, also from the conference bureau, for her excellent administrative skills. She was responsible for daily administrative matters including mailing lists, letters and faxes and receipt of manuscripts.

We look forward to the 9th international meeting in Salzburg in 1997.

*Milton Taylor
Department of Biology
Indiana University
Bloomington, IN 47405

Amrick Sahota
Department of Medical and Molecular Genetics
Indiana University School of Medicine
Indianapolis, IN 46202

---

*Contact Editor

CONTENTS

REVIEW

REGULATION OF PURINE AND PYRIMIDINE METABOLISM

CLINICAL

GOUT

CANCER TREATMENT

NUCLEOTIDE METABOLISM AND CANCER

ISCHEMIA

DIAGNOSIS AND TREATMENT OF INHERITED DISEASES

GENE THERAPY

ADENOSINE REGULATORY AGENTS

ENZYMES OF PURINE METABOLISM

ENZYMES OF PYRIMIDINE METABOLISM

GENE REGULATION

ENZYME STRUCTURE AND PURIFICATION

MOLECULAR BIOLOGY AND GENE STRUCTURE

DIFFERENTATION

NUCLEOSIDE TRANSPORT

CYTOKINES

METHODS

DIAGNOSIS AND TREATMENT OF INBORN ERRORS OF PURINE AND PYRIMIDINE METABOLISM: AN OVERVIEW

H. Anne Simmonds

Purine Research Laboratory, UMDS Guy's Hospital, London, GB

INTRODUCTION

There have been many exciting developments in this field since the 1st International Symposium was held in Tel Aviv in 1973. Such progress is all the more remarkable considering that the metabolic basis for the first such genetic defects, xanthinuria and hereditary oroticaciduria, was established as recently as 1959 and by 1973 the number had increased only to include complete or partial deficiency of hypoxanthine-guanine phosphoribosyltransferase: HPRT) and superactivity of its companion X-linked disorder, phosphoribosylpyrophosphate synthetase (PRPS). Both owed their recognition then, as now, to the fact that they can present as gout, kidney stones or hyperuricaemia (1). Allopurinol had made its appearance as a uric acid lowering drug for treating hyperuricaemia, and the mechanisms of the altered renal urate handling in the so-called 'primary gout' of the middle aged male were being hotly debated.

In the intervening years there has been an exponential increase in the number of such disorders recognised (2), as the Proceedings of subsequent Symposia testify, and at the time of this 8th International meeting they now total 22. Knowledge of their metabolic basis has grown apace, but such recent description means that many are little known in the clinic or laboratory. This diagnostic problem is compounded by the fact that they cover a broad spectrum of illnesses (1-9), sometimes with multiple forms of presentation, as in the Lesch-Nyhan syndrome (LNS) often identified only after presentation with gout or acute renal failure in a child/adolescent institutionalised for 'cerebral palsy' (3). They may be the cause of unexplained immunodeficiency or anaemia, or gout in children and adolescents, and neurological deficits varying from muscle weakness, autistic behaviour, delayed development, deafness, epilepsy, fitting, self mutilation, choreoathetosis and inability to walk or talk. Hyper/hypotonicity, microcephaly, mental retardation and premature death are frequent accompaniments. They can be the unexpected cause of intolerance to therapy with purine and pyrimidine analogues (3).

The history of the first recognition of these disorders is one of skill (e.g similar symptoms in siblings suggesting a genetic disorder: HPRT, PRPS), serendipity (e.g.by workers in a different discipline: adenosine deaminase (ADA),purine nucleoside phosphorylase (PNP), or adenylosuccinase (ASA) deficiency), or occasionally the refusal of parents, or patients, to accept a (non)diagnosis (eg adenine phosphoribosyltransferase (APRT) deficiency). As additional cases have been recognised.a considerable genetic heterogeneity has become evident, which means that these disorders can present at birth leading to early death, or with milder forms up to the 80's (1-10).

DIAGNOSIS IN THE CLINIC

From the above summary it is evident that diagnosis is not always easy. Let us look at some of the problems encountered.

Acute renal failure (ARF) is a frequent presentation Many disorders are associated

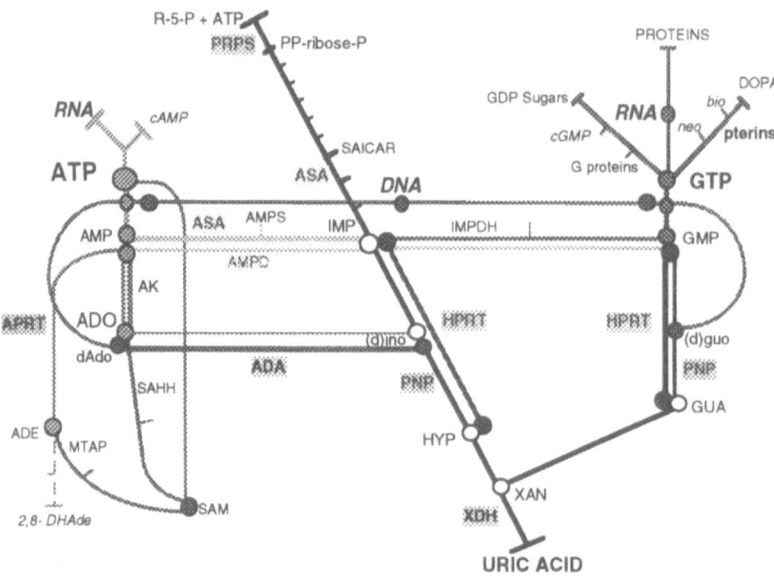

Figure 1. Pathways of purine metabolism in human cells. Abbreviations for the enzymes catalysing the different steps are listed in the text.

with renal complications due to the poor solubility of purines (2,3).HPRT deficiency is a prime example and xanthine oxidase (XDH) deficiency, previously considered a benign disorder of adults presenting with kidney stones, is also now being identified in infants presenting in ARF, whilst APRT deficiency, originally described in children, often accompanied by ARF, has now been found in adults up to the 8th decade .

Factors such as renal handling may be important . For example, patients with LNS overproduce purines (1,3), but may have normal plasma uric acids because of the high uric acid clearance in children. Simple clues in such cases have been reports of crystals on the diaper (LNS, partial HPRT, APRT, XDH and UMPS deficiencies, and PRPS superactivity), or on the tip of the penis (1). Renal ultrasound gave the first hint of the underlying cause in all five of the above disorders in infants presenting in ARF, sometimes in coma (3). Symptoms and signs, such as colic or haematuria present from birth have often been overlooked.

Hyperuricaemia, sometimes renal failure or gout, is also a feature of familial juvenile hyperuricaemic nephropathy (FJHN), which unusually affects young women as well as men, is not associated with uric acid overproduction, but rather gross underexcretion. Untreated it has led to progressive renal disease, dialysis, transplantation or death in the 30's (3). The genetic basis is unknown, but FJHN must be excluded in young men, women and children with gout, hyperuricaemia or familial renal disease (2 and McBride et al this volume).

Genetic heterogeneity also exists in neurological disorders. This is the second problem for diagnosticians. HPRT deficiency is characterised by the severe LNS at one end of the spectrum and adolescent gout at the other (4), with varying degrees of neurological involvement in between (1). The same is true of two defects of the *de novo* synthetic pathway (1,2), the latest developments in which are described in detail at this Symposium. The first, PRPS, characterised at one extreme by a variety of neurological deficits including abnormal facies, sometimes inherited nerve deafness, contrasts with presentation only with gout or stones in early adulthood is discussed by Dr Becker. PRPS, unlike HPRT deficiency, may also manifest in the carrier female and should be suspected in any seemingly X-linked defect where the mother presents with gout and may be deaf (2). The second disorder, the recessive adenylosuccinase (ASA) deficiency, is described by Dr Van den Berghe, and manifests with varying degrees of psychomotor delay which likewise makes recognition difficult - especially since the autism in the index case (5), has not been a consistent feature.

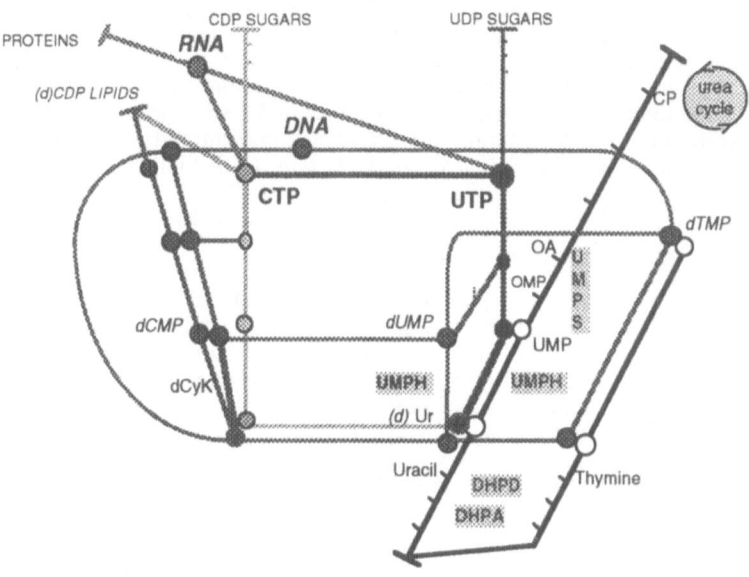

Figure 2. Pathways of pyrimidine metabolism in human cells. Abbreviations for the enzymes catalysing the different steps are listed in the text.

Neonatal fitting can be symptomatic of purine or pyrimidine disorders. Combined xanthine oxidase slphite oxidase (XDH/SO) deficiency, is characterised generally by severe neurological deficits, including neonatal fitting, ocular lens dislocation and death in the first year, but a teenager was reported (2) where, as in late presenters with PNP deficiency the disorder may be missed because of significant uric acid in plasma and urine.

Neonatal fitting is also a form of presentation in the pyrimidine disorder, dihyropyrimidinase (DHPA) deficiency (2,6), associated with gross microcephaly and spastic quadraplegia. Neurological abnormalities in the other pyrimidine defects include, in the companion disorder, dihydropyrimidine dehydrogenase (DHPD) deficiency, which also presents neonatally, (2,6), epilepsy, microcephaly, dysmorphic features and mental retardation; whilst growth retardation and inability to sit unaided are found in the *de novo* synthetic disorder hereditary oroticaciduria - uridine monophosphate synthase (UMPS) deficiency (1,2).

Failure to thrive is an early warning sign, especially in the immunodeficiency disorders Failure to thrive and recurrent life-threatening infections, should evoke suspicion of the recessive genetic immunodeficiency disorders, ADA deficiency, and if there are also associated neurological abnormalities, PNP deficiency (especially in consanguineous kindreds) (2,7,8). ADA deficiency generally appears neonatally, associated with vomiting, diarrhoea, generalised candidiasis, severe combined immunodeficiency (SCID), and early death if untreated. Later presenters have milder symptoms, the oldest reported case until now being 8 years (9). The finding of ADA deficiency in two HIV-negative sisters in their thirties (10 and this volume) presenting with respiratory dysfunction and CD4+ T-lymphocytopenia, a clinical syndrome resembling 'non-HIV AIDS', confirms that ADA deficiency must also be suspected at any age.

The immunological problems in PNP deficiency are milder (1,8). The disorder frequently presents first with neurological abnormalities, commonly spastic diplegia, similar to HPRT deficiency. Although HPRT is not defective, the enzyme cannot function without the substrates normally provided by PNP (2). Consequently, despite the history of recurrent infections - ultimately the cause of death at around five years - PNP deficiency may not be suspected. Less severe cases present later, with mild to absent neurological abnormalities (2 and this volume).

Anaemia is frequently indicative of pyrimidine defects. Intractable megaloblastic anaemia is the more characteristic presentation of UMPS deficiency, but infections and immunodeficiency may also be found (1,2). Non-spherocytic haemolytic anaemia with basophilic stippling is the presentation in uridine monophosphate hydrolase (UMPH) deficiency - commonly known as pyrimidine 5'- nucleotidase deficiency (1). Another disorder associated with abnormal pyrimidine nucleotide metabolism and presenting with chronic haemolytic anaemia is a putative deficiency of the enzyme CDP choline phosphotransferase (2).

Purine and pyrimidine disorders can be the basis for adverse responses to analogue therapy

Severe neurotoxicity has been reported in patients treated with 5 - flourouracil for common neoplasms, subsequently found to be DHPD deficient (2,6). A similar toxicity could be predicted in DHPA deficiency. The use of azathioprine for transplant immunosuppression in patients with thiopurine methyltransferase (TPMT) deficiency, purine 5'-nucleotidase deficiency and in an undiagnosed case of XDH deficiency, has also had potentially catastrophic consequences (2,3). ITPase deficiency is a seemingly benign but relatively common mutation in humans (2) If, as speculated, it serves a function in degrading rogue nucleotides, then ITPase deficiency might also be an unappreciated cause of analogue nucleotide toxicity.

LABORATORY DIAGNOSIS

As with genetic metabolic disorders generally (1), laboratory diagnosis is based on the presence or absence of specific metabolites and/or altered enzyme activity, but in the case of the purine and pyrimidine disorders laboratory diagnosis is not easy due to numerous potential pitfalls. In rare instances, as discussed below, altered renal handling or altered metabolites alone (eg where the enzyme is expressed only in the liver), or (where no abnormal metabolites accumulate) abnormal cellular nucleotides, may provide the only clue.

The first paper on the application of HPLC, now widely used instead of TLC or radioactive methods,to monitor the characteristic abnormal metabolites accumulating in cells, body fluids and/or altered enzyme activity, was also presented in Tel Aviv. There has been a similar exponential increase in other diagnostic methods, which include in-line Photodiode Array and radiodetection, capillary electrophoresis (CE) and molecular biology techniques (11). These methods may also be applied to prenatal diagnosis.

Diagnosis from specific metabolites in body fluids

Correct laboratory diagnosis depends on a number of factors. Normal ranges must be established for local populations relative to age and sex and the method used.This is very important for uric acid, since dietary purine intake varies in different countries (2,3).The clinical condition can render a common yardstick meaningless (e.g.- urine uric acid/creatinine ratio in.ARF). A keen awareness of the numerous pitfalls is vital as follows:

A full list of diet, drugs and therapeutic manipulations must be obtained. Lack of knowledge of prior treatment (IVP or blood transfusion) could lead to misdiagnosis, especially by HPLC if monitoring only at one wavelength. (ie 254nm).In some systems hypaque has a retention time similar to hypoxanthine, methylated xanthines to deoxyadenosine and paracetamol metabolites, septrin, acyclovir etc. co-elute with endogenous purines such as uric acid or adenine. In-line diode-array detection and enzyme peak shift are vital for confirmation.

. Blood transfusion - not always communicated to the investigator - can produce essentially normal RBC lysate activity and spurious activity can take up to 6 months to disappear (2). Characteristic blood or urine metabolites eg deoxyadenosine or deoxyinosine, may likewise disappear when the low but detectable red cell dATP is the only abnormal finding in ADA deficiency and plasma and urine uric acid will be normal.in PNP deficiency.

Acid lability is a particular problem. Degradation in the cationic systems used led to deoxyadenosine being mistaken for adenine as the abnormal metabolite excreted in ADA deficiency (2). Nucleosides can be degraded rapidly to bases or uric acid by bacterial contamination, as has occurred in PNP and XDH deficiency, disorders where uric acid is normally undetectable.

Specific metabolites may accumulate secondary to non purine or pyrimidine disorders. Elevated urinary uracil, together with thymine, is the hall-mark of DHPD and DHPA deficiency (2,6), but elevated uracil (without thymine) is also characteristic of carriers for ornithine carbamoyltransferase (OCT) deficiency (2), and is sometimes a better guide than orotic acid . Abnormally elevatied uracil requires scrutiny, since on rare occasions pseudouridine (normally excreted with a constancy equivalent to creatinine) is undetectable and is replaced by a comparable amount of uracil. The absence of pseudouridine, is the vital clue .

Intact erythrocytes can be a useful guide to diagnosis

Characteristic erythrocyte nucleotide patterns are found in some pyrimidine and purine disorders, even in the absence of detectable abnormalities in plasma or urine. They are not only useful in diagnosis (2), including prenatal diagnosis, but also in determining prognosis. and recently in monitoring PEG-ADA and gene therapy in ADA deficiency (9,10). They may also give some clue as to the mechanism by which the abnormal gene product produces disease (2). Incubation of radiolabelled substrates with red cells from freshly drawn blood has been informative in establishing the phenotype in HPRT deficiency (2). Red cells in LNS show no incorporation of radiolabelled hypoxanthine into IMP, but cells of patients with milder to absent neurological abnormalities showed a corresponding increase in the amount of IMP formed.

Two types of APRT deficiency have been recognised since the independent description of the defect in London and Paris in the mid-70's (2). Type 1 mainly in Caucasians with no detectable erythrocyte lysate activity and type II, found only in Japan to date, with heterozygote lysate levels but undetectable activity in intact cells (1,2,). Dr Kamatani (this volume) describes their latest results in the more than 100 patients now identified in Japan alone.

TREATMENT OF GENETIC PURINE AND PYRIMIDINE DISORDERS

Intensive studies at the molecular level have delineated a variety of novel mutations in many of these disorders (1,7,8). These include deletions, substitutions, missense mutations and splicing defects. It has been proposed that tissue specific variations in splicing efficiency may ameliorate disease severity and thereby explain the absence of symptoms, or their late onset, in some disorders (1,7). It was originally hoped that DNA analysis would demonstrate a useful correlation between genotype and phenotype in LNS (4), but no such association was identified in a study of a large number of patients with partial or complete HPRT deficiency (1).

Thus, despite the detailed knowledge now available from the above molecular studies, the sad fact is that we still have no clear understanding of the metabolic basis for the associated neurological abnormalities in patients with aberrant PRPS, or the other purine enzyme deficiencies HPRT, PNP, ASA and XDH/SO, or the pyrimidine defects DHPD and DHPA. Only a handful of purine and pyrimidine disorders respond to treatment and these can be monitored by the decrease in the concentration of the specific metabolite concerned.

Disorders associated with hyperuricaemia: Allopurinol has been extremely useful in reducing raised uric acid levels, but must be used with care in overproducers because of the poor solubility of the xanthine which accumulates and can lead to kidney stones or xanthine nephropathy (3). Likewise accumulation of oxypurinol can have severe side effects, e.g.bone marrow depression. Consequently, allopurinol should be reduced in severe renal failure with an even further reduction in patients also receiving azathioprine immunosuppression. Early introduction of allopurinol is particularly important in treating the hyperuricaemia in patients with FJHN. since studies in some 68 patients for up to 20 years indicate that allopurinol appears to prevent or ameliorate the hitherto rapid progression to end stage renal disease (3).

Neurological disorders:As mentioned above the majority of the neurological disorders listed are as yet untreatable and this problem must be the subject of active collaborative research between investigators in the field. Hereditary oroticaciduria is the one exception (1). The clinical manifestations respond to uridine, but not uracil therapy, demonstrating an important difference in salvage pathways between purine and pyrimidine metabolism. The first surviving patient has now been treated with uridine for more than 30 years. Uridine also corrects the megaloblastic anaemia characteristic of this disorder.

Immunodeficiency disorders The treatment of choice in ADA and PNP deficiency is bone marrow transplantation (BMT) with cells from an HLA-genotypically identical donor, but only 2 of the 5 cases of PNP deficiency who have received BMT are still alive (8). The success rate using haploidentical-related donors in ADA deficiency is also poor (7) -around 50%. PEG-ADA has been used successfully in up to 40 patients (9) but the treatment is lifelong and the cost prohibitive. Somatic cell gene therapy has been developed and the first human gene implants have been performed in several ADA deficient children; all have initially or eventually been given PEG-ADA (9). An increase in circulating T cells has resulted and lymphocyte ADA has increased 10 fold (this volume) Success in this disorder could pave the way in others.

DIAGNOSIS IN THE YEAR 2000+

Methods based on specific metabolite or enzyme assay, are rapid and inexpensive and the only way to identify new defects and are thus unlikely to be superseded totally by molecular biology. .Gene probes will only be informative in recognised kindreds - not in the 30% who represent new mutations. CE may become important (11). It has a higher sensitivity and smaller sample requirement than HPLC. Developments in LC-MS, LC-NMR and MS with MALDI-TOF and electrospray systems may give definitive structural information.

WHAT HAVE WE LEARNT FROM PURINE AND PYRIMIDINE DISORDERS?

The potential to apply knowledge derived from these rare disorders to the more common killers of mankind is a rapidly developing field, particularly in heart disease, organ transplantation and the treatment of malignancies. The inherited immunodeficiencies have expanded our knowledge about the controls vital to the immune system and have suggested ways it may be manipulated to the benefit of the host, and detriment of the invader (1,2,7-10).

We are approaching the possibility of xenograft transplantation. The immune problems are being overcome, but the physiological consequences ignored. Studies indicating the proximal tubule brush-border urate transporter is lacking in pig kidney (12) have implications for their use in human transplantation, particularly for the handling of drugs in common use.

Devising treament for the neurological problems in these devastating disorders must be another challenge. Such knowledge could then be usefully applied to the more common neurological disorders.

It is predicted that by the year 2000 there will be a widespread (re)recognition of the importance of purine and pyrimidine metabolism and a swing back to this neglected area

REFERENCES

1.CR Scriver, AL Beaudet, WS Sly, D Valle, eds. The Metabolic Basis of Inherited Disease.McGraw-Hill, New York 1989; 6th edition (1989).
2. H.A. Simmonds. Chapter 8. Purine and pyrimidine disorders, in: The Inherited Metabolic Diseases, J.B. Holton, ed. Churchill Livingstone, Edinburgh; 2nd Edition: (1993)
3. J.S.Cameron, F.Moro and H.A.Simmonds. Gout, uric acid and purine metabolism in paediatric nephrology: Nephrology Review. Pediatr Nephrol7:105(1993).
4. R.W.E.Watts. Defects of tetrahydrobipopterin synthesis and their possible relationship to a disorder of purine metabolism (the Lesch-Nhyan syndrome). Adv Enz Reg 1985;23:25-58.
5. J.Jaeken and G.Van den Berghe. An infantile autistic syndrome characterised by the presence of succinylpurines in body fluids. Lancet 1984;2:1058-1061.
6.A.H. van Gennip, S. Busch, L.Elzinga, A.E.M. Stroomer, A.van Cruchten, E.G. Scholten and N.G.G.M. Abeling. Application of simple chromatographic methods for the diagnosis of defects in pyrimidine degradation. Clin Chem 39:380 (1993)
7. R.Hirschorn. Overview of biochemical abnormalities and molecular genetics od adenosine deaminase deficiency. Pediatr. Res. 33:S35 (1993)
8. M.L.Markert Purine nucleoside phosphorylase deficiency. Immunodeficiency Reviews 3:45 (1991).
9.M.S. Hershfield, S. Chaffee and R.U. Sorensen Enzyme replacement therapy with polyethylene glycol-adenosine deaminase and adenosine deaminase deficiency. Pediatr. Res. 33:S42 (1993)
10. CL Shovlin, ADB,Webster LD Fairbanks, H.A. Simmonds, S.Deacock, R. Lechler, I. Roberts and JMB Hughes. Adult presentation of inherited adenosine deaminase deficiency. Lancet 341:1472 (1993)
11.T. Grune, G.A..Ross,H.Schmidt, W. Siems, and D. Perrett,. Optimized separation of purine bases and nucleosides in human cord plasma by capillary zone electrophoresis.J Chromatogr 636:105 (1993)
12. D Werner, F. Martinez and F Roch-Ramel. Urate and p-amino hippurate transport in the brush border membrane of the pig kidney. J Pharmacol Exp Ther 252:792 (1990).

DETERMINATION OF DIHYDROPYRIMIDINE DEHYDROGENASE (DPD) IN FIBROBLASTS OF A DPD DEFICIENT PEDIATRIC PATIENT AND FAMILY MEMBERS USING A POLYCLONAL ANTIBODY TO HUMAN DPD

R.B. Diasio,[1] A.B.P. Van Kuilenburg,[2] Z. Lu,[1] R. Zhang,[1] H.Van Lenthe,[2] H.D. Bakker,[2] and A.H. Van Gennip[2]

[1]Department of Pharmacology, Division of Clinical Pharmacology, Comprehensive Cancer Center, University of Alabama at Birmingham, Birmingham, Al 35294, USA
[2]Laboratory of Metabolic Diseases and Procreation, Academic Medical Center, Amsterdam, Netherlands

INTRODUCTION

Dihydropyrimidine dehydrogenase (EC 1.3.1.2, DPD) is the initial rate-limiting enzyme in pyrimidine catabolism, which catalyzes the following reaction.

$$\text{Pyrimidine} + \text{NADPH} \rightleftharpoons \text{dihydropyrimidine} + \text{NADP}^+$$

The biological significance of DPD has been demonstrated in several previous studies.[1-3] This enzyme catalyzes the first reaction in the three-step catabolic pathway which converts uracil into ß-alanine; this is the only pathway for the biosynthesis of ß-alanine in mammalian tissues.[1] ß-Alanine has been suggested to be involved in several metabolic and neurotransmitter functions.[2] This enzyme has a critical role in cancer chemotherapy with fluoropyrimidine antimetabolite drugs.[3,4] More than 85% of administered fluorouracil (FUra), one of the most frequently used anticancer drugs, is catabolized by this enzyme.[4] Previous studies have shown a correlation between DPD activity and FUra pharmacokinetics.[4-7] It has also been suggested that the anticancer efficacy of FUra and its drug-associated host toxicity is related to DPD activity. Several studies have demonstrated that DPD follows a circadian pattern in cancer patients[6] and in experimental animals.[8]

More recently, the significance of this enzyme has been investigated in patients with DPD deficiency.[9-16] Our laboratories have previously described DPD deficiency in pediatric patients where affected patients have severe developmental abnormalities[16] and more recently in cancer patients where the manifestations are pharmacogenetic (i.e., individuals are phenotypically "normal" until exposed therapeutically to 5-fluorouracil).[10-13] Since our initial demonstration of genetic deficiency of DPD in humans[10], several clinical studies[11-14] have shown that patients with DPD deficiency experience severe toxicity during fluoropyrimidine administration which requires cessation of chemotherapy. Pharmacogenetic and pharmacoepidemiologic studies have suggested that the frequency of this genetic defect is greater than previously recognized.[13] More recently, human liver DPD has been purified to homogeneity and a polyclonal antibody generated and shown to be specific for human DPD.[17]

In previous studies from our laboratories, DPD activity was typically assessed by methods to directly measure enzyme activity.[6,10-13,16] The availability of purified human liver DPD and a polyclonal antibody to human DPD[17] has made a potential alternative method available for determining DPD activity in patients. This method has already proven useful in demonstrating DPD deficiency in adult cancer patients.[17,18] The present study was undertaken to determine the usefulness of this polyclonal antibody in quantitation of DPD in pediatric patients.

MATERIALS AND METHODS

A double-blind study design was used. DPD activities were quantitated in Amsterdam, The Netherlands. Antibody studies were conducted on the same samples in Birmingham, Alabama, U.S.A.

DPD Assay

A male 3 year old child with developmental and mental retardation was identified to be DPD deficient using a previously reported method[16] for measuring DPD activity in fibroblasts. DPD activities from his family members, parents and two brothers, and controls were also determined.

Immunoblot Analysis (Western Blot)

The primary antibody used in the present study was the purified rabbit polyclonal antibody generated against human liver DPD.[13,17] A 7% SDS-PAGE was performed using the freshly prepared fibroblast supernatant from the patient, family members, and controls. SDS-PAGE was carried out in a 1.0 mm thick, 7% (w/v) polyacrylamide gel containing 0.375 M Tris-HCl (pH 8.8) and 0.1% SDS as described previously. Electrophoresis was performed at a constant current of 30 mA for 45 min at 25°C. The proteins were transferred from the gel to a nitrocellulose filter following the method described previously.[17] Following incubation overnight at 4°C with the primary antibody (diluted 1:2000) in a 120 mM borate-saline solution containing 1% (w/v) BSA, pH 8.5, the nitrocellulose filter was washed with borate-saline containing 0.5% (v/v) Tween 20 and then incubated with a secondary, alkaline phosphatase-labeled goat anti-rabbit antibody. The location of immunoreactive proteins on the nitrocellulose filter was detected in a 0.1 M sodium carbonate buffer, pH 9.5, containing 30 mg nitro blue tetrazoliumand 15 mg 5-bromo-4-chloro-3-indolyl phosphate toluidine salt.

RESULTS AND DISCUSSION

DPD Activity

DPD activities from the patient and family members are shown in Table 1, together with demographic data for each individual. DPD activity was undetected (lower than the sensitivity limit of the method). The patient's eleven year old brother was observed to have DPD activity in the range of normal controls. DPD activities from the mother and father as well as the thirteen year old brother were found to be intermediate between the patient's level (undetectable) and the normal control level. These data suggested that the patient had complete DPD deficiency and the parents and one brother had partial deficiency. The distribution of DPD activities in this family is consistent with an autosomal recessive pattern of inheritance as reported previously.[10-13,15,16]

Table 1 DPD Activity in Patient and Family Members

Name	Sex	Relation	Age	DPD Activity (nmol/hr/mg protein)	
				1st Assay	2nd Assay
BRDB	M	Patient	3	< 0.06	< 0.03
TB	F	Mother	33	0.14	0.08
CB	M	Father	39	0.11	0.11
FFB	M	Brother	11	0.94	1.06
RCB	M	Brother	13	0.28	0.18

Western Blot Analysis

Figure 1 illustrates the amount of DPD protein for each individual and control detected by the antibody. The density of the DPD protein band in the fibroblast cytosol was shown to correlate with DPD activities shown in Table 1.

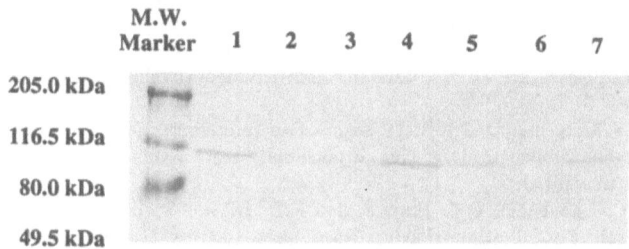

Figure 1 Immunoblot analysis of fibroblast DPD from deficient patient, family members, and normal subjects. For lanes 1-6, each lane contains 50 µg of crude cytosol. Lane 7 contains 25 µg of crude cytosol. Lane 1: Normal control; Lane 2: Mother; Lane 3: Father; Lane 4: Brother 1(FFB); Lane 5: Brother 2 (RCB); Lane 6: Patient; and Lane 7: Normal control.

The sample from the affected patient had no detectable DPD protein on Western blot analysis. The sample from the thirteen year old brother with normal DPD activity had a single band corresponding to control samples in location and density of known DPD protein. Samples from the other family members (mother, father, and eleven year old brother) with intermediate DPD activity also had a single band but with an intensity intermediate between control (normal DPD activity) and the patient (undetectable).

As reported earlier, this antibody generated from purified human liver DPD has been demonstrated to be useful in quantitation of DPD from various tissues including human liver[13, 18,19] and peripheral blood mononuclear cells[18] from cancer patients as well as normal individuals. The present study is the first report of the use of this antibody in pediatric patients.This is also the first time the antibody has been used for detecting DPD in fibroblasts.In summary, this study demonstrates that this polyclonal antibody developed for use in cancer patients is useful in the pediatric population in identifying both affected patients (homozygotes) and carriers (heterozygotes) with this form of DPD deficiency.

ACKNOWLEDGEMENTS

This study was supported by USPHS NIH/NCI grants CA 40530 and CA 62164.

REFERENCES

1. T. W. Traut, and S. Loechel, Pyrimidine catabolism: individual characterization of the three sequential enzymes with a new assay, *Biochemistry* 23: 2533-2539 (1984).
2. F. Zafra, M.C. Aragon, F. Valdiviesco, and C. Gimenez, ß-Alanine transport into plasma membrane vesicles derived from rat brain synaptosomes,*Neurochem. Res.* 9: 695-707 (1984).
3. R.B. Diasio and B.E. Harris, Clinical pharmacology of 5-fluorouracil, *Clin. Pharmacokinet. 16*: 215-237, 1989.
4. G.C. Daher, B. E. Harris, and R.B. Diasio, Metabolism of pyrimidine analogues and their nucleosides, *Pharmacol. Ther.* 48:189-222 (1990).
5. G.D. Heggie, J.-P.Sommadossi, D.S. Cross,W.J. Huster, and R.B. Diasio, Clinical Pharmacokinetics of 5-fluorouracil and its metabolites in plasma, urine, and bile, *Cancer Res.* 47:2203-2206 (1987).
6. B.E. Harris, R. Song, S.-J. Soong, and R.B. Diasio, Relationship between dihydropyrimidine dehydrogenase activity and plasma 5-fluorouracil levels with evidence for circadian variation of enzyme activity and plasma drug levels in cancer patients receiving 5-fluorouracil by protracted continuous infusion, *Cancer Res.* 50:197-201 (1990).
7. R.A. Fleming, G. Milano, A. Thyss,M.-C. Etienne, N. Renee, M. Schneider, and F. Demard, Correlation between dihydropyrimidine dehydrogenase activity in peripheral mononuclear cells and systemic clearance of fluorouracil in cancer patients, *Cancer Res.* 52: 2899-2902 (1992).
8. R. Zhang, Z. Lu, T. Liu, S.-J. Soong, and R.B. Diasio, Relationship between circadian-dependent toxicity of 5-fluorodeoxyuridine and circadian rhythms of pyrimidine enzymes; Possible relevance to fluoropyrimidine chemotherapy, *Cancer Res.* 53: 2816-2822 (1993).
9. M. Tuchman, J.S. Stoeckeler,D.T. Kiang, R.F. O'Dea, M.L. Rammaraine, and B.L. Mirkin, Familial pyrimidinemia and pyrimidinuria associated with severe fluorouracil toxicity, *New Engl. J. Med.* 313:245-249 (1985).
10. R.B. Diasio, T.L. Beavers, and J.T. Carpenter, Familial deficiency of dihydropyrimidine dehydrogenase, *J. Clin. Invest .* 81:47-51 (1988).
11. B.E. Harris, J.T. Carpenter, and R.B. Diasio, Severe 5-fluorouracil toxicity secondary to dihydro-pyrimidine dehydrogenase deficiency: a potentially more common pharmacogenetic syndrome, *Cancer* 68: 499-501 (1991).
12. A.P. Lyss, R.C. Lilenbaum, B.E. Harries, and R.B. Diasio, Severe 5-fluorouracil toxicity in a patient with decreased dihydropyrimidine dehydrogenase activity, *Cancer Invest.* 11: 239-240 (1993).
13. Z. Lu, R. Zhang, R.B. Diasio, Dihydropyrimidine dehydrogenase activity in human peripheral blood mononuclear cells and liver: population characteristics, newly identified deficient patients, and clinical implication in 5-fluorouracil chemotherapy, *Cancer Res.* 53: 5433-5438 (1993).
14. P. Houyau, C. Gay, E. Chatelut, P.Canal, H. Roche, and G. Milano, Severe fluorouracil toxicity in a patient with dihydropyrimidine dehydrogenase deficiency,*J . Nat. Cancer Inst.* 85: 1602-1603 (1993).
15. K. ward, M.J. Henderson, H.A. Simmonds, J.A. Duley, and P.M. Davies, Dihydropyrimidinuria presenting in childhood with severe developmental retardation, *in:* "Molecular Genetics, Biochemistry and Clinical Aspects of Inherited Disorders of Purine and Pyrimidine Metabolism", U. Gresser, ed.,pp. 165-167, Springer-Verlag, Berlin Heidelberg (1993).
16. M. Brocksted, C. Jakobs, L.M.E. Smith, A.H. van Gennip, and R. Berger, A new case of dihydropyrimidine dehydrogenase deficiency, *J. Inher. Metab. Dis.* 13: 121-124 (1990).
17. Z. Lu, R. Zhang, and R. B Diasio, Purification and characterization of dihydropyrimidine dehydrogenase from human liver, *J. Biol.Chem.* 267:17102-17109 (1992).
18. R. Zhang, Z. Lu, and R. B. Diasio, Quantitation of dihydropyrimidine dehydrogenase (DPD) using a polyclonal antibody to human liver DPD, *Proc. Am. Assoc. Cancer. Res.* 35:200 (1994).
19. Z. Lu, R. Zhang, and Diasio, R. B. Genetic polymorphism of dihydropyrimidine dehydrogenase (DPD), the key enzyme in 5-fluorouracil (FUra) catabolism, *Clin. Pharm. Ther.* 54: 180 (1994).

THE EFFECT OF PARTIAL HEPATECTOMY ON BLOOD PURINE LEVELS IN RATS AND PATIENTS

Makoto Usami, Koji Furuchi, Hiroshi Kasahara, Seiji Haji, George Kotani, Atsunori Iso, Kai Sun, Enmei Sou, Jyang-hua Zheng, Kazuya Sakata, Kyosuke Ohta, Taichi Kanamaru, and Yoichi Saitoh

First Department of Surgery, Kobe University School of Medicine, 7-5-2 Kusunoki-cho, Chuo-ku, 650, Kobe, Japan

INTRODUCTION

Purine metabolism is a key metabolic process after partial hepatectomy in the synthesis of DNA and RNA, and in the energy metabolism for hepatic regeneration[1]. However it is not evaluated in clinical situations due to the difficulty in measuring changes of purine levels in the liver itself. In order to get a basic understanding of purine metabolism in clinical hepatectomy cases, blood samples are used for analysis in this study. Also, the liver is the main nondietary source of purines in mammalian cells[2] and changes in blood level may affect purine metabolism in various tissues.

It is the purpose of this study to evaluate changes in purine metabolism by measuring blood purine levels in patients and rats after partial hepatectomy.

MATERIALS AND METHODS

Animal experiment consisted of male Wistar rats with 70% partial hepatectomy. Hepatic regeneration ratio was calculated. Labelling index (LI) of hepatocytes was measured by histoimmunochemical staining using bromo-deoxyuridine and nucleotides in the liver was determined by enzymatic analysis.

Patient population studied were 15 hepatectomized patients for hepatic cancer or hemangioma, (H) group, including 7 cirrhotic cases (CL) and 8 non-cirrhotic cases (NCL), and 19 nonhepatectomized patients ((NH) group) without cirrhosis of the liver who have undergone gastrectomy for gastric cancer. Diagnosis was confirmed by histological studies.

Heparinized venous blood was obtained before the operation, and 1, 3, 7 and 14 days after the operation (POD) following overnight fasting in patients and rats. Whole blood, diluted with PBS, were mixed with cold 20% perchloric acid and centrifuged to obtain an acid-soluble fraction, then stored at -80°C. Nucleotides in blood, and nucleosides and nucleic bases in plasma and urine were measured according to the method developed by Stocchi and Yamamoto using HPLC[3,4]. An Inertsil ODS-2 column (GL Science, Japan) for nucleotide measurement and a Shimazu STR-ODS-M column (Shimazu Japan) for nucleoside and nucleobase measurement were used. Plasma and urine uric acid were measured using uricase-POD method[5]. All data were expressed as mean ± standard deviation and Student's t test was used to compare two samples.

Purine and Pyrimidine Metabolism in Man VIII, Edited by
A. Sahota and M. Taylor, Plenum Press, New York, 1995

RESULTS

In the animal experiment, LI peak value appeared at POD 1 and regenerated hepatic mass reached 2 to 3 fold on POD 7 after partial hepatectomy, indicating rapid hepatic regeneration in rats. Liver purine nucleotide levels decreased after the operation and recovered gradually (Figure 1).

In the patient group, none of the patients developed lethal complications and there were no statistically significant differences between the three groups in distribution of sex, age, operation time and blood loss, except for operative procedure and preoperatively complication of cirrhotic liver .

Purine nucleotides level in the blood, mainly in red blood cells, was 7-800 nmol/ml and formed the majority of blood purine and pyrimidine in the three patient groups. ATP formed about 85% of purine nucleotides. AMP, GMP, inosine, adenosine and guanosine were not detected in the assays. Plasma xanthine level and xanthine/hypoxanthine ratio were statistically higher in the CL group than the NH group ($p < 0.01$, Table).

Table 1   Concentration of purine in plasma and whole blood (nmol/ml)

Groups	NH(19)[a]	H(15)	
		NCL(8)	CL(7)
Purine Nucleotide	785 ± 152[b]	742 ± 216	744 ± 107
ATP	639 ± 141	587 ± 199	636 ± 81
Xanthine	2.30 ± 0.65	2.41 ± 0.75	3.53 ± 1.28*
Hypoxanthine	6.50 ± 4.50	4.25 ± 4.04	3.54 ± 2.01
X/H Ratio	0.46 ± 0.22	1.03 ± 0.92	1.27 ± 0.57*
Uric Acid	37.7 ± 18.6	35.1 ± 20.1	32.8 ± 11.2

HN=nonhepatectomized group; H= hepatectomized group; NCL=non cirrhotic patients after partial hepatectomy; CL= cirrhotic patients after partial hepatectomy; X/H=xanthine/hypoxanthine. [a]Numbers in parenthesis, number of patients examined. [b] Values are mean ± SD. *$p < 0.01$ vs. NH, Student's t test.

Whole blood nucleotide levels decreased in both the H and the NH groups after operation and their levels did not recover to the preoperative value even on POD 14. Plasma xanthine level decreased on POD 1 and recovered by POD 7 in the H group (Figure 2, upper). Plasma hypoxanthine level in the H group decreased on POD 1 and recovered to their preoperative value by POD 3 ($p < 0.05$, Figure 2, lower). In the H group, the ratio of xanthine to hypoxanthine increased abruptly on POD 1, decreased to it's preoperative level by POD 3 and then maintained a constant level, but did not change in the NH group ($p < 0.05$-0.01, Figure 3, upper). There was no relationship among these changes, operation time and blood loss. Partial hepatectomized rats showed the same changes in the blood xanthine/hypoxanthine ratio. Plasma uric acid level decreased until POD 3, then recovered in the H group ($p < 0.05$) and was higher in the NH group than in the H group (Figure 3, lower). The amount of urinary excretion in purine catabolites increased after the operation, and those of xanthine and hypoxanthine in the H group were less than the NH group without statistical significance.

DISCUSSION

Preoperative observations revealed that plasma purine nucleosides and nucleobase contents were less than 1/100 of blood ATP level. Purine nucleoside was not detected but purine bases were detected in the plasma. These blood purine levels in patients are comparable to the data obtained in normal rats[6].

The results obtained from rats and patients clearly indicate a marked increase in the ratio of plasma xanthine to hypoxanthine on POD 1, followed by a rapid recovery to its preoperative level by POD 3 after partial hepatectomy. Rapid changes in ATP metabolism after the operation have been considered as the metabolic response to various surgical stress, including operative procedure itself, tissue damage, ischemia, reperfusion and blood

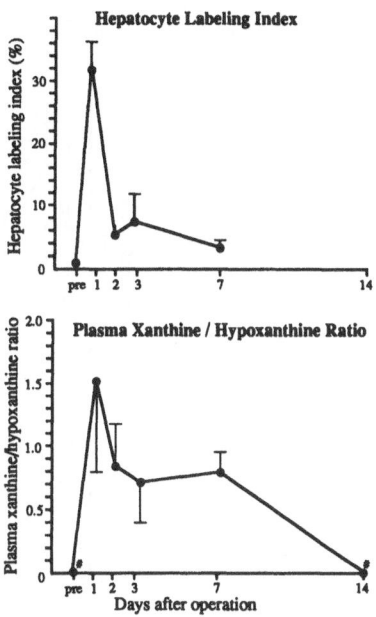

Figure 1. Hepatocyte DNA synthesis and plasma purine base level in rats after 70% partial hepatectomy. Mean ± SD, #: levels below the measurable limit.

transfusion[7], but there was no relation between the ratio of xanthine and hypoxanthine on POD 1 and the operation method, operation time and blood loss. It is suggested that partial hepatectomy itself results in significant elevation of plasma xanthine/hypoxanthine ratio. These observations after partial hepatectomy in rats and patients were exactly the same.

Hepatic purine nucleotide level decreased in regenerating rat liver due to both the acceleration of ATP-utilizing reactions in response to metabolic overload and the increase in DNA and RNA synthesis in the regenerating liver. Purine nucleosides and nucleobases are utilized as a source for salvage purine synthesis in the regenerating liver[2]. The increased xanthine/hypoxanthine ratio is considered in two different mechanisms; a decreased hypoxanthine level due to its utilization by the salvage pathway for purine nucleotide synthesis and an increased xanthine oxidase activity in purine catabolism, which converts hypoxanthine to xanthine and then xanthine to uric acid. Our data, decreased plasma and urine uric acid level in hepatectomized patients, supports the former mechanism, indicating the utilization of hypoxanthine for salvage synthesis. Hypoxanthine is the major source for purine salvage pathway and is trapped into the cytoplasm of animal cells in a short time[2].

Preoperative changes in xanthine/hypoxanthine ratio in cirrhotic patients indicates xanthine oxidase level changes. Xanthine oxidase is an intracellular enzyme in human, confined mainly to the liver and the liver is the only source of serum xanthine oxidase[8]. There is no report of an increase in serum xanthine oxidase activity in cirrhotic patients. Also increased xanthine oxidase activity after partial hepatectomy has not been established[7]. Further study is required to investigate the increased xanthine oxidase activity in relation to hepatic surgery.

These changes in blood purine metabolism are important for liver regeneration but also important for whole body metabolism. Since nucleosides and nucleobases are supplied to various tissues by the liver, it is postulated that tissue purine and pyrimidine metabolism changes in response to hepatic purine and pyrimidine utilization. It suggests of the possibility of intravenous nucleosides supplementation. A mixture solution containing nucleosides and a nucleotide has been reported to improve protein metabolism after partial hepatectomy in rats[9] and to enhance hepatic DNA synthesis[10]. Detailed kinetic study is required for the analysis of purine and pyrimidine metabolism in various organs in relation to changes in blood purine and pyrimidine levels in the future.

In summary, blood purine and pyrimidine levels in patients change after hepatectomy. Significant decrease in hypoxanthine and uridine levels, and the increase in the ratio of

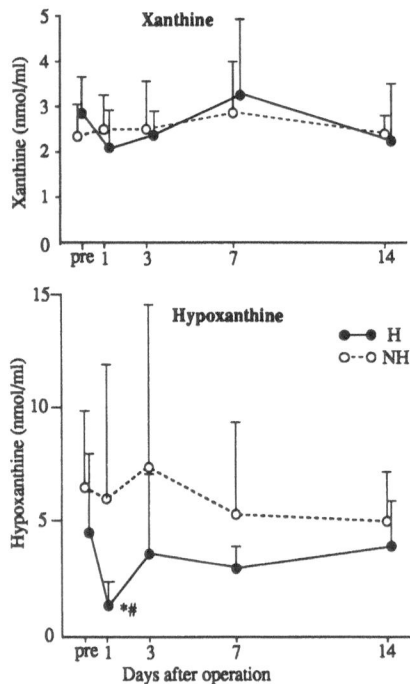

Figure 2. Plasma purine nucleobase level after operation in patients. Mean ± SD, #p<0.05 vs. NH, *p<0.05 vs. pre.

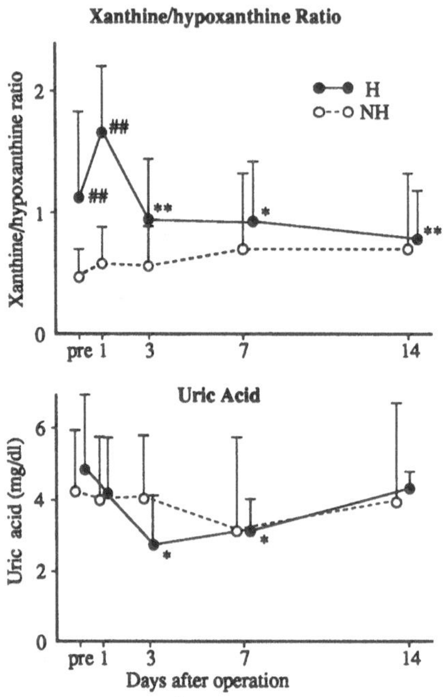

Figure 3. Plasma xanthine/hypoxanthine ratio and uric acid level after operation in patients. Mean ± SD, #p<0.05, ##p<0.01 vs. NH, *p<0.05, **p<0.01 vs. day 1.

xanthine/hypoxanthine indicates increased salvage synthesis of purine and pyrimidine in the regenerating liver. Also, it is suggested that measurement of blood purine level is important in the management of surgical metabolism.

REFERENCES

1. M. Itakura, T. Maeda, M. Tsuchiya, and K. Yamashita, Increased rate of de novo purine synthesis and its mechanism in regenerating liver, *Am J Physiol* . 251:G585 (1986).
2. A.W. Murry, The biological significance of purine salvage, *Ann Rev Biochem* 40: 811 (1971).
3. V. Stocchi, L. Cucchiarine, F. Canestrari, M. Piacentini, and G. Fornani, A very fast ion-pair reversed-phase HPLC method for the separation of the most significant nucleotides and their degradation products in human red blood cells, *Anal Biochem.* 167:181 (1987).
4. T. Yamamoto, Y. Moriwaki, S. Takahashi, T. Hada, and K. Higashino, Separation of hypoxanthine and xanthine from pyrazinamide and its metabolites in plasma and urine by high-performance liquid chromatography, *J Chromatogr.* 382:270 (1986).
5. P. Kabasakalia, S. Kalliney, and A. Westcott, Enzymatic blood glucose determination by colorimetry of N, N-diethylailine-4-aminoantipyrine, *Clin Chem.* 20:606 (1974).
6. M. Usami, K. Furuchi, M. Ogino, J. Kotani, A. Iso, K. Oh , I. Yasuda, E. Sou, T. Kanamaru, H. Kasahara, Y. Saitoh, H. Yokoyama, and T. Yumisashi, Changes in purine and pyrimidine metabolism after partial hepatectomy in rats (in Japanese), *Jap J Surg Metab Nutr.* 27:35 (1993).
7. T. Kurokawa, T. Nonami, K. Kuroe, M. Satake, A. Harada, A. Nakano, and H. Takagi, Nucleotide metabolism in remnant rat liver after major hepatic resection (in Japanese), *J Jap Surg Soc.* 9:994 (1990).
8. S.H. Giler, O. Sperling, S. Brosh, I. Urca, and A. De Vries, Serum xanthine oxidase in jaundice, *Clin Chim Acta.* 63:37 (1975).
9. S. Ogoshi, M. Iwasa, T. Yonezawa, and T. Tamiya, Effect of nucleotide and nucleoside mixture on rats given total parenteral nutrition after 70% hepatectomy, *J Parenter Enteral Nutr.* 9:339 (1985).
10. H. Ohyanagi, S. Nishimatu, Y. Kanbara, M. Usami, and Y. Saitoh, Effect of nucleosides and a nucleotide on DNA and RNA syntheses by the salvage and de novo pathway in primary monolayer cultures of hepatocytes and hepatoma cells, *J Parenter Enteral Nutr.* 13:51 (1989).

ABNORMAL PURINE AND PYRIMIDINE METABOLISM IN INHERITED SUPERACTIVITY OF PRPP SYNTHETASE

Claude Bory,[1] Christiane Chantin,[1] Roselyne Boulieu[2]

[1]Laboratoire d'Etude des Maladies Métaboliques
Hopital Debrousse, 29 rue soeur Bouvier,
69322 Lyon Cedex 05 France
[2]Laboratoire de Pharmacie Clinique
Institut des Sciences Pharmaceutiques et Biologiques
8 Avenue Rockefeller 69373 Lyon France

INTRODUCTION

Superactivity of phosphoribosylpyrophosphate synthetase (PRPPS), the enzyme catalyzing synthesis of the purine regulatory substrate, phosphoribosylpyrophosphate, (PRPP) from adenosine triphosphate (ATP) and ribose 5-phosphate, is an inherited disorder in which excessive enzyme activity is associated with uric acid overproduction and gout[1]. Erythrocytes and cultured fibroblasts from affected individuals show increases in PRPP concentration. The metabolic basis for the increase enzyme activity of PRPPS has been fully studied[2-4] whereas data on alteration in nucleotide and nucleoside profiles are scare[5]. In this study the biochemical abnormalities in a patient with PRPPS superactivity was investigated.

MATERIALS AND METHODS

Case report

The patient was a 6-year-old boy. PRPPS was diagnosed at 1 year of age. The child was hypotonic and presented neurodevelopmental impairement including sensorineural deafness. He was treated by allopurinol.

Assay of nucleotides, nucleosides and bases

Nucleotides, nucleosides and bases were measured by high performance liquid chromatographic methods[6-8] using a Chromatem 800 instrument (Touzart et Matignon Paris France). In addition peak identification and quantification were confirmed by enzymatic peak shift and peak purity was controlled using a photodiode array detector (SPD-M6A Shimadzu Touzart et Matignon Paris France). Nucleotides in erythrocytes were separated on a weak exchanger (Hypersil APS, 3 µm) using a pH and a phosphate buffer gradient elution. Nucleoside and base separations in plasma were performed on a reversed phase (Hypersil

ODS 3 µm) using a methanol gradient in the phosphate buffer elution. Blood samples were immediately centrifuged to prevent metabolic changes[9]. Plasma and erythrocytes were rapidly deproteinized with perchloric acid.

RESULTS AND DISCUSSION

Nucleotide levels in erythrocytes from the patient with PRPPS superactivity and from age-matched children are given in table 1.This table shows, that in the patient, high levels of GTP, UDPG, NAD and particularly IMP were found compared to values obtained from healthy subjects. IMP was the most raised nucleotide level and its concentration was approximatively 20 times higher than the normal range. In the mother red blood cells, only IMP level was elevated.

Table 1. Nucleotide levels in erythrocytes from the patient, his mother and healthy controls.

Subjects	Erythrocyte	Nucleotide	levels	$\mu mol.l^{-1}$		packed	cells	
	ATP	ADP	AMP	GDP	GTP	IMP	UDPG	NAD
patient	1900	151	16	51	180	130	58	164
control children n=5	1750 ± 200	135 ± 15	18 ± 4	30 ± 5	85 ± 20	7 ± 2	25 ± 5	61 ± 13
patient's mother	1207	150	25	38	75	19	28	75
control adults n=25	1525 ± 85	114 ± 13	20 ± 5	30 ± 5	84 ± 18	4.5 ± 1.5	26 ± 5	75 ± 16

Peak identification and quantification were confirmed using 5' nucleotidase for IMP and GTP and using PNP for inosine. The control of peak purity for GTP gave a coefficient of similarity of 0.9999.

The increased levels of NAD and UDPG suggest that purine and pyrimidine metabolic disorder are involved in PRPPS superactivity. An increase in NAD and UDPG concentrations may be related to the elevated PRPP concentration in the erythrocytes[10], however it must be noted that a depletion of NAD has been reported in a patient[5] whereas UDPG level was raised. A depletion of GTP was also found in this patient.

With regard to nucleosides and bases,the results are given in table 2. In plasma and also in urine, levels of hypoxanthine, uric acid and inosine were elevated whereas xanthine level was in the normal range. An increase in xanthine concentration was only observed after allopurinol therapy. Moreover a marked increase of uridine was found in plasma. In the mother only a slight increase of uric acid was found in plasma and urine.

Table 2. Nucleosides and purine bases in plasma and urine from the patient, his mother and heatly controls.

Subjects	Uridine	Inosine	Uric acid		Hypoxanthine		Xanthine	
	plasma $\mu mol.l^{-1}$	plasma $\mu mol.l^{-1}$	plasma $\mu mol.l^{-1}$	urine mmol $.24h^{-1}$	plasma $\mu mol.l^{-1}$	urine μmol $.24h^{-1}$ $.kg^{-1}$	plasma $\mu mol.l^{-1}$	urine μmol $.24h^{-1}$ $.kg^{-1}$
patient	12	2.1	718	2.8	13.3	9.6	1.2	2.5
control children n=5	4 ± 2	0.4 ± 0.1	220 ± 100	1.1 ± 0.9	2.5 ± 1.0	1.3 ± 1.3	1.4 ± 0.7	2.2 ± 1.6
						μmol $.24h^{-1}$		μmol $.24h^{-1}$
patient's mother	4,6	0,3	518	5.5	1.6	80	0.7	64
control adults n=20	4.3 ± 1.6	0.3 ± 0.1	255 ± 105	2.8 ± 1.3	2.5 ± 1.0	48 ± 26	1.4 ± 0.7	68 ± 42

In conclusion, altered concentrations of NAD, UDPG, GTP and IMP in red blood cells associated with a rise of uric acid and hypoxanthine levels in plasma and urine should be considered as biochemical characteristics of PRPP superactivity. Thus, the measurement of these purine and pyrimidine metabolites represents a useful tool for the diagnosis of this inherited purine metabolic disorder.

REFERENCES

1. M.A. Becker, L.J. Meyer and J.E. Seegmiller, Purine overproduction in man associated with increased phosphoribosylpyrophosphate synthetase activity, *Science (Wash. DC)* 179:1123 (1973).
2. M.A. Becker, M.J. Losman, A.L. Rosenberg, I. Mehlam, D.J. Levinson, and E.W. Holmes, Phosphoribosylpyrophosphate synthetase superactivity. A study of five patients with catalytic defects in the enzyme, *Arthritis Rheum.* 7:880 (1986).
3. M.J. Losman, D. Rimon, M. Kim, and M.A. Becker, Selective expression of phosphoribosylpyrophosphate synthetase superactivity in human lymphoblast lines, *J. Clin. Invest.* 76:1657 (1985).
4. M.A. Becker, M.J. Losman, and M. Kim, Mechanisms of accelerated purine nucleotide synthesis in human fibroblasts with superactive phosphoribosylpyrophosphate synthetases, *J. Biol. Chem.* 262:5596 (1987).
5. H.A. Simmonds, D.R. Webster, J. Wilson, S. Lingham, An X-linked syndrome characterised by hyperuricaemia, deafness, and neurodevelopmental abnormalities, *Lancet*, 10 juillet (1982).
6. R. Boulieu, C. Bory, and C. Gonnet, Liquid chromatographic measurement of purine nucleotides in blood cells, *Clin. Chem.* 31:727 (1985)
7. R. Boulieu, C. Bory, and C. Gonnet, High performance liquid chromatograhic method for the analysis of purine and pyrimidine bases, ribonucleosides and deoxyribonucleosides in biological fluids, *J. Chromatogr.* 339:380 (1985).
8. R. Boulieu, C. Bory, P. Baltassat,and C. Gonnet, High performance liquid chromatographic determination of hypoxanthine and xanthine in biological fluids, *J. Chromatogr.* 233:131 (1982).

9. R. Boulieu, C. Bory, P. Baltassat, and C. Gonnet, Hypoxanthine and xanthine levels determined by HPLC in plasma erythrocyte and urine samples from healthy subjects. The problem of hypoxanthine level evolution as a function of time, *Anal. Biochem.* 129:398 (1983).
10. L.D. Fairbanks, H.A. Simmonds, and D.R. Webster, Use of intact erythrocytes in the diagnosis of inherited purine and pyrimidine disorders, *J. Inher. Metab. Dis.* 10:174 (1987).

ACCELERATED PURINE BASE SALVAGE -
A POSSIBLE CAUSE OF ELEVATED NUCLEOTIDE POOL IN THE ERYTHROCYTES OF PATIENTS WITH URAEMIA

Maciej Marlewski, Ryszard T. Smolenski, Julian Swierczynski,
Boleslaw Rutkowski#, John A.Duley*, H.Anne Simmonds*,
Mariusz M.Zydowo

Department of Biochemistry and #Department of Kidney Diseases
Academic Medical School, Gdansk, Poland
*Purine Research Laboratory, U.M.D.S.
Guy`s Hospital, London, U.K.

INTRODUCTION

Erythrocytes from patients with uraemia invariably contain markedly increased ATP concentration[1]. Other effects of uraemia on erythrocyte metabolism are: elevation of GTP, ITP and 2.3-DPG concentration[1,2], increased utilisation of glucose, higher lactate production or decreased Na/K dependent membrane ATP-ase activity[3,4]. There is no effect of haemodialysis on ATP in uraemic erythrocytes. Only renal transplantation or hypophosphatemic drug therapy causes reversal of this abnormality[5,6]. The haematological status of the uraemic patients is always very poor, which is a consequence both of slow regeneration of erythrocytes and the reduction of its half-life. Two possibilities are currently considered to explain the elevated ATP concentration. The first is the regulating effect of inorganic phosphate on nucleotide metabolism, which activates glycolysis and the pathways of purine synthesis and reutilisation and inhibits purine degradation[4,7-9]. The metabolic acidosis accompanying uraemia increases the influx of inorganic phosphate into the erythrocyte as well as is amplifying its effect[10]. The second reason could be the preponderance of younger red blood cells (RBC) in uraemic blood, which are known to contain higher concentrations of ATP[11,12].

To obtain further information concerning the nature of the ATP abnormality in the uraemic erythrocytes, we compared the rate of adenine nucleotide synthesis from adenine and adenosine in uraemic and normal red blood cells. Adenine or adenosine are the only substrates for synthesis of adenine nucleotides in erythrocytes. Furthermore we evaluated the age distribution in the RBC population from normal and uraemic patients, to explain the role of young erythrocytes as a source of higher ATP concentration.

METHODS

Freshly drawn blood from healthy subjects (n=4) and from patients with chronic renal failure (n=4), was added to tubes containing lithium heparin as an anticoagulant. Subsequently, tubes were centrifuged, plasma was removed and RBC were washed 3 times, twice in saline and finally in HEPES buffered Krebs-Ringer solution, containing 125 mM NaCl, 2.7 mM KCl, 1.2 mM $MgCl_2$, 1.2 mM KH_2PO_4, 5 mM glucose and 10mM HEPES adjusted to pH 7.4. An inhibitor of adenosine deaminase - erythro-9(-2-hydroxy-3-nonyl)adenine (EHNA) was added to the last washing medium at 15 µM concentration. Then the isolated RBC were diluted to a 20% haematocrit and preincubated at 37°C for 30 min. After preincubation, [8-^{14}C] adenine or adenosine were added at a final concentration of 10 µM. Incubations were terminated by addition of 1.3 M perchloric acid (PCA), and extracts were neutralised with 3M K_3PO_4. Purine nucleotides, nucleosides and bases were analysed in the erythrocyte extracts, using the method described[13] on an automated Merck-Hitachi HPLC system with a reversed-phase Hypersil 3 µm ODS 150 x 4.6 mm column. In-line radiodetection was conducted in addition to standard UV recording. To establish the age distribution of RBC we used Percoll density gradient centrifugation with the washed red cells from normal (n=4) as well as from uraemic patients (n=5). The discontinuous gradient consisting of 5 steps of increasing density in prepared tubes, was made as described[14]. After adjusting the packed cell volume (PCV) of washed erythrocytes to 50 %, RBC were layered on the top of the Percoll gradient. Centrifugation was carried out in a Sorvall RC-25 at 1000 x g for 10 min at 24°C. After centrifugation the RBC fractions were transferred to tubes, washed and quantified by evaluation of the percentage distribution of PCV.

RESULTS

The rate of adenine incorporation into the adenylate pool (Fig. 1) was more than two times greater in the erythrocytes from patients with uraemia (0.81±0.11 nmol/min/ml RBC), than the rate in healthy subjects (0.35±0.06 nmol/min/ml RBC).The rate of adenosine incorporation (Fig. 2) under normal conditions was similar in uraemic patients (12.1±0.8 nmol/min/ml RBC) and normal subjects (10.1±1.3 nmol/min/ml RBC).

The percentage distribution of erythrocytes age (Fig. 3) on Percoll density gradient showed only a slightly higher proportion of younger erythrocytes from uraemic blood as compared to normal blood.

DISCUSSION

This study demonstrates accelerated incorporation of adenine but not adenosine into the adenylate pool in uraemic erythrocytes. This higher flux through adenine phosphoribosyltransferase (APRT) could explain the elevation of adenine nucleotides in uraemic erythrocytes as has been suggested previously[11]. However, in contrast to the other report, we did not observe a shift in the age of erythrocytes sufficient to cause such a significant elevation in ATP. Moreover ATP content, analysed in different age fractions of uraemic erythrocytes was still markedly elevated, even in the oldest fraction. Thus it seems to be more likely that metabolic effects play a more significant role in acceleration of adenine reutilisation and consequent elevation in ATP.

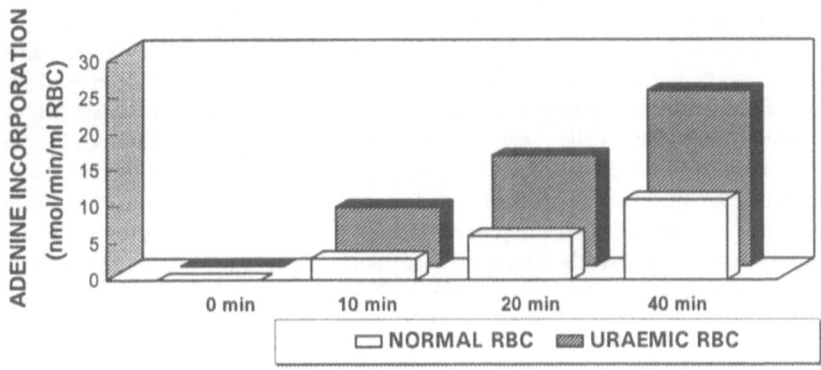

Figure 1. Incorporation of adenine into nucleotide pool of erythrocytes of patients with uraemia and normal subjects. Values represent mean of 4 experiments.

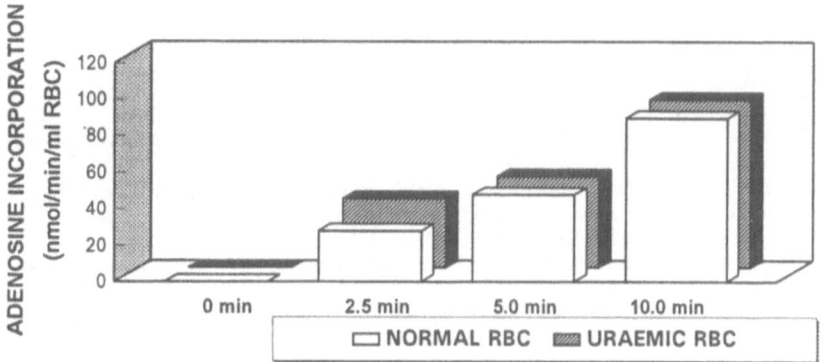

Figure 2. Incorporation of adenosine into nucleotide pool of erythrocytes of patients with uraemia and normal subjects. Values represent mean of 4 experiments.

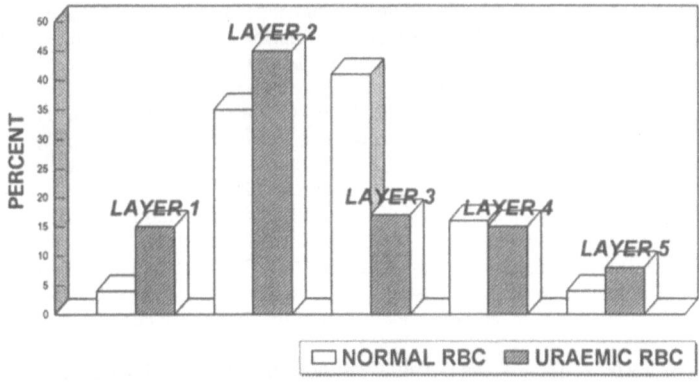

Figure 3. Age distribution of RBC of patients with uraemia and healthy subjects obtained using Percoll density gradient centrifugation. Layer nr 1 with lowest density contain the youngest RBC fraction. Values represent mean of 4-5 experiments.

21

The regulating role of pH and inorganic phosphate on purine metabolism is well known and could play a key role in the turnover of adenine moiety. This is of special significance since hyperphosphatemia and the metabolic acidosis is invariably observed in uraemia. Elevated plasma phosphate, especially in combination with acidosis, which shifts phosphate into the cells can raise intracellular phosphate and stimulate phosphoribosylpyrophosphate (PRPP) synthesis[15]. This in turn can accelerate adenine reutilisation. Our previous data documented accelerated adenylate catabolism in uraemic erythrocytes[16]. Combined with the present results, it suggests that erythrocyte turnover of adenine nucleotides is greatly accelerated in renal failure with excess of synthesis over breakdown, giving steady state elevation in ATP.

ACKNOWLEDGEMENTS

This study was supported by the Polish Committee for Scientific Research, Special Trustees of Guy's Hospital and Collaborative Research Grant from NATO (CRG 921365).

REFERENCES

1. G.A. Hurt and A. Chanutin, Organic phosphate compounds of erythrocytes from individuals with uremia, *J. Lab. Clin. Med.* 64:675(1964).
2. M.A. Mansell, J. Allsop, M.E. North, R.J. Simmonds, R.A. Harkness, and W.E. Watts, Effect of renal failure on erythrocyte purine nucleotide, nucleoside and base concentrations and some related enzyme activities, *Clin. Sci.* 61:757(1981).
3. H.J. Kramer, D. Gospodinov, and F. Kruck, Inhibition of transport ATP-ase by uraemic toxins, *Proc. Eur. Dial. Transpl. Assoc.* 9:521(1972).
4. M.A. Lichtman and D.R. Miller, Erythrocyte glycolysis, 2,3-diphosphoglycerate and adenosine triphosphate concentration in uremic subjects: Relationship to extracellular phosphate concentration, *J. Lab. Clin. Med.* 76:267(1970).
5. A.S.M. Rejman, M.A. Mansell, A.J. Grimes, and A.M. Joekes, Rapid correction of red-cell nucleotide abnormalities following successful renal transplantation, *Brit. J. Haem.* 61:433(1985).
6. M.A. Lichtman, D.R. Miller, and R.B. Freeman, Erythrocyte adenosine triphosphate depletion during hypophosphatemia in uremic subject, *New Eng. J. Med.* 280:240(1969).
7. M.A. Johnson, K. Tekkanat, S.P. Schmaltz, and I.H. Fox, Adenosine triphosphate turnover in humans. Decreased degradation during relative hyperphosphatemia, *J. Clin. Invest.* 84:990(1989).
8. M.A. Lichtman and M.S. Murphy, Red cell adenosine triphosphate in hypoproliferative anemia with and without chronic renal disease: relationship to hemoglobin deficit and plasma inorganic phosphate, in: "Blood Cells 1", Anonymous., ed., Springer-Verlag, Berlin, pp. 467-484 (1975).
9. E.M. Warrendorf and D. Rubinstein, The elevation of adenosine triphosphate levels in human erythrocytes, *Blood* 42:637(1973).
10. P.A. Berman, D.A. Black, L. Human, and E.H. Harley, Oxipurine cycle in human erythrocytes regulated by pH, inorganic phosphate and oxygen, *J. Clin. Invest.* 82:980(1988).
11. H.J. Becher, H.J. Weise, U. Volkermann, and P. Schollmeyer, Enhanced purine nucleotide synthesis in erythrocytes of uremic patients, *Klin. Wochenschr.* 58:1243(1980).
12. A.M. Shojania, L.G. Israels, and A. Zipursky, The relationship of adenosine triphosphate concentration to erythrocyte aging, *J. Lab. Clin. Med.* 71:41(1968).
13. R.T. Smolenski, D.R. Lachno, S.J.M. Ledingham, and M.H. Yacoub, Determination of sixteen nucleotides, nucleosides and bases using high-performance liquid chromatography and its application to the study of purine metabolism in hearts for transplantation, *J. Chromatogr.* 527:414(1990).
14. V. Micheli, C. Ricci, A. Taddeo, and R. Gili, Centrifugal fractionation of human erythrocytes according to age: comparison between ficoll and percoll density gradients, *Quad. Sclavo Diagn.* 21:236(1985).
15. P.A. Berman and L. Human, Regulation of 5-phosphorybosyl 1-pyrophosphate and of hypoxanthine uptake and release in human erythrocytes by oxypurine cycling, *J. Biol. Chem* 265:6562(1990).
16. M. Marlewski, R.T. Smolenski, J. Swierczynski, B. Rutkowski, and M.M. Zydowo, Adenylate catabolism in erythrocytes of uraemic patients, *Adv. Exp. Med. Biol.* 309B:349(1991).

QUANTITATION OF TK1 mRNA IN PATIENTS WITH CHRONIC LYMPHATIC LEUKEMIA

Tina Kristensen, Helle Kock Jensen and Birgitte Munch-Petersen

Department of Life Sciences and Chemistry, Roskilde University,
DK-4000 Roskilde, Denmark

INTRODUCTION

Thymidine kinase is an enzyme in the pyrimidine salvage pathway that, with ATP as co-substrate, catalyzes the phosphorylation of deoxythymidine to deoxythymidine monophosphate (dTMP) which is subsequently converted to dTTP and utilized for DNA synthesis.

In mammalian cells there are two thymidine kinases (TK), the constitutively expressed TK2, and the S-phase specific TK1 which is only present in dividing cells[1].

Lymphocytes from patients with chronic lymphatic leukemia (CLL) are non-dividing and it is therefore plausible that the low TK activity in these cells almost exclusively is due to TK2. However, a thymidine kinase with similar enzyme kinetic pattern as that observed with TK1 from lymphocytes stimulated to growth by the mitogene phytohemagglutinin has been reported[2]. Since TK1 expression is tightly regulated thoughout the cell cycle with transcriptional, translational as well as post-translational regulatory mechanism[3], the occurrence of TK1 in non-dividing CLL cells may be due to a change in the control of the cell cycle regulated expression of the TK1 gene.

To investigate the transcriptional expression of TK1 mRNA in CLL cells, we have measured the level of TK1 mRNA with the competitive polymerase chain reaction (competitive PCR), and compared this mRNA level with the TK enzyme activity. Surprisingly, we have found that the ratio of TK1 mRNA/TK activity in lymphocytes from CLL patients was about 60-400 fold higher than in lymphocytes from healthy persons.

METHODS

Lymphocytes from peripheral blood from 6 healthy persons and from 5 patients with untreated CLL were isolated by the Ficoll-Isopaque technique.
Lymphocytes from healthy persons were stimulated to growth by PHA in RPMI 1640 medium supplemented with 10% fetal calf serum, 20 μg/ml PHA and 20 μg/ml penicillin/streptomycin at a concentration of 10^6 cells pr ml in 5% CO_2 at 37°C. The lymphocytes were divided in portions of 5 x 10^6 cells, and in each portion, TK activity was determined by the DE-81 paper method as described[4] and total protein was determined by the Bradford assay[5]. Total RNA was isolated with the guanidine thiocyanate method[6], transcribed to cDNA and quantitated by the competitive PCR method[7]. TK1 cDNA, taken as representative for TK1 mRNA, was co-amplified with a dilution series of competitor DNA. Exon 1 and 2 with intron from the TK1 gene served as competitor DNA and exon 1 and 2 of the TK1 gene as the cDNA fragment to be quantitated. The fragments were amplified using a pair of primers identical to those reported by Lipson and Baserga[8]. The sizes of the resulting fragments was 138 bp with cDNA as template and 248 bp with competitor DNA as template. The relative amounts of cDNA versus competitor DNA were measured by scanning of ethidium-bromide stained gels. Because the starting concentration of the competitor DNA was known, the amount of cDNA (in grams) in the sample could be estimated as that amount of competitor DNA where equal intensities of the two amplification products were obtained. The number of TK1 cDNA copies was calculated from the amount of cDNA, by dividing the amount of cDNA with the molecular weight of 1 copy of the 138 bp cDNA fragment.

The amplification was performed in a Perkin-Elmer/Cetus Thermal Cycler according to the following program: denaturation for 1 min at 95°C, annealing for 1 min at 60°C and polymerization for 1 min at 72°C, for 35 cycles.

RESULTS AND DISCUSSION

Table 1 shows the ratio of TK1 mRNA copies and TK activity in non-dividing lymphocytes from 6 healthy persons and in lymphocytes from 5 patients with CLL. As seen, the ratio TK1 mRNA copies/TK activity in CLL cells is 60 to 400 fold higher than in non-dividing lymphocytes. The TK activity in CLL cells is of a magnitude as expected for non-dividing cells, while the expression of TK1 mRNA is very high and in the range of the TK1 mRNA level in PHA stimulated healthy donor lymphocytes. In these experiments the TK1 mRNA level is 3-98 x 10^6 copies/mg protein (results are not shown).

The detection limit in the assay is around 6 x 10^4 copies of TK1 mRNA/mg protein or 0.006 copies/cell. Below this level, a 248 bp amplification product, interferes with the competitive PCR. This is probably a result of traces of DNA or non-spliced RNA in the RNA preparation. The results indicate that there, as expected, is no TK1 mRNA in non-dividing lymphocytes from healthy persons.

Due to the high TK1 mRNA level in non-dividing CLL cells it was of importance to clarify whether the dominating TK in CLL cells was TK1 or TK2, using the characteristic differences in phosphate donor specificity towards ATP and CTP. Both enzymes can utilize ATP, but only TK2 is capable of utilizing CTP[9]. The relative TK activity with CTP as phosphate donor was expressed as % of activity with ATP as phosphate donor. PHA-stimulated lymphocytes showed a 85-90% decrease in relative activity, while non-dividing lymphocytes from healthy persons and lymphocytes from CLL patients showed a 7-30% decrease. The conclusion is that the enzyme in CLL cells is the same as in non-dividing lymphocytes from healthy persons, namely TK2.

Table 1. Ratio of TK1 mRNA copies and TK activity.

	TK1 mRNA copies x 10^6/mg protein	TK activity Units/mg protein	TK1 mRNA copies x 10^6/TK activity
Non-dividing lymphocytes			
1	< 0.06	0.009	< 6.7
2	< 0.06	0.013	< 4.6
3	0.21	0.013	16.2
4	0.06	0.008	7.5
5	< 0.06	0.009	< 6.7
6	< 0.06	0.016	< 3.8
Lymphocytes from CLL patients			
1	10.3	0.008	1287
2	7.4	0.006	1233
3	22.7	0.013	1746
4	15.2	0.005	3040
5	6.1	0.006	1016

The ratio between TK1 mRNA and TK activity as estimated in non-dividing lymphocytes from 6 donors and in lymphocytes from 5 patients with CLL. The numbers refer to the individual donors and patients.
1 unit is the amount of enzyme that phosphorylate 1 nmol substrate per minute.

The occurence of a high level of TK1 mRNA without concomittant expression of TK1 enzyme activity may indicate that CLL cells have an abnormal regulation of the cell-cyclus regulated TK1. The regulations mechanism are not fully understood, but several investigations have shown that the changes in TK1 mRNA during cell cycle can not fully account for the rise in TK activity. Translational and post-translational modifications may contribute to the regulation of TK1. Chang and Huang[10] have demonstrated that seryl residues of the TK1 polypeptide are phosphorylated in cycling HL-60 cells. An increasing phosphorylation of the polypeptide was followed by an increase in enzyme activity, during the cell cycle. Another post-translational mechanism has been reported by Kauffman and Kelly[11]. They have shown that amino acid residues near the C-terminal end are responsible for degradation of thymidine kinase protein in the G_2 and M phase, and that mutations in this part of the gene allow expression in G_0 cells.

It is possible that a post-translational mechanism serve as a secondary back-up system for the regulation of TK. This may explain why we can measure a high TK1 mRNA level but no TK1 activity.

REFERENCES

1. D.L. Coppock and A.B. Pardee, Control of thymidine kinase mRNA during cell cycle. Molecular and cellular biology. 7:2925 (1987).
2. B. Munch-Petersen and G. Tyrsted, Thymidine kinase isoenzymes in human acute and chronic lymphatic leukemia. Leukemia Res. 10:637 (1986).
3. J.L. Sherley and T.J. Kelly, Regulation of human thymidine kinase during the cell cycle. J. Biol. Chem. 263:375 (1988b).
4. B. Munch-Petersen, Differences in the kinetic properties of thymidine kinase isoenzymes in unstimulated and phytohemagglutinin-stimulated human lymphocytes. Mol. Cell. Biochem. 64:173 (1984).
5. M.M. Bradford, A rapid and sensitive method for the quantitation of microgram quanties of protein utilizing the principle of protein-dye binding. Anal.Biochem. 72:248 (1976).
6. P. Chomczynski and N. Sacchi, Single-step method of RNA isolation by acid guanidinium thiocyanate-phenol-chloroform extraction. Anal. Biochem. 162:156 (1987).
7. G. Gilliand, S. Perrin and F. Bunn, in: PCR Protocols. A guide to methods and applications, pp 60-69. Ed. by Innis M. A, Gelfand D. H., Sninsky J. J & White T. J. Academic Press, INC, San Diego ET (1990).
8. K.E. Lipson and R. Baserga, Transcriptional activity of the human thymidine kinase gene determined by a method using the polymerase chain reaction and a intron-specific probe. Proc. natl. Acad. Sci. USA 86:9774, (1989)
9. P.H. Ellims, M.B. Van der Weyden, G. Medley, Thymidine kinase isoenzymes in human malignant lymphoma. Cancer Res. 41:691 (1981)
10. Z-F. Chang and D-Y. Huang, The regulation of thymidine kinase in HL-60 human promyelo leukemia cells. J. Biol. Chem. 268:1266 (1993)
11. M.G. Kauffman and T.J. Kelly, Cell cycle regulation of thymidine kinase: residues near the carboxyl terminus are essential for the specific degradation of the enzyme at mitosis. Mol. Cell. Biol. 11:2538 (1991)

EFFECT OF HYPOURICEMIC AGENTS ON SERUM CARBOHYDRATE-DEFICIENT TRANSFERRIN IN GOUTY PATIENTS

Kiyoko Kaneko[1], Shin Fujimori[2], Hisashi Yamanaka[3], and Ieo Akaoka[2]

[1]Central Laboratory of Analytical Biochemistry and [2]Department of Internal Medicine, Teikyo University School of Medicine,
[3] Institute of Rheumatology, Tokyo Women's Medical College,
Tokyo, Japan

INTRODUCTION

Carbohydrate-deficient transferrin (CDT) was demonstrated to be a reliable biochemical marker of alcohol abuse by Stibler et al [1,2]. CDT is currently regarded as the most accurate biological indicator of recent, heavy alcohol consumption. Alcohol consumption has been reported to be closely related to hyperuricemia and gout[3]. The measurement of CDT in gouty patients is considered to be helpful on their management of hyperuricemia.

In this study, we examined the effect of administered hypouricemic agents on serum level of CDT in gouty patients. As we have previously reported that the metabolism of allopurinol was influenced after acute ethanol intake in normal subjects[4], the plasma level of oxipurinol with long-term administration of allopurinol was also determined and compared with the amount of alcohol consumption.

MATERIALS AND METHODS

Subjects

Seventy three male outpatients with gout, aged 35-88 years, were examined. Information on their alcohol consumption was obtained from personal questionnaires. According to their alcohol consumption, these patients were divided into heavy drinking patients with alcohol intake $\geq$ 80g/day, social drinking patients with alcohol intake 40-80g/day, and abstinent patients with alcohol intake < 40g/day. Allopurinol 100 mg was used by 14, allopurinol 200 mg by 26, probenecid or benzbromarone by 11, these combination therapy by 16, and no medication by 6. Renal function of these patients was in the normal range (serum creatinine levels < 1.3 mg/dl).

Determination of CDT in Serum

Isoelectric Focusing Electrophoresis followed by Direct Immunofixation (IEF/DI). Serum (10 μl) was iron-saturated with 40 μl of 1 mM Fe^{3+} citrate. According to the method described in the previous literature[5], analytical IEF was performed twice in polyacrylamide gel with a pH gradient of 3-9. Direct immunofixation (DI) was carried out with anti-human transferrin antibodies. The gel was washed, stained with Coomassie brilliant blue, and analyzed using transmission at 590 nm.

Micro Anion Exchange Chromatography followed by Radioimmunoassay (MAEC/RIA). According to the method of Stibler H[6], iron-saturated serum was passed through microcolumns with DEAE-Sephacel. The effluent was analyzed in duplicate using a human transferrin RIA kit.

Determination of Oxipurinol

Plasma was deproteinized by passage through ultrafiltration tubes. Oxipurinol levels in the filtrate were measured using HPLC with a prepacked ODS column [4].

Statistical Procedure

Statistical analysis was carried out using the two-tailed Student's t-test and Welch's procedure when variance was not equal.

RESULTS

Comparison between IEF/DI and MAEC/RIA

Separation and determination of CDT was carried out by IEF/DI and by MAEC/RIA. CDT/total transferrin determined by IEF/DI in gouty patients ranged from 0.47% to 4.71% (n=73). CDT contents determined by MAEC/RIA ranged from 7.3 μg/ml to 72.3 μg/ml (n=59). As shown in Figure 1, significant correlation was found between the value obtained by IEF/DI and that obtained by MAEC/RIA (correlation coefficient : r = 0.66, P<0.01).

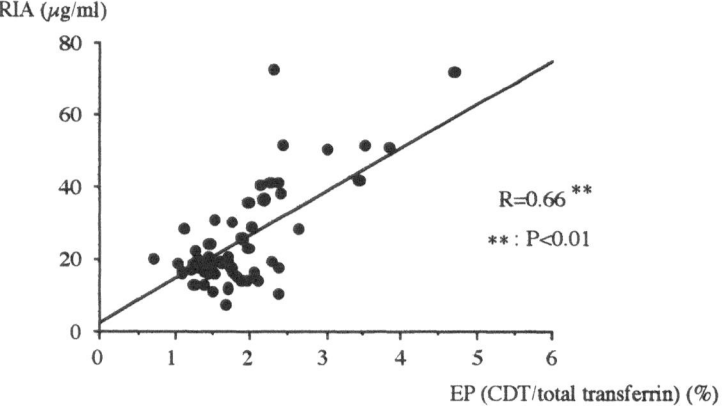

Figure 1 Comparison between electrophoresis and radioimmunoassay

Serum Level of CDT in Gouty Patients

CDT/total transferrin in heavy drinking patients (alcohol intake ≥ 80g/day), social drinking patients (alcohol intake 40-80g/day), and abstinent patients (alcohol intake < 40g/day) was $2.7 \pm 0.2\%$ (n=17), $1.8 \pm 0.1\%$ (n=23), and $1.5 \pm 0.1\%$ (n=35), respectively. In heavy drinking patients, serum CDT was significantly higher than those in abstinent patients and in social drinking patients (both : p<0.001). Cutoff point was calculated 2.08%, using the mean+2SD of abstinent patients [5].

Most of these gouty patients were on long-term therapy with hypouricemic agents. With any hypouricemic agents, average CDT/total transferrin in heavy drinking patients was higher than that in abstinent patients. In the patients taking allopurinol 200 mg, CDT/total transferrin in heavy drinking, social drinking, and abstinent patients were 2.5 ± 0.1% (n=6), 1.7 ± 0.2% (n=10), and 1.4 ± 0.1% (n=10), respectively (Table 1). CDT in heavy drinking patients with allopurinol 200 mg was significantly elevated (P<0.001) compared with abstinent patients.

Oxipurinol Level in Gouty Patients with Administration of Allopurinol

Oxipurinol is reported to be cleared slowly via the kidneys with a serum half-life of 14 to 26 hours in subjects with normal renal function [8]. Oxipurinol levels in patients receiving allopurinol were shown in Table 1. In heavy drinking patients taking allopurinol 200 mg, oxipurinol level was 43.4 ± 2.7 μM (n=6) which was significantly decreased compared with the level of 62.9 ± 6.4 μM (n=10) in abstinent patients (P<0.05).

Table 1. Serum level of CDT and plasma level of oxipurinol

	allopurinol 100mg			allopurinol 200mg		
	heavy drinking (n=2)	social drinking (n=4)	abstinent (n=8)	heavy drinking (n=6)	social drinking (n=10)	abstinent (n=10)
Carbohydrate-Deficient Transferrin (CDT) (%)						
	3.4±1.4	1.8±0.2	1.5±0.1	2.5±0.1[***]	1.7±0.2	1.4±0.1
Oxipurinol (μM)						
	24.3±1.7	30.1±7.9	36.6±4.6	43.4±2.7[*]	51.1±4.2	62.9±6.4

* : significantly decreased compared with abstinent patients with allopurinol 200mg (P < 0.05)
*** : significantly elevated compared with abstinent patients with allopurinol 200mg (P < 0.001)

Serum Levels of CDT, GGT, AST and ALT in Gouty Patients

Table 2 shows CDT, GGT, AST and ALT in gouty patients. In patients with administration of allopurinol, only CDT was significantly elevated in heavy drinking patients. In patients with other hypouricemic agents, CDT, GGT, AST and ALT were all elevated in heavy drinking patients compared with abstinent patients (CDT and GGT : P<0.001, AST : P<0.05, ALT : P<0.01).

Table 2. Serum levels of CDT, GGT, AST and ALT

	allopurinol only			other hypouricemic agents		
	heavy drinking (n=7)	social drinking (n=15)	abstinent (n=18)	heavy drinking (n=9)	social drinking (n=8)	abstinent (n=16)
CDT (%)						
	2.7±0.4[*§]	1.8±0.1[*]	1.4±0.1	2.7 ±0.2[***§]	1.9±0.3	1.5±0.1
GGT (γ-GTP) (IU/L)						
	76.9±10.9	59.8±8.9	40.1±10.9	107.6±13.2[***]	103.4±30.0	38.4±9.2
AST (GOT) (IU/L)						
	24.9±2.6	24.9±1.7	22.5±1.1	33.4±4.3[*]	25.0±3.9	22.2±1.6
ALT (GPT) (IU/L)						
	26.9±3.7	24.8±2.7	22.9±1.9	36.8±5.0[**]	27.5±4.7	22.1±2.5

* P< 0.05, ** P< 0.01, *** P< 0.001 vs. abstinent patients with same hypouricemic agents
§ P< 0.05 vs. social drinking patients with same hypouricemic agents

DISCUSSION

The elevation of serum CDT in heavy drinking subjects was previously described in alcoholics [1,5,6], in patients with alcoholic or non-alcoholic liver disease [9,10], and in different ethnic populations [2]. Increment of CDT levels was also demonstrated in gouty patients with recent heavy drinking [7].

Since most of gouty patients are on long-term therapy with hypouricemic agents, the effect of these medicines on serum CDT level was examined in this study. Average of serum CDT were all elevated in heavy drinking patients (ethanol intake $\geq$ 80g /day) with allopurinol 100mg, allopurinol 200mg, benzbromarone 50mg or these combination. It was considered that internal used medicines did not change serum CDT level.

We have demonstrated in our previous study that the plasma level of oxipurinol was reduced after acute ethanol intake in normal subjects [4]. It has been reported that the average alcohol intake in gouty patients is higher than that of controls [11]. In clinical, there should be many gouty patients who are receiving chronic allopurinol therapy and with heavy drinking. The plasma level of oxipurinol was significantly decreased in heavy drinking patients compared with abstinent patients. Decrease of plasma oxipurinol concentration would affect the long-term pharmacological effects of allopurinol. In patients taking allopurinol, it is considered that the amount of ethanol intake should be checked carefully in order to evaluate the pharmacological effect of allopurinol.

In conclusion, internal used hypouricemic agents do not change CDT level in gouty patients and the measurement of CDT would be helpful in the treatment of gout, especially with administration of allopurinol.

REFERENCES

1. H. Stibler, Carbohydrate-deficient transferrin in serum: a new marker of potentially harmful alcohol consumption reviewed, Clin Chem 37: 2029 (1991)
2. U.J. Behrens, T.M. Worner, L.F. Braly, et al, Carbohydrate-deficient transferrin, a marker for chronic alcohol consumption in different ethnic populations, Alcoh Clin Exp Res 12: 427 (1988)
3. T. Gibson, A.V. Rodgers, H.A. Simmonds, et al, A controlled study of diet in patients with gout, Ann Rheum Dis 42:123 (1983)
4. K, Kaneko, S. Fujimori, I. Ishizuka, et al, Effect of ethanol on metabolism of the hypouricemic agents allopurinol and benzbromarone, Clin Chim Acta 193: 181 (1990)
5. H. Stibler and O. Sydow, Quantitative estimation of abnormal microheterogeneity of serum transferrin in alcoholics, Pharmacol Biochem Behav 13: 47 (1980)
6. H. Stibler, S. Borg, aand M. Joustra, Micro anion exchange chromatography of carbohydrate-deficient transferrin in serum in relation to alcohol consumption, Alcoh Clin Exp Res 10: 535 (1986)
7. K, Kaneko, S. Fujimori, H. Yamanaka, et al, Serum level of carbohydrate-deficient transferrin in gouty patients, (in press)
8. G.B. Elion, T.F. Yu, A.B. Gutman, et al, Renal clearance of oxipurinol, the chief metabolite of allopurinol, Am J Med 45: 69 (1968)
9. L.M. Fletcher, I. Kwoh-Gain, E.E. Powell, et al, Markers of chronic alcohol ingestion in patients with nonalcoholic steatohepatitis: an aid to diagnosis, Hepatol 13: 455 (1991)
10. Y. Xin, A.S. Rosman, J.M. Lasker, et al, Measurement of carbohydrate- deficient transferrin by isoelectric focusing/western blotting and by micro anion exchange chromatography/radioimmunoassay: comparison of diagnostic accuracy, Alcoh Alcoh 27: 425 (1992)
11. C.R. Sharpe, A case-control study of alcohol consumption and drinking behavior in patients with acute gout, Can Med Assoc J 131: 56 (1984)

RENAL HEMODYNAMIC IN FAMILIAL NEPHROPATHY ASSOCIATED WITH HYPERURICEMIA (FNAH)

Felícitas A. Mateos and Juan G. Puig, on behalf of the Spanish Group for the Study of FNAH

Divisions of Clinical Biochemistry, Internal Medicine, Nephology, Pediatrics, Radiology and Pathology, "La Paz" Hospital, Universidad Autónoma, Madrid, Spain

INTRODUCTION

Renal vasoconstriction is accompanied by markedly reduced urinary uric acid excretion and hyperuricemia. FNAH (McKusik 16200) is an autosomal dominant disease characterized by severe uric acid underexcretion, usually preceding renal insufficiency (1). Because the pathogenetic role of increased serum urate concentration is controversial, we examined renal hemodynamics in patients with FNAH to assess whether renal vasoconstriction mediates the disturbance of uric acid metabolism.

PATIENTS AND METHODS

Five Spanish families with FNAH were included in this study (Figure 1). Family members were defined as affected if they had any of the following conditions: hyperuricemia, gout, arterial hypertension, renal insufficiency and small kidney size. Hyperuricemia was defined as a serum urate concentration above 7.0 mg/dL (417 µmol/L) in men and above 6.5 mg/dL (387 µmol/L) in women on a self-selected diet. Gout was diagnosed according to the American Rheumatism Association criteria. Renal insufficiency was defined as a serum creatinine concentration above 133 µmol/L. Kidney size was measured by means of intravenous urography and considered to be diminished if the renal index was below the mean minus two SD with respect to previously reported normal values for age and sex (2). Patients were admitted to "La Paz" University Hospital. The studies were performed according to a previously described protocol (3). The glomerular filtration rate (GFR) and the effective renal plasma flow (RPF) were calculated from the rates of clearance of inulin and p-aminohippuric acid, respectively. Filtration fraction (FF) was calculated by dividing the GFR by the RPF. Renal blood flow (RBF) was estimated by dividing the effective renal plasma flow by one minus the hematocrit. Renal vascular resistance (RVR) was estimated by dividing the calulated mean arterial pressure by the RBF. Uric acid and creatinine in plasma and urine were measured in a multichannel autoanalyzer (Hitachi 737, Boehringer Mannheim, Germany). Relation among selected variables was assessed by the Pearson's correlation coefficient. The BMDP program was used to analyze the data. A P value of 0.05 or less was considered to indicate statistical significance.

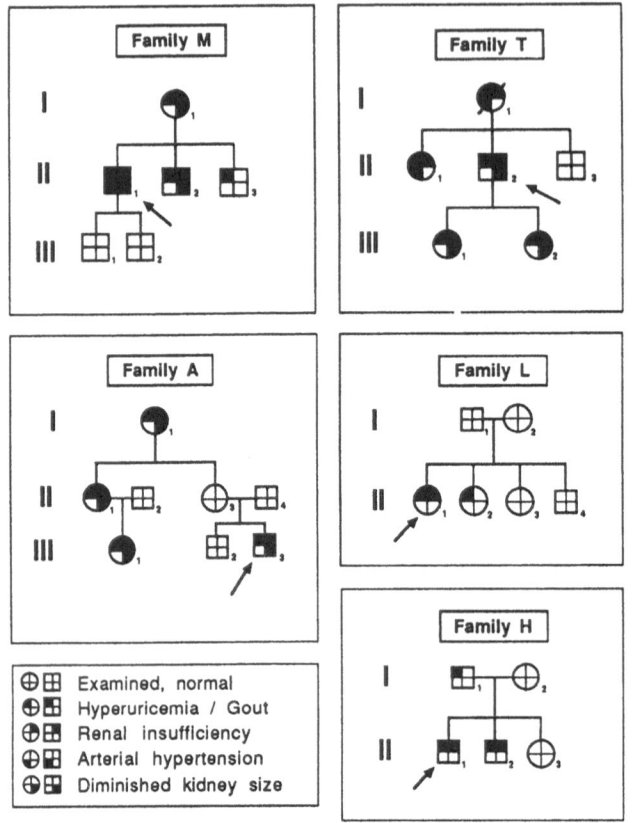

Figure 1. Pedigree of five families with hereditary nephropathy associated with hyperuricemia or gout (FNAH). Roman numerals denote generations; squares denote men and circles women; the slash denotes a dead subject; symbols with quadrants denote persons examined; shaded squares denote pathological features at initial evaluation.

RESULTS

Eighteen patients belonging to five different families were diagnosed as having FNAH (Figure 1). The median age of the patients was 28 years (range, 12 to 87 years). At initial evaluation and among these 18 patients, 15 (83%) showed a serum creatinine concentration equal to or above 115 μmol/L (1.3 mg/dl). The clinical characteristics and renal hemodynamic parameters in 18 subjects belonging to these 5 families in whom purine metabolism under a purine restricted diet and renal hemodynamic could be investigated are shown in Table 1. In 5 subjects [II-3(M), II-3(A), I-1(L), II-3(L) and II-4(L)] serum creatinine and urate concentrations were normal, as was blood pressure and kidney size. Thus, on clincal grounds, these subjects were not affected. In addition, they showed normal or borderline creatinine and inulin clearances. However, patient II-3(M) had gout and among the remaining 4 subjects two were parents of affected siblings [II-3(A) and I-1(L)]. In these two patients the only abnormal findings were a reduced RPF and increased RVR (Table 1). Among the 18 studied subjects RVR was found to be increased in 16 (mean, 25.2; range, 9.3 to 62.6; normal 7.6±0.6 mm Hg/dL/min/1.73 m2), being within normal limits in II-3(L) and II-4(L). RPF was significantly related to serum urate levels (r=-0.523; p=0.038) and to the fractional excretion of uric acid (r=-0.531; p=0.034). Moreover, RVR was directly related to serum urate concentrations (r=0.506; p=0.046). Subjects II-3(A) and I-1(L) showed hyperuricemia and decreased 24-h uric acid excretion after 4 and 21 months of follow-up, respectively.

Table 1. Clinical characteristics and renal hemodynamic in 18 subjects belonging to 5 Spanish families with familial nephropathy associated with hyperuricemia.

Patient (Family)	Age (yrs)	Sex	Serum[1] creatinine (mg/dL)	Serum urate (mg/dL)	Fractional excretion of urate (%)	Gout	High blood pressure	Reduced kidney size	Inulin clerance (mL/min/1.73m²)	Renal plasma flow (mL/min/1.73m²)	Renal vascular resistance (mmHg/mL/min/1.73m²)
I-1 (M)	59	F	1.9	6.5	6.5	-	-	+	41	182	37.5
II-1 (M)	34	M	2.4	7.1	5.6	+	+	+	40	178	48.4
II-2 (M)	33	M	2.4	6.8	5.1	+	+	+	43	251	28.7
II-3 (M)	28	M	1.2	6.3	4.2	+	-	-	104	529	9.3
II-2 (T)	46	M	2.9	9.8	3.2	+	+	+	32	142	62.6
III-1 (T)	15	F	2.1	8.0	5.2	-	-	+	57	186	24.9
III-2 (T)	12	F	2.0	9.0	3.5	-	-	+	82	407	22.5
II-1 (A)	53	F	2.7	6.3	6.5	+	-	+	28	278	30.2
II-3 (A)	52	F	1.0	4.6	6.1	-	-	ND	76	430	19.4
III-1 (A)	22	F	2.1	6.3	5.0	-	-	+	29	374	21.3
III-3 (A)	20	M	1.5	7.1	2.2	+	-	+	56	478	21.6
I-1 (L)	47	M	0.9	4.9	3.6	-	-	-	79	386	18.5
II-1 (L)	17	F	1.5	7.8	2.6	-	-	+	56	200	17.7
II-2 (L)	15	F	1.1	7.8	1.8	-	-	-	51	474	17.3
II-3 (L)	14	F	0.8	5.0	6.8	-	-	-	113	593	7.5
II-4 (L)	13	M	0.8	4.9	5.9	-	-	-	98	540	7.7
II-1 (H)	17	M	1.3	7.3	2.1	-	-	+	79	493	11.6
II-2 (H)	15	M	1.3	7.1	2.2	-	-	+	74	396	12.4

[1]Normal upper limits for these subjects age are: serum creatinine, 1.2 mg/dL; serum urate, 7.0 (males, M) and 6.5 mg/dL (females, F); Fractional excretion of urate (males), 5.0 and (females), 7.0 mg/dL; Inulin clearance, 80 mL/min/1.73 m²; renal vascular resistance, 8.8 mm Hg/mL/min/1.73 m². The normal lower limit for renal plasma flow is 500 mL/min/1.73 m². ND, not determined; a positive sign (+) denotes present and a negative sign (-) denotes absent.

DISCUSSION

This study shows that a reduced RPF and renal vasoconstriction appeared to be the primary pathogenic events in FNAH. The disturbance of renal hemodinamic may explain uric acid underexcretion and increased serum urate concentrations in this hereditary nephropathy. An increased serum urate concentration appears not to be the primary event since some obligate affected subjects did not show hyperuricemia under a purine-restricted diet. Uric acid underexcretion could not be entirely attributed to renal insufficiency since it was more pronounced than in patients with chronic renal failure and similarly reduced GFR (4,5). The fractional excretion of uric acid appeared to be more severely diminished in patients with FNAH than in gouty patients with normal GFR (6) and in patients with gout and renal failure due to hypertension and atherosclerosis (3). Thus a severe uric acid underexcretion is a prominent characteristic of FNAH. The finding of a markedly increased renal vascular resistance in our patients suggests that a severe disruption of renal hemodynamics may be of primary pathogenetic importance. Serum urate was directly related to RVR and inversely related to RPF. In patients with essential hypertension decreased renal blood flow usually precedes a decreased GFR and the increase in renal vascular resistance is directly related to the increased serum urate levels commonly observed in these patients (7). Renal blood flow is a major determinant of uric acid excretion since in normotensive subjects renal plasma flow and urate excretion decrease during infusion of pressor doses of angiotensin and norepinephrine despite unchanged GFR (8). We conclude that a disturbance of renal hemodynamic appears to be of primary pathogenic relevance in FNAH and may explain the severe disturbance in uric acid metabolism.

ACKNOWLEDGEMENTS

Investigators of the Spanish Group for the Study of FNAH are: Clinical Biochemistry, F. Mateos, A. Herranz, I. Martínez, and R. Torres; Internal Medicine, JG. Puig, and MªE. Miranda; Nephrology, L. Espinosa, and J. Martínez Ara; Radiology, T. Berrocal, and C. Prieto; Pathology, Mª Luz Picazo. We are indebted to T. Ramos, Javier Díaz, and Mª Paz Canencia for valuable technical assistance; to Rosario Madero for the statistical analysis; to Erik Lundin for editorial assistance. Supported by grants from Caja de Madrid and Fondo de Investigación Sanitaria (FIS, 94/0218), Spain.

REFERENCES

1. McKusick VA. Mendelian Inheritance in Man: Catalogs of Autosomal Dominant, Autosomal Recessive and X-linked Phenotypes. 9th ed. Baltimore; Johns Hopkins University Press; 1990:648.
2. Friedenberg MJ, Walz BJ, McAlister WH, Locksmith JP, Gallagher TL. Roentgen size of normal kidneys. Radiology 1965;84:1022-30.
3. Puig JG, Miranda MªE, Mateos ME, et al. Hereditary nephropathy associated with hyperuricemia and gout. Arch Intern Med 1993;153:357-65.
4. Steele TH, Rieselbach RE. The contribution of residual nephrons within the chronically diseased kidney to urate homeostasis in man. Am J Med. 1967;43:876-886.58.
5. Garyfallos A, Magoula I, Tsapas G. Evaluation of the renal mechanisms for urate homeostasis in uremic patients by probenecid and pyrazinamide test. Nephron 1987;46:273-80.
6. Puig JG, Mateos FA, Jiménez ML, Ramos TH. Renal excretion of hypoxanthine and xanthine in primary gout. Am J Med 1988;85:533-37.
7. Messerli FH, Frohlich ED, Dreslinski GR, Suarez DH, Aristimuno GG. Serum uric acid in essential hypertension: An indicator of renal vascular involvement. Ann Intern Med 1980;93:817-21.
8. Ferris TF, Gorden P. Effect of angiotensin and norepinephrine upon urate clearance in man. Am J Med 1968;44:359-65.

RENAL URATE HYPOEXCRETION IN POLYNESIAN WOMEN IS NOT AS SEVERE AS IN UNITED KINGDOM (UK) WOMEN WITH FAMILIAL JUVENILE HYPERURICAEMIC NEPHROPATHY (FJHN)

MB McBride#, HA Simmonds#, PJ Hatfield¶, R Graham§, J McCaskey*, M Jackson**

Purine Research Laboratory#, UMDS Guy's Hospital, London UK; Departments of Renal Medicine¶, Clinical Chemistry**, Dietetic Services*, Wellington Hospital; Mitsubishi Health Centre§, Porirua, NZ

INTRODUCTION

Polynesians have a predisposition to hyperuricaemia and gout. A number of risk factors have been identified including obesity, hyperglycaemia and hypertriglyceridaemia, however renal disease is not one of the associated risk factors. Studies in healthy Maori men[1] established that the factor which put them at risk for hyperuricaemia and gout was an extremely reduced fractional clearance of uric acid, FEur (uric acid clearance factored by creatinine clearance x 100%)[1]. Asymptomatic hyperuricaemia was identified in 23% of healthy Maori men, with a mean FEur (3.9%) lower than in the UK gouty male (5.4%). Even normouricaemic Maori males had a low FEur (4.9%) compared with UK males (8.1%).

Gout is rare in European women[2,3]. Renal disease is equally rare today in gouty men, whether European or Maori. This makes the many kindreds in the UK with FJHN - a dominant disorder affecting young men and women equally - even more unusual. The hallmark of FJHN is a grossly reduced FEur, which may preceed the onset of renal disease. The purpose of this study was to determine whether Polynesian women have a similarly reduced FEur to that in Maori men and how this compared with healthy UK women and those with FJHN.

SUBJECTS AND METHODS

Forty-three Polynesian women, (age 19 to 58 years) volunteered to participate in the study. The data for healthy UK women are drawn from previous publications[2], detailed studies in healthy subjects, and unaffected relatives in FJHN kindreds. Blood and a 24 hour urine was collected for the measurement of creatinine and uric acid clearances, as described[2]. Clearances were corrected for body surface area. Statistical analyses were performed using Student's t test and the Analysis of Variance (ANOVA).

Purine and Pyrimidine Metabolism in Man VIII, Edited by
A. Sahota and M. Taylor, Plenum Press, New York, 1995

RESULTS

Two of the 43 women (aged 46 and 58), had both a personal history and a family history of gout; in total 18 women had a family history of gout. Hyperuricaemia, (plasma uric acid of $\geq$300 μmol/l), was found in 19 asymptomatic women (45%), while 22 (51%) were normouricaemic.   The mean FEur in the hyperuricaemic Polynesian group (6.7 $\pm$ 1.5) was significantly lower than in the normouricaemic women (9.7 $\pm$ 1.9), (t = 5.7; p<0.001).   The FEur in normouricaemic Polynesian women was also significantly lower (t = 3.89; p<0.001) than that found in normouricaemic UK women (12.8 $\pm$ 2.9), (table 1). This finding resembles that in Maori men (table 2).   Plasma creatinine was within the normal range in all the Polynesians investigated.   Mean creatinine clearances ranged from 83.0 to 91.7 ml/min corrected to 1.73m^2, and did not differ significantly between the hyperuricaemic and normouricaemic groups (t = 0.24; p<0.001), indicating normal renal function in these women.

The mean FEur in the FJHN women, predominantly asymptomatic hyperuricaemic, was significantly lower than in asymptomatic hyperuricaemic Polynesian women (t = 3.33; p<0.005). The remarkable finding was the absence of any sex difference in FEur between young men and women with FJHN (t =0), the FEur in FJHN being lower even than in the middle-aged UK gouty male and the Polynesian women, whether hyperuricaemic or normouricaemic.

TABLE 1 : WOMEN

	POLYNESIAN			**UK**	
	Normouricaemia	Asymptomatic Hyperuricaemia	Gout	Normouricaemia	FJHN
Number	22	19	2	14	20
Plasma UA μmol/l	252 ±25	386 ±76	395	200 ±37	450 ±135
FEur %	9.7 ±1.9	6.7 ±1.5	6.5	12.8 ±2.9	5.1 ±1.5
Creatinine Clearance ml/min/1.73m^2	90.4 ±19.0	91.7 ±14.2	83.0	87.3 ±17.0	58.3 ±25.9
Ponderal Index	12.0 ±0.7	11.5 ±0.8	11.8	12.8 ±1.0	12.6 ±0.9

Results are expressed as the mean $\pm$ standard deviation

TABLE 2 : MEN

	POLYNESIAN			UK		
	Normouricaemia	Asymptomatic Hyperuricaemia	Gout	Normouricaemia	Gout	FJHN
Number	79	26	10	51	51	23
Plasma UA μmol/l	380 ± 40	480 ± 30	520 ± 100	270 ± 60	390 ± 60	402 ± 110
FEur %	4.9 ± 1.5	3.9 ± 1.4	3.6 ± 1.3	8.1 ± 3.2	5.4 ± 1.7	5.1 ± 1.6
Creatinine Clearance ml/min/1.73m^2	113.0 ± 22.0	106.0 ± 28.0	95.0 ± 27.0	102.5	92.6	81.4 ± 33.3
Ponderal Index	12.2 ± 0.7	11.9 ± 0.7	11.4 ± 0.9	12.5 ± 0.6	12.2 ± 0.6	12.9 ± 1.1

Results are expressed as the mean $\pm$ standard deviation

DISCUSSION

These studies demonstrate that normouricaemic Polynesian women resemble Maori men in also having a reduced FEur compared with their healthy UK counterparts. The mean FEur in the asymptomatic hyperuricaemic Polynesian women was even lower than in the normouricaemic group. The Polynesian women, both normouricaemic and asymptomatic hyperuricaemic, had a higher FEur than that reported for Maori men[1] confirming the same sex difference in FEur reported for Caucasians[2]. The prevalence of asymptomatic hyperuricaemia in Polynesian women was extremely high (45%), higher even than that reported for asymptomatic Maori men (23%)[1]. These studies confirmed that Polynesian women also have a tendency to obesity and habitually consume a high purine intake, (data not shown), which coupled with the low FEur explains the prevalence of gout in this race

The other interesting finding in this study was that the reduced FEur in the asymptomatic hyperuriceamic Polynesian women was not as low as that found in young UK women with FJHN, and unlike the latter was not associated with renal disease. Moreover, these young women with FJHN had the same reduced FEur as FJHN men, i.e. there was no sex difference, and they were slender, not obese. The low FEur in FJHN is all the more remarkable considering that reduced renal function is normally associated with an increase in FEur.

The reduced FEur in these Polynesian women supports the hypothesis that indigenous Pacific races share a similar genetic defect in renal urate handling, and indicate that the reduced FEur in FJHN has a different genetic basis.

ACKNOWLEDGMENTS

This study was supported by the Arthritis and Rheumatism Council of the United Kingdom. We would also like to thank the Polynesian volunteers and the nursing staff at Mitsubishi Motors, Porirua and the Renal Unit, Wellington Hospital, for the collaboration which made the study possible.

REFERENCES

1. T. Gibson, R. Waterworth, P. Hatfield, G. Robinson, and K. Bremner, Hyperuricaemia, gout and kidney function in young New Zealand Maori men, Br. J. Rheumatol. 23:276 (1984).

2. J.S. Cameron, F. Moro, H.A. Simmonds, Gout, uric acid and purine metabolism in paediatric nephrology, Review, Ped. Nephrol. 7:105 (1993).

3. J.G. Puig, A.D. Michan, M.L. Jiminez, C.P. de Ayala, F.A. Mateos, C.F. Capitan, E. de Miguel, J.B. Gijon, Female gout; clinical spectrum and uric acid metabolism, Arch. Int. Med. 151:726 (1991).

RENAL CLEARANCES OF PURINE BASES AND OXYPURINOL DURING GLUCOSE INFUSION

Yuji Moriwaki, Tetsuya Yamamoto, Sumio Takahashi, Yumiko Nasako, Toshikazu Hada, and Kazuya Higashino

Third Department of Internal Medicine, Hyogo College of Medicine
Mukogawa-cho 1-1, Nishinomiya, Hyogo 663, Japan

INTRODUCTION

Glucose infusion causes increased excretion of uric acid[1,2], suggesting a functional relationship between the renal handling of uric acid and glucose. Furthermore, some functional association between the renal transport of uric acid, oxypurines and oxypurinol have been suggested[3,4]. However, the effect of glucose infusion on the renal transport of oxypurines and oxypurinol, an allopurinol metabolite, has not yet been investigated, to our knowledge. Therefore, we examined the response of renal clearances of purine bases and oxypurinol to intravenous glucose infusion.

SUBJECTS AND METHODS

Subjects

The subjects were five males aged from 27- to 39-year-old, who presented normal laboratory and physical examinations.

Experimental protocol

Six hours prior to the study, 300mg of allopurinol was administered orally. Following a 1-hour control period, 500 mL of 1.1 M glucose solution, or same volume of 1.1 M mannitol solution (after at least a two-week interval), was infused intravenously during an hour. Urine samples were collected every hour. Blood were drawn at the mid-point of each 1 hour period and plasma was immediately separated for analysis of purine bases, oxypurinol, creatinine and glucose.

Analysis of samples

The plasma and urinary concentrations of uric acid were measured by an enzymatic spectrophotometric assay. The plasma and urinary concentrations of oxypurines and oxipurinol were measured by the modified method of Yamamoto et al[5], using high-performance liquid chromatography (HPLC). The fractional clearance of purine compunds and oxipurinol was calculated from the following formula: clearance of the substance/creatinine clearance. Plasma and urinary concentrations of creatinine were measured with an autoanalyzer. Plasma and urinary glucose were determined by the enzymatic method.

Statistical analyses

Data were expressed as the mean$\pm$SD. The observed differences were evaluated using Student's t test and a p value of less than 0.05 was considered statistically significant.

RESULTS

Effect of glucose infusion on plasma concentration, urinary excretion of glucose and plasma IRI

Both plasma concentration and urinary excretion of glucose were significantly increased during glucose infusion (50 ± 3mmol/L vs 241 ± 28mmol/L and below the detection limit vs 8.4 ± 15.2g/hr, p<0.01 respectively).

Effect of glucose and mannitol infusion on purine bases, oxypurinol and albumin in plasma

Plasma concentration of oxypurinol (Pox) was significantly decreased during glucose infusion when compared with that during pre-infusion period ($21.5\pm2.4\mu$mol/L vs $19.3\pm0.9\mu$mol/L, p<0.05). However, plasma concentrations of uric acid (Pua), hypoxanthine (Phx) and xanthine (Px) were not different from those in the pre-infusion period (Table 1).

Table 1. Effect of glucose infusion on plasma concentrations of albumin, purine bases and oxypurinol

	(1)	(2)	p value
Albumin (g/L)	47.6 ± 2.9	41.8 ± 2.0	p<0.01
Pua (mmol/L)	0.28 ± 0.04	0.28 ± 0.04	NS
Phx (μmol/L)	2.5 ± 0.5	2.5 ± 0.6	NS
Px (μmol/L)	7.5 ± 2.7	6.9 ± 2.5	NS
Pox (μmol/L)	21.5 ± 2.4	19.3 ± 0.9	p<0.05

Values are expressed as mean $\pm$SD.(1): before glucose infusion (2): during glucose infusion NS denotes not significant.

Pua and Pox were significantly decreased during mannitol infusion (0.28 ± 0.01mmol/L vs 0.25 ± 0.01mmol/L and $23.2\pm3.9\mu$mol/L vs $18.8\pm3.8\mu$mol/L, p<0.05 respectively). However, Phx and Px were not different between the two periods. Plasma albumin levels were significantly decreased both during glucose and mannitol infusion.

Effect of glucose and mannitol infusion on purine bases and oxypurinol in urine

Urinary excretions of uric acid (Uua) and oxypurinol (Uox) were significantly increased during glucose infusion (0.12 ± 0.03mM/hr vs 0.25 ± 0.03mM/hr, $p<0.01$ and $22.7\pm6.6\mu$M/hr vs $30.2\pm4.8\mu$M/hr, $p<0.05$). Urinary excretions of hypoxanthine (Uhx) and xanthine (Ux) were also increased, however the increases were not statistically significant (Table 2).

Table 2. Effect of glucose infusion on urinary excretions of purine bases and oxypurinol

	(1)	(2)	
Uua (mM/hr)	0.12 ± 0.03	0.25 ± 0.03	$p<0.01$
Uhx (μM/hr)	21.6 ± 5.6	21.9 ± 5.4	NS
Ux (μM/hr)	33.3 ± 6.8	37.5 ± 3.2	NS
Uox (μM/hr)	22.7 ± 6.6	30.2 ± 4.8	$p<0.05$
Urine volume(mL)	150.0 ± 24.0	354.5 ± 106.4	$p<0.01$

Values are expressed as mean $\pm$SD. (1), (2) and NS are the same as in Table 1.

In contrast, Uua, Uhx, Ux and Uox were not changed by mannitol infusion.

Effect of glucose and mannitol infusion on fractional clearances of purine bases and oxypurinol

Fractional clearances of uric acid (Fua), xanthine (Fx) and oxypurinol (Fox) were significantly increased during glucose infusion (0.06 ± 0.02 vs 0.11 ± 0.01, $p<0.01$, 0.51 ± 0.14 vs 0.75 ± 0.15, $p<0.05$, 0.15 ± 0.06 vs 0.22 ± 0.06, $p<0.05$). In contrast, fractional clearance of hypoxanthine (Fhx) was comparable to that in the pre-infusion period (Table 3).

Table 3. Effect of glucose infusion on fractional clearances of purine bases and oxypurinol

	(1)	(2)	
Fua	0.06 ± 0.02	0.11 ± 0.01	$p<0.01$
Fhx	0.93 ± 0.29	1.12 ± 0.23	NS
Fx	0.51 ± 0.14	0.75 ± 0.15	$p<0.05$
Fox	0.15 ± 0.06	0.22 ± 0.06	$p<0.05$

Values are expressed as mean $\pm$SD. (1), (2) and NS are the same as in Table 1.

Fua, Fhx, Fx and Fox were not changed by mannitol infusion.

DISCUSSION

In the present study, we administered 500 mL of 1.1 M glucose solution to 5 normal subjects after 300 mg administration of allopurinol, and demonstrated that glucose decreased the plasma concentration of oxypurinol, while increasing the fractional clearances of uric acid, xanthine and oxypurinol. Since osmotic diuresis may contribute to the increase in the fractional clearances of these substances, we administered 500 mL of 1.1 M mannitol solution to investigate the effect of osmotic diuresis on the renal clearances of these purine compounds and oxypurinol. However, there were no significant changes in the fractional clearances of purine bases and oxypurinol, suggesting that the increase in the fractional clearances of these compounds seems to be related to glycosuria and/or hyperglycemia rather than osmotic effect. As for uric acid, uricosuric response was observed only with the appearance of glycosuria[6]. Moreover, subjects with renal glycosuria showed an increase in uric acid excretion in spite of normal plasma glucose level[6]. These results support the contention that uricosuric response is related to urinary glucose concentration. It is possible that similar transport mechanism may work on the renal transport of oxypurines and oxypurinol during glucose infusion, since pyrazinamide which markedly suppress the urinary excretion of uric acid decreases the fractional clearances of oxypurines and oxypurinol[3] and in contrast probenecid increases the fractional clearances of oxypurines and oxypurinol as well as uric acid[3]. This being supposition, further examination is required to elucidate the precise mechanism of the increased fractional clearances of purine bases and oxypurinol during glucose infusion.

Oxypurinol is converted from allopurinol and is also a potent xanthine oxidase inhibitor. The present study also indicated that the biological half-life of oxypurinol is decreased by glucose infusion. Therefore, glycosuria and/or hyperglycemia-induced decrease in the biological half-life of oxypurinol must be taken into consideration when gouty subjects with uncontrolled diabetes mellitus are treated with allopurinol.

REFERENCES

1. R.W.Bonsnes, E.S.Dana: On the increased uric acid clearance following the intravenous infusion of hypertonic glucose solutions. *J Clin Invest* 25:386(1946).
2. P.J.Christensen, O.R.Steenstrup: Uric acid excretion with increasing plasma glucose concentration (pregnant and non-pregnant cases). *Scand J Clin Lab Invest* 10:182 (1958).
3. T.Yamamoto, Y.Moriwaki, S.Takahashi, T.Hada, K.Higashino: Renal excretion of purine bases- Effects of probenecid, benzbromarone and pyrazinamide. *Nephron* 48:116(1988).
4. K.Kaneko, S.Fujimori, H.Itoh: Renal handling of hypoxanthine and xanthine in normal subjects and in four cases of idiopathic renal hypouricemia. *J Rheumatol* 15:325(1988).
5. T.Yamamoto, Y.Moriwaki, S.Takahashi, T.Hada, K.Higashino: Separation of hypoxan thine and xanthine from pyrazinamide and its metabolites in plasma and urine by high-performance liquid chromatography. *J Chromatogr* 382:270(1986).
6. G.Boner, R.E.Rieselbach: Effect of glucose upon reabsorptive transport of urate by the kidney. *Adv Exp Med Biol* 41B:781(1977).

IN VITRO AND *IN VIVO* STUDY ON THE CONVERSION OF ALLOPURINOL AND PYRAZINAMIDE

Yumiko Nasako, Tetsuya Yamamoto, Yuji Moriwaki, Sumio Takahashi, Zenta Tsutsumi, Toshikazu Hada and Kazuya Higashino

Third Department of Internal Medicine, Hyogo College of Medicine, Mukogawa-cho 1-1, Nishinomiya, Hyogo 663, Japan

INTRODUCTION

Allopurinol is widely used in the treatment of gout. The main metabolic pathway of allopurinol is its oxidation to oxypurinol. Pyrazinamide is an antituberculous drug. One of the main metabolic pathways of pyrazinamide is its oxidation to 5-hydroxypyrazinamide. The oxidation of allopurinol and pyrazinamide is known to be attributed to xanthine oxidase[1,2]. However, it has been suggested that aldehyde oxidase also participates in the oxidation of these agents[3,4]. Therefore, we investigated whether aldehyde oxidase plays a role in the oxidation of allopurinol and pyrazinamide by *in vivo* and *in vitro* studies.

MATERIALS AND METHODS

Chemicals

Pyrazinamide was kindly provided by Sankyo Pharmaceutical Company (Tokyo, Japan). BOF-4272[5] (xanthine oxidase inhibitor) was kindly provided by Otsuka Pharmaceutical Factory, Inc. (Tokushima, Japan). 5-hydroxypyrazinamide was obtained by the method described previously[6]. Other reagents were of analytical grade.

Methods

In vitro study. Purification of aldehyde oxidase from rat liver and measurement of its activity were performed according to the previously described methods[7,8]. The oxidation of allopurinol or pyrazinamide was determined by measuring the formation of oxypurinol or 5-hydroxypyrazinamide, respectively with or without benzamidine or BOF-4272. The assay mixture contained 50mM potassium phosphate buffer, pH 9.0, 2.4mM allopurinol and $25 \mu l$ of enzyme preparation in a final volume of $150 \mu l$ with or without $10 \mu M$ BOF-4272 or 10mM benzamidine. When pyrazinamide was used as the substrate, the assay mixture contained 5.4mM pyrazinamide instead of allopurinol in the reaction mixture described above. The reaction was terminated after 120 minutes incubation at 37℃ by the addition of 20% HClO₄, then neutralized with 1M K₂CO₃. The precipitated protein was removed by centrifugation and filtered through Chromatodisc 4A (0.2 μ m). The oxidation products of allopurinol and pyrazinamide, oxypurinol and 5-hydroxypyrazinamide, were measured by HPLC[9] (high-performance liquid chromatography).

In vivo **study.** Catheterization was performed in Wistar male rats as described previously[10] and then physiological saline containing 4% mannitol and inulin was intravenously administered from the femoral vein at 3ml/hour. To investigate the effect of BOF-4272 on allopurinol and pyrazinamide metabolism, 40 minutes after the beginning of the infusion of mannitol and inulin, BOF-4272 (30mg/kg weight) was intravenously administered to 6 rats (BOF-4272-treated rats) as a solution of 10mg/ml BOF-4272 in physiological saline and then 10 minutes later, BOF-4272 (30mg/kg weight) was administered again. Immediately after the second administration of BOF-4272, allopurinol (14.7 μ mol/kg weight) or pyrazinamide (650 μ mol/kg weight) was administered as a solution of 7.4 μ mol/ml allopurinol or 163 μ mol/ml pyrazinamide in physiological saline to 3 rats, respectively. At 3, 6, 9, 15, 45, 75 and 105 minutes after the administration of allopurinol or pyrazinamide, 0.3ml of blood was drawn from the femoral artery with a heparinized syringe. In addition, after the administration of allopurinol or pyrazinamide, urine was collected 4 times every 30 minutes. The control study was performed in 3 rats (the controls for BOF-treated rats) respectively, in the same method as described above except the administration of only physiological saline instead of physiological saline containing BOF-4272. Allopurinol , oxypurinol, pyrazinamide and its metabolites in both plasma and urine were detertmined by HPLC[6,9,11] . The concentration of inulin in plasma and urine was determined as descrived previously[10]. The activity of xanthine oxidase + xanthine dehydrogenase was measured by a reversed-phase HPLC as described previously [9].

RESULTS

In vitro study

Aldehyde oxidase purified from rat liver was homogeneous as judged by PAGE (Polyacrylamide gel electorophoresis). The estimated molecular weight of aldehyde oxidase was 270kd. When compared against molecular weight standards run simultaneously on thin layer SDS-PAGE (Sodium dodecyl sulfate-polyacrylamide gel electorophoresis), a molecular weight of approximately 135kd was estimated for this protein. Specific activity of the purified aldehyde oxidase was about 120 fold as that of the cytosolic aldehyde oxidase. Five mM benzamidine inhibited aldehyde oxidase activity about 90%.However, neither 320 μ M BOF-4272 nor 2.5×10^{-4}M allopurinol had effect on the activity of aldehyde oxidase. During incubation aldehyde oxidase converted allopurinol to oxypurinol and pyrazinamide to 5-hydroxypyrazinamide. Benzamidine inhibited the formation of oxypurinol and 5-hydroxypyrazinamide, but BOF-4272 did not.

In vivo study

The plasma concentration of oxypurinol in the BOF-4272-treated rats (N=3) after the administration of allopurinol was 0.14±0.03 μ g/ml, 0.19±0.05 μ g/ml, 0.17±0.02 μ g/m,0.17±0.02 μ g/ml, 0.19±0.03 μ g/ml, 0.16±0.03 μ g/ml and 0.18±0.04 μ g/ml (mean±S.E.) at 3, 6, 9, 15, 45, 75 and 105 minutes, respectively, while that plasma concentration of oxypurinol was not changed among those periods. The plasma concentration of oxypurinol in the control rats (N=3) after the administration of allopurinol was $0.24 \pm 0.17 \mu$ g/ml, $0.89 \pm 0.09 \mu$ g/ml, $0.69 \pm 0.03 \mu$ g/ml, $0.58 \pm 0.03 \mu$ g/ml, $0.45 \pm 0.06 \mu$ g/ml, $0.34 \pm 0.04 \mu$ g/ml and $0.26 \pm 0.02 \mu$ g/ml (mean $\pm$ S.E.) at 3, 6 ,9, 15, 45, 75 and 105 minutes, respectively. The plasma concentrations of oxypurinol after the administration of allopurinol was markedly decreased in the BOF-4272-treated rats, as compared with that in the control rats. The plasma concentration of 5-hydroxypyrazinamide in the BOF-4272-treated rats (N=3) after the administration of pyrazinamide was 0.04 ± 0.02 μ g/ml, $0.15 \pm 0.03 \mu$ g/ml, $0.73 \pm 0.12 \mu$ g/ml, $1.05 \pm 0.15 \mu$ g/ml, $1.35 \pm 0.09 \mu$ g/ml and $1.63 \pm 0.17 \mu$ g/ml (mean $\pm$ S.E.) at 3, 6, 15, 45, 75 and 105 minutes, respectively. In contrast, the plasma concentration of 5-hydroxypyrazinamide in the control rats (N=3) after the administration of pyrazinamide was $2.26 \pm 0.4 \mu$ g/ml, $4.54 \pm 0.16 \mu$ g/ml, 6.1 ± 0.23 μ g/ml, $7.76 \pm 0.81 \mu$ g/ml, $8.12 \pm 1.29 \mu$ g/ml and $7.96 \pm 1.71 \mu$ g/ml (mean $\pm$ S.E.) at 3, 6, 15, 45, 75 and 105 minutes, respectively. The plasma concentration of 5-hydroxypyrazinamide after the administration of pyrazinamide was markedly decreased in the BOF-4272-treated rats, as compared with the control rats. Total urinary excretion of oxypurinol in the BOF-4272-treated rats after the administration of allopurinol was $3.58 \pm 0.27 \mu$ g/ml (mean $\pm$ S.E.), compared to $15.6 \pm 0.8 \mu$ g/ml (mean $\pm$ S.E.) in the control rats. The urinary excretion of oxypurinol after the administration of allopurinol was decreased in the BOF-4272-treated rats,as compared with that in the control rats. Total urinary excretion of 5-hydroxypyrzinamide in the BOF-4272-treated rats after the administration of pyrazinamide was $46.3 \pm 13.4 \mu$ g/ml (mean $\pm$ S.E.), in the control rats was 808 ± 241.3 μ g/ml (mean $\pm$ S.E.). The urinary excretion of 5-hydroxypyrazinamide was also decreased in the BOF-4272-treated rats, as compared with that in the control rats.

DISCUSSION

Allopurinol and pyrazinamide are oxidized to oxypurinol and 5-hydroxy-pyrazinamide, respectively, by xanthine oxidase[1,2,12]. In the previous studies, it has been suggested that aldehyde oxidase may oxidize allopurinol to oxypurinol. Therefore, to examine the oxidation of allopurinol and pyrazinamide by aldehyde oxidase, we conducted an *in vitro* study. The *in vitro* study demonstrated that aldehyde oxidase oxidized both allopurinol and pyrazinamide, suggesting that aldehyde oxidase may play a role in the oxidation of allopurinol to oxypurinol and that of pyrazinamide to 5-hydroxypyrazinamide along with xanthine dehydrogenase. However, the relative contribution of xanthine dehydrogenase or aldehyde oxidase to the oxidation of these two agents is not yet determined*in vivo*. In the present study, we demonstrated that BOF-4272 inhibited the activity of xanthine oxidase as described previously[5] but inhibited neither the activity of aldehyde oxidase nor HGPRT *in vitro*. Therefore, using BOF-4272, we investigated the contribution of xanthine dehydrogenase to the oxidation of allopurinol and pyrazinamide *in vivo* and demonstrated that BOF-4272 almost inhibited the oxidation of allopurinol to oxypurinol, and that of pyrazinamide to 5-hydroxypyrazinamide. These results strongly suggest that xanthine dehydrogenase is the most important enzyme in the conversion of allopurinol to oxypurinol, and pyrazinamide to 5-hydroxypyrazinamide, respectively, and that aldehyde oxidase plays a minor role in the oxidation of allopurinol to oxypurinol and that of pyrazinamide to 5-hydroxypyrazinamide *in vivo*.

REFERENCES

1. G.B.Elion. Allopurinol and other inhibitors of urate synthesis. *In: Uric Acid*

(Eds.W.N.Kelley and I.M.Weiner),pp.485.Springer,Berlin(1978).

2. G.A.C.Murrell and W.G.Rapeport.Clinical pharmacokinetics of allopurinol.*Clin pharmacokinet* 11:343(1986).

3. D.G.Johns.Human liver aldehyde oxidase: differential inhibition of oxidation of changed and unchanged substrates.*J Clin Invest* 46:1492(1967).

4. T.A.Krenitzky,S.M.Neil,G.B.Elion andG.H.Hitchings.A comparison of the specificities of xanthine oxidase and aldehyde oxidase.*Arch Biochem Biophys* 150:585(1972).

5. S.Sato,K.Tatsumi,T.Nishino.A novel xanthine dehydrogenase inhibitor(BOF-4272).*Adv Exp Med Biol* 309A:135(1991).

6. T.Yamamoto,Y.Moriwaki,S.Takahashi,T.Hada and K.Higashino.Rapid and simultaneous determination of pyrazinamide and its major metabolites in human plasma by high-performance liquid chromatography.*J Chromatogr* 413:342 (1987).

7. J.G.P.Stell,A.J.Warne and C.Lee-Woolley.Purification of rabbit liver aldehyd oxidase by affinity chromatography on Benzamidine Sepharose6B. *J.Chromatogr* 475:363(1989).

8. J.Kurth and A.Kubiciel.Method zur photometrischen Bestimmung der Aktivität von Aldehydoxydase.*Biomed Biochim Acta* 43:1223(1984).

9. T.Yamamoto,Y.Moriwaki,S.Takahashi,T.Hada and K.Higashino.Separation of hypoxanthine and xanthine from pyrazinamide and its metabolites in plasma and urine by high-performance liquid chromatography.*J Chromatogr* 382:270(1986).

10. Y.Yonetani and K.Iwaki.Effects of uricosuric drugs and diuretics on uric acid excretion in oxonate-treated rats.*Jpn J Pharmacol* 33:947(1983).

11. T.Yamamoto,Y.Moriwaki,S.Takahashi,T.Hada and K.Higashino.Study of the metabolism of pyrazinamide using a high-performance liquid chromatographic analysis of urine samples.*Anal Biochem* 160:346(1987).

12. T.Yamamoto,Y.Moriwaki,S.Takahashi,T.Hada and K.Higashino.In vitro conversion of pyrazinamide into 5-hydroxypyrazinamide and that of pyrazinoic acid into 5-hydroxypyrazinoic acid by xanthine oxidase from human liver. *Biochem Pharmacol* 36:3317(1987).

FREE OXYPURINES IN PLASMA AND URINE OF GOUT PATIENTS BEFORE AND AFTER A PURINE-FREE DIET

B.Porcelli, D.Vannoni, R.Leoncini, M.Pizzichini, R.Pagani, and E.Marinello

Institute of Biochemistry and Enzymology
University of Siena, Italy

INTRODUCTION

The metabolism of uric acid and its levels in plasma and urine of gout patients have been extensively investigated[1]. Much less is known about the behavior of its precursors, hypoxanthine and xanthine in plasma and urine of the same patients. Reports in the literature are few and contradictory[2,3]. Gutman et al.[4] report no changes in plasma concentration and urinary excretion of oxypurines in gouty patients. McBurney and Gibson[5] observed a non significant increase in plasma levels of hypoxanthine and xanthine in gout patients, with a high dispersion of values and greater variation of data within a group than between groups.

Urinary oxypurine excretion in gout patients has been studied by different methods in a very limited patient population[2,6]. Only a few recent studies have used modern techniques[5-7].

In this study, we examined the behavior of oxypurines (hypoxanthine and xanthine) in plasma and urine of normal subjects and gout patients, under basal conditions and after a purine-free diet. The assay technique used was high performance liquid chromatography (HPLC). Specific differences between the two groups of subjects are discussed.

METHODS

Patients and controls

Control subjects (12 males and 13 females) and 28 male gout patients (38 to 66 years of age), who had not taken allopurinol in the 15 days immediately preceding the experiment, were chosen for our experiment. Male and female controls (20-50 years of age) were chosen because a previous study of ours showed no difference in plasma (unpublished data) and urinary oxypurines in relation to sex[8]. The gout patients had a one- to 25- year history of the disease (mean: 13 years). All patients were hyperproducers, according to the criterion[9] of high plasma levels and high excretion of uric acid even after a purine-free diet. None had

Address for correspondence: Prof.Enrico Marinello, Istituto di Biochimica e di Enzimologia, Università di Siena, Pian dei Mantellini, 44, 53100 SIENA Italy, Tel. +39-577-298026, FAX +39-577-298057

acute attacks during the experimental period.

All gout patients were hospitalized in the Institute of Rheumatology of the University of Siena; the controls worked in our Institute. The activity levels of all subjects were standardized. No controls had liver, kidney or metabolic disorders. No subject was more than 30% overweight. In no case were they on corticosteroids or other drugs.

All subjects were given a free, balanced diet of not more than 2500 calories/day (60% carbohydrates, 15% protein, 25% fat) with no purine restriction. After one week hospitalization, 8 control subjects and 22 gout patients were kept on a purine-free diet. The purine-free diet was isocaloric with previous food intake and had the same composition of basic components, with a low purine level (75 mg%) and 20% fat. Purine-rich foods were rigorously eliminated, such as meat, fish, salads, eggs, sausages, legumes, chocolate, alcohol, tea, coffee and Coca Cola; noodles, bread, cheese, dairy products, fruit and vegetables were allowed. Body weight was constant throughout the study period.

Preparation of plasma and urine samples

Heparinized blood samples for plasma oxypurines and uric acid were taken at 8 a.m. after overnight fasting and bed rest. They were rapidly chilled and centrifuged at 4°C (3000 rpm, for 15 minutes): an equal volume of 10% TCA was added for deproteinization. Clear supernatant was extracted 5 times with ethyl ether, until pH was neutral.

Twenty-four-hour urine samples (pH ranging between 4.5-5.5) were collected for oxypurine and uric acid assay: the urine samples were diluted 5 times and ultrafiltered through 0.22 μm Sartorius membranes. Plasma and urine were stored at -80°C until assay.

Determination of oxypurines and uric acid in plasma and urine

The HPLC apparatus was a Beckman model 332 equipped with UV detector at 254 nm and a 5 μm, 25 cm x 4.6 mm Supelcosil C18 column (Supelco Inc. Bellefonte, Pa., U.S.A.). 20 μl of neutralized plasma samples, or diluted urine samples, were injected for analysis. For plasma samples, the procedure of Hartwick et al.[10] was used with methanol instead of methanol-water (80:20); elution was carried out with KH_2PO_4 buffer (0.01 M, pH 5.5) with a linear gradient of 0-25% methanol, for 30 min, at a flow rate of 1 ml/min. For urine samples, isocratic elution was carried out with 0.01 M KH_2PO_4 buffer (pH 5.5) at a flow rate of 1 ml/min[11].

Urinary uric acid was determined with the Boehringer Biochemie kit.

Statistical analysis of the results was carried out by conventional one-way analysis of variance.

Determination of clearances

Creatinine was determined in blood and urine with the Sclavo kit.

Creatinine clearance was calculated by the well-known formula and expressed in ml/min. Uric acid, hypoxanthine and xanthine clearances were also calculated, by analogy, with the same formula, and expressed in ml/min.

Statistical analysis of the results was carried out by conventional one-way analysis of variance.

RESULTS

In Figure 1 we report individual values of plasma oxypurines and uric acid in control subjects and gout patients.

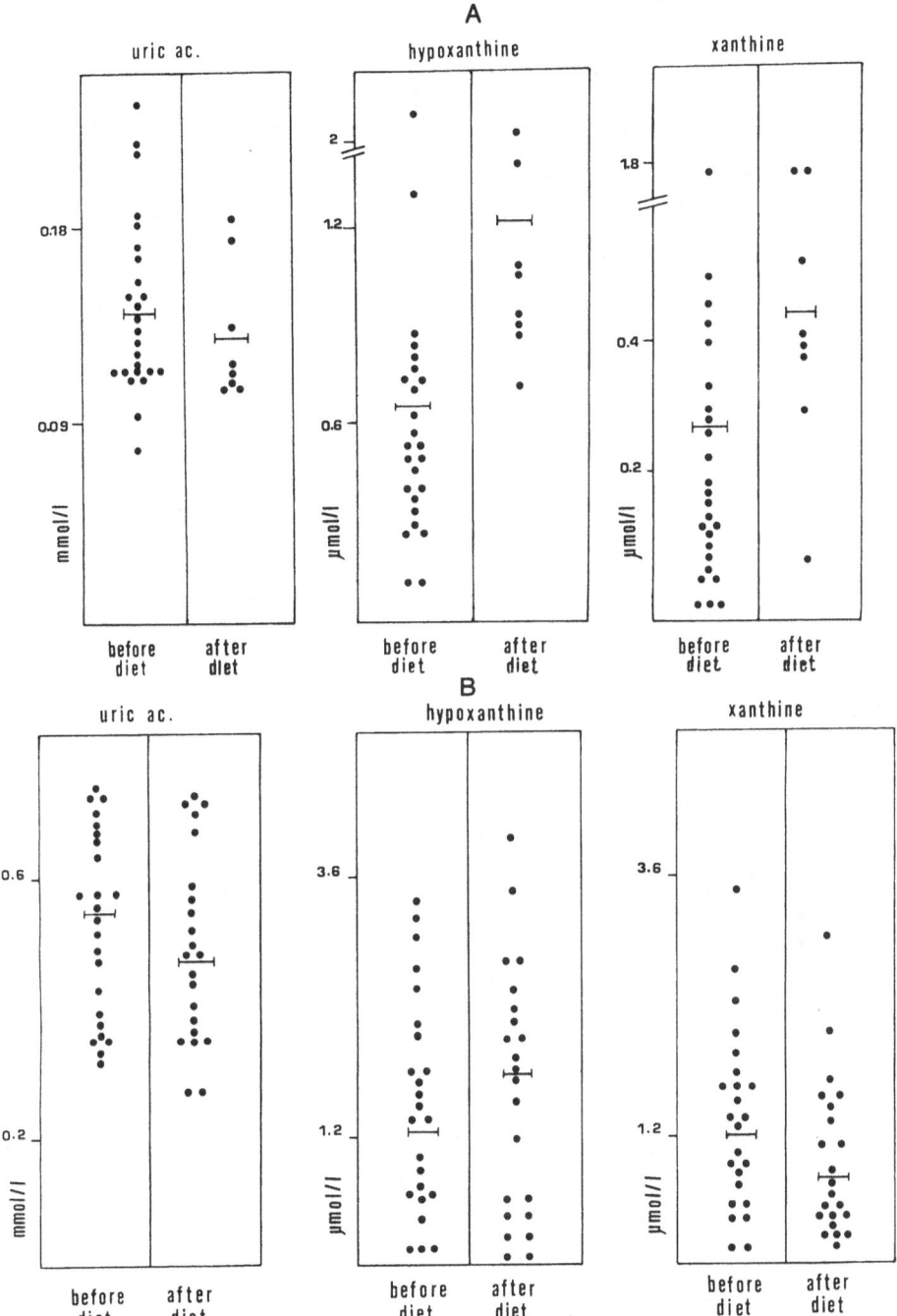

Figure 1. Individual values of plasma oxypurines and uric acid in control subjects (A) and gout patients (B) before and after purine-free diet.

Plasma basal levels of uric acid and oxypurines were remarkably and significantly higher in gout patients. After the purine-free diet, oxypurines were much increased in controls, while a decreasing trend was observed in gout patients for xanthine. Uric acid remained practically unchanged in both groups. After the diet, urinary oxypurines were substantially unchanged in normal subjects; in gout patients, xanthine and uric acid showed a decreasing trend (Figure 2).

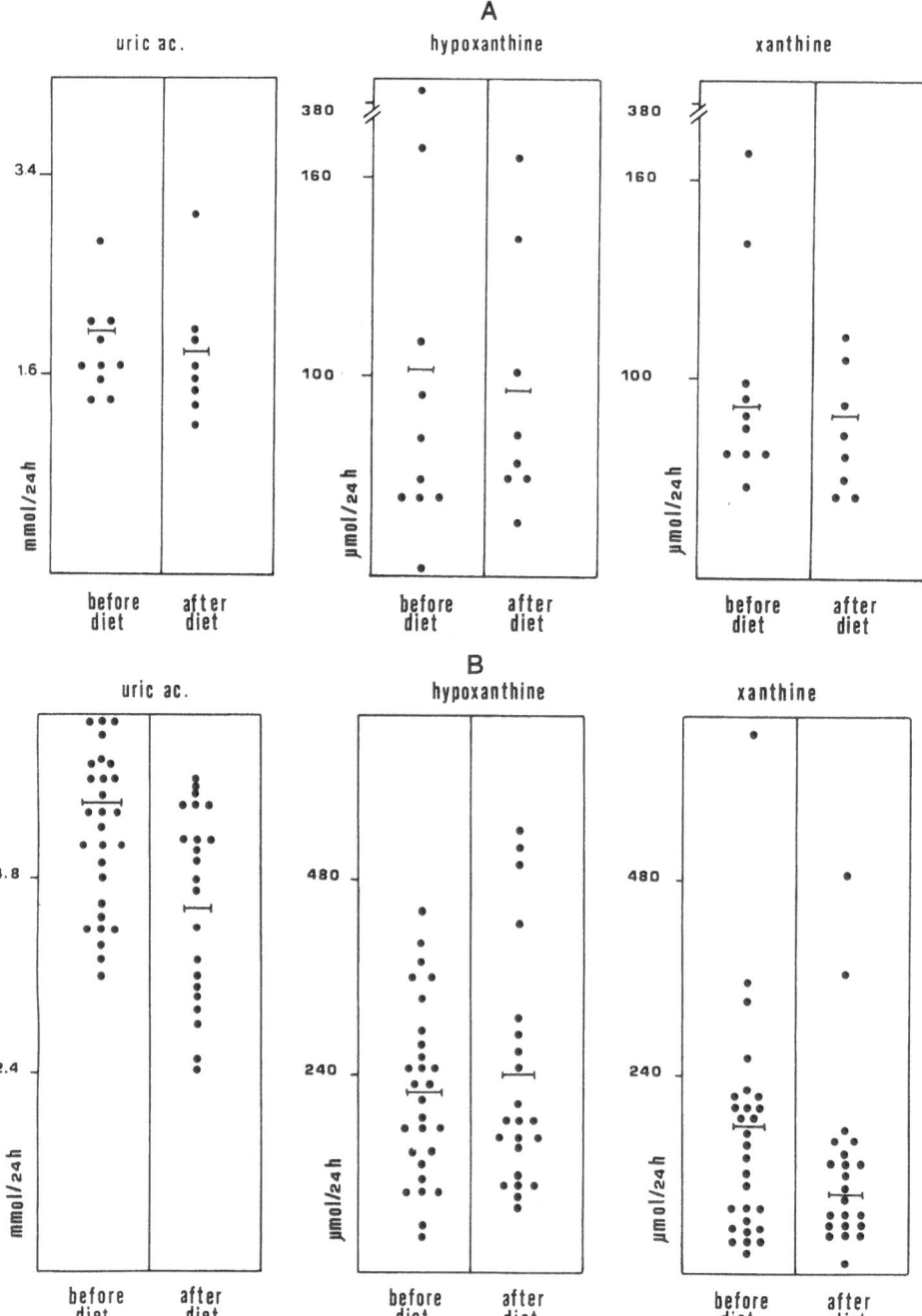

Figure 2. Individual values of urine oxypurines and uric acid in control subjects (A) and gout patients (B) before and after purine-free diet.

In Table 1 we report the clearances of creatinine, hypoxanthine, xanthine and uric acid in control subjects and gout patients, before and after diet.

Table 1. Clearance of creatinine, hypoxanthine, xanthine and uric acid (ml/min) in control subjects and in gout patients, before and after diet. In brackets, the number of cases. The values are expressed as mean ± SE.

		Cl.creatinine	Cl.hypoxanthine	Cl.xanthine	Cl.uric acid
Controls:					
- before diet	(10)	95.57 ± 4.1	136.19 ± 21.9	604.0 ± 107.1	11.35 ± 2.3
- after diet	(8)	71.97 ± 5.1	79.12 ± 18.3[1]	282.32 ± 98.5[1]	10.10 ± 1.59
Gout patients:					
- before diet	(24)	111.9 ± 19.7	40.3 ± 8.2[1]	233.60 ± 128.9[1]	7.40 ± 1.4
- after diet	(22)	81.80 ± 8.3	55.2 ± 6.3	166.50 ± 81.4	6.40 ± 0.70[1]

Variations ($p \leq 0.01$) with respect to: [1] control subjects before diet.

Basal clearance of uric acid was depressed in gout patients, as already reported in the literature[12,13]. Hypoxanthine and xanthine clearance decreased in the same group ($p \leq 0.01$). After the purine-free diet, the clearances of all compounds (creatinine, hypoxanthine, xanthine and uric acid) were lower, not always significantly, in normal subjects. A further decrease also occurred in the gout patients, but not for hypoxanthine.

DISCUSSION

A. Increased plasma and urinary levels of uric acid in gout patients have hitherto been regarded as hallmark of the disease. The present findings demonstrate that plasma and urinary concentrations of hypoxanthine and xanthine are also remarkably higher in gout patients. Similar results have been reported by Puig et al.[14].

B. The behavior of plasma oxypurines in normal subjects after the purine-free diet may be due to various factors: 1) increased nucleotide catabolism affecting the production of free purines; 2) decreased xanthine-oxidase activity affecting the transformation of oxypurines to uric acid (the enzyme is diet-sensitive[15], increasing after purine administration and could therefore decrease in their absence); 3) the ability of the body to save certain physiologically important compounds, for instance, sodium and magnesium[16,17], by renal or hormonal regulation, during dietary restriction.

Oxypurines values were remarkably high in gout patients under basal conditions, reaching the levels observed in controls after the diet. The tendency to spare such metabolites does not seem necessary in gout patients: the oxypurines have probably reached a limit value and the equilibrium is difficult to shift.

C. The main variations observed in normal subjects and gout patients could also be due to changes in renal clearance. The clearance of uric acid and oxypurines was reduced in normal subjects after the diet. The clearance of uric acid has long been known to be low in gout patients[13]. Decreased clearance of oxypurines in gout has only recently been observed and studied[14].

The present study confirms the results of Puig et al.[14] of lower clearance not only of uric acid but also of hypoxanthine and xanthine in all gout patients. It also extends this conclusion to other conditions, namely before and after the purine-free diet.

We conclude underlining the main results of our research: increased plasma levels and urinary excretion of oxypurines in gout patients. Their peculiar pattern before and after a purine-free diet are a new and unexpected result obtained by the use, for the first time, of HPLC. These results open a new chapter in the study of gout, providing evidence in favour of the hypothesis that an intrinsic and subtle renal defect is the underlying cause of primary gout in many patients, and that such defect involves not only uric acid but also oxypurines.

REFERENCES

1. J.B.Wyngaarden and W.N.Kelley, Epidemiology of hyper-uricaemia and gout, *in:* "Gout and hyperuricaemia", J.B.Wyngaarden, W.N.Kelley, Academic Press, New York (1976).
2. A.B.Gutman, T.F.Yii and B.Weissman, The concept of secondary gout: relation to purine metabolism in polycytemia and myeloid metaplasia, *Trans.Assoc.Am. Physician* 69:229 (1956).
3. W.S.Adams, F.Davis and M.Nakatani, Purine and pyrimidine excretion in normal and leukemic subjects, *Am.J.Med.* 28:726 (1960).
4. A.B.Gutman and F.F.Yu, A three component system for regulation of renal excretion of uric acid in man, *Trans.Ass.Am.Phys.* 74:353 (1961).
5. A.McBurney and T.Gibson, Reverse phase partition HPLC for determination of plasma purines and pyrimidines in subjects with gout and renal failure, *Clin.Chim.Acta* 102:19 (1980).
6. H.A.Simmond, Urinary excretion of purines, pyrimidines and pyrazolopyrimidines in patients treated with allopurinol or oxypurinol, *Clin.Chim.Acta*; 23:353 (1969).
7. J.L.Chabard, C.Lartigne-Mattei, F.Verdone, J.Petit and J.A.Berger, Mass fragmentographic determination of xanthine and hypoxanthine in biological fluids, *Chromatograph* 221:9 (1980).
8. D.Vannoni, B.Porcelli, R.Leoncini, L.Terzuoli, M.Pizzichini, A.Di Stefano, E.Marinello and R.Marcolongo, The excretion of oxypurines in normal subjects, *Biom. & Pharmacother.* 43:513 (1989).
9. J.B.Wyngaarden and W.N.Kelley, Uric acid overproduction in gout, *in:* "Gout and hyperuricaemia", J.B.Wyngaarden, W.N.Kelley, Academic Press, New York (1976).
10. R.A.Hartwick, A.M.Krstulovic and P.R.Brown, Identification and quantitation of nucleosides, bases and other UV-absorbing compounds in serum, using reversed phase high performance liquid chromatography, *J.Chrom.*, 186:659 (1979).
11. R.Bolieu, C.Bory, P.Baltassat and C.Gonnet, Simultaneous determination of allopurinol, oxypurinol, hypoxanthine and xanthine in biological fluids by high performance liquid chromatography, *J.Chrom.* 307:469 (1984).
12. A.B.Gutman and F.F.Yu F.F., Renal function in gout with a commentary on the renal regulation of urate excretion, and the role of the kidney in the pathogenesis of gout, *Am.J.Med.* 23:600 (1957).
13. M.L.Smith and J.T.Scott, Uric acid clearance in patients with gout and normal subjects, *Ann.Rheum.Dis.* 30:285 (1971).
14. J.G.Puig, F.A.Mateos, M.L.Jimenez, T.Ramos, M.C.Capitàn and A.A.Gil, Impaired renal excretion of hypoxanthine and xanthine in primary gout, *in:* "Purine and pyrimidine metabolism in man VI", New York, Plenum Press (1989).
15. A.Carcassi, R.Marcolongo and E.Marinello, Liver xanthine oxidase in gouty patients, *Arthritis Rheum.* 12:17 (1969).
16. J.Bourgoignie, J.P.Pennel and A.J.Jacob, Sodium metabolism and volume regulation, *in:* "Current Nephrology", Houghton Mifflin, Boston (1979).
17. M.G.Fitzgerald and P.Forman, An experimental study of magnesium deficiency in man, *Clin.Sci.* 15:635 (1965).

ANALYSIS OF THE GENOTYPES FOR ALDEHYDE DEHYDROGENASE 2 IN JAPANESE PATIENTS WITH PRIMARY GOUT

Hisashi Yamanaka, Naoyuki Kamatani, Masayuki Hakoda, Chihiro Terai, Ryuji Kawaguchi* and Sadao Kashiwazaki

Institute of Rheumatology, Tokyo Women's Medical College
KS bldg., 9-12 Wakamatsu-cho, Shinjuku-ku, Tokyo, 162, Japan
*SRL laboratories
51 Komiya-cho, Hachi-Oji, Tokyo, 192, Japan

SUMMARY

Alcoholic ingestion is one of the major factors for increasing serum uric acid levels. Genotypes of aldehyde dehydrogenase 2 (ALDH2, E.C.1.2.1.3), which regulates the sensitivity of an individual to ethanol, were determined in Japanese patients with gout and control subjects by allele specific oligonucleotide hybridization using PCR amplified gene. The most common allele $ALDH2*1$ codes for normal ALDH2 activity, while the less common allele $ALDH2*2$ codes for a lower enzyme activity. The frequency of homozygotes of $ALDH2*2$ was significantly lower in patients with gout than those with rheumatoid arthritis or a normal population. Plasma and urinary hypoxanthine levels were strikingly increased after ethanol drinking in homozygotes for $ALDH2*1$ but not in heterozygotes for $ALDH2*1/ALDH2*2$, indicated extensive purine nucleotide degradation in homozygote for $ALDH2*1$. These data indicated that alcohol ingestion may not be the requisite factor but is deeply involved in the pathogenesis of gout and hyperuricemia.

INTRODUCTION

Alcohol ingestion is considered to be one of the major factors in the pathogenesis of gout and hyperuricemia . Ethanol increases the production of uric acid by the degradation of purine nucleotides[1], and decreases the renal clearance of uric acid as a result of the accumulation of lactate[2]. Among the alleles of aldehyde dehydrogenase 2 (ALDH, E.C.1.2.1.3), an $ALDH2*2$, which is associated with the phenotype of the lower enzyme activity, is widely distributed in Mongoloid population including Japanese, but not in the Caucasoid population[3-7]. On alcohol ingestion, those individuals who are homozygous for $ALDH2*2$ show exaggerated acute responses to ethanol including facial flushing and palpitation, because of the rapid accumulation of acetaldehyde. Heterozygotes for *ALDH2*1/ALDH2*2* also show these responses to ethanol but less severely[3-7].

To know how ethanol intake is involved in the pathogenesis of gout and hyperuricemia, we have investigated the prevalence of the genotypes of *ALDH2* in Japanese patients with primary gout, and the extent of ethanol-induced purine nucleotide degradation was evaluated in individuals with three different types of ALDH2 genotypes.

MATERIALS AND METHODS

Genomic DNA was separated from peripheral blood mononuclear cells from 180 patients with gout and 39 patients with rheumatoid arthritis (RA). Portion of genomic DNA for aldehyde dehydrogenase was amplified by the PCR method using primers as previously described[5,7] and hybridized with allele specific oligonucleotide (ASO) of *ALDH2*1* (normal genotype) and *ALDH2*2* (mutated genotype).

Ten healthy male volunteers, including five homozygotes for *ALDH2*1*, four heterozygotes for *ALDH2*1/ALDH2*2* and two homozygotes for *ALDH2*2* drank 60 ml of Scottish Whiskey (ethanol 25.8 g), and venous blood and urine were obtained from these individuals serially. Plasma and urinary hypoxanthine, xanthine and uric acid were determined by HPLC, using a reversed phase column.

RESULTS

Frequency of genotype of ALDH2 in patients with gout:

Frequency of the genotypes of *ALDH2* in patients with gout was compared with those with RA and with a control group in Japanese population in a previous study[6].

Among 180 patients with gout, only four was homozygous for *ALDH2*2*, and the frequency of *ALDH2*2/ALDH2*2* genotype was significantly less in patients with gout compared to those with RA ($P < 0.001$) or the previous data for normal population ($P < 0.005$). Similarly, homozygotes with the genotype of *ALDH2*1* were more frequent in patients with gout than RA patients or normal subjects ($P < 0.005$) (Figure-1).

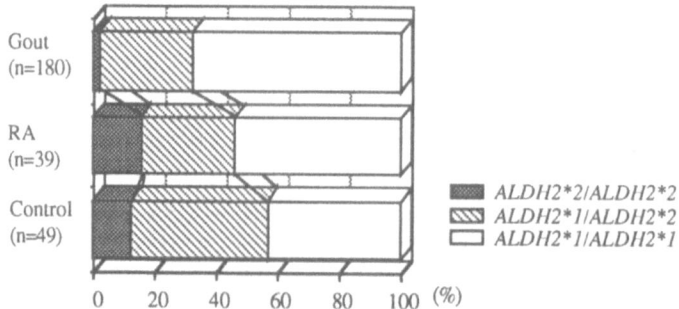

Figure 1. Frequency of *ALDH2* genotypes in Japanese population

Purine nucleotide degradation by alcohol drinking:

On 25.8 g ethanol drinking, blood acetaldehyde concentration were elevated highest in homozygotes for *ALDH2*2* .

Plasma hypoxanthine and urinary excretion were significantly higher in homozygotes for *ALDH2*1* than in heterozygotes for *ALDH2*1/ALDH2*2* (p < 0.001). Neither of plasma nor urinary hypoxanthine was elevated in homozygotes for *ALDH2*2* (Figure-2).

In homozygotes for *ALDH2*1* , greater elevation of plasma and urinary hypoxanthine was observed in those with daily alcohol drinking habit than those without daily drinking. On the contrary, in hererozygotes for *ALDH2*1/ALDH2*2* , almost no elevation of plasma and urinary hypoxanthine was observed even in those with daily drinking habit (data not shown).

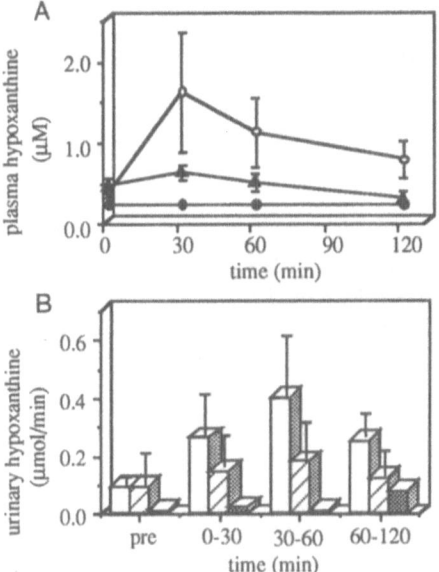

Figure 2. Plasma levels and urinary excretion of hypoxanthine on ethanol loading test.
Plasma level (Panel A) and urinary excretion (Panel B) of hypoxanthine were demonstrated in individuals with homozygotes for *ALDH2*1* (-○-,□), *ALDH2*2* (-●-,■), and heterozygotes for *ALDH2*1/ALDH2*2* (-▲-,▨).

DISCUSSION

The mechanism by which alcohol ingestion induces the purine nucleotide degradation has been well investigated. Rapid metabolism of ethanol through acetate to acetyl-CoA consumes considerable amounts of cellular ATP, which is a well known inducer of the degradation of purine nucleotides[8], and as a consequence, uric acid production increases.

ALDH2 is the key enzyme of ethanol metabolism in humans, and was shown to be polymorphic. The homozygotes of *ALDH2*2* , the genotype associated with a lower ALDH activity, are present in approximately 10 % of the Japanese and Asian populations, but extremely rare in Caucasoid[6]. The individuals homozygous for *ALDH2*2* are quite sensitive to ethanol, and cannot have daily drinking custom. Heterozygotes with genotypes of *ALDH2*1/ALDH2*2* also accumulate acetaldehyde in lesser amount, and they usually do not drink alcohol beverages so much, if any. However, in these people, daily drinking may induce ALDH2 enzyme and make them resistant to alcohol.

Thus, the *ALDH2*2* gene is likely to protect people from some diseases in which alcohol ingestion is closely involved in the pathogenesis. Indeed, strikingly low frequencies of *ALDH2*2* allele are reported in patients with alcoholic liver disease and chronic alcoholism[9].

If alcoholic consumption is a major factor in the pathogenesis of gout, then the prevalence of the *ALDH2*2/ALDH2*2* genotype in patients with gout should be extremely low. Our results showed that the frequency of *ALDH2*2* allele is significantly lower in patients with primary gout compared to control groups. Thus, alcohol ingestion may not be a requisite factor but is deeply involved in the pathogenesis of gout and hyperuricemia.

When loading ethanol, those homozygotes for *ALDH2*1* but not those with *ALDH2*2* induced purine nucleotide degradation. This phenomenon was prominent in individuals with daily alcohol drinking behavior. Thus, together with the data that indicated that the majority of Japanese patients with gout are homozygotes for *ALDH2*1*, individuals with homozygous for *ALDH2*1* are susceptible for gout and hyperuricemia, because they can readily induce purine nucleotide degradation on ethanol metabolism and furthermore, can drink a lot of alcoholic beverages.

Gout and hyperuricemia have been considered as a polygene disease whose phenotype is influenced by environmental factors, and alcohol drinking is only one of them. We have provided additional evidence for a critical role of alcohol in the development of gout and hyperuricemia from the molecular data.

ACKNOWLEDGMENTS

Authors thank to Ms. Tomoko Tsuchiya and Ms. Sanae Otsuka for excellent technical assistance. Portion of this research has been supported by a grant from Japan Gout Research Foundation, 1991.

REFERENCES

1. Faller, J. and Fox, I.H.: Ethanol-induced hyperuricemia. Evidence for increased urate production by activation of adenine nucleotide turnover. N. Eng. J. Med. 307:1598 (1982).

2. Lieber, C.S., Jones, D.P., Losowsky, M.S., Davidson, C.S.: Interaction of uric acid and ethanol metabolism in man. J. Clin. Invest. 41:1863 (1962).

3. Harada, S.S., Misawa, S., Agarwal, D.P., and Goedde, H.W. : Liver aldehyde dehydrogenase in Japanese-isozyme variation and its possible role in alcohol intoxication. Am. J. Hum. Genet. 32:8 (1980).

4. Yoshida, A., Haung, I-Y. and Ikawa, M. Molecular abnormality of an inactive aldehyde dehydrogenase variant commonly found in Orientals. Proc. Natl. Acad. Sci. U.S.A.. 81:258 (1984).

5. Crabb, D.W., Edenberg, H.J., Bosron, W.F., and Li, T-K. Genotypes for aldehyde dehydrogenase deficiency and alcohol sensitivity. J Clin Invest 83:314 (1989).

6. Shibuya, A. and Yoshida, A. Frequency of the atypical aldehyde dehydrogenase-2 gene (ALDH2/2) in Japanese and Caucasians. Am. J. Hum. Genet. 43:741 (1988).

7. Xu, Y., Carr, L.G., Bosron, W.F., Li, T-K and Edenberg, H. Genotyping of human alcohol dehydrogenases at the ADH2 and ADH3 loci following DNA sequence amplification. Genomics 2:209 (1988).

8. Fox, I.H., Pallela, T.D., and Kelley, W.N.: Hyperuricemia: a marker for cell energy crisis. N Engl. J Med 317:111 (1987).

9. Shibuya, A. and Yoshida, A.: Genotypes of alcohol-metabolizing enzymes in Japanese with alcohol liver disease: A strong association of the usual Caucasian-type aldehyde dehydrogenase gene (ALDH1/2) with the disease. Am J Hum Genet 43:744 (1988).

HYPOURICEMIA AND AN INCREASED CLEARANCE OF URIC ACID ARE OBSERVED IN LIVER DISEASES ?

A. Pelatti, C. P. Quaratino, C. D' Amario, R. Tentarelli,
G. Riario Sforza and A. Giacomello

Institute of Medical Physiopathology, University of Chieti and
Ospedale Civile di Pescara
Italy

Hypouricemia has been noted in patients with hepatocellular disease.[1-5] The hypouricemia may be chiefly due to induced tubular defects but reduced production may also play a role. In the presence of liver disease either a uricemic substance (perhaps bilirubin) accumulates or a substance that enhances uric acid reabsorption is not produced. The purpose of the present paper was to determine the prevalence of low serum urate levels in liver diseases and to evaluate which degree of hepatocellular damage should be present in order to observe this metabolic alteration.

Patients and methods

To answer these questions an epidemiological study on the correlations between serum urate, fractional urate excretion (FE) [(urate clearance / creatinine clearance) x100], total serum bilirubin levels and serum ALT (alanine aminotransferase) activity was carried out .

During a few months period serum urate, creatinine , total bilirubin levels, serum ALT activity and urinary creatinine and uric acid concentrations were determined in all outpatients from a Hospital of an Italian region with a high prevalence of liver diseases. 991 outpatients were studied (512 women).

Furthermore, serial measurements (at least 6) of these variables were carried out in patients with liver diseases in order to determine if in the same subject an increase in serum ALT activity or in serum bilirubin levels was associated with a decrease of serum urate concentration and (or) to an increase of fractional urate excretion.

18 patients were studied (7 females); 7 had acute and 9 had chronic hepatitis; 2 subjects were cirrhotic.

Purine and Pyrimidine Metabolism in Man VIII, Edited by
A. Sahota and M. Taylor, Plenum Press, New York, 1995

MALES

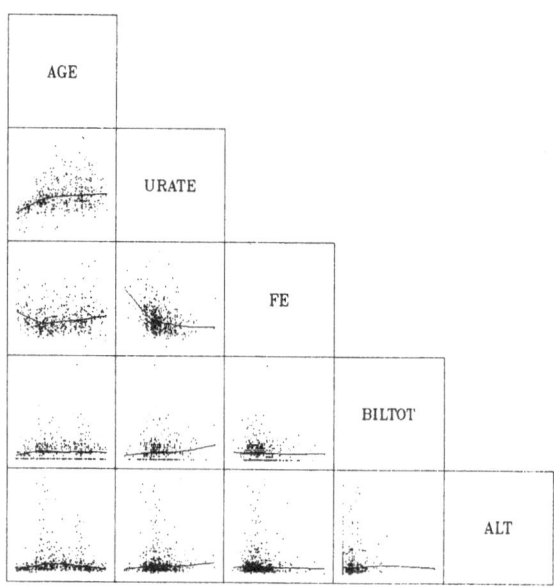

FEMALES

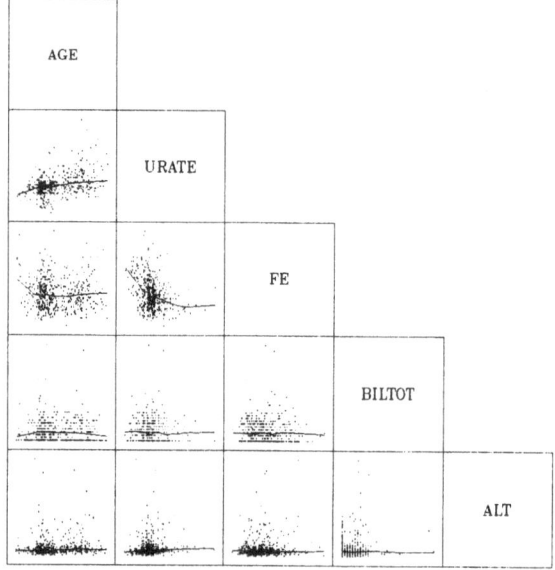

Figure 1. Scatterplot matrices of the measured variables

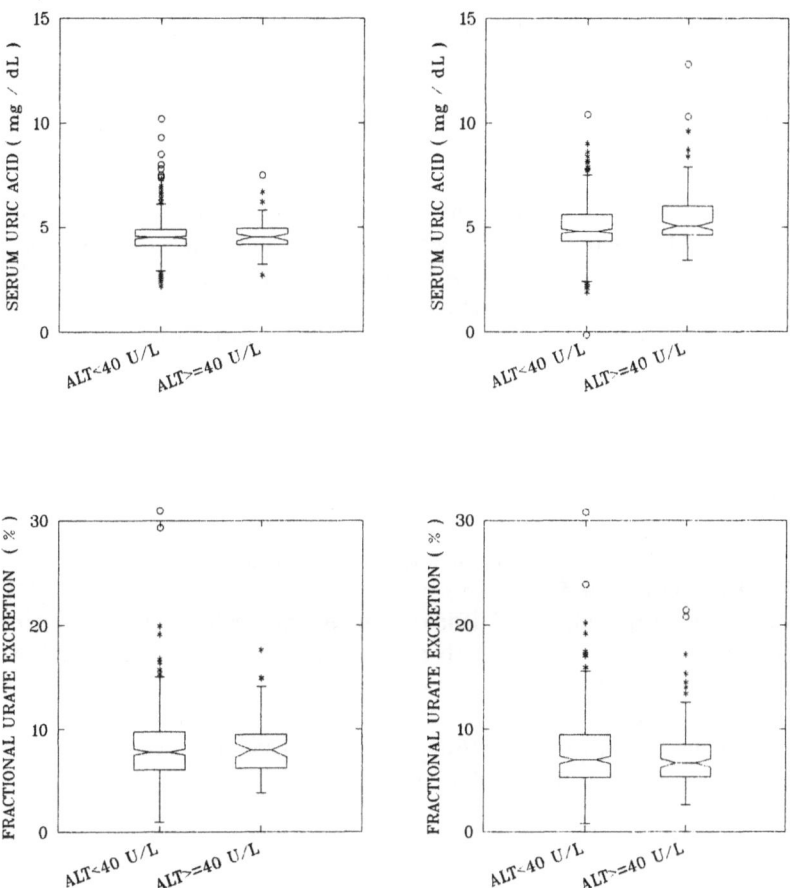

Figure 2. Grouped notched box plots of the examined variables

Results and Discussion

Figure 1. shows the scatterplots of the main variables measured in outpatients. No negative significant correlation was observed between serum ALT activity or total serum bilirubin and serum urate levels.

124 men and 57 women had ALT serum activity higher than 40 U/L (range 41-718 U/L). In these subjects serum urate level were lower then those obtained in subjects with ALT activities in the normal range (<=40 U/L) (Figure 2). Furthermore, as shown in Figure 2, fractional urate excretion was comparable in the two groups of subjects. In both sexes no significant correlation was observed between fractional urate excretion and total serum bilirubin (range : 0.1- 5.7 mg/dL).

In patients with acute hepatitis an inverse relationship between serum ALT or total serum bilirubin was evident only in one subject in which serum ALT activity reached very high values (10.000 U/L). In this subject serum total bilirubin ranged from 0.1 to 8 mg/dL. In the other subjects with an ALT activity lower then 6.000 U/L and with serum total bilirubin values up to 16 mg/dL this trend was not clearly evident.

In the patients with chronic hepatitis and with cirrhosis no negative correlation between serum urate and ALT activity or total serum bilirubin could be observed. In these patients ALT activity was lower than 300 U/L and total serum bilirubin concentration was lower than 7.5 mg/dL. These results lead to the conclusion that a decrease in serum urate levels cannot be frequently observed in liver disorders and perhaps occurs only in the presence of very severe hepatocellular damage.

References

1. H.Ullmann , Zur Frage der Harnsaureausscheidung in Urin bei Ikteruskranken, Klin Wochenschr 2 : 2174-2175,(1923).
2. G. Pasero, G. Masini , L' ipouricemia negli itteri colurici. Minerva Med. 49 : 3155 - 3158, (1958).
3. R. Matz , J. Christodoulou , N.Vianna , at al , Renal tubular dysfunction associated with alcoholism and liver disease. NY State J. Med. 69 : 1312 - 1314, (1969).
4. M.F. Michelis, P.C. Warms, R.D. Fusco, at al , Hypouricemia and hyperuricosuria in: Laennec cirrhosis. Arch. Intern. Med.134 : 681 - 683 , (1974).
5. L. Schlosstein, I. Kippen, R. Bluestone, et al , Association between hypouricemia and jaundice. An Rheum .Dis. 33: 308 - 312, (1974)

THE INFLUENCE OF ALLOPURINOL ON RENAL DETERIORATION IN FAMILIAL NEPROPATHY ASSOCIATED WITH HYPERURICEMIA (FNAH)

María Eugenia Miranda on behalf of the Spanish Group for the Study of FNAH

Divisions of Internal Medicine, Clinical Biochemistry, Nephrology, Pediatrics, Radiology and Pathology, "La Paz" Hospital, Universidad Autónoma, Madrid, Spain

INTRODUCTION

The kidney has been traditionally considered as one of the two primary target organs in gout (1), but long-term follow-up studies have failed to show a deleterious effect of gout on renal function (2). FNAH is characterized by hyperuricemia or gout, usually appearing at an early age and often in both sexes, associated with progressive renal failure (3-17). The pathogenesis of this syndrome is controversial. Some authors believe that renal deterioration is the result of sustained hyperuricemia because in some patients allopurinol halted the progression of renal insufficiency (8,16). Most authors, however, have documented the progression of renal disease despite allopurinol therapy (7,10,11,14,15,17). We examined the influence of allopurinol on the evolution of renal disease in patients with FNAH.

PATIENTS AND METHODS

Three Spanish families with FNAH were included in this prospective study (Figure 1). The proband of Family M, Patient II-1, was referred in 1987 to "La Paz" University Hospital for evaluation of hyperuricemia and arterial hypertension. He knew that his mother (patient I-1) and his two brothers (patient II-2 and patient II-3) had increased serum urate levels. The proposita of Family T (patient II-2) and Family A (patient III-1) were referred to us in 1988 and 1992, respectively, because of hyperuricemia and renal insufficiency. Clinical findings prompted the study of other family members. Twelve patients with FNAH belonging to these 3 families were followed-up. Routine hematological determinations, biochemical parameters, urinalysis, and blood and urinary urate and creatinine concentrations were assayed by standard laboratory techniques. One hypertensive patient (patient II-1 of Family M) and one patient who developed arterial hypertension during follow-up (patient II-2 of Family T) received a hyposodic diet and enalapril. Five patients with gout arthritis and one asymptomatic female patient (patient II-3, Family A) were treated with allopurinol (300 mg/day) for 29-patient-years (Group A). Five patients with no articular symptoms and one female patient with gout did not receive allopurinol therapy for 15-patient-years (Group B). Blood pressure, and serum and urinary concentrations of urate and creatinine were determined in each patient every 6 months.

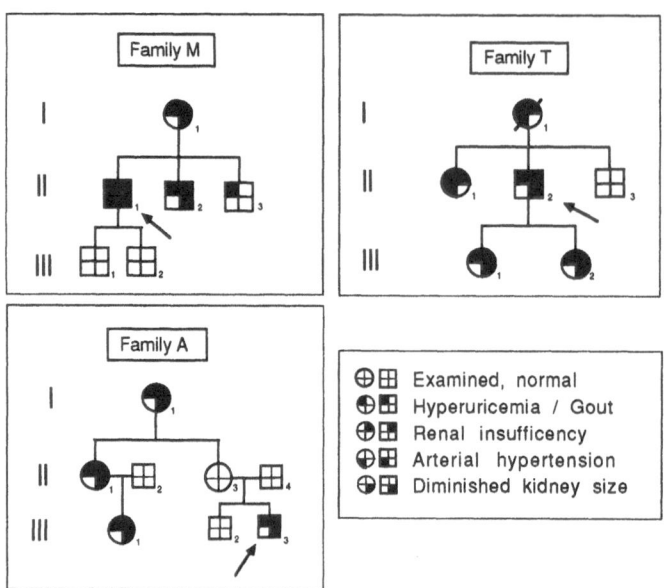

Figure 1. Pedigree of three families with familial nephropathy associated with hyperuricemia or gout (FNAH). Roman numerals denote generations; squares denote men and circles denote women; shaded squares denote pathological features at initial evaluation and arrow, proposita. The slash indicates a dead subject.

RESULTS

The mean age of the patients in Group A was 29 years (range, 15 to 46 years) and in Group B, 48 years (range, 12 to 87 years) (Table 1). Low-sodium diet and enalapril normalized blood pressure (below 140/85 mm Hg) in hypertensive patients. Mean baseline serum urate concentrations were similar in both groups (7.5 vs 7.2 mg/dL). Mean serum urate concentration diminished in allopurinol-treated patients (Group A) to 4.6 mg/dL (P=0.028), whereas it remained elevated in Group B (7.9 mg/dL). Mean baseline serum creatinine level was 2.0 mg/dL in Group A and 1.4 mg/dL (P=0.200) in Group B. Serum creatinine increased to 4.5 mg/dL (range, 1.6 to 8.1 mg/dL; P=0.027) in allopurinol-treated patients. Similarly, serum creatinine increased to 2.1 mg/dL (range, 1.0 to 3.0 mg/dL; P=0.028) in patients not treated with allopurinol (Table 1, Figure 2). The mean increase in serum creatinine levels did not significantly differ between groups (Group A, 93% vs Group B, 54%; P=0.810).

Individual follow-up data shows that allopurinol therapy does not influence the course of renal disease in FNAH. At initial evaluation patient II-3 of Family M had gout but normal serum creatinine levels (0.9 to 1.2 mg/dL) and GFR (104 ml/min/1.73 m2). His serum urate concentration was maintained below 5.0 mg/dL with allopurinol. Four-years later, serum creatinine increased to 1.5 to 1.7 mg/dL and GFR decreased to 72 ml/min/1.73 m2. Patient II-2 of Family M and Patient II-2 of Family T had gout, renal insufficiency, diminished kidney sizes but normal blood pressure when first examined. Four and three years later, respectively, and while being on allopurinol therapy, both developed arterial hypertension and evidenced a decline in renal function (Figure 2). Two sisters aged 15 and 12 years (patients III-1 and III-2 of Family T) had baseline serum creatinine concentrations of 1.7 and 1.4 mg/dL, respectively. Only the older patient was treated with allopurinol which normalized serum urate levels (3.7 mg/dL). Serum urate remained elevated in the younger (10.5 mg/dL). After 5 years of follow-up, serum creatinine was 2.0 mg/dL in both patients. In summary, despite control of gout and hyperuricemia, creatinine clearance decreased in all patients. The rate of decline in

Table 1. Clinical and laboratory data in the studied patients.

Patient	Age	Sex	Follow-up	Serum urate (mg/dL)		Serum creatinine (mg/dL)	
(Family)	(yrs)		(yrs)	Initial	Final	Initial	Final
GROUP A, patients treated with allopurinol							
II-1(M)	34	M	6	7.1	5.6	2.4	8.1
II-2(M)	33	M	6	6.8	3.8	2.4	5.7
II-3(M)	28	M	6	6.3	3.7	1.2	1.7
II-2(T)	46	M	5	9.8	6.1	2.9	5.8
III-1(T)	15	F	5	8.0	3.7	1.7	2.0
III-3(A)	20	M	1	7.1	4.4	1.3	1.6
GROUP B, patients nontreated with allopurinol							
I-1(M)	59	F	6	7.4	6.6	1.9	2.7
III-2(T)	12	F	5	8.9	10.5	1.4	2.0
I-1(A)	87	F	1	6.3	7.1	1.3	1.6
II-1(A)	53	F	1	8.5	8.5	1.3	1.6
II-3(A)	52	F	1	4.6	6.9	0.8	1.0
III-1(A)	22	F	1	7.7	7.7	1.6	2.1

creatinine clearance ranged from -0.17 mL/month (Patient II-1, Family M) to -0.67 mL/month (Patient II-3, Family M).

DISCUSSION

Chronic hyperuricemia has long been considered deleterious for the kidney, and according to some authors FNAH represents a severe form of gouty nephropathy (8,9,16). However, three lines of evidence argue against this hypothesis. First, the concept of gouty nephropathy as a complicating renal disease secondary to increased serum urate levels has vanished (18). Second, urate crystal deposits have been thought to be a distinctive feature of gouty nephropathy (1) but despite intensive search, with the exception of one case (9), in no other previously reported patients have uric acid crystals been encountered (4,10,11,17). Third, the results of our follow-up studies indicate that renal deterioration similarly progresses in allopurinol-treated and untreated patients with FNAH. We conclude that this syndrome is a distinctive interstitial nephropathy presumably inherited as an autosomal dominant trait; it may progress to renal failure and is not halted nor prevented by an effective control of hyperuricemia. Thus, allopurinol does not appear to influence the evolution of renal disease in this familial syndrome.

ACKNOWLEDGEMENTS

Investigators of the Spanish Group for the Study of FNAH are: Internal Medicine, JG. Puig, and MªE. Miranda; Clinical Biochemistry, F. Mateos, A. Herranz, I. Martínez, and R. Torres; Nephrology, L. Espinosa, and J. Martínez Ara; Radiology, T. Berrocal, and C. Prieto; Pathology, Mª Luz Picazo. We are indebted to the Clinical Research Unit nursing staff and the dietetic staff for excellent patient care; to T. Ramos, Javier Díaz, and Mª Paz Canencia for valuable technical assistance; to Rosario Madero for the statistical analysis; to Erik Lundin for assistance in preparing the manuscript. Supported by grants from Caja de Madrid and Fondo de Investigación Sanitaria (FIS, 94/0218), Spain.

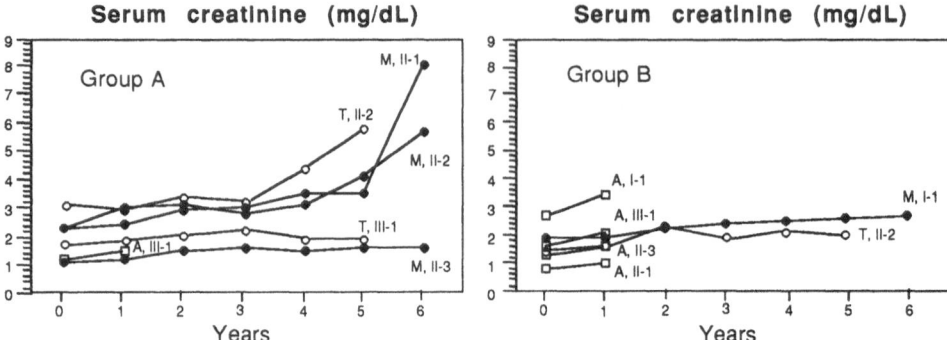

Figure 2. Serum creatinine over time in patients with FNAH treated (Group A) and untreated (Group B) with allopurinol.

REFERENCES

1. Talbott JH, Terplan KL. The kidney in gout. Medicine (Balt) 1960; 39:405-7.
2. Berger L, Yü T-F. Renal function in gout. IV. An analysis of 524 gouty subjects including long-term follow-up studies. Am J Med 1975; 59:605-13.
3. Duncan H, Dixon AS. Gout, familial hyperuricemia, and renal disease. Q J Med 1960; 29:127-35.
4. Rosenbloom FM, Kelley WN, Carr AA, Seegmiller JE. Familial nephropathy and gout in a kindred (abstract). Clin Res 1967; 15:270.
5. Treadwell BLJ. Juvenile gout. Ann Rheum Dis 1971; 30:279-84.
6. Van Goor W, Kooiker CJ, Mees EJD. An unusual form of renal disease associated with gout and hypertension. J Clin Path 1971; 24:354-9.
7. Massari PU, Hsu CH, Barnes RV, Fox IH, Gikas PW, Weller JM. Familial hyperuricemia and renal disease. Arch Intern Med 1980; 140:680-4.
8. Simmonds HA, Warren DJ, Cameron JS, Potter CF, Farebrother DA. Familial gout and renal failure in young women. Clin Nephrol 1980; 14:176-82.
9. Farebrother DA, Pincott JR, Simmonds HA, Warren DJ, Dillon MJ, Cameron JS. Uric acid crystal-induced nephropathy: Evidence for a specific renal lesion in a gouty family. J Pathol 1981; 135:159-68.
10. Richmond JM, Kincaid-Smith P, Whitworth JA, Becker GJ. Familial urate nephropathy. Clin Nephrol 1981; 16:163-8.
11. Leumann EP, Wegmann W. Familial nephropathy with hyperuricemia and gout. Nephron 1983; 34:51-7.
12. Fleckenstein JL, Simkin PA. Corrected clearance identifies underexcretion of uric acid in a gouty kindred. Adv Exp Med Biol 1986; 195A:295-7.
13. Calabrese G, Simmonds HA, Cameron JS, Davies PM. Precocious familial gout with reduced fractional urate clearance and normal purine enzymes. Q J Med 1990; 75:441-50.
14. Murakami T, Kawakami H, Nakatsuka K, Jojima K, Nohno H, Matsuzaki H. Underexcretory-type hyperuricemia, disproportionate to the reduced glomerular filtration rate in two boys with mild proteinuria. Nephron 1990; 56:439-42.
15. Yokota N, Yamanaka H, Yamamoto Y, Fujimoto S, Eto T, Tanaka K. Autosomal dominant transmission of gouty arthritis with renal disease in a large Japanese family. Ann Rheum Dis 1991; 50:108-11.
16. Moro F, Ogg C, Simmonds HA, et al. Familial juvenile gouty nephropathy with renal urate hypoexcretion preceding renal damage. Clin Nephrol. 1991;35:263-69.
17. Puig JG, Miranda MªE, Mateos FA, Picazo ML, Jiménez ML, Calvin TS, Gil AA. Hereditary nephropathy associated with hyperuricemia and gout. Arch Intern Med 1993; 153: 357-65.
18. Beck LH. Requiem for gouty nephropathy. Kidney Int 1986; 30:280-7.

ULTRASOUND IMAGING AND COLOUR DOPPLER STUDIES IN FAMILIAL NEPHROPATHY ASSOCIATED WITH HYPERURICEMIA (FNAH)

Consuelo Prieto, and Teresa Berrocal on behalf of the Spanish Group for the Study of FNAH

Divisions of Radiology, Internal Medicine, Clinical Biochemistry, Nephrology, and Pathology, "La Paz" Hospital, Universidad Autónoma, Madrid, Spain

INTRODUCTION

FNAH is an inherited disease of unknown pathogenesis. By means of clearance techniques we have demonstrated that diminished renal plasma flow and increased renal vascular resistance may be primary pathogenic events in FNAH (1). Ultrasonography and two-dimension echo-color-doppler (2D ECD) are two methods that have been applied to the study of renal morphology and the intrarenal vascular bed, respectively. Sonography is routinely used for the study of kidney size (2). In addition, sonography allows the assessment of the parenchymal thickness and the cortico-medular echogenicity (3). Two-dimensional ultrasound imaging and blood flow detection with color doppler are accepted methods to quantify intrarenal vascular resistance (4). In this study we have examined renal morphology and the intrarenal vascular bed in patients with FNAH.

SUBJECTS AND METHODS

Twenty-four subjects (10 males and 14 females aged 5 to 86 years) belonging to 5 Spanish families with FNAH were studied. Fifteen subjects were considered to be affected because they were related to patients with FNAH and showed at least one of the main four characteristics of the disease: hiperuricemia or gout, arterial hypertension, renal insufficiency and decreased kidney size. Nine subjects were healthy members of these 5 families. The studies were performed after all subjects fasted for at least 12 hours. Real time ultrasonography was performed with a 3.5 MHz sectorial transducer. Four morphological parameters were evaluated: kidney size, parenchimal thickness, echogenicity and cortico-medular differentiation. Longitudinal and transversal standard scans of both kidneys were obtained. Kidney size was measured from the longitudinal and antero-posterior diameters. The observed values were compared to those previously obtained in normal subjects (2). Parenchymal thickness was equated to the distance from the capsula to the most external part of the renal sinus at both poles and the medial third (3). The kidney cortical echogenicity was compared to the liver echogenicity and was considered normal if the former was lower (3). Normally, the renal medula shows a lower echogenicity than the renal cortex. Thus, nonvisualization of the pyramids suggests parenchymal involvement. 2D ECD served to assess the vascular map of the kidneys. An

Table 1. Ultrasonography and 2D echo-color-doppler parameters in subjects belonging to five families with FNAH.

Patient (Family)	Age (yrs)	Clinically affected[1]	Kidney size Right	Left	Parenchimal thickness	Parenchimal echogenicity	Vascular map	Resistance index (RI)	Pulsatility index (PI)	Velocity blood pressure index (VBI)
Normal values			10-12x4.0-4.5		1.1-1.4	N	N	<0.65	<1.1	<2.74
I-1 (M)	59	Yes	8.1x3.0	7.5x3.2	0.4-0.8	high	poor	0.67	1.3	3.03
II-1 (M)	34	Yes	9.0x2.7	7.5x2.7	0.5-0.9	high	poor	0.59	1.2	4.08
II-2 (M)	33	Yes	8.9x3.5	8.7x3.7	0.5-1.1	high	poor	0.76	2.2	5.93
II-3 (M)	28	Yes	9.5x3.2	10.5x4.0	0.4-1.4	high	poor	0.60	1.2	3.76
III-2 (M)	10	No	9.0x3.5	10x3.7	1.1-1.4	N	N	0.57	0.8	2.32
III-1 (M)	5	No	8.5x3.5	9.0x3.7	1.1-1.4	N	N	0.70	1.4	2.24
III-1 (T)	15	Yes	8.4x2.8	8.5x3.3	0.5-1.2	high	poor	0.73	1.4	4.81
III-2 (T)	12	Yes	8.0x2.8	8.5x3.8	0.6-1.5	high	NC	0.68	1.2	3.33
I-1 (A)	86	Yes	6.1x2.6	6.7x3.0	0.4-1.0	high	NC	0.80	2.1	ND
II-1 (A)	53	Yes	8.7x2.8	8.4x3.7	0.4-1.1	high	NC	0.74	2.3	4.11
II-2 (A)	53	No	10.0x4.7	11.1x4.9	1.1-1.5	N	N	0.55	0.9	2.71
III-1 (A)	22	Yes	9.5x2.6	9.0x4.7	0.5-2.0	high	NC	0.69	1.4	3.15
III-3 (A)	20	Yes	9.2x3.5	9.4x4.7	0.6-1.8	high	NC	0.61	1.0	1.96
I-1 (L)	47	No	12.0x5.0	12.2x5.0	1.0-1.5	N	N	0.64	1.1	ND
I-2 (L)	43	No	10.6x4.5	11.4x5.0	0.8-1.4	N	N	0.64	1.5	4.65
II-1 (L)	17	Yes	9.7x3.0	9.5x3.5	0.4-1.5	N	N	0.68	1.5	3.46
II-2 (L)	15	Yes	10.5x3.0	10.4x3.6	1.1-1.5	N	N	0.62	1.2	2.75
II-3 (L)	14	No	10.0x4.0	10.0x4.5	1.1-1.6	N	N	0.62	1.1	2.43
II-4 (L)	13	No	9.0x3.5	9.2x3.7	1.1-1.5	N	N	0.63	1.1	2.36
I-1 (H)	46	Yes	11.0x4.5	11.2x4.5	1.0-1.4	N	N	0.62	1.2	4.46
I-2 (H)	44	No	10.3x3.7	11.0x3.7	1.0-1.4	N	N	0.65	1.1	1.91
II-1 (H)	17	Yes	10.0x3.5	9.4x4.5	1.0-1.4	N	N	0.62	1.1	1.69
II-2 (H)	15	Yes	9.6x3.7	9.4x4.0	1.1-1.3	N	N	0.62	1.3	2.17

[1]A subject was defined as clinically affected if showing hyperuricemia or gout, renal insufficiency, arterial hypertension or diminished kidney size. N, denotes normal; ND, not determined; NC, not conclusive. Kidney size and parenchimal thickness are given in centimeters.

electrocardiogram and blood pressure were obtained simultaneously (4). The hemodynamic study obtained with pulsed doppler is based on the spectral analysis of doppler curves (4). Quantification of the vascular resistance was obtained by calculating the resistance index (RI=Vsyst-Vdiast/Vdiast; being Vsyst, systolic velocity and Vdiast, diastolic velocity) and the pulsatility index (PI=Vsyst-Vdiast/mean syst. and diast. velocities). Vsyst and Vdiast were measured in at least two interlobar arteries of each kidney (4-6). The velocity blood pressure index (VBI) was equated to SBP-DBP/MAP; being SBP, systolic blood pressure; DBP, diastolic blood pressure; and MAP, mean arterial pressure (6). All measures were obtained in a blinded manner. To assess the clinical usefulness of the different parameters obtained by pulsed doppler two by two tables were constructed and sensibility, specificity and predictive values were calculated.

RESULTS

Among the 15 affected subjects with FNAH, 13 (87%) showed a decreased kidney size (Table 1). Patients with severe renal insufficiency had markedly diminished kidney size. Kidney size was normal in two patients (II-2, Family L and I-1, Family H) (see family pedigrees elsewhere in this volume) with asymptomatic hyperuricemia. In contrast, kidney size was diminished in subject II-3 (Family A) with normal serum urate levels, renal function and blood pressure. This obligate carrier showed during follow-up hyperuricemia and decreased inulin clearance and renal plasma flow. Eleven patients (73%) evidenced some degree of parenchimal atrophy, focal in 6 and diffuse in 5. An increased kidney echogenicity and/or loss of cortico-medular differentiation was found in 10 patients (67%). Parenchimal thickness, echogenicity and cortico-medular differentiation were normal in non-affected subjects.

By means of 2D ECD five (36%) patients clearly showed a poor vascular map (Table 1). In 6 subjects it could not be established whether the vascular map was normal or poor (not conclusive). An increase of RI, PI and VBI was documented in 8 (53%),12 (87%), and 10 (71%) of the patients, respectively. Among unaffected subjects, RI, PI and VBI were normal in 7 (78%), 5 (56%) and 5 (63%), respectively. The RI showed a low sensibility and a high specificity. The PI and the VBI had a similar and high sensibility and specificity with an accuracy of 71 and 80%, respectively.

DISCUSSION

This study shows that ultrasonography is highly sensitive in the detection of kidney abnormalities in patients with FNAH. In addition, a certain parallelism was evident

Table 2. Sensibility, specificity and predictive values of different parameters obtained by means of two-dimension echo-color-doppler in patients with FNAH.

	Renal vascular map	Resistance index (RI)	Pulsatility index (PI)	Velocity blood pressure index (VBI)
Sensibility	50[1]	53	87	79
Especificity	100	100	71	83
Positive predictive value	100	100	82	92
Negative predictive value	55	50	87	63
Accuracy	65	68	71	80

[1]Data are in percentages.

between the disease severity and the ultrasonographic abnormalities. All patients with two main characteristics of the disease showed diminished kidney size, parenchimal atrophy and abnormal echogenicity. In only two patients with mild hyperuricemia was the ultrasonography study within normal limits.

Among the four parameters examined with 2D ECD (vascular map, RI, PI and VBI) the least sensitive was the appearance of the vascular map since this image may be influenced by several counfounding variables such as obesity, meteorism or patient collaboration. However, it facilitated the localization of the vessels that were further studied with pulsed doppler (5). By means of clearance studies we showed that patients with FNAH have markedly increased intrarenal vascular resistance (1). In fact, two obligate carriers with no evidence of the disease showed an increased vascular resistance as the only abnormal parameter of renal function. These two subjects evidenced normal RI, but one patient (subject I-1 of Family H) showed both increased PI and VBI when examined with 2D ECD. The low sensibility of the RI may be due to the fact that this index only evaluates the maximal systolic and minimal diastolic blood velocities and does not take into account the blood velocity variations during the cardiac cycle (5). In contrast, the PI varies proportionally to the peripheral vascular resistance. However, PI might also depend on several confounding variables such as the artery diameter, blood viscosity and blood pressure. These physiological variables led Bardelli et al (6) to define the BPI which represents a "dimensionless" ratio between pulse pressure and mean arterial pressure. Since in this study the predictive values of PI and VBI were similar, we propose the use of the PI as a working parameter to assess intrarenal vascular resistance. We conclude that (a) ultrasonography is the method of choice to examine kidney morphology in FNAH, (b) the most relevant ultrasonographic findings were: decreased kidney size, parenchimal atrophy and increased cortical echogenicity with loss of cortico-medular differentiation, (c) PI and VBI were markedly elevated in patients with FNAH, and (d) although not pathognomonic, a disturbance of vascular resistance is a common abnormality in patients with FNAH and confirms previous findings with clearance techniques.

ACKNOWLEDGEMENTS

We are indebted to Dr. I. Pastor for his valuable technical assitance, to Pilar Martínez and Victorina Riomoros for their excellent patient care and to Dr. I. Al-Assir and Erik Lundin for assistance in preparing the manuscript. Investigators of the Spanish Group for the Study of FNAH are: Radiology, T. Berrocal, and C. Prieto; Internal Medicine, JG. Puig, and MªE. Miranda; Nephrology, L. Espinosa, and J. Martínez Ara; Clinical Biochemistry, F. Mateos, A. Herranz, I. Martínez, and R. Torres; Pathology, Mª Luz Picazo. Supported by grants from Caja de Madrid and Fondo de Investigación Sanitaria (FIS, 94/0218), Spain.

REFERENCES

1. Puig JG, Miranda MªE, Mateos FA, et al. Hereditary nephropathy associated with hyperruricemia and gout.Arch Intern Med 1993; 153:357-65.
2. Brandt TD, Neiman HL, Dragowski MJ, et al. Ultrasound assessment of normal renal dimensions. J Ultrasound Med 1982; 1:49
3. Hricak H, Cruz C, Romanski R, et al. Renal parenchymal disease: Sonographic-histologic correlation. Radiology 1982; : 141-4.
4. Mastorakou I, Robbins M, Bywaters T. Resistance and pulsatility Doppler indices: how accurately do they reflect changes in renal vascular resistance. Br J Radiol 1993; 66:577-80.
5. Mostbeck G, Gossinger H, Mallek R, et al. Effect of heart rate on Doppler measurements of resistive index in renal arteries. Radiology 1990; 175:511-3.
6. Bardelli M, Jensen G, Volkmann R, Caidahl K, Aurell M. Experimental variations in renovascular resistance in normal man as detected by means of ultrasound. Eur J Clin Invest 1992; 22:619-24.

PURINE METABOLISM IN FEMALE PATIENTS WITH PRIMARY GOUT

Juan G. Puig,[1] Felícitas A. Mateos,[2] Mª.E. Miranda,[1] Rosa J. Torres,[2] Eugenio de Miguel,[3] Carlos Pérez de Ayala[3] and Antonio Gil[1]

Divisions of Internal Medicine,[1] Clinical Biochemistry,[2] and Rheumatology,[3] "La Paz" Hospital, Universidad Autónoma, Madrid, Spain

INTRODUCTION

Uncontrolled studies have shown that mean serum urate levels are higher in female patients with gout than in gouty male patients (1-5) and that daily uric acid excretion is similar (4) or lower (3,5) in the former than in the latter. These findings prompted the hypothesis that female patients with primary gout have a more defective tubular transport of uric acid than gouty male patients (3). In this prospective study, carried out under controlled conditions, we examined purine metabolism in female patients with primary gout. In addition, we investigated whether there are sex-related differences in purine metabolism among patients with primary gout.

PATIENTS AND METHODS

Fifty female patients diagnosed as having had gouty arthritis between May 1980 and September 1993 and who were followed up by the internal medicine and rheumatology divisions at "La Paz" University Hospital, Madrid, Spain, were selected for the study. The clinical characteristics of these patients have been previously described (9). For inclusion in the study, patients had to fulfill the following criteria: (a) demonstration of negatively birefringent crystals obtained by aspiration from synovial fluid or from soft-tissue deposits (tophus); (b) absence of secondary causes of gout; (c) serum creatinine concentration equal to or below 133 μmol/L (1.5 mg/dL); and (d) physical status stable with no medications that could influence uric acid metabolism.

Patients and normal females matched for age (within ± 2 years), race, and body mass index (within ± 1.0 Kg/m^2) were admitted to the Clinical Research Unit. Informed consent was obtained from all subjects. All medications were discontinued 4 weeks before the study, except for colchicine (0.5 mg/d), which was given as prophylaxis against acute gout arthritis. All subjects were placed on a weight-maintenance, isoenergy, purine-restricted diet (less than 75 mg/d of purines) with a protein content of 10% to 15% for 5 days before and throughout the studies. On the sixth to the tenth hospital day purine metabolism was examined by determination of plasma and 24-hour urinary uric acid, hypoxanthine and xanthine concentrations, and quantification of urinary radioactivity after the infusion of tracer doses of [8-^{14}C]adenine to radiolabel the adenine nucleotide pool. Twenty-four-hour urine samples were collected daily under toluene (3 mL) during the ensuing 5 days for measurements of creatinine, purine concentrations, and excretion of radioactivity. Venous blood was obtained at the end of 24-hour urine collections after

Purine and Pyrimidine Metabolism in Man VIII, Edited by
A. Sahota and M. Taylor, Plenum Press, New York, 1995

overnight fast and rest, according to the method previously described (6). The means of three determinations for plasma purines and five for urinary purines were used. The results of plasma and urinary purine measurements in female patients with primary gout were also compared with those obtained in male patients with primary gout, matched for age (within ± 2 years), race, and body mass index (within ± 1.0 Kg/m^2), and who were subjected to an identical protocol to assess purine metabolism.

Uric acid and creatinine in plasma and urine were measured in a multichannel autoanalyzer (Hitachi 737, Hitachi Ltd, Tokyo, Japan). Plasma and urinary hypoxanthine and xanthine were determined by high-performance liquid chromatography (6). Standard formulas were used to calculate the clearances of uric acid, hypoxanthine and xanthine and the fractional excretion of purines. Results are expressed as means and SEs, or 95% confidence intervals as indicated. The significance of the differences were assessed by means of the Mann-Whitney rank-sum test. Spearman rank coefficients were calculated to test the degree of linear association between selected variables. All tests were two tailed; a P value of 0.05 or less was considered to indicate a statistically significant difference.

RESULTS

Among the 50 women diagnosed as having had gouty arthritis, 37 patients did not meet the inclusion criteria: 5 patients did not have crystal-proved gout, 4 patients had secondary causes that might have contributed to the development of gout, 23 patients had mild renal insufficiency, and 5 patients were receiving diuretics for heart failure. In addition, 2 patients were unavailable for the study and one patient refused to participate. These exclusions left 10 patients with primary gout for further study. Their mean age was 63.8 years (range, 43 to 82 years) and all were caucasian. Gout was diagnosed in 8 patients after the onset of menopause. When gout was diagnosed none of the patients was taking medications known to influence uric acid metabolism (eg. diuretics). The articular features of gout were similar to those previously reported in 37 female patients with gout (3). Seven patients had osteoarthritis, 5 obesity, 5 arterial hypertension, 5 hypercholesterolemia, 2 alcoholism and 1 diabetes mellitus. The enzyme activities adenine phosphoribosyltransferase, hipoxanthine-guanine phosphoribosyltransferase and phosphoribosylpyrophosphate synthetase in dialyzed hemolysates were normal in all the patients (data not shown).

The results of glomerular filtration rate and purine metabolism were compared to those obtained in 10 healthy caucasian females (mean age, 64.0 years, range 42 to 84 years) of a similar age and body mass index (Table 1). Mean creatinine clearance was significantly lower in gouty female patients than in controls (P=0.017). Marked differences in purine metabolism were observed between female patients with primary gout and normal females (Table 1). The mean plasma urate concentration was significantly higher in female patients with primary gout than in controls (P=0.0024). The mean difference in plasma urate concentration was 147 µmol/L (95% confidence intervals [CI], 116 to 178 µmol/L). In contrast, daily urinary uric acid excretion was similar in both groups (P=0.212). The mean difference in urinary uric acid excretion was 0.42 mmol/d per 1.73 m^2 (95% CI, 0.12 to 0.72 mmol/d per 1.73 m^2). Female patients with primary gout evidenced significantly increased mean plasma hypoxanthine and xanthine concentrations and significantly decreased mean 24-hour urinary hypoxanthine and xanthine excretion rates (P<0.05) as compared with control females. The renal clearances of uric acid, hypoxanthine and xanthine and the fractional excretion of these purines were markedly lower (P<0.05) in gouty female patients than in normal females. The results of plasma and urinary purines in female patients with gout could be attributed to their lower creatinine clearances as compared to control females. To assess this hypothesis, patients and controls were matched according to their creatinine clearance (± 5 mL/min/1.73 m^2). The mean creatinine clearance in 6 gouty patients and 6 normal females was 86.7 and 87.4 mL/min/1.73 m^2, respectively (P=0.948). The significant differences in plasma and urinary purines between both groups were similar to those shown in Table 1, except for plasma and urinary xanthine concentrations which became non-significant (data not shown). Cumulative excretion of urinary radioactivity after [8-14C]adenine infusion was similar in 8 female patients with gout and in 5 normal females (Table 1).

Table 1. Purine Metabolism in Female Patients with Primary Gout, Normal Females and Male Patients with Primary Gout and Uric Acid Underexcretion.

	Female Gout (n = 10)	Normal Females (n = 10)	Male Gout (n = 20)
Age, y	63.8±3.6[1]	64.0±3.7	63.6±2.5
Body Mass Index (Kg/m^2)	28.8±1.7	27.6±1.5	28.0±0.6
PLASMA			
Creatinine (μmol/L)	85.7±5.1[2]	62.6±3.9	94.1±4.9
Urate (μmol/L)	415±19[2]	268±25	432±18
Hypoxanthine (μmol/L)	4.6±0.7[3]	2.3±0.4	4.0±0.5
Xanthine (μmol/L)	1.4±0.2[3]	0.8±0.1	1.1±0.1
URINE			
Creatinine (mmol/d per 1.73 m^2)	8.96±0.66	8.04±0.62	11.60±0.54[4]
Uric acid (mmol/d per 1.73 m^2)	2.15±0.21	2.57±0.22	2.29±0.14
Uric acid/creatinine (mmol/mmol)	0.25±0.03	0.33±0.04	0.20±0.01
Hypoxanthine (μmol/ g creatinine)	17±2[2]	38±4	19±3
Xanthine (μmol/ g creatinine)	14±4[3]	30±4	12±2
CLEARANCES			
Creatinine (mL/min per 1.73 m^2)	73.8±5.3[3]	99.0±6.4	91.5±4.7[d]
Uric acid (mL/min per 1.73 m^2)	3.7±0.4[2]	7.0±0.7	3.9±0.3
Hypoxanthine (mL/min per 1.73 m^2)	4.4±1.2[2]	10.3±1.7	5.2±0.8
Xanthine (mL/min per 1.73 m^2)	9.7±2.9[3]	26.6±6.9	10.0±1.2
FRACTIONAL EXCRETION (%)			
Uric acid	5.0±0.5[2]	7.1±0.5	4.3±0.3
Hypoxanthine	5.9±2.2[3]	10.5±1.4	6.1±1.2
Xanthine	13.1±4.6[3]	26.8±5.7	11.2±1.6
^{14}C-excretion[5] (% of the dose given)	5.24±0.61	5.46±0.66	4.90±0.62

[1]Values are means±SEs.

[2]P<0.01 versus normal females.

[3]P<0.05 versus normal females.

[4]P<0.05 versus female patients with primary gout.

[5]Determined in 8 female patients with primary gout, 5 normal females and 10 male patients with primary gout.

Twenty caucasian male patients with primary gout (mean age 63.6 years, range 41 to 80 years) were studied to assess sex-related differences in purine metabolism. Male patients with primary gout were of a similar age and had a body mass index similar to gouty female patients (Table 1). Mean plasma creatinine concentration was similar in both groups but mean 24-hour creatinine excretion was significantly higher in male than in female patients (P=0.011). Thus, mean creatinine clearance was significantly higher in male patients with primary gout than in gouty female patients (P=0.026). Purine metabolism showed a similar pattern in female and male patients with primary gout: increased plasma urate, hypoxanthine and xanthine concentrations and markedly diminished urinary excretion of hypoxanthine and xanthine, mean uric acid excretion rate being similar to controls. Uric acid, hypoxanthine and xanthine clearances and the fractional excretion of these purines were similarly diminished in female and male patients with primary gout. Plasma urate concentrations in the patients were inversely associated with the fractional excretion of uric acid (r=-0.520; P=0.003). Similarly, plasma hypoxanthine and xanthine levels were inversely related to the fractional excretion of hypoxanthine (r=-0.555; P=0.002) and xanthine (r=-0.384; P=0.040), respectively. However, no significant associations were found between the renal excretion of uric acid, hypoxanthine and xanthine in female and male patients with primary gout.

DISCUSSION

Previous studies have emphasized sex-related differences in the clinical features and uric acid metabolism of patients with gout (2,3,5,7,8). However, these studies included female patients with secondary gout or diuretic-induced gout. In addition, uric acid metabolism was examined under uncontrolled conditions. In our study, confunding variables that may influence purine metabolism, such as age, race, body mass index, exercise, drugs and renal insufficiency, were carefully controlled. The increased and relatively diminished concentrations of uric acid, hypoxanthine and xanthine in plasma and urine, respectively, suggest a tubular dysfunction for purine excretion in female patients with primary gout. Conditions associated with increased plasma oxypurine concentrations are accompanied by an enhanced urinary excretion of uric acid, hypoxanthine and xanthine (6). In this study, gouty female patients evidenced mean plasma purine concentrations almost twice and mean urinary purine excretion rates about half compared to those of normal women. These results most likely indicate that the renal excretion of uric acid, hypoxanthine and xanthine is severely impaired in female patients with primary gout. The second aim of this study was to determine whether the magnitude of the tubular dysfunction for purine excretion in female gout was similar to that documented in gouty male patients (6). No significant sex-related differences in plasma or urinary purines could be documented. The similar disturbance in purine metabolism in female and male patients with primary gout indicates a severe renal purine underexcretion relative to their increased plasma purine levels. The significant inverse association between plasma urate, hypoxanthine and xanthine and the fractional excretion of these purines provides further support for this concept.

The observation of an impaired renal excretion of uric acid, hypoxanthine and xanthine in female gout, being of a similar magnitude to that documented in male gout, has specific implications. First, it suggests a common tubular dysfunction for the excretion of purine compounds in primary gout. Second, it extends the hypothesis that an intrinsic renal disease may be the underlying cause of primary gout in most patients.

ACKNOWLEDGEMENTS

We are indebted to the Clinical Research Unit nursing staff and the dietetic staff for excellent patient care; to Teresa Ramos, MD, Javier Díaz, and Mª Paz Canencia for valuable technical assistance; to Rosario Madero for the statistical analysis; to Erik Lundin for assistance in preparing the manuscript; and to Julio Ortíz Vázquez, MD, PhD, for his continued support and guidance. Supported by grants from Caja de Madrid and Fondo de Investigación Sanitaria (FIS, 93/0622), Spain.

REFERENCES

1. Kuzell WC, Schaffarzick RW, Naugler WE, et al. Some observations on 520 gouty patients. J Chron Dis 1955; 2:645-69.
2. Meyes OL, Monteagudo FSE. A comparison of gout in men and women. A 10-year experience. S Afr Med J 1986; 70:721-23.
3. Puig JG, Michán AD, Jiménez ML, et al. Female Gout. Clinical spectrum and uric acid metabolism. Arch Intern Med 1991; 151:726-32.
4. Borg EJT, Rasker JJ. Gout in the elderly, a separate entity? Ann Rheum Dis 1987; 46:72-6.
5. Delbarre F, Braun S, Saint-Georges-Chaumet F. La goutte fémenine. Analyse de quarante observations. Sem Hop Paris 1967; 44:623-33.
6. Puig JG, Mateos FA, Jiménez ML, Ramos TH. Renal excretion of hypoxanthine and xanthine in primary gout. Am J Med 1988; 85:533-7.
7. Turner RE, Frank MJ, Van Ausdal D, Bollet AJ. Some aspects of the epidemiology of gout: Sex and race incidence. Arch Intern Med 1960; 106:400-4.
8. Lally EW, Ho G Jr, Kaplan SR. The clinical spectrum of gouty arthritis in women. Arch Intern Med 1986; 146:2221-5.

FAMILIAL JUVENILE HYPERURICAEMIC NEPHROPATHY IN

ADOLESCENTS

I.Sebesta[1], J.Krijt[1], K.Pavelka[2], J.Maly[3], H.A.Simmonds[4], and M.B.McBride[4]

Centre of Metabolic Disorders, Department of Clinical Biochemistry, Charles University[1], Research Rheumatology Institute[2], Institute of Clinical and Experimental Medicine[3], Prague, Czech Republic, Purine Research Laboratory UMDS, Guy's Hospital[4], London, UK

INTRODUCTION

Familial juvenile hyperuricaemic nephropathy (FJHN; McKusick 16200) is a disorder characterized by progressive renal disease and gout, affecting young people of either sex equally. Gout and hypertension are inconsistent features. There are two biochemical markers of this disorder. The first is hyperuricaemia disproportionate to the degree of renal dysfunction; the second is a grossly reduced clearance of uric acid relative to creatinine, disproportionate to age, sex and degree of renal failure[1]. This condition was first noted in 1960 by Duncan and Dixon[2]. So far more than 43 kindreds have been described in the literature[1]. The mode of inheritance is autosomal dominant. The genetic defect remains undefined.

The purpose of this paper is to describe new kindred that came to attention through early onset of gout associated with renal disease in young woman.

PATIENTS

Case 1: J.P., born 1974, female. This patient, the propositus was referred for metabolic study because of early onset of characteristic gouty attack of her right big toe at the age of 17 years. Further investigation revealed hyperuricaemia, chronic renal insufficiency. Renal echography showed small kidney size.

Case 2: J.P., born 1977, male. The apparently healthy younger brother of case 1 was investigated because of the family history. He had no gouty attacks. Renal biopsy following purine metabolic study showed the tubulointerstitial nephropathy without evidence of urate deposition.

Purine and Pyrimidine Metabolism in Man VIII, Edited by
A. Sahota and M. Taylor, Plenum Press, New York, 1995

Case 3: J.P., born 1952, male. The father of case 1 and 2 suffered repeated attacks of gout affecting the big toe since the age of 24 years. Chronic renal insufficiency was found. Renal ultrasound revealed small kidney size.

METHODS

Detailed purine metabolic studies were performed in the three subjects and endogenous purines and purine enzyme activities were investigated in lysed erythrocytes as described[3]. Uric acid excretion was evaluated from the ratio of the urine uric acid concentration relative to creatinine concentration (UA/Cr) in morning samples. Uric acid excretion rate (ER) was determined from the product of urinary uric acid (μmol/l) and plasma creatinine (μmol/l) divided by the urinary creatinine (μmol/l).

RESULTS

The biochemical investigation revealed elevated plasma uric acid (UA) and grossly reduced fractional clearance of uric acid FE_{UA} in all three subjects. Uric acid excretion expressed in terms of creatinine excretion (UA/Cr) and uric acid excretion rate (ER) were normal or low in all instances. Normal activity of hypoxanthine phosphoribosyltransferase (HPRT) in lysed erythrocytes was found in all patients (table 1).

Table 1. Laboratory investigations in father and his two children with FJHN.

case (age, years)	urinary UA/Cr	ER of UA	FE_{UA} %	Plasma UA μmol/l	Cr	HPRT nmol/ mgHb/h
1. 19	0.14	18.6	4.4	422	140	156
2. 16	0.14	12.5	2.6	478	87	164
control adolescents (m,f)	0.6	18-47	12-26	120-360	40-70	80-160
3. 41	0.14	20.0	3.1	635	136	135
control adults (m)	0.4	18-47	6-10	120-420	76-120	80-160

Overproduction of uric acid was excluded as the genetic basis of the gout and hyperuricaemia. This was confirmed by the normal activity of HPRT in lysed erythrocytes, low FE_{UA}, and the fact that levels of the other minor purine bases in urine were all within the normal range (results not shown).

Table 2. Major clinical and laboratory findings in FJHN, of use in selective screening for this disorder.

- progressive renal disease

- early onset (childhood, adolescence)

- affecting both sexes

- strong family history

- gout

       - hyperuricaemia

       - grossly reduced fractional uric acid clearance

DISCUSSION

The pathogenesis of this disorder remains unknown. The possible relationships between the low FE_{UA}, hyperuricaemia and renal failure were reported[1,4]. Also the finding of disruption of renal hemodynamics may be of primary pathogenetic importance[5]. A recent report revealed strong evidence against linkage between FJHN and regions of the loci for autosomal dominant polycystic kidney disease type I[6]. Therapy with allopurinol does seem to ameliorate the progression of renal damage[7] but the successful outcome depends on renal function at the start of allopurinol[1]. These observations render early diagnosis even more important. The condition may be missed and a diagnosis of "familial nephritis" made until other characteristic features becomes evident and purine metabolic investigations performed. Also in older FJHN patients when progressive renal disease, accompained with hypertension is evident, hyperuriceamia could be wrongly interpreted as the secondary consequence of other renal disease. Renal biopsy in almost all described cases of FJHN did not revealed urate crystals. However, the absence of crystals does not necessarily exclude their original action as the causative agent for nonspecific interstitial nephropathy seen in this disorder[1].

Unlike other purine and pyrimidine disorders familial juvenile hypeuricaemic nephropathy has no characteristic metabolite, which could be detected in body fluids. The diagnosis is based on complex clinical evaluation, together with family history and detailed purine metabolic studies. It is necessary to establish more data about the pathogenesis of this disorder. The target symptoms are shown in table 2. The coexistence of these features should suggest to FJHN. But it must be taken to account that clinical gout could be an inconsistent feature[1].

Detailed evaluation of uric acid levels in blood and urine is important as the renal hypoexcretion of urate in this condition is generally extreme. Other conditions that may cause hyperuricaemia such as Barters syndrome, Alport syndrome, medullary cystic disease, lead nephropathy, chronic haemolytic anaemia, the use of drugs altering uric acid production or excretion and other causes of renal failure must be ruled out.

It is also necessary to exclude genetic overproduction of uric acid as the basis of the gout and hyperuricaemia.

The recent finding of a second family by our department suggests that this disorder could be more prevalent in the Czech population than is currently recognized. The results underline the fact that attacks of gout together with renal impairment in young people of either sex, is a clear indication for the detailed purine investigations.

REFERENCES

1. J.S.Cameron, F.Moro, and H.A.Simmonds, Nephrology review: Gout, uric acid purine metabolism in paediatric nephrology. *Pediatr Nephrol.* 7:105 (1993).
2. H.Duncan, A.StJ, Gout, familial hyperuricaemia and renal disease,*Q J Med.* 29:127 (1960).
3. H.A.Simmonds, J.A.Duley, and P.M.Davies, Analysis of purines and pyrimidines in blood, urine and other physiological fluids, *in*: "Techniques in diagnostic human biochemical genetics. A laboratory manual," F.Hommes, ed., Wiley- Liss, New-York (1991).
4. J.S.Cameron, F.Moro, and H.A.Simmonds, *Adv Exp Med Biol.* What is the pathogenesis of familial gouty nephropathy? 309A:185 (1991).
5. M.E.Miranda, J.G.Puig, F.A.Mateos, and J.O.Vazquez, Increased renal vascular resistance in familial nephropathy associated with hyperuricaemia or gout, *PWS.* 15:F11 (1993).
6. F.Moro, I.Noam, J.S.Cameron, H.A.Simmonds, M.B.McBride, C.G.P. Mathew, C.S. Ogg, J.G.Puig, M.E. Miranda, and F.A.Mateos, Mapping the gene for familial juvenile hyperuricaemic nephropathy, *PWS.* 15:F23 (1993).
7. F.Moro, H.A.Simmonds, M.B.McBride, J.S.Cameron, D.G.Williams,and C.S.Ogg, Does allopurinol ameliorate progression in familial juvenile gouty nephropathy (FGHN)? *Adv Exp Med Biol.* 309A:199 (1991).
8. D.A.Farebrother, J.R.Pincott, H.A.Simmonds, D.J.Warren, M.J.Dillon, J.S.Cameron, Uric acid crystal-induce nephropathy: evidence for a specific renal lesion in a gouty family. *J Pathol.*135:159 (1981).

CORRELATION OF SERUM URIC ACID, CHOLESTEROL AND TRIGLYCERIDE LEVELS

N. Di Sciascio°, C.P. Quaratino, C. Rucci, P. Ciaglia,
M.N. Mariani, and A. Giacomello
Institute of Medical Physiopathology, University of Chieti and
Renzetti Hospital°, Lanciano
Italy

There is an apparent high incidence of hyperuricemia in patients selected for diabetes, however epidemiological studies have not demonstrated any relation between serum urate and blood glucose concentration.[1] A correlation of serum urate to serum cholesterol and/or triglyceride levels has been described but also disclaimed.[2-3]

The purpose of the present paper was to study the correlations between glucose, cholesterol, triglycerides and uric acid.

Patients and methods

During a few months period serum urate, cholesterol, creatinine, glucose and triglycerides levels were determined in adult patients aged 21-70 years from a big Hospital of Central Italy. 3224 patients (1639 males) were examined. 1795 subjects (872 males) were outpatients. Uric acid, cholesterol, creatinine, glucose and triglycerides were determined enzymatically using an automated analyzer (Monarch, Instrumentation Laboratory).

Results and Discussion

In a previous study[4] we have seen that in adult males with serum creatinine concentration <= 1.3mg/dL serum uric acid levels do not vary significantly among different age groups or between out and in patients.

In adult females with serum creatinine concentration <= 1.2 mg/dL at least two groups of patients must be considered in order to not have a significant difference in mean serum urate values among different age groups or between out and in patients. These groups were : premenopausal (age <= 46) and postmenopausal (age >= 54) women. Pearson correlation matrix of the variables considered in the three groups of patients are shown in table 1.

In good agreement with previous studies no significant correlation between serum urate and glucose levels was observed in all groups. The correlation between serum urate and triglyceride levels was significant in all groups, while that between serum urate and cholesterol can be considered significant only in males. In conclusion a significant correlation of serum uric acid and triglycerides levels has been observed in both sexes.

A weak correlation of serum uric acid and cholesterol levels is observed. However this correlation appears to be significant only in man.

Purine and Pyrimidine Metabolism in Man VIII, Edited by
A. Sahota and M. Taylor, Plenum Press, New York, 1995

Table 1. Pearson Correlations of Serum Urate with examined variables

----

SEX		VARIABLES	
MALES	S_CHOLESTEROL	S_TRIGLYCERIDES	S_GLUCOSE
	0.200	0.313	-0.020
	p*= 0.000	p= 0.000	p= 1.000
FEMALES			
Premenopausal	0.112	0.292	0.070
	p= 0.052	p= 0.000	p= 0.617
Postmenopausal	0.030	0.185	-0.002
	p= 1.000	p= 0.000	p= 1.000

----

* 'p' are Bonferroni-adjusted probabilities

References

1. J. B. Herman, F. W. Mount, J.H. Medallie, at al. : Diabetes prevalence and serum uric acid : Observations among 10.000 men in a survey of ischemic heart desease in Israel. Diabetes 16: 858-868, (1967).
2. T. Gibson, R. Grahame : Gout and hyperlipidaemia. Ann Rheum Dis 33: 298-303, (1974)
3. B. S. Gathof, M. A. Schreiber, U. Gresser, I. Kamilli, N. Zollner, Importance of the confounding factors age and sex in the correlation of serum uric acid, colesterol and triglyceride levels in : " Purine and Pyrimidine metabolism in man VII", A. Harkness. G. B. Elion and N. Zollner eds, Plenum Press; New York (1991).
4. C.P. Quaratino, N. Di Sciascio, C. Rucci , et al. The normal range of serum urate levels and of fractional urate excretion. This Book.

SERUM CONCENTRATION OF Lp(a) IN PATIENTS WITH PRIMARY GOUT

Sumio Takahashi, Tetsuya Yamamoto, Yuji Moriwaki, Zenta Tsutsumi, and Kazuya Higashino

Third Department of Internal Medicine, Hyogo College of Medicine
Mukogawa-cho 1-1, Nishinomiya, Hyogo 663, Japan

INTRODUCTION

Many studies have demonstrated that Lp(a), a plasma lipoprotein similar to LDL, is associated with the development of atherosclerotic diseases[1,2]. However, there have been few studies on serum Lp(a) concentration in patients with gout who are frequently complicated with atherosclerotic diseases. Therefore, we measured the serum concentration of Lp(a) in patients with gout to investigate whether Lp(a) contributes to the high incidence of atherosclerotic diseases in gout.

SUBJECTS AND METHODS

Subjects

The subjects were 143 male patients with primary gout who visisted Gout Clinic during the past 4 years and 143 normal subjects who were recruited from applicants for annual health check-up and gave normal physical and laboratory examinations including 75 g OGTT. All the patients fulfilled the criteria on primary gout as outlined by the ARA[3]. After informed consent was obtained, any medication which may affect serum lipid metabolism was withheld at least 6 months before the study. Subjects who had diseases known to cause hyperlipidemia such as diabetes mellitus, renal, hepatic or endocrine diseases were excluded from this study. Information concerning alcohol consumption was obtained by questionnaire on frequency, quantity and type and converted to gram ethanol ingested per day.

Analysis of samples

Blood samples were obtained after an overnight fast. Serum Lp(a) was determined by enzyme-linked immunosorbent assay using a commercially available kit. Serum total cholesterol, triglyceride and phospholipid were measured by enzymatic methods. High-density lipoprotein cholesterol(HDL-C) was measured by the heparin calcium precipitation method[4]. Serum apolipoproteins were determined by a single radial immunodiffusion method.

Purine and Pyrimidine Metabolism in Man VIII, Edited by
A. Sahota and M. Taylor, Plenum Press, New York, 1995

Statistical analyses

Data were expressed as the mean±SE. Mann-Whitney U test was used to compare the Lp(a) values in subjects with gout and controls because of skewed distribution of Lp(a) concentrations. Comparison of Lp(a) distribution frequency between gout and control was made by the Chi-square test. Other data were evaluated by Student's t test for significance. Multiple regression analysis was used to assess the association among serum Lp(a) level and other variables.

RESULTS

Clinical features of the subjects

Serum uric acid and systolic blood pressure were significantly higher in patients than in controls(p<0.01), while BMI, alcohol consumption, serum ALT, AST ,γ-GTP, FBS and creatinine values were not different between subjects with gout and controls.

Serum lipids and apolipoproteins

Serum Lp(a), triglyceride, apolipoprotein B, C-II, C-III and E levels were significantly higher in patients with gout than in normal controls, while HDL-C level was significantly lower in patients with gout than in controls. Serum cholesterol level was not different between the two groups (Table 1).

Distribution and concentration of serum Lp(a)

The distribution of Lp(a) levels in patients with gout was different from that in controls (x_4^2=16.254, p<0.01). The incidence of high Lp(a) level (more than 20mg/dl) was significantly higher in the patient group(56/143, 39.1% vs 32/143, 22.4%, p<0.05). Multiple regression analysis disclosed that serum Lp(a) was not related to all variables tested except for age(Table 2).

Table 1. Serum concentrations of Lp(a), lipids and apoliporoteins of the subjects

	Gout(n=143)	Control(n=143)	p value
Lp(a)(mg/dL)	21.0±1.5	13.1±1.0	p<0.01
T-cholesterol(mg/dL)	211.3±3.2	208.9±3.1	NS
Triglyceride(mg/dL)	180.2±11.6	156.7±12.8	p<0.05
Phospholipid(mg/dL)	228.3±3.9	235.8±3.5	NS
HDL-C(mg/dL)	47.2±1.1	51.0±1.2	p<0.05
Apo A-I(mg/dL)	125.9±2.0	129.0±1.8	NS
Apo A-II(mg/dL)	35.5±0.7	32.6±0.5	NS
Apo B(mg/dL)	102.6±2.0	90.5±2.0	p<0.01
Apo C-II(mg/dL)	5.0±0.2	4.2±0.2	p<0.01
Apo C-III(mg/dL)	13.3±0.5	11.5±0.5	p<0.01
Apo E(mg/dL)	5.9±0.2	5.0±0.2	p<0.01

Table 2. Multiple regression analysis of serum Lp(a) on selected independent variables in 143 gouty patients

Variable	Partial regression coefficient	Standard regression coefficient	Partial correlation coefficient	Partial F value	P value
Age	0.4237	0.3269	0.2498	8.715	p<0.01
SBP	-0.1291	-0.1121	-0.0739	0.7186	NS
DBP	-0.0308	-0.0166	-0.0120	0.0204	NS
BMI	0.6369	0.0933	0.0876	1.0127	NS
Alcohol intake	-0.0725	-0.0858	-0.0763	0.7673	NS
γ-GTP	0.0394	0.1532	0.1275	2.1640	NS
ALT	-0.0852	-0.0399	-0.0213	0.0597	NS
AST	-0.0718	-0.0572	-0.0298	0.1162	NS
S-Cr	5.2949	0.0586	0.0553	0.4012	NS
Uric acid	-1.6719	-0.1097	-0.1120	1.6650	NS
FBS	-0.1680	-0.1174	-0.1147	1.7459	NS

Abbreviations are as follows: SBP, systolic blood pressure; DBP, diastolic blood pressure; ALT, alanine aminotransferase; AST, aspartate aminotransferase; S-Cr, serum creatinine; FBS, fasting blood sugar

DISCUSSION

Atherosclerotic diseases such as ischemic heart disease have become an important cause of death in patients with gout. However, it remains undetermined whether or not hyperuricemia per se is an independent risk factor for atherosclerosis. On the other hand, it is well known that hypercholesterolemia, low HDL-cholesterol level, hypertriglyceridemia, increased serum apolipoprotein B and Lp(a) levels are important risk factors for atherosclerosis and some of these lipid abnormalities, especially hypertriglyceridemia is frequently seen in gout. The present study confirmed the earlier observations that serum triglyceride level is higher, while HDL-cholesterol is lower in patients with gout. However, the most intriguing result was an increase in Lp(a) level in patients with gout. In addition, frequency distribution of Lp(a) in gout was significantly different from that in controls (Frequency of Lp(a) below 10mg/dl was low. In contrast, Lp(a) level above 40mg/dl was high in patients with gout compared with controls), although skewed distribution of Lp(a) was noted in both groups as previously described. Previous studies indicated that Lp(a) level was higher in patients with ischemic heart disease[5] and Lp(a) level was related to the degree of coronary artery stenosis[6]. Therefore, an increase in serum Lp(a) level may contribute to the development of atheroslerosis in primary gout together with low HDL-C, and high triglyceride and apo B levels. Lp(a) consists of 7 phenotypes(F, B, S_1, S_2, S_3, S_4, O), and although close relationship between phenotype of Lp(a) and Lp(a) concentration has been indicated in previous studies, phenotype of Lp(a) was not determined this being beyond the purpose of the present study.

Since Lp(a) is not correlated with other risk factors (lipids, blood pressure and smoking *etc*) for atherosclerosis, it seems to be an independent risk factor except for age for atherosclerosis in gout.

In conclusion, since lipid abnormalities including elevated Lp(a) concentration may accelerate atherosclerosis in patients with gout, it is important to measure serum lipids, Lp(a) and apolipoproteins routinely in patients with gout, and to lower Lp(a) with drug therapy to prevent the development of atherosclerotic complications.

REFERENCES

1. G.G.Rhoads, G.Dahlen, K.Berg, N.E.Morton, A.L.Dannenberg: Lp(a) lipoprotein as a risk factor for myocardial infarction. *J Am Med Assoc* 256:2540(1986).
2. A.Murai, T.Miyahara, N.Fujimoto, M.Matsuda, M.Kameyama: Lp(a) lipoprotein as a risk factor for coronary heart disease and cerebral infarction. *Atherosclerosis* 59:199(1986).
3. S.L.Wallace, H.Robinson, A.T.Masi, J.L.Decker, D.J.MaCarty, T.F.Yu: Preliminary crite ria for the classification of the acute arthritis of primary gout. *Arthritis Rheum* 20:895 (1977).
4. A.Noma, K.Nezu-Nakayama, M.Kita, H.Okabe: Simultaneous determination of serum cholesterol in high- and low- density lipoproteins with use of heparin, Ca^{2+}, and an anion-exchange resin. *Clin Chem* 24:1504(1978).
5. H.F.Hoff, G.J. Beck, C.I.Skibinski, G.Jurgens, J.O'Neil, J.Kremer, B.Lytle: Serum Lp(a) level as a predictor of vein graft stenosis after coronary artery bypass surgery in patients. *Circulation* 77:1238(1988).
6. G.H. Dahlen, J.R.Guyton, M.Attar, J.A.Farmer, J.A.Kautz, A.M.Gotto: Association of level of lipoprotein Lp(a), plasma lipids, and other lipoproteins with coronary artery disease documented by angiography. *Circulation* 74:758(1986).

STUDY ON LIPOPROTEIN LIPASE AND HEPATIC TRIGLYCERIDE LIPASE ACTIVITIES IN PATIENTS WITH GOUT

Zenta Tsutsumi, Sumio Takahashi, Tetsuya Yamamoto, Yuji Moriwaki, Yumiko Nasako, Keisai Hiroishi, Toshikazu Hada, and Kazuya Higashino

Third Department of Internal Medicine, Hyogo College of Medicine Mukogawa-cho 1-1, Nishinomiya, Hyogo 663, Japan

INTRODUCTION

Hypertriglyceridemia is frequently observed in patients with gout whose first cause of death in Japan is coronary heart disease. However, its mechanism still remains undetermined. Until recently hypertriglyceridemia has not been accepted as an independent risk factor for coronary atherosclerosis, and it has often been ignored in clinical practice. However, recent epidemiological studies showed that hypertriglyceridemia might indeed be a risk factor for coronary atherosclerosis[1]. To clarify the etiology of hypertriglyceridemia in patients with gout, the relationship between postheparin lipolytic enzyme activity, lipoprotein lipase (LPL), hepatic triglyceride lipase (HTGL), and serum lipids was investigated.

SUBJECTS AND METHODS

Subjects

Age-matched 20 male patients with primary gout and 20 healthy male adults with normal lipid levels (serum triglyceride (TG) <150 mg/dl, and total cholesterol (TC) <250 mg/dl) were included in the study. There were no blood relatives among these subjects, and they were all free of hepatic, renal, or endocrine diseases, and none took drugs which might affect lipid metabolism. All subjects underwent 75 g OGTT , and those who showed diabetic pattern according to the WHO diagnostic criteria were excluded from the study, as well as those with severe obesity (BMI >29). Apolipoprotein CII, a cofactor of LPL, was not abnormally low in any subject.

Sample collection

Venous blood was obtained after an overnight fast. Lipoproteins were isolated from fresh serum by sequential ultracentrifugation[2]. Postheparin plasma (PHP) was collected as follows: Venous blood was collected in a tube containing EDTA-2Na 10 min after intravenous injecton of heparin (30 U/ body weight, Novo-heparin 1000 U/ ml, Novo Nordisk A/S, Denmark), then plasma was immediately prepared by low speed centrifugation at 4°C and stored at -40°C until used.

Lipid and apolipoprotein analysis

The concentrations of TG, TC, HDL-C in serum and isolated lipoproteins were measured using enzymatical kits as follows: Triglyceride T-test Wako, Nescot TC kit, HDL-C kit (Wako Pure Chem Co. Osaka, Japan). Serum apolipoprotein concentrations were measured by single radial immunodiffusion method using the respective apoprotein plate kits (Daiichi Chem Co. Tokyo, Japan).

Assay of LPL and HTGL in PHP

Enzyme activities were determined by the method as published elsewhere[3] with an emulsion of tri[9,10-^{3}H]olein in gum arabic as the substrate. LPL activity was determined after the inhibition of HTGL with 50 mM of sodium dodecyl sulfate (SDS), and HTGL activity after the inhibition of LPL with 1 M of sodium chloride. Both lipase activities were expressed as FFA formed per hour and 1 unit of the enzyme activity was defined as µg FFA/ h / ml PHP.

Statistical analysis

Data were expressed as the mean ± S.E. The observed differences between the groups were analyzed by Student's t-test. A p value below 0.05 was considered statistically significant.

RESULTS

Serum TG level was significantly higher in the patients with gout than in the controls (202±25 mg/dl vs 122±20 mg/dl, $p<0.05$) as described previously[4,5]. Serum LDL-TG was also significantly higher in the patients with gout than in the controls (41.5±2.4 mg/dl vs 36.0±1.8 mg/dl, $p<0.01$).

HTGL activity was significantly lower in the patients with gout compared to the controls as we had expected (14.5±1.6 U vs 18.3±2.3 U, $p<0.01$) but LPL activity was not different between the patients with gout and the controls (Table 1).

Table 1. Clinical and laboratory findings of the subjects.

		Gout (n=20)		Control (n=20)		
		mean	SE	mean	SE	Significance
Age	(years)	43.3 ±	2.5	44.1 ±	2.2	ND
Uric acid	(mg/dl)	9.3 ±	0.4	5.7 ±	0.2	p<0.01
BMI	(kg/m²)	23.6 ±	0.7	25.1 ±	0.5	ND
TG	(mg/dl)	202 ±	25	122 ±	20	p<0.05
TC	(mg/dl)	216 ±	10	224 ±	9	ND
HDL-C	(mg/dl)	48 ±	3	48 ±	2	ND
LDL-TG	(mg/dl)	41.5 ±	2.4	36 ±	1.8	p<0.05
LDL-TG/apoB		0.66 ±	0.03	0.57 ±	0.02	p<0.01
LPL activity	(U)	5.8 ±	0.7	7.2 ±	0.6	ND
HTGL activity	(U)	14.5 ±	1.6	18.3 ±	2.3	p<0.01

In addition, it was demonstrated that HTGL activity was negatively correlated with LDL-TG in the patients with gout (r=-0.54, p<0.01) and in the controls (r=-0.55, p<0.01) (Fig.1).

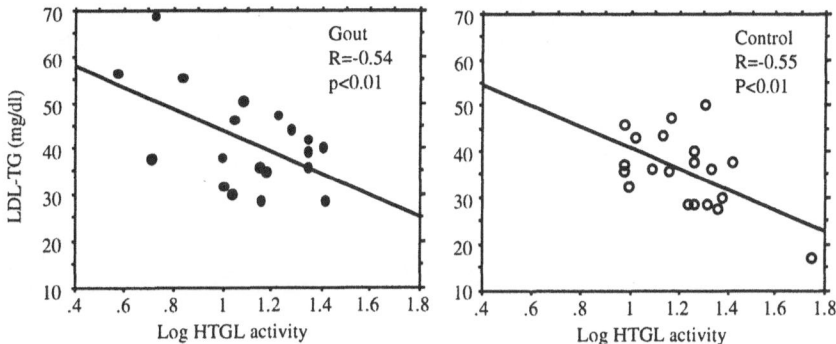

Fig.1. Relationship between HTGL activity and LDL-TG in patients with gout and control subjects

DISCUSSION

LPL acts at the first step in the degradation of TG. It hydrolyzes TG in chylomicrons and VLDL on vascular endothelial cells, while HTGL acts in hydrolyzing TG in the other lipoproteins (IDL, LDL, HDL) on hepatic sinusoidal endothelial cells[6]. Thus, it is implied that lower activities of LPL and HTGL contribute to hypertriglyceridemia in the patients with gout. In the present study, HTGL activity was significantly lower in the patients with gout as we had expected. However, LPL activity was not different between the patients with gout and the controls. These results suggested that the elevated serum LDL-TG level may be caused by the lower activity of HTGL but the elevated serum TG level is not ascribed to the activity of LPL.

It has been reported that patients with coronary heart disease have lower activity of HTGL than normal controls[7,8], although the etiology of low HTGL activity is not determined. Therefore, it is important to clarify the underlying mechanism reducing HTGL activity in the patients with gout.

REFERENCES

1. W.P. Castelli, The Triglyceride issue : A view from Framingham. *Am. Heart J.*112: 432 (1986).
2. F.T. Hatch, R.S. Lees, Practical methods for plasma lipoprotein analysis. *Adv. Lipid Res.* 6:1 (1968).
3. M.L. Baginski, W.V. Brown, A new method for the measurement of lipoprotein lipase in postheparin plasma using sodium dodecyl sulfate for the inactivation of hepatic triglyceride lipase. *J. Lipid Res.* 20: 548 (1979).
4. T. Kodama, T. Murase, H. Itakura, Y. Akanuma, F. Takaku, Y. Nishida, Postheparin plasma lipoprotein lipase and hepatic triglyceride lipase activities in patients with primary asymptomatic gout. *Clin.Chem.* 29:2124 (1983).
5. Y. Nishida, T. Miyamoto, T. Kodama, M. Okazaki, I. Hara, Postheparin plasma lipoprotein lipase and hepatic triglyceride lipase activities in gout. *Adv. Exp. Med.Biol.* 165A:137 (1984).
6. A. Bensadown A, Lipoprotein lipase. *Ann. Rev. Nutr.* 11:217 (1991).
7. P.H. Groot, W.A. van Stiphout, X.H. Krauss, H. Jansen, A. van Tol, E. van Ramshorst, S.O. Chin, S.R. Cresswell, L. Havekers, Postprandial lipoprotein metabolism in normolipidemic men with and without coronary artery disease. *Arterioscler.Thromb.* 11:653(1991).
8. P. Tornvel, K. Karpe, L.A. Carlson, A. Hamstren, Relationships of low density lipoprotein subfractions to angiographically defined coronary artery disease in young survivors of myocardial infarction. *Atherosclerosis* 90:67 (1991).

BENZBROMARONE AND FENOFIBRATE ARE LIPID LOWERING AND URICOSURIC: A POSSIBLE KEY TO METABOLIC SYNDROME?

Ursula Gresser, Birgit S. Gathof, and Manfred Gross

Purine Research Laboratory
Medizinische Poliklinik,
University of Munich
D-80336 München, Germany

INTRODUCTION

Patients with hyperlipidemia often suffer from hyperuricemia and conversely. This is proven in many studies, e. g. our laboratory studied the levels of uric acid, cholesterol and triglycerides in German blood donors (Gresser et al., 1990; Gathof et al., 1991). 68 % of the men and 57 % of the women showed cholesterol levels of 200 mg/dl and more. Levels above 250 mg/dl were found in 25 % of the men and 15 % of the women. Hyperuricemia occured in 28.6 % of men and 2.6 % of women. There was a statistically significant correlation ($p<0.001$) between the level of uric acid and cholesterol respectively triglycerides. After elimination of the factor age the correlation coefficients dropped, especially in women, but still remained significant.

Fenofibrate (FF) is a fibric acid derivate, which is metabolized to a number of compounds in humans. Hydrolysis to fenofibric acid, the active principal metabolite, occurs immediately after absorption. The principal excretion pathway for the metabolites is renal excretion (Balfour et al., 1990). Evidence for an uricosuric effect of FF was found in some studies. In 10 healthy volunteers FF (300 mg) and benzbromarone (BB; 100 mg) both showed an increase of uric acid clearance; FF lead to a doubling, BB to a quadruple of uric acid clearance compared to the initial value (Desager et al., 1980). In 19 out-patients with hyperlipidemia FF (300 mg/day for one year) lead to a decrease of plasma uric acid of 12 % in normo-uricemic patients to 30 % in hyperuricemic patients (Harvengt et al., 1980). In 17 out-patients with hyperlipidemia FF (200 mg/day) lead to a decrease of plasma uric acid of average 20 % (Steinmetz et al., 1981). 1988 Bastow et al. compared the effects of placebo, FF ($3 \cdot 100$ mg/day) and bezafibrate ($3 \cdot 200$ mg/day) in 10 patients with hyperlipidemia. Placebo and bezafibrate showed no uricosuric effect. FF lead to a decrease of plasma uric acid of average 20 % and an increase of uric acid clearance of average 30 %.

Purine and Pyrimidine Metabolism in Man VIII, Edited by
A. Sahota and M. Taylor, Plenum Press, New York, 1995

Most of these studies were performed to measure the lipid lowering effect of FF and therefore give no detailed informations on extent and course of time of the uricosuric effect of FF. Aim of our study was to get detailed informations on these points in comparison to a well-known uricosuric drug.

METHODS

7 healthy volunteers were given a single dose of 100 mg BB or 200 mg micronized FF at Saturday 8.00 a.m. after a run-in phase with constant purine intake of 3 days. Between the experiments with BB and FF there was an interval of 6 weeks. We measured plasma uric acid, renal uric acid excretion, triglycerides, total cholesterol, HDL-cholesterol and LDL-cholesterol (see figures 1 to 4).

RESULTS

Additional renal uric acid excretion (AUC) was 376 mg after BB and 204 mg (= 54 %) after FF within 14 hours after drug ingestion (figure 1). Both drugs caused a decrease in triglycerides (BB 15.5; FF 21.2 %) and cholesterol (BB 9.1; FF 3.0 %), ba-

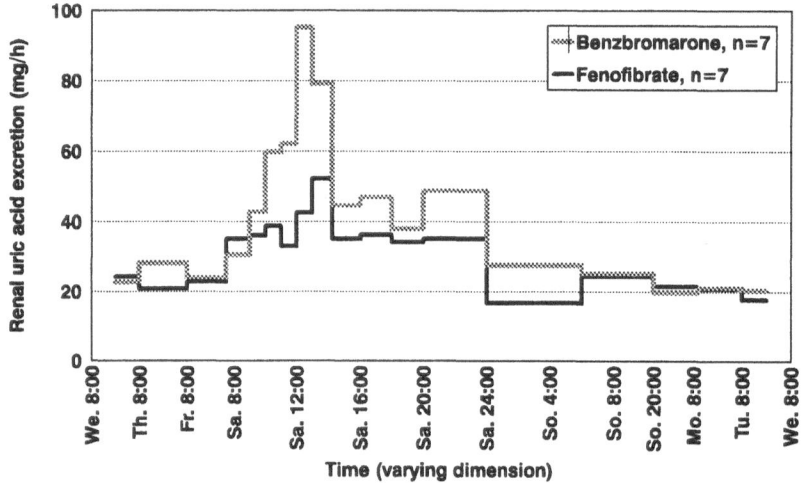

Figure 1. Renal uric acid excretion after a single dose of 100 mg BB or 200 mg FF in 7 healthy volunteers (drug application at Sa. 8:00 a.m.).

sed on a decrease in LDL-cholesterol (BB 17.5; FF 6.3 %). After a single dose of 100 mg BB plasma cholesterol decreases from 176 mg/dl to 160 mg/dl two days after drug application (after 200 mg FF from 168 mg/dl to 163 mg/dl). A single dose of 100 mg BB at least has a comparable hypolipidemic effect compared to 200 mg FF (figures 2 and 3).

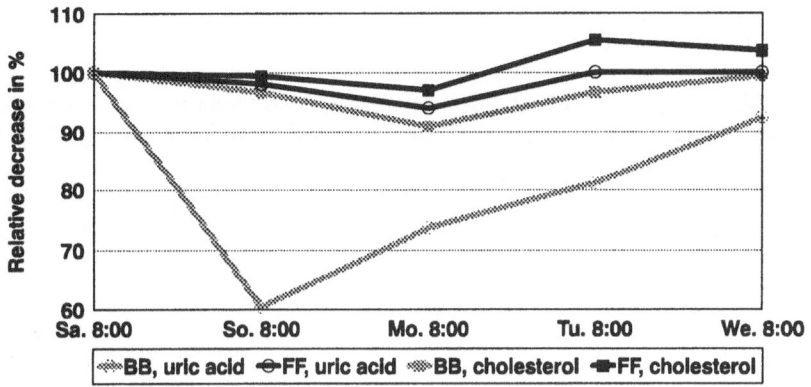

Figure 2. Relative decrease of plasma uric acid and cholesterol after a single dose of 100 mg BB or 200 mg FF in 7 healthy volunteers (drug application at Sa. 8:00 a.m.).

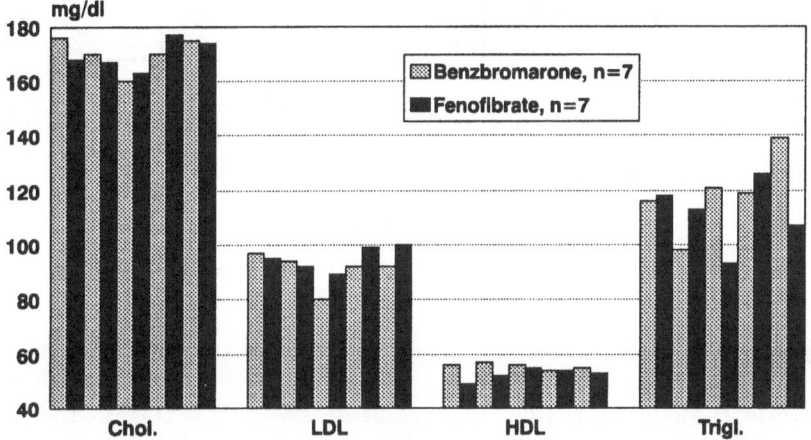

Figure 3. Cholesterol, LDL-cholesterol, HDL-cholesterol and triglycerides after a single dose of 100 mg BB or 200 mg FF in 7 healthy volunteers (each pair of columns marks one day at 8:00 a.m., Sa. to We., drug application at Sa. 8:00 a.m.).

In 1991 we reported our first results on a biliary-intestinal-biliary recirculation of BB after oral application (Gresser et al., 1991). In this study we could confirm this finding. Volunteer O. A. took part in both studies and showed the same behaviour with respect to renal uric acid excretion: renal uric acid excretion showed two peaks instead of a normally occuring exponential elimination (figure 4).

CONCLUSIONS

Our results on the uricosuric effect of FF confirm the observations described in the literature. The lipid-lowering effect of BB, as shown by our data, is not yet described.

The cause of the common occurance of hyperuricemia and hyperlipidemia is still unknown. Hyperuricemia and hyperlipidemia are part of the so-called metabolic syndrome (obesity, hypertension, hyperlipidemia, hyperuricemia, diabetes mellitus). The cause of this syndrome is still unknown. Interestingly we found two substances, which act on two or more of these symptoms: BB is uricosuric and lipid-lowering, FF is lipid-lowering, uricosuric and has some effect on glucose metabolism and on platelet aggregation (Balfour et al., 1990). With its combined lipid lowering and uricosuric properties FF and BB will help to shed light on the pathophysiology of the correlation between hyperlipidemias and hyperuricemia.

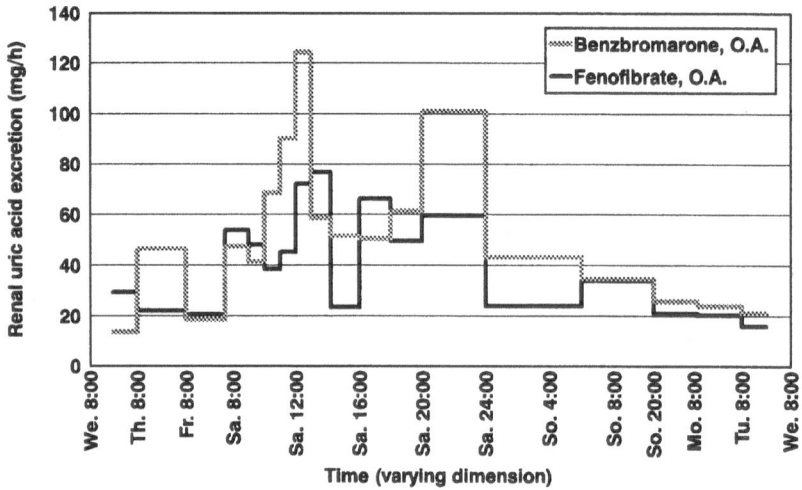

Figure 4. Renal uric acid excretion after a single dose of 100 mg BB or 200 mg FF (drug application at Sa. 8:00 a.m.) in a healthy volunteer (O.A.).

REFERENCES

J. A. Balfour, D. Mc Tawish, and R. C. Heel, Fenofibrate: a review of its pharmacodynamic and pharmacokinetic properties and therapeutic use in dyslipidemia, *Drugs* 40: 260-290 (1990)

M. D. Bastow, P. N. Durrington, and M. Ishola, Hypertriglyceridemia and hyperuricemia: effects of two fibric acid derivates (bezafibrate and fenofibrate) in a double-blind, placebo controlled trial, *Metabolism* 37: 217-220 (1988)

J. P. Desager, R. Hulhoven, and C. Harvengt, Uricosuric effect of fenofibrate in healthy volunteers, *J Clin Pharmacol* 10: 560-564 (1980)

B. S. Gathof, M. A. Schreiber, U. Gresser et al., Importance of the confounding factors age and sex in the correlation of serum uric acid, cholesterol and triglyceride levels, *Adv Exp Med Biol* 309A: 231-234 (1991)

U. Gresser, H. Empl, I. Kamilli et al., Pharmacokinetics of benzbromarone: Evidence for a biliary-intestinal-biliary recirculation. *Adv Exp Med Biol* 309A: 147-150 (1991)

U. Gresser, B. S. Gathof, and N. Zöllner, Uric acid levels in Southern Germany 1989. A comparison with studies from 1962, 1971 and 1984, *Klin Wochenschr* 68: 1222-1228 (1990)

C. Harvengt, F. Heller, and J. P. Desager, Hypolipidemic and hypouricemic action of fenofibrate in various types of hyperlipoproteinemias, *Artery* 7 (1): 73-82 (1980)

J. Steinmetz, C. Morin, E. Panek et al., Biological variations in hyperlipidemic children and adolescents, treated with fenofibrate, *Clinica Chimica Acta* 112: 43-53 (1981)

THE NORMAL RANGE OF SERUM URATE LEVELS AND OF FRACTIONAL URATE EXCRETION

C. P. Quaratino, N.Di Sciascio, C.Rucci, P.Ciaglia and A.Giacomello

Institute of Medical Physiopathology, University of Chieti and
Renzetti Hospital Lanciano
Italy

In the present paper an epidemiological study on serum urate concentration and on fractional urate excretion (FE) [(urate clearance-creatinine clearance) x 100] has been carried out .

Patients and methods

During a few months period urate and creatinine concentrations were determined in serum and in urine of all adult patients aged 21-70 years from a big Hospital of Central Italy. 3224 patients (1639 males) were examined. 1795 subjects (872 males) were outpatients.

Uric acid and creatinine were determined enzymatically using an automated analyzer (Monarch , Instrumentation Laboratory).

Results and Discussion

Sex and age specific means of serum urate and FE values in the population studied are shown in Figure 1. Only males with serum creatinine <= 1.3 mg/dL and females with serum creatinine <= 1.2 mg/dL were included in the study. Serum urate levels were higher in men while FE was higher in women. In males serum uric acid concentrations and FE did not vary significantly among different age groups or between division (out and in-patients).

The same results for urate were obtained by other investigators[1] In females serum urate concentration was dependent on age. In particular, outpatients aged 54-70 years and inpatients aged 47-70 years had higher urate values then outpatients aged 21-35 years and inpatients aged 21-46 years. This result was probably related to the rise in serum urate levels observed during menopause[2] Within each group serum urate concentrations were not significantly dependent on age. Although FE was indipendent on age, mean values of this variable obtained in outpatients were significantly higher than those obtained in inpatients (p< 0.01).

The significance of this difference disappeared when FE values in the 5-95 percentiles range were considered, only subjects with serum creatinine values in the reference interval of the laboratory method employed (0.6-1 mg/dL) were included and when patients were divided into age groups (21-50 and 51-70 years). Descriptive nonparametric statistics of serum urate concentration and FE are reported in Table 1.

Purine and Pyrimidine Metabolism in Man VIII, Edited by
A. Sahota and M. Taylor, Plenum Press, New York, 1995

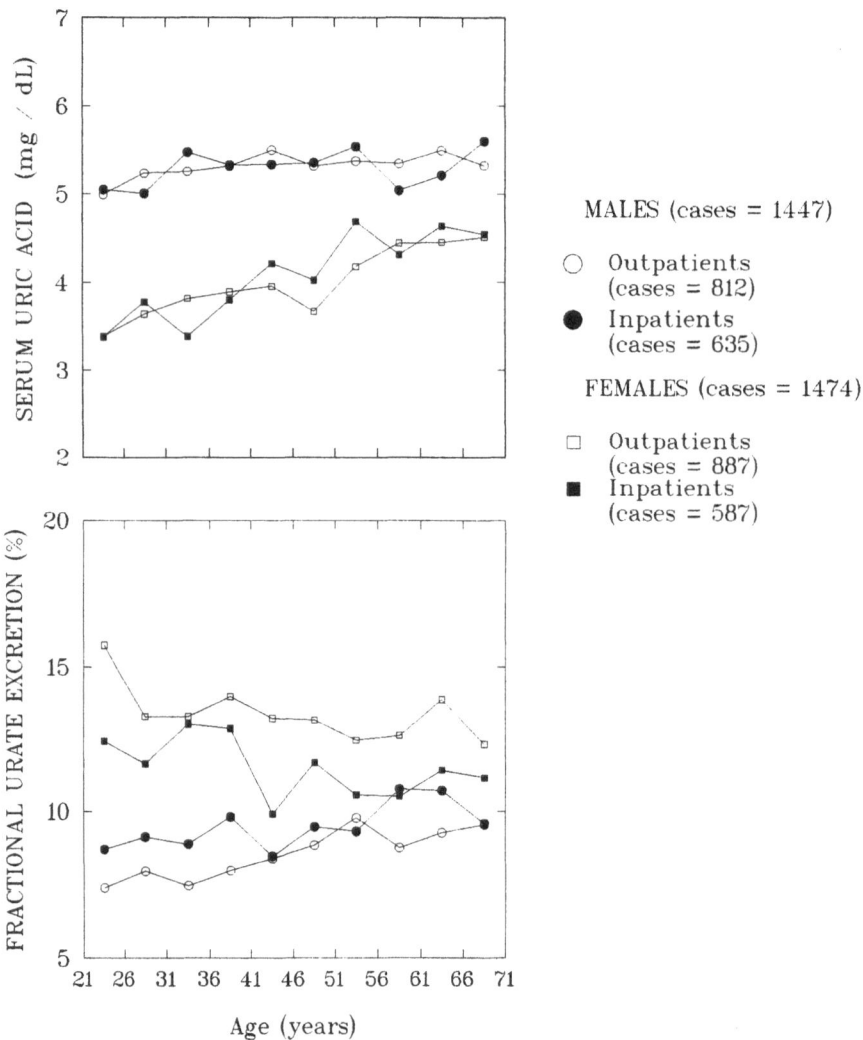

MALES (cases = 1447)

○ Outpatients
  (cases = 812)
● Inpatients
  (cases = 635)

FEMALES (cases = 1474)

□ Outpatients
  (cases = 887)
■ Inpatients
  (cases = 587)

Figure 1. Category plots of means

Parametric statistics for each group considered are reported in Table 2. In the youngest and in the oldest women age groups no significant difference in mean serum urate levels were observed between inpatients and outpatients.

Obtained results can be summarized as follows: in men serum urate and FE values are not dependent on age. The reference values (Mean ± 2SD) are 2.91-9.04 mg/dL for urate and 3.80-18.20 % for FE.

In women at least two groups of subjects must be considered: premenopausal women with serum urate levels ranging from 1.96 to 6.70 mg/dL and FE ranging from 4.79 to 24.7 % and postmenopausal women with urate levels ranging from 2.35 to 7.71 mg/dL and FE values ranging from 4.62 to 23.17 %.

Table 1. Descriptive nonparametric statistics

SEX	VARIABLE	DIVISION	AGE	CASES	MIN	MAX	25°	75°	MEDIAN
MALES									
	URATE	Total Cases	21-70	1447	1.5	19.4	4.3	6.2	5.1
	FE	Total Cases	21-70	1447	0.13	99.05	5.9	11.1	8.2
FEMALES									
	URATE	Outpatients	21-53	418	0.9	20.6	3.0	4.2	3.6
		Outpatients	54-70	469	1.1	11.5	3.5	5.1	4.2
		Inpatients	21-46	251	1.2	9.2	2.8	4.2	3.5
		Inpatients	47-70	336	1.4	20.4	3.5	5.1	4.2
	FE	Outpatients	21-70	887	0.18	258.6	8.7	15.6	11.8
		Inpatients	21-70	587	0.10	72.2	7.1	14.4	10.4

Table 2. Descriptive parametric statistics

SEX	VARIABLE	DIVISION	AGE	CASES	DISTRIB	MEAN	SD	SKEW.	KURT.
MALES									
	URATE	Total Cases	21-70	1447	Log-norm.	0.71	0.12	-0.12	1.15
	FE	Total Cases	21-70	1339	Log-norm.	0.92	0.17	-0.08	-0.55
FEMALES									
	URATE	Outpatients	21-53	418	Log-norm.	0.56	0.13	0.24	3.99
		Outpatients	54-70	469	Log-norm.	0.63	0.13	-0.27	0.95
		Inpatients	21-46	251	Log-norm.	0.54	0.14	-0.33	0.65-
		Inpatients	47-70	336	Log-norm	0.63	0.13	0.18	2.04
	FE	Total Cases	21-70	416	Log-norm	1.04	0.18	-0.30	-0.23
		Total Cases	21-70	430	Log-norm	1.01	0.17	-0.40	-0.17

References

1. U.Gresser and B.Gathof Epidemiology of Hyperuricaemia, in: "Urate Deposition in Man and its Clinical Consequences", U. Gresser , N. Zollner , Eds. Springer-Verlag, Berlin(1991).
2. B.Wyngaarden, M.D. William N. Kelley,Epidemiology of Hyperurcemia and Gout in: "Gout and Hyperuricemia", M.D., Grune & Stratton, New York (1976).

THERAPY RELATED DISTURBANCES
IN NUCLEOTIDES IN CANCER CELLS

Godefridus J. Peters

Department of Oncology
Free University Hospital
PO Box 7057
1007 MB Amsterdam
the Netherlands

INTRODUCTION

Ribonucleotides (NTP) and deoxyribonucleotides (dNTP) form the end products of the purine and pyrimidine synthetic pathways, both the *de novo* and the salvage pathways. They form the direct substrates for RNA and DNA synthesis. Nucleotide pools in the cell are the result of a dynamic equilibrium of synthesis and consumption. Besides their essential role in nucleic acid synthesis, ribonucleotides are essential for a number of other functions in the cell such as a source for energy supply (ATP), nucleotide sugar synthesis (UTP, CTP and GTP), lipid biosynthesis (CTP), signal transduction (ATP, GTP as part of the GTP binding protein, cyclic nucleotides such as cAMP and cGMP), protein elongation, tubulin-binding and ras-oncogene (all GTP), phosphate donors for kinase catalyzed reactions (mostly ATP, but other nucleotides can also act as a phosphate donor), allosteric activators and feedback regulators of several synthetic pathways, co-enzyme in several synthetic reactions such folyl-polyglutamate synthetase and phosphoribosylpyrophosphate (PRPP) synthetase. Besides being a substrate for DNA synthesis, dNTP play an essential role in DNA repair and as feedback regulators of a number of enzymes. *E.g.* dTTP is an inhibitor of dCMP deaminase [1], and of thymidine kinase [2]; dATP is a potent feedback inhibitor of ribonucleotide reductase [3,4], inhibiting the synthesis of all deoxyribonucleoside diphosphates, dGTP that of dGDP, dUDP and dCDP, dTTP that of dUDP and dCDP, while ATP is an activator of this enzyme, and dTTP an activator of the reduction of GDP to dGDP and dGTP of that of ADP to dADP. Thus, any perturbation of the synthesis or use of both NTP and dNTP can have major effects on cellular functioning. Thus measurement of nucleotide pools in cells during exposure to any drug and especially to drugs directed against synthesis of (deoxy)ribonucleotides can give essential information on the mechanism of action of these drugs.

NORMAL NUCLEOTIDES IN CELLS AND TISSUES

Measurement of nucleotides

For the measurement of normal nucleotides in cells and tissues a number of different methods have been described. Currently HPLC is one of the most commonly used procedures. The initial methods consisted of the use of an anion-exchange column from which the nucleotides were eluted using a gradient with increasing salt concentrations [5]. Numerous variations in columns, slope of the gradients, pH, ionic strength and other variations have been described, usually to answer specific questions, such as measurement of all nucleotides, that of nucleotide sugars, pyridine nucleotides, analog nucleotides such as ara-CTP, FUTP, dF-dCTP, thionucleotides, etc. Other more recent applications of HPLC consist of ion-pair reversed chromatography, again these methods were usually developed to answer specific questions [6]. More recently capillary zone electrophoresis has been applied to measure the concentration of nucleotides; this method combines speed, selectivity and sensitivity, but specific (expensive) apparatus and expertise is still required at this moment [7]. Another advantage of this method is the possibility for not only the measurement of nucleotides, but also of that of nucleosides and bases. It should be mentioned that this can also be achieved with some of the ion-pair reversed phase systems.

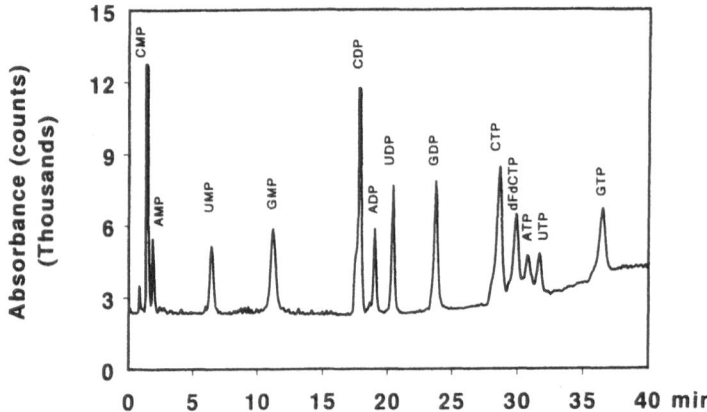

Fig. 1: Typical example of a nucleotide separation using a gradient with increasing salt concentrations with a Partisil SAX column connected to a photodiode array detector set at 254 nm.

Localization of nucleotides

What are normal nucleotide pools in a cell and what do they mean? In order to answer that question one should realize that nucleotide pools in the cell are always an equilibrium between synthesis and use. Besides that nucleotides in the cell are localized in several compartments [8,9,10], while the synthesis of various nucleotides can also occur via different pathways, the so-called channelling. Compartments which have classically be associated with compartmentation are the cytoplasm and various cellular organelles, such mitochondria and the nucleus [10]. In addition to that nucleotides can be bound to proteins, such as GTP to tubulin, ADP to actin and ATP to hemoglobin. Channelling of (deoxy)nucleotide synthesis has been described for various pathways, such as the difference between *de novo* and salvage pathways, and the synthesis of UTP. Pels Rijcken et al [13] demonstrated that in rat hepatocytes double-labelling with ^{14}C-orotic acid and ^{3}H-uridine, resulted in a higher ^{14}C/^{3}H ratio in UDP-sugars than in RNA, indicating that nucleotide synthesis via the *de novo* pathway is channelled to UDP-sugars and that via the salvage pathway is channelled to RNA. The ratio for free cytoplasmic UTP pools was intermediate, indicating an overflow from each pool. Sometimes these 'channelling' pathways are related with the compartments, such dNTP synthesis in the nucleus. It was demonstrated in permeabilized cells that NTP's are incorporated in DNA faster than the direct precursors dNTP's [8], although this hypothesis was challenged by others demonstrating that in intact cells the dCMP in the DNA originating from CDP was diluted to some extent by dCMP coming from deoxycytidine [12]. Thus although sufficient evidence of different dNTP pools in the various cellular compartments exist, evidence for channelling is not (yet) completely convincing.

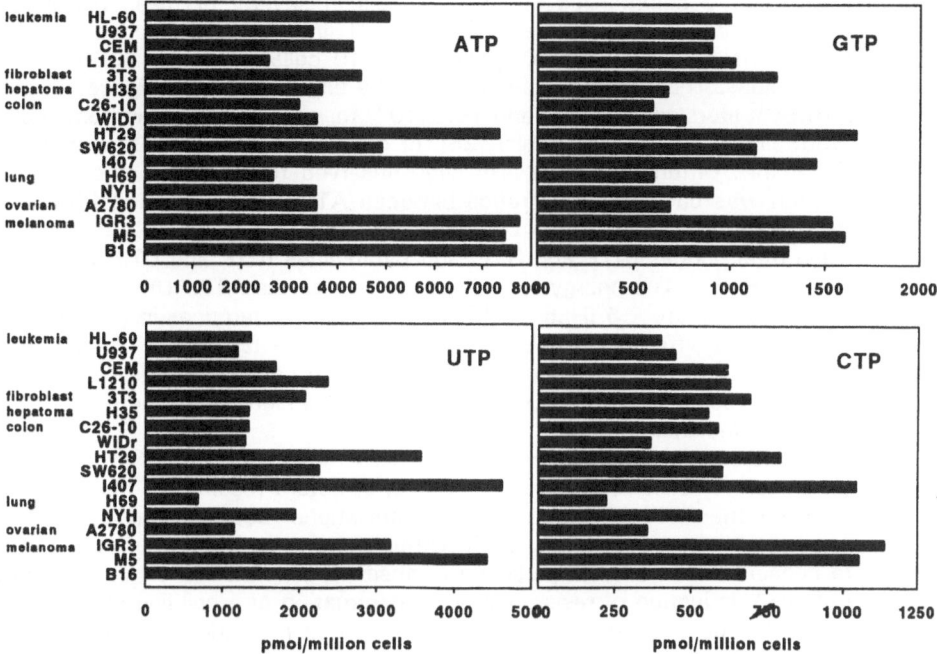

Fig. 2: Nucleotide concentrations in several cell lines from murine (L1210, 3T3, B16), rat (H35) and human origin. Values are means of at least 3 until 27 separate experiments, each performed in duplicate as reported in [17-26].

Normal nucleotide pools

Nucleotide pools have been measured by numerous groups in many different systems. However, many of these studies have been limited to easily accessible cells such as cultured leukemic lymphoblasts or circulating blood cells, both from healthy persons [13-15] and from patients with some form of leukemia [16]. Measurements of nucleotides in solid tissue (both normal and tumor) are scarce due to several technical limitations. Measurements of nucleotides in cell lines derived from solid tumors are more frequent. However, for studies on both leukemic and solid tumor cell lines, measurements are usually limited to one or a few cell lines for the specific purpose of that study. Thus it is difficult to get a good impression of "normal" nucleotide pools in cancer cells, or even to conclude that anything such as "normal" nucleotide pools exists in cancer cells. In the course of more than a decade [17-26], we measured nucleotide pools in several cell lines from different histological origins from various species (human, rat, mouse). Cell lines obtained from colorectal cancer, breast cancer, ovarian cancer, leukemia, melanoma, and derived from normal tissues such as fibroblasts have been analyzed. In addition to that nucleotide pools from several normal tissues (lymphocytes, bone marrow cells, liver, brain) and from tumors (colon, melanoma, leukemia) have been analyzed. For the cell lines an approximate similar procedure has been used throughout the years; harvest cells by trypsinization (for solid tumor cell lines) two days after plating in order to assure that cells were in an exponential growth phase. Some of these cell lines have been assayed several times in this period and similar pools have been observed, despite changing culture stocks (cell lines are restarted from liquid nitrogen stock after a certain number of passages), and different analytical procedures, such manual operated HPLC pumps, detectors and recorders [17,18], compared to a currently automated procedure consisting of autosamplers, computer controlled gradients, sensitive photodiode array detectors coupled with a computerized data-acquisition system [26]. Fig. 2 summarizes nucleotide pools in several of these cell lines, demonstrating a large difference between the cell lines in all nucleotide pools, varying for ATP from 2500 pmol/10^6 cells in L1210 cells to almost 8000 in the melanoma cell lines. These differences could not be explained by differences in cell size (e.g. $10.5x^{-13}$ liters/cell for CEM and $15.2x10^{-13}$ for murine colon carcinoma cells C26-10, with almost similar ATP pools) [26], or growth rates (e.g. WiDr and HT29 both about 24 hr). Differences have also been observed for the other nucleotides, but the pattern was different; thus ratios between ATP and GTP varied from 2.5 (L1210) to 5.9 (B16); that between UTP and CTP from 2.4 (H35) to 4.5 (HT29); that between total purines (ATP and GTP) and pyrimidines (UTP and CTP) from 1.2 (L1210) to 3.4 (HL60). The energy ratio (here defined as the ATP/ADP ratio) varied from 2.4 (L1210; H35) to 9.5 (HL60), although no major differences in growth rate were observed. This difference was also not related to the fact that leukemic cells are in principle easy to harvest since for other leukemic cell lines ratios between 3 and 5 were observed, while also for solid tumor cell lines high values were observed (e.g. 8.4 for IGR3 melanoma cells).

Nucleotide concentrations in cell lines are very much dependent on the culture conditions. Culture of cells in a three-dimensional structure resembling that of solid tumors, reduced the ATP, UTP and CTP concentrations, but not that of GTP in A2780 ovarian carcinoma cell line [25]. Other culture conditions affecting the nucleotide concentrations are the presence or absence of folates in culture medium (data not shown). Induction of resistance or transformation of a cell line also causes changes in nucleotide concentration, e.g. transformation of murine fibroblasts NIH-3T3 with the either ras or trk oncogene significantly affected nucleotide pools [24].

In solid tumors values for nucleotide concentrations are dependent on more variables than in cell culture. This is partly due to potential problems during removal of the tissue and subsequent homogenization and extraction. Other interfering factors are the heterogeneity of tumors, with more or less differentiated parts, contamination with normal tissue, vascularization, but also the presence of quiescent and necrotic parts, which by definition would have a lower ATP/ADP ratio. Careful rapid removal of the tumor (under anesthesia, e.g. ether, nembutal), rapid freezing (e.g. liquid nitrogen), processing under nucleotide preserving conditions (e.g. pulverization of the frozen tissue) immediately followed by extraction in e.g. ice-cold trichloroacetic acid or perchloric acid [19,26,27], will enable to make comparisons between drug-effects on a similar tumor. However, even under these conditions we have observed large differences between several tumors with a different histological origin [26].

Measurement of nucleotides in human samples is even more difficult, especially in tissues. Current phosphor-NMR techniques enable the measurement of ribonucleotides, but only specialized institutes have the equipment and expertise [28].

DEPLETION OF RIBONUCLEOTIDES

It is evident that drugs directed against one of the purine and pyrimidine synthetic pathways can cause a depletion of these nucleotides. Numerous inhibitors have been synthesized for both the purine and pyrimidine de novo nucleotide synthesis but not for the major salvage enzymes, uridine-cytidine kinase and hypoxanthine-guanine and adenine phosphoribosyltransferases (HGPRT and APRT) for the pyrimidine and purine ribonucleotides, respectively. A major disadvantage for any of the inhibitors of both the purine and pyrimidine de novo ribonucleotide synthesis is the ample availability of salvage precursors, in plasma and tissues; e.g. the uridine concentration in plasma is 5-10 μM [19,22,23] and even higher in several tissues [19,29]; the plasma concentration of the major precursor for purine nucleotide synthesis, hypoxanthine, is sometimes even higher. These compounds are derived either from the diet or from degradation of nucleic acids. Thus, at this moment no inhibitor of either the purine or the pyrimidine de novo synthesis is in extensive clinical use for treatment of cancer as a single agent [30]. This does not mean that these agents do not or will not have a clinical application. These agents either have a therapeutic application in combination with other agents (see below) or for the treatment of other diseases such as parasitic infections; these organisms depend almost entirely on their *de novo* pathways for nucleotide synthesis [31]. The circulating nucleosides and bases will than protect the normal host tissues.

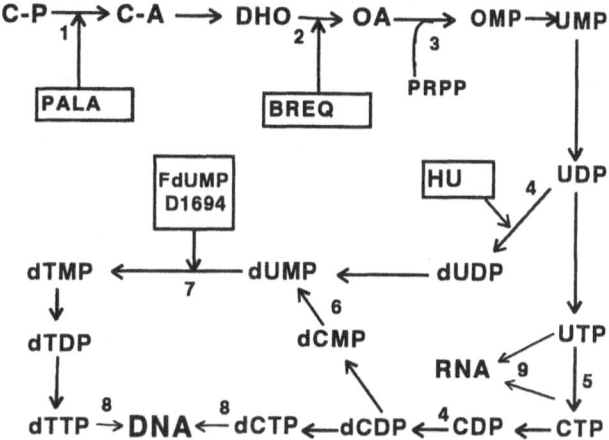

Fig. 3: Schematic representation of the pyrimidine *de novo* nucleotide synthesis, with the targets for several antimetabolites used in clinical practice. C-P, carbamyl-phosphate; C-A, carbamyl-aspartate; DHO, dihydroorotic acid; OA, orotic acid; Breq, Brequinar; HU, Hydroxyurea. Enzymes; 1, aspartate transcarbamylase; 2, DHO-dehydrogenase; 3, orotate phosphoribosyltransferase (OPRT); 4, ribonucleotide reductase; 5, CTP synthetase; 6, dCMP deaminase; 7, thymidylate synthase; 8, DNA polymerase; and 9, RNA polymerase.

Pyrimidine ribonucleotides

A now classical inhibitor of the pyrimidine *de novo* nucleotide synthesis is *N*-phosphon-acetyl-L-aspartate (PALA) an inhibitor of the second enzyme of the *de novo* pathway, aspartate transcarbamylase (ATC) (Fig. 4). In *in vitro* and *in vivo* model systems PALA depleted both ribo- and deoxyribonucleotide pools [18,32,33]; in *in vitro* models it could be clearly demonstrated that the depletion of pyrimidine ribonucleotides was accompanied by a similar depletion of the deoxyribonucleotides [18] (Fig. 4); in *in vivo* models nucleotide measurements were usually limited to the ribonucleotides [32,33]. Another more recent inhibitor of the pyrimidine *de novo* nucleotide synthesis, Brequinar, is a potent inhibitor of the mitochondrial enzyme dihydroorotate dehydrogenase (DHO-DH) [34,35]. Also for this compound a similar depletion of the pyrimidine ribo- and deoxynucleotide pools could be demonstrated [20]. Interestingly, the concentrations of the purine ribonucleotides and more pronounced of the deoxyribonucleotides, increased [20] (Fig. 4). This is possibly related to a deregulation of the nucleotide synthesis due to a lack of feedback inhibition and a larger availability of substrates, *e.g.* PRPP has a larger availability following inhibition of the pyrimidine *de novo* pathway; while a deregulation of ribonucleotide reductase will also occur.

From a biochemical point of view these inhibitors are of potential interest, since application of these compounds to patients in a clinical phase I trial, provided one of the few possibilities to determine the importance of the *de novo* pathway to ribonucleotide synthesis, *i.e.* how essential is this pathway for nucleotide synthesis compared to the salvage pathway [23]. It is known that a rare inherited form of a deficiency of orotate phosphoribosyltransferase (OPRT), causes orotic aciduria leading to megaloblastic anemia [36]; thus it would be expected that these agents would also cause some hematological side effects. Thus, in a clinical Phase I trial with Brequinar we have observed that increase of the dose led to a more pronounced and prolonged inhibition of lymphocytic DHO-DH and a larger and longer depletion of plasma uridine concentration [23]. At the maximum tolerated dose (MTD) in patients a 80-90% inhibition of lymphocytic DHO-DH, with a similar depletion of plasma uridine was associated with a Grade 3 lymphocytopenia and a Grade 3 gastrointestinal toxicity; pyrimidine ribonucleotide pools were depleted similarly. These results were comparable to the changes observed in mice after administration of Brequinar at the MTD [22].

In the above mentioned studies it was observed that both PALA and Brequinar can cause a depletion of pyrimidine NTP [18,20,22,32,33] and an elevation of the PRPP availability when administered at a lower non-toxic dose [32,37]. However, these effects were sufficient to potentiate the efficacy of 5-fluorouracil (5FU) [32]. This can be due to 1) enhanced metabolism due to a larger availability of PRPP, 2) enhanced incorporation into RNA, due a lack of competition by UTP, 3) enhanced inhibition of thymidylate synthase due to a lower concentration of dUMP. For both PALA and Brequinar is has been shown that they can enhance the therapeutic efficacy of 5FU in preclinical models, both *in vitro* and in animals [32,38-40], while for PALA it has been demonstrated that a combination with 5FU can results in a higher response rate that with monotherapy with 5FU [41]. These combinations are currently evaluated in randomized Phase III trials [41].

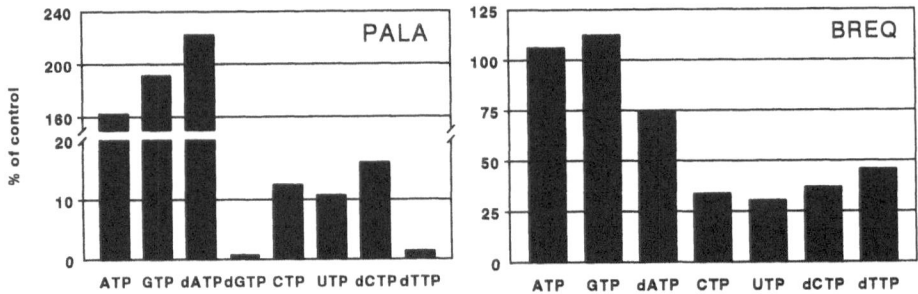

Fig. 4: Effect of PALA and Brequinar (Breq) on (deoxy)ribonucleotide pools. Values are percentages of concentrations in control cells harvested at the same time. IGR3 melanoma cells were exposed for 4 hr to 1 mM PALA [18], while L1210 leukemia cells were exposed to 25 μM Brequinar for 2 hr [20].

Purine ribonucleotides

A number of commonly used antimetabolites have a secondary effect on purine *de novo* ribonucleotide synthesis, *e.g.* (methyl)thionucleotides and polyglutamates of methotrexate (MTX). However, this effect is not the major mechanism by which these agents exert their antiproliferative effect. Currently several classes of new purine *de novo* synthesis inhibitors are in clinical development, Tiazofurin, a pyridine nucleotide analog, which (after activation) is a potent inhibitor of IMP-dehydrogenase (IMP-DH) [42,43], and dideazatetrahydrofolate (DDA-THF, Lometrexol), a folate analog, which is a potent inhibitor of glycinamide ribonucleotide formyltransformylase (GARFT) [44]. Tiazofurin selectively depletes (deoxy)GTP pools while DDA-THF depletes all pools. The action of both compounds is prevented by a large availability of purine bases. For the clinical application of Tiazofurin in the treatment of chronic myeloblastic leukemia [43], the compound is combined with allopurinol, which through inhibition of xanthine oxidase, causes an increase of hypoxanthine which competitively inhibits phosphoribosylation of guanine [43]. Inhibition of IMP-DH and the subsequent depletion of GTP in blast cells of patients correlated with the response of patients to treatment [43]. *In vitro* studies on Lometrexol are compromised by high hypoxanthine concentrations in serum, leading to a large variation in IC50 values [45]. Clinical studies are not only complicated by this high hypoxanthine concentration but also by a delayed toxicity possibly caused by a long retention of polyglutamates of DDA-THF [46]. This toxicity could be prevented by administration of either folic acid or leucovorin. Thus, at this moment the use of purine *de novo* synthesis inhibitors is complicated by a small therapeutic window, due to the above-mentioned factors.

DEPLETION OF DEOXYRIBONUCLEOTIDES

Synthesis of deoxyribonucleotides from ribonucleotides is mediated by ribonucleotide reductase. This four-substrate enzyme is, as mentioned above, very tightly regulated by its end products, the dNTP's and by ATP [3,4]. dNTP synthesis, is however, not only dependent on ribonucleotide reductase but also on the activity of several deoxynucleoside kinases (thymidine kinases, deoxycytidine kinase, deoxyguanosine kinase), which do not have the substrate specificity as suggested by their name [47], by thymidylate synthase (TS), which specifically catalyzes the rate-limiting step in dTTP synthesis, and by dCMP deaminase, which is one of the sources for dUMP, the substrate for TS. dNTP synthesis can thus by affected at a number of different sites; inhibitors of ribonucleotide reductase such as hydroxyurea will cause a depletion of all dNTP, while inhibitors of TS will specifically deplete dTTP. In order to facilitate the understanding of the complicated regulation of dNTP synthesis and the effect of antimetabolites on this Jackson [1] described a model in which a number of effects produced by antimetabolites could be explained in terms of dNTP metabolism.

Depletion of dTTP

Selective depletion of dTTP can be mediated by inhibition of TS, the rate-limiting enzyme in *de novo* thymidylate synthesis. FdUMP, the active nucleotide of 5FU, is one of the best described inhibitors of this enzyme [41,48,49]. FdUMP inhibits TS by the formation of a stable ternary complex in the presence of the co-substrate for this reaction, 5,10-methylene-tetrahydrofolate (CH_2-THF). Modulation of the anti-tumor activity of 5FU is based on an additional supply of CH_2-THF by administration of its precursor leucovorin. *In vitro* studies with more than 72 cell lines showed a potentiation of more than 1.5-fold in about 60% of these lines [48], associated with a more pronounced and more prolonged inhibition of TS. In animal model systems co-administration of leucovorin and 5FU prevented the 5FU induced overexpression of TS, which was possibly responsible for the resistance to 5FU of that model; leucovorin enhanced the antitumor activity of 5FU [50]. In patients we demonstrated that 5FU induced inhibition of TS was related with the response to TS and that leucovorin could enhance the 5FU induced TS-inhibition [51]; also others demonstrated that inhibition of TS is of major importance in the development of clinical resistance to 5FU [52]. These studies clearly demonstrate that inhibition of TS, leading to depletion of dTTP (Fig. 5), is a major therapeutic target [48].

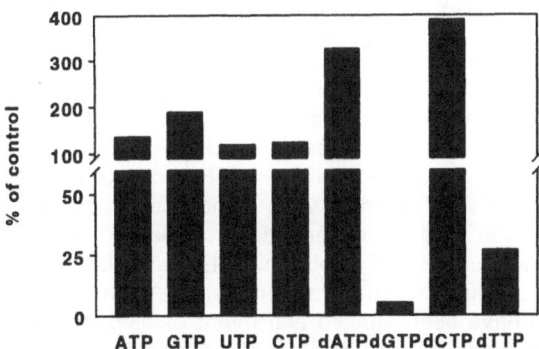

Fig. 5: Effect of 5FU on (deoxy)ribonucleotide pools of IGR3 melanoma cells. Cells were exposed to 20 µM 5FU for 20 hr [18].

A new generation of TS-inhibitors has recently entered clinical phase I-III studies. These compounds are analogs of the folate substrate for TS, CH_2-THF, and act by competition with this folate. ZD-1694 (Tomudex) has now completed clinical Phase II trials with a 27% response rate in colorectal cancer, similar to that for the combination of leucovorin and 5FU [53]. Other compounds of interest are LY231514 from Eli Lilly, BW1843U89 from Burroughs Welcome, and AG331 and AG337 from Agouron [48]. The latter compounds have been synthesized based on the three-dimensional structure of TS. All compounds are characterized by inhibition of TS leading to a depletion of dTTP [54]. The extent and retention of depletion are related with the extent and persistence of the growth inhibitory effect, *e.g.* the polyglutamable ZD-1694 will retain its growth-inhibitory after removal from the medium [54], while the non-polyglutamable AG331 and AG337 will not [55]. An additional effect of TS-inhibitors besides depletion of dTTP, is a disturbance of other dNTP pools, including an elevation of dATP [56, 18], leading to an imbalance of dNTP, which is possibly responsible for the ultimate effect of these compounds.

SYNTHESIS OF ANALOG NUCLEOTIDES

A number antimetabolites exert their activity by conversion to a nucleotide. Enzymes from purine and pyrimidine metabolism catalyze these reaction [30]. Examples are HGPRT for 6-mercaptopurine [57], deoxycytidine kinase for ara-C, chlorodeoxyadenosine and 2'2'-difluorodeoxycytidine (Gemcitabine, dFdC) [47], thymidine kinase for 5-fluoro-2'-deoxyuridine [48], OPRT for 5FU [58]. Since normal nucleotides are potent allosteric regulators for these enzymes [47,48], it is evident that synthesis of analog nucleotides and normal nucleotides will affect each other. Jackson [59] also developed a computer simulation model to predict inhibition of one pathway by an antimetabolite and the subsequent inhibition of RNA and DNA synthesis. This demonstrated that a similar inhibition of two different enzymes in the same synthetic pathway would have different effects on RNA synthesis. It is thus not only important that a compound is a potent inhibitor of just an enzyme but a potent inhibitor of a specific enzyme.

Synthesis and action of Gemcitabine-triphosphate

An example of a compound exerting a number of different actions leading to a self-potentiation is Gemcitabine (dFdC). This compound is activated by the action of deoxycytidine kinase [47], and induced resistance to dFdC has been associated with a deficiency of deoxycytidine kinase leading to a depletion of dF-dCTP, the activated dFdC nucleotide [60]. dF-dCTP will be incorporated into DNA [61,62] and into RNA [62]. Accumulation and retention of dF-dCTP has been associated with the growth-inhibitory effects of dFdC [26]; a major difference with another deoxycytidine analog ara-C is the longer intracellular half-life of dF-dCTP compared to ara-CTP [63]. In a primer extension assay Huang et al [61] demonstrated that dF-dCTP permitted the incorporation of one additional nucleotide but inhibited incorporation of more dNTP. dF-dCTP has a number of additional actions on (deoxy)ribonucleotide metabolism, leading to a depletion of dCTP [64], due to inhibition of ribonucleotide reductase by dF-dCDP [64]. In addition to that it has been shown that dF-dCTP is a potent inhibitor of dCMP deaminase, preventing the deamination of the parent compound dFdC [65]. Besides disturbance in the dNTP pools, Gemcitabine also affects normal ribonucleotide pools. In several solid tumor cell lines we demonstrated a time-, concentration- and cell line dependent depletion of CTP accompanied by a simultaneous increase in UTP pools, indicating an inhibition of CTP synthetase [26]. This complicated mechanism of action, which seems to vary between the model systems which have been studied and might explain the potent therapeutic efficacy observed in the clinic against non-small cell lung cancer, ovarian cancer and breast cancer [66].

INDIRECT EFFECTS OF dNTP IMBALANCE; RELATION WITH CELL KILL

As mentioned above a number of antimetabolites induce pronounced disturbances in ribo- and/or deoxyribonucleotide pools. The aim of treatment with antimetabolites is to selectively kill the cancer cells. The question can thus be reformulated to; when is an imbalance of (deoxy)ribonucleotides lethal for the cell? Should one increase or decrease these pools (or just of one dNTP) and to what extent? The direct effect of an imbalance in dNTP pools can considered to be 1) an inhibition of DNA synthesis, 2) a depressed repair, 3) an increased mutagenesis, 4) a disruption in DNA proofreading. Snyder [67] demonstrated that an imbalance in dNTP can inhibit UV induced excision/repair in a fibroblast system. In the presence of normal dNTP concentrations the basal level of DNA strand breaks is very low and pyrimidine dimer removal amounts to 25%; however, an increased dATP/dTTP ratio caused a 26-fold increase in DNA single strand breaks and removal of pyrimidine dimers was less than 10%. Meuth [68] demonstrated that an excess of dCTP in the absence of dTTP increased the mutation rate in a mutated CHO model system to 69% due to a mismatch during proofreading of the DNA strands; repletion of dTTP reduced the mutation rate to 6%. Thus it is clear that decreased DNA repair can lead to severe damage to the DNA ultimately leading to cell death. Also Mattano et al [69] demonstrated that an imbalance in dNTP pools can cause an increase in the mutations rate as studied at the HGPRT locus. This type of studies has been summarized by Kunz [70]. It is however not clear what is the relation between this imbalance in dNTP pools and cell death, although it has been demonstrated for several antimetabolites (*e.g.* chlorodeoxyadenosine, Gemcitabine, fludarabine, 5-fluorodeoxyuridine and other specific TS inhibitors) that they can cause apoptosis [71-73].

Yoshioka [74] postulated that the imbalance in dNTP caused by inhibition of TS leading to a depletion of dTTP and an increase in dATP, would lead to the activation of the endonuclease gene, the gene products would than attack the DNA to cause double strand breaks, ultimately leading to cell death. Fisher et al [73] demonstrated that dTTP depletion and induction of strand breaks preceded the onset of apoptosis, and postulated that enhanced expression of *bcl*-2 would reduce the rate of apoptosis. Also other cellular genes such as c-*myc* and *p53* may determine the ability of cell to engage apoptosis.

Application of a dNTP imbalance

How can we use this knowledge? It is clear that an imbalance of dNTP affects DNA repair. Thus the antitumor activity of agents for which it is known that enhanced DNA repair is a resistance mechanism, may be potentiated by an imbalance of dNTP. Thus, it was tested whether a PALA induced depletion of pyrimidine NTP and dNTP, associated with enhanced dATP (see *e.g.* Fig. 4), would enhance the activity of cisplatin [75]. Mice bearing advanced colon tumors were treated at their MTD with cisplatin in combination with a non-toxic ineffective dose of PALA; the antitumor activity of this combination (median doubling time of the tumors 26 days) was significantly higher than that of cisplatin alone (median doubling time of 10 days) [75]. Another example is a combination aimed at modulation of dNTP pools and of poly (ADP) ribose polymerase. This enzyme is chromatin bound, synthesizes (ADP)ribose from NAD and is involved in regulation of DNA repair. A combination of 6-aminonucleotide (which inhibited synthesis of NAD, leading to inhibition of poly(ADP)ribose polymerase), PALA and 6-methylmercaptopurine (also interfering with poly(ADP) synthesis) resulted in a significantly better antitumor activity (>60% partial responses) than any double combination of these compounds (not more than 20% partial responses), or than any of the single agents (no responses), in a murine breast carcinoma tumor [32]. It can thus be concluded that a better knowledge of substrate specificity of deoxynucleotide analogs for DNA polymerases [76], of DNA repair mechanism in the cancer cells and the role of dNTP imbalance in this will enable to develop more rationale combinations for chemotherapy of tumors.

CONCLUSIONS

It can be concluded that in order to enable optimal cell growth cellular nucleotide concentrations have a large so-called "safety window". Thus in order to inhibit cellular proliferation nucleotide pools have to be depleted considerably, possibly more than 90%, depending on the nucleotides. It should be considered that ATP concentrations are approximately 3 mM which is much higher than the Km for many reactions in which ATP is involved as a substrate or an activator. Similar conclusions can be drawn for the other NTP. For specific pyrimidine *de novo* synthesis inhibitors it can be concluded that a depletion of the ribonucleotides will be reflected by a similar decrease in the corresponding dNTP. An imbalance in dNTP can significantly affect DNA repair and replication. Monitoring changes in normal cellular nucleotide pools (or markers for that in plasma) and/or the accumulation of an analog nucleotide *e.g.* dF-CTP for gemcitabine, can serve as a marker for response/toxicity.

ACKNOWLEDGEMENTS

I thank P. Noordhuis and G. Veerman for their cooperation in collecting data reviewed in this manuscript.

REFERENCES

1   Jackson RC (1984). A kinetic model of regulation of the deoxyribonucleoside triphosphate pool composition. Pharmac. Ther. 26; 279-301.
2   Munch-Petersen B, Tyrsted G (1988). Thymidine kinase in human leukemia. Expression of the lymphoblastic isozyme in three patients with acute myelocytic leukemia. Leukemia Res. 12; 173-178.
3   Hunting D, Henderson JF (1982). Models of the regulation of ribonucleotide reductase and their evaluation in intact mammalian cells. CRC Crit. Rev. Biochem. 13; 325-348.
4   Reichard P (1987). regulation of deoxyribotide synthesis. Biochemistry 26; 3245-3248.
5   Perret D, Herbert KE, Morris G, Simmmonds HA (1989). Optimised conditions for the routine HPLC separation of nucleotides in cell extracts. Adv. Exp. Med. Biol. 253B; 463-468
6   Werner A (1991). Analysis of nucleotides, nucleosides, nucleobases in cells by ion-pair reversed-phase HPLC. Chromatographia 31; 401-410.
7   Perret D, Ross G (1991). Capillary electrophoresis for the analysis of cellular nucleotides. Adv. Exp. Med. Biol. 309B; 1-5.
8   Mathews CK, Slabaugh MB (1986). Eukaryotic DNA metabolism. Are deoxyribonucleotides channelled to replication sites? Exp Cell Res 162; 285-295
9   Moyer JD, Henderson JF (1985). Compartmentation if intracellular nucleotides in mammalian cells. CRC Crit. Rev. Biochem 19; 45-61.
10   Bestwick RK, Moffett GL, Mathews CK (1982). Selective expansion of mitochondrial nucleoside triphosphate pools in antimetabolite-treated HeLa cells. J Biol Chem 257; 9300-9304.
11   Pels Rijcken WR, Overdijk B, Van den Eijnden DH, Ferwerda W (1993). Pyrimidine nucleotide metabolism in rat hepatocytes: evidence for compartmentation of nucleotide pools. Biochem J 293; 207-213.
12   Chiba P, Bacon PE, Cory JG (1984) Biochem Biophys Res Comm 128; 345
13   Peters GJ, Veerkamp JH (1983). Purine and pyrimidine metabolism in peripheral blood lymphocytes. Int. J. Biochem. 15: 115-123.
14   De Abreu RA, Peters GJ, Bakkeren JAJM & Veerkamp JH (1985).Discrepancies in ribonucleotide concentrations in human lymphocytes isolated from heparinized and defibrinized blood. Clin. Chim. Acta 145: 349-355.
15   Peters GJ, De Abreu RA, Oosterhof A & Veerkamp JH (1983). Concentration of nucleotides and deoxynucleotides in peripheral and stimulated mammalian lymphocytes. Effects of adenosine and deoxyadenosine. Biochim. Biophys. Acta. 759: 7-15.
16   De Korte D, Haverkort WA, Roos D, Behrendt H, Van Gennip AH (1986). Imbalance in ribonucleotide pools of lymphoid cells from acute lymphoblastic leukemia. Leuk Res 10; 389-396
17   Peters GJ, Laurensse E, Lankelma J, Leyva A, Pinedo HM (1984). Separation of several 5-fluorouracil metabolites in various melanoma cell lines. Evidence for the synthesis of 5-fluorouracil-nucleotide sugars. Eur. J. Cancer Clin. Oncol. 20: 1425-1431.

18   Peters GJ, Laurensse E, Leyva A, Lankelma J, Pinedo HM (1986). Sensitivity of human, murine and rat cells to 5-fluorouracil and 5'deoxy-5-fluorouridine in relation to drug metabolizing enzymes. Cancer Res. 46: 20-28.

19   Peters GJ, Van Groeningen CJ, Laurensse E, Lankelma J, Leyva A, Pinedo HM (1987). Uridine-induced hypothermia in mice and rats in relation to plasma and tissue levels of uridine and its metabolites. Cancer Chemother. Pharmacol. 20: 101-108.

20   Schwartsmann G, Peters GJ, Laurensse E, De Waal FC, Loonen AH, Leyva A & Pinedo HM (1988). DUP-785 (NSC 368390): Schedule-dependency of growth-inhibitory and anti-pyrimidine effects. Biochem. Pharmacol. 37: 3257-3266

21   Pels Rijcken WR, Telleman F, Peters GJ, Ferwerda W (1989). Incorporation of 5-fluorouracil into nucleotide sugars and the effect on glycoconjugates in rat hepatoma cells and hepatocytes. Adv. Exp. Med. Biol. 253B: 313-320.

22   Peters GJ, Nadal JC, Laurensse EJ, De Kant E, Pinedo HM (1990). Retention of in vivo antipyrimidine effects of Brequinar sodium (DUP-785; NSC368390) in murine liver, bone marrow and colon cancer. Biochem. Pharmacol. 39: 135-144.

23   Peters GJ, Schwartsmann G, Nadal JC, Laurensse EJ, Van Groeningen CJ, Van der Vijgh WJF, Pinedo HM (1990). In vivo inhibition of the pyrimidine de novo enzyme dihydroorotic acid dehydrogenase by Brequinar Sodium (DUP-785; NSC 368390) in mice and patients. Cancer Res. 50; 4644-4649.

24   Peters GJ, Wets M, Keepers YPAM, Oskam R, Van Ark-Otte, Noordhuis P, Smid K, Pinedo HM (1993). Transformation of mouse fibroblasts with the oncogenes H-ras or trk is associated with pronounced changes in drug sensitivity and metabolism. Int. J. Cancer 54; 450-455.

25   Pizao PE, Peters GJ, Van Ark-Otte J, Smets LA, Smitskamp-Wilms E, Winograd B, Pinedo HM, Giaccone G (1993). Cytotoxic effects of anticancer agents on subconfluent and multilayered postconfluent cultures. Eur. J. Cancer 29A; 1566-1573.

26   Ruiz van Haperen VWT, Veerman G, Boven E, Noordhuis, Vermorken JB, Peters GJ (1994). Schedule-dependence of sensitivity to 2',2'-difluorodeoxycytidine (Gemcitabine) in relation to accumulation and retention of its triphosphate in solid tumor cell lines and solid tumors. Biochem. Pharmacol., in press

27   Jackson RC, Lui MS, Boritzki TJ, Morris HP, Weber G (1980). Purine and pyrimidine nucleotide patterns of normal, differentiating, and regenerating liver and of hepatomas in rats. Cancer Res. 40; 1286-1291.

28   Malet-Martino M-C, Martino R (1991). uses and limitations of nuclear magnetic resonance spectroscopy in clinical pharmacokinetics. Clin Pharmacokin 20; 337-349.

29   Darnowski JW, Handschumacher RE (1989) Enhancement of fluorouracil therapy by the manipulation of tissue uridine pools. Pharmacol Therap 41: 381-392.

30   Peters GJ, Beijnen JH (1994). Purine and pyrimidine metabolism; still a black box? Pharm World Sci 16; 37-38.

31   Krungkai J, Krungkai SR, Phakanont K (1992). Antimalarial activity of orotate analogs that inhibit dihydroorotase and dihydroorotate dehydrogenase. Biochem Pharmacol 43; 1295-1301.

32   Martin DS (1987) Biochemical modulation: perspectives and objectives. In: KR Harrap and TA Connors (eds). In: Proceedings 8th Bristol-Myers Symposium on Cancer Research - New Avenues in Developmental Cancer Chemotherapy, pp 113-162, Academic Press, London

33   Van Laar JAM, Durrani FA, Rustum YM (1993). Antitumor Activity of the Weekly Intravenous Push Schedule of 5- Fluoro-2'-Deoxyuridine +/- N-Phosphonacetyl-L-aspartate in Mice Bearing Advanced Colon Carcinoma 26. Cancer Research 53, 1560- 1564.

34   Peters GJ, Sharma SL, Laurensse E, Pinedo HM (1987). Inhibition of pyrimidine de novo synthesis by DUP-785 (NSC 368390). Invest. New Drugs 5: 235-244.

35   Chen S-F, Ruben RL and Dexter DL (1986). Mechanism of action of the novel anticancer agent 6-fluoro-2-(2'-fluoro-1,1'-biphenyl-4-yl)-3-methyl-4-quinolinecarboxylic   acid   sodium   salt   (NSC 368390): Inhibition of de novo pyrimidine nucleotide biosynthesis, Cancer Res 46: 5014-5019.

36   Simmonds HA (1994). When and how does one search for inborn errrors of purine and pyrimiidne metabolism Pharm World Sci 16; 139-148.

37   Peters GJ, Laurensse, E, Leyva, A & Pinedo, HM (1985). The concentration of 5-phosphoribosyl-1-pyrophosphate in monolayer cells and the effect of various pyrimidine antimetabolites. Int. J. Biochem. 17: 95-99.

38   Peters GJ, Kraal I, Pinedo HM (1992). In vitro and in vivo studies on the combination of Brequinar sodium (DUP-895; NSC 368390) with 5-fluorouracil; effects of uridine. Brit J Cancer 65; 229-233.

39   Pizzorno G, Wiegand RA, Lentz SK, Handschumacher RE (1992). Brequinar Potentiates 5-Fluorouracil Antitumor Activity in a Murine Model Colon 38 Tumor by Tissue-specific Modulation of Uridine Nucleotide Pools. Cancer Research 52: 1660-1665.

40   Chen T-L, Erlichman C (1992). Biochemical modulation of 5-fluorouracil with or without leucovorin by a low dose of brequinar in MGH-U1 cells. Cancer Chemother Pharmacol 20; 370-376.

41   Peters GJ, Van Groeningen CJ (1991). Clinical relevance of biochemical modulation of 5-fluorouracil. Ann. Oncology 2; 469-480.

42   Geyssen GJ, Pieters R, Veerman AJP, Pinedo HM, Peters GJ (1991). Do inhibitors of IMP dehydrogenase have a future in cancer chemotherapy? Int. J. Purine Pyrimidine Res. 2; 17-26.

43 Weber G, Hata Y, Prajda N (1994). Role of differentiation inducers in the action of purine antimetabolites. Pharm World Sci 16; 77-83.

44 Beardsley GP, Moroson BA, Taylor EC, Moran RG (1989) A new folate antimetabolite, 5,10-dideaza-5,6,7,8-tetrahydrofolate is a potent inhibitor of de novo purine synthesis. J Biol Chem 264: 328-333.

45 Van der Laan BFAM, Jansen G, Kathmann GAM, Westerhof GR, Schornagel JH, Hordijk GJ (1992). In vitro activity of novel antifolates againsthuman squamous carcinoma cell lines of the head and neck with inherent resistance to methotrexate. Int J Cancer 51; 909-914.

46 Ray MS, Muggia FM, Leichman CG, Grunberg SM, Nelson RL, Dyke RW, Moran RG (1993). Phase I study of (6R)-5,10-dideazatetrahydrofolate; a folate antimetabolite inhibitory to de novo purine synthesis. J Natl Cancer Inst 85; 1154-1159.

47 Ruiz van Haperen VWT, Peters GJ (1994). New targets for pyrimidine antimetabolites for the treatment of solid tumours. 2. Deoxycytidine kinase. Pharm. World Sci. 16; 104-112.

48 Van der Wilt CL, Peters GJ (1994). New targets for pyrimidine antimetabolites for the treatment of solid tumours. 1. Thymidylate synthase. Pharm. World Sci. 16; 84-103.

49 Danenberg PV (1977) Thymidylate synthetase - a target enzyme in cancer chemotherapy. Biochimica et Biophysica Acta 473: 73-92.

50 Van der Wilt CL, Smid K, Pinedo HM, Peters GJ (1992). Elevation of thymidylate synthase following 5-fluorouracil treatment is prevented by addition of leucovorin in murine colon tumors. Cancer Res. 52; 4922-4928.

51 Peters GJ, Van der Wilt CL, Van Groeningen CJ, Meijer S, Smid K, Pinedo HM (1994). Thymidylate synthase inhibition after administration of 5-fluorouracil with or without leucovorin; implications for treatment with 5-fluorouracil. J. Clin. Oncol., in press

52 Swain SM, Lippman ME, Egan EF, Drake JC, Steinberg SM, Allegra CJ (1989) Fluorouracil and high-dose leucovorin in previously treated patients with metastatic breast cancer. Journal of Clinical Oncology 7 890-899.

53 Zalcberg J, Cunningham D, Van Cutsem E, et al (1994). Good antitumor activity of the new thymidylate synthase inhibitor Tomudex (ZD 1694) in colorectal cancer. Ann Oncol 5 (supp 5), 133

54 Jackman AL, Taylor GA, Gibson W, Kimbell R, Brown M, Calvert AH, Judson IR, Hughes LR (1991). ICI D1694, a quinazoline antifolate thymidylate synthase inhibitor that is a potent inhibitor of L1210 tumor cell growth in vitro and in vivo: a new agent for clinical study. Cancer Res. 51; 5579-5586.

55 Jackson RC, Johnston AL, Shetty BV, Varney MD, Webber S, Webber SE (1993). Molecular design of thymidylate synthase inhibitors. Proc. Am. Ass. Cancer Res. 34; 566-567.

56 Kwok JBJ, Tattersall MHN (1992). DNA fragmentation, dATP pool elevation and potentiation of antifolate cytotoxicity in L1210 cells by hypoxanthine. Br. J. Cancer 65; 503-508.

57 Pinkel D (1993). Intravenous Mercaptopurine: Life Begins at 40. Journal of Clinical Oncology 11, 1826-1831.

58 Peters GJ, Van Groeningen CJ, Laurensse EJ, Pinedo HM (1991). A comparison of 5-fluorouracil metabolism in human colorectal cancer and colon mucosa. Cancer 68; 1903-1909.

59 Jackson RC (1987). Computer simulation of the effects of antimetabolites on metabolic pathways. Connors (eds). In: Proceedings 8th Bristol-Myers Symposium on Cancer Research - New Avenues in Developmental Cancer Chemotherapy, pp 3-35, Academic Press, London

60 Ruiz van Haperen VWT, Veerman G, Eriksson S, Boven E, Stegmann APA, Hermsen M,in Developmental Cancer Chemotherapy, pp 113-162, Academic Press, London Vermorken JB, Pinedo HM, Peters GJ (1994). Development and characterization of a 2′,2′-difluorodeoxycytidine-resistant variant of the human ovarian cancer cell line A2780. Cancer Res., in press

61 Huang P, Chubb S, Hertel LW, Grindey GB and Plunkett W (1991) Action of 2′,2′-Difluorodeoxycytidine on DNA Synthesis. Cancer Research 51 6110-6117

62 Ruiz Van Haperen VWT, Veerman G, Vermorken JB, Peters GJ (1993). 2′,2′-Difluoro-deoxycytidine (gemcitabine) incorporation into RNA and DNA from tumour cell lines. Biochem. Pharmacol. 46; 762-766.

63 Heinemann V, Hertel LW, Grindey GB and Plunkett W (1988) Comparison of the cellular pharmacokinetics and toxicity of 2′,2′-difluorodeoxycytidine and 1-ß-D-arabinofuranosylcytosine. Cancer Research 48 4024-4031

64 Gandhi V, Plunkett W (1990). Modulatory activity of 2′,2′-difluorodeoxycytidine on the phosphorylation and cytotoxicity of arabinosyl nucleosides. Cancer Res 50; 3675-3680.

65 Xu Y-Z, Plunkett W (1992). Modulation of deoxycytidylate deaminase in intact human leukemia cells. Biochem Pharmacol 44; 1819-1827.

66 Lund B, Kristjansen PEG, Hansen HH (1993). Clinical and preclinical activity of 2′,2′-difluorodeoxycytidine (gemcitabine). Cancer Treat. Rev. 19, 45-55.

67 Snyder RD (1988). Consequences of depletion of cellular deoxyribonucleoside triphosphate pools on the excision-repair process in cultured human fibroblasts. Mutation Res. 200; 193-199.

68 Meuth M (1989). The molecular basis of mutations induced by deoxyribonucleoside triphosphate pool imbalances in mammalian cells. Exp. Cell Res 181; 305-316.

69 Mattano SS, Palella TD, Mitchell BS (1990). Mutations induced at the hypoxanthine-guanine phosphoribosyltransferase locus of human T-lymphoblasts by perturbations of purine

deoxyribonucleoside triphosphate pools. Cancer Res <u>50</u>; 4566-4571.

70   Kunz BA (1988). Mutagenesis and deoxyribonucleotide pool imbalance. Mutation Res <u>200</u>; 133-147.

71   Huang P, Plunkett W. A quantitative assay for fragmented DNA in apoptotic cells. Anal. Biochem. 207, 1992, 163-167.

72   Robertsen LE, Chubb S, Meyn RE, Story M, Ford R, Hittelman WN, Plunkett W (1993). Induction of apoptotic cell death in chronic lymphocytic leukemia by 2-chloro-2'-deoxyadenosine and 9-β-D-arabinosyl-2-fluoroadenine. Blood <u>81</u>; 143-150.

73   Fisher TC, Miller AE, Gregory CD, Jackman AL, Aherne GW, Hertley JA, Dive C, Hickman JA (1993). *bcl*-2 Modulation of apoptosis induced by anticancer agents: resistance to thymidylate stress is independent of classical resistance pathways. Cancer Res. <u>53</u>; 3321-3326.

74   Yoshioka A, Tanaka S, Hiraoka O, Koyama Y, Hirota Y, Ayusawa D, Seno T, Garrett C, Wataya Y (1987). Deoxyribonucleoside triphosphate imbalance. 5-Fluorodeoxyuridine-induced DNA double strand breaks in mouse FM3A cells and the mechanism of cell death. J Biol Chem <u>262</u>; 8235-8241.

75   Van Laar JAM, Mayhew E, Durrani FA, Peters GJ, Rustum YM (1993). Enhancement of the antitumor activity of 5-fluoro-2'-deoxyuridine and cis-platinum by *N*-(phosphonacetyl)-*L*-aspartate in the murine colon 26 carcinoma. Pharmacy World & Science <u>15</u> (sup F), F29

76   Wright GE, Brown NC (1990). Deoxyribonucleotide analogs as inhibitors and substrates of DNA polymerases. Pharmac. Ther. <u>47</u>; 447-497.

COMPARISON OF 5-FLUORO-2'-DEOXYURIDINE AND 5-FLUOROURACIL

IN THE TREATMENT OF MURINE COLON CANCER;

EFFECTS ON THYMIDYLATE SYNTHASE

Clasina L. van der Wilt[1], Jan A.M. van Laar[1,2], Kees Smid[1],
Youcef M. Rustum[2], Godefridus J. Peters[1]

[1] Dept. of Medical Oncology, Free University Hospital, PO Box 7057,
1007 MB Amsterdam, The Netherlands
[2] Dept. of Experimental Therapeutics, Roswell Park Memorial Institute, Elm
and Carlton Streets, Buffalo, NY 14263, USA

INTRODUCTION

The fluoropyrimidines 5-fluoro-2'-deoxyuridine (FdUrd) and 5-fluorouracil (FUra) are both applied in the treatment of patients with colorectal cancer. These agents exert their action after metabolization to the nucleotide level (Fig. 1). The nucleotides 5-fluorouridine triphosphate (FUTP) and 5-fluoro-2'-deoxyuridine triphosphate (FdUTP) may mediate antiproliferative effects by their incorporation into cellular RNA and DNA, respectively. Anabolism to 5-fluorodeoxyuridine monophosphate (FdUMP) causes an inhibition of the key-enzyme of the pyrimidine de novo synthesis, thymidylate synthase (TS)[1,2]. FdUrd's main mechanism of action is thought to be inhibition of TS[3], while the action of FUra is incorporation into RNA and TS inhibition in different magnitudes, depending on the schedule[4,5].

FdUrd is usually given as a continuous infusion. We have demonstrated in the treatment of a murine Colon 26 tumor (here designed Colon 26-B), that with i.v. push administration of FdUrd at its maximum tolerated dose (MTD), greater therapeutic activity and selectivity can be achieved[6]. To delineate the mechanisms associated with the therapeutic superiority of FdUrd over FUra, the effect of treatment on TS was studied up to 1 week post drug treatment.

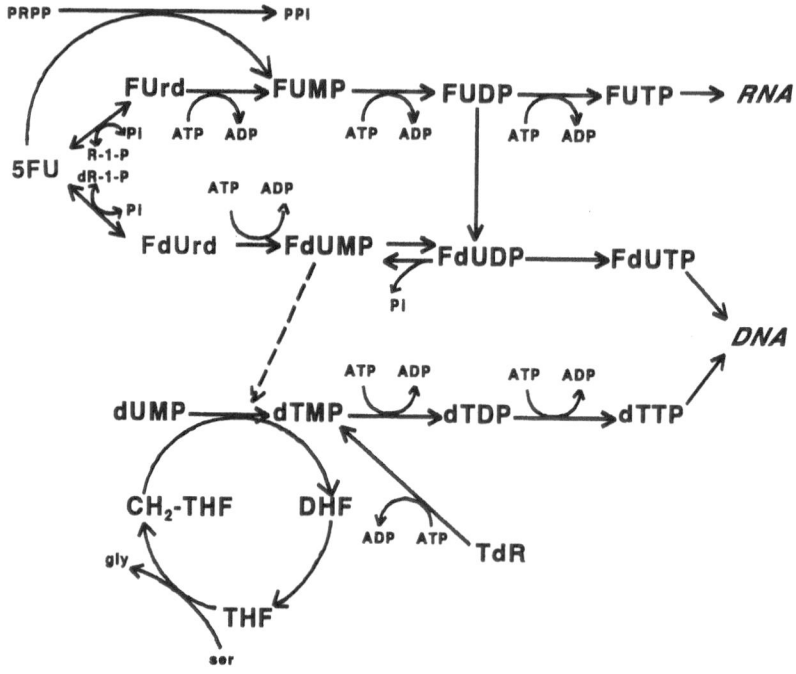

Figure. 1. Metabolism of FUra and FdUrd

MATERIAL AND METHODS

Female Balb/c mice (Harlan/Olac, Zeist, The Netherlands) of 6-7 weeks old were kept 6 per cage with water and food *ad libitum*. Colon 26-B tumor[6,7] fragments were inoculated in both flanks of the mice under slight ether anaesthesia. Ten days after transplantation when the tumors had reached a size of about 200 mm², drugs were administered. FUra (80 mg/kg) and FdUrd (400 mg/kg) were given as a bolus i.p. injection. Three weekly courses were administered and antitumor activity was evaluated by calliper measurements of the tumor volume (length x width x height x 0.5) at least three times a week.

Tumors for biochemical measurements were removed at 2, 8, 24, 48, 72 h and 1 week (=168 h) after a single treatment. Tissues were removed and immediately frozen in liquid nitrogen and kept at -80°C until assayed. Tumors from untreated mice served as controls. Frozen tumors were pulverized with a microdismembrator and the still frozen powder was suspended in a TRIS-buffer (200 mM TRIS, 15 mM CMP, 100 mM NaF, 20 mM ß-mercaptoethanol) at a concentration of 0.25 g/ml[8]. Each sample was split in 5 parts for the different assays.

FUra concentrations in the tumor were measured after extraction and derivatization by gas chromatography coupled to mass spectrometry with [^{15}N,^{15}N]-FUra as an internal standard[9]. Intratumoral levels of FdUMP were determined using the isotope-dilution assay with *Lactobacillus casei* TS. [6-^{3}H]-FdUMP was used as the radioactive substrate[10].

The inhibition of TS catalytic activity was measured by the conversion of [5-^{3}H]-dUMP into dTMP and ^{3}H$_2$O. Binding of [6-^{3}H]-FdUMP to TS gave an indication about the number of free binding sites of TS after treatment. Details on both assays have been described elsewhere[10]. Protein content of the tumor homogenates was evaluated with the Bio-Rad assay.

RESULTS

Antitumor activity of FUra (80 mg/kg) and FdUrd (200 and 400 mg/kg) is shown in Fig. 2. Furthermore in comparing FdUrd with FUra by the i.v. push weekly schedule at their MTDs (80 and 400 mg/kg, respectively), tumor doubling time was 30 *vs* 7 days, tumor volume of treated/untreated animals was 0.023 *vs* 0.26 and complete tumor regression was 70% *vs* 0% for FdUrd and FUra, respectively[6]. More recently FdUrd (400 mg/kg, weekly x 3) resulted in 20-30% complete tumor regression, but this was still significantly different from the results with FUra (80 mg/kg, weekly x 3). The antitumor activity of FdUrd (200 mg/kg, weekly x 3) was intermediate.

Evaluation of the biochemical effects of FUra and FdUrd treatment related to TS in murine Colon 26-B tumors were evaluated. Eight hours after administration of 400 mg/kg FdUrd the concentrations of FUra in the tumors were 2.5-fold higher than the levels measured in tumors from mice that received 80 mg/kg FUra (Fig. 3). At later time points tumor FUra concentrations were comparable for both drugs.

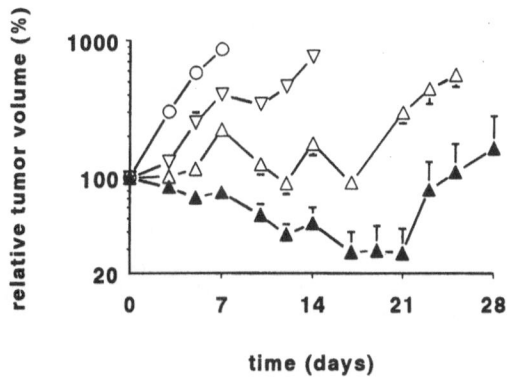

Figure 2.   Antitumor effect of FUra (80 mg/kg)(-▽-) and FdUrd (200 mg/kg) (-△-) and FdUrd (400 mg/kg)(-▲-) against Colon 26-B compared to control (-o-). Therapy was given on day 0, 7 and 14. Values are means ± SD of 8-12 tumors. Data from Van Laar *et al*[6].

Intratumoral free FdUMP levels were 3-fold higher at 2 h post FdUrd administration compared to FUra administration. At 48 h post treatment free FdUMP was below detection limit of the assay (10 fmol/mg wet weight (ww)) (data not shown).

The number of free binding sites for ³H-FdUMP decreased from 103 fmol/mg ww to 30 and 20 fmol/mg ww at 2 h post administration for FUra and FdUrd, respectively (Fig. 4). Values remained low for 72 h, but 1 week post FUra treatment the number of FdUMP binding sites had recovered and was actually higher than control (153 fmol/mg ww). FdUMP binding to TS in tumors from FdUrd treated mice was still low (17 fmol/mg ww) after 1 week.

The catalytic activity of TS measured with [5-³H]-dUMP was 5536 pmol/h/mg protein in control tumors. Maximal inhibition of the activity to 10% of the control value was obtained 8 h post FdUrd and 48 h post FUra treatment. One week post FUra treatment the catalytic activity had recovered and was slightly higher than control (1.3x), while in the FdUrd tumors TS activity was still inhibited at the level of 25% of control (data not shown).

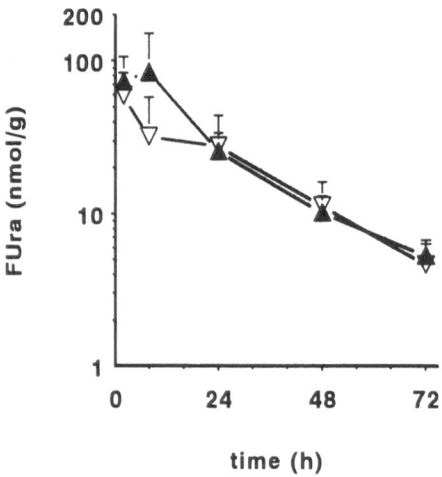

Figure 3. FUra concentrations in tumor tissue measured at different time points after FUra (80 mg/kg) (-▽-) and FdUrd (400 mg/kg) (-▲-) administration. Values are means ± SD of 3-5 tumors. Small SD's are within the symbol.

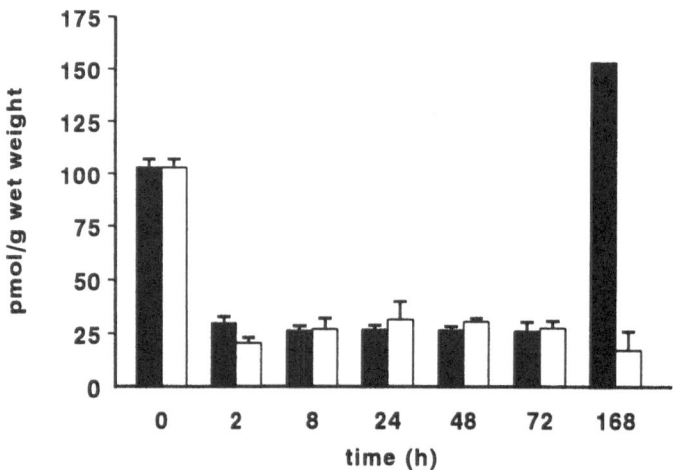

Figure 4. FdUMP binding to TS in Colon 26-B tumors at different time points after administration of 80 mg/kg FUra (closed bars) or 400 mg/kg FdUrd (open bars). Values are means of at least 3 tumors ± SD. Statistics: values at 2 h post treatment were significantly different ($p < 0.01$, Mann Whitney U test)

DISCUSSION

The high therapeutic efficacy of FdUrd *vs* FUra at equitoxic dose in Colon 26-B can possibly be explained on the basis of a long term difference in retention of TS inhibition. After one week TS in tumors from mice treated with FdUrd was still inhibited in contrast to tumors from mice treated with FUra.

The 400 mg/kg FdUrd dose was higher than the FUra dose (80 mg/kg) and initially this led to higher FUra and free FdUMP levels in the tumor resulting in a significant lower TS catalytic activity and FdUMP binding to TS at 2 h after treatment. This rapid inhibition of TS seemed to be an advantage of the FdUrd therapy, however after 24 h and up to 72 h similar values for TS inhibition concerning FdUrd and FUra treatment were observed. It seemed unlikely that this small difference in TS inhibition at an early time point could explain the superiority of FdUrd antitumor effect over the effect of FUra. Moreover also at a lower dose of FdUrd (200 mg/kg, weekly x 3), which was about equimolar to 80 mg/kg FUra antitumor activity was significantly higher for FdUrd therapy compared to FUra therapy[6]. This dose FdUrd would likely give lower initial FUra and FdUMP levels affecting the rapid inhibition of TS at 2 and 8 h post treatment. The retention effect of FdUrd on TS inhibition seemed more important. The difference with FUra effects was larger.

The FUra induced increase of TS seems to be comparable to that shown for a related Colon 26 tumor line, which was maintained in the laboratory in Amsterdam for several years (now designated as Colon 26-A)[10]. Although the Colon 26-B, which was maintained in Buffalo and the Colon 26-A share many characteristics, some striking differences have evolved during the years. For example Colon 26-A causes cachexia and is sensitive to modulation of FUra antitumor activity by folinic acid, while Colon 26-B lacks these properties. TS catalytic activity and FdUMP binding of Colon 26-A and Colon 26-B are in the same range and the increase of TS levels that occurred after bolus i.p. FUra administration could be caused by the same mechanism. Increase of TS was also shown in Colon 26-A after continuous infusion of FUra[11]. It is very interesting that the increase of TS did not occur 1 week after FdUrd treatment. The metabolism of FdUrd and FUra seems to be very similar and after 24 h FUra and free FdUMP levels were comparable. It seems unlikely that the initial difference in these levels was responsible for the increase of TS observed after one week. High FdUMP levels, such as observed after FdUrd would cause a more severe disturbance of the regulation of TS mRNA translation[12] than lower levels corresponding to FUra administration. Further investigations at RNA and protein level might reveal why FUra caused an increase of TS and FdUrd did not. So far this has only been investigated in an *in vitro* system[12], more complicating factors may play a role *in vivo*.

The antiproliferative effect of FdUrd is thought to be mediated mainly by TS inhibition, while FUra might also act at the level of RNA. We found a prolonged TS inhibition for FdUrd treatment, which could explain its high antitumor activity against Colon 26-B. the increase of TS that occurred especially after FUra therapy could not be clarified yet, but might be related to an effect at the RNA level.

ACKNOWLEDGEMENT

This study was supported by the Dutch Cancer Society (grant IKA 92-88).

REFERENCES

1. Y.M. Rustum. Modulation of fluoropyrimidines by leucovorin: rational and status. *J. Surg. Oncol.* suppl. 2:116-123 (1991).

2. H.M. Pinedo, G.J. Peters. 5-Fluorouracil: Biochemistry and pharmacology. *J. Clin. Oncol.* 6:1653-1664 (1988).

3. Z.G. Zhang, A. Harstrick, Y.M. Rustum. Modulation of fluoropyrimidines: role of dose and schedule of leucovorin administration. *Sem. Oncol.* 19 (suppl. 3):10-15 (1992).

4. C. Aschele, A. Sobrero, M.A. Faderan, J.R. Bertino. Novel mechanism(s) of resistance to 5-fluorouracil in human colon cancer (HCT-8) sublines following exposure to two different clinically relevant dose schedules. *Cancer Res.* 52:1855-1864 (1992).

5. R.J. Langenbach, P.V. Danenberg, C. Heidelberger. Thymidylate synthetase: mechanism of inhibition by 5-fluor-2'-deoxyuridylate. *Biochem. Biophys. Res. Commun.* 48:1565-1571 (1972).

6. J.A.M. Van Laar, F.A. Durrani, Y.M. Rustum. Antitumor activity of the weekly push schedule of 5-fluoro-2'-deoxyuridine $\pm$ N-phosphonacetyl-L-aspartate in mice bearing advanced colon carcinoma 26. *Cancer Res.* 53:1560-1564 (1993).

7. T.H. Corbett, D.P. Griswold, B.J. Roberts, J.C. Peckham, F.M. Schabel. Tumor induction relationship in development of transplantable cancer of the colon in mice for chemotherapy assays, with a note on carcinogen structure. *Cancer Res.* 35:2434-2439 (1975).

8. G.J. Peters, E Laurensse, A. Leyva, H.M. Pinedo. Tissue homogenization using a microdismembrator for the measurement of enzyme activities. *Clin. Chim. Acta* 158:193-198 (1986).

9. G.J. Peters, J. Lankelma, R.M. Kok, P. Noordhuis, C.J. Van Groeningen, C.L. Van der Wilt, S. Meyer, H.M. Pinedo. Prolonged retention of high concentrations of 5-fluorouracil in human and murine tumors as compared with plasma. *Cancer Chemother. Pharmacol.* 31:269-276 (1993).

10. C.L. Van der Wilt, H.M. Pinedo, K. Smid, G.J. Peters. Elevation of thymidylate synthase following 5-fluorouracil treatment is prevented by the addition of leucovorin in murine colon tumors. *Cancer Res.* 52:4922-4928 (1992).

11. G.J. Peters, G. Codacci-Pisanelli, C.L. Van der Wilt, J.A.M. Van Laar, K. Smid, P. Noordhuis, H.M. Pinedo. Comparison of continuous infusions and bolus injections of 5-fluorouracil with or without leucovorin; implications for inhibition of thymidylate synthase, *in:* "Novel Approaches to Selective Treatments of Human Solid Tumors: Laboratory and Clinical Correlation", Y.M. Rustum, ed., Plenum Publishing Company, New York. pp. 9-20 (1993).

12. E. Chu, D.M. Koeller, P.G. Johnston, S. Zinn, C.J. Allegra. Regulation of thymidylate synthase in human colon cancer cells treated with 5-fluorouracil and interferon-γ. *Mol. Pharmacol.* 43:527-533 (1993).

INTRACELLULAR PHARMACOLOGY AND BIOCHEMISTRY OF METHOTREXATE AND 6-MERCAPTOPURINE IN CHILDHOOD ACUTE LYMPHOBLASTIC LEUKEMIA

C.W. Keuzenkamp-Jansen, J.P.M. Bökkerink, J.M.F. Trijbels, M.A.H. v.d. Heijden, R.A. De Abreu

Center for Pediatric Oncology, University Hospital, POB 9101, 6500 HB Nijmegen, The Netherlands

INTRODUCTION

After treatment with methotrexate (MTX), 6-mercaptopurine (6MP) demonstrated a synergistic effect and an enhanced incorporation into DNA and RNA in human lymphoblastic cell lines[1].

Based on these results a randomized clinical trial was started in cooperation with the Dutch Childhood Leukemia Study Group to investigate whether the synergism in vitro can be translated to the clinical setting[2,3]. In the present study the adminstration of low daily oral doses of 6MP is compared with high intravenous doses (HD-6MP) during MTX infusions.

The aim of this part of the study is to investigate the intracellular pharmacology and biochemistry of HD-6MP infusions. It is of special interest to investigate the intracellular compartment and not only plasma, because 6MP is a prodrug and exerts its cytotoxic effect after extensive intracellular metabolism.

6MP is an analogue of hypoxanthine and in the first step of the anabolic pathway 6MP is converted by hypoxanthine guaninephosphoribosyltransferase into thioinosine monophosphate (tIMP). tIMP can be converted by two pathways and both routes have cytotoxic effects in vitro. Whether both routes are important in vivo is not yet known.

First, tIMP can be converted by IMP dehydrogenase (IMPDH) into thioguanosine monophosphate (tGMP) and, after further phosphorylation, thioguanine nucleotides can be incorporated into DNA and RNA, which causes cytotoxicity[4]. Second, tIMP can be methylated by thiopurine methyltransferase (TPMT) into methyl-tIMP (MetIMP), which is a strong inhibitor of the purine de novo synthesis[5].

We investigated the metabolism of 6MP in peripheral mononuclear cells (pMNC) and in red blood cells (RBC) and the effects of MTX and 6MP on the purine and pyrimidine nucleotide pools in pMNC of patients with acute lymphoblastic leukemia (ALL).

PATIENTS AND METHODS

All patients were treated according to the protocol of the Dutch Childhood Leukemia Study Group (DCLSG-ALL-8) and were in complete remission before they entered the study. So, at that time no malignant lymphoblasts were present in the peripheral blood.

All patients received 4 courses with HD-MTX infusions (5 g/m^2.24 h) in 8 weeks and a standard leucovorin rescue, which dosage was increased or prolonged when the MTX excretion was delayed. In addition, all children received 6MP. Group 1 received 6MP in a low oral dose during 8 weeks (25 mg/m^2 daily). Group 2 received 6MP intravenously in a high dose immediately after the MTX infusion (1300 mg/m^2). The cumulative dose of 6MP in the intravenous group is 3.7 times that in the oral group.

All patients were studied during 4 courses. Blood was sampled before and 24, 28, 44, 48, 52 and 72 hours after start of the MTX infusion. PMNC were isolated from 10 patients in the oral and from 10 in the intravenous group. From 7 patients in each group RBC were also obtained.

PMNC were isolated from defibrinized blood by a ficoll isolation procedure and contaminating RBC were removed by a NH_4Cl shock[6]. RBC were isolated from heparinized blood. RBC and pMNC were deproteinized with perchloric acid (final concentration 0.4 M) and neutralized with K_2HPO_4. This extraction procedure yielded good recoveries for MetIMP and the thionucleotides in pMNC and for MetIMP in RBC. Measurement of thionucleotides in RBC was not possible with this extraction procedure, due to poor recoveries.

MetIMP and the thionucleotides were measured with HPLC according to the method described[7].

RESULTS

MetIMP levels in pMNC of the intravenous group are indicated in Figure 1. They increased slowly during the 6MP infusion from 24 till 48 hours. After termination of the 6MP infusion the MetIMP levels still increased to a level of 35 $pmol/10^6$ pMNC (sem 10).

MetIMP was not detectable in pMNC of the oral group (detection limit 20 pmol). Thioinosine and thioguanine nucleotides were not detectable in any of the pMNC, neither in the oral group, nor in the intravenous group (detection limit 45 pmol).

The levels of the endogenous purine and pyrimidine nucleotides in pMNC of the oral and intravenous group were within normal ranges[8] and did not change during the courses.

The MetIMP levels in RBC of patients in the intravenous group (data not shown) gradually increased during the infusion and remained at a constant level of 1300 $pmol/10^8$ RBC (sem 220) during the first 24 hours after the end of the infusion. At the start of the next course, two weeks later, MetIMP was still measurable in RBC of these patients and these levels were comparable with those reached in the oral group.

The MetIMP levels in RBC of patients in the oral group are indicated in Figure 2. These levels were about 35 $pmol/10^8$ RBC (sem 4). The MetIMP levels in the oral group decreased with about 50 % after cessation of the MTX infusion and started to increase after 44 hours. The leucovorin rescue was started at 42 hours.

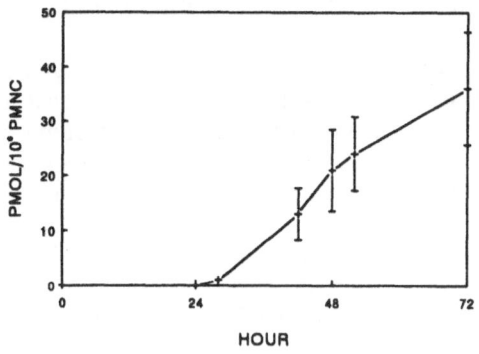

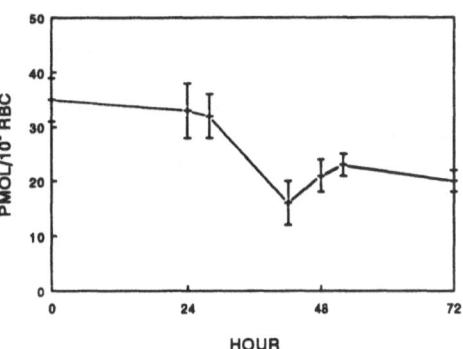

Figure 1. MetIMP in pmol/10⁶ PMNC of the intravenous group.

Figure 2. MetIMP in pmol/10⁸ RBC of the oral group.

DISCUSSION

Methylation is an important pathway for 6MP in vivo, especially after HD-6MP. This is reflected by methylmercaptopurine and methylmercaptpurine riboside in plasma[2], but also by MetIMP in RBC and pMNC. Whether the methylation pathway is a cytotoxic or a detoxification pathway in vivo is not yet known.

TIMP is converted by IMPDH into tGMP and by TPMT into MetIMP. We measured IMPDH and TPMT activity in pMNC of 11 patients and found much higher activities for IMPDH compared to TPMT. Nevertheless, thioguanine nucleotides (tGN) could not be detected in pMNC. This suggests that tGN are immediately incorporated into DNA and RNA and that this pathway can be important although its metabolites are not detectable.

The formation of MetIMP contributes to the cytotoxic activity of 6MP in vivo by inhibition of the purine de novo synthesis. Although, with respect to the purine metabolism, pMNC are a better model compared to RBC for the target cells of antileukemic therapy, the malignant lymphoblast, they still are biochemically different from their malignant counterpart. PMNC do not have an active purine de novo synthesis, which means that the cytotoxic effects of MetIMP after HD-6MP cannot be studied in these cells, as was demonstrated in our study by normal levels of purine and pyrimidine nucleotides.

Low levels of tGN in RBC of patients during prolonged maintenance treatment of ALL have been correlated with an increased risk of relapse[9]. Moreover, toxicity of low dose 6MP is greater in patients with a TPMT deficiency and with high tGN levels in RBC[10,11]. These data suggest that the formation of tGN is responsible for the treatment response and toxicity and that methylation is a detoxification pathway. However, we have to keep in mind that these studies were performed in RBC and that RBC do not have DNA synthesis and are not the target cells.

The first results of another study during the maintenance treatment of ALL showed that equal doses of 6-thioguanine or 6MP resulted in 5 times higher tGN levels in RBC of patients treated with 6-thioguanine, but that leucopenia was comparable in both groups[12].

These data suggest that cytotoxicity is not reflected by tGN levels in RBC and gives an indication that 6MP acts in vivo not only by the formation of tGN.

The MetIMP levels in RBC of the oral 6MP group decreased with about 50% after cessation of the MTX infusion. After start of the leucovorin rescue they started to increase slightly. A possible explanation for this phenomenon might be the inhibition of dihydrofolate reductase (DHFR) by MTX which can result in a decreased formation and usage of the universal methyldonor S-adenosyl methionine. Leucovorin circumvents the MTX effects on DHFR.

Concerning the relevance of methylation of 6MP, which is an important metabolic pathway in vivo, (HD-)6MP might influence the metabolism of S-adenosyl methionine, which can have implications for the cytotoxic effects of 6MP. This needs further investigation.

ACKNOWLEDGEMENTS

This project is financially supported by the Dutch Cancer Society (NUKC-92-79).

REFERENCES

1.   J.P.M. Bökkerink, M.A.H. Bakker, T.W. Hulscher, et al, Purine de novo synthesis as the basis of synergism of methotrexate and 6-mercaptopurine in human malignant lymphoblasts of different lineages. *Biochem. Pharmacol.* 37:2321 (1988).

2.   C.W. Keuzenkamp-Jansen, J.P.M. Bökkerink, R.A. De Abreu, et al, Pharmacokinetics of 24 hours intravenous high-dose 6-mercaptopurine after 24 hours intravenous methotrexate in childhood acte lymphoblastic leukemia. *Med. Ped. Oncol.*, 21:539 (1993).

3.   J.P.M. Bökkerink, C.W. Keuzenkamp-Jansen, New use of an old drug: clinical application and pharmacokinetics of intravenous 6-mercaptopurine. In: R. Riccardi, J. Borsi (eds), The Role of Clinical Pharmacology in Pediatric Oncology. Rome: Tipolitografia Feroce, 105 (1992).

4.   D.M. Tidd, A.R.P. Paterson, A biochemical mechanism for the delayed cytotoxic reaction of 6-mercaptopurine. *Cancer Res.* 34:738 (1974).

5.   M.H.J. Vogt, E.H. Stet, R.A. De Abreu, et al, The importance of methylthio-IMP for methylmercaptopurine ribonucleoside cytotoxicity in Molt F4 human malignant T-lymphoblasts, *Biochem. Biophys. Acta* 1181:189 (1993).

6.   R.A. De Abreu, G.J. Peters, J.A.J.M. Bakkeren, et al, Discrepancies in ribonucleotide concentrations in human lymphocytes isolated from heparinized and defibrinized blood. *Clin. Chim. Acta* 145:349 (1985)

7.   E.H. Stet, R.A. De Abreu, Y.P.G. Janssen, et al, A Biochemical basis for synergism of 6-mercaptopurine and mycophenolic acid in Molt F4 human malignant T-lymphoblastic cell line, *Biochim. Biophys. Acta* 1180:277 (1993).

8.   D. de Korte, W.A. Haverkort, A.H. van Gennip, et al, Nucleotide profiles of normal human blood cells determined by high-performance liquid chromatography. *Analyt. Biochem.* 147:197 (1985).

9.   L. Lennard, J.S. Lilleyman, Variable mercaptopurine metabolism and treatment outcome in childhood lymphoblastic leukemia. *Leukemia, J. Clin. Oncol.* 7:1816 (1989).

10.   W.E. Evans, M. Horner, Y. Qin Chu, et al, Altered mercaptopurine metabolism, toxic effects and dosage requirement in a TPMT deficient child with acute lymphoblastic leukemia, *J. Pediatr.* 119:985 (1991).

11.   L. Lennard, J.A. Van Loon, J.S. Lilleyman, et al, Thiopurine pharmacogenetics in leukemia: correlation of erythrocyte TPMT activity and 6-thioguanine nucleotide concentrations. *Clin Pharmacol. Ther.* 41:18 (1987).

12.   N. Erb, I. Ekem, D. Harms, et al, Differing metabolism of 6-thioguanine versus 6-mercaptopurine in maintenance treatment of childhood acute lymphoblastic leukemia. *Ann. Hematol.* 68 (suppl 1) A93 (1994).

MODULATION OF ARABINOSYLCYTOSINE METABOLISM DURING

LEUKEMIA THERAPY

Varsha Gandhi[1], Elihu Estey[2], and William Plunkett[1]

[1]Department of Clinical Investigation
[2]Department of Hematology
The University of Texas
M.D. Anderson Cancer Center
1515 Holcombe Boulevard
Houston, TX 77030

INTRODUCTION

Knowledge of the pharmacokinetics of a drug is essential to the optimal design of the dose and schedule of chemotherapy protocols. As an extension, an understanding of the mechanism of drug action is necessary to construct the optimal strategy for combination chemotherapy. A nucleoside antimetabolite such as arabinosylcytosine (ara-C) is a pro-drug that must enter cells and be phosphorylated to the ara-C triphosphate (ara-CTP) before it can elicit biologic activity. Thus, knowledge of the pharmacokinetics of the triphosphate in target cells and an understanding of the mechanisms by which this active form of the drug act are indispensable to the rational design of treatment protocols.

OPTIMIZING ara-CTP ACCUMULATION

The recognition that clinical response of acute leukemias strongly correlates with the pharmacokinetics of ara-CTP has emphasized the importance of strategies that enhance ara-CTP accumulation and retention in leukemia cells.[1,2] Initial questions focused on accumulation of ara-CTP by leukemia cells when ara-C was infused as a single agent. The fact that accumulation remained linear for more than 1 hour after the end of a 2-hour infusion of high-dose ara-C[3,4] suggested that the plasma ara-C concentrations (60 to 100 μM) produced by infusion rates of 1.5 to 3 $g/m^2/h$ exceeded those needed to maintain the maximal rate of ara-CTP accumulation and that this function was saturated at such plasma ara-C levels. This hypothesis was confirmed when serial ara-C infusions of 3 g/m^2 over 2 hours and then over 1 hour showed that the higher infusion rate yielded less than half the ara-CTP accumulated by the 2-hour infusion.[5,6] Furthermore, with infusion of 3 g/m² over 4 hours, the ara-CTP AUC was more than twice that of the 2-hour infusion in the same patient. These findings are consistent with the hypothesis that ara-CTP accumulation was saturated by the plasma ara-C concentrations achieved by high-dose ara-C infusions.

To determine optimal ara-C infusion rates, patients were infused with serial doses of ara-C at two different dose rates and the ability of circulating leukemia cells to accumulate ara-CTP was measured.[5,6] In the initial cohort of patients, infusion of 3 g/m² over 2 hours was followed 12 hours later by infusion of 1 g/m² over 2 hours. Circulating leukemia cells had equal ara-CTP AUC values after each dose. Assuming that infusion of 1 g/m² produced ara-CTP AUC values equal to those of 3 g/m², in subsequent patients the initial dose was decreased to 1 g/m² over 2 hours and the second dose was set at 0.5 g/m². Again the AUC values in the first and second infusions were similar, indicating that 0.5 g/m² also saturated ara-CTP accumulation. A stepwise decrease in the second, lower ara-C dose to 0.4 g/m² and finally to 0.3 g/m² administered over 2 hours was then tested. The ara-CTP AUC ratio, determined by comparing the ara-CTP AUC on the higher dose with that of the lower dose, gave values significantly > 1 at 0.4 ($P < .05$) and 0.3 g/m² ($P < .01$). These investigations showed that an ara-C dose rate of 0.5 g/m² administered over 2 hours provided adequate plasma ara-C concentrations (≤ 10 μM) to maximize ara-CTP accumulation. Thus, this intermediate-dose ara-C regimen has the potential of producing equivalent responses but with lower toxic effects than produced by conventional high-dose ara-C.[7,8] Different strategies are required to further enhance ara-CTP accumulation and retention in the circulating leukemia cells during therapy.[9]

MODULATION OF ARA-C METABOLISM

Both biochemical and biological modulation of ara-C metabolism have been described and achieved *in vitro* by several agents.[9-15] During therapy, however, studies are restricted to fludarabine, granulocyte-macrophage colony stimulating factor (GM-CSF), and granulocyte colony stimulating factor (G-CSF). The essential element of these investigations was a comparison of ara-CTP accumulation and elimination in the circulating leukemia cells of an individual before and after infusion of the modulator. Hence, as a control, it becomes necessary to compare ara-CTP accumulation when 2 ara-C doses were infused without any modulator.[16] Additionally, this comparison provides a frame of reference for these investigations because it determines the variability in the cellular pharmacokinetics of ara-CTP in the circulating leukemia blasts of patients who received treatment with 2 doses of ara-C alone. Five patients with acute leukemias received 3 g/m² of ara-C infused i.v. over 2 hours. The interval between the 2 doses of ara-C was 12 hours. Samples of peripheral blood were obtained at 1-2 hour intervals for 24 hours starting with the first dose of ara-C. Mononuclear cells were isolated and cellular nucleotides were extracted by the perchloric acid method and separated using anion-exchange high-pressure liquid chromatography.[17] The total intracellular exposure of ara-CTP was calculated as the area under the curve (AUC) of ara-CTP accumulation and elimination.[3,18] In this calculation the area under the linear accumulation phase was summed with that under monoexponential elimination phase. The ara-CTP AUC values were expressed in units of μM-hr.

Comparison of the rate of ara-CTP accumulation demonstrated that although there were substantial differences in the rates of accumulation of ara-CTP among patients, there was no significant change in the ara-CTP accumulation determined after serial ara-C infusions in the cells of each patient. Similarly ara-CTP elimination rates were not significantly different in each patient during serial infusions of ara-C. Determinations of intracellular ara-CTP AUC values in the cells of patients revealed both increases and decreases, the AUC values after the second dose of ara-C differed by only 0.94-fold (Table 1).[16]

Table 1. Effect of modulators on ara-CTP AUC in leukemia cells during therapy.

Modulator	Patients (n)	# of Patients AUC Ratio* > 1.2	Median* Ratio	Ref #
None	5	0	0.94	16
Fludarabine	10	9	1.83	21
GM-CSF	4	0	0.71	24
G-CSF	4	0	0.48	27

*AUC Ratio = $\dfrac{\text{ara-CTP AUC post modulator}}{\text{ara-CTP AUC pre modulator}}$

Given the complexity associated with the repeated sampling required to construct the pharmacokinetic profiles in the cells of each patient and the likely possibility of ongoing cellular damage, the level of variation seen in the cells of the patients who received ara-C alone was considered to be remarkably small. Hence only a difference in the AUC of ara-CTP which is more than or less than 20% of initial ara-CTP AUC is taken as a significant augmentation or decrease in the AUC due to the modulation strategy.

BIOCHEMICAL MODULATION OF ARA-C

The intracellullar accumulation of ara-CTP is a multistep process. The first step, phosphorylation of ara-C to its monophosphate by deoxycytidine kinase (dCyd kinase), is rate-limiting (Reviewed in Ref. 9). Our previous studies demonstrated that fludarabine increases the activity of this crucial enzyme, leading to a higher rate of ara-CTP accumulation and consequently resulting in greater AUC of ara-CTP in K562 cells.[19] These studies were extended to lymphocytes from patients with chronic lymphocytic leukemia (CLL) who received fludarabine therapy; these leukemic lymphocytes were isolated before and after therapy and were incubated *in vitro* with 100 μM ara-C for 2 hours. Accumulation of ara-CTP was potentiated 1.7-fold in lymphocytes obtained after fludarabine therapy.[20] Based on these results, the combination of fludarabine and ara-C was evaluated for biochemical modulation of ara-C metabolism in patients with acute myelogenous leukemia (AML).[21] Ara-C was infused for 2 to 6 hours at a dose rate that maximizes ara-CTP accumulation (0.5 g/m^2/h). Fludarabine (30 mg/m^2 over 30 minutes) was administered 20 hours later followed by a second, identical dose of ara-C at 24 hours when the concentration of F-ara-ATP was maximal in leukemia cells. Comparison of ara-CTP pharmacokinetics during therapy showed that the ara-CTP AUC increased by a median of 1.8-fold (range, 1.6 to 2.4; $P = .004$) in AML blasts after fludarabine infusion (Table 1). The median plasma ara-C concentrations, the levels of its deamination product arabinosyluracil, and the rate of ara-CTP elimination from circulating blasts were not affected by fludarabine infusion. However, the rate of ara-CTP accumulation in leukemia cells after fludarabine infusion increased by a median 2.0-fold in circulating AML blasts.[21] These studies clearly show that protocols based on biochemical and pharmacological rationales can successfully modulate ara-C metabolism in leukemia cells during treatment.

BIOLOGICAL MODULATION OF ARA-C

The hematopoietic growth factors such as GM-CSF, G-CSF, and interleukin-3 (IL-3) are responsible for the survival, proliferation, and differentiation of hematopoietic cells. Two major strategies have guided their use: first, reduction of hematologic toxicities associated with many chemotherapeutic agents; second, combinations with ara-C to sensitize leukemia cells to such cytotoxic chemotherapy and thereby enhance clinical response. The latter approach is based on the fact that recombinant growth factors such as G-CSF, GM-CSF, and IL-3 might stimulate the proliferation of leukemia progenitors.[22] Therapeutic advantage may be gained by transient stimulation of leukemia followed by treatment with a cell-cycle-specific drug, such as ara-C. *In vitro* studies indicate that growth regulatory molecules might also selectively alter the ratio of ara-CTP to dCTP and thereby affect the therapeutic index.[23]

A protocol was designed to test the possibility that GM-CSF might augment ara-CTP accumulation in circulating leukemia blasts.[24] Because of the heterogeneity of ara-CTP pharmacology among patients and the role of host factors in the cellular response to GM-CSF, it was important to compare *in vivo* ara-CTP metabolism in myeloblasts before and after GM-CSF in individuals. Patients received 1 g/m^2 ara-C over 2 hours; 24 hours after the start of administration of ara-C, 125 $\mu g/m^2$ of GM-CSF was infused intravenously over 6 hours daily for 3 days. Twenty-four hours after the start of the third day of GM-CSF administration, a second identical dose of ara-C was infused. Comparison of ara-CTP pharmacokinetics before and after GM-CSF treatment showed that the rate of ara-CTP accumulation in circulating blasts did not change in one patient; it was reduced to 70% in two and 30% in one of the pre-GM-CSF ara-CTP AUC values (Table 1).[24] Because serial infusions of ara-C without GM-CSF resulted in similar ara-CTP pharmacokinetics, these results suggest that GM-CSF may not act by increasing ara-CTP accumulation in circulating cells.[16] The results obtained with GM-CSF were contrary to those obtained during *in vitro* incubations; therefore it became essential to study biological modulations using other hematopoietic growth factors. The fact that G-CSF but not GM-CSF or IL-3 resulted in sensitizing AML blasts to ara-C[25,26] suggested a study design in which G-CSF was used as a modulator. The treatment plan stipulated that patients with AML receive a 2 g/m^2 dose of ara-C infused i.v. over 4 hours. At 24 hours, patients received a 6-hour infusion of 400 $\mu g/m^2$ G-CSF. At 48 hours, ara-C dose was repeated. Comparison of ara-CTP pharmacokinetics in circulating AML cells of patients on this regimen demonstrated that the AUC of ara-CTP did not increase in any patients after G-CSF infusion. In fact, the AUC of ara-CTP accumulation in these patients was decreased by a median of 48% (range, 30-80%) after G-CSF infusion (Table 1). Consistent with these *in vivo* investigations, *ex vivo* ara-CTP accumulation was decreased in the AML blasts after G-CSF infusion.[27]

CONCLUSION

In the absence of any other effective agents for the treatment of AML, there is a need to improve upon the available effective nucleoside analog such as ara-C. The growth factors such as GM-CSF and G-CSF did not effectively modulate ara-CTP accumulation in leukemia cells during therapy, but biochemical modulation by fludarabine was consistently demonstrated in several patients and several disease types[21,28]. This opens up a new arena of several agents such as chlorodeoxyadenosine and gemcitabine to be tested in combination. Secondly, the use of a nucleoside analog as biochemical modulators adds a new cytotoxic metabolite in the circulating leukemia blasts of patients treated with combination regimen. The effect of the combination of 2 cytotoxic metabolites on other cellular and molecular events warrants investigation.

ACKNOWLEDGEMENTS

This work was supported in part by grants CA32839, CA55164, and CA57629 form the National Cancer Institute, Department of Health and Human Services. The authors are grateful to Lidia Vogelsang for her expert assistance in typing this manuscript.

REFERENCES

1. Kantarjian, H.M., Estey, E.H., Plunkett, W., et al, Phase I-II clinical and pharmacologic studies of high-dose cytosine arabinoside in refractory leukemia. *Am J Med* 81:387 (1986).
2. Estey, E., Plunkett, W., Dixon, D.O., et al, Variables predicting response to high dose cytosine arabinoside therapy with refractory acute leukemia. *Leukemia* 1:580 (1987).
3. Liliemark, J.O., Plunkett, W., Dixon, D.O., Relationship of 1-β-D-arabinofuranosylcytosine in plasma to 1-β-D-arabinofuranosylcytosine 5'-triphosphate levels in leukemia cells during treatment with high-dose 1-β-D-arabinofuranosylcytosine. *Cancer Res* 45:5952 (1985).
4. Plunkett, W., Iacoboni, S., Estey, E., et al, Pharmacologically directed ara-C therapy for refractory leukemia. *Semin Oncol* 12[suppl 3]:20 (1985).
5. Plunkett, W., Liliemark, J.O., Estey, E., et al, Saturation of ara-CTP accumulation during high-dose ara-C therapy: Pharmacologic rationale for intermediate-dose ara-C. *Semin Oncol* 14[suppl 1]:159 (1987).
6. Plunkett, W., Liliemark, J.O., Adams, T.M., et al, Saturation of 1-β-D-arabinofuranosylcytosine 5'-triphosphate accumulation in leukemia cells during high-dose 1-β-D-arabinofuranosylcytosine therapy. *Cancer Res* 47:3005 (1987).
7. Estey, E., Plunkett, W., Kantarjian H., et al, Treatment of relapsed or refractory AML with intermediate-dose arabinosylcytosine (ara-C): Confirmation of the importance of ara-C triphosphate formation in mediating response to ara-C. *Leuk Lymph* 10[suppl]:115 (1993).
8. Hiddemann, W., Cytosine arabinoside in the treatment of acute myeloid leukemia: The role and place of high-dose regimens. *Ann Hematol* 62:119 (1991).
9. Grant, S., Biochemical modulation of cytosine arabinoside. *Pharmacol Ther* 28:29 (1990).
10. Zittoun, R., Zittoun, J., Marquet, J., et al, Modulation of 1-β-D-arabinofuranosylcytosine metabolism by thymidine in human acute leukemia. *Cancer Res* 45:5186 (1985).
11. Barlogie, B., Plunkett, W., Raber, M., et al, *In vivo* cellular kinetic and pharmacological studies of 1-β-D-arabinofuranosylcytosine and 3-deazauridine chemotherapy for relapsing acute leukemia. *Cancer Res* 41:1227 (1981).
12. Shackney, S.E., Ford, S.S., Occhipinti, S.J., et al, Schedule optimization of hydroxyurea and 1-β-D-arabinofuranosylcytosine in sarcoma 180 *in vitro*. *Cancer Res* 42:4339 (1982).
13. Cannistra, S.A., Groshek, P., Griffin, J.D., Granulocyte-macrophage colony-stimulating factor enhances the cytotoxic effects of cytosine arabinoside in acute myeloblastic leukemia and in the myeloid blast crisis phase of chronic myeloid leukemia. *Leukemia* 3:328 (1989).
14. Tanaka, M., Recombinant GM-CSF modulates the metabolism of cytosine arabinoside in leukemic cells in bone marrow. *Leukemia Res* 17:585 (1993).
15. Bhalla, K., Holladay, C., Arlin, Z., et al, Treatment with interleukin-3 plus granulocyte-macrophage colony-stimulating factors improves the selectivity of ara-C *in vitro* against acute myeloid leukemia blasts. *Blood* 78:2674 (1991).
16. Plunkett, W., Nowak, B., Keating, M.J., Effect of amsacrine on ara-CTP cellular pharmacology in human leukemia cells during high-dose cytarabine therapy. *Cancer Treat Rep* 71:479 (1987).

17. Plunkett, W., Hug, V., Keating M.J., et al, Quantitation of 1-β-D-arabinofuranosylcytosine 5'-triphosphate in the leukemic cells from bone marrow and peripheral blood of patients receiving 1-β-D-arabinofuranosylcytosine therapy. *Cancer Res* 40:488 (1980).
18. Shewach, D.S., Plunkett, W., Correlation of cytotoxicity with total intracellular exposure to 9-β-D-arabinofuranosyladenine-5'-triphosphate. *Cancer Res* 42:3637 (1982).
19. Gandhi, V., Plunkett, W., Modulation of arabinosyl nucleoside metabolism by arabinosylnucleotides in human leukemia cells. *Cancer Res* 48:329 (1988).
20. Gandhi, V., Nowak, B., Keating, M.J., et al, Modulation of arabinosylcytosine metabolism by arabinosyl-2-fluoroadenine in lymphocytes from patient with chronic lymphocytic leukemia: Implications for combination therapy. *Blood* 74:2070 (1989).
21. Gandhi, V., Estey, E., Keating, M.J., et al, Fludarabine potentiates metabolism of cytarabine in patients with acute myelogenous leukemia during therapy. *J Clin Oncol* 11:116 (1993).
22. Tafuri, A., Andreeff, M., Kinetic rationale for cytokine-induced recruitment of myeloblastic leukemia followed by cycle-specific chemotherapy *in vitro*. *Leukemia* 4:826 (1990).
23. Bhalla, K., Birkhofer, M., Arlin, Z., et al, Effect of recombinant GM-CSF on the metabolism of cytosine arabinoside in normal and leukemic human bone marrow cells. *Leukemia* 2:810 (1988).
24. Gandhi, V., Du, M., Kantarjian, H.M. et al, Effect of granulocyte-macrophage colony-stimulating factor on the metabolism of arabinosylcytosine triphosphate in blasts during therapy of patients with chronic myelogenous leukemia. *Leukemia* 8:1463, 1994.
25. Koistinen, P., Wang, C., Yang, G.S., et al, OCI/AML-4 an acute myeloblastic leukemia cell line: Regulation and response to cytosine arabinoside. *Leukemia* 5:704 (1991).
26. Koistinen, P., Wang, C., Curtis, J.E., et al, Granulocyte macrophage colony-stimulating factor and interleukin-3 protect leukemic blasts cells from ara-C toxicity. *Leukemia* 5:789 (1991).
27. Gandhi, V., Du, M., Nowak, B., et al, Granulocyte colony-stimulating factor modulates the metabolism of fludarabine in leukemia blasts during therapy. *Proc Am Assoc Cancer Res* 35:210 (1994).
28. Gandhi, V., Kemena, A., Keating, M.J., et al, Fludarabine infusion potentiates arabinosylcytosine metabolism in lymphocytes of patients with chronic lymphocytic leukemia. *Cancer Res* 52:897 (1992).

A NEW MECHANISM OF TOXICITY OF
2-CHLORODEOXYADENOSINE (2CdA)

Krystyna Fabianowska-Majewska,[1] Tadeusz J.Wasiak,[1]
Krzysztof Warzocha,[2] Maciej Marlewski,[3] Lynette Fairbanks,[4]
Ryszard T. Smoleński,[3] John Duley,[4] Anne Simmonds[4]

[1]Department of General Chemistry, Medical University of Łódź,
Łódź 90-131, Poland,
[2]2nd Clinic of Internal Medicine, Medical University of Łódź,
Łódź, Poland
[3]Department of Biochemistry, Medical School in Gdańsk, Gdańsk, Poland
[4]Purine Research Laboratory, UMDS Guy's Hospital, London SE1 9RT, UK

INTRODUCTION

2-chloro-2'-deoxyadenosine (2CdA), a new anticancer drug has been introduced as therapy against some kinds of leukemia, i.e. hairy cell leukemia and chronic lymphocytic leukemia (CLL) . According to published data this drug induces cell apoptosis[1] and the mechanism of 2CdA cytotoxicity leads *via* phosphorylated derivatives to inhibition of DNA synthesis which involves ribonucleotide reductase, DNA polymerases and DNA repair [2].

The possibilities for the metabolism of 2CdA are limited. This drug can be phosphorylated to the mononucleotide by deoxycytidine (dCyt) kinase, or by deoxyguanosine kinase in tissues with a deficiency of dCyt kinase[3]. 2CdA is not deaminated by adenosine deaminase (ADA)[2] but can be catabolised by phosphorolytic cleavage to 2-chloroadenine and ribose-1-phosphate, this reaction being catalysed by methylthioadenosine (MTA) phosphorylase[4].

Our *in vitro* studies on the molecular mechanism of 2CdA toxicity complement both the successful treatment of human gliomas with 2CdA (Dept.of Neurophysiology, Warsaw) and the positive response in patients with CLL plus hemolysis (2nd Clinic of Internal Medicine, Łódź). We have previously reported the effect of several adenosine (Ado) and deoxyadenosine (dAdo) analogues on the activity of enzymes which are important for the metabolism and toxicity mechanisms of dAdo (i.e.S-adenosylhomocysteine (SAH) hydrolase and MTA phosphorylase). The present studies have investigated the effect of 2CdA on these enzymes to ascertain whether the cytotoxicity of 2CdA involves only perturbation of DNA synthesis

Purine and Pyrimidine Metabolism in Man VIII, Edited by
A. Sahota and M. Taylor, Plenum Press, New York, 1995

by 2CdA triphosphate. The question arises whether 2CdA, as an agent resistant to cellular deamination, might mimic the ADA-deficient state, and whether 2CdA simulated the action of dAdo plus dCF with all its mechanisms of toxicity, i.e.also including inhibition of SAH hydrolase and hence inhibition of transmethylation reactions.

MATERIAL AND METHODS

Chemicals. Both labelled and cold substrates were products of the Sigma Chemical Co., 2-chloro-2'-deoxyadenosine was a kind gift from Dr.P.Grieb (Dept.of Neurophysiology, PAN, Warsaw, Poland)

Enzyme asay. MTA phosphorylase activity was assayed in the phosphorolysis direction by HPLC[5] or by a spectrophotometric method in which formation of adenine from MTA was determined by measurement of the absorbance increase at 305 nm in the presence of an excess of xanthine oxidase which converts the adenine formed into 2,8-dihydroxyadenine.
SAH hydrolase was determined in the SAH synthesis direction according to either HPLC[5] or by radiochemical TLC analysis. ADA was also determined using radiochemical TLC analysis. Adenine release from dAdo or Ado was also determined by a radiochemical method using reversed phase HPLC with in-line radiodetection[6]. In all experiments in which 2CdA was used as an inhibitor, its concentration was double the substrate concentration.

RESULTS AND DISCUSSION

The inhibitory effect of 2CdA on the activities of some purine pathway enzymes in lysed erythrocytes is summarized in Table 1. The results show that 2CdA slightly inactivates both SAH hydrolase and ADA. The result for ADA is not in agreement with Piro's data[7]. Although it is reported that 2CdA is not an inhibitor of ADA, the inhibitory effect of 2CdA may vary with the cell type.

Our *in vitro* studies of ADA activity of erythrocyte lysates were stimulated by unexpected results of clinical studies with 2CdA in patients with advanced CLL plus hemolytic anaemia and included determination of activity of enzymes of the purine pathway (i.e. ADA, PNP, 5'-N, MTA phosphorylase and SAH hydrolase). Fig.1 shows the activities of patient ADA and SAH hydrolase before and after treatment with 2CdA. The dramatic decrease in activity of both ADA and SAH hydrolase (by 70% and 60%, respectively) was observed after 7 days of treatment with 2CdA at a dose of 0.1 mg/kg/d. It is worth mentioning that the activity of ADA in untreated patients with CLL was above the normal range, and that the monitored decrease in enzyme activities were accompanied by a positive response to therapy. Activities of the remaining enzymes were quite stable (data not shown).

The present studies *in vitro* demonstrate that 2CdA does not inhibit MTA but that adenine release from dAdo or Ado in lysed or intact erythrocytes (RBC) is blocked by up to 70% by 2CdA. This different influence of 2CdA on phosphorolytic cleavage can be explained by the fact that 2CdA is a substrate for MTA phosphorylase, the specific activity with both the drug and adenosine being of the same order, although approximately 10-fold less than towards the natural substrate MTA (Table 2). One explanation for this is that 2CdA can be competitive for Ado, but not for MTA at the active site of the enzyme, and this effect of 2CdA could also contribute to the frequently observed decline in ATP level[8]. This depletion of the ATP pool should be greater for tissues with MTA phosphorylase deficiency e.g. the malignant leukemia cell line CEM, the murine leukemia cell L1210, or some types of gliomas[9]. This may also be the reason for the observed high sensitivity to 2CdA. Investigation of the possible routes of

Table 1.Influence of 2-chloro-2'-deoxyadenosine on activity of enzymes of the purine pathway

Enzyme	% of inhibition
SAH hydrolase both partial purified from human liver and of lysate RBC	25 - 30
MTA phosphorylase both partial purified from human liver and of lysate RBC	0
Adenine release from Ado or dAdo by lysate RBC or intact RBC	65 - 70
ADA of lysate RBC	30 - 35

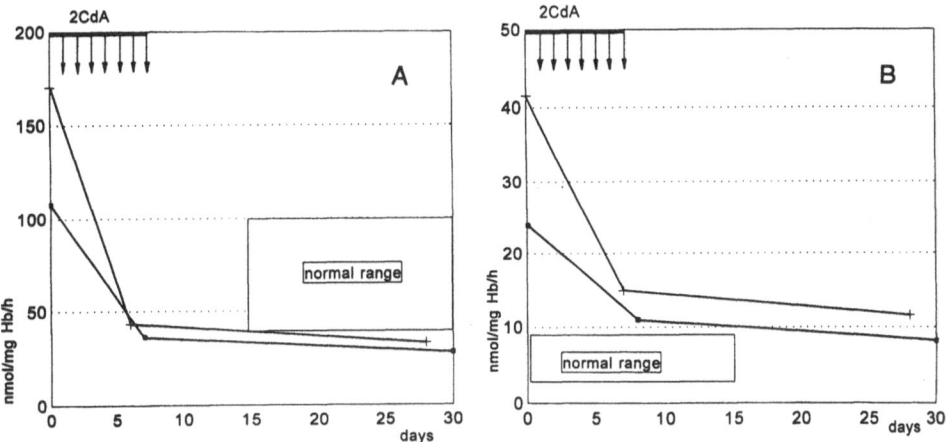

Figure 1. ADA (A) and SAH hydrolase (B) activities in erythrocyte lysates of patients with CLL treated with 2CdA (———·———— I patient, ———+——— II patient)

Table 2. Substrate activity of MTA phosphorylase from human liver

Substrate	MTA phosphorylase activity (μmol/mg of protein/h)
MTA	5.5 ± 0.4[1]
5'-Iodo-5'-deoxyadenosine	12.8 ± 0.9
Adenosine	0.69 ± 0.05
2-chloro-2'deoxyadenosine	0.6 ± 0.04
2-chloro-5'-O-methyl-2'-deoxyadenosine	0.63 ± 0.05

[1]Each value represents the mean $\pm$ SEM for four experiments

metabolism of dAdo in human cells[6] suggests that dAdo metabolism in the presence of 2CdA is "arrested" in cells due to the blocking of activities of ADA and SAH hydrolase and adenine release (Fig.2). According to this hypothesis, the "arrest" of dAdo may induce, *via* inhibition of SAH hydrolase, perturbation of S-adenosylmethionine dependent transmethylation reactions and depletion of the energy capability of the cell. In this way, the cytotoxic effect of 2CdA would simulate that caused by dAdo plus dCF.

This proposal would be in agreement with Carrera's studies in which the effect of 2CdA is compared with the effect of dAdo and dCF[8].

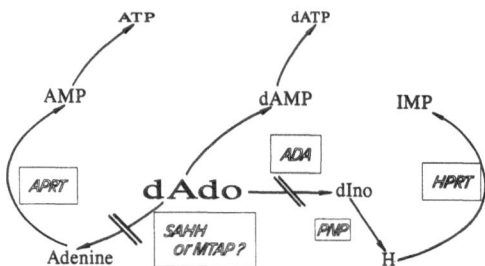

Figure 2. The effect of 2CdA on metabolism of deoxyadenosine

Acknowledgments

This work was partly supported by Medical University of Łódź grant No 502-11-91(58).

REFERENCES

1. L.E.Robertson, S.Chub, R.E.Meyn, M.Story, R.Ford, W.N.H.Hittelman, W.Plunkett, Induction of apoptosis cell death in chronic lymphocytic leukemia by 2 chloro-2,-deoxyadenosine and 9-ß-D-arabinosyl-2-fluoroadenine, *Blood* 81:143 (1993)
2. P.A.Bonnet, R.K.Robins, Modulation of leukocyte genetic expression by novel purine nucleoside analogues. A new approach to antitumor and antiviral agents, *J.Med.Chem.* 36:635 (1993)
3. L.Wang, A.Karlsson, E.S.J.Arnen, S.Eriksson, Substrate specificity of mitochondrial 2'-deoxyguanosine kinase. Efficient phosphorylation of 2 chlorodeoxyadenosine, *J.Biol.Chem.*268:22847 (1993)
4. K.Fabianowska-Majewska, J,Duley, L.Fairbanks, T.Wasiak, A.Simmonds, Substrate specificity of methylthioadenosine phosphorylase from human liver, *Acta Biochimica Polonica* (in press)
5. K.Fabianowska-Majewska, J.Duley, A.Simmonds, Novel, anti-viral analogues of adenosine and activity of S-adenosylhomocysteine hydrolase from human liver, *Biochem.Pharmacol.*(in press)
6. R.T.Smoleński, K.Fabianowska-Majewska, C.Montero, J.A.Duley, L.D.Fairbanks, M.Marlewski, A.Simmonds, A novel route of ATP synthesis, *Biochem.Pharmacol.* 43:2053 (1992)
7. L.D.Piro, C.J.Carrera, E.Beutler, D.A.Carson, 2-chlorodeoxyadenosine: an effective new agent for the treatment of chronic lymphocytic leukemia, *Blood* 72:1069 (1988)
8. C.J.Carrera, C.Terai, M.Lotz, J.G.Curd, L.D.Piro, E.Beutler, D.A.Carson, Potent toxicity of 2-chlorodeoxyadenosine towards human monocytes *in vitro* and *in vivo,* *J.Clin.Invest.* 86:1480 (1990)
9. T.Nobori,J.G.Karras, F.D.Ragione, T.A.Waltz, P.C.Chen, D.A.Carson, Absence of methylthioadenosine phosphorylase in human gliomas, *Cancer Res.* 51:3193 (1991)

THE INFLUENCE OF 2-CHLORODEOXYADENOSINE
(2-CdA) ALONE AND IN COMBINATION WITH
CYCLOPHOSPHAMIDE OR METHOTREXATE ON
NORMAL AND LEUKEMIC HEMATOPOIESIS

Joanna Góra-Tybor[1], Tadeusz Robak[2]

[1] Department of Pharmacology, Medical University of Lodz
Muszyńskiego 1, 90-150 Lodz
[2] 2nd Clinic of Internal Medicine, Medical University of Lodz
Pabianicka 62, 93-513 Lodz

2-Chlorodeoxyadenosine (2-CdA) is a new halogenated purine analog, very active in different lymphoproliferative disorders (1). Recent data have demonstrated that this agent is also active in myeloid malignacies, both in acute myeloid leukemia (AML) and in chronic myeloid leukemia (CML) resistant to conventional treatment (2, 3). Until now, there have been no clinical reports on the therapeutic effect of the combination of 2-CdA with other cytotoxic drugs.

The present study was designed to investigate the influence of 2-CdA and cyclophosphamide (CY) or methotrexate (MTX) on murine leukemias L1210 and P388. We have chosen CY and MTX because of their wide application and effectiveness in the treatment of hematological malignancies both as single agents and in combination regiments. In the second part of our experiments we investigated the influence of 2-CdA on normal and CML granulocyte-macrophage progenitor cells (CFU-GM) as well as on AML clonogenic blasts (CFU-L) in cultures in vitro.

MATERIALS AND METHODS

In vivo assay

2-CdA produced according to the method described by Kazimierczuk et al., (4) was kindly supplied by the Foundation of Development of Diagnosis and Therapy (Warsaw) as dry powder, On day 0, 10^4 of L1210 and 10^6 of P388 leukemic cells were implanted i.p. into DBA/2 mice. Beginning on day 1 the mice received i.p. injection of 2-CdA alone or in combination with CY (Boehringer Ingelheim) or MTX (Lederle, USA).

In vitro assay

Specimens of peripheral blood or bone marrow were collected from 10 patients with CML in chronic phase and 10 patients with AML. None had received cytotoxic drugs before the collection of the blood or marrow for experiments. Normal bone marrow was collected from 10 hematologically normal patient. The assay for CFU-GM was based on the method described by Iscove et al. and modified by Hibbin et al. (5) Our method for AML CFU-L was based on the method described by Buick et al. (6) At the same time the mononuclear cells were plated with 2-CdA at the concentration of 20nM/L, 40nM/L and 60nM/L. All cultures were incubated at 37°C in the atmosphere of 5% CO_2 in air and examined with an inverted microscope. "Late" CFU-GM colonies as well as CFU-L colonies were scored on day 7 and "early" CFU-GM on day 14.

Statistical analysis

Statistical analysis was performed using Student's t-test. Results were considered significant when $p < 0.05$.

RESULTS AND DISCUSSION

The survival time of mice with leukemia L1210 treated with 2-CdA at doses 20 mg/kg and 35 mg/kg was longer than that of control groups. 2-CdA at the same doses caused only slight prolongation of life of mice with leukemia P388. MST of mice with both kinds of leukemias receiving 2-CdA at dose 50 mg/kg was shorter than that of the control group. The mice with leukemia L1210 or P388 treated with 2-CdA and CY lived significantly longer than the animals treated with these agents separately. The addition of MTX to 2-CdA caused a very limited increase in the survival time of the treated mice with both leukemias, as compared with 2-CdA or MTX singly treated groups. (Table 1)

Table 1. The influence of 2-CdA alone and in combination with CY or MTX on the MST of mice with leukemia L1210 and P388

Therapeutics		MST[e](days)		$x \pm SD^f$		ILS[g](%)	
2CDA[a] (mg/kg)	second drug	L1210	P388	L1210	P388	L1210	P388
0[b]	0	10.0	9.0	10.0 ± 0.0	9.0 ± 0.63		
50	0	7.0*	8.0	6.8 ± 0.40*	7.8 ± 0.75	−30*	−12
35	0	15.5*	13.0	12.5 ± 5.08*	13.0 ± 0.89	55*	44
20	0	13.5*	11.0	14.0 ± 1.26*	10.8 ± 1.32	35*	22
35	MTX[c]	17.0*	14.5*	16.66 ± 0.51*	14.5 ± 1.87*	70*	61*
0	MTX	15.0*	12.0	15.16 ± 0.75*	12.0 ± 0.63	50*	33
35	CY[d]	29.0#	36.5#	29.83 ± 4.7#	36.1 ± 6.73#	190#	305#
0	CY	19.0*	17.5*	20.5 ± 0.54*	17.2 ± 2.71	90*	94*

Explanation: [a] Treatment once daily, for 5 days, i.p. − [b] 0.9% sol. of sodium chloride, given once daily, for 5 days, i.p. − [c] Treatment, single i.p. injection on the 1st day of experiment, 10 mg/kg − [d] Treatment, single, i.p. injection on the 1st day of experiment, 75 mg/kg − [e] Median survival time, six mice were used per group − [f] Mean values and standard deviations − [g] ILS = [(MST of the treated group/MST of the control group)−1] × 100% − * statistical significance as compared with the control group ($p < 0.05$) − # Statistical significance as compared with the control group and the groups treated with single drug.

The high degree of therapeutic synergy between 2-CdA and CY, revealed in our experiments, may be of great clinical importance since both agents have high antineoplastic activity against chronic lymphocytic leukemia and non Hodgin lymphomas (1).

UTILIZATION OF PURINE AND PYRIMIDINE IN HUMAN GASTRIC CANCER CELLS (KATO III): EFFECT OF A NUCLEOTIDE AND NUCLEOSIDES MIXTURE (OG-VI) SOLUTION ON PROLIFERATION WITH COADMINISTRATION OF 5-FLUOROURACIL

Makoto Usami, Jian-hua Zheng, Jianping Wang, Ichiro Yasuda, Seiji Haji, George Kotani, Atsunori Iso, Kai Sun, Kazuya Sakata, Kyosuke Ohta, Taichi Kanamaru, Hiroshi Kasahara, and Yoichi Saitoh

First Department of Surgery, Kobe University School of Medicine, 7-5-2 Kusunoki-cho, Chuo-ku, Kobe, 650, Japan

INTRODUCTION

Purine and pyrimidine are essential for cellular proliferation[1,2,3]. But, purine and pyrimidine are synthesized from glutamine in de novo synthesis pathway and cultured cancer cells do not require them in the medium. This is the reason that utilization of nucleosides by human cancer cells and their changes with 5 fluorouracil (5FU), the fluorinated analogue of uracil which is used in chemotherapy for various tumors[2,3], has not been reported. In this study, as a source of purine and pyrimidine, a solution containing a mixture of nucleosides and a nucleotide (OG-VI), developed recently as a nutritional supplement[4], was used in the medium. It consists of purines, guanosine 5'-mono-phosphate (5'-GMP) and inosine, and pyrimidines, cytidine, uridine, and thymidine (TdR). It is known to increases hepatic regeneration and improve protein metabolism after partial hepatectomy in rats[5,6], and increase DNA synthesis in primary cultured hepatocytes[7].

The purpose of this study is to evaluate the utilization of purine and pyrimidine by human cancer cells, the direct effect of OG-VI on cellular growth and their modulation on 5FU in an in vitro and in vivo study.

MATERIALS AND METHODS

KATO III human gastric cancer cells, 2×10^4 cells/ml suspension, (Immuno-biological Research Institute, Japan), William's-E medium (Flow Laboratories, Scotland) with 100 U/ml of penicillin G (Meiji Seika Kaisya, Japan) 100 µg/ml of kanamycine (Meiji), 5FU (Kyowa Hakko Kogyo Co., Japan) and 10% fetal calf serum (GIBCO, NY) were used. OG-VI, composed of 30 mM inosine, 30 mM 5'-GMP, 30 mM cytidine, 23 mM uridine and 7.4 mM TdR, and TdR were kind gifts from Otsuka Pharmaceutical Co., Tokushima, Japan. OG-VI or TdR were added at various concentration, with or without 5FU at concentrations ranging from 0.01 to 1.0 µg/ml. The cells were incubated at 37 °C in 5% CO_2 and 100% humidified atmosphere for 72 hours. The concentration of purine and pyrimidine in the medium before and after the incubation were measured by high-performance liquid chromatography (HPLC) with a TSK gel ODS-80 TM column (Shimazu, Japan) and LC-6AD system (Shimazu)[7]. The consumption rate of purine and pyrimidine was calculated as follows: (Conc. pre - Conc. post)/Conc. pre x 100(%).

Tumor growth of Yoshida sarcoma (SRL Hachioji Lab., Tokyo, Japan), 1×10^6 cells inoculated subcutaneously, was evaluated in male Donryu rats after 7 days of treatments, under nucleotide deficient diet (AIN-B, Oriental Co., Japan), with or without intragastric administration of 5FU (10-20 mg/kg/day) and intraperitoneal administration of OG-VI

Purine and Pyrimidine Metabolism in Man VIII, Edited by
A. Sahota and M. Taylor, Plenum Press, New York, 1995

solution (2.8-14 ml/kg/day). Each result was taken as the mean value of a series of 5-12 experiments carried out and expressed as mean ± standard deviation (SD). Statistical analysis was based upon Student's t test.

RESULTS

The consumption rate of purine (inosine and 5'-GMP) was greater than that of pyrimidine (cytidine, uridine and TdR) without 5FU (Figure 1.). Xanthine and hypoxanthine in the medium, the catabolic products of purine nucleotide, increased after the incubation. Consumption of purine and pyrimidine was greater in the higher concentration ranges. By the addition of 0.1 μg/ml 5FU, the consumption rate of purine decreased, but that of pyrimidine increased (Table, $p<0.05$). The consumption rate of TdR was 16.7% with OG-VI and 54.4% with TdR alone. Further more, the consumption rate of TdR in the TdR group with 5FU was up by 72.2%. As to the purine metabolism, the production of xanthine and hypoxanthine was less in the TdR group than those in the OG-VI group.

Addition of OG-VI did not enhance the cellular proliferation in all concentration ranges examined. Furthermore, OG-VI decreased cell growth to 84% of the control in higher concentration range, 1-10% of OG-VI ($p<0.05$). 5FU decreased cell growth in a concentration dependent fashion. Addition of OG-VI enhanced the inhibitory effect of 5FU in all 5FU concentration ranges ($p<0.05$).

In vivo tests revealed the same result as cell culture showing the enhanced inhibition of tumor growth without major complication (Fig.2). The effect of OG-VI mixture was stronger than the effect of each component of OG-VI alone ($p<0.05$). Enhancive effect of OG-VI on 5FU action was more clear than in vitro test ($p<0.01$).

Table Consumption rate of purine and pyrimidine in the medium
(KATO III, 5FU 0.1 μg/ml, OG-VI 1/10)

	5FU(-)	5FU(+)
inosine	66.7 ± 9.8	33.5 ± 3.0*
5' GMP	97.5 ± 9.8	58.3 ± 13.6*
cytidine	11.0 ± 12.2	16.8 ± 9.8
uridine	0	13.7 ± 9.0
thymidine	16.7 ± 17.9	33.7 ± 8.4*

Consumption rate= (Cpre - Cpost)/ Cpre x 100(%), mean ± SD, *$p<0.05$ vs 5FU(-)

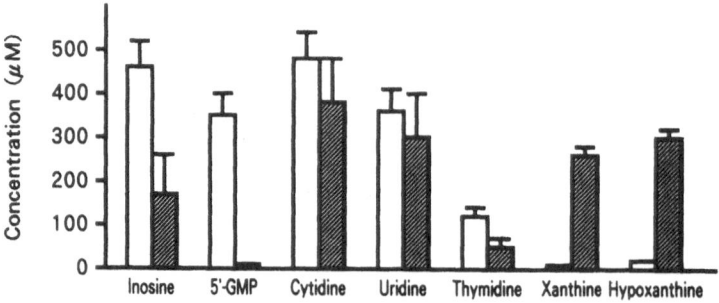

Figure 1. Changes of purine and pyrimidine concentration in the medium, before and 3 days after incubation (OG-VI 1/100). Mean ± SD, open square; before, closed square; after incubation.

DISCUSSION

The consumption rate of purine and pyrimidine in the medium was different and may imply their competitive utilization in salvage synthesis pathway of cancer cells. TdR utilization was more and purine catabolism was less in the TdR group than in the OG-VI group. This result indicates that TdR was used in place of other pyrimidines. It suggests that nutrients in OG-VI are better balanced than TdR alone for purine and pyrimidine metabolism in cancer cells, and that they may influence cellullar growth. No enhancement of cellular proliferation was observed by the addition of OG-VI within the broad concentration ranges (1/10-1/1000), but inhibition was observed in higher concentration. This result is in agreement with the report that TdR produces cytokinetic arrest of cellular progression through S phase at mM concentrations[1].

As to the consumption of purine and pyrimidine under 5FU, it decreased the usage of purine and increased the usage of pyrimidine. The difference in utilization can be explained by the inhibition effect of 5FU on both thymidylate synthetase and phosphoribosyl pyrophosphate amidotransferase in the de novo pyrimidine synthesis pathway and then increased salvage pyrimidine synthesis[2,6,8]. Therefore, the addition of 5FU increases the consumption of uridine and cytidine. The decreased amount of purine consumption observed is postulated to be the result of decreased cellular activity secondary to the metabolic disturbances of pyrimidine by 5FU.

These results of purine and pyrimidine utilization in cell culture strongly suggest the modulation of biochemical activity 5FU by OG-VI. When the concentration of OG-VI in the culture medium was in the range of 1/10-1/100 dilution of the original concentration, there was an obvious augmentation on the inhibitive effect of 5FU on KATO III cell proliferation. The anti-tumor effect under the combination of 0.1 μg/ml of 5FU with OG-VI is equal to that of 1.0 μg/ml of 5FU. This result suggests the possibility of decreasing the dosage and consequently decreasing the side effect of 5FU without changing the anti-tumor effect of 5FU by the combined administration of OG-VI. Also, the in vivo results revealed that OG-VI by itself did not influence tumor growth, but instead enhanced the tumor growth inhibitory effect of 5FU without major complication. The result of this biochemical modulation is in agreement with Spiegelman et al.[9], who reported that TdR pretreatment can cause a three to five fold increase in the amount of 5FU incorporated at equivalent RNA synthesis. However, the mechanism of biochemical modulation by the addition of different materials differ. Inosine, GMP and TdR increases cytotoxicity of 5FU[2,3], but uridine decreases the effect[8]. The result of in vivo comparison among OG-VI and its components on tumor growth showed the advantage of the mixture solution in biochemical modulation.

It is possible that OG-VI is effective in alleviating the side effects of 5FU and also in cellular immunity modulation, as observed in experiments using yeast RNA diet[10]. Further in vivo study is required to evaluate the effect of combined administration of 5FU and OG-VI has on cellular growth in different cancer cells and its toxicity.

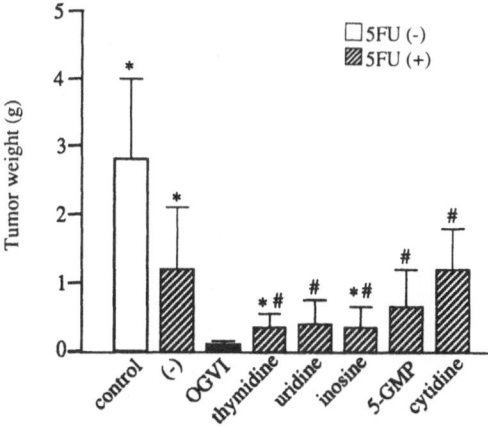

Figure 2. Effect of OG-VI and its components on tumor proliferation with 5FU (in vivo). n=6 - 8, mean ± SD, *p<0.05 vs 5FU(+), #p<0.05 vs 5FU + OG-VI.

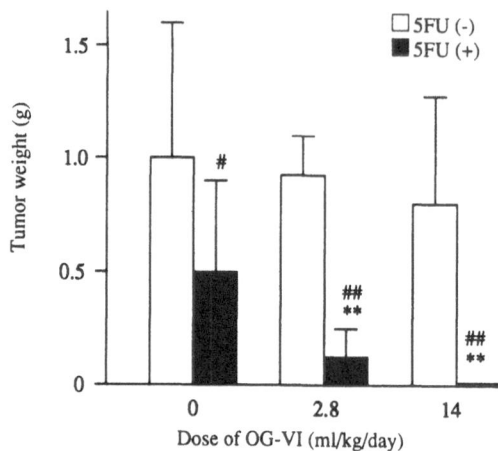

Figure 3. Effect of OG-VI on tumor proliferation with 5FU (in vivo). n=12, mean ± SD, *p<0.01 vs 5FU(+) and OG-VI(-), #p<0.05, ##p<0.01 vs 5FU(-)

REFERENCES

1. D. Bootsma, L. Budke, and O. Vos, Studies on synchronous division of tissue culture cells initiated by excess thymidine, *Exptl.Cell Res.* 33:301 (1964).
2. C. Engelbrecht, I. Ljungquist, L. Lewan, and T. Yanner, Modulation of 5-fluorouracil metabolism by thymidine: In vivo and vitro studies on RNA-directed effects in rat liver and hepatoma, *Biochem Pharmacol.* 33:745 (1984).
3. M. Iigo, N. Ando, A. Hoshi, and K. Kuretani, Effect of pyrimidines, purines and their nucleosides on antitumor activity of 5-fluorouracil against L-1210 leukemia, *J Pharmacobiodyn.* 5:515 (1982).
4. S. Ogoshi, M. Iwasa, T. Yonezawa, and T. Tamiya, Effect of nucleotide and nucleoside mixture on rats given total parenteral nutrition after 70% hepatectomy, *JParentalEnteralNutr,* 9:339 (1985).
5. S. Ogoshi, M. Iwasa, S. Mizobuchi, Effect of nucleoside and nucleotide mixture on protein metabolism in rats given total parenteral nutrition after 70% hepatectomy. *in:* "Nutritional Support in Organ Failure", T. Tanaka, and A. Okada, ed., Elsevier, Amsterdam (1990).
6. M. Usami, H. Ohyanagi, K. Furuchi, M. Ogino, I. Yasuda, J. Kotani, J.P. Wang, K. Sun, T. Kanamaru, H. Nomura, Y. Saitoh, H. Yokoyama, and K. Takahashi, Augmentation of hepatic regeneration after partial hepatectomy in rats by nucleosides and a nucleotide mixture (in Japanese), *Jpn J Surg Metal Nutr.* 26:187 (1992).
7. H. Ohyanagi, S. Nishimatu, Y. Kanbara, M. Usami, and Y. Saitoh, Effect of nucleosides and a nucleotide on DNA and RNA syntheses by the salvage and de novo pathway in primary monolayer cultures of hepatocytes and hepatoma cells, *J. Parenter. Enteral. Nutr.* 13:51 (1989).
8. L.D. Nord, R.L. Stolfi, and D.S. Martin, Biochemical modulation of 5-fluorouracil with leucovorin or delayed uridine rescue, *Biochem. Pharamcol.* 43:2543 (1992).
9. S. Spiegelman, R. Nayak, R. Sawyer, R. Stolfi, and D. Martin, Potentiation of the anti-tumor activity of 5-Fu by thymidine and its correlation with the formation of (5-Fu) RNA, *Cancer* 45:1129 (1980).
10. C.T. Van Buren, A.D. Kulkarmi, W.C. Fanslow, and F.B. Rudolph, Dietary nucleotides, a requirement for helper/inducer T lymphocytes, *Transplantation.* 40:694 (1985).

INHIBITION OF THE PROLIFERATION OF HUMAN CANCER CELLS IN-VITRO BY SUBSTRATE-ANALOGOUS INHIBITORS OF dUTPASE

Petra Zalud[1], Wolfgang O. Wachs[1], Per Olof Nyman[2], and
Michael M. Zeppezauer[1]

[1]12.4. Biochemie, Universität des Saarlandes
PO Box 151150, D-66041 Saarbrücken, Germany

[2]Biochemistry, Chemical Center, PO Box 124
S-22100 Lund, Sweden

INTRODUCTION

2'-Deoxyuridine-5'-triphosphate-nucleotidohydrolase (dUTPase E.C. 3.6.1.23) serves two major functions in nucleotide metabolism, namely the production of dUMP, the immediate precursor of thymidine nucleotides, and the control of the intracellular dUTP level, thus minimising incorporation of uracil into DNA.

It has long been known that impairment of dUTPase synthesis causes a profound increase in the production of small replication fragments (Tye et al. 1977, Hochhauser & Weiss 1978, Kornberg 1980). Therefore it has been noticed by several authors that dUTPase seems to represent a perfect target for the development of substances suitable for the control of cell proliferation (Beck et al. 1986, McIntosh et al. 1992).

We have synthesised several substrate-analogous inhibitors of dUTPase and tested their efficiency with dUTPase from E. coli as a suitable model system which is 35% homologous to human dUTPase (McIntosh et al. 1992) and which has been overexpressed (Hoffmann et al. 1987). Inter alia, replacement of the α-ß oxygen bridge in dUTP by $-CH_2-$ (M-DUTP) or other suitable groups should prevent hydrolysis of the substrate , and attachment of lipophilic groups, e.g. benzyl groups (BM-dUTP) should increase the membrane permeability.

The inhibitors were tested in vitro on human cancer cell lines and on fibroblasts serving as non-transformed proliferating controls. As will be shown below a selective response of all cancer cell lines was observed, the effect depends on the chemical structure of the inhibitors and is time- and dose-dependent.

Purine and Pyrimidine Metabolism in Man VIII, Edited by
A. Sahota and M. Taylor, Plenum Press, New York, 1995

MATERIALS AND METHODS

2'- Deoxy- 5'- uridinyl- [hydroxythiophosphinyl- oxo- hydroxyphosphinyl- oxo]-phosphonic acid (ASD) was prepared according to the procedure of Ludwig & Eckstein (1989). The syntheses of 2'- deoxy- 5'- uridinyl-[hydroxyphosphinyl- methylenyl-hydroxyphosphinyl- oxo]- phosphonic acid (M-dUTP), benzyl- [2'- deoxy- 5'- uridinyl-hydroxyphosphinyl- methylenyl- hydroxyphosphinyl- oxo]- phosphonate (BM-dUTP) and dibenzyl- [2'- deoxy- 5'- uridinyl- hydroxyphosphinyl-methylenyl-hydroxyphosphinyl- oxo]- phosphonate (DBM-dUTP) were performed by standard procedures using transient protecting group methodology (Zeppezauer et al. 1993) and will be published elsewhere.

dUTPase from E. coli was overexpressed and purified according to Hoffmann et al. (1987). Enzyme kinetic measurements were performed by monitoring liberated protons in a spectrophotometric assay (Zalud et al., in preparation).

The human lymphoma cell line OH 77 and the human melanoma cell line EG 463 were established at the Department of Genetics, University of Lund (Sweden), the human glioma cell line U 373 and the fast-growing fibroblast line CRL 1635 (newborn foreskin, human) were obtained from the American Type Culture Collection (ATCC), Rockville Maryland, USA. The slow-growing fibroblasts were a gift from Dr. H.P. Leinenbach from our laboratory. The cell lines were grown under sterile conditions in Mc Coy's medium in the absence of penicillin and streptomycin, completed with 1% of glutamine and 10% of Fetal Calf Serum. The cell lines were grown in the presence or absence of dUTPase inhibitors as described in the Results section. Vitality was determined by the MTT-test (Mosman 1983).

RESULTS AND DISCUSSION

M-dUTP and BM-dUTP were found to be slow-reacting substrates (competitive inhibitors) of E. coli dUTPase with apparent K_I-values of 2.5 $\pm$ 0.3 μM and 0.3 $\pm$ 0.02 μM, respectively. The α-thio derivative ASD also acts inhibitory, but owing to its complex kinetics no data are presented. Due to the high degree of homology between the E. coli and the human enzyme in general and the very narrow substrate specificity in particular we believe that the inhibitors may interact with both enzymes with comparable strength.

This is corroborated by the cytostatic and cytotoxic effect of the inhibitors on the human cancer cell lines. As shown in figure 1 the growth of the lymphoma cell line OH 77 is significantly retarded by DBM-dUTP, the effect of which is comparable to that of methotrexate. A very strong cytostatic effect is shown by M-dUTP which reduces the growth of the cancer cells to 10% of the control. A cytotoxic effect is shown by ASD and most pronounced by BM-dUTP which reduce the growth rate to -10% and -43% of the control, respectively. In comparison, under the given conditions, fast-growing fibroblasts are almost unaffected by methotrexate (MTX) and most dUTPase inhibitors except BM-dUTP. The selectivity of the cytotoxic effect of the latter compound is shown in a dose-response curve comparing the cell line OH 77 with slow-growing fibroblasts. It is apparent from figure 2 that it is possible to select a concentration range where the effect is highly cytotoxic against the cancer cells whereas it is cytostatic against the fibroblasts. Similar data were obtained with the melanoma cell line EG 463 and the glioma cell line U 373 (data not shown). The time course of the cytotoxic effect was investigated for BM-dUTP. It was found that the inhibitor's effect is observable within a short time span and significant differences are found within 3 hours under the conditions employed (i.e. 5 μg/ml). However, for the sake of statistics, the cytotoxicity tests were routinely run during a time interval of 48 hours.

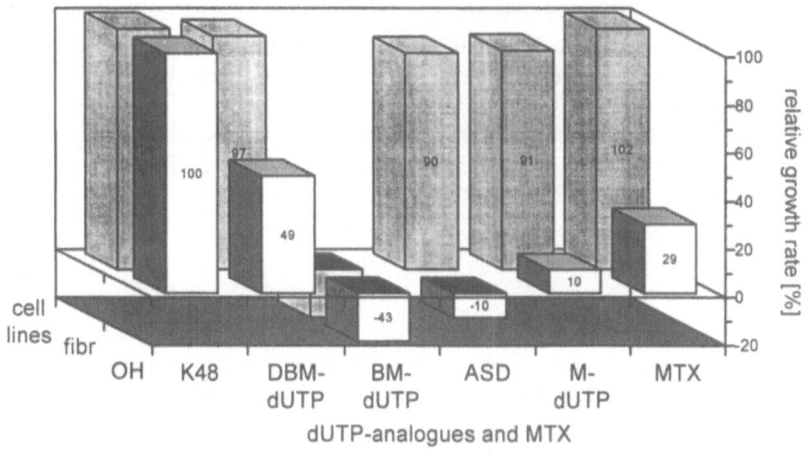

Figure 1. Selective response of the human cancer cell line OH 77 compared to fast-growing human fibroblasts upon incubation with novel dUTP-analogs and methotrexate (MTX) in-vitro. Final concentration of antimetabolites (DBM-dUTP, BM-dUTP, ASD, M-dUTP and MTX) in the culture medium was 10 µg/ml. K48: control experiment, i.e. cultivation under standard conditions. Incubation time and duration of the cell experiments was 48 hours.

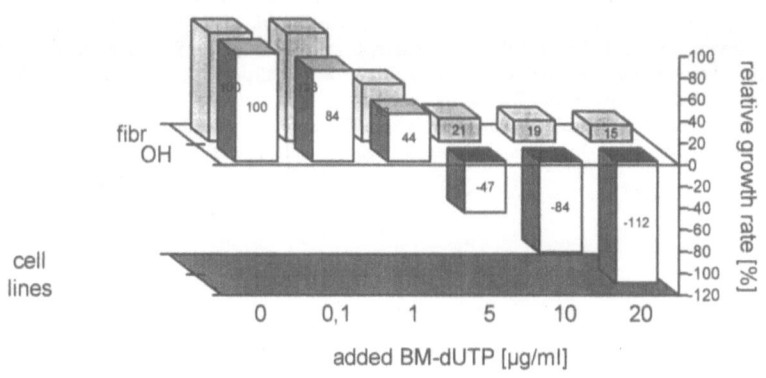

Figure 2. Selective response of the human cancer cell line OH 77 compared to slow-growing human fibroblasts dependent on the BM-dUTP-concentration. The incubation time was 48 hours.

CONCLUSIONS

The data obtained so far allow the following tentative conclusions:

1. All dUTP analogs seem to be sufficiently membrane-permeable in order to exert cytostatic/cytotoxic activity.
2. The selective effect on cancer cells corresponds to the proposed model of cell death as a consequence of DNA fragmentation due to excessive uracil incorporation.
3. The comparison of cancer cells, fast and slow-growing fibroblasts suggests a correlation between the cytostatic action of the dUTPase inhibitors and the proliferation rate of the target cells.

REFERENCES

Beck, W.R., Wright, G.E., Nusbaum, N.J., Chang, J.D., and Isselbacher, E.M., 1986, Enhancement of methotrexate cytotoxicity by uracil analogues that inhibit deoxyuridine triphosphate nucleotidohydrolase (dUTPase) activity, *Adv. Exp. Med. Biol.* 195B:97.

Hochhauser, S.J. and Weiss, B., 1978, Escherichia coli mutants deficient in deoxyuridine triphosphatase, *J. Bacteriol.* 134:157.

Hoffmann, I., Widström, J., Zeppezauer, M., and Nyman, P.O., 1987, Overproduction and large scale preparation of deoxyuridine triphosphate nucleotidohydrolase from Escherichia coli, *Eur. J. Biochem.* 164:45.

Kornberg, A., 1980, "DNA replication", W.H Freeman & Co., San Francisco, USA.

Ludwig, J. and Eckstein, F., 1989, Rapid and efficient synthesis of nucleoside 5'-O-(1-thiotriphosphates), 5'-triphosphates and 2',3'-cyclophosphorothioates using 2-chloro-4H-1,3,2-benzodioxaphosphorin-4-one, *J. Org. Chem.* 54:631.

McIntosh, .M., Ager, D.D., Gadsden, M.H., and Haynes, R.H., 1992, Human dUTP pyrophatase: cDNA sequence and potential biological importance of the enzyme, *Proc. Natl. Acad. Sci. USA* 89:8020.

Mosman T., 1983, Rapid colorimetric assay for cellular growth and survival: application to proliferation and cytotoxic assays, *J. Immunol. Meth.* 65:55.

Tye, B.K., Nyman, P.O., Lehman, I.R., Hochhauser, S., and Weiss, B., 1977, Transient accumulation of Okazaki fragments as a result of uracil incorporation into nascent DNA, *Proc. Natl. Acad. Sci. USA* 74:154.

Zeppezauer,M., Zalud, P., Wachs, W.O., and Nyman, P.O., 1993, "Membrangängiger Wirkstoff zur Störung der DNA-Biosynthese", German Patent Application P 4341 161.4; Dec 2, 1993.

SYNERGISTIC INTERACTION BETWEEN CISPLATIN AND GEMCITABINE

IN OVARIAN AND COLON CANCER CELL LINES

André M. Bergman, Veronique W.T. Ruiz van Haperen, Gijsbert Veerman, Catharina M. Kuiper, Godefridus J. Peters

Dept. Medical Oncology
Free University Hospital
PO BOX 7057
1007 MB Amsterdam
The Netherlands

INTRODUCTION

Cisplatin (cis-diammine dichloroplatinum, CDDP) is an established square planar coordination compound, which is effective against ovarian cancer (1). The antitumour activity of CDDP is the result of formation of adducts within the DNA (2). Resistance against CDDP treatment may be related to an increased DNA repair. Like CDDP, Gemcitabine (2',2'-difluorodeoxycytidine, dFdC) is active against human ovarian carcinoma (3,4). After entering the cell, dFdC requires activation catalyzed by deoxycytidine kinase (dCK), and can be inactivated by the action of deoxycytidine deaminase (dCDA)(5). The active metabolite dFdCTP can be incorporated into both DNA and RNA (6). For both 1-ß-D-arabinofuranosylcytosine (ara-C) and 2'-deoxy-5-azacytidine (DAC) two other deoxycytidine analogues, synergy with CDDP has already been described (7,8). The next logical step was to combine dFdC with CDDP. Both agents have a different mechanism of action. A combination is also attractive from a clinical point of view, since both drugs have different side effects. We tested this combination in different concentrations and schedules in the human ovarian cancer cell line A2780, its 50 fold CDDP resistant variant ADDP (9), its 150,000 fold dFdC resistant variant AG6000 (10), and in the murine colon cancer cell line C26-10, which has an inherent lower sensitivity to both drugs compared to A2780 (11). These results were related to both dCK and dCDA activities and the effect of CDDP on dFdCTP accumulation.

METHODS

Growth inhibition tests: the four cell lines were exposed to CDDP and dFdC as separate agents and a combination of both. Cells were plated in 96 wells plates in different densities, depending on their doubling time (11). dFdC was added at final concentrations ranging from 4.10^{-11} M to 2.10^{-6} M and CDDP from 2.10^{-8} M to 10^{-3} M. The concentrations in the combination had the same range as the separate agents in a dFdC:CDDP ratio of 500. The cells were exposed for 1, 4, 24 and 72 hours, and cultured until 72 hours. The combination studies were evaluated by the sulforhodamine B (SRB) assay and expressed relative to control growth without drugs, which was set at 100% (12,13). Synergism was determined by using isobolograms and the multiple effect analysis (14). A combination index >1 indicates synergism and <1 indicates antagonism.

Enzyme assays: Since dCK reveals biphasic kinetics (15) the assay was performed with ^{3}H-dCyd as a substrate at 2 μM and 230 μM deoxycytidine in the presence of thymidine to block dCyd phosphorylation by thymidine kinase. Substrate and product were separated on a TLC plate, as described previously (16). The dCDA assay was performed with non-tritiated dCyd as a substrate and analyzed by HPLC (16).

DFdCTP accumulation: the cells were exposed for 4 hours to two concentrations of dFdC 1.10^{-7} M or 1.10^{-6} M with or without CDDP in a concentration of 2.10^{-5} M or 2.10^{-4} M respectively. Nucleotides were extracted and analyzed by HPLC as described (11).

RESULTS

Growth inhibition tests: A2780 is the most sensitive cell line for both agents. Both AG6000 and ADDP have higher IC-50 values for dFdC and CDDP, respectively, as a result of their induced resistance. C26-10 was also less sensitive to both agents (Tab.1). Furthermore, it appeared that AG6000 and ADDP were cross resistant, AG6000 about 4-fold to CDDP and ADDP about 40-fold to dFdC, depending on scheduling. It is remarkable that the IC-50 value of ADDP for dFdC is higher at 72 hours than at 24 hours of incubation, this in contrast to the wild-type A2780 and other cell lines (11).

Table 1.   IC-50 values of dFdC and CDDP in the human ovarian cancer cell lines A2780, ADDP, AG6000, and in the murine colon cancer cell line C26-10.

Time	A2780		ADDP		AG6000		C26-10	
dFdC (nM)								
1H	59.0	± 5.8	540.0	±110.9	>1000		>1000	
4H	15.8	± 2.0	128.8	± 44.7	>1000		238.0	±48.0
24H	3.0	± 0.6	157.3	± 24.4	>1000		22.0	±4.0
72H	1.6	± 0.2	515.6	±115.7	50.5	±20.2	6.0	±1.0
CDDP (μM)								
1H	12.6	± 2.2	460.0	±155.6	54.2	±12.4	18.7	± 7.9
4H	3.3	± 0.5	152.0	± 45.7	17.1	± 4.0	5.9	± 1.7
24H	0.7	± 0.1	71.1	± 13.4	4.7	± 0.8	2.4	± 0.7
72H	0.6	± 0.1	50.1	± 9.3	4.7	± 1.0	2.6	± 0.8

Values are means ± SEM of 3-12 experiments. Cells were exposed for 1, 4, 24, 72 hours.

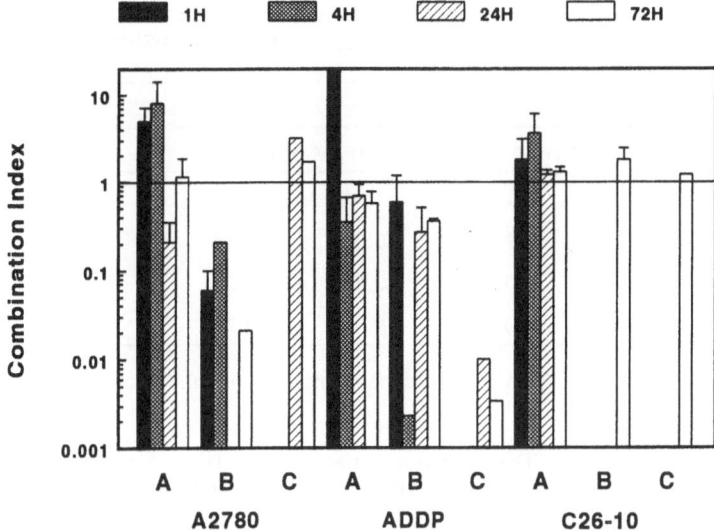

Fig 1. Combination indices (CI) for the human ovarian cancer cell lines A2780, ADDP and the murine colon cancer cell line C26-10, exposed to dFdC and CDDP in different schedules. Values are means ± SEM of 2-5 experiments. Cells were exposed for 1,4,24 and 72 hours. A: Simultaneous exposure to CDDP and dFdC; B: CDDP exposure 4 H before start dFdC exposure; C: CDDP exposure 4 H after start dFdC exposure. (CI > 1 = antagonism, CI < 1 = synergism).

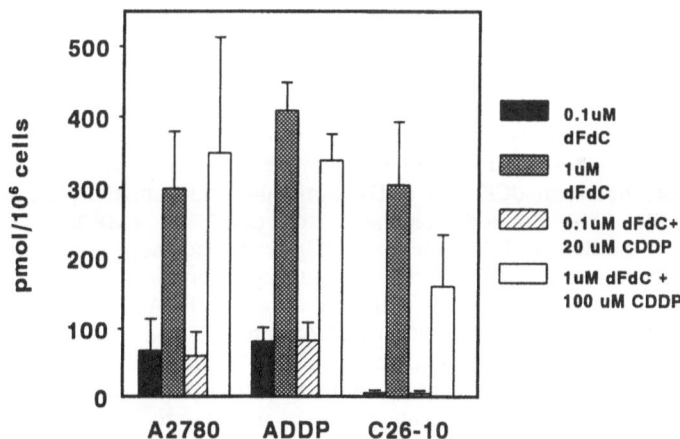

Fig 2. Accumulation of dFdC in A2780, ADDP and C26-10 after a 4 hr exposure to dFdC with or without CDDP. Values are means ± SEM of 4 seperate measurements.

Simultaneous exposure to the combination of dFdC and CDDP was synergistic in A2780 at an exposure time of 24 hours, but in ADDP also at exposure times of 4, 24 and 72 hours (Fig.1). All combinations were antagonistic for C26-10 cells. In AG6000 growth inhibition was due to CDDP alone (dFdC was ineffective). Combinations were not only tested at simultaneous exposure, but also at sequential exposure to CDDP for 4 hours followed by simultaneous exposure to both drugs for various periods. A 4 hour preincubation with CDDP was synergistic in A2780 and ADDP cells, but not for C26-10. A 4 hour pre-incubation with dFdC followed by simultaneous exposure to both drugs for various periods was only synergistic in ADDP.

Table 2.   dCK and dCDA activities of the human ovarian cancer cell line A2780, its CDDP resistant variant ADDP and its dFdC resistant variant AG6000.

cell line	dCK activity						dCDA activity		
	2 μM dCyd			230 μM dCyd					
A2780	0.200	±	0.1	1.42	±	0.63	0.12	±	0.04
ADDP	0.490	±	0.1	0.63	±	0.15	13.42	±	0.40
AG6000	0.001	±	0.0	0.16	±	0.03	0.03	±	0.01
C26-10	0.250	±	0.02	2.30	±	0.78	2.15	±	0.18

Activity values are means ± SEM of 3-10 experiments in nmol/hr/10^6 cells.

DFdCTP accumulation: CDDP did not enhance dFdCTP accumulation in A2780 at both concentrations (Fig.2). In C26-10 cells CDDP seemed to decrease dFdCTP accumulation.

Enzyme assays: despite its cross-resistance to dFdC, ADDP has a higher dCK activity than A2780 at the low deoxycytidine concentration, the dCDA activity is higher in ADDP cells (Tab.2). In AG6000 the dCK and dCDA activities are very low

DISCUSSION

In this paper we show that synergism of the combination of dFdC and CDDP is schedule and time dependent. Preincubation with CDDP was synergistic in both ADDP and A2780. DFdC followed by CDDP was not synergistic in A2780 but highly synergistic in ADDP.

Since the dCK activity in ADDP is higher than in the parent cell line A2780, the cross resistance to dFdC is not related with a lower dCK activity. The high dCDA might explain the insensitivity to dFdC, although earlier it was reported that no correlation exits between dCK and dCDA activities and sensitivity to deoxycytidine analogues (16). The induced resistance to dFdC in AG6000, was already demonstrated to be due to dCK deficiency (10), and associated with an aberrant dCK mRNA amplicon as shown by PCR analysis and the absence of dCK protein (10).

Accumulation of dFdCTP was comparable in both A2780 and ADDP cells, although ADDP cells have a very high dCDA activity. The observation that CDDP tended to decrease dFdCTP accumulation in C26-10 cells, might be related to the observed antagonism. The lack of an effect of CDDP on dFdCTP accumulation indicates that the mechanism for synergism is unlikely to be related to enhanced dFdCTP accumulation. This suggests that the mechanism of interaction between CDDP and dFdC is also unlikely to be related to dFdC uptake, although it was found in a combination of DAC and CDDP, that CDDP decreased the cellular uptake of DAC (17), but DNA platination was increased. Thus it seems more likely that dFdC affects DNA repair of damage caused by CDDP. It has been demonstrated that ara-C markedly delayed the recovery of DNA synthesis inhibited by CDDP (8). Future research should reveal wheter dFdC affects repair of DNA damage caused by platination or whether another mechanism of synergy exists.

ACKNOWLEDGEMENTS

This research was supported by a grant from the Dutch Cancer Society IKA-VU 90-19 and by Eli Lilly, Nieuwegein, the Netherlands

REFERENCES

1. F. Muggia, Introduction: Cisplatin update, *Semin. Oncol*, 18(1-Suppl.3): 1-4(1991).
2. J.J. Roberts and A.J. Thomson, The mechanism of action of antitumour platinum compounds, *Prog. Nucleic Acid Res. Mol. Biol*, 22: 71-133 (1979).
3. B. Lund, P.E. Kristjansen, H.H. Hansen, Clinical and preclinical activity of 2',2'-difluorodeoxycytidine (gemcitabine), *Cancer Treat. Rev.* 19(1): 45-55 (1993).
4. E. Boven, H. Schipper, C.A. Erkelens, S.A. Hatty, H.M. Pinedo, The influence of the schedule and the dose of gemcitabine on the anti-tumour efficacy in experimental human cancer, *Br. J. Cancer*, 68(1): 52-56 (1993).
5. V. Heinemann, Y.Z. Xu, S. Chubb, A. Sen, L.W. Hertel, G.B. Grindey, W. Plunkett, Cellular elimination of 2',2'-difluorodeoxycytidine-5'-triphosphate: a mechanism of self potentiation, *Cancer Res*, 52(3): 533-539 (1992).
6. V.W.T. Ruiz van Haperen, G. Veerman, J.B. Vermorken, G.J. Peters , 2',2'-Difluoro-deoxycytidine (gemcitabine) incorporation into RNA and DNA from tumour cell lines, *Biochem. Pharmacol*, 46: 762-766 (1993).
7. P. Frost, J.L. Abbruzzese, B. Hunt, D. Lee, M. Ellis, Synergistic cytotoxicity using 2'-deoxy-5-azacytidine and cisplatin or 4-hydroperoxycyclophosphamide with human tumor cells, *Cancer Res*, 50: 4572 (1990).
8. R.J. Fram, N. Robichaud, S.D. Bishov, J.M. Wilson, Interactions of cis-diamminedichloroplatinum(II) with 1-ß-D-arabinofuranosylcytosine in LoVo colon carcinoma cells, *Cancer Res*, 47: 3360 (1987).
9. B.T. Hill, S.A. Shellard, L.K. Hasking, W.C. Dempke, A.M. Fichtinger-Schepman, T. Tone, K. Scanlon, Characterization of a cisplatin-resistant human ovarian carcinoma cell line expressing cross-resistance to 5-fluorouracil but collateral sensitivity to methotrexate, *Cancer Res*, 52(11): 3110-3118 (1992).
10. V.W.T. Ruiz van Haperen, G. Veerman, S. Eriksson, E. Boven, A.P.A. Stegmann, M. Hermsen, J.B. Vermorken, H.M. Pinedo, G.J. Peters, Development and characterization of a 2',2'-difluorodeoxycytidine-resistant variant of the human ovarian cell line A2780. *Cancer Res*, in press.
11. V.W.T Ruiz van Haperen, G. Veerman, E. Boven, P. Noordhuis, J.B. Vermorken, G.J. Peters, Schedule dependence of sensitivity to 2',2'-difluorodeoxycytidine (gemcitabine) in relation to accumulation and retention of its triphosphate in solid tumour cell lines and solid tumours, *Biochem. Pharmacol.*, in press.
12. Y.P. Keepers, P.E. Pizao, G.J. Peters, J. Van Ark-Otte, B. Winograd, H.M. Pinedo, Comparison of the sulforhodamine B protein and tetrazolium (MTT) assays for in vitro chemosensitivity testing, *Eur. J. Cancer* 27: 897-900 (1991).
13. G.J. Peters, M. Wets, Y.P.A.M. Keepers, R. Oskam, J. Van Ark-Otte, P. Noordhuis, K. Smid, H.M. Pinedo, Transformation of mouse fibroblasts with the oncogenes *H-ras* or *trk* is associated with pronounced changes in drug sensitivity and metabolism, *Int. J. Cancer*, 54: 450-455 (1993).
14. T.C. Chou and P. Talahay, Applications of the median-effect principle for the assessment of low-dose risk of carcinogens and for the quantitation of synergism and antagonism of chemotherapeutic agents, New Avenues in Developmental Cancer Chemotherapy (Harrap KR, Connors TA, eds.). New York: Academic Press, 37 (1987).
15. V.W.T. Ruiz van Haperen and G.J. Peters. New targets for pyrimidine antimetabolites for the treatment of solid tumours. Part II; deoxycytidine kinase, *Pharm. World Sci.* 16 (1994).
16. V.W.T. Ruiz van Haperen, G. Veerman, B.J.M. Braakhuis, J.B. Vermorken, E. Boven, A. Leyva, G.J. Peters, Deoxycytidine kinase and deoxycytidine deaminase activities in human xenografts, *Eur. J. Cancer* 29A: 2132-2137 (1993).
17. J.A. Ellerhorst, P. Frost, J.L. Abbruzzese, R.A. Newman, Y. Chernajovsky, 2'-deoxy-5-azacytidine increases binding of cisplatin to DNA by a mechanism independent of DNA hypomethylation, *Br. J. Cancer*, 67: 209-215 (1993).

ANTIMETABOLITES REDUCE THE ACTIVITIES OF ENZYMES WITH SHORT HALF-LIVES IN ADDITION TO INHIBITING THEIR SPECIFIC TARGETS

George Weber*, Noemi Prajda**, and Radhey L. Singhal

Laboratory for Experimental Oncology
Indiana University School of Medicine
Indianapolis, IN 46202-5200, USA

INTRODUCTION

De novo and Salvage Activity and Enzyme Pattern-Targeted Chemotherapy

In evolution two pathways emerged for the biosynthesis of pyrimidine and purine nucleotides. The *de novo* biosynthetic pathway assembles nucleotides from small building blocks. By contrast, the salvage pathways provide mechanisms to recycle nucleosides and bases from the dead cells of tissues and from the blood stream[1]. In some lower organisms only one of these pathways operates[2]. However, in mammalians both *de novo* and salvage pathways function in all tissues. It has been recognized in the past 10 years that the activities of salvage enzymes in each of the biosynthetic segments of metabolism are markedly higher than those of the rate-limiting enzymes of *de novo* biosynthesis[1].

In cancer cells the operation of the *de novo* and salvage pathways poses a therapeutic problem. The antimetabolites in clinical use are targeted against key enzymes of the *de novo* pathway; however, the high salvage capacity could circumvent the block provided by the antimetabolite[1]. Enzyme pattern-targeted chemotherapy has been suggested to overcome the circumvention activity of salvage[1]. Evidence has been provided that combined inhibition of *de novo* and salvage pathways can provide a synergistic chemotherapeutic impact[1,3,4].

Failure of chemotherapy frequently may be due to the high salvage pathway activities and the emerging resistance. The observation that antimetabolite therapy does have at least partial success (probably depending on the extent of the functioning salvage) suggested to us that the antimetabolites might have an effect on the activity or amount of salvage enzymes[5,6]. Our investigations of the past 7 years revealed that antimetabolites, in addition to attacking their specific targets, also lead to a decrease in activities of non-targeted, unrelated enzymes[5-7].

---

*To whom correspondence and reprint requests should be addressed.
**Permanent address, National Oncological Institute, Budapest, Hungary.

Purine and Pyrimidine Metabolism in Man VIII, Edited by
A. Sahota and M. Taylor, Plenum Press, New York, 1995

Non-Targeted Action of Antimetabolites - Enzymes with Short Half-Lives

Studies on the action of antimetabolites (tiazofurin, methotrexate (MTX), acivicin, and 5-fluorouracil (5-FU)) revealed that these drugs, in addition to inhibiting or inactivating their specific enzymic targets, also caused the decrease in the activities and amounts in other enzymic activities unrelated to the primary antimetabolite action[5-7]. We recognized the principle that accounts for the decrease in non-targeted enzymic activities: those enzymes decrease in activity which have short half-lives[5-7]. When the antimetabolites inhibit or decrease macromolecular biosynthesis, the enzymes with the shortest half-lives decreased the most rapidly. Some of these are salvage enzymes, e.g. thymidine (TdR) and deoxycytidine (CdR) kinases. This insight provides a deeper understanding of antimetabolite action. Furthermore, there is an opportunity to employ more effective and synergistic enzyme pattern-targeted chemotherapy and more efficient protection against bone marrow damage.

Initial Observation with Tiazofurin and Acivicin: Inhibition of Hepatic Target Enzymes and a Reduction of TdR Kinase Activity

In 1988 we showed that when rats were treated with tiazofurin the hepatic activity of the target enzyme, IMP dehydrogenase, was inhibited; however, there was also a decrease in the activity of the non-targeted enzyme, TdR kinase[7]. We also observed in animals treated with the antiglutamine agent, acivicin, that there occurred the expected inhibition in the liver activities of the five purine and pyrimidine glutamine utilizing enzymes, but there was also a reduction in the activity of the non-targeted enzyme, TdR kinase[7].

New Light on the Mechanism of Action of Methotrexate

The primary target of the antifolate, MTX, was identified as dihydrofolate reductase (DHFR)[8]. The inhibition of DHFR in pyrimidine metabolism results in a decrease in the conversion of dUMP to dTMP, a reaction catalyzed by dTMP synthase. It is thought that reducing the activity of the dTMP synthase reaction accounts for the decrease in thymidylate levels. This leads to limiting of the concentration of dTTP and reducing DNA biosynthesis through an imbalance of the dNTP pools. However, inhibition of the dTMP synthase activity only blocks the *de novo* biosynthesis of dTMP. But dTMP is also produced by TdR kinase activity which salvages TdR. When anticancer agents destroy neoplastic cells, TdR is liberated which can be recycled by TdR kinase activity. Moreover, we showed that in all examined human and murine cancer cells the activity from the salvage enzyme, TdR kinase, is orders of magnitude higher than that of the enzyme of the *de novo* pathway, dTMP synthase[6]. There is TdR in rat plasma (1.3 ± 0.3 μM) and in liver (0.3 ± 0.03 μM)[9] and more should become available when cancer cells are destroyed by MTX. Thus, the antipyrimidine cytotoxicity of MTX is not accounted for by inhibition of the dTMP synthase reaction because the high TdR kinase capacity could circumvent this block[1,3-6]. The salvage circumvention hypothesis[1] is supported by the well-known ability of TdR to rescue when added to human or murine tissue culture cells and in patients treated with MTX[3,4,10].

Our recent investigations provided new light on the action of MTX and other antimetabolites. We demonstrated that injection of rats with MTX caused a decrease in the activity and amount of TdR kinase in bone marrow, liver and transplanted hepatoma[5]. Dose-response studies showed that not only the activity of TdR kinase but also that of dTMP synthase was decreased by MTX[5]. A time course of MTX impact revealed a sharp decline in the activities of dTMP synthase and TdR kinase in the bone marrow. The reduction in enzymic activities did not merely reflect a decrease in bone marrow cellularity because the enzymic activities decreased to 50% of controls when cellularity had not yet declined by 10%[5,6].

Unifying Principle: Degradation Rates of Pyrimidine and Purine Enzymes

In searching for a unifying principle in the decrease in activities of target and non-target enzymes in antimetabolite treatment we tested the hypothesis that MTX and other antimetabolites might have curtailed enzyme biosynthetic processes which would cause a degradation of the enzymes that have short half-lives. To answer this question we studied the effect of an inhibitor of protein biosynthesis, cycloheximide.

Cycloheximide, when injected into rats in a dose of 15 mg/kg, markedly curtails protein biosynthesis, permitting measurement of degradation rates of enzymes, largely unopposed by cellular protein biosynthesis[5,6]. Our studies indicated that among the hepatoma enzymes studied, TdR and CdR kinases, had short half-lives (0.8 and 0.6 h), but other enzymes examined in purine and pyrimidine metabolism exhibited longer half-lives[5,6] (Table 1).

Table 1. Effect of cycloheximide on $t_{1/2}$ of enzyme activities in rat bone marrow

Enzymes	EC numbers	$t_{1/2}$: h
GMP reductase	1.6.6.8	4.3
Adenine phosphoribosyltransferase	2.4.2.7	3.2
Guanine phosphoribosyltransferase	2.4.2.8	3.0
IMP dehydrogenase	1.1.1.205	1.6
dTMP synthase	2.1.1.45	1.4
GMP synthase	6.3.5.2	1.2
TdR kinase	2.7.1.21	0.8
CdR kinase	2.7.1.74	0.6
Phosphatidylinositol 4-kinase	2.7.1.67	0.12
Phosphatidylinositol 4,5 kinase	2.7.1.68	0.12

A single cycloheximide i.p. injection (15 mg/kg) was given to ACI/N rats and groups of animals were killed at 0.5, 1, 2, 4, 6, 8 and 12 h after treatment.

Tiazofurin: Impact on Enzymic Programs

Tiazofurin, a C-nucleoside, is a pro-drug which when converted in cells to the active metabolite, TAD, strongly inhibits the activity of IMP DH, the rate-limiting enzyme of *de novo* GTP biosynthesis. This leads to a decrease in the concentrations of GTP and dGTP[11]. Tiazofurin infused to patients with chronic granulocytic leukemia in blast crisis yielded therapeutic responses in 77% of cases[12]. Tiazofurin has an impact on 4 targets. 1. Chemotherapy. Through inhibition of IMP DH activity and the reduction of guanylate concentration in the blast cells, chemotherapy is achieved as shown by the tendency to increased uric acid production. With concurrent administration of allopurinol, the production of uric acid decreased and there was an increased level of serum hypoxanthine which competitively inhibits the activity of guanine phosphoribosyltransferase (GPRT), yielding decreased guanine salvage and a synergistic decrease in the GTP concentration of the blast cells. 2. Differentiation. Tiazofurin was particularly effective in inducing differentiation in patients[13]. In HL-60 and K562 cells tiazofurin also decreased GTP concentration resulting in induced differentiation; addition of guanosine abrogated differentiation by restoring GTP concentration through guanine salvage[14,15]. 3. Down-regulation of oncogenes. In tiazofurin-treated patients IMP DH activity and subsequently GTP concentration decreased in the blast cells accompanied by down-regulation of the expression of ras and myc oncogenes[16]. Then the blast cells differentiated. This agrees with observations in K562 cells[15]. 4. Reduction of activities of non-targeted enzymes with short half-lives. Our recent investigations revealed that the overall impact of tiazofurin entails a decrease in the activities and amounts of enzymes with rapid decay rates. We observed that after a single tiazofurin injection, resulting in a steep decrease of IMP DH activity in rat bone marrow, there was also a rapid decline in the activities of TdR and CdR kinases and of GPRT[17]. Current studies also revealed that tiazofurin caused a rapid decrease in the activity (amount) of key enzymes in the signal transduction biosynthetic pathway[18].

The Signal Transduction Pathway and Tiazofurin

A major signal transduction mechanism operates through the utilization of phosphatidylinositol (PI) by subsequent phosphorylation by PI kinase and PIP kinase, yielding PIP_2. PIP_2 is the substrate of PIP_2 phosphodiesterase which produces the second messengers, IP_3 and diacylglycerol (DAG). We recently showed that the activities of PI kinase and PIP kinase and the concentrations of the end product, IP_3, were increased in human and murine cancer cells[19]. Enzymes of particular significance in metabolic processes frequently have short half-lives because, in part at least, their role is to appear promptly when there is expression of a vital program and to decline rapidly when the function is completed. Using cycloheximide treatment in the rat, we showed that the bone marrow PI kinase and PIP kinase activities had $t_{1/2} = 0.12$ h. This was the shortest degradation rate in comparison with 8 enzymes of purine and pyrimidine biosynthesis with $t_{1/2} = 0.6$ to 4.3 h (Table 1). In liver and hepatoma PI and PIP kinase activities decreased less rapidly ($t_{1/2} = 5$ to 13 h) (not shown).

Since PI kinase[20] and PIP kinase[19] activities are markedly increased in hepatomas in presence of similar enzyme decay rates in normal liver and hepatoma, therefore the increased activity (amount) of these signal transduction enzymes may be attributed to increased biosynthesis. A similar situation is documented for carbamoyl-phosphate synthase II activity which is elevated 9-fold in hepatoma 3924A[21]. The liver and hepatoma enzymes have similar decay rates but the 9-fold elevated enzymic activity is paralleled by a 9-fold increase in the immunoprecipitable enzyme amount and the increased incorporation of amino acids into the enzyme. There is also an elevated mRNA production for this enzyme[21].

The presently discovered new action of tiazofurin in reducing the activities of PI and PIP kinases might explain, in part at least, the observation that tiazofurin down-regulates the signal transduction pathway in the nuclei of Friend cells[22]. Down-regulation of the signal transduction pathway, in turn, might play a role in the remissions and induced differentiation observed with tiazofurin treatment in leukemic patients[12,13].

Conclusions: Targeted and Non-Targeted Actions of Anti-Cancer Drugs

Examination of the impact of tiazofurin, acivicin, MTX and 5-FU on the enzymic program of the rat bone marrow revealed that in addition to the well-known specific enzymic targets of the antimetabolites there is also a reduction in the activity of non-target enzymes because they have short half-lives. It is assumed that these enzyme activities (amounts) decrease because of the curtailment of DNA and RNA biosynthesis. Our conclusion is supported by results showing that taxol which does not act primarily through inhibition of nucleic acid biosynthesis had no effect on the activities of any of the enzymes studied in the bone marrow[6].

It is important to keep in mind that tiazofurin and other anti-metabolites do not inhibit the activities of the non-target enzymes, e.g., TdR kinase, but they act through curtailing macromolecular biosynthesis, leading to a decrease in the amounts of enzymes of rapid degradation rates[5,6]. It is also relevant that although these anti-metabolites cause a reduction in the amounts of target and salvage enzymes some activities remain in the tissues. Therefore, it is necessary to discover and utilize inhibitors for the salvage enzymes to provide synergistic action with blockers of the *de novo* pathways, as suggested by the approaches of enzyme pattern-targeted chemotherapy.

ACKNOWLEDGMENTS

This work was supported by Fogarty International Res. Collab. Award TW00031 to G.W. and by OTKA Grant 3024 to N.P.

REFERENCES

1. G. Weber, Biochemical strategy of cancer cells and the design of chemotherapy: G. H. A. Clowes Lecture, *Cancer Res.* 43:3466 (1983).
2. T. Aoki, Initial steps of *de novo* pyrimidine nucleotide biosynthesis in parasites and mammalian tissues: purification, regulation, adaptation, and evolution, *Jpn. J. Parasitology* 43:1 (1994).
3. G. Weber, S. Ichikawa, M. Nagai, and Y. Natsumeda, Azidothymidine inhibition of thymidine kinase and synergistic cytotoxicity with methotrexate and 5-fluorouracil in rat hepatoma and human colon cancer cells, *Cancer Commun.* 2:129 (1990).
4. G. Weber, M. Nagai, N. Prajda, H. Nakamura, T. Szekeres, and E. Olah, AZT: a biochemical response modifier of methotrexate and 5-fluorouracil cytotoxicity in human ovarian and pancreatic carcinoma cells, *Cancer Commun.* 3:127 (1991).
5. M. Abonyi, N. Prajda, Y. Hata, H. Nakamura, and G. Weber, Methotrexate decreases thymidine kinase activity, *Biochem. Biophys. Res. Commun.* 187:522 (1992).
6. G. Weber, and N. Prajda, Targeted and non-targeted actions of anti-cancer drugs, *Advan. Enzyme Regul.* 34:71 (1994).
7. N. Prajda, Y. Natsumeda, T. Ikegami, M.A. Reardon, S. Szondy, Y. Hashimoto, J. Emrani, and G. Weber, Enzymic programs of rat bone marrow and the impact of acivicin and tiazofurin, *Biochem. Pharmacol.* 37:875 (1988).
8. M.J. Osborn, M. Freeman, and F.M. Huennekens, Inhibition of dihydrofolate reductase by aminopterin and amethopterin, *Proc. Soc. Exptl. Biol. Med.* 97:429 (1958).
9. K. Pillwein, H.N. Jayaram, and G. Weber, Effect of ischemia on nucleosides and bases in rat liver and hepatoma 3924A, *Cancer Res.* 47:3092 (1987).
10. C.J. Allegra, Antifolates, *in*: "Cancer Chemotherapy: Principles and Practice," B.A. Chabner and J.M. Collins, eds., J.B. Lippincott Co., Philadelphia (1972).

11. M.S. Lui, M.A. Faderan, J.J. Liepnieks, Y. Natsumeda, E. Olah, H.N. Jayaram, and G. Weber, Modulation of IMP dehydrogenase activity and guanylate metabolism by tiazofurin (2-ß-D-ribofuranosylthiazole-4-carboxamide), *J. Biol. Chem.* 259: 5078 (1984).

12. G. Weber, IMP dehydrogenase and GTP as targets in human leukemia treatment, *Advan. Exptl. Med. Biol.* 309B:287 (1991).

13. G. Tricot, H.N. Jayaram, G. Weber, and R. Hoffman, Tiazofurin: biological effects and clinical uses, *Intl. J. Cell Cloning* 8:161 (1990).

14. D. Wright, A role for guanine ribonucleotides in the regulation of myeloid cell maturation, *Blood* 69:334 (1987).

15. E. Olah, Y. Natsumeda, T. Ikegami, Z. Kote, M. Horanyi, J. Szelenyi, E. Paulik, T. Kremmer, S.R. Hollan, J. Sugar, and G. Weber, Induction of erythroid differentiation and modulation of gene expression by tiazofurin in K-562 cells produced by inhibitors of inosine 5'-phosphate dehydrogenase, *Cancer Res.* 46:2314 (1986).

16. G. Weber, M. Nagai, Y. Natsumeda, J.N. Eble, H.N. Jayaram, E. Paulik, W. Zhen, R. Hoffman, and G. Tricot, Tiazofurin down-regulates expression of c-Ki-*ras* oncogene in a leukemic patient, *Cancer Commun.* 3:61 (1991).

17. N. Prajda, Y. Hata, M. Abonyi, R.L. Singhal, and G. Weber, Sequential impact of tiazofurin and ribavirin on enzymic program of the bone marrow, *Cancer Res.* 53:5982 (1993).

18. R.L. Singhal, Y.A. Yeh, K.Y. Look, G.W. Sledge, Jr., and G. Weber, Coordinated increase in activities of the signal transduction enzymes PI kinase and PIP kinase in human cancer cells, *Life Sci.*, accepted for publication (1994).

19. R.L. Singhal, N. Prajda, Y.A. Yeh, and G. Weber, 1-Phosphatidylinositol 4-phosphate 5-kinase (EC 2.7.1.68): A proliferation- and malignancy-linked signal transduction enzyme, *Cancer Res.* Nov. issue, 54: (1994).

20. M.T. Rizzo, and G. Weber, 1-Phosphatidylinositol 4-kinase (EC 2.7.1.67): An enzyme linked with proliferation and malignancy, *Cancer Res.* 54:2611 (1994).

21. G. Weber, and M.A. Reardon, Regulation of carbamoyl-phosphate synthase II, *Advan. Enzyme Regul.* 25:65 (1986).

22. A.M. Billi, L. Cocco, A.M. Martelli, R.S. Gilmour, and G. Weber, Tiazofurin-induced changes in inositol lipid cycle in nuclei of Friend erythroleukemia cells, *Biochem. Biophys. Res. Commun.* 195:8 (1993).

INHIBITION OF MURINE AMIDO PHOSPHORIBOSYLTRANSFERASE BY FOLATE DERIVATIVES

Sarah L. Schoettle, and Richard I. Christopherson

Department of Biochemistry
University of Sydney
Sydney NSW 2006
Australia

INTRODUCTION

Amido phosphoribosyltransferase (PRTase) catalyses the first reaction of the pathway for *de novo* biosynthesis of purine nucleotides:

$$\overset{1}{\text{P-Rib-PP}} \to \overset{2}{\text{PRA}} \to \overset{3}{\text{GAR}} \to \overset{4}{\text{FGAR}} \to \overset{5}{\text{FGAM}} \to \overset{6}{\text{AIR}} \to \overset{7}{\text{CAIR}} \to \text{SAICAR} \to \overset{8}{\text{AICAR}} \to \overset{9}{}$$

$$\overset{10}{\text{FAICAR}} \to \text{IMP} \tag{1}$$

The reaction involves transfer of the γ-amido nitrogen of Gln to P-Rib-PP with the displacement of pyrophosphate:

$$\text{P-Rib-PP} + \text{Gln} \xrightarrow{\text{Mg}^{2+}} \text{PRA} + \text{Glu} + \text{PPi} \tag{2}$$

Amido PRTase conforms to an ordered sequential mechanism, P-Rib-PP binds first with positive cooperativity followed by Gln without cooperativity (Holmes, 1980). The enzyme is subject to feed-back inhibition by the end-prodcuts of the pathway, AMP, GMP and IMP and Holmes (1980) proposed that these 6-amino and 6-oxo nucleotides bind at two distinct allosteric sites giving synergistic inhibition. These data suggest that amido PRTase and hence the flux through the *de novo* purine pathway are regulated by a fine balance between the cellular concentrations of P-Rib-PP (56 μM; Sant *et al.*, 1992), GMP (38 μM) and IMP (57 μM; Sant *et al.*, 1989). Holmes *et al.* (1973) showed by sucrose density gradient centrifugation that partially purified human amido PRTase is found as a mixture of two molecular forms with the smaller form (M_r 133,000, $s_{20,w} = 5.9$) predominating over the larger one (M_r 270,000, $s_{20,w} = 10.0$). Most of the small form was converted to the large form by 5 mM AMP and as the concentration of AMP increased, amido PRTase activity decreased in concert with the proportion of the small form remaining, consistent with the small form being active. Conversely, 10 mM P-Rib-PP converted most of the amido PRTase to the small (active) form. Experiments with pure amido PRTase from pigeon liver suggested that the small form is a dimer and the large form is a tetramer (Rowe and Wyngaarden, 1968). In support of this conclusion, Zhou *et al.* (1990) calculated a subunit molecular weight of 55 kDa from the cDNA sequence for amido

Purine and Pyrimidine Metabolism in Man VIII, Edited by
A. Sahota and M. Taylor, Plenum Press, New York, 1995

PRTase from chicken liver and a molecualr weight of 210 kDA was determined for the tetramer from sedimentation and diffusion measurements with pure chicken liver amido PRTase (Hartman, 1963).

We have recently discovered that the pentaglutamyl derivative of dihydrofolate (DHF-Glu$_5$) is a potent non-competitive inhibitor or amido PRTase with a dissociation constant of 3.4 μM for interaction with the enzyme-glutamine complex (Sant et al., 1992). Such potent inhibition provided an explanation of how the antifolate drug, methotrexate, induced inhibition of the de novo purine pathway. Inhibition of dihydrofolate reductase by methotrexate results in accumulation of DHF polyglutamates to 20% of the total folate pool (Allegra et al., 1986) and the DHF polyglutamates inhibit amido PRTase. Sant et al. (1992) also found that amido PRTase is subject to direct, potent inhibition by the antifolate, piritrexim (PTX), although DHF polyglutamates would also accumulate in PTX-treated cells. In this chapter, we have investigated the effect of PTX on the aggregation state of amido PRTase, and the kinetic mechanism of the potent inhibition of amido PRTase by PTX has also been determined.

EXPERIMENTAL PROCEDURES

5-Phosphoribosyl-1-pyrophosphate (P-Rib-PP), L-glutamine, ATP, AMP, IMP, GMP and the marker proteins alcohol dehydrogenase, lactate dehydrogenase and catalase were obtained from the Sigma Chemical Company, USA. Piritrexim (PTX) was a kind gift from Burroughs Wellcome Company, USA. L-[U-^{14}C]glutamine (0.447 mM, 224 Ci/mol) was purchased from DuPont Ltd, USA and [1-^{14}C]glycine (180 mM, 56 Ci/mol) was from Amersham International plc, UK.

Purification of Amido PRTase

Mouse L1210 leukaemia cells were grown in RPMI 1640 medium supplemented with 10% (v/v) foetal calf serum. Cells were harvested in late exponential phase (9x10^5 cells/ml), lysed by sonication and amido PRTase was purified by fractionation with streptomycin sulfate (3.5% w/v) and ammonium sulfate (40-60% of saturation) followed by election from a column of DEAE-Sephacel with a linear gradient of 0-300 mM KCl in 20 mM K.Hepes (pH 7.3), 1.0 mM dithiothreitol. The final purification factor was 12.7-fold with a yield of 41%.

Assay of Amido PRTase

For the purification of amido PRTase and analysis of its states of aggregation by density gradient centrifugation, enzymic activity was determined by formation of [^{14}C]Glu from [^{14}C]Gln (Sant et al., 1992). For enzyme kinetic experiments, a more sensitive assay was used where the unstable 5-phosphoribosylamine (PRA) formed was immediately converted to [^{14}C]glycineamide ribotide (GAR) by excess recombinant GAR synthetase (Schendel et al., 1988; Shen et al., 1990). Assay mixtures contained in a total volume of 25 μl: 50 mM K.Hepes (pH 7.2), 1.0 mM MgCl$_2$, P-Rib-PP, Gln, 800 μM [1-^{14}C]glycine (20 Ci/mol), 500 μM MgATP, pure GAR synthetase (1.2 μg protein) and amido PRTase. Saturating concentrations of P-Rib-PP (500 μM) and Gln (3.0 mM) were used unless indicated otherwise. [^{14}C]GAR formed was isolated by thin layer chromatography as described elsewhere (S.L. Schoettle and R.I. Christopherson, manuscript submitted).

Sucrose Density Gradient Centrifugation

Linear sucrose gradients (10-25% w/v, 11.6 ml) contained 20 mM K.Hepes (pH 7.3), 1.0 mM MgCl$_2$, 1.0 mM dithiothreitol and effectors as required. Samples for analysis (200 μl) contained amido PRTase (0.5 mg protein), alcohol dehydrogenase (1 mg), lactate dehydrogenase (150 μg), catalase (0.5 mg) and effectors as required. Sedimentation profiles for amido PRTase were developed by centrifugation at 40,000 rpm in a Beckman SW 41 Ti rotor for 24 h at 4°C. Gradients were fractionated and amido PRTase and the three marker enzymes were assayed. Sedimentation coefficients ($s_{20,w}$) used for the standard proteins were: alcohol dehydrogenase 5.1 S, lactate dehydrogenase 7.0 S, and catalase 11.3 S.

RESULTS AND DISCUSSION

Molecular Forms of Amido PRTase

In agreement with Holmes *et al.* (1973), sedimentation of amido PRTase through sucrose gradients in the presence of AMP or P-Rib-PP induced two distinct forms of the enzyme. In the presence of 5 mM AMP, amido PRTase sedimentated as a tetramer ($s_{20,w}$ = 10.2±0.4 S) while 250 μM P-Rib-PP induced formation of a dimer ($s_{20,w}$ = 6.7±0.3 S). GMP (5 mM) had a similar effect to AMP and the activity peaks with these three ligands were symmetrical and clearly defined. In the absence of effectors, amido PRTase sedimented over a broader range with a peak giving an apparent $s_{20,w}$ value of 9.3±0.6 S. The activity profile consisted of a mixture of dimer with a predominant tetramer, perhaps with dynamic equilibrium existing between these two species during sedimentation. When amido PRTase was sedimented in the presence of 100 μM PTX, the sedimentation rate was similar to the active, dimeric form of the enzyme. This dimeric form of the enzyme ($s_{20,w}$ = 7.2±0.5 S) was inactive and an activity profile could only be obtained after delution of the PTX into the assay mixture. Thus amido PRTase can exist in three forms: as an inactive dimer ($s_{20,w}$ = 7.2±0.5 S) in the presence of PTX, an active dimer ($s_{20,w}$ = 6.7±0.3 S) with P-Rib-PP, or an inactive tetramer ($s_{20,w}$ = 10.2±0.4 S) with AMP (GMP or IMP). The latter two forms of the enzyme were previously identified by Holmes *et al.* (1973) with amido PRTase from human placenta. Transition from the inactive tetramer to the inactive dimer induced by increasing concentrations of PTX was consistent with a dissociation constant of 6.0 μM for PTX acting as a non-competitive inhibitor of the enzyme (Sant *et al.*, 1992).

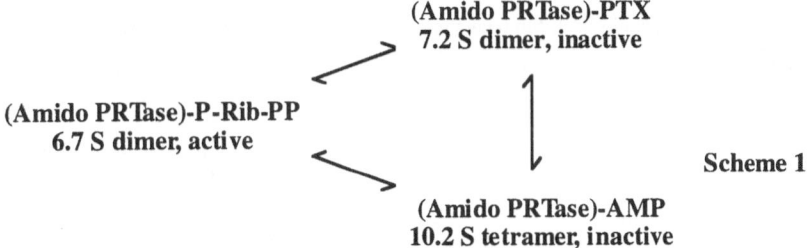

(Amido PRTase)-PTX
7.2 S dimer, inactive

(Amido PRTase)-P-Rib-PP
6.7 S dimer, active

(Amido PRTase)-AMP
10.2 S tetramer, inactive

Scheme 1

Inhibition of Amido PRTase

The pentaglutamyl derivative of dihydrofolate (DHF-Glu$_5$) is a potent non-competitive inhibitor of amido PRTase (K_i = 3.4 μM, Sant *et al.* 1992). The data obtained were consistent with a model where all enzyme species are in rapid equilibrium, all enzyme-inhibitor complexes are catalytically inactive, amido PRTase (E) is present as the active or inactive dimer and forms the indicated complexes with folate derivatives (I) and P-Rib-PP (S) (Scheme 2). Inhibition patterns for PTX with P-Rib-PP or Gln as the varied substrate indicated that PTX is a non-competitive inhibitor of amido PRTase bound with positive cooperativity, like P-Rib-PP. The dissociation constant for binding of the first molecule of PTX is approximately 66 μM and the second molecule is bound more tightly with a lower value of 12 μM. Thus in the presence of PTX, most enzyme-inhibitor complex would be IEI due to the positive cooperativity, this form of the enzyme would be equivalent to the 7.2 S form characterised by ultracentrifugation (Scheme 1). It is assumed that the natural effector of amido PRTase, DHF-Glu$_5$, would involve formation of the inactive 7.2 S dimer of the enzyme like PTX (Scheme 1). In normal cells, inhibition of amido PRTase by accumulated DHF polyglutamates could stop *de novo* purine biosynthesis when there is insufficient N^{10}-formyl tetrahydrofolate to convert GAR→FGAR and AICAR→FAICAR.

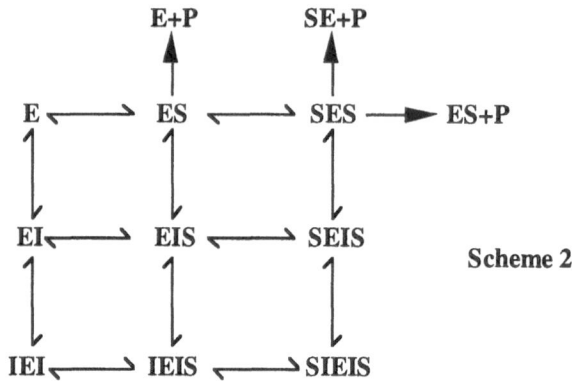

Scheme 2

Acknowledgements

We thank Leesa B. Crisp for technical assistance. This research was supported by project grant 930019 from the National Health and Medical Research Council of Australia.

REFERENCES

Allegra, C.J. Fine, R.L., Drake, J.C. and Chabner, B.A., 1986, The effect of methotrexate on intracellular folate pools in human MCF-7 breast cancer cells, *J. Biol. Chem.*, 261:6478.

Hartman, S.C., 1963, Phosphoribosyl pyrophosphate amidotransferase: purification and general catalytic properties, *J. Biol. Chem.*, 238:3024.

Holmes, E.W., 1980, Kinetic, physical, and regulatory properties of amido phosphoribosyl transferase, *Adv. Enz. Reg.*, 19:215.

Holmes, E.W., Wyngaarden, J.B., and Kelley, W.N., 1973, Human glutamine phosphoribosylpyrophosphate amidotransferase: two molecular forms interconvertible by purine nucleotides and phosphoribosylpyrophosphate., *J. Biol. Chem.*, 248:6035.

Rowe, P.B., and Wyngaarden, J.B., 1968, Glutamine phosphoribosyl pyrophosphate amidotransferase: purification, sub-structure, amino acid composition and absorption spectra, *J. Biol.. Chem.*, 243:6373.

Sant, M.E., Lyons, S.D., Phillips, L., and Christopherson, R.I., 1992, Antifolates induce inhibition of amido phosphoribosyltransferase in leukaemia cells, *J. Biol. Chem.*, 267:11038.

Sant, M.E., Poiner, A., Harsanyi, M.C., Lyons, S.D., and Christopherson, R.I., 1989, Chromatographic analysis of purine precursors in mouse L1210 leukaemia, *Anal. Biochem.*, 182:121.

Schendel, F.J., Cheng, Y.S., Otvos, J.D., Wehrli, S., and Stubbe, J., 1988, Characterisation and chemical properties of phosphoribosylamine, an unstable intermediate in the *de novo* purine biosynthetic pathway, *Biochemistry*, 27:2614.

Shen, Y. Rudolph, J., Stern, M., and Stubbe, J., 1990, Glycinamide ribonucleotide synthetase from *Escherichia coli*: cloning, overproduction, sequencing, isolation, and characterisation, 29:218.

Zhou, G., Dixon, J.E., and Zalkin, H., 1990, Cloning and expression of avian glutamine phosphoribosylpyrophosphate amidotransferase, *J. Biol. Chem.*, 265:21152.

INOSINE 5'-MONOPHOSPHATE DEHYDROGENASE AS A CHEMOTHERAPEUTIC TARGET

Trevor J. Franklin[1], Gwynneth Edwards[2] and Philip Hedge[1]

[1]Cancer Research Department, Zeneca Pharmaceuticals, Alderley Park, Macclesfield, SK10 4TG, England
[2]Department of Cell and Structural Biology, University of Manchester, Oxford Rd., Manchester, M13 9PT, England

INTRODUCTION

Inosine 5'-monophosphate, EC 1.1.1.205 (IMPDH) catalyzes the conversion of IMP to XMP utilizing NAD as a proton acceptor. Its role in catalyzing the rate determining step in the biosynthesis of GTP[1] gives IMPDH a position of central importance in cellular activity because of the myriad activities of GTP in biosynthesis and cellular regulation. The activity of IMPDH is much higher in proliferating tissues, both normal and malignant[2,3] suggesting that the salvage of guanine by hypoxanthine-guanine phosphoribosyltransferase is probably inadequate to satisfy the requirements of dividing cells for guanine nucleotides.

Several nucleoside analogues, including ribavirin, tiazofurin, bredinin (mizoribine), oxanosine and benzamide riboside, which are metabolized to nucleotide derivatives, inhibit intracellular IMPDH. However, the derived nucleotides may not be entirely specific for IMPDH. Tiazofurin and benzamide riboside are metabolized to analogues of NAD and act as competitive inhibitors of the cofactor[4,5]. The tiazofurin metabolite also inhibits several other dehydrogenases[6] although it is unclear whether this contributes to the biological actions of tiazofurin. The most specific inhibitor of IMPDH is the antibiotic mycophenolic acid (MPA) which is unique in directly inhibiting the enzyme without prior metabolism[7]. An ester pro-drug of MPA, 'Mofetil', has remarkable immunosuppressive properties, combined with low toxicity, in both experimental animals and human patients[8]. Progress towards defining the molecular biology of IMPDH[9] has provided an added impetus to the discovery of novel inhibitors of IMPDH with several potential applications in medicine.

PROPERTIES AND REGULATION OF IMPDH

Eleven cDNA clones for IMPDH from both eukaryotic and prokaryotic species have been sequenced (Fig.1). Although there is only approximately 26% amino acid sequence identity between human IMPDH and that of the protozoal parasite, *Tritrichomonas foetus*, the amino acids flanking Cys-331 of the human enzyme are extensively conserved throughout the species and Cys-331 has been located at the binding site for IMPDH[10]. The two isoforms of human IMPDH have about 83% amino acid sequence identity[9]. The Type II enzyme is upregulated, at both mRNA and protein levels, in malignant cells[11]. Dayton et al[12]., reported that both isoforms are upregulated in normal human T cells stimulated to divide by phorbol myristate acetate or ionomycin. Glesne et al.,[13] found that the level of IMPDH gene expression is inversely related the cellular concentration guanine nucleotides by a post-transcriptional event in the nucleus. Since both isoforms of IMPDH have closely similar kinetic properties[14], the physiological rationale for the expression of both enzymes in the same cells is not obvious.

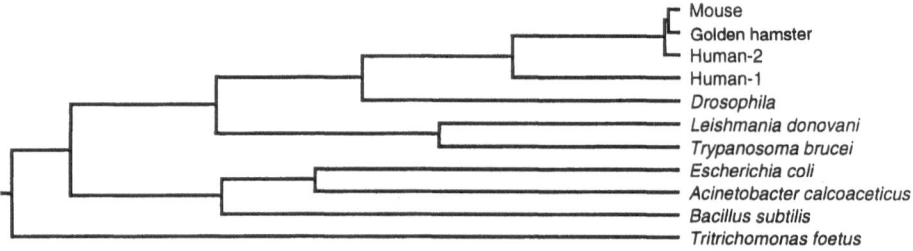

Fig. 1. Phylogenetic tree of IMP dehydrogenases using Clustal method with PAM250 residue weight table

Isolation of IMPDH

Natural forms of IMPDH have been isolated from mammalian tissues[2,15], bacteria[16] and protozoal parasites[17,18]. The mammalian preparations contain a mixture of both isoforms. Publication of the cDNA sequences for the human Type I and Type II isoforms[9] facilitated the production of the recombinant enzymes. We cloned full length cDNAs for Types I and II from a Clontech human placental library into the vector pBluescript II KS+ and transferred them into the Trp expression vector pICI117 for expression in a *guaB* ⁻ mutant of *E. coli* (HT12) which is defective for IMPDH. Types I and II IMPDH were isolated from cell lysates by IMP-Sepharose affinity chromatography[19]. The expression of the Type II enzyme was maximal during stationary phase and reached up to 30% of the total soluble protein of the cells. Whilst this work was in progress Carr et al.,[14] described the production of both human isoforms also in *guaB* ⁻ *E. coli* under the control of a hybrid trp/lac promoter and also achieved an expression level of 30% of total soluble protein.

Stability of IMPDH

Elucidation of the X-ray structure of IMPDH would be of considerable interest. Stable solutions of the enzyme at concentrations of at least 10mg/ml are required for crystallization trials. Unfortunately purified IMPDH is notoriously unstable and we have found that the recombinant Type II enzyme is no exception. At pH7.4 the enzyme rapidly precipitates from solutions of concentrations greater than ~2mg/ml. Carr et al.,[14] found that both isoforms could be retained in solution at concentrations up to 30mg/ml in 0.25M KCl. If the activity of the enzymes is maintained this type of preparation may be suitable for crystallization trials.

EFFECTS OF INHIBITION OF IMPDH ON BIOLOGICAL ACTIVITY

Cultured cells

Inhibitors of IMPDH have three major effects on proliferating cells in culture: 1. Cellular GTP is depleted by up to 80% of the normal level without loss of cellular viability[20].

2. Cell division is arrested at the G_1/S interphase or in early S phase[21]. (3) Differentiation of malignant cells is frequently observed[22,23,24]. The effects of specific inhibitors of IMPDH on cells, including inhibition of cell division and the induction of differentiation, are reversed by the addition of guanine or guanosine to the culture medium, which restores GTP levels to normal via the purine salvage pathway.

The reduction in GTP resulting from the inhibition of IMPDH also affects signal transduction events involving GTP. Franklin and Twose[25] found that the cyclic AMP response of C6 glioma cells to β-adrenergic stimulation was impaired by pretreatment of the cells with mycophenolic acid, an effect which was reversed by the addition of guanine to the cultures. Hata et al.,[26] showed that tiazofurin decreases the intracellular concentration of the ras-GTP complex by more that 50% in K562 cells. The biological significance of this on the functioning of the ras-dependent signalling cascade remains to be seen.

Another potentially important result of the inhibition of IMPDH is a reduction in the GTP-dependent expression of high-mannose oligosaccharides and the mannosylation of some membrane glycoproteins on the surface of activated human lymphocytes[27]. The glycosylation of the integrin VLA4 is also inhibited. These effects may contribute to the immunosuppressive action of mycophenolic acid

Antitumor activity

Although nucleoside analogues which are metabolized to nucleotide inhibitors of IMPDH *in vivo* inhibit tumor growth, the pleiotropic biochemical actions of such compounds makes it difficult to define the contribution of inhibition of IMPDH to the antitumor effects. The first convincing evidence that specific inhibition of this enzyme suppresses tumor growth was obtained with MPA which is effective against a wide range of rodent tumors[28,29]. Disappointingly, the compound was ineffective against human malignant disease, probably because of its rapid clearance as the inactive glucuronide[30]. MPA suppresses psoriasis, possibly because of the high β-glucuronidase activity of psoriatic plaques[31]. Tiazofurin has useful activity in patients with blast cell crisis in late-stage chronic myeloid leukemia[32] and its use in advanced ovarian carcinoma is apparently under consideration[33].

The inhibition of IMPDH therefore appears to be valid target for cancer chemotherapy, although an agent with an appropriate combination of potency, specificity and sustained pharmacokinetics remains to be discovered. The ability of inhibitors of IMPDH to induce differentiation in tumor cells, in addition to the arrest of mitosis, may prove to be a valuable feature of this type of agent.

Immunosuppression

The immunosuppressive activity of MPA was first reported many years ago[34] but not exploited clinically at the time. More recently, Allison and coworkers have developed a morpholinoethyl ester of MPA (Mofetil) which improves the bioavailability of MPA *in vivo*[35] and which is a safe and effective immunosuppressive agent in animals and man[35]. The compound arrests and even reverses acute graft rejection[35]. In a similar development, the antibiotic bredinin, which is metabolized to a competitive antagonist of IMP, also prevents the rejection of renal allografts[36].

Mofetil is also beneficial in refractory rheumatoid arthritis[37], possibly due to its immunosuppressive action.

Antiviral Activity

Ribavirin inhibits a wide range of RNA and DNA viruses *in vitro* and has clinically useful activity against respiratory syncytial virus infection[38] . The antiviral action of ribavirin in cultured cells is probably due to inhibition of host cell IMPDH although other mechanisms may also contribute[38]. Inhibition of IMPDH also underlies the antiviral action of the markedly more potent compound 5-ethynyl-1-β-D-ribofuranosylimidazole-4-carboxamide (Eicar)[39].

Adverse Effects of Inhibiting IMPDH

MPA has surprisingly low systemic toxicity in patients compared with other antimetabolites. The bone marrow is unaffected and drug-induced alopecia has not been reported[31,37,]. The most common side-effects are gastrointestinal irritation and a higher than normal incidence of infections, although these have not posed serious clinical problems[31,37]. The toxicity of tiazofurin in a Phase I trial in cancer patients[40] may be due to the action of the active metabolite on other NAD-dependent dehydrogenases at high doses of tiazofurin.

We were concerned that the depletion of cellular GTP caused by the inhibition of IMPDH might impair physiological and pharmacological responses to neuro-endocrine effectors dependent upon GTP-mediated signal tranduction. However, administration of MPA to rats at doses which markedly depress GTP in a dividing tissue, such as spleen, had no effect on GTP in a non-dividing tissue (heart), presumably because the net consumption of guanine nucleotides is lower in the latter[41]. MPA had no effect on such GTP-mediated activities as blood pressure, heart rate, gastrointestinal motility or histamine-provoked stimulation of gastric acid secretion[41] .

CONCLUSIONS

IMPDH appears to be a valid target for cancer chemotherapy although none of the currently available inhibitors is ideal for this application. A preferred agent should be highly selective for the major isoform of IMPDH in malignant cells, and possess the appropriate pharmacokinetic properties to ensure sustained inhibition. Clinical experience with Mofetil indicates that inhibition of IMPDH provides safe and effective immunsuppression with applications in organ grafting and the treatment of autoimmune disease such as rheumatoid arthritis. Clinically useful antiviral activity can be obtained with inhibitors of host cell IMPDH. Finally, the considerable sequence differences amongst IMPDH variants from different species suggests that selective inhibition of microbial forms of the enzyme might also yield novel antimicrobial drugs.

REFERENCES

1. G Weber, H. Nakamura, Y. Natsumeda, T. Szekeres and M. Nagai, Regulation of GTP synthesis. *Advan. Enzyme Regul.* 32: 57 (1992).
2. R.C. Jackson, H.P. Morris and G. Weber, Partial purification, properties and regulation of inosine 5'-monophosphate dehydrogenase in normal and malignant rat tissues, *Biochem. J.* 166: 7 (1977)
3. R.C. Jackson, G. Weber and H.P. Morris, IMP dehydrogenase, an enzyme linked with proliferation and malignancy. *Nature* 256: 331 (1975).
4. D.A. Cooney H.N. Jarayam, G. Gebeyehu, C.R. Betts, J.A. Kelley, V.E.Marquez and D.G. Johns, The conversion of 2-β-D-ribofuranosylthiazole-4-carboxamide to an anlogue of NAD with potent IMP dehydrogenase inhibitory properties. *Biochem. Pharmcol.* 31: 2133 (1982).

5. K. Gharehbaghi, K.D. Paull, J.A. Kelley, J.J. Barchi, J.E. Marquez, D.A. Cooney, A. Monks, D. Scudiero, K. Krohn and H.N. Jarayam, Cytotoxicity and characterization of an active metabolite of benzamide riboside,a novel inhibitor of IMP dehydrogenase. *Int.J.Cancer* 56:892 (1994)

6. B.M. Goldstein, J.E. Bell and V.E. Marquez, Dehydrogenase binding by tiazofurin anabolites, *J.Med. Chem.* 33:1123 (1990).

7. T.J. Franklin and J.M. Cook, The inhibition of nucleic acid synthesis by mycophenolic acid. *Biochem.J.* 113: 515 (1969).

8. H.W. Sollinger, M.H. Deierhol, F.O.Belzer, A.G. Diethelm and R.S.Kauffman, RS-61443 - a phase I clinical trial and pilot rescue study, *Transplantation* 53: 428 (1992).

9. Y. Natsumeda, S.Ohno, H. Kawasaki, Y. Konno, G. Weber and K. Suzuki, Two distinct cDNAs for human IMP dehydrogenase. *J.Biol.Chem.* 265:5292 (1990)

10. L.C. Antonino, K. Straub and J.C. Wu, Probing the active site of human IMP dehydrogenase using halogenated purine riboside 5'-monophosphate and covalent modification reagents, *Biochemistry* 33:1760 (1994)

11. Y.Konno, Y. Natsumeda, M. Nagai, Y.Yamaji, S. Ohno, K. Suzuki and G. Weber, Expression of human IMP dehydrogenase Types I and II in *Escherichia coli* and distribution in human normal and leukemic cell lines, *J.Biol.Chem.* 266:506 (1991).

12. J.S. Dayton, T. Lindsten, C.B. Thompson and B.S. Mitchell, Effects of human T lymphocyte activation on inosine monophosphate dehydrogenase expression, *J. Immunol. 152*:984 (1994).

13. D.A. Glesne, F.R. Collart and E. Huberman, Regulation of IMP dehydrogenase gene expression by its end products, guanine nucleotides, *Mol.Cell.Biol.* 11:5417 (1991).

14. S.F. Carr, E. Papp, J.C.Wu and Y. Natsumeda, Characterization of human Type I and Type II IMP dehydrogenase, *J.Biol.Chem.* 268:27286 (1993).

15. S.D. Hodges, E. Fung, B.S. Renaux and F.F. Snyder, Increased activity, amount and altered kinetic properties of IMP dehydrogenase from mycophenolic acid-resistant neuroblastoma cells, *J.Biol.Chem.* 264: 18137 (1989).

16. H.J Gilbert, C.R. Lowe and W.T. Drabble, Inosine 5'-monophosphate dehydrogenase in *Escherichia coli*, *Biochem.J.* 183:481 (1979)

17. D.J. Hupe, B. Azzolina and N.D. Behrens, IMP dehydrogenase from the intracellular parasite *Eimeria tenella* and its inhibition by mycophenolic acid, *J.Biol.Chem.* 261:8363 (1986)

18. L. Hedstrom and C.C. Wang, Mycophenolic acid and thiazole adenine dinucleotide inhibition of *Tritrichomonas foetus* inosine 5'-monophosphate dehydrogenase: implication on enzyme mechanism, *Biochemistry*, 29:849 (1990).

19. T. Ikegami, Y. Natsumeda and G. Weber, Purification of IMP dehydrogenase from rat hepatoma 3924A, *Life Sciences* 40:2277 (1987).

20. M.B. Cohen, J. Maybaum and W. Sadee, Guanine nucleotide depletion and toxicity in mouse T lymphoma (S-49) cells, *J.Biol.Chem.* 256:8713 (1981).

21. T.J. Franklin and V.N. Jacobs, In preparation, (1994).

22. O. Itoh, S. Kuroiwa, S.Atsumi, K. Umezawa, T.Takeuchi and M. Hori, Induction by guanosine analogue oxanosine of reversion toward the normal phenotype of K-*ras*- transformed rat kidney cells, *Cancer Res.* 49:996 (1989).

23. F.R. Collart and E. Huberman, Expression of IMP dehydrogenase in differentiating HL-60 cells, *Blood* 75:570 (1990).

24. K. Kiguchi, F.R.Collart, C. Henning-Chubb and E. Huberman, Induction of cell differentiation in melanoma cells by inhibitors of IMP dehydrogenase: altered patterns of IMP dehydrogenase expression and activity, *Cell Growth and Differentiation* 1:259 (1990).

25. T.J. Franklin and P.A. Twose, Reduction in β-adrenergic response of culture glioma cells following depletion of intracellular GTP, *Eur.J.Biochem.* 77:113 (1977).

26. Y. Hata, Y. Natsumeda and G. Weber, Tiazofurin decreases *Ras*-GTP complex in K562 cells, *Oncol.Res.* 5:161 (1993).

27. A.C. Allison, W.J. Kowalski, C.J. Muller, R.V. Waters and E.M. Egui, Mycophenolic acid and brequinar, inhibitors of purine and pyrimidine synthesis, block the glycosylation of adhesion molecules, *Transplant.Proc.* 25, Suppl.2:67 (1993).

28. S.B. Carter, T.J. Franklin, D.F. Jones, B.J. Leonard, S.D. Mills, R.W. Turner and W.B. Turner, Mycophenolic acid; an anti-cancer compound with unusual properties, *Nature* 223:848 (1969).

29. M.J. Sweeney, K. Gerzon, P.N. Harris, R.E. Holmes, G.A. Poore and R.H. Williams, Experimental antitumor activity and preclinical toxicology of mycophenolic acid, *Cancer Res.* 32:1795 (1972).

30. D.S. Platt, personal communication.

31. W.W. Epinette, M.D. Cohen, C.M. Cohen, E.L. Jones and M.C. Greist, Mycophenolic acid for psoriasis, *J.Am. Acad. Dermatol.*17:962 (1987).

32. G.J. Tricot, H.N. Jarayam, E. Lapis, Y. Natsumeda, C.R. Nichols, P. Kneebone, N. Heerema, G. Weber and R. Hoffman, Biochemically directed therapy of leukemia with tiazofurin, a selective blocker of inosine 5'-monophosphate dehydrogenase activity, *Cancer Res.* 49: 3696 (1989).

33. Anon. *Scrip* 1915: 11 (1994).

34. S.B. Carter, Pharmaceutical compositions containing mycophenolic acid or a salt or ester thereof.Anti-tumour and immunosuppressive activity, *UK Patent #* 1,157,100 (1967).

35. A.C. Allison and E.M. Egui, Mycophenolate mofetil, a rationally designed immunosuppressive drug, *Clinical Transplantation* 7:96 (1993).

36. T. Osakabe, H. Uchida, Y. Masaki, K. Sato, Y.Nakayama, M. Ohkubo, K. Kumano, T. Endo, K. Watanabe and K. Aso, Studies on immunosuppression with low-dose cyclosporine combined with mizoribine in experimental and clinical cadaveric renal allotransplantation, *Transplant. Proc.* 21:1598 (1989).

37. R. Goldblum, Therapy of rheumatoid arthritis with mycophenolate mofetil, *Clin.Exp. Rheumatol.* 11 (Suppl .8): S117 (1993).

38. H-J. Liao and V. Stollar, Reversal of the antiviral activity of ribavirin against Sindbis virus in *Ae. albopictus* mosquito cells, *Antiviral Res.* 22:285 (1993)

39. J. Balzarini, A. Karlsson, L. Wang, C. Bohman, K. Horska, I. Votruba, A. Fridland, A. Van Aerschot, P. Herdewijn and E. De Clercq, Eicar (5-ethynyl -1- β - D-ribofuranosylimidazole - 4 - carboxamide). A novel potent inhibitor of inosinate dehydrogenase activity and guanylate biosynthesis, *J. Biol.Chem.* 268:24591 (1993).

40. D.L. Trump, K.D. Tutsch, J.M. Koeller and D.C. Tormey, Phase I clinical study with pharmacokinetic analysis of 2 - β - D- ribofuranosylthiazole - 4 - carboxamide 9NSC 286193) administered as a five-day infusion, *Cancer Res.* 45:2853 (1985)

41. T.J. Franklin and W.P. Morris, Pharmacodynamics of the inhibition of GTP synthesis *in vivo* by mycophenolic acid, In press: *Advan.Enzyme Regul.* 34 (1994)

PHARMACOKINETICS/PHARMACODYNAMICS OF CI-1000, A PURINE NUCLEOSIDE PHOSPHORYLASE (PNP) INHIBITOR, IN RATS AND MONKEYS

Hussein Hallak, Amy Hayes, Mi Dong, Richard Gilbertsen, and
Robert Guttendorf

Parke-Davis Pharmaceutical Research
Warner-Lambert Company
Ann Arbor, MI 48105

INTRODUCTION

Purine nucleoside phosphorylase (PNP) is a purine salvage enzyme that catalyzes the reversible phosphorylation of guanine and hypoxanthine-based nucleosides to their respective purine bases. Patients with homozygous deficiency in PNP have markedly impaired T cell function with normal to elevated B cell function. Because of the sparing of B cell function in PNP deficiency, inhibitors of PNP are purported to have the potential to be T cell selective immunosuppressive agents with application to a wide variety of clinical settings.[1]

CI-1000 is a potent PNP inhibitor currently being developed for the treatment of autoimmune diseases including psoriasis and rheumatoid arthritis.[2] The work described here gives an overview of the pharmacokinetics and bioavailability of CI-1000 in rat and monkey. It also describes pharmacokinetic-pharmacodynamic studies in rat and monkey, including computer modeling which helped provide a better understanding of this new class of therapeutic agents.

METHODS

Study Design

Rat Pharmacokinetics-Bioavailability Study. The absolute bioavailability of CI-1000 was determined in fasted male Wistar rats. Rats (6/dose) were given single 15 mg/kg oral or intravenous doses of CI-1000. Serial plasma samples were collected up to 24 hours postdose. CI-1000 plasma concentrations were measured using a validated HPLC method.

Monkey Pharmacokinetics-Bioavailability Study. A single dose study was performed in fasted male and female cynomolgus monkeys. Two monkeys per sex received single 5 mg/kg intravenous and oral doses of CI-1000 in a 2-way crossover study design with 4 weeks between treatments. Serial plasma samples were collected and assayed for CI-1000 concentration.

Rat Pharmacokinetic-Pharmacodynamic Study. In this study rats (n=8/time point) were given an oral gavage dose of CI-1000 (5 mg/kg). Plasma samples were collected by serial sacrifice at 1, 3, 6, and 24 hours.

Monkey Toxicokinetic-Pharmacodynamic Study. In this monkey toxicokinetic study CI-1000 was administered orally (4/sex/group) at doses of 0, 10, 20, 40 mg/kg. On Day 15, blood samples were obtained from each monkey predose and at, 2 and 4 hours postdose. The relationship between plasma CI-1000 concentration and pharmacologic response was explored using plasma inosine levels as an indicator of PNP inhibition. The pharmacokinetic-pharmacodynamic model used for data analysis could not fit the 10 and 20 mg/kg dose levels because the majority of animals had no detectable CI-1000 concentration at predose on Day 15. However the model was able to fit data for 6 of 8 animals at the 40 mg/kg dose level.

Plasma CI-1000 Analysis

CI-1000 was extracted from plasma using aminopropyl (NH_2) solid phase extraction columns. The samples were injected on a C18 column (4.6x250 mm) and were detected with a UV detector. This HPLC assay was validated according to internal SOP requirements.

Plasma Inosine Analysis

The methodology for assessing the effects of PNP inhibitors on plasma nucleosides has been described previously. [3-5] Plasma inosine concentrations were analyzed by HPLC.[4]

Data Analysis

Noncompartmental pharmacokinetic analysis was conducted using the validated computer program PSP (ver. 2). Compartmental pharmacokinetic analysis was conducted using Topfit (ver. 2). The relationship between plasma CI-1000 concentration and pharmacologic response was explored using plasma inosine levels as an indicator of PNP inhibition. Topfit was used to fit pharmacokinetic-pharmacodynamic parameters. and for modeling concentration vs. time, effect vs. time and the effect vs. concentration profiles for CI-1000.

RESULTS

Pharmacokinetic Bioavailability Studies

CI-1000 plasma concentrations declined in a biexponential manner after IV administration in both rats (Figure 1A) and Monkeys (Figure 1B). CI-1000 was rapidly absorbed following oral administration, with an absolute oral bioavailability of 76% in rat and 48% in monkey (Table 1). In monkey, no gender difference was observed in the pharmacokinetics after intravenous or oral dosing.

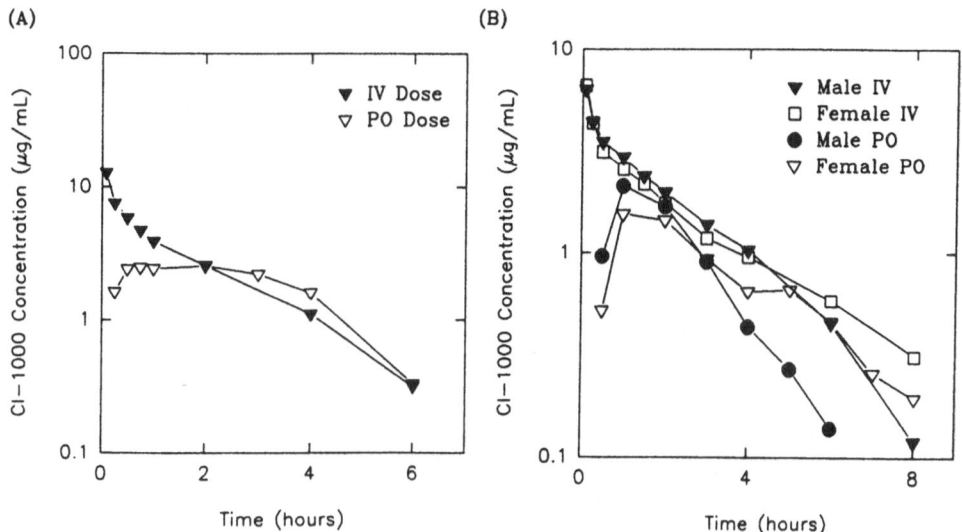

Figure 1. Concentration-time profiles for CI-1000 following IV or PO administration in (a) rat and (b) monkey.

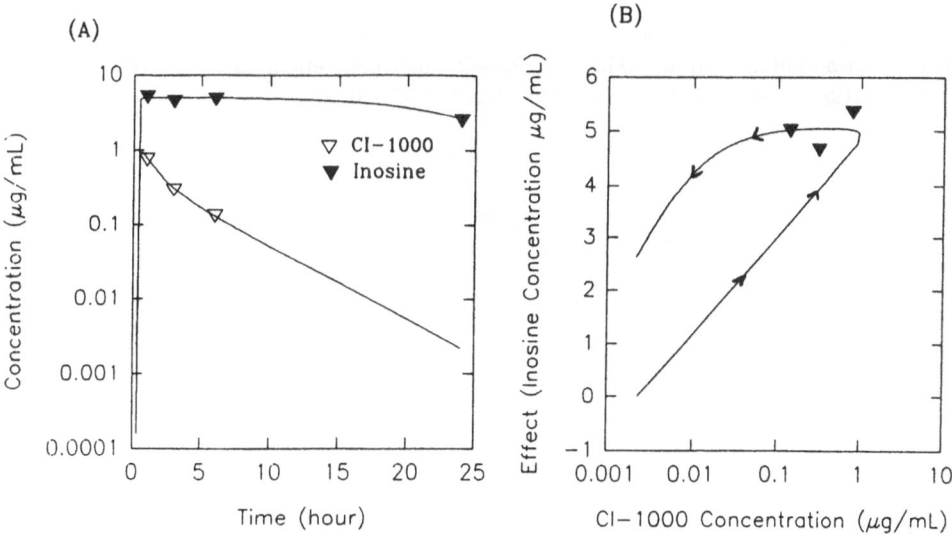

Figure 2. (A) Effect and CI-1000 concentration vs. time. (B) Effect vs. CI-1000 following a single 5 mg/kg dose in rats (symbols are experimental data points, —— best fit lines from model).

Table 1. Mean (± SD) CI-1000 pharmacokinetic parameters for the PO and IV dose as determined by noncompartmental analysis.

Dose Route	Cmax (μg/mL)	Tmax (hour)	T1/2 (hour)	$AUC_{(0-\infty)}$ (μg.hr/mL)	F
		Rat Noncompartmental Analysis			
IV	----	----	1.52 (± 0.409)	14.6 (± 3.35)	
Oral	2.90 (± 0.506)	1.67 (± 1.32)	1.30 (± 0.412)	11.1 (± 1.37)	76%
		Monkey Noncompartmental Analysis			
IV	----	----	1.88 (± 0.101)	11.4 (± 3.11)	
Oral	1.87 (± 0.691)	1.13 (± 1.629)	1.23 (± 0.386)	5.57 (± 2.12)	48%

Rat Pharmacokinetic-Pharmacodynamic Study

A two compartment oral absorption model with an added effect compartment was used to fit concentration-time and effect-time profiles for CI-1000 in rats (Figure 2A). Various pharmacodynamic models were investigated, with a basic Emax model (E=Emax*C / Kd+C) returning an excellent fit. Using the results of the pharmacokinetic-pharmacodynamic study, a computer simulation was conducted to better characterize the effect-concentration profile. The relationship between effect and CI-1000 plasma concentration shows a counter-clockwise hysteresis (Figure 2B).

Monkey Toxicokinetic-Pharmacodynamic Study

A one compartment oral absorption model with an added effect compartment provided the best fit for concentration-time and effect-time profiles for CI-1000 in monkeys (Figure 3A). The observed effect was best described by a basic Emax model. A computer simulation was conducted to better characterize the effect-concentration profile. The relationship between effect and CI-1000 plasma concentration in monkeys shows a counter-clockwise hysteresis (Figure 3B).

DISCUSSION

CI-1000 has a 1-2 hr half life in rats and monkey. The oral bioavailability is 76% and 48% in rat and monkeys respectively.

The basic Emax model returned an excellent fit for effect data both in rat and monkey. The effect-concentration relationship revealed a counter-clockwise hysteresis which is usually observed when the site of action is associated with a peripheral compartment or when an active metabolite exerts an agonistic activity to the drug's effect. In rat the time of predicted maximum effect coincided with the plasma CI-1000 tmax and the effect persisted beyond the time of last measurable CI-1000 concentration. However, in monkey the time of predicted maximum effect was slower than the tmax for CI-1000 and the effect declined simultaneously with CI-1000 concentration.

In summary, CI-1000 pharmacokinetics and bioavailability were determined in rat and monkey and pharmacodynamic analysis reveals that the plasma CI-1000 concentrations are predictive of PNP inhibition.

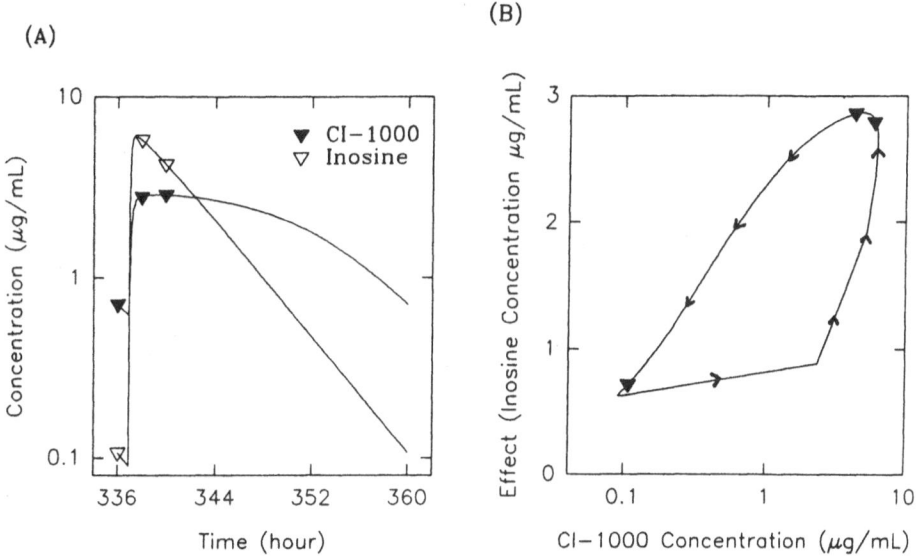

Figure 3. (A) Effect and CI-1000 concentration vs. time. (B) Effect vs. CI-1000 in a representative monkey (#809 Female) on day 15 of a 4 week toxicokinetic study. Dose = 40 mg/kg/day, (symbols are experimental data points, —— best fit lines from model).

REFERENCES

1. R.B. Gilbertsen and J.C. Sircar, Enzyme cascades: Purine metabolism and immunosuppression, *in*: "Comprehensive Medicinal Chemistry," C. Hansch, P.G. Sammes, and J.B. Taylor, eds., Pergamon Press, New York (1990).

2. R.B. Gilbertsen, U. Josyula, J.C. Sircar, M.K. Dong, W. Wu, D.J. Wilburn, and M.C. Conroy, Comparative in vitro and in vivo activities of two 9-deazaguanine analog inhibitors of purine nucleoside phosphorylase, CI-972 and PD 141955, *Biochem. Pharm.* 44:996 (1992).

3. R.B. Gilbertsen and M.K. Dong, Effects of 8-aminoguanosine, an inhibitor of purine nucleoside phosphorylase, on plasma nucleosides in Wistar rats, *Annals New York Acad. Sci.* 451:313 (1985).

4. R.B. Gilbertsen, M.K. Dong, and L.M. Kossarek, Aspects of the purine nucleoside phosphorylase (PNP) deficient state produced in normal rats following oral administration of 8-amino-9-(2-thienylmethyl)guanine (PD 119,229), a novel inhibitor of purine nucleoside phosphorylase, *Agents and Actions* 22:379 (1987).

5. M.K. Dong, M.E. Scott, D.J Schrier, M.J. Suto, J.C. Sircar, A. Black, T. Chang and R.B. Gilbertsen, The biochemistry and pharmacology of PD 116124 (8-amino-2'-nordeoxyguanosine), an inhibitor of purine nucleoside phosphorylase (PNP), *J. Pharmacol. Exp. Ther.* 260: 319 (1992).

BLOCKADE OF NUCLEOSIDE DEGRADATION IN MONKEY WHOLE BLOOD IN VITRO BY CI-1000, A PURINE NUCLEOSIDE PHOSPHORYLASE (PNP) INHIBITOR

Richard B. Gilbertsen and Mi K. Dong

Immunopathology, Experimental Therapeutics Dept.
Parke-Davis Pharmaceutical Research, Warner-Lambert Co.
Ann Arbor, MI 48105

INTRODUCTION

Humans with congenital homozygous deficiency of the purine degradative enzyme purine nucleoside phosphorylase (PNP, E.C. 2.4.2.1) exhibit profound impairment in the T cell, but not B cell, arm of the immune system.[1] Because of the importance of PNP in T cell development and/or function, it has been reasoned that PNP inhibitors might provide novel therapy for human disorders in which T cell function is central to the disease pathogenesis. Such an agent would, in theory, be B cell-sparing and relatively selective for replicating T cells, and hence non-inhibitory for quiescent (both memory and naive) T cells. After several years of pursuing this objective, we have identified a novel and highly potent inhibitor of PNP, CI-1000 (PD 141955-2; 2-amino-3,5-dihydro-7-(3-thienylmethyl)-4H-pyrrolo[3,2-d]-pyrimidin-4-one HCl; 9-deaza-9-(3-thienylmethyl)guanine; BCX-5*).[2] The Ki of CI-1000 for human erythrocyte PNP is 67 nM,[2] while the IC$_{50}$ using calf spleen PNP is reported to be 80 nM.[3] When tested concomitantly with 2'-deoxyguanosine (GdR), CI-1000 potently inhibits ^{3}H-TdR uptake by MOLT-4 and CEM T cell lymphoblasts, but not B cell lymphoblasts,[2,4] produces concentration-dependent inhibition of the human mixed lymphocyte reaction (MLR),[5] and dose-dependently elevates dGTP both in T lymphoblasts and in the MLR.[2,4,5] CI-1000 also exerts potent growth inhibition of the human cutaneous T cell lymphoma cell line MyLa via a mechanism that does not appear to involve accumulation of dGTP.[6] In rats following oral administration, CI-1000 causes dose-related elevation of plasma nucleosides over the range of 0.5 to 150 mg/kg.[2] Plasma inosine levels in rats following a single oral dose of CI-1000 approximate the lower levels found in PNP-deficient children, and persist despite declining CI-1000 concentrations in plasma.[1,2] When added to human or rat whole blood that had been spiked with GdR or inosine (HxR), CI-1000 markedly reduced the rate of purine nucleoside degradation compared with

---

*Designation of BioCryst Pharmaceuticals, Inc.

untreated whole blood.[7] The studies described here were undertaken to determine the extent of activity of CI-1000 in a comparable system using whole blood from cynomolgus monkeys. Studies in monkeys were prompted by the need to construct an integrated hypothesis incorporating the pharmacokinetic activity of CI-1000 with pharmacodynamic effect (nucleoside elevation) and toxicologic and immunotoxicologic findings in monkeys, so as to provide a model to assist in the evaluation of CI-1000 in human trials.

MATERIALS AND METHODS

Blood Collection and Processing

Blood was collected in EDTA-treated vacutainer tubes from normal cynomolgus monkeys. Specimens of whole blood were spiked in 5-mL glass tubes at ambient temperature (20-24° C) with 50 µL HxR (Aldrich Chemical Co., Milwaukee, WI) or GdR (Sigma Chemical Co., St. Louis, MO), each at 2.5 µg/mL whole blood, mixed, and incubated for varying periods at ambient temperature. Ambient temperature was selected over 37°C because nucleoside breakdown is slower and more quantifiable at ambient temperature, and to enable comparison with other PNP inhibitors and results using blood from other species. Spiking with exogenous nucleosides was necessary because endogenous PNP substrates are below the limits of detection of the HPLC system (0.02 µM). CI-1000 (1.0 µg/mL) was added simultaneously with HxR or GdR and nucleoside degradation was followed as a function of time. Care was taken to insure that samples were processed rapidly at 0-4°C after completion of the ambient temperature incubation period, which was defined as being stopped when the tubes were placed on ice. Base line data, referred to as time-zero, were obtained from blood to which nucleosides had been added, the tubes briefly mixed, and then immediately placed on ice. Tubes were promptly transferred to a refrigerated centrifuge and centrifuged at 1200 x g for 15 min at 4°C. Clarified monkey plasma was removed and again centrifuged at 1200 x g for 25 min at 4°C through a Centriflo® (Amicon Corp., Lexington, MA) ultrafiltration membrane cone to remove plasma macromolecules.

Nucleoside Analysis and Calculations

The ultrafiltrate was analyzed by HPLC as described previously.[8,9] In brief, the ultrafiltrate was chromatographed on a Microsorb C18 column with a mobile phase of 4 mM potassium phosphate buffer containing 0.75% methanol, pH 5.75, using a detection wavelength of 254 nm. Nucleoside half-lives were calculated using a validated pharmacokinetic statistical package (PSP #27) for determining half-lives of test substances in the Department of Pharmacokinetics and Drug Metabolism, Parke-Davis Pharmaceutical Research, Warner-Lambert Co.

RESULTS AND DISCUSSION

GdR was degraded so quickly in untreated cynomolgus monkey whole blood at ambient temperature that 48% of the exogenous GdR had been degraded by the time-zero time point. The half-life of GdR was calculated to be 1.2 min, but this was extended to approximately 17.8 min upon coaddition of 1.0 µg/mL of CI-1000 (Figure 1, Table 1). The half-life of HxR in untreated monkey whole blood was too short to be calculated (91% of the spiked HxR was degraded at the time-zero time point). CI-1000 retarded HxR breakdown, extending the half-life to >39.8 min. A structural analog of CI-1000, CI-972

[PD 126547; 2,6-diamino-3,5-dihydro-7-(3-thienylmethyl)-4*H*-pyrrolo[3,2-*d*]pyrimidin-4-one],[10,11] was tested in parallel at 1 μg/mL and had substantially less effect on the breakdown of GdR and HxR than did CI-1000 (Figure 1). As was observed here using monkey blood, CI-972 has been reported to be minimally effective in retarding nucleoside breakdown in human whole blood, although it is quite capable of doing so in rat blood.[7] In contrast, CI-1000 is highly effective at blocking degradation of nucleosides in rat blood as well as in human blood (Table 1).[7]

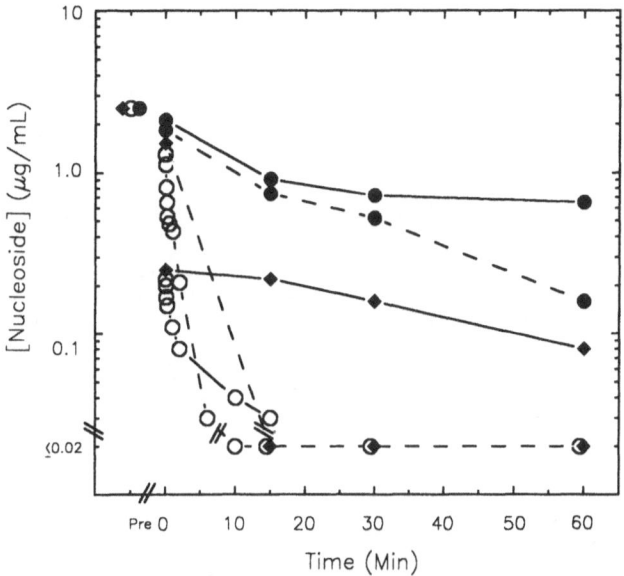

Figure 1. CI-1000 blockade of PNP substrate degradation in cynomolgus monkey whole blood in vitro. EDTA-treated blood was spiked with HxR (solid lines) or GdR (dashed lines) and treated with vehicle (○), CI-1000 (1 μg/mL, ●), or CI-972 (1 μg/mL, (♦), and breakdown of nucleosides followed as a function of time.

SUMMARY

Purine nucleosides HxR or GdR (2.5 μg/mL blood) were added to EDTA-treated cynomolgus monkey whole blood in vitro, alone or with the PNP inhibitor CI-1000 (1 μg/mL), mixed, and the concentration of nucleosides remaining in plasma followed as a function of time. The half-lives of GdR and HxR in control blood were 1.2 and < 1 min, respectively, and were extended to 17.8 and 39.8 min, respectively, by coaddition of CI-1000. In contrast, a structural analog of CI-1000, CI-972, when tested in parallel at 1 μg/mL, had markedly less effect on the breakdown of either nucleoside. The ability of CI-1000 to retard nucleoside breakdown in blood in vitro may be a predictor of in vivo activity, and can be viewed as an early and essential biochemical consequence of PNP inhibition culminating in immunosuppression.

Table 1. Nucleoside half-lives in spiked whole blood treated with CI-1000 in vitro.

Species	Nucleoside[a]	Control	CI-1000 Concentration (µg/mL)		
			0.3	1.0	3.0
Monkey	GdR	1.2 min[b]	---[c]	17.8 min (14.8)[d]	---
	HxR	<1 min[e]	---	39.8 min (>39.8)	---
Rat[f]	GdR	17.7 min	>60 min (>3.4)	>60 min (>3.4)	>60 min (>3.4)
Human[f]	GdR	12.0 ± 1.4 sec[g]	---	>60 min[h] (>300)	---
	HxR	15.4 ± 2.4 sec[g]	---	57.9 min[h] (226)	---

[a] Nucleosides measured were 2'-deoxyguanosine (GdR) and inosine (HxR).
[b] 48% of the GdR was metabolized at the time-zero time point.
[c] ---, not determined.
[d] Values in parentheses are the fold increase in nucleoside half-life compared with control.
[e] The half-life of HxR in monkey blood could not be calculated due to its extremely rapid rate of disappearance (91% of the HxR had been metabolized at the time-zero time point).
[f] Data from reference 7.
[g] Mean ± SEM from three studies.
[h] Mean from two studies.

REFERENCES

1. N.M. Kredich and M.S. Hershfield, Immunodeficiency diseases caused by adenosine deaminase deficiency and purine nucleoside phosphorylase deficiency, *in*: "The Metabolic Basis of Inherited Disease," J.A. Stanbury, ed., 6th Edition, pp. 1045-1075, McGraw-Hill Information Services Company, New York, (1989).

2. R.B. Gilbertsen, U. Josyula, J.C. Sircar, M.K. Dong, W. Wu, D.J. Wilburn, and M.C. Conroy, Comparative in vitro and in vivo activities of two 9-deazaguanine analog inhibitors of purine nucleoside phosphorylase, CI-972 and PD 141955, *Biochem. Pharm.* 44:996 (1992).

3. S.E. Ealick, Y.S. Babu, C.E. Bugg, M.D. Erion, W.C. Guida, J.A. Montgomery, and J.A. Secrist, III, Application of crystallographic and modeling methods in the design of purine nucleoside phosphorylase inhibitors, *Proc. Natl. Acad. Sci. USA* 88:11540 (1991).

4. R.B. Gilbertsen, M.K. Dong, U. Josyula, J.C. Sircar, D.J. Wilburn, and M.C. Conroy, Activities of two 9-deazaguanine analog inhibitors of purine nucleoside phosphorylase, CI-972 and PD 141955, in vitro and in vivo, *Annals New York Acad. Sci.* 685:248 (1993).

5. D.J. Wilburn, M.K. Dong, and R.B. Gilbertsen, PD 141955 and CI-972: 9-deazaguanine analog inhibitors of purine nucleoside phosphorylase (PNP). I. Suppression of the human mixed lymphocyte reaction (MLR), *Agents and Actions* 39:C96 (1993).

6. W-H Boehncke, R.B. Gilbertsen, J. Hemmer, and W. Sterry, Evidence for a pathway independent from 2'-deoxyguanosine and reversible by IL-2 by which purine nucleoside phosphorylase inhibitors block T cell proliferation, *Scand. J. Immunol.* 39:327 (1994).

7. M.K. Dong and R.B. Gilbertsen, PD 141955 and CI-972: 9-deazaguanine analog inhibitors of purine nucleoside phosphorylase (PNP). II. Effects on nucleoside catabolism in human and rat blood in vitro, *Agents and Actions* 39:C99 (1993).

8. R.B. Gilbertsen and M.K. Dong, Effects of 8-aminoguanosine, an inhibitor of purine nucleoside phosphorylase, on plasma nucleosides in Wistar rats, *Annals New York Acad. Sci.* 451:313 (1985).

9. R.B. Gilbertsen, M.K. Dong, and L.M. Kossarek, Aspects of the purine nucleoside phosphorylase (PNP) deficient state produced in normal rats following oral administration of 8-amino-9-(2-thienylmethyl)guanine (PD 119,229), a novel inhibitor of purine nucleoside phosphorylase, *Agents and Actions* 22:379 (1987).

10. R.B. Gilbertsen, M.K. Dong, L.M. Kossarek, J.C. Sircar, C.R. Kostlan, and M.C. Conroy, Selective in vitro inhibition of human MOLT-4 T lymphoblasts by the novel purine nucleoside phosphorylase inhibitor, CI-972, *Biochem. Biophys. Res. Comm.* 178:1351 (1991).

11. J.C. Sircar, C.R. Kostlan, R.B. Gilbertsen, M.K. Bennett, M.K. Dong, and W.J. Cetenko, Inhibitors of human purine nucleoside phosphorylase. Synthesis of pyrrolo[3,2-*d*]pyrimidines, a new class of purine nucleoside phosphorylase inhibitors as potentially T-cell selective immunosuppressive agents. Description of 2,6-diamino-3,5-dihydro-7-(3-thienylmethyl)-4*H*-pyrrolo[3,2-*d*]pyrimidin-4-one, *J. Med. Chem.* 35:1605 (1992).

CHANGES IN MONKEY PLASMA PURINES INDUCED BY REPEATED DOSES OF CI-1000, A NOVEL INHIBITOR OF PURINE NUCLEOSIDE PHOSPHORYLASE

Donald G. Robertson, Ellen. R. Urda, Michael R. Bleavins, and Narendra D. Lalwani

Deptartment of Pathology and Experimental Toxicology, Parke-Davis Pharmaceutical Research Division, Warner-Lambert Co., Ann Arbor, MI

INTRODUCTION

CI-1000, a 9-deazaguanine analog, is an orally active and reversible inhibitor of purine nucleoside phosphorylase (PNP). *In vitro*, CI-1000 inhibits human T-lymphoblast replication and the mixed lymphocyte response, as well as facilitating the intracellular accumulation of deoxyguanosine triphosphate (Gilbertsen *et al.*, 1992; Dong and Gilbertsen, 1993; Wilburn *et al.* 1993). These changes are consistent with inhibition of PNP (Osborne and Scott, 1983; Fairbanks *et al.*, 1990; Gilbertsen and Sircar, 1990) and the drug is approximately 10-fold more potent than a structurally related compound (Gilbertsen *et al.*, 1991a, 1991b, and 1992).

Purine nucleoside phosphorylase deficient patients exhibit elevated plasma concentrations of inosine and guanosine (Stoop *et al.*, 1977; Staal *et al.*, 1980; Rijksen *et al.*, 1987) related to altered metabolism in the absence of significant PNP activity. Elevated purine concentrations have been demonstrated in rats (Gilbertsen and Dong, 1985) and dogs (Frederick *et al.*, 1985; Osborne *et al*, 1986) following chemotherapeutic inhibition of PNP, but these changes have not been examined in non-human primates or humans. In this study we report the that oral CI-1000 administration produces plasma purine elevations in the cynomolgus monkey, consistent with inhibition of PNP.

MATERIALS AND METHODS

CI-1000 (4H-pyrrolo:3,2-d:pyrimidin-4-one,2-amino-3, 5-dihydro-7-(3-thienylmethyl), monohydrochloride, monohydrate) was synthesized by the Department of Chemistry, at Parke-Davis Pharmaceutical Research Division. The nuclear magnetic resonance, infrared, and mass spectra were consistent with the structure of the compound and the purity of the

Purine and Pyrimidine Metabolism in Man VIII, Edited by
A. Sahota and M. Taylor, Plenum Press, New York, 1995

compound used was >99% as determined by HPLC and elemental analysis.

Thirty-two adult feral cynomolgus monkeys (*Macaca fasicularis*) were administered single daily doses of CI-1000 suspended in 0.5% aqueous methylcellulose at doses of 0, 10, 20 and 40 mg/kg (4 monkeys/sex/group). On Day 15, blood samples were obtained prior to dose and 2 and 4 hours post-dose. Samples were divided and prepared for analysis of plasma drug levels and plasma purines. The method for determination of plasma drug level was as described elsewhere in this volume (Hallak et. al., 1994). Blood collected for purine determinations was immediately placed in an ice water bath and plasma prepared in a refrigerated centrifuge as soon as feasible (< 30 min). Care was taken to avoid hemolysis of the blood samples. Plasma ultrafiltrates were prepared by centrifugation (4° C) using Ultrafree-MC (Millipore) ultrafilters. Ultrafiltrates were immediately placed in a refrigerated Waters UltraWisp 712 autosampler. Chromatographic separation of inosine, guanosine and hypoxanthine was achieved using a gradient elution technique (Assenza and Brown, 1980) employing a Waters 600E Multisolvent delivery system and a Waters μBondapak C-18 column. Detection and quantitation of eluted peaks was performed with a Waters 991 Photodiode array detector monitoring absorbance 254 nm with subsequent integration of peak area. Identification of purine peaks separated by the chromatographic procedure was confirmed by co-elution with pure standards, spectral analysis (220 - 400 nm), and by enzymatic reaction (disappearance of inosine and guanosine and increase in hypoxanthine after incubation of plasma with purified PNP). Standard curves made from pure standards of each purine (Sigma) were used to establish concentration of purines in each ultrafiltrate.

RESULTS

Combined sex plasma drug level data are reported in Table 1.

Table 1. Mean (± SEM) combined sex plasma CI-1000 concentrations (μg/mL)

| Time | Dose (mg/kg) | | | |
	0	10	20	40
Predose	BLQ[1]	BLQ	BLQ	0.21 ± 0.05
2 hours	BLQ	1.33 ± 0.30	2.87 ± 0.45	5.82 ± 0.27
4 hours	BLQ	0.62 ± 0.14	1.79 ± 0.37	4.79 ± 0.19

[1] BLQ = Below level of quantitation

Quantitation of monkey plasma purines revealed pronounced inter-animal variation. However, it was evident that CI-1000 produced profound changes in measured purines. Within 2 hours of dose inosine was increased 3, 5 and 5 fold and guanosine was increased 9, 12 and 14 fold at 10, 20 and 40 mg/kg respectively (Figure 1). Corresponding decreases in hypoxanthine levels were noted at 2 hours with 7 of 8 animals having no measurable hypoxanthine in the 40 mg/kg group. At the 10 and 20 mg/kg doses, inosine and guanosine appeared to return towards predose levels correlating with plasma drug levels. However, despite a slight decrease in plasma drug levels between 2 and 4 hours, both purines remained elevated at 4 hours in the 40 mg/kg group. In addition, 0 time hypoxanthine levels (24 hours after the 14th dose) for the 40 mg/kg group were decreased relative to the control, 10 and 20 mg/kg groups suggesting a prolonged biologic effect at the high dose.

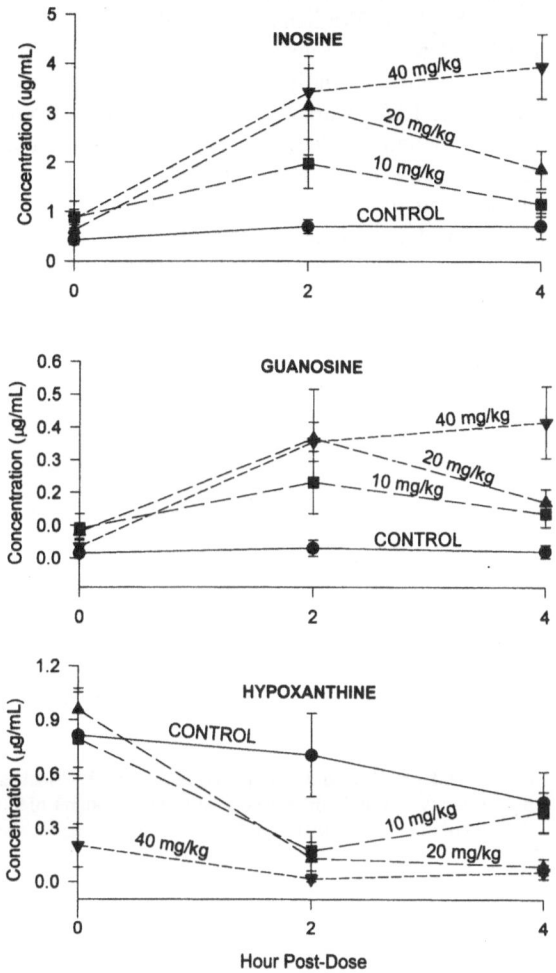

Figure 1. Changes in combined sex mean plasma purine levels produced by CI-1000. Each point represents the mean ± SEM of 8 animals (4 male/4 female).

DISCUSSION

Despite individual animal variation, it was evident that repeated doses of CI-1000 had profound effects on monkey plasma purines. Accumulation of inosine and guanosine and depletion of hypoxanthine are consistent with the proposed mechanism of action, PNP inhibition. To our knowledge this is the first report of efficacy of a PNP inhibitor in a primate model. Effects were generally dose related and the magnitude of changes at 10 mg/kg suggested that significant results may be obtained at even lower doses. The peak activity observed at 2 hours post-dose for the 10 and 20 mg/kg groups is consistent with the T_{max} of the compound. Effects at 40 mg/kg suggested longer duration of biologic effects that did not strictly correlate to plasma drug levels.

Children congenitally deficient in the enzyme purine nucleoside phosphorylase (PNP) have profoundly impaired T-lymphocyte function, yet retain normal or elevated B-lymphocyte activity (Giblett, 1975; Kredich and Hershfield, 1989). In contrast, patients with inherited deficiencies of adenosine deaminase exhibit depression of both T- and B-lymphocyte activity, with T-cell function inhibited to a somewhat greater extent (Giblett *et al.*, 1972; Pickering *et al.*, 1974; Carson *et al.* 1978). The strong selectivity toward the cell-mediated immune function, while preserving humoral immunologic components, suggests that compounds capable of inhibiting PNP activity may provide a means of abrogating abnormal or exaggerated T-lymphocyte function in disease. Potential therapeutic applications of CI-1000 include rheumatoid arthritis, psoriasis, Type I diabetes, systemic lupus erythrematosis, multiple sclerosis, and T-cell leukemias (Sircar and Gilbertsen, 1988). The results reported here indicate that CI-1000 should serve as a valuable tool for assessing the activity of PNP inhibitors as therapeutic immunomodulators.

REFERENCES

Assenza, S.P., and Brown P.R., 1980, Comparison of high-performance liquid chromatographic serum profiles of humans and dogs. *J. Chromatog.* 181:169-176.

Carson, D.A., Kaye, J., and Seegmiller J.E., 1978, Differential sensitivity of human leukemic T cell lines and B cell lines to growth inhibition by deoxyadenosine. *J. Immunol.* 121:1726-1731.

Dong. M.K., and Gilbertsen R.B., 1993, PD 141955 and CI-972: 9-Deazaguanine analog inhibitors of purine nucleoside phosphorylase (PNP). II. Effects on nucleoside catabolism in human and rat blood *in vitro*. *Agents Act.* 39:C99-C101.

Fairbanks, L.D., Taddeo, A., Duley, J.A., and Simmonds, H.A., 1990, Mechanisms of deoxyguanosine lymphotoxicity. Human thymocytes, but not peripheral blood lymphocytes accumulate deoxyGTP in conditions simulating purine nucleoside phosphorylase deficiency. *J. Immunol.* 144:485-491.

Frederick, D.L.,Epstein, R.B., and Benear, J., 1985, Pharmacokinetics of 8-aminoguanosine. *Clin. Chem.* 31:939.

Giblet, E.R., Anderson, J.E., Cohen, F., Pollara, B., and Meuwissen, H.J., 1972, Adenosine-deaminase deficiency in two patients with impaired cellular immunity. *Lancet* 2:1067-1069.

Giblett, E.R. (1975). Nucleoside-phosphorylase deficiency in a child with severely defective T-cell immunity and normal B-cell immunity. *Lancet* 1:1010-1013.

Gilbertsen , R.B., Dong, M.K., Kossarek, L.M., Sircar, J.C., Kostlan, C.R., and Conroy, M.C., 1991a, Selective *in vitro* inhibition of human MOLT-4 lymphoblasts by the novel purine nucleoside phosphorylase inhibitor, CI-972. Biochem. Biophys. Res. Comm. 178:1351-1358.

Gilbertsen , R.B., Dong, M.K., Wilburn, D.J., Kossarek, L.M., Sircar, J.C., Kostlan, C.R., and Conroy, M.C., Biochemical and pharmacologic properties of CI-972, a novel 9-deazaguanine analog purine nucleoside phosphorylase (PNP) inhibitor, *in*: "Purine and Pyrimidine Metabolism in Man VII". R.A. Harkness, G.B. Elion, and N. Zollner, editors. Plenum Press, New York, (1991b)

Gilbertsen, R.B. and Dong, M.K., 1985, Effects of 8-aminoguanosine, an inhibitor of purine nucleoside phosphorylase, on plasma nucleosides in Wistar rats. *Ann. N.Y. Acad. Sci.* 451:313-314.

Gilbertsen , R.B., U. Josyula, J.C. Sircar, M.K. Dong, W. Wu, D.J. Wilburn, and M.C. Conroy (1992). Comparative *in vitro* and *in vivo* activities of two 9-deazaguanine analog inhibitors of purine nucleoside phosphorylase, CI-972 and PD 141955. Biochem Pharmacol. 44:996-999.

Gilbertsen, R.B. and Sircar, J.C., Enzyme cascades: Purine metabolism and immunosuppression. *in*: " Comprehensive Medicinal Chemistry, Volume 2", C. Hansch, P.G. Sammes, and J.B. Taylor, editors. Pergamon Press, Oxford, (1990).

Hallak, H., Gilbertsen, R., Dong, M., and Guttendorf R., 1994, Pharmakokinetic/pharmacodynamics of CI-1000, a purine nucleoside phosphorylase (PNP) inhibitor, in rat and monkey. This Volume

Kredich, N.M. and Hershfield, M.S., Immunodeficiency diseases caused by adenosine deaminase deficiency and purine nucleoside phosphorylase deficiency. *in*: "The Metabolic Basis of Inherited Disease, 6th Edition". J.B. Stanbury, J.B. Wyngaarden, and D.S. Fredrickson, editors, McGraw-Hill, New York (1989).

Osborne, W.R.A. and Scott C.R., 1983, The metabolism of deoxyguanosine and guanosine in human B and T lymphoblasts. A role for deoxyguanosine kinase activity in the selective T cell defect associated with purine nucleoside phosphorylase deficiency. *Biochemistry* 214:711-718.

Osborne, W.R.A., Deeg, H.J., and Slichter, S.J., 1986, A canine model of induced purine nucleoside phosphorylase deficiency. *Clin. Exp. Immunol.* 66:166-172.

Pickering, R.J., Pollara, B., and Meuwissen, H.J., 1974, .Meeting Report: Workshop on severe combined immunological deficiency and adenosine deaminase deficiency. *Clin. Immunol. Immunopath.* 3:301-303.

Rijksen, G., Kuis, W., Wadman, S.K., Spaapen, L.J.M., Duran, M., Voorbrood, B.S., Staal, G.E.J., Stoop, J.W., and Zegers, B.J.M., 1987, A new case of purine nucleoside phosphorylase deficiency: Enzymologic, clinical, and immunologic characteristics. *Pediatr. Res.* 21:137-141.

Sircar, J.C. and Gilbertsen, R.B., 1988, Purine nucleoside phosphorylase (PNP) inhibitors: Potentially selective immunosuppressive agents. *Drugs of the Future* 13:653-668.

Sircar, J.C. and Gilbertsen, R.B., 1990, Selective immunosuppressive agents. *Drug News Perspect.* 3:213-217.

Staal, G.E., van Stoop, J.W., Zegers, B.J.M., Siegenbeek van Heukelom, L.H., van der Vlist, M.J.M., Wadman, S.K., and Martin, D.W., 1980, Erythrocyte metabolism in purine nucleoside phosphorylase deficiency after enzyme replacement therapy by infusion of erythrocytes. *J. Clin. Invest.* 65:103-108.

Stoop, J.W., Zegers, B.J.M., Hendricks, G.F.M., Siegenbeek van Heukelom, L.H., Staal, G.E.J., de Bree, P.K., Wadman, S.K., and Ballieux R.E., 1977, Purine nucleoside phosphorylase deficiency associated with selective cellular immunodeficiency. *N. Engl. J. Med.* 296:651-655.

Wilburn, D.J., Dong, M.K., and Gilbertsen R.B., 1993, PD 141955 and CI-972: 9-Deazaguanine analog inhibitors of purine nucleoside phosphorylase (PNP). I. Suppression of the human mixed lymphocyte reaction (MLR). *Agents Act.* 39:C96-C98.

INHIBITORS OF THYMIDYLATE SYNTHASE AND GLYCINAMIDE RIBONUCLEOTIDE TRANSFORMYLASE

Robert C. Jackson

Agouron Pharmaceuticals, Inc.
San Diego, CA 92121

INTRODUCTION

The antipyrimidine drug, 5-fluorouracil (5-FU), as its active metabolite, FdUMP, is a potent inhibitor of thymidylate synthase (TS), and this is generally considered to be its primary mechanism of action (Pratt et al., 1994). Drug combinations including 5-FU continue to be the primary treatment for metastatic colon carcinoma. This tumor is relatively refractory to DNA cross-linking agents, and because it usually expresses the multidrug resistance (MDR) phenotype is unresponsive to anthracyclines and vinca alkaloids. Although 5-FU is not subject to these limitations, it has other disadvantages: it requires intracellular activation to FdUMP, and when it inhibits TS, the resulting accumulation of deoxyuridylate (dUMP) tends to compete with the FdUMP for binding to TS. Because 5-FU has other biochemical effects, most notably incorporation into RNA, it is difficult to determine which of its therapeutic and toxic effects are a result of TS inhibition. In addition, 5-FU has rather unfavorable pharmacokinetics, with a terminal plasma half-life in humans of 10 to 20 minutes.

ANTIFOLATE THYMIDYLATE SYNTHASE INHIBITORS

The folate analogue, 10-propargyl-5,8-dideazafolate (CB3717) was designed by Jones et al.(1991) to provide an alternative means of inhibiting TS. Because it binds to TS at the folate cofactor site, it is not affected by dUMP accumulation. The early results with CB3717 have been reviewed by Jackman et al.(1985). It showed clinical activity, but serious toxicity to liver and kidney which was attributed to its limited solubility. These results led to a program at ICI Pharmaceuticals (now Zeneca) and the Institute of Cancer Research that culminated in tomudex (D-1694) (Jackman et al., 1991), a more soluble compound. Tomudex is a less potent TS inhibitor than CB3717, but a more potent inhibitor of cell growth. Unlike CB3717, which enters cells slowly, and probably primarily through the folic acid binding protein (FABP), tomudex is a good substrate for the reduced folate carrier (RFC); it is also a better substrate for folylpolyglutamate synthetase (FPGS), which results in extensive intracellular retention. Tomudex has shown clinical activity in colorectal and breast carcinomas (Cunningham et al., 1994).

Grindey et al. (1992) described LY231514, a pyrrolopyrimidine TS inhibitor that had IC_{50} against CCRF-CEM cells of 16 nM, and had experimental antitumor activity in mice at doses from 12.5 to 200 mg/kg, without renal or liver toxicity. LY231514 is currently in phase I clinical trials.

Another structurally novel TS inhibitor, BW1843U89, has been described by Duch et al. (1993). This compound is notable for its high inhibitory potency (K_i of 90 pM for TS, and IC50 values for several cell lines of < 1 nM). It is a good substrate for the RFC, and also for FPGS, which converts it to the diglutamate, but not to higher polyglutamates.

GLYCINAMIDE RIBONUCLEOTIDE TRANSFORMYLASE INHIBITORS

The first selective inhibitor of glycinamide ribonucleotide transformylase (GART) was lomotrexol (DDATHF), described by Beardsley et al. (1989). This classical antifolate had K_i for GART of 60 nM, and an IC_{50} against CCRF-CEM cells of about 8 nM. It was efficiently polygluamylated by FPGS (K_m of 9.7 micromolar), and was a potent ligand to the FABP (Kd of 0.3 nM). Lomotrexol had broad-spectrum preclinical anticancer activity, and gave responses in the early clinical trials (Moran, 1992).

A clinical problem with lomotrexol was delayed toxicity, including leukopenia, anemia, and anorexia. Grindey et al. (1991) showed that folate depletion increased the toxicity of lomotrexol to mice by 100-fold. Conversely, they showed that low doses of folic acid in the drinking water could protect the mice from toxicity without abrogating the antitumor activity. They proposed that this behavior was associated with the tight binding of lomotrexol to FABP. Based upon these results, a phase I clinical trial of lomotrexol with folic acid is being conducted.

Smith et al. (1992) reported studies with another GART inhibitor, 5-deaza-acyclotetrahydrofolate (5-DACTHF). They concluded that, although 5-DACTHF killed cells to a lesser extent, and more slowly, than a TS inhibitor, it was nevertheless cytotoxic and could trigger apoptosis in treated cells. Moran's laboratory (Smith et al., 1993), studied lomotrexol in human colon carcinoma cells, and found that cytotoxicity was slower, and lesser in extent than for tomudex. By analogy with the effects of dihydrofolate reductase inhibitors, in which the antipurine effect tends to diminish the cytotoxicity caused by the antithymidylate effect, it is possible that the initial G1/S arrest caused by GART inhibitors is not cytotoxic, and that cells only begin to die when they have been in this arrested state for several days, or, alternatively, when they escape from the G1 block and continue into S-phase.

In addition to being cytostatic and cytotoxic, lomotrexol was recently shown to stimulate cytodifferentiation in HL-60 promyelocytic leukemia cells (Sokoloski et al., 1993). By metabolite replacement experiments, this effect was specifically attributed to depletion of GTP caused by the lomotrexol.

Habeck et al. (1994) recently described two novel GART inhibitors in which the p-aminobenzoyl moiety of lomotrexol has been replaced by a 2',5'-thiophene (LY254155) and a 2',5'-furan, (LY222306) respectively. Both compounds were tight-binding GART inhibitors, even as the monoglutamates. The binding affinity of LY222306 to FABP was 350-fold weaker than that of lomotrexol, which should make it possible to test the hypothesis that the delayed toxicity of lomotrexol is associated with its binding to FABP.

Based upon the X-ray crystal structures of the *E.coli* GART (Almassy et al., 1992), and of the GART domain of the human trifunctional enzyme (Almassy et al., unpublished), a number of novel GART inhibitors have been designed. An example is AG-2007, N-[5-(3-[(2,6-diamino-4(3H)-oxopyrimidin-5-yl)thio]propyl)thieno-2-yl]-L-glutamic acid (Varney , 1993). Like 5-DACTHF, these compounds are substituted pyrimidines. Members of this series have low nanomolar inhibition of GART, are cytotoxic in the low to mid-nanomolar range, and have in vivo antitumor activity. Some of the novel compounds designed from the GART crystal structure have greatly decreased binding to FABP.

CONCLUSIONS

Extensive experience has now been gained with a wide range of anticancer TS and GART inhibitors. Factors that influence cellular pharmacokinetics include transport by RFC, and substrate activity for FPGS. Binding to FABP is believed to contribute to delayed toxicity. Nonclassical TS inhibitors, which do not utilize RFC or FPGS, may nonetheless have high-level antitumor activity. However, they are not retained tightly in the target cell, so that high doses and prolonged schedules are necessary. It is too early to tell whether these compounds will avoid delayed toxicity problems. An interesting example of a novel nonclassical antifolate is ZD9331, which is transported but not polyglutamated, and should therefore help in a dissection of the importance of these factors for activity, potency, and toxicity.

It is often stated that antimetabolites kill cell cells by starving them of macromolecule precursors, that they are indiscriminately active against all rapidly dividing cells, and that they possess no inherent anticancer selectivity. Recent experience with

1843U89 has good in vivo activity against the GC3TK⁻ colon carcinoma xenograft, and is being developed for clinical trial.

Nonclassical Thymidylate Synthase Inhibitors

The compounds described above, CB3717, tomudex, LY231514, and 1843U89, are classical antifolates, i.e. structural analogues of folic acid. In particular, they contain a glutamate moiety. As such they are transported into cells on the RFC (with the exception of CB3717), and they are subject to polyglutamylation. These properties contribute to favorable cellular pharmacokinetics of classical antifolates, and since tumor cells tend to have high levels of RFC and FPGS, they contribute to antitumor selectivity. However, the classical antifolate structure may present two potential liabilities: it confers the potential for high level accumulation in certain tissues (e.g. liver) that may result in cumulative toxicity, and deletion or mutation of either RFC or FPGS in tumor cells may result in acquired drug resistance.

For these reasons, it is of interest to explore the pharmacology and clinical spectrum of nonclassical TS inhibitors that do not contain a glutamate moiety. Based on the x-ray crystal structure of TS, several chemical series of lipophilic TS inhibitors were designed, and two compounds were advanced into clinical trials. AG-331 (Varney et al., 1992) is a benzindole that acts as a subnanomolar inhibitor of TS, and an micromolar inhibitor of cell division. Though AG-331 binds to DNA, its cytostatic effect was blocked by thymidine, suggesting that TS inhibition was its primary site of action in L1210 cells.

AG-337 is a small molecule (MW 296) 5-pyridyl-substituted quinazoline that is also highly lipophilic but (as the dihydrochloride salt) water soluble. It is an 11 nM TS inhibitor, and is growth inhibitory at about one micromolar.

Both AG-331 and AG-337 are in late phase I clinical trial. AG-331 has not yet reached a maximum tolerated dose. The dose of AG-337 has been escalated to 960 mg/m2 per day, for 5 days by continuous infusion. This dose gives myelosuppression, mucositis, and diarrhea. AG-337 was shown to cause elevation of plasma deoxyuridine, suggesting that it was effectively inhibiting TS (Calvete et al., 1994). The two compounds differed in their pharmacokinetic properties, with AG-337 showing a short terminal half-life (2 to 3 hr) while that of AG-331 was about 24 hr. It is clear that both compounds, despite being potent TS inhibitors, have low *in vivo* potency, and require prolonged administration to elicit toxicity.

A novel nonclassical TS inhibitor, ZD9331, was recently described by Jackman et al. (1994). ZD9331 is a subnanomolar TS inhibitor that is transported by the RFC, but which is not polyglutamylated. When given on prolonged schedules of administration it had high-level in vivo antitumor activity.

Thymidylate Synthase: Recent Developments

In addition to the development of five TS inhibitors to clinical trial, there have been a number of interesting preclinical developments. Evidence continues to accumulate to support uracil misincorporation as a primary mechanism of cytotoxicity of TS inhibitors. Yin et al. (1993) showed that LY231514 and AG-331 caused DNA strand breaks in nascent DNA, and tomudex also caused breaks in mature DNA. Another study supporting the importance of uracil misincorporation was the observation by Canman et al. (1994) that high level expression of dUTPase in tumor cells caused resistance to FUdR.

Chu and colleagues (1993) demonstrated that synthesis of TS protein is regulated at the posttranscriptional level, and that human TS possesses a specific binding site for its own mRNA. Since the accumulation of total TS protein occurs in cells treated with TS inhibitors, an understanding of this process may suggest ways of downregulating TS synthesis.

A number of studies in the past two years have suggested that the way in which a thymidylate-starved cell responds may depend upon the status of its cell cycle checkpoint proteins. Lowe et al. (1993) proposed that p53-dependent apoptosis modulates the cytotoxicity of anticancer agents, including 5-fluorouracil. Houghton et al. (1994) proposed that wild-type p53 caused cell cycle arrest in thymineless conditions, and thus protected cells. Fisher et al., (1993) examined the relation of apoptosis, triggered by treatment with TS inhibitors, to expression of bcl-2, and found that bcl-2 expression protected cells from cytotoxicity induced by TS inhibitors.

novel antifolates has shown that the situation is more interesting: antimetabolites kill cells by creating irreversible metabolic imbalances, to which the cell may respond by entering apoptosis. Antimetabolites may have effective antitumor activity against slowly proliferating cells, such as moderately differentiated colon carcinomas. Finally, our rapidly evolving knowledge of cell cycle checkpoint proteins suggests that a cell's status with respect to proteins such as p53 (whether normal or mutant) may influence whether an antimetabolite will be cytostatic or cytotoxic to a particular cell. The convergence of anticancer drug design with a deeper knowledge of control of the cell cycle should guide the development of new, more effective antimetabolite drugs.

REFERENCES

Almassy, R.J., Janson, C.A., Kan, C.C., and Hostomska, Z., 1992, Structues of apo and complexed *Escherichia coli* glycinamide ribonucleotide transformylase, Proc. Natl. Acad.Sci. USA 89: 6114.

Beardsley, G.P., Moroson, B.A., Taylor, E.C., and Moran, R.G., 1989, A new folate antimetabolite, 5,10-dideaza-5,6,7,8-tetrahydrofolate is a potent inhibitor of *de novo* purine synthesis. J. Biol. Chem. 264: 328.

Calvete, J.A., Balmanno, K., Taylor, G.A., Rafi, I., Newell, D.R., Lind, M.R. and Calvert, A.H., 1994. Preclinical and clinical studies of prolonged administration of the novel thymidylate synthase inhibitor, AG337. Proc.AACR 35: 306.

Canman, C.E., Radany, E.H., Parsels, L.A., Davis, M.A., Lawrence, T.S. and Maybaum, J. 1994, Induction of resistance to fluorodeoxyuridine cytotoxicity and DNA damage in human tumor cells by expression of *Escherichia coli* deoxyuridine triphosphatase. Cancer Res. 54: 2296.

Chu, E., Voeller, D., Koeller, D.M., Drake, J.C., Takimoto, C.H., Maley G.F. et al., 1993, Identification of an RNA binding site for human thymidylate synthase. Proc. Natl. Acad. Sci. USA 90: 517.

Cunningham, D., Zalcberg, J., François, E., Van Cutsem, E., Schornagel, J.H. et al., 1994, Tomudex (ZD1694) a new thymidylate synthase inhibitor with good antitumor activity in advanced colorectal cancer. Proc.ASCO 13: 199.

Fisher, T.C., Milner, A.E., Gregory, C.D., Jackman, A.L., Aherne, G.W., et al., 1993. bcl-2 Modulation of apoptosis induced by anticancer drugs: resistance to thymidylate stress is independent of classical resistance pathways. Cancer Res. 53: 3321.

Grindey, G.B., Alati, T., and Shih, C., 1991, Reversal of the toxicity but not the antitumor activity of lomotrexol by folic acid. Proc.AACR 32: 324.

Grindey, G.B., Shih, C., Barnett, C.J., Pearce, H.L., Engelhardt, J.A., Todd, G.C. et al., 1992, LY231514, a novel pyrrolopyrimidine antifolate that inhibits thymidylate synthase. Proc.AACR 33: 411.

Habeck, L.L., Leitner, T.A., Shackelford, K.A., Gossett, L.S., Schultz,R.M., et al., 1994, A novel class of monoglutamated antifolates exhibits tight-binding inhibition of human glycinamide ribonucleotide formyltransferase and potent activity against solid tumors. Cancer Res. 54: 1021.

Houghton, J.A., Harwood, F.G., and Houghton, P.J., 1994, Commitment to thymineless death is influenced by cell cycle control processes. Proc.AACR 35: 316.

Jackman, A.L., Jones, T.R., and Calvert, A.H., 1985, Thymidylate synthetase inhibitors: experimental and clinical aspects, in:"Experimental and Clinical Progress in Cancer Chemotherapy" (ed. by F. M. Muggia), p. 155. Martinus Nijhoff, Boston.

Jackman, A.L., Taylor, G.A.,Gibson, W., Kimbell, R., Brown, M. et al., 1991, ICI D1694, a quinazoline antifolate thymidylate synthase inhibitor that is a potent inhibitor of L1210 tumor cell growth in vitro and in vivo. Cancer Res. 51: 5579 .

Jackman, A.L., Aherne, G.W., Kimbell, R., Brunton, L., Hardcastle, A., Wardleworth, J.W. et al., 1994, ZD9331, a non-polyglutamatable quinazoline thymidylate synthase inhibitor. Proc.AACR 35: 301.

Jones, T.R., Calvert, A.H., Jackman, A.L., Brown, S.J., Jones, M.,and K.R. Harrap, 1981,A potent antitumour quinazoline inhibitor of thymidylate synthetase: synthesis, biological properties, and therapeutic results in mice. Eur. J. Cancer 17: 11.

Lowe, S.W., Ruley, H.E., Jacks, T., and Housman, D.E.,1993, p53-Dependent apoptosis modulates the cytotoxicity of anticancer agents. Cell 74: 957.

Moran, R.G., 1992, Folate antimetabolites inhibitory to *de novo* purine synthesis, in: "New Drugs, Concepts and Results in Cancer Chemotherapy," F.M. Muggia, ed., p. 65, Kluwer Academic Publishers, Boston.

Pratt, W.B., Ruddon, R.W.,Ensminger, W.D., and Maybaum, J., 1994, "The Anticancer Drugs," Oxford University Press.

Smith, G.K., Duch, D.S., Dev, I.K., and Kaufmann, S.H., 1992. Metabolic effects and kill of human T-cell leukemia by 5-deazaacyclotetrahydrofolate, a specific inhibitor of glycineamide ribonucleotide transformylase. Cancer Res. 52: 4895.

Smith, S.G., Lehman, N.L., and Moran, R.G., 1993, Cytotoxicity of antifolate inhibitors of thymidylate and purine synthesis to WiDr colonic carcinoma cells. Cancer Res. 53: 5697.

Varney, M.D., Marzoni, G.P., Palmer, C.L., Deal, J.G., Webber, S.,et al., 1992, Crystal structure-based design and synthesis of benz[cd]indole-containing inhibitors of thymidylate synthase. J. Med. Chem. 35: 663.

Webber, S.E., Bleckman, T.M., Attard, J., Deal, J.G., Kathardekar, V., et al., 1993, Design of thymidylate synthase inhibitors using protein crystal structures: the synthesis and biological evaluation of a novel class of 5-substituted quinazolines. J. Med. Chem. 36: 733.

Varney,M.D., 1993, Antiproliferative substituted 5-thiapyrimidinone compounds, US Patent Application Serial No. 07/991,259.

Yin, M.B., Schober, C., Arredonto, M.A., and Rustum, Y.M., 1993, Effects of specific thymidylate synthase inhibitors, ICI D1694, LY231514 and AG-331, on fragmentation of mature and nascent DNA. Proc.AACR 34: 274.

THE ANTITUMOUR ACTIVITY OF ZD9331, A NON-POLYGLUTAMATABLE QUINAZOLINE THYMIDYLATE SYNTHASE INHIBITOR

Ann L. Jackman, Rosemary Kimbell, Melody Brown, Lisa Brunton, Kenneth R. Harrap, *J. Michael Wardleworth, and *F. Thomas Boyle

The CRC Centre for Cancer Therapeutics, Block E, The Institute of Cancer Research, Sutton, Surrey, UK and *ZENECA Pharmaceuticals, Alderley Park, Macclesfield, Cheshire, UK

INTRODUCTION

Thymidylate synthase (TS) is regarded as a good target for the development of quinazoline (folate-based) anticancer agents. The quinazoline-based TS inhibitor, D1694 or Tomudex (trade mark of Zeneca PLC) demonstrated exciting activity against colorectal tumours in the worldwide Phase II[1] clinical studies and is currently in Phase III study for this tumour type. Tomudex is an excellent substrate for folylpolyglutamate synthetase (FPGS) and is almost completely metabolised to polyglutamate forms (mainly tetra and pentaglutamates) that are not readily effluxed from the cell[2]. Prolonged intracellular drug retention is thus a feature of this drug. This has the advantage of showing antitumour activity by bolus administration and the current clinical protocol is a 15min infusion once every 3 weeks. The polyglutamates of Tomudex are ~60-fold more active than the parent drug as TS inhibitors[2]. Polyglutamation is thus a requirement for antitumour activity and may offer some tumour selectivity. However a mechanism of acquired and possibly intrinsic resistance to Tomudex is the failure of cells to polyglutamate the drug[3]. In order to broaden the spectrum of tumours responsive to TS inhibition by antifolates we synthesised and evaluated compounds unable to undergo such metabolism. As drug retention is not a general feature of this class of compound, infusion protocols were employed to evaluate their activity in mice[4]. Since TS inhibition is only achieved during drug administration this offers a high degree of control over the length of time that DNA synthesis is inhibited. Prevention of polyglutamation is possible through modification of the quinazoline ring (7-methylation)[5] or the glutamate residue. Compound design also focused on developing TS inhibitors with potency at least equal to the polyglutamates of Tomudex (Ki tetraglu = 1nM). Acid-containing (water-soluble) and highly lipophilic analogues were synthesised but we concentrated on those that had high water-solubility and used the reduced-folate/MTX carrier (RFC) for cell entry. One such compound, ZD9331, was chosen as the most promising compound with activity against a number of experimental tumours (murine and human xenografts) in mice[6,7,8]. The in vitro activity of ZD9331 is summarised below and, where appropriate, is compared with other TS inhibitors of known biochemical profile. Thus

N^{10}-propargyl-5,8-dideazafolic acid (CB3717), the prototype quinazoline TS inhibitor[9], uses the RFC poorly and is slowly polyglutamated when compared with ICI 198583[10] or Tomudex[2]. AG337, represents a newer class of lipophilic quinazolines discovered by Agouron Pharmaceuticals that neither uses the RFC nor is a substrate for FPGS[11].

RESULTS AND DISCUSSION

ZD9331 is an analogue of 2-desamino-2-CH_3-N^{10}-propargyl-5,8-dideazafolic acid (ICI 198583) a water-soluble, potent TS inhibitor ($Ki = 10nM$)[10]. A 25-fold improvement in TS inhibition resulted from substituting the 2' carbon of the benzoyl ring with fluorine, methylating C7 of the quinazoline, and replacing the glutamate γ-carboxyl with tetrazole. Either C7-methyl or the γ-tetrazole alone will prevent polyglutamation but the combination led to superior TS inhibition and cytotoxicity (Table 1). Improved in vivo antitumour activity was observed with this combination of modifications[4,6].

Table 1: In vitro activity of ZD9331 and related compounds

ZM No. (ICI)	R (C7)	X (2')	Y	L1210 TS Ki, nM	L1210 IC_{50}, μM	W1L2 IC_{50}, μM	FPGS Km, μM
198583	H	H	COOH	10	0.15	0.058	*43
207478	CH_3	H	COOH	~4	0.21	0.15	*non-sub
214888	CH_3	F	COOH	~6	0.08	0.07	-
ZD9331	CH_3	F	tetrazole	0.44	0.024	0.0073	+non-sub
CB3717			COOH	3 (glu$_4$ ~0.05)	5.0	2.6	*40
D1694			COOH	60 (glu$_4$ = 1)	0.0088	0.0046	*1.3
AG337				70	0.70	0.78	non-sub

TS was partially purified and assayed by ^{3}H release as previously described[2]. Mouse L1210 and human W1L2 cells were grown in suspension with continuous exposure to compounds for 48 and 72hrs respectively, as previously described[2]. *mouse liver FPGS, Moran et al[5] ; +L1210 FPGS.

Experiments suggest that ZD9331 predominantly uses the RFC for entry into the cell. ZD9331 inhibits the transport of ^{3}H methotrexate (MTX) into mouse L1210 and human W1L2 cells with a Ki of ~1μM (Table 2). This is similar to that measured for Tomudex and ICI 198583 and significantly better than CB3717. Internalisation of ZD9331 by the RFC is supported by the high level of cross-resistance in the L1210 variant cell line (L1210:1565) which has a greatly impaired ability to transport MTX into the cell and is thus highly resistant to MTX (Table 3).

The use of resistant cell lines (described in reference 2) has also confirmed TS as the intracellular target for ZD9331. The W1L2:C1 cell line has acquired resistance to ICI 198583 through TS gene amplification and is cross-resistant to ZD9331 and all the other quinazoline-based TS inhibitors but sensitive to the dihydrofolate reductase inhibitor, MTX. An L1210 resistant line, L1210:R^{D1694} is unable to polyglutamate Tomudex or other antifolates such as MTX or ICI 198583 that are substrates for FPGS. As expected ZD9331, and those glutamate analogues possessing a C7-methyl, displayed good activity in this resistant cell line.

Confirmation that ZD9331 was not retained in cells, once drug was removed from the extracellular environment, would provide clear evidence of a difference from polyglutamated drugs such as Tomudex. It would also allow control over the time that TS is inhibited. One semi-quantitative but rapid assay to indicate how rapidly TS activity recovers after drug removal is the whole cell TS assay where the conversion of 5-^{3}H deoxyuridine to ^{3}H$_2$0 is used as a measure of TS activity. When W1L2 cells were treated with 0.1-10μM ZD9331 they had barely detectable TS activity after 1hr. Cells exposed to drug for 4hrs, then washed and resuspended in drug-free medium for 4hrs had TS activity even higher than control cells. This was in contrast to cells treated with Tomudex where activity had not recovered at the same time. Direct measurement of dUMP and TTP pools by radio-immunoassay has confirmed this difference between ZD9331 and Tomudex (Aherne et al, accompanying paper).

Table 2: Affinity of ZD9331 and related quinazoline TS inhibitors for the reduced-folate/MTX cell membrane carrier (RFC)

	Inhibition of ^{3}H MTX transport, Ki (μM)*	
	L1210 (mouse)	W1L2 (human)
MTX	4.2	2.1
198583	2.7	1.1
207478	4.5	1.3
214888	5.0	1.2
ZD9331	1.6	0.72
CB3717	41	19
AG337	>100	>100
D1694	2.6	0.66

One ml logarithmically growing L1210 or W1L2 cells (resuspended at ~4 x 10^6/ml) were incubated with 0.5μM ^{3}H MTX plus varying concentrations of MTX or quinazolines for 5min at 37^0 in HBSS transport buffer. Cells were washed x 3 with ice-cold HBSS[5] by centrifugation and the cell pellet counted for radioactivity. Cellular ^{3}H MTX was estimated and the data was fitted to the competitive inhibition equation by non-linear least squares analysis and values for Ki obtained. *Mean of two experiments.

SUMMARY

ZD9331 is a very potent, quinazoline-based TS inhibitor, which although having structural similarities with classical TS inhibitors, has biochemical features in common with the non-classical inhibitors of this enzyme. Water-solubility, transport into the cell via the RFC but lack of intracellular polyglutamation give this drug its hybrid properties. The 150-fold improvement in TS inhibitory activity over Tomudex predominantly accounts for equivalent cytotoxic activity, when given by continuous exposure. The drug exerts its effect only while it is present extracellularly. This feature, together with rapid plasma clearance in mice, should give the control over inhibition of DNA synthesis desired from this class of agent. ZD9331 is active in a Tomudex-resistant cell line unable to polyglutamate classical antifolates. Pre-clinical in vivo studies are ongoing to define more closely the toxicology and administration protocols appropriate to ZD9331.

Table 3: Cross-resistance studies for ZD9331 and related quinazoline TS inhibitors

	W1L2:C1 TS+++	L1210:1565 RFC ---	L1210:R^{D1694} polyglu --- (small RFC-)
	Resistance factors (IC$_{50}$ resistant cells/IC$_{50}$ wild-type cells)		
MTX	0.4	90	2*
198583	27,000	40	18
207478	>670	62	3
214888	>1400	97	5
ZD9331	2300	50	3
CB3717	>100	1	3
AG337	32	2	2
D1694	~22,000	90	>11,000

* ratio of 15 under 6hr drug exposure conditions. W1L2:C1 = overexpression of TS 200-fold; L1210:1565 = impaired RFC ; L1210:R^{D1694} = polyglutamation defect. Cells were treated as described above.

REFERENCES

1. J. Zalcberg, D. Cunningham, E. Van Cutsem, E. Francois, J.H. Schornagel, A. Adenis, M. Green, H. Starkhammer, and M. Azab, Annals of Oncology, 5: supl 5, (1994).
2. A.L. Jackman, G.A. Taylor, W. Gibson, R. Kimbell, M. Brown, A.H. Calvert, I.R. Judson, and L.R. Hughes, Cancer Res 51: 5579-5586 (1991).
3. A.L. Jackman, L.R. Kelland, M. Brown, W. Gibson, R. Kimbell, W. Aherne, and I.R. Judson, Proc Amer Assoc Cancer Res 33: 406 (1992).
4. T.C. Stephens, M.N. Smith, S.E. Waterman, M.L. McCloskey, A.L. Jackman, and F.T. Boyle, in: Advances in Experimental Medicine and Biology , Vol. 338 (Ed. Ayling J) pp 589-592, Plenum Press, New York (1993).
5. P.C. Sanghani, A.L. Jackman, V.R. Evans, T. Thornton, L. Hughes, A.H. Calvert, and R.G. Moran, Mol Pharmacol, 45: 341-351 (1994).
6. T.C. Stephens, M.N. Smith, M.L. McCloskey, S.E. Waterman, A.J. Gwyne, B.E. Valcaccia, A.L. Jackman, J.M. Wardleworth, F.T. Boyle, Proc Amer Assoc Cancer Res. 35:305 (1994).
7. J.M. Wardleworth, F.T. Boyle, R.J. Barker, L.F. Hennequin, S.J. Pegg, T.C. Stephens, R. Kimbell, M. Brown and A.L. Jackman, Annals of Oncology, 5: suppl 6 (1994).
8. A.L. Jackman, G.W. Aherne, R. Kimbell, L. Brunton, A. Hardcastle, J.M. Wardleworth, T.C. Stephens and F.T. Boyle, Proc Amer Assoc Cancer Res. 35: 301 (1994).
9. T.R. Jones, A.H. Calvert, A.L. Jackman, S.J. Brown, M. Jones, and K.R. Harrap, Eur J Cancer, 17: 11-19 (1981).
10. A.L. Jackman, D.R. Newell, W. Gibson, D.I. Jodrell, G.A. Taylor, J.A. Bishop, L.R. Hughes, and A.H. Calvert, Biochem Pharmacol 42: 1885-1895 (1991).
11. S.E. Webber, T.M. Bleckman , J. Attard and 18 other authors, J Med Chem 36:733-746 (1993).

CROSS-RESISTANCE TO THYMIDYLATE SYNTHASE INHIBITORS IN P-GLYCOPROTEIN AND NON-P-GLYCOPROTEIN CELL LINES

B. van Triest, F. Telleman, H.M. Pinedo,
C.L. van der Wilt, G.J. Peters

Free University Hospital
Department of Medical Oncology
po box 7057
1007 MB Amsterdam
The Netherlands

INTRODUCTION

The fluoropyrimidines have been used for the treatment of patients with solid tumors of the breast, head and neck, and gastrointestinal tract. Although the response to treatment is not impressive 5-fluorouracil (5FU) is part of the standard protocols in advanced colorectal cancer[1,2]. The mechanism of action of 5FU is rather complicated and may consist of: 1. incorporation into DNA; 2. incorporation into RNA; and 3. inhibition of thymidylate synthase (TS) leading to inhibition of DNA synthesis due to lack of thymidylate formation. Various mechanisms of resistance against 5FU have been described; some are associated with alterations in the 5FU metabolism, some with inhibition of TS (e.g. relative intracellular folate depletion) and mutations or increased TS levels. Chu et al. found cross-resistance to 5FU in Multi Drug Resistant (MDR) human breast and human colon cell lines which showed a significant overexpression of TS[3]. Zhang et al. found cross-resistance to 5FU and 5-fluoro-2'-deoxy-uridine in MCF7/Adr cells caused by increased level of TS[4]. In these studies only cell lines with MDR due to overexpression of P-glycoprotein (P-gp) have been studied; however, no study has been performed on possible cross-resistance in cell lines with a MDR-phenotype not due to an overexpression of P-gp. We selected a panel of three cell lines, the wild type SW1573 and two doxorubicin (DOX) resistant cell lines; a MDR variant resistant to DOX due to P-gp overexpression (SW1573/2R160) and a subline resistant to DOX but with no P-gp overexpression (SW1573/2R120). In these cell lines we determined whether they exhibit a cross-resistance to 5FU.

We also determined a possible cross-resistance to several TS-inhibitors with different structural properties; ZD1694 (Tomudex) an antifolate dependent on transport via the reduced folate carrier and a good substrate for folylpolyglutamate synthetase; and AG337 a lipophilic compound, transported by passive diffusion which can not be polyglutamylated.

MATERIALS AND METHODS

SW1573 is a squamous cell carcinoma cell line of the lung. The two resistant sublines were selected by culture in increasing doxorubicin concentrations[5,6,7]. Cells were cultured under standard conditions. All experiments were carried out after a drug free period of at least 7 days.

Chemosensitivity was determined using the Sulforhodamine B (SRB) assay as described previously[8,9]. Drug exposure was 72 hr. The IC_{50} was the concentration that is equal with half-maximal growth of the control cells based on the difference of optical density values at day 0 and day 4 of drug exposure. The IC_{50} concentrations were based on at least three separate experiments.

TS assays (FdUMP binding assay and TS catalytic assay) were performed as described previously[10,11,12]. The procedure for immunohistochemical staining for TS has been described elsewhere[13]. Western blotting of the TS protein was performed according to a standard method[14]. The TS antibody was a generous gift of Dr. G.W. Aherne, Institute of Cancer Research, Sutton, Surrey, United Kingdom.

Flow-cytometry to determine the cell-cycle distribution was performed essentially according to Vindelov and Christensen[15].

RESULTS

Growth inhibition studies

Growth inhibition experiments confirmed the cross-resistance pattern for DOX previously observed with the MTT-assay[5]. In the 2R120 a collateral sensitivity was observed for 5FU and AG337 but a cross-resistance for ZD1694. 2R160 cells showed cross-resistance to all drugs.

Table 1. Growth inhibitory effect of 5FU, folate-based TS inhibitors and DOX against non-P-gp (2R120) and P-gp (2R160) cell lines.

	Cell lines		
	SW1573 IC_{50}	2R120 RF	2R160 RF
5FU	3.65 ±0.66	0.4	2.4
ZD1694	0.036±0.003	8.8	18.8
AG337	4.5 ±0.46	0.7	17
DOX	0.069±0.012	8.9	73

The cell lines were exposed to 5FU, ZD1694, AG337 and DOX for 72 hr. IC_{50} values (μM) are means ± SE from at least 3 experiments performed in triplicate. Resistance Factor (RF) = $IC_{50resist}/IC_{50wild\ type}$ determined per experiment.

TS assays

Since TS is a common target for the three drugs we determined whether TS expression could explain these results. TS activity as measured by the FdUMP binding assay showed a slight increase of 1.5 fold in the 2R120 but a decrease of 0.7 in the 2R160 compared to the parental cell line SW1573 (Table 2).

The TS catalytic activity (conversion of dUMP to TMP) did not show significant changes. Evaluation of the expression of TS protein using immunohistochemical staining (IHC) revealed contrasting results, a high expression was found in the parental cell line and the 2R160 and a low expression in the 2R120 cell line.

Since both 5FU and ZD1694 have been reported to induce overexpression of TS[3, 16] we determined whether this phenomenon might contribute to the observed resistance pattern. Treatment with 5FU showed an increase in TS protein level only in the 2R120 (Table 3); but ZD1694 induced an increase in all cell lines. AG337, however, decreased the TS protein level in the parental cell line, while in the 2R120 the TS protein level was increased.

Table 2. TS levels in non-P-gp (2R120) and P-gp (2R160) cell lines.

Cell line	FdUMP binding assay[a]	TS catalytic assay[b]	IHC
SW1573	328 ± 146	209 ± 4	+ + +
2R120	477 ± 159	253 ± 63	+
2R160	237 ± 109	202 ± 44	+ + +

Results are means $\pm$ SE of three separate experiments. a = FdUMP binding assay: fmol/10^6 cells. b = TS catalytic assay: pmol ^{3}H-H$_2$O/hr/10^6 cells.

Table 3. Western immunoblot analysis in a non-P-gp (2R120) and P-gp (2R160) cell lines after treatment with 5FU and TS-inhibitors.

	Cell line		
	SW1573	2R120	2R160
5FU	$100\% \pm 3$	$155\% \pm 38$	$111\% \pm 50$
ZD1694	$143\% \pm 16$	$336\% \pm 44$	$148\% \pm 24$
AG337	$70\% \pm 12$	$189\% \pm 39$	$96\% \pm 13$

Cells were exposed at their IC$_{50}$ for 24 hr, thereafter Western blotting was performed. The amount of protein was related to that of the untreated cells. Quantitation was performed by densitometric scanning. Values are means $\pm$ SE.

Cell cycle distribution

Since TS is a cell cycle related protein, we determined the cell-cycle distribution of untreated and treated cells. These experiments showed a similar S-phase distribution in all three cell lines. Treatment with ZD1694 resulted in a decrease in S-phase cell percentage in SW1573 and 2R120; but an increase in the 2R160. Treatment with 5FU showed a different pattern, decrease of S-phase cell percentage in all cell lines compared with the untreated cell lines.

Table 5. Cell cycle distribution of non-P-gp (2R120) and P-gp (2R160) cell lines before and after treatment with 5FU and ZD1694.

| | % cells in S-phase | | |
	SW1573	2R120	2R160
control	34%	38%	34%
5FU	14%	23%	30%
ZD1694	27%	28%	47%

Cells were treated for 24 hr at their IC_{50} concentration. Thereafter cell cycle distribution was determined.

DISCUSSION

In this paper we demonstrated a cross-resistance to 5FU and to two TS-inhibitors in cell lines with a P-gp mediated MDR, however in the non-P-gp MDR cell line only cross-resistance was observed to ZD1694 but collateral sensitivity to 5FU and AG337. Elevation of the TS enzyme expression as observed in the studies of Chu et al[4], was not found.

The cross-resistance shown in the P-gp cell line (2R160) and the collateral sensitivity as well as cross-resistance in the non-P-gp cell line (2R120) for TS-inhibitors was striking. The compounds being tested have different characteristics; lipophilic (AG337) or non-lipophilic (ZD1694) with consequences for the mechanism of transport across the cell membrane. To exclude the influence of the P-gp drug efflux pump experiments were performed with verapamil which can inhibit competitively the P-gp mediated efflux. In our study reversal of the antifolate cross-resistance by verapamil could not be established (data not shown).

The 2R120 cell line showed cross-resistance to ZD1694 but collateral sensitivity to AG337. This is suggestive for a polyglutamylation defect because only ZD1694 has to be polyglutamylated to exert its intracellular activity. Since a defect of polyglytamylation has been described as a mechanism of resistance to antifolates, we will determine the polyglutamylation pattern in all cell lines.

Both 5FU and the folate-based TS inhibitors increased the TS expression in the 2R120 cell line, but in the parental and 2R160 only ZD1694 increased the expression of TS-protein. In order to determine whether this is related to alterations in translation or transcription of the TS gene we will determine the mRNA expression as described previously[17]. Gene amplification is thought to be a mechanism of drug resistance for TS-dependent chemotherapeutic agents[18]. Such studies will give evidence whether the differences between the cell lines will be at (post)transcriptional level, such as has been described by Chu et al.[3] and Keyomarsi et al.[16].

The regulation of TS enzyme is cell cycle dependent and the enzyme is mainly active in the S-phase[19]. Our cell cycle distribution analysis showed a comparable amount of cells in the S-phase in the untreated SW1573, 2R120 and 2R160. So, the selective treatment with doxorubicin in the 2R120 and the 2R160 did not influence the cell cycle distribution. Treatment with 5FU and ZD1694 showed a different cell cycle distribytion pattern indicating that the mechanisms of action of both drugs are different.

We can conclude that a different pattern in collateral sensitivity and cross-resistance to 5FU and antifolate TS-inhibitors could be observed in non-P-gp (SW1573-/2R120) and P-gp (SW1573/2R160) cell lines.

ACKNOWLEDGMENTS

This project was supported by a grant from the Netherlands Cancer Society, grant number VU-IKA 93-627.

REFERENCES

1. HM Pinedo and GJ Peters, Fluorouracil: biochemistry and pharmacology, J. Clin. Oncol. 6:1653-1664 (1988).
2. GJ Peters, CJ Van Groeningen, Clinical relevance of biochemical modulation of 5-fluorouracil, Ann.Oncology 2:469-480 (1991).
3. E Chu, JC Drake, DM Koeller et al., Induction of thymidylate synthase associated with multidrug resistance in human breast and colon cancer cell lines. Mol. Pharm. 39:136-143 (1991).
4. ZG Zhang, A Harstrick and YM Rustum, Mechanisms of resistance to fluoropyrimidines, Sem. in Oncol. 19; suppl 3:4-9 (1992).
5. CM Kuiper, HJ Broxterman, F Baas et al., Drug transport variants without P-glycoprotein overexpression from a human squamous lung cancer cell line after selection with doxorubicin, J. Cell. Pharm. 1:35-41 (1990).
6. F Baas, APM Jongsma, HJ Broxterman et al., Non-P-glycoprotein mediated mechanism for multidrug resistance precedes P-glycoprotein expression during in vitro selection for doxorubicin resistance in a human lung cancer cell line, Cancer Res. 50:5392-5398 (1990).
7. HG Keizer, GJ Schuurhuis, HJ Broxterman et al., Correlation of multidrug resistance with decreased drug accumulation, altered subcellular drug distribution, and increased P-glycoprotein expression in cultured SW-1573 human lung tumor cells, Cancer Res. 49:2988-2993 (1989).
8. P Skehan, R Storeng, D Scudiero et al. New colorimetric cytotoxicity assay for anticancer-drug screening, J Nath Cancer Inst 82:107-112 (1990).
9. YP Keepers, PE Pizao, GJ Peters et al., Comparison of the sulforhodamine B protein and tetrazolium (MTT) assays for in vitro chemosensitivity testing, Eur J Cancer 27:897-900 (1991).
10. GJ Peters, E Laurensse, A Leyva et al., Sensitivity of human, murine and rat cells to 5-fluorouracil and 5'-deoxy-5-fluorouridine in relation to drug metabolizing enzymzes, Cancer Res. 46:20-28 (1986).
11. CL Van der Wilt, HM Pinedo, K Smid et al., Elevation of TS following 5-fluorouracil treatment is prevented by the addition of leucovorin in murine colon tumors, Cancer Res. 52:4922-4928 (1992).
12. CL Van der Wilt, GWM Visser, BJM Braakhuis et al., In vitro antitumour activity of cis- and trans-5-fluoro-5,6-hydro-6-alkoxy-uracils; effect on thymidylate synthesis, Brit J Cancer 38:702-707 (1993).
13. CL Van der Wilt, K Smid, GW Aherne et al., Evaluation of Immunohistochemical Staining and Activity of Thymidylate Synthase in Cell Lines.In: "Chemistry and Biology of Pteridines and Folates". (Eds. J.E. Ayling, M.G. Nair, C.M. Baugh). Plenum Publishing Corporation, New York. Adv. Exp. Med. Biol. 338; 605-608 (1993).
14. SJ Freemantle, GW Aherne, A Hardcastle et al., Increases in TS protein levels measured using newly developed antibodies, Proc. Am. Assoc. Cancer Res. 32:360 (1991).
15. LL Vindelov and IJ Christensen, A review of techniques and results obtained in one laboratory by an integrated system of methods designed for routine clinical flow cytometric DNA analysis, Cytometry 11:753 (1990).
16. K Keyomarsi, J Samet, G Molnar et al., The thymidylate synthase inhibitor, ICI D1694, overcomes translational detainment of the enzyme, J. Biol. Chem. 268:15142-15149 (1993).
17. SJ Freemantle, J Lunec, CL Van der Wilt et al., Polymerase chain reaction quantification of thymidylate synthase (TS) mRNA levels in tumour samples of patients recently treated with 5-fluorouracil (5-FU) $\pm$ leucovorin (LV). Proc. Amer. Assoc. Cancer Res. 34; 348 (Abstract 2076) (1993).
18. M Kashani-Sabet, JJ Rossi, Y Lu et al., Detection of drug resistance in human tumors by in vitro enzymatic amplification, Cancer Res 48:5775-5778 (1988).
19. LG Navelgund, C Rossana, AJ Muench, LF Johnson, Cell cycle regulation of thymidylate synthase gene expression in cultured mouse fibroblasts, J. Biol. Chem. 255:7386-7390 (1980).

NUCLEOTIDE METABOLISM: MODE OF ACTION OF THIOPURINES IN LEUKEMIA

Ronney A. De Abreu

Center for Pediatric Oncology S.E. Netherlands
Department of Pediatrics
University Hospital Nijmegen, St. Radboud
PO Box 9101, 6500 HB Nijmegen, The Netherlands

INTRODUCTION

Thiopurines are active in a variety of lymphoid and myeloid leukemia and represent an important component of the systemic treatment of these malignancies[1-4]. Oral 6-mercaptopurine (6MP) in combination with methotrexate (MTX) forms the cornerstone of the maintenance therapy of children with acute lymphoblastic leukemia (ALL)[4] and thioguanine (6TG) is used in regiments of acute myeloblastic leukemia and several other malignancies[5,6]. The enhanced efficacy of combination of 6MP and MTX on survival of children with acute lymphoblastic leukemia has been empirically demonstrated. However, absorption of 6MP after oral dose is incomplete and variable[7] and very little is known about the distribution or clearance of the metabolites of these drugs, and about the optimal concentration and duration of exposure required for cytotoxicity in vivo.

To elucidate the metabolic routes leading to cytotoxicity and to search for better routes of administration, prolonged exposure to thiopurines were studied with respect to cell kinetic parameters, purine nucleotide and phosphoribosyl pyrophosphate levels, DNA and RNA synthesis, formation of thiopurine metabolites and incorporation of thioguanine nucleotides into newly synthesized DNA and RNA.

MECHANISMS OF THIOPURINES INVOLVED IN CELLULAR CYTOTOXICITY

Effects on DNA and RNA level

The biochemical mechanisms of action of 6MP and 6TG are closely related. 6MP and 6TG are initially activated by hypoxanthine-guanine phosphoribosyltransferase to 6-thioinosine monophosphate (t-IMP) and 6-thioguanosine monophosphate (t-GMP), respectively (scheme 1). T-IMP is further converted by the purine interconversion via 6-thioxanthosine monophosphate (t-XMP) to t-GMP. The enzymes catalyzing these metabolic conversions are IMP dehydrogenase and GMP synthetase, respectively. Subsequently, t-GMP is converted further by GMP and GDP kinases and ribonucleotide reductase, to form 6-thioguanosine triphosphate (t-GTP) and 6-thiodeoxyguanosine triphosphate (t-dGTP), which are incorporated into RNA and DNA, respectively. At present it is thought that the primary mechanisms involved in cytotoxicity of malignancies caused by thiopurines, is incorporation of t-dGTP into DNA. T-dTGP incorporation into DNA induces DNA damage, such as single strand breaks, DNA-protein cross-links, interstrand cross-links and sister chromatid exchanges[8-13]. These effects on DNA result in delayed

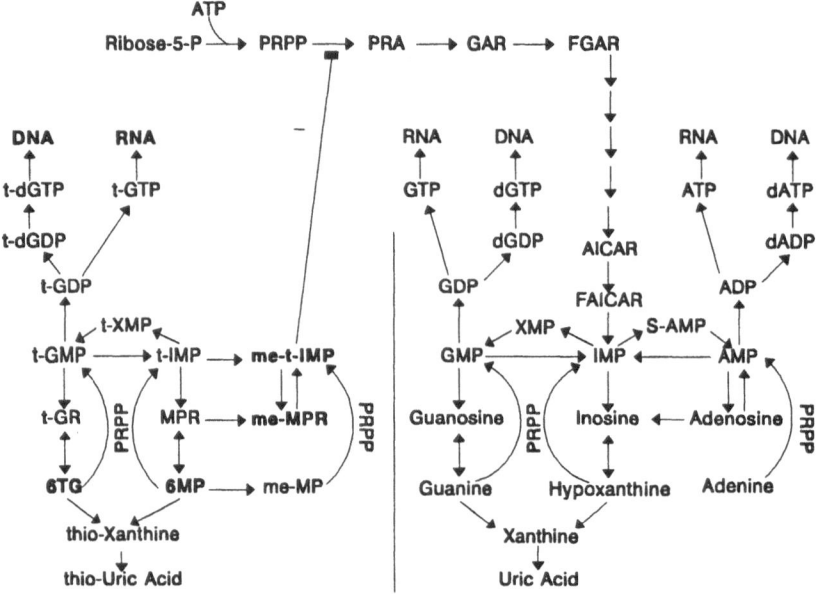

Scheme 1: Metabolism of thiopurines

cytotoxicity in most tumor cell lines, which is reflected by growth arrest of tumor cells in G_2 + M phase of the cell cycle[14]. Incorporation of 6MP-nucleotides into DNA was 5 times as high as incorporation into RNA of MOLT F4 cells[14]. This is consistent with the observation that the incorporation of 6TG in place of guanine proceeds readily in DNA synthesis reactions catalyzed by mammalian and bacterial polymerases, while t-dG-containing DNA functions as a poor template[15]. Moreover, kinetic studies showed inhibition of RNA polymerase activities by 6-thio-ITP, which could be reversed by increasing concentrations of GTP for its binding sites. These data suggest that inhibition of RNA transcription by 6-thio-ITP may be considered as one of the mechanisms of cytotoxic action of 6MP in tumor cells[16].

Inhibition of purine de novo synthesis

A second route for cytotoxicity is methylation of 6MP metabolites by thiopurine methyltransferase into methyl-mercaptopurines (scheme 1). This conversion is S-adenosyl-L-methionine dependent[17]. Methyl-thioinosine monophosphate (me-t-IMP), formed by this conversion, is a strong inhibitor of purine de novo synthesis (PDNS). Inhibition of PDNS occurs at the enzyme PRPP amidotransferase, catalyzing conversion of phosphoribosylpyrophosphate (PRPP) into phosphoribosylamine ribonucleotide[18]. T-IMP can also inhibit PDNS, but me-t-IMP is the most effective of the two.

Inhibition of PDNS can provoke cytotoxicity by several mechanisms. We studied prolonged in vitro exposure of human lymphoblastic T-cells MOLT F4 to $2\mu M$ and 10 μM 6MP[14]. Exposure to $2\mu M$ and 10 μM 6MP resulted in a rapid inhibition of PDNS by increased levels of me-t-IMP[14]. Subsequently, the inhibition of PDNS resulted in increased levels of PRPP and decreased purine nucleotide concentrations. Exposure to the low dose of 6MP resulted in partially inhibited cell growth and clonal growth, while cell viability was hardly effected. DNA synthesis was decreased, associated with an increasing delay of cells in S phase. Incorporation of thioguanine nucleotides into newly synthesized DNA resulted in an increasing arrest of cells in $G_2 + M$ phase. RNA synthesis was initially decreased, but recovered partially at 48 hrs, at this point the purine nucleotide concentrations were recovering also. Exposure to the high dose of 6MP resulted in early cytotoxicity at 24 hrs, associated a larger pool of me-t-IMP and lower levels of purine ribonucleotides as compared to the low dose of 6MP. A more severe delay of cells in S phase was associated with an inhibition of DNA synthesis to 14% of untreated cells within the first 24 hrs, and an arrest in $G_2 + M$ phase. Further increasing levels of me-t-IMP caused an arrest of cells and late cytotoxicity in S phase at 48 hrs, preventing further progression into G2 + M phase.

These data suggest that inhibition of PDNS by me-t-IMP is an important mechanism involved in 6MP cytotoxicity. Leukemic lymphoblasts, with an active PDNS, have a high need of purine nucleotides for cell growth and cell proliferation. With high concentrations of 6MP, the large pool of me-t-IMP causes severe reduction of endogenous nucleotide concentrations in lymphoblasts. The increased PRPP, resulting from inhibition of PDNS induces increase of pyrimidine biosynthesis[19-20]. These effects of me-tIMP were also observed when MOLT F4 cells were exposed to several concentrations of 6-methylmercaptopurine riboside (me-MPR)[21]. Me-MPR is converted to me-t-IMP by adenosine kinase. In this study, PRPP increase was maximal at intracellular me-t-IMP concentrations of approximately 500 pmol/10^6 viable cells. At higher intracellular concentrations of me-t-IMP, PRPP synthesis was inhibited due to severe depletion of ATP. ATP is a co-substrate for the PRPP synthetase dependent conversion of ribose-5-phosphate to PRPP. Depleted purine nucleotide pools, together with increased pyrimidine nucleotide pools may lead to unbalanced cell growth.

Depletion of purine nucleotides can be circumvented by amidoimidazole carboxamide ribonucleoside (AICAR) and by hypoxanthine, inosine and adenosine. Reversal of cytotoxicity of 6MP and me-MPR was demonstrated in Molt F4 cells[22,23]. AICAR as well as hypoxanthine, inosine and adenosine were able to partially prevent 6MP and me-MPR cytotoxicity. Under these conditions, me-t-IMP formation decreased. Moreover these data suggest that depletion of adenine nucleotides is the main cause for me-t-IMP cytotoxicity.

Depletion of S-adenosyl-L-methionine

Recently, depletion of intracellular S-adenosyl-L-methionine (SAM) concentrations and increase of intracellular S-adenosylhomocysteine (SAH) and methionine concentrations was observed after exposure of Molt F4 cells to 6MP and Me-MPR[23]. The decreased SAM/SAH ratio correlated well with the decrease of ATP observed under these conditions, whereas the increase of intracellular methionine correlated inversely with the decrease of ATP. ATP is required as co-substrate for conversion of methionine to SAM by methionine adenosyltransferase. In proliferating cells, depletion of SAM induced by 6MP and Me-MPR can result in deregulation of several cellular processes, since SAM is an universal methyl donor and is involved in methylation of several macromolecules, e.g. DNA and RNA. Ultimately, the altered methylation state of these macromolecules may lead to cell death.

NEW STRATEGIES FOR TREATMENT WITH THIOPURINES

Despite the convincing evidence from the in vitro studies that thiopurine methylation is involved in 6MP cytotoxicity, conflicting ideas still exist with respect to its clinical relevance. Levels of thiopurine methyltransferase (TPMT) activity in human tissue are controlled by a common genetic polymorphism[24]. On a genetic basis wide differences exist, 88.6% of humans have a high, 11.1% have intermediate and 0.3% have very low activity. Lennard[25] observed a negative correlation between 6TG and TPMT activity in erythrocytes from children with ALL on long-term oral 6MP therapy (full dose therapy of 75 mg/m^2). Children below the group median had higher TPMT activities and a higher subsequent relapse rate. Fifty of the 105 long-term survivors of ALL - no longer on treatment - had been treated with "gentle" low-dose protocols and this subgroup contained an excess of children with lower TPMT activities. From this, she concluded that TPMT activity may be a substantial regulator of the cytotoxic effect of 6MP and that this effect in turn could be important in influencing the outcome of therapy for childhood ALL. In this view, methylation is considered to be a detoxification pathway for 6MP. On the other hand, however, it was reported that the concentration of 6MP required to induce 50% inhibition of DNA synthesis in phytohemagglutinin stimulated peripheral blood lymphocytes was higher in subjects with a genetically low TPMT activity[26]. This seems to indicate that methylated metabolites of 6MP are important for cytotoxicity. So, the clinical relevance of the methylation route for 6MP cytotoxicity is not yet solved. It may be that intracellular me-t-IMP concentrations in the lymphoblasts of children receiving oral 6MP are to low to inhibit PDNS, even in patients with high TPMT activity. This may be due to the poor bioavailility of oral 6MP, with plasma 6MP concentrations of approximately 20 - 160 nM[7]. In this case the methylation route may indeed be considered as a detoxification route. However, in a preliminary study in 9 children with ALL receiving an intravenous infusion of 6MP during 12-24 hours at a dose rate of 50 mg/m^2/hour a mean steady-state plasma concentration of 6MP was measured of 6,2 μM, with less than two-fold difference between the highest and lowest plasma concentrations[27]. Under these conditions enough me-t-IMP may be formed to inhibited PDNS in blast cells in vivo.

As mentioned, during maintenance treatment of ALL both methotrexate and 6MP are administered orally in a low dose. However, combination treatment of intravenous methotrexate (MTX) and intravenous 6MP can be very beneficial. Biochemical evidence for potential synergism of combination of these drugs was demonstrated in vitro on inhibition of PDNS, DNA synthesis, RNA synthesis, growth and clonal growth and on incorporation of labeled 6MP-nucleotide incorporation into DNA and RNA[28-30]. In these studies concentrations of 6MP and MTX were used, which are easily obtained during intravenous administration of these drugs. First studies using both drugs intravenously, have shown good results[31].

Pharmacological and intracellular studies, comparing intravenous administration of 6MP with low dose oral administration may provide more insight in the therapeutic efficacy of 6MP and the role of thiopurine methylation in this respect. Such a study is undertaken at the moment in the Dutch Childhood Leukemia Study Group during protocol M of the DCLSG-ALL-8 protocol, using intravenous MTX and 6MP during consolidation therapy.

ACKNOWLEDGEMENTS

Work in the laboratory of the author was funded by the Dutch Cancer Society.

REFERENCES

1. J.H. Burchenal, M.L. Murphy, R.R. Elison, M.P. Sykes, T.C.Tan, L.A. Leone, D.A. Kanofsky, H.W. Craven, H.W. Dargeon and C.P. Roads, Clinical evaluation of new antimetabolite, 6-mercaptopurine intreatment of leukemia and allied diseases. *Blood* 8:965 (1953).
2. S. Zimm, J.H. Collins, R. Riccardi, D. O'Neill, P.K. Narang, B. Chabner and D.G. Poplack, Variable bioavailability of oral mercaptopurine. Is maintenance chemotherapy in acute lymphoblastic leukemia being optimally delivered? *N. Engl. J. Med.* 308:1005 (1983).
3. A.M. Mauer Therapy of acute lymphoblastic leukemia in childhood. Blood 50:1 (1980).
4. J.V. Simone and G. Rivera. Management of acute leukemia. *In*: Clinical pediatric oncology. W.W. Sutow, D.J. Fernbach and T.J. Vietti, eds., CV Mosby, Sint Louis (1986).
5. L.W. Brox, L.L. Birkett and A. Belch, Clinical pharmacology of oral thioguanine in acute myelogenous leukemia. *Cancer Chemother. Pharmacol.* 6:35 (1981).
6. J.C. Britell, C.G. Moertel, L.K. Kvols, M.J. O'Connell, J. Rubin and A.J. Schutt, Phase II trial of IV 6-thioguanine in advanced colorectal carcinoma. *Cancer Treat. Rep.* 65:909 (1981).
7. T.J. Schouten. R.A. De Abreu, C.H.M.M. De Bruyn, E. Van der Kleijn, M.J.M. Oosterbaan, E.D.A.M. Schretlen and G.A.M. de Vaan, 6-Mercaptopurine: Pharmacokinetics in animals and preliminary results in children. *In*: Purine Metabolism in Man IV, part B; Biochemical, Immunological and Cancer Research. C.H.M.M. De Bruyn, H. A. Simmonds and M.M. Müller, eds., Plenum Press, New York and London (1984).
8. J. Maubaum and H.G. Mandel, Differential chromatid damage induced by t-thioguanine in CHO cells. *Exp. Cell. Res.* 135:465 (1981)
9. J. Maubaum and H.G. Mandel, Unilateral chromatid damage: a new basis for 6-thioguanine cytotoxicity. *Cancer Res.* 43:3852 (1983).
10. N.T. Christie, S. Drake, R.E. Meyn and J.A. Nelson, 6-Thioguanine-induced DNA damage as a determinant of cytotoxicity in cultured chinese hamster ovary cells. *Cancer Res.* 44:3665 (1984).
11. J.M. Covey, M. D'Incalci and W. Kohn, Production of DNA-protein crosslinks (DPC) by 6-thioguanine (TG) and 2'-deoxy-6-thioguanosine (TGdR) in L1210 cells in vitro. Proc. Am. Assoc. *Cancer Res.* 27:17 (1986).
12. B.F. Pan and J.A. Nelson, Characterization of the DNA damage in 6-thioguanine-treated cells. *Biochem. Pharmacol.* 40:1063 (1990).
13. W.J. Bodell, Molecular dosimetry of sisterchromatid exchange induction in 9L cells treated with 6-thioguanine. *Mutagen.* 6:175 (1991).
14. J.P. Bökkerink, E.H. Stet, R.A. De Abreu, F.J. Damen, T.W. Hulscher, M.A. Bakker and J.M. van Baal, 6-Mercaptopurine: cytotoxicity and biochemical pharmacology in human malignant T-lymphoblasts. *Biochem. Pharmacol.* 45:1455 (1993).

15. Y.H. Ling. J.Y. Chan, K.L. Beatle and J.A. Nelson, Consequences of 6-thioguanine incorporation into DNA on polymerase, ligase, and endonuclease reactions. *Mol. Pharmacol.* 42:802 (1992)
16. R.T. Kawata, L-F Chuang, C.A. Holmberg et al., Inhibition of human lymphoma DNA-dependent RNA polymerases activity by 6-mercaptopurine ribonucleoside triphosphate. *Cancer Res.* 43:3655 (1983).
17. R.M. Weinshilboum, Methylation pharmacogenetics: thiopurine methyltransferase as a model system. *Xenobiotica* 22:1055 (1992).
18. L.L. Bennett Jr. and P.W. Allan, Formation and significance of 6-methylmercaptopurine ribonucleotide as a metabolite of 6-mercaptopurine. *Cancer Res.* 31:152 (1971).
19. J.A. Sokoloski and A.C. Sartorelli, Inhibition of HL-60 promyelocytic leukemia by 6-mercaptopurine ribonucleoside. *Cancer Res.* 47:6283 (1987).
20. M.B. Cohen and W. Sadee, Contribution of the depletions of guanine and adenine nucleotides to the toxicity of purine starvation in mouse T-lymphoma cells. *Cancer Res.* 43:1587 (1983).
21. M.H.J. Vogt, E.H. Stet, R.A. De Abreu, J.P.M. Bökkerink, L.H.J. Lambooy and F.J.M. Trijbels, The importance of methylthio-IMP for methylmercaptopurine ribonucleoside (MeMPR) cytotoxicity in Molt F4 human malignant T-lymphoblasts. *Biochem. Biophys. Acta* 1181:189 (1993).
22. E.H. Stet, R.A. De Abreu, J.P.M. Bökkerink, T.M. Vogels, L.H. Lambooy, F.J. Trijbels and R.C. Trueworthy, Reversal of 6-mercaptopurine and 6-methylmercaptopurine ribonucleoside cytotoxicity by amidoimidazole carboxamide ribonucleiside in Molt F4 human malignant T-lymphoblasts. *Biochem. Pharmacol.* 46:547 (1993).
23. E.H. Stet. The relevance of methylation for thiopurine cytotoxicity. Thesis, University Hospital Nijmegen, The Netherlands (1994).
24. J.A. Van Loon and R.M. Weinshilboum, Thiopurine methyltransferase isozymes in human renal tissue. *Drug Met. Dispos.* 18:632 (1990).
25. L. Lennard, J.S. Lillyman, J. van Loon and R.M. Weinshilboum, Genetic variation in response to 6-mercaptopurine for childhood acute lymphoblastic leukemia. *Lancet* 336:225 (1990).
26. J.A. van Loon and R.M. Weinshilboum, Thiopurine pharmacokinetics in leukemia: Correlation of erythrocyte thiopurine methyltransferase and 6-thioguanine nucleotide concentration. *Clin. Pharmacol. Ther.* 41:18 (1987).
27. S. Zimm, L. Ettinger, J. Holcenberg et al., 6-Mercaptopurine administratered as a prolonged intravenous (IV) infusion: plasma and CSF pharmacokinetics. *Proc. Am. Soc. Clin. Oncol.* 3:37 (1984).
28. J.P.M. Bökkerink, M.A.H. Bakker, T.W. Hulscher, R.A. De Abreu and E.D.A.M. Schretlen, Purine de novo synthesis as the basis of synergism of methotrexate and 6-mercaptopurine in human malignant lymphoblasts of different lineages. *Biochem. Pharmacol.* 37: 2321 (1988).
29. R.A. De Abreu, F. van Strien, L.H. Lambooy and J.P. Bökkerink, Synergistic Interaction of methotrexate and 6-mercaptopurine in human derived malignant T-ALL and CALA⁺ cell lines. *Adv. Exp. Med. Biol.* 309A: 87 (1991).
30. J.P. Bökkerink, R.A. De Abreu, E.H. Stet and F.J. Damen, Cell-kinetic and biochemical pharmacology of methotrexate and 6-mercaptopurine in human malignant T-lymphoblasts. *Klin. Pädiatr.* 204:293 (1992).
31. B. Camitta, B. Leventhal, S. Lauer, J.J. Shuster, S. Adair, J. Casper, C. Civin, M. Graham, D. Mahoney, L. Munoz, G. Kiefer and B. Kamen. Intermediate-dose intravenous methotrexate and mercaptopurine therapy for non-T, non-B acute lymphocytic leukemia of childhood: a Pediatric Oncology Group study. *J. Clin. Oncol.* 7:1539 (1889).

MITOCHONDRIAL VERSUS CYTOSOLIC ACTIVITIES OF DEOXYRIBONUCLEOSIDE SALVAGE ENZYMES

Johan C.F. Söderlund and Elias S.J. Arnér

Medical Nobel Institute for Biochemistry
Department of Medical Biochemistry and Biophysics
Karolinska Institutet
S-171 77 Stockholm, Sweden

INTRODUCTION

The four major mammalian anabolic enzymes in the salvage pathway of deoxyribonucleosides are deoxycytidine kinase (dCK), the two thymidine kinase isoenzymes (TK1 and TK2) and deoxyguanosine kinase (dGK). These enzymes were recently cloned and/or homogeneously purified and thereby thoroughly characterized (reviewed by Arnér, 1993; Arnér and Eriksson, 1994). dCK and TK1 are both considered to be cytosolic, whereas dGK is regarded to have a mitochondrial location. The subcellular location of TK2, however, is not fully clarified, as discussed below.

The substrate specificities of the enzymes overlap, with both TK1 and TK2 phosphorylating Thd, TK2 and dCK phosphorylating dCyd and both dCK and dGK phosphorylating purine deoxyribonucleosides. With nucleoside analogs, all four enzymes show individual patterns of specificity, but with a certain overlap for selected analogs (Eriksson et al., 1991; Kierdaszuk et al., 1992; Munch-Petersen et al., 1991; Wang et al., 1993).

The subcellular distribution of the salvage enzymes has consequences for the metabolism of their natural substrates as well as that of chemotherapeutically used nucleoside analogs. It is also of interest to clarify in view of the mitochondrial toxicity seen in therapy with nucleoside analogs, such as that described for anti-HIV dideoxynucleosides (Chen et al., 1991).

To address the subcellular distribution of salvage enzyme activities, we have used crude extracts of cytosolic and mitochondrial fractions two mutant human T-lymphoblastic cell lines (CCRF-CEM cells) that are dCK$^-$ and TK1$^-$, respectively. The dCK$^-$ CEM$_{ddC}$50 cells were selected as described earlier (Hershfield et al., 1982; Ullman et al., 1988) by resistance to ddC and the TK1$^-$ CEM$_{BrdU}$ cells are resistant to BrdU. To use crude extracts retains the possibility of detecting additional salvage enzyme activities besides the four well characterized enzymes and, moreover, if the results correlate well to the kinetic characteristics of the pure enzymes this becomes one way of verifying that data to be reasonably valid also in the complex mixture of a crude cell extract.

MATERIALS AND METHODS

Materials and cell lines

Radiolabeled nucleosides were from Amersham (U.K.) or Moravek Biochemicals (Ca., U.S.A.) whereas unlabeled substrates came from Sigma (Mo., U.S.A.) except CdA which was provided by Dr. Z. Kazimierczuk (The foundation for the development of diagnostics and therapy, Warsaw, Poland). The dCK$^-$ CEM$_{ddC}$50 and TK1$^-$ CEM$_{BrdU}$ cell lines were originally developed by Dr. B. Ullman (Dept. of Biochemistry and Molecular Biology, University of Oregon, Portland, Or., U.S.A.) and were kindly provided by Dr. S. Eriksson (Dept. of Veterinary Medical Chemistry, Biomedical Center, Uppsala, Sweden).

Enzyme assays

Enzyme activity was measured radiochemically as described earlier (Arnér et al., 1992). Shortly, substrates were [5-^{3}H]dCyd (10, 300 and 600 μM), [8-^{3}H]CdA (50 μM), [8-^{3}H]AraG (20 μM), [methyl-^{3}H]Thd (10 μM)

Purine and Pyrimidine Metabolism in Man VIII, Edited by
A. Sahota and M. Taylor, Plenum Press, New York, 1995

and [5-^{3}H]ddC (600 μM) with specific activities of about 500-1500 cpm/pmole. Assays were performed with 10-40 μg of protein at 37°C in 40-75 μl of a reaction mixture containing 50 mM Tris-HCl (pH 7.6 in all assays except for Thd when pH was 8.0), 5 mM MgCl$_2$, 5 mM ATP, 2 mM DTT and 10 mM NaF. With dCyd as substrate, 25 μM THU was also included. At four time points within 60 min., aliquots of 10-15 μl were spotted on Whatman DE-81 filter discs, which were washed in ammonium formate. Thereafter filter-bound nucleotide products could be determined using a scintillation counter.

Preparation of cytosolic and mitochondrial protein extracts

Mitochondria were isolated from about 900 x 10^6 exponentially grown cells, using the protocol described by Tapper and coworkers (Tapper et al. 1983), with the following alterations. The supernatant (8 ml) from the centrifugation (13500 rpm for 15 min.) of the nuclei-free cell homogenate was recovered as the fraction containing cytosolic proteins and was upon addition of 90 μl 0.5 M benzamidine, 90 μl 50 mM PMSF, 450 μl 1.0 M Tris-HCl pH 7.6, 45 μl Nonidet P-40, and 18 μl 1.0 M DTT frozen in 1,5 ml aliquots and stored at -70 °C until analyzed. The mitochondria were resuspended in 7 ml of mitochondrial storage buffer (30 mM Tris-HCl pH 8.0, 0,5 mM EDTA pH 8,1, 0.25 M sucrose) and were after collection from the subsequent sucrose gradient isolated by centrifugation (13500 rpm for 15 min.). Hereafter the mitochondrial pellet was resuspended in 3 ml of 10 mM MgCl$_2$, 1 mM EDTA pH 8.1, 7,5 % glycerol, 1 mM DTT, 350 mM NaCl, 0,5 mM PMSF, 0.5 % Triton X-100, and 10 mM Tris-HCl, pH 8.0. The mitochondria were then disrupted in a small glass homogenizer, and after centrifugation at 45.000 rpm for 60 min. the supernatant, containing extracted mitochondrial proteins, was kept at -70°C in 250 μl aliquots until analyzed.

RESULTS

Figure 1 shows the CdA and AraG phosphorylation in cytosolic and mitochondrial fractions of crude extracts from the dCK⁻ CEM$_{ddC50}$ and the TK1⁻ CEM$_{BrdU}$ cell lines. No CdA or AraG phosphorylation could be detected in the cytosol of the dCK⁻ cells, whereas the cytosol of the TK1⁻ cells displayed a high CdA phosphorylation, which was completely inhibited by excess dCyd. Like in the CEM$_{ddC50}$ cytosol, no AraG phosphorylation could be found in that of the CEM$_{BrdU}$ cells. Both cell lines displayed a significant CdA phosphorylation in the mitochondrial fractions and this CdA phosphorylation was not inhibited by excess dCyd. In the same mitochondrial extracts an additional AraG phosphorylation was found.

Using a concentration of 10 μM, dCyd phosphorylation was significant only in the cytosol of the TK1⁻ cells (Figure 2). After increasing the concentration to 300 μM or 600 μM dCyd, however, phosphorylation could be detected in the cytosol of the dCK⁻ cells as well as in mitochondria of both cell lines. Moreover, with 600 μM dCyd this increase in dCyd phosphorylation was superimposed on the phosphorylation seen with 10 μM and 300 μM dCyd in the cytosol of the TK1⁻ cells (Figure 2).

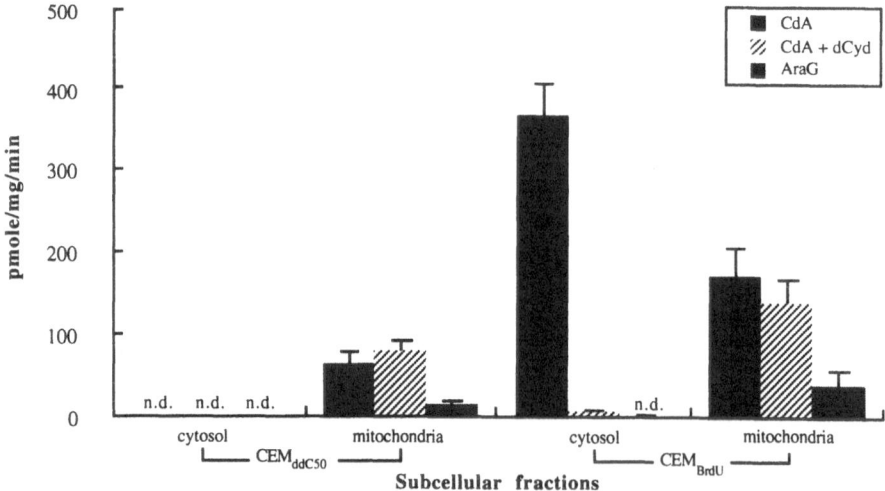

Figure 1. CdA and AraG phosphorylation in cytosolic and mitochondrial fractions of crude extracts from the dCK⁻ CEM$_{ddC50}$ and the TK1⁻ CEM$_{BrdU}$ cell lines.

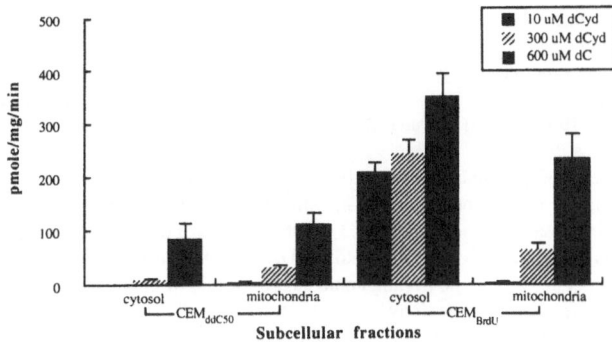

Figure 2. Cytosolic and mitochondrial phosphorylation of dCyd.

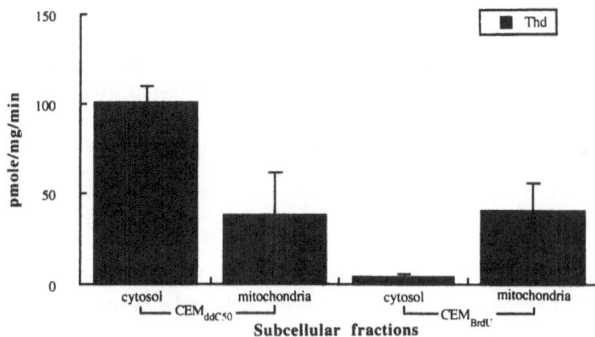

Figure 3. Thd phosphorylation in mitochondria and cytosol.

A Thd phosphorylation of about 40 pmole/mg/min was found in the mitochondria of both the dCK⁻ and the TK1⁻ cells (Figure 3). Twice this specific activity was seen in the cytosol of the dCK⁻ cells, whereas the cytosol of the TK1⁻ cells displayed a very low Thd phosphorylation (Figure 3).

When measuring ddC phosphorylation, this could only be detected in the cytosol of the TK1⁻ cells (171 ± 65 pmole/mg/min) but not in the cytosol of the dCK⁻ cells nor in any of the two mitochondrial extracts.

DISCUSSION

Based on the findings of this study we conclude that dGK activity was solely found in the mitochondrion, while TK2 activity was found in both the cytosol and the mitochondrion. It was recently shown that dGK phosphorylates CdA without being inhibited by dCyd (Wang et al., 1993). The CdA phosphorylation in the cytosolic fraction of the TK1⁻ cell line was inhibited by dCyd, whereas no cytosolic phosphorylation at all could be found in the dCK⁻ cells (Figure 1). This demonstrates that dCK is the sole CdA phosphorylating enzyme in the cytosol, in agreement with earlier studies showing dCK to be important for mediating CdA cytotoxicity (Carson et al., 1980). In addition, however, the mitochondria of both cell lines displayed a CdA phosphorylation unaffected by addition of excess dCyd (Figure 1). This activity was found earlier and can readily be explained by dGK-mediated CdA phosphorylation (Wang et al., 1993). The lack of cytosolic CdA phosphorylation in the dCK⁻ cell line (Figure 1) shows that dGK is a truly mitochondrial enzyme. This is also supported by the pattern of AraG phosphorylation (Figure 1) since AraG is a good substrate for dGK but not dCK (Wang et al., 1993).

The mitochondrial fractions of both cell lines as well as the cytosolic fraction of the dCK⁻ cells all show the same profile of dCyd phosphorylation, with increased phosphorylation upon increasing concentrations of dCyd up to 600 μM. We suggest that this dCyd phosphorylation is due to TK2 activity, as it is well in agreement with the dCyd phosphorylation by purified TK2 that has a K_m value around 40 μM for dCyd (Munch-Petersen et al., 1991). The specific activity was about the same in both the cytosol and the mitochondria of the dCK⁻ cells, indicating that TK2 is quite freely distributed to both the cytosolic and the mitochondrial compartments. This was found earlier, but was suggested to be due to leakage of TK2 from the mitochondria due to the subfractionation procedure (Jansson et al., 1992). The cytosolic activity of TK2 implied here, however, can not easily be explained by a leakage artifact as in that case also dGK activity should have been found in the cytosolic fractions, which was not the case.

A high Thd phosphorylation was found in the cytosol of the dCK⁻ cells, but not in the cytosol of the TK1⁻ cells, reflecting the TK1-derived Thd phosphorylation (Figure 3). In addition, a significant Thd phosphorylation was found in the mitochondria of both cell lines, demonstrating the mitochondrial presence of TK2. The TK2-derived Thd phosphorylation was readily detected in the mitochondria 10 μM Thd (Figure 3), in contrast to the TK2-derived phosphorylation of dCyd, where substrate concentrations of 300 μM and above were needed in order to detect activity (Figure 2). This is in agreement with the lower (0.3 μM) K_m value of TK2 for Thd (Munch-Petersen et al., 1991), which is considerably lower than the K_m value of TK2 for dCyd (as discussed above).

Interestingly, the dCyd phosphorylation in the cytosol of the dCK⁻ cells suggested a presence of TK2 in the cytosol as well as in the mitochondria (Figure 2), while the Thd phosphorylation in the cytosol of the TK1⁻ cells was lower than in the mitochondria (Figure 3). If this reflects a decrease of cytosolic TK2 concomitant with loss of TK1 upon development of resistance to BrdU, which is phosphorylated by both purified TK1 and purified TK2 (Eriksson et al., 1991), the finding could suggest a role of the cytosolic TK2, but not of the mitochondrial TK2, in mediating BrdU cytotoxicity.

We also found that ddC seems only to be phosphorylated by dCK, indicating that side effects of ddC due to mitochondrial toxicity are independent of intra-mitochondrial phosphorylation. This was also suggested by Chen and Cheng (1992) who showed that ddC entered the mitochondrion from the cytosol on the nucleotide level.

In conclusion, our results can readily be interpreted with the kinetic data from earlier studies with the cloned and/or purified salvage enzymes, and show no evident signs of additional kinase activities in the cytosol or mitochondria for the substrates used in this study. Further experiments using mitochondrial and cytosolic subfractionation for studying the phosphorylation of additional nucleoside analogs, utilizing other triphosphates than ATP as phosphate donors or determine mitochondrial versus cytosolic 5'-nucleotidase activities should be of great interest.

ACKNOWLEDGMENTS

This study was supported by the Swedish Society of Medicine, the Swedish Cancer Society (project nr 961 and contract 942026) and the Medical Faculty of the Karolinska Institute. Generous support from Drs Staffan Eriksson and Arne Holmgren is also gratefully acknowledged.

REFERENCES

Arnér ESJ, Eriksson S., 1994, Expression and properties of mammalian deoxyribonucleoside kinases. Pharm Ther; in preparation.

Arnér ESJ, Spasokoukotskaja T, Eriksson S., 1992, Selective Assays for Thymidine kinase 1 and 2 and Deoxycytidine Kinase and Their Activities in Extracts from Human Cells and Tissues. Biochem. Biophys. Res. Comm.;188:712-718.

Arnér ESJ., 1993, Studies of the salvage and metabolism of deoxyribonucleosides in human cells and tissue [Thesis]. Karolinska Institute, Stockholm, ISBN 91-628-0984-9.

Carson DA, Wasson DB, Kaye J, et al., 1980, Deoxycytidine kinase-mediated toxicity of deoxyadenosine analogs toward malignant human lymphoblasts in vitro and toward murine L1210 leukemia in vivo. Proc Natl Acad Sci U S A;77(11):6865-9.

Chen CH, Cheng YC., 1992, The role of cytoplasmic deoxycytidine kinase in the mitochondrial effects of the antihuman immunodeficiency virus compound, 2',3'-dideoxycytidine. J Biol Chem;267(5):2856-9.

Chen CH, Vazquez Padua M, Cheng YC., 1991, Effect of anti-human immunodeficiency virus nucleoside analogs on mitochondrial DNA and its implication for delayed toxicity. Mol Pharmacol;39(5):625-8.

Eriksson S, Kierdaszuk B, Munch-Petersen B, Öberg B, Johansson NG., 1991, Comparison of the substrate specificities of human thymidine kinase 1 and 2 and deoxycytidine kinase toward antiviral and cytostatic nucleoside analogs. Biochem Biophys Res Commun;176(2):586-592.

Hershfield MS, Fetter JE, Small WC, et al., 1982, Effects of mutational loss of adenosine kinase and deoxycytidine kinase on deoxyATP accumulation and deoxyadenosine toxicity in cultured CEM human T-lymphoblastoid cells. J Biol Chem;257:6380-6386.

Kierdaszuk B, Bohman C, Ullman B, Eriksson S., 1992, Substrate specificity of human deoxycytidine kinase toward antiviral 2',3'-dideoxynucleoside analogs. Biochem Pharmacol;43(2):197-206.

Munch-Petersen B, Cloos L, Tyrsted G, Eriksson S., 1991, Diverging substrate specificity of pure human thymidine kinases 1 and 2 against antiviral dideoxynucleosides. J Biol Chem;266(14):9032-9038.

Tapper DP, Van Etten RA, Clayton DA., 1983, Isolation of mitochondrial DNA and RNA and cloning of the mitochondrial genome. Methods Enzymol.;97:426-434.

Ullman B, Coons T, Rockwell S, McCartan K. Genetic analysis of 2',3'-dideoxycytidine incorporation into cultured human T lymphoblasts. J Biol Chem 1988;263(25):12391-6.

Wang L, Karlsson A, Arnér ESJ, Eriksson S., 1993, Substrate specificity of mitochondrial 2'-deoxyguanosine kinase: Efficient phosphorylation of 2-chlorodeoxyadenosine (CdA). J. Biol. Chem.;268:22847-228852.

DOES LOW-DOSE ALLOPURINOL, WITH AZATHIOPRINE, CYCLOSPORIN AND PREDNISOLONE, IMPROVE RENAL TRANSPLANT IMMUNOSUPPRESSION?

PR Chocair[1], JA Duley[2], JS Cameron[2], S Arap[1], L Ianhez[1], E Sabbaga[1], HA Simmonds

[1] Unidade de Transplante Renal, Hospital das Clinicas, Universidade de Sao Paulo, Brazil
[2] Purine Research Laboratory, UMDS Guy's Hospital, London, UK

INTRODUCTION

This communication reports the results of two studies aimed at improving the immunosuppressive efficacy of azathioprine in renal transplant patients. Although azathioprine has been used in transplantation for over 30 years, the mechanism of its cytotoxicity is not yet completely defined. It is generally accepted that its principal mode of action involves rapid conversion in vivo to 6-mercaptopurine (6MP), which is further metabolised via 6-thioinosinic acid to cytotoxic thioguanine nucleotides (1). The efficacy of 6MP is reduced by catabolic pathways, one of them being via xanthine oxidase (XO), which oxidises 6MP to thiouric acid. Thus the therapeutic combination of azathioprine with allopurinol, which inhibits the XO path, is usually contra-indicated, having been associated with bone marrow toxicity (2). The importance of this catabolic route is illustrated by the severe myelotoxicity reported in an XO-deficient patient treated with azathioprine (3). Azathioprine and 6MP catabolism may also occur via aldehyde oxidase (AO) and thiopurine methyltransferase (TPMT) (Figure 1).

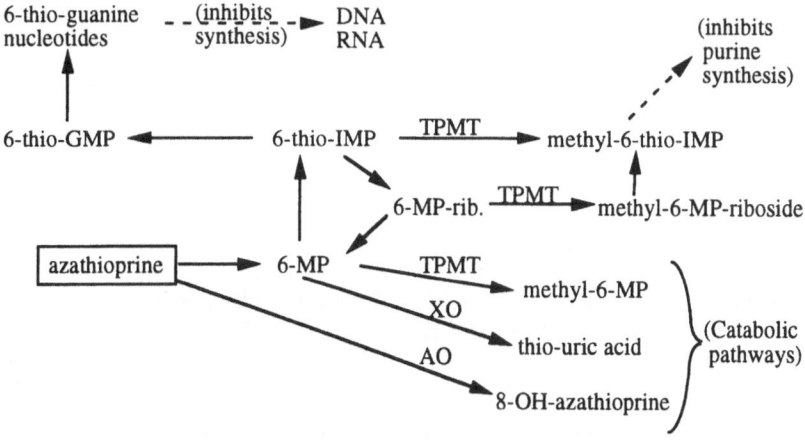

Figure 1. Azathioprine metabolic pathways.

Episodes of early acute rejection are common following cadaver renal transplantation, even since the advent of cyclosporin. Since the high cost of cyclosporin may inhibit its usage, the rationale behind our studies was to investigate whether small doses of allopurinol, in association with conventional immunosuppression using 'triple therapy' (cyclosporin + prednisolone + azathioprine), might improve the efficacy of azathioprine therapy without inducing the myelotoxicity associated with a normal allopurinol dosage. The first study was conducted during 1992/93, with two patient groups selected as clinically similar. The second study was conducted during 1993/94, as a 'double blind' prospective trial on randomly selected patients.

PATIENTS

Study 1: Twelve cadaver renal transplant recipients received low-dose allopurinol (25 mg, alternative days) plus 'triple therapy'. These were compared with fifteen matching case control subjects for whom clinical criteria were identical, i.e. adult cadaver recipients on normal 'triple therapy' only commencing the day of surgery, and with comparable immunological status. There was one retransplanted patient in the treated group and none in the controls, while one control had panel–reactive antibodies higher than 20%.

Study 2: Thirty two patients were randomly assigned into two groups. One group of sixteen cadaver renal transplant recipients received low-dose allopurinol (25 mg once a day) plus 'triple therapy'. Note that this allopurinol dose was low, but double that of the first study. The second group of sixteen patients received 'triple therapy' alone.

Tissue typing was not considered an issue, as both groups received cadaver kidneys.

Table 1. Immunosuppressive therapy and general parameters of controls and treated patients.

	STUDY 1		STUDY 2	
	Control[1] (n=15)	Treated (n=12)	Control (n=16)	Treated (n=16)
Azathioprine (mg/kg)	1.84+0.06	1.33+0.18	1.7+0.36	1.3+0.43
Prednisone (mg/kg)	0.17+0.01	0.21+0.03	0.18+0.03	0.19+0.05
Cyclosporin (mg/kg)	4.4+0.25	4.7+0.28	4.4+1.35	4.5+0.96
Allopurinol (mg)	0	25 alt.	0	25 o.d.
Age (years)	40.6+3.1	40+2.8	43.5+14.3	41.6+11.0
Male/Female	6/9	9/3	6/10	9/7
Dialytic age (months)	45.5+8	39.6+9˙	32.6+30.4	41.6+41.5
Follow-up (months)	4.2+0.22	3.6+0.15	¶	¶

[1] All values are expressed as mean±SEM; Student's T-test used to compare groups;
¶ follow-up presently approx 3 months and continuing.

All patients were immunosuppressed by 'triple therapy': cyclosporin was given orally twice daily post-operatively, regardless of graft function, at 8 mg/kg/day for the first five days, reducing to about 4 mg/kg/day, the cyclosporin concentrations in whole blood being maintained around 100 µmol/L. Methylprednisolone was given before transplantation, with hydrocortisone after operation, followed by prednisone which was tapered off progressively. Acute rejection episodes (diagnosed on clinical grounds with histological confirmation) were treated for three days with intravenous methylprednisolone (0.5–1 g/day). Azathioprine was started at 5 mg/kg/day for three days, then reduced to about 2 mg/kg/day, according to the peripheral white blood cell count (WBC). There were no significant differences between the groups with respect to transfusions, antihypertensive therapy or diuretics. ATG, ALG or OKT3 were not used. Allopurinol was given post–operatively to treated-group patients when the plasma creatinine had fallen below 220 µmol/L (mean: 10th post-operative day), and maintained for about 3 months.

Brazil has a very heterogeneous racial mix. The majority of patients (approx 75%) were 'white', there were 2 Brazilian Japanese, and the remainder were 'black'. There were no significant differences between the racial distributions of the treated and control groups.

RESULTS

Study 1: The results were surprising (Table 2): only one rejection episode occurred in the allopurinol-treated group, compared with 11 of the 15 case controls who experienced rejections. As expected, it was necessary to reduce the azathioprine dose in the treated group (mean 1.3 versus 1.8mg/kg in controls), depending on WBC count. A total of 7 patients (3 controls and 4 allopurinol-treated) required a further reduction in azathioprine dose because of mild leucopenia. Haemoglobin was slightly but significantly lower in the treated group.

Only one patient (from treated group) experienced serious leucopenia (2200 WBC/mm^3), 15 days after starting allopurinol. Allopurinol and azathioprine were stopped, and ten days later the WBC had returned to the normal range: the patient was then given the usual dose of azathioprine, but not allopurinol, with no subsequent leucopenia. Plasma oxypurinol, measured in 4 treated patients, ranged from 1 to 11 μmol/L. Serum creatinines and plasma uric acids were not different in both groups by the end of the study.

Study 2: Two patients, one in each group, showed humoral rejection and were excluded from the analysis (leaving 15 in each group). The result for cellular rejections was again highly significant, only 4 patients experiencing rejection episodes following allopurinol therapy, compared with 11 patients in the control group. Fuller results for other parameters are being analysed, however hemoglobins in the treated group were again reduced.

Table 2. Clinical results for controls and treated patients[1].

	STUDY 1		STUDY 2	
	Control[1] (n=15)	Treated (n=12)	Control (n=15)	Treated (n=15)
Acute rejection	11	1*	11	4**
2 or more crises	7	0*	8	1*
Serum creatinine[2]	97.3±3.5	98±8.8	110+35	115+31
Hemoglobin (g%)	12.6±0.38	10.8±0.52*	na	na
WBC (/mm^3)	6240±349	5590±520	na	na
Infections incl. CMV	3	5	3	5
Death	0	0	1	0
Plasma oxypurinol	-	6 (1-11)	-	-
Blood cyclosporin	145±10	99±13*	196+116	152+43

[1]Values are expressed as mean±SEM (range in brackets); Fisher's exact test used to compare groups (*= signif less than 0.01, **= signif less than 0.025); [2]mean of the last 3 controls; oxypurinol, cyclosporin, creatinine, uric acid shown as umol/L; na = not available at present.

DISCUSSION

These two studies arose out of a previous interest in the relevance of azathioprine metabolism, via the body's purine pathways, to its immunosuppressive efficacy (4). In both of the studies reported here, there was an impressive reduction of acute rejections in the treated groups after the introduction of allopurinol, compared with the control groups.

In the absence of allopurinol, the prevalence of rejections was more than 70% in both studies, which accords well with the general experience in Sao Paulo. This rejection rate is higher than that usually observed in European countries, but demonstrates the racial differences in outcome that have been well-documented, for example, among black renal graft recipients in the U.S.A. Allopurinol treatment reduced the rejection rate to less than 20% overall. This improved outcome in the treated groups is particularly intriguing, considering that it was observed in cadaver kidney recipients receiving low-dose azathioprine.

These results are particularly important when taking into account the strong evidence that acute rejections are devastating events. This is not only because they are the most significant cause of graft loss in the first year following transplantation, but also because they are predictive of chronic rejection, even when the acute rejection phase is thought to have been adequately treated (5).

The use of allopurinol is usually contra-indicated with azathioprine therapy, yet with careful adjustment of the azathioprine dose, the bone marrow tolerated this low-dose regime well. The level of haemoglobin was noticeably higher in the controls presumably because of slight bone marrow depression by allopurinol, even at low dose.

The significantly improved immunosuppression could relate to the potentiation of azathioprine, by inhibition 6MP catabolism via xanthine oxidase and/or aldehyde oxidase (6). However, simply increasing the azathioprine dose has not been found clinically to benefit immunosuppression.

An alternative explanation is that the inhibition of pyrimidine biosynthesis by the ribotide of oxypurinol (the active metabolite of allopurinol), concomitant with the purine metabolic effects of 6MP, is the mechanism behind the seemingly synergistic effect. Additionally, allopurinol is reported to have an important suppressive effect on the production of oxygen free–radicals under both ischemic and inflammatory conditions (7). The involvement of these radicals in tissue injury is well known in a variety of situations, including renal transplants.

It has been reported recently that cyclosporin provides a long-term renal graft survival superior to azathioprine (8). However, cyclosporin is very expensive compared with azathioprine, inhibiting its use in Brazil and other developing countries. The incorporation of both drugs into 'triple therapy' has provided an effective and cost-efficient means of immunosuppression: the inclusion of allopurinol in the post-operative period appears to provide additional and excellent protection against acute rejection.

REFERENCES

1. W.N. Kelley, F.M. Rosenbloom, J.E. Seegmiller, Effects of azathioprine on purine synthesis in clinical disorders of purine metabolism, *J. Clin. Invest..* 46: 1518 (1967).
2. V.G. Raman, V.L. Sharman, H.A. Lee, Azathioprine and allopurinol: a potentially dangerous combination, *J. Intern. Med.* 228: 69 (1990).
3. G. Hillebrand, S. Reiter, Hypourikämie – ein differentialdiagnostisches Problem, *Internist* 32: 226 (1991).
4. P.R. Chocair, J.A. Duley, H.A. Simmonds, J.S. Cameron, The importance of thiopurine methyltransferase activity for the use of azathioprine in transplant recipients. *Transplantation* 53: 1051 (1992).
5. G.P. Bassadonna, A.J. Matas, K.J. Gillingham, W.D. Payne, D.L. Dunn, D.E.R. Sutherland, P.F. Gores, R.W.G. Gruessner, J.S. Najarian, Early versus late acute renal allograft rejection: impact on chronic rejection, *Transplantation* 55: 993 (1993).
6. A.H. Chalmers, P.R. Knight, M.R. Atkinson, 6–Thiopurines as substrates of purine oxidases: a pathway for conversion of azathioprine into 6–thiouric acid without release of 6–mercaptopurine, *Aust. J. Exp. Med. Sci.* 47: 263 (1969).
7. K.A. Nath, Reactive oxygen species and renal injury, *in:* "International Yearbook of Nephrology", V.E. Andreucci, L.G. Fine, eds., Kluwer Academic, N.Y. (1991).
8. J.W. Slaton, K.A. Kropp, J.S. Jhunjhunwala, S.H. Selman, Cyclosporin versus azathioprine: a 5-year followup of 200 consecutive cadaver renal transplant recipients, *J. Urol.* 151: 582 (1994).

PURINE NUCLEOTIDE INTERCONVERSION ENZYMES IN CHILDHOOD LEUKEMIA: RELATION WITH CELL CYCLE AND CLINICAL OUTCOME

D.C. van Oostveen[1], R. Pieters[1*], G.J. Peters[2], A.J.P. Veerman[1]

Department of Pediatrics[1] and Oncology[2]
Free University Hospital
POB 7057, 1007MB Amsterdam
The Netherlands
(*corresponding author)

INTRODUCTION

Many authors have studied enzymes of the purine metabolism in relation to subtypes of acute lymphoblastic leukemia leukemia (ALL). Recently we found that low activities of adenosine deaminase (ADA), purine nucleoside phosphorylase (PNP) and hypoxantine-guanine phosphoribosyltransferase (HGPRT) correlated with a poorer prognosis in childhood ALL.[1,2] A small subset of children with a very high ecto-5'nucleotidase (ecto-5'NT) expression on their leukemic cells also had a relatively poorer prognosis.[3,4] The hypothesis that the prognostic value of these enzymes could be explained by a relation with thiopurine resistance was not confirmed.[1,2,4]

In lymphoblasts PNP and HGPRT have a purine salvage function. Some studies showed that activities of several salvage enzymes are lower in quiescent cells compared to those of proliferating cells[5-9]. This suggests that ALL cells with low activities of purine salvage enzymes have a lower therapeutic index compared to normal hematopoietic cells which might perhaps explain the prognostic relevance of the enzyme activities. Therefore we investigated the relation between the percentage of cells in S phase and the activities of ADA, PNP, HGPRT and ecto-5'NT in children with precursor B-lineage ALL.

MATERIALS AND METHODS

Bone marrow or peripheral blood cells of 37 children with CD10[+] precursor B-ALL were studied. Activities of ADA, PNP, ecto-5'NT and HGPRT were determined as described earlier.[1,2,4] The cell cycle distribution was determined with the FACS scan and the percentage of cells calculated as described elsewhere.[10]

RESULTS

The ADA activity showed a large variation between the patients ranging from 30 to more than 300 nmol/hr/10^6 cells. When ADA activity was plotted against the percentage of cells in S phase no significant correlation was observed.

For PNP activity an almost similar variation in enzyme activity was observed as for ADA activity (fig. 1). The PNP activity was significantly ($p<0.05$) related to the percentage of cells in the S phase. The Pearson correlation coefficient was 0.40.

For HGPRT the variation in enzyme activity was less than for ADA and PNP and ranged from 3.48 to 20.85 nmol/hr/10^6 cells (fig. 2) The relation between HGPRT and S phase was statistically significant ($p<0.05$). The correlation coefficients was 0.41.

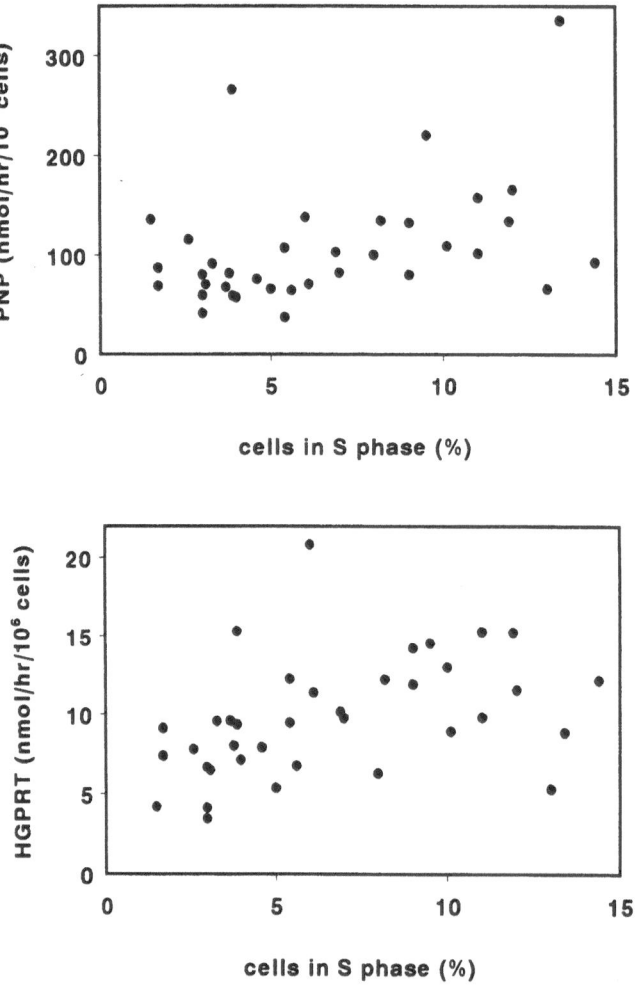

Fig. 1 and 2. Relation between PNP (fig 1) and HGPRT (fig 2) activities and percentage of cells in the S phase of the cell cycle in childhood precursor B-ALL.

A large variation in ecto-5'NT activity from not detectable to more than 80 nmol/hr/10^6 was found. A weak and negative correlation between eco-5'NT and cells in S phase was observed. The correlation coefficient was -0.21 and not statistically significant.

DISCUSSION

Part of the changes in enzyme activities associated with the progression of cells from the G_1 to the S phase are due to the increase in cell size and cell mass during proliferation. However, cell cycle dependent changes in the rates of DNA and RNA synthesis are also of great importance.[7] Müller et al.[5] showed in neoplastic cells an enhanced uptake of purine bases and a higher HGPRT activity compared with normal cells. In phytohemagglutin stimulated peripheral blood lymphocytes an increase in both PNP and HGPRT activity was observed compared to resting lymphocytes[6]. Several authors showed that the activities of HGPRT and PNP increased during the progression of the cell cycle or proliferation.[6,11,12]

In line with these results we found a statistically significant positive correlation between the percentage of ALL cells in S phase and the activities of both PNP and HGPRT. These relations were however weak with a correlation coefficient of 0.4. Earlier we demonstrated that low HGPRT and PNP activities were associated with a poorer prognosis in children with ALL.[1,2] So a leukemic cell population with a low HGPRT or PNP activity might represent a relative quiescent cell population with a lower therapeutic index compared to normal, not malignant hemopoietic cells.

Contradictive results have been published about the relationship between 5'NT activity and phase of the cell cycle or proliferation state.[13-16] A high ecto-5'NT expression on the leukemic cells of children with cALL and pre B ALL was related to a relatively poorer prognosis.[2,3] In the present study, we found no relation between ecto-5'NT activity and percentage of leukemic cells in the S phase. So, these results do not support the hypothesis that high ecto-5'NT activities correspond with a relatively slowly proliferating cell population. In lymphyocytes stimulated to proliferation with mitogens no significant increase in ADA compared to non-stimulated lympgocytes was observed.[7-9] In the present study there was no relation between ADA activity and the S phase.

CONCLUSIONS

A significant but weak relation was found between HGPRT and PNP activity on one hand and the percentage of cells in the S phase on the other hand. These findings might contribute to the explanation of our earlier reports on the relationship between low HGPRT and PNP activities and a poor prognosis: quiescent malignant cell populations with low activities of purine salvage pathway enzymes might have a low therapeutic index compared to normal hematopoietic cells in vivo.

REFERENCES

1. R. Pieters, D.R. Huismans, A.H. Loonen, G.J. Peters, K. Hählen, A. van der Does-van den Berg, E.R. van Wering, A.J.P. Veerman. Adenosine deaminase and

purine nucleoside phosphorylase in childhood lymphoblastic leukemia: Relation with differentiation stage, in vitro drug resistance and clinical prognosis. Leukemia 6:375 (1992).

2. R. Pieters, D.R. Huismans, A.H. Loonen, G.J. Peters, K. Hählen, A. van der Does-van den Berg, E.R. van Wering, A.J.P. Veerman. Hypoxantine-guanine phosphoribosyltransferase in childhood leukemia: relation with immunophenotype, in vitro drug resistance and clinical prognosis. Int J Cancer 51:213 (1992).

3. A.J.P. Veerman, P.H.G. Hogeman, C.H. van Zantwijk, P.D. Bezemer. Prognostic value of 5-nucleotidase in acute lymphoblastic leukemia with the common-ALL phenotype. Leuk Res 6:1227 (1985).

4. R. Pieters, D.R. Huismans, A.H. Loonen, G.J. Peters, K. Hählen, A. van der Does-van den Berg, E.R. van Wering, A.J.P. Veerman. Relation of 5'-nucleotidase and phosphatase activities with immunophenotype, drug resistance and clinical prognosis in childhood leukemia. Leuk Res 16:873 (1992).

5. M.M. Müller, M. Kraup, P. Chiba, H. Rumpold. Regulation of purine uptake in normal and neoplastic cells. Adv Enzyme Regul 21:239 (1983).

6. J. Hordern, J.F. Henderson. Comparison of purine and pyrimidine metabolism in G_1 and S phases of HeLa and Chinese hamster ovary cells. Can J Biochem 60:422 (1982).

7. F.F. Snyder, J. Mendelsohn, J.E. Seegmiller. Adenosine metabolism in phytohemagglutigin-stimulated human lymphocytes. J Clin Invest 58:654 (1976).

8. T. Hovi, J.F. Smyth, A.C. Allison, S.C. Williams. Role of adenosine in lymphocyte proliferation. Clin Exp Immunol 23:395 (1976).

9. G.J. Peters, J.H. Veerkamp. Purine and pyrimide metabolism in peripheral blood lymphocytes. Int J Biochem 15:115 (1983).

10. L.A. Smets, E. Mulder, F.C. de Waal, F.J. Cleton, J. Blok. Early responses to chemotherapy detected by pulse cytophotometry. Br J Cancer 34:153 (1976).

11. F.F. Snyder, J.F. Henderson, S.C. Kim, A.R.P. Paterson, L.W. Brox. Purine nucleotide metabolism and nucleotide pool sizes in synchronized lymphoma L5178Y cells. Cancer Res 33:2425 (1973).

12. H.R. Zielke, Su-Chen Hong, J.W. Littlefield. Different timing of increases in activity of four X-chromosome linked enzymes during the cell cycle of synchronized human lymphoblasts. Exp Cell Res 97:427 (1976).

13. J. Turnay, N. Olmo, G. Risse, K. von der Mark, M.A. Lizarbe. 5'-nucleotidase activity in cultured cell lines. Effect of different assay conditions and correlation with cell proliferation. In Vitro Cell Dev Biol 25:1055 (1989).

14. L. Lelièvre, B. Prigent, A. Paraf. Contact inhibition- Plasma membranes enzymatic activities in cultured cell lines. Biochem Biophys Res Commun 45:637 (1971).

15. H.B. Bosmann. Cellular membranes: Membrane marker enzyme activities in synchronized mouse leukemic cells L5178Y. Biochim Biophys Acta 203:256 (1970).

16. A.S. Sun, J.F. Holland, T. Ohnuma, M. Slankard-Chahinian. 5'-nucleotidase activity in permanent human lymphoid cell lines. Implication for cell proliferation and aging in vitro. Biochim Biophys Acta 714:530 (1982).

BEHAVIOR OF ENZYMES INVOLVED IN PURINE NUCLEOTIDE METABOLISM IN TUMORS

L.Lorenzini,[2] A.De Martino,[2] W.Testi,[2] F.Sorbellini,[2] E.Dispensa,[3]
A.Tabucchi,[1] F.Carlucci,[1] and F.Rosi[1]

[1]Institute of Biochemistry and Enzymology
[2]Institute of General Surgery
 University of Siena, Italy
[3]Division of Haematology
 Hospital of Siena, Italy

INTRODUCTION

Alterations in nucleotide metabolism have been observed in all experimental tumors. The "biosynthetic enzymes" and, specifically, phosphoribosylamidotransferase, which is the "key biosynthetic enzyme", are clearly enhanced in experimental tumors, whereas "catabolic enzymes" - adenylate deaminase, 5'-nucleotidase, purine nucleoside-phosphorylase and, specifically, xanthine oxidase (the "key catabolic enzyme") - are decreased. The phosphoribosylamidotransferase/xanthine oxidase ratio increases with a characteristic pattern and is related to the aggressivity and invasiveness of the tumor[1,2].

We have taken into consideration the enzymes at the "inosinic branch point" of the de novo synthesis: IMP-dehydrogenase (EC 1.1.1.205); GMP-synthetase (EC 6.3.5.2); AMP-succinate (AMP-S) synthetase (EC 6.3.4.4); AMP-succinate (AMP-S) lyase (EC 4.3.2.2). The determinations were carried out in lymphocyte extracts from normal and leukemic subjects as well as in the intestinal mucosa of healthy controls and colon-rectal cancer bearing patients.

The importance of this study arises from the fact that IMP occupies a central position in purine metabolism, acting as the common precursor for adenine and guanine and also leading to the purine degradative pathway. The control of such a branch point is strictly maintained by a combination of positive and negative feedback effects in normal conditions[3] and is dramatically lost when the cell undergoes malignant transformation.

Address for correspondence: Prof. Enrico Marinello, Istituto di Biochimica e di Enzimologia, Università di Siena, Pian dei Mantellini, 44, 53100 SIENA Italy, Tel. +39-577-298026, FAX +39-577-298057

EXPERIMENTALS

Chemicals

The nucleotides used as chromatographic standards were obtained from Sigma Chemical Corp. (St.Louis, MO, USA). All radiochemicals were obtained from Amersham International (Amersham, U.K.). Lymphocyte Separation Medium and phosphate-buffered saline were purchased from Flow Laboratories (Ayrshire, U.K.). All other chemicals were of analytical grade and were obtained from Merck (Darmstadt, Germany). DEAE-cellulose disks (DE-81) were from Whatman.

Subjects

Lymphocyte extract determinations were performed from samples taken from 8 normal and 8 leukemic subjects. The patients, aged from 47 to 75 years, had been affected by B-cell chronic lymphocytic leukemia (B-CLL) for a mean of 2.7 years. The diagnosis was based on lymphocyte number in peripheral blood (betwwen 22,000-425,000/mm^3), bone marrow examination, and lymphocyte typification by cyto-fluorimetry.

None of the patients with colon-rectal cancer had a history or clinical evidence of kidney disease; urea and creatinine levels were normal and urine sediment did not present any pathological findings. All patients had primitive neoplastic disease and had undergone surgery.

Preparation of lymphocytes

All blood samples for enzymatic determinations were drawn in the morning (8 a.m.) from patients who had fasted: 60 ml of blood taken from normal subjects was usually sufficient for all determinations. In the case of leukemic patients, volumes proportional to the counts were used. Lymphocytes were isolated using Lymphocytes Separation Medium[4] and then suspended in 0.29 M sucrose containing 10 mM Tris (pH 7). In all cases, the final concentration of the cells was 140,000/μl. Cells were counted in a Delcon cell counter. Lymphocyte extracts for all enzymatic determinations were prepared by sonication, and lysis was controlled microscopically. Extracts were spun down at 300 xg, and pellets were discarded. The supernatant (1 ml) obtained after sonication was sometimes treated with Norit A (150 μl of Norite A at 50% in H$_2$O) to eliminate endogenous nucleotides, which can interfere with enzymatic activities. In other cases, the lymphocyte supernatants were dialyzed for 2 hours against four changes of the same buffer.

Determination of the enzymes

The radiochemical method with labelled substrates was used for determination of all enzymes (except AMP-S lyase). Separation of the substrates or of labelled products was done by HPLC. Some of the following methods, as well as some chromatographic conditions, have already been described in the literature. They have been improved upon in our laboratory or have been adapted for the determination of enzymes in lymphoid cells. We specifically followed the indications of Green and Smith[5] for the separation of nucleotides and those of Welch and Rudolph[6] and Barankiewicz and Cohen[7] for the determination of the enzymes involved in nucleotide metabolism in lymphocyte extracts. In most cases, the separation of nucleotides used as substrates or formed during incubation was carried out by HPLC. A Beckman model 332 chromatograph equipped with a UV detector at 254 nm was used. Column, gradients and other details are described for each case. A 20-μl sample was always injected for HPLC analysis by overloading the injection loop with 50 µl. Unless otherwise indicated, deproteinization was carried out with a final 0.21 N $HClO_4$ concentration and neutralization with a final 0.22 N KOH concentration. Neutralized supernatants were frozen and stored at -70°C until HPLC analysis was performed. The identification of nucleotides was carried out by comparison with the retention time and coelution of internal standards. Their quantification was done by determining the radio-activity of the peak collected in a scintillation vial.

Enzymes of the inosinate branch point

1. Enzymes involved in the formation of AMP.
AMP-S synthetase (EC 6.3.4.4). The formation of AMP-S was followed in incubation mixtures as already reported by Schultz and Lowenstein[8]. Incubation was carried out at 37°C. The AMP-succinate formed was analyzed by HPLC according to Vannoni et al.[9] but with the flow increased to 1.5 ml/min.
AMP-S lyase (EC 4.3.2.2). The incubation mixtures were prepared essentially according to Jackson et al.[10]. The formation of AMP was analyzed by HPLC as in the case of AMP-S synthetase.
2. Enzymes involved in the formation of GMP.
IMP dehydrogenase (EC 1.1.1.205). The enzyme was measured by following the formation of XMP. We modified the assay mixture elaborated by Jackson et al.[11] for determination in rat tissues. IMP and XMP were separated by HPLC according to Greene and Smith[5].
GMP-synthetase (EC 6.3.5.2). We modified the mixture reported by Welch and Rudolph[6] and followed the formation of ^{14}C-glutamic acid using DEAE-cellulose disks. The difference in the glutamic acid formed in the presence or the absence of the cells (spontaneous hydrolysis of glutamine) represented the enzyme activity.

Enzymatic determinations of the intestinal mucosa

After surgical removal the tissues were kept in ice for up to 3 h at which time they were frozen and stored at -70°C. The mucosa was homogenized in four volumes of buffer using a Turrax Homogenizer (four cycles of 60 sec., with cooling at 0-4°C inbetween). The homogenization buffer was 0.29 M sucrose, pH 7.8, containing 10 mM Tris and 5 mM dithiothreitol (DTT). The homogenate underwent centrifugation at 50,000 xg for 1 h at 4°C. The supernatant was treated with Norite A and then dialyzed for 2 hours against four changes of the homogenization buffer.
Supernatant enzymatic activities were immediately determined after dialysis, as already reported[12].

Determination of enzyme activity

Enzyme activity was expressed as nmol/h/10^6 cells in the case of lymphocyte extracts and as nmol/h/mg protein in the case of colon-rectal mucosa. Suitable controls were run to ascertain that activity was linear with time and proportional to cells and protein concentration, in all cases.

Statistical analysis

Data were analyzed using the conventional one-way analysis for variance. Only p values < 0.05 were considered significant.

RESULTS

The determination of the enzymes at the inosinic branch point on lymphocyte extracts was carried out on supernatants as such, after treatment with Norite and after dialysis (Table 1).

Table 1. Enzymes of the inosinc branch point in peripheral blood lymphocytes from healthy subjects and B-CLL patients.

ENZYMES	CONTROLS	PATIENTS
IMP dehydrogenase	0.000	0.170
GMP synthetase	0.056	0.221
AMP-S synthetase	0.000	0.000
AMP-S lyase	0.000	9.598

The values, reported as $nmol/h/10^6$ cells, represent a typical experiment.
Extracts were treated with Norite A.

Figure 1 shows an HPLC chromatogram for the radiochemical determination of the IMP-dehydrogenase.

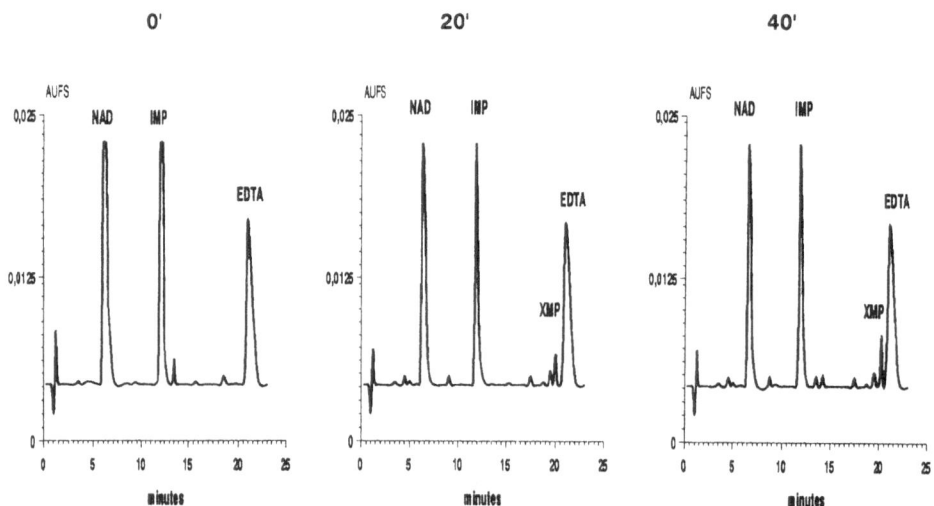

Figure 1. Determination of IMP dehydrogenase in a patient affected by B-CLL. The analysis was carried out on supernatants after treatment with Norite. HPLC chromatograms at 0', 20' and 40' of incubation are reported. The peaks of NAD, IMP, XMP and EDTA are indicated. Elution was carried out as reported in Experimentals.

From our experiments it is evident that:
1) the enzymes of the inosinic branch point are not detectable in normal lymphocytes even after treatment with Norite (except GMP synthetase);

2) the same enzymes are evident in the leukemic cells after Norite treatment (except AMP-S synthetase).

The results of the experiments on colon-rectal mucosa are reported in Table 2.

In analogy with the data reported by Reed et al.[13], in the case of breast and prostate malignancies, AMP-S lyase exhibits a tendency to increase in cancer mucosae.

Table 2. Enzymatic activities at the inosinic branch point in colon-rectal mucosa from healthy subjects and cancer-bearing patients.

ENZYMES	CONTROLS	PATIENTS
IMP-dehydrogenase	1.44	1.99
GMP-synthetase	34.84	22.35
AMP-S synthetase	51.76	41.82
AMP-S lyase	64.09	179.18

The activities, in nmol/h/mg protein, represent a typical experiment. Extracts were dialyzed.

Although our study on the enzymes at the inosinic branch point is preliminary and does not permit to draw a definite conclusion, some results are interesting:
1) they can be detected only under special conditions (tumors) and after special treatments;
2) it is doubtful if they are active in normal or only in tumor cells;
3) they can be useful in monitoring the tumor disease, from a diagnostic and prognostic point of view (this must be further investigated);
4) they can be useful in studying its pathogenesis.

The behavior of the enzymes at the "inosinic branch point" is specifically interesting, because the enzyme activity is evident only after Norite and after dialysis. Since Norite eliminates all nucleotides, it is evident that effectors - other than purine nucleotides - normally repress the above quoted activities.

Some similarities between the behavior of the enzymes at the branch point in leukemia and colon-rectal cancer indicate that the reprogramming of gene expression for these enzymes is in positive correlation with the neoplastic transformation.

ACKNOWLEDGEMENTS

Thanks are due to Progetto Finalizzato A.C.R.O. for financial supports.

REFERENCES

1. N.Prajda, N.Katunuma, H.P.Morris and G.Weber, Imbalance of purine metabolism in hepatomas of different growth rates, as expressed in behavior of glutamine phosphoribosyl-pyrophosphate amido transferase (EC 2.4.2.14), *Cancer Res.* 35:3061 (1975).
2. N.Prajda, H.P.Morris and A.Weber, Imbalance of purine metabolism in hepatomas of different growth rates as expressed in behavior of xanthine oxidase (EC 1.2.3.2), *Cancer Res.* 36:4639 (1976).
3. S.C.Hartman, *Metabolic Pathways* 4:1 (1970).
4. A.Boyum, Isolation of leukocytes from human blood, *Scand.J.Clin.Lab.Invest.*, Suppl.97:31 (1968).
5. S.V.Greene and J.M.Smith, High-performance liquid chromatographic detection of XMP as a basis for an improved IMP dehydrogenase assay. *J.Chromatogr.* 343:160 (1985).
6. M.M.Welch and F.N.Rudolph, Regulation of purine biosynthesis and interconversion in the chick, *J.Biol.Chem.* 257:13253 (1982).
7. J.Barankiewicz and A.Cohen A., Purine nucleotide metabolism in phytoemagglutinin-induced human T lymphocytes, *Arch.Biochem.Biophys.* 258:167 (1987).

8. V.Schultz V. and J.M.Lowenstein, Purine nucleotide cycle: evidence for the occurrence of the cycle in brain, *J.Biol.Chem.* 251:485 (1976).

9. D.Vannoni, B.Porcelli, A.Tabucchi, R.Leoncini, E.Marinello, Metabolismo dell'acido adenilosuccinico nel fegato di ratto, Communication to 34° Congr.Naz.SIB, Padova, 2-4 ottobre 1988.

10. R.C.Jackson, H.P.Morris and G.Weber, Increased adenylosuccinase activity in hepatomas and kidney tumors, *Life Science* 18:1043 (1976).

11. R.C.Jackson, H.P.Morris and G.Weber, Partial purification, properties and regulation of inosine 5'-phosphate dehydrogenase in normal and malignant rat tissue, *Biochem.J.* 166:1 (1977).

12. A.Tabucchi, F.Carlucci, A.De Martino and E.Marinello, Determinazione degli enzimi del punto d'incrocio inosinico in mucose colon-rettali tumorali, submitted to *Riv.It.Biol.Med.* (1994).

13. V.L.Reed, O.D.Mack and L.D.Smith, Adenylosuccinate lyase as an indicator of breast and prostate malignancies: a preliminary report, *Clin.Biochem.* 29:349 (1987).

PURINE AND PYRIMIDINE METABOLISMS IN HUMAN GLIOMAS, MELANOMAS AND COLON CARCINOMAS XENOGRAFTS, RELATION TO THEIR CYTOGENETIC PROFILE

V. Bardot[1], A.M. Dutrillaux[2], J. Beaumatin[1], D. Lefrançois[2], F. Apiou[2], B. Dutrillaux[2], and C. Luccioni[1]

[1]Commissariat à l'Energie Atomique, DSV/DPTE/LCG
BP 6 99265 Fontenay aux Roses cedex, France
[2]URA 620 CNRS, Section Biologie, Institut Curie
26 rue d'Ulm, 75231 Paris cedex 05, France

INTRODUCTION

Cytogenetic analyses performed on human solid tumors have shown, for some of them characteristic patterns of chromosomes anomalies. To understand the meaning of recurrent chromosomal imbalances, it was postulated that they could be related to specific metabolic modifications. This hypothesis was first tested on colorectal carcinomas[1, 2, 3] and verified on gliomas, by comparing tumor samples to corresponding normal tissue[4, 5]. In an attempt to precise the role of chromosomal aberrations, it appeared of interest to develop a metabolic study of three tumors, presenting either similar or distinct cytogenetic patterns, in order to compare their metabolic patterns.

Cytogenetic studies have been largely performed on gliomas and melanomas[6-11] and these two tumors present rather similar alterations, affecting principally chromosomes carrying genes encoding for enzymes involved in purine metabolism. Colorectal carcinomas are characterized by chromosomal abnormalities suggesting imbalances of genes involved in pyrimidine metabolism[1, 2, 12].

Grafted tumors on nude mice, widely used for metabolic and therapeutic studies, represent an interesting source of partially "purified" tumor tissue, because devoid of human non-cancerous cells. They also permit to amplify tumoral material[13, 14].

The activities of ten enzymes involved in purine metabolism and of two enzymes implicated in pyrimidine metabolism were measured on human gliomas, melanomas and colon carcinomas xenografts, previously karyotyped, at different passages on nude mice. For the salvage synthesis of purine nucleotides, adenosine kinase (ADK), adenine phosphoribosyltransferase (APRT), hypoxanthine phosphoribosyltransferase (HPRT), methylthioadenosine phosphorylase (MTAP) and adenylate kinase (AK) were studied. The activities of two enzymes involved in the purine *de novo* synthesis, inosine monophosphate dehydrogenase (IMPDH) and adenylosuccinate lyase (ADSL), were measured. For purine catabolism, adenosine deaminase (ADA), adenosine monophosphate deaminase (AMPD) and nucleoside phosphorylase (NP) were studied. In parallel, the activities of thymidine kinase (TK) and thymidylate synthase (TS), involved in pyrimidine metabolism, were measured on the same samples.

MATERIAL AND METHODS

The metabolic study was developped on 6 glioblastomas, 5 malignant cutaneous melanomas and 12 colon carcinomas growing as nude mouse xenografts, at different passages. Tumor fragments of 0.5 cm were subcutaneously transplanted on the flank of anesthetized Swiss nude mice, under aseptic conditions. Serial transplantations and sampling for metabolic studies were realized for each passage on three mice, when the grafts were still in exponential growth phase and reached 1 cm diameter.

Enzyme assays for purine and pyrimidine metabolisms were performed on 23 samples of glioblastomas corresponding to 3 to 5 passages (ranging from P1 to P16) on nude mice for each tumor, on 19 samples of melanomas, corresponding to 3 to 4 passages for each primary tumor (P3 to P16) and on 28 samples of colon carcinomas, corresponding to 1 to 4 passages for each primary tumor (P1 to P29). Enzyme assays and cytogenetic studies were performed as previously described.

The purine metabolism, indicating the 10 enzymes studied and the chromosomal localization of the genes coding for these enzymes is schematized in figure 1. The reactions catalyzed by TK and TS and the chromosomal localization of the corresponding genes are schematized in figure 2.

A statistical test of Mann-Whitney was performed to compare enzyme activities measured in the three tumor types.

RESULTS AND DISCUSSION

For the 10 enzymes involved in the purine metabolism, the mean activities measured in glioblastomas, melanomas and colon carcinomas xenografts, are shown in figure 3. The mean activities for TK and TS measured on the same samples are shown in figure 5. For the three tumor types, as indicated by the large standard variations, there were high inter-tumoral variations in activities for all the enzymes studied. There were also slight variations in enzyme activities from passage to passage on nude mice, especially for colon carcinomas xenografts; nevertheless, they did not correspond to any evolution of enzyme activities with time.

Activities are expressed in 10^{-9} mole/h/mg protein for ADK, IMPDH, TK and TS, in 10^{-8} mole/h/mg protein for APRT, HPRT, MTAP and AMPD, in 10^{-7} mole/h/mg protein for ADSL and ADA, and in 10^{-6} mole/h/mg protein for AK and NP.

A statistical test of Mann-Whitney was performed to compare enzyme activities measured in the three tumor types.

Purine metabolism

Our results, summarized in figure 3, show a low activity of purine metabolism in glioblastomas and melanomas when compared to colorectal carcinomas, especially for the enzymes involved in the salvage synthesis and in the catabolism, since higher ADK ($p<0.01$), AK ($p<0.01$), MTAP ($p<0.03$), ADA ($p<0.01$), AMPD ($p<0.03$) and NP ($p<0.01$) activities are observed in the last tumor type. As indicated by their very low IMPDH and ADSL activities, colon carcinomas are also characterized by a low rate of purine *de novo* synthesis, by comparison to the two other tumor types, especially melanomas. These results are in agreement with previously published data[15-17].

As shown in figure 3, comparable activities are observed between glioblastomas and melanomas, especially for the enzymes ADK, HPRT, MTAP, AK, IMPDH and ADA, suggesting a similar purine metabolic pattern in these two tumor types, by comparison to colon carcinomas.There are however marked differences for ADSL ($p<0.01$) and APRT ($p=0.01$) activities, which are higher in melanomas than in glioblastomas, suggesting a more active adenine nucleotides synthesis in melanomas. Another difference is the very low NP ($p<0.01$) activity in melanomas when compared to glioblastomas.

Cytogenetic studies performed on human glioblastomas and melanomas have shown

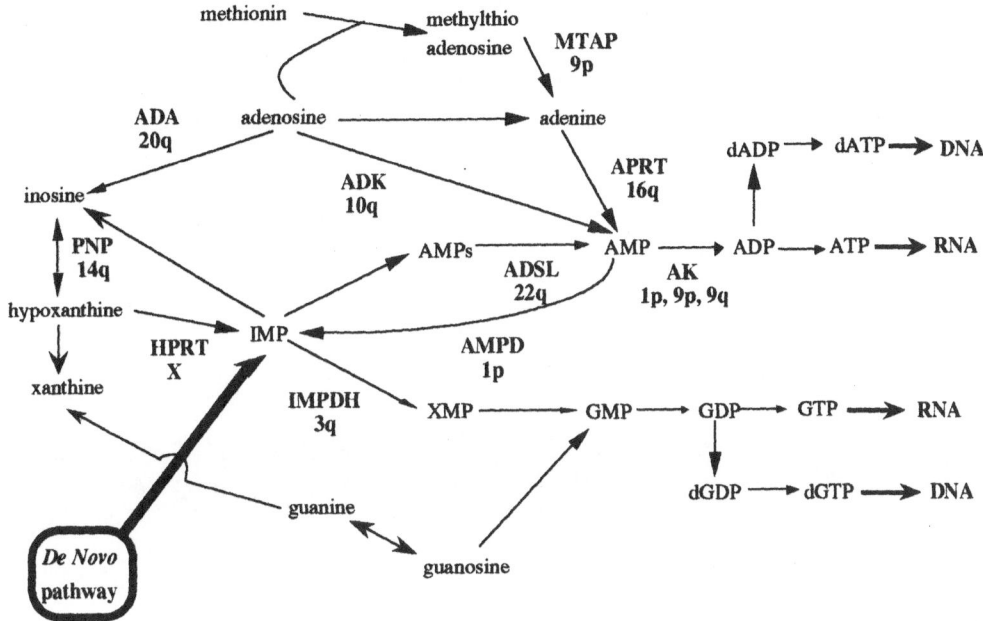

Figure 1: Purine metabolism: enzymes studied and chromosomal
localization of the corresponding genes

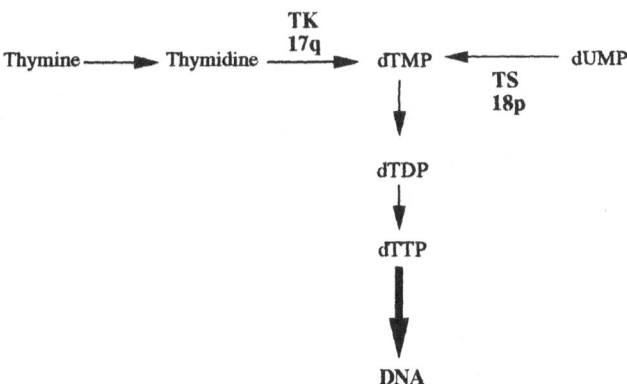

Figure 2: Pyrimidine metabolism: thymidine kinase and thymidylate synthase
and chromosomal localization of the corresponding genes

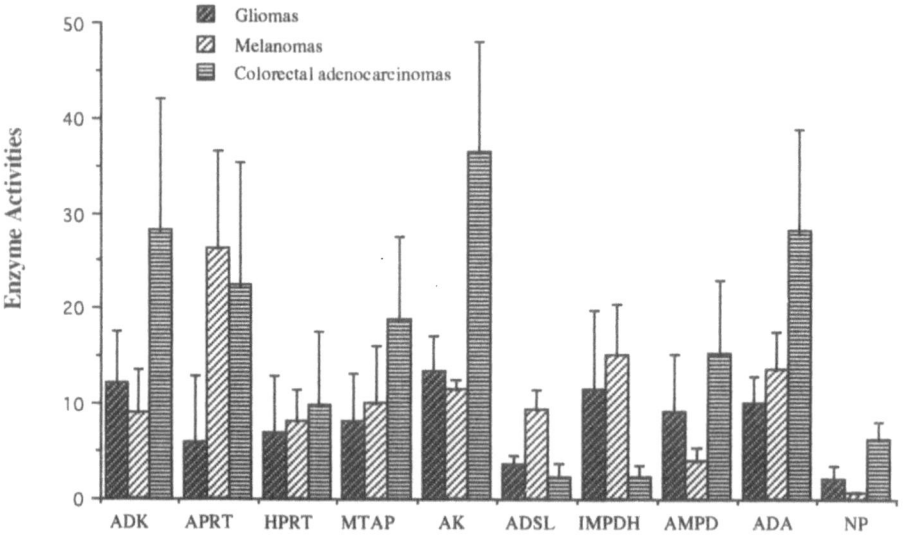

Figure 3: Purine metabolic data in glioblastomas, melanomas and colon carcinomas xenografts

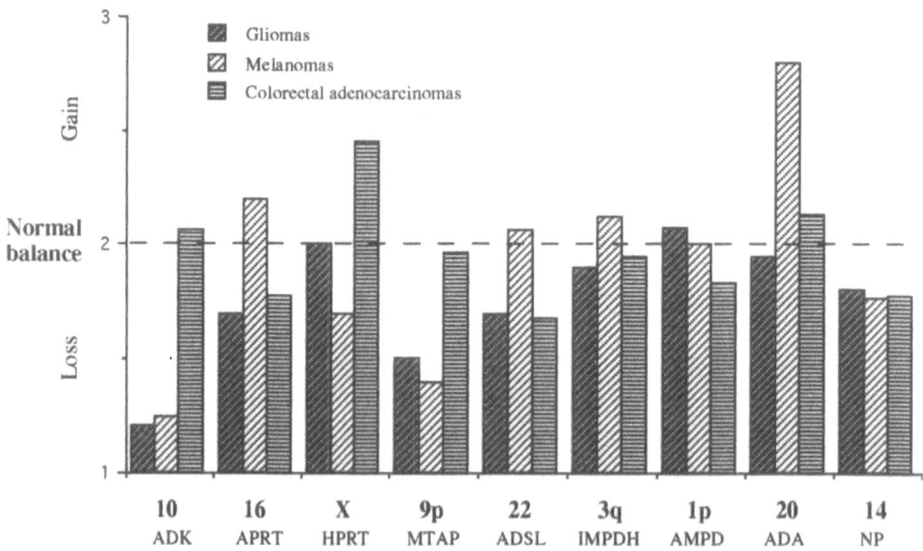

Figure 4: Average chromosome number for glioblastomas, melanomas and colon carcinomas xenografts

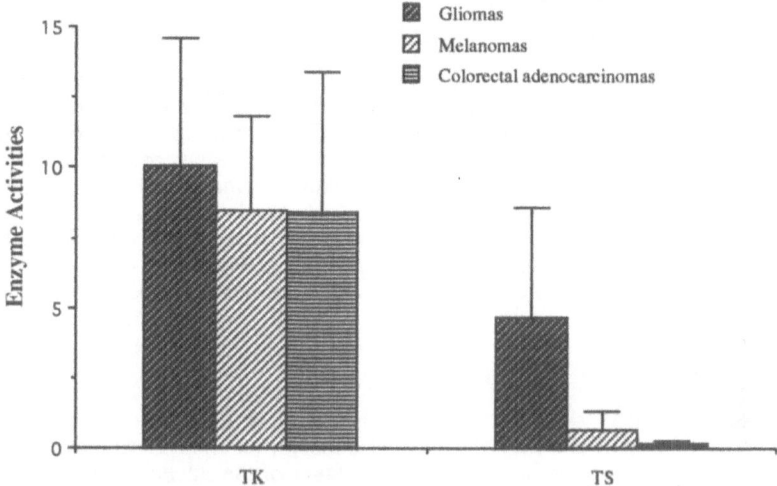

Figure 5: Pyrimidine metabolic data in glioblastomas, melanomas and colon carcinomas xenografts

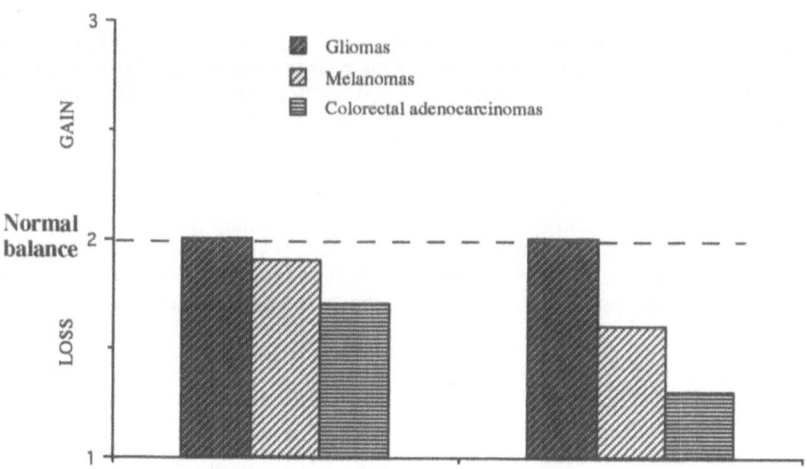

Figure 6: Average chromosome number for glioblastomas, melanomas and colon carcinomas xenografts

essentially numerical chromosomes abnormalities[6-11], affecting principally chromosomes carrying genes coding for enzymes involved in purine metabolism[4, 5]. These abnormalities are most frequently losses or deletions of chromosomes 10 for ADK, 9p for MTAP, 22 for ADSL and 1p for AMPD or occasionally losses or deletions of chromosomes 9q for AK, 16 for APRT and 14 for NP. Except for chromosomes 14 and 1p, these chromosomes are not deficient in human colorectal cancers. The cytogenetic results are summarized for the three tumor types on figure 4. Our results are in agreement with previously published data, excepted for chromosomes 22 and 20 in melanomas and chromosome 22 in colon carcinomas[9-11, 1, 12].

As suggested by cytogenetic data, the activities of most of the enzymes involved in purine metabolism are very low in glioblastomas and melanomas, when compared to colorectal tumors, especially ADK, MTAP, AK, AMPD and NP activities. More, the purine metabolic patterns are similar for the two tumors presenting a similar pattern of chromosome disturbances. The higher APRT and ADSL activities in melanomas, when compared to glioblastomas, may be partly explained by the fact that in the 5 melanomas xenografts studied, the chromosomes 16 and 22 were more frequently gained than deleted. However, when tumors are considered case by case, the relation between low enzyme activities and chromosomes losses or deletions is not direct and does not correspond to a simple gene dosage effect.

Pyrimidine metabolism

Our results (figure 5), show TK and TS activities lower than those measured for purine enzymes, except for ADK and IMPDH. They also display an imbalance between salvage and *de novo* syntheses in the three tumors studied, in favor of the salvage pathway, as indicated by the high TK activity in the three tumor types. TK activity is quite similar in glioblastomas, melanomas and colorectal carcinomas. However, large variations are observed for TS activity among the three tumor types, with the highest activity in glioblastomas. Colorectal carcinomas are characterized by the highest TK/TS ratio (54.1), by comparison to melanomas (12.9) and glioblastomas (2.1).

According to previous cytogenetic studies, the chromosome 17q carrying gene encoding for TK is frequently duplicated and the chromosome 18, carrying gene encoding for TS, frequently deleted in human colon carcinomas[1, 2, 12]. These two abnormalities are rarely observed in glioblastomas and melanomas. Our results concerning chromosomes 17q and 18 are summarized in figure 6. In spite of this correlation, the relationship between enzyme activity and chromosome number is not direct, as for purine metabolism.

As suggested by cytogenetic data, glioblastomas and melanomas present a similar pattern of activities for enzymes involved in purine metabolism, with few exceptions, when compared to colon carcinomas. For purine and pyrimidine, metabolic profiles roughly parallel cytogenetic profiles. However, the relation between deletions and low enzyme activities or between gains and higher enzyme activities is not very often direct when tumors are considered case by case and does not correspond to a simple gene dosage effect.

REFERENCES

1. C. Luccioni, M. Muleris, L. Sabatier, and B. Dutrillaux, Chromosomal and enzymatic patterns provide evidence for two types of human colon cancers with abnormal nucleotides metabolism, *Mutation Res.* 200: 55-62 (1988).

2. V. Bardot, C. Luccioni, D. Lefrançois, M. Muleris, and B. Dutrillaux, Activity of thymidylate synthase, thymidine kinase and galactokinase in primary and xenografted human colorectal cancers in relation to their chromosomal patterns, *Int. J. Cancer* 47: 670-674 (1991).

3. A. Bravard, C. Luccioni, M. Muleris, D. Lefrançois, and B. Dutrillaux, Relationships between UMPK and PGD activities and deletions of chromosome 1p in colorectal cancers, *Cancer Genet. Cytogenet.* 56: 45-56 (1991).

4. V. Bardot, A.M. Dutrillaux, C. Luccioni *et al.*, Anomalies chromosomiques et métabolisme de l'adénine dans les tumeurs gliales humaines, *Rev Neurol. (Paris)* 148: 408-416 (1992).

5. V. Bardot, A.M. Dutrillaux, J.Y. Delattre, F. Vega, M. Poisson, B. Dutrillaux, and C. Luccioni, Purine and pyrimidine metabolism in human gliomas, in relation to chromosomal aberrations, *Br. J. Cancer* (In Press).

6. S.H. Bigner, J. Mark, and D.D. Bigner, Cytogenetic of human brain tumors, *Cancer Genet. Cytogenet.* 47: 141-154 (1990).

7. S.H. Bigner, J. Mark, P.C. Burger, M.S. Mahaley, D.E. Bullard, L.H. Muhlbaier, and D.D. Bigner, Specific chromosomal abnormalities in malignant gliomas, *Cancer Res.* 48: 405-411 (1988).

8. G. Thiel, T. Losanowa, D. Kintzel *et al.*, Karyotypes in 90 human gliomas, *Cancer Genet. Cytogenet.* 58: 109-120 (1992).

9. J. Limon, P. Dal Cin, S.N.J. Sait, C. Karakousis, and A.A. Sandberg, Chromosome changes in metastatic human melanoma, *Cancer Genet. Cytogenet.* 30: 201-211 (1988).

10. P.C. Nowell, Cytogenetic of tumor progression, *Cancer* 65: 2172-2177 (1990).

11. G. Balaban, M. Herlyn, D. Guerry, R. Bartelo, H. Koprowski, W.H. Clark, and P.C. Nowell, Cytogenetics of human malignant melanoma and premalignant lesions, *Cancer Genet. Cytogenet.* 11: 429-439 (1984).

12. B. Dutrillaux, and M. Muleris, Induction of increased salvage pathways of nucleotide synthesis by dosage effect due to chromosome imbalances may be fundamental in carcinogenesis: the example of colorectal carcinoma, *Ann. Genet.* 29: 11-15 (1986).

13. F. Plénat, A. Duprez, Chimérisme tissulaire homme-souris nude: caractérisation des populations cellulaires par hybridation *in situ*, *C. R. Acad. Sci. Paris* 310: 53-59 (1990).

14. D. Lefrançois, S. Olschwang, O. Delattre, M. Muleris, A.M. Dutrillaux, G. Thomas, and B. Dutrillaux, Preservation of chromosomal and DNA characteristics of human colorectal carcinomas after passages in nude mice, *Int. J. Cancer* 44: 871-878 (1989).

15. G.W. Crabtree, D.L. Dexter, J.D. Stoeckler, T.M. Savarese, L.Y. Ghoda, *et al.*, Activities of purine-metabolizing enzymes in human colon carcinoma cell lines and xenograft tumors, *Biochemical Pharmacology* 30: 793-798 (1981).

16. Y. Natsumeda, N. Prajda, J.P. Donohue, J.L. Glover, and G. Weber, Enzymic capacities of purine *de novo* and salvage pathways for nucleotide synthesis in normal and neoplastic tissues, *Cancer Res.* 44: 2475-2479 (1984).

17. Y. Natsumeda, M.S. Lui, J. Emrani, M.A. Faderan, M.A. Reardon, J.N. Eble, J.L. Glover, and G. Weber, Purine enzymology of human colon carcinomas, *Cancer Res.* 45: 2556-2559 (1985).

ECTO-5'-NUCLEOTIDASE ACTIVITY IN LYMPHOCYTES FROM HEALTHY AND LEUKEMIA PATIENTS

A.B.Agostiñho,[1] F.Rosi,[1] F.Carlucci,[1] R.Pagani,[1] M.Pizzichini,[1] E.Marinello,[1] P.Galieni,[2] E.Dispensa,[2] and R.Leoncini[1]

[1]Institute of Biochemistry and Enzymology
 University of Siena, Italy
[2]Division of Haematology
 Hospital of Siena, Italy

INTRODUCTION

Purine nucleotide catabolism involves different enzymes and can be represented as follows:

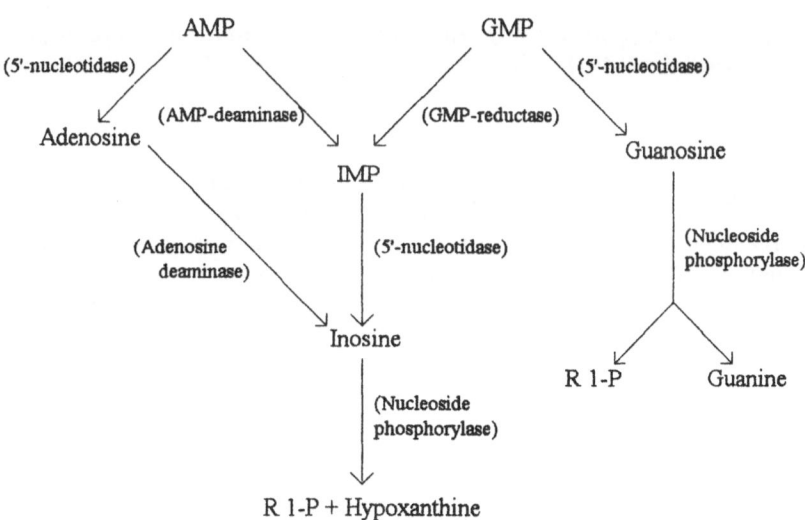

It is evident that 5'-nucleotidase plays a very important role in the degradation of purine nucleotides, transforming them into the corresponding nucleosides. A 5'-nucleotidase,

Address for correspondence: Prof. Enrico Marinello, Institute of Biochemistry and Enzymology, University of Siena, Pian dei Mantellini, 44, 53100 SIENA Italy. Tel. +39-577-298026, FAX +39-577-298057

existing in different forms, acting on pyrimidine nucleotides, has been described in erythrocytes[1]. Different 5'-nucleotidases have been described in rat liver and human placenta, and their primary structure has been investigated[2,3]. On the surface of lymphocytes, two forms have been reported[4], one resistant to cleavage by phosphatidylinositol-specific phospholipase C (e-N), and the other soluble and susceptible to cleavage (e-N$_s$).

The function of ecto-5'-nucleotidase is presumably to regulate intracellular penetration of nucleotides as nucleosides, and also the formation of adenosine, which has specific receptors in lymphocytes[5].

In the cytosol of the same cells, two isoforms of 5'-nucleotidase have been described. These cytosolic enzymes are both soluble and have similar properties, but can be differentiated on the basis of their preferential affinities for AMP (cytoplasmic 5'-nucleotidase I, c-N-I) and IMP (cytoplasmic 5'-nucleotidase II, c-N-II)[6].

The relationship between the internal and external forms is not well known. Stanley et al.[7] regarded them as the same enzyme which shuttles between the cytosol and the cell surface. The same authors described "internalization" in the liver and hypothesized it in lymphocyte extracts, but other studies do not support their hypothesis[6].

The aim of the present paper is to describe the behavior of the ecto 5'-nucleotidase in normal subjects and in B-chronic lymphocytic leukemia patients.

MATERIALS AND METHODS

Chemicals

Adenosine 5'-monophosphate, adenosine, inosine and hypoxanthine used as chromatographic standards were obtained from Sigma Chemical Corp. (St.Louis, MO, USA). ^{14}C-AMP was obtained from Amersham International (Amersham, U.K.). Lymphocyte Separation Medium and phosphate-buffered saline were purchased from Flow Laboratorie (Ayrshire, U.K.). M-450 Pan B (CD19) and M-450 Pan T (CD2) Dynabeads were obtained from Unipath Oxoid, Oslo, Norway. All other chemicals were of analytical grade from Merck (Darmstadt, Germany).

Patients

The subjects were 5 healthy volunteers and 5 leukemia patients. The patients, aged 52 to 82 years, had B-cell chronic lymphocytic leukemia (B-CLL). Diagnosis was based on lymphocyte count in peripheral blood (22,000-435,000/mm^3), bone marrow examination, and lymphocyte typing (cytofluorimetry, immunocytochemistry).

Preparation of human peripheral blood lymphocytes

All blood samples for enzyme determinations were drawn in the morning (8 a.m.) after overnight fasting; 50 ml of blood from healthy subjects was usually sufficient for all determinations. In leukemia patients, volumes proportional to the counts were used. Lymphocytes were isolated using Lymphocyte Separation Medium[8] and suspended in 0.29 M sucrose containing 10 mM Tris-HCl (pH 7.4). B and T cells were isolated from peripheral blood lymphocytes (PBL) using Dynabeads (CD2 for T lymphocytes and CD19 for B lymphocytes)[9]. The cells were counted in a Delcon cell counter.

Ecto 5'-nucleotidase assay

Determination of ecto 5'-nucleotidase activity was carried out by a radiochemical method coupled with HPLC. Incubation mixtures contained 50 mM Tris-HCl (pH 7.4), 10 mM MgCl$_2$, 0.25 mM ^{14}C-AMP (specific activity 12 μCi/μmol) and 6x10^6 cells in a total volume of 400 μl. Incubation was carried out for 5, 10, 15, 20 and 30 minutes. Unless otherwise indicated, deproteinization was performed with a final HClO$_4$ concentration of 0.21 N and neutralization with a final KOH concentration of 0.22 N. Neutralized supernatants were frozen and stored at -20°C until HPLC analysis.

Enzyme activity was determined by evaluating the disappearance of ^{14}C-AMP during the incubation, and the formation of the catabolic compounds: ^{14}C-adenosine, ^{14}C-inosine and ^{14}C-hypoxanthine. Separation of nucleotides, nucleosides and nucleobases was carried out by HPLC using a Beckman model 110A chromatograph equipped with a UV detector set at 254 nm. We used a Supelcosil LC-18 column (loop of volume 20 μl) with isocratic elution with NH$_4$H$_2$PO$_4$/CH$_3$OH (10:1) buffer at a flow rate of 1 ml/min.

Identification of substrate and products was carried out by comparison of retention times and coelution with internal standards. Quantification was performed by determining the radioactivity of the peak collected in a scintillation vial. Enzyme activity was expressed in nmol/h/10^6 cells. Suitable controls were run to ascertain that activity was linear with time and proportional to cell number (Fig.1).

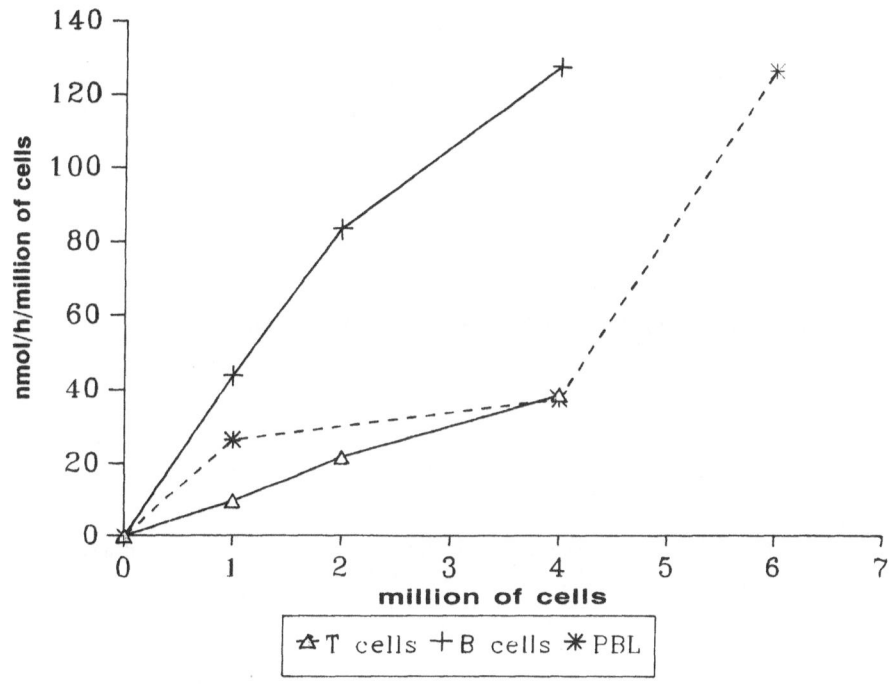

Figure 1. Relationship between enzyme activity and number of cells.

The data was analyzed using one-way analysis of variance. P values <0.05 were considered to be significant.

RESULTS

The results are reported in Table 1.

Table 1. Ecto-5'-Nucleotidase activity in PBL, and B and T lymphocytes from healthy subjects and leukemia patients.

Subjects	PBL nmol/h/10^6 cells*	B cells nmol/h/10^6 cells*	T cells nmol/h/10^6 cells*
Healthy	20.96 ± 2.76 (n = 5)	97.87 ± 45.67 (n = 5)	10.42 ± 4.18 (n = 5)
Leukemic	3.19 ± 1.10 (n = 5)	2.40 ± 1.10 (n = 4)	2.13 ± 1.62 (n = 2)

* Ecto 5'-nucleotidase activity is given as mean ± standard error. n = number of cases.
Significance level: P ≤ 0.05.

DISCUSSION

The results show a remarkable decrease in ecto 5'-nucleotidase activity in the cells of leukemia patients. It will be interesting to ascertain whether this is due to:
1) a different primary structure of the protein;
2) lower expression of the enzyme, due to decreased synthesis of mRNA;
3) rapid internalization of ecto-5'-nucleotidase in lymphocytes as described by some authors.
Irrespective of the mechanism of this decrease, the determination of 5'-nucleotidase activity is useful for the determination of cell maturity[10], and, therefore, will be useful in the diagnosis and monitoring of the disease.

ACKNOWLEDGEMENTS

Thanks are due to Progetto Finalizzato A.C.R.O. for financial supports.

REFERENCES

1. F.Bontemps, Van Den Berghe and H.G.Hers, 5'-Nucleotidase activities in human erythrocytes: identification of a purine 5'-nucleotidase stimulated by ATP and glycerate 2,3-biphosphate, *Biochem.J.* 250:687 (1988).
2. Y.Misumi, S.Ogata, S.Hirose and Y.Ikehara, Primary structure of rat liver 5'-nucleotidase deduced from the cDNA, *The Journal of Biological Chemistry* 265(4):2178 (1990).
3. Y.Misumi, S.Ogata, K.Ohkubo, S.Hirose and Y.Ikehara, Primary structure of human placental 5'-nucleotidase and identification of the glycolipid anchor in the mature form, *Eur.J.Biochem.* 191:563 (1990).
4. M.R.Klemens, W.R.Sherman, N.J.Holmberg, J.M.Ruedi, M.G.Low and L.F.Thompson, Characterization of soluble vs membrane-bound human placental 5'-nucleotidase, *Biochemical and Biophysical Research Communications* 172(3):1371 (1990).
5. G.Marone, S.Vigorita, M.Triggiani and M.Condorelli, Adenosine receptors on human lymphocytes, *Purine and Pyrimidine Metabolism in Man*, IV, 195B:7 (1986).
6. H.Zimmermann, 5'-Nucleotidase: molecular structure and functional aspects, *Biochem.J.* 285:345 (1992).
7. K.K.Stanley, M.R.Edwards and J.P.Luzio, Subcellular distribution and movement of 5'-nucleotidase in rat cells, *Biochem.J.* 186:59 (1980).
8. A.Böyum, Isolation of leukocytes from blood and bone marrow, *Scand.J.Clin.Lab.Invest.* Suppl. 97:31 (1968).
9. J.E.Brinchmann, F.Vartdal, G.Gaudernack, G.Markussen, S.Funderud, J.Ugelstad and E.Thorsby, Direct immunomagnetic quantification of lymphocyte subsets in blood, *Clin.Exp.Immunol.* 71:182 (1988).
10. W.Guttenson, E.Thiel and S.Buschette, Ecto-5'-nucleotidase as a leukemia marker, *Purine and Pyrimidine Metabolism in Man*, IV, 165B:249 (1984).

ENDO-5'-NUCLEOTIDASE ACTIVITY IN LYMPHOCYTES FROM HEALTHY SUBJECTS AND LEUKEMIA PATIENTS

F.Rosi,[1] A.B.Agostiñho,[1] A.Tabucchi,[1] R.Pagani,[1] M.Pizzichini,[1] E.Marinello,[1] R.Leoncini,[1] P.Galieni,[2] and E.Dispensa[2]

[1]Institute of Biochemistry and Enzymology
 University of Siena, Italy
[2]Division of Haematology
 Hospital of Siena, Italy

INTRODUCTION

The dephosphorylation of nucleoside 5'-monophosphates is catalyzed by 5'-nucleotidase (EC 3.1.3.5). This reaction is considered to be the first committed and irreversible step in the purine nucleotide degradation pathway in many animal and plant tissues. Its behaviour differs from one type of cell to another and seems to be linked to cell maturation in B and T lymphocytes[1].

It is interesting that the enzyme occurs at the cell surface and in the cytosol of lymphocytes and two forms have been described in both cases[2]. The properties of the internal enzymes are different: one form (c-N-II) has a strong preference for IMP, the other soluble form (c-N-I) for AMP. C-N-I is activated by ADP but not by ATP, whereas c-N-II is activated by ATP and less well, by ADP. C-N-I shows absolute dependence on Mg^{2+} ions.

Intracellular c-N-I and c-N-II enzymes are involved in the catabolism of AMP, IMP and other mononucleotides, producing the corresponding nucleoside. They are thought to be of particular importance in the hydrolysis of AMP and IMP in situations of low energy charge.It has never been possible to demonstrate if the internal enzymes are specific forms or the "internalization" of the surface enzyme. The properties and specificity of the four isoenzymes are different. The internalization and shuttle of the two forms has been described in the liver[3] and postulated in lymphocytes[4].

The biological role of the two internal isoenzymes is still unclear, and even less is known about their behaviour under pathological conditions. This prompted us to study the behaviour of the internal 5'-nucleotidases in normal lymphocytes and those of B-cell chronic lymphocytic leukemia (B-CLL) patients.

Address for correspondence: Prof. Enrico Marinello, Istituto di Biochimica e di Enzimologia, Università di Siena, Pian dei Mantellini, 44, 53100 SIENA Italy, Tel. +39-577-298026, FAX +39-577-298057

MATERIALS AND METHODS

Patients

The subjects were 10 healthy volunteers and 10 leukemia patients. Enzyme determinations were carried out in total lymphocytes. The patients, aged 52 to 82 years, had B-cell chronic lymphocytic leukemia (B-CLL). Diagnosis was based on lymphocyte count in peripheral blood (22,000-435,000/mm³), bone marrow examination, and lymphocyte typing (cytofluorimetry, immunocytochemistry).

Chemicals

The nucleotides used as chromatographic standards were obtained from Sigma Chemical Corp. (St.Louis, MO, USA). All radiochemicals were obtained from Amersham International (Amersham, U.K.). Lymphocyte Separation Medium and phosphate-buffered saline were purchased from Flow Laboratories (Ayrshire, U.K.). M-450 Pan B (CD 19) and M-450 Pan T (CD 2) Dynabeads were obtained from Unipath Oxoid, Oslo, Norway. All other chemicals were of analytical grade from Merck (Darmstadt, Germany).

Preparation of lymphocytes

All blood samples for enzyme determinations were drawn in the morning (8 a.m.) after overnight fasting; 50 ml of blood from healthy subjects was usually sufficient for all determinations. In the leukemia patients, volumes proportional to the counts were used. Peripheral blood lymphocytes (PBL) were isolated using Lymphocyte Separation Medium[5] and then suspended in 0.29 M sucrose containing 10 mM Tris (pH 7.4). B and T lymphocytes were isolated using M-450 Pan B (CD 19) and M-450 Pan T (CD 2) Dynabeads respectively[6], and counted in a Delcon cell counter. Lymphocyte extracts for all enzyme determinations were prepared by sonication, and lysis was controlled microscopically. Extracts were spun down at 300 xg, and the pellets were discarded. The supernatant obtained after sonication was treated with Norit A (150 µl of Norit A at 50% in H_2O) to eliminate endogenous nucleotides which can interfere with enzyme activities.

Endo-5'-nucleotidases assay

Endo-5'-nucleotidase assay in total PBL, B and T cells was carried out by a radiochemical method, coupled with HPLC. Six million cells were incubated at 37°C with 50 mM Tris-HCl (pH 7.4), 10 mM $MgCl_2$, 0.25 mM ^{14}C-AMP (specific activity 12 µCi/µmol) or 100 mM Tris-HCl (pH 7.4), 10 mM $MgCl_2$, 0.54 mM ^{14}C-IMP (specific activity 1.625 µCi/µmol), 3 mM ATP, 500 mM NaCl (total volume 400 µl). After 5, 10, 15, 20 and 30 minutes of incubation, 60 µl aliquots of incubation mixture were deproteinized with $HClO_4$ at a final concentration of 0.21 N, and neutralized with KOH at a final concentration of 0.22 N. Neutralized supernatants were frozen and stored at -20°C until HPLC analysis.

The enzyme activities were determined by evaluating the disappearance of ^{14}C-AMP or ^{14}C-IMP during incubation, and the formation of the corresponding nucleosides and bases. In the case of c-N-I, we followed the formation of all catabolic compounds: ^{14}C-adenosine, ^{14}C-inosine and ^{14}C-hypoxanthine; for c-N-II, we evaluated ^{14}C-inosine and ^{14}C-hypoxanthine.

Separation of nucleotides, nucleosides and bases was carried out by HPLC. A Beckman model 110 A chromatograph equipped with UV detector at 254 nm, Supelcosil LC-18 column and isocratic elution were used according to the procedure of Harmsen et al.[7]. Samples injected for HPLC analysis were of 20-µl. Identification of the substrate and reaction products was carried out by comparison of retention times and coelution of internal standards.Quantification was performed by determining the radioactivity of the peak collected in a scintillation vial.

Protein determination

The protein content of the extracts was evaluated according to Bradford[8] using bovine albumin as standard.

Determination of enzyme activity

Enzyme activity was expressed in $nmol/h/10^6$ cells. Suitable controls were run to ascertain that activity was linear with time and proportional to cell number and protein concentration.

Statistical analysis

The data was analyzed by conventional one-way analysis of variance. Only P values < 0.05 were considered significant.

RESULTS

We report a typical chromatogram of the HPLC determination of c-N-I in peripheral blood lymphocytes of a normal subject (Figure 1).

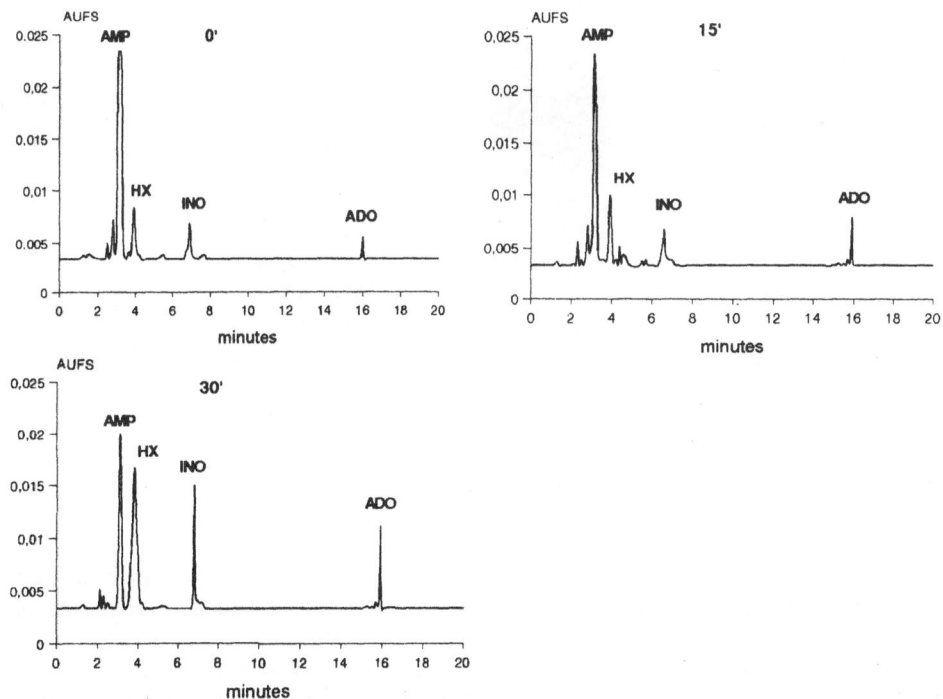

Figure 1. Determination of c-N-I in a normal subject, showing HPLC chromatograms after 15 and 30 minutes of incubation. Peaks of AMP, hypoxanthine (HX), inosine (INO) and adenosine (ADO) are indicated. Elution was carried out as reported in Materials and Methods.

The results are summarized in the Table 1.

Table 1. C-N-I and c-N-II activity in PBL from normal subjects and leukemia patients. The activity is expressed in nmol/h/10^6 cells as mean ± S.E. n = number of cases.

	Controls n = 10	B-CLL n = 10
c-N-I	19.36 ± 4.61	2.82 ± 0.77*
c-N-II	15.10 ± 2.08	4.48 ± 1.07*

* $p < 0.05$

Protein content is shown in Figure 2.

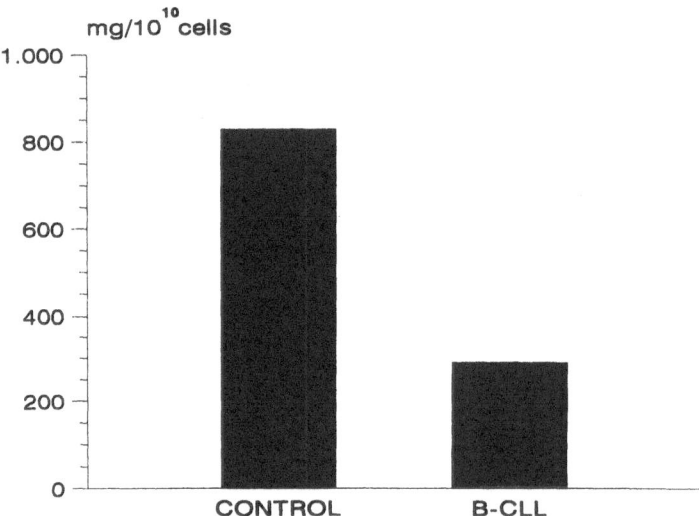

Figure 2. Protein content in lymphocytes of normal subjects and leukemia patients, determined according to Bradford[8]. P values <0.05 were considered significant.

DISCUSSION

These results show a remarkable decrease in both forms of the endo 5'-nucleotidase, with c-N-II showing less reduction than c-N-I in total lymphocytes. Enzyme determinations in B and T cells were performed in too few cases to permit sure conclusions.

The decrease in enzyme activity in leukemia cells is not solely linked to the decrease in protein content, since it is more significant. The lower activity is therefore specific to the disease. It will be interesting to ascertain whether it is due to:

1) effectors, such as the disappearance of an activator or the appearance of an inhibitor;
2) lower expression of the isoenzymes due to decreased synthesis of mRNA;

3) a different primary structure of the protein.

The ratio between surface and intracellular forms observed in PBL (e.g. 2.30 in a normal subject and 2.03 in B-CLL patient)[9] indicates that there is no relationship between the decrease in these forms. The simultaneous decrease therefore does not seem to be due to internalization of the ecto form or to different distribution of the forms.

ACKNOWLEDGEMENTS

Thanks are due to Progetto Finalizzato A.C.R.O. for financial supports.

REFERENCES

1. M.T.C.Kramers, D.Cotovsky, R.Foa, M.Cherchi and D.A.G. Galton, 5-Nucleotidase activity in leukaemic lymphocytes, *Biomedicine* 25:363 (1976).
2. H.Zimmermann, 5'-nucleotidase: molecular structure and functional aspects, *Biochem.J.* 285:345 (1992).
3. K.K.Stanley, M.R.Edwards and S.P.Luzio, Subcellular distribution and movement of 5'-nucleotidase in rat cells, *Biochem.J.* 186:59 (1980).
4. K.K.Stanley, M.R.Edwards and J.P.Luzio, Rapid internalization of plasma-membrane 5'-nucleotidase in rat spleen lymphocytes in response to rabbit anti-(rat liver 5'-nucleotidase) serum, *Biochem.Soc.Trans.* 7:1023 (1979).
5. A.Böyum, Isolation of leukocytes from blood and bone marrow, *Scand.J.Clin.Lab.Invest*, Suppl. 97:31 (1968).
6. J.E.Brinchmann, F.Vartdal, G.Gaudernack, G.Markussen, S.Funderud, J.Ugelstad and E.Thorsby, Direct immunomagnetic quantification of lymphocyte subsets in blood, *Clin.Exp.Immunol.* 71:182 (1988).
7. E.Harmsen, J.W.De Jong and P.W.Serruys, Hypoxanthine production by ischemic heart demonstrated by high-pressure liquid chromatography of blood purine nucleosides and oxypurines, *Clin,Chim.Acta* 115:73 (1981).
8. M.M.Bradford, A rapid and sensitive method for the quantitation of microgram quantities of protein utilizing the principles of protein-dye-binding, *Anal.Biochem.* 72:248 (1976).
9. A.B.Agostinho, F.Rosi, Unpublished results.

PURINE NUCLEOTIDE CONTENT IN THE LEUKOCYTES OF LEUKEMIA PATIENTS

F.Carlucci,[1] A.Tabucchi,[1] E.Consolmagno,[1] E.Dispensa,[2] P.Galieni,[2] R.Pagani,[1] M.Pizzichini,[1] R.Leoncini,[1] and E.Marinello[1]

[1]Institute of Biochemistry and Enzymology
 University of Siena, Italy
[2]Division of Haematology
 Hospital of Siena, Italy

INTRODUCTION

Purine nucleotides play an important role in cell life: they are precursors of RNA and DNA, and act as coenzymes and condensers of energy. The evaluation of their metabolism under physiological and pathological conditions is therefore of great interest.

The metabolism of purine nucleotides can be studied by: 1) evaluation of intracellular content; 2) the kinetics of incorporation of labelled precursors (^{14}C-formate) and 3) assaying the activity of related enzymes[1-3]. The first parameter is the main approach in this type of research, since the quantification of the nucleotide pool is an interesting tool for the in vivo study of purine metabolism. It could be interesting to follow this parameter in two apparently different types of blood cells, the mono- and polymorphonuclear cells, because both are derived from the same "staminal cell", changes in which underlie the disease leukemia. If the staminal cell is altered, biochemical changes may be transmitted to both mono- and to polymorphonuclear cells of leukemia patients.

We therefore studied the nucleotide content of mono- and polymorphonuclear cells from patients with B-chronic lymphocytic leukemia (B-CLL), acute myeloid leukemia (AML) or chronic myeloid leukemia (CML).

EXPERIMENTAL

Chemicals

The nucleotides employed as standards were from Sigma (St. Louis, MO, USA). PBS was purchased from Flow Laboratories (Ayrshire, UK). All chemical reagents were of

Address for correspondence: Prof. Enrico Marinello, Istituto di Biochimica e di Enzimologia, Università di Siena, Pian dei Mantellini, 44, 53100 SIENA Italy, Tel. +39-577-298026, FAX +39-577-298057

---

analytical grade from Merck (Darmstadt, Germany). HPLC solvents were filtered through 0.45 μm membrane filters (Sartorius, Göttingen, Germany) before use.

Healthy subjects and patients

The healthy subjects were 4 males and 4 females, aged 50 to 81 years, admitted at the Division of Haematology, Hospital of Siena, for minor non-haematological complaints. The leukemia patients were hospitalized in the same Department and were so grouped:
    a) 10 subjects with B-chronic lymphocytic leukemia (B-CLL).
    b) 5 subjects with acute myeloid leukemia (AML).
    c) 10 patients with chronic myeloid leukemia (CML).

Blood sampling and preparation of leukocytes

For the purification of mononuclear (MN) and polymorphonuclear (PMN) leukocytes from human blood, we used the procedure of Ferrante[4], which enables pure neutrophils and lymphocytes to be obtained with a satisfactory yield. Venous blood (60 ml) was drawn into preservative-free heparin. 3.5 ml of fresh blood was carefully layered over 3 ml Mono-Poly-Resolving Medium (d = 1.114 g/ml, Flow Laboratories) and centrifuged at 600 x g for 60 min at room temperature. Differential migration during centrifugation resulted in two distinct cell bands and a red cell pellet. The top leukocyte band was 94-98% mononuclear and the lower, 96-99% polymorphonuclear. The cells were recovered from the two bands and washed with phosphate buffered saline (pH 7.4), as already described[3].

Determination of nucleotides

For the analysis of purine nucleotides 2×10^7 cells were pelleted, the supernatants removed and the nucleotides dissolved by treating the cell pellets with 400 μl ice cold 0.4 M perchloric acid. The extracts were left at 0°C for 15 min, mixed and centrifuged for 10 min at 8,000 x g at 4°C. The supernatants were separated from the protein pellets, neutralized with potassium carbonate and centrifuged to remove the potassium perchlorate.

All details of storage, injection of samples, HPLC and the nucleotide determination are reported in previous papers[1-3]. The nucleotides were separated by HPLC, reading at 254 nm. Satisfactory resolution of the principal standard nucleotides (NAD, AMP, IMP, GMP, ADP, GDP, ATP and GTP) was obtained. The method was linear for all nucleotides from 10 pmol to 10 nmol. Run-to-run and day-to-day precision was satisfactory and CV values were always < 5%. Pyrimidine nucleotides and XMP were not considered because they were not constant and reproducible in the chromatograms.

Determination of cell energy and viability parameters

According to the literature[1-8], ATP/ADP, GTP/GDP ratios and EC represent cell energy, viability and metabolic state. The EC for adenine and guanine nucleotides is given by the ratios ½ADP+ATP / AMP+ADP+ATP and ½GDP+GTP / GMP+GDP+GTP, respectively. The total adenine and guanine nucleotides ratio, A/G, is an inverse index of proliferation rate. All these parameters were calculated by us on the basis of nucleotide content.

Statistical analysis of data

Statistical evaluation was performed by conventional one-way analysis of variance, taking p values < 0.05 as significant.

RESULTS AND DISCUSSION

Figure 1 shows a typical purine nucleotide pattern of PMN from healthy subjects (A), patients with B-CLL (B) and patients with acute and chronic myeloid leukemia, AML (C) and CML (D).

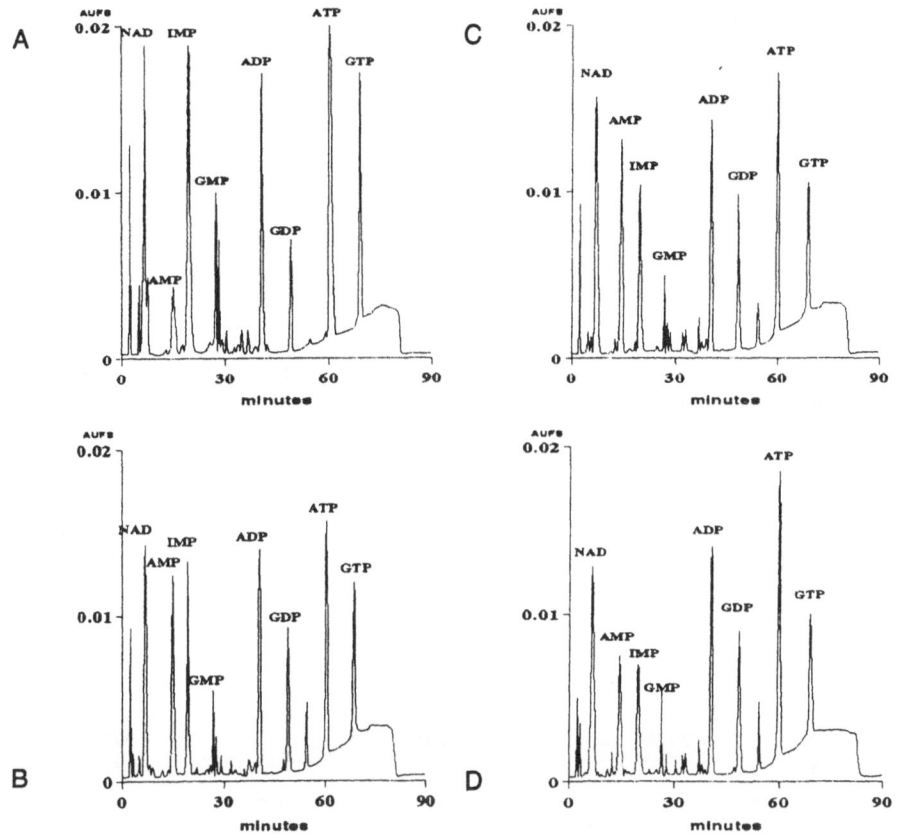

Figure 1. HPLC-UV profile of nucleotides from PMN cells of A. healthy control; B. B-CLL patients; C.AML patient; D. CML patient. For chromatographic conditions, see Materials and Methods.

The results are reported in Tables 1, 2 and 3, which show the following variations in purine nucleotide content:

a) B-CLL:
in MN cells:	IMP increased; di- and triphosphates decreased;
in PMN cells:	AMP was enhanced;
	NAD,IMP,GMP,ATP and GTP were depressed;

b) AML:
in MN cells:	IMP showed an increasing trend;
	NAD, di- and triphosphates were reduced;
in PMN cells:	AMP was enhanced;
	NAD,IMP,ATP and GTP were depressed;

c) CML:
in MN cells:	AMP and ATP increased;
in PMN cells:	IMP decreased;
	AMP was significantly higher;
	ATP and GTP were depressed.

239

Table 1. Levels of purine nucleotides in mono- and polymorphonuclear cells from healthy subjects and patients with B-CLL:

| | Controls | | B.CLL | |
	MN	PMN	MN	PMN
NAD	242 ± 98	289 ± 93	415 ± 111	189 ± 52
AMP	53 ± 21	62 ± 31	98 ± 38	193* ± 81
IMP	98 ± 58	530 ± 211	311* ± 91	215 ± 97
GMP	74 ± 15	110 ± 42	48 ± 16	24* ± 15
ADP	720 ± 111	589 ± 164	314* ± 58	615 ± 103
GDP	212 ± 58	214 ± 68	72* ± 32	188 ± 32
ATP	1,115 ± 349	1,347 ± 252	539* ± 115	589* ± 128
GTP	402 ± 81	392 ± 82	91* ± 24	214* ± 51

Values, in pmol/10^6 cells, are the means ± S.E. of 8 cases.
* $p < 0.05$ with respect to the corresponding cells of healthy subjects.

Table 2. Levels of purine nucleotides in mononuclear cells from healthy subjects and patients with acute and chronic myeloid leukemia.

	Controls n = 8	AML n = 5	CML n = 10
NAD	242 ± 98	35* ± 14	282 ± 99
AMP	53 ± 21	74 ± 31	235 ± 98
IMP	98 ± 58	163 ± 83	141 ± 26
GMP	74 ± 15	35 ± 10	109 ± 28
ADP	720 ± 111	232* ± 58	785 ± 110
GDP	212 ± 58	53* ± 16	173 ± 22
ATP	1,115 ± 349	575* ± 300	1,320 ± 477
GTP	402 ± 81	213* ± 94	299 ± 99

Values (means ± S.E.) are expressed in pmol/10^6 cells. n = number of cases.
* $p < 0.05$ with respect to cells of healthy subjects.

Table 3. Levels of purine nucleotides in polymorphonuclear cells from healthy subjects and patients with acute and chronic myeloid leukemia.

	Controls n = 8	AML n = 5	CML n = 10
NAD	289 ± 93	72 ± 18	104 ± 64
AMP	62 ± 31	105 ± 29	199* ± 59
IMP	530 ± 211	68* ± 22	260* ± 93
GMP	110 ± 42	38 ± 11	48 ± 25
ADP	589 ± 164	431 ± 56	442 ± 88
GDP	214 ± 68	105 ± 18	113 ± 2.3
ATP	1,347 ± 252	552* ± 130	442* ± 123
GTP	392 ± 82	142 ± 23*	185* ± 67

Values, in pmol/10^6 cells, represent means ± S.E.; n = number of cases.
* $p < 0.05$ with respect to controls.

Regarding the ratios calculated on the basis of purine nucleotide content, no variation was observed for the energy charge in any form of leukemia. The most interesting variations concerned the tri-phosphate/di-phosphate ratios, which showed a decreasing trend in mono- and polymorphonuclear cells in CLL. In myeloid leukemia we observed a marked increase of these ratios in mononuclear cells during the acute phase of the disease, as can be calculated by the Tables of purine nucleotide content.

The results of our study can be summarized as follows:

1) In MN and PMN cells of patients with B-CLL, AML and CML, many interesting changes in purine nucleotide content were observed.

2) In order to understand the mechanism of these changes, a study of the kinetics (incorporation of labelled precursors) and enzyme activities is required. An increase, for instance, may be due to increased synthesis or reduced breakdown.

3) Many variations are common to both mono- and polymorphonuclear cells, which is in line with the hypothesis that occur in the staminal precursor cell.

ACKNOWLEDGEMENTS

Thanks are due to Progetto Finalizzato A.C.R.O. for financial supports.

REFERENCES

1. D.De Korte, W.A.Haverkort, A.H.Van Gennip and D.Roos, Nucleotide profile of normal human blood cells determination by high performance liquid chromatography, *Anal.Biochem.* 147:197 (1985).
2. Y.M.T.Marijnen, D.De Korte, W.A.Haverkort, E.J.S.Del Breejen, A.H.Van Gennip and D.Roos, Studies on the incorporation of precursors into purine and pyrimidine via de novo and salvage pathways in normal lymphocytes and in lymphoblastic cell lines, *Biochim. Biophys.Acta* 1012:148 (1989).
3. E.Marinello, R.Pagani, F.Carlucci, M.Molinelli, P.Valerio and A.Tabucchi, The purine nucleotide content in normal lymphocytes, *Biochem.Soc.Trans.* 19:347S (1991).
4. A.Ferrante and Y.H.Thong, Optimal conditions for simultaneous purification of mononuclear and poly- morphonuclear leukocytes from human peripheral blood by the hypaque-ficoll method, *J.Immunol. Methods* 36:109 (1980)
5. G.J.Peters, R.A.De Abreu, A.Oosterhof and J.H.Veerkampf, Concentration of nucleotides and deoxy- nucleotides in peripheral and phytohemagglutinin-stimulated mammalian lymphocytes, *Biochim. Biophys.Acta* 759:7 (1983).
6. L.J.M.Spaapen, I.G.M.Scharenberg, B.J.M.Zegers, G.T.Rijkers, M.Duran and S.K.Wadman, Intracellular purine and pyrimidine pools of human T and B lymphocytes, *Adv.Exp.Med.Biol.* 195A:567 (1985).
7. R.A.De Abreu, J.M.Van Baal, J.A.J.M.Bakkeren, C.H.N.M.De Bruyn and E.D.A.M.Schretlen, A high performance liquid chromatography assay for identification and quantitation of nucleotides in lymphocytes and malignant lymphoblasts, *J.Chromatogr.* 227:45 (1982).
8. A.Goday, H.A.Simmonds, D.R.Webster, R.J.Levinsky, A.R.Watson and A.V.Hoffbrand, Importance of platelet-free preparations for evaluating lymphocyte nucleotide levels in inherited or acquired immunodeficiency syndromes, *Clin.Sci.* 65:635 (1983).
9. R.A.De Abreu, G.J.Peters and J.H.Veerkampf, Concentration of nucleotides in peripheral blood lympho- cytes of various mammalian species, *Adv.Exp.Med.Biol.* 165B:125 (1984).

AMP-DEAMINASE FROM HUMAN UTERINE MUSCLE NEOPLASM (LEIOMYOMA)

Krystian Kaletha[1], Zygmunt Chodorowski[2], Gabriela Nagel-Starczynowska[1], Tomasz Gos[3], and Wiesław Makarewicz[1].

[1] Department of Biochemistry
[2] Department of Internal Medicine
[3] Department of Forensic Medicine
  Medical University of Gdańsk
  80-211 Gdańsk, Poland

INTRODUCTION

Leiomyomas are common, benign tumors which may occur at any location containing smooth muscle cells (1). Most of these tumors originate in the female genital tracts, and in particular, in the uterus. Leiomyomas are relatively avascular and tightly compacted tumors of smooth muscle cells and fibroblasts. Usually they are surrounded by a pseudocapsule of alveolar tissue and have two relatively large nourishing vessels. The majority of Leiomyomas are of normal karyotype, but their cytogenic profile is strinkingly similar to lipomas (2).

The experimental studies reported that the specific activity of enzymes involved in carbohydrate metabolism (hexokinase, phosphofruktokinase, lactate dehydrogenase and glucogen phosphorylase) is not different in leiomyoma tissues when compared with normal adjacent myometrium. The same was true for the specific activity of enzymes involved in tricarboxylic acid cycle (succinate dehydrogenase), but significant differences were detected in the specific activity of enzymes involved in fatty acid oxidation (hydroxy acetyl-CoA dehydrogenase) (2).

In the present study, kinetic and regulatory properties of AMP-deaminase (EC 3.5.4.6) isolated from human uterine benign tumors (Leiomyomas) were described and compared with these of the enzyme isolated from adjacent, unaffected myometrium.

MATERIALS AND METHODS

Human uterus with well circumscribed, rounded, firm masses of Leiomyoma tumors was taken during hysterectomy of 62 year old female. After peeling out the tumors from the adjacent myometrium, the 5 g weighing samples of the tumor and unaffected, normal muscle were washed, homogenized (in three volumes (v/w) of the extraction buffer - 0.089 M phosphate buffer, pH 6.5, containing 0.18 M KCl and 1 mM thioethanol), and chromatographied on the phosphocellulose column, as described previously (3). The most active fractions of AMP-deaminase (second activity peak on Fig. 1) were pooled, and used for kinetic studies.

Purine and Pyrimidine Metabolism in Man VIII, Edited by
A. Sahota and M. Taylor, Plenum Press, New York, 1995

Activity of AMP-deaminase was estimated according to the phenol-hypochlorite method of Chaney and Marbach (4). The incubation medium in the final volume of 0.5 ml, contained 0.1 M potassium-succinate buffer, pH 6.5, various concentrations of the substrate (AMP) and effectors (ADP, ATP and orthophosphate).

The kinetic parameters of the reaction ($S_{0.5}$, V_{max} and n_H) were calculated, as described previously (3).

RESULTS AND DISCUSSION

The activity of AMP-deaminase in the tissue extract prepared from human uterine Leiomyoma tumors amounted about 0.015 μmoles/min per mg of extractable protein, and was roughly 30 per cent higher than that found in the adjacent, unaffected myometrium.

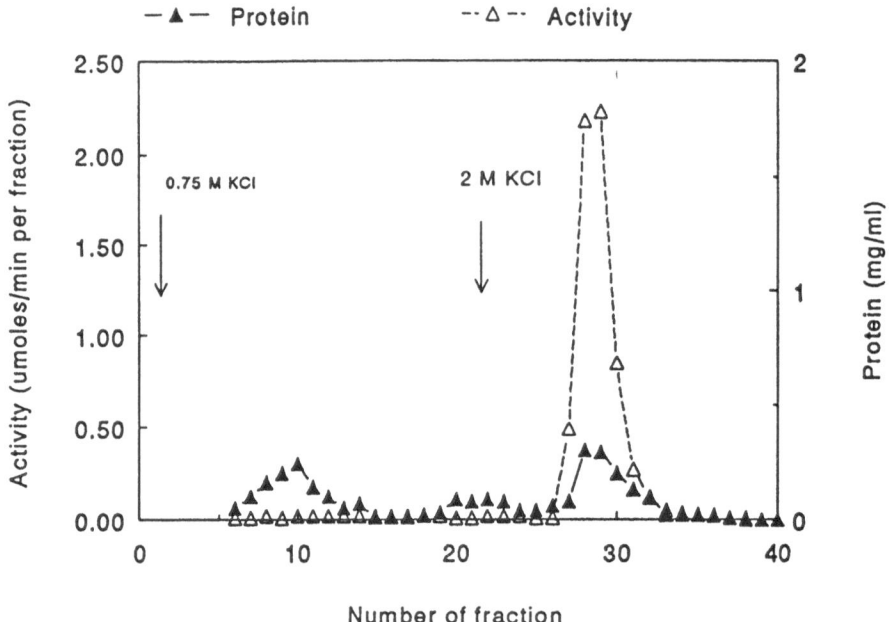

Fig.1. Phosphocellulose elution profile of AMP-deaminase from human uterine neoplasm benign tumor (Leiomyoma).

Figure 1 presents elution profile of AMP-deaminase adsorbed from the Leiomyoma tumors extract. As may be seen from the Figure, the enzyme eluted in the form of two well separated activities, indicating on the existence of two, chromatographically different forms of AMP-deaminase. The elution profile of the enzyme adsorbed from the unaffected myometrium (not shown on Figure 1) was very similar (3).

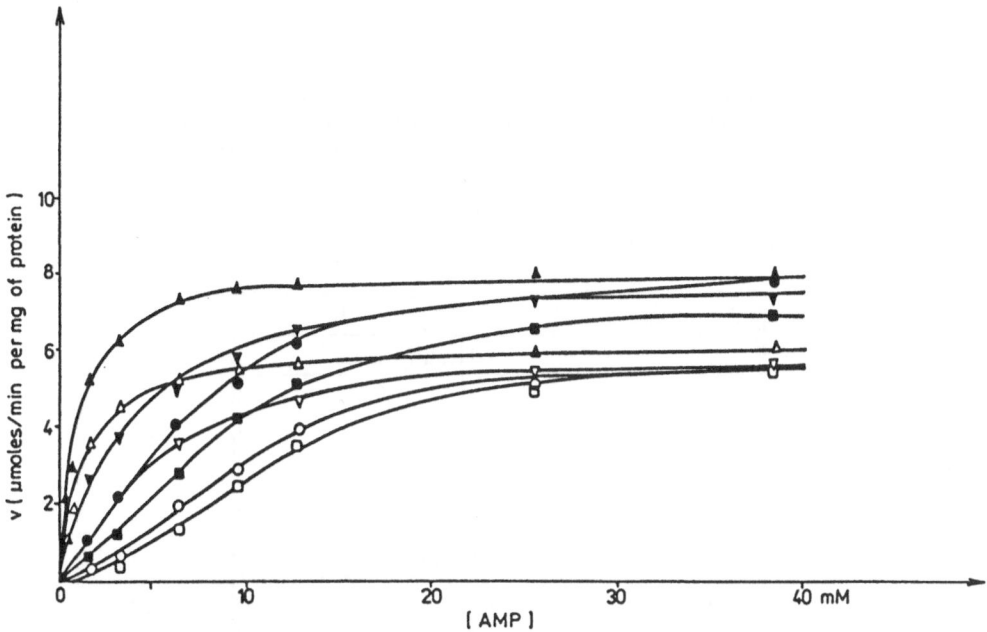

Fig.2. Effect of substrate concentration on velocity of the reaction catalysed by AMP-deaminase isolated from healthy (open symbols) and neoplasmatically changed (closed symbols) human uterine smooth muscle. The reaction was measured in the absence (O,●) or in the presence of 1 mM ATP (Δ,▲) , 1 mM ADP (∇,▼) or 2.5 mM orthophosphate (□,■).

Figure 2 presents a set of kinetic substrate saturation curves produced by AMP-deaminases isolated from the normal and diseased uterine muscle tissue, while measured at optimal pH 6.5, in the absence (control conditions) and in the presence of three important regulatory ligands. As may be seen from this figure, in the absence of the effectors tested, the enzyme, independently on its tissue derivation, manifested a similar, sigmoid-shaped (n_H = 1.7 and 1.5) substrate saturation kinetic profile. 1 mM ADP, and especially 1 mM ATP activated potently the enzyme, hyperbolizing nearly completely the sigmoidal shape of the control saturation curves. The inhibitory effect of 2.5 mM orthophosphate was much less distinctly depicted.

Table 1 compares the values of the kinetic parameters calculated for the reaction catalysed by the two AMP-deaminases studied, even in the absence or in the presence of regulatory ligands. As it may be seen from the Table, none of the regulatory ligands tested influenced markedly the control values of the maximal velocity of the reaction. The only object of the enzyme activity modulation, exerted by the effectors tested were the values of the K0.5 parameter. AMP-deaminase from the tumor tissue was more susceptible for adenylate nucleotide activation than the enzyme from the unaffected myometrium, but simultaneously less susceptible for orthophosphate inhibition.

Table 1. The effect of some important regulatory ligands on the $S_{0.5}$ and V_{MAX} values of the reaction catalysed by AMP-deaminase from healthy and neoplasmatically changed human uterine smooth muscle. Assay conditions: 0.05 M succinate-KOH buffer (pH 6.5).

Effector added	$S_{0.5}$	V_{MAX}	h
Healthy tissue			
Control	9.5	100	1.7
ATP (1mM)	1.3	105	1.1
ADP (1mM	4.4	98	1.4
P_i	11.6	102	1.9
Tumor tissue			
Control	6.7	100	1.5
ATP (1mM)	1.1	102	1.1
ADP (1mM	3.5	90	1.2
P_i	8.4	88	1.6

S.D.<1

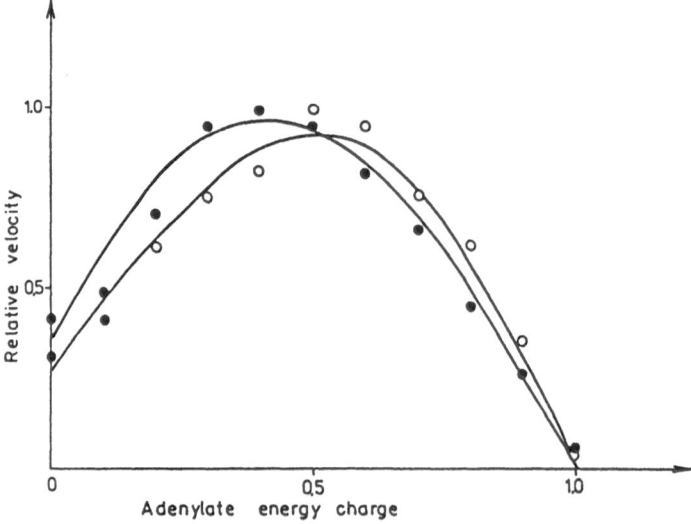

Fig.3. Effect of adenylate energy charge variation on velocity of the reaction catalysed by AMP-deaminase from healthy (open symbols) and neoplasmatically changed (closed symbols) human uterine smooth muscle. Measurements conditions: pH 6.5, 2.5 mM concentration of adenylate nucleotide pool.

Figure 3 demonstrates activity changes manifested by the two AMP-deaminases studied in the response to variation of the value of adenylate energy charge, as induced by the 2.5 mM adenylate nucleotide pool concentration. As it may be seen from the figure, the response curves produced by the two AMP-deaminases studied, had a similar, bell-shaped profile with an optimum at 0.4 and 0.5 for the enzyme extracted from the tumor-affected and unaffected myometrium, respectively.

The observed changes of AMP-deaminase properties (increase in specific activity, changes in sensitivity for allosteric ligands influence) may indicate on the increased capacity of the tumor transformed smooth muscle cell for deamination of AMP to IMP - the precursor of GMP, an essential nucleotide for DNA biosynthesis (5).

REFERENCES

1. F.T. Kraus, Female genitalia, in: "Anderson's Pathology", J.M. Kissane , ed., The C.V. Mosby Company, St.Louis - Toronto -Princeton (1985).
2. M. Koutsilieris, Pathophysiology of uterine leiomyomas, Biochem.Cell Biol. 70: 273-278 (1992).
3. G. Nagel-Starczynowska, G. Nowak, and K. Kaletha, Purification and properties of AMP-deaminase from human uterine smooth muscle, Biochim. Biophys. Acta 1073: 470-473 (1991).
4. A.L. Chaney, and E.P. Marbach, Modified reagents for determination of urea and ammonia, Clin. Chem. 8: 130-135 (1962).
5. R.C. Jackson, H.P. Morris, and G. Weber, Enzymes of the purine ribonucleotide cycle in rat hepatomas and kidney tumors. Cancer Res. 37: 3057-3065 (1977).

REGULATION OF GENE EXPRESSION OF ADENOSINE DEAMINASE, PURINE NUCLEOSIDE PHOSPHORYLASE AND TERMINAL DEOXYNUCLEOTIDYL TRANSFERASE BY DEXAMETHASONE AND cAMP IN HUMAN LEUKEMIC CELLS

Fernández-Mejia, C[1]., Peralta-Zaragoza, 0[2]., Cerezo-Roman, J.,[2] Navarro-Duque, C.,[2] Barrera-Rodríguez, R.,[3] Martínez-Valdez, H., and Madrid-Marina, V.[2]*

1. Unidad de Genética de la Nutrición. Instituto Nacional de Pediatría
2. * To whom all correspondance should be addressed. Department of Immunopathology Research Center for Infectious Diseases. Instituto Nacional de Salud Pública. Av. Universidad No. 655. Cuernavaca, 62508, Morelos México
3. Department of Biochemistry. Instituto Nacional de Enfermedades Respiratorias

INTRODUCTION

Inherited deficiencies of enzymes in the purine catabolic pathway, 5'-Nucleotidase (5'-NT), Adenosine Deaminase (ADA), Purine Nucleoside Phosphorylase (PNP) have been associated with states of immunodeficiency.[1-3] These enzymes change during lymphoid cell differentiation, ADA activity is higher in immature cells, while PNP and 5'-NT activities are lower.[4] As the lymphoid cells differentiation proceeds, ADA activity decrease, and PNP and 5'-NT activities increase, which has established the importance of purine metabolism in the immune system.[5] On the other hand, Terminal Deoxynucleotidyl Transferase (TdT) is DNA pol, which catalyse the addition of deoxynucleotides in absence of a template.[6] TdT activity is restricted to immature lymphoid cells.[7] It has been proposed that TdT plays an important role in the generation of diversity of immunoglobulins and T-cell antigen receptor molecules during gene rearrangements.[8,9] These observations led the postulation these enzymes may play a role in lymphocyte differentiation,[10] and led to the idea that in early lymphoid cells, 5'-NT, ADA and PNP regulate intracellular deoxynucleoside triphosphate pool, which are substrates of TdT.[11] Thus, the expression of ADA, PNP and TdT genes may be coordinately regulated during lymphoid cell differentiation. We shown that PKC activation regulates the gene expression of ADA, PNP and TdT in normal lymphoid and leukemic cells.[12-13] To further test which other intra-cellular signals regulate the expression of ADA, PNP and TdT in undifferentiated malignant cells, we studied the role of cAMP and Dexamethasone in the regulation of gene expression of these enzymes.

MATERIAL AND METHODS

Materials. All reagents were purchased from Sigma Chemical (St. Louis, MO), or same quality. **Cells.** Human lymphoblastoid B-cells NALM-6, and HYON, and T-cell lines PEER, and JURKAT, acute lymphoblastic leukemia cell lines, were a gift of Dr. Amos Cohen, from The Hospital for Sick Children, Toronto, Canada.

RNA isolation, Probes, Quantitation of Enzyme Activities and Protein.   All procedures were done as previously reported.[13]

Slot Blot and Densitometry Scanning of mRNA Levels.   After 4 hs incubation with 0.5 mM Dibutyryl-cAMP and 0.1 μM Dexamethasone, the cell lines were harvested by centrifugation and total RNA was extracted as described above. 10 ug of total RNA were spotted onto nitrocellulose, using a Schleicher and Schuell slot blotter apparatus and following manufactures instructions. Autoradiograms were scanned in a Beckman DU-40 spectrophotometer. The value of the cell sample incubated with no additions was normalized to 1.0 unit of relative absorvance, and other experimental points were calculated as the mRNA ratio (cAMP/control). The results are representative of three or four experiments of each cell line.

RESULTS AND DISCUSSION

The highest activities of ADA and TdT in lymphoid cells occur in immature lymphocytes and decrease during the course of lymphocyte maduration, while the activity of PNP is lower and increases during this process. It is therefore expected that TdT, ADA, and PNP expression may be coordinately regulated during lymphocyte differentiation. Recently, we have shown that the expression of TdT, ADA, and PNP mRNA are coordinately regulated by phorbol esters in human thymocytes.[12] In addition, we reported that phorbol esters show the same pattern of regulation of gene expression in human leukemic cells.[13] Thus, we investigate whether other signals coordinately regulate the gene expression of ADA, TdT, and PNP in undifferentiated malignant cells. For that end, we have used an *in vitro* system to study the regulation of gene expression in acute lymphoblastic leukemia cells of T-cells and pre-B-cell lineages, using pharmacological manipulations of intracellular signals.

TABLE I: EFFECT OF cAMP OR DEXAMETHASONE ON THE mRNA LEVELS
OF ADA, PNP, AND TdT IN HUMAN LEUKEMIC CELL LINES.

	ADA	PNP	TDT
I. T-Cell line			
cAMP			
PEER	↓ ↓	↑ ↑ ↑	↓
JURKAT	↓	↑ ↑	↓
Dexamethasone			
PEER	↓ ↓ ↓	↑ ↑	↓ ↓
JURKAT	↓	↑ ↑	↓
II. Pre-B cell line			
cAMP			
HYON	↓	↑ ↑	↓ ↓
NALM-6	↓ ↓	↑	↓ ↓ ↓
Dexamethasone			
HYON	↓ ↓	↑ ↑	↓ ↓
NALM-6	↓	↑	↓ ↓

TABLE I. Total RNA was extracted, slot blotted, and hybridized with appropiate radioactive cDNA probes. Autoradiograms were scanned. The changes produced by Dibutyryl-cAMP (0.5 mM) and Dexamethasone (0.1 μM) on ADA, PNP, and TdT mRNA levels in human leukemic cell lines are expressed in fold of changes with respect to control without treatment. One arrow denotes weak change, Two arrows moderate and , Three arrows strong change.

Dibutyryl-cAMP and dexamethasone effects on TdT, ADA, and PNP expression in human leukemic cell of T- and B-lineage were analyzed by slot blot. We chose to study HYON and NALM-6 as a representative of B cell lineage, and PEER and JURKAT as a representative of T cell lineage. While the treatment with Dibutyryl-cAMP and Dexamethasone decreased the mRNA levels of ADA, and TdT in both B and T cell lines, Dibutyryl-cAMP and Dexamethasone caused an increased of PNP mRNA levels. The analysis of ADA, PNP, and TdT mRNA levels in all leukemic cell lines of B and T lineage treated with Dibutyryl-cAMP and dexamethasone is summarized in TABLE I.

To investigate whether the changes produced by Dexamethasone and Dibutyryl-cAMP on specific mRNA levels are followed by comparable effects on enzyme activities, we compared the enzymatic activities of ADA, and PNP in cell-free extracts from the control and Dexamethasone and Dibutyryl-cAMP treated cell cultures. The data in the FIGURE 1 show that the activity of ADA decreased with dexamethasone in 60% at 18 hs, while that with Dibutyryl-cAMP the activity decreased in 40% at 12 hs. At 12 hs the activity of PNP increased in 140% and 120% with Dexamethasone and Dibutyryl-cAMP respectively, and its activity decreased on 18 hs at 125% and 115% of control for each case. This results correlate well with the effects of Dexamethasone and Dibutyryl-cAMP on ADA and PNP mRNA levels (FIGURE 1 and TABLE I).

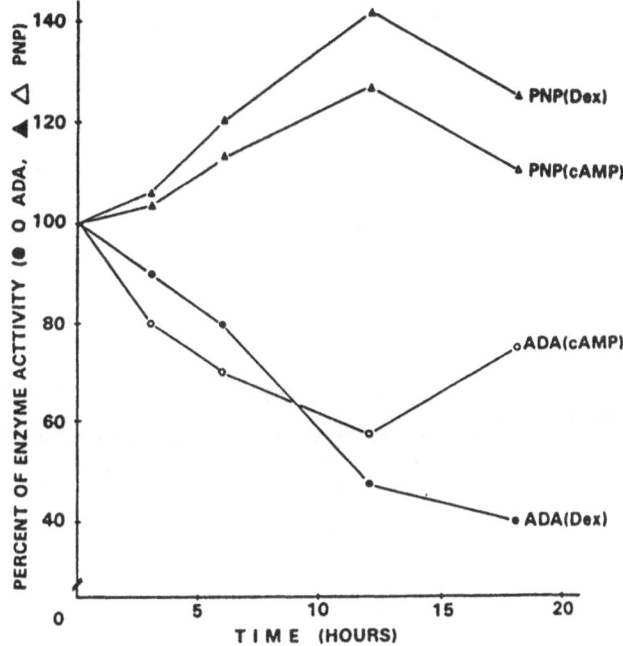

FIGURE 1. ADA, PNP, AND TdT ACTIVITIES IN DEXAMETHASONE- AND DIBUTYRYL-cAMP-TREATED CELLS. Cells (1 x 10^6/ml) of the pre-B-cell line HYON were incubated in presence or in absence of Dibutyryl-cAMP (0.5 mM) and Dexamethasone (0.1 μM). At the indicated times, samples were taken and the enzyme activities were determinated and activities are represented in percentage of change with respect to control.

The expression of ADA and TdT has been previously studied in human normal and leukemic cells. For example, cAMP induces the expression of TdT in immature B leukemic cells.[14] Furthermore, an increase of TdT expression by cAMP has been observed in normal and transformed thymocytes,[15] and after treatment by thymosin and thymopoietin.[16] Also, the ADA mRNA expression during differentiation of human promyelocytic leukemia cells (HL-60) was found to decrease following phorbol ester treatment.[17] In our lab, we have previously reported similar pattern of regulation of TdT, ADA, and PNP mRNA expression, to those reported here, in normal human thymocytes and in acute lymphoblastic leukemia cells induced by phorbol esters.[12-13]

The changes induced by Dibutyryl-cAMP and Dexamethasone in the mRNA levels of ADA, PNP, and TdT, and their enzyme activities in human leukemic cells mimic the changes in the activities of these enzymes in developing T- and B-lymphocytes during differentiation *in vivo*, suggesting a role for PKA activation and glucocorticoids in the regulation of ADA, PNP and TdT gene expression during lymphoid cell differentiation. Since it has been proposed that purine enzyme activities regulates intracellular deoxynucleotides pool, substrates of TdT,[11] and that the maintenance of an adecuate deoxynucleotide pool by adenosine deaminase is critical to at least two stages of thymocyte differentiation;[18] thus, our data indicate that the coordinate regulation of ADA, PNP, and TdT gene expression by these intracellular signals, is an crucial step for lymphocyte development.

ACKNOWLEDGEMENTS

This work was partially supported by Consejo Nacional de Ciencias y Tecnología (CONACYT) of México. Dr. Madrid-Marina is a recipient of a Molecular Biology Scholarship from Secretaría de Salubridad of México, throughout Sabritas S.A. de C.V.

REFERENCES

1). Giblett, E.R., Anderson, J.E., Cohen, F., Pollara, B., and Meuwissen, H.J Adenosine Deaminase   Deficiency in two Patients with Severely Impaired Immunity. Lancet 2:1067-1069, 1972.

2). Giblett, E.R., Amman. A.J., Wara, D.W., Sandman, R., and Diamond, L.K. Nucleoside   Phosphorylase Deficiency in a Child with Severely Defective T-cell Immunity. Lancet 1:1010-1013, 1975.

3). Edwards, N.L., Magilavy, D., Cassidy, J., and Fox, LH. Lymphocyte 5'-NT Deficiency in Agammaglobulinemia. Science 201:628-630, 1978.

4). Martin, D.W.Jr., and Gelfand, E.W. Biochemistry of Immunodevelopment Diseases. Ann. Rev.   Biochem. 50:845-877, 1978.

5). Hoffbrand, A.V., Drexsler, H.G., Ganeshaguru, K., Piga, A. and Wickremasinghe, R. Biochemical Aspects of Acute Leukaemia. Clin. in Haematol. 15:669-694, 1986.

6). Bollum, F.J. Terminal Deoxynucleotidyl Transferase as a Hematopoietic Cell Marker. Blood 54:1203-1215, 1976.

7). Bollum, F. Terminal Deoxynucleotidyl Transferase. In The Enzymes. (Boyer, P.D. ed.) Acad. Press New York. Vol. 11 p. 145-177, 1974.

8). Desiderio, S.V., Yancopoulos, G.D., Paskind, M., Thomas, E., Boss, M.A., Landau, N., Alt, F.W.   and Baltimore, D. Insertion of N-regions into Heavy-chain Genes is Correlated with the expression of Terminal Deoxynucleotidyl Transferase in B cells. Nature 311:752, 1984.

9). Landau, N.R., Schatz, D.G., Rosa, M., and Baltimore, D. Increased Frequency of N-region Insertion in a Murine pre-B cell line Infected with Terminal Deoxynucleotidyl Transferase Retroviral Expression Vector. Mol. Cell. Biol. 7:3237, 1987.

10). Ma, D.D.F., Sylvestrowicz, T., Janossy, G., and Hoffbrand, A.V. The Role of Purine Metabolic Enzymes and TdT in Intrathymic T-cell Differentiation. Immunol Today 4:65-69, 1983.

11). Cohen, A., Barankiewicz, J., Lederman, H.M., and Gelfand, E.W. Purine and Pyrimidine Metabolism in Human T-lymphocytes. J. Biol. Chem. 258:12334, 1983.

12). Martínez-Valdez, H., and Cohen, A. Coordinate Regulation of Adenosine Deaminase, Purine Nucleoside Phosphorylase, and Terminal Deoxynucleotidyl Transferase mRNA Levels by Phorbol Esters in Human Thymocytes. Proc. Natl. Acad. Sci. USA. 85:6900, 1988.

13). Madrid-Marina, V., Martínez-Valdez, H., and Cohen, A. Phorbol Ester Induce Changes in Adenosine Deaminase, Purine Nucleoside Phosphorylase, and Terminal deoxynucleotidyl Transferase mRNA in Human Leukemic Cells. Cancer Res. 50:2891-2894, 1990.

14). Siden, E.J., Gifford, A., Baltimore, D. Cyclic AMP induced terminal deoxynucleotidyl tranferase in immature B cell leukemia lines. J. Immunol. 135:1518-1522, 1985.

15). Pazmino, N.H., Ihle, J.N., and Goldstein, A.L. Induction in vivo and in vitro terminal deoxynocleotidyl transferase by thymosin in bone marrow cells from athymic mice. J. Exp. Med. 147:708, 1978.

16). Goldschneider, I., Ahmed, A., Bollum, F.J., and Goldstein A.L. induction of TdT and Lyt antigens with thymosin. Proc. Natl. Acad. Sci. USA 78:2469- , 1981.

17). Berkvens, Th.M., Schoute, F., van Ormondt, H., Meera Khan, P., et al. Adenosine deaminase mRNA expression is regulated transcriptionally during differentiation of HL-60 cells. Nucl. Ac. Res. 15:6575-6587, 1987.18).

18). Doherty, P.J., Pan, S., Mulloy, J.C., Thompson, E., Thorner, P., Barankiewicz, J., Roifman, C.M., and Cohen, A. Adenosine deaminase and thymocyte maturation. Scand. J. Immunol. 33:405-410, 1991.

ALTERED SUBSTRATE AND INHIBITOR SPECIFICITY OF PURIFIED

HUMAN ADULT THYMIDINE KINASES (TK2) FROM LEUKEMIC CELLS

Birgitte Munch-Petersen[1], Christof Völker[2], Lisbet Cloos[3], Reinhold
Hofbauer[2], Børge Thing Mortensen[4], Gerda Tyrsted[3]

[1]Institute of Life Sciences and Chemistry, Roskilde University,
 DK 4000, Denmark
[2]Wiener Biozentrum, University of Vienna, A-1030, Austria
[3]Panum Institute, Copenhagen University, DK 2200 N, Denmark
[4]Department of Hematology, Rigshospitalet, DK 2100 O, Denmark

INTRODUCTION

The two thymidine kinases in mammalian cells, TK1 and TK2, are important for the
balanced supply of deoxypyrimidine nucleotides for DNA replication and repair[1,2,3]. TK1
is cell cycle regulated and has the most restricted substrate specificity[4,5]. TK2 is
constitutively expressed and differs substantially from the cell cycle regulated TK1 regarding kinetic reaction mechanisms and substrate specificity[6,7]. The inhibition with TTP is
cooperative with TK1, but non-cooperative and competitive with TK2. The substrate
kinetics with ATP are positive cooperative with TK1, but non-cooperative with TK2.
Finally, thymidine is phosphorylated by TK2 with a characteristic, negative cooperative
mechanism.

The specificity of the two kinases towards natural substrates and nucleoside analogs
is substantially different. Thus, deoxycytidine (dCyd) was not a substrate for TK1,
whereas TK2 phoshorylated dCyd and thymidine (dThd) to the same extent. Arabinosyl-
dThd (Ara-T) was a good substrate and azidothymidine (AZT) a poor substrate, whereas
the reverse was found with TK1[5,7,8]. 5-fluorodeoxyuridine was a good substrate for both
TK1 and TK2[7].

These results were obtained with TK2 from human leukemic spleen and lymphocytes
from healthy donors. Recently, however, we have found different substrate and inhibitor
specificity with human TK2 purified from acute monocytoid bone marrow cells, and from
two cell lines, promyelocytic HL60 cells, and acute myelomonocytic ML-2 cells.

Purine and Pyrimidine Metabolism in Man VIII, Edited by
A. Sahota and M. Taylor, Plenum Press, New York, 1995

MATERIALS AND METHODS

Cells

Lymphocytes were isolated from peripheral blood from healthy volunteer by the Ficoll-Isopaque technique[6].

AMOL cells was isolated by the Ficoll-Isopaque technique from 200 ml bone-marrow aspirate taken from a 21 year old female with acute monocytoid leukemia (AMOL), type M4-M5. The human cell lines, ML2, an acute monocytic leukemic cell line type M4, and HL60, a promyelocytic cell line, were obtained from the Human Tissue Culture Collection and cultured in RPMI+20% FCS at the laboratories at University of Vienna and Rigshospitalet, Copenhagen, respectively.

Cell extracts were prepared by freeze-thawing cells suspended in hypotonic buffer (50 mM tris-HCL, PH 7.6 containing 2 mM dithiotreitl, 20% glycerol, 5 mM benzamidine, 0.5 mM phenylmethylsulfonyl fluoride) and centrifugating the homogenates at 100 000 g.

The specific activities of thymidine kinase activity, as determined in the desalted ammoniumsulfate precipitation were 0.06, 0.12, 12.8 and 2.6 nmol/min/mg in the lymphocytes, AMOL, ML2 and HL60 cells, respectively.

Purification of TK2

TK2 was purified and concentrated as previously described[7]. Briefly, supernatants from streptomycin sulfate precipitated cell extracts were precipitated with 65% ammonium sulfate, desalted on G-25, separated from TK1 by DEAE chromatography, and purified on 3'-dTMP-Sepharose. Thymidine was removed and the enzyme concentrated by hydroxyl apatite chromatography on a small column (10 x 15 mm).

Both the subunit and native molecular mass of TK2 from human lymphocytes (LY-TK2) was 30 KD. The amounts of purified TK2 from the leukemic cells was too limited for protein and molecular mass determinations.

Enzyme assays

The enzyme activity was determined with the DEAE-cellulose 81 filter paper method from initial velocity measurements using the standard assay previously described[7]. The radioactive substrates were from Amersham, UK and the radioactive analogs from Moravek. Biochemicals Inc. Brea.

RESULTS AND DISCUSSION

The substrate specificity of TK2 purified from leukemic cells and quiescent lymphocytes is shown in fig. 1. LY-TK2 had essentially the same properties as previously found with TK2 from human leukemic spleen[7,8] (The % in () denote the relative activity compared to that with dThd): dCyd was phosphorylated to the same extent as dThd; AZT was a poor substrate (3-4%); Ara-T a good substrate (50%) and CTP was an efficient phosphate donor (40%).

With TK2 from the leukemic cells, surprising differences were found. AMOL-TK2 did not phosphorylate AZT (0%). ML2-TK2 behaved as LY-TK2 regarding Ara-T and CTP, but was also capable of efficient phosphorylation of AZT (30%). HL60-TK2 had almost lost the ability to phosphorylate Ara-T (2%) and dCyd (< 1%), and to use CTP as phosphate donor (5%), whereas AZT was efficiently phosphorylated (30%).

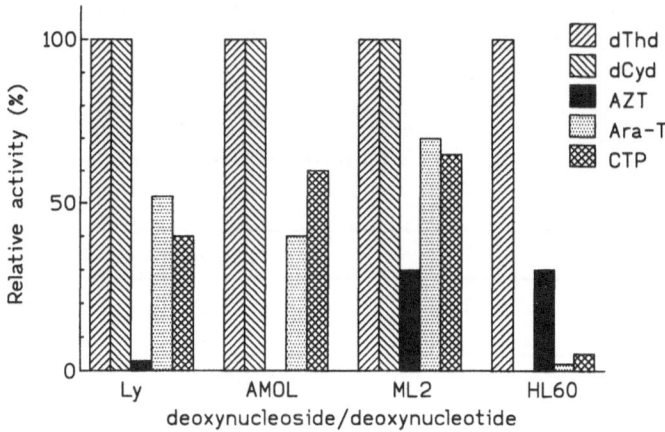

Figure 1. The phosphorylation capacity was measured at standard assay conditions with 10 μM of the indicated radiolabelled compound.

Likewise, the inhibitor specificities LY-TK2 and the leukemic TK2's differed markedly. AMOL-TK2 and ML2-TK2 were more sensitive to TTP with Ki values 28 and 14 fold, respectively, compared to LY-TK2. On the other hand, HL60-TK2 was more resistant to inhibition with AZT and FdUrd with Ki values 28 and 20 fold higher compared to LY-TK2. Especially the high Ki value with AZT was surprising considering the efficient phosphorylation of this substrate by HL60-TK2.

The altered substrate specificities and inhibitor sensitivities were unexpected and surprising. But they indicated that the TK2's from the leukemic cells may be variants of the TK2 found in normal lymphocytes. That the observed deviations may be caused by contamination of the purified leukemic TK2's with TK1 is unlikely for the following two reasons. Firstly, two different procedures were used to separate TK1 and TK2, DEAE chromatography before and hydroxyl-apatite chromatography after the dThd-affinity step. TK1 does not bind to any of these materials. Secondly, all TK2's displayed negative cooperative kinetics with dThd and competitive inhibition kinetics with TTP. Even minor contamination with TK1 would alter this pattern.

The cellular significance of TK2 is yet unclear. TK2 is frequently referred to as the mitochondrial TK and probably serves to maintain the dCTP and dTTP level in these organels. Recent results indicate, hovever, that at least 50% of TK2 is cytosolic[7,9]. The usual proportion of TK1 and TK2 activity is in the range of 20:1 in dividing cells, and 1:100 in quiescent cells. These ratios were also found with the quiescent lymphocytes and the dividing HL-60 and ML2 cells. In the AMOL cells, the ratio of TK2 to TK1 was 2:1. This ratio was much higher than expected considering the presence of proliferating cells in the bone marrow cells.

The occurrence of TK2 with altered properties in leukemic cells may be due to changes in post-translational regulatory mechanisms or more unlikely, changes at the genetic level. Whatever the explanation can be, TK2 with altered substrate specificities in malignant cells such as found for monocytic cells and HL60 cells in the present work may open a possibility for selectively impairing cellular and mitochondrial DNA synthesis and repair in leukemic cells using drugs specific for the altered TK2.

Table 1. Inhibitor specificity of TK2 purified from human lymphocytes and leukemic cells.

	Inhibitor[1]		
	TTP K_i (μM)	AZT K_i (μM)	FdUrd K_i (μM)
Lymphocytes	7	0.7	4
AMOL	0.25	0.95	3
ML2	0.5	1	2
HL60	2.6	20	86

[1]The K_i values were calculated from plots of the reciprocal initial velocity, versus inhibitor concentration at 1 and 10 μM dThd.

REFERENCES

1. P. Reichard, Interactions between deoxyribonucleotide and DNA synthesis. Ann. Rev. Biochem. 57:349 (1988).
2. V. Bianchi and L. Celotti, Accuracy of UV-induced DNA repair in V79 cells with imbalance of deoxynucleotide pools. Mutat. Res. 146:277 (1985).
3. M. Meuth, The molecular basis of mutations induced by deoxyribonucleoside triphosphate pool imbalances in mammalian cells. Exp. Cell Res. 181:305 (1989).
4. J.L. Sherley and T.J. Kelly, Regulation of human thymidine kinase during the cell cycle. J.Biol. Chem. 263:8350 (1988).
5. S. Eriksson, B. Kierdaszuk, B. Munch-Petersen, B. Öberg and N.G. Johansson, Comparison of the substrate specificities of human thymidine kinase 1 and 2 and deoxycytidine kinase toward antiviral and cytostatic nucleoside analogs. Biochem. Biophys. Res. Commun. 176:586 (1991).
6. B. Munch-Petersen, Differences in the kinetic properties of thymidine kinase isoenzymes in unstimulated and phytohemagglutinin-stimulated human lymphocytes. Mol. Cell. Biochem. 64:173 (1984).
7. B. Munch-Petersen, L. Cloos, G. Tyrsted and S. Eriksson, Diverging substrate specificity of pure human thymidine kinases 1 and 2 against antiviral dideoxynucleosides. J. Biol. Chem. 266:9032 (1991).
8. E.S.J. Arnér, T. Spasokukotskaja and S. Eriksson, Selective assays for thymidine kinase 1 and 2 and deoxycytidine kinase and their activities in extracts from human cells and tissues. Biochem. Biophys. Res. Commun. 188:712 (1992).
9. L.M. Wang, G.L. Kucera and R.L. Capizzi, Purification and Characterization of Deoxycytidine Kinase from Acute Myeloid Leukemia Cell Mitochondria. Biochim. Biophys. Acta 1202:309 (1993).

ELEVATED RATIO BETWEEN DEOXYCYTIDINE KINASE AND THYMIDINE KINASE 2 IN CLL LYMPHOCYTES COMPARED TO CONTROL CELLS

[1]Svend Erik Nielsen, [2]Birgitte Munch-Petersen and
[1]Johannes Mejer

[1]Department of Haematology and Oncology, Roskilde County Hospital, DK-4000 Roskilde, Denmark
[2]Institute of Life Sciences and Chemistry, Roskilde University Centre, DK-4000 Roskilde, Denmark

INTRODUCTION

Earlier studies comparing the phosporylation of Cytosine arabinoside (Ara-C) with deoxycytidine (dcyd) in leucocytes from patients with acute myeloid leukaemia (AML), chronic myeloid leukaemia (CML) and chronic lymphatic leukaemia (CLL) showed that the phosphorylation of ara-C was twice as high in the malignant cells than in the controls while there was no differences in the deamination of the two substrates (1,2). An obvious explanation for the results was hard to give because purification of the involved enzymes was not carried out at time of the findings. Now both purification and substrate specificity of deoxycytidine kinase and thymidine kinase 2 has been done (3,4).

Deoxycytidine kinase, dCK, is the rate limiting enzyme which catalyses both the initial step of the phosphorylation of dcyd to deoxycytidine triphosphate (dCTP) and of many deoxynucleoside analogs (e.g. Ara-C, C-dA)(4,5,6). The dCK activity is feed back inhibited by dCTP and by some of the phosphorylated analogs (e.g. ara-CTP).

Recently thymidine kinase 2, TK2, has been purified and it showed some similarities to dCK. It can phosphorylate a broad spectrum of deoxynucleosides and it has dCK activity. TK2 can not phosphorylate Ara-C (3).

In this study we have examined the relation between the dcyd phosphorylation carried out by dCK and by TK2 (dCK/TK2-dcyd) in lymphocytes from patients with CLL and from healthy donors and the results may give an explanation on the earlier findings.

METHODS

Blood samples were obtained from healthy donors and from patients with untreated chronic lymphatic leukaemia (CLL). Lym-phocytes were isolated by Isopaque-Ficoll density gradient centrifugation. The cells were sonificated in hypotonic buffer and centrifugated at 10000G.

The supernatant was precipitated with Streptomycin sulphate to a final concentration of 0.7 % and centrifugated.

The supernatant was fractionated with ammonium sulphate in two steps (20% and 60%). The solution was desalted by passage over a Sephadex G-25 column. The desalted proteins were chromatographed on a DEAE sepharose column with a linear gradient of KCl of 0-0.25M.The activity of TK2 and dCK was assayed by the DE-81 paper method. For precise method description see ref. 2

RESULTS

In the control group the dCK/TK2-dcyd ratio was between 7 and 15 with one as high as 36 (table 1).
In the CLL lymphocytes the dCK/TK2-dcyd ratio was elevated in all but one case where the value was 11. The average was about 3.5 fold elevated compared to the controls, but there were large deviations.

Table 1. Measurement of the total activity of dCK and TK2: TK2-dthd:thymidine kinase activity of TK2. TK2-dcyd: deoxycytidine kinase. n.d.:not determined.

pmol substrate phosphorylated/minute

No.	TK1	TK2-dthd	TK2-dcyd	dCK	dCK/TK2-dcyd
K1	0	3.5	2.9	37.0	13
K2	0	24.9	20.7	749.0	36
K3	0	26.0	48.0	700.0	15
K4	9	73.0	37.1	502.0	14
K5	4	42.0	36.7	337.0	9
K6	n.d	n.d.	73.7	563.0	8
K7	0	69.0	48.4	328.0	7
CLL1	0	51.3	32.0	725.0	23
CLL2	0	12.0	8.8	1443.0	164
CLL3	0	13.0	14.5	160.0	11
CLL4	5	11.0	9.7	688.0	70
CLL5	1	64.3	27.0	1599.0	59
CLL6	0	62.6	59.0	2018.0	34
CLL7	0	52.0	33.4	2087.0	62

There was no differences in the thymidine kinase activity of TK2, TK2-dthd, in the two groups, and thymidine kinase 1 could hardly be measured which was expected because it is cell cycle regulated.

DISCUSSION

The differences in the dCK/TK2-dcyd ratio found in the two groups may give an explanation to the previous findings in AML, CML, CLL leucocytes where the ratio between the ara-C and the dcyd phosphorylation was twice as high as in the controls. The same studies showed no differences in the ratio between ara-C and dcyd deamination.

Our explanation to the results is the following: ara-C is a substrate for dCK but not for TK2, which means that the relative high dCK activity in the leukaemic cells will support the phosphorylation of ara-C, compared with the normal cells. When dCK in vivo is occupied with the ara-C phosphorylation dCyd may primarily be phosphorylated by TK2. A relatively lower level of TK2 in the malignant cells will result in a higher ara-C/dcyd ratio of phosphorylation.

REFERENCES

1. Mejer, J., Enzymatic studies on possible improvement of cytosine arabinoside treatment. Scand. J. Lab. Invest. 42, p. 401 (1982).

2. Mejer, J., and Nygaard P. Cytosine arabinoside phosphorylation and deaminationin acute myeloblastic leukaemia cells. Leukaemia Research Vol. 2 No. 2 p. 127 (1977).

3. Munch-Petersen, B., Cloos, L., Tyrsted, G., and Eriksson, S., Diverging substrate specificity of pure human thymidine kinases 1 and 2 against antiviral dideoxynucleosides. The Journal of Biological Chemistry, vol. 266, no. 14, p. 9032 (1990).

4. Eriksson, E., Kierdaszuk, B., Munch-Petersen, B., Öberg, B, and Johansson, N.G., Comparison of the substrate specificities of human thymidine kinase 1 and 2 and deoxycytidine kinase toward antiviral and cytostatic nucleoside analogs. Biochemical and biophysical research communications, vol. 176, no. 2, p. 586 (1991).

5. Plunkett, W., Huang, P, Gandhi, V. Metabolism and action of Fludarabine Phosphate. Seminars in Oncology, vol 17, No 5, Suppl 8, p 3 (1990).

6. Gandhi, V., plunkett, W. Modulatory Activity of2',2'-diflourodeoxycytidine on the phosphorylation and cytotoxicity of Arabinosyl Nucleosides. Cancer Research 50, p. 3675 (1990).

Abbriviations: F.Ara-A, Fludarabine; C-dA,2chloro-2'-deoxyadenosin.

THE ROLES OF URIDINE-CYTIDINE KINASE AND CTP SYNTHETASE IN THE SYNTHESIS OF CTP IN MALIGNANT HUMAN T-LYMPHOCYTIC CELLS [1]

A. André van den Berg, Henk van Lenthe, André B.P. van Kuilenburg
and Albert H. van Gennip

Academic Medical Center
P.O. Box 22700
F-0-223, Meibergdreef 9
1100 DE, Amsterdam, The Netherlands

INTRODUCTION

In human T-lymphocytic cells pyrimidine ribonucleotides are predominantly synthesized via the salvage pathways [1, 2]. Cytosine ribonucleotides can be synthesized by salvage of cytidine or through the salvage of uridine and subsequent conversion of UTP into CTP by the action of CTP synthetase (E.C. 6.3.4.2.). Uridine and cytidine are competing substrates for uridine/cytidine kinase (urd/cyd kinase, EC 2.7.1.48) [3]. Before, we demonstrated increased CTP pools in human lymphocytic leukemia blasts and MOLT-3 T-ALL cell-line cells as well as in proliferating normal human T-lymphocytic cells in comparison with resting human peripheral blood lymphocytes [1, 4]. At physiological concentrations of uridine and cytidine, in the MOLT-3 cells this increase correlates with an increased activity of CTP synthetase [4]. For benign proliferating human T lymphocytes a similar increase in cytosine ribonucleotides as in MOLT-3 cells was found. However, CTP synthetase contributes less to this pool compared to the enzyme in the MOLT-3 cells [4, 5]. Therefore, we hypothesized that cytidine is more avidly converted into CTP in the normal T lymphocytes, compared to the MOLT-3 cells at physiological concentrations of uridine and cytidine.

The CTP pool both regulates the activity of the urd/cyd kinase, depending on the amount of UTP [6], and regulates the activity of CTP synthetase, probably through a mechanism of allosteric feedback [7]. Because CTP synthetase contributes more to the increased CTP pools in the MOLT-3 cells independent of the growth rate and nucleic acids synthesis activity, compared to normal T cells [4], we wondered if the enzyme CTP synthetase in MOLT-3 cells was to some extent defective in the allosteric feedback control. On the other hand, it might be that the affinities of the urd/cyd kinase towards its substrates is dissimilar in the normal T lymphocytes and MOLT-3 cells. Therefore, we have now investigated the relationships between the sizes of the CTP and UTP pools and their ratio, and the activities of CTP synthetase as well as the activity of urd/cyd kinase in MOLT-3 cells versus proliferating normal human T cells.

MATERIALS AND METHODS

Nucleotides used as chromatographic standards were obtained from Sigma Chemical Corp. (St. Louis, MO, USA). [14C]uridine was obtained from Amersham International (Amersham,

---

[1]This study was supported by grant 89-01 from the Dutch Foundation for Paediatric Research S.K.K.

UK). All other chemicals were of analytical grade and were obtained from Merck (Darmstadt, Germany). Proliferating MOLT-3 cells and growth-induced normal human T lymphocytes were obtained, cultured, and cells were subjected to a pulse-chase study of the pyrimidine ribonucleotide metabolism as described elsewhere [4]. In the present study we emphasize on the synthesis of CTP from uridine and cytidine. Briefly, 2 - 4 μM of [^{14}C]uridine was given to cells in the presence of < 0.5 μM, 2 μM, 4 μM or 10 μM cytidine. After one hour (pulse), the nucleotides were extracted. The remaining cells were centrifuged and resuspended in fresh medium of 37 °C. Further extracts were made 2, 4 and 8 hours after the application of [^{14}C]uridine (chase). Extracts were analysed for content of nucleotides and of incorporated [^{14}C]uridine by HPLC and on-line radiodetection [1]. Statistical significance was analysed with the double sided students' T-test.

RESULTS

Amounts of UTP and CTP and their ratios at various concentrations of cytidine in malignant and benign proliferating human T-lymphocytic cells

As Fig. 1 shows, at a physiological concentration of cytidine (< 0.5 μM ; present in the fetal calf serum) and after addition of 2 μM cytidine the CTP pools were similar in MOLT-3 cells. After addition of 4 μM of cytidine no increase in CTP was observed yet, whereas a decrease in the UTP over CTP ratio was observed. After addition of 10 μM of cytidine an increase in CTP was observed, with a decrease in the UTP over CTP ratio. The CTP pools increased and the UTP over CTP ratio decreased in normal human T lymphoblasts after addition of 10 μM cytidine in a quite similar fashion as in the MOLT-3 cells. So, no significant differences were observed in the relative increases of the CTP and UTP pools between the MOLT-3 cells and the normal human T lymphocytes after increasing the concentration of cytidine to supra-physiological levels.

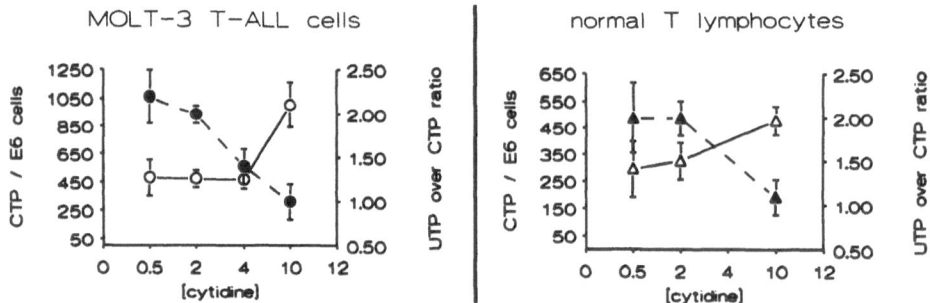

Figure 1. The amounts of CTP and the ratio of UTP over CTP in malignant and benign proliferating human T-lymphocytic cells at various concentrations of cytidine. Solid symbols on right y-axis, ratios of UTP over CTP; open symbols on left y-axis, CTP pools in pmole per million cells. Note the difference in left y-axes and the similarity of right y-axes in either panel.

Incorporation of [^{14}C]uridine into pyrimidine nucleotides and nucleic acids at various concentrations of cytidine

Table 1 gives the amounts of [^{14}C]uridine incorporated into nucleotides or nucleic acids after one hour labeling with [^{14}C]uridine in the presence of various concentrations of cytidine. In case the concentration of cytidine was increased from 0.5 μM to 2 μM, in the normal human T-lymphocytic cells an increase in the amounts of [^{14}C]uridine incorporated into nucleotides and nucleic acids was observed. In the MOLT-3 cells the incorporation was decreased as soon as the concentration of cytidine was increased to 4 μM or 10 μM. In the normal T lymphocytes a similar decrease (to 22 % of the value at 0.5 μM) was observed as in the MOLT-3 cells (decrease to 17 %) when 10 μM of cytidine was present.

Table 1. Incorporation of [^{14}C]uridine into nucleotides and nucleic acids of one million malignant or benign proliferating human T-lymphocytic cells at various concentrations of cytidine for one hour. The values in table 1 give the mean of three (0.5 μM and 2 μM), two (4 μM) or one experiment (10 μM) $\pm$ S.D.

	MOLT-3		T-lympho's	
	Nucleic acids	Nucleotides	Nucleic acids	Nucleotides
0.5 CYD	436 $\pm$ 28	247 $\pm$ 34	115 $\pm$ 48	53 $\pm$ 11
2 CYD	423 $\pm$ 84	280 $\pm$ 18	136 $\pm$ 32	92 $\pm$ 20
4 CYD	189 $\pm$ 19	208 $\pm$ 40	n.d.	n.d.
10 CYD	69	69	18	19

The contribution of CTP synthetase to the synthesis of CTP at various concentrations of cytidine in malignant and benign proliferating human T-lymphocytic cells

In MOLT-3 cells the conversion of UTP into CTP was decreased when the concentration of cytidine was increased from 0.5 μM to 2 μM (Table 2). In contrast, in the normal T lymphocytes the contribution of CTP synthetase to the synthesis of CTP was unchanged when the concentration was increased from 0.5 to 2 μM of cytidine. For the MOLT-3 cells, a further increase in cytidine with 4 μM or 10 μM resulted in a further decrease in the activity of CTP synthetase. In the benign proliferating human T-lymphocytic cells the contribution of CTP synthetase to the synthesis of CTP was decreased to less then one percent in case the concentration of cytidine was increased to 10 μM.

The relation between the increases in the size of the CTP pool and the decrease in the activity of CTP synthetase is similar in the malignant MOLT-3 human T-lymphocytic cells and the benign proliferating human T-lymphocytic cells. The results show that the activity of CTP synthetase in the malignant MOLT-3 human T lymphoblasts is still subject to feedback inhibition of the enzyme by its product, just as it is in the proliferating normal human T-lymphocytic cells.

Table 2. The contribution of cytidine salvage and CTP synthetase to the synthesis of CTP in benign and malignant human T-lymphocytic cells at various concentrations of cytidine. The values in the table give the mean ^{14}C-CTP over ^{14}C-UTP ratio in the steady-state phase of pulse-chase experiments of three (0.5 μM and 2 μM) or one experiment $\pm$ S.D. and represent the contribution of CTP synthetase to the synthesis of CTP (cf. [5]).

	MOLT-3	T-lympho's
0.5 CYD	0.66 $\pm$ 0.16	0.35 $\pm$ 0.10
2 CYD	0.38 $\pm$ 0.16	0.37 $\pm$ 0.06
4 CYD	0.08	n.d.
10 CYD	0.03	< 0.01

DISCUSSION

The sizes of the UTP and CTP pools, and the rate of incorporation of [^{14}C]uridine into nucleotides and nucleic acids are increased in MOLT-3 cells compared to the proliferating normal human T lymphocytes in the presence of 0.5 μM or 2 μM of cytidine. With respect to the pool sizes this is a reflection of the state of tetraploidy of the MOLT-3 cells [8], and with respect to the incorporation a combination of the former fact and of the apparent preferential uptake of uridine over cytidine in the MOLT-3 cells.

Recently, we demonstrated that the increased activity of CTP synthetase in proliferating MOLT-3 cells, compared to proliferating normal T lymphocytes, is independent of the rate of synthesis of pyrimidine ribonucleotides and nucleic acids [4]. Since, excessive production of CTP in cells with a defective feedback of CTP on the activity of CTP synthetase has been reported [6] we hypothesized that a similar phenomenon could underly the increased activity of CTP synthetase in MOLT-3 cells. However, because an increase in the concentration of cytidine with 10 μM resulted in an increase in the intracellular CTP pools and a concomitant decrease in the activity of CTP synthetase in a similar fashion in both MOLT-3 cells and in normal T lymphocytes, we regard the feedback of CTP on CTP synthetase in MOLT-3 cells as intact.

As soon as the concentration of cytidine was increased with 2 μM, the contributions of uridine and cytidine salvage to the synthesis of CTP in MOLT-3 cells were similar to that observed in the proliferating normal T lymphocytes (Table 2). Moreover, in the T lymphocytes, but not in the MOLT-3 cells the incorporation of [^{14}C]uridine was enhanced when the concentration of cytidine was increased (Table 1). Since uridine and cytidine are competing substrates for the same enzyme [3] this must imply that the total activity of this enzyme in the proliferating normal T lymphocytes was increased to a larger extent than in the proliferating MOLT-3 cells. So, the activity of the urd/cyd kinase in the presence of various concentrations of its competing substrates uridine and cytidine is different in MOLT-3 cells and normal T lymphocytes.

We regard the increased activity of CTP synthetase in the MOLT-3 cells as directly correlated to a more efficient salvage of uridine at physiological concentrations of uridine and cytidine. So, the fact that CTP synthetase makes a larger contribution to the CTP pool in MOLT-3 cells, in comparison with normal human T lymphocytes seems to be related to altered kinetic properties of the urd/cyd kinase, rather than to altered properties of the CTP synthetase.

We conclude that at physiological concentrations of uridine and cytidine, the re-utilisation of uridine is a preferred route in the synthesis of CTP for MOLT-3 cells whereas in proliferating normal human T lymphocytes CTP is largely synthesized through re-utilisation of cytidine. This difference in salvage of pyrimidine ribonucleosides may be exploited for selective chemotherapy.

REFERENCES

1. D. De Korte. Ribonucleotide metabolism of leukemic cells. Thesis, Kanters B.V., Alblasserdam (1987).
2. R.I. Christopherson, and S.D. Lyons. Potent inhibitors of de novo pyrimidine and purine biosynthesis as chemotherapeutic agents. *Med. Res. Rev.* 10: 505-548 (1990).
3. A.S. Liacouras, and E.P. Anderson. Uridine-cytidine kinase. III. Competition between uridine and cytidine for a single enzyme. *Mol. Cell. Biochem.* 17: 141-146 (1977).
4. A.A. Van den Berg, H. Van Lenthe, S. Busch, D. de Korte, D. Roos, A.B.P. Van Kuilenburg, and A.H. Van Gennip. Evidence for transformation-related increase in CTP synthetase activity in situ in human lymphoblastic leukemia. *Eur. J. Biochemistry* 216: 161-167 (1993).
5. A.A. van den Berg, H. van Lenthe, S. Busch, D. de Korte, A.B.P. van Kuilenburg, and A.H. van Gennip. The roles of uridine-cytidine kinase and CTP synthetase in the synthesis of CTP in malignant human T-lymphocytic cells. *Leukemia*: in press (1994).
6. N. Cheng, R.C. Payne, and T.W. Traut. Regulation of uridine kinase: Evidence for a regulatory site. *J. Biol. Chem.* 261: 13006-13012 (1986).
7. Aronow B, and Ullman B. *In situ* regulation of mammalian CTP synthetase by allosteric inhibition. *J. Biol. Chem.* 262: 5106-5112 (1987).
8. J.M. Greenberg, R. Gonzalez-Sarmiento, D.C. Arthur, C.W. Wilkowski, B.J. Streifel, and J.H. Kersey. Immunophenotypic and cytogenetic analysis of Molt-3 and Molt-4: human T-lymphoid cell lines with rearrangement of chromosome 7. *Blood* 72: 1755-1760 (1988).

METABOLISM OF UDP-N-ACETYL-HEXOSES AND UDP-HEXOSES IN

NORMAL HUMAN T-LYMPHOCYTES AND MOLT-3 T-LEUKEMIA CELLS [1]

A. André van den Berg, Henk van Lenthe and Albert H. van Gennip

Academic Medical Center
P.O. Box 22700
F-0-223, Meibergdreef 9
1100 DE, Amsterdam, The Netherlands

INTRODUCTION

UDP sugars serve as cosubstrates in the synthesis of glycolipids, glycoproteins, glycosami-noglycans, etcetera. [1]. Knowledge of their metabolism of in normal versus malignant cells could be used to develop anti-cancer drugs on a rational basis. Peripheral blood cells from patients with acute lymphoblastic leukemia (ALL) contain increased amounts of UDP-sugar compounds with a decreased ratio of UDP-N-acetyl-hexoses (UDP-HexNAc) to UDP-hexoses (UDP-Hex) as compared to peripheral blood leukocytes from healthy donors (PBL) [2]. However, changes in pools of metabolites do not necessarily reflect changes in their steady-state turnover. Recently, we reported on a pulse-chase system designed to study the synthesis of pyrimidine compounds [3]. In this system the MOLT-3 T-ALL cell line which like freshly isolated human leukemia cells, has increased UDP-sugar pools with a decreased UDP-HexNAc to UDP-Hex ratio and which can be induced to differentiate, is used as a model to perform comparative studies with PBL or growth-stimulated normal human T lymphocytes.

MATERIALS AND METHODS

Nucleotides used as chromatographic standards were obtained from Sigma Chemical Corp. (St. Louis, MO, USA). [14C]uridine was obtained from Amersham International (Amersham, UK). All other chemicals were of analytical grade and were obtained from Merck (Darmstadt, Germany). PBL, proliferating MOLT-3 cells and differentiated MOLT-3 cells and growth-induced normal human T lymphocytes were obtained, cultured, and cells were subjected to a pulse-chase study of the pyrimidine ribonucleotide metabolism as described elsewhere [3] with emphasis on the synthesis of UDP sugars. Briefly, 2 - 4 μM of [14C]uridine was added to cell samples and after one hour (pulse), the nucleotides were extracted. The remaining cells were centrifuged and resuspended in fresh medium of 37 °C. Further extracts were made 2, 4 and 8 hours after application of [14C]Urd (chase). Extracts were analysed for content of nucleotides and of incorporated [14C]Urd by HPLC and on-line radiodetection [2]. Statistical significance was analysed with the double sided students' T-test.

RESULTS

Amounts of pyrimidine nucleotides, UDP sugars and their ratios

Proliferating MOLT-3 cells and proliferating normal human T lymphocytes with similar population doubling times (27-30 hours) contained similar amounts of pyrimidine ribonucleotides per μg of protein (Table 1). Both cell types contained much more pyrimidine nucleotides than

---

[1]This study was supported by grant 89-01 from the Dutch Foundation for Paediatric Research S.K.K.

did PBL. Non-proliferating, differentiated MOLT-3 cells contained about as much pyrimidine nucleotides per μg of protein as did proliferating MOLT-3 cells and thus far more compared to the PBL. The amounts of UDP-sugar compounds were slightly increased in proliferating MOLT-3 and in differentiated MOLT-3 cells compared to proliferating T-lymphocytes (p < 0.0001) and all three forementioned cell types had increased amounts of UDP-sugars with a decreased UDP-HexNAc to UDP-Hex ratio compared to PBL (Table 1). However, expressed as a percentage of the total amount of pyrimidine nucleotides (including UDP sugars) the total amount of all UDP sugars was similar in all the types of cells studied. Only the UDP-HexNAc pool in the differentiated MOLT-3 cells had decreased compared to proliferating MOLT-3 cells, leading to a decreased (p < 0.01) UDP-HexNAc to UDP-Hex ratio (Table 1).

Table 1.

	[a]Peripheral blood lymphocytes (n=19)	Proliferating T lymphocytes (n=5)	Proliferating MOLT-3 cells (n=6)	Differentiated MOLT-3 cells (n=7)
uracil nucleotides	121 ± 11	560 ± 150	1038 ± 317	834 ± 338
cytosine nucleotides	21 ± 5	298 ± 104	482 ± 128	371 ± 136
pmole pyrimidine nucleotides / μg protein	3 ± 0.5	10 ± 4	15 ± 4	14 ± 4
UDP-HexNAc*	41 ± 7	182 ± 28	407 ± 33	250 ± 54
synthesis UDP-HexNAc	n.d.	0.57 ± 0.19	0.28 ± 0.20	0.40 ± 0.08
UDP-Hex*	21 ± 8	165 ± 19	286 ± 51	271 ± 109
synthesis UDP-Hex	n.d.	1.19 ± 0.20	1.18 ± 0.47	1.13 ± 0.14
UDP-HexNAc : UDP-Hex	2.0 ± 0.3	1.1 ± 0.2	1.5 ± 0.2	1.0 ± 0.2
UDP sugars / total pyrimidine nucleotides * 100	30 ± 5	28 ± 8	28 ± 4	27 ± 5

* UDP-HexNAc ; UDP-GlcNAC and/or UDP-GalNAc, UDP-Hex ; UDP-Glc and/or UDP-Gal

Synthesis of UDP-HexNAc and UDP-Hex pools

The amounts of [^{14}C]uridine incorporated into UDP sugars do not directly give information about the synthesis rate of the UDP sugars, because the amounts of [^{14}C]uridine incorporated into UDP sugars depend on the specific activity of UTP. Therefore we measured the specific activities of UTP, UDP-HexNAc and UDP-Hex and calculated the synthesis rate of UDP-HexNAc and UDP-Hex.

The specific activities of the UTP, the UDP-HexNAc and the UDP-Hex pools for the various types of cells are given in Fig. 1. From the mean specific activity of the UTP pool in the first hour (pulse) we calculated the amounts of UDP-sugars synthesized within this one hour (radioactivity in a UDP-sugar pool divided by half the specific activity of UTP as determined after one our) (Table 1). Furthermore, Fig. 1 shows that the *relative* turnovers of the UDP-HexNAc (Fig. 1A) and UDP-Hex pools (Fig. 1B) are similar in proliferating normal human T lymphocytes, in proliferating MOLT-3 cells and non-proliferating differentiated MOLT-3 cells. The specific activities of both types of UDP-bound sugars were decreased in differentiated MOLT-3 cells compared to proliferating MOLT-3 cells or proliferating normal human T-lymphocytes. This corresponded with a decreased specific activity of UTP, which in turn was probably caused by a decreased uptake of [^{14}C]uridine in the differentiated MOLT-3 cells. So, although the amount of UDP-HexNAc has decreased in differentiated MOLT-3 cells (Table 1 part A), the turnover of UDP-HexNAc is similar as compared to proliferating MOLT-3 cells (Table 1 part B). Taken together the figures in Table 1 and the graphs in Fig. 1 show that the relative turnover of both the UDP-HexNAc and UDP-Hex pools in the various types of human T lymphocytic cells studied showed no gross differences.

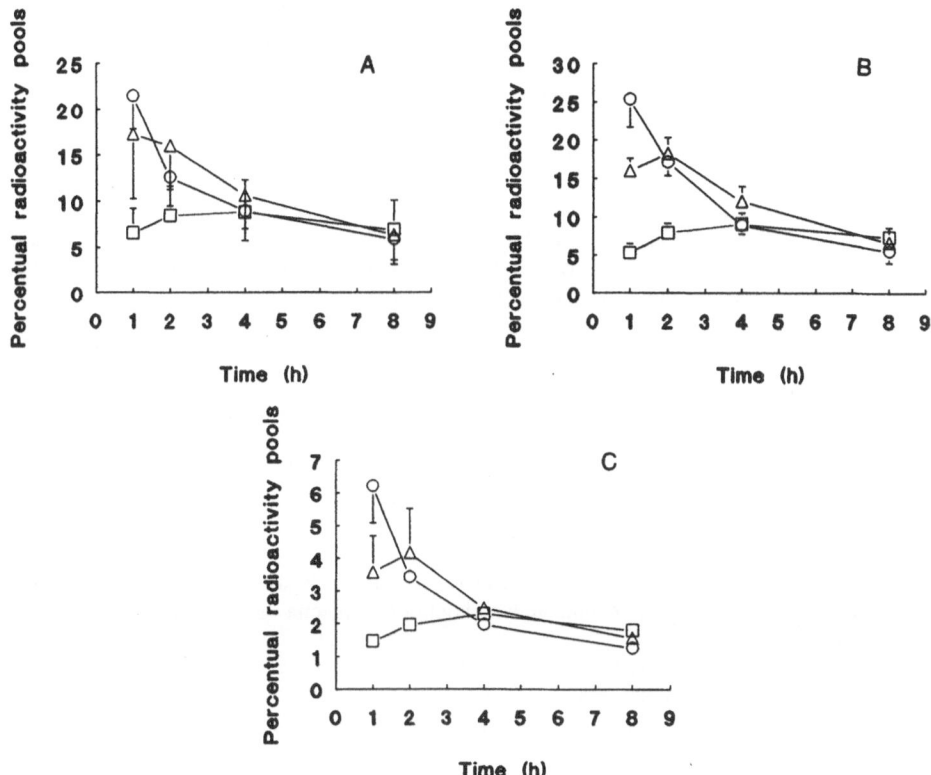

Figure 1. Specific activities of UTP (○), UDP-HexNAc (□) and UDP-Hex (△) during pulse-chase experiments. Chart A, proliferating normal human T lymphocytes; chart B, proliferating MOLT-3 cells; chart C, differentiated MOLT-3 cells. Notice the differences of the ordinates.

DISCUSSION

The increased amounts of pyrimidine nucleotides, UDP sugars and the increased incorporation of [¹⁴C]uridine into UTP and UDP sugars in proliferating MOLT-3 cells compared with proliferating T lymphocytes where both cell types had a comparable growth rate may be explained by the tetraploid character of the MOLT-3 cell line [4]. Because we have not observed changes in the relative turnover of UDP sugars in differentiated MOLT-3 cells, we hypothesize that the changes in glycosylation of plasmamembrane constituents of human MOLT-3 leukemia cells during differentiation [5] are not caused by changes in the metabolism of UDP-sugar compounds but originate from a different utilisation of (acetyl-amino-) sugars attached to UDP.

Our results show that the increased amounts of UDP sugars and the decreased UDP-HexNAc to UDP-Hex ratio in MOLT-3 cells compared to PBL are not fully determined by the proliferative state of the MOLT-3 cells, and are not solely caused by the immature state of the MOLT-3 cells either. Also, the relative turnover of UDP sugars of MOLT-3 cells was not subject to changes in the proliferative or maturation state. Therefore, differences in the UDP-sugar metabolism that have been observed in MOLT-3 cells in comparison with PBL seem to be associated with the malignantly transformed nature of the MOLT-3 cells. A similar phenomenon has been described for HL-60 myeloid leukemia cells [2]. As leukocytes from ALL patients show similar features it is likely that these features are related to the transformed character of patient ALL cells as well. However, because a similar metabolism of UDP sugars has been observed in proliferating benign T cells, the development of a *specific* anti-T cell leukemia drugs on basis of the UDP sugar metabolism seems to offer little perspective.

On the other hand, aberrant glycosylation of several molecules of human T-ALL leukemic cells has been demonstrated [6, 9], and the presence of aberrant oligo-saccharide side chains on these molecules has been suggested to be associated with the malignant capacities (invasiveness, escaping immune surveillance) of the human leukemia cells involved [6, 9]. Therefore, it could be that changing these antigenic (saccharide related) characteristics via the introduction of analogue sugars into the saccharide side chains will appear to be a succesfull strategy to diminish the evil potential of leukemic cells. Moreover, for HL-60 human leukemia cells the activities of certain sugar transferases involved in the glycosylation of gangliosides have a decisive role in directing the differentiation process of these cells into either granulocytic or monocytic cells [7]. So, perhaps it is possible to induce differentiation of leukemic cells through changing the activity of specific transferases via manipulation of the concentration of UDP sugars. These are strategies that await further exploration.

REFERENCES

1. H.R. Mahler, E.H. Cordes. Biological Chemistry (second edition). Harper and Row publishers New York, Evanston, San Fransisco, London (1971).
2. D. De Korte. Ribonucleotide metabolism of leukemic cells. Thesis, Kanters B.V., Alblasserdam (1987).
3. A.A. Van den Berg, H. Van Lenthe, S. Busch, D. de Korte, D. Roos, A.B.P. Van Kuilenburg, and A.H. Van Gennip. Evidence for transformation-related increase in CTP synthetase activity in situ in human lymphoblastic leukemia. *Eur. J. Biochemistry* 216: 161-167 (1993).
4. J.M. Greenberg, R. Gonzalez-Sarmiento, D.C. Arthur, C.W. Wilkowski, B.J. Streifel, and J.H. Kersey. Immunophenotypic and cytogenetic analysis of Molt-3 and Molt-4: human T-lymphoid cell lines with rearrangement of chromosome 7. *Blood* 72: 1755-1760 (1988).
5. M. Akashi, F. Takaku, H. Nojiri, Y. Miura, Y. Nagai, and M. Saito. Neutral and sialosyl glycosphingolipid composition and metabolism of human T lymphoblastic cell line MOLT-3 cells: distinctive changes as markers specific for their differentiation. *Blood* 72: 469-479 (1988).
6. M. Petrini, E. Pelosi-Testa, N.M. Sposi, G. Mastroberardino, A. Camagna, L. Bottero, F. Mavilio, U. Testa, and C. Peschle Constitutive expression and abnormal glycosylation of transferrin receptor in acute T-cell leukemia. *Cancer Res.* 49: 6989-6996 (1989).
7. M. Nakamura, A. Tsunoda, K. Sakoe, J. Gu, A. Nishikawa, N. Tanigushi, and M. Saito. Total metabolic flow of glycosphingolipid biosynthesis is regulated by UDP-GlcNAc:lacto-sylceramide beta 1--:3N-acetylglucosaminyltransferase and CMP-NeuAc:lactosylceramide alpha 2--:3 sialyltransferase in human haemapoietic cell line HL-60 during differentiation. *J. Biol. Chem.* 267: 23507-23514 (1992).
8. M.E. Perlman, D.G. Davis, S.A. Gabel, and R.E. London. Uridine diphospho sugars and related hexose phosphates in the liver of hexosamine-treated rats: identification using 31P-[1H] two-dimensional NMR with HOHAHA relay. *Biochemistry* 29: 4318-4325 (1990)
9. L.A. Smets, and W.P. Van Beek. Carbohydrates of the tumor cell surface. *Biochim. et Biophys. Acta* 738: 237-249 (1984).

PATTERN OF METHYLATED PURINE BASES IN URINE OF CANCER PATIENTS. ANALYSIS BY MASS SPECTROMETRY

L.Lorenzini,[1] A.De Martino,[1] W.Testi,[1] F.Sorbellini,[1] S.Catinella,[2] P.Traldi,[2] E.Marinello,[3] and B.Porcelli,[3]

[1]Institute of General Surgery
[3]Institute of Biochemistry and Enzymology
 University of Siena, Italy
[2]CNR, Area di Ricerca
 Padova, Italy

INTRODUCTION

Methylated purine bases are constituents of both RNA and DNA. tRNA contains the highest number of such compounds. The methyl groups play different roles: they avoid "base pairing", they allow codon identification, they recognize the enzyme aminoacyl tRNA synthetase, and they participate in the aminoacid transfer from tRNA to ribosomes[1].

Methyl bases have long been known to be present in normal urine, and to change under pathological conditions, such as leukemia, tumors and immunodeficiencies[2,3]. Their presence in urine has always been regarded as an index of tRNA turnover[4]. The first methylated purine bases to be demonstrated in urine by Weissmann[5] were 1-methylxanthine, 7-methylxanthine and 1,7 dimethylxanthine. Later research by Park et al.[3], led to the identification of other methylated bases, such as 8-hydroxy-7-methylguanine, 1-methylguanine, 7-methylguanine and 1-methylhypoxanthine. The experimental procedure for separation and identification of all these compounds in urine was based on purification of urine by ion-exchange chromatography, precipitation of the bases with $AgNO_3$ and further separation by paper chromatography, thin layer chromatography[6], gas-chromatography[7] and HPLC techniques[8-10].

All these studies have been focused on the identification or determination of a single methylated base or a small class of such compounds (such as 7-methylguanine[11]) and exogenous methylated xanthines[9,10]. The screening of the whole set of methylated purine bases would be an interesting advance that we attempted in the present study. Two different approaches were employed to cross-check the results. The first was based on off-line measurements by HPLC and mass spectrometry; the second was performed by direct analysis of urinary extracts by Tandem Mass Spectrometry, the hyphenated method with the highest informing power[12]. The compounds were been monitored in urinary extracts of healthy subjects and colorectal tumor patients.

Address for correspondence: Prof.Enrico Marinello, Istituto di Biochimica e di Enzimologia, Università di Siena, Pian dei Mantellini, 44, 53100 SIENA Italy, Tel.+39-577-298026, FAX +39-577-298057

METHODS

Clinical data

We examined 6 healthy control subjects and 6 patients with colorectal cancer: none had a history or clinical evidence of kidney disease or anomalies of purine metabolism; urea and creatinine levels were normal and urine sediment showed no pathological signs. All patients had primitive neoplasia and had undergone surgery (explorative or palliative 32%, radical 68%).

Purification of methylated bases from 24 h urine

Methylated bases were obtained from 24 h urine, by the Weissmann procedure[5], which is not reported here for the sake of brevity. We started from 200 ml urine. At the end of Weissmann procedure[5], we obtained 5 ml of acqueous urine extract.

Materials

Guanase and xanthine oxidase (20 U/ml) were purchased from Boerhinger Mannheim, GmbH (Germany). Hypoxanthine, xanthine, 1-methylhypoxanthine, guanine, 7-methyl-guanine, N_2-methylguanine, adenine, 1-methyladenine, 2-methyladenine and 3-methyl-adenine standards were purchased from Sigma (St.Louis, Mo, USA). Standards of 1-methyl uric acid, 3-methyl uric acid, 7-methyl uric acid, 9-methyl uric acid, 1-methylxanthine, 3-methylxanthine, 7-methylxanthine, 1.7-dimethylxanthine, 8-methylxanthine, 9-methyl-xanthine, theophylline, theobromine, caffeine, 1-methylguanine, 3-methylguanine and 9-methylguanine, were purchased from Fluka (Buks, Switzerland).

Identification of urinary purine bases

Purine bases were identified by two procedures: a) high performance liquid chromatography and b) tandem mass spectrometry.

a) High performance liquid chromatography. A Beckman mod.332 chromatograph with an UV detector at 254 nm, a Supelcosil C18 5μm column (250x4.6 mm) and Supelcosil precolumn 5μm (20x4.6 mm) were utilized. 200 ml of urine samples were eluted for centimes with a 0.01 M KH_2PO_4 buffer (pH 5.5) up to 32 min and from 32 min with a linear gradient between 0-30% methanol for 30 min. The flow rate was 1.5 ml/min. The used UV detector was a Shimadzu UV 160.

Identification of purine bases after HPLC separation was based on retention time, coelution with internal standard and enzyme treatment with guanase and xanthine oxidase.

b) Mass spectrometry. The mass spectrometry measurements of the fractions obtained by preparatory HPLC were performed with an VG ZAB2F (Altricham, UK) double focus, reverse geometry instrument, operating in EI conditions (70 eV, 200 μA). High energy collision experiments were obtained by colliding 8KeV ions with nitrogen in the collision cell placed in the second field-free region. The pressure in the cell was such to reduce the main beam intensity to 40% of its usual value. Accurate mass measurements were performed by the peak matching technique at 10,000 resolving power (10% valley definition).

Mass spectrometry measurements of whole urinary extract were performed introducing 2 μl of acqueous solution containing 3 μg/μl of urine extract in an ion trap mass spectrometer (ITMS, Finnigan MAT, St.José, CA). Collision experiments were carried out by selection of parent ions using the two-step procedure[13], at a qz value of 0.3. A supplementary AC voltage ("tickle" voltage), applied to the two end-caps was used for ion excitation. Typical values of tickle time were in the range 150-200 mV and 15-35 ms respectively. Helium (10^{-4} torr) was used as a buffer gas.

RESULTS

Figure 1 shows the different analytical procedures used for the identification of the purine bases.

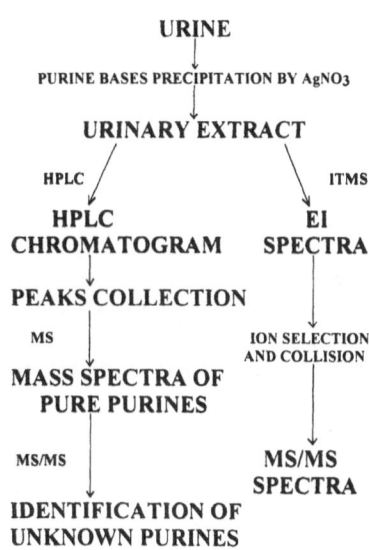

Figure 1. Analytical procedures used for the identification of the purine bases.

Table 1 reports the list of components analyzed in urine extracts and identified by HPLC.

Table 1. List of components of urinary extracts identified by HPLC

1	3-METHYL URIC ACID	10	7-METHYLGUANINE
2	1-METHYL URIC ACID	11	N_2-METHYLGUANINE
3	9-METHYL URIC ACID	12	9-METHYLGUANINE
4	7-METHYL URIC ACID	13	7-METHYLXANTHINE
5	HYPOXANTHINE	14	3-METHYLXANTHINE
6	XANTHINE	15	1-METHYLXANTHINE
7	3-METHYLGUANINE	16	THEOBROMINE
8	1-METHYLGUANINE	17	1,7-DIMETHYLXANTHINE
9	1-METHYLHYPOXANTHINE		

Table 2 reports the list of components analyzed in urine extracts and identified by Mass Spectrometry.

Table 2. List of components of urinary extracts identified by Mass spectrometry

1	3-METHYL URIC ACID	14	N$_2$-METHYLGUANINE
2	1-METHYL URIC ACID	15	9-METHYLGUANINE
3	9-METHYL URIC ACID	16	7-METHYLXANTHINE
4	7-METHYL URIC ACID	17	9-METHYLXANTHINE
5	HYPOXANTHINE	18	3-METHYLXANTHINE
6	XANTHINE	19	1-METHYLXANTHINE
7	3-METHYLGUANINE	20	1-METHYLADENINE
8	3-METHYLADENINE	21	THEOBROMINE
9	8-METHYLXANTHINE	22	2-METHYLADENINE
10	1-METHYLGUANINE	23	GUANINE
11	1-METHYLHYPOXANTHINE	24	CAFFEINE
12	ADENINE	25	THEOPHYLLINE
13	7-METHYLGUANINE	26	1,7-DIMETHYLXANTHINE

DISCUSSION

The following points emerge from our results:

1) HPLC and tandem mass spectrometry enable detection of a great number of methylated purine bases in urine;

2) some of them (2-methyladenine, 8-methylxanthine, 9-methylxanthine) have never previously been detected in urine;

3) tandem mass spectrometry permits the identification of a large number of compounds in a very short time;

4) tandem mass spectrometry is therefore a suitable method for the analysis of methyl purine bases in cancer patients and other subjects.

ACKNOWLEDGEMENTS

Thanks are due to Progetto Finalizzato A.C.R.O. for financial supports.

REFERENCES

1. E.Borek and S.J.Kerr, Atypical transfer RNA's and their origin in neoplastic cells, *Adv.Cancer Res.* 15:163 (1972).
2. W.S.Adams, F.Davis and M.Nakatani, Purine and pyrimidine excretion in normal and leukemic subjects, *Am.J.Med.* 28:726 (1960).
3. R.W.Park, J.F.Holland and A.Jenkins, Urinary purines in leukemia, *Canc.Res.* 22:469 (1962).
4. E.Borek, B.S.Baliga, C.W.Gehrke, K.C.Kuo, S.Belman, W.Trole and T.P.Waalkes, High turnover rate of transfer RNA in tumor tissue, *Cancer Res.* 37:3362 (1977).
5. B.Weissmann, P.A.Bromberg and A.B.Gutman, The purine bases of human urine, *J.Biol.Chem.* 224:407 (1957).
6. A.H.Van Gennip, J.Grift, E.J.Van Bree-Blom, D.Ketting and S.K Wadman, Urinary excretion of methylated purines in man and in the rat after the administration of theophylline, *Clin.Chim.Acta* 86:7 (1978).
7. C.W.Gehrke and D.B.Lakings, Gas-liquid chromatography of the purine and pyrimidine bases, *J.Chrom.* 61:45 (1971).

8. H.Topp, G.Sander and G.Heller-Schoch, A high performance liquid chromatography method for the determination of pseudouridine and uric acid in native human urine and ultrafiltrated serum, *Anal.Bioch.* 150:153 (1985).

9. A.Wahllander, E.Renner and G.Karlaganis, High-performance liquid chromatographic determination of dimethylxanthine metabolites of caffeine in human plasma, *J.Chrom.* 338:369 (1985).

10. N.R.Scott, J.Chakraborty and V.Marks, Determination of urinary chromatography. A comparative study of a direct injection and an ion pair extration procedure, *J.Chrom.* 375:321 (1986).

11. H.Yuki, T.Yajma, H.Kawasaki and A.Jamaji, Analysis of 7-Methylguanine in deoxyribonucleic acids by high performance liquid chromatography, *Anal.Bioch.* 97:203 (1979).

12. R.A.Yost, MS/MS tandem mass spectrometry, *Spectra* 9:2 (1983).

13. C.E.Ardanaz, J.Kavka, F.Guidugli, P.Traldi and V.Vettori, The ion trap mass spectrometer in ion structure studies. The keys of M-H+ ions from Chalcone, *Rap.Comm.Mass Spectrom.* 5:5 (1991).

CYTOTOXICITY OF DEOXYCOFORMYCIN ON HUMAN COLON CARCINOMA CELL LINES

M. Camici,[1] M. Turriani,[1] G. Turchi,[2] M.G. Tozzi,[3] J. Cos,[4] C. Alemany,[4] V. Noe,[4] and C.J. Ciudad[4]

[1]Dipartimento di Fisiologia e Biochimica, Università di Pisa, Italy
[2]European Centre for the Validation of Alternative Methods (ECVAM), Ispra, Italy
[3]Istituto di Chimica Biologica, Università di Sassari, Italy
[4]Unitat de Bioquimica, Facultat de Farmacia, Universitat de Barcelona, Spain

INTRODUCTION

Deoxycoformycin (DCF), a powerful inhibitor of adenosine deaminase[1], by virtue of its immunosuppresive action[2], has found clinical application for the treatment of several types of lymphatic leukemia[3,4]. In fact, the prevention of the catabolism of deoxyAdo to deoxyIno, results both in the accumulation of intracellular deoxyATP[5], an inhibitor of ribonucleotide reductase, and in the depletion of ATP[6]. Indeed, it has been reported that human T lymphocyte cells are very sensitive to the cytotoxic action of deoxyAdo, and that their sensitivity to DCF is higher than B lymphocyte cells probably as a result of a reduced cytosolic 5'deoxynucletidase activity, with a concomitant preservation of the deoxyATP pool[7]. While many studies have been performed on the cytotoxicity of DCF on human lymphocytes and several reports describe its possible mechanism of action in this type of cells[8,9], the effect of this drug on different mammalian cells has not been thoroughly examined so far, also on account of the reported inefficacy of DCF on the growth of cultured cells[10]. Human colon carcinoma cultured cells are a model for the study of the responsiveness of this solid tumor to possible chemotherapic agents. In view of the involvement of the purine salvage enzymes in the activation of purine analogs, the knowledge of their program in tumor cells might be of great help not only for the understanding of the analog mechanism of action, but also for the design of specific pro-drugs. In this line is the present study on the cytotoxicity of DCF in combination with deoxyAdo on both two human colon carcinoma (LOVO and HT29) and a chinese hamster ovary (CHO K-1) cell lines. The different sensitivity of the three cell lines to DCF is well related to their distinct purine salvage enzyme pattern, thus supporting, in a general point of view, the enzymatic approach to the choice of a suitable chemotherapic agent.

EXPERIMENTAL PROCEDURES

Cells were grown as a monolayer in Ham's F-12 medium with 7% foetal calf serum and antibiotics. When cells were in logarithmic phase of growth the monolayers were washed twice with PBS and the cells were scraped with a rubber policeman, collected and centrifuged at 800xg for 3 min. The pellets were further washed with PBS and stored at -70°C. The pellets were then resuspended in 3 volumes of 50 mM Tris-HCl, pH 7.4, subjected to sonication and centrifuged at 100.000xg for 1 hr. The supernatant was used for the determination of the enzyme activities. Adenosine deaminase, adenosine kinase, AMP deaminase, and 5'-nucleotidase with 5'-AMP or 5'-deoxyAMP as substrates were assayed essentially according to Cappiello et al [11], except that 5 μM

DCF was added to the assay for adenosine kinase and 5'-nucleotidase activities. In order to test the cytotoxicity, one thousand cells were plated in 35 mm dishes and incubated in the absence or in the presence of the effectors until formation of visible colonies (usually one week). Then, colonies were stained with crystal violet and counted. The incorporation of deoxyAdo was assessed by incubation of the cells at initial density ranging between 15 and 31 x 10^3 cells/cm^2 for various periods of time with DCF and dypiridamole. Control cultures contained neither additives.

RESULTS AND DISCUSSION

The addition of DCF alone, up to a concentration of 10^{-6}M, a value significantly higher than the K_i of adenosine deaminase for DCF[1] was not cytotoxic for CHO, LOVO and HT29 cells growing in Ham's F-12 nutrient medium (data not shown), thus indicating that, in these conditions, the adenosine deaminase activity is not essential for the growth of these cells. However, when deoxyadenosine was also added to the culture medium, DCF became a powerful inhibitor of cell growth (Fig.1). The concentration of deoxyAdo required in combination with 10^{-8}M DCF to attain 50% inhibition of the cell growth was 25, 50 and 70 μM for CHO, LOVO and HT29 cell lines, respectively, thus indicating a lower sensitivity of the latter two cell lines to the treatment.

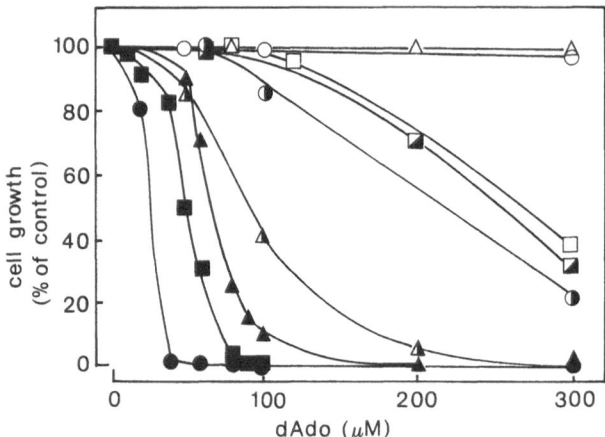

Figure 1. Effect of DCF (10^{-8}M) on the growth of CHO K-1 (circle), LOVO (square), and HT29 (triangle) cells in the presence of increasing concentrations of deoxyAdo; alone (filled symbols), with 10^{-5}M 5'amino5'deoxyAdo (open symbols), with 10^{-6}M dypiridamole (half filled symbols). Results are expressed as percent of the control.

Indeed, the inhibition of adenosine deaminase by DCF, results in an enhancement of the intracellular level of deoxyAdo which, on one hand, is itself an inhibitor of S-adenosylhomocysteine hydrolase, with subsequent inhibition of S-adenosylmethione dependent methylation reactions[12], on the other hand, by successive phosphorylation steps, is converted to deoxyATP, a potent inhibitor of ribonucleotide reductase[13].

Several enzyme activities are involved both in the conversion of deoxyAdo to the cytotoxic agent (activation) and in the transformation of deoxyAdo and/or its products to a non cytotoxic compound (inactivation). Therefore, the cellular enzyme levels should be conceivably related to the different sensitivity exhibited by the three cell lines. The highest level of adenosine deaminase activity was found in CHO K-1 cells (Fig.2) and this is related to their greater sensitivity to the cytotoxic action of deoxyAdo (Fig.1): in fact these cells must be more protected against a compound, which is potentially more cytotoxic for them than for LOVO and HT29. Since the first step towards the formation of deoxyATP is the phosphorylation of deoxyAdo, the fact that CHO cells are the most sensitive is in agreement with their significantly enhanced adenosine kinase activity; this enzyme activity, together with deoxycytidine kinase is in fact responsible for the phosphorylation of deoxyAdo. The involvement of adenosine kinase in the determination of the cytotoxic action of DCF is further supported by the increased resistance of the three cell lines when 10^{-5}M 5'amino5'deoxyAdo, an inhibitor of adenosine kinase, was added to the culture medium (Fig.1). On the other hand, when 50 μM deoxycytidine was added to the culture medium, the cytotoxity of DCF

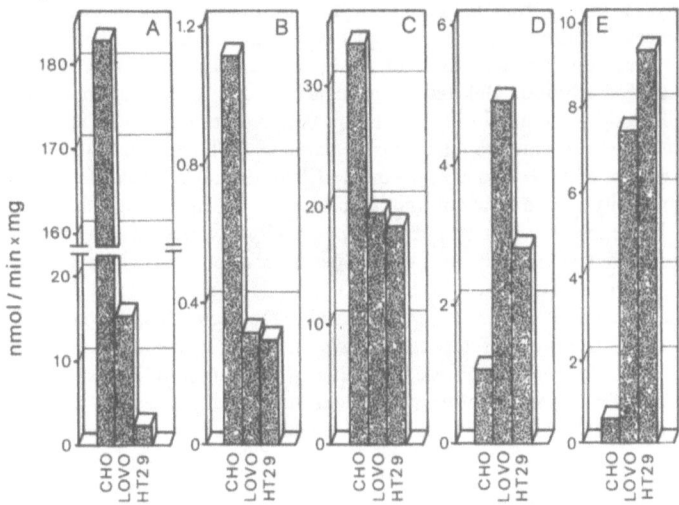

Figure 2. Levels of enzyme activities involved in the metabolism of deoxyAdo. A: adenosine deaminase; B: adenosine kinase; C: AMP deaminase; D: 5'AMP dephosphorylating activity; E: 5'deoxyAMP dephosphorylating activity. Each bar represents the mean of the values measured in the crude extracts obtained by at least seven distinct cell cultures.

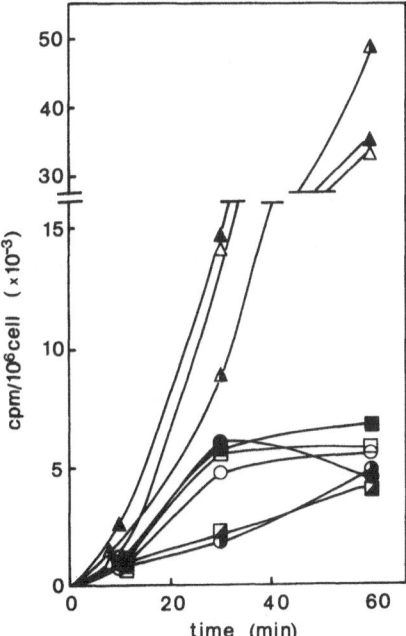

Figure 3. Effect of DCF and dypiridamole on the incorporation of [14C]deoxyAdo (specific activity 9 µCi/µmol) into total cell material. CHO K-1: circle; LOVO: square; HT29: triangle. DCF was added 1 hr before deoxyAdo and/or dypiridamole. Filled symbols: deoxyAdo (40 µM); Open symbols: DCF (10^{-8}M) and deoxyAdo; Half filled symbols: DCF, deoxyAdo, and dypiridamole (10^{-6}M). The experiments were carried out in duplicate cultures.

was unaffected. This seems to indicate that in vivo, at least in these types of cells, deoxyAdo is not phosphorylated to a significant extent by deoxycytidine kinase, and that the major deoxyAdo phosphorylating activity is exerted by adenosine kinase. Relevant differences among AMP deaminase, 5'AMP and 5'deoxyAMP dephosphorylating activities were found in the three cell lines (Fig.2). In fact AMP deaminase activity is higher in the most sensitive cells, and this contrasts the assessment that the "deactivating" enzymes should be more abundant in the most resistant cells. However, deoxyAMP is a poor substrate for AMP deaminase[14] and therefore the involvement of this enzyme as a detoxifying activity seeems to be unlikely. The lower level of 5'AMP and 5'deoxyAMP dephosphorylating activities in CHO cells is a further indication that the phosphorylation of deoxyAdo plays a crucial role in the mechanism of cytotoxicity. Indeed, a consequence of a diminished nucleoside monophosphate dephosphorylation is a greater formation of deoxyATP.

The observed different sensitivity of the three cell lines could be also related to a different cellular uptake of deoxyAdo. Our results indicate that the incorporation of deoxyAdo was not inhibited significantly in the total cell material by the presence of DCF (Fig.3). Furtermore, HT29 cells which are the less sensitive to the cytotoxic action of DCF and deoxyAdo, are able to take up a much greater amount of deoxyAdo as compared to CHO and LOVO cells. This is in accordance with the lower cytotoxicity of deoxyAdo in this type of cells which, through mechanisms only in part elucidated in this report, are able to buffer its potentially toxic effect. The addition to the growth medium of dypiridamole, an inhibitor of nucleoside transport[15], which reduced the uptake of deoxyAdo (Fig.3), diminished the cytotoxicity of the deoxynucleoside in combination with DCF (Fig.1), and the extent of this effect was proportional to the inhibition of deoxyAdo transport.

In conclusion, the growth of mammalian cells, including human colon carcinoma cell lines, can be affected by the presence of DCF in combination with deoxyAdo; furthermore, the extent of this effect is related to their distinct purine salvage enzyme pattern, indicating that the study of purine metabolism in tumor cells can be of great help for a better approach to purine analog treatment.

Acknowledgments

This work was supported by grants from CNR, "Progetto Bilaterale", and FISss (#92/0775).

REFERENCES

1. R.P. Agarwal, T. Spector, and R.E. Parks Jr, Tight-binding inhibitors-IV. Inhibition of adenosine deaminase by various inhibitors, *Biochem. Pharmacol.* 26:359 (1977).
2. R.I. Glazer, Adenosine deaminase inhibitors: their role in chemotherapy and immunosuppression, *Cancer Chemother. Pharmacol.* 4:227 (1980).
3. B.D. Cheson, New chemotherapeutic agents for the treatment of low-grade non-Hodgin's lymphomas, *Seminars in Oncology* 20:96 (1993).
4. A. Saven, and L.D. Piro, The newer purine analogs. Significant therapeutic advance in the management of lymphoid malignancies, *Cancer* 72:3470 (1993).
5. C.M. Smith, and J.F. Henderson, Deoxyadenosine triphosphate accumulation in erythrocytes of deoxycoformycin-treated mice, *Biochem. Pharmacol.* 31:1545 (1982).
6. H.A. Simmonds, R.J. Levinsky, D. Perrett, and D.R. Webster, Reciprocal relationships between erythrocyte ATP and deoxyATP levels in inherited adenosine deaminase deficiency, *Biochem. Pharmacol.* 31:947 (1982).
7. T. Iizasa, F. Takeuchi, Z. Honda, Y. Nishida, N. Kamatani, and T. Miyamoto, Differential composition of cytosol 5'-nucleotidases between T and B lymphoblasts, *Biochim. Biophys. Acta* 882:228 (1986).
8. M.S. Hershfield, J.E. Fetter, W.C. Small, A.S. Bagnara, S.R. Williams, B. Ullman, D.W. Martin Jr., D.B. Wasson, and D.A. Carson, Effects of mutational loss of adenosine kinase and deoxycytidine kinase on deoxyATP accumulation and deoxyadenosine toxicity in cultured CEM human T-lymphoblastoid cells, *J. Biol. Chem.* 257:6380 (1982).
9. M.S. Hershfield, and N.M. Kredich, Resistance of an adenosine kinase-deficient human lymphoblastoid cell line to effects of deoxyadenosine on growth, S-adenoylhocysteine hydrolase inactivation, and deoxyATP accumulation, *Proc. Natl. Acad. Sci. USA* 77:4292 (1980).
10. P. Rowland III, J. Pfeilsticker, and P.A. Hoffee, Adenosine deaminase gene amplification in deoxycoformycin-resistant mammalian cells, *Arch. Biochem. Biophys.* 239:396 (1985).
11. M. Cappiello, D. Barsacchi, A. Del Corso, M.G. Tozzi, M. Camici, U. Mura, and P.L. Ipata, Purine salvage as a metabolite and energy saving mechanism in the ocular lens, *Curr.Eye Res.* 11:435 (1992).
12. C.E. Cass, M. Selner, P.J. Ferguson, and J.R. Phillips, Effects of 2'-deoxyadenosine, 9-β-D-arabino furanosyladenine, and related compounds on S-adenoyl-L-homocysteine hydrolase activity in synchronous and asynchronous cultured cells, *Cancer Res.* 42:4991 (1982).
13. I.H. Fox, and W.N. Kelley, The role of adenosine and 2'-deoxyadenosinein mammalian cells, *Ann. Rev. Biochem.*, 47:655 (1978).
14. S.-L., Yun, and C.H. Suelter, Human erythrocyte 5'-AMP aminohydrolase. Purification and characterization, *J. Biol. Chem.* 253:404 (1978).
15. P.G.W. Plagemann, and J. Erbe, The deoxynucleoside transport systems of cultured Novikoff rat hepatoma cells, *J. Cell. Physiol.* 83: 337 (1974).

IMBALANCE BETWEEN THE PYRIMIDINE RIBONUCLEOTIDE POOLS IN RAT RHABDOMYOSARCOMA R1 CELLS

Robbert J. Slingerland, André B.P. Van Kuilenburg, Jeroen Bodlaender, Henk Van Lenthe, Edith Kreuk, P.A. Voûte, Albert H. Van Gennip

Academic Medical Center
University of Amsterdam
Departments of Paediatrics and Clinical Chemistry
PO BOX 22700
1100 DE AMSTERDAM, The Netherlands

INTRODUCTION

Imbalances between ribonucleotide pools are at least partly associated with transformation in pheochromocytoma[1] and in leukemic cells. In leukemic cells, imbalances in pyrimidine ribonucleotide pools were shown to be caused by altered activities of 'key-enzymes' such as CTP-synthetase[2]. We investigated whether such ribonucleotide imbalances also exist in rhabdomyosarcoma R1 cells in comparison to their normal counterparts. We used CPEC (NSC 375575) to study the involvement of CTP-synthetase and/or uridine-cytidine kinase in causing a pyrimidine ribonucleotide imbalance. CPEC is a carbocyclic analogue of cytidine in which the ribofuranose moiety is replaced by a cyclopentenyl ring[3]. An antineoplastic activity of CPEC has been shown against several tumors[4-7]. CPEC acts as a substrate inhibitor for uridine-cytidine kinase[8]. Intracellularly, CPEC is converted to its 5'triphosphate (CPE-CTP), which inhibits the conversion of UTP to CTP by CTP-synthetase causing a depletion of the endogenous CTP pools.

MATERIALS AND METHODS

The rat rhabdomyosarcoma R1 cell line was propagated continuously in Dulbecco's Modified Eagles Medium (DMEM) (Gibco Laboratories, Paisley, Scotland) supplemented with 2 mM L-glutamine (Flow laboratories, Irvine, UK), 50 I.U./ml penicillin, 50 μg/ml streptomycin (Imperial, UK) and 10%(v/v) fetal calf serum (FCS)(Gibco Laboratories). Loosely capped culture flasks (Costar Corp., Cambridge, MA, USA) were incubated at 37 °C, in humidified (96%) air with 5% CO_2. Cells were passaged once a week and maintained in logarithmic growth phase. Cultures were persistently free of mycoplasma (tested with Genprobe, ICN, UK). Rat L6 myoblasts were cultured in a similar way as described for the R1 cells except the antibioticum used was gentamicine at a concentration of 50 μg/ml.

To extract ribonucleotides from R1 cells, medium was removed and the cells were gently washed two times with ice cold phosphate-buffered saline (PBS; 140 mM NaCl, 9.2 mM Na_2HPO_4, 1.3 mM NaH_2PO_4, pH 7.4). After removal of PBS, ice-cold PCA (0.4 M) was added to the flask and the flask was incubated for 10 min on ice with intermittently scraping of the culture surface with a cell scraper (Costar corp.). The suspension was removed and centrifuged at 10000 x g for 3 min at 4 °C. The supernatant was neutralized with K_2CO_3 and used for analysis of the ribonucleotides. Analysis of the ribonucleotides was performed by HPLC as described elsewhere[9]. Protein content of the remaining cell-pellet was measured after dissolving protein with 0.2 M NaOH[10].

Cell-cycle distribution was measured by a two step immunofluorescent detection of incorporated bromo-2'-deoxyuridine (BrdU) and staining of nuclei with propidium iodide[2]. Cell-cycle analyses were performed by flow-cytometric analyses of red and green fluorescence of 10000 cells.

A modified (MTT) assay (Boehringer Mannheim GmbH Biochemica, Germany) was used to determine the sensitivity of the rhabdomyosarcoma R1 cells to CPEC (NSC 375575, Drug Synthesis & Chemistry Branch, Developmental Therapeutics Program, Division of Cancer Treatment, National Cancer Institute, NC, USA). A cell suspension was serially diluted to concentrations varying from 0.2×10^5 - 1.25×10^3 cells/ml and 100 μl of these solutions were plated into 96-well microtiter plates and incubated at 37 °C. Twenty-four hours later, 50 μl medium was removed and CPEC was added to each well at various concentrations in replicates of 4. Cells were exposed to CPEC (100 μl final volume in well) for 72 hr and the numbers of surviving cells were subsequently quantitated using the previously described MTT assay[11]. Additionally, cells were counted under a phase-contrast microscope.

As a marker of differentiation creatine kinase activity was determined (Boehringer Mannheim GmbH Biochemica). Cells were harvested by trypsinization and centrifuged at 200 x g. The pellet was resuspended in 100 μl 0.5 M cysteine. This suspension was frozen. After thawing, the suspension was 3 x 30 (s) sonicated with a "Vibrocell" (Sonics & Materials inc., Danbury, USA) with a cooling period at 0 °C of 50 (s) in between. This suspension was used for determination of creatine kinase activity.

RESULTS

The uracil(U)/cytosine(C) ribonucleotide balance in rat rhabdomyosarcoma R1 cells was 2.2 ± 0.5 (mean ± S.D., n=3). In the rat myoblast cell-line L6 the U/C balance was 4.7 ± 0.08 (n=3). In contrast, the adenine (A)/guanine(G) ribonucleotide ratios were comparable for both cell types: the A/G ratio was 5.3 ± 0.5 (n=3) in R1 cells and 6.3 ± 0.9 (n=3) in L6 cells.

To investigate the influence of plausible factors causing the imbalance, R1 cells with a different cell-cycle distribution (83% ± 5% G0/G1- phase, 6% ± 5% S-phase, 11% ± 6% G2/M-phase, n=7) were studied. These cells were obtained by inducing growth limitation by culturing R1 cells to confluency. In comparison to exponential growing cells (53% ± 9% G0/G1-phase, 40% ± 10% S-phase, 9% ± 2% G2/Mphase, n=9) no change was observed in the U/C ratio and the A/G ratio in the cells with the altered cell-cycle distribution and a significantly reduced growth rate.

L6 myoblasts were differentiated with insuline (0.85 μM). Differentiated L6 cells (20 % were large myotubes) had a creatine kinase activity of 0.45 x E-2 units/mg protein whereas undifferentiated L6 cells and R1 cells had a ten times lower activity. The U/C as well as the A/G ribonucleotide ratio in the differentiated L6 cells remained unchanged.

The cytotoxicity of CPEC, a carbocyclic analogue of cytidine towards the rhabdomyosarcoma R1 cells and the L6 cells was determined with a colorimetric assay based on the reduction of tetrazolium. In the presence of 10 percent fetal calf serum in the culturing medium CPEC had an IC_{50} value of 1 μM for R1 cells after 3 days incubation, whereas the L6 cells were less sensitive (IC_{10} of CPEC was 2 μM after 3 days incubation).

DISCUSSION

Neither the cell-cycle distribution, nor the proliferation rate or the degree of differentiation seems to influence ribonucleotide ratios in R1/L6 cells. We postulate that the imbalance between uracil and cytosine ribonucleotide pools in R1 rhabdomyosarcoma cells compared to L6 myoblasts is at least partly associated with transformation.

This imbalance between pyrimidine pools in rhabdomyosarcoma R1 cells in comparison with L6 myoblasts may have been induced by different activities of either deaminases or Urd/Cyd kinase or CTP-synthetase or any combination of these enzymes. So far, incubation for 3 days with 2 μM CPEC, inhibitor of Urd/Cyd kinase and CTP-synthetase, revealed that the imbalance between the pyrimidine ribonucleotide pools in R1 cells can be used to kill selectively R1 cells vs. L6 cells.

Acknowledgements

This study was supported by a grant from the *Dutch Foundation for Pediatric Cancer Research S.K.K.* grant no. 90-02 and by a grant of the *Maurits and Anna de Kock Foundation*. We thank Dr. Johnson of the National Cancer Institute, USA for kindly providing CPEC.

REFERENCES

1. R.J. Slingerland, J.M. Bodlaender, H. Van Lenthe, A.B.P. Van Kuilenburg, and A.H. Van Gennip, Imbalance between the pyrimidine ribonucleotide pools of rat pheochromocytoma PC-12 cells, Clin. Chem. Enzym. Comms. 5:315(1993).
2. A.A. Van den Berg, H. Van Lenthe, S. Busch, D. De Korte, D. Roos, A.B.P. Van Kuilenburg, and A.H. Van Gennip, Evidence for transformation-related increase in CTP-synthetase activity in situ in human lymphoblastic leukemia, Eur. J. Biochem. 216:161 (1993).
3. V.E. Marquez, MU-I. Lim, S.P. Treanor, J. Plowman, M.A. Priest, A. Markovac, M.S. Khan, B. Kaskar, and J.S. Driscoll, Cyclopentenylcytosine. A carbocyclic nucleoside with antitumor and antiviral properties, J. Med. Chem., 31: 1687 (1988).
4. H. Zhang, D.A. Cooney, M. H. Zhang, G. Ahlowalia, H. Ford, Jr., and D.G. Johns, Resistance to cyclopentenyl cytosine in murine leukemia L1210 cells, Cancer Res., 53: 5714 (1993).
5. R.I. Glazer, M.B. Cohen, K.D. Hartman, M.C. Knode, MU-I. Lim, and V.E. Marquez, Induction of differentiation in the human promyelocytic leukemia cell line HL-60 by the cyclopentenyl analogue of cytidine, Biochem. Pharmac. 35:1841 (1986).
6. L.K. Yee, C.J. Allegra, J.B. Trepel, and J.L. Grem, Metabolism and RNA incorporation of cyclopentenyl cytosine in human colorectal cancer cells, Biochem. Pharmac. 43: 1587 (1992).
7. H. Ford, Jr., D.A. Cooney, G.S. Ahluwalia, Z. Hao, M.E. Rommel, L. Hicks, K.A. Dobyns, J.E.

Tomazewski, and D.G. Johns, Cellular pharmacology of cyclopentenyl cytosine in Molt-4 lymphoblasts, Cancer Res. 51:3733 (1991).

8. G.J. Kang, D.A. Cooney, J.D. Moyer, J.A. Kelley, H.-Y. Kim, V.E. Marquez, and D.G. Johns, Cyclopentenylcytosine triphosphate. Formation and inhibition of CTP synthetase, J. Biol. Chem. 264:713 (1989).

9. D. De Korte, W.A. Haverkort, D.Roos, and A.H. Van Gennip, Anion-exchange high performance liquid chromatography method for the quantitation of nucleotides in human blood cells, Clin. Chim. Acta 148:185 (1985).

10. P.K. Smith, R.I. Krohn, G.T. Hermanson, A.K. Mallia, F.H. Gartner, and M.D. Provenzano, Measurement of protein using bicinchoninic acid, Anal. Biochem. 150: 76 (1985).

11. T. Mosmann, Rapid colorimetric assay for cellular growth and survival: application to proliferation and cytotoxicity assays, J. Immunol. Meth. 65:55 (1983).

COMPARTMENTATION OF RIBONUCLEOTIDES IN PC-12 CELLS: FREE AND PROTEIN BOUND RIBONUCLEOTIDES

Robbert J. Slingerland, André B.P. Van Kuilenburg, Jeroen Bodlaender, Henk Van Lenthe, Edith Kreuk, P.A. Voûte, Albert H. Van Gennip

Academic Medical Center
University of Amsterdam
Departments of Paediatrics and Clinical Chemistry
PO BOX 22700
1100 DE AMSTERDAM, The Netherlands

INTRODUCTION

Pheochromocytoma is a cancer derived from the neural crest, mainly occurring at adult age. Rat pheochromocytoma PC-12 cells have an imbalance between pyrimidine ribonucleotide pools in comparison with adrenal medulla tissue[1]. No imbalance has been noticed between the ribonucleotide pools of the purine pathway. In contrast, leukemic cells have an imbalance between the total ribonucleotide pools within both pathways. Inhibition of 'key-enzymes' of the purine and pyrimidine pathway, resulting in a temporary restoration of the balance between the ribonucleotide pools, induces differentiation of leukemic cells[2]. So far, the mechanism underlying this differentiation proces is unknown. *Mycophenolic acid* (MPA) and tiazofurin inhibit inosine-monophosphate dehydrogenase, a 'key-enzyme' of the purine pathway. Tiazofurin induced a trans-differentiation in the neural crest derived neuroblastoma cells[3].

Our aim is to investigate whether ribonucleotides in PC-12 cells are differently *compartmented* compared to their normal counter-parts, the adrenal medulla cells. We focussed as a first part of this study on the ribonucleotide fraction bound to protein in the pheochromocytoma PC-12 cells. An imbalance between *compartmented* purine ribonucleotide pools may exist in PC-12 cells in comparison to normal cells and if so, be used as a target for drugs which possess a selectivity towards these compartments. Furthermore this may provide information about the mechanism inducing (trans-)differentiation as has been noticed in neuroblastoma and other cells.

Purine and Pyrimidine Metabolism in Man VIII, Edited by
A. Sahota and M. Taylor, Plenum Press, New York, 1995

MATERIALS AND METHODS

The rat pheochromocytoma PC-12 cell line was propagated continuously in DMEM/HAMF-12 (Gibco Laboratories, Paisley, Scotland) supplemented with 2 mM L-glutamine (Flow laboratories, Irvine, UK), 50 I.U./ml penicillin, 50 μg/ml streptomycin (Imperial, UK), 10% (v/v) heat-inactivated horse serum and 10% (v/v) fetal calf serum (FCS)-(Gibco Laboratories). Loosely capped culture flasks (Costar Corp., Cambridge, MA, USA) were incubated at 37 °C, in humidified (96%) air with 5% CO_2. Cells were passaged once a week and maintained in a logarithmic growth phase. Cultures were persistently free of mycoplasma (tested with Gen-probe, ICN, UK).

Exponentially growing PC-12 cells on poly-D-lysine growth-layers were used for quantification of the free (ethanol/EDTA) ribonucleotide pool vs. the ribonucleotide pool bound to protein. Medium was removed and cells were washed twice with ice-cold PBS (without Mg^{2+} or Cl^{-}). Subsequently, the flasks were put on extra cold ice (prepared with NaCl and ethanol -70 °C) and the cells were extracted with ethanol/EDTA (5 mM EDTA and 65% (v/v) ethanol) cooled at -20 °C for 5 min[4]. The supernatant was removed and checked for protein. The extraction of free ribonucleotides was repeated. Remaining cell pellets were washed with PBS at 0 °C.

To release the ribonucleotides bound to protein, remaining cell pellets were extracted with perchloric acid (0.4 M) for 10 min. The solution was removed and centrifuged for 3 min at 10.000 x g. The supernatant was removed and stored at - 20 °C for further analysis. Protein content of the remaining cell-pellet was measured after dissolving protein with 0.2 M NaOH[5].

One day prior to experiments with ^{14}C-guanosine, cells were grown in a chemically defined medium consisting of DMEM/HAMF-12 with 0.5% Human Serum Albumine (Boehringer Mannhein GmbH Biochemica, Germany), 1% N_2-bottenstein supplement (Gibco), 2 mM L-glutamine, 12 μM hypoxanthine, 6 μM uridine, 3 μM cytidine (Sigma Chemical Company, St. Louis, MO, USA). Moreover, these cells were also plated (for 48 hr) on poly-D-lysine (Sigma) coated flasks (20 μg/cm^2).

15 min before starting labeling with 1 μM ^{14}C-guanosine (0 - 60 min.), cells were incubated with 10 μM *mycophenolic acid* to inhibit the flux from hypoxanthine into guanine ribonucleotides.

During 60 min, samples were collected for quantification of the free and protein-bound ribonucleotide pools and the amount of radiolabel in the various guanine ribonucleotide pools. Separation was performed as described above.

The ethanol/EDTA supernatant and the perchloric acid supernatant have been analysed by HPLC as described elsewhere[5]. Radioactivity was determined on-line with a RAMONA detector (radiolabeled guanine ribonucleotide pools). Furthermore, samples were analysed with a ß-counter.

RESULTS

Investigation of the distribution of purine ribonucleotides between the free fraction and the protein-bound fraction revealed that approximately 20 percent of the guanine ribonucleotide pools was bound to protein whereas about 4.5 percent of the adenine pools was bound to protein. High ATP/ADP and GTP/GDP ratios (ca 11 and 7 resp.) were observed in whole cell nucleotides (only perchloric acid extraction), whereas the ribonucleotides bound to protein had an ATP/ADP ratio close to 2.5 and GTP/GDP ratio between 1 and 2. Pretreatment of PC-12 cells with *mycophenolic* acid and subsequent labeling of the *compartmented* guanine-ribonucleotide pools with ^{14}C-guanosine showed the presence of metabolically different guanine ribonucleotide pools. Especially the free GDP-pool (ethanol/EDTA extract) had a ca 45 % lower specific activity (2.1 % $\pm$ 0.5 % of free GDP was labeled) then the protein-bound GDP in the perchloric extract (3.8 % $\pm$ 0.2 % of GDP bound to protein was labeled) after a 65 min incubation period with ^{14}C-guanosine (Figure 1).

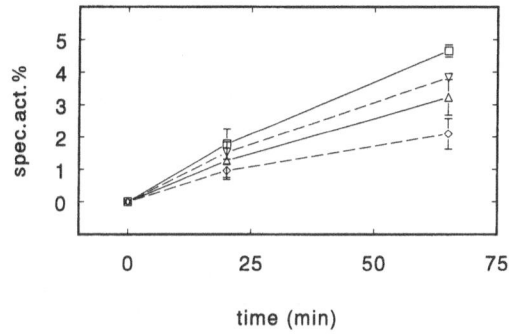

Figure 1. (^{14}C)-guanosine labeling of PC-12 cells. Specific activities of the "free" and "bound GTP and GDP pools were followed during 65 min (n=3). Specific activity % is expressed as (^{14}C-G*P/tot G*P)*100. (-□- = free GTP-pool, -△- = bound GTP-pool, -◇- = free GDP-pool, -▽- bound GDP-pool).

DISCUSSION

We report here that ca 20 % of the guanine ribonucleotides in rat pheochromocytoma PC-12 cells are physically compartmented by binding to proteins. Although many authors have addressed the problem of *compartmentation* of nucleotides[4] when studying the metabolism of the purine and pyrimidine nucleotides, actual measurements on cells are still limited. According to our knowledge, no complete data are available of cells derived from neural crest tissue. Recently, the adenine and guanine nucleotide content of Triton X-100 extracted cytoskeletal proteins from PC-12 was described[7]. Analysis of the adenine fraction bound to actin is complicated due to the fact that a high proportion of ATP can be very rapidly hydrolyzed to ADP and AMP[8]. In order to prevent break-down of ribonucleotides, taxol was added by Angelastro to PC-12 cells in order to stabilize the binding of nucleotides to microtubuli prior to extraction. An evaluation of the metabolic activity of these pools was omitted. The fraction of the guanine ribonucleotides bound to protein described here is in line with that reported by Angelastro for these cells. Also the determined GTP/GDP ratio of the fraction bound to protein is close to those described for cytoskeletal proteins extracted with Triton X-100. The disadvantage of the use of taxol is that these cells have to be manipulated with taxol before the proportion of nucleotides bound to e.g. actine can be measured. Sampling at early time-points without disturbing the nucleotide metabolism is difficult during metabolic labeling studies in which taxol is used for stabalization of nucleotides. In order to prevent the break-down of ribonucleotides we prefer to extract at low temperature.

After pretreatment of PC-12 cells with *mycophenolic acid*, labeling with ^{14}C-guanosine revealed differences in specific activities (^{14}C labeled guanine ribonucleotides/total guanine ribonucleotides) in the various guanine ribonucleotide pools. Based on the higher specific activity of GDP pool bound to protein compared to free GDP we postulate that guanosine is preferably channeled to proteins under the conditions tested.

Currently, we are investigating which and to what extent metabolically different *compartments* are influenced by *mycophenolic acid*, a known (trans-)differentiation inducing agent. This research may provide more detailed information about the mechanism underlying the differentiation proces as has been noticed in leukemic cells and the trans-differentiation in neuroblastoma cells.

Acknowledgements

This study was supported by a grant from the *Dutch Foundation for Pediatric Cancer Research S.K.K.* grant no. 90-02 and by a grant from the *Maurits and Anna de Kock Foundation*.

REFERENCES

1. R.J. Slingerland, J.M. Bodlaender, H. Van Lenthe, A.B.P. Van Kuilenburg, and A.H. Van Gennip, Imbalance between the pyrimidine ribonucleotide pools of rat pheochromocytoma PC-12 cells, Clin. Chem. Enzym. Comms 5:315(1993).
2. A.A. Van den Berg, H. Van Lenthe, S. Busch, D. De Korte, D. Roos, A.B.P. Van Kuilenburg, and A.H. Van Gennip, Evidence for transformation-related increase in CTP-synthetase activity in situ in human lymphoblastic leukemia, Eur. J. Biochem. 216:161 (1993).
3. K. Pillwien, K. Schuchter, G. Ressmann, K. Gharebaghi, A. Knoflach, B. Cermak, H.N. Jayaram, S. M. Szalay, T. Szekeres, and P. Chiba, Cytotoxicity, differentiating activity and metabolism of tiazofurin in human neuroblastoma cells, Int. J. Cancer 55:92 (1993)
4. J.L. Daniel, Ilean R. Molish, and H. Holmsen, Radioactive labeling of the adenine nucleotide pool of cells as a method to distinguish among intracellular compartments, Biochim. Biophys. Acta 632:444 (1980).
5. P.K. Smith, R.I. Krohn, G.T. Hermanson, A.K. Mallia, F.H. Gartner, and M.D. Provenzano, Measurement of protein using bicinchoninic acid, Anal. Biochem. 150: 76 (1985).
6. D. De Korte, Y.M.T. Marijnen, W.A. Haverkort, A.H. Van Gennip, and D. Roos, Sensitive on-line radioactivity measurement with a heterogeneous flow cell: application to HPLC-separated ribonucleotides in lymphoid cells, J. Chromatogr 415:383 (1987).
7. J.M. Angelastro, and L. Purich, Adenine and guanine nucleotide content of triton-extracted cytoskeletal fractions of nonmuscle cells, Anal. Biochem. 204:47 (1992).
8. J.L. Daniel, L. Robkin, I.R. Molish, and J. Holmsen, Determination of the ADP concentration available to participate in energy metabolism in an actin-rich cell, the platelet, J. Biol. Chem. 254:7870 (1979).

THE EFFECT OF ISCHEMIC PRECONDITIONING ON NUCLEOTIDE METABOLISM AND FUNCTION OF RAT HEART AFTER PROLONGED COLD STORAGE

Hitoshi Ogino, Ryszard T. Smolenski, Anne-Marie L. Seymour and Magdi H. Yacoub

Department of Cardiothoracic Surgery, NHLI at Harefield Hospital, Harefield, Middx UB9 6JH, U.K.

INTRODUCTION

Experimental data suggests that brief periods of myocardial ischemia followed by reperfusion induced by coronary artery occlusion or global ischemia enhances myocardial tolerance to subsequent, more prolonged ischemic episodes (Ischemic Preconditioning)[1]. However, the effect of ischemic preconditioning on the contractility and metabolism of the heart subjected to prolonged cold storage induced by a single use of cold cardioplegia as in clinical heart transplantation has not been determined. In this study we evaluated the effect of 5 min ischemia and 10 min reperfusion on the recovery of myocardial mechanical function and metabolism following 6 hours of 4°C preservation using an isolated rat heart preparation.

MATERIAL AND METHODS

Wistar rats weighing 300 to 350 g were used. Hearts were perfused in the Langendorff mode with Krebs-Henseleit bicarbonate buffer at 37°C at a constant perfusion pressure of 100 cm of water. Functional assessment was performed using a balloon catheter inserted into the left ventricle. The balloon was loaded in stepwise manner with water and systolic and diastolic pressures recorded. These values were used to calculate pressure-volume relation[2,3]. The coronary effluent was collected during reperfusion to evaluate purine catabolite release. Nucleotide content was evaluated in hearts freeze-clamped at the end of experiment and extracted with perchloric acid. For metabolic determinations chromatographic procedure described previously was used[4]. Results are expressed as mean ±S.E.M.. Student's unpaired t test was used to assess the level of statistical significance of difference between IP and C group. The experimental protocol for groups subjected to ischemic preconditioning (IP) and control (C) is shown below:

Purine and Pyrimidine Metabolism in Man VIII, Edited by
A. Sahota and M. Taylor, Plenum Press, New York, 1995

Table 1.

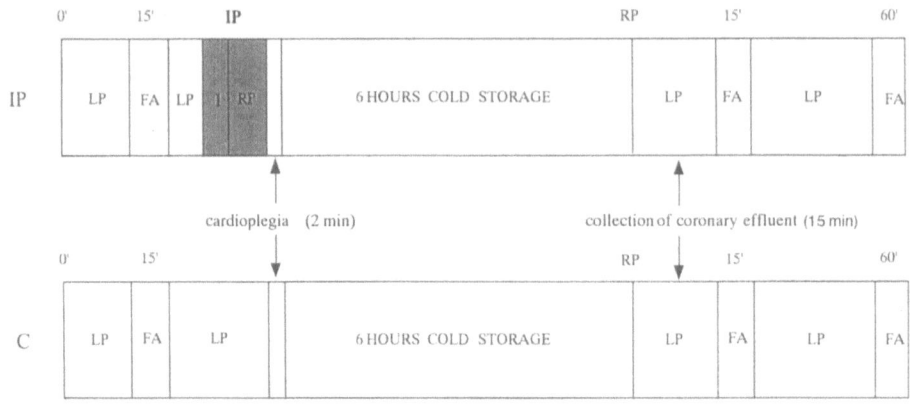

LP : perfusion in the Langendorff mode,   FA : mechanical functional assessment

I : ischemia,   RP : reperfusion,   IP : ischemic preconditioning

RESULTS

Mechanical Function

No significant difference was observed in peak systolic pressure-volume (P-V) and diastolic P-V relationship for IP and C prior to preservation. In the postpreservation phase, no significant difference in the slope values of systolic P-V curves was shown. End diastolic P-V relation which indicate chamber stiffness show significantly sharper slope for C group both at 15 and 60 min of reperfusion (Fig. 1). Functional assessment by comparing the recovery of left ventricular developed pressure (LVDP) showed no significant difference between two groups. Thus, ischemic preconditioning does not exert a strong positive effect on the mechanical function after prolonged cold preservation.

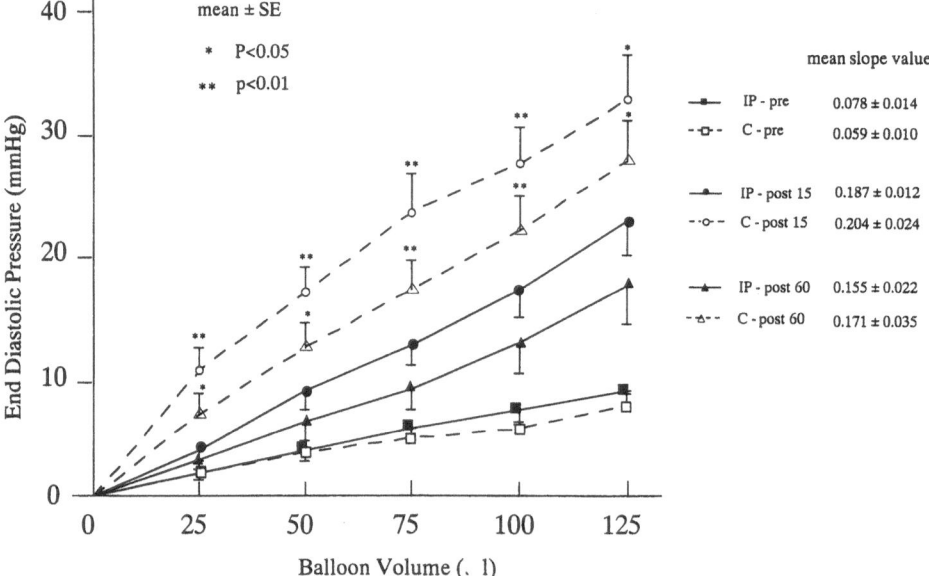

Figure 1. End diastolic pressure-volume relationship of the rat heart before (pre) and after (post) 6 hours of hypothermic ischemia. C- control group, IP- ischemic preconditioning.

Metabolism

Initial values (before preservation) of ATP concentration were 24.3±0.4 nmol/mg dry weight. After preservation, at the end of protocol ATP concentration was 19.7±0.4 nmol/mg dry weight in IP compared to 18.5±0.8 nmol/mg dry weight in C, which was not significantly different. There was no significant differences in the tissue concentrations of the other metabolites. Concentration of purine catabolites released from the heart during the first 15 min reperfusion was significantly lower in IP (0.61±0.06 μmol) than in C (0.85±0.09 μmol).

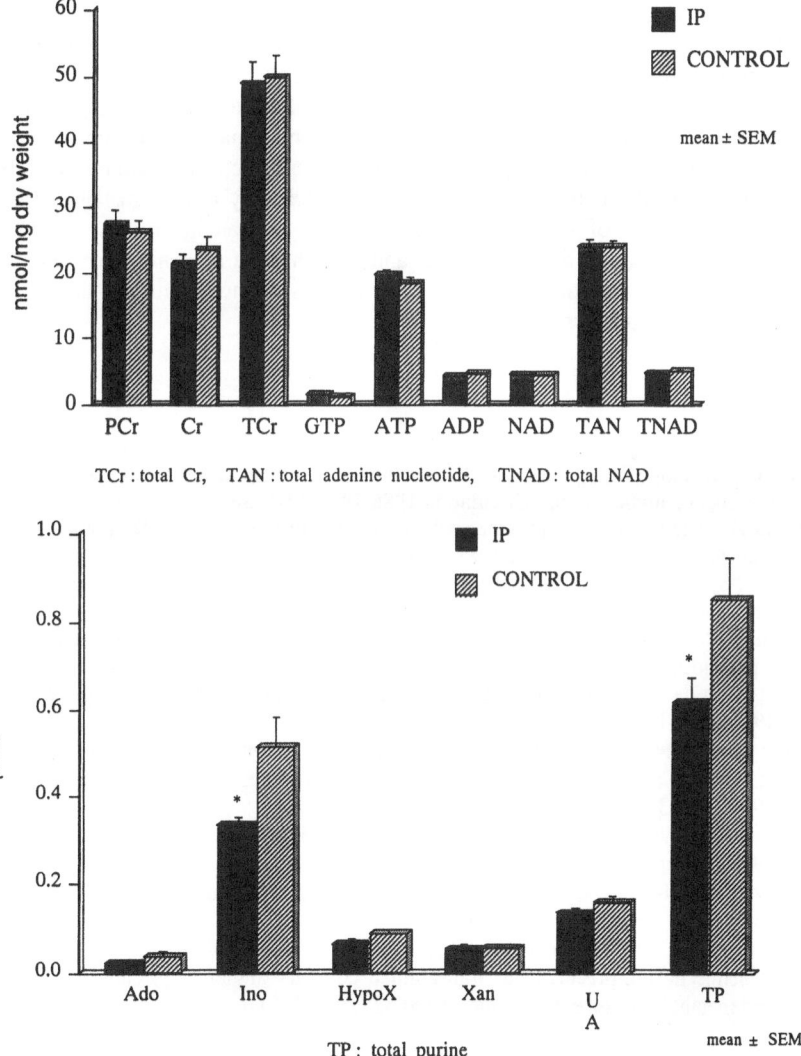

Figure 2. Myocardial concentration of nucleotide and creatine metabolites in the heart (A) and the release of purine catabolites from the heart (B) after prolonged hypothermic ischemia in control and preconditioning (IP) group.

DISCUSSION

Since Reimer and Murry[1] first reported the effect of ischemic preconditionig is reducing lethal cell injury in ischemic myocardium, many experimental data have suggested that brief periods of myocardial ischemia followed by reperfusion enhance myocardial tolerance to subsequent, more prolonged ischemic episodes[5-7]. Moreover, clinical evidences during PTCA or coronary artery bypass have shown that short ischemia similar to preconditioning protect the heart from subsequent prolonged coronary occlusion during the procedures and can avoid myocardial infarction or severe myocardial dysfunction[8]. However, in cardiac surgery this technique has not been used for prolonged myocardial preservation induced by cardioplegia.

Our study evaluate both nucleotide metabolism and functional consequences of ischemic preconditioning with routine cardioplegic arrest followed by hypothermic preservation. However, ischemic preconditioning exerted only moderate effect on the recovery of mechanical function. There was also no significant difference in ATP concentration at the end of reperfusion, although the total loss of purine catabolite was slightly less in the preconditioning group than the control.

There are many mechanisms which may contribute to this relatively small effect of preconditioning. Cardioplegic arrest or hypothermia may share similar protective mechanisms to those involved in preconditioning. Another consideration may be the short time of protection produced by preconditioning which may not be sustained throughout the prolonged period of cardioplegic preservation. In conclusion, preconditioning as additional procedure to cardioplegic arrest and hypothermia does not exert any additional beneficial effect as can be observed with other models neither by functional improvement nor by metabolic parameters.

REFERENCES

1. Murry CE, Jennings RB, Reimer KA. Preconditioning with ischemia: a delay of lethal cell injury in ischemic myocardium. Circulation 1986;74:1124-1136
2. Sagawa K. The end-systolic pressure-volume relation of the ventricle: Definition, modification and clinical use. Circulation 1981;63:1223-1227
3. Mankad PS, and Yacoub MH. Systolic and diastolic function of both ventricles after prolonged cardioplegic arrest. Ann Thorac Surg 1993;55:933-939
4. Smolenski RT., Lachno DR., Ledingham SJM., and Yacoub MH. Determination of sixteen nucleotides, nucleosides and bases using high-pefomance liquid chromatograghy and its application to the study of purine metabolism in heart for transplantation. J Chromatogr 1990;527:414-420
5. Murry CE, Richard VJ, Reimer KA, Jennings RB. Ischemic preconditioning slows energy metabolism and delay ultrastructual damage during a sustained ischemic episode. Circ Res 1990;66:913-931
6. Downey JM, Jordan M. Preconditioning limits infact size in rabbit. Circulation 1989;80(Suppl.I):238
7. Walker DM, Yellon DM Ischemic preconditioning: from mechanisms to exploitation Cardiovasc Res 1992;26:734-739
8. Deutsch E, Berger M, Kussmal WG, Hirshfeld JW, Herrman HC, LAskey WK Adaptation to ischemic during precutaneous transluminal coronary angioplasty: clinical, hemodynamic, and metabolic feature. Circulation 1990;82: 2044-2051

DOES ADENOSINE DEAMINASE INHIBITION PROTECT ISCHEMIC MYOCARDIUM?

Kim P. Gallagher, Thomas B. McClanahan, Bradley J. Martin,
Lori J. Saganek, Diane P. Ignasiak, Thomas E. Mertz,
David G.L. Van Wylen*, and Jakob Vinten-Johansen†

Parke-Davis Pharmaceutical Division
Warner Lambert Co.
Ann Arbor, MI 48105
*State University of New York
Buffalo, NY 14215
†The Bowman Gray Medical School
Winston-Salem, NC

INTRODUCTION

The protective effects of adenosine in myocardial ischemia are well established (1,2). Exogenous adenosine, however, can produce hypotension, bradycardia, and AV block. Since ischemic myocardium itself is a significant endogenous source of adenosine, the rationale has developed that inhibition of adenosine catabolism to inosine with adenosine deaminase (ADA) inhibition could augment or sustain local adenosine levels during ischemic conditions. The elevated adenosine levels would be largely restricted to the ischemic area, minimizing potentially adverse side effects. The phrase "site and event specific intervention" has evolved to describe this appealing concept and ADA inhibitors such as erythro-9-(2-hydroxy-3-nonyl) adenine (EHNA) and 2-deoxycoformycin (pentostatin) have been used to test it in experimental models of myocardial ischemia.

RESULTS

Significant benefit has been demonstrated in studies focused on functional recovery (reversible injury) after global myocardial ischemia. In isolated, buffer perfused rabbit hearts exposed to moderately hypothermic (34º C) global ischemia for two hours, for example, pentostatin (1 μmol/L) enriched cardioplegia significantly improved functional recovery (3). Although ATP levels were depressed to the same degree in control and pentostatin treated groups during ischemia, there was modest but significantly better repletion of ATP levels in the pentostatin treated group after reperfusion. Since adenosine levels were also modestly but signficantly higher in the pentostatin treated animals during global ischemia, the data suggested that even slight augmentation of local adenosine levels can ameliorate the functional consequences of global ischemia. A number of other studies in isolated hearts have provided similar results, supporting the conclusion that modification of adenosine metabolism during global ischemia leads to improved functional recovery (4,5).

A particularly rigorous test of this conclusion was performed recently in dogs supported on cardiopulmonary bypass (6). The aorta was crossclamped for 30 minutes to render the heart globally ischemic without myocardial protection. Then blood cardioplegia was administered and global ischemia was maintained for an additional 60 minutes. In animals pretreated with pentostatin (0.2 mg/kg), myocardial dialysate adenosine levels (measured with the microdialysis technique which provides an index of interstitial adenosine concentration) were elevated approximately 120-fold on the average during global ischemia. In animals with myocardial protection from blood cardioplegia alone and in animals with pentostatin added to the blood cardioplegia, microdialysate measurements of adenosine increased only two to five-fold. Left ventricular function (measured as preload recruitable stroke work or with the end-systolic pressure volume relationship after removing the aortic crossclamp and weaning the animals from cardiopulmonary bypass) recovered completely in the pentostatin pretreatment group but was signficantly depressed in the other two groups. Thus, pretreatment with an inhibitor of ADA led to striking accumulation of interstitial adenosine during global ischemia and was associated with significantly better functional recovery after reperfusion.

Equally striking results on functional recovery have been obtained in studies on dogs with global myocardial ischemia when the ADA inhibitor EHNA was combined with the nucleoside transport inhibitor p-nitrobenzylthioinosine NBMPR (7). In terms of regional myocardial dysfunction, less striking but statistically significant reductions in mechanical stunning were observed in dogs treated with EHNA (8), intracoronary EHNA plus NBMPR (9), or pentostatin (10) before 15 minute occlusions of the LAD artery.

In studies focused on myocardial infarction (irreversible injury), however, beneficial results have not been obtained consistently. In open-chest, anesthetized dogs, for example, pentostatin (0.2 mg/kg) dramatically increased interstitial adenosine levels during 60 minute coronary occlusions but did not reduce infarct size which averaged approximately 35% of the region at risk (RAR) in control and treated groups (11).

In anesthetized rabbits, preliminary results with pentostatin suggested that infarct size could be reduced when the animals were pretreated with this potent ADA inhibitor, but our early results may have been influenced by mildly hypothermic body temperatures. In subsequent experiments, normothermia (38-39°C) was rigorously sustained and no significant change in infarct size was detected across a range of pentostatin doses (0.02-2.0 mg/kg). Similarly negative findings were obtained in rats pretreated with pentostatin 24 hours before 90 minute coronary occlusions were performed and infarct size was determined (12).

In anesthetized, open-chest Micropigs®, dramatic elevations of adenosine during 60 minute coronary occlusions were produced with pentostatin, but in normothermic animals (37°C), control (31±4% of RAR) and pentostatin treated infarcts (23±4% of RAR) were not significantly different. Only when ADA inhibition with pentostatin was combined with mild hypothermia (35°C) in pigs, were signficant reductions in infarct size evident in pentostatin treated (16±3% of RAR) compared to control (28±4% of RAR) animals.

CONCLUSION

The rationale underlying studies on the potential therapeutic utility of ADA inhibition in myocardial ischemia was based on the assumption that augmenting local adenosine levels would protect the myocardium through direct and/or indirect effects (1). The protection so provided would be "site and event specific" thereby avoiding the potentially adverse side effects associated with exogenous administration of adenosine.

The data indicate that ADA inhibition with EHNA or pentostatin leads to very striking elevations in adenosine within ischemic areas that are largely restricted to the ischemic areas, fulfilling the criteria of site and event specificity. The data also suggest that elevating local adenosine levels during ischemia is beneficial when the ischemic injury is evaluated in terms of functional recovery. When the principal endpoint is infarction (irreversible injury), however, the data suggest that even dramatic elevations of local adenosine levels during ischemia are not protective. Since the intensity of ischemia required to produce infarction is more severe and/or prolonged than that required to produce mechanical stunning, the potential therapeutic utility of augmenting adenosine levels during ischemia with ADA inhibition may be restricted to circumstances in which ischemia is not severe or prolonged enough to kill cardiomyocytes.

Since the available data demonstrate that raising local adenosine levels alone does not reduce infarct size, we speculate that it may be necessary to combine ADA inhibition with interventions such as hypothermia to significantly modify infarct size, a possibility supported by our observations in pigs that infarct size reduction was achievable when mildly hypothermic pigs were pretreated with pentostatin. Additional investigation will be required to pursue this possibility.

REFERENCES

1. Ely SW, Berne RM: Protective effects of adenosine in myocardial ischemia. Circulation 85: 893-904, 1992.

2. Engler RL: Adenosine. The signal of life? Circulation 84: 951-954, 1991.

3. Bolling SF, Bies LE, Bove EL, Gallagher KP. Augmenting intracellular adenosine improves myocardial recovery. J Thorac Cardiovasc Surg 99: 469-474, 1990.

4. Sandhu GS, Burrier AC, Janero DR: Adenosine deaminase inhibitors attenuate ischemic injury and preserve energy balance in isolated guinea pig hearts. Am J Physiol 265 (Heart Circ Physiol 34): H1249-H1256, 1993.

5. Van Belle H: Nucleoside transport inhibition: a therapeutic approach to cardioprotection via adenosine? Cardiovasc Res 27: 68-76, 1993.

6. Hudspeth DA, Williams MW, Zhao Z-Q, Sato H, Nakanishi K, McGee DS, Hammon JW, Jr, Vinten-Johansen J, Van Wylen DGL: Pretreatment pentostatin augments interstitial fluid adenosine and prevents postischemic dysfunction in canine hearts protected with blood cardioplegia. Ann Thorac Surg (In press).

7. Abd-Elfattah AS, Ding M, Dyke CM, Wechsler AS: Protection of the stunned myocardium. Selective nucleoside transport blocker administered after 20 minutes of ischemia augments recovery of functional recovery. Circulation 88[part 2]: 336-343, 1993.

8. Dorheim TA, Hoffman A, Van Wylen DGL, Mentzer RM, Jr: Enhanced interstitial fluid adenosine attenuates myocardial stunning. Surgery 110: 136-145, 1991.

9. Zughaib ME, Abd-Elfattah AS, Jeroudi MO, Sun J-Z, Sekili S, Tang X-L, Bolli R: Augmentation of endogenous adenosine attenuates myocardial stunning independently of coronary flow or hemodynamic effects. Circulation 88 [part 1]:2359-2369, 1993.

10. McClanahan TB, Ignasiak DP, Martin BJ, Mertz TE, Gallagher KP: Inhibition of adenosine deaminase with pentostatin attenuates myocardial stunning in dogs (abstract). Circulation 88 [part 2]: I-187, 1993.

11. Van Wylen DGL: Inhibition of adenosine deaminase with pentostatin augments interstitial fluid adenosine during regional myocardial ischemia (abstract). Circulation 88 [part 2]: I-432, 1993.

12. Li Y, Kloner RA: Adenosine deaminase inhibition is not cardioprotective in the rat. Am Heart J 126: 1293-1298, 1993.

IS HYPOXANTHINE A USEFUL MARKER OF PERINATAL HYPOXIA ?

Tilman Grune[1], Reiner Mueller[1], Manuela Jakstadt[1], Heike Schmidt[2], and Werner G. Siems[3]

[1] Clinics of Physical Therapie and Rehabilitation, Medical Faculty (Charite), Humboldt University, Schumannstr. 20/21, D-10098 Berlin, Germany
[2] Children Hospital, D-78224 Singen, Germany
[3] Herzog-Julius Hospital, D-38655 Bad Harzburg, Germany

INTRODUCTION

Perinatal hypoxia is one of the major causes of morbidity and mortality of newborns and was therefore intensively studied (O'Connor et al., 1981; Saugstadt and Gluck, 1982; Ripalda et al., 1989; Gloeckner and Kretschmar, 1991). Each kind of oxygen deficiency is combined with an accelerated purine degradation and therefore an accelerated formation of purine degradation products. On the other hand the final purine degradation could lead to an increased formation of reduced oxygen species via the xanthine oxidase reaction. Some of these reduced oxygen species are able to damage cellular structures and components, e.g. proteins, nucleic acids and lipids. The cascade of oxidation of polyunsaturated fatty acids via reactive oxygen species includes the formation of a number of aldehydes as secondary lipid peroxidation products. The quantitative most important among these is malondialdehyde (MDA).

The aim of this study was the determination of final purine degradation products, as hypoxanthine and uric acid, and the simultaneous measurement of parameters of free radical action in umbilical cord blood of newborns to estimate both - the purine nucleotide breakdown and the lipid peroxidation - as possible markers of hypoxia.

MATERIALS AND METHODS

Samples of umbilical cord blood were taken from more than 100 newborns born in the Department of Neonatology, Medical Faculty of Humboldt University Berlin. The basic clinical data are presented in Table 1. With respect to the pH of umbilical arterial blood, the Apgar score and the gestational age the babies were divided into 5 groups. An Apgar score less than 8 assessed 5 min after birth was taken as an indicator of asphyxia and therefore of perinatal hypoxia.

Blood was taken during the birth and immediately serum was separated from blood cells. Samples of serum and blood were stored in liquid nitrogen until biochemical determinations.

Hypoxanthine and uric acid was determined as published earlier by capillary electrophoresis and reversed phase HPLC (Grune et al., 1991; Grune et al., 1993).

Products of lipid peroxidation were determined after addition of the radical scavenger BHT to prevent further lipid peroxidation. Malondialdehyde was determined after modification with thiobarbituric acid using an HPLC method according to Wong (1987). Glutathione status was determined as described by Siems (1994).

Table 1. Clinical data of the newborns studied.

group	A	B	C	D	E
number of newborns	49	12	14	15	19
birth weight (g)	3407±478	3522±305	3402±459	2419±556	1751±723
gestational age (weeks)	39.0±1.2	40.3±1.3	39.9±1.0	35.8±0.9	32.1±3.1
pH of umbilical cord blood	7.3±0.05	7.1±0.06	7.2±0.10	7.3±0.04	7.3±0.09
Apgar score	9.4±0.6	8.2±1.3	6.8±0.4	8.8±0.7	6.5±1.1

Values are given as mean ± S.D. Group A, full term, healthy newborns; group B, full term newborns with acidosis (pH<7.20); group C, full term babies with asphyxia (5 min Apgar score< 8); group D, preterm, healthy infants (gestational age <36 weeks); and group E, preterm babies suffering from asphyxia.

RESULTS AND DISCUSSION

The results of measurement of reduced and oxidized glutathione, malondialdehyde and oxopurines are presented in tables 2 and 3. If one compares the groups A and D, which represent the healthy term and preterm babies, one can remark a significantly decreased level of reduced glutathione and malondialdehyde in the case of group D. Whereas the levels of oxidized glutathione, hypoxanthine and uric acid remain uneffected by the gestational age. We found a clear increase of the reduced glutathione and the lipid peroxidation product malondialdehyde dependent upon the gestational age (data not shown). The decreased level of reduced glutathione in preterm newborns in comparison to term newborns is in agreement with the findings of Smith (1993). There are a number of reports about an increase of antioxidants at the end of the pregnancy. This may be an adaptation to the increased levels of polyunsaturated fatty acids as reported by Hamosh (1987).

Table 2.   Serum concentrations of hypoxanthine and uric acid in term and preterm newborns.

	Hypoxanthine (nmoles/ml)	Uric acid (nmoles/ml)
Group A	10.97 ± 0.65	306.9 ± 24.0
Group B	10.99 ± 1.06	403.4 ± 43.4 +
Group C	13.40 ± 1.67	377.4 ± 54.0
Group D	9.65 ± 0.69	304.8 ± 49.5
Group E	15.34 ± 1.75 *	480.6 ± 42.3 *

Values are given as mean ± S.E.M. Statistical differences are representented as $p < 0.05$ in comparison to group A (+) or group D (*) using the Wilcoxon test.

The accumulation of polyunsaturated fatty acid may result in an increased lipid peroxidation, due to the accumulation of substrates. This possibility was demonstrated by higher levels of malondialdehyde in term newborns compared with those concentrations in preterm babies (Table 3).

However we could not find any differences in the levels of hypoxanthine and uric acid that were dependent upon the gestational age. The levels of hypoxanthine remain also uneffected by acidosis, whereas the uric acid concentration in the serum shows an increase in comparison to the control group. Only a minor increase of hypoxanthine levels was found following asphyxia both in term and preterm newborns, more distinctly in preterm infants.

In previous studies an increase of hypoxanthine levels was found due to perinatal hypoxia, but also a wide overlapping of the hypoxanthine concentrations in healthy and hypoxic newborns was reported. Interestingly there were increased levels of uric acid in asphyxic preterm newborns.

Table 3.   Concentrations of malondialdehyde (MDA) in serum of umbilical cord blood, levels of reduced glutathione (GSH) in red blood cells and the glutathione ratio (2GSSG/ [2GSSG+GSH]) as indicators of oxidative stress.

	MDA (nmoles/ml serum)	GSH (mmoles/l cells)	glutathione ratio
Group A	1.80±0.01	2.18±0.11	2.40±0.21
Group B	3.94±0.66	1.13±0.15	4.04±0.79
Group C	3.62±0.48	1.24+0.08	3.54±0.80
Group D	0.62±0.07	1.13±0.06	3.96±0.72
Group E	3.33±0.39	0.63±0.04	8.87±1.67

Values are given as mean±S.E.M. Statistical differences are representented as $p < 0.05$ in comparison to group A (+) or group D (*) using the Wilcoxon test.

These higher uric acid levels may be due to an increased purine degradation and a following increased flux through the xanthine oxidoreductase.

Summarizing, one can conclude that parameters of lipid peroxidation, as malondialdehyde, and parameters of purine degradation, mainly hypoxanthine, reflect pathophysiological processes during perinatal hypoxia. Malondialdehyde is a more sensitive parameter but it is less specific and therefore more dependent on gestational age, pH of umbilical blood and asphyxia. Whereas hypoxanthine shows a more specific increase due to peinatal hypoxia, but with the disadvantage of a broad overlapping in the concentration range of healthy and asphyctic newborns.

ACKNOWLEDGMENT

This study was supported by a grant from the Bundesministerium fuer Forschung und Technologie (FP Risikoneugeborenes), Bonn, Germany.

REFERENCES

Gloeckner, R., and Kretschmar, M., 1991, Perinatal glutathione levels in liver and brain of rats from large and small litters, Biol. Neonate 60:236.

Grune, T., Siems, W., Gerber, G., Tikhonov, Y.V., Pimenov, A.M., andToguzov, R.T., 1991, Changes of nucleotide patterns in liver, muscle and blood during the growth of Ehrlich ascites cells, J. Chromatogr. 563:53.

Grune, T., Ross, G.A., Schmidt, H., Siems, W., and Perrett, D., 1993,Optimized separation of purine bases and nucleosides in human cord plasma by capillary zone electrophoresis. J. Chromatogr. 636:105.

Hamosh, M., 1987, Lipid metabolism in premature infants. Biol. Neonate 52:50.

O'Connor, M.C., Harkness, R.A., Simmonds, R.J., and Hytten, F.E., 1981,The measurement of hypoxanthine, xanthine inosine and uridine in umbilical cord bloob and fetal scalp blood samples as a measure of fetal hypoxia, Br. J. Obstet. Gynaecol. 88:381.

Ripalda, M.J., Rudolph, N., and Wong, S.L., 1989, Developmental patternsof antioxidant defense mechanisms in human erythrocytes. Pediatr. Res. 26:366.

Saugstadt, O.D., and Gluck, L., 1982, Plasma hypoxanthine levels in newborn infant. J. Perinat. Med. 10:266.

Siems, W., van Kuijk, F.J.G.M., Maas, R., and Brenke, R., 1994, Uricacid and glutathione levels during short-term whole body cold exposure, Free Rad. Biol. Med. 16:299.

Smith, C.V., Hansen, T.N., Martin, N.E., McMicken, H.W., and Elliott,S.J., 1993, Oxidant stress responses in premature infants during exposure to hyperoxia. Pediatr. Res. 34:360.

Wong, S.H.J. Knight, J.A., Zaharia, O., and Leach, C.N., 1987, Lipoperoxides in plasma as measured by liquid-chromatographic separation of malondialdehyde-thiobarbituric acid adduct, Clin. Chem. 33:214.

ENDOTHELIAL CELLS' RESPONSES TO HYPOXIA AND REPERFUSION

Andrea Griesmacher[1], Andreas Windischbauer[1], and Mathias M. Müller[2]

[1]Dept. of Cardiothoracic Surgery, University Hospital Vienna, A-1090 Vienna, Austria
[2]Dept. of Laboratory Diagnostics, Kaiser Franz Josef Hospital Vienna, A-1100 Vienna, Austria

INTRODUCTION

The potential sources of toxic reactive oxygen species (ROS) generated during hypoxia and reoxygenation are prostaglandin biosynthesis, mitochondrial electron transport systems, purine catabolism by means of xanthine oxidase and infiltration by phagocytes. These ROS are reported to play a crucial role in tissue damage [1]. This study was aimed at investigating the metabolism in human umbilical venous endothelial cells (HUVECs) under hypoxic/hyperoxic conditions, thus simulating hypoxia/reoxygenation. Cellular concentration of ATP and creatine phosphate (CP), the formation of lipid peroxidation products (LPO) as well as the ratio between reduced (GSH) and oxidised glutathione (GSSG) were measured.

MATERIALS AND METHODS

Cell Culture

Endothelial cells (HUVECs) were isolated from human umbilical veins and cultivated according to a standard procedure [2]. After reaching confluence the cells were seeded into precoated 6 well culture plates. The cells were identified as endothelial cells by the typical "cobblestone" contact-inhibited morphology [3] and by factor VIII (FVIII: vWF) staining [4].

Exposure of Cells to Hypoxia and Reoxygenation

A degassed phosphate buffered saline, which had been perfused with 100 % N_2 for 20 min, was used for the incubation. The O_2 content was approx. 0 - 0.5%. The culture plates were placed into a modified incubator which allowed handling under low oxygen tension (1-2% O_2). The cell culture plates were incubated with the degassed buffer and placed into an absolutely airtight chamber filled with N_2 for 3 or 24 h. At the end of the incubation time one group (hypoxia group) was used to determine the metabolic effects of hypoxia. The second group (hypoxia/reperfusion group) was exposed for 60 min to phosphate buffered saline (perfused with 100% O_2 for 20 min), thus simulating reoxygenation.

Control experiments were performed with air-equilibrated phosphate buffer (normoxia group) with and without "subsequent reperfusion".

Determination of High Energy Phosphates

HUVECs were lysed with 0.5 mol/l perchloric acid. The lysates were neutralised by adding 4 mol/l K_2HPO_4 and analysed for ATP and CP. CP was determined after conversion to ATP using creatine kinase, with ADP as substrate and N-acetylcysteine as activator. ATP was measured by means of bioluminescence [5].

Determination of LPO

LPO were determined as thiobarbituric acid reactive substances [6].

Determination of GSH and GSSG

The simultaneous photometrical determination of GSH and GSSG was carried out using specific enzymes [7].

Statistical Analysis

To compare control values with those obtained during the incubation experiments a matched pairs t test was used. $p < 0.001$ was taken to indicate a significant difference.

RESULTS

Three hours of hypoxia had no influence on ATP levels, whereas CP levels decreased by 23.9 % compared to controls (Table 1). Subsequent hyperoxic treatment for 60 min resulted in restoration of CP, whereas ATP remained stable. 24 h of hypoxia exhibited a pronounced effect on the content of high energy phosphates. Both ATP and CP levels decreased dramatically (Table 2). A 60 min period of reoxygenation following 24 h of hypoxia slightly increased the ATP content compared to hypoxia alone. In

Table 1: Intracellular levels of ATP, CP, LPO, GSH and GSSG after 3 h of hypoxia and 3 h of hypoxia followed by 60 min of reoxygenation

	N (3 h)	H (3 h)	N/R (3 h / 60 min)	H/R (3 h/ 60 min)
ATP(nmol/ 10^6cells)	10.8 ± 0.3	11.0 ± 0.5	11.9 ± 0.4	10.4 ± 0.5
CP (nmol/ 10^6cells)	20.5 ± 1.4	$15.6 \pm 2.3*$	20.9 ± 1.8	22.1 ± 2.1
LPO (pmol/ 10^6cells)	32 ± 4	36 ± 5	$135 \pm 5*$	$143 \pm 21*$
GSH (nmol/ 10^6cells)	16.0 ± 1.3	$27.2 \pm 2.1*$	$21.4 \pm 2.5*$	$17.4 \pm 1.9*$
GSSG (nmol/10^6cells)	0.3 ± 0.1	$5.9 \pm 1.1*$	$4.1 \pm 0.7*$	$7.1 \pm 1.4*$

* = statistically significant compared with normoxia group ($p < 0.001$)

Table 2: Intracellular levels of ATP, CP, LPO, GSH and GSSG after 24 h of hypoxia and 24 h of hypoxia followed by 60 min of reoxygenation

	N (24 h)	H (24 h)	N/R (24 h/60 min)	H/R (24 h/60 min)
ATP (nmol/ 10^6 cells)	10.6 ± 0.6	4.9 ± 0.3*	11.7 ± 0.5	6.2 ± 0.5*
CP (nmol/ 10^6 cells)	19.8 ± 2.1	9.6 ± 1.0*	21.2 ± 1.5	5.3 ± 1.5*
LPO (pmol/ 10^6 cells)	35 ± 2	46 ± 5	140 ± 20*	143 ± 21*
GSH (nmol/ 10^6 cells)	16.0 ± 1.3	not detected	21.4 ± 2.5*	not detected
GSSG (nmol/ 10^6 cells)	0.3 ± 0.1	9.9 ± 1.9*	4.5 ± 0.8*	12.1 ± 2.4*

* = statistically significant compared with normoxia group ($p < 0.001$)

contrast, the CP content decreased further. No significant differences in ATP and CP content were detectable after 60 min of reoxygenation without hypoxic pretreatment in comparison to control experiments. Under normoxic conditions the ratio between GSSG and GSH shifted from 1:100 to 1:4.5 after 3 h of hypoxia, subsequent reoxygenation led to a further increase in GSSG (GSSG:GSH = 1:2.4). After 24 h of hypoxia no intracellular GSH could be detected. The content of LPO was nearly unaffected during hypoxia, whereas reoxygenation led to an enormous increase.

DISCUSSION

The aim of this study was to determine cellular metabolic changes after different periods of hypoxia or normoxia followed by reoxygenation. Furthermore, the LPO formation as an index for oxygen derived free radical activity was investigated. This was of special interest, since human endothelial cells do not contain measurable activities of xanthine oxidase under normoxic, hypoxic and hyperoxic conditions (8), which is in contrast to endothelial cells of many other species. Long periods of hypoxia (3 h, 24 h) were necessary to influence the cellular ATP and CP. In contrast to physiological conditions cultured endothelial cells are not exposed to shear stress, thus requiring a lower level of metabolic activity to maintain their function. Reoxygenation following a 3 h hypoxic period induced increased consumption of reduced glutathione, a further increase in oxidized glutathione and the restoration of total high energy phosphates. A different situation occurred when 24 h of hypoxia was followed by reperfusion. No reduced glutathione could be detected and the total high energy phosphate content was dramatically reduced. Recently, we could show that increasing concentrations of hydrogen peroxide lead to an increased content of high energy phosphates followed by a breakdown (9). Therefore injury during reoxygenation is probably due to enhanced radical production.

SUMMARY

Our results suggest that longer periods of hypoxia lead to a deficiency of high energy phosphates. Reoxygenation leads to the formation of ROS, irrespectively of the duration of hypoxia. It might be concluded that a diminished level of intracellular high energy phosphates upon hypoxia followed by oxidative stress plays a key role in the reperfusion associated cellular dysfunctions.

REFERENCES

1. J.L. Zweier, P. Kuppusamy, and G.A. Lutty, Measurement of endothelial cell free radical generation: evidence for a central mechanism of free radical injury in post ischemic tissues, Proc. Natl. Acad. Sci., USA 85, 4046 (1988).

2. E.A. Jaffe, R.L. Nachman, C.G. Becker, and C.R. Minck, Culture of human endothelial cells derived from umbilical veins, J. Clin. Invest. 52: 2745 (1973).

3. C.C. Haudenschild, R.S. Cotran, M.A Gimbrone, and J. Folkman, Fine structure of vascular endothelium in culture, J. Ultrastruct. Res. 50: 611 (1975).

4. E.A. Jaffe, L.W. Hoyer, and R.L. Nachman, Synthesis of antihaemophilic factor antigen by cultured human endothelial cells, J. Clin. Invest. 52: 2757 (1973).

5. B.L. Strehler, Adenosin - 5´- triphosphat und Kreatininphosphat - Bestimmung mit Luciferase, in: "Methoden der enzymatischen Analyse", H.U. Bergmeyer, ed., Academic Press, New York (1974).

6. H. Ohkawa, N. Ohishi, and K. Yagi, Assay for lipid peroxides in animal tissues by thiobarbituric acid reaction. Anal. Biochem. 95, 351-355 (1979).

7. E. Bernet, and U. Bergmeyer, Glutathione, in: "Methoden der Enzymatischen Analyse", H.U. Bergmeyer, ed., Academic Press, New York (1974).

8. K.O. Raivio, and T.K. Aalto, Endothelial cells and ischemia-reperfusion (I/R) injury, Intern. J. Purine Pyrimidine Res. 2, 79 (1991).

9. A. Griesmacher, G. Weigel, I. Schimke, A. Windischbauer, and M.M. Müller, The H_2O_2 induced effects on purine metabolism in human endothelial cells, Free Rad. Biol. Med. 15, 603 (1993).

INHIBITION OF LIPID PEROXIDATION BY PURINES AND ANALOGUES OF PURINE

Y. Nishida

Health Administration Center, Tokyo University of Foreign Studies
4-51-21, Nishigahara, Kita-ku, Tokyo, Japan

INTRODUCTION

The importance of oxygen radicals in tissue injury is widely recognized[1]. Mutation, cancer, inflammation, and ageing are thought to be caused by oxygen radicals. It has also been confirmed that lipids are peroxidized by oxygen radicals leading to the damaging effects on cellular constituents. Mickel & Horbar[2] reported that peroxidized lipids affect platelet aggregation. The concentration of lipid peroxide in the arterial wall and the severity of atherosclerosis are positively correlated[3]. These findings suggest a possible role of peroxidized lipids in the pathogenesis of atherosclerosis.

Recently, uric acid has been suggested as an important physiological antioxidant against oxidative injury[4,5,6,7,8]. This protective role of uric acid is proposed as an important factor in the prevention of ageing and cancer. However, uric acid is an insoluble agent. High serum urate levels result in gout or kidney damage.

In the present experiment, the antioxidant effect of purine analogues and 5-amino-4-imidazole carboxamide (AICA) compounds has been studied.

MATERIALS AND METHODS

Purines, methylated analogues of purines and AICA compounds were purchased from Sigma Chemical Corp.

Human blood was obtained from healthy donors and collected into heparinized tubes. Blood was centrifuged at 2000 g for 20 min at 0°C and the plasma and buffy coats were removed by aspiration. Erythrocytes were washed with isotonic phosphate buffered saline, pH 7.4, and were lysed in cold 10 mM phosphate buffer, pH 7.4. Haemolysate was centrifuged at 25000 g for 30 min at 0°C and then washed with 10 mM phosphate buffer until they were no longer pink.

To measure peroxidation of lipids in erythrocyte membranes, a thiobarbituric acid method was used[9]. Erythrocyte membranes containing 3 mg protein were suspended in 0.9 mL of 10 mM phosphate buffer with or without various drugs. Reactions were started by addition of 0.1 mL of 10 mM hydrogen peroxide and were incubated at 37°C for 15 min. To generate ozone, a silent electric discharge apparatus (Nipponmicron, Tokyo) was used. The ozone/oxygen mixture was bubbled through a 1 mL sample solution for 30 s. Ozone delivery was 40 µmol min^{-1} with a gas flow rate of 1L min^{-1}.

Reactions were stopped by addition of 0.5 mL of 20% trichloroacetic acid. After centrifugation at 2000 g for 10 min, the supernatant (1 mL) was added to 0.5 mL of 1% thiobarbituric acid solution in 0.05 M NaOH heated in a boiling water bath for 15 min and allowed to cool. The absorbance was measured at 535 nm. Protein content was estimated by the method of Lowry et al [10], using crystalline bovine serum albumin as standard.

RESULTS AND DISCUSSION

Treatment of erythrocyte membranes with hydrogen peroxide or ozone resulted in an increase in thiobarbituric acid-reactive material. Several purine analogues, purine bases and AICA compounds suppressed the lipid peroxidation induced by hydrogen peroxide and or ozone in human erythrocyte membranes (Table 1). 1,3-Dimethyluric acid and 1,3,7-trimethyluric acid were shown to have high potency in prevention of lipid peroxidation, although the relationship between the relative potencies of the various derivatives and structures remains unclear.

As shown in Fig. 1. lipid peroxidation in human erythrocyte membranes were inhibited by 1,3-dimethyluric acid in a dose-dependent manner. Fig. 2 shows the inhibitory effect of 1,3-dimethyluric acid on lipid peroxidation by varying concentrations of hydrogen peroxide or ozone. The antioxidant effects of 200 μM 1,3-dimethyluric acid were more potent at relatively low concentrations of hydrogen peroxide or ozone, suggestng this agent has antioxidant effect in-vivo.

Previous experiments in-vitro have shown that uric acid protects erythrocytes against damage by singlet oxygen or t-butylhydroxide[4], inhibits lipid peroxidation [11,12,13,14,15,16,17] and protects against oxidant damage to DNA [18]. Uric acid also scavenges hydroxyl radicals [19] and protects certain enzymes against inactivation by reactive radical species [20]. Moorhouse et al[21] reported that allopurinol and oxypurinol were the effective hydroxyl radical scavengers. However, the inhibitory

Table 1. Inhibitory effect of purines and AICA compounds on hydrogen peroxide-(1 mM) or ozone-(20 μmol) induced lipid peroxidation in human erythrocyte membranes. Results are expressed as the mean±s.d. of triplicate experiments.

Additions (500 μM)	Inhibition (%)	
	H_2O_2	O_2
Uric acid	52.5±3.4	32.7±4.8
1-Methyluric acid	69.2±14.8	43.8±5.5
3-Methyluric acid	60.2±13.8	32.7±9.8
7-Methyluric acid	65.3±11.9	14.6±3.3
9-Methyluric acid	55.4±15.6	35.1±10.3
1,3-Dimethyluric acid	87.4±3.6	63.9±5.7
1,7-Dimethyluric acid	34.9±4.6	30.1±5.2
1,3,7-Trimethyluric acid	76.1±14.0	49.0±13.1
Hypoxanthine	6.9±4.1	9.2±1.2
Xanthine	11.1±3.0	8.2±2.1
3-Methylxanthine	0	5.8±4.2
7-Methylxanthine	8.9±5.3	0
8-Methylxanthine	11.9±4.2	22.9±2.5
1,3-Dimethylxanthine (theophylline)	0	0
1,3,7-Trimethylxanthine (caffeine)	15.4±2.7	0
Allopurinol	11.4±9.1	16.5±4.2
Oxypurinol	1.1±0.8	9.2±0.8
Adenine	25.8±18.4	10.2±4.9
1-Methyladenine	0	0
2-Methyladenine	41.5±15.7	0
Guanine	16.9±4.3	4.1±3.6
7-Methylguanine	0	0
5-Amino-4-imidazole carboxamide	74.4±7.8	ND
5-Amino-4-imidazole carboxamide riboside	28.0±7.5	ND

ND: Not determined

effects of uric acid, allopurinol and oxypurinol on lipid peroxidation in the present experiment were relatively weak. In addition, uric acid, 1-methyluric acid, allopurinol and oxypurinol were very insoluble.

1,3-Dimethyluric acid and 1,3,7-trimethyluric acid are very soluble and have little cytotoxicity (data not shown). These methyluric acid may be useful as antioxidants.

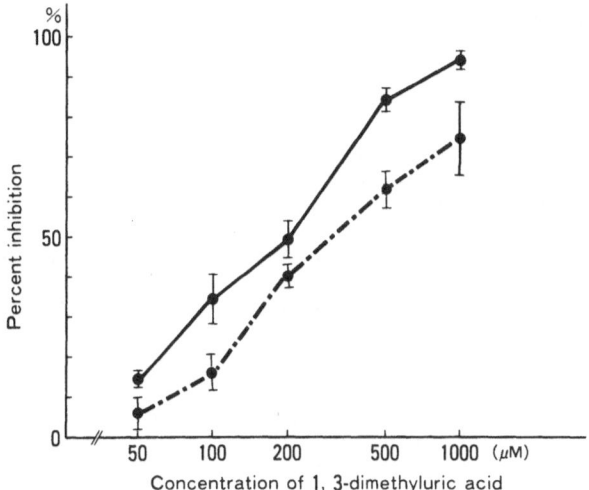

Fig. 1. Inhibitory effects of varying doses of 1,3-dimethyluric acid on lipid peroxidation by a fixed standard dose of hydrogen peroxide (●——●) and or ozone (●–·–●). Results are expressed as the mean ± s.d. of triplicate experiments.

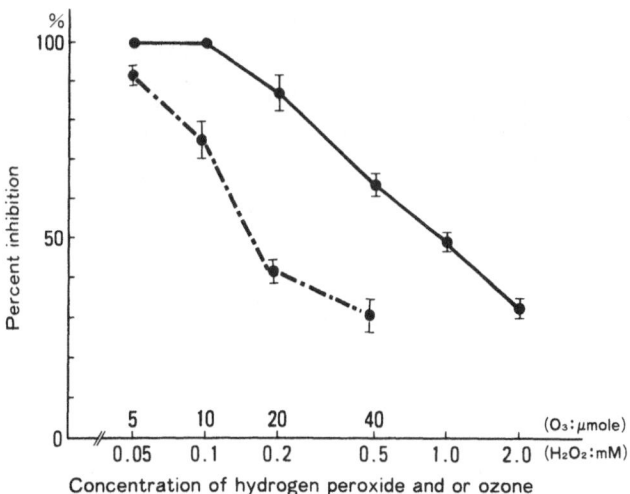

Fig. 2. Inhibitory effect of 1,3-dimethyluric acid (200 μM) on lipid peroxidation by varying concentrations of hydrogen peroxide (●——●) and or ozone (●–·–●). Results are expressed as the mean ± s.d. of triplicate experiments.

REFERENCES

(1) J.L. Marx, Oxygen free radicals linked to many disease. *Science.* 235:529 (1987).

(2) H.S. Mickel, J. Horbar, The effect of peroxidized arachidonic acid upon human platelet aggregation. *Lipids* 9:68 (1973).

(3) J. Clavind, S. Hartmann, J. Clemmensen, K.E. Jensen, H. Dam, Studies on the role of lipoperoxides in human pathology. II. The presence of peroxidized lipids in the atherosclerotic aorta. *Acta Pathol. Microbiol. Scand.* 30:1 (1952).

(4) B.N. Ames, R. Cathcart, E. Schwiers, P. Hochstein, Uric acid provides an antioxidant defense in humans against oxidant and radical-caused aging and cancer: hypothesis. *Proc. Natl. Acad. Sci. USA* 78:6858 (1981).

(5) K.J.A. Davies, A. Sevanian, S.F. Muakkassah, P. Hochstein, Uric acid-iron ion complexes. A new aspect of the antioxidant functions of uric acid. *Biochem. J.* 235:747 (1986).

(6) T.Z. Liu, Hyperuricemia associated with sickle-cell anemia may compensate for chronic antioxidant deficiency. *Clin. Chem.* 32:560 (1986).

(7) D.D. Wayner, G.W. Burton, K.U. Ingold, L.R. Barclay, S.J. Locke, The relative contributions of vitamin E, urate, ascorbate and proteins to the total peroxyl radical-trapping antioxidant activity of human blood plasma. *Biochim. Biophys. Acta* 924:408 (1987).

(8) M.G. Simic, S.V. Jovanovic, Antioxidation mechanisms of uric acid. *J. Am. Chem. Soc.* 111:5778 (1989).

(9) M. Yoshikawa, S. Hirai, Lipid peroxide formation in the brain of aging rats. *J. Gerontol.* 22:162 (1976).

(10) O.H. Lowry, N.J. Rosebrough, A.L. Farr, R.J. Randall, Protein measurement with the folin phenol regent. *J. Biol. Chem.* 193:265 (1951).

(11) S. Matsushita, F. Ibuki, A. Aoki, Chemical reactivity of the nucleic acid bases. I. Antioxidative ability of the nucleic acids and their related substances on the oxidation of unsaturated fatty acids. *Arch. Biochem. Biophys.* 102:446 (1963).

(12) R.C. Smith, L. Lawing, antioxidant activity of uric acid and 3-*N*-ribosyluric acid with unsaturated fatty acids and erythrocyte membranes. *Arch. biochem. Biophys.* 223:166 (1983).

(13) J. Meadows, R.C. Smith, Uric acid protection of nucleobases from ozone-induced degradation. *Arch. biochem. Biophys.* 246:838 (1986).

(14) J. Meadows, R.C. Smith, Uric acid protects erythrocytes from ozone-induced charges. *Environ. Res.* 43:410 (1987).

(15) J. Meadows, R.C. Smith, J. Reeves, Uric acid protects membranes and linolenic acid from ozone-induced oxidation. *Biochem. biophys. Res. Commun.* 137:536 (1986).

(16) E. Niki, M. Saito, Y. Yoshikawa, Y. Yamamoto, Y. Kamiya, Oxidation of lipids. XII Inhibition of oxidation of soybean phosphatidylcholine and methyl linoleate in aqueous dispersions by uric acid. *Bull. Chem. Soc. Jpn.* 59:471 (1986).

(17) G. Healing, C.J. Green, S. Simpkin, B.J. Fuller, J. Lunec, Inhibition of lipid peroxidation in renal tissue: in vitro studies with desferrioxamine, mannitol and uric acid. *Cryo-Letters* 10:7 (1989).

(18) A.M. Cohen, R.E. Aberdroth, P. Hochstein, Inhibition of free radical-induced DNA damage by uric acid. *FEBS. Lett,* 174:147 (1984).

(19) K.J. Kittridge, R.L. Wilson, Uric acid substantially enhances the free radical-induced inactivation of alcohol dehydrogenase. *FEBS. Lett.* 170:162 (1984).

(20) O.I. Aruoma, B. Halliwell, Inactivation of α_1-antiproteinase by hydroxyl radicals. The effect of uric acid. *FEBS. Lett.* 244:76 (1989).

(21) P.C. Moorhouse, M. Grootveld, B. Halliwell, J.E. Quinlan, J.M.C. Gutteridge, Allopurinol and oxypurinol are hydroxyl radical scavengers. *FEBS. Lett.* 213-23 (1987).

A STUDY OF ANOXIA IN RAT HEPATOCYTES

Anke Schwendel[†], Hermann-Georg Holzhütter[†],
Tilman Grune[§], and Werner Siems[‡]

[†]Institute of Biochemistry, Humboldt University,
 Berlin, Germany
[§]Department of Biochemistry and Molecular Biology,
 Medical College, Albany, NY
[‡]Herzog-Julius-Hospital, Bad Harzburg, Germany

INTRODUCTION

Pathophysiology of ischemia/reperfusion or hypoxia/reoxygenation is a focal point of both basic and clinical research.

The investigation of metabolic regulation requires the knowledge about substrate flux rates in intact cells and in cells under pathological conditions. Computer-based mathematical modeling has been established as an efficient technique for analysing the regulation of metabolic systems. The purine metabolism was first modeled by Franco and Canela[1] and later by Bartel and Holzhütter[2]. Both groups used sophisticated mathematical expressions for the enzymatic rate laws. But the values of most kinetic parameters were obtained from *in vitro* experiments.

In this paper a mathematical model for the simulation of experiments under anoxic conditions is presented. For the first time the starting point of modeling is given by metabolic concentrations and flux rates determined from *in vivo* tracer experiments. These values were measured in the metabolic steady state. The theoretical background for the estimation of metabolic flux rates has been described previously[3].

EXPERIMENTAL METHODS

Cell preparation. Hepatocytes were prepared from starved male Wistar rats according to the method of Berry and Friend[4] with modifications described by van den Berghe et al.[5].

Cell incubation. The incubation medium was a Krebs-Henseleit-Bicarbonate-Solution with 2.5 mM $CaCl_2$, 20 mM HEPES and 1% HSA. The cells were incubated at 37°C with the gas phase O_2/CO_2 (19:1). The cell viability was determined by cell staining with trypan blue. The cell concentration was adjusted to a cytokrit of 2% (v/v).

Purine analysis. Nucleotides were determined by isocratic ion-pair reversed-phase HPLC. The buffer contained 10 mM $NH_4H_2PO_4$, 2 mM tetrabutylammonium phosphate and 15% acetonitrile. Nucleosides and nucleobases were analysed in the isocratic reversed-phase mode using a buffer containing 50 mM KH_2PO_4 and 1 mM tetrabutylammonium phosphate. A $4\mu M$ C_{18} Nova Pak cartrige was used. The determination of radioactivity is described by Grune et al.[6]

The experiments under anoxic conditions were carried out in a similar way, but with N_2/CO_2 (19:1) substituting O_2/CO_2 in the gas phase.

MODEL DESCRIPTION

The enzyme reactions of the purine metabolism considered in the mathematical model, are shown in Figure 1. The concentrations of the purine metabolites and other

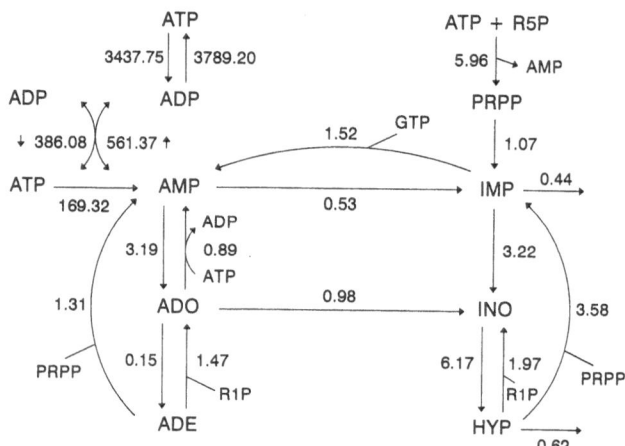

Figure 1. Reactions of purine metabolism considered in the model; values of the *in vivo* flux rates from tracer experiments in μmol/l·cells/min.

compounds used for modeling are listed in Table 1. The concentration of R5P is used instead of R1P as a simplification. The flux rate v_j of an enzyme reaction depends on the metabolite concentrations x_i

$$v_j = k_j \cdot \prod_i x_i^{-c_{ij}} \ , \tag{1}$$

TABLE 1. Concentrations of purine metabolites in isolated rat hepatocytes in metabolic steady state from tracer experiments and other compounds taken from literature.

Metabolite		Concentration in μmol/l cells
ATP		1940.0 ± 192
ADP		611.0 ± 33
AMP		224.0 ± 25
GTP		385.0 ± 37
IMP		160.0 ± 44
ADO	adenosine	11.2 ± 28
INO	inosine	69.4 ± 14
ADE	adenine	150.0 ± 29
HYP	hypoxanthine	45.9 ± 6
P_i	inorganic phosphate	5500[7]
NAD		5[8]
R5P	ribose-5-phosphate	38[9]
PRPP		5.6[10]

where i sums over all substrates of reaction j, c_{ij} is the corresponding element of the stoichiometric matrix and k_j is a proportionality factor. In this model all enzymes are considered as non-equilibrium ones. The temporal evolution of the metabolite concentrations can be described by a system of coupled ordinary differential equations[11]

$$\frac{dx_i}{dt} = \sum_{j=1}^{m} c_{ij} \cdot v_j(x_1, \ldots, x_n) \; ; \quad i = 1, \ldots, n \; , \tag{2}$$

with m and n being the number of reactions and metabolites, respectively. The system of differential equations governing the reaction scheme depicted in Figure 1 reads

$$\dot{x}_{\text{ADE}} = v_{\text{ADO}\to\text{ADE}} - v_{\text{ADE}\to\text{ADO}} - v_{\text{ADE}\to\text{AMP}}$$

$$\dot{x}_{\text{ADO}} = v_{\text{ADE}\to\text{ADO}} - v_{\text{ADO}\to\text{ADE}} - v_{\text{ADO}\to\text{AMP}} + v_{\text{AMP}\to\text{ADO}} - v_{\text{ADO}\to\text{INO}}$$

$$\dot{x}_{\text{AMP}} = v_{\text{ADO}\to\text{AMP}} - v_{\text{AMP}\to\text{ADO}} + v_{\text{ADE}\to\text{AMP}} + v_{\text{ADP}\to\text{AMP},\text{ATP}}$$
$$- v_{\text{ATP},\text{AMP}\to\text{ADP}} - v_{\text{AMP}\to\text{IMP}} + v_{\text{IMP}\to\text{AMP}} + v_{\text{R5P}\to\text{PRPP}} + v_{\text{ATP}\to\text{AMP}}$$

$$\dot{x}_{\text{ADP}} = v_{\text{ADO}\to\text{AMP}} - 2v_{\text{ADP}\to\text{AMP},\text{ATP}} + 2v_{\text{AMP},\text{ATP}\to\text{ADP}} - v_{\text{ADP}\to\text{ATP}} + v_{\text{ATP}\to\text{ADP}}$$

$$\dot{x}_{\text{ATP}} = v_{\text{ADP}\to\text{ATP},\text{AMP}} - v_{\text{ADO}\to\text{AMP}} - v_{\text{ATP},\text{AMP}\to\text{ADP}} + v_{\text{ADP}\to\text{ATP}}$$
$$- v_{\text{ATP}\to\text{ADP}} - v_{\text{R5P}\to\text{PRPP}} - v_{\text{ATP}\to\text{AMP}}$$

$$\dot{x}_{\text{IMP}} = v_{\text{AMP}\to\text{IMP}} - v_{\text{IMP}\to\text{AMP}} + v_{\text{PRPP}\to\text{IMP}} - v_{\text{IMP}\to\text{INO}}$$
$$+ v_{\text{HYP}\to\text{IMP}} - v_{\text{IMP}\to}$$

$$\dot{x}_{\text{INO}} = v_{\text{IMP}\to\text{INO}} + v_{\text{ADO}\to\text{INO}} - v_{\text{INO}\to\text{HYP}} + v_{\text{HYP}\to\text{INO}}$$

$$\dot{x}_{\text{HYP}} = v_{\text{INO}\to\text{HYP}} - v_{\text{HYP}\to\text{INO}} - v_{\text{HYP}\to\text{IMP}} - v_{\text{HYP}\to}$$

$$\dot{x}_{\text{PRPP}} = v_{\text{R5P}\to\text{PRPP}} - v_{\text{ADE}\to\text{AMP}} - v_{\text{HYP}\to\text{IMP}} + v_{\text{PRPP}\to\text{IMP}}$$

The time-dependent change of the GTP concentration has to be determined separately. This is accomplished by an interpolation based on experimental data[12].

RESULTS AND DISCUSSION

The time-dependent changes of the concentrations in the adenine nucleotides during inhibition of ATP synthesis resulting from mathematical modeling are plotted in Figure 2. Additionally, the experimental data for ATP, ADP, and AMP during anoxia[7,12] are depicted.

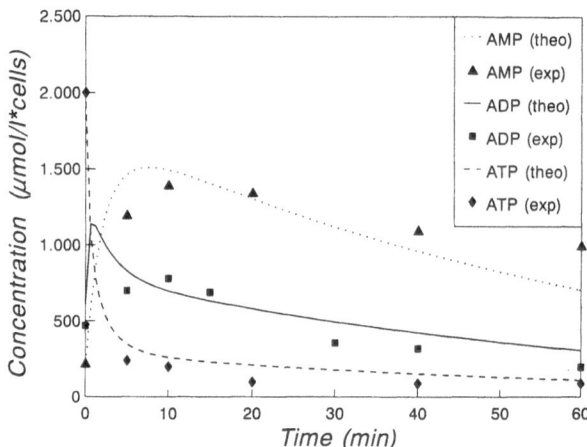

Figure 2. Changes of the adenine nucleotide concentrations during anoxia.

From equation (1) it follows, that with changing metabolite concentrations (anoxia) the substrate flux rates will change, and vice versa. A rapid decrease of the ATP content was observed, followed by an increase of the concentrations of purine degradation products. The latter is due to a significant increase of the flux rates related to the enzymes of AMP degradation (5'-nucleotidase (AMP) and AMP deaminase) after 10 minutes of anoxia. But, during the next 50 minutes there is a slow decrease of the above mentioned fluxes.

The results of our investigations are in good agreement with earlier, more sophisticated mathematical models, and with experimental studies. Hence, simple enzymatic rate laws allows for a realistic simulation of metabolic pathways even under pathological conditions. Most important is a reliable set of experimental data describing the steady state regime and pathological effects. All data has to be obtained under comparable conditions, e.g. cell type, incubation. Then the *in vivo* flux rates from the steady state tracer experiments and the steady state concentrations can be used as starting point for further simulations of different regimes, like anoxia, reoxygenation, drug influence etc.

REFERENCES

1. R. Franco and E.I. Canela, Computer simulation of purine metabolism, *Eur. J. Biochem.*, 144:305 (1984)

2. T. Bartel and H.-G. Holzhütter, Mathematical modelling of the purine metabolism of the rat liver, *Biochim. Biophys. Acta*, 1035:331 (1990)

3. H.-G. Holzhütter and A. Schwendel, Estimation of enzymatic flux rates from kinetic isotope experiments, *in:* "Modern Trends in Biothermokinetics", Plenum Press, New York

4. M.N. Berry and D.S. Friend, High-yield proparation of isolated rat liver parenchymal cells, *J. Cell Biol.*, 43:506 (1969)

5. G. van den Berghe, F. Bontemps, and H.G. Hers., Purine catabolism in isolated rat hepatocytes, *Biochem. J.*, 188:913 (1980)

6. T. Grune, W.G. Siems, G. Gerber, and R. Uhlig, Determination of the ultraviolet absorbance and radioactivity of purine compounds separated by high-performance liquid chromatography, *J. Chromatogr.*, 553:193 (1991)

7. M.F. Vincent, G. van den Berghe, and H.G. Hers, The pathway of adenine nucleotide catabolism and its control in isolated rat hepatocytes subjected to anoxia, *Biochem. J.*, 202:117 (1982)

8. T. Bartel, "Ein Mathematisches Modell des Purinnukleotidmetabolismus der Rattenleber zur Untersuchung des hypoxischen Stoffwechsels und seiner pharmakologischen Beeinflußbarkeit",thesis, Humboldt University, Berlin (1988)

9. M.F. Vincent, G. van den Berghe, and H.G. Hers, Increase in phosphoribosyl pyrophosphate induced by ATP and Pi depletion in hepatocytes, *FASEB J.*, 3:1862 (1989)

10. G. Weber, Y. Natsumeda, M.S. Lui, M.A. Faderan, J.J. Liepnieks, and W.L. Elliott, Control of enzymic programs and nucleotide pattern in cancer cells by acivicin and tiazofurin, *in:* "Adv. Enzyme Regulation", G. Weber, ed., Pergamon Press, Oxford (1984)

11. R. Heinrich, S.M. Rapoport, and T.A. Rapoport, Metabolic regulation and mathematical models, *Prog. Biophys. Mol. Biol.*, 32:1 (1977)

12. G. Gerber, W. Siems, and A. Werner, Purine nucleotide degradation and free radical generation in the hypoxic liver, *in:* "Membran Lipid Oxidation, Vol. III", C. Vigo-Pelfrey, ed., CRC Press, Boca Raton (1990)

ADENINE NUCLEOTIDE LEVELS AND LIPID PEROXIDATION AT HYPOXIA AND REOXYGENATION IN DIFFERENT CELL TYPES

Werner G. Siems,[1] Tilman Grune,[2] and Anke Schwendel[3]

[1] Herzog-Julius Hospital for Rheumatology and Orthopaedics, D-38655 Bad Harzburg, Germany
[2] Department of Biochemistry & Molecular Biology, Albany Medical College, Albany, NY 12208, USA,
[3] Institute of Biochemistry, Medical Faculty, Humboldt University, D-10115 Berlin, Germany

INTRODUCTION

Posthypoxic/postischemic reoxygenation plays a major role in the pathogenesis of diseases, such as myocardial infarction and stroke. The causal chain of pathobiochemical changes including the evaluation of different causal components for cell injury during oxygen deficiency and reoxygenation/ reperfusion has not been clarified yet.
Without any doubt two of the most important causal components of reoxygenation/reperfusion injury following hypoxia/ischemia are 1.) the depletion and restoration of energy-rich nucleotides and 2.) the formation of oxygen free radicals, leading to lipid peroxidation. Furthermore, the xanthine oxidoreductase reaction steps combine these two causal components of reoxygenation injury.
It was suggested that endothelial cells play a key role in pathophysiology of reperfusion syndrome. Therefore, in this study five different cell types with priority for endothelial/epithelial cells were used for comparative studies on depletion of energy-rich nucleotides during oxygen deficiency and on their restoration during reoxygenation period. In parallel the accumulation of lipid peroxidation products such as 4-hydroxynonenal (HNE) and malondialdehyde (MDA) was measured. As cell types were used: Renal tubular cells, aortic endothelial cells (AEC), brain endothelial cells (BEC), hepatocytes and small intestine.

MATERIAL AND METHODS

Isolation and cultivation of aortic endothelial cells and brain endothelial cells. Bovine aortic endothelial cells (AEC)

Purine and Pyrimidine Metabolism in Man VIII, Edited by
A. Sahota and M. Taylor, Plenum Press, New York, 1995

were cultivated in Eagle MEM with 10% FCS, 2 mM glutamine without antibiotics at 37°C in presence of 5% CO_2/95% air. Subcultivation was performed twice a week with 0.25% trypsin/0.25% EDTA in PBS.

Brain endothelial cells (BEC) were isolated according to a modified method of Bowman et al. (1981), Audus and Borchardt (1986) and Mischek et al. (1989). 6-8 brains (pig) fresh from slaughter were placed in PBS, washed, covered with 70% ethanol and flamed. Brains were minced after discarding the meninges and mixed for 2-3 hours at 37°C, 60 rpm with dispase II from B. polymyxa (5 mg/ml). Density gradient centrifugation was performed with 15% dextran solution (molecular weight 500,000) at 5800g and at 4°C for 10 min. The pellet was dissolved in PBS (with Ca and Mg) in presence of collagenase/dispase (from Achromobacter ioghages/Bacillus polymyxa (1 mg/ml). Mixing was performed for 1.5 - 2 hours at 37°C and 10 rpm. Than a centrifugation step at 1000 g, 4°C for 10 min was performed. The pellet was dissolved in M 199 with 20 % FCS, 2 mM glutamine, 100 I.E. penicilline, 100 mg/ml streptomycine, 2.5 mg amphotericine/ml. Cells were cultivated in 24 well-plates, covered for 2 hours with gelatine before plating. Cells were cultivated with 5% CO_2/95% air, at 37°C and 95% relative humidity for 10 days. Every other day cells were washed with PBS at 37°C for removal of erythrocytes and covered again with medium.

Preparation of human renal tubular cells. Tubular cells were prepared from healthy parts of human cancerous kidneys under sterile conditions. The cell suspensions were flushed through sieves of 60, 140 and 400 mesh. The cells, lying on the 400 mesh-sieve were washed with 50 ml of HEPES-buffer (pH 5.5) and centrifuged at 250 g for 3 min. The cell-material was incubated in a cell culture flask flushed with 0.2 % gelatine in a CO_2 incubator (37°C). The incubation medium was RPMI-1640 fluent medium supplemented with 10% human blood serum, penicillin, streptomycin, amphotericin B, glutamine and epidermal growth factor (EGF). After 4 days the cell suspension was washed with HEPES buffer, centrifuged and supplied with new incubation medium. The cells were reaped with a medium of 0.05% trypsine and 0.02% EDTA solved in HEPES buffer. Cell suspensions with a number of 10^6 cells/ml were used for the experiments.

Preparation of hepatocytes. Male Wistar H-strain rats with a body weight of about 190 g were used. Hepatocytes were prepared according to Berry and Friend (1969) with modifications. Starved rats (24 h) were anaesthetized by an intraperitoneal injection of pentobarbital (30 mg/kg b.w.). After an initial perfusion of the liver as described in van den Berghe et al. (1980) cell isolation was performed by a retrograde recirculating perfusion with a collagenase-containing solution (50 mg collagenase/100 ml of buffer). As perfusion medium Krebs-Henseleit bicarbonate buffer gassed with carbogen was used.

Anoxia and reoxygenation of AEC and BEC. Experiments were carried out with confluent monolayers of endothelial cells on 6-well-plates in a shaking water bath (rotating incubator GFL, 37°C, 25 rpm). AEC were used at the 4th day of cultivation (between 16th to 20th passage). BEC were used at the 10th day of cultivation. For the experiments the cells were washed twice with PBS without glucose. 2 ml PBS was added to each of the 6 wells. Plates were gassed for 2 hours hypoxia with Nelson gas (95% N_2 and 5% carbon dioxide) and during reoxygenation with carbogen (95% oxygen and 5% carbon dioxide).

Anoxia and reoxygenation of human renal tubular cells. The cells were incubated in HEPES-buffer (10^6 cells/ml) at 37°C under continuous shaking (50 rotations per minute). Anoxia was induced by replacement of carbogen by Nelson gas. The duration of anoxia was 60 min, followed by 30 min of reoxygenation. Controls were incubated under normoxic conditions during the whole experiment.

Anoxia and reoxygenation of rat hepatocytes. The cell concentration was adjusted to a cytocrit of 2% (v/v; 1% cytocrit corresponds to $1.68 + 0.11 * 10^6$ cells/ml). The incubation medium was Krebs-Henseleit buffer. The cells were incubated at 37°C under continuous shaking (50 rpm). The anoxia was induced by replacement of carbogen by Nelson gas. The hepatocytes from each animal were used for one sample which was incubated under normoxic conditions (control) and samples incubated for 60 min under anoxic conditions and for further 30 min under reoxygenation conditions. In the case of hepatocyte experiments the effects of xanthine oxidase inhibitor oxipurinol (1 mM and 20 uM) were investigated, too.

Ischemia and reperfusion of small intestine. Male Wistar rats of 285 +/- 20 g body weight were used. The animals were anaesthetized initially with diethylether, followed by the application of pentobarbital. The animals were heparinized. The abdomen of the rats was opened and after taking off initial value samples of the middle jejunum for the measurement of biochemical parameters the superior mesenteric artery was ligated with a microvascular clamp (time zero of the ischemia). After 30 and 60 min of the ischemia the next tissue samples were taken. After 60 min the microvascular clamp was removed finishing the ischemic period and starting the reperfusion period of the intestine. During the reperfusion period further tissue samples were taken for analyses. The samples were immediately transfered into fluid nitrogen.

Nucleotide determination. Samples were deproteinized with ice cold 60 g/l perchloric acid, centrifuged at 1200 g for 10 min and neutralized with potassium carbonate. After a second centrifugation the supernatant was stored at -20°C. Ten or twenty ul of supernatants were analysed by HPLC. HPLC equipment of Waters Assoc., Milford, MA was used consisting of a M510 pump, Rheodyne injector, variable wavelength detector (adjusted to 254 nm), 740 Data Module and a 640-system controller. 5um Novapak C_{18} cartridges (100 mm * 8 mm i.d.) with a Z-module compression system were used. Purine nucleotides were determined with an isocratic ion pair reversed phase HPLC system (Grune et al., 1991). The eluent contained 10 mmol/l $NH_4H_2PO_4$, 2 mmol/l tetrabutylammonium phosphate and 20 % acetonitrile. The flow-rate was 2 ml/min.

Determination of lipid peroxidation products. The detection of 4-hydroxynonenal (HNE) was performed according to Esterbauer et al. (1982) and Poli et al. (1985). The TBA-RS concentration was determined by means of the method of Uchigama and Mihara (1978). In the experiments with AEC, BEC and renal tubular cells the MDA specific method of Wong et al. (1987) was used: addition of phosphoric acid, thiobarbituric acid to the biological sample, boiling for 60 min, cooling in ice bath, immediately before HPLC analysis addition of an equal volume of 1 M NaOH, centrifugation. The HPLC equipment consists of a Waters 510 pump. A Shimadzu fluorescence detector RF-530 (525/550 nm) and integrator C-R6A were used. The column was a Supelcosil 150 * 4 mm LC-18-S (5um). As eluent a 50 mM potassium phosphate buffer solution pH 6.8 with 40% methanol was used.

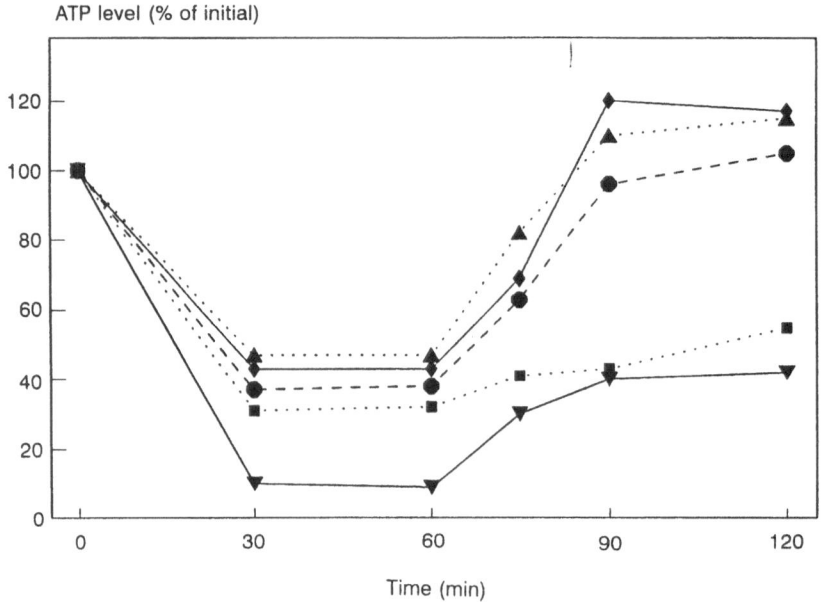

Figure 1. ATP concentration as % of initial value during 60 min hypoxia and following reoxygenation.
The symbols are: ♦ aortic endothelial cells; ▲ brain endothelial cells; ● renal tubular cells; ▼ hepatocytes and ■ small intestine.

RESULTS

In all five cell types which were investigated a rapid decrease of ATP level during hypoxia was observed, but the new steady state level of ATP as well as the velocity of ATP restoration were different. Fig. 1 demonstrates the changes of ATP level given as given as % of initial.
The cellular ATP levels decreased under hypoxic conditions to the following new steady state concentrations: about 10% of the initial value in hepatocytes, 30% in small intestinal cells, 37 to 39% in renal tubular cells, 43% in aortic endothelial cells (AEC) and to 47% in brain endothelial cells (BEC). In Fig. 1 is additionally demonstrated, that the restoration of ATP during the reoxygenation period was faster in the following range: BEC>AEC>renal tubular cells>small intestine>hepatocytes. In small intestine after 1h of oxygen deficiency only a very slow ATP restoration could be measured.

In freshly prepared hepatocytes ATP restoration was only possible if the duration of hypoxia did not exceed 30 min. In all five cell types a free radical induced lipid peroxidation during the first minutes of posthypoxic reoxygenation was measured (not documented here). The extent of MDA and HNE accumulation was low only in small intestine if compared with the other cell types used. In AEC, BEC, renal tubular cells and hepatocytes the peaks of MDA levels were between 2 and 5 nmol/mg protein and the peaks of HNE levels were between 0.07 and 0.4 nmol/mg protein.

Mathematical modelling with the aim of calculation of flux rates of purine nucleotide metabolism during anoxia and reoxygenation - studied in hepatocyte suspensions - shows drastic increases of catabolic pathways and decreases of nucleotide syntheses during anoxic period and normalization of flux rates during reoxygenation (Table 1).

Table 1. Calculation of flux rates of purine metabolism during oxygen deficiency and posthypoxic reoxygenation in hepatocytes. Values are given as nmoles/ml cells/min.

Reaction step	Normoxia	30 min anoxia	60min reoxygenation
HPRT	3.58	0.43	2.96
APRT	1.31	0.09	0.61
IMP nucleotidase	3.22	8.52	3.69
AMP deaminase	0.53	13.68	0.39

DISCUSSION

In comparative studies on hypoxia/ reoxygenation in five different cell types a rapid decrease of ATP during hypoxia was observed. Degree of ATP loss correlated inversely to velocity of ATP restoration during posthypoxic reoxygenation. In all five cell types a free radical induced lipid peroxidation during the first minutes of reoxygenation was demonstrated. Prerequisites for increased posthypoxic free radical formation are the accumulation of hypoxanthine during hypoxia and the presence of oxidase form of xanthine oxidoreductase (Grune et al., 1993; Gerber et al., 1991; Siems et al., 1986, 1991, 1993). Posthypoxic ATP restoration was not influenced by MDA and HNE levels which were about equal in AEC, BEC, renal tubular cells and hepatocytes.

It is suggested that the ATP recovery depends rather on metabolic adaptation of different cell types to hypoxic conditions than to the extent of posthypoxic radical formation and lipid peroxidation. Cell cultivations seem to be more resistant against ATP loss than freshly prepared cell suspensions or solid tissues. another interpretation may be: endothelial cells (AEC and BEC) which are continuously exposed to changing oxygen pressure and high free radical formation are better adapted to resist reperfusion injury and nucleotide depletion than other cell types.

REFERENCES

Audus, K.L., and Borchardt, R.T., 1986, Characteristics of the large neutral amino acid transport system of bovine brain microvessel endothelial cells, J. Neurochem. 47:484.

Berry, M.N., and Friend, D.S.,1969, High-yield preparation of isolated rat liver cells, J. Cell. Biol. 43:506.

Bowman, P.D., Betz, A.L., and Ar, D., 1981, Primary culture of capillary endothelium from rat brain, In Vitro 17:353.

Esterbauer, H., Cheeseman, K.H., Dianzani, M.U., Poli, G., and Slater, T.F., 1982, Separation and characterization of the aldehydic products of lipid peroxidation stimulated by ADP-Fe2+ in rat liver microsomes, Biochem. J. 208:129.

Gerber, G., Siems, W., and Werner, A., 1991, Purine nucleotide degradation and free radical generation in the hypoxic liver, in: "Membrane Lipid Oxidation" Vol.1, C. Vigo-Pelfrey, ed., CRC Press, Inc., Boca Raton,FL, pp.116-140.

Grune, T., Siems, W., Gerber, G., Tikhonov, Y.V., Pimenov, A.M., and Toguzov, R.T., Changes of nucleotide patterns in liver, muscle and blood during the growth of Ehrlich ascites cells, J. Chromatogr. 563:53.

Grune, T., Siems, W.G., and Schneider, W., 1993, Accumulation of aldehydic lipid peroxidation products during postanoxic reoxygenation of isolated rat hepatocytes, Free Rad. Biol. Med. 15:125.

Mischek, U., Meyer, J., and Galla, H.-J., 1989, Characteri-zation of g-GT-activity of cultured endothelial cells from porcine brain capillaries, Cell Tissue Res. 256:221.

Poli, G., Dianzani, M., Cheeseman, K.H., Slater, T.F., Lang, J., and Esterbauer, H., 1985, Separation and characterization of the aldehydic products of lipid peroxidation stumulated by carbon tetrachloride or ADP-iron in isolated rat hepatocytes and rat liver microsomal suspension, Biochem. J. 227:629.

Siems, W.G., Grune, T., Zollner, H., and Esterbauer, H., 1993, Formation and metabolism of the lipid peroxidation product 4-hydroxynonenal in liver and small intestine, in: "Free Radicals: From Basic Science to Medicine", G. Poli, E. Albano, and M.U. Dianzani, eds., Birkhaeuser Verlag, Basel, pp. 89-101.

Siems, W., Kowalewski, J., David, H., Grune, T., and Bimmler, M., 1991, Discrepancy between biochemical mormalisation and morphological recovery of jejunal mucosa during postischemic reperfusion in presence of the xanthine oxidase inhibitor oxypurinol, Cell. Molec. Biol. 37:213.

Siems, W., Schmidt, H., Mueller, M., Henke, W., and Gerber, G., 1986, H_2O_2 formation during nucleotide degradation in the hypoxic rat liver: A quantitative approach, Free Rad. Res. Comms. 1:289.

Uchiyama, M., and Mihara, M., 1978, Determination of malonaldehyde precursor in tissues by thiobarbituric acid test, Anal. Biochem. 86:271.

van den Berghe, G., Bontemps, F., and Hers, H.-G., 1980, Purine catabolism in isolated rat hepatocytes, Biochem. J. 188:913.

Wong, S.H.J., Knight, J.A., Hopfer, S.M., Zaharia, O., Leach, C.N., and Sunderman, F.W., 1987, Lipoperoxides in plasma measured by liquid chromatographic separation of malon-dialdehyde-thiobarbituric acid adduct, Clin.Chem. 33:214.

PURINE AND PYRIMIDINE CATABOLITE PRODUCTION IN THE POSTISCHEMIC RAT HEART - EFFECT OF ADENOSINE SUPPLY DURING REPERFUSION

Ryszard T. Smolenski[1], H. Anne Simmonds[2], and David J. Chambers[3]

[1]Department of Biochemistry, Academic Medical School, Gdansk, Poland.
[2]Purine Research Laboratories and [3]Cardiac Surgical Research,
Rayne Institute, UMDS, Guy's & St. Thomas' Hospital, London, U.K.

INTRODUCTION

Marked reduction in purine catabolite production is invariably observed in the course of repeated ischemic events[1-3]. Both beneficial and deleterious effects of this phenomenon should be taken into consideration. Inhibition of nucleotide breakdown protects the nucleotide pool particularly ATP in the heart but at the expense of the reduction of the release of endogenous protective metabolite - adenosine (ADO). Exogenous ADO exerts a significant reviving effect on the heart when supplied during reperfusion[4-6] which may be partially mediated by substitution for impaired release of endogenous ADO. Exogenous ADO is also the most efficiently used substrate for resynthesis of adenine nucleotides. To obtain further information concerning the mechanism producing beneficial effect of ADO supplied during reperfusion we investigated the effects of three concentrations of ADO (optimal for vasoactive, metabolic and cardioplegic effect) on mechanical recovery, myocardial nucleotides and purine and pyrimidine catabolite production in Langendorff perfused rat heart.

MATERIALS AND METHODS

Male Wistar rats weighing 200-250 g were used for this study. Hearts were perfused in Langendorff mode with Krebs-Henseleit buffer solution at constant pressure (100 cm H_2O) at 37°C. Cardiac function was monitored by measuring developed tension using a pressure transducer connected to the apex of the heart. Hearts were paced at 350 beats per min. All hearts were perfused for a total of 75 min. After 20 min of stabilization period hearts were subjected to 30 sec of global ischemia followed by reperfusion. At 30 min hearts were subjected to 10 min of global (37°C) ischemia with subsequent reperfusion with or without supply of ADO during the first 15 min of reperfusion. ADO was added to the perfusion buffer at concentrations of 1 μM, 30 μM or 1 mM and the no ADO group was reperfused with standard buffer. At 70 min of perfusion a further 30 sec ischemia was applied. A control series of hearts without 10 min ischemia (no 10 min ischemia group) was subjected to three 30 sec periods of ischemia at 20, 45, and 70 min of perfusion. At the end of perfusion hearts were frozen with a clamp precooled in liquid nitrogen. In all groups coronary effluent was collected for analysis of nucleotide catabolites by HPLC[7,8]. Values are presented as means ±S.E.M.. Statistical comparison was performed with analysis of variance or with paired Student t-test. $p<0.05$ was considered significant difference.

RESULTS AND DISCUSSION

Highest recovery of developed tension (DT) occurred with 1 μM and 30 μM ADO (72±3% and 72%±5 ($p<0.05$) of preischemic DT, respectively) compared to 53±5% and 63±5% in control and 1 mM Ado respectively. Tissue content of nucleotide triphosphates is presented in Table 1.

Table 1. Final myocardial content of ATP, GTP, UTP and CTP in no 10 min ischemia control and in the hearts subjected to 10 min of ischemia, supplied with ADO at differnt concentration in the first 15 min of reperfusion.

	ATP	GTP	UTP	CTP
	μmol/g dry wt			
no 10 min isch	20.7±0.7	0.99±0.05	1.21±0.05	0.26±0.01
no ADO	15.2±0.6*	0.74±0.03*	0.78±0.06*	0.22±0.01
1 μM ADO	16.4±0.4*	0.82±0.03*	0.84±0.04*	0.24±0.01
30 μM ADO	18.9±0.5*#	0.79±0.03*	0.88±0.05*	0.24±0.01
1 mM ADO	17.2±0.4*	0.86±0.03*	0.86±0.03*	0.25±0.02

Value are mean±S.E.M. (n=6-9). *$p<0.05$ in comparison with no 10 min ischemia.
#$p<0.05$ in comparison with 1 μM and 1 mM ADO

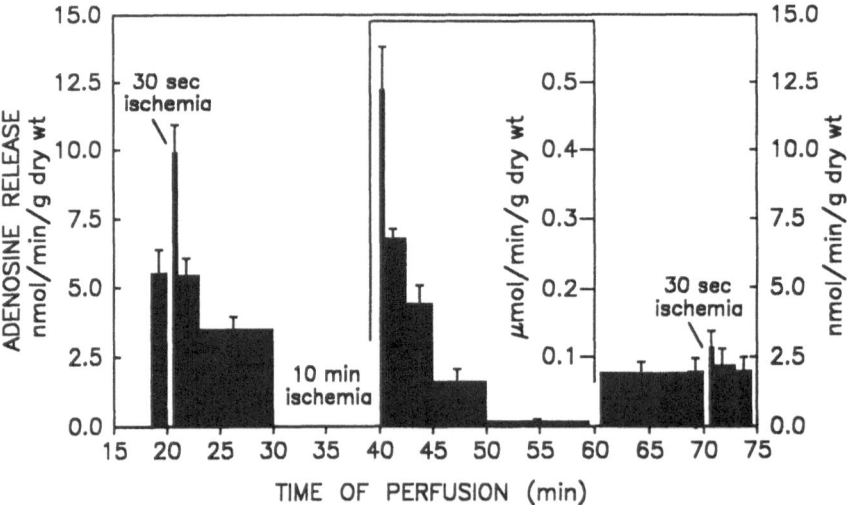

Figure 1. Release of ADO from rat heart subjected to 10 min and two 30 sec ischemic events

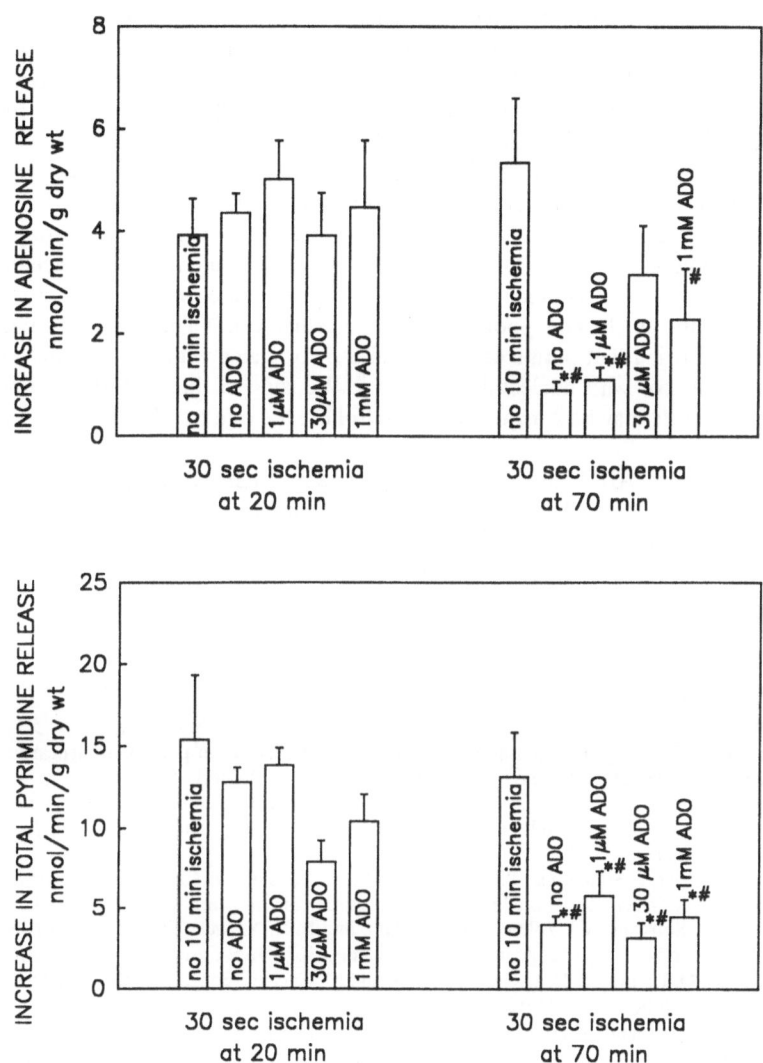

Figure 2. Increases in release of ADO and pyrimidine catabolites from rat heart subjected to 30 sec ischemia at different time of perfusion. #$p<0.05$ in comparison with 30 sec ischemia at 20 min. *$p<0.05$ in comparison with no 10 min ischemia.

ATP significantly increased with 30 μM ADO in comparison to control while these differences were not significant at 1 μM and 1 mM ADO. Concentration of GTP, UTP and CTP was similar in all hearts subjected to 10 min ischemia. Proportion of ATP to other triphosphate nucleotides was thus selectively enhanced by supply of 30 μM ADO during reperfusion.

Preischemic coronary flow was restored after 30 sec of reperfusion in hearts treated with 1 and 30 μM ADO while more than 2 min were necessary without or with 1 mM ADO. Reactive hyperaemia after 30 sec ischemia was significantly enhanced by 1 or 30 μM ADO infusion, while this was not observed without or with 1 mM ADO. Endogenous ADO, purine and pyrimidine release to coronary effluent increased immediately after the first 30 sec ischemia after 20 min of perfusion (Fig. 1,2). This increase was reduced to 30-50% after 30 sec ischemia at 70 min of perfusion in controls, with 1 μM or 1 mM ADO. However, increase in endogenous ADO and purines was restored while pyrimidine release remained reduced after 30 sec ischemia at 70 min of perfusion when 30 μM ADO was supplied during reperfusion. In the group not subjected to 10 min ischemia group increases of purines, pyrimidines and ADO after each 30 sec of ischemia remained unchanged. It is likely therefore that ADO and other nucleotide catabolite release from the heart is dependent on the nucleotide pool size in myocardial cells. This contrasts with earlier observations showing that purine catabolite release correlates with the ATP utilization rate in the heart and not with nucleotide pool size[9]. We suggest that in certain situations the nucleotide pool size may be a major factor controlling the rate of ADO and other catabolite production.

In summary, ADO supplied during reperfusion improves mechanical performance already at its vasoactive 1 μM concentration which restores coronary flow in early reperfusion. However, 30 μM ADO while maintaining functional and vasoactive effects, also elevated ATP and consequently enhanced endogenous adenosine production, which may provide additional benefits for in vivo reperfused heart.

REFERENCES

1. K.A. Reimer, C.E. Murry, I. Yamasawa, and R.B. Jennings, Four brief periods of myocardial ischemia cause no cumulative ATP loss or necrosis, *Am. J. Physiol.* 251:H1306(1986).
2. K.J. Henrichs, H. Matsuoka, and J. Schaper, Influence of repetitive coronary oclusions on myocardial adenine nucleosides, high energy phosphate and ultrastructure, *Basic. Res. Cardiol.* 82:557(1987).
3. R.T. Smolenski, H.A. Simmonds, P.B. Garlick, G.E. Veen, and D.J. Chambers, Depressed adenosine and total purine catabolite production in the postischemic rat heart, *Cardioscience* 4:235(1993).
4. M. Amrani, R. Shirvani, N.J. Allen, S. Ledingham, and M.H. Yacoub, Enhancement of low coronary reflow improves postischemic myocardial function, *J. Thorac. Cardiovasc. Surg.* 104:1375(1992).
5. S. Ledingham, O. Katayama, D. Lachno, N. Patel, and M. Yacoub, Beneficial effect of adenosine during reperfusion following prolonged cardioplegic arrest, *Cardiovasc. Res.* 24:247(1990).
6. C.J. Pitarys, R. Virmani, H.D. Vildibill, E.K. Jackson, and M.B. Forman, Reduction of myocardial reperfusion injury by intravenous adenosine administered during early reperfusion phase period, *Circulation* 83:237(1991).
7. R.T. Smolenski, D.R. Lachno, S.J.M. Ledingham, and M.H. Yacoub, Determination of sixteen nucleotides, nucleosides and bases using high-performance liquid chromatography and its application to the study of purine metabolism in hearts for transplantation, *J. Chromatogr.* 527:414(1990).
8. C. Montero, R.T. Smolenski, J.A. Duley, and H.A. Simmonds, S-adenosylmethionine increases erythrocyte ATP in vitro by a route independent of adenosine kinase, *Biochem. Pharmacol.* 40:2617(1990).
9. R. Zucchi, G. Yu, S. Ronca Testoni, M. Mariani, and G. Ronca, Energy metabolism in myocardial stunning, *J. Mol. Cell Cardiol.* 24:1237(1992).

THE AMP - ADENOSINE CYCLE IS ACTIVE DURING NORMOXIA AND IMPAIRED IN ATP DEPLETION IN ISOLATED RABBIT CARDIOMYOCYTES

Daniel R. Wagner, Françoise Bontemps,
and Georges van den Berghe

Laboratory of Physiological Chemistry,
International Institute of Cellular and Molecular
Pathology, and University of Louvain Medical School,
B-1200 Brussels

INTRODUCTION

Adenosine (Ado), a breakdown product of adenine nucleotides, is a potent coronary vasodilator, which has been proposed to regulate coronary circulation according to myocardial oxygen demand (1). Recycling of Ado into AMP has been demonstrated to play a crucial role in the regulation of the concentration of Ado in isolated rat hepatocytes (2,3). In the heart, however, the AMP - Ado substrate cycle has been considered either inactive (4), or operating at a very low rate (5), except for a recent study in normoxic guinea pig heart (6).

In the present study, the existence and the role of a substrate cycle between AMP and Ado have been investigated in isolated rabbit cardiomyocytes, in control conditions and during ATP depletion.

METHODS

Rabbit cardiomyocytes were isolated by collagenase perfusion (7), and incubated in HEPES-buffered Krebs-Henseleit solution at 37°C, pH 7.4, 100% oxygen, in control conditions and in ATP depletion achieved by inhibiting glycolysis with 5 mM iodoacetate. After isolation, more than 90% of the cardiac myocytes were viable as assessed by trypan blue exclusion, and 75 - 90% of the cells were rod shaped. Contamination of the preparation with endothelial cells was ruled out by measuring hypoxanthine, a catabolite which is only formed in the latter cells and not in myocytes. Inosine (Ino) is the terminal purine catabolite in cardiomyocytes due to the absence of nucleoside phosphorylase in these cells (8). Additional details are given in ref. (9).

To study Ado metabolism, iodotubercidin 2 μM (ITu), an inhibitor of Ado kinase, and deoxycoformycin 2 μM (dCF), an inhibitor of Ado deaminase, were used. It had been verified in preliminary studies that these concentrations inhibited by 94% and 99% the phosphorylation and deamination, respectively, of ^{14}C labeled Ado by the cardiomyocytes.

RESULTS AND DISCUSSION

Adenine nucleotides in control conditions

In control conditions, the contents of ATP, ADP, and AMP (expressed in nmol/mg Lowry protein) were 28.9±1.2, 5.4±0.3 and 4.3±0.3, respectively (n = 12). The adenine nucleotides remained stable over the 60 min incubation and were not significantly affected by the addition of either 2 μM ITu, 2 μM dCF, or the combination of both inhibitors.

Accumulation of Ado in control conditions

In the absence of inhibitors, Ado accumulated in the cardiomyocyte suspensions at the rate of 4±1 pmol/min per 10^6 cells (n = 6) (Table 1). Addition of dCF had no influence on the accumulation of Ado. This indicates that little or no deamination of Ado to Ino proceeds in rabbit cardiomyocytes in control conditions. Addition of ITu provoked a 13-fold increase in Ado accumulation, to 53±16 pmol/min per 10^6 cells (n = 6) (p<0.05). This proves that Ado is normally recycled into AMP. Addition of both dCF and ITu increased Ado accumulation to 210±32 pmol/min per 10^6 cells. This value reflects the "absolute" rate of dephosphorylation of AMP. It reached 20-fold the rate of accumulation of Ado and Ino in the absence of inhibitors, and indicates that 95% of Ado produced by dephosphorylation of AMP is normally recycled into AMP.

Adenine nucleotides in ATP depletion

After addition of 5 mM iodoacetate, ATP decreased in a biphasic way. During the first 10 min, ATP decreased by approx. 15% to a plateau which was maintained for 10 min. Thereafter, ATP linearly declined up to complete depletion at 60 min. ADP increased slightly at 40 min, and decreased thereafter. AMP and IMP increased strikingly during the first 5 min, to 3.9±0.5 and 6.8±2.3 nmol/mg Lowry protein, respectively. Later on, AMP remained at a plateau while IMP continued to increase progressively.

Ado accumulation in ATP depletion

In the absence of dCF and ITu, Ado accumulated during the first 10 min at a rate of 461 ± 92 pmol/min per 10^6 cells to a plateau which was maintained during the following 10 min; thereafter Ado accumulated linearly. The overall production of Ado reached 335 ± 63 pmol/min per 10^6 cells over 60 min (Table 1). Addition of dCF increased Ado accumulation 2-fold during the first 40 min, and 4-fold thereafter ($p < 0.01$), indicating that Ado is deaminated to Ino in ATP depletion. Addition of ITu increased Ado accumulation 2-fold during the first 10 min ($p < 0.05$). Thereafter ITu had no effect. This indicates that recycling of Ado occurs during the first 10 min and is suppressed later on. Addition of dCF and ITu increased Ado accumulation to 814 ± 16 pmol/min per 10^6 cells. This is a 4-fold increase compared to the rate of Ado accumulation in control conditions with dCF and ITu, and reflects the rate of dephosphorylation of the elevated concentration of AMP.

Table 1. Accumulation rates of Ado (pmol/min per 10^6 cells)

	Control	Iodoacetate
No addition	4 ± 1	335 ± 63
dCF	5 ± 5	743 ± 123**
ITu	53 ± 18*	504 ± 66
dCF + ITu	210 ± 32***	814 ± 16***

Means of 3 to 6 experiments $\pm$ SEM; *$p < 0.05$, **$p < 0.01$, ***$p < 0.005$ vs. control

CONCLUSIONS

Under control conditions, isolated rabbit cardiomyocytes continuously form Ado by dephosphorylation of AMP, but 95% thereof is recycled. This recycling allows the cells to maintain low, non vasodilatory concentrations of Ado. During ATP depletion, owing to the elevation of AMP, the rate of formation of Ado increases 4-fold, and initially 50% thereof is recycled. Later on recycling is suppressed. As first proposed by Arch and Newsholme (10), dependence of recycling of Ado on the concentration of ATP, AMP and Ado, provides a sensitive mechanism for elevation of Ado upon ATP depletion.

REFERENCES

1. Berne RM. The role of adenosine in the regulation of coronary blood flow. *Circ Res* 47: 807-813 (1980)
2. Bontemps F, Van den Berghe G, Hers HG. Evidence for a substrate cycle between AMP and adenosine in isolated hepatocytes. *Proc Natl Acad Sci USA* 80: 2829-2833 (1983)
3. Bontemps F, Vincent MF, Van den Berghe G. Mechanisms of elevation of adenosine levels in anoxic hepatocytes. *Biochem J* 290: 671-677 (1993)
4. Newby AC, Holmquist CA, Illingworth J, Pearson JD. The control of adenosine concentration in polymorphonuclear leukocytes, cultured heart cells and isolated perfused heart from the rat. *Biochem J* 217: 317-323 (1983)
5. Achterberg PW, Stroeve RJ, De Jong JW. Myocardial adenosine cycling rates during control conditions and under conditions of stimulated purine release. *Biochem J* 235: 13-17 (1986)
6. Kroll K, Decking UKM, Dreikorn K, Schrader J. Rapid turnover of the AMP-adenosine metabolic cycle in the guinea pig heart. *Circ Res* 73: 846-856 (1993)
7. Altschuld RA, Gamelin LM, Kelley RE, Lambert MR, Apel LE, Brierley GP. Degradation and resynthesis of adenine nucleotides in adult rat heart myocytes. *J Biol Chem* 262: 13527-13533 (1987)
8. Rubio VR, Wiedmeier T, Berne RM. Nucleoside phosphorylase: localization and role in the myocardial distribution of purines. *Am J Physiol* 222: 550-555 (1972)
9. Wagner DR, Bontemps F, Van den Berghe G. Existence and role of substrate cycling between AMP and adenosine in isolated rabbit cardiomyocytes under control conditions and in ATP depletion. *Circulation* (in press)
10. Arch JRS, Newsholme EA. The control of the metabolism and the hormonal role of adenosine. *Essays in Biochem* 14: 82-123 (1978)

SEARCH FOR THE MECHANISMS OF HIGH INCIDENCE OF APRT DEFICIENCY AMONG JAPANESE

Naoyuki Kamatani, Hisashi Yamanaka, Masayuki Hakoda, Chihiro Terai and Sadao Kashiwazaki

Institute of Rheumatology, Tokyo Women's Medical College, Tokyo, Japan

INTRODUCTION

The incidence of 2,8-dihydroxyadenine urolithiasis or adenine phosphoribosyltransferase (APRT) deficiency is apparently higher among Japanese than other ethnic groups (1). Thus, our laboratory alone has so far diagnosed more than 100 subjects with homozygous APRT deficiency as well as many heterozygotes. The incidence of heterozygotes among the Japanese has been estimated to be 0.47-1.2% (2, 3). In the present study, we attempted to search for the mechanisms of the high incidence of APRT deficiency among this ethnic group by analyzing the data from molecular studies using samples from these patients.

SUBJECTS AND METHODS

Subjects: Some patients with homozygous APRT deficiency came to our out-patient department (Tokyo Women's Medical College). Heparinized blood was drawn from them and mononuclear cells were separated. From most of the patients, however, blood samples were drawn in other hospitals and sent to our laboratory for the diagnosis. In most of the subjects, the possibility of the enzyme deficiency was suspected because they had excreted 2,8-dihydroxyadenine stones into urine or the surgically removed urinary stones from them contained 2,8-dihydroxyadenine. In a few subjects, however, the presence of spherical crystals in urinary sediments had lead laboratory technicians to suspect the enzyme deficiency. In some others, the blood samples were tested because they were family members of individuals with homozygous APRT deficiency.
Diagnosis of homozygotes: The homozygotes were confirmed to be homozygously deficient by the diagnostic technique using peripheral blood T-cell culture. They were cultured with phytohemagglutinin, interleukin 2 and 6-methylpurine and tested wether they were resistant (homozygotes) or sensitive (not homozygotes) to the adenine analog. Details of the procedure were described elsewhere (4).

Diagnosis of heterozygotes: The diagnosis of heterozygotes was done by cloning peripheral blood T-cells in the presence of 2,6-diaminopurine in the culture. Details were described elsewhere (3).

RESULTS AND DISCUSSION

Mutations: By the analysis of genomic DNA from these patients, we found 4 different mutations responsible for the deficiency; i.e. a missense base substitution at codon 136, a nonsense base substitution at codon 98, 4-base pair insertion in exon 3, and a gross gene change (5). The missense base substitution at codon 136 (designated *APRT*J* allele) (6) was always associated with a partial APRT deficiency (type II APRT deficiency). All other mutations were associated with complete APRT deficiency (designated *APRT*Q0* allele). When a homozygote had both *APRT*J* and *APRT*Q0* alleles (compound heterozygotes) (7), they exhibited a partial APRT deficiency (type II deficiency).

Statistics of the genotypes and phenotypes: Numbers of families and individuals with homozygous APRT deficiency were listed in Table 1. The total number of homozygotes diagnosed in our laboratory is 106 as of May 14, 1994. Table 1 include 104 homozygous subjects whose genotypes were determined. As shown, 79 individuals from 64 families were type II deficiency while 25 individuals from 22 families were type I deficiency. Thus, as to the numbers of the subjects, 24% and 76% were type I and II deficiencies, respectively. Although all type I homozygotes had only *APRT*Q0* alleles, some type II homozygotes had both *APRT*J* and *APRT*Q0* alleles. In two individuals from two families, it was not completely sure whether their genotypes were *APRT*J/APRT*J* or *APRT*J/APRT*Q0* (Table 1). The genotypes of 64 subjects was *APRT*J/APRT*J*, while that of 13 patients was *APRT*J/APRT*Q0*.

Table 1. Statistics of homozygous APRT deficiency diagnosed in our laboratory.

Genotype	Families	Individuals	Phenotype
*APRT*J/APRT*J*	51	64	
*APRT*J/APRT*Q0*	11	13	type II
J/J or *J/Q0*	2	2	(64 families, 79 Individuals)
*APRT*Q0/APRT*Q0*	22	25	type I
Total	86	104	

Actual and expected ratios of genotypes: The ratios of genotypes in Table 1 suggests a shift from Hardy-Weinberg's equilibrium. Thus, if the frequencies of the alleles *APRT*J* and *APRT*Q0* are p and q, and the population is in Hardy-Weinberg's equilibrium, then the frequencies of genotypes *APRT*J/APRT*J*, *APRT*J/APRT*Q0* and *APRT*Q0/APRT*Q0* should be p^2, $2pq$ and q^2, respectively. The values p and q calculated from the number of the subjects in Table 1, are p= 0.69r and q=0.31r (r is the frequency of all the defective APRT alleles causing APRT deficiency). The expected ratios of the genotypes calculated from these values are shown in Table 2. Comparison of these data with the actual ratios of the genotypes suggest that there is a shift from the Hardy-Weinberg's equilibrium. The actual ratio of the number of compound heterozygotes (*APRT*J/APRT*Q0*) is much lower than expected as compared with those of the homozygotes. The reason for this shift from the Hardy-Weinberg's equilibrium is unclear. There may be some uneven distribution of APRT*J and APRT*Q0 alleles among Japanese. If there is a shift from the random mating as to the defective APRT alleles due to the uneven geographical distribution, then the increase in the relative ratios of the homozygotes would be expected.

Another possibility is that there are differences in the penetrances for different genotypes. If the penetrance for APRT*J/APRT*Q0 is lower than those for the homozygotes, then the ratio of symptomatic compound heterozygotes would be relatively low. In fact, there is evidence for the difference in the severity of symptoms between different genotypes. Thus, the ages at the diagnosis were compared among 60 propositi. Those were 31.4 ± 19.5 (n=37), 33.3 ± 28.1 (n=7) and 24.1 ± 17.6 (n=16) for propositi with genotypes of APRT*J/APRT*J, APRT*J/APRT*Q0 and APRT*Q0/APRT*Q0, respectively. These data suggest that the subjects with type II deficiency may develop symptoms at elder ages than the type I deficiency patients. The penetrance for each genotype may be similarly different. Accurate determination of the frequencies of the defective alleles in the population would be a good test to examine these possibilities as the reason for the shift from the Hardy-Weinberg's equilibrium.

Table 2. Actual and expected ratios of genotypes

Genotypes	Expected ratios (%)	Actual ratios (%)
APRT*J/APRT*J	47.8	62.7
APRT*J/APRT*Q0	42.7	12.7
APRT*Q0/APRT*Q0	9.5	24.5
Total	100	100

Statistics of mutations: Statistics of mutations for 141 separate alleles obtained from different families were shown in Table 3. Four different mutations explained 97% of the defective APRT alleles among the Japanese. The most frequent mutational allele (APRT*J) explained 68% of all the defective alleles, while the second and third common mutations explained 21 and 7%, respectively. In 3% of the alleles, we have not succeeded to identify mutations responsible for the deficiency.

There is evidence that the origin of the mutational alleles with the same sequence in Table 3 is single but not multiple. Thus, no patients with the type II APRT deficiency have been identified in western countries, thereby indicating that the frequencies of the APRT*J allele in non-Japanese populations are very low, if present at all, and that frequent recurrence of this mutation is unlikely. When we examined in vivo mutations at the human APRT locus in somatic T and B cells, mutations causing APRT*J alleles were not observed (8, 9). These data also indicate that this type of mutation is not common in human DNA. Furthermore, the data of the linkage disequilibrium between the germline mutations in Table 3 and intragenic and extragenic polymorphic sites were consistent with the hypothesis that mutational alleles with the same sequence originated from a single ancestral gene (10).

Since each common mutation in Table 3 is likely to have originated from a single mutational event, random factors may have made a great difference in the frequency of defective alleles. The frequency of the heterozygotes of APRT deficiency among the Japanese was estimated to be 1.2% (2). Therefore, the mutant allele frequency should be 0.6%. Since the most common *APRT*J* alleles explain 68% of the total mutational alleles, the mutant allele frequency would be 0.6 x 0.32 = 0.192 (%) if the most common mutation had not occurred. Therefore, the removal of the most common mutational allele would greatly reduce the defective allele frequency among the Japanese. From Table2, we can calculate the expected homozygote frequency in the population with no *APRT*J* mutant alleles. Thus, only subjects with the genotype of *APRT*Q0/APRT*Q0* should be symptomatic and the homozygote frequency would be 24.5% of the actual value. Taken together, the high apparent incidence of APRT deficiency among the Japanese may be explained simply by the presence of the *APRT*J* alleles. This mutation occurred in a single ancestor of Japanese and the frequency may have increased by a random genetic drift. Although there remains a possibility that the mechanism of overdominance is involved in the process of the increase in the allele frequency, the random genetic drift is likely to be the major mechanism.

Table 3. Statistics of mutations in defective alleles (Modified from 5).

Allele	Mutation	No	Percentage
*APRT*J*	ATG to ACG at codon 136	96	68
	TGG to TGA at codon 98	30	21
*APRT*Q0*	4bp insertion in exon 3	10	7
	gross alteration	1	1
	undefined	4	3
	Total	141	100

REFERENCES

1. H.A. Simmonds, A.S. Sahota and K.J. Van Acker, Adenine phosphoribosyltransferase deficiency and 2,8-dihydroxyadenine lithiasis, *in*:"Metabolic Basis of Inherited Disease," 6th ed. C.R. Scriver, A.L. Beaudet, W.S. Sly and D. Valle, ed., McGraw-Hill, New York (1989) p1029.
2. N. Kamatani, T. Sonoda and K. Nishioka, Distribution of the patients with 2,8-dihydroxyadenine urolithiasis and adenine phosphoribosyltransferase deficiency in Japan, *J. Urol.* 140:1470 (1988)
3. M. Hakoda, H. Yamanaka, N. Kamatani, and N. Kamatani, Diagnosis of heterozygous states for adenine phosphoribosyltransferase deficiency based on detection of in vivo somatic mutants in blood T cells: Application to screening of heterozygotes, *Am. J. Hum. Genet.* 48:522 (1991)
4. N. Kamatani, F. Takeuchi, Y. Nishida, H. Yamanaka, K. Nishioka, K. Tatara, S. Fujimori, K. Kaneko, I. Akaoka and Y. Tofuku, Severe impairment in adenine metabolism with a partial deficiency of adenine phosphoribosyltransferase, *Metabolism* 34:164 (1985)
5. N. Kamatani, M. Hakoda, S. Otsuka, H. Yoshikawa and S. Kashiwazaki, Only three mutations account for almost all defective alleles causing adenine phosphoribosyltransferase deficiency in Japanese patients, *J. Clin. Invest.* 90:131(1992)
6. N. Kamatani, C. Terai, S. Kuroshima, K. Nishioka and K. Mikanagi, Genetic and clinical studies on 19 families with adenine phosphoribosyltransferase deficiencies. *Hum. Genet.* 75:163 (1987)
7. T. Ishidate, S. Igarashi and N. Kamatani, Pseudodominant transmission of an autosomal recessive disease, adenine phosphoribosyltransferase deficiency, *J. Pediatr.* 118:90 (1991)
8. N. Kamatani, S. Kuroshima, C. Terai, K. Kawai, K. Mikanagi and K. Nishioka, Selection of human cells having two different types of mutations in individual cells (genetic/artificial mutants)-Application to the diagnosis of the heterozygous state for a type of adenine phosphoribosyltransferase deficiency, *Hum. Genet.* 76:148 (1987)
9. J. Chen, A. Sahota, G.F. Martin, M. Hakoda, N. Kamatani, P.J. Stambrook and J.A. Tischfield, Analysis of germline and in vivo somatic mutations in the human adenine phosphoribosyltransferase gene - mutational hot spots at the intron-4 splice donor site and at codon-87, *Mutat. Res.* 287:217 (1993)
10. N. Kamatani, S. Kuroshima, M. Hakoda, T.D. Palella and Y. Hidaka, Crossovers within a short DNA sequence indicate a long evolutionary history of *APRT*J* mutation. *Hum. Genet.* 85:600 (1990)

GENETIC AND CLINICAL HETEROGENEITY IN HYPOXANTHINE
PHOSPHORIBOSYLTRANSFERASE DEFICIENCIES

Renate Burgemeister[1], W. Gutensohn[1], G. Van den Berghe[2] and J. Jaeken[2]

[1]Institute of Anthropology and Human Genetics, University of Munich, Germany
[2]International Institute of Cellular and Molecular Pathology, Brussels, Belgium
[3]Department of Pediatrics, University of Leuven, Belgium

INTRODUCTION

Complete deficiency of HPRT causes the Lesch-Nyhan syndrome (LNS) which is characterized by hyperuricemia, mental retardation, choreoathetosis, and compulsive self-mutilation. Partial deficiency of HPRT leads to a severe form of gout and nephrolithiasis. In contrast to the Lesch-Nyhan syndrome it has been proposed to designate this as Kelley-Seegmiller syndrome, which by this definition should never include autoaggression. In single cases, however, it even seems difficult to make a clearcut decision between the two syndromes. Autoaggressive behaviour may either develop more slowly and with a later onset than in the classical Lesch-Nyhan syndrome and may thus escape an early postnatal diagnosis.

In the following we will describe mutations on the molecular level in four nonrelated families, including two classical Lesch-Nyhan patients and two different borderline cases as mentioned above.

MATERIALS AND METHODS

Sample Preparation And Enzyme Assays

Erythrocyte lysates were prepared and the HPRT enzyme assays were performed by previously published procedures (Burgemeister and Gutensohn 1989).

Amplification Of Genomic DNA

With genomic DNA as template exons 2 through 9 of the HPRT gene including more or less extended flanking intron sequences were amplified by PCR using the set of primers and the conditions published by Gibbs et al. (1990). For the amplification of exon 1 plus flanking sequences the following protocol had to be chosen: Forward primer as given by Gibbs et al. (1990), reverse primer corresponding to positions 1845 - 1827 of the genomic sequence (Edwards et al. 1990). The PCR was performed with 100 ng of genomic DNA, 250 ng of each of the primers, 2% dimethylformamide and 2 U of

Vent(exo-) polymerase (New England Biolabs). PCR amplified DNA was prepared for direct sequencing by isopropanol precipitation.

Single strand conformational polymorphism (SSCP)

This was done with the PCR amplified exons. For exon 2 the forward and the sequencing primers of Gibbs et al. (1990) were used resulting in a 379 bp fragment. For exon 3 and exon 7 foreward and reverse primers of Gibbs et al. (1990) were used. Restriction enzyme digestion with AluI for exon 3 leads to 5 fragments. The 345 bp fragment shows the mutation. Digestion of the exon 7 PCR product with restriction enzymes AvaII and SspI results in 2 fragments. The mutation is found in the 364 bp fragment.

SSCP analysis was performed by electrophoresis on vertical 15% polacrylamide gels. A 3 μl aliquot of amplified exons was mixed with 5 μl of formamide, denatured at 95°C for 5 min, and immediately cooled on ice for 3 min. For optimal resolution electrophoresis was run at 100 V for 28 hours. Subsequently the gels were silver-stained.

Sequencing

For direct sequencing of exons 2 through 9 the sequencing primers indicated by Gibbs et al. (1990) were used, for exon 1 a forward primer corresponding to positions 1634 - 1651 of the genomic sequence. The PCR amplified DNA was denatured at 95°C in the presence of the appropriate sequencing primer and immediately cooled in a dry ice bath.

RESULTS

In the patients IJ, KM and GS the sizes of the PCR amplified exons corresponded to those of normal controls excluding a major deletion or duplication as the underlying mutation. With the genomic DNA of patient P-JG it was only possible to amplify exons 1 and 2, which is an indication for a deletion of the gene which comprises exons 3 to 9.

The mutations in three of the patients were finally detected by sequencing and the results are shown in Fig. 1.

The mutation in family J. is a single base (G) insertion in exon 3 within a poly-G stretch in codons 68-70. The resulting frame shift leads to a stop codon two codons downstream and on the protein level to an early chain termination. The control panel for this family (Fig. 1) shows the sequence of patient IJ's sister. It is evident that she could be excluded as a carrier.

The mutation in patient GS is a point mutation in exon 7. In codon 169 a C to T transition converts an arginine into a stop codon resulting in a late chain termination during protein synthesis. This mutation is already described as HPRT$_{North Mymms}$ (Davidson et al. 1991; Tarlé et al. 1991). The description of the same mutation in still another family (Marcus et al. 1992) would suggest that we are dealing with a hot spot of mutation in this CpG sequence. In contrast to the report of Marcus et al. (1992) our patient GS (now at age 10) has not developed the typical autoaggressive behavior. So we have to consider some clinical heterogeneity even with the same mutation.

In family M. the A to T transversion in exon 2 converts codon 41 from ATT to TTT resulting in an exchange of isoleucine to phenylalanine on the protein level. This mutation is hitherto undescribed and we propose to designate it HPRT$_{Isar}$.

As can be seen in Fig. 1 the mothers of patients IJ and KM are clearly heterozygous, whereas the mother in family S. shows the normal pattern. The mutation in GS must have arisen de novo. In this three families no additional mutation was found in the other exons including the flanking intron sequences.

Single strand conformational polymorphism (SSCP) analysis was performed as a confirmation of the sequence data in families J., S. and M. and was performed prospectively in patient P-JG and his family. The results are summarized in Figs. 2+3. Resolution of the SSCP analysis was optimal in our hands when the DNA fragments introduced were 300 - 400 bp in length. When the amplified exons including flanking intron sequences were much longer than that, they were cut by appropriate restriction enyzmes. This results in somewhat more complex band patterns, however, Fig. 2 only shows the informative region of the gels. SSCP analysis of patient P-JG shows a normal band pattern for exon 1 (data not shown), a different pattern for exon 2 compared to control (Figure 3) and no bands for exons 3 to 9.

The mother, grandmother and the sister of his grandmother show only control bandpatterns for all nine exons. The sequence data of Fig. 1 are fully confirmed by the SSCP analysis in Fig. 2: A mutant pattern for the three patients, a heterozygous pattern for the mothers in families J. (exon 3) and M. (exon 2), and a control pattern for the mother of GS (exon 7).

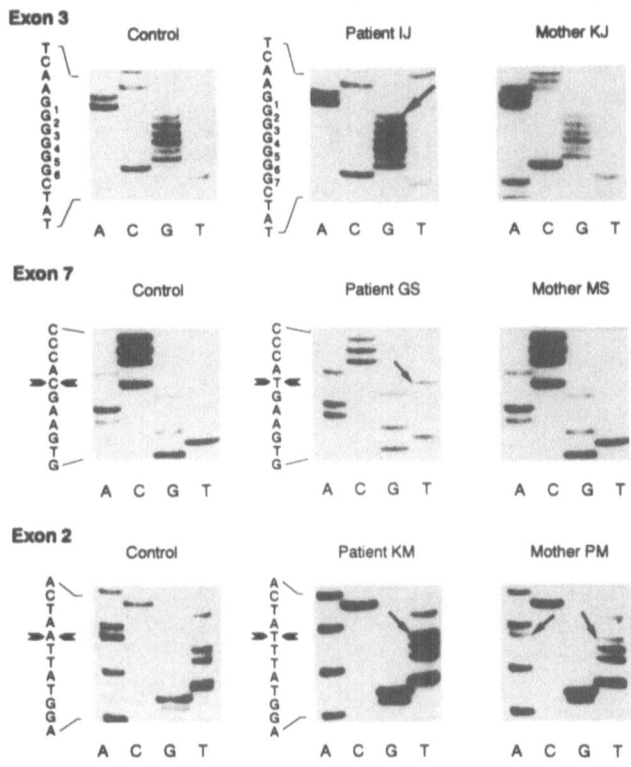

Figure 1: Sequencing gels

DISCUSSION

Four different mutations at the HPRT locus were detected in four unrelated patients with a complete or partial deficiency of the HPRT enzyme. Two of these mutations (the ones shown here in families J. and S.) have already been described in the literature. The insertion of a single G in an oligo stretch in exon 3 (as in family J.) has been interpreted as being due to replication slippage (Gibbs et al. 1989; Tarlé et al. 1991). The mutation in exon 7 (in family S.) is the typical methylation mediated C to T transition in a CpG sequence and has been designated as $\text{HPRT}_{\text{North Mymms}}$ (Davidson et al. 1991; Tarlé et al. 1991). Marcus et al. (1992) have described another family with this mutation where they observed nonrandom X-inactivation in several heterozygous females. The authors point to the importance of genomic amplification and sequencing for carrier detection in such a situation. Since our study is based exclusively on genomic amplification and sequencing, we are confident that the normal sequence seen in the mother of family S. is

Exon 3
       345 bp

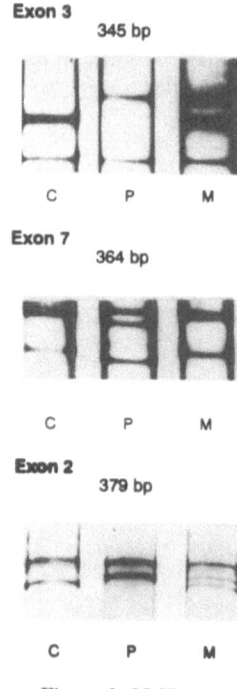

Exon 7
       364 bp

Exon 2
       379 bp

Figure 2: SSCP gels
C=control
P=patient
M=mother

Exon 2   **379 bp**

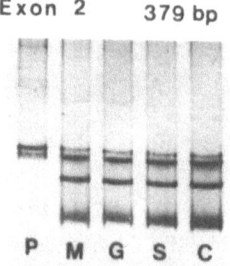

Figure 3: SSCP gel
C=control
P=patient
M=mother
G=grandmother
S= sister of G

334

not due to an experimental error, and that the mutation in GS is in fact spontaneous. Since both mutations (families J. and S.), especially the one in exon 7, have now been described in several unrelated families, it can be assumed that we are dealing with hot spots of mutation.

The cases studied here demonstrate, once again, the genetic and clinical heterogeneity of HPRT deficiencies. In addition there is reason to believe that even with the same mutation some discrepancies in the clinical expression might be observed. In the family reported by Marcus et al. (1992) with the Arg to Stop mutation in exon 7 the two index patients have developed selfmutilatory symptoms gradually from 3 years of age and now with age 18 and 16 resp. exhibit the complete form of the Lesch-Nyhan syndrome. Our patient GS with the same mutation was 7 years old when we first diagnosed his enzyme deficiency, and now with age 10 still does not bite.

Patient KM showed at age 4 years the onset of massive autoaggression with biting of lips. It is remarkable that inspite of a residual acitivity of HPRT (2.7% of normal) the patient has gradually developed the full clinical picture of the Lesch-Nyhan syndrome including selfmutilation.

ACKNOWLEDGEMENTS

The financial support of the following institutions is gratefully acknowledged: Bundesministerium für Forschung und Technologie (No. SVG 0706509); Deutsche Forschungsgemeinschaft (No. Gu123/17-1).

REFERENCES

Burgemeister R, Gutensohn W (1989), Normal ranges of the activities of ten different enzymes in 100 independent preparations of chorionic villi. Comparison of specimens from induced abortions, biopsies, and cultured cells, *Prenat Diagn* 9:195-204

Davidson BL, Tarlé SA, Palella TD, Kelley WN (1989), Molecular basis of hypoxanthine-guanine phosphoribosyl-transferase deficiency in ten subjects determined by direct sequencing of amplified transcripts, *J Clin Invest* 84:342-346

Davidson BL, Tarlé SA, van Antwerp M, Gibbs RA, Watts RWE, Kelley WN, Palella TD (1991), Identification of 17 independent mutations responsible for human hypoxanthine-guanine phosphoribosyltransferase (HPRT) deficiency, *Am J Hum Gen* 48:951-958

Edwards A, Voss H, Rice P, Civitello A, Stegemann J, Schwager C, Zimmermann J, Erfle H, Caskey CT, Ansorge W (1990), Automated DNA sequencing of the human HPRT locus, *Genomics* 6:593-608

Gibbs RA, Nguyen PN, McBride LJ, Koepf SM, Caskey TC (1989), Identification of mutations leading to the Lesch-Nyhan-syndrome by automated direct DNA sequencing of in vitro amplified cDNA, *Proc Natl Acad Sci (USA)* 86:1919-1923

Gibbs RA, Nguyen PN, Edwards A, Civitello AB, Caskey CT (1990), Multiplex DNA deletion detection and exon sequencing of the hypoxanthine phosphoribosyltransferase gene in Lesch-Nyhan families, *Genomics* 7:235-244

Marcus S, Steen AM, Andersson B, Lambert B, Kristoffersson U, Francke U (1992), Mutation analysis and prenatal diagnosis in a Lesch-Nyhan family showing non-random X-inactivation interfering with carrier detection tests, *Hum Genet* 89:395-400

Tarlé SA, Davidson BL, Wu VC, Zidar FJ, Seegmiller JE, Kelley WN, Palella TD (1991), Determination of the mutations responsible for the Lesch-Nyhan syndrome in 17 subjects, *Genomics* 10:499-501

MOLECULAR MECHANISMS OF THE SECOND FEMALE LESCH-NYHAN PATIENT

Yasukazu Yamada,[1] Haruko Goto,[1] Takeo Yukawa,[2] Hirofumi Akazawa,[2] and Nobuaki Ogasawara[1]

[1]Department of Genetics, Institute for Developmental Research,
 Aichi Prefectural Colony, Kasugai, Aichi 480-03, Japan
[2]Ehime Disabled Children's Hospital, Matsuyama, Ehime 790, Japan

INTRODUCTION

Lesch-Nyhan syndrome is an X linked recessive disorder characterized by hyperuricemia, physical and mental retardation, choreoathetosis, and compulsive self-mutilation.[1] This disease is associated with the absence of activity of an enzyme involved in purine metabolism, namely hypoxanthine guanine phosphoribosyltransferase (HPRT, EC 2.4.2.8).[2] HPRT locus is on long arm of X chromosome, q26 and because of inability of reproduction in Lesch-Nyhan patient, this disorder occurs only in males. We have, however, two unusual cases of female Lesch-Nyhan patients.

The first female patient reported in 1982,[3] has a normal karyotype, 46XX and mRNA⁻ phenotype.[4] The molecular mechanisms of the first female Lesch-Nyhan patient are (1) a total maternal HPRT gene deletion and (2) a gene inactivation due to the methylation at the active copy specific *Hpa* II site on the first intron of the paternal HPRT gene.[4,5] The second case of a female with the syndrome characterized by hyperuricemia, decreased HPRT activity and characteristic central nervous system symptoms has been reported in 1992.[6] In this study, we present the molecular mechanisms of the second female Lesch-Nyhan patient.

MATERIALS AND METHODS

Subject. The patient is now 11 years of age and has normal female external genitalia and karyotype, 46XX.[6] Her erythrocyte HPRT activity was about 5-10 % of the control and adenine phosphoribosyltransferase (APRT) activity was increased about 2-fold compared with the control value. The patient is mRNA⁺ phenotype. B-lymphoblastoid cell lines from the patient and her mother were established and maintained as described previously.[4]

Gene analysis. All the methods of HPRT gene analysis, identification of the genomic mutation and the altered mRNA, were described previously.[7] DNA sequences were determined according to the simplified direct sequencing method[7] and were recorded into a personal computer and analyzed with a software of gene analysis, GENETYX version 9.0 (SDC, Japan).

RESULTS AND DISCUSSION

The multiplex amplification from genomic DNA revealed no differences in product sizes between the patient and normal control. By direct sequencing of all nine amplified exons, a single nucleotide substitution of C to T on the exon 3 has been identified in one allele of the HPRT genes. Sequence analysis of cDNA synthesized by RT-PCR from B-lymphoblasts of the patient showed that the majority had the identified mutation, since the sequencing pattern was similar to the homozygote. The substitution results in the nonsense mutation, CGA (Arg) to TGA (stop) at codon 51, and should produce premature enzyme protein consisted of 50 amino acids (Figure 1). But, other mutation was not detected on all the region sequenced. The *Xho* I restriction site (CTCGAG) is lost in the mutant allele (CTTGAG) (Figure 1). DNA fragments including exon 3 were amplified from genomic DNAs of the patient and the mother, and the fragments were then digested with *Xho* I. The results of the analysis showed that the mother had only normal HPRT genes. Sequence analysis of the mother's DNA fragment were also demonstrated that the mother had not the identified mutation. Therefore, the identified mutation must be *de novo* mutation occurred in mother's or father's germ cell.

The identified mutation is in the CpG context and thus may be indicative of a hot spot for mutation in HPRT gene.[8] The same mutation had been reported previously in four male Lesch-Nyhan patients.[9-12] In both two Japanese cases, HPRT$_{Fujimi}$ and HPRT$_{Kanagawa}$, their respective mothers were normal, similar to this study, indicating *de novo* mutation of maternal HPRT gene. The data in this study and previous studies supported that the C at nucleotide position 151 was a hot spot for mutation in HPRT gene.

```
(A) genomic DNA

Normal
          (intron 2)*(exon 3)         Xho I
5'-ttttatttctgtag.GACTGAACGTCTTGCTCGAGATGTGATGAAG-3'
   ************* *************** **************
5'-ttttatttctgtag.GACTGAACGTCTTGCTCGAGATGTGATGAAG-3'
                                   T
Patient                       (C/T heterozygous)

(B) cDNA

Normal
 127 GAC.AGG.ACT.GAA.CGT.CTT.GCT.CGA.GAT.GTG.ATG 162
  44 Met-Asp-Arg-Thr-Glu-Arg-Leu-Arg-Asp-Val-Met  54
     *** *** *** *** *** *** ***
 127 GAC.AGG.ACT.GAA.CGT.CTT.GCT.TGA.GAT.GTG.ATG 162
  44 Met-Asp-Arg-Thr-Glu-Arg-Leu-stop             50
Patient
```

Figure 1. Direct sequencing analysis of genomic DNA (A) and cDNA (B) from the second female Lesch-Nyhan Patient.

Table 1. Digestion by *Xho* I of recombinant clones inserted HPRT cDNA fragments from the patient.

Clone	Number (%)
+ *Xho* I site	2 (8.3)
− *Xho* I site	22 (91.7)
Total	24 (100)

Direct sequence analysis and *Xho* I digestion analysis of cDNA showed that the majority had the identified mutation. However, a low activity of HPRT was detected in the patient's erythrocyte. cDNA fragments were cloned into pUC vector and then digested by *Xho* I, to test the existence of normal mRNA (Table 1). In 24 pUC recombinants inserted the HPRT cDNA, only 2 recombinants were digested by *Xho* I but 22 recombinants were not. Sequencing the entire coding region of 2 clones, which have not the identified mutation, showed no other mutation.

The cloning experiments demonstrated the existence of about 1/10 amount of cDNA with no mutation. The decreased activity of HPRT in erythrocyte (5-10 % of the control value) might be reflected by this small amount of normal mRNA. The existence of small amount of normal mRNA might be due to the regulatory defect, such as the decrease of expression and the enhancement of mRNA turnover, or predominant inactivation of normal X chromosome. The analyses of the 3'-flanking region of coding region were carried out, because the enhanced mRNA turnover was reported in the presence of mutation in this region. However, no change was found in this region. The analysis of the promoter region is now in progress.

CONCLUSION

The molecular mechanisms of the second female Lesch-Nyhan patient are (1) nonsense mutation at codon 51 on one of alleles, and (2) the decreased mRNA expression from the other allele, or predominant inactivation of X chromosome carrying normal HPRT gene.

ACKNOWLEDGMENTS

This work was supported by a Gout Research Foundation grant, an Intractable Disease grant from the Ministry of Health and Welfare of Japan, and a grant from the Ministry of Education, Culture and Science of Japan.

REFERENCES

1. M. Lesch, and W.L. Nyhan, A familial disorder of uric acid metabolism and central nervous system function., *Am. J. Med.* 36:561 (1964).
2. J.E. Seegmiller, F.M. Rosenbloom, and W.N. Kelley, Enzyme defect associated with a sex-linked human neurological disorder and excessive purine synthesis., *Science* 155:1682 (1967).
3. K. Hara, S. Kashiwamata, N. Ogasawara, H. Ohishi, R. Natsume, T. Yamanaka, S. Hakamada, S. Miyazaki, and K. Watanabe, A female case of the Lesch-Nyhan syndrome., *Tohoku J. Exp. Med.* 137:275 (1982).

4. N. Ogasawara, J.T. Stout, H. Goto, S. Sonta, A. Matsumoto, and C.T. Caskey, Molecular analysis of a female Lesch-Nyhan patient., *J. Clin. Invest.* 84:1024 (1989).

5. N. Ogasawara, Y. Yamada, and H. Goto, HPRT gene mutations in a female Lesch-Nyhan patient., *Adv. Exp. Med. Biol.* 309B:109 (1991).

6. T. Yukawa, H. Akazawa, Y. Miyake, Y. Takahashi, H. Nagao, and E. Takeda, A female patient with Lesch-Nyhan syndrome., *Developmental Med. Child Neurol.* 34:554 (1992).

7. Y. Yamada, H. Goto, K. Suzumori, R. Adachi, and N. Ogasawara, Molecular analysis of five independent Japanese mutant genes responsible for hypoxanthine guanine phosphoribosyltransferase (HPRT) deficiency., *Hum. Genet.* 90:379 (1992).

8. D.G. Sculley, P.A. Dawson, B.T. Emmerson, and R.B. Gordon, A review of the molecular basis of hypoxanthine-guanine phosphoribosyltransferase (HPRT) deficiency., *Hum. Genet.* 90:195 (1992).

9. S. Fujimori, N. Kamatani, Y. Nishida, N. Ogasawara, and I. Akaoka, Hypoxanthine guanine phosphoribosyltransferase deficiency: nucleotide substitution causing Lesch-Nyhan syndrome identified for the first time among Japanese. *Hum. Genet.* 84:483 (1990).

10. B.L. Davidson, S.A. Tarle, M. van Antwerp, D.A. Gibbs, R.W.E. Watts, W.N. Kelley, and T.D. Palella, Identification of 17 independent mutations responsible for human hypoxanthine-guanine phosphoribosyl transferase (HPRT) deficiency., *Am. J. Hum. Genet.* 48:951 (1991).

11. S.A. Tarle, B.L. Davidson, V.C. Wu, F.J. Zidar, J.E. Seegmiller, W.N. Kelley, and T.D. Palella Determination of the mutation responsible for the Lesch-Nyhan syndrome in 17 subjects., *Genomics* 10:499 (1991).

12. S. Fujimori, Genetic diagnosis and therapy of Lesch-Nyhan syndrome., *Purine and Pyrimidine Metabolism* 17:158 (1993).

CLINICAL SYMPTOMS OF PATIENTS WITH PARTIAL HPRT DEFICIENCY

Birgit S. Gathof, Daniela Jurgens, Ursula Gresser

Purine Research Laboratory
Medizinische Poliklinik
University of Munich
Pettenkoferstr. 8a
80336 München, Germany

INTRODUCTION

Since the first description of partial Hypoxanthine-Guanine-Phosphoribosyl-transferase (HPRT) deficiency in 1967 by Kelley, Seegmiller and coworkers a large number of cases has been described. The clinical picture has been quite variable, reaching from asymptomatic hyperuricemia to very severe disease with arthritis, nephrolithiasis and neurological symptoms. In contrast to classical Lesch-Nyhan syndrome selfmutilation is not observed in patients with partial HPRT deficiency.

In the past different authors have tried to correlate clinical symptoms and enzyme activities. This proved to be difficult due to too few cases reported (Greene, 1972) or different methods in determining enzyme activities (Page et al., 1988). This report gives a brief review on three patients with partial HPRT deficiency treated in our department. The second part contains preliminary results of our study (Gathof et al. in preparation) that summarizes the symptoms and clinical course of the patients with partial HPRT deficiency in relation to enzyme activities and results of molecular genetic studies.

PATIENTS

In the purine research unit three patients with partial HPRT deficiency have been studied and followed over a period of 3 to 30 years. The data on these patients are given in table 1. A detailed report on patient I.V. has been published previously (Kamilli et al., 1993). No neurological symptoms were observed in the three patients.

Table 1. Clinical data of three patients with partial HPRT deficiency treated in the purine research unit.

patient	I.V., *1931	R.Sch., *1962	C.S., *1963
first manifestation	17 y: tophaceous gout	8 y: nephrolithiasis	2.5 y: nephrolithiasis
course of the disease	gouty attacks, chronic gout and recurrent renal colics	11-29 y: recurrent renal colics	21-28 y: recurrent renal colics
course of therapy	since consequent treatment free from symptoms, impaired renal function remained stable, good compliance	free from symptoms with uric acid lowering therapy, compliance problems	after a period without symptoms recurrent episodes of nephrolithiasis due to dietary and compliance problems
age at diagnosis	47 y	29 y	28 y
enzyme acitivity	4-7 %	5-8%	2-4%
additional diseases	pheochromocytoma, erythrocytosis	-	-

METHODS

In addition to the case reports of the patients treated in our unit, we collected all data available on other patients with partial HPRT deficiency published.

All reports on patients with partial HPRT deficiency, and also patients without detectable HPRT activity in lysed erythrocytes who did not show self mutilation, were included in this study.

The different parameters noted were: age at diagnosis, at occurance of the different symptoms (arthritis, tophi, hyperuricemia, nephrolithiasis, pyelonephritis, acute renal failure, impaired creatinine clearence, hypertension, epilepsy, spasticity, mental retardation), enzyme activity compared to the normal values of the report, method of determination of enzyme activities, course of the disease.

We examined the relationship between age and first onset of the clinical symptoms. Also the correlation of enzyme activity and age at onset of the different clinical symptoms was evaluated.

RESULTS

A total of 95 case reports with partial HPRT deficiency was selected. Arthritis and/or tophi were described in 49 patients; in 65 signs of renal involvent, in 12 hypertension and in 15 neurological symptoms were described at any time during the course of disease reported.

The mean age at first occurance of the different symptoms is shown in figure 1. Neurological symptoms as well as hematuria and nephrolithiasis were reported in an earlier age than arthritis.

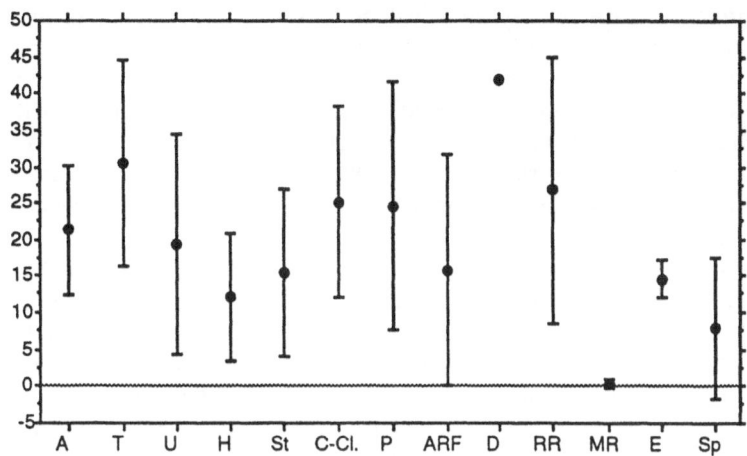

Figure 1. Mean age at onset of the different symptoms in 95 patients with partial HPRT deficiency reported in the literature. Abbreviations: A=arthritis, T=tophi, U=hyperuricemia, St=nephrolithiasis, H=hematuria, P=pyelonephritis, ARF=acute renal failure, D=dialysis, RR=hypertension, MR=mental retardation, E=epilepsy. The bars indicate the standard deviation.

The possible relation between the age at the onset of the different symptoms and the relative enzyme activity was evaluated. There was no correlation.

DISCUSSION

Different attempts have been made to subclassify patients with HPRT deficiency. One problem which occurs in comparing cases in the literature is the inhomogeneity of tests for HPRT activity. According to some authors only studies in intact cells give reliable results (Page and Nyhan 1988). Page and Nyhan proposed a classification according to enzyme activites in lysed erythrocytes which differentiates: classic Lesch-Nyhan syndrome with enzyme activity below 1.5%, neurologically affected patients with enzyme activity of up to 10% patients with the symptoms of hyperuricemia with 10 to 80% enzyme activity.

More than 80% of the cases reported have been examined using lysed erythrocytes. Almost every laboratory reported a different normal value. This allows comparison of relative values only.

In our patients with enzyme acitivities of 4 to 8% in lysed erythocytes only symptoms of hyperuricemia (gout in 1 of 3 and nephrolithiasis in 3 of 3) were observed.

In the literature study only one of the neurologically affected patients showed enzyme activites above 10%. There was a number of patients with enzyme activities below 1% who did not selfmutilate, and also many with enzyme activities below 10% without neurological symptoms. This is in good agreement with the observations in our patients.

According to Page and Nyhan (1988) the severity of symptoms of hyperuricemia was not dependent upon the level of enzyme activity. This is confirmed by the results of our literature research. It is also illustrated by several family studies where the propositus shows different symptoms of partial HPRT deficiency, whereas a brother with the same level of HPRT activity remains more or less asymptomatic.

Within the last years several studies were undertaken to determine a possible relationship between the site of the mutation and the severity of the clinical symptoms (Sage-Peterson et al. 1992, Scully et al. 1992). Generally large deletions seem to cause Lesch-Nyhan Syndrome, whereas point mutations especially those at the aminoacid terminal predominantly seem to cause partial deficiency. Other factors like diet, fluid intake, other genetic or environmental influences might be responsible for the different clinical picture caused by the same mutation in different members of the same family. This difference in clinical picture which is observed for the same mutations remains subject to further studies as well as the clinical follow up and search for the individual mutation in each new patient and his family.

Acknowledgement

We thank Prof. Dr. Martin Schreiber, Institute for Medical Statistics, University of Munich, for his advice.

REFERENCES

Greene M.L, 1972, Clinical features of patients with the "partial" deficiency of the X-linked uricaciduria enzyme, Arch Intern Med 130: 193-198

Kamilli, I., Gresser, U., Gathof, B., Gröbner, W., 1993, Partial HGPRT-deficiency, pheochromocytoma and erythrocytosis. J Inherit Metab Dis 16: 484-485

Kelley, W.N., Rosenbloom, F.M., Henderson J.F., Seegmiller, J.E., 1967, A specific enzyme defect in gout associated with overproduction of uric acid. Proc Natl Acad Sci 57: 1736-1739

Page, T., Nyhan W., 1988, The spectrum of HPRT deficiency: An update. Adv Exp Med Biol 253 A:129-133

Sage-Peterson, K., Chambers, J., Page, T., Jones, O.W., Nyhan, W.L., 1992, Characterization of mutations on phenotypic variants of hypoxanthine phosphoribosyltransferase deficieny, Hum Mol Gen 1: 427-432

Sculley, G.D., Dawson, P.A., Emmerson, B.T., Gordon, R.B., 1992, A review of the molecular basis of hypoxanthine/guanine phosphoribosyltransferase (HPRT) deficiency, Hum Genet 90: 195-207

NORMAL HYPOXANTHINE AND AMMONIA RELEASE FROM WORKING MUSCLE IN PARTIAL HPRT DEFICIENCY

Manfred Gross, Birgit S. Gathof, Ursula Gresser

Medizinische Poliklinik
University of München
Pettenkoferstraße 8a
80336 München, Germany

INTRODUCTION

Partial deficiency of HPRT (EC 2.4.2.8.) is one of the enzyme defects causing hyperuricemia and gout. The most obvious explanation for this phenomenon is the reduced activity of the salvage pathway with increased degradation of hypoxanthine and xanthine into uric acid. Since the salvage reactions consume PRPP, HPRT deficiency results in an accumulation of PRPP which further activates the purine nucleotide synthesis de novo and ultimately increases uric acid formation.

A recent study by Hisatome et al.[1] in which two related patients were exercised using a semi-ischemic forearm test reported evidence for the excess release of hypoxanthine from exercising muscle in patients with partial HPRT deficiency. In these patients, the increase in hypoxanthine level was much higher and the increase in ammonia level lower than that found in control subjects. Since the increase in ammonia level was lower than that found in control subjects, the increased hypoxanthine levels were attributed to a reduced activity of the salvage pathway due to HPRT deficiency and not to an increased degradation of purine nucleotides. This increased exercise-induced hypoxanthine formation was considered to contribute to the hyperuricemia in HPRT deficiency.

These findings, however, are surprising since in working skeletal muscle, most of the IMP produced during exercise is converted into adenylosuccinate and not degraded into inosine and hypoxanthine because of the high activity of adenylosuccinate synthetase[5] and low activities of 5´-nucleotidases and purine nucleoside phosphorylase[6,7]. The vast majority of the adenine nucleotides degraded during vigorous exercise is therefore converted into the metabolites of the purine nucleotide cycle (AMP, IMP and adenylosuccinate)[2]. Only a small portion of degraded adenine nucleotide is converted into inosine and hypoxanthine which is released from working muscle[3,4]. Furthermore, the activity of the HPRT is rather low in skeletal muscle[7]. Therefore, a partial defect in HPRT activity does not seem likely to affect the hypoxanthine release from working muscle in a significant way.

Aerobic or semi-ischemic exercise tests of the forearm as used by Hisatome et al.[1] result in much lower formations of lactate, ammonia and hypoxanthine production in comparison to the results under nonischemic conditions[8]. More reliable results can be expected from ischemic exercise tests resulting in both a stronger activation of glycolysis and a more pronounced degradation of purine nucleotides.

METHODS

An ischemic exercise test was performed on the dominant arm of patients with partial HPRT deficiency and control subjects. Prior to exercise, the subjects rested for one hour. Baseline blood samples were drawn. Ischemia was produced by inflating a sphygmomanometer cuff on the proximal arm to 200 mm Hg (all subjects had normal blood pressure). During ischemia, the subjects squeezed another sphygmomanometer cuff with maximum force at a rate of 1 grip every 2 seconds until exhaustion occurred (after 1 - 2 minutes). Blood samples were drawn immediately after exercise and the inflated cuff was released afterwards. Additional blood samples were taken from the patients at several points in time up to 20 minutes after exercise. In control subjects, no blood samples were taken later than 10 minutes after exercise since significant changes were not observed in the patients after 10 minutes.

The control subjects were healthy men of average size and normal weight, with ages of 34 (M.R.), 41 (H.L.), 55 (H.C.) and 64 (E.G.) years. The three male patients with partial HPRT deficiency were also of normal size and weight, age 30 years (C.S.), 62 years (I.V.), and 31 years (R.Sch.). They had been treated with 300 mg allopurinol daily for a number of years. The exercise tests were performed in the early afternoon, 6-8 hours after both the last intake of allopurinol and breakfast.

All blood samples were analyzed for lactate (enzymatic reaction, Sigma, procedure no. 826-UV), ammonia (Kodak ektachem 500 analyzer system), and hypoxanthine (HPLC with a diode array detection system, Beckman System Gold).

RESULTS

In control subjects, the ischemic exercise test resulted in an increase in plasma hypoxanthine levels between 4.4 - 30.0 µmol/l (mean 19.8 µmol/l). There is a correlation between the increase in hypoxanthine level and both lactate levels and ammonia levels (Table 1). Subject M.R., with the smallest increase in hypoxanthine plasma level, had both the smallest increase in lactate level and in ammonia level, indicating that he had not exercised vigorously. The mean initial plasma level in hypoxanthine in control subjects was 1.7 µmol/l (range 1.5 - 1.9 µmol/l).

In contrast, the initial hypoxanthine plasma levels were much higher in patients (mean 7.7 µmol/l, range 4.2 - 10.5 µmol/l) than in controls. The mean exercise-induced increase in hypoxanthine plasma level was 24.7 µmol/l (range 22.4 - 28.5 µmol/l), the data of all patients were in normal range. The time course of hypoxanthine plasma level for all subjects is illustrated in Figure 1.

The increase in the serum lactate level (mean 3.8 mmol/l) and in the ammonia level (mean 122 µmol/l) in patients was within the normal range except for a high ammonia increase in patient R.Sch. (Table I). The ratio of the increase in ammonia and lactate levels was within the normal range (> 0.007) for both patients and control subjects.

Table 1. Hypoxanthine plasma level prior to exercise, exercise-induced maximum increases in plasma or serum levels of hypoxanthine, lactate and ammonia, and ratio of ammonia and lactate increase in 3 patients with partial HPRT deficiency and 4 healthy control subjects.

subject	Hypoxanthine initial plasma level [µM]	Hypoxanthine increase [µM]	Lactate increase [mM]	Ammonia increase [µM]	Ratio of ammonia and lactate increase
Patients with partial HPRT deficiency					
C.S.	4.2	28.5	3.8	77	0.020
I.V.	10.5	22.4	3.1	45	0.015
R.Sch.	8.3	23.3	4.6	245	0.053
Control subjects					
H.C.	1.5	14.8	3.9	63	0.016
E.G.	1.6	30.0	4.7	117	0.025
H.L.	1.9	29.8	4.4	58	0.013
M.R.	1.6	4.4	3.7	37	0.010

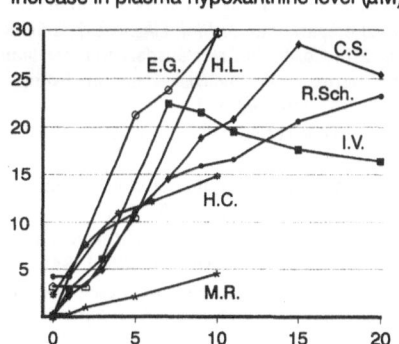

Increase in plasma hypoxanthine level (µM)

Figure 1. Time course of hypoxanthine plasma level in 3 patients with partial HPRT deficiency and 4 healthy control subjects.

DISCUSSION

The ischemic exercise test used in this study resulted in a much higher increase in hypoxanthine plasma levels than did the semi-ischemic protocol as used by Hisatome et al.[1]. Based on the greater production of hypoxanthine, effects of the partial HPRT deficiency are expected to be more pronounced in our experiment.

The initial hypoxanthine plasma level was more than 4 times higher in the patients. It is known that patients with partial HPRT deficiency have higher hypoxanthine levels in their body fluids because of higher de novo synthesis of purine nucleotides[10,11]. Furthermore, the patients were treated with 300 mg of allopurinol daily which is also known to increase hypoxanthine plasma levels as a result of the reduced conversion into xanthine and uric acid[12].

The increase in hypoxanthine plasma levels in three partial HPRT deficient patients was within the range of data obtained in healthy control subjects. These findings support our hypothesis that the exercise-induced hypoxanthine production in skeletal muscle is not different between these patients and the control subjects and is not a mechanisms accounting for hyperuricemia and gout found in these patients.

Acknowledgments: We are grateful to Prof. Dr. Roscher, Haunersche Kinderklinik, University of Munich, for providing the ammonia levels.

REFERENCES

1. I. Hisatome I, H. Kitamura H, M. Saito et al., Excess release of hypoxanthine from exercising muscle in two gout patients with partial HGPRTase deficiency: Lack of ammonium release. *Am. J. Med.* 90:533-535 (1991).
2. J.M. Lowenstein, Ammonia production in muscle and other tissues: the purine nucleotide cycle. *Physiol. Review* 52:382-414 (1972).
3. M.N. Goodman, J.M. Lowenstein, The purine nucleotide cycle. Studies of ammonia production by skeletal muscle in situ and in perfused preparations. *J. Biol. Chem.* 252:5054-5060 (1977).
4. R.A. Harkness, R.J. Simmonds, S.B. Coade, Purine transport and metabolism in man: the effect of exercise on concentrations of purine bases, nucleosides and nucleotides in plasma, urine, leucocytes and erythrocytes. *Clin. Sci.* 64:333-340 (1983).
5. V. Schultz, J.M. Lowenstein, Purine nucleotide cycle. Evidence for the occurrence of the cycle in brain. *J. Biol. Chem.* 251:485-492 (1976).
6. G. Schopf, M. Havel, R. Fasol, M.M. Müller, Enzyme activities of purine catabolism and salvage in human muscle tissue. *Adv. Exp. Med. Biol.* 195B:507-509 (1986).
7. Jacobs AEM, Oosterhof A, Veerkamp JH. Purine and pyrimidine metabolism in human muscle and cultured muscle cells. *Biochim. Biophys. Acta* 970:130-136 (1988).
8. V.H. Patterson, K.K. Kaiser, M.H. Brooke MH, Exercising muscle does not produce hypoxanthine in adenylate deaminase deficiency. *Neurology* 33:784-786 (1983).
9. W.N. Fishbein, Human myoadenylate deaminase deficiency. *Adv. Exp. Med. Biol.* 165A:77-84 (1984).
10. J.G. Puig, M.L. Jiménez, F.A. Mateos, I.H. Fox, Adenine nucleotide turnover in hypoxanthine-guanine phosphoribosyl-transferase deficiency: evidence for an increased contribution of purine biosynthesis de novo. *Metabolism* 38:410-418 (1989).
11. J.G. Puig, F.A. Mateos, M.L. Jiménez, T.H. Ramos, Renal excretion of hypoxanthine and xanthine in primary gout. *Am. J. Med.* 85:533-537 (1988).
12. C. Lartigue-Mattei, J.L. Chabard, J.M. Ristori et al., Kinetics of allopurinol and its metabolite oxypurinol after oral administration of allopurinol alone or associated with benzbromarone in man. Simultaneous assay of hypoxanthine and xanthine by gas chromatography-mass spectrometry. *Fundam. Clin. Pharmacol.* 5:621-633 (1991).

ALTERED PYRIDINE METABOLISM IN THE ERYTHROCYTES OF A MENTALLY RETARDED INFANT WITH PARTIAL HPRT DEFICIENCY

V.Micheli, M.Pescaglini, M.Rocchigiani, S.Sestini, G.Jacomelli, G.Hayek*, G. Pompucci

Dip. Biologia Molecolare - Università di Siena
*Rip. Neuropsichiatria infantile, USL 30 - Siena, Italia

INTRODUCTION

Alterations in the erythrocyte NAD(P) concentration have been reported in inherited defects of purine metabolism, such as PNP and HPRT deficiency, and phosphoribosylpyrophosphate synthetase (PRPS) superactivity[1]. The neurological disturbances associated with these disorders, as well as the utilization of nicotinate (NA) or nicotinamide (NAm) for the treatment of psychotic states, schizophrenia and depression[2] and the findings on the role of NAD in the synaptic modulation[3], suggested a correlations between neurological disorders and pyridine nucleotides.

Present study provides further informations on pyridine nucleotide involvement in HPRT deficiency, confirming our previous findings in this disorder[4].

PATIENTS AND METHODS

The propositus (P) was a mentally retarded child, aged 18 months, presenting with severe mental retardation and recurrent convulsive attacks. His brother (B), aged 8 years, presented with autistic trait, ataxia and recurrent convulsive attacks. Both children were treated with antiepilectic drugs; they both bore renal cysts. 17 age-matched children with no metabolic disorders were studied as controls. Blood samples were obtained as part of the patients' normal treatment programme. Washed erythrocytes and plasma were isolated by centrifugation of fresh heparinized blood. An unique RP-HPLC-linked method, coupled or not with radiodetection, was employed, using a System Gold apparatus (Beckman, Berkeley, CA) for all the determinations, as described elsewhere[5]. Nucleotide concentration was measured in erythrocyte perchloric extracts[5]. The incorporation rate of precursors into the respective pyridine or purine nucleotides was determined by incubation of erythrocytes in a PRPP generating medium containing 0.2 mM [^{14}C]NA (plus glutamine) or [^{14}C]NAm (6h at 37°C) or [^{14}C]adenine (Ade) or [^{14}C]hypoxanthine (Hyp) (1h at 37°C)[6]; perchloric extracts of the incubation suspensions were analysed by HPLC. The activity of the following purine and pyridine enzymes was measured in erythrocyte lysates: HPRT, APRT, ADA, PNP, PRPS; NA-phosphoribosyltransferase (NAPRT), NA- and NAm-mononucleotide adenylyltransferases (NAMN-AT and NMN-AT). Nucleoside and base content was measured in plasma perchloric extracts.

Purine and Pyrimidine Metabolism in Man VIII, Edited by
A. Sahota and M. Taylor, Plenum Press, New York, 1995

RESULTS

Nucleotide concentrations in red blood cell extracts (Tab.1) were within the normal range except NAD, raised to twofold normal in the propositus, and slightly above controls in his brother. Plasma uric acid concentration was within normal range for both (165 and 197 µM, controls ranging 246±62). HPRT activity in the propositus' erythrocyte lysate was very low, about 17% of controls, while APRT was twofold normal, and NAPRT and NMN-AT activities were much higher than controls. All the enzyme activities tested were within the normal range in the brother (Tab. 2).

Table 1- Erythrocyte nucleotide concentration (µM ± S.D.)

	IMP	AMP	ADP	ATP	GDP	GTP	NAD	NADP	UDP-S
controls	4±	14±	159±	1226±	16±	78±	53±	38±	32±
	3	7	34	164	6	25	12	8	22
P	3	8	111	1348	15	103	134	34	28
B	2	6	126	1325	10	78	72	46	10

Table 2- Enzyme activities in erythrocyte lysates (nmol/h · mgHb^{-1})

	HPRT	APRT	ADA	PNP	PRPS	NAPRT	NMN-AT	NAMN-AT
controls	136±	22±	91±	4523±	46±	0.64±	0.37±	0.36±
	27	5	21	758	6	0.33	0.11	0.10
P	24	52	106	5254	35	1.94	0.55	0.31
B	112	19	138	4871	35	0.99	0.38	0.28

In the propositus' intact erythrocytes , the conversion of labeled hypoxanthine into IMP was about 30% of controls, while that of adenine into its nucleotides was higher than normal; nucleotide production from both NA and NAm was higher than controls. In the brother, NA conversion into nucleotides was also increased, though at a lesser extent (Tab. 3).

Table 3- [^{14}C] Labeled precursors (0,2mM) incorporation into the respective nucleotides.

| | nmol/h/mlRBC | | nmol/6h/mlRBC | |
	ADENINE	HYPOXANTHINE	NICOTINATE	NICOTINAMIDE
controls	170± 34	154± 32	329± 54	11± 4
P	235	52	515	31
B	169	215	468	13

HPRT activity measured in the erythrocyte lysates of the children's mother was in the lower range of controls (data not shown). Studies in progress on the sequence of genomic DNA are revealing a point mutation leading to the substitution of a single aminoacid in the propositus[7].

CONCLUSIONS

Two mentally retarded brothers have been examined. The younger child (propositus) showed very low HPRT and increased APRT, NAPRT and NMN-AT activities in erythrocyte lysates, low Hyp conversion and increased Ade, NA and NAm conversion into the respective nucleotides by intact erythrocytes, and increased NAD erythrocyte concentration. The elder brother did not show any alteration of purine metabolism, but slightly higher NAD concentration and increased NA conversion into nucleotides were found. Plasma uric acid concentration was normal, also in the HPRT deficient child, in agreement with other reports on HPRT deficient children of the same age[8].

Present data in the HPRT deficient subject confirm the hypothesis of enhanced NAD synthesis in this disorder, supported by the increase of lysate NAPRT activity, as previously reported[9], of NMN-AT activity, and by the raised pyridine nucleotide production from radiolabeled precursors by intact erythrocytes. Both the deamidated and the amidated pathways of NAD synthesis seem to be involved. Neurological disorders are an unusual finding in partial HPRT deficiency[10]. The neurological manifestations in the two brothers examined have no apparent connections, except in some NAD related abnormalities. We have recently reported alterations in pyridine nucleotide synthesis in some subjects affected by Rett syndrome, and increased NAPRT activity in three more mentally retarded children[11]. The contribution of alterations of pyridine metabolism to the clinical features is not known, nevertheless our data suggest possible association with neurological and behavioral impairments; the relationship between the children's parents (first cousins) might suggest unknown inherited factors.

ACKNOWLEDGEMENTS

This study was supported by the Italian M.U.R.S.T (60% and 40% funds).

REFERENCES

1. H.A.Simmonds, D.R.Webster, J.Wilson, S.Lingham. An X-linked syndrome characterized by hyperuricaemia, deafness and neurodevelopmental abnormalities. Lancet 2:68 (1982).
2. V.Micheli, H.A.Simmonds, C.Ricci. Regulation of NAD synthesis in erythrocytes of patients with hypoxanthine-guanine phosphoribosyltransferase deficiency and a patient with phosphoribosylpyrophosphate synthetase superactivity. Clin. Sci. 78:239 (1990).
3. W.Blom, G.B.Van den Berg, J.G.M Huijmans, J.A.R.Sanders-Woudstra, Successful nicotinamide treatment in an autosomal dominant behavioral and psychiatric disorder, J Inter Metab Dis. 8, 2:107 (1985).
4. C.H.V.Hoyle, Pharmacological activity of adenine dinucleotides in the periphery: possible receptor classes and transmitter function. Gen Pharmac. 21:827 (1990).
5. V.Micheli, H.A.Simmonds, M.Bari, G.Pompucci. HPLC determination of oxidized and reduced pyridine coenzymes in human erythrocytes. Clin Chim Acta.220:1 (1993.)
6. V.Micheli, S.Sestini, C.Ricci. Purine and pyridine nucleotide production in human erythrocytes. Arch Biochem Biophys. 244:454 (1986).
7. M.Rocchigiani, paper in preparation
8. H.A.Simmonds, Purine and pyrimidine disorders, in: "The inherited metabolic diseases", J.B.Holton, ed. Churchill Livingstone, Edimburgh (1987)
9. M.Pescaglini, V.Micheli, H.A.Simmonds, M.Rocchigiani, G.Pompucci. Nicotinic acid phosphoribosyltransferase activity in human erythrocytes: studies using a new HPLC-linked method. Clin. Chim. Acta, in press.
10. I.Kamilli and U.Gresser, The clinical aspects of HGPRT deficiency, in "Molecular genetics,, biochemistry and clinical aspects of inherited disorders of purine and pyrimidine metabolism" , U. Gresser Ed., Springer Verlag, Berlin Heidelberg (1993).
11. V.Micheli, M.Pescaglini, M.Rocchigiani, S.Sestini, G.Jacomelli, G. Hayek*, G. Pompucci. Alterations of pyridine metabolism in the erythrocytes of children with neurological disorders. IBST.4:105 (1993)

TREATMENT OF LESCH-NYHAN SYNDROME WITH AICAR

T. Page, B. Barshop, A.L. Yu, and W.L. Nyhan

Department of Pediatrics, 0609
University of California, San Diego
La Jolla, CA 92093

Lesch-Nyhan syndrome is neurological disorder caused by the complete deficiency of the purine salvage enzyme hypoxanthine phosphoribosyltransferase (HPRT) (1). Patients display spasticity, choreoathetosis, and aggressive, self-injurious behavior; megaloblastic anemia is also commonly associated with these patients. The connection between the enzyme defect and the symptoms is unknown, but the role of HPRT in purine production suggests a possible purine nucleotide deficiency. Metabolic therapies designed to correct this supposed deficiency, including adenine, ribose, inosine, and guanosine, have not been successful (1); however, none of these compounds has been demonstrated _in vitro_ to be capable of replacing hypoxanthine as a source of purine nucleotides. The compound 5-aminoimidazole-4-carboxamide ribonucleoside (AICAR) has been shown to produce quantities of adenine and guanine nucleotides in HPRT$^-$ cells equal to those produced by hypoxanthine in normal cells (2). AICAR is therefore theoretically capable of compensating cells for the lack of HPRT

Based on this rationale, treatment of Lesch-Nyhan patient with AICAR was undertaken. Patient MS was a 16 year old classic Lesch-Nyhan patient with megaloblastic anemia (MCV 110) and megaloblastic bone marrow. He was given 30 mg/kg/day AICAR orally for four days followed by 100 mg/kg/day for four days, along with 400 mg/day allopurinol. AICAR, uric acid, and oxypurines were determined in the plasma and urine as described previously (3). No improvement was noted in his neurological, behavioral, or hematological status. Plasma AICAR remained <2 uM, and no changes in plasma uric acid or oxypurines were noted. Nor was any AICAR detected in the urine; urinary uric acid and oxypurines were also unchanged. These results are in agreement with the manufacturer's estimate of <5% oral bioavailability of AICAR in humans.

Patient CW was a 29 year old man with classic Lesch-Nyhan syndrome, megaloblastic anemia (MCV 102) and megaloblastic bone marrow. He received daily infusions of AICAR as shown in Fig. 1.

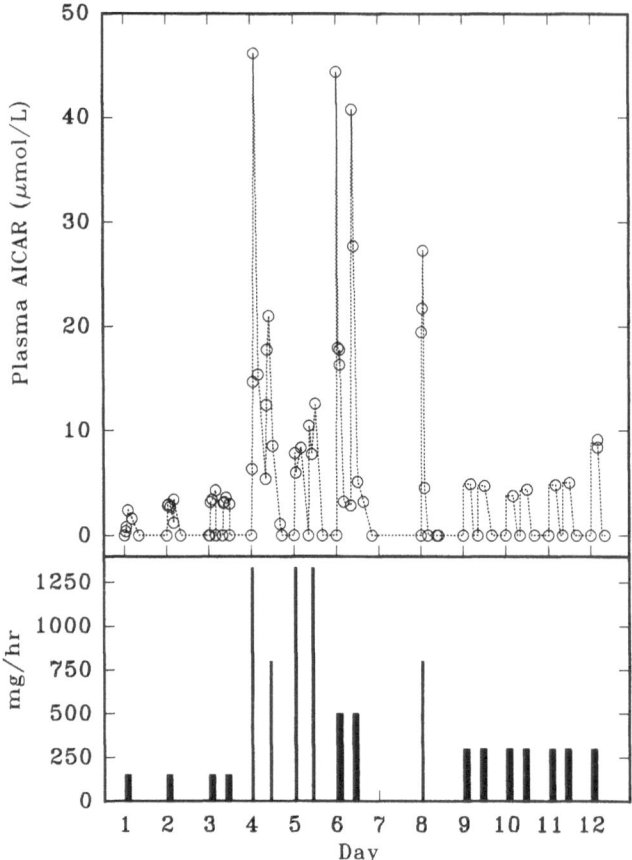

Figure 1. Plasma concentration of AICAR (upper panel) and AICAR infusion (lower panel). The width of the bars in the lower panel is proportional to the length of infusion.

Plasma and urinary AICAR, uric acid, and oxypurines were monitored as before. During the treatment regimen CW also received 400 mg/day allopurinol. Plasma AICAR varied between 0 and 50 uM, as shown in Fig. 1. Except for peaks and troughs corresponding to AICAR infusion and allopurinol ingestion, plasma uric acid, xanthine, and hypoxanthine were relatively unchanged, as seen in Fig 2. Total purine excretion, shown in Fig 3, increased only moderately during the course of therapy. A spinal tap was performed before the start of therapy, and again on day 12 immediately following infusion.

Despite the maintenance of a relatively high plasma level of AICAR, no improvement was noted in the patient's behavioral, neurological, or hematological status, although increased

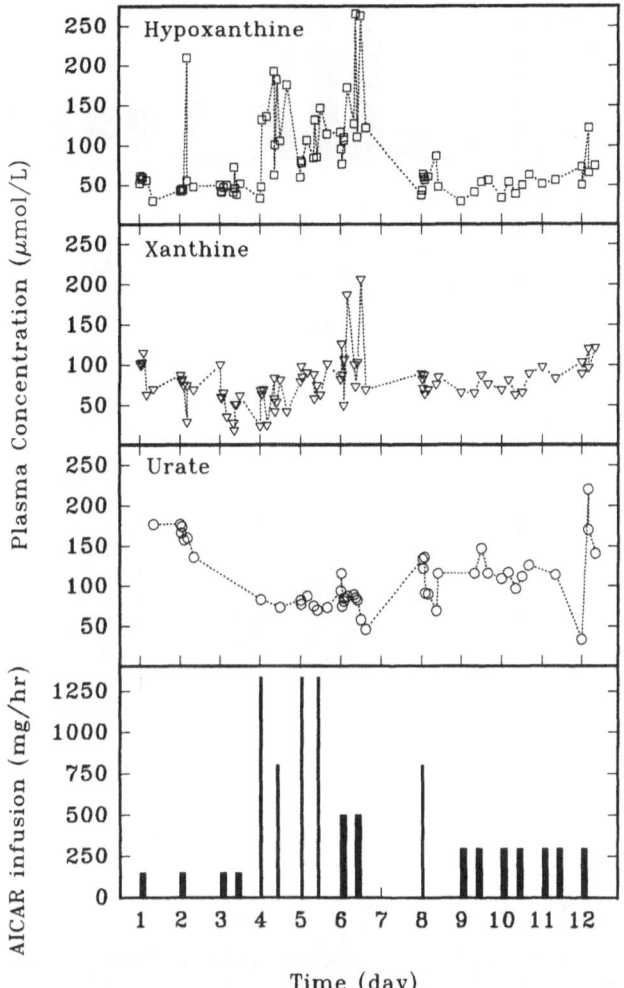

Figure 2. Plasma concentration of hypoxanthine (top panel), xanthine (second panel), and uric acid (third panel) during AICAR infusion (lower panel).

somnolence was noted. Significantly, no AICAR was detected in CSF before or after therapy; since the tap was performed immediately following the infusion, this suggests poor entry of AICAR into the CSF rather than rapid clearance. The relatively low dose of AICAR was chosen to prevent possible formation of stones, but plasma oxypurines remained largely unchanged. Total purine excretion increased only slightly, indicating that the amount of AICAR infused was small compared to endogenous purine

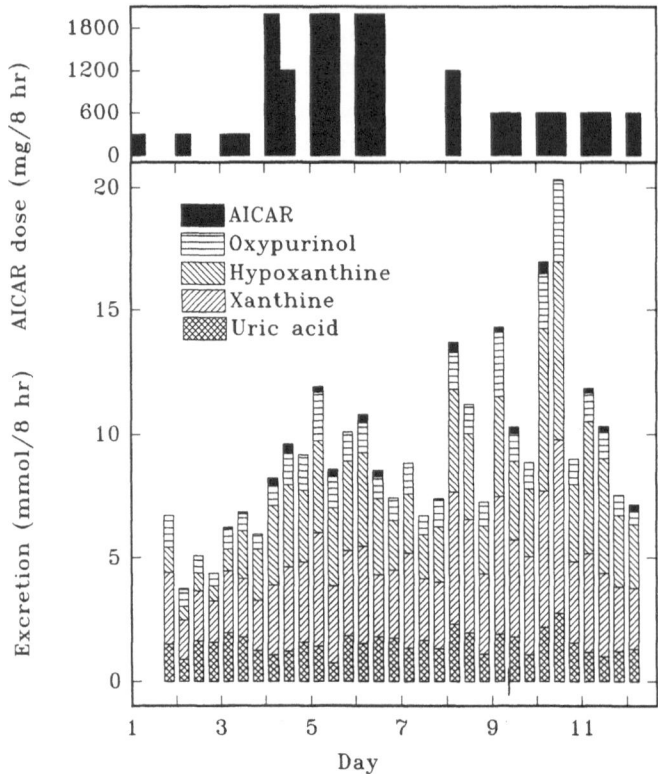

Figure 3. Purine excretion during AICAR infusion. AICAR infusion is shown in the upper panel, purine excretion in the lower panel.

production. These findings suggest that a larger dose consisting of longer infusions could have been employed safely.

REFERENCES

1. Kelley, W.N., and Wyngaarden, J.B., 1983, Clinical syndromes associated with hypoxanthine-guanine phosphoribosyltransferase deficiency, in, "The Metabolic Basis of Inherited Disease," J.B. Stanbury, J.B. Wyngaarden, D.S. Fredrickson, et al (eds), McGraw-Hill, New York.

2. Page, T., 1989, Purine nucleotide production in normal and HPRT$^-$ cells, Int. J. Biochem. 21:1377.

3. Page, T. Bakay, B., Nissinen, E., and Nyhan, W.L., 1981. Hypoxanthine-guanine phosphoribosyltransferase variants: correlation of clinical phenotype with enzyme activity. J. Inher. Metab. Dis. 4:203.

PLASMA AND URINARY OXYPURINES IN LESCH-NYHAN PATIENT AFTER ALLOPURINOL TREATMENT

G.Roscioni,[1] M.A.Farnetani,[1] R.Pagani,[2] M.Pizzichini,[2] E.Marinello,[2] and B.Porcelli [2]

[1]Institute of Pediatrics
[2]Institute of Biochemistry and Enzymology
 University of Siena, Italy

INTRODUCTION

In 1967, Seegmiller, Rosenbloom and Kelley described a virtually complete deficiency of an enzyme of purine metabolism, hypoxanthine-guanine phosphoribosyltransferase (HGPRT), in erythrocyte lysates from three children and in cultured skin fibroblasts from a fourth patient with the Lesch-Nyhan syndrome[1]. This disorder is characterized by choreoathetosis, striking growth and mental retardation, spasticity, self-mutilation, and marked hyperuricemia, with excessive uric acid production and uric acid crystalluria[2]. The enzyme defect was subsequently confirmed in other tissues, as well as in cultured skin fibroblasts and erythrocytes from many similarly affected subjects[3].

In addition, adolescent and adult patients have now been described with a "partial" rather than a "complete" deficiency of HGPRT[4,5]. These patients usually present with uric acid calculi or gouty arthritis and do not have the devastating neurological and behavioral features characteristic of children with the complete enzyme defect[4].

Hypoxanthine-guanine phosphoribosyltransferase activity

Hypoxanthine-guanine phosphoribosyltransferase (HGPRT) catalyzes the conversion of hypoxanthine to inosinic acid and guanine to guanylic acid in the presence of phosphoribosylpyrophosphate[6]. The natural purine base xanthine, as well as several purine analogs including 6-mercaptopurine, allopurinol, 8-azaguanine, and 6-thioguanine, are also substrates[7].

Patients with the Lesch-Nyhan syndrome have levels of HGPRT activity ranging from undetectable to 5 percent of normal[1,3,8].

Patients with the partial enzyme defect show levels of HGPRT activity in erythrocytes ranging from 0.01 to nearly 30 percent of normal[5]. Allopurinol is widely employed in the therapy of the disease, showing striking effects.

Address for correspondence: Prof. Enrico Marinello, Istituto di Biochimica e di Enzimologia, Università di Siena, Pian dei Mantellini, 44, 53100 SIENA Italy, Tel. +39-577-298026, FAX +39-577-298057

The effects of allopurinol in patients lacking HGPRT activity differ in several distinctive ways from its effects in normal subjects: (a) it does not have its usual inhibitory effect on purine biosynthesis de novo[9]; (b) it does not reduce erythrocyte PP-ribose-P concentrations[10]; (c) a normally occurring ribonucleoside derivative of allopurinol does not appear in the urine[11]; and (d) the drug produces an even more striking inhibitory effect on xanthine oxidase[4]. All these observations can be attributed to the deficiency in HGPRT activity.

In most patients with normal HGPRT activity the decrease in uric acid excretion produced by allopurinol is not accompanied by a stechiometric increase in excretion of the immediate precursors of uric acid, hypoxanthine and xanthine.

The failure of allopurinol to reduce purine synthesis de novo in HGPRT-deficient patients results in an increase in excretion of hypoxanthine and xanthine which roughly equals the decrease in uric acid excretion. Although hypoxanthine usually predominates in the urine under these conditions, the excretion of xanthine can be high enough to exceed solubility levels.

Here we report a case of Lesch-Nyhan syndrome, in which allopurinol also seems to have an effect on renal clearance of oxypurines and uric acid.

MATERIALS AND METHODS

Clinical aspects of the patient

The patient we have considered is a 7 year-old child; he is the second child of non-consanguineous parents; his older brother is healthy. Family history regarding hyperuricemia, gout, kidney or neurologic disorders was negative.

The patient was born at term after an uneventful pregnancy and normal delivery. Birth weight was 4.350 kg. He was fine until the fourth month of age, when he developed renal insufficiency.

At 6 months of age, because of progressive serum uric acid increase, allopurinol therapy was started in a dosage of 4 mg/kg/day plus urine alkalinization with oral sodium bicarbonate.

HGPRT activity was completely absent in erythrocytes and fibroblasts.

When admitted to our hospital physical examination revealed mild generalized hypotonia, normal deep tendon reflexes, mild neuromotor retardation, but afterwards the patient showed a progressive psychomotor injury with piramidal and extrapiramidal signs and self mutilation.

The dose of allopurinol was progressively increased up to 11.5 mg/kg/day and a normalization of serum and urine uric acid was observed.

At 5 years of age, the child presented complex partial seizures with secondary generalization and phenobarbital therapy was started. In this period the patient eliminated brick red kidney stones consisting of uric acid and xanthine.

In this last two years, period in which we have analyzed the effect of allopurinol on patient, he was on allopurinol for two months receiving either 200 mg/day, then 150 mg/day, and, subsequently, 100 mg/day.

Biochemical determinations

1) HGPRT was determined by the method of Cartier et al.[12].

2) Oxypurines and uric acid were determined in the blood by HPLC; elution was carried out with KH_2PO_4 buffer 0.01M, pH 5.5, with a linear gradient of 0-25% methanol for 30 min, at a flow rate of 1 ml/min.

3) Oxypurines in urine were determined always by HPLC, with an isocratic elution carried out with KH_2PO_4 buffer 0.01M, pH 5.5, at a flow rate of 1 ml/min.

4) Urinary uric acid was determined with the Boehringer Biochemie kit, according to Fossati et al.[13].

5) Urinary clearances were calculated by the well-known formula:

$$\frac{\text{mg}/100 \text{ ml urine} \times \text{ml 24h urine}}{\text{mg}/100 \text{ ml plasma} \times 1440}$$

and expressed in ml/min.

RESULTS

Table 1 reports plasma values of hypoxanthine, xanthine, uric acid, HX/U, X/U and X/HX at the different moments of the therapy, when different doses of allopurinol were given.

Table 1. Plasma values of oxypurines (μmol/l), uric acid (mg/dl), HX/U, X/U and X/HX (μmol/l) at different moments of allopurinol therapy.

Date	Allopurinol mg/day	Uric acid	Hypoxanthine	Xanthine	HX/U	X/U	X/HX
13.7.92	200	2.8	0.06	0.04	0.37	0.25	0.6
21.9.92	150	4.4	0.01	0.004	0.053	0.015	0.28
30.11.92	100	6.0	18	23	51	67	1.3
13.1.93	100	5.3	33	38	107	125	1.16
22.2.93	100	5.5	31	44	97	137	1.41

Table 2 reports the same parameters in urine, under the same conditions.

Uricemia and uricuria were low (2.8 mg/dl and 128 mg/24h) with 200 mg/day of allopurinol. Plasma levels of oxypurines were very much reduced but urinary excretion of oxypurines was very high. After the dose of allopurinol was reduced to 100 mg/day (5 mg/kg/day) the patient eliminated again yellowish kidney stones. Oxypurines were very high in plasma (the increase was most evident for xanthine) and urinary excretion dropped.

Table 2. Urinary values of oxypurines (μmol/24h), uric acid (mg/24h), HX/U, X/U and X/HX (μmol/24h) at different moments of allopurinol therapy,

Date	Allopurinol mg/day	Uric acid	Hypoxanthine	Xanthine	HX/U	X/U	X/HX
13.7.92	200	128	2,500	2,000	3.2	2.6	0.80
21.9.92	150	88	2,209	1,581	4.2	3.0	0.70
30.11.92	100	-	1,560	1,500	-	-	0.96
13.1.93	100	136	1,310	1,282	1.6	1.6	0.98
22.2.93	100	185	1,315	1,422	1.2	1.3	1.08

It is evident that the urinary excretion of both bases was lower and similar when allopurinol was reduced.

Table 3 reports the clearance values for uric acid, hypoxanthine and xanthine, which all decreased under the same conditions; the decrease was most evident for xanthine.

Table 3. Clearance of uric acid, hypoxanthine and xanthine (ml/min) at different moments of allopurinol therapy.

Date	Allopurinol mg/day	Uric acid	Cl.Hypoxanthine	Cl.Xanthine
13.7.92	200	3.2	29,058	34,941
21.9.92	150	1.4	11,126	276,218
30.11.92	100	-	60.2	44.3
13.1.93	100	1.8	27.4	23.0
22.2.93	100	2.3	29.4	22.4

DISCUSSION

From our data it is evident that a patient with Lesch-Nyhan syndrome shows a response to allopurinol which varies according to the dose administered:

a) hypoxanthine and xanthine are very low in plasma and very high in urine, with high doses of allopurinol; under these conditions the clearance of both oxypurines was very incredible high;

b) when the dose of allopurinol was reduced to 100 mg/day, higher plasma levels and lower urinary excretion of oxypurines were observed, together with decreased renal clearance. The urinary excretion bore a perfectly inverse relation to plasma levels.

While the higher plasma levels are in agreement with the fact that allopurinol is known not to reduce de novo synthesis in Lesch-Nyhan patients, these observations may be due to an effect of allopurinol on the renal clearance of oxypurines, a hitherto unsuspected mechanism of the drug.

REFERENCES

1. J.E.Seegmiller, R.M.Rosenbloom and W.N.Kelley, An enzyme defect associated with a sex-linked human neurological disorder and excessive purine synthesis, *Science* 155:1682 (1967).
2. M.Lesch and W.L.Nyhan, A familial disorder of uric acid metabolism and central nervous system function, *Am.J.Med.* 36:561 (1964).
3. W.N.Kelley, Hypoxanthine-guanine phosphoribosyltransferase deficiency in the Lesch-Nyhan syndrome and gout, *Fed.Proc.* 27:1047 (1968).
4. W.N.Kelley, M.L.Greene, J.F.Rosenbloom et al., Hypoxanthine-guanine phosphoribosyltransferase deficiency in gout, *Ann.Intern.Med.* 70:155 (1969).
5. F.Delbarre, P.Cartier, C.Auscher et al., Gouttes enzymopathiques: Dyspurines par deficit en hypo-xanthine-guanine phosphoribosyltransferase frequence et caracteres cliniques de l' anenzymose. *Presse Med.* 78:729 (1970).
6. A.Kornberg, I.Liberman and E.S.Simms, Enzymatic synthesis of purine nucleotides, *J.Biol.Chem.* 215:417 (1955).
7. T.A.Krenitsky, R.Papaionnou and G.B.Elion, Human hypoxanthine phosphoribosyltransferase. I. Purification, properties and specificity, *J.Biol.Chem.* 244:1263 (1969).
8. L.B.Sorensen, Mechanism of excessive purine biosynthesis in hypoxanthine-guanine phosphoribosyl-transferase deficiency, *J.Clin.Invest.* 49:968 (1970).
9. W.N.Kelley, F.M.Rosenbloom, J.Miller et al., An enzymatic basis for variation in response to allopurinol, *N.Engl.J.Med.* 278:287 (1968).
10. I.H.Fox, J.B.Wyngaarden and W.N.Kelley, Depletion of erythrocyte phosphoribosylpyrophosphate in man. A newly observed effect of allopurinol, *N.Engl.J.Med.* 283:1177 (1970).
11. L.Sweetman, Urinary and cerebrospinal fluid oxypurine levels and allopurinol metabolism in the Lesch-Nyhan syndrome. *Fed.Proc.* 27, 1055 (1967).
12. P.Cartier and M.Hamet, Les activités purine-phosphoribosyl transférasiques des globules ranges humains. Technique de dosage, *Clin.Chim.Acta* 20:205 (1968).

13. P.Fossati, L.Prencipe and G.Berti, Use of 3,5-dichloro-2-hydroxy benzene sulfonic acid-4 amino phenazone chromogenic system indirect enzymatic assay of uric acid on serum and urine, *Clin. Chem.* 26:227 (1980).

ADENYLOSUCCINATE LYASE DEFICIENCY: AN UPDATE

Georges Van den Berghe[1], Françoise Van den Bergh[1], M. Françoise Vincent[1] and Jaak Jaeken[2]

[1]Laboratory of Physiological Chemistry, International Institute of Cellular and Molecular Pathology, and University of Louvain Medical School, B-1200 Brussels
[2]Department of Pediatrics, University of Leuven, Belgium

INTRODUCTION

Adenylosuccinate lyase (adenylosuccinase, ASase) catalyzes both the conversion of succinylaminoimidazole carboxamide (SAICAR) into AICAR in the *de novo* pathway of purine synthesis, and that of adenylosuccinate (S-AMP) into AMP in the conversion of IMP into AMP. The deficiency of ASase, the first enzyme deficiency reported in man on the purine biosynthetic pathway, was discovered in 1984 (1). The hallmark of the defect is the accumulation in body fluids, particularly cerebrospinal fluid and urine, of two normally undetectable compounds, SAICAriboside and succinyladenosine (S-Ado). These succinyl-purines are the products of the dephosphorylation of the two substrates of ASase by cytosolic 5'-nucleotidase(s) (2).

Severe psychomotor retardation is the principal symptom in 11 of the 12 patients diagnosed hitherto in Belgium and The Netherlands. The majority of these children also have epilepsy. Autistic features (failure to make eye-to-eye contact, repetitive behavior, temper tantrums) are found in about half of the patients. Some affected children display in addition profound growth retardation, associated with muscular wasting. Strikingly, one patient, a girl, is only slightly retarded. Whereas ASase deficiency with profound mental retardation is often referred to as type I, this variant is termed type II (3). In Type I, body fluid S-Ado/SAICAriboside ratios are between 1 and 2. In type II, they reach 4 to 5, owing to an approx. 4-fold higher concentration of S-Ado. The marked clinical heterogeneity justifies systematic screening for ASase deficiency in unexplained psychomotor retardation and neurological disease. For this purpose, a modified Bratton-Marshall test (4), performed on urine, appears most practical. More recently, a point mutation resulting in a Ser[413]Pro substitution has been evidenced in the ASase gene of a type I family (5).

In this presentation, data will be reviewed that *(i)* indicate that the wide clinical spectrum of ASase deficiency is caused by different enzyme lesions, and *(ii)* suggest that the mental retardation caused by the deficiency is more likely due to the accumulation of the succinylpurines, than to a deficient synthesis of purine nucleotides.

METHODS

Skin fibroblasts were cultured by standardized techniques in MEM to which 5 mM glutamine, 10 % fetal calf serum and 100 mg/ml kanamycin were added (6). For assay of ASase (7), cells were trypsinized, washed with ice-cold PBS, resuspended in 20 mM Tris-Cl buffer, pH 7.4, 1 mM DTT, and supernatants prepared by homogenization followed by centrifugation. For measurements of the incorporation of [14C]formate (0.2 mM for 7 h) or of [14C]hypoxanthine (20 µM for 20 min), cell monolayers or suspensions, respectively, were incubated in medium containing dialyzed fetal calf serum (8). After extraction with 10 % perchloric acid and neutralization, radioactivity in purine nucleotides was determined after separation by TLC or HPLC.

RESULTS AND DISCUSSION

Studies of fibroblast adenylosuccinase

In ASase-deficient lower organisms (9,10) and Chinese hamster ovary fibroblasts (11), the activities of the enzyme with both substrates are always lost in parallel. The higher S-Ado/SAICAriboside ratios in type II raised the possibility of a more profound deficiency of the activity of ASase with S-AMP as compared to SAICAR. To verify this hypothesis, studies were performed in cultured skin fibroblasts, which were known to express partially the enzyme defect (3).

As shown in Table 1, in control fibroblasts the activity of ASase with S-AMP was 1.6-fold higher than with SAICAR. In fibroblasts of type I patients, the activities of ASase with both S-AMP and SAICAR were about 30 % of normal, so that the ratio of both activities was barely modified. In contrast, in fibroblasts of the type II patient, the activity with S-AMP was only 3 % of normal, whereas that with SAICAR was 30 % of normal. If also present in other tissues, this non-parallel deficiency provides an explanation for the higher concentration of S-Ado in type II.

Table 1. Activities of adenylosuccinase is skin fibroblasts[1]

	with S-AMP	with SAICAR	Ratio
Controls (n = 4)	1.40 ± 0.08	0.90 ± 0.12	1.6
Patients			
Type I (n = 5)	0.47 ± 0.04	0.25 ± 0.01	1.9
Type II	0.04	0.25	0.2

[1]nmol/min per mg of protein

Additional studies (not illustrated) showed that in both types of ASase-deficient fibroblasts, Km of the enzyme for SAICAR was increased 3- to 4-fold. Type I ASase was retarded as compared to the normal enzyme on Mono-Q, an anion exchanger, whereas the type II enzyme was not. Control and Type I ASase were not influenced, but Type II ASase was potently inhibited by KCl, other anions, and nucleoside triphosphates.

Taken together, these results indicate modifications of ASase that are markedly different in type I and in type II. In type I, the changes result in decreased stability of the enzyme (5,12). The non-parallel loss of the activity of the enzyme in type II suggests a change in the active site which affects more the catalysis of S-AMP than that of SAICAR. One possibility is the addition of one or more positive charges, which would render the enzyme more sensitive to inhibitory anions, including nucleoside triphosphates.

Studies of nucleotide synthesis

Incorporation of [14C]formate proceeds at two steps along the *de novo* pathway of purine synthesis, a first located prior to ASase, and a second immediately thereafter. In control fibroblasts, total incorporation of 0.2 mM [14C]formate into purine nucleotides reached 11.4 ± 0.9 pmol/min per 10^6 cells (n = 3). After 7 h of incubation, 90 % of the radioactivity was found in the adenine nucleotides, 10 % in the guanine nucleotides, and no radioactivity was detectable in SAICAR and S-AMP (Table 2). In type I fibroblasts, the rate of incorporation of formate was about 30 % higher, namely 15.2 pmol/min per 10^6 cells (n = 2), but distribution of the radioactivity was similar. In the type II fibroblasts, total incorporation was similar to that in type I, slightly less radioactivity was incorporated into the adenine nucleotides, but both SAICAR and S-AMP were detectable. These results indicate that the partial deficiency of fibroblast ASase, characteristic of type I, does not hamper purine nucleotide synthesis. In type II, in contrast, flux through the biosynthetic pathway is hampered, although far from suppressed, notwithstanding the profound deficiency of ASase with S-AMP.

Table 2. Synthesis of [14C]purine nucleotides in skin fibroblasts

	Control	Type I	Type II
From [14C]formate			
Total nucleotides[1]	11.4 ± 0.9	15.2	14.5
Adenine nucleotides[2]	90	86	76
Guanine nucleotides[2]	10	14	12
SAICAR[2]	n.d.	n.d.	7
S-AMP[2]	n.d.	n.d.	5
From [14C]hypoxanthine			
Total nucleotides[1]	22.3 ± 5.4	22.2 ± 7.3	20.4
IMP[2]	2	5	10
Adenine nucleotides[2]	88	89	24
Guanine nucleotides[2]	10	6	6
S-AMP[2]	n.d.	n.d.	59

[1]pmol/min per 10^6 cells; [2] percentage of total; n.d., not detectable

Incorporation of [14C]hypoxanthine occurs at the level of hypoxanthine-guanine phosphoribosyltransferase, and the resulting IMP is converted into guanine and adenine nucleotides, the latter via ASase. Rates of incorporation of 20 μM [^{14}C]hypoxanthine into total purine nucleotides were similar in control, type I, and type II fibroblasts (Table 2). Similarly to [^{14}C]formate, most [^{14}C]hypoxanthine was incorporated into adenine nucleotides after 20 min, both in control and in type I fibroblasts, confirming that partial ASase deficiency does not hamper flux through the enzyme step. In contrast, in type II fibroblasts, incorporation of [^{14}C]hypoxanthine resulted in marked build-up of [^{14}C]S-AMP. Nevertheless, adenylate synthesis remained possible.

These findings are in agreement with the observation that normal purine nucleotide concentrations were measured in ASase-deficient tissues (2). They also accord with the fact that the activity of normal ASase is several-fold higher than that of the rate-limiting steps of purine synthesis. Taken together, our studies suggest that the pathophysiology of the disorder is mediated by accumulation of SAICAR and S-AMP, and/or of their dephosphorylated derivatives, SAICAriboside and S-Ado, rather than by deficiency of nucleotide synthesis. The observation of a strikingly less severe psychomotor retardation in the type II patient, with similar SAICAriboside levels but higher S-Ado/SAICAriboside ratios, suggests that SAICAriboside is the offending compound, and that S-Ado could protect against its toxic effects.

REFERENCES

1. Jaeken J, Van den Berghe G. An infantile autistic syndrome characterised by the presence of succinylpurines in body fluids. *Lancet* 2: 1058-56, 1984.
2. Van den Berghe G, Jaeken J. Adenylosuccinase deficiency. *Adv Exp Med Biol* 195A: 27-33, 1986.
3. Jaeken J, Wadman SK, Duran M, van Sprang FJ, Beemer FA, Holl RA, Theunissen PM, de Cock P, van den Bergh F, Vincent MF, van den Berghe G. Adenylosuccinase deficiency : an inborn error of purine nucleotide synthesis. *Eur J Pediatr* 148: 126-31, 1988.
4. Laikind PK, Seegmiller JE, Gruber HE. Detection of 5'-phosphoribosyl-4-(N-succinyl-carboxamide)-5-aminoimidazole in urine by use of the Bratton-Marshall reaction: identification of patients deficient in adenylosuccinate lyase activity. *Anal Biochem* 156: 81-90, 1986
5. Stone RL, Aimi J, Barshop BA, Jaeken J, Van den Berghe G, Zalkin H, Dixon JE. A mutation in adenylosuccinate lyase associated with mental retardation and autistic features. *Nature Genetics* 1: 59-63, 1992.
6. Van den Bergh F, Vincent MF, Jaeken J, Van den Berghe G. Residual adenylosuccinase activities in fibroblasts of adenylosuccinase-deficient children: parallel deficiency with adenylosuccinate and succinyl-AICAR in profoundly retarded patients and non-parallel deficiency in a mildly retarded girl. *J Inher Metab Dis* 16: 415-24, 1993.
7. Van den Bergh F, Vincent MF, Jaeken J, Van den Berghe G. Radiochemical assay of adenylosuccinase. Demonstration of parallel loss of activity toward both adenylosuccinate and succinylamidazole carboxamide ribotide in liver of patients with the enzyme defect. *Analyt Biochem* 193: 287-91, 1991.
8. Van den Bergh F, Vincent MF, Jaeken J, Van den Berghe G. Functional studies in fibroblasts of adenylosuccinase-deficient children. *J Inher Metab Dis* 16: 425-34, 1993.
9. Giles NH, Partridge CWH, Nelson JH. The genetic control of adenylosuccinase in *Neurosopora Crassa. Proc Natl Acad Sci USA* 43: 826-34, 1957.
10. Gollub EG, Gots JS. Purine metabolism in bacteria. VI. Accumulations by mutants lacking adenylosuccinase. *J Bacteriol* 78: 320-25, 1959.
11. Patterson D. Biochemical genetics of Chinese hamster cell mutants with deviant purine metabolism. IV. Isolation of a mutant which accumulates adenylosuccinic acid and succinylaminoimidazole carboxamide ribotide. *Somat Cell Genet* 2: 189-203, 1976.
12. Laikind PK, Gruber HE, Jansen I, Miller L, Hoffer M, Seegmiller JE, Willis RC, Jaeken J, Van den Berghe G. Purine biosynthesis in Chinese hamster cell mutants and human fibroblasts partially deficient in adenylosuccinate lyase. *Adv Exp Med Biol* 195B: 363-69, 1986.

ADENYLOSUCCINATE LYASE DEFICIENCY IN A CZECH GIRL AND TWO SIBLINGS

J. Krijt[1], I.Sebesta[1], A.Svehlakova[2], A.Zumrova[3] and J.Zeman[1]

[1]Centre for Metabolic Disorders, Charles University, Prague
[2]Children's Hospital, Zlin, Czech Republic
[3]Children's Hospital Motol, Prague, Czech Republic

INTRODUCTION

Adenylosuccinase deficiency (McKusick 10305) is a recently described genetic defect of purine nucleotide metabolism. Adenylosuccinase (adenylosuccinate lyase, ASase, E.C.4.3.2.2.) catalyses the eighth step in the ten-step *de novo* purine synthesis and the second step in the formation of AMP from IMP. Affected children are normal at birth, but psychomotor retardation, the principal symptom of this defect becomes evident in the first years of life. The other, inconstant, clinical features are epilepsy, involuntary movement, agitation, hypotonia, autistic features, muscle wasting and growth failure.

The biochemical marker of this disease is the accumulation in the urine, cerebrospinal fluid (CSF) and plasma of two normally undetectable compounds: succinyladenosine (S-Ado) and succinylaminoimidazol carboxamid riboside (SAICAr)[1,2,3] . These are the products of the dephosphorylation of succinylaminoimidazol carboxamid ribotide (SAICAR) and adenylosuccinate (S-AMP) by cytoplasmic 5'-nucleotidase.

An autosomal recessive mode of inheritance was observed[2]. Only thirteen patients with this disorder have been reported to date. Here we report three additional patients with putative ASase deficiency of Czech origin.

PATIENTS

Case 1

This 4-year-age girl is the first child of unrelated parents. She was born in term after an uncomplicated pregnancy with birth weight 3200 g and length 50 cm. Early postnatal adaptation was uneventful, but apneic pauses occurred frequently from the third day of life followed by myoclonic convulsions. There was no growth retardation or muscular wasting, she was able to eat, but no progress in her psychomotor development was observed. She is not able to sit or walk till now and her mental development corresponds to severe oligophrenia. She is quadriplegic and has no eye contact. Regardless of various antiepileptic therapy EEG showed epileptic discharges. Computed tomography of the brain revealed severe cortical and periventricular atrophy.

Case 2

6-year-old girl of unrelated parents was born in term after the second uncomplicated pregnancy with birth weight 3550 g and length 50 cm. Her parents and older brother are healthy. Early postnatal development was normal, but mild hypotonia was observed in the fourth month of life and severe psychomotor retardation developed. She was standing alone at one year of age and she was walking after the third year. She is not able to speak till now. Neurological examination revealed central hypotonic syndrome, cereberal signs and movement hyperactivity. EEG examination showed generalized epileptic discharges. Psychologic investigation revealed severe erethitic form of idiocy. Changes of her mood and temper tantrums occur frequently. Autistic features were not evident.

Case 3

5-year-old younger brother of case 2 was born in term after the third uncomplicated pregnancy with birth weight 3200 g and length 50 cm. Early postnatal development was uneventful, but hypotonia and psychomotor retardation was observed in the second half of the first year. He started to walk in two years but never started to speak. Neurological examination revealed central hypotonia and hyperactivity. Clinical course is identical with his sister.

METHODS

We used screening TLC method (Pauly reagent detection) according to de Bree[4]. Two separate HPLC methods were used for quantification of succinylpurines in body fluids.

1) Reversed phase chromatography using gradient elution with ammonium acetate buffers according to Morris et.al.[5]

2) Anion-exchange chromatography on amino phase column using linear gradient elution from 0.007 mol/l KH_2PO_4 pH 4.0 to 0.125 mol/l KH_2PO_4 pH 4.5 in 20 minutes.

Concentrations of succinylpurines were calculated from their molar extinction coefficients at 268 nm (S-Ado [19.2×10^3]; SAICAr [13.2×10^3]). Mild acid hydrolysis of CSF samples was performed in 0.5 M HCl at 80°C for 1 hour.

RESULTS

TLC screening for SAICAr was positive in urines of all three patients (blue spot of SAICAr, R_f= 0.35). This finding was confirmed by HPLC quantification of succinylpurines in body fluids (Table 1). These data are consistent with ASase deficiency as classified by Jaeken et.al.[1]

Table 1. Concentrations of succinylpurines in body fluids

Cases	Urine		Plasma		Cerebrospinal fluid		
	S-Ado	SAICAr	S-Ado	SAICAr	S-Ado	SAICAr	S-Ado/
	(mmol/mol Creat.)		(μmol/l)		(μmol/l)		SAICAr
1	94	91	5.2	7	126	147	0.9
2	110	45	4.2	2	283	127	2.2
3	89	27	4.2	2.4	260	125	2.1

DISCUSSION

Although we did not perform the enzymatic assays in these patients the presence of succinylpurines in body fluids is highly suggestive of ASase deficiency.

Clinical features of the described patients correspond to nonspecific clinical presentation of this disease, without any characteristic sign. These three new cases emphasize the need for further systematic screening for this disorder in patients with unexplained neurological disease and psychomotor retardation.

Jaeken et al.[3] classifies the ASase deficient patients in four groups according to their clinical picture and CSF succinylpurines concentrations. Our patients resemble type I classification, although no autistic features are evident.

The pathogenesis of this disease is not known. A hypothesis exists "that SAICAr is the offending compound and that S-Ado could protect against its toxic effects[2]". This hypothesis is supported by our findings in all three patients. SAICAr concentrations in CSF are almost the same in all three patients. However the CSF S-Ado/SAICAr ratios in less severe retarded siblings are significantly higher due to higher S-Ado concentrations. This supports the hypothesis of protective role of S-Ado.

Analyzing twelve CSF control samples we found a peak with retention times and UV spectra identical to S-Ado in both HPLC systems. The peak disappeared after mild acid hydrolysis of control CSF samples, similarly to S-Ado peak in ASase deficient patients. We suggest that S-Ado is a normal component of cerebrospinal fluid (range 2.3-5.1 μmol/l, n=12).

REFERENCES

1. Jaeken J,Van den Berghe G.,An infantile autistic syndrome characterized by the presence of succinyl purines in body fluids, *Lancet* 2:1058 (1984).
2. Jaeken J, Wadman SK, Duran M, van Sprang FJ, Beemer FA, Holl RA, Theunissen PM, deCock P,Van den Berghe F, Vincent MF, Van den Berghe G.,Adenylosuccinase deficiency: an inborn error of purine nucleotide synthesis. *Eur J Pediatr* 148:126 (1988).
3. Jaeken J, Casaer P, De Cock P, Van den Berghe G., The clinical aspects of ASase deficiency,*in:* "Molecular Genetics, Biochemistry and Clinical Aspects of Inherited Disorders of Purine and Pyrimidine Metabolism" Gresser,ed.,Springer Verlag,Berlin, Heidelberg 1993

4. de Bree Pk, Wadman SK, Duran M, Fabery de Jonge H. Diagnosis of inherited adenylosuccinase by thin-layer chromatography of urinary imidazoles and by automated cation exchange chromatography of purines. *Clin Chim Acta*156:279 (1986).
5. Morris GS,Simmonds HA,Davies PM,Use of biological fluids for the rapid diagnosis of potencially lethal inherited disorders of human purine and pyrimidine metabolism,*Biomed Chromatogr.* 1:109 (1986).

ANOMALOUS RESPONSE TO INTRAVENOUS FRUCTOSE TOLERANCE TEST IN A CASE OF DEFICIT OF ADENYLOSUCCINATE LYASE

C.Salerno,[1] C.Crifò,[2] E.Capuozzo,[2] and O.Giardini[3]

[1]Department of Human Biopathology,
[2]Department of Biochemical Sciences,
[3]Institute of Paediatrics,
 University of Roma La Sapienza, Italy

INTRODUCTION

Adenylosuccinate lyase (adenylosuccinase, EC 4.3.2.2) deficiency is an autosomal recessive disorder that has been diagnosed hitherto in 13 children (nine females and four males) from nine families: seven with one patient, one with two patients, and one with four patients.[1] They belong to five nationalities: three families are Belgian, three are Dutch, one Italian, one Moroccan, and one Turkish. In the last two families there is consanguinity between the parents. Affected children are normal at birth, but psychomotor retardation, the principal symptom of the defect, becomes evident in the first two years. Autistic features and epilepsy are often present. Sometimes, severe growth failure and muscular wasting have been observed.[2]

The inherited defect is characterized by the presence in body fluids of the products of the dephosphorylation, by cytosolic 5'-nucleotidase,[1] of the two substrates of adenylosuccinate lyase, namely succinylaminoimidazole carboxamide riboside (SAICA riboside) and succinyladenosine (S-Ado), that are normally undetectable. Measurements of the activity of adenylosuccinate lyase in the biopsies showed that the degree of enzyme deficiency is variable in the patients and not generalized in all the tissues.[3]

The enzyme catalyses two steps in the purine pathway: the formation of aminoimidazole carboxamide ribotide (AICAR) from succinylaminoimidazole carboxamide ribotide (SAICAR) along the *de novo* synthesis and the formation of adenosine monophosphate (AMP) from adenylosuccinate (S-AMP) in the conversion of inosine monophosphate (IMP) into adenine nucleotides. Both reactions involve the cleavage of a succinyl group, yielding fumarate. Adenylosuccinate lyase forms, together with adenylosuccinate synthetase and AMP deaminase, a functional unit (the purine nucleotide cycle), which has been shown to operate in intact muscle and brain at rates sufficient to account for the rates of ammonia production by these tissues.[4,5] This cycle is thought to be also important for the production of Krebs cycle intermediates and for the removal of AMP, in order to pull the adenylate kinase reaction in the direction of formation of ATP.[6]

Purine and Pyrimidine Metabolism in Man VIII, Edited by
A. Sahota and M. Taylor, Plenum Press, New York, 1995

In this work we studied whether, owing to the impairment of the purine nucleotide cycle, the deficit of adenylosuccinate lyase caused an anomalous response to the intravenous fructose tolerance test: This analysis is commonly employed to assess fructose toxicity in children with hereditary fructose intolerance, but it is well known that fructose administration markedly enhances degradation of adenine nucleotides also in normal subjects.[7] This is explained by a release of the physiological inhibition of AMP deaminase, due to the decrease in concentration of inorganic phosphate and GTP during fructose metabolism.[8]

CASE REPORT

The patient (now 10 years old) is the second daughter of unrelated Italian parents. A few months after birth, frequent crying attacks, motor restlessness and hypertonicity were noticed. Eye contact was difficult, while reaction to auditory stimuli was exaggerated. Psychomotor development was delayed (she was standing alone at 15 months and walking after 2 years; up to now, she speaks unintelligibly in short sentences with an incorrect structure and seems to understand very little of what is said to them). At age of 5 years, generalized tonic-clonic convulsions were observed for the first time. There is an intermittent external strabismus with crisis of convergence spasm. Striking bouts of extreme agitation, especially involving arms and legs, are common. Growth and head circumference are normal.

The patient excretes approximately 130 nmol S-Ado/min in the urine, while S-Ado concentration in plasma is about 3 μM. The molar ratio of S-Ado to SAICA riboside is about 1.2. Serum urate concentration is 2-4 mg/dl. The activity of adenylosuccinate lyase in mixed peripheral blood lymphocytes is reduced to about 30%, while the hemolysate activity is 35-40% of normal. The erythrocyte enzyme shows an anomalous heat denaturation curve and it is extremely unstable with time after partial purification. Substrate affinity of the partially-purified enzyme is not modified as compared to controls.

RESULTS AND DISCUSSION

The injection of a single intravenous test dose of 200 mg fructose / kg body weight (20% solution) in our patient caused a rapid fall, first of serum phosphate and urate and then of glucose, and a rise of magnesium (Figure 1).

The change in serum phosphate concentration did not markedly differ from what was observed in some of the controls and it could be attributed to the conversion of fructose to fructose-1-phosphate and to the rephosphorylation of ADP in mitochondria. By contrast, the increase of serum magnesium was significantly above the control levels, most probably reflecting a loss of ATP. Indeed, ATP is a strong Mg^{2+} chelator and a drop in its concentration is likely to result in the release of cellular magnesium ions.[9] The depletion of ATP could be also suggested by the slow decrease of blood glucose. Fructose-induced hypoglycemia (exceptionally observed in healthy subjects[10]) is generally attributed to the impairment of gluconeogenesis that, in turn, seems to depend, at least in perfused rat liver,[11] on the concentration of ATP up to the physiological level.

As far as the link between the deficit of adenylosuccinate lyase and the decrease in ATP is concerned, we suggest that this can be found in the impairment of the purine nucleotide cycle due to the inherited disease, preventing IMP reutilization. Indeed, an increase in IMP concentration was already observed even in normal liver tissue upon

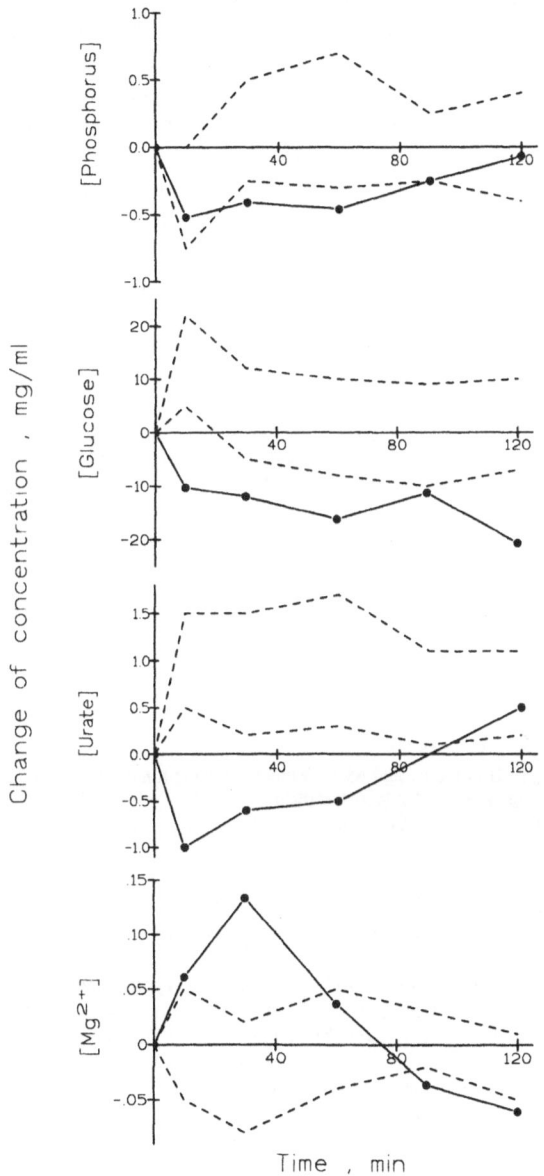

Figure 1. Intravenous fructose tolerance test (200 mg/kg; 20% solution) in the patient with deficit of adenylosuccinate lyase. <u>Solid lines:</u> time course of the concentration change of serum analytes upon fructose infusion. <u>Broken lines:</u> means ± 1 SD of 16 control children.

fructose administration.[12] An approximately three-fold increase of IMP content in biopsies of liver and kidney from two patients with deficit of adenylosuccinate lyase was reported, but this finding was not confirmed in another patient.[13]

The decrease in serum urate level upon fructose loading is somewhat unexpected, taking into account the well known hyperuricemic effect of fructose not only in patients with hereditary fructose intolerance but also in normal children.[7] Since urinary excretion of uric acid, expressed as the creatinine ratio, was constant (data not shown), a decrease in purine biosynthesis was suggested. A proposal is that inhibition of the *de novo* pathway could result from both the lack of ATP and the accumulation of IMP, which behaves as an allosteric inhibitor of purine biosynthesis.[14]

Sporadic depletion of intracellular ATP could contribute to the impairment of the neurological function in subjects with inherited deficiency of adenylosuccinate lyase. Alteration of the energy-generating systems could also account for the cerebral hypotrophy and the marked decrease of brain glucose utilization, observed in some of these patients.[2,15]

REFERENCES

1. J. Jaeken and G. Van den Berghe, An infantile autistic syndrome characterized by the presence of succinylpurines in body fluids, *Lancet* 2:1058 (1984).
2. J. Jaeken, S.K. Wadman, M. Duran, F.J. Van Sprang, F.A. Beemer, R.A. Holl, P.M. Theunissen, P. De Cock, F. Van den Bergh, M.F. Vincent, and G. Van den Berghe, Adenylosuccinase deficiency: an inborn error of purine nucleotide synthesis, *Eur. J. Pediatr.* 148:126 (1988).
3. J. Jaeken, P. Casaer, P. De Cock, and G. Van den Berghe, The clinical aspects of ASase deficiency, *in*: "Molecular Genetics, Biochemistry and Clinical Aspects of Inherited Disorders of Purine and Pyrimidine Metabolism ", U. Gresser, ed., Springer-Verlag, Berlin (1993).
4. K. Tornheim and J. Lowenstein, The purine nucleotide cycle: the production of ammonia from aspartate by extracts of rat skeletal muscle, *J. Biol. Chem.* 247:162 (1972).
5. V. Schultz and J. Lowenstein, Purine nucleotide cycle: evidence for the occurrence of the cycle in brain, *J. Biol. Chem.* 253:1938 (1978).
6. G. Van den Berghe, F. Bontemps, and M.F. Vincent, Purine nucleotide cycle, molecular defects and therapy, *Adv. Exp. Med. Biol.* 309B:281 (1991).
7. B. Steinmann and R. Gitzelmann, The diagnosis of hereditary fructose intolerance, *Helv. Paediatr. Acta* 36:297 (1981).
8. G. Van den Berghe, M. Bronfman, R. Vanneste, and H.G. Hers, The mechanism of adenosine triphosphate depletion in the liver after a load of fructose, *Biochem. J.* 162:601 (1977).
9. P.H. Maempaa, Fructose-induced alterations in liver polysome profiles and Mg levels, *FEBS Lett.* 24:37 (1972).
10. R. Schartz, H. Gamsu, P.B. Mulligan, S.H. Reisner, S.H. Wybregt, and M. Cornblath, Transient intolerance to exogenous fructose in the newborn, *J. Clin. Invest.* 43:433 (1964).
11. J. Wilkening, J. Nowack, and K. Decker, The dependence of glucose formation from lactate on the adenosine triphosphate content in isolated perfused rat liver, *Biochim. Biophys. Acta* 392:299 (1975).
12. H.F. Woods, L.V. Eggleston, and H.A. Krebs, The cause of hepatic accumulation of fructose 1-phosphate on fructose loading, *Biochem. J.* 119:501 (1970).
13. G. Van den Berghe and J. Jaeken, Adenylosuccinase deficiency, *Adv. Exp. Med. Biol.* 195A:27 (1986).
14. T.D. Palella and I.H. Fox, Hyperuricemia and gout, *in*: "The Metabolic Basis of Inherited Disease", J.B. Stanbury et al., eds., McGraw-Hill, New York (1989).
15. A.G. De Volder, J. Jaeken, G. Van den Berghe, A. Bol, C. Michel, M. Cogneau, and A.M. Goffinet, Regional brain glucose utilization in adenylosuccinase-deficient patients measured by positron emission tomography, *Pediatr. Res.* 24:238 (1988).

EFFECT OF ALLOPURINOL ON THE XANTHINURIA IN A PATIENT WITH MOLYBDENUM COFACTOR DEFICIENCY

Albert H. van Gennip[1], Hanna Mandel[2], Lida E.M. Stroomer[1] and Arno G. van Cruchten[1]

[1]Academic Medical Center, University of Amsterdam,
Depts. of Pediat. and Clin. Chem., P.O. Box 22700,
1100 DE Amsterdam, The Netherlands
[2]Rambam Medical Center, Dept. of Pediat., Haifa, Israel

INTRODUCTION

In humans molybdenum-cofactor deficiency (MCF) leads to the combined deficient activities of sulphite oxidase (SO), aldehyde oxidase (AO) and xanthine dehydrogenase (XDH). Patients present from birth on with frequent severe seizures refractory to anticonvulsants[1] and abnormal sulphur and purine metabolites in their body fluids[2,3]. The deficiency of SO causes the severe neurological manifestations as they also occur in patients with isolated SO, but not in patients with isolated XDH deficiency or patients with the double defect of XDH and AO[4]. The deficiency of XDH leads to xanthinuria and some patients even present with xanthine calculi.

We diagnosed a two month-old boy with MCF deficiency who already had urinary calculi. An allopurinol loading test was performed in order to investigate whether the xanthine to hypoxanthine ratio could be altered in favour of the more soluble hypoxanthine as has been described for some patients with isolated XDH deficiency. The parents were also loaded for comparison.

Since it has been suggested that AO from rabbit liver or rat liver is able to convert hypoxanthine into xanthine[5,6] we were interested in a possible role of AO in the 'in vivo' oxidation of hypoxanthine in humans. Therefore, we compared the results of allopurinol loading on purine metabolism in the patient with the combined AO and XDH deficiency (and SO deficiency) with the results obtained in three patients with isolated XDH deficiency.

MATERIALS AND METHODS

Allopurinol was administered orally in a dose of 10 mg/kg. b.w. in patient B.S. with MCF deficiency, his parents and three patients with isolated XDH deficiency. Urine was collected during 24 hrs before and 24 hrs after the load. Before and during the loading experiments, all individuals were on a free diet.

Amino acids were analyzed by cation-exchange chromatography with lithium citrate buffers and post-column ninhydrin detection using an amino acid analyzer. Sulphite was detected with a dip-stick test (Mercko-quant 10013).

Sulphite, thiosulphate and sulphate were quantified by anion-exchange chromatography with a high performance liquid chromatography (HPLC) system combined with a conductometric detector[7]. Purine analysis was performed by bidim. thin-layer chromatography and HPLC[8].

RESULTS

Metabolic investigation of a 24 hrs urine of patient B.S. revealed the metabolite pattern characteristic for MCF deficiency (see Table 1). The pattern was comparable with the patterns found in two other patients with MCF deficiency. The diagnosis was confirmed at the enzyme level by determination of the SO activity in the patient's fibroblasts.

The results of allopurinol loading in patient B.S. were compared with his parents and three patients with isolated XDH deficiency (see Table 2). In the parents a substantial part of allopurinol was converted into oxipurinol, but in patient B.S. no oxipurinol was formed. The patients with isolated XDH deficiency also had oxipurinol in their urines but in much smaller amounts than the parents of B.S. Patient B.S. and patient XDHD-3 excreted large amounts of allopurinol and allopurinolribonucleoside, the parents of B.S. and patient XDHD-1 excreted moderate amounts of these substances and in the urine of patient XDHD-2 these substances were not detected at all. Oxipurinolribonucleoside was found in very small amounts in the urines from the parents of B.S and the urines from two patients with isolated XDH deficiency, but not

Table 1. Abnormal urinary metabolites and enzyme activities in patient B.S. Data of two other patients with MCF deficiency and of controls are given for comparison.

biochemical findings	patients			controls
	K.N.	J.K.	B.S.	
urinary metabolites[1]				
S-sulphocysteine	171	338	169	< 10
taurine	1261	466	931	< 200
cystine	7	trace	trace	> 30
sulphite	24	66	280	trace
thiosulphate	53	263	210	< 9
sulphate	186	857	730	> 1500
xanthine	991	228	326	< 28
hypoxanthine	88	46	11	< 8
enzyme activities				
SO	n.d.[2]	n.d.[3]	n.d.[4]	85;103[3]; > 1[4]
XDH	n.a.	0.08[3]	n.a.	7.1; 12.1[3]

[1] μmol/mmol creatinine; n.d. = not detectable; n.a. = not available
[2] in cultured fibroblasts by J.L. Johnson, U.S.A.
[3] in liver tissue and [4] in cultured fibroblasts by C. Dorche, France

Table 2. Results of allopurinol loading in patient B.S. with MCF deficiency and his parents in comparison with three patients with isolated XDH deficiency.

Subject	No	hyp	xan	x/h	uri	all	alo	oxi	oxo
B.S.	(1)	28	247	8.8	37	-	-	-	-
	(2)	49	382	7.8	4	376	431	n.d.	n.d.
Mo	(1)	4	4	1.0	518	-	-	-	-
	(2)	31	109	3.5	329	35	29	133	trace
Fa	(1)	6	4	0.7	347	-	-	-	-
	(2)	21	59	2.8	187	18	16	74	trace
XDHD-1	(1)	11	155	14.1	n.d.	-	-	-	-
	(2)	22	71	3.2	n.d.	37	135	2	n.d.
XDHD-2	(1)	24	151	6.2	10	-	-	-	-
	(2)	37	76	2.0	0.9	n.d.	n.d.	1.3	0.5
XDHD-3	(1)	46	745	16.2	33	-	-	-	-
	(2)	93	524	5.6	n.d.	362	337	trace	trace

Allopurinol load: 10 mg/kg b.w. orally. No (1) and (2): 24 hrs urines collected before (1) or after (2) loading. Hyp, hypoxanthine; xan, xanthine; x/h, xanthine to hypoxanthine ratio; uri, uric acid; all, allopurinol; alo, allopurinolribonucleoside; oxi, oxipurinol; oxo, oxipurinolribonucleotide.

in the third XDH deficient patient and patient B.S.

Urinary hypoxanthine appeared to have increased after loading in all subjects. Xanthine had increased in patient B.S. and his parents, but it had decreased in the patients with isolated XDH deficiency. The xanthine to hypoxanthine ratio was hardly changed in patient B.S., it was increased in his parents and it was substantially decreased in the patients with XDH deficiency. The excretion of uric acid was decreased in all subjects, except in one who did not excrete uric acid, neither before nor after allopurinol loading.

DISCUSSION

MCF deficiency is an inherited disorder of both purine and sulphur metabolism as is demonstrated in Table 1. Our patient B.S. presented already shortly after birth with urinary xanthine calculi. A beneficial effect of allopurinol on xanthine-stone formation has been described in patients with isolated XDH deficiency, but the underlying mechanism has not been fully elucidated[9]. As shown in Table 2, allopurinol had a substantial effect on the urinary X/H ratio in the patients with isolated XDH deficiency, but hardly any effect in the patient with MCF deficiency. In this patient xanthine had increased instead of decreased. Moreover, oxipurinol was only found in the urines from the patients with the isolated XDH defect, but not in the patient with the combined defects of XDH, AO and SO. Therefore we conclude that in patients with isolated XDH deficiency AO plays a role in the conversion of hypoxanthine into xanthine and of allopurinol into oxipurinol. The latter finding has been reported recently[4]. The increase of xanthine in the urine from the MCF deficient patient and decrease in the urines of the isolated XDH-deficient patients after allopurinol loading can also be explained by the activity of AO in the patients with isolated XDH deficiency. In the latter patients inhibition of AO by allopurinol prevents the AO-catalysed conversion of hypoxanthine into xanthine. In patients with isolated XDH deficiency as well as in the patient with MCF deficiency hypoxanthine and guanine may have increased by the reduced availability of phosphoribosylpyrophosphate (PRPP) for the formation of their nucleotides due to the competitive formation of allopurinolribonucleotide. The guanine after conversion into xanthine may have caused an increase of xanthine, which is only manifested in the urine from the MCF deficient patient, in who there was no reduction of xanthine by inhibition of AO by allopurinol.

The decrease of the uric acid excretion in the parents of B.S. is expected because of inhibition of the conversion of xanthine into uric acid. In patient B.S. with MCF deficiency and in the XDH-deficient patients the decrease my be caused by some very low rest activity of XDH.

REFERENCES

1. H.M.J. Slot, W.C.G. Overweg-Plandsoen, H.D. Bakker, N.G.G.M. Abeling, P. Tamminga, P.G. Barth and A.H. van Gennip, Molybdenum cofactor deficiency: an easily missed cause of neonatal convulsions. Neuropediatrics 24:139-142 (1993).
2. J.L. Johnson and S.K. Wadman, Molybdenum cofactor deficiency, in: "The Metabolic Basis of Inherited Disease," C.R. Scriver, A.L. Beaudet, W.S. Sly and D. Valle, eds., McGraw-Hill, New York, pp 1463-1475 (1989).
3. A.H. van Gennip, A. Stroomer, W.C.G. Plandsoen and N.G.G.M. Abeling, The effect of molybdenum cofactor deficiency on the purine pattern of cerebrospinal fluid, J. Inher. Metab. Dis. 14:364-366 (1991).
4. S. Reiter, H.A. Simmonds, N. Zöllner, S.L. Braun and M. Knedel, Demonstration of a combined deficiency of xanthine oxidase and aldehyde oxidase in xanthinuric patients not forming oxipurinol, Clin. Chim. Acta 187: 221-234 (1990).
5. W.W. Hall and T.A. Krenitsky, Aldehyde oxidase from rabbit liver: specificity toward purines and their analogs, Arch. Biochem. Biophys. 15:36-46 (1986)
6. T. Yamamoto, Y. Moriwaka, M. Suda, Y. Nasako, S. Takahashi, K. Hiroishi, T. Nakano, T. Hada and K. Higashino, Effect of BOF-4272 on the oxidation of allopurinol and pyrazinamide 'in vivo', Biochem. Pharmacol. 46:2277-2284(1993)
7. Dionex Corporation, Installation instructions and troubleshooting guide for the Ionpac AG9-SC guard Column and Ionpac AS9-SC analytical column, Document No. 034656-01, Dionex Corporation (1992)
8. A.H. van Gennip, Screening for inborn errors of purine and pyrimidine metabolism by bidimensional TLC and HPLC, in: "Handbook of Chromatography Nucleic Acids and Related Compounds," A.M. Krstulovic, Ed., CRC Press Inc., Boca Raton, pp 221-245 (1987).
9. E.W. Holmes and J.B. Wijngaarden, Hereditary xanthinuria, in: "The Metabolic Basis of Inherited Disease", C.R. Scriver, A.L. Beaudet, W.S. Sly and D. Valle, eds., McGraw-Hill, New York, pp 1085-1094 (1989).

DIFFERENTIAL DIAGNOSIS OF MAIN RHEUMATIC DISEASES IN MAN

M. L. Sorgi, C. P. Quaratino, C. Rucci, P. Ciaglia, A. Zoppini
and A. Giacomello

Institute of Medical Physiopathology, University of Chieti
and Institute of Rheumatology, University of Rome
Italy

The purpose of the present study was to determine which of the routine laboratory variables are most useful in the differential diagnosis of main rheumatic diseases in man.

During a twenty year period 2045 males were admitted to the Institute of Rheumatology of the University of Rome. The most prevalent diseases were: rheumatoid arthritis (RA) (22.1%), gout (G) (11.2%), osteoarthritis (OA) (10.3%), ankylosing spondylitis (AS) (8.4%) and psoriatic arthritis (PA) (5.4%). For each diagnostic group patient samples were randomly selected and the following laboratory variables in all subjects were determined : fasting serum glucose (S_GLU), blood urea nitrogen (BUN), serum urate (S_UA), erythrocyte sedimentation rate (ESR) (Westergren) and Rheumatoid Factor Latex test (RF).

Age and number of radiological examinations performed on admission were also considered. A discriminant analysis[1] was performed in order to determine which variables are the best predictors of the diagnosis.

Results and Discussion

Categorized notched box plots[2] of the continuous variables considered are shown in Figure 1. Age was significantly lower is AS than in other diseases perhaps owing to the earlier age at onset of the disease. The oldest patients were those with OA.

The highest number of radiological examinations on admission were found in PA and in RA, while the lowest number was found in OA patients. Serum urate concentration was significantly higher in gout than in other diseases. Patients with OA had serum urate levels higher than those with AS, PA and RA. ESR was significantly lower in OA than in other diseases. BUN and glucose were the variables that contributed least to the discrimination between the diagnostic groups.

Relationships between the categoric variable RA test and the diagnostic groups are shown in Table 1, where the significance tests are also reported.

Pearson correlation matrix of the continuous independent variables is shown in Table 2.

A summary of the discriminant function analysis where all variables are included in the model is shown in Table 3.

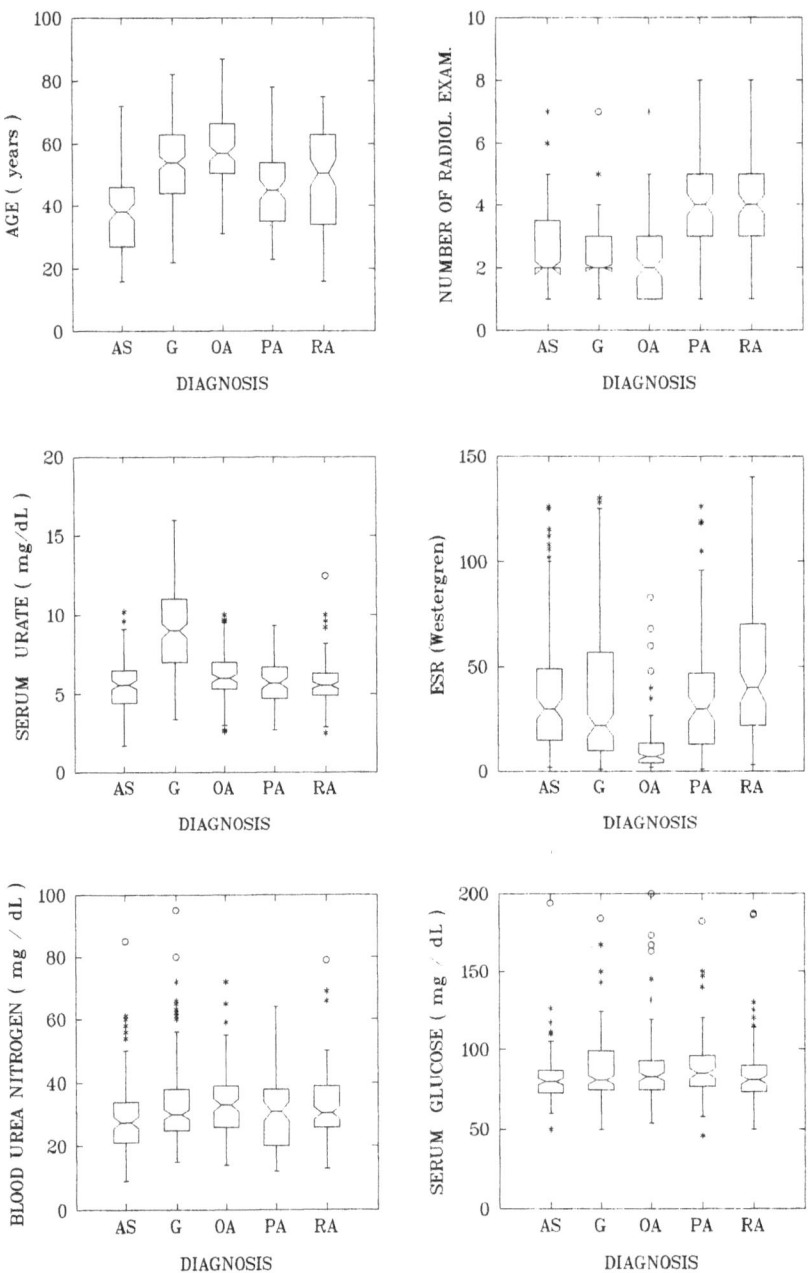

Figure 1. : Grouped Notched box plots of the examined variables

Table 1. Table of RA Test (rows) by Diagnosis (columns)

Frequencies	AS	G	OA	PA	RA	TOTAL
+	3	21	6	9	66	105
-	123	173	113	81	42	532
TOTAL	126	194	119	90	108	637

TEST STATISTIC	VALUE	DF	PROB.
PEARSON CHI-SQUARE	194.024	5	0.000
LIKELIHOOD RATIO CHI-SQUARE	159.832	5	0.000

Table 2. Person Correlation Matrix

VARIABLE	AGE	ESR	S_UA	BUN	S_GLU	NMESRX
AGE	1.000	.0795	.0909	.1639	.1908	-.0577
	p= ---	p=.045	p=.022	p=.000	p=.000	p=.146
ESR	.0795	1.000	-.0022	.1238	.0402	.2044
	p=.045	p= ---	p=.956	p=.002	p=.312	p=.000
S_UA	.0909	-.0022	1.000	.1217	-.0693	-.1772
	p=.022	p=.956	p= ---	p=.002	p=.081	p=.000
BUN	.1639	.1238	.1217	1.000	.0080	-.0700
	p=.000	p=.002	p=.002	p= ---	p=.840	p=.077
S_GLU	.1908	.0402	-.0693	.0080	1.000	.0465
	p=.000	p=.312	p=.081	p=.840	p= ---	p=.241
NMESRX	-.0577	.2044	-.1772	-.0700	.0465	1.000
	p=.146	p=.000	p=.000	p=.077	p=.241	p= ---

Table 3. Discriminant Function Analysis

N= 637	Wilks' Lambda	Partial Lambda	F	p-level	Toler	1-Toler.
AGE	.303	.764	48.26	.000	.891	.109
NMESRX	.287	.806	37.53	.000	.978	.022
S_UA	.360	.642	87.22	.000	.975	.024
ESR	.262	.884	20.60	.000	.926	.073
BUN	.233	.993	1.11	.352	.954	.045
S_GLU	.232	.997	.393	.814	.960	.040
RATEST	.313	.739	55.17	.000	.966	.034

Wilks' Lambda : .23151   approx. F (28.2258)= 40.370   p< 0.0000

As expected the lowest F values obtained were for the variables glucose and BUN that were removed from the models. No variable was highly redundant as evidenced by the high tolerance values obtained. Two models were considered ; in the first one all the considered variables useful in the prediction of diagnostic group memberships were included (model 1) while in the second model only the useful routine laboratory variables were considered (model 2).

Classification matrices of the two models are shown in Table 4. A remarkable drop in the percentage of patients with PA correctly classified was observed in the second model. This drop can be attributed, for the most part, to the importance of the variable number of radiological examinations performed on admission in the differential diagnosis between PA and other diseases.

In conclusion, serum urate concentration, ESR and RA latex test are routine laboratory examinations very useful in the differential diagnosis of the main rheumatic diseases in man.

However, clinical data are necessary to perform the correct diagnosis and in particular, to discriminate PA from other rheumatic diseases.

Table 4. Classification matrices of models

Model 1.
(Rows : observed classifications; Columns : predicted classifications)

GROUP	Percent Correct	AS $p=.198$	G $p=.304$	OA $p=.187$	PA $p=.141$	RA $p=.169$
AS	68.25	86	11	11	15	3
G	68.56	15	133	26	5	15
OA	68.91	7	18	82	7	5
PA	48.89	24	7	6	44	9
RA	60.18	17	4	7	15	65
Total	64.36	149	173	132	86	97

Model 2.
(Rows : observed classifications ; Columns : predicted classifications)

GROUP	Percent Correct	AS $p=.198$	G $p=.304$	OA $p=.187$	PA $p=.141$	RA $p=.169$
AS	57.14	72	16	35	0	3
G	66.49	19	129	26	0	20
OA	62.18	17	22	74	0	6
PA	0.00	43	16	22	0	9
RA	60.18	29	3	11	0	65
Total	53.37	180	186	168	0	103

REFERENCES

1. B.J. Winer, Statistical principles in experimental design, 2nd.Ed.
   New York : McGraw-Hill.(1971).
2. R. Mc Gill , J. W.Tukey and W.A. Larsen, Variations of box plots.
   The American Statistician, 32, 12-16. (1978).

DIHYDROPYRIMIDINURIA: THE FIRST CASE IN JAPAN

Satoru Ohba,[1] Kiyoshi Kidouchi,[2] Satoshi Sumi,[3] Masayuki Imaeda,[2] Naohito Takeda,[4] Hideo Yoshizumi,[4] Akira Tatematsu,[4] Kyoko Kodama,[5] Katsumi Yamanaka,[5] Masanori Kobayashi[1] and Yoshiro Wada[1]

[1]Department of Pediatrics, Nagoya City Univ. Med. Sch.,
[2]Department of Pediatrics, Nagoya City Higashi General Hospital,
[3]Department of Pediatrics, Nagoya Child Welfare Center,
[4]Faculty of Pharmacy, Meijo Univ.,
[5]Nagoya City Health Research Institute, Nagoya, Japan

INTRODUCTION

Dihydropyrimidinuria (McKusick 222748) is a recently described disorder of pyrimidine metabolism that presents neurological symptoms different in degree. Only two cases have been reported to date (Duran et al, 1991; Henderson et al, 1993). The patients with dihydropyrimidinuria excrete large amount of dihydrouracil and dihydrothymine, and moderate amount of uracil and thymine in urine. Therefore this disease is thought to be caused by a deficiency of dihydropyrimidine amidohydrolase (DHPase; EC 3.5.2.2), the second step of pyrimidine base catabolism. The first case, reported by Duran et al (1991), was hospitalized for convulsion and disturbed consciousness at the age of 8 weeks, but whose subsequent development had been normal. The second case reported by Henderson et al (1993) presented severe developmental delay. These two patients were discovered by the urinary gas chromatography mass spectrometry (GC-MS) analysis for the neurological sick children. We report here another case of dihydropyrimidinuria which is the first case in Japan and probably the third worldwide. We discovered her by using high-performance liquid chromatography (HPLC) at the mass screening program.

CASE REPORT

Index patient Y.S. is the first child of healthy non-consanguineous Japanese parents. Pregnancy and delivery were uneventful. She fed on formula for 11 months with normal weaning foods after 6 months of age without taking any medicine. She happened to be discovered in a screening program for inborn errors of pyrimidine metabolism that was operated as a pilot study.

MATERIALS AND METHODS

Screening for Inborn Errors of Pyrimidine Metabolism by HPLC

Dried urine samples on filter papers were collected from 2237 healthy infants with informed consent. Urine constituents were extracted by 5% methanol and were quantified for pyrimidines, i.e. orotic acid(OA), pseudouridine(PsU), dihydrouracil (DHU) and uracil(URA), by column switching high performance liquid chromatography (HPLC) that is based on the previously described report (Ohba et al., 1991) except for monitoring at UV 210 and 280nm absorption. Creatinine concentrations were measured by using an autoanalyser with micro plate based on the method of Folin-Wu.

Quantification of Pyrimidines in Urine and Serum by HPLC

Single voided urine and serum samples were collected from the patient at the age of 8 months and from her parents. They were frozen and stored at −20°C until analysis. Just before analysis, urine samples were passed through a 0.20-μm membrane Millipore filter (Millipore, Japan) to remove particle matter, and serum samples were passed through other Millipore filter with a smaller pore membrane to remove protein bigger than mw 10000 dalton. DHU and URA were quantified by the same method as mentioned above. Dihydrothymine (DHT) and thymine (THY) were quantified by HPLC based on the previously reported method (Kidouchi et al., 1991) except for monitoring at UV 210 and 280nm.

Gas Chromatography Mass Spectrometry (GC-MS)

A 100 μl of urine sample from the patient was passed through column switching HPLC, and the 1.6 ml of eluate containing of respective dihydropyrimidine fraction was collected. To eliminate SO_4^{2-} ion from the fraction, the eluate was passed through the disposable cation-exchange column (TOYOPACK IC-SPM cartridge: Tosoh Inc., Tokyo), which was previously changed to Ba^{2+}type by 10 ml of $0.1M$ $BaCl_2$ followed by 10 ml of water. SO_4^{2-} ion and Ba^{2+} ion banded together into $BaSO_4$ and were eliminated by adhesion to the column. The passed liquid was evaporated to dryness. The residue was trimethylsilylated with pyridine (10μl) and N,O-Bis trimethylsilyl trifluoroacetamide (BSTFA) (20 μl) containing 1% trimethylchlorosilane (TMCS) at 100°C for 60 min. A Shimadzu GC-9A gas chromatograph combined with a double-focusing mass spectrometer (Shimadzu 9020-DF) was used. The GC-MS conditions were as follows: column temperature program, from 100°C to 280°C at 5°C/min; electron-impact (EI) ionization energy, 70eV; chemical-ionization (CI) gas, isobutane.

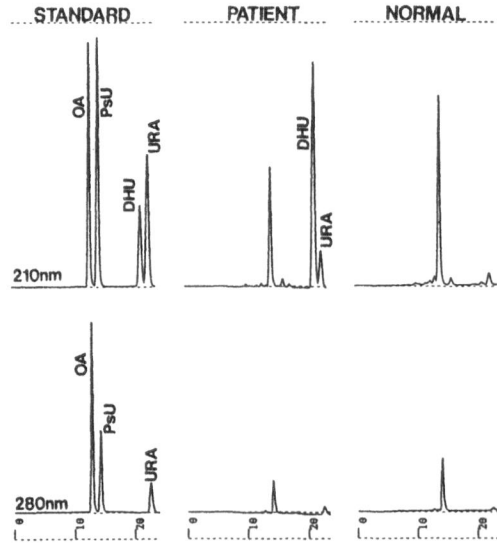

Figure 1. Chromatogram of the column switching HPLC for OA, PsU, DHU and URA from both a 500μM of standard mixture (left), and urine samples of a patient (center) and a normal infant (right). Detection wave length: (upper) 210nm; (lower) 280nm.

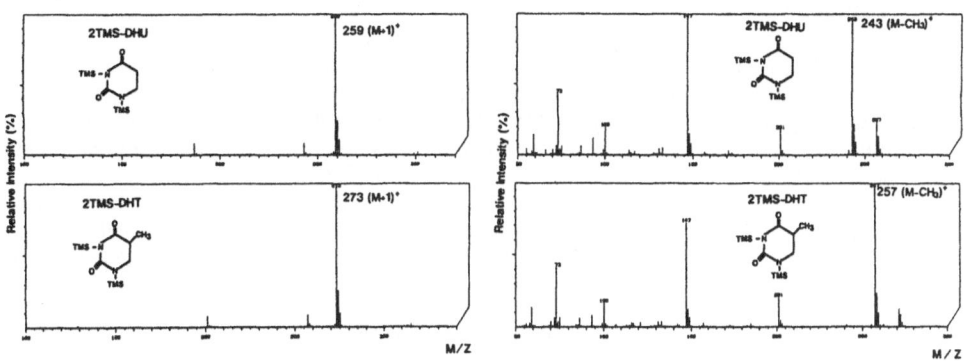

Figure 2. CI mass spectra (left) and EI mass spectra (right) of trimethylsilylated urinary compounds .

Table 1. Pyrimidine metabolite concentrations in urines (μmol/ mmol creatinine) and sera (μmol/l) from a patient with dihydropyrimidinuria and her parents

	DHU	DHT	URA	THY	PsU	ß-Ala	ß-AIB
<Urine>							
patient Y.S.	626.1	451.1	46.9	12.9	83.8	n.d.	n.d.
mother	4.1	n.d.	9.6	n.d.	32.6	n.d.	9.7
father	2.8	n.d.	5.7	n.d.	29.5	n.d.	n.d.
controls(infant)	12.3	trace	15.2	trace	102.6	15>	200>
(n=16)(±S.D.)	±6.3		±10.1		±26.6		
controls(adult)	3.0	trace	6.4	trace	23.2	15>	200>
(n=12)(±S.D.)	±1.2		±3.0		±3.5		
<Serum>							
patient Y.S.	24.2	23.6	n.d.	n.d.	n.d.	n.d.	n.d.
mother	n.d.	n.d.	n.d.	n.d.	n.d.	n.d.	n.d.
father	n.d.	n.d.	n.d.	n.d.	n.d.	n.d.	n.d.
control (infant) (n=1)	n.d.	n.d.	n.d.	n.d.	n.d.	n.d.	n.d.

DHU:dihydrouracil, DHT:dihydrothymine, URA:uracil, THY:thymine, PsU: pseudouridine, ß-Ala:ß-alanine, ß-AIB:ß-aminoisobutyric acid, n.d.:not detected

RESULTS

As to the screening program, we detected one case with a normal PsU peak and an extremely high DHU peak in this pilot study. We confirmed the result after the reexamination.

We gained freshly single voided urine from that case and determined pyrimidines by the column switching HPLC (Table 1). Chromatograms of the column switching HPLC for DHU and URA portion are shown in Figure 1. A remarkable DHU peak and moderately high URA peak are seen in this chromatogram, but are not seen from normal urine. Chromatogram for DHT of a patient urine displayed the prominent peak as well (data not shown).

GC-MS was performed for further identification of dihydropyrimidine. We collected eluate of DHU portion and DHT portion from the HPLC respectively and analyzed by GC-MS. Left side panels of Figure 2 showed Chemical ionization (CI) mass spectra of trimethylsilyl (TMS) derivatives of urinary compounds. These showed a base peak ion at m/z are 259 and 273, respectively. Right side panels of Figure 2 showed electron-impact ionization (EI) mass spectra of the same derivatives. These spectra were the same as those of authentic standards (data not shown).

Table 1 shows pyrimidine metabolites concentrations which revealed extremely high concentration of DHU and DHT, with moderately increased amounts of URA and THY in patient's urine. Blood pyrimidine analysis of the patient revealed excess of DHU, and DHT. The examination of urine and blood for her parents did not reveal abnormal quantities of the dihydropyrimidines.

DISCUSSION

Identification of DHU or DHT excreted in this disorder has been difficult by HPLC, because the maximum UV absorbance of both compounds is adjacent to one another at about 210nm, and furthermore monitoring at UV 210nm cannot rule out almost all of substances including target substances. However, we made it possible to determine these compounds (DHU and DHT) by the special technique with a column switching HPLC. We used on-lined 2 kinds of column for the analysis. Firstly, the sample was analyzed by an octadecylsilane-bonded (ODS) column. Secondly, one portion from ODS column was fractionated to the second column (cation exchange column) and pure dihydropyrimidine peak appeared. And we confirmed the peak as dihydropyrimidine by GC-MS.

We detected one case who excreted extremely large amount of DHU and DHT, and moderate amount of URA and THY, among 2237 apparently healthy infants. In our patient, the PsU level was within the normal value, and ß-Ala and ß-AIB were not detected. Moreover, the serum level of dihydropyrimidines was high and was not influenced by bacteria. These results indicate that the cause of this disorder might be the enzyme defect of DHPase (EC 3.5.2.2), an enzyme that catalyzes the hydrolysis of both dihydropyrimidines.

Two cases have been reported to date. We report here the third case of dihydro-pyrimidinuria worldwide. The first case (Duran et al. 1991) manifested transient convulsion and disturbed consciousness at the age of 8 weeks. The second case (Henderson et al. 1993) presented prolonged severe developmental delay. Our case showed no neurological symptoms until now, the age of 11 months. These results indicate this disease has heterogeneity like another pyrimidine catabolism disorder, dihydropyrimidine dehydrogenase (DPD) deficiency. However special care should be taken in an administration of cancer drugs such as 5-fluorouracil, which are not normally metabolized and accumulate in these patients.

In this disorder, results of laboratory examination are all normal except for the pyrimidine metabolites which are difficult for identification. We established a simple method for determination of dihydropyrimidines by using a column switching HPLC and discovered one case among 2237 healthy infants.

REFERENCES

Duran M, Rovers P, de Bree PK, Schreuder CH, Beukenhorst H, Dorland L and Berger R, 1991, Dihydropyrimidinuria: a new inborn error of pyrimidine metabolism. *J Inher Metab Dis* 14: 367-370

Henderson MJ, Ward K, Simmonds HA, Duley JA and Davies PM, 1993, Dihydropyrimidinase deficiency presenting in infancy with severe developmental delay. *J Inher Metab Dis* 16: 574-576

Kidouchi K, Nakamura C, Katoh T, Kibe T, Ohba S and Wada Y, 1991, Automated quantitative analysis for orotidine and uridine/thymine in urine by high-performance liquid chromatography with column switching. *Adv Exp Med Biol* 309B: 31-34

Ohba S, Kidouchi K, Katoh T, Kibe T, Kobayashi M and Wada Y, 1991, Automated determination of orotic acid, uracil and pseudouridine in urine by high-performance liquid chromatography with column switching. *J Chromatogr* 568: 325-332

NORMAL URIC ACID CONCENTRATIONS IN A PURINE NUCLEOSIDE PHOSPHORYLASE (PNP) DEFICIENT CHILD PRESENTING WITH SEVERE CHICKEN POX, POSSIBLE IMMUNODEFICIENCY AND DEVELOPMENTAL DELAY

Richard J Hallett[#], Stephen M Cronin[#], Gareth Morgan[*], John A Duley[¶], Lynette D Fairbanks[¶], H Anne Simmonds[¶]

[#]Dept of Paediatrics, St Mary's Hospital, Portsmouth; [*]Molecular Immunology Unit, Institute of Child Health and [¶]Purine Research Laboratory, UMDS Guy's Hospital, London, UK

INTRODUCTION

Purine nucleoside phosphorylase (PNP) catalyses the degradation of purine nucleosides and deoxynucleosides to the corresponding bases hypoxanthine and guanine. PNP deficiency is one of the rarest inherited purine disorders. Approximately 40 patients have been reported, 60% of whom have had neurological disorders, with developmental delay evident from birth and generally preceding the onset of immunodeficiency (reviewed in 1). There is a history of recurrent infection dating from infancy, with severe T cell dysfunction. B cell function may be normal initially, but declines with time. Most have died.

Metabolic markers in early presenters are the absence of uric acid in body fluids and its replacement by inosine, guanosine and their deoxyanalogues (Table 1). deoxy GTP (dGTP) is elevated in erythrocytes, associated with severe GTP depletion. There is considerable heterogeneity, both in the age of presentation and the immunological and metabolic parameters.

CASE HISTORY

This male child was first investigated at the age of 13 months due to delay in motor development. He was found to have mild cerebral palsy, but no cause was found. At 2 years he was admitted with severe chicken pox, laryngeal stridor and right middle lobe consolidation, requiring intensive care. Immunological investigations revealed low IgG1, low lymphocyte count, poor T cell function and CD4 deficiency, but he was HIV negative (Table 1). The association of developmental delay, with immunodeficiency suggested PNP deficiency, but plasma uric acid concentration by routine laboratory testing was normal.

Because of the similarity of these findings to a previous case of PNP deficiency (2) who had received a prior blood transfusion (Table 2), the child was referred for purine studies, although blood transfusion was later excluded. The analytical methods used have been reported in detail in previous publications (2).

RESULTS

Uric acid concentration in plasma was just below the bottom of the normal range for children- (130-230µmol/l) and the urine concentration on a creatinine basis was also within the normal range for children when tested by a specific enzymic method (Table 2). However, the

TABLE 1   Immunological investigations at presentation

Immunoglobulins (g/l)		IgG 5.2	IgA 0.9	IgM 2.1
Control age 2-3 yrs		(3.7-15.8)	(0.3-1.3)	(0.1-0.65)
IgG subclasses	IgG1 3.51 (4.3-9.8)	IgG2 0.87 (0.3-3.9)	IgG3 0.18 (0.1-0.8)	IgG4 0.23 (0.1-0.65)

"IgG 1 is low"

	10^9/l
Lymphocyte count	1.2
T lymphocytes (CD3)	0.56
T helper (CD4)	0.12
T supressor (CD8)	0.42
B lymphocytes (CD19)	0.19
NK cells (CD 58)	0.01

"CD4 deficiency"

urine contained elevated concentrations of the normally undetectable nucleosides, inosine and guanosine and their deoxy analogues, as well as elevated concentrations of hypoxanthine and xanthine (data not shown).The latter two bases are normally undetectable in PNP deficiency - even in our transfused case (2). Total urinary purine excretion on a creatinine basis however was elevated, consistent with the purine overproduction characteristic of PNP deficiency. Intact and lysed erythrocytes repeatedly showed low but detectable PNP activity (Table 2), whilst APRT activity (normally elevated) was essentially within the normal (range 16-32). The parents' erythrocytes exhibited heterozygote activity (data not shown). Intact fibroblasts showed low residual activity, but no significant lysate activity. Consistent with the presence of residual PNP activity in the erythrocytes was the absence of detectable dGTP (Table 2) and the fact that GTP concentrations were not depleted (normal for children 48-84μmol/l).

The patient showed steady improvement following immediate treatment with intravenous (IV) acyclovir and antibacterial agents. Subsequent therapy has included prophylactic antimicrobials and three weekly infusions of IV immunoglobulin. Bone marrow transplant is planned.

DISCUSSION

Repeated investigations of PNP activity in erythrocytes from this child have demonstrated the same low residual activity, thereby excluding spurious elevation due to a prior (undisclosed) blood transfusion, the explanation in our previous case (2). The biochemical results in the present child are consistent with a diagnosis of 'partial'PNP deficiency, but the clinical findings differ from those reported in two siblings with 'partial' PNP deficiency (3) who also had normal uric acid concentrations in plasma and urine, but a milder immune disorder, growth retardation only in the elder and the complete absence of neurological abnormalities in the younger who also presented later (6 years).

The lymphotoxicity in PNP deficiency has been related to the inability to degrade deoxyguanosine with resultant accumulation of dGTP in T cells which inhibits ribonucleotide reductase and hence cell division (1,4), although the toxicity to resting T cells is unclear (1). This dGTP accumulation is normally mirrored by the elevated dGTP concentrations in the erythrocytes of severe cases (Table 2). The neurological deficits have been considered an indirect consequence of the defect - in the absence of PNP the next step in the salvage cycle catalysed by hypoxanthine-guanine phosphoribosyltransferase (HPRT) is inoperable due to lack of substrate (2). The low erythrocyte GTP concentrations normally found in PNP deficiency (Table 2), are characteristic also of the Lesch-Nyhan syndrome (complete HPRT deficiency) and it has been postulated that the neurological deficits in both disorders may reflect

Table 2   Biochemical data from PNP deficient patients studied in London, GB

Patient	1	2	3	4	5*	Present case
Sex	m	f	m	m	m	m
Age (yrs)	birth	4	11/12	4 5/12	6	2
Plasma (μmol/l)						
uric acid	4	<1	<1	<1	50*	109
inosine	40	71	49	62	16	9
guanosine	8	14	13	23	1	<1
deoxyinosine	5	9	6	19	6	<1
deoxyguanosine	5	6	5	14	1	<1
Urine (mmol/l)						
uric acid	<1	T	T	T	1.4*	0.83
inosine	1.94	1.65	1.44	3.02	6.4	0.46
guanosine	0.63	0.58	0.74	1.01	4.8	0.10
deoxyinosine	0.43	0.49	0.41	0.99	5.2	0.26
deoxyguanosine	.31	0.37	0.33	0.75	3.0	0.09
Erythrocyte						
nucleotides (μmol/l)						
ATP	1140	1012	1569	1784	1716*	2024
GTP	6	6	5	9	152*#	46
dGTP§	7	4	4	7	<1	<1
enzymes (nmol/h/mg/Hb)						
PNP	0	0	0	0	580*	44
APRT	45	49	42	70	36*	33

*T = trace (<0.01)   *blood transfusion # = spuriously elevated due to ribavirin therapy (2) § = normally <.01*

a similar inability of the brain to sustain GTP at concentrations compatible with normal physiological function (2).

The comparative severity of the immunological involvement in the present case, coupled with the presence of neurological abnormalities, despite the detectable PNP activity in erythrocytes, the absence of erythrocyte dGTP accumulation or GTP depletion, is not consistent with the above hypothesis. The present findings suggest the existence of an unusual tissue specificity of the enzyme in this patient. This is being examined by isozyme analysis. There was no history of consanguinity - a feature of other kindreds (1, 2) - although the HLA typing showed a remarkable similarity.

In summary there are three lessons to be learnt from this new case. First, the presence of uric acid alone should not be used to exclude PNP deficiency in an immunodeficient child. A full metabolic screen should be undertaken. Second, neurological abnormalities in a child presenting with immunodeficiency should alert the clinician to the possibility of PNP deficiency. Third, structural features may be important in controlling PNP activity in different tissues and thus the relationship between genotype and phenotype.

REFERENCES

1. L.M. Markert. Purine nucleoside phosphorylase deficiency. Immunodeficiency Reviews 3:45 (1991).
2. G. Morgan, S. Strobel, C. Montero, J.A. Duley, P.M. Davies and H.A. Simmonds. Raised IMP-dehydrogenase activity in the erythrocytes of a case of purine nucleoside phosphorylase deficiency. Adv Exp Med Biol 309B:297 (1991).
3. W.D. Biggar, E.R. Giblett, R.L. Ozere and B.D. Grover. A new form of purine nucleoside phosphorylase in two brothers with defective T-cell function. J Pediatrics 92:354 (1977).
4. L.D.Fairbanks, A.Taddeo, J.A.Duley and H.A.Simmonds. Mechanisms of deoxyguanosine toxicity: Human thymocytes but not peripheral blood lymphocytes accumulate deoxy GTP in conditions simulating PNP deficiency. J Immunol 144:485 (1990).

BIOCHEMICAL AND IMMUNOLOGICAL STATUS FOLLOWING GENE THERAPY AND PEG-ADA THERAPY FOR ADENOSINE DEAMINASE (ADA) DEFICIENCY

Lynette D. Fairbanks¶, H. Anne Simmonds¶, Peter M. Hoogerbrugge§, Victor W. van Beusechem§, Dinko Valerio§, Anne Moseley§, Roland J. Levinsky#, Hubert B. Gaspar#, and Gareth Morgan#

#Molecular Immunology Unit, Institute of Child Health and
¶Purine Research Laboratory, UMDS Guy's Hospital, London, GB;
§TNO Biological Laboratories & Introgene, Rijswijk, The Netherlands

INTRODUCTION

A deficiency of the enzyme adenosine deaminase (ADA) has been established as the cause of the immunodeficiency in 25-30% of patients with inherited severe combined immunodeficiency (SCID). ADA is essential for the catabolism of deoxyadenosine (dAdo) and a deficiency leads to the accumulation of dAdo with subsequent conversion to the equally lymphotoxic metabolite, deoxy ATP (dATP), accompanied by ATP depletion in severe cases (1-3). This is one of the commonest genetic purine disorders. Some 33 patients have been identified by us alone, 60% of whom have presented in the first weeks of life (reviewed in 1,2). In the small number of children presenting later there is a history of recurrent infection with severe T cell dysfunction, but total lymphocyte numbers and B cell function may be normal initially, when the defect may not be suspected (1). ADA deficiency has been diagnosed recently in two sisters in their 30's, originally categorised as 'non HIV' AIDS (4, and this volume), indicating the same genetic heterogeneity common to other purine disorders.

Untreated, the defect is invariably fatal. Many modalities were tried initially without success (1,2). The cure rates achieved with HLA identical bone marrow transplants are very good, but mismatched haploidentical transplantation has been less successful. Enzyme replacement therapy with polyethylene glycol ADA (PEG-ADA) was introduced in 1986 and has provided sustained benefit in up to 40 cases (5). The success of bone marrow transplantation has made ADA deficiency a prime target for somatic gene therapy and it was thus the first genetic metabolic human disorder in which this technique has been performed. However, the clinical efficacy of this treatment in the patients so far reported has been difficult to interpret because the protocol required simultaneous administration of PEG-ADA (6). This paper presents data in the first such case to be treated in Great Britain (GB).

CASE HISTORY

This 1.5 year old female was the second child of non-consanguineous parents. She was investigated and found to be ADA deficient at birth because the elder male sibling had been diagnosed as ADA deficient shortly before death from disseminated adenoviral infection at 2yrs of age. The immunological and biochemical parameters in the baby (Tables 1 and 2) were

Table 1   Immunological parameters at birth and following gene therapy

Age	Gene Therapy			PEG-ADA	
	2/52	9/12	12/12	17/12	21/12
Haematology					
WCC	1.8	3.0	1.4	3.7	3.2
Lymphocytes	0.9	0.72	0.3	1.46	1.4
T-cell subsets %					
CD3	89	82	82	74	85
CD19	4	2	1	7	5
Immunoglobulin production (on IVIG)					
IgG	8.3	15.7	18	10.4	15.24
IgA	<0.03	0.19	0.17	0.23	0.28
IgM	0.08	0.36	0.29	0.48	0.51
Isohaemagglutinins (Gp 0 +ve)	anti A-ve anti B 1:4	anti A 1:1 anti B 1:1	Absent	anti A 1:2 anti B 1:1	anti A 1:8 anti B 1:2

consistent with the milder form of the defect which explained the later presentation in the first child.Both had low but not absent lymphocyte count and preserved T cell function, the main abnormality in this case being a low B cell count as indicated by the CD19%.(Table 1) Careful measures were taken to prevent infection, including daily prophylactic Septrin, Acyclovir and Itroconazole, plus 2 weekly IV immunoglobulin. Apart from an episode of RSV bronchiolitis at 7 months (treated with Ribavirin) there have been no documented infections.

No histocompatible donor was available and because of the poor success rate using haploidentical bone marrow transplant (<50%), and the child's relatively preserved immune function and good clinical status, official consent was obtained to use the protocol for transfection without concomitant use of PEG-ADA (7). At 9 months of age, CD 34 +ve bone marrow progenitor cells were selected and transfected using a modified retroviral vector carrying the ADA gene. Transduced cells were then reinfused into the patient.

RESULTS AND DISCUSSION

The dATP concentrations in the erythrocytes of this child in the first month of life (Table 2) were much lower (520μmol./l) than in 12 previous untransfused cases with neonatal onset (mean dATP 1783μmol./l) and lower than in the majority of cases with late onset (4m-2y:mean dATP 557μmol./l).As in the other later presenters ATP concentrations were normal in this child, and low but detectable ADA activity was present in unfractionated lymphocytes,. consistent with the milder form of the defect.

Following infusion of transfected bone marrow progenitor cells bone marrow and peripheral blood have been analysed regularly, but early evidence of the presence of the gene in peripheral blood at 2 months has not been a consistent finding subsequently.. PCR analysis of bone marrow at 6 months showed the gene was stably integrated, but it was not possible to ascertain either the proportion of cells transfected or the level of expression..Studies at twelve months have demonstrated an increment in lymphocyte ADA activity (Table 2) and investigations to determine which cell population has the transfected gene are in progress..

Immunology and purine nucleotide metabolism studies carried out at weekly intervals over the four months post transfection showed no significant change in either dATP concentrations or immunological parameters (Tables 1, 2, Fig 1) At this point, because of the known deterioration in immunity in the sibling at this age, PEG-ADA therapy was commenced.

Table 2   Biochemical data at birth and following gene therapy

Patient Age		Erythrocytes (µmol/l)		(nmol/h)	Plasma µmol/ml/h	Lymphocytes (pmol/10⁶ cells)		(nmol/h)#
		ATP	dATP§	SAHH	ADA	ATP	dATP§	ADA
Born			*27/6/92*					
	2/52	1888	520	0.19		3625	115	7
	9/12	1690	816	0.2		3590	254	6
Gene Therapy (GT)			*18/3/93*					
	1yr	1649	580	0.39		3682	252	21
PEG-ADA			*27/7/93*					
	15/12	1634	92	4.6	44	3878	<1	507
	21/12	1949	25	4	49	3473	<1	438
Control (mean or range)								
		1570	*<1*	*3-12*		*3070*	*<1*	*(1162-4500)*

ADA deficient children			
(n)	Age onset		
12	<2/12	813	1783
Range		(527-1025)	(1121-2478)
6	4/12 - 2yr	1602	557
Range		(761-1888)	(304-912)

 # enzyme activity either in nmol/h/mg haemoglobin (RBC) or protein (lymphocytes)

Clinically the child has progressed well since then and showed normal development. An immediate response to PEG-ADA was evident from the rapid rise in B cell numbers and the steady reduction in dATP which resulted (Fig 1). However, the concentrations over the past six months have rarely fallen below 20µmol./l and values at one year post therapy have remained around this level, despite acceptable plasma ADA concentrations. (Fig 1). Six months after starting PEG-ADA lymphocyte numbers started to fall again, as did the biochemical parameters, suggesting the formation of antibodies or increased drug clearance. The problem was overcome by increasing the dose from the initial 250u/week to 500u/week (55u/kg/week)

The principal mechanism of lymphotoxicity in ADA deficiency has been related to the inability to degrade dAdo with accumulation of dATP in T and B cells which inhibits ribonucleotide reductase and hence cell division (1,4). The toxicity to resting T cells may relate to the accumulation of dAdo itself, which inhibits S-adenosylhomocysteine hydrolase (SAHH) and hence restricts vital cellular methylation reactions (1). It is noteworthy that erythrocyte SAHH activity returned rapidly from virtually undetectable values to activities within the normal range following PEG-ADA, but this was not accompanied by an equally rapid fall in dATP concentrations. The accumulation of dATP in erythrocytes has been used as a marker for the severity of the defect and also to determine the efficacy of therapy with PEG-ADA (4,5) It is now being used together with immunological parameters, as a guide to progress in gene therapy (6).

In summary, one year following gene therapy the clinical condition of this child is very good and she has been at home for six months with no illnesses. There is limited evidence, both from PCR data and the increased ADA activity in peripheral blood lymphocytes that the transfected gene has been expressed.She is being monitored on a regular basis for further evidence of gene integration and expression, or any adverse effects of gene therapy, such as wild type virus formation. At present the changes are not sufficiently dramatic to allow the withdrawl of PEG-ADA.

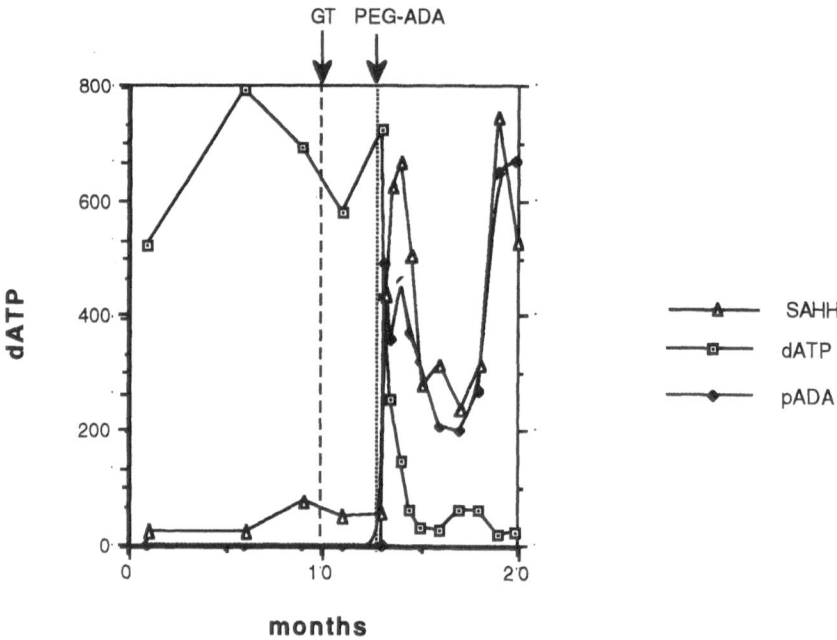

Figure 1 Response to therapy from March 1993 to March 1994.

REFERENCES

1.G.Morgan, R.J.Levinsky, K.Hugh-Jones, L.D.Fairbanks, G.S.Morris and H.A.Simmonds Heterogeneity of biochemical, clinical and immunological parameters in severe combined immunodeficiency due to adenosine deaminase deficiency. Clin exp Immunol 70:491(1987)
2.N.M.Kredich and M.S.Hershfield Adenosine deaminase deficiency and purine nucleoside phosphorylase deficiency. Chapter 40, in The Metabolic Basis of Inherited Disease,.C.R. Scriver, A.L. Beaudet, W.S. Sly, D. Valle, eds., McGraw-Hill, New York, 6th edition (1989).
3. A. Goday, H.A. Simmonds, G.S. Morris and L.D. Fairbanks. Human B lymphocytes and thymocytes but not peripheral blood mononuclear cells accumulate high dATP levels in conditions simulating ADA deficiency. Biochem Pharmacol 34: 3561(1985)
4. C.L. Shovlin, A.D.B. Webster, L.D. Fairbanks, H.A. Simmonds, S. Deacock, R. Lechler, I. Roberts and J.M.B. Hughes Adult presentation of inherited adenosine deaminase deficiency. Lancet 341:1472 (1993).
5.M.S. Hershfield, S. Chaffee and R.U Sorensen. Enzyme replacement therapy with polyethylene glycol-adenosine deaminase and adenosine deaminase deficiency. Pediatr. Res. 33:S42 (1993)
6.R.M.Blaese et al Treatment of SCID due to ADA deficiency with autologous lymphocytes transduced with the ADA gene. Human Gene Therapy 4:521 (1993).
7.V.W. van Beusechem, A. Kukler, P.J. Heidt and D. Valerio. Long-term expression of human adenosine deaminase in rhesus monkeys transplanted with retrovirus-infected bone-marrow cells. Proc. Natl Acad. Sci USA 89:7640 (1992).

GENETIC MANIPULATION OF HEMATOPOIETIC STEM CELLS IN ADENOSINE DEAMINASE DEFICIENCY

David A. Williams[1,2] and Thomas Moritz[1]

[1]Herman B Wells Center for Pediatric Research
James Whitcomb Riley Hospital for Children
[2]Howard Hughes Medical Institute
Indiana University
Indiana University School of Medicine
Indianapolis, IN 46202-5225

INTRODUCTION

Increasing use of molecular biologic methods, specifically gene transfer technology, has made genetic therapy a potential therapeutic approach for some human diseases. Somatic gene therapy is the attempt at correction of a disease by the transfer and expression of a normal genetic sequence in somatic cells which are deficient in the expression or function of the homologous gene. Retroviral vectors, which have been developed by a number of investigators over the last decade, can efficiently transfer new genetic material into a variety of mammalian cells. Lifelong correction of several severe genetic diseases of the bone marrow should be possible by the efficient transfer of new genetic sequences into reconstituting hematopoietic stem cells, those primitive cells in the bone marrow responsible for long term and multilineage hematopoiesis *in vivo*.

Genetic Correction of Adenosine Deaminase Deficiency

Adenosine deaminase (ADA) plays a key role in purine metabolism and genetic deficiency of this enzyme is associated with 30% of cases of severe combined immunodeficiency (SCID) (Kredich and Hershfield, 1983). SCID due to ADA deficiency is generally fatal, if untreated; inherited as an autosomal recessive disease; and phenotypically characterized as a lack of both cell-mediated and humoral immunity. SCID due to ADA deficiency has been considered by many investigators as an appropriate model for somatic gene therapy due to several disease characteristics. These characteristics include: a severe and predictable phenotype; the ability to correct the disease with allogeneic bone marrow transplantation demonstrating that the disease is expressed in progeny of hematopoietic stem cells; the nature of the abnormality (abscence or low expression of a single polypeptide normally present in all cells); and the fact that low levels of enzyme may be adequate to correct the phenotype *in vivo*.

An understanding of some aspects of the retroviral life cycle is essential to appreciate some of the current limitations of retroviral vector applications to treatment of human diseases. A detailed review of the use of recombinant retroviral vectors has recently been published (Apperley and Williams, 1990; Miller, 1992; Anderson, 1992). After entry of the recombinant viral genome into the infected cell and generation of proviral DNA by reverse transcriptase, integration of the double stranded proviral structure requires cell division. This aspect of the retroviral life cycle is problematic for genetic modification of primitive hematopoietic stem cells, since the majority of these cells are normally in G_0 of the cell cycle.

Analysis of recombinant retroviral vectors expressing the ADA cDNA has been accomplished in our and several other laboratories over the past decade. A recombinant retroviral vector which expresses the human ADA cDNA from an internal human phosphoglycerate kinase (PGK) promoter has been useful for expression of the transferred sequence *in vivo* in murine bone marrow recipients (Williams et al., 1986). This simplified vector design does not rely on dominant selectable markers, such as neo phosphotransferase (Neo) for selection in infected cells, since some evidence suggests that expression of two genes has deleterious effects on the level and stability of expression *in vivo* (Apperley et al., 1989). Transduction of murine hematopoietic stem cells is accomplished using a modification described by Dick et al. (1985) of the method originally described in 1984 (Williams et al., 1984). Primitive hematopoietic stem cells are induced to enter cell cycle *in vivo* following 5-fluorouracil-induced ablation. These cells are collected and infected by cocultivation with a retroviral packaging cell line producing high titer recombinant PGK-hADA virus. Cells harvested after a 48 hour infection are infused into lethally-irradiated syngeneic recipients using standard murine bone marrow transplantation protocols.

Analysis of expression of the transferred human ADA sequence has been accomplished by harvesting primary hematopoietic tissues from recipient mice 4-6 months after transplantation (Lim et al., 1987). Electrophoretic separation of the human and murine ADA isozyme on cellulose acetate and *in situ* staining for enzyme activity provides a simple and semi-quantitative assay for the expression of the introduced sequence. Expression of the endogenous murine enzyme provides a convenient internal control for comparison. Using this protocol and the simplified PGK-hADA vector, expression in all primary hematopoietic lineages, including T-cells and thymus was easily detected after full hematopoietic reconstitution. In some mice, the level of expression of human ADA was equal to or greater than endogenous murine ADA.

Subsequent modification of this infection protocol by Luskey et al improved these results (Luskey et al., 1992). After the harvest of murine bone marrow cells following 5-FU treatment, a 48 hour prestimulation with recombinant growth factors prior to infection by cocultivation was utilized. Use of IL-6 and stem cell factor (SCF) led to transduction of hematopoietic stem cells at relatively high frequency, since all transplant recipient mice were documented to express human ADA after reconstitution in three consecutive experiments. In summary, our laboratory and other investigators have used recombinant retroviral vectors to transduce murine hematopoietic stem cells at a frequency which allows consistent expression of new genetic sequences in mice *in vivo*.

Gene transfer into human hematopoietic cells

The translation of retroviral technology from murine to large animal hematopoietic cells has been more problematic. Current dilemmas in the use of retroviral vectors for transduction of hematopoietic stem cells in large animals include: inefficiency of infection or transduction of primitive (reconstituting) hematopoietic stem cells; lack of stable expression of the introduced genetic sequences using some vectors; and for some applications, inability to precisely control the level of expression of the introduced sequences. The latter point is particularly important for future attempts at treating hemoglobinopathies since precise control of globin expression would be required for corrective treatment.

Our laboratory and other investigators have developed several approaches to improve gene transfer efficiency into primitive hematopoietic cells of large animal species (Table 1).

Table 1. Approaches to improved gene transfer efficiency into primitive human (primate) hematopoietic cells.

---

1) Manipulate the microenvironment of hematopoietic cells during infection

       a) use of multiple growth stimulating proteins
       b) alteration in growth factor presentation
       c) presentation of adhesive ligands

2) Use of alternative sources for reconstituting hematopoietic stem cells

       a) mobilized peripheral blood stem/progenitor cells
       b) human umbilical vein stem/progenitor cells

These include both manipulation of the microenvironment of the hematopoietic cells during infection and use of alternative sources for reconstituting hematopoietic stem cells. The use of umbilical cord stem and progenitor cells as an alternative target for genetic modification will be further detailed here. Cord blood contains a high concentration of committed hematopoietic progenitor cells (Broxmeyer et al., 1989). Previous transplant experience in humans demonstrates that a single cord blood sample achieves successful bone marrow reconstitution in children implying that cord blood contains reconstituting hematopoietic stem cells (Gluckman et al., 1989). Some data from *in vitro* cultures suggest that a single cord blood sample might be capable of engrafting an adult, although this has never been attempted *in vivo*. Initial experiments in our laboratory sought to determine if cord blood stem/progenitor cells were capable of genetic modification using retroviral vectors. For these experiments two recombinant retroviral vectors were utilized (figure 1). A low titer vector expressing the neo phosphotransferase selectable marker from an internal thymidine kinase promoter (N_2/ZipTkNeo) was used. The titer of this vector, 2×10^5/ml, was determined by formation G418-resistant colonies (Moritz et al., 1993). A high titer vector expressing the *murine* adenosine deaminase (mADA) cDNA from an internal phosphoglycerate kinase (PGK) promoter was also utilized. The titer of this vector was determined by the expression of the murine ADA protein in human cells as analyzed by protein gel electrophoresis and *in situ* staining of enzyme activity (Moritz et al., 1993).

Infection of human cord blood and bone marrow cells was carried out by pre-stimulation of hematopoietic cells in IL-6 and stem cell factor (SCF) for 48 hours followed by infection using cocultivation of the target cells with viral packaging cells producing either of the two vectors for 48 hours (Moritz et al., 1993). After infection, cells were plated in short term progenitor assays or in long term culture-initiating cell (LTC-IC) assays. Analysis of gene transfer was accomplished by determining the percentage of colonies resistant to G418 (N_2/ZipTkNeo) or the percentage of colonies which expressed murine ADA (PGK-mADA).

Using the low titer N_2/ZipTkNeo virus, infection of cord blood progenitor cells was consistently 2-3 fold more efficient than adult bone marrow cells. The increased efficiency of gene transduction was noted in both mixed/erythroid progenitor-derived (CFU-Mix/BFU-E) colonies and myeloid progenitor-derived (CFU-GM) colonies. The increased transfer efficiency was also evident using this low titer virus when more primitive LTC-IC derived colonies was examined. In this assay, progenitor colonies derived from 5 week old long term cultures are analyzed. Over 2-fold more LTC-IC derived from cord blood were transduced compared to adult bone marrow.

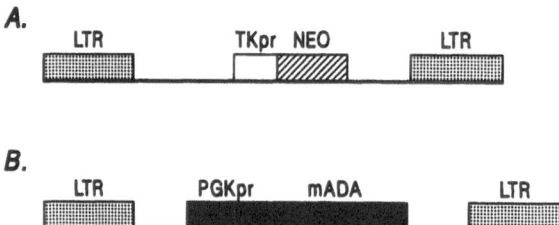

Figure 1. Schematic structure of retroviral vectors. A. N_2/Zip TKNeo. B. PGK-mADA. LTR, long terminal repeats; TKpr, herpes thymidine kinase promoter; NEO, neo phosphotransferase; PGKpr, phosphoglycerate kinase promoter; mADA, murine adenosine deaminase cDNA.

The differences in gene transfer efficiency was also noted with the high titer PGK-mADA virus. Using this vector 100% of progenitor cells from cord blood were consistently transduced (Table 2). In addition 75-100% of LTC-IC derived colonies (Table 2) were transduced and the level of expression of the introduced mADA cDNA was consistently 2-4 fold higher than the endogenous human ADA expression (figure 2).

Table 2. Infection efficiency of cord blood LTC-ICs using PGK-mADA vector

Week in Culture

exp. #	protocol	day 0[1]	week 5[2]
1	co culture	10/10[3]	7/11
2	co culture	12/12	11/11

[1]Committed progenitors
[2]LTC-IC-derived progenitors
[3]number of mADA expressing colonies/total colonies analyzed

In summary, umbilical cord blood stem/progenitor cells appear to be capable of genetic modification using recombinant retroviral vectors. These cells have already been shown to be an alternative source of transplantable stem cells in pediatric transplant studies. Together with the increasing ability to perform pre-natal diagnosis, these observations suggest that umbilical cord blood stem/progenitor cells may be useful targets for genetic manipulation for treatment of severe genetic diseases of childhood.

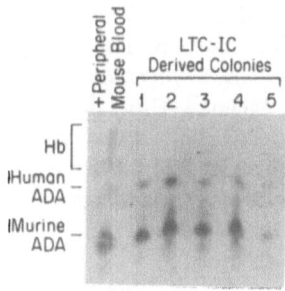

Figure 2. Expression of murine ADA in LTC-IC-derived colonies. Five individual LTC-IC-derived colonies analyzed for expression of introduced murine ADA after electrophoretic separation from endogenous human ADA. Hb, hemoglobin; +peripheral mouse blood, sample of blood from mouse expressing both murine and human ADA as positive control.

ACKNOWLEDGEMENTS

We thank D. Giarla for manuscript preparation. This work is supported by National Institutes of Health grant RO1 HL46528 and National Cancer Institute Program Project Grant PO1 CA59348. T. Moritz is supported by grant #300 402 635/2 from the German Cancer Aid/Mildred Scheel Foundation. Bone marrow donations are supported by the Riley Cancer Research for Children.

REFERENCES

Anderson, W.F., 1992, Human gene therapy. *Science* 256:808.

Apperley, J.F., Lim, B., Orkin, S.H., and Williams, D.A., 1989, Gene transfer of human adenosine deaminase in mice: effect if selectable marker sequences on long term expression. *Exp. Hematol.* 17:484.

Apperley, J.F. and Williams, D.A., 1990, Gene therapy: current status and future directions. *Br. J. Haematol.* 75:148.

Broxmeyer, H.E., Douglas, G.W., Hangoc, G., Cooper, S., Bad, J., English, D., Arny, M., Thomas, L., and Boyse, F.A., 1989, Human umbilical cord blood as a potential source of transplantable hematopoietic stem/progenitor cells. *Proc. Natl. Acad. Sci. , USA* 86:3828.

Dick, J.E., Magli, M.C., Huszar, D., Phillips, R.A., and Bernstein, A., 1985, Introduction of a selectable gene in primitive stem cells capable of long-term reconstitution of the hematopoietic system of w/wᵛ mice. *Cell* 42:71.

Gluckman, E., Broxmeyer, H.A., Auerbach, A.D., Friedman, S., Douglas, G.W., Devergie, A., Esperou, H., Thierry, D., Socie, G., Lehn, P., Cooper, S., English, D., Kurtzberg, J., Bard, J., and Boyse, E.A., 1989, Hematopoietic reconstituttion in a patient with Fanconi's anemia by means of umbilical-cord blood from an HLA-identical sibling. *N. Eng. J. Med.* 321:1174.

Kredich, N.M. and Hershfield, M.S. (1983). Diseases caused by adenosine deaminase deficiency and purine nucleoside phosphorylase deficiency. In The Metabolic Basis of Inherited Disease. J.B. Stanbury, J.B. Wyngaarden, D.S. Fredrickson, J.L. Goldstein, and M.S. Brown, eds. (New York: McGraw-Hill), pp. 1157-1183.

Lim, B., Williams, D.A., and Orkin, S.H., 1987, Retrovirus-mediated gene transfer of human adenosine deaminase: Expression of functional enzyme in murine hematopoietic stem cells *in vivo*. *Mol. Cell. Biol.* 7:3459.

Luskey, B.D., Rosenblatt, M., Zsebo, K., and Williams, D.A., 1992, Stem cell factor, IL-3 and IL-6 promote retroviral-mediated gene transfer into murine hematopoietic stem cells. *Blood* 80:396.

Miller, D.A., 1992, Human gene therapy comes of age. *Nature* 357:455.

Moritz, T., Keller, D.C., and Williams, D.A., 1993, Human cord blood cells as targets for gene transfer: Potential use in genetic therapies of severe combined immunodeficiency disease. *J. Exp. Med.* 178:529.

Williams, D.A., Lemischka, I.R., Nathan, D.G., and Mulligan, R.C., 1984, Introduction of new genetic material into pluripotent stem cells of the mouse. *Nature* 310:476.

Williams, D.A., Orkin, S.H., and Mulligan, R.C., 1986, Retrovirus-mediated transfer of human adenosine deaminase gene sequences into cells in culture and into murine hematopoietic cells *in vivo*. *Proc. Natl. Acad. Sci. , USA* 83:2566.

A1 ADENOSINE RECEPTORS IN HUMAN NEUTROPHILS: ELECTRON MICROSCOPE LOCALIZATION USING A COLLOIDAL CHA-GOLD-ALBUMIN PROBE

Claudia Martini[1], Umberto Montali[2], Laura Giusti[1], Marcello Fiorini[1], Alessandra Falleni[3], Vittorio Gremigni[3], Antonio Lucacchini[1]

[1] Istituto Policattedra di Discipline Biologiche
[2] Istituto di Chimica Biologica
[3] Dipartimento di Biomedicina Sperimentale, Infettiva e Pubblica
  Università di Pisa, 56126 Pisa, Italy

INTRODUCTION

Adenosine is an ubiquitous nucleoside that mediates several important physiological effects on numerous tissues and cells, including cells involved in immune and inflammatory reactions[1]. These effects are mediated by the binding of adenosine to either of two receptor subtypes, which have been pharmacologically classified as A_1 and A_2[2]. Adenosine modulates a variety of human polymorphonuclear leukocyte (PMN) functions acting via PMN surface adenosine receptors as demonstrated by functional evidence using adenosine analogues[3,4,5,6,7]. The direct labeling of A_1 adenosine receptors using potent adenosine analogues and the localization of these receptors on PMN by electron microscopy are important tools to understand the effects of adenosine modulation. Previously we used a ferritin labelled probe to visualize the A_1 adenosine receptors in guinea pig synaptoneurosome preparations[8].

In this study we labeled A_1 adenosine receptors using ^{3}H-^{6}N-cyclohexyladenosine ([^{3}H]CHA) as specific ligand and synthesized a colloidal CHA-gold-albumin with the goal of obtaining a specific marker that could bind to the A_1 receptors and localize them in human neutrophils by electron microscopy.

MATERIALS AND METHODS

[^{3}H]CHA was from New England Nuclear (specific activity 30.2 Ci/mmol); 20 nm colloidal gold conjugate bovine serum albumin was from Bio-Cell Research Laboratories, all other compounds, salts and reagents were from standard commercial sources.

Purine and Pyrimidine Metabolism in Man VIII, Edited by
A. Sahota and M. Taylor, Plenum Press, New York, 1995

Synthesis of the probe

Oxidized CHA preparation was performed essentially as described by Gilhman[9]: 3 mM CHA and 3 mM metaperiodate cold solutions in water, pH 6.0, were mixed together and the reaction was allowed to proceed for 1 hour at 0 °C in the dark. Colloidal gold-albumin (2.3 nmoles) was dialyzed overnight against 50 mM Tris-HCl buffer pH 7.4 (T), and incubated with oxidized CHA (40 nmoles) in 0.25 ml of T buffer for 3 hours at 25 °C. For preparation of a stable oxidized CHA-gold-albumin complex, the Schiff bases, formed by the reaction between the aldheidic functions of the oxidized CHA and the primary amines present in albumin, were reduced by addition of $NaBH_4$ at a final concentration of 80 mM in the incubation mixture. The reduction was allowed to proceed for 60 min. at 25 °C and the reaction mixture was dyalized exhaustively against the T buffer. The substitution (nmoles of CHA / nmoles of colloidal gold-albumin) was determined by carrying out a parallel preparation in which labelled CHA was used.

Neutrophil preparation

Neutrophils were obtained essentially as described by Boyum[10] using a mixture of sodium metrizoate 9.6% (w/v) and polysaccharide 5.6% (w/v) (Lymphoprep, Nycomed Pharma, Oslo, Norway).

Healthy volunteer blood was collected into tubes containing anticoagulant (heparin), and was diluted by addition of an equal volume of plasma substitute (Emagel, Behring) for 40 min. to obtain a preliminary sedimentation of erythrocytes.

The supernatant was carefully layered over an equal volume of Lymphoprep and was centrifuged at 800 x g for 20 min. at room temperature (approximately 20 °C) in a swing-out rotor.

After centrifugation, neutrophils were sedimented to the bottom of the tube. The contamination of erythrocytes was removed by using a lysing solution containing 1% ammonium oxalate and 0.01% sodium azide. This procedure allowed to obtain a 95±2% of neutrophil population with few erythrocytes or platelets.

For membrane preparations, cells were washed with physiological solution (0.9% NaCl), centrifuged at 48000 x g for 20 min. at 4 °C, homogenized in 20 vol. of 10 mM Tris-HCl pH 7.7, containing 2 mM $MgCl_2$ (T_2 buffer) and protease inhibitors as previously described[11], and then centrifuged at 48000 x g for 20 min. at 4 °C. The pellet was then treated with adenosine deaminase (2 U.I./ml) for 60 min. at 37 °C to remove endogenous adenosine, centrifuged at 48000 x g for 20 min. at 4 °C and used in the binding assay.

For electron microscope localization studies, neutrophil pellets were washed twice with 20 mM Hepes-Tris buffer pH 7.5 containing 118 mM NaCl, 4.7 mM KCl, 1.18 mM $MgCl_2$, 2.5 mM $CaCl_2$ and 10 mM D-Glucose (T_1 buffer).

A_1 adenosine receptor binding assay

[3H]CHA binding assay was performed by incubating aliquots of membrane fractions of human neutrophils (0.25 mg of protein) for 45 min. at 25 °C in 0.5 ml of T_2 buffer containing 2 nM [3H]CHA. Specific binding was determined by measuring the difference in membrane-bound radioactivity either in the presence or in the absence of 17 μM of R-phenylisopropyladenosine (R-PIA). Saturation analysis of [3H]CHA binding sites was performed on the membrane receptor preparations using 1.2-30 nM [3H]CHA.

The concentration of colloidal CHA-gold-albumin that inhibits specific [3H]CHA binding by 50% (IC_{50}) was determined by log-probit analysis with six concentrations of the compound, each performed in duplicate. The Ki value was calculated from IC_{50} value using the Cheng and Prusoff equation[12].

Electron Microscope Localization

Neutrophils (approximately 11×10^6 cells) were incubated with colloidal CHA-gold-albumin (1 μM) either in the absence (total binding) or in the presence (non specific binding) of 160 μM R-PIA in T_1 buffer for 12 hours at 4 °C. After incubation, the samples were centrifuged for 1 min. in an Eppendorf centrifuge and rinsed three times with T_1 buffer. Pellets were cut in small pieces (1 mm^3) and fixed with 2.5% glutaraldehyde in 0.1 M cacodylate buffer pH 7.2, for 1 hour at 4 °C. After rinsing in the same buffer, specimens were postfixed in 1% cacodylate buffered osmium tetroxide for 2 hr at room temperature. Following dehydration in a graded series of ethanol, specimens were briefly transferred to propylene oxide and embedded in Epon-Araldite. Ultrathin sections (50-60 nm thick) were cut with a Reichert-Jung Ultratome E equipped with a diamond knife, placed on formvar-carbon coated copper grids (200 mesh), stained with uranyl acetate and lead citrate and observed with a JEOL 100 SX electron microscope. Only neutrophils which appeared intact and roundish were selected for this investigation. First a micrograph of each cell was taken at 5,000 magnification for the determination of the circumference. Then the magnification was increased to 20,000 for the reckoning of gold particles on the plasma membrane. The number of gold particles was referred to unit of plasma membrane (1 μm). Approximately the same length of plasma membrane was analyzed in both CHA-gold-albumin and CHA-gold-albumin plus R-PIA specimens.

RESULTS AND DISCUSSION

Functional data[4,7] demonstrated that human neutrophils have two different classes of adenosine receptors (A_1 and A_2). The occupancy of A_1 adenosine receptors promotes chemotaxis and neutrophil adherence to the endothelium, whereas the occupancy of A_2 adenosine receptors inhibits neutrophil adherence to the endothelium and the generation of reactive oxygen species (e.g. O^{2-} and H_2O_2). In the present work we demonstrated directly the presence of A_1 adenosine receptors by the use of tritiated agonist [^{3}H]CHA. The binding was specific and saturable. Scatchard analysis of the saturation binding data (1.2-30 nM) indicated the presence of a single class of binding sites with a maximum number (Vmax) of 27 ± 2 fmol/mg protein and a dissociation constant (Kd) of 6 ± 0.3 nM. Therefore, the localization of A_1 adenosine receptors on neutrophils was important to understand the effects of adenosine action. This study demonstrated the electron microscope localization of A_1 adenosine receptors in human neutrophils. In a previous paper we used a ferritin labelled probe as a tool to visualize the A_1 adenosine receptors in guinea pig synaptoneurosome preparations[8]; however, the quality of ferritin as high resolution marker was suboptimal at the electron microscope level. Many advantages of colloidal gold over other markers, as the easy conjugation of proteins to colloidal gold particles, the high electron density of gold particles, and the time sparing of the gold method, led us to obtain a CHA-gold-albumin complex. For this purpose CHA was covalently bound to 20 nm gold-albumin, after oxidation with periodate of the cis-vicinal hydroxyl groups of ribose moiety. The aldheidic function of the periodate oxidized CHA may react with primary amines present in albumin to form Schiff bases. These are reduced with NaBH$_4$ to form stable secondary amines. In these conditions the substitution was approximately 1 nmol of CHA / 2.5 nmol of albumin. The ability of CHA-gold-albumin to retain a biological activity was evaluated by assaying the ability of increasing amounts of CHA-gold-albumin to compete with the binding of 2 nM [^{3}H]CHA with neutrophil membrane preparations as described in materials and methods. The probe inhibited ^{3}H-CHA binding with Ki of 11.5 ± 1.7 nM compared with the Ki value of CHA (3.0 ± 0.26 nM); gold-albumin complexes were unable to compete. Ultrastructurally, the CHA-gold-albumin complexes were visualized as electron dense gold particles (20 nm) localized on the cell surface. For this purpose, CHA-gold-albumin probe was incubated with neutrophils either with or without R-PIA, then samples were examined with the electron microscope.

Neutrophils from untreated specimens were roundish (9-10 μm in diameter) and had a polilobed nucleus. The heterochromatin was characteristically massed on the inner nuclear membrane (Figure 1). In neutrophils of CHA-gold-albumin treated specimens gold particles were distributed along the plasma membrane usually as single particles (Figure 2) and rarely in small aggregates (Figure 3).

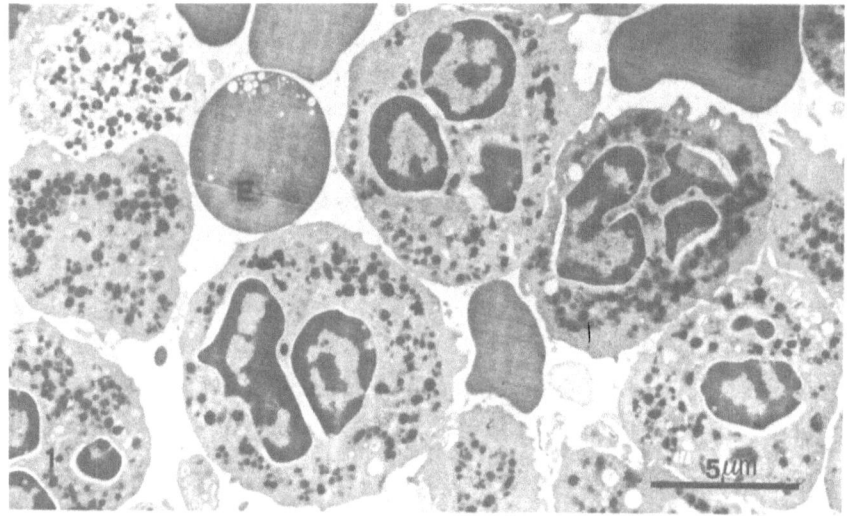

Figure 1. Cluster of untreated neutrophils. E= erythrocyte.

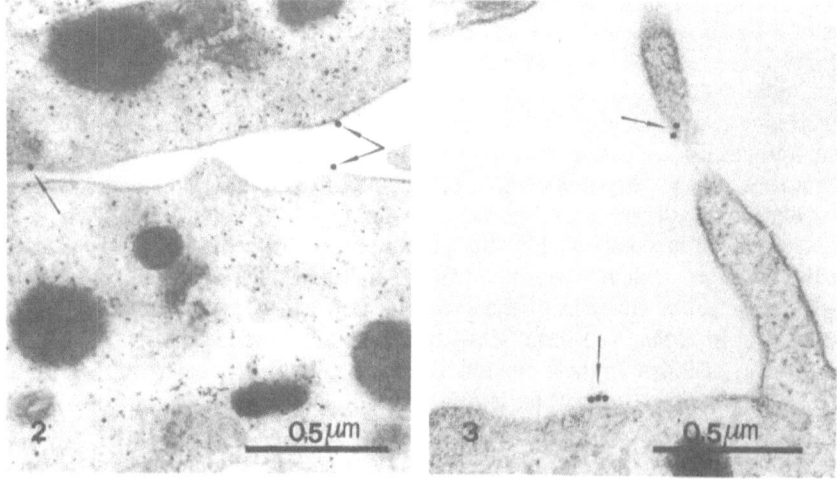

Figure 2. CHA-gold-albumin treated neutrophils. Single gold particles (arrows) on the plasma membrane of two adjacent cells. **Figure 3.** CHA-gold-albumin treated neutrophils. Two small aggregates of gold particles (arrows) on the plasma membrane of a cell.

404

The examination of 100 neutrophils from CHA-gold-albumin treated specimens revealed the presence of 62 gold particles on a total of 2510 μm of plasma membrane with an average of 0.03±0.05 gold particles per μm of membrane (total binding). The examination of 100 neutrophils from CHA-gold-albumin plus R-PIA treated specimens revealed the presence of 8 gold particles on a total of 2347 μm of plasma membrane with an average of 0.004±0.01 gold particles per μm of membrane (non specific binding). The difference between total and non specific binding was significant ($p<0.002$). These findings suggest that CHA bound to gold is an useful tool for the localization of A_1 adenosine receptors on the neutrophil plasma membrane. The use of this probe in a temperature controlled system could also elucidate the internalization mechanism of the receptor and its intracellular transport.

REFERENCES

1. J.L. Daval, A. Nehlig, and F. Nicolas, Physiological and pharmacological properties of adenosine: therapeutic implications, *Life Sciences*, 49:1435 (1991).
2. C. D. Londos, M. F. Cooper, and J. Wolff, Subclasses of external adenosine receptors, *Proc. Natl. Acad. Sci. USA*, 77,2551 (1988).
3. J.E. Salmon and B.M. Cronstein, Fc gamma receptor-mediated functions in neutrophils are modulated by adenosine receptor occupancy. A_1 receptors are stimulatory and A_2 receptors are inhibitory, *J.Immunol.*, 145:2235 (1990).
4. B.N. Cronstein, L. Daguma, D. Nichols, A.J. Hutchison, and M. Williams, The adenosine/neutrophil paradox resolved: human neutrophils posses both A_1 and A_2 receptors that promote chemotaxis and inhibit O^{2-} generation, respectively, *J.Clin.Invest.*, 85:1150 (1990).
5. B.N. Cronstein, E.D. Rosenstein, S.B. Kramer, G. Weissmann, and R. Hirschhorn, Adenosine: a physiologic modulator of superoxide anion generation by human neutrophils. Adenosine acts via an A_2 receptor on human neutrophils, *J.Immunol.*, 135:1366 (1985).
6. G. Marone, R. Petracca, and S. Vigorita, Adenosine receptors on human inflammatory cells, *Int. Arch. Allergy Appl. Immunol.*, 77:259,(1985).
7. B.N. Cronstein, R.I. Levin, M. Philips, R. Hirschhorn, S.B. Abramson, and G. Weissmann, Neutrophil adherence to endothelium is enhanced via adenosine A_1 receptors and inhibited via adenosine A_2 receptors, *J.Immunol.*, 148:2201 (1992).
8. C. Martini, U. Montali, L. Giusti, G. Giannaccini, A. Falleni, V. Gremigni, and A. Lucacchini, A ferritin-cycloexyladenosine probe for electron microscope localization of A_1 adenosine receptor, *in* "Purine and Pyrimidine metabolism in man VII", R.A.Harkness et al. eds, p.451, Plenum Press, New York (1991).
9. P.T. Gilham , The covalent binding of nucleotides, polinucleotides and nucleic acid to cellulose, *Methods Enzymol.*, 21:191 (1971).
10. A. Boyum, Isolation of mononuclear cells and granulocytes from human blood. Isolation of mononuclear cells by centrifugation and sedimentation at 1xg, *Scand. J. Clin. Lab. Invest.*, 21:77, (1968).
11. C. Martini, G. Giannaccini, A. Lucacchini, L.Bazzichi, A. Soletti, and M.L. Ciompi, A_1 and A_2 adenosine receptors in synovial cells from patients with rheumatic diseases, *in* "Purine and Pyrimidine metabolism in man VI", Mikanagi K., Nishioka and KelleyW.N. eds, p.429, Plenum Publishing Corp. New York (1989).
12. Y.C. Cheng, and W.H. Prusoff, Relation between the inhibition constant Ki and the concentration of inhibitor which causes fifty per cent inhibition (IC_{50}) of an enzymatic reaction, *Biochem.Pharmacol.*, 22:3099, (1973).

A_2 ADENOSINE RECEPTORS IN NEUTROPHILS FROM HEALTHY VOLUNTEERS AND PATIENTS WITH RHEUMATOID ARTHRITIS

Claudia Martini,[1] Paolo Tacchi,[1] Laura Bazzichi,[2] Marcello Fiorini,[1] Franca Bondi,[2] Maria Laura Ciompi,[2] Antonio Lucacchini [1]

[1] Istituto Policattedra di Discipline Biologiche
[2] Istituto Patologia Speciale Medica
Università di Pisa, 56100 Pisa, Italy

INTRODUCTION

At physiological concentrations, adenosine can modulate a variety of biological activities by engaging specific cells surface receptors, termed A_1 and A_2, with different affinity for adenosine and adenosine analogues[1]. Engagement of A_2 adenosine receptors induces an increase in cAMP levels in several cells types, in contrast stimulation of A_1 receptors causes opposite effects [2]. It has been shown that adenosine and its analogues inhibit O_2 generation [3,4], phagocytosis and adherence by occupancy of specific adenosine A_2 receptors, while the occupancy of A_1 adenosine receptors enhance chemiotaxis [5], phagocytosis and adherence [6]. In general, activation of adenosine receptors on leukocytes reduces immune and inflammatory responses [7]. Therefore, it may be suggested that release of adenosine is one mechanism by which normal cells protect themselves from activated neutrophils. Since it is possible that a decreased adenosine receptor functions are implicated in diseases like rheumatic pathologies characterised by an excess of inflammation.

In the present report we described characteristics of adenosine binding sites on human neutrophils from healthy volunteers and rheumatic patients afflicted with rheumatoid arthritis by using [^{3}H]N ethylcarboxamidoadenosine (NECA) and [^{3}H] 2-p-(2-carboxyethyl) phenethylamino 5' N-ethylcarboxamidoadenosine (CGS 21680) as ligands.

MATERIALS AND METHODS

[^{3}H]-NECA (specific activity 18 Ci/mmol), [^{3}H]-CGS 21680 (specific activity 42.6 Ci/mmol) were from New England Nuclear, NECA, R-PIA (N^6-(R)phenylisopropyladenosine), CPA (N-cyclopentyladenosine), were from Sigma. All other chemicals were from standard supplies .

Human neutrophils were isolated from heparinized whole blood of healthy donors (170 cc) and patients with Rheumatoid Arthritis (150 cc). Granulocytes are obtained using a mixture of 9.6% (w/v) sodium metrizoate and 5.6% (w/v) polysaccharide (Lymphoprep, Nycomed Pharma, Oslo Norway) [8]. Bloods were collected into tubes containing anticoagulant (heparin), and were diluted by addition of an equal volume of plasma's substitute (Emagel, Behring) for 40 minutes to obtain a preliminary erythrocytes sedimentation. The supernatant was carefully layered over an equal volume of Lymphoprep and was centrifuged at 800 x g for 20 minutes at room temperature (approximately 20 °C) in a swing-out rotor. After centrifugation, granulocytes were sedimented to the bottom of the tube. The contaminating erythrocytes were lysed by 1% ammonium oxalate and 0.01% sodium azide solution. The cells, which are known to be essentially neutrophils (95 ± 2), were washed with physiological solution, centrifuged at 48000 x g at 4 °C for 20 min. and homogenised in 20 vol. of 10 mM Tris-HCl pH 7.4 containing 10 mM $MgCl_2$ and protease inhibitors [9]; than centrifuged at 48000 x g at 4 °C for 20 min.. The resulting pellet was treated with adenosine deaminase (2 U.I./ml) at 37 °C for 60 min. to remove endogenous adenosine, centrifuged at 48000 x g. at 4 °C for 20 min and used in the binding assay.

^{3}H-NECA binding assay was performed as previously described [10]. Non-specific binding was defined in the presence of 100 µM NECA or 100 µM R-PIA. The assay included 50 nM CPA, a selective A_1 agonist, to block pharmacologically ^{3}H-NECA binding to A_1 sites.

^{3}H-CGS 21680 binding to human neutrophils membranes was measured essentially as described by Martini et al.[10]. In brief membranes (0.2-0.3 mg protein) were incubated with 21 nM ^{3}H-CGS 21680 in a final volume of 0.5 ml of 50 mM/ 10 mM Mg Cl$_2$ buffer at pH 7.5 for 90 min. at 25 °C. The samples were filtered into Whatman GF/C glass fiber filters under vacuum. After washing 3x5 ml of Tris-buffer, the radioactivity was counted in 4 ml of Ready-Safe Beckman scintillation cocktail. The specific binding was obtained in the presence of 50 µM CGS 21680.

Saturation studies were performed by incubating membranes with increasing concentrations of tritiated ligands ^{3}H-NECA (8-350 nM) and ^{3}H-CGS 21680 (5-140 nM). The maximum number of binding sites (Bmax) and affinity constant (Kd) were calculated by Scatchard analysis using EBDA, Ligand program written by G.A. McPherson[11].

RESULTS

^{3}H-NECA specific binding, obtained in the presence of 100 µM NECA and 50 nM CPA, to human normal and rheumatoid neutrophil membrane preparations was saturable. The Bmax values were 489 ± 30 fmol/mg protein and 210 ± 20 fmol/mg for normal and rheumatoid membranes respectively. A significant difference was also found for the dissociation constant (Kd=150 ±6 nM for control neutrophil membranes versus 39 ± 2 nM for rheumatoid neutrophil membranes) (figure 1). To evaluate the contribution of the A_2-like component to ^{3}H-NECA binding, the assay was carried out in the presence of 100 µM R-PIA, a compound able to discriminate between A_2 adenosine receptor and A2-like binding protein whose function is unknown [12].

Under these conditions, ^{3}H-NECA specific binding yielded a single straight line in both membrane preparations, indicating a single population of specific binding sites.

The Bmax values were 39 ± 5 fmol/mg protein for normal neutrophil membrane preparations and 40 ± 6 fmol/mg protein for rheumatoid neutrophils membranes. On the contrary, a significant difference was found with regard to the dissociation constant (Kd = 51 ± 4.2 nM for control membranes versus 22 ± 1.9 nM for patients membranes).(see figure 2). These data suggested that ^{3}H-NECA specific binding obtained in the presence of 100 µM NECA labeled essentially A2-like binding protein. To specifically label A_2 adenosine receptors in human neutrophil membranes, we used ^{3}H-CGS 21680, a selective ligand for A_2 receptors [13]. The saturation analysis of the ^{3}H-CGS 21680 specific binding to controls and patients neutrophil membranes are showed in figure 2.

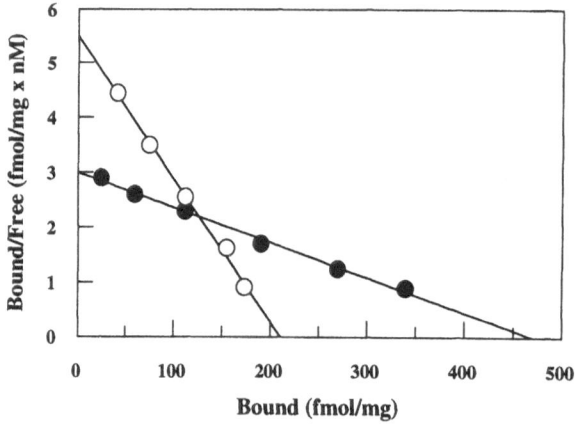

Figure 1. Representative Scatchard analysis of specific ^{3}H-NECA bound of neutrophils membranes from healthy volunteers (filled circles) and patients with Rheumatoid Arthritis (open circles) Membrane preparations were incubated in presence and absence of unlabelled NECA (100 µM)

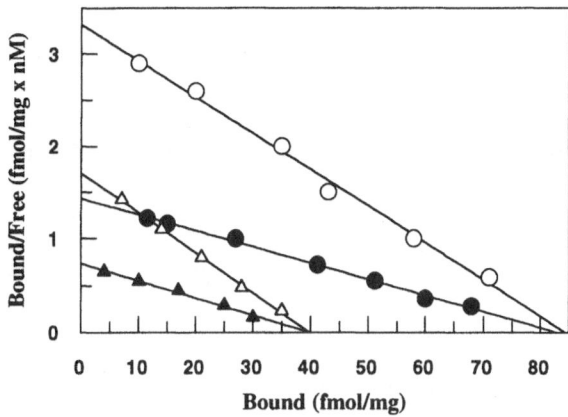

Figure 2. Representative Scatchard analysis of specific ^{3}H-NECA bound of neutrophils membranes from healthy volunteers (filled triangles) and patients with Rheumatoid Arthritis (open triangles). Membrane preparations were incubated in presence and absence of unlabelled R-PIA (100 μM). ^{3}H-CGS 21680 Scatchard plot of neutrophils membranes from healthy volunteers (filled circles) and patients with Rheumatoid Arthritis (open circles). Membrane preparations were incubated in presence and absence of unlabelled CGS 21680 (50 μM)

No variations of the Bmax values of A_2 adenosine receptors were observed (Bmax= 83 ± 7.2 fmol/mg versus 84 ± 8.1 fmol/mg); on the contrary the Kd was significantly increased in patients membranes (volunteer's Kd=58 ± 4.9 nM and patients Kd=25 ± 2.1 nM).

DISCUSSION

Rheumatoid arthritis is characterized by an inappropriate, unlimited chronic inflammation causing the disruption of the affected articulations. In synovial fluid 75-80% of cellular population is represented by PMNs with multilobate aspects, a great number of cytoplasmic vacuoles and broken up nuclei. These cells release in synovial fluid powerful enzymes, such as cathepsin, collagenase, elastase, acid protease or oxygen radicals, playing therefore a crucial role in the articular damage. Consequently the mechanisms ruling PMNs' activity and function appear to be very important. A powerful endogenous modulator of PMNs' function is adenosine. This purine mediator binds to PMNs' membrane receptors increasing their chemotaxis, but (at higher concentrations) inhibiting their production of superoxide anions, particularly in reply to such a stimulation as that of C_{5a}, FMLP and Zymosan particles. As a consequence to a pathogenic noxa, adenosine increase in ischemic tissues and plays its role acting on lymphocytes (T suppressor decrease and T helper increase)[14] and ruling PMNs' reply through A_1, A_2 adenosine and perhaps A_2-like binding sites.

In the present report we describe the presence of A_2 adenosine receptors in neutrophils from healthy volunteers and from patients afflicted with rheumatoid arthritis. We used for our studies two different radiolabeled adenosine analogues ^{3}H-NECA and ^{3}H-CGS 21680.

As shown in table 1, ^{3}H-NECA and ^{3}H-CGS 21680 binding characteristics (Bmax and Kd) were significantly different in neutrophils membrane preparation from healthy volunteers with respect to patients affected by rheumatoid arthritis. An increase of A_2 adenosine receptor affinity was evident in rheumatoid arthritis patients. On the contrary, no significant differences of the Bmax values were observed. A dramatic decrease of A_2-like binding sites was evident in rheumatoid neutrophils membranes. A tentative explanation of these results might be that if adenosine, released during the inflammatory process, acts with a mechanism inhibiting the inflammation, an increase of A_2 adenosine receptor affinity could be a further attempt to decrease the chronic inflammatory reply.

Table 1. Bmax and Kd values for A2 receptors in healthy volunteers and patients with Rheumatoid Arthritis.

	[^{3}H]-NECA		[^{3}H]-NECA		[^{3}H]-CGS 21680	
Displacers	NECA		R-PIA		CGS 21680	
	Healthy Volunteers	Patients	Healthy Volunteers	Patients	Healthy Volunteers	Patients
Bmax	489±30	210±20	39±5	40±6	83±7.2	84±8.1
Kd	150±6	39±2	51±4.2	22±1.9	58±4.9	25±2.1

REFERENCES

1. J.L. Daval, A. Nehlig and F. Nicolas, Physiological and pharmacological properties of adenosine: therapeutic implications, *Life Sciences* 49:1435-1453 (1991).
2. D.M. van Calker, M. Muller and B.Hamprecht, Adenosine regulates via two different types of receptors, the accumulation of cyclic AMP in cultured brain cells, *J.Neurochem.*, 33, 999 (1979).
3. P.A. Roberts, A.C. Newby, M.B. Hallet. and A.K. Campbell, Inhibition by adenosine of reactive oxygen metabolite production by human polymorphonuclear leucocytes, *Biochem. J.*, 277, 669 (1985).
4. B.N. Cronstein, E.D. Rosenstein, S.B. Kramer, G. Weissmann and R. Hirschhorn, Adenosine: a physiologic modulator of superoxide anion generation by human neutrophils. Adenosine acts via A_2 receptor on human neutrophils, *J. Immunol.*,135:1366.(1985).
5. F.R. Rose, R. Hirschorn, G. Weissmann and B.N. Cronstein, Adenosine promote neutrophils chemiotaxis, *J. Exp. Med.*, 167,1186 (1988).
6. B.N. Cronstein, Adenosine is an autacoid of inflammation: effect of adenosine on neutrophil function, *in:*"Role of adenosine and adenine nucleotides in the biological system", Shoichi Imai & Mikio Nakazawa, eds., Elsevier Science Publishers BV, New York (1991).
7. B.N. Cronstein,R.I. Levin,J. Belanoff,G. Weissmann and R.Hirschhorn, Adenosine: an endogenous inhibitor of neutrophil-mediated injury to endothelial cells, *J. Clin. Invest.*, 78:760 (1986).
8. A.Boyum , Isolation of monuclear cells and granulocytes from human blood. Isolation of mononuclear cells by centrifugation and sedimentation at 1 x g, *Scand. J. Clin. Lab. Invest.*, 21:77.(1968).
9. C. Martini, G. Giannaccini, A. Lucacchini, L. Bazzichi, A. Soletti, and M.L. Ciompi, A_1 and A_2 adenosine receptors in synovial cells from patients with rheumatic diseases, *in:* "Purine and Pyrimidine Metabolism in Man VI, Mikanagy K., Nishioka and Kelley W.N. eds., p.429, Plenum Publishing Corp. New York, (1989)..
10. C. Martini, S Di Sacco, P. Tacchi, L.Bazzichi, A.Soletti, F. Bondi, M.L. Ciompi, A. Lucacchini, A_2 Adenosine receptors in neutrophils from healthy volunteers and patients with rheumatic disease, *in:* "Purine and Pyrimidine Metabolism in Man VII, Part A, R. A. et al. eds., p.459, Plenum Press, New York (1991)
11. G.A. McPherson, Kinetic, EBDA, Ligand, Lowry. A collection of radioligand binding analysis programs, Elsevier, Cambridge.
12. S. Zolnierowicz, C.Work, K. Hutchison and I.H. Fox, Partial separation of platelet and placental adenosine receptors from adenosine A_2 -like binding protein. *Mol.Pharmacol.*, 37:554-559.
13. M.F. Jarvis, Schulz R., A.J. Hutchinson, U.H. Do, M.A. Sills and M. Williams, 3H-CGS 21680, a selective A_2 adenosine receptor agonist directly labels A_2 receptors in rat brain, *J. Pharm. Exper. Ther.*,251, 888-893.
14. R.E. Birsch, A.K. Rosenthal and S.H Palmer, Pharmacological modification of immunoregulatory T lymphocytes. II. Modification of T lymphocytes.II. Modification of T lymphocytes cell surface characters, *Clin. Exp. Immunol.* 48, 231-238 (1982).

THE ANTIINFLAMMATORY EFFECTS OF METHOTREXATE ARE MEDIATED BY ADENOSINE

Bruce N. Cronstein, Dwight Naime, Edward Ostad

Department of Medicine, Division of Rheumatology
New York University Medical Center
550 First Ave.
New York, New York   10016

INTRODUCTION

The introduction of low-dose, intermittent methotrexate was the single most important advance made in the treatment of Rheumatoid Arthritis during the past decade. Although methotrexate was initially introduced for the treatment of RA as a folate antagonist that inhibits synthesis of purines and pyrimidines necessary for cellular proliferation there is little evidence that the therapeutic effects of methotrexate result from diminished proliferation of the cells of inflammation. We originally proposed a novel mechanism of action of methotrexate[1]. Previous studies had indicated that methotrexate is taken up by cells and polyglutamated[2]. Even low concentrations of the polyglutamated form of methotrexate, in combination with a modest methotrexate-induced increase in dihydrofolate polyglutamate, inhibit an intermediate step in the de novo pathway of purine biosynthesis, methylation of 5-aminoimidazolecarboxamidoribonucleotide, AICAR[3-5]. The intracellular accumulation of AICAR leads, by a mechanism which is only incompletely understood, to increased release of adenosine into the extracellular space[6,7]. Adenosine, acting at cell surface receptors, is a potentially potent antiinflammatory agent by virtue of its capacity to inhibit neutrophil, macrophage and lymphocyte function[8]. We have previously reported in vitro studies[1] and others have presented evidence from in vivo studies that lend support to this hypothesis although in the in vivo studies high doses of methotrexate were applied directly to an inflamed site[9]. We therefore tested the hypothesis that the antiinflammatory effects of low dose, weekly methotrexate result from the selective inhibition of AICAR transformylase with the intracellular accumulation of AICAR and increased release of the antiinflammatory autocoid adenosine into the extracellular milieu using an IN VIVO model of inflammation.

METHODS

In the murine air pouch model of inflammation a pouch is formed on the back of a mouse by injection of 3cc of air subcutaneously three times a week for three weeks. Inflammation was induced in the air pouch by the injection of carrageenan. After four hours the animals were sacrificed, the pouch exudates collected, the leukocytes counted and adenosine concentration measured by reverse phase HPLC. In some animals spleens were collected and splenocyte lysates were assayed for their AICAR content by reverse phase, anion exchange HPLC. Treatment of the mice during the induction of the pouch consisted of a single weekly intraperitoneal injection of methotrexate or an identical volume of saline[10].

RESULTS

As in patients with Rheumatoid Arthritis, methotrexate is antiinflammatory in the murine air pouch model of inflammation. There was a dose-dependent reduction of the number of leukocytes which accumulated in the inflamed air pouches of animals treated with methotrexate. The concentration of methotrexate at which 50% of maximal effect was observed, IC_{50}, in these experiments was 0.08 mg/kg/week and the maximal antiinflammatory effect was achieved with a dose of 0.1 mg/kg/week, a dose equivalent to 7mg/week in a 70 kg individual (Figure 1). Thus, low dose, weekly methotrexate is antiinflammatory in this murine model at doses and intervals comparable to those used to treat patients with Rheumatoid Arthritis.

To determine whether low dose weekly methotrexate therapy inhibits AICAR transformylase we examined the intracellular AICAR concentration in splenocytes. If methotrexate were acting as a non-specific folate antagonist we would expect that cellular AICAR concentrations in mice treated with methotrexate would be either unchanged or decreased from those in untreated mice since synthesis of AICAR is folate-dependent. Instead we found that methotrexate treatment increases intracellular AICAR concentrations. The AICAR concentration in control, saline-treated mice was 26.5 ± 10pmol/10^6 splenocytes. Weekly treatment of mice with 0.5mg/kg/week of methotrexate significantly increased splenocyte AICAR concentration to 72.4 ± 16pmol/10^6 splenocytes (P<0.02, N=6). We conclude from these experiments that low dose, weekly methotrexate treatment selectively inhibits AICAR transformylase and leads to intracellular AICAR accumulation.

We would predict that increased AICAR concentrations would lead to higher extracellular adenosine concentrations, particularly at inflamed or injured sites. We therefore measured adenosine concentration in the inflamed air pouch exudates. We found that methotrexate treatment increases adenosine concentration in the air pouch exudate. The mean adenosine concentration in control animals, 570 ± 90nM, a concentration that is almost twice the concentration of adenosine measured in whole blood in humans. We found, as predicted, that treatment of animals with methotrexate almost doubled the concentration of adenosine in the air pouch exudate to 1110 ± 190nM (P<0.008, N=16).

To determine whether the adenosine released into the air pouches of methotrexate-treated mice mediated the antiinflammatory effect of methotrexate we eliminated the extracellular adenosine by injection of adenosine deaminase or ADA, an enzyme which converts adenosine to its inactive metabolite inosine, into the pouch with the carrageenan. Adenosine deaminase reverses, partially, the antiinflammatory effect of methotrexate treatment in this model. In these experiments that 4×10^6 leukocytes accumulated in pouches of control mice and weekly methotrexate treatment diminished leukocyte accumulation to approximately 1.2×10^6 leukocytes per pouch. Injection of ADA into the pouch did not significantly increase the number of leukocytes that accumulated in controls but partially reversed the antiinflammatory effect observed in the methotrexate treated animals (Figure 2).

The physiological effects of adenosine are mediated by three major classes of receptors, A1, A_2 and A3. These receptors can be distinguished, in part, by using specific

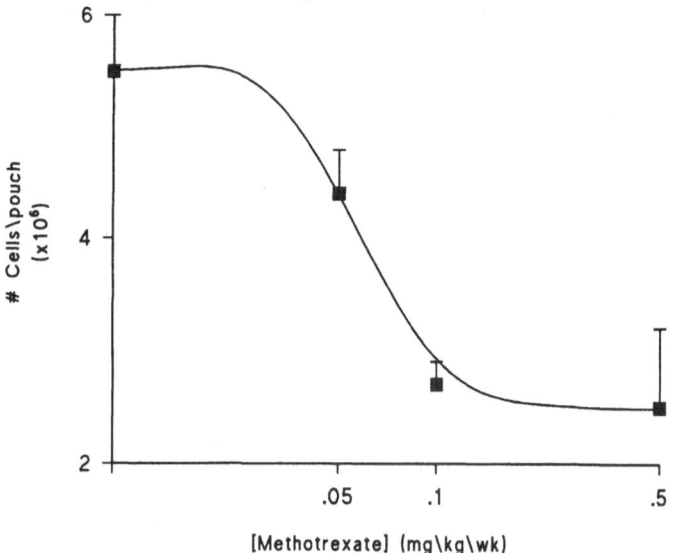

Figure 1.   Methotrexate was given to the mice by intraperitoneal injection at the indicated doses for 3-4 weeks during induction of the air pouch. The air pouch was injected with carrageenan (2%wt/v), exudates were harvested 4hrs. later and the cells counted.   Each point represents the mean (±SEM) of cell counts from three mice. (Reprinted with permission from the Journal of Clinical Investigation).

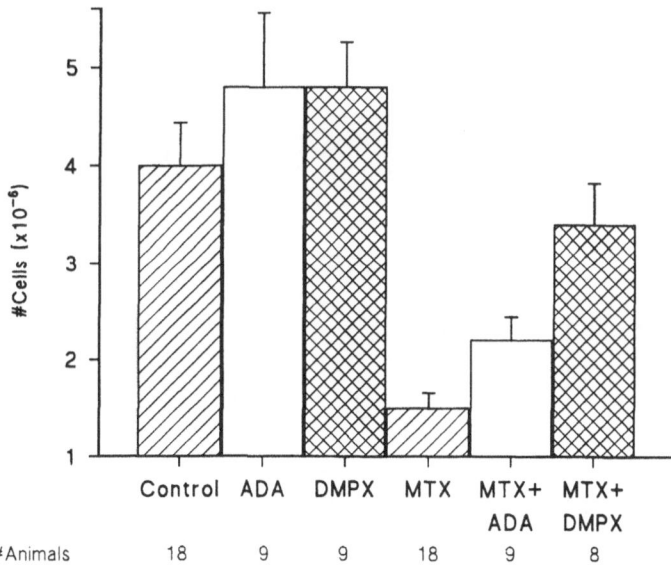

Figure 2.   Adenosine deaminase (ADA, 0.15IU/ml) and DMPX (0.5mg/kg/pouch) reverse the antiinflammatory effects of methotrexate. Mice were treated with saline or methotrexate for 3-4 weeks before inflammation was induced in the air pouch. Shown are the means (±SEM) of the number of cells that accumulated in the pouch exudates. Neither ADA nor DMPX significantly affected the number of cells that accumulated in the pouch of saline-treated animals but both ADA ($p<0.006$) and DMPX ($p<0.001$) reversed the antiinflammatory effects of methotrexate treatment. (Reprinted with permission from the Journal of Clinical Investigation).

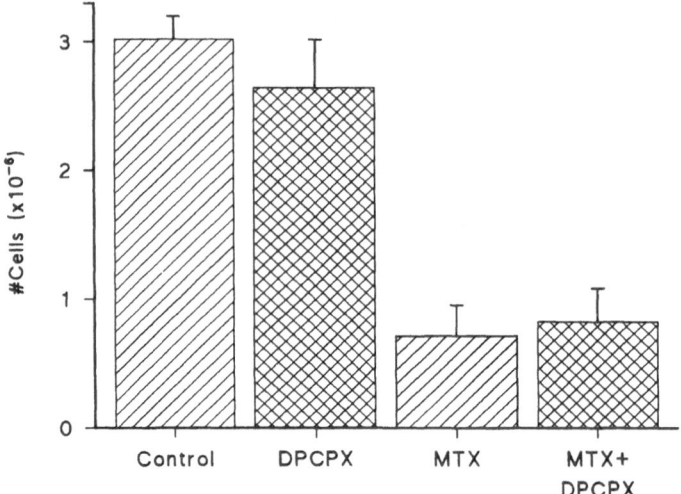

Figure 3.   The adenosine A_1 antagonist DPCPX (2mg/kg/pouch) does not reverse the antiinflammatory effects of methotrexate. (Reprinted with permission from the Journal of Clinical Investigation).

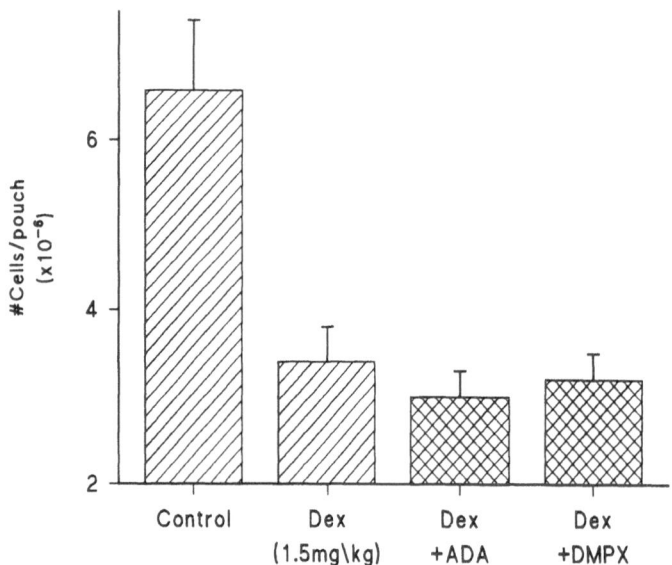

Figure 4.   Neither ADA (0.15IU/ml) nor DMPX (0.5mg/kg/pouch) reverses the antiinflammatory effect of dexamethasone (1.5mg/kg, intraperitoneally). (Reprinted with permission from the Journal of Clinical Investigation).

agonists and antagonists. To further confirm that adenosine, acting at specific receptors, mediates the antiinflammatory effects of methotrexate treatment we injected specific adenosine receptor antagonists into the air pouch with the inflammatory stimulus. We found that an adenosine A_2 receptor antagonist, DMPX, reverses the antiinflammatory effect of methotrexate. DMPX did not significantly affect the number of leukocytes that accumulated in saline treated animals but completely reversed the antiinflammatory effect of methotrexate treatment (Figure 2). In contrast, an adenosine A_1 antagonist, DPCPX, does not reverse the effect of methotrexate on inflammation in the air pouch model. Control leukocyte accumulation was 3 million, and methotrexate reduced leukocyte accumulation to 0.9million in these experiments. DPCPX did not affect leukocyte accumulation in control animals or in methotrexate-treated animals (Figure 3). Thus, the increased adenosine released at inflamed sites in methotrexate-treated animals inhibits inflammation by binding to A_2 receptors.

To further confirm that the reversal by ADA and DMPX ,of methotrexate's antiinflammatory effects was specifically due to elimination or antagonism of adenosine we studied the effect of these agents on inflammation in animals treated with dexamethasone, a glucocorticoid which inhibits inflammation by a mechanism which does not involve adenosine. We found that neither adenosine deaminase nor the adenosine A_2 antagonist DMPX reverses the antiinflammatory effect of dexamethasone. 6.5 million leukocytes accumulated in the air pouches of control animals. Dexamethasone reduced leukocyte accumulation to 3.8 million. Neither ADA nor DMPX affected leukocyte accumulation in dexamethasone treated animals (Figure 4).

SUMMARY AND CONCLUSION

In summary, intermittent, low dose methotrexate treatment is: 1.) antiinflammatory in the murine air pouch model of inflammation; 2.) selectively increases intracellular AICAR concentration; 3.) increases adenosine concentration in an inflammatory exudate; and, 4.) inhibits leukocyte accumulation at an inflamed site by a mechanism that is specifically reversed by adenosine deaminase and the adenosine A_2 receptor antagonist DMPX but not the A_1 antagonist DPCPX.

In conclusion, we have demonstrated a novel mechanism for the antiinflammatory action of methotrexate; methotrexate is a nonsteroidal antiinflammatory agent that acts by promoting the release of adenosine which engages A_2 receptors on inflammatory cells.

REFERENCES

1. Cronstein BN, Eberle MA, Gruber HE, Levin RI: Methotrexate inhibits neutrophil function by stimulating adenosine release from connective tissue cells. Proc Natl Acad Sci USA 88:2441, 1991

2. Chabner BA, Allegra CJ, Curt GA, Clendeninn NJ, Baram J, Koizumi S, Drake JC, Jolivet J: Polyglutamation of methotrexate. Is methotrexate a prodrug?. J Clin Invest 76:907, 1985

3. Allegra CJ, Drake JC, Jolivet J, Chabner BA: Inhibition of phosphoribosylaminoimidazolecarboxamide transformylase by methotrexate and dihydrofolic acid polyglutamates.. Proc Natl Acad Sci USA 82:4881, 1985

4. Allegra CJ, Hoang K, Yeh GC, Drake JC, Baram J: Evidence for direct inhibition of de

novo purine synthesis in human MCF-7 breast cells as a principal mode of metabolic inhibition by methotrexate.. J Biol Chem 262:13520, 1987

5. Baggott JE, Vaughn WH, Hudson BB: Inhibition of 5-aminoimidazole-4-carboxamide ribotide transformylase, adenosine deaminase and 5'-adenylate deaminase by polyglutamates of methotrexate and oxidized folates and by 5-aminoimidazole-4-carboxamide riboside and ribotide.. Biochem J 236:193, 1986

6. Barankiewicz J, Ronlov G, Jimenez R, Gruber HE: Selective adenosine release from human B but not T lymphoid cell line. J Biol Chem 265:15738, 1990

7. Barankiewicz J, Jimenez R, Ronlov G, Magill M, Gruber HE: Alteration of purine metabolism by AICA-riboside in human B lymphoblasts. Archives of Biochemistry & Biophysics 282:377, 1990

8. Cronstein BN: Adenosine, an endogenous anti-inflammatory agent.. J Appl Physiol 76:5, 1994

9. Asako H, Wolf RE, Granger DN: Leukocyte adherence in rat mesenteric venules: effects of adenosine and methotrexate.. Gastroenterology 104:31, 1993

10. Cronstein BN, Naime D, Ostad E: The antiinflammatory mechanism of methotrexate: increased adenosine release at inflamed sites diminishes leukocyte accumulation in an in vivo model of inflammation.. J Clin Invest 92:2675, 1993

ENDOGENOUS ADENOSINE FORMATION CAN REGULATE HUMAN

NEUTROPHIL FUNCTION

Jerzy Barankiewicz, Roland Jimenez, Jon Uyesaka, Elisabeth Colmerauer,
Gary S. Firestein

Gensia, Inc., 9360 Towne Centre Drive, San Diego, CA 92121, U.S.A.

Neutrophils, in addition to their role in host defense, can cause injury to normal tissues during inflammatory processes.   Oxygen radicals and secreted proteases, in particular, are responsible for some aspects of neutrophil-mediated injury to endothelial cells and cardiomyocytes. A variety of neutrophil functions, including adhesion and reactive oxygen species production, are inhibited by adenosine (Ado) (Cronstein, 1991 and Cronstein, et al., 1992).   Furthermore, inhibition of neutrophil adhesion by adenosine regulating agents like acadesine and adenosine kinase (AK) inhibitors (Firestein, et al., 1994) appears to be mediated by Ado, since it is reversed by the addition of adenosine deaminase (ADA) or Ado receptor antagonists. Although Ado and adenine nucleotides can be released at inflammatory sites during platelet aggregation or from endothelial cells during ischemic stress conditions, little is known about Ado formation by human neutrophils.   To determine if neutrophils can serve as an endogenous Ado source and thereby provide an autocrine stimulus, we evaluated purine metabolism and Ado formation in human neutrophils.

INTRACELLULAR ADENOSINE FORMATION

Ado formation and release was studied in resting and fMLP-activated human neutrophils that had been incubated with specific inhibitors of ADA and AK (2'-deoxycoformycin (DCF) or 5-iodotubercidin (ITU), respectively). Ado production was also studied in neutrophils that were permitted to bind to cytokine-activated endothelial cells. Ado release from neutrophils $(1x10^7)$ was studied after stimulation with 10^{-7} M fMLP in the presence or absence of 20 μM DCF or 20 μM ITU. Ado, Ino and Hyp were measured in the supernatants after 60 min by HPLC. To study Ado release during adhesion, fMLP-stimulated neutrophils containing radiolabelled adenine nucleotides were incubated with unlabelled IL-1β-stimulated human aortic endothelial cells. Ado release was also measured when IL-1β-stimulated endothelial cells containing radiolabelled adenine nucleotides were incubated with unlabelled fMLP-stimulated neutrophils. Radiolabelled Ado was measured in the supernatants using Kodak cellulose TLC.

No Ado release was detected from resting neutrophils but abundant Ado was present in the culture supernatants after activation by fMLP (Table 1). There was a dramatic increase in Ado release if cells were activated in the presence of 20 μM DCF or ITU. During neutrophil adhesion to endothelial cells, both neutrophils and endothelial cells released Ado (Table 1).

Table 1. Adenosine release from human neutrophils and endothelial cells.

		Neutrophils only*		Adhesion**	
Resting	fMLP	fMLP+DCF	fMLP+ITU	Neutrophils	Endothelial cells
<0.1*	8.3±3*	86.3±2	88.8±8	16.6±2	17.6±5

* pmoles/hr/10^6 neutrophils
** pmoles/5 min/10^6 neutrophils or endothelial cells during adhesion

In additional studies on resting neutrophils, acceleration of intracellular ATP degradation by 5.5 mM 2-deoxyglucose caused rapid degradation of ATP to Ino, Hyp and Ado catabolites. Ado levels increased continuously but represented only a small portion of the catabolites. No ADP or AMP accumulation was observed during 60 minutes incubation.

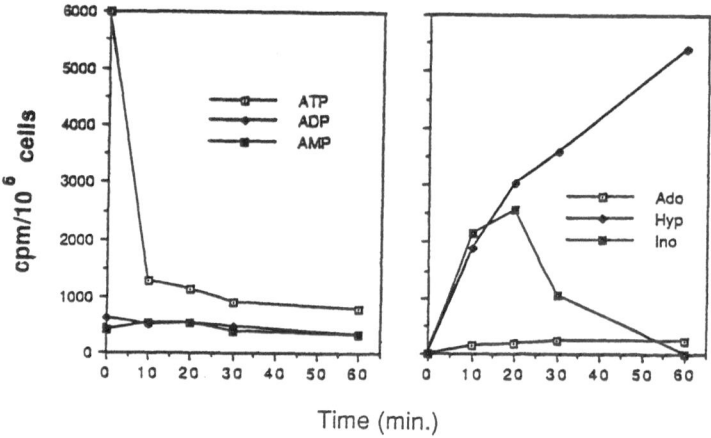

Figure 1. Intracellular ATP degradation in human blood neutrophils in experimental stress conditions. Cells (2x10^7/ml) were incubated for 60 min with radiolabelled [2,8-^{3}H] Ado. Unincorporated Ado was then removed and cells were treated with 5.5 mM 2-deoxyglucose in glucose free RPMI medium. Nucleotides were analyzed in cell extracts and nucleosides and bases in incubation medium using TLC. A = ATP, ADP, and AMP levels; B = adenosine, hypoxanthine, and inosine levels.

EXTRACELLULAR ATP DEGRADATION

Human neutrophils rapidly degraded extracellular ATP and ADP, indicating high levels of ectoATPase and ectoADPase on their membrane (Figure 2). AMP accumulated in the supernatants when exogenous ATP was degraded. Extracellular Ado formation was not observed, indicating minimal ecto5'-nucleotidase activity. Indeed, when human neutrophils were incubated with exogenous AMP, no degradation occurred after 60 min (Figure 2). Activation of human blood neutrophils with 10^{-7} M fMLP had no effects on ectonucleotidase activity (data not shown).

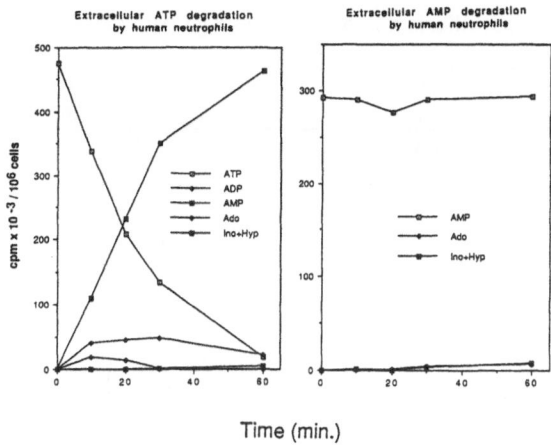

Figure 2. Neutrophils express ectoATPase and ectoADPase. Cells (1x10⁶/0.1 ml) were incubated with 5 μM [8-¹⁴C] ATP or AMP for 60 min. After incubation, cells were removed by centrifugation and nucleotides, nucleosides and bases were analyzed in the incubation medium using TLC.

ADENINE NUCLEOTIDE RELEASE

Human neutrophils released adenine nucleotides, especially if they were activation with 10^{-7} M fMLP. AMP represented the largest pool of nucleotides found in incubation medium and increased about 10-fold after stimulation, whereas ADP and ATP only increased 2-3-fold (Table 2). Although adenine nucleotides were released from neutrophils and might act on P_2 receptors, they were not a source of Ado because neutrophils lacked ecto-5'NT activity.

Table 2. Adenine nucleotide release from human blood neutrophils.

Additions	AMP	ADP	ATP
None	205±43*	123±5	100±51
fMLP	2248±271	290±29	457±58

*cpm/10⁶ cells

Cells (2.5x10⁷) were preincubated with [2,8-³H] Ado for 60 min to radiolabel intracellular adenine nucleotides. After removing unincorporated radiolabelled Ado, cells were incubated for 60 min in the presence or absence of 10^{-7}M fMLP. Nucleotides were evaluated in incubation medium using TLC.

PURINE METABOLISM

To study purine biosynthetic pathways in neutrophils, cells were incubated with 1 μM radiolabelled glycine, formate, adenine, adenosine, hypoxanthine and guanine for 60 min. Incorporation of radioactivity into nucleotides was evaluated in cell extracts using TLC. Human neutrophils did not incorporate radioactivity into the purine nucleotide pool from radiolabelled glycine, formate, adenine, hypoxanthine or guanine (Table 3). Only Ado was incorporated into purine nucleotides indicating that neutrophils have active (AK). This was confirmed by the observation that acadesine (AICA-riboside) was phosphorylated almost exclusively to ZMP (Figure 3). Activation of neutrophils with 10^{-7} M fMLP increased acadesine incorporation into nucleotides by about 10% (data not shown). These data indicate that neutrophils have limited purine biosynthesis de novo and salvage pathways.

Table 3. Purine biosynthesis de novo and salvage pathways in human neutrophils.

Substrate	Initial substrate (cpm/10^6 cells)	Nucleotides (cmp/10^6 cells	(% substrate)
Glycine	101,400	182	0.18
Formate	82,400	59	0.07
Adenine	138,800	702	0.51
Adenosine	369,700	10,520	2.85
Hypoxanthine	184,000	442	0.24
Guanine	222,500	776	0.35

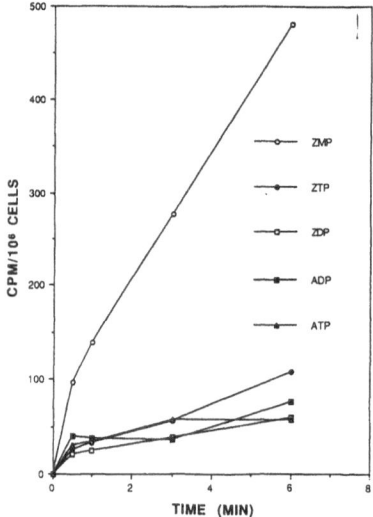

Figure 3. Acadesine metabolism in human neutrophils. 1×10^7 neutrophils were incubated with 50 µM radiolabelled acadesine. Incorporation of radioactivity into nucleotides was determined using TLC.

NUCLEOSIDE TRANSPORT

Ado transport into the neutrophils was moderately sensitive to NBMPR, since 500 nM NBMPR decreased Ado uptake by only 48% (Table 4). Activation of cells with 10^{-7} fMLP had no effect on nucleoside (Ado or formycin B) transport (data not shown).

Table 4. Effect of NBMPR on Ado transport in human neutrophils.

NBMPR (nM)	Nucleoside influx (% inhibition)
1	19±7
10	34±9
500	48±5

Transport of 0.5 µM [2,8-^{3}H] Ado was evaluated during a 10 sec assay followed by centrifugation of cells through a mineral oil layer.

In conclusion, human neutrophils formed Ado from intracellular nucleotides but not from the extracellular pool. Neutrophil activation increased Ado release and the inhibition of ADA or AK elevated Ado release further. Ado formed from neutrophil-derived AMP can only occur after hydrolysis by ecto5′-NT on other cells since neutrophils appeared to lack this enzyme. Finally, acadesine was a poor substrate for nucleotide synthesis but accumulated in neutrophils as ZMP.

REFERENCES

1. Cronstein, B.N., Purines and inflammation: Neutrophils possess P_1 and P_2 purine receptors, *in*: "Adenosine and Adenine Nucleotides as Regulators of Cellular Function," J.W. Phillis, ed., CRC Press, Boca Raton (1991).
2. Cronstein, B.N., Levin R.I., Philips, M., Hirschhorn, R., Abramson, S.B., and Weissmann, G., Neutrophil adherence to endothelium is enhanced via adenosine A1 receptors and inhibited via adenosine A2 receptors, *J. Immunol.* 148:2201 (1992).
3. Firestein, G.S., Bullough, D.A., Erion, M.D., Jimenez, R., Ramirez-Weinhouse, M., Barankiewicz, J., Smith., C.W., Gruber, H.E., and Mullane, K.M., Inhibition of neutrophil adhesion by adenosine and an adenosine kinase inhibitor: The role of selectins, submitted for publication.

ACADESINE: PRECLINICAL OVERVIEW

Harry E. Gruber

Gensia, Inc., San Diego, California 92121

Acadesine (aica riboside) is the first in a new class of drugs termed adenosine regulating agents (ARAs). Acadesine and other drugs in this class elevate adenosine in a site and event specific manner at times and sites of pathology where there is net ATP breakdown but not in healthy settings or tissues. Acadesine and its 5' monophosphate (ZMP) elevates extracellular adenosine by reportedly interacting with a number of enzymes in the adenosine regulating pathways, especially during times of net ATP catabolism (figure 1). The localized increase in adenosine is efficacious in a variety of clinical settings through the natural protective actions of adenosine. Of note, with the extremely short half-life of adenosine (< 1 second in human blood) the beneficial effects can be achieved without the deleterious effects of systemically acting adenosine, such as hypotension and bradycardia. Adenosine has numerous protective effects, especially in combating the pathologic sequelae of ischemia. The enhanced adenosine caused by acadesine treatment is directly protective to cardiomyocytes, inhibits aggregation of platelets, is anti-inflammatory and enhances local blood flow by causing smooth muscle vasodilation. With these numerous protective effects, acadesine was developed as a novel cardioprotective drug for use in settings of myocardial ischemia and reperfusion.

A variety of studies have demonstrated the cardioprotective properties of acadesine. In a model of simulated ischemia and reperfusion performed by Gruver, et al[1] on isolated rat cardiomyocytes, acadesine prevented the hypercontraction indicative of cell death and this benefit was reversed by the adenosine receptor antagonist, 8 sulfophenyltheophyline (8 SPT), confirming the adenosinergic mechanism of protection. In an isolated guinea pig heart model of ischemia and reperfusion performed by Bullough, et al[2] (figure 2) acadesine improved post ischemic function (LVDP), an effect reversed by adenosine deaminase (ADA) or by 8SPT. Kitakaze, et al[3] demonstrated that acadesine could prevent myocardial stunning in a dog model of brief ischemia and reperfusion, an effect reversed by 8PT (figure 3). In a related model of chronic angina in the dog, Young and Mullane[4] reported acadesine could prevent the deterioration in myocardial function in animals experiencing multiple cycles of accelerated cardiac pacing and recovery. Recently, Zhao, et al[5] reported a reduction in infarct size in acadesine treated rabbits compared to vehicle controls following 30 minutes of ischemia and 2 hours of reperfusion.

Acadesine has been demonstrated to have antithrombotic effects. In human blood, acadesine inhibits platelet aggregation, an effect reversed by ADA[6]. Montag, et al[7] demonstrated that acadesine could prevent cyclic flow reductions (CFRs) caused by endothelial injury and subsequent platelet thrombosis and that these benefits were abrogated by 8 SPT. Henry, et al[8] reported acadesine prevents reocclusion following tPA treatment in a canine model of coronary artery occlusion. In a photocoagulation induced carotid artery injury model of embolic stroke in rats, acadesine reduced the number of platelet emboli in the brain[9]. Acadesine's beneficial effects on regional blood flow during myocardial ischemia was demonstrated by Gruber, et al[10]. In this same study the elevation of adenosine was documented in the venous effluent draining the ischemic heart (figure 4) but not in mixed venous blood. Acadesine treated animals had no alterations in blood pressure or heart rate, consistent with only a localized effect of adenosine. The treated animals also had an inhibition of neutrophil trapping in the ischemic tissue consistent with the anti-inflammatory effects of adenosine.

The numerous beneficial properties of acadesine in settings of myocardial ischemia prompted clinical research into acadesine's use in the treatment of cardiac ischemia and reperfusion. Coronary artery bypass (CABG) surgery is an especially appealing clinical setting to evaluate acadesine as all patients undergo myocardial ischemia and reperfusion and a significant percent have subsequent myocardial infarcts and strokes. Galiñanes, et al[11] demonstrated that acadesine preserved myocardial function in isolated perfused rat hearts when added to cold cardioplegic solution. Acadesine preservation of myocardial function was also shown by Bolling, et al[12] in a canine model of cardioplegic arrest with cardiopulmonary bypass in the dog (figure 5). Vinten-Johansen, et al[13] performed a confirmatory study of acadesine's cardioprotective effects that more thoroughly evaluated myocardial function using pressure volume loop measurements. Galiñanes further demonstrated[14] that the protective effects of acadesine salvages the myocardium rather than merely increasing the rate of recovery, by documenting improved rat cardiac function after 4 hours of cardioplegic arrest and 24 hours of heart transplantation using acadesine (figure 6). In summary, these animal studies suggest that acadesine may be an ideal drug to develop for the clinical settings of myocardial ischemia and reperfusion, especially for preventing the adverse cardiovascular events which occur during CABG surgery.

ACKNOWLEDGEMENTS

Figures 4 and 6 are reproduced with permission from the American Heart Association. Figure 5 is reproduced with permission from Elsevier.

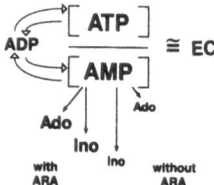

Figure 1.   ARAs increase adenosine generation from AMP during net ATP catabolism.

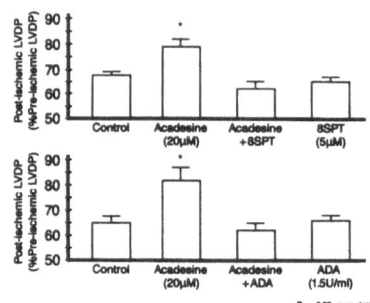

Figure 2.   Acadesine's protection from ischemia/reperfusion injury in isolated guinea pig hearts is reversed by ADA or 8SPT[2].

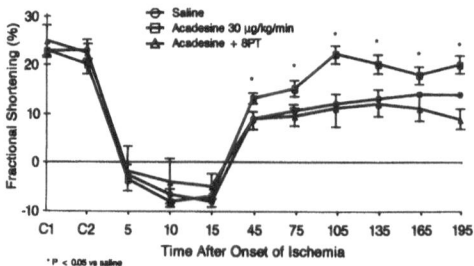

Figure 3.   Acadesine's protection from canine myocardial stunning is reversed by 8PT[3].

424

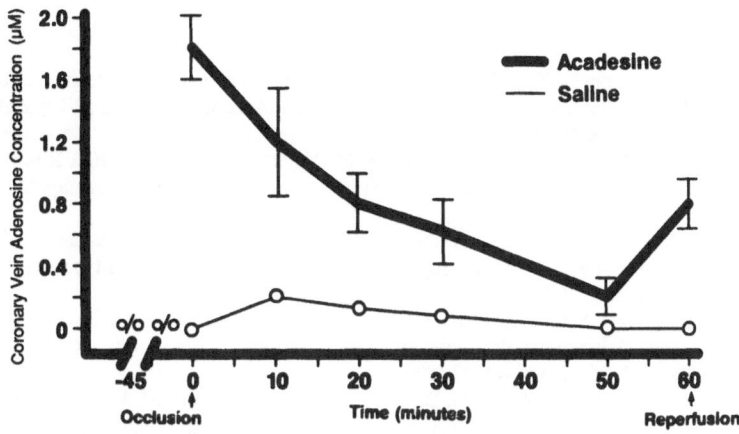

Figure 4.   Acadesine enhances coronary vein adenosine concentrations during canine myocardial ischemia[10].

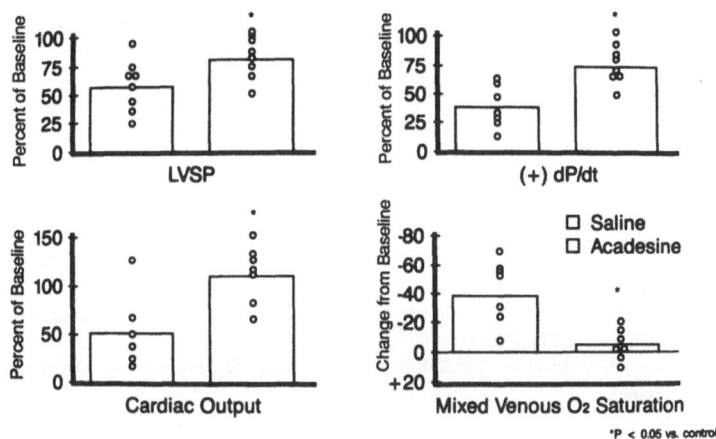

Figure 5.   Acadesine protects myocardial function in a canine model of cardiopulmonary bypass[12].

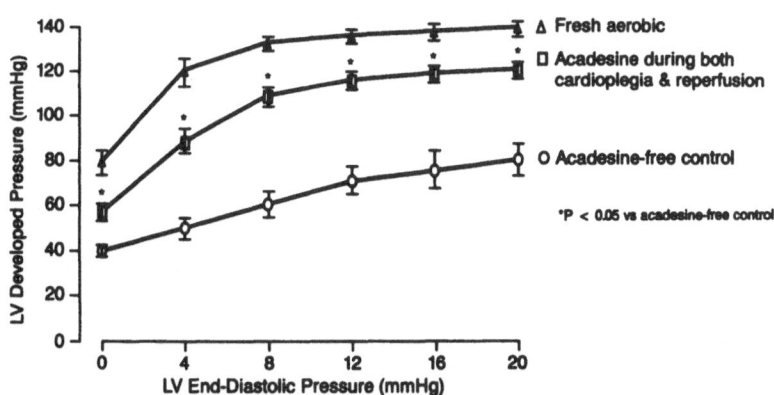

Figure 6.   Rat cardiac transplant model demonstrates acadesine preserves myocardial function at 24 hrs. of reperfusion rather than just delaying injury[14].

REFERENCES

1. E.J. Gruver, D. Toupin, T.W. Smith, J.D. Marsh. "Acadesine Improves Tolerance to Ischemic In jury in Rat Adult Cardiac Myocytes by an Adenosine Receptor-Mediated Process", *Circulation*, Pg. I-541 (1993).

2. D. Bullough, M. Fox, S. Potter, K. Metzner, M. Young, K. Mullane. "AICA Riboside Protects Myocardial Tissue From Reperfusion Injury and Oxidant-Induced Damage", *FASEB*, Vol. 5, No. 5, Pg. A1436 (1991).

3. M. Hori, M. Kitakaze and S. Takashima. "AICA-Riboside (5-amino-4-imidazole carboxamide riboside 100), a Novel Adenosine Potentiator, Attenuates Myocardial Stunning." *Circulation*, Vol 82, No. 4, Pg. 466; (Suppl. III), (1990).

4. M.A. Young, and K.M. Mullane. "Progressive Cardiac Dysfunction With Repeated Pacing-Induced Ischemia: Protection by AICA-Riboside". *Am J. Physiol.* 261 (Heart Circ. Physiol. 30):H1570-H1577 (1991).

5. Z. Zhao, D.S. McGee, K. Nakanishi, D.A. Hudspeth, M.W. Williams, J. Vinten-Johansen, D. G.L. Van Wylen. "Acadesine Reduces Infarct Size Without Augmenting Interstitial Adenosine Levels During Coronary Occlusion in the Rabbit". *Circulation*, Vol. 88, No. 4, Part II, Pg. I-432, (1993).

6. D.A. Bullough, C. Zhang, A. Montag, K.M. Mullane, and M.A. Young. "Adenosine-Mediated Inhibition of Platelet Aggregation by Acadesine: A Novel Antithrombotic Mechanism In Vitro and In Vivo". *Submitted for publication.*

7. A. Montag, D. Bullough, K. Mullane, M.A. Young. "Enhanced Antithrombotic Efficacy of Locally-Produced vs. Systemically Administered Adenosine in a Canine Model of Unstable Angina". *J. Mol. & Cell. Cardio.*, Vol. 25 (Suppl. III); S.5.

8. C.A. Henry, M.A. Young, C. Zhang, D.A. Bullough, and K. Mullane. "Adenosine Release From Red Cells Mediates Inhibition of Platelet Aggregation by Acadesine and Delays Post-Thrombolytic Reocclusion in Dogs". *Circulation*, Vol. 84, No. 4, Pg. 247; (Suppl. II), (1991).

9. L.P. Miller, W.D. Dietrich, R. Prado, M.K. Dewanjee and H. Gruber. "Acadesine Reduces Indium Labeled Platelet Deposition in Rat Brain Following Photothrombosis of the Common Carotid Artery". *Soc. Neurosci.*, Vol. 19, Pg. 321 (1993).

10. H.E. Gruber, M.E. Hoffer, D.R. McAllister, P.K. Laikind, T.A. Lane, G. W. Schmid-Schoenbein, and R. L. Engler. "Increased Adenosine Concentration in Blood From Ischemic Myocardium by AICA Riboside: Effects on Flow, Granulocytes, and Injury". *Circulation* 80:5, 1400-1411 (1989).

11. M. Galiñanes, K.M Mullane, D. Bullough, D.J. Hearse. "Acadesine and Myocardial Protection: Studies of Time of Administration and Dose-Response Relations in the Rat. *Circulation* 86:598-608 (1992).

12. S.F. Bolling, M.A. Groh, A.M. Mattson, R.A. Grinage, and K.P. Gallagher. "Acadesine (AICA-riboside) Improves Postischemic Cardiac Recovery". *Ann. Thorac. Surg.* 54:93-98 (1992).

13. J. Vinten-Johansen, K. Nakanishi, Z.Q. Zhao, D.S. McGee, P. Tan. "Acadesine Improves Surgical Myocardial Protection With Blood Cardioplegia in Ischemically Injured Canine Hearts". *Circulation* 88:350-358 (1993).

14. M. Galiñanes, D. Bullough, K.M.Mullane, D.J. Hearse. "Sustained Protection by Acadesine Against Ischemia- and Reperfusion-Induced Injury: Studies in the Transplanted Rat Heart". *Circulation* 86:589-597 (1992).

REGULATION OF ENDOGENOUS ADENOSINE LEVELS IN THE CNS: POTENTIAL FOR THERAPY IN STROKE, EPILEPSY AND PAIN

Alan C. Foster, Leonard P. Miller, James B. Wiesner

Department of CNS Pharmacology, Gensia, Inc., 9360 Towne Centre Drive, San Diego, CA 92121, U.S.A.

INTRODUCTION

In the central nervous system (CNS), adenosine is proposed to play a role as a neuromodulator. One function is to provide negative feedback in response to increased activation of neurons. These effects are of importance for the "normal" physiological functioning of the CNS, but may have greater relevance during pathophysiological events. This chapter will review the evidence which shows that endogenous adenosine increases in the extracellular compartment in the CNS in response to pathological events, the mechanisms by which adenosine can exert beneficial effects which counteract the pathology, and the approaches taken by Gensia to develop drugs which can harness this natural protective effect of adenosine.

ADENOSINE RECEPTOR SUB-TYPES IN THE CNS

Four adenosine receptors - A_1, A_{2A}, A_{2B}, and A_3 - have been identified from pharmacological and molecular cloning experiments (1,2). These all belong to the family of receptors which are linked to their effectors through G-proteins. The A_1 receptor sub-type is the most abundant in the mammalian CNS, having high density in the cerebral cortex, hippocampus, cerebellum and spinal cord. A_1 receptors are present on both neurons and glia, and have been demonstrated to be strongly associated with both pre- and post-synaptic elements of the glutamate neurotransmitter system. Through their associated G-proteins, A_1 receptors inhibit adenylate cyclase, activate K^+ channels and inhibit Ca^{++} channels. A_{2A} receptors are located primarily in the basal ganglia, and are associated particularly with areas of dopaminergic innervation such as the caudate nucleus, nucleus accumbens and olfactory tubercle. A_{2A} receptors increase adenylate cyclase activity. A_{2B} receptors have a restricted location in the CNS being present in the pars tuberalis of the pituitary and in the spinal cord and have also been detected on astrocytes. Like the A_{2A} receptors, A_{2B} receptors increase the activity of adenylate cyclase. A_3 receptors inhibit the activity of adenylate cyclase, but little else is known of their location or function in the CNS.

PHYSIOLOGICAL ROLES FOR ADENOSINE IN THE CNS

Adenosine is not regarded as a classical neurotransmitter substance, but is termed a "neuromodulator" which, although not directly involved in neuronal signalling, can influence neuronal activity (3). The existence of an adenosine inhibitory "tone" in the CNS is suggested by the observation that adenosine receptor antagonists which penetrate the blood-brain barrier, e.g. theophylline and caffeine, act as stimulant drugs. This inhibitory tone may have several different functions depending on the brain region and adenosine receptor sub-types involved. The enhanced locomotor activity caused by adenosine receptor antagonists appears to be mediated by inhibition of A_{2A} receptors located in the basal ganglia (4). This may suggest that adenosine can control locomotion through functional inhibition of the dopamine system. A role for adenosine in regulating arousal has been suggested, based on its ability to alter the firing patterns of cholinergic neurons in the basal forebrain through activation of post-synaptic A_1 receptors which modify ion channel function (5). In the hippocampus, a brain region associated with learning and memory, local adenosine release appears to mediate paired-pulse depression, a mechanism which may be important for organizing the processing of neuronal information. In this case adenosine acts over a millisecond timeframe and is released upon activation of excitatory synapses to provide a local inhibition of excitatory neurotransmission via the A_1 receptor (6). A role for adenosine in the control of pain responses is suggested by the findings that an inhibitory post-synaptic potential (ipsp) detected in spinal cord neurons which transmit pain information is mediated by adenosine. In this case a nucleotide (probably ATP) is released from primary afferent fibres and is degraded extracellularly by the enzyme ecto-5'-nucleotidase to form adenosine which inhibits the neurons via post-synaptic A_1 receptors linked to ATP-sensitive K^+ channels (7).

These studies suggest diverse physiological roles for endogenous adenosine in the CNS, but many questions remain. Is a constant level of extracellular adenosine present which modifies neuronal function, or as in the example of the hippocampus, is adenosine released locally in response to neuronal activity? Does this vary between brain regions and neuronal types? What is the source of extracellular adenosine, i.e release from intracellular stores via the nucleoside transporters, or extracellular degradation from a nucleotide? What are the roles played by glial cells, which possess A_1 and A_2 receptor sub-types?

ROLES OF ADENOSINE IN PATHOLOGICAL STATES IN THE CNS

Extracellular concentrations of adenosine are increased in a variety of pathological situations in the CNS. Measurements using intracerebral microdialysis indicate that during cerebral ischemia extracellular adenosine is elevated by as much as 30 times basal values (8). This is probably a result of the breakdown of ATP when the supply of energy substrates is reduced, but could also be evoked by the increased activity of excitatory neurotransmitters, particularly glutamate, the levels of which also rise in ischemic CNS tissue (9). Excessive activation of neurons also leads to increased extracellular adenosine. Tissue levels of adenosine are elevated within seconds of seizure activity (10). Microdialysis experiments in human hippocampus have indicated that extracellular adenosine increases by several-fold during temporal lobe seizures (11). The elevation of extracellular adenosine during cerebral ischemia or seizure activity may reflect an attempt by brain tissue to counteract the pathology. Adenosine can act at its receptors to inhibit the glutamate neurotransmitter system whose overactivity is largely responsible for the degeneration and excessive activation of neurons which occur in cerebral ischemia and epilepsy (12). Adenosine acts at presynaptic A_1 receptors to inhibit glutamate release and at post-synaptic A_1 receptors to open K^+ channels and reduce neuronal excitation (13). In the case of cerebral ischemia, local release of adenosine in the vicinity of cerebral arterioles can activate A_2 receptors on smooth muscle to produce vasodilation and improve blood flow, and can activate the same receptor sub-type on platelets and neutrophils to inhibit their function and reduce clot formation and deleterious inflammatory responses.

In the spinal cord, activation of adenosine receptors can inhibit pain responses (14). One line of evidence indicates that morphine, through activation of opiate receptors, can provoke the release of endogenous adenosine in the spinal cord, and that part of the analgesic actions of morphine is mediated by this mechanism, since adenosine receptor antagonists have been shown to inhibit opiate-mediated effects in animal models of acute pain (15).

THERAPEUTIC STRATEGIES BASED ON ADENOSINE

The evidence that adenosine has beneficial effects which can counteract pathological mechanisms occurring in stroke, seizure and pain suggests that therapeutic strategies based on adenosine may lead to novel and effective treatments for these CNS disorders. Adenosine receptor agonists have been shown to be neuroprotective in animal models of stroke (16), are potent anticonvulsants in several animal seizure models (17) and provide analgesia in animal models of pain (14). However, clinical development of these agents has been prevented by side effects which occur at or below the effective neuroprotective, anticonvulsant or analgesic doses due to indiscriminate activation of adenosine receptors throughout the body. For example, A_1 receptor agonists produce profound hypotension, bradycardia, hypothermia and sedation at doses which are anticonvulsant (18).

An alternative approach is to find compounds which can enhance the extracellular levels of endogenous adenosine when it is released during pathological conditions. Gensia is developing a class of compounds called adenosine regulating agents (ARAs) which are targeted at enzymes responsible for the metabolism of adenosine and achieve their therapeutic effects by increasing the local concentration of adenosine. These enzymes are particularly active during certain pathological events where ATP levels fall and adenosine is produced. As a result, the ARAs selectively increase adenosine during, and at the site of, pathological events. Combined with the fact that adenosine has a very short half-life in the circulation, this site and event specificity allows ARAs to effectively harness the benefits of adenosine without the often-encountered systemic side effects.

ARAs are currently under investigation for their effectiveness in animal models of stroke, seizure, and pain. Results obtained in two animal models relevant to stroke are described below.

ADENOSINE REGULATING AGENTS IN STROKE

A_1 receptor agonists, adenosine transport inhibitors and inhibitors of adenosine deaminase have been reported to reduce neuronal degeneration in animal models of focal cerebral ischemia (16). We have tested an analog of 5-amino-4-imidazole carboxamide riboside (acadesine; Protara™), the prototype ARA, in a rat model of focal stroke with reperfusion (19). In this model a nylon suture was inserted via the internal carotid artery to occlude the middle cerebral artery (MCA) for a period of 105 minutes followed by suture removal and reperfusion for 24 hours. The endpoints used were motor function of the animals at 24 hours and volume of infarction in the cerebral cortex and basal ganglia as measured by the 2,3,5-triphenyltetrazolium method. 5-Amino-1-[β-D-5-phosphoribofuranosyl]-imidazole-1-N-[4-nitrophenyl)-methyl]-carboxamide (GP-1-668) was given as an iv infusion (12.5, 25, 50 or 100 µg/kg/min) for 24 hours, starting 30 min prior to MCA occlusion. A dose dependent reduction in both infarct size and motor deficits was observed up to 50 µg/kg/min which reached a maximum of approximately 50%. At 100 µg/kg/min, no significant protective effect was apparent. The magnitude of infarct size reduction obtained with GP-1-668 was equivalent to that seen with any pharmacological approach in a focal stroke model. These data indicate the potential of ARAs as neuroprotective agents.

Acadesine was found to reduce the incidence of stroke following coronary artery bypass graft surgery by 89% in a phase III clinical trial. In models of cardiovascular function, acadesine has been shown to enhance the extracellular levels of adenosine in ischemic, but not non-ischemic tissue, to improve collateral flow in ischemic tissue and inhibit adherence of neutrophils (20). These intravascular effects of acadesine may be relevant for the observed reduction in the incidence of stroke. To investigate these mechanisms, acadesine has been tested in an animal model of embolus deposition in the brain (21). A thrombus was generated in the carotid artery by shining a laser onto the exposed artery in rats injected with rose bengal. The platelets were prelabelled with radiolabelled indium, and autoradiograms prepared of brain sections to assess the deposition of platelet emboli which break off from the thrombus and lodge in the cerebral circulation. Acadesine dose-dependently reduced the number of platelet emboli found in the cerebral vessels, over a dose range which achieved the plasma drug concentration targeted in the clinical study. These data indicate the potential benefits of enhancing the intravascular effects of adenosine for cerebrovascular disease.

CLOSING REMARKS

Therapeutic agents targeted at enhancing the natural protective effects of adenosine constitute a novel approach to the treatment of CNS disorders. By enhancing the intravascular and neuroprotective effects of adenosine, ARAs have the potential to be effective in cerebrovascular diseases. Similar approaches are being taken to identify the optimal ARA approach to provide novel anticonvulsant and analgesic drugs.

REFERENCES

1. K.A. Jacobson, P.J.M. van Galen and M. Williams, Adenosine receptors: pharmacology, structure-activity relationships, and therapeutic potential, *J. Med. Chem.* 35:407 (1992).
2. A.M. Curruthers and J.R. Fozard, Adenosine A_3 receptors: two into one won't go, *Trends Pharmacol. Sci.* 14:290 (1993).
3. S.H. Snyder, Adenosine as a neuromodulator, *Ann. Rev. Neurosci.* 8:103 (1985).
4. R.A. Barraco, K.A. Martens, M. Parizon and H.J. Normile, Adenosine A_{2A} receptors in the nucleus accumbens mediate locomotor depression, *Brain Res. Bull.* 31:397 (1993).

5. D.G Rainne, H.C.R. Grunze, R.W. McCarley and R.W. Greene, Adenosine inhibition of mesopontine cholinergic neurons: implications for EEG arousal, *Science* 263:689 (1994).

6. J.B. Mitchell, C.R. Lupica and T.V. Dunwiddie, Activity-dependent release of endogenous adenosine modulates synaptic responses in the rat hippocampus, *J. Neurosci.* 13:3439 (1993).

7. M.W. Salter, Y. de Koninck and J.L. Henry, ATP-sensitive K^+ channels mediate an IPSP in dorsal horn neurones elicited by sensory stimulation, *Synapse* 11:214 (1992).

8. L. Hillered, A. Hallström, S. Segersvärd, L. Persson and U. Ungerstedt, Dynamics of extracellular metabolites in the striatum after middle cerebral artery occlusion in the rat monitored by intracerebral dialysis, *J. Cereb. Blood Flow Metab.* 9:607 (1989).

9. H. Benveniste, J. Drejer, A. Schousboe and N.H. Diemer, Elevation of extracellular concentrations of glutamate and aspartate in rat hippocampus during transient cerebral ischaemia monitored by intracerebral microdialysis, *J. Neurochem.* 43:1369 (1984).

10. H.R. Winn, J.E. Welsh, R. Rubio and R.M. Berne, Changes in brain adenosine during bicuculline-induced seizures in rats. Effects of hypoxia and altered systemic blood pressure, *Circ. Res.* 47:568 (1980).

11. M.J. During and D.D. Spencer, Adenosine: a potential mediator of seizure arrest and postictal refractoriness, *Ann. Neurol.* 32:618 (1992).

12. R. Schwarcz and B.S. Meldrum, Excitatory amino acid antagonists provide a therapeutic approach to neurological disorders, *Lancet* 2:140 (1985).

13. L.P. Miller and C. Hsu, Therapeutic potential for adenosine receptor activation in ischemic brain injury, *J. Neurotrauma* 9:563 (1992).

14. J. Sawynok and M.I. Sweeney, The role of purines in nociception, *Neuroscience* 32:557 (1989).

15. J. Sawynok, M.I. Sweeney and T.D. White, Adenosine release may mediate spinal analgesia by morphine, *Trends Pharmacol. Sci.* 10:186 (1989).

16. K.A. Rudolphi, P. Schubert, F.E. Parkinson and B.B. Fredholm, Adenosine and brain ischemia, *Cereb. Brain Metab. Rev.* 4:346 (1992).

17. M. Druganow, Adenosine: the brain's natural anticonvulsant, *Trends Pharmacol. Sci.* 7:128 (1986).

18. T. H. Dunwiddie and T. Worth, Sedative and anticonvulsant effects of adenosine analogs in mouse and rat, *J. Pharmacol. Exp. Therap.* 220:70 (1982).

19. L.P. Miller, P.C. Chiang, S. Carriedo, K. Metzner and A.C. Foster, The adenosine regulating agent, GP-1-668, reduces infarct volume and neurological deficits in a rat model of focal ischemia with reperfusion, *Abs. Stroke Council 19th Int. Joint Conf. Stroke and Cereb. Circ.* p21 (1994).

20. H.E. Gruber, M.E. Hoffer, D.R. Mcallister, P.K. Laikind, T.A. Lane, G.W. Schmid-Schoenberg and R.L. Engler, Increased adenosine concentration in blood from ischemic myocardium by AICA riboside. Effects on flow, granulocytes and injury. *Circulation* 80:1400 (1989).

21. L.P. Miller, W.D. Dietrich, R. Prado, M.K. Dewanjee and H. Gruber, Acadesine reduces indium labeled platelet deposition in rat brain following photothrombosis of the common carotid artery, *Abs. Soc. Neurosci.* 19: 673.1 (1993).

ENZYMES OF ADENOSINE METABOLISM IN THE HEART, CARDIOMYOCYTES AND ENDOTHELIUM

Zdzislaw Kochan, Ryszard T. Smolenski, Anne-Marie L. Seymour and Magdi H. Yacoub

Magnetic Resonance Spectroscopy Group
National Heart and Lung Institute
Harefield Hospital
Harefield UB9 6JH, Middx, U.K.

INTRODUCTION

Adenine nucleotide metabolism in the heart has important physiological implications for the regulation of the high energy phosphate pool and the production of adenosine, a "retaliatory" metabolite of the heart (Newby, 1984). The extent of operation of metabolic pathways within various cell types in a single organ may differ substantially. Two cell types are of particular importance in the heart; cardiomyocytes which represent the majority of weight of the heart and the endothelium contributing only 3% to the weight of heart (Gerlach et al., 1985), but strategically localized in the vessel wall.

The purpose of this study was to analyze the activities of enzymes of purine metabolism in the human heart, rat heart, rat cardiomyocytes and cultured human endothelial cells. Evaluation of the distribution of enzyme activities within various cellular compartments can provide important information on the potential contribution of different pathways to the overall metabolism of the myocardium. The distribution of these activities in the heart has been evaluated and, in addition, differences between human and rat heart have been detected.

MATERIALS AND METHODS

Samples of human myocardium were collected from the left ventricle of explanted heart during heart or heart-lung transplantations, donor heart not used for transplantation or papillary muscle removed during mitral valve replacement (n=6). Rat hearts were rapidly removed from male Wistar rats (n=5). Cardiac myocytes (n=6) were isolated using a collagenase perfusion method previously described (Smolenski et al., 1991). Endothelial cell cultures were prepared from umbilical cord veins by collagenase digestion and were plated on 25 cm² plastic culture flasks (Jaffe et al., 1973). Confluent cultures were passaged and cells were used for enzyme assays after the second to third passage (n=4).

Human or rat heart tissue or rat cardiomyocytes were homogenized at the ratio of 9 ml of buffer (150 mM KCl, 20 mM TRIS, 1 mM EDTA, 1 mM Dithiothreitol, pH 7.0) per 1 g of wet weight of tissue. To obtain endothelial homogenates, the culture medium was removed from culture flasks, and cells rinsed with 0.9 % NaCl. 0.1 ml of homogenization buffer was added to the flask and cells were scraped. Cell suspensions from 6-7 culture flasks were pooled and subjected to homogenization. The following enzyme activities were determined: adenosine kinase (AK), adenosine deaminase (ADA), S-adenosylhomocysteine hydrolase (SAHH), purine nucleoside phosphorylase (PNP), AMP deaminase (AMPD), membrane 5'nucleotidase (M5'N), AMP specific (AC5'N) and IMP specific (IC5'N) cytosolic 5'nucleotidases. Samples were analysed for conversion of substrates into products by reversed-phase HPLC (Smolenski et al., 1990).

RESULTS AND DISCUSSION

As can be seen in Table 1, enzyme activities in the human heart were generally lower than in the rat heart with the exception of AC5'N and IC5'N where activities were equivalent. The difference was most marked in the case of AMPD (9-fold lower in human heart) and M5'N (6 fold lower in the human heart). The results presented here are in agreement with an earlier comparative study between human and rat heart (Meghji et al., 1988). There is also a good correlation between the low AMPD/IC5'N activity ratio in human heart observed here and the lack of IMP accumulation in human cardiomyocytes subjected to metabolic stress (Smolenski et al., 1992). In contrast, the high ratio of AMPD/IC5'N activities observed in the rat heart was consistent with the previous findings (Altshuld et al, 1987) showing the ability of the rat heart or cardiomyocytes to accumulate IMP. Although the majority of human samples were taken from explanted hearts or papillary muscles affected by disease process or hypertrophy, three determinations were performed on normal human myocardium collected from a donor human heart not used for transplantation and two explanted hearts from heart and lung transplantation.

Rat cardiomyocyte enzyme activities were comparable to those measured in the whole rat heart with the exception of ADA (6-fold lower) and PNP (16-fold lower). Endothelial activities were notably different to the whole human heart in the case of SAHH (9-fold higher) and PNP (16-fold higher). Taken together, these results suggest a predominantly endothelial location for ADA, PNP and SAHH.

Our data on the cellular distribution of ADA are in agreement with previous findings in the rat heart (Bowditch et al., 1986, Dow et al., 1987, De Jong et al., 1991) or guinea pig heart (Dendorfer et al., 1987). PNP on the other hand was found to be almost totally absent in rat cardiomyocytes by Bowditch et al. (1986) but present at a similar level of activity in ventricular homogenates by De Jong et al. (1991). The difference in the latter study was the use of one day old cultured cardiomyocytes as compared to fresh myocytes used in this study and by Bowditch et al, (1986). The culturing of cells may allow time for re-expression of PNP activity which could be reduced during the isolation procedure. However, culturing may also lead to contamination by other types of cells or to enhanced expression of activities which are normally present at low levels. Histochemical data show a predominantly endothelial location of PNP and an absence in cardiomyocytes (Rubio and Berne, 1980, Borgers and Thone, 1992).

Table 1. Purine enzymes activities in the human and rat heart, endothelium and cardiomiocytes.

Enzyme	Human heart	Rat heart	Rat cardiomiocytes (μmol/min/g wet weight)	Human endothelium
Adenosine kinase	0.14 ±0.02	0.33 ±0.06	0.30 ±0.03	0.22 ±0.06
Adenosine deaminase	0.46 ±0.03	1.62 ±0.23	0.20 ±0.04	1.36 ±0.29
SAH hydrolase	0.0023 ±0.0012	0.0057 ±0.0009	0.0039 ±0.0004	0.0207 ±0.0018
Purine nucleoside phosph.	0.43 ±0.08	1.45 ±0.25	0.15 ±0.008	6.67 ±0.51
AMP deaminase	0.41 ±0.05	3.77 ±0.19	2.80 ±0.30	1.97 ±0.16
Membrane 5'nucleotidase	1.75 ±0.12	7.52 ±0.56	5.40 ±1.10	6.08 ±0.68
Cytosolic 5'nucleotidase AMP-specific	0.11 ±0.02	0.17 ±0.04	0.08 ±0.009	0.27 ±0.1
Cytosolic 5'nucleotidase IMP-specific	0.21 ±0.04	0.27 ±0.02	0.085 ±0.008	0.17 ±0.01

Values represent mean ± S.E.M (n=4-6)

Our finding that SAHH activity is greater in endothelium is in agreement with observations of Mistry and Drummond (1986), whereas other authors suggested a more uniform distribution (Borst et al., 1992). Species differences could be a possible explanation for this discrepancy. The potentially greater capacity of the SAHH pathway in endothelial cells is very interesting as the product of this pathway– adenosine– is an important physiological metabolite in the heart (Newby, 1984). SAHH is a significant source of adenosine in the perfused rat heart under normoxic conditions (Achterberg et al., 1989, Deussen et al., 1989). Data obtained after selective radiolabelling of the endothelial nucleotide pool in the perfused rat heart during normoxia has shown that a similar proportion of the adenosine released originates from endothelium (Deussen et al., 1986, Bardenheuer et al., 1987, Kroll et al., 1987). It is thus possible that this adenosine originates from the endothelial SAHH pathway.

Distribution of other adenosine metabolising enzymes seems to be more uniform in cardiomyocytes and endothelium, and differences in activities may reflect species differences. This may be the case for E5'N which has been shown to be present histochemically in both cardiomyocytes and vascular cells in the rat heart but only in the vasculature of the human heart (Borgers and Thone, 1992). Our results for M5'N activity support these findings. AK activity was found to be uniformly distributed in the heart, despite the high rate and specificity of incorporation of exogenous adenosine into the endothelium (Gerlach et al., 1985). It appears therefore that the efficiency of endothelial incorporation of adenosine arises predominantly from its location rather than its enzyme content.

In conclusion, adenosine metabolism is particularly active in endothelial cells. In addition to the exclusive location of the enzymes of adenosine catabolic pathway, these cells also show significant synthetic capacity by intra and extracellular dephosphorylation of AMP and especially by SAH hydrolysis.

ACKNOWLEDGMENT

The authors thank Rhoda McDouall for her excellent guidance in the preparation of the endothelial cell cultures. This study was supported by The British Heart Foundation (Grant no 91/167) and The Royal Society.

REFERENCES

Achterberg PW, De Tombe PP, Harmsen E, de Jong JW (1985) Myocardial S-adenosylhomocysteine hydrolase is important for adenosine production during normoxia. Biochim Biophys Acta 840:393-400

Bardenheuer H, Whelton B, Sparks HV (1987) Adenosine release by the isolated guinea pig heart in response to isoproterenol, acetylocholine, and acidosis: the minimal role of vascular endothelium. Circulation Res 61:594-600

Borgers M, Thone F (1992) Species differences in adenosine metabolic sites in the heart. Histochem J 24:445-452

Borst MM, Deussen A, Schrader J (1992) S-adenosylhomocysteine hydrolase activity in human myocardium. Cardiovasc Res 26:143-147

Bowditch J, Brown AK, Dow JW (1986) Accumulation and salvage of adenosine and inosine by isolated mature cardiac myocytes. Biochim Biophys Acta 844:119-128

de Jong JW, Keijzer E, Huizer T, Schoutsen B (1990) Ischemic nucleotide breakdown increases during cardiac development due to drop in adenosine anabolism/catabolism ratio. J Mol Cell Cardiol 22:1065-1070

Dendorfer A, Lauk S, Schaff A, Nees S (1987) New insights into the mechanism of myocardial adenosine formation. In: Becker BF, Gerlach E (eds) Topics and Perspectives in Adenosine Research. Springer-Verlag, Berlin, Heidelberg, pp 170-185

Deussen A, Moser G, Schrader J (1986) Contribution of coronary endothelial cells to cardiac adenosine production. Pflugers Arch 406:608-614

Deussen A, Lloyd HGE, Schrader J (1989) Contribution of S-adenosylhomocysteine to cardiac adenosine formation. J Mol Cell Cardiol 21:773-782

Dow JW, Bowditch J, Nigdikar SV, Brown AK (1987) Salvage mechanisms for regeneration of adenosine triphosphate in rat cardiac myocytes. Cardiovasc Res 21:188-196

Gerlach E, Nees S, Becker F (1985) The vascular endothelium: a survey of some newly evolving biochemical and physiological features. Basic Res Cardiol 80:459-474

Jaffe EA, Nachman RL, Becker CG, Minick CR (1973) Culture of human endothelial cells derived from umbilical veins. Identification by morphologic and immunologic criteria. J Clin Invest 52:2745-2756

Kroll K, Schrader J, Piper HM, Henrich M (1987) Release of adenosine and cyclic AMP from coronary endothelium in isolated guinea pig hearts: relation to coronary flow. Circ Res 60:659-665

Meghji P, Middleton KM, Newby AC (1988) Absolute rates of adenosine formation during ischaemia in rat and pigeon hearts. Biochem J 249:695-673

Mistry G, Drummond GI (1986) Adenosine metabolism in microvessels from heart and brain. J Mol Cell Cardiol 18:13-22

Newby AC (1984) Adenosine and the concept of retaliatory metabolites. Trends in Biochem Sci 9:42-44

Rubio R, Berne RM (1980) Localisation of purine and pyrimidine nucleoside phosphorylases in heart, kidney and liver. Am J Physiol 239:H721-H730

Smolenski RT, Lachno DR, Ledingham SJM, Yacoub MH (1990) Determination of sixteen nucleotides, nucleosides and bases using high-performance liquid chromatography and its application to the study of purine metabolism in hearts for transplantation. J Chromatogr 527:414-420

Smolenski RT, Schrader J, de Groot H, Deussen A (1991) Oxygen partial pressure and free intracellular adenosine of isolated cardiomyocytes. Am J Physiol 260:C708-C714

Smolenski RT, Suitters A, Yacoub MH (1992) Adenine nucleotide catabolism and adenosine formation in isolated human cardiomyocytes. J Mol Cell Cardiol 24:91-96

ADENOSINE UPTAKE AND METABOLISM IN HUMAN ENDOTHELIAL CELLS

Ryszard T. Smolenski, Zdzislaw Kochan, Rhoda McDouall,
Anne-Marie L. Seymour and Magdi H. Yacoub

M.R.S. Group, National Heart and Lung Institute, Harefield Hospital,
Harefield, U.K.
Department of Biochemistry, Academic Medical School,
Gdansk, Poland

INTRODUCTION

The handling of adenosine by the endothelium is crucial for the effects of this nucleoside in the vascular system. Strategic location of endothelial cells at the vessel wall border indicate its important role in the turnover of adenosine. High endothelial activity of ecto 5'nucleotidase[1] implies predominant role in the extracellular formation of adenosine from secreted nucleotides. However, the capacity to release intracellularly produced adenosine, phosphorylation of extracellular adenosine and to some extent adenosine transit between interstitial and intravascular space depend on the membrane transport capacity in this type of cells.

Adenosine transport in endothelium of experimental animals is well characterized[2,3]. However, little data is available concerning human cells. In addition, few studies allow the direct comparison of adenosine transport capacity between endothelium and cardiomyocytes or evaluate metabolic pathways of transported adenosine. The aims of this study were thus twofold: to characterize adenosine uptake and metabolism in human endothelium and to compare its kinetics with adenosine uptake in isolated cardiomyocytes evaluated using similar methodology.

MATERIALS AND METHODS

Confluent cultures of human umbilical cord endothelial cells were used to study kinetics of adenosine uptake together with adenosine metabolic pathways by human endothelium. Cells were prepared and cultured according to method of Jaffe[4,5] and passaged 2-3 times. Before experimentation, incubation medium was removed and incubation flasks were washed with Hank's Balanced Salt Solution (HBSS). After 30 min preincubation in HBSS at 37°C, cultures were briefly exposed to [2,8-³H] adenosine at concentrations varying from 0.3 to 1000 μM with [U-¹⁴C] sucrose as a marker of extracellular space. Immediately after exposure (for blank samples) or after defined incubation time, the incubation medium was quickly removed and cells were extracted with 0.6 M perchloric acid. Adenosine uptake was linear for up to 2 min from 0.3 to 10 μM concentrations and up to 30 sec from 30 to 1000 μM concentrations. Cardiomyocytes were isolated using collagenase perfusion technique[6]. Cardiomyocyte suspension was exposed to [2,8-³H] adenosine/[U-¹⁴C] sucrose mixture and immediately separated from medium (blank samples) or incubated for preset time (0.25-1 min) before separation. Radioactivity of ¹⁴C and ³H was counted both in the medium and in cell extracts using a dual channel scintillation counter and used to calculate quantity of transported adenosine. Cell extracts and medium were also subjected to HPLC[7] coupled to fraction collector or with on line radiodetection to calculate the profile of adenosine metabolism.

RESULTS

Data presented in Fig. 1 suggest the existence of two components of adenosine transport: a saturable component with half saturation constant ($S_{0.5}$) at 3 μM and maximal rate 700 pmol/min/mg protein and non saturable component with the slope $y = 6.0x + 801$ pmol/min/mg prot..

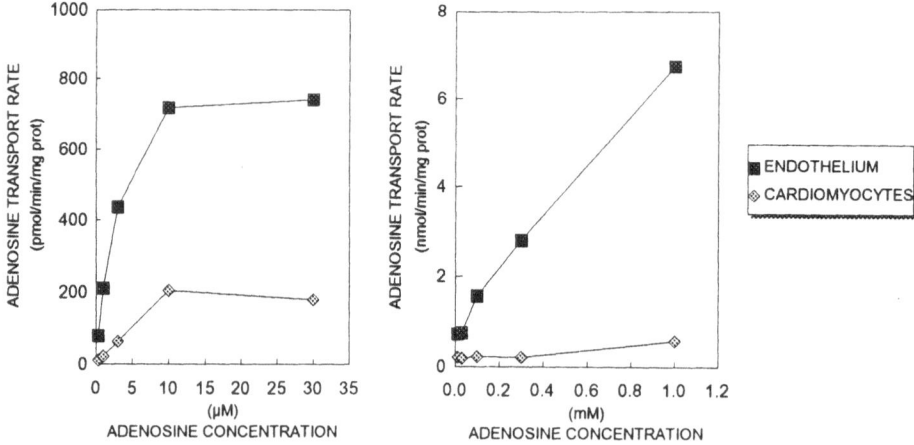

Figure 1. Adenosine transport rate in endothelium and cardiomyocytes at low (left) and high (right) adenosine concentration.

As may be seen from Fig. 2 adenosine was predominantly phosphorylated to ATP at concentration 0.3 and 3 μM. However, little accumulation of intracellular adenosine was also shown. At higher concentrations phosphorylation, was almost completely inhibited and accumulation of intracellular adenosine with its breakdown products inosine and hypoxanthine were observed. Separate experiments with [8-^{14}C] adenosine (not shown) demonstrated no further breakdown to xanthine and uric acid.

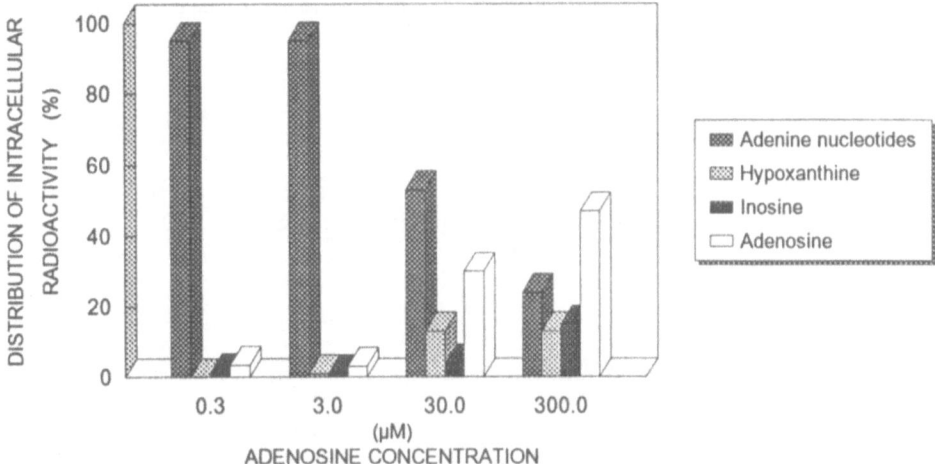

Figure 2. Cellular metabolism of adenosine in endothelial cells.

Experiments on adenosine transport with rat cardiomyocytes (Fig. 1) demonstrated much lower rates for both saturable and non saturable component than in endothelial cells. The concentration dependence of incorporation into ATP was generally similar to that in endothelium with rapid incorporation of adenosine into ATP at concentrations below 30 μM and complete inhibition of phosphorylation above that level. However, there was less inosine and an absence of hypoxanthine accumulation. A value of $S_{0.5}$ in cardiomyocytes was the same as in endothelium for high affinity adenosine transport component.

DISCUSSION

Kinetic parameters of low $S_{0.5}$ adenosine transport component in human umbilical vein endothelium were found to be similar to those reported for animal endothelium. For pig aortic endothelial cells and porcine aortic and pulmonary endothelium, $S_{0.5}$ values were equal to 3 μM[2,3]. The maximal rates of high affinity component are also equivalent to the values demonstrated here. However, the kinetics of adenosine transport at concentrations above 30 μM observed here fit a non saturable linear relation. This may represent simple diffusion or an initial phase of low affinity component as suggested in previous studies[2,3] which seems to have higher $S_{0.5}$ values than found in animal endothelium. In general, the kinetics and degree of expression of adenosine transport proteins in human endothelium seems to be similar to values previously found in animals.

Comparing adenosine transport rates in cardiomyocytes and endothelium lower rate of both high affinity and non saturable (low affinity ?) component in cardiomyocytes is very clear. $S_{0.5}$ value for high affinity component in cardiomyocytes was found to be similar to endothelial. There were few points in which metabolic pattern of transported adenosine in cardiomyocytes and endothelium differed. Unlike cardiomyocytes there was some accumulation of intracellular adenosine at low concentrations in endothelium most likely due to differences in transport rate versus adenosine kinase activity. Another difference is the greater breakdown of adenosine in endothelium observed at 30 and 1000 µM adenosine concentrations which is the result of higher endothelial activities of adenosine deaminase and purine nucleoside phosphorylase.

In conclusion, endothelial cells must thus play a vital role in adenosine metabolism not only because of their strategic location but also due to large capacity of membrane adenosine transport and rapid metabolism.

ACKNOWLEDGMENTS

This study was supported by the British Heart Foundation (Grant No: 91/167) and by the Polish Committee for Scientific Research (Grant No: 4 S 402 016 04)

REFERENCES

1. G. Mistry and G.I. Drummond, Adenosine metabolism in microvessels from heart and brain, *J. Mol. Cell Cardiol.* 18:13(1986).
2. J.D. Pearson, J.S. Carleton, A. Hutchings, and J.L. Gordon, Uptake and metabolism of adenosine by pig aortic and smooth-muscle cells in culture, *Biochem. J.* 170:265(1978).
3. Y. Dieterle, C. Ody, A. Ehrensberger, H. Stalder, and A.F. Junod, Metabolism and uptake of adenosine triphosphate and adenosine by porcine aortic and pulmonary endothelial cells and fibroblasts in culture, *Circ. Res.* 42:869(1978).
4. E.A. Jaffe, R.L. Nachman, C.G. Becker, and C.R. Minick, Culture of human endothelial cells derived from umbilical veins. Identification by morphologic and immunologic criteria, *J. Clin. Invest.* 52:2745(1973).
5. R.T. Smolenski, Z. Kochan, R. McDouall, C. Page, A-M.L. Seymour, and M.H. Yacoub, Endothelial nucleotide catabolism and adenosine production, *Cardiovasc. Res.* 28:100(1994).
6. R.T. Smolenski, J. Schrader, H. de Groot, and A. Deussen, Oxygen partial pressure and free intracellular adenosine of isolated cardiomyocytes, *Am. J. Physiol.* 260:C708(1991).
7. R.T. Smolenski, D.R. Lachno, S.J.M. Ledingham, and M.H. Yacoub, Determination of sixteen nucleotides, nucleosides and bases using high-performance liquid chromatography and its application to the study of purine metabolism in hearts for transplantation, *J. Chromatogr.* 527:414(1990).

SAH-HYDROLASE ACTIVITY IN HAEMOLYSATE AND INTACT ERYTHROCYTES; THE EFFECT OF ADENOSINE ANALOGUES

Krystyna Fabianowska-Majewska, Ryszard T.Smolenski#,
Maciej Marlewski#, John A.Duley* and H.Anne Simmonds*

Dept General Chemistry, Medical University Lodz, Poland
Dept of Biochemistry#, Academic Medical School, Gdansk, Poland
Purine Research Laboratory*, UMDS, Guys Hospital, London, UK.

INTRODUCTION

S-adenosylhomocysteine hydrolase (SAHH) is essential for S-adenosylmethionine (SAM) mediated transmethylations because of its role in the removal of SAH, a direct transmethylation product. This reaction is an important target for anticancer and immunosuppressive chemotherapy using nucleoside analogues[1,2]. In this study, the effect of 18 different nucleoside analogues on SAHH activity was evaluated in haemolysates and intact erythrocytes.

METHODS

The SAHH activity was measured in the synthetic direction in the haemolysates. as well as in intact erythrocytes, using [8-^{14}C] adenosine and L-homocysteine as the substrates. Freshly drawn blood was added to heparinised tubes, centrifuged to remove the plasma and washed twice in saline. To obtain haemolysates, washed erythrocytes (100 μl alliquots) were suspended in 500 μl distilled water and frozen and thawed twice, followed by centrifugation to remove erythrocyte membranes. For the assay, 25 μl of lysate was added to 80 μl of incubation buffer. Various nucleoside analogues were added at 1.25 mM concentrations. This mixture were preincubated for 30 min at 37°C. The reaction was started by addition of 20 μl of [8-^{14}C] adenosine and L-homocysteine solution and the incubation was continued for the next 40 min. Final concentrations in the incubation medium were as follows: 26 mM NaH_2PO_4 (pH 7.4), 0.8 mM EDTA. 0.8 mM dithiothreitol, 10 μM deoxycoformycin (dCf) an adenosine deaminase inhibitor, 20 μM iodotubercidine (Itu), an adenosine kinase inhibitor, 6 mM L-homocysteine and 0.5 mM adenosine.

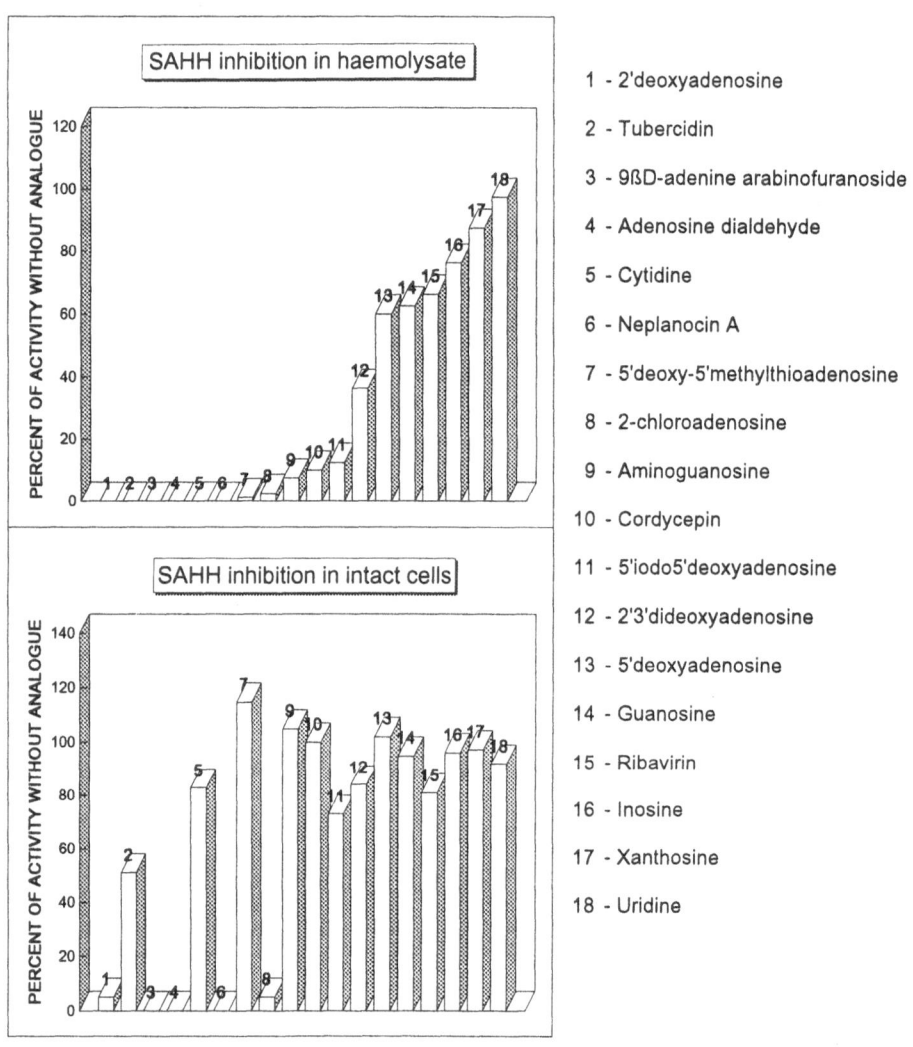

Figure 1. Comparison of the effect of nucleoside analogues on SAH hydrolase activity in the haemolysate and in the intact cells. Each value rapresent the average of at least 3 experiments.

The inhibitors dCf and Itu were added to minimise the flux of adenosine through the phosphorylation and deamination pathways. Reactions were terminated by addition of 25 μl of 40% TCA. Conversion of adenosine into SAH was determined by reverse-phase HPLC on an automated Milipore-Waters system using UV and radiodetection[3]. Intact cells assays of SAHH activity were based on determination of SAH accumulation in the cells incubated in the presence of adenosine and L-homocysteine thiolactone[4]. Isolated and washed erythrocytes were added to the incubation medium - Earls Balanced Salt Solution (EBSS), containing 5.6 mM glucose, 18 mM H_3PO_4 as well as 10 μM dCf and 20 μM Itu. Nucleoside analogues were present in the medium at 1.25 mM concentration. After 30 min preincubation at 37°C, [8-^{14}C] adenosine and L-homocysteine thiolactone mixture (concentrations 0.2 mM and 0.4 mM respectively) was added and incubation was continued for 40 min. Termination of incubation as well as determination of conversion of adenosine into SAH were performed as described above.

RESULTS AND DISCUSSION

As may be seen from Figure 1 complete inhibition of SAHH activity in the haemolysates was obtained with 2'deoxyadenosine, 9ßD-adenine arabinofuranoside, 2-chloroadenosine, adenosine dialdehyde, 5'iodo-5'deoxyadenosine, tubercidin, cytidine, 5'deoxy-5'methylthioadenosine, 2-aminoguanosine, cordycepin and neplanocin-A. Partial inhibition was found using 2'3'dideoxyadenosine, ribavirin, guanosine and 5'deoxyadenosine while uridine, xanthine and inosine were without any significant influence. In the intact cells only 2'deoxyadenosine, 9ßD-adenine arabinofuranoside, 2-chloroadenosine, adenosine dialdehyde, and neplanocin-A completely inhibited SAHH, while 5iodo-5deoxyadenosine and tubercidin shows a small effect. Other compounds exhibited no effect. The results presented here showed a substantial discrepancy for the effect of SAHH inhibitors between lysates and intact cells. A number of effective inhibitors of the enzyme in erythrocytes lysate were ineffective in intact cells, possibly as a consequence of intracellular metabolism or membrane transport effects.

ACKNOWLEDGMENT

This study was supported by Collaborative Research Grant from NATO (CRG 921365).

REFERENCES

1. N.J. Prakash, G.F. Davis, E.T. Jarvi, M.L. Edwards, J.R. McCarthy, and T.L. Bowlin, Antiretroviral activity of mechanism-based irreversible inhibitors of S-adenosylhomocysteine hydrolase, *Life Sci.* 50:1425(1992).
2. R.T. Smolenski, C. Montero, J. Duley, and H.A. Simmonds, Effects of adenosine analogues on ATP concentration in human erythrocytes. Further evidence for a route independent of adenosine kinase, *Biochem. Pharmacol.* 42:1767(1991).
3. R.T. Smolenski, K. Fabianowska Majewska, C. Montero, J.A. Duley, Fairbanks, M. Marlewski, and H.A. Simmonds, A novel route of ATP synthesis, *Biochem. Pharmacol.* 43:2053(1992).
4. A. Deussen, M. Borst, K. Kroll, and J. Schrader, Formation of S-adenosylhomocysteine in the heart. II: A sensitive index for regional myocardial underperfusion, *Circ. Res.* 63:250(1988).

CHEMOTHERAPEUTIC OPTIONS IN HIV INFECTION

Susan Cox, Sarah Palmer, Kajsa Aperia, and Britta Wahren

Virology Dept
Swedish Institute for Infectious Disease Control
Karolinska Institute
S 105 21 Stockholm
Sweden

INTRODUCTION

Human immunodeficiency virus, (HIV), the causative agent of Acquired Immunodeficiency Syndrome (AIDS), can be inhibited in vitro by several different dideoxynucleoside analogues (Sommadossi, 1993). 3'-Azido-3'-deoxythymidine (AZT), an analogue of thymidine, is the most commonly used dideoxynucleoside for treatment of HIV infection and AIDS. Recently, 2',3'-dideoxyinosine (ddI) has also become available for clinical use. Both AZT and ddI are nucleoside analogues which must be phosphorylated intracellularly before inhibition of the reverse transcriptase (RT) of HIV by the 5'-triphosphate of the drug (Furman et al., 1986; Matthes et al., 1988; Cox and Harmenberg, 1990). AZT is phosphorylated by the thymidine salvage pathway, while ddI is phosphorylated first by a 5'-nucleotidase, and then, after conversion to ddA monoposphate, by adenosine-metabolising enzymes (fig. 1).

One of the major problems associated with clinical treatment of HIV infection with these drugs is the development of antiviral resistance. Prolonged treatment with either AZT or ddI has been shown to result in the development of specific mutations in the RT of HIV associated with drug resistance (Larder et al., 1989; Larder and Kemp, 1989; Rooke et al., 1989; Land et al., 1990; St. Clair et al., 1991; McLeod et al., 1992; Wahlberg et al., 1992). To overcome this, the use of combinations of antiviral drugs has been suggested.

Combination chemotherapy involves the use of two or more antiviral drugs in combination with the aim of both increasing the antiviral activity of the combination while decreasing the risk for development of resistance (Yarchoan et al., 1990; Cox, 1993). Several combinations of anti-retroviral drugs, including AZT and ddI (Dornsife et al., 1991) have been shown to be synergistic when combined, and therefore have potential as improved therapies for HIV infection (Johnson et al., 1989; Koshida et al., 1989a, 1989b; Harmenberg et al., 1990; Johnson et al., 1991; Kong et al., 1991; Eron et al., 1992; Cox, 1993; Cox et al., 1993a). The mechanism of the synergistic effects are, however, in most cases unknown. The RT from HIV is not synergistically inhibited by combinations of drug triphosphates (Harmenberg et al., 1990; White et al., 1993).

Purine and Pyrimidine Metabolism in Man VIII, Edited by
A. Sahota and M. Taylor, Plenum Press, New York, 1995

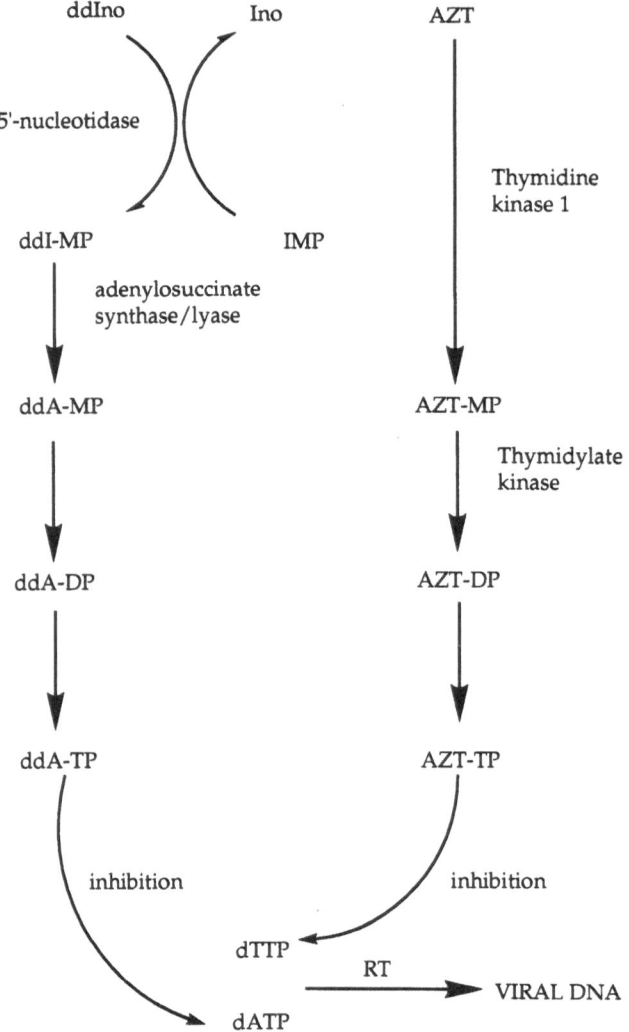

Figure 1. Intracellular metabolism of AZT and ddI.

Here, we will report on the antiviral activity against different primary HIV isolates, including AZT-resistant isolates, of AZT and ddI, both alone and in combination. The metabolism of the individual drugs and the combination in lymphocytes, and a possible mechanism for the synergy will be described.

MATERIALS AND METHODS

Cells and viruses

Peripheral blood mononuclear cells (PBMC) from HIV seronegative healthy blood donors were isolated by centrifugation on Ficoll-Hypaque and cultured in RPMI medium with 10% heat inactivated foetal calf serum and antibiotics, and stimulated for 2 days with 2.5 µg/ml phytohaemagglutinin (Sigma Chemical Co., St. Louis, USA). Primary isolates of HIV-1 were made by co-cultivation of patient PBMC with phytohaemagglutinin-stimulated PBMC from healthy seronegative donors in RPMI 1640 medium with 10% foetal calf serum, 5 units/ml interleukin-2 (Amersham, Amersham, UK) and antibiotics. Isolates were either used directly, or passaged once on blood donor PBMC, and were never cultivated in cell lines. Isolates were used fresh or frozen at -70° C for later use.

Patients

Primary HIV-1 isolates were from seropositive individuals at various stages of disease who had been treated with AZT for various lengths of time. Viral isolations were performed to check for resistance to AZT and cross-resistance to ddI.

Antiviral sensitivity determination

Sensitivity of primary isolates was determined in PBMC in microtitre plates. 50 000 PBMC were placed in each well with the virus isolate at two different dilutions, usually 1:20 and 1:100. Several dilutions of AZT or ddI dissolved in medium were added. The drugs were also tested in combination at a range of concentrations with a constant ratio between the two drugs. The ratios used were determined by the relative 50% inhibitory concentrations (IC_{50}s) of the two drugs in the combination, according to the median effect method of assessing drug interactions (Chou and Talalay, 1983; Harmenberg et al., 1990). All drugs used were a kind gift from Prof. Bo Öberg, Medivir, Huddinge, Sweden. Each plate also contained four five-fold titrations of the virus to determine the 50% infective dose in tissue culture ($TCID_{50}$). A $TCID_{50}$ of between 10-50 was used to evaluate drug sensitivity. The plates were incubated at 37° C in a humidified atmosphere of 5% CO_2 in air. After 7 days incubation, 20 µl of lysis buffer (phosphate buffered saline with 0.5% bovine serum albumin, 0.05% Tween 20 and 5% Triton X100) were added to the plates. The resulting cell lysate was then tested in an HIV-1 p24 ELISA (Koshida et al., 1989a; Sundqvist et al., 1989). Control wells were first titrated and then diluted to give appropriate positive controls and minimum background.

Analysis of results

The 50% inhibitory concentrations (IC_{50}) for the drugs alone and the different combinations were calculated graphically using regression analysis by comparison to viral growth in the absence of drugs. The results were analyzed for synergy using a computer programme following the median effect method (Chou and Talalay, 1983; Harmenberg et al., 1990). The programme calculates the slope of the dose-effect curve, and the combination index (CI) for the combination. A CI of 1 indicates addition, while a CI of <1, synergy, and >1, antagonism, respectively. A deviation of 15% in the CI was allowed before deducing synergy, in accordance with the deviation in the raw data.

Chemicals

3'-Azido 3'-deoxythymidine, (methyl-^{3}H, 14Ci mmol^{-1}, 99.9% pure) and 2',3'-dideoxyinosine, (2',3'-^{3}H(N), 42Ci mmol^{-1}, 98.3% pure) were purchased from Moravek Biochemicals Inc. (Brea, CA, USA).

Metabolic studies

For metabolism studies, CEM cells (a human CD4+ lymphoblastoid cell line) were grown at 37°C in a humidified atmosphere of 5% CO_2 in air in RMPI 1640 medium containing 10% heat-inactivated foetal calf serum, penicillin, streptomycin, and pyruvate.

Cells (6-8 x 10^5 ml^{-1}) were incubated with a clinically relevant concentration of 1μM ^{3}H-AZT (specific activity 440000 dpm nmol^{-1}), or 1μM ^{3}H-ddI (specific activity 1760000 dpm nmol^{-1}) or both ^{3}H-AZT and ^{3}H-ddI (1μM each, specific activity of each 1760000 dpm nmol^{-1}). After a 20h incubation at 37°C, cells were harvested by centrifugation at 1000 g for 10 min and extracted overnight with 60% cold methanol (Palmer and Cox, 1994).

HPLC analysis

Prior to analysis by high-performance liquid chromatography (HPLC) the ribonucleotides in each sample were degraded by periodate oxidation. All HPLC analyses were done on a chromatograph from Gilson Medical Electronics Inc. equipped with a dual wavelength spectrophotometer programmed to measure absorbance at 254 and 280 nm. All integration was performed on a Macintosh Classic II computer using RAININ Dynamax HPLC method manager software. Nucleotides were separated on a strong anion exchange column (Partisil 10 SAX, 250 x 4.6 mm I.D.; Whatman, Clifton, NJ, USA) using a phosphate buffer gradient elution. The two buffers used were 0.02 M potassium phosphate (pH 3.5, buffer A) and 0.80 M potassium phosphate (pH 3.5, buffer B). The gradient was completed in four steps. Step 1 was 100% buffer A for five minutes. Step 2 was a 30 minute linear increase in buffer B to 100%. Step 3 was a constant isocratic 20 min. delivery of 100% buffer B. In Step 4, buffer B was decreased to 0% in 10 minutes. A constant flow of 1.3 ml/min was used throughout. During the gradient, thirty second fractions were collected to which 10 ml of scintillation fluid was added and radioactivity determined by a Beckman LS 1801 scintillation counter.

Peaks were identified by retention time on the chromatograph, ratio of absorbance at 254 nm to that at 280 nm, and comparison with known standards. Recoveries were calculated by peak area comparisons of injected samples to those of pure dNTP standards with known concentrations.

Statistical analysis

Results were analysed for statistical significance using the Spearman rank correlation, Mann Whitney test and one-way ANOVA.

RESULTS

Antiviral activity of AZT and ddI

The antiviral activity of AZT and ddI for a range of primary isolates of HIV is shown in table 1. The sensitivity to AZT varied more than five hundred fold, and that for ddI more than two hundred fold. 45% of the isolates had an IC$_{50}$ for AZT of <0.1μM, while 55% had an IC$_{50}$ >0.1μM, and were designated AZT resistant. There was a significant correlation between the IC$_{50}$s for AZT and ddI ($p < 0.001$), indicating that resistance to AZT was associated with reduced susceptibility to ddI (fig. 2).

Table 1. Antiviral activity of AZT and ddI for 300 different primary isolates of HIV.

	IC$_{50}$ (μM) for AZT	IC$_{50}$ (μM) for ddI
Range	0.016 - 9.6	0.16 - 44.0
Median	0.2	2.6
Mean ± SE	0.89 ± 0.08	4.4 ± 0.31

Antiviral activity of the combination of AZT and ddI

The antiviral activity of the combination of AZT and ddI was tested against select virus isolates, and the results analysed by the median effect method. Figure 3 shows the combination indices for the combination of AZT and ddI for different viral isolates. Isolates A and B, which are from untreated HIV patients and are sensitive to AZT (IC$_{50}$s of 0.02 and 0.01 μM, respectively) show synergy with the combination of AZT and ddI, as shown by combination indices of <1. Isolates C and D, however, which are from patients treated with AZT for more than two years, and which show resistance to AZT (IC$_{50}$s of 4.2 and 0.33 μM, respectively) do not show synergy with the combination, but instead antagonism is seen, as shown by combination indices of greater than 1.

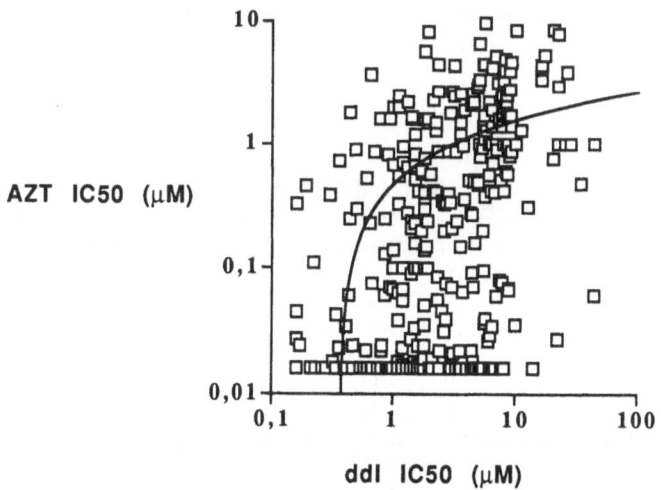

Figure 2. Correlation between the IC$_{50}$s for AZT and ddI for the primary isolates tested.

Metabolism of AZT and ddI in lymphocytes

The phosphorylation of AZT and ddI, alone and in combination in lymphocytes, is shown in table 2. AZT was phosphorylated to a greater extent than ddI, and the level of AZT triphosphate was nearly one hundred times higher than that of ddA triphosphate (the active form of ddI). The level of AZT triphosphate formed in the presence of ddI was similar to that formed when AZT was used alone, whereas the level of ddA triphosphate was doubled when the drug was combined with AZT compared to that when ddI was used alone.

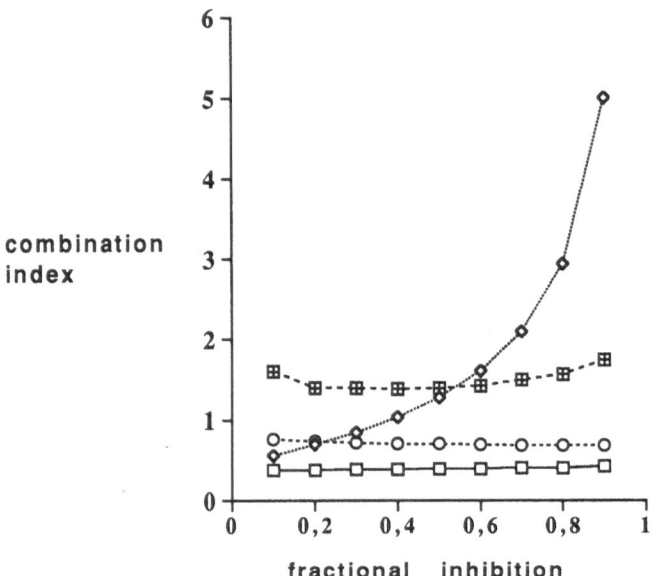

Figure 3. Effect of the combination of AZT and ddI on two AZT-sensitive primary isolates (squares and circles) and two AZT-resistant primary isolates (triangles and filled squares). Combination indices of <1 indicate synergy, while those of >1 indicate antagonism.

DISCUSSION

The antiviral sensitivity to AZT and ddI of more than 300 primary isolates of HIV from patients treated with AZT was measured in PBMCs. The sensitivity of the isolates to AZT varied greatly. Isolates from untreated persons had an IC_{50} for AZT of <0.1 μM, in accordance with other studies on the susceptibility of wild-type HIV-1 to AZT (Mitsuya et al., 1985; Mitsuya and Broder, 1986). More than half of the isolates tested had an IC_{50} for AZT of >0.1μM, and can therefore be classed as resistant to AZT. This indicates that resistance to AZT is common in Swedish patients treated with AZT.

The sensitivity of the isolates to ddI also varied, and a significant correlation was found between the IC_{50}s for AZT and ddI. This indicates that development of resistance to AZT is associated with a reduction in the sensitivity to ddI. A similar such correlation has also been shown for dideoxycytidine (Cox et al., 1993b), another antiviral dideoxynucleoside, but not, however, for Foscarnet, a pyrophosphate analogue (Cox et al., 1993b).

Table 2. Phosphorylation of AZT, ddI and a 1:1 combination
in human lymphocytes

Metabolite	Amount formed (pmol/10^6 cells) in the presence of		
	AZT(1µM)	ddI (1µM)	AZT + ddI (1µM each)
AZT-MP	71.6 ± 1.8	-	ND[1]
AZT-DP	0.87 ± 0.05	-	ND
AZT-TP	2.33 ± 0.28	-	2.29 ± 0.11
ddI-MP	-	1.5 ± 0.04	ND
ddA-DP	-	0.084 ± 0.01	ND
ddA-TP	-	0.029 ± 0.002	0.063 ± 0.002***

[1]ND, not determined. *** $p = 0.001$ compared to ddI alone (one-way ANOVA).

The antiviral activity of the combination of AZT and ddI was also assessed for four primary isolates of AZT, two sensitive to AZT and two resistant. The two sensitive isolates were synergistically inhibited by the combination of AZT and ddI, in agreement with previous results (Dornsife et al., 1991). However, the two AZT-resistant isolates did not show synergy with the combination of AZT and ddI, but instead showed addition or antagonism. This indicates that the synergistic effect of the combination of AZT and ddI may be lost upon development of resistance to AZT. This synergistic combination may therefore be best used early on in treatment, before the development of resistance, rather than after resistance has developed. After resistance has developed, switching to another drug might be advantageous (Kahn et al., 1992).

The mechanism behind the synergy between AZT and ddI was also investigated. It was previously shown that the RT from HIV was not synergistically inhibited by combinations of the drug triphosphates (White et al., 1993), in agreement with our previous studies on another synergistic drug combination, that of AZT and 3'-fluorothymidine (Harmenberg et al., 1990). These studies indicate that the synergistic effect between the two drugs does not lie at the level of the RT per se. Instead, we found that the phosphorylation of ddI was increased when the drug was combined with AZT. The amount of ddA triphosphate formed (the active form of ddI) doubled when ddI was combined with AZT, compared to the separate drug. The level of AZT triphosphate formed, however, was unchanged, whether the drug was used alone or combined with ddI. These studies provide an explanation for the antiviral synergy seen between AZT and ddI. Increased formation of ddA triphosphate in the presence of AZT would lead to increased inhibition of RT. This effect would be especially noticeable since the level of ddA triphosphate formed in cells was found to be low.

In conclusion, both AZT and ddI show antiviral activity against primary isolates of HIV in vitro. Resistance to AZT is common among patients treated with AZT, and is associated with reduced susceptibility to ddI. The combination of AZT and ddI shows synergistic inhibition of AZT-sensitive primary HIV isolates, but not of AZT-resistant ones. Increased phosphorylation of ddI in the presence of AZT might explain the synergy seen between the two drugs.

ACKNOWLEDGEMENTS

These studies were supported by grants from Stiftelsen Läkare mot AIDS forskningsfond and the Swedish Society of Medicine.

REFERENCES

Chou, T.-C. and Talalay, P. ,1983, Analysis of combined drug effects, *Trends in Pharmacol. Sci.* 4: 450.

Cox, S. ,1993, Combination chemotherapy in HIV disease, *Antiviral News* 1(9): 132.

Cox, S., Albert, J., et al. ,1993a, Synergistic inhibition of primary isolates of HIV by combinations of fluorothymidine and dideoxyinosine., *Antiviral Chem. Chemother.* 4(4): 241.

Cox, S., Aperia, K., et al. ,1993b, Cross-resistance between AZT, ddI, and other antiretroviral drugs in primary isolates of HIV-1., *Antiviral Chem. Chemother.* 5(1): 7.

Cox, S. and Harmenberg, J. ,1990, Comparison of the intracellular metabolism of 3'-azido-3'-deoxythymidine and 3'-fluoro-3'-deoxythymidine in lymphocytes in the presence of 5-fluoro-2'-deoxyuridine., *Antiviral Chem. Chemother.* 1(2): 155.

Dornsife, R. E., St. Clair, M. H., et al. ,1991, Anti-human immunodeficiency virus synergism by Zidovudine (3'-azidothymidine) and didanosine (dideoxyinosine) contrasts with their additive inhibition of normal human marrow progenitor cells., *Antimicrob. Agents Chemother.* 35(2): 322.

Eron, J., Johnson, V., et al. ,1992, Synergistic inhibition of replication of human immunodeficiency virus type 1, including that of a zidovudine resistant isolate, by zidovudine and 2',3'-dideoxycytidine in vitro., *Antimicrob. Agents Chemother.* 36: 1559.

Furman, P., Fyfe, J., et al. ,1986, Phosphorylation of 3'-azido-3'-deoxythymidine and selective interaction of the 5'-triphosphate with human immunodeficiency virus reverse transcriptase, *Proc. Natl. Acad. Sci. USA* 83: 8333.

Harmenberg, J., Åkesson Johansson, A., et al. ,1990, Synergistic inhibition of human immunodeficiency virus replication in vitro by combinations of 3'-azido-3'-deoxythymidine and 3'-fluoro-3'-deoxythymidine, *AIDS Res. Hum. Retroviruses* 6: 1197.

Johnson, V., Merrill, D., et al. ,1991, Two drug combinations of zidovudine, didanosine and recombinant interferon alpha A inhibit replication of zidovudine resistant human immunodeficiency virus type 1 synergistically in vitro., *J. Infect. Dis.* 164: 646.

Johnson, V., Walker, B., et al. ,1989, Synergistic inhibition of human immunodeficiency virus type 1 and type 2 replication in vitro by castanospermine and 3'-azido-3'-deoxythymidine, *Antimicrob. Agents Chemother.* 33: 53.

Kahn, J., Lagakos, S., et al. ,1992, A controlled trial comparing continued zidovudine with didanosine in human immunodeficiency virus infection, *N. Engl. J. Med.* 327: 581.

Kong, X.-B., Zhu, Q.-Y., et al. ,1991, Synergistic inhibition of human immunodeficiency virus type 1 replication in vitro by two drug and three drug combinations of 3'-azido-3'-deoxythymidine, phosphonoformate, and 2',3'-dideoxythymidine., *Antimicrob. Agents Chemother.* 35: 2003.

Koshida, R., Cox, S., et al. ,1989a, Structure activity relationships of fluorinated nucleoside analogues and their synergistic effect in combination with phosphonoformate against HIV 1, *Antimicrob. Agents Chemother.* 33: 2083.

Koshida, R., Vrang, L., et al. ,1989b, Inhibition of human immunodeficiency virus in vitro by combinations of 3'-azido-3'- deoxythymidine and foscarnet, *Antimicrob. Agents Chemother.* 33: 778.

Land, S., Treloar, G., et al. ,1990, Decreased in vitro susceptibility to zidovudine of HIV isolates obtained from patients with AIDS., *J. Infect. Dis.* 161: 326.

Larder, B., Darby, G., et al. ,1989, HIV with reduced sensitivity to zidovudine (AZT) isolated during prolonged therapy, *Science* 243: 1731.

Larder, B. and Kemp, S. ,1989, Multiple mutations in HIV 1 reverse transcriptase confer high level resistance to zidovudine (AZT), *Science* 246: 1155.

Matthes, E., Lehmann, C., et al. ,1988, Phosphorylation, anti-HIV activity and cytotoxicity of 3'-fluoro-thymidine, *Biochem. Biophys. Res. Commun.* 153: 825.

McLeod, G., McGrath, J., et al. ,1992, Didanosine and zidovudine resistance patterns in clinical isolates of human immunodeficiency virus type 1 as determined by a replication endpoint concentration assay., *Antimicrob. Agents Chemother.* 36: 920.

Mitsuya, H. and Broder, S. ,1986, Inhibition of the in vitro infectivity and cytopathic effect of human T-lymphotrophic virus type III/lymphadenopathy associated virus (HTLV-III/LAV) by 2',3'-dideoxynucleosides., *Proc. Natl. Acad. Sci. USA* 83: 1911.

Mitsuya, H., Weinhold, K. J., et al. ,1985, 3'-Azido-3'-deoxythymidine (BW A509U): an antiviral agent that inhibits the infectivity and cytopathic effect of human T-lymphotropic virus type III/lymphadenopathy associated virus in vitro., *Proc. Natl. Acad. Sci. USA* 82: 7096.

Palmer, S. and Cox, S. ,1994, Comparison of extraction methods for the high performance liquid chromatography analysis of cellular deoxynucleotides, *J. Chromatogr.* in press:

Rooke, R., Tremblay, M., et al. ,1989, Isolation of drug resistant variants of HIV 1 from patients on long term zidovudine therapy, *AIDS* 3: 411.

Sommadossi, J.-P. ,1993, Nucleoside analogues: similarities and differences., *Clin. Inf. Dis.* 16(Suppl 1): S7.

St. Clair, M., Martin, J., et al. ,1991, Resistance to ddI and sensitivity to AZT induced by a mutation in HIV-1 reverse transcriptase., *Science* 253: 1557.

Sundqvist, V.-A., Albert, J., et al. ,1989, Human immunodeficiency virus type 1 p24 production and antigenic variation in tissue culture of isolates with various growth characteristics., *J. Med. Virol.* 29: 170.

Wahlberg, J., Albert, J., et al. ,1992, Dynamic changes in HIV-1 quasispecies from azidothymidine (AZT) treated patients., *FASEB* 6: 2843.

White, E., Parker, W., et al. ,1993, Lack of synergy in the inhibition of HIV-1 reverse transcriptase by combinations of the 5'-triphosphates of various anti-HIV nucleoside analogues, *Antiviral Res.* 22: 295.

Yarchoan, R., Mitsuya, H., et al. ,1990, Strategies for the combination therapy of HIV infection., *J. Acq. Immune Defic. Syndromes* 3(suppl 2): S99.

CHANGES OF PURINE NUCLEOTIDE CONTENT IN LYMPHOCYTES FROM PATIENTS INFECTED WITH HUMAN IMMUNODEFICIENCY VIRUS

C.A.Boggiano,[2] F.Carlucci,[1] A.Tabucchi,[1] R.Pagani,[1] E.Marinello,[1] M.Pizzichini,[1] and R.Leoncini[1]

[1]Institute of Biochemistry and Enzymology
  University of Siena
[2]Institute of Infectious Diseases
  Hospital of Siena; Italy

INTRODUCTION

Few reports have been published on changes in purine catabolic enzymes during HIV-1 infection. Indications in the literature suggest that purine nucleotide metabolism is altered in virus infected cells. Some authors[1,2] report decreased total 5'-N activity in peripheral blood lymphocytes (PBL) from AIDS patients, while others find normal ecto-5'-N in T cells, but markledy decreased activity in B cells.

The accumulation of unintegrated viral DNA[3], altered levels of cyclic nucleotides in infected cells[4] and an increase of ADA in the lymphocytes of AIDS patients[5,] all indicate that the penetration of an RNA-dependent virus, like HIV-1, involves changes in purine nucleotide metabolism[3].

This question is of great interest, due to its theoretical importance and many practical applications. The evaluation of intracellular nucleotide content is a useful approach to the study of purine nucleotide metabolism since it gives an overview of the reactions that regulate the synthesis and degradation of purine nucleotides. We analyzed the purine nucleotide content of peripheral blood lymphocytes (PBL), in T, non-T, T4 and T8 cells from healthy subjects and asymptomatic and symptomatic patients with HIV-1 infection.

MATERIALS AND METHODS

Chemicals

The nucleotides used as standards were from Sigma (St.Louis, MO, USA). PBS was purchased from Flow Laboratories (Ayrshire, UK). All chemical reagents were of analytical grade from Merck (Darmstadt, Germany). HPLC solvents were filtered through 0.45 μm

---

Address for correspondence: Prof. Enrico Marinello, Istituto di Biochimica e di Enzimologia, Università di Siena. Pian dei Mantellini. 44. 53100 SIENA Italv. Tel. +39-577-298026. FAX +39-577-298057

Purine and Pyrimidine Metabolism in Man VIII, Edited by
A. Sahota and M. Taylor, Plenum Press, New York, 1995

membrane filters (Sartorius, Göttingen, Germany) before use.

Control subjects and patients

The control subjects were 15 male and 7 female blood donors, 27 to 50 years of age, on a normal balanced diet and taking normal exercise. The HIV-1 patients were 14 males and 10 females, 28 to 52 years of age (mean age 32 years), divided according to the Walter Reed classification. None of them had Kaposi sarcoma. Subjects in WR stage 1-2 were grouped as asymptomatic, and those in WR stage 5-6 as symptomatic.

Erythrocytes, white blood cells, $CD4^+$, $CD8^+$, platelet counts and hemoglobin were measured. These parameters were normal, in all subjects including patients. The $CD4^+/CD8^+$ ratio (which was about 1 in control subjects), fell to values included between 0.37 and 0.72 in asymptomatic HIV-1 patients, and between 0 and 0.16 in symptomatic patients.

Human peripheral blood lymphocytes (PBL)

Peripheral blood samples (15-30 ml) were taken from controls and HIV-1 seropositive patients. The lymphocytes were isolated from whole blood and then washed as previously described[6]. No contamination by platelets or erythrocytes was evident on microscope examination. The pellet was suspended in PBS to a final concentration of 40×10^6 cells/ml. Subpopulations of lymphocytes were prepared by an immunomagnetic procedure. T-cells were separated from non-T-cells using M-450 Pan T Dynabeads (Dynal, Oslo, Norway), magnetizable polystyrene beads coated with monoclonal antibody specific for CD2 antigen of human lymphocytes. The rosetted cells were separated by holding the tube in a magnetic device for 180 s. The unbound cells, designated as non-T-cells, were decanted and collected by centrifugation.

T4 and T8 cells were separated directly from PBLs using M-450 CD4 and M-450 CD8 Dynabeads respectively. After separation, all cells were washed three times with PBS containing 0.1% BSA and then resuspended in PBS, at a concentration of 5×10^6/ml.

Assessment of cell viability

The cells were counted after staining with Trypan blue to assess the percentage of viable cells.

Purine nucleotide determination

The preparation of perchloric extracts for the determination of nucleotides, the details of storage and injection of samples, HPLC and the determinations are reported in previous papers. The chromatographic conditions are given in Table 1 [8,9,10].

The nucleotides were separated by HPLC, reading at 254 nm. Satisfactory resolution of the principal standard nucleotides (NAD, AMP, IMP, GMP, ADP, GDP, ATP and GTP) was obtained. The method was linear for all nucleotides from 10 pmol to 10 nmol. Run-to-run and day-to-day precision was satisfactory and CV values were always < 5%. Pyrimidine nucleotides and XMP were not considered because they were not constant and reproducible in the chromatograms.

Statistical analysis of data

Statistical evaluation was carried out by conventional one-way analysis of variance and p values < 0.05 were considered significant.

Table 1. Chromatographic conditions

Column:	Whatman Partisil 10 SAX 5 μm (250x4.6 mm)
Flow rate:	1.7 ml/min
Buffers:	
(A) Low ionic strength	7.5 mM $NH_4H_2PO_4$, 2% (v/v) CH_3CN (pH 3.80)
(B) High ionic strength	750.0 mM $NH_4H_2PO_4$, 2% (v/v) CH_3CN (pH 4.92)
Gradient:	14.4 min 0.1-1,5% B
	9.6 min 1.5-12% B
	12.0 min 12.0-17.0% B
	33.6 min 17.0-56.0% B
	7.2 min 56% B
	2.4 min 56-0.1% B
Equilibration:	9.6 min 0.1% B
Total analysis time:	88.8 min

RESULTS

The results are reported in the following Figure and Tables.

Table 2 shows the results obtained for PBL.

Table 3 shows the data obtained with T and non-T lymphocytes of controls and patients.

Table 4 shows the results for T4 and T8 cells.

Table 2. Purine nucleotide content of peripheral blood lymphocytes from control subjects and HIV-1 infected patients.

	Control (n = 15)	Asymptomatic (n = 5)	Symptomatic (n = 6)
NAD	264± 80	276 ± 68	350* ± 130
AMP	23 ± 6	31 ± 10	22 ± 6
IMP	28 ± 8	96* ± 15	53 ± 10
GMP	10 ± 3	18 ± 6	17 ± 5
ADP	409 ± 67	209* ± 40	570*± 120
GDP	57 ± 9	35* ± 6	102* ± 30
ATP	886± 107	540* ± 62	1,377* ± 260
GTP	173 ± 9	132* ± 15	342* ± 78

Values are mean ± S.E. in pmol/10^6 cells. For experimental conditions, see Materials and Methods. In brackets, the number of cases. * $p < 0.05$ with respect to control subjects.

Table 3. Purine nucleotide content of T and non-T cells from control subjects and HIV-1 infected patients.

	Control		Asymptomatic		Symptomatic	
	T n=4	non-T n=4	T n=3	non-T n=3	T n=4	non-T n=4
NAD	162 ± 37	1367 ± 99	80 ± 10	120* ± 83	84 ± 30	200* ± 45
AMP	9 ± 4	233 ± 124	4 ± 2	20* ± 7	133 ± 30	40 ± 11
IMP	92 ± 27	469 ± 158	75 ± 33	35* ± 11	83 ± 32	30*± 10
GMP	22 ± 3	135 ± 20	25 ± 8	22* ± 6	28 ± 5	21* ± 6
ADP	345 ± 160	3,796 ± 217	150 ± 49	110* ± 24	230 ± 43	300* ± 75
GDP	46 ± 9	471 ± 25	38 ± 16	63* ± 12	78 ± 18	38* ± 13
ATP	787 ± 273	4,400 ± 238	560 ± 185	400* ± 70	940 ± 114	502*±113
GTP	110 ± 12	668 ± 47	125 ± 20	90* ± 23	203 ± 35	180* ± 34

Values are means ± S.E. in pmol/10^6 cells. For cell preparation and experimental conditions, see Materials and Methods. n = number of cases. * $p < 0.05$ with respect to corresponding cells from control subjects.

It is evident that there were changes in purine nucleotide content in all patients examined. The pattern obtained with PBL is very difficult to interpret; the data becomes clearer when we observe the nucleotide content of T and non-T cells. First, the purine nucleotide content of non-T cells from control subjects was strikingly higher than that of T-cells.

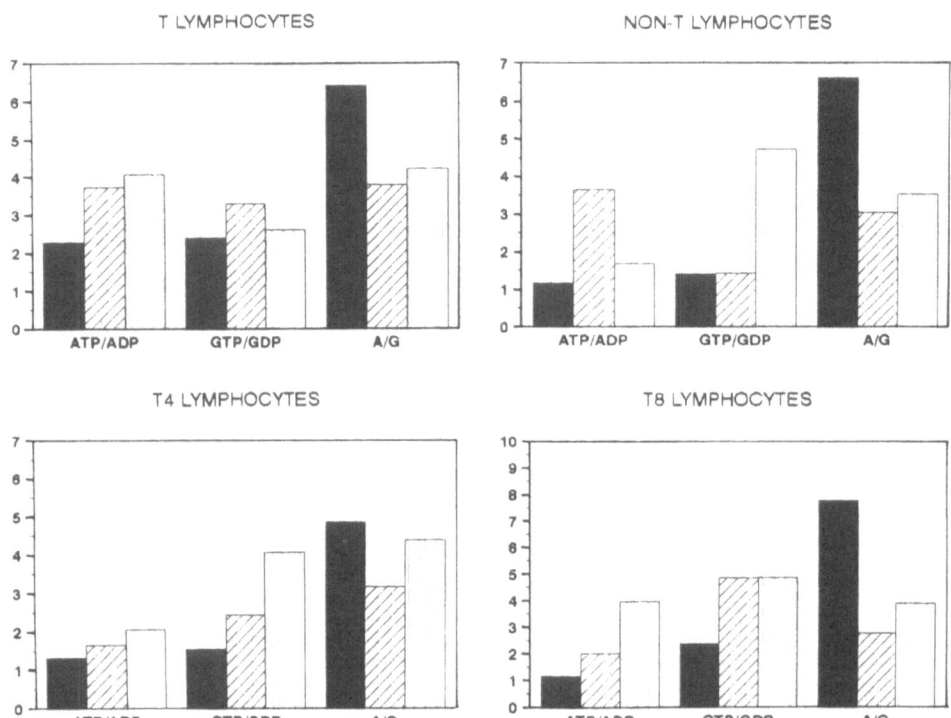

Figure 1. Some important ratios between purine nucleotides in lymphocyte subpopulations from controls and patients.

Secondly, the purine nucleotide patterns in lymphocytes of control subjects were different from those of HIV-infected persons. These differences were much more evident when non-T-cells from the three groups were compared: a dramatic general decrease in purine nucleotides was observed upon sero-conversion, and to a lesser extent with ATP and GTP (Table 3). In contrast, only minor changes in purine nucleotides were seen in separated T-cells, with a marked increase in AMP, GDP and GTP in symptomatic patients.

Table 4. Purine nucleotide content of T4 and T8 cells from control subjects and HIV-1 infected patients.

	Healthy		Asymptomatic		Symptomatic	
	T4	T8	T4	T8	T4	T8
	n=3	n=3	n=3	n=3	n=3	n=3
NAD	52 ± 30	32 ± 8	252* ± 92	185 ± 52	110 ± 32	242 ± 58
AMP	22 ± 8	76 ± 25	155* ± 89	38 ± 18	34 ± 11	58 ± 15
IMP	220 ± 87	29 ± 15	42 ± 6	56 ± 38	317 ± 117	68 ± 32
GMP	30 ± 5	16 ± 6	189* ±62	18 ± 16	63 ± 27	32 ± 17
ADP	378 ± 40	400 ± 138	812* ± 112	340 ± 92	998* ± 157	282 ± 58
GDP	61 ± 24	31 ± 11	157* ± 38	62 ± 18	127 ± 49	58 ± 24
ATP	506 ± 144	463± 92	1,340 ± 339	682 ± 110	2,064* ± 327	1,114* ± 311
GTP	95 ± 23	74 ± 13	382 ± 79	301 ± 80	516* ± 62	282 ± 85

Values are means ± S.E. in pmol/10^6 cells. For cell preparation and experimental conditions, see Materials and Methods. n = number of cases. * $p < 0.05$ with respect to corresponding cells from control subjects.

In T4 cells most of the nucleotides (NAD, AMP, GMP, ADP, GDP, ATP and GTP) were higher in asymptomatic subjects than in healthy controls. In the later stages of the infection, the values of AMP returned to normal, and ATP further increased. IMP was lower in asymptomatic than in symptomatic patients.

In separated T8 cells only NAD, ATP and GTP increased after seroconversion with a further enhancement of ATP as the disease progressed.

It is widely accepted[8,10,11] that ATP/ADP, GTP/GDP and EC represent cell energy, viability and metabolic state. The EC for adenine and guanine nucleotides is given by the ratios ½ADP+ATP/AMP+ADP+ATP and ½GDP+GTP/GMP+GDP+GTP, respectively. The total adenine and guanine nucleotide ratio A/G is an inverse index of proliferation rate. All these parameters were calculated by us on the basis of the nucleotide content (Fig.1).

The Figure 1 shows dramatic increases in parameters like the ATP/ADP and GTP/GDP ratios and a decrease in the A/G ratio upon seroconversion, especially in non-T cells. These changes are an indication of the activation of these cells in energy and metabolic state[13], and are so large as to constitute an indicator of viral infection for risk groups. Although preliminary, these results show the utility of research on purine nucleotide metabolism in HIV-1 virus infected lymphocytes.

The low levels of purine nucleotides observed in non-T-cells of HIV-infected subjects are probably due to 'activation' of lymphocytes, known to occur in this disease[12]. An increase in mRNA has always been observed during lymphocyte 'activation'[13,14,15]. Formation of mRNA presumably occurs at the expense of nucleotides.

Further study is needed to elucidate the cause and significance of decreased levels of purine nucleotides in non-T-cells of HIV-infected subjects, and the slight increases observed in T-cells. Full interpretation of our observations will require the evaluation of RNA, assay of the enzyme activities involved in purine nucleotide metabolism and kinetic data such as the incorporation of ^{14}C-formate and ^{14}C-glycine into nucleotides.

REFERENCES

. 1. J.I.Murray, D.W.Bywaters, J.M.Reben, P.W.Mansell and E.M.Hersh, Decreased 5' nucleotidase activity in suppressor (OKT8) T-lymphocytes from homosexuals with AIDS related complex: nonassociation with enhanced deoxynucleoside toxicity, *Clin.Immunol.Immunopathol.* 42:10 (1987).
2. J.I.Murray, J.M.Reuben, C.G.Munn, P.W.A.Mansell, G.R.Newell and E.M.Hersh, Decreased 5'-nucleotidase activity in lymphocytes from asymptomatic sexually active homosexual men and patients with acquired deficiency syndrome, *Blood* 64:1016 (1984).
3. A.S.Fauci, The human immunodeficiency virus: infectivity and mechanism of pathogenesis, *Science* 239:617 (1988).
4. M.Nokta and R.Pollard, Human immunodeficiency virus infection: association with altered intracellular levels of cAMP and cGMP in MT-4 cells, *Virology* 181:211 (1991).
5. L.D.Christensen, M.Svenson, P.Nygaard, V.Andersen and V.Faber, Decreased B lymphocyte ecto-5' nucleotidase and increased adenosine deaminase in mononuclear cells from patients infected with human immunodeficiency virus, *APMIS* 96:882 (1988).
6. A.Tabucchi, R.Leoncini, R.Pagani, M.Pizzichini, L.Terzuoli, D.Vannoni, B.Porcelli, E.Marinello and E.Dispensa, Some aspects of purine nucleotide metabolism in lymphocytes of B-CLL, *Tumori* 77:112 (1991).
7. D.De Korte, W.A.Haverkort, A.H.Van Gennip and S.Roos, Nucleotide profile of normal human blood cells determination by high performance liquid chromatography, *Anal.Biochem.* 147:197 (1985).
8. Y.M.T.Marijnen, D.De Korte, W.A.Haverkort, E.J.S.Den Breejen, A.H.Van Gennip and S.Roos, Studies on the incorporation of precursors into purine and pyrimidine nucleotides via de novo and salvage pathways in normal lymphocytes and in lymphoblastic cell lines, *Biochim. Biophys.Acta* 1012:148 (1989).
9. E.Marinello, R.Pagani, F.Carlucci, M.Molinelli, P.Valerio and A.Tabucchi, The purine nucleotide content in normal lymphocytes, *Biochem.Soc.Trans.* 19:347S (1991).
10. G.J.Peters, R.A.De Abreu, A.Oosterhof and J.H.Veerkampf, Concentration of nucleotides and deoxynucleotides in peripheral and phytohemagglutinin-stimulated mammalian lymphocytes, *Biochim.Biophys.Acta* 759:7 (1983).
11. L.J.M.Spaapen, I.G.M.Scharenberg, B.J.M.Zegers, G.T.Rijkers, M.Duran and S.K.Wadman, Intracellular purine and pyrimidine pools of human T and B lymphocytes, *Adv.Exp.Med.Biol.* 195A:567 (1985).
12. J.I.Murray, K.C.Loftin, C.G.Munn, M.R.Reuben, P.W.Mansell and E.M.Hersh, Elevated adenosine deaminase and purine nucleoside phosphorylase activity in peripheral blood lymphocytes from patients with the acquired immunodeficiency syndrome, *Blood* 6:1318 (1985).
13. H.L.Cooper, Studies on RNA metabolism during lymphocyte activation, *Transplant.Rev.* 11:3 (1972).
14. J.Bernheim and I.Mendelsohn, DNA synthesis and proliferation of human lymphocytes in vitro. Characterization of the DNA newly synthesized after phytohemagglutinin stimulation, *J.Immunol.* 120(3):963 (1978).
15. M.Somasundaran and H.L.Robinson, Unexpectedly high levels of HIV-1 RNA and protein synthesis in a cytocidal infection, *Science* 242:1554 (1988).

ANTI-RETROVIRAL AND PHARMACOLOGICAL PROPERTIES OF 9-(2-PHOS-PHONYLMETHOXYETHYL)ADENINE (PMEA)

J. Balzarini

Rega Institute for Medical Research
Minderbroedersstraat 10
B-3000 Leuven, Belgium

INTRODUCTION

Acyclic nucleoside phosphonates (ANP) represent a structural class of compounds that contain a phosphonate group linked to an acyclic (alkyl) side chain of purine or pyrimidine bases.[1-3] Due to the unusual direct linkage between the phosphor atom of the phosphonate moiety and a carbon atom of the acyclic side chain, the ANP derivatives are resistant to phosphorolytic cleavage by cellular esterases. Therefore, ANP derivatives are both enzymatically and chemically stable and are taken up by the cells in an unaltered intact form. Several subclasses of ANP can be considered (Fig. 1): (i) HPMP derivatives[1,4] [prototype compound: HPMPA, (S)-9-(3-hydroxy-2-phosphonylmethyl-propyl)adenine], (ii) PME derivatives[1,4,5] [prototype compound: PMEA, 9-(2-phosphonylmethoxyethyl)adenine], (iii) PMP derivatives[6,7] [prototype compounds: PMPA, (R)-9-(2-phosphonylmethoxypropyl)adenine and FPMPA, (S)-9-(3-fluoro-2-phos-phonylmethoxypropyl)adenine]. Each subclass of ANP is endowed with a specific and characteristic antiviral activity spectrum.

ANTIVIRAL ACTIVITY OF ACYCLIC NUCLEOSIDE PHOSPHONATES

(S)-HPMPA is inhibitory to a broad spectrum of DNA viruses in cell culture, including herpesviruses [i.e. herpes simplex virus type 1 (HSV-1) and type 2 (HSV-2), varicella-zoster virus (VZV), cytomegalovirus (CMV), Epstein-Barr virus (EBV), human herpes virus type 6 (HHV-6)], adenoviruses, poxviruses and iridoviruses (i.e. African swine fever virus). (S)-HPMPA is inactive against RNA viruses, and shows poor, if any, inhibitory activity against retroviruses [i.e. human immunodeficiency virus type 1 (HIV-1) and type 2 (HIV-2) and simian immunodeficiency virus (SIV)]. It has a slight activity against Moloney murine sarcoma virus (MSV).[1,4].

PMEA displays also an antiviral activity against herpesviruses, although it should be mentioned that its activity against these different groups of viruses is not so pronounced than observed for (S)-HPMPA.[1,4] Also, PMEA is inactive against other DNA viruses such as adenoviruses and iridoviruses. However, in contrast with (S)-HPMPA, PMEA shows pronounced *in vitro* activity against hepadnaviruses (i.e. human and duck hepatitis B viruses),[8-10] and it also has a pronounced inhibitory activity against a panel of retroviruses, including HIV-1, HIV-2, SIV, feline immunodeficiency virus (FIV), MSV, Rausher and Friend murine leukemia virus (MLV) and murine LP-BM5 retrovirus complex.[10-19]

Among the PMP derivatives, both prototype compounds (R)-PMPA and (S)-FPMPA are endowed with potent antiretroviral activity[6,7] and inhibitory potential against hepatitis B virus in cell culture.[9,10] In contrast with PMEA, they are devoid of any

PMEG and (S)-HPMPG, phosphonoformate (PFA, foscarnet), and purine nucleoside analogues such as tubercidine and araA. Moreover, uptake studies with radiolabeled PMEA revealed that PMEA was taken up at an at least 20-fold lower efficiency by the L1210/PMEA-1 cell line compared to wild-type L1210/0 cells. Thus, treatment of murine L1210 cells by escalating PMEA concentrations had selected for a cell line that is affected at the level of its uptake mechanism for several ANP's. This cell line is now subject to further biochemical characterization to reveal the molecular basis of its markedly decreased uptake capacity for PMEA and PMEDAP.

When the metabolism of radiolabeled PMEA was studied in human T-lymphocyte MT-4 or CEM cells, only the parent compound (PMEA), and its monophosphorylated (PMEAp) and diphosphorylated (PMEApp) derivative could be detected. Neither deamination to the hypoxanthine derivative PMEHx, nor hydrolysis to free adenine could be detected.[23,24] As a rule, the amount of phosphorylated metabolites of PMEA in MT-4 cells represented ~ 10% of that of the parent compound in the intracellular compartment.[23] In human lung fibroblast MRC-5 cells, additional metabolites of PMEA could be detected, but were not further characterized.[24] PMEApp, that represents the intracellular active species for anti-retroviral activity, was found to have an intracellular half-life of approximately 16-18 hrs.[23]

CHARACTERIZATION OF PMEA PHOSPHORYLATING ENZYMES

AMP kinase from rabbit muscle and several other kinases, including GMP kinase, nucleoside 5'-monophosphate kinase, nucleoside diphosphate kinase, pyruvate kinase, were unable to catalyse the phosphorylation of PMEA. However, Merta and coworkers characterized an enzyme from murine leukemia L1210 cells that was able to convert PMEA to its mono- and diphosphorylated form in the presence of ATP.[25] The enzyme responsible for the conversion of PMEA to eventually PMEApp was found to be an AMP kinase. PMEAp appeared first, followed by PMEApp. The same sequence of reactions could be shown for (S)-HPMPA and (S)-FPMPA but not for the (R)-enantiomer of FPMPA, which was not recognized by the enzyme as a substrate.[25] The latter data are in agreement with the lack of anti-retroviral activity of the (R)-FPMPA derivative, and the assumption that the ANP's are antivirally active as their diphosphate derivatives.

Another enzyme has also been found to recognize ANP derivatives as a substrate. Indeed, PRPP synthetase from either E.coli (Sigma Chemical Co., St. Louis, Mo),[26,27] rat liver (kind gift of Dr. M. Tatibana, Chiba, Japan) and human erythrocytes (kind gift of Dr. M.A.. Becker, Chicago, USA) was able to convert PMEA, (S)-HPMPA, (S)-FPMPA and (R)-PMPA to their corresponding diphosphate derivatives in a one-step reaction. However, it should be mentioned that the K_m values of these ANP derivatives were 2.5- to 8-fold higher, and their V_{max} values 150- to 500-fold lower than the K_m and V_{max} for the natural substrate AMP.[26,27] It should also be mentioned that the equilibrium of this PRPP synthetase reaction is in favor of the formation of AMP + PRPP. Therefore, PMEA and related analogues are not very efficiently converted to their diphosphate derivatives by PRPP synthetase. It is currently unclear if PRPP synthetase plays a physiologically relevant role in converting PMEA to its active metabolite in the intact cells.

OTHER ENZYMES THAT MAY INTERACT WITH ACYCLIC NUCLEOSIDE PHOSPHONATES

The ANP may also interact with other enzymes than AMP kinase, PRPP synthetase and their target enzyme reverse transcriptase or cellular DNA polymerases. Herpes simplex virus type 1 (HSV-1) ribonucleotide reductase (RR), in particular conversion of CDP to dCDP, was found to be strongly inhibited by PMEApp and (S)-HPMPApp.[28] Conversion of ADP and GDP to their respective dADP and dGDP congeners by HSV-1 RR proved less sensitive to the inhibitory effect of these compounds.[3,28] It is not clear if the ANP derivatives markedly inhibit cellular RR in the intact cell system. It was also

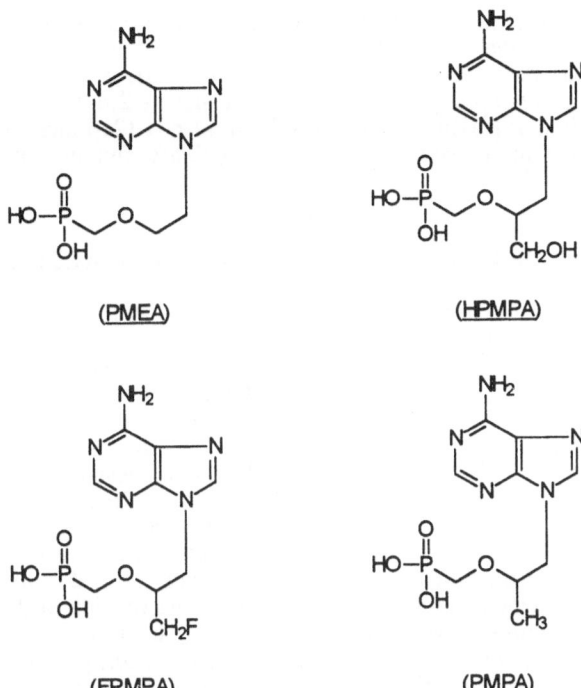

Figure 1. Structural formulae of PMEA [9-(2-phosphonylmethoxyethyl)adenine], HPMPA [9-(3-hydroxy-2-phosphonylmethoxypropyl)adenine], FPMPA [9-(3-fluoro-2-phosphonylmethoxypropyl)adenine] and PMPA [9-(2-phosphonylmethoxypropyl)adenine].

marked anti-herpesvirus activity.[7] Thus, minor structural changes in the alifatic side chain of the ANP analogues may have a major impact on the antiviral selectivity spectrum of these test compounds.

PMEA has also been evaluated on its anti-retroviral activity in several animal models incuding MSV infection in newborn NMRI mice, MLV infection in adult NMRI mice,[20] LP-BM5 retrovirus complex infection in Balb/c mice,[15] SIV infection in Rhesus monkeys[11] and FIV infection in cats.[13] In all retrovirus animal models, PMEA proved markedly effective in delaying the onset of retrovirus-induced disease and/or prevention of the retrovirus infection and related symptoms. Thus, PMEA proved not only to be a selective anti-retroviral agent *in vitro*, but it showed also pronounced anti-retroviral activity in a panel of retrovirus models *in vivo*. PMEA is now subject to clinical Phase I/II trials in the USA.

PHARMACOLOGICAL PROPERTIES OF PMEA

Uptake and metabolism of PMEA has been studied in several human and murine cell lines.[21,22] Not surprisingly, PMEA uptake was not mediated by the nucleobase or nucleoside carrier transport system (i.e. passive carrier-mediated transport). Indeed, uptake of PMEA was insensitive to inhibitors of nucleoside transport (i.e. NBMPR and dipyridamole).[21,22] However, the influx of PMEA was found to require energy, to be sodium-dependent and temperature-sensitive. It has been suggested that PMEA uptake is mediated by an endocytosis-like process.[21] Very recently, we selected a PMEA-resistant L1210 cell line that was 300-fold less sensitive to the cytostatic effects of PMEA [50%-cytostatic concentration (CC_{50}): 1.5 μM and $\geq$ 500 μM, respectively]. The PMEA-resistant cell line proved markedly cross-resistant to the closely related 2,6-diaminopurine derivative PMEDAP, and to PMEoA (9-[2-(phosphonylmethoxy)alkoxy]adenine (BRL 47923)) and to D4API (9-[(2R,5R)-2,5-dihydro-5-(phosphonomethoxy)-2-furanyl]adenine (BMS-181167-02)), but not to (*S*)-HPMPA, (*S*)-HPMPDAP, the guanine derivatives

reported that L1210 purine nucleoside phosphorylase (PNPase)-catalysed hydrolysis of inosine was inhibited by the parental PMEA and (S)-HPMPA derivatives, but also by their monophosphorylated PMEAp and (S)-HPMPAp derivatives in a cell-free enzyme assay.[3,29] However, we recently found that PMEA and its 2,6-diaminopurine derivative PMEDAP did not inhibit [^{3}H]inosine hydrolysis in intact L1210 and CEM cells, using a newly established tritium release assay to detect PNPase and inosinate dehydrogenase activity in intact cells.[30]

TARGET ENZYME OF PMEA FOR ANTI-RETROVIRAL ACTION

The diphosphorylated derivatives of PMEA and (S)-FPMPA are potent and selective inhibitors of HIV-1 and MLV reverse transcriptases.[6,23,31] Inhibition is competitive with respect to the natural substrate dATP. The K_i/K_m values of PMEApp and FPMPApp are 0.01 and 0.002, respectively, which means that these compounds have a much stronger affinity for HIV-1 RT than the natural substrate. Moreover, the ratio of IC$_{50}$ (50% inhibitory concentration) of PMEApp for HIV-1 or AMV RT *versus* DNA polymerase α is ~ 40. Thus, the ANP diphosphate derivatives should be considered as selective anti-RT agents. Beside the competitive inhibitory activity of the ANP diphosphates against RT, these compounds also serve as efficient alternative substrates for the enzyme to become incorporated into the growing DNA chain and therefore may also act as DNA chain terminators.[3,6,23] Preincubation of template.primer with RT and PMEApp results in a progressive loss of enzyme activity, presumably due to inactivation of the template.primer by terminal incorporation of PMEA, thereby preventing elongation of the primer by the RNA-dependent DNA polymerase activity of RT. It has been postulated that the DNA chain terminating potential of the PME and FPMP derivatives is a prerequisite for efficient anti-retroviral action of the test compounds. This latter premise is in agreement with the findings that (S)-HPMPApp did not cause an instant DNA chain terminating activity (presumably due to the presence of an hydroxyl group available for DNA chain elongation) and has no pronounced anti-retroviral activity.

ANTIMETABOLIC PROPERTIES OF PMEA

At subtoxic to moderately toxic concentrations, PMEA causes a slight but significant accumulation of intracellular NTP pools. This phenomenon may be suggestive for an inhibition of cellular RR by one or several PMEA metabolites. Also, preliminary data indicate that PMEA may cause a decrease of the endogenous dTTP pools in CEM cells (Balzarini & Karlsson, unpublished). The latter observation is in agreement with our most recent findings that PMEA causes a marked stimulation of radiolabeled dThd uptake and phosphorylation to its 5'-triphosphate metabolites in CEM cells (~ 4- to 5-fold) and L1210 cells (~ 20- to 35-fold). These data let us to hypothesize that inhibition of UDP reductase by PMEA metabolites (presumably PMEAp and/or PMEApp) cause a drop of endogenous dTTP pools which in turn causes a feedback stimulation of cytosol TK activity resulting in enhanced phosphorylation of dThd. The validity of this hypothesis has still to be proven. However, the decreased dTTP pools that occur under PMEA treatment of cell cultures seem not (solely) to contribute to the cytotoxic effects of the compound since toxicity of PMEA could not be reversed by co-administration with exogenous dThd or dCyd.

CONCLUSION

The anti(retro)viral properties of PMEA in cell culture and in several animal models indicate that PMEA may become an interesting candidate drug for the treatment of HIV-infected patients. Its mechanism of anti-retroviral action is primarily directed to the retroviral reverse transcriptase. The diphosphate metabolite of PMEA acts both as a potent competitive inhibitor of the RT reaction and as a strong DNA chain terminator. AMP kinase and PRPP synthetase do recognize PMEA and related analogues as a (poor) substrate for phosphorylation, but their physiological role in the activation of PMEA in the intact cells is currently unclear.

REFERENCES

1.   E. De Clercq, A. Holy, I. Rosenberg, T. Sakuma, J. Balzarini, and P.C. Maudgal, A novel selective broad-spectrum anti-DNA virus agent, *Nature* 323:464 (1986).

2.   A. Holy, and I. Rosenberg, Synthesis of 9-(2-phosphonylmethoxyethyl)adenine and related compounds, *Collect. Czech. Chem. Commun.* 52:2801 (1987).

3.   A. Holy, I. Votruba, A. Merta, J. Cerny, J. Vesely, J. Vlach, K. Sedivá, I. Rosenberg, M. Otmar, H. Hrebabecky, M. Travnicek, V. Vonka, R. Snoeck, and E. De Clercq, Acyclic nucleotide analogues: synthesis, antiviral activity and inhibitory effects on some cellular and virus-encoded enzymes in vitro, *Antiviral Res.* 13:295 (1990).

4.   E. De Clercq, T. Sakuma, M. Baba, R. Pauwels, J. Balzarini, I. Rosenberg, and A. Holy, Antiviral activity of phosphonylmethoxyalkyl derivatives of purine and pyrimidines, *Antiviral Res.* 8:261 (1987).

5.   R. Pauwels, J. Balzarini, D. Schols, M. Baba, J. Desmyter, I. Rosenberg, A. Holy, and E. De Clercq, Phosphonylmethoxyethyl purine derivatives: a new class of anti-human immunodeficiency virus agents, *Antimicrob. Agents Chemother.* 32:1025 (1988).

6.   J. Balzarini, A. Holy, J. Jindrich, H. Dvorakova, Z. Hao, R. Snoeck, P. Herdewijn, D.G. Johns, and E. De Clercq, 9-[(2RS)-3-fluoro-2-phosphonylmethoxypropyl] derivatives of purines: a class of highly selective antiretroviral agents *in vitro* and *in vivo*, *Proc. Natl. Acad. Sci. USA* 88:4961 (1991).

7.   J. Balzarini, A. Holy, J. Jindrich, L. Naesens, R. Snoeck, D. Schols, and E. De Clercq, Differential antiherpesvirus and antiretrovirus effects of the (S) and (R) enantiomers of acyclic nucleoside phosphonates: potent and selective *in vitro* and *in vivo* antiretrovirus activities of (R)-9-(2-phosphonomethoxypropyl)-2,6-diaminopurine, *Antimicrob. Agents Chemother.* 37:332 (1993).

8.   T. Yokota, K. Konno, S. Shigeta, A. Holy, J. Balzarini, and E. De Clercq, Inhibitory effects of acyclic nucleoside phosphonate analogues on hepatitis B virus DNA synthesis in HB611 cells, *Antiviral Chem. Chemother.* 5:57 (1994).

9.   R.A. Heijtink, G.A. De Wilde, J. Kruining, L. Berk, A. Holy, J. Balzarini, E. De Clercq, and S.W. Schalm, Antiviral activity of 9-(2-phosphonylmethoxyethyl)adenine (PMEA) on human and duck hepatitis B virus infection, *Antiviral Res.* 21:141 (1993).

10. J. Balzarini, J. Kruining, R. Heijtink, and E. De Clercq, Comparative anti-retrovirus and anti-hepadnavirus activity of three different classes of nucleoside phosphonate derivatives, *Antiviral. Chem. Chemother.*, in press (1994).

11. J. Balzarini, L. Naesens, J. Slachmuylders, H. Niphuis, I. Rosenberg, A. Holy, H. Schellekens, and E. De Clercq, 9-(2-Phosphonylmethoxyethyl)adenine (PMEA) effectively inhibits simian immunodeficiency virus (SIV) infection in Rhesus monkeys, *AIDS* 5:21 (1991).

12. H. Egberink, M. Borst, H. Niphuis, J. Balzarini, H. Neu, H. Schellekens, E. De Clercq, M. Horzinek, and M. Koolen, Suppression of feline immunodeficiency virus infection *in vivo* by 9-(2-phosphonomethoxyethyl)adenine, *Proc. Natl. Acad. Sci. USA* 87:3087 (1990).

13. K. Hartmann, A. Donath, B. Beer, H.F. Egberink, M.C. Horzinek, H. Lutz, G. Hoffmann-Fezer, I. Thum, and S. Thefeld, Use of two virustatica (AZT, PMEA) in the treatment of FIV and FeLV seropositve cats with clinical symptoms, *Vet. Immunol. Immunopathol.* 35:167 (1993).

14. K. Hartmann, J. Balzarini, J. Higgins, E. De Clercq, and N.C. Pedersen, *In vitro* activity of acyclic nucleoside phosphonate derivatives against feline immunodeficiency virus in Crandell feline kidney cells and feline peripheal blood lymphocytes, *Antiviral Chem. Chemother.* 5:13 (1994).

15. J.D. Gangemi, R.M. Cozens, E. De Clercq, J. Balzarini, and H.-K. Hochkeppel, 9-(2-Phosphonylmethoxyethyl)adenine in the treatment of murine acquired immunodeficiency disease and opportunistic herpes simplex virus infections, *Antimicrob. Agents Chemother.* 33:1864 (1989).

16. H. Thormar, J. Balzarini, A. Holy, J. Jindrich, I. Rosenberg, Z. Debyser, J. Desmyter, and E. De Clercq, Inhibition of visna virus replication by 2',3'-dideoxynucleosides and acyclic nucleoside phosphonate analogues, *Antimicrob. Agents Chemother.* 37:2540 (1993).

17. J. Balzarini, L. Naesens, P. Herdewijn, I. Rosenberg, A. Holy, R. Pauwels, M. Baba, D.G. Johns, and E. De Clercq, Marked *in vivo* antiretrovirus activity of 9-(2-phosphonylmethoxyethyl)adenine, a selective anti-human immunodeficiency virus agent, *Proc. Natl. Acad. Sci. USA* 86:332 (1989).

18. L. Naesens, J. Balzarini, A. Holy, I. Rosenberg, and E. De Clercq, Antiretroviral efficacy of 9-(2-phosphonylmethoxyethyl)adenine and 9-(2-phosphonylmethoxyethyl)-2,6-diaminopurine in mice infected with Friend leukemia virus (abstract), UCLA Symposia on Molecular Biology and Cellular Biology - HIV and AIDS: Pathogenesis, Therapy and Vaccine, Keystone, Colorado, USA, March 31-April 6, 1990, *J. Cell. Biochem.* 14D (Suppl.): 140 (1990).

19. E.A. Hoover, J.P. Ebner, N.S. Zeidner, and J.I. Mullins, Early therapy of feline leukemia virus infection (FeLV-FAIDS) with 9-(2-phosphonylmethoxyethyl)adenine (PMEA), *Antiviral Res.* 16:77 (1991).

20. L. Naesens, J. Balzarini, and E. De Clercq, Single-dose administration of 9-(2-phosphonylmethoxyethyl)adenine (PMEA) and 9-(2-phosphonylmethoxyethyl)-2,6-diaminopurine (PMEDAP) in the prophylaxis of retrovirus infection in vivo, *Antiviral Res.* 16:53 (1991).

21. G. Palú, S. Stefanelli, M. Rassu, C. Parolin, J. Balzarini, and E. De Clercq, Cellular uptake of phosphonylmethoxyalkylpurine derivatives, *Antiviral Res.* 16:115 (1991).

22. K.L. Prus, E.L. Hill, and M.N. Ellis, The transport of 9-(2-phosphonylmethoxyethyl)adenine (PMEA) into Vero cells, Fourth Int. Conf. Antiviral Res., New Orleans, Louisiana, USA, April 21-26, 1991, *Antiviral Res.* Suppl. 1:144, abstract no. 188 (1991).

23. J. Balzarini, Z. Hao, P. Herdewijn, D.G. Johns, and E. De Clercq, Intracellular metabolism and mechanism of anti-retrovirus action of 9-(2-phosphonylmethoxyethyl)adenine, a potent anti-human immunodeficiency virus compound, *Proc. Natl. Acad. Sci. USA* 88:1499 (1991).

24. J.J. Bronson, H.-T. Ho, H. De Boeck, K. Woods, I. Ghazzouli, J.C. Martin, and M.J.M. Hitchcock, Biochemical pharmacology of acyclic nucleotide analogues, *Ann. N.Y. Acad. Sci.* 616:398 (1990).

25. A. Merta, I. Votruba, J. Jindrich, A. Holy, T. Cihlar, I. Rosenberg, M. Otmar, and H.Y. Tchaou, Phosphorylation of 9-(2-phosphonomethoxyethyl)adenine and 9-(S)-(3-hydroxy-2-phosphonomethoxypropyl)adenine by AMP (dAMP) kinase from L1210 cells, *Biochem. Pharmacol.* 44:2067 (1992).

26. J. Balzarini, and E. De Clercq, 5-Phosphoribosyl 1-pyrophosphate synthetase converts the acyclic nucleoside phosphonates 9-(3-hydroxy-2-phosphonylmethoxypropyl)adenine and 9-(2-phosphonylmethoxyethyl)adenine directly to their antivirally active diphosphate derivatives, *J. Biol. Chem.* 266:8686 (1991).

27. J. Balzarini, and E. De Clercq, Conversion of acyclic nucleoside phosphonates to their diphosphate derivatives by 5-phosphoribosyl-1-pyrophosphate (PRPP) synthetase, *in*: Purine and Pyrimidine Metabolism in Man VII, Part A", R.A. Harkness, G. Elion, and N. Zöllner, eds., Plenum Press, New York, p. 29 (1991).

28. J. Cerny, I. Votruba, V. Vonka, I. Rosenberg, M. Otmar, and A. Holy, Phosphonylmethyl ethers of acyclic nucleoside analogues: inhibitors of HSV-1 induced ribonucleotide reductase, *Antiviral Res.* 13:253 (1990).

29. K. Sedivá, A.V. Ananiev, I. Votruba, A. Holy, and I. Rosenberg, Inhibition of purine nucleoside phosphorylase by phosphonylmethoxyalkyl analogues of nucleotides, *Int. J. Purine Pyrimidine Res.* 2:35-39 (1991).

30. J. Balzarini, and E. De Clercq, Assay method for monitoring the inhibitory effects of antimetabolites on the activity of inosinate dehydrogenase in intact human CEM lymphocytes, *Biochem. J.* 287:785 (1992).

31. I. Votruba, M. Trávnícek, I. Rosenberg, M. Otmar, A. Merta, H. Hrebabecky, and A. Holy, Inhibition of avian myeloblastosis virus reverse transcriptase by diphosphates of acyclic phosphonylmethyl nucleotide analogues, *Antiviral Res.* 13:287 (1990).

PROPERTIES OF MITOCHONDRIAL DNA METABOLISING ENZYMES; IMPLICATIONS FOR CHEMOTHERAPY

Staffan Eriksson[1], Baoji Xu[2] and David A. Clayton[2]

[1]Department of Veterinary Medical Chemistry, Swedish
University of Agricultural Sciences, The Biomedical Center,
Box 575, S-75123, Uppsala, Sweden
[2]Department of Developmental Biology,
The Beckman Center, Stanford University School of
Medicine, Stanford, California

INTRODUCTION

Antiviral chemotherapy against HIV infection is based on the use of dideoxynucleoside analogs such as Zidovudine (3′-azido-deoxythymidine, AZT), dideoxycytidne (ddC) and dideoxyinosine, that are inhibitors of HIV reverse transcriptase. Although these drugs give clinical improvement to patients with AIDS they all give severe side effects[1]. On prolonged treatment with dideoxynucleosides a substantial number of patients show myophaty, pheripheral neurophaty and pancreatitis and these symptoms are most likely related to mitochondrial dysfunction[2,3]. Cell culture studies as well as experiments with isolated mitochondria have shown that AZT, ddC and some other dideoxynucleosides cause defects in miotchondrial DNA synthesis, which subsequently will lead to delayed cell toxicity. Recently, several patients participating in a clinical trial of a new anti-hepatitis B drug (2′deoxy-2′fluoro-arabinosyl 5-iodo)uracil (FIAU) tragically died due to drug induced liver failure. There were indications of damaged mitochondria in these patients and it is likely that also FIAU is toxic due to its interference with mitochondrial DNA synthesis[2,3].

The underlying mechanism for the mitochondrial side effects of nucleosides therapy has been attributed to the broad substrate specefcity of mitochondrial(mt) DNA polymerase, enabling it to efficiently incorporate several nucleotide analogs. A prerequisite for this mechanism is the formation of sufficient mitochondrial pools of phosphorylated analogs, that can serve as precursors for mtDNA synthesis. Some properties of the mitochondrial

oligonucleotide substrates. Yeast mt DNA polymerase is the product of the nuclear MIP1 gene and it is the first mt-polymerase which has been cloned and sequenced[4]. The gene cods for a protein of 145 kDa with sequence similarities to eukaryotic polymerases as well as reverse transcriptases. Yeast mtDNA polymerase show many biochemical similarities with other mt polymerases isolated from higher eucaryots. The goal with the approach here is to provide biochemical information that migth help improving future design of nucleoside therapeutics.

MATERIALS AND METHODS

Materials: Pure recombinant HIV-RT was a gift from T. Unge (Dept of Mol. Biol. The Biomedical Center, Uppsala, Sweden). The oligonucleotide primer-templates were as described[5].

Purification of yeast mtDNA polymersase. The enzyme was purifies from yeast cells harboring the MIP1 overpression plasmid as described[6]. Cells were disrupted in a buffer containing 25 mM Tris-Cl (pH 7.9), 20% glycerol, 1mM DTT, 5 mM EDTA, 1 mM PMSF, 2 mM benzamidine, 1 mg/ml of E64, and 2 mg/ml each of pepstatin A, leupeptin, and antipain] plus 0.1% Triton X-100 and 250 mM KCl. Yeast mtDNA polymerase was purified 300-fold by DEAE Sephacel-, Phosphocellulose 11-, Heparin Sepharose- and Butyl Sepharose chromatography. SDS gelelectrophoresis showed a single 135 kDa polypeptide in the fractions from the Butyl Sepharose step.

DNA polymerase Assay: To prepare the primer templates for the running and standing start reactions, primer oligonucleotides were labeled with T4 polynucleotide kinase at the 5´-end, gel purified and annealed to the respective templates as described[5]. Reactions were performed for 10 min at 37 $^{\circ}$ C with 3-5 ng DNA polymerase, 0.1 pmole template-primer and varying concentrations of nucleoside triphosphates. The products were analysed by 19 % polyacrylamide urea gel electrophoresis which were subsequently exposed to X-ray film. The films were scanned and the percentage of product band formed as related to totally loaded radioactivity determined. The kinetic constants were calculated using a reiterative curve fitting program for the Michaelis- Menten equation.

RESULTS AND DISCUSSION

Yeast cells contain a very low level of mtDNA polymerase. In order to get a good source for the purification of the yeast mtDNA polymerase the coding region of the MIP1 gene was cloned into a high copy expression vector and put under the control of a strong and inducible Gal1 promoter. Yeast mtDNA polymerase was purified to homogeneity from whole cell extract by a simple four-step purification procedure. It was important to include a mixture of protease inhibitors in the buffers to isolate a 135 kDa mtDNA polymerase with a reasonable yield[6].

The main aim of the study was to determine the kinetic properties of mtDNA polymerase with pharmacologically important dideoxynucleotides. We used an assay system testing the capacity of the polymerase to incorporate a nucleoside triphosphate either as first (standing start template) or the third nucleotide (running start template) in a defined primer-template. The properties of human alfa and beta polymerases were recently described by Copeland et al with the same assay system[5].

The elongation of the primer by the polymerase in the standing start assay was saturated at nM concentration of nucleotide substrates and the addition of the other deoxynucleotides led to synthesis of a complete 36 mer, demonstrating a high processivity of the mtDNA polymerase. When 3'-AZTTP was used as substrate there was minimal incorporation even at high concentrations, while 3'-FTTP and ddTTP served as a relatively efficient substrates (Table 1). Another standing start primer-template specific for cytidine nucleotide incorporation was used to determine the efficiency of dCTP and ddCTP substrates. A series of experiments of this type were performed and the gel results quantified by densitometry so that the apparent K_m and V_{max} values for these substrate could be determined (Table 1).

Table 1. Apperent kinetic constants for nucleoside triphosphates with standing start primer- templates for mtDNA polymerase and HIV-RT.

	dTTP	3'-FTTP	ddTTP	AZTTP	dCTP	ddCTP
mtDNA pol:						
App. K_m (nM)	10 ± 3	52 ± 16	76 ± 3	>6000	5.4 ± 0.5	20 ± 9
App. V_{max}	7.1 ± 1	7.7 ± 0.2	3.6 ± 0.4	0.8	7.7 ± 0.2	4.4 ± 0.7
Rel	1	0.21	0.66		1	0.15
HIV-RT:						
App. K_m (nM)	5.4 ± 1	60 ± 4	24 ± 2	11 ± 2	4.4 ± 0.4	6.1 ± 2
App. V_{max}	11 ± 0.6	7 ± 3	6.1 ± 1.4	8 ± 0.9	10 ± 0.8	8.4 ± 0
Rel	1	0.06	0.12	0.36	1	0.59

A similar set of reactions were carried out with pure recombinant HIV-RT and the kinetic constants observed with mtDNA polymerase and HIV-RT were overall very similar with K_m values for dTTP or dCTP between 5-10 nM and with ddCTP, ddTTP and 3'-FTTP about 2-10 fold less efficient as substrates. The exception being that 3'-AZTTP was a poor substrate for mtDNA polymerase while it was almost as efficient as dTTP in case of HIV-RT (Table 1).

With the running start synthesis the gel assay results were somewhat more complicated but the results were qualitatively similar to those reported above. In this case AZTTP (K_m= 40 nM) was also incorporated but approximately 30 fold less efficient than the other nucleotide analogs. Copeland et al[5] showed with the same assay that pure human alfa polymerase was capable of incorpoation of AZTTP, but only with high (more than 100 uM) concentration and thus the nuclear enzyme discriminates very efficiently against dideoxynucleotides substrates.

The results presented here support the hypothesis that the capacity of mtDNA polymerase to incorporate dideoxynucletides is the underlying mechanism for the mitochondrial defects observed during anti-HIV treatment with this type of analogs. The synthesis of deoxynucleotides for mtDNA replication occurs at least in mammalian cells both via deoxyribonucleotide translocation across the mt membranes as well as via salvage of deoxyribonucleosides. The details of this metabolism is still not known.

TK2 and dGK are most likely responsible for supplying mtDNA precursors in resting animal cells with very low cellular deoxyribonucleotide pools. It is in these type of tissues (i e muscles and nerve cells) that interference of mtDNA replication lead to functional mt defects in case of nucleoside analog treatments. TK2 and dGK have relatively recently been completely purified. They are dimers or in case of TK2 possibly a monomer of 29 kDa and 28 kDa, respectively[7,8]. TK2 is a pyrimidine deoxynucleoside kinase that can phosphorylate thymine, uracil and cytidine containing nucleosides, even with substantial 5-substitutions. In case of sugar modifications TK2 can accept 2'- and to a much lower extent 3'-modifcations[7,9]. FIAU is also a very efficient substrate (Wang, J. and Eriksson, S. in preparation). dGK is a purine nucleoside kinase that can phosphorylate guanine, hypoxanthine and adenine nucleosides with minor 2- or 7- modifications and it accepts arabinosylsugars and other 2'-substitutions but not 3'-modifications or acyclic compounds[8].

In order to make prediction about the capacity of a certain analog to block mtDNA replication it is thus necessary to know to what extent the compound can be phosphorylated and transported into the mt in the cells of interest as well as if it is capabale of inhibiting mtDNA polymerase. Thus, there is a need for further basic studies on the mechanism of mtDNA replication and mtDNA precursor synthesis in order to design antiviral drugs, which will not interfer with mtDNA metabolism.

Acknowledgment

This work was supported by grants to D.C. from the National Institute of General Medical Sciences and to S.E from The Swedish Medical Research Councel, The Royal Society of Swedish Physiciens, The Medical Faculty of the Karolinska Institute, and Swedish University of Agricultural Sciences.

REFERENCES

1. H. Mitsuya and S. Broder. Inhibition of the in vitro infectivity and cytopathic effect of human T-lymphotrophic virus type III lymphadenopathy-associated virus (HTLV-III/LAV) by 2',3'-dideoxynucleosides. *Proc .Natl. Acad. Sci. USA* 8:1911 (1986).

2. C.-H. Chen and Y.-C. Cheng. Effect of anti-human immunodeficiency virus nucleoside analogs on mitochondrial DNA and implication for delayed toxicity. *Mol. Pharmacol.* 39:625 (1991).

3. W.B. Parker and Y-C. Cheng. Mitochondrial toxicity of antiviral analogs. *J. NIH Res.* 6:57 (1994).

4. F. Foury. Cloning and sequencing of the nuclear gene MIP1 encoding the catalytic subunit of yeast mitochondrial DNA polymerase. *J. Biol. Chem.* 264:20552 (1989).

5. Copeland C.W., Chen M.S. and Wang T. Human DNA polymerase a and ß are able to incorporate anti-HIV deoxynucleosides into DNA. *J. Biol. Chem.* 267:21459 (1992).

6. S. Eriksson, B. Xu and D. A. Clayton. Efficient incorporation of anti-HIV deoxynucleotides by yeast mitochondrial DNA polymerase. Submitted.

7. B. Munch-Petersen , L. Cloos, G. Tyrsted and S. Eriksson. Diverging substrate specificity of pure human thymdine kinase 1 and 2 against antiviral dideoxynucleosides. *J. Biol. Chem.* 266:9032 (1991).

8. L. Wang, A. Karlsson, S. J. Arnér and S. Eriksson. Substrate specificity of mitochondrial 2´-deoxyguanosine kinase. Efficient phosphorylation of 2-chlorodeoxyadenosine. *J. Biol. Chem.* 268:22847 (1993).

9. S. Eriksson, B. Kierdaszuk, B. Munch-Petersen, B. Öberg and N.G. Johansson. Comparison of the substrate specificities of human thymidine kinase 1 and 2 and deoxycytidine kinase toward antiviral and cytostatic nucleoside analogs. *Biochem. Biophys. Res. Commun.* 176:586 (1991).

ADENOSINE DEAMINASE (ADA) DEFICIENCY AS THE UNEXPECTED CAUSE OF CD4+ T-LYMPHOCYTOPENIA IN TWO HIV-NEGATIVE ADULT FEMALE SIBLINGS

Lynette D.Fairbanks[¶], H.Anne Simmonds[¶], A.David B.Webster[≠], Claire L.Shovlin[§] and J.Michael B.Hughes[§]

[¶]Purine Research Laboratory, UMDS Guy's Hospital, London, GB;
[≠]Dept of Clinical Immunology, Royal Free Hospital, London, GB;
[§]Respiratory Medicine, Royal Postgraduate Medical School, Hammersmith Hospital, London GB

INTRODUCTION

Adenosine deaminase (ADA) deficiency results in the syndrome of severe combined immunodeficiency (SCID). The effects on the immune system result almost exclusively from the build up of one of the substrates, deoxyadenosine (dAdo), derived from the rapid DNA turnover in haematopoietic cells (1-5). dAdo is phosphorylated to the deoxyribonucleotide (dATP) in erythrocytes, thymic cells and B cells. The chief mechanism of lymphotoxicity is considered to relate to dATP inhibition of ribonucleotide reductase and hence DNA synthesis (1-3). Loss of erythrocyte S-adenosylhomocysteine hydrolase (SAHH) activity relates directly to dAdo build up (2).

A considerable heterogeneity in age of presentation and biochemical parameters has been found (1,2). The presence of low level ADA activity in white cells in late presenters, capable of metabolising some of this toxic dAdo, explains the milder symptoms and late-onset of the life-threatening consequences which have led to death in early presenters. Available forms of therapy include bone marrow transplantation, enzyme replacement with packed erythrocytes or PEG-ADA, and recently somatic cell gene replacement therapy (1,2,4,5). Until now the oldest published case was first diagnosed at 15 years of age (2). This paper reports biochemical studies in two sisters who presented with chronic chest disease in their teens and were first identified as ADA deficient in their thirties (4,6,7).

CASE HISTORY

Case 1 was originally diagnosed as asthmatic and had 2 episodes of pneumonia in childhood. She remained well until her twenties when recurrent chest infections developed. She had 2 successful pregnancies (Fig 1), but developed deteriorating lung function during the second. By 1990 she required continuous antibiotic therapy, salbutamol and beclomethasome inhalers, intermittent courses of oral steroids and had widespread viral warts with recurrent oral and vaginal candidiasis.

Case 2, the elder sister of case 1, has 1 child. She also remained well until her late teens, when she developed a resistant idiopathic thrombocytopaenic purpura. By the early twenties she had asthma, recurrent chest infections, vaginal and oral candidiasis, and widespread viral warts. By 1992 there was clinical and radiological evidence of extensive lung damage.

When investigated both were lymphopaenic with severely depressed numbers of CD4+ T cells (Table 1), functional in vitro T-cell defects and mildly deranged serum immunoglobulin levels. Both were HIV negative (4). Such similar abnormalities in two siblings suggested a genetic aetiology and led to investigations for a defect of purine metabolism.

Table 1 Immunological parameters at presentation and after 12 months on PEG-ADA

	Pre PEG-ADA		1 year post PEG-ADA	
	Case 1	Case 2	Case 1	Case 2
Lymphocytes (x 10^9/l)				
T cells				
$CD3^+$ (0.8-2.5)	0.2	0.35	0.60	2.23
$CD4^+$ (0.4-1.5)	0.08	0.05	0.28	0.49
$CD8^+$ (0.2-1.1)	0.15	0.20	0.38	1.82
B cells				
CD19+ (0.14-0.3)	<0.01	0.01	0.03	0.05
Immunoglobulins (g/l)				
IgG (5-16)	15.2	14.4	10.0	25.0
IgG1 (3.2-10.2)	>5.9	14.6	8.7	>12.0
IgG2 (1.2-6.0)	0.73	0.5	0.9	0.7
IgA (1.2-4.2)	0.96	0.4	0.8	0.7
IgM (0.5-4.25)	0.56	0.3	0.7	0.8

IgG Antibodies to:
Pneumovax (pneumococcal polysaccharide) (titre)

Pre vaccination	52	28	20	30
Post vaccination	nd	23	20	25

Tetanus toxoid

Pre vaccination	2560	29	2349	30
Post vaccination	nd	384	nd	nd

RESULTS

Table 2 lists the concentrations of ATP and dATP found in the erythrocytes compared with the mean values in erythrocytes of ADA deficient children studied by us, the oldest patient hitherto in our experience being 2yrs of age. This demonstrates that the highest dATP concentrations were found in the earliest presenters, along with a concomitant ATP depletion.

Table 2 Biochemical data at presentation and one year post PEG-ADA

		Erythrocyte		Lymphocyte	Plasma
Patient	Age	ATP	dATP	ADA	ADA
			(µmol/l)	(nmol/h/mg)	(µmol/ml/h)
Case 1	34yr	1686	234	68	
Case 2	35yr	1424	105	55	
12 months post PEG ADA					
Case 1	35yr	1092	1	nd	30-50
Case 2	36yr	1469	<1	nd	30-60
ADA deficient children (n)					
12	<2/12	813	1783		
range		(527-1025)	(1121-2478)		
6	4/12 - 2yr	1602	557		
range		(761-1888)	(304-912)		

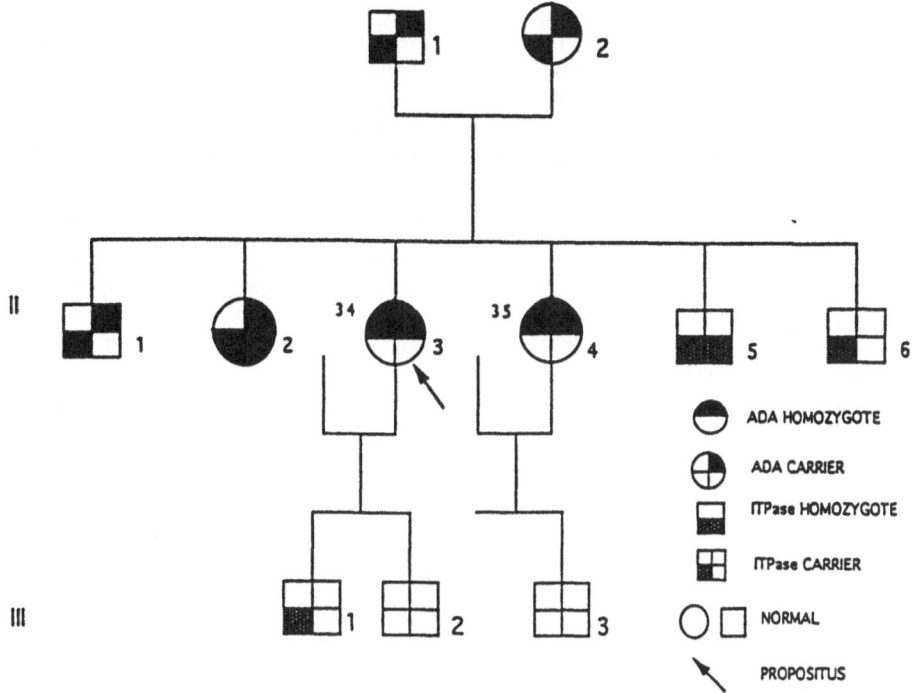

Figure 1 Family tree showing inheritance of ADA deficiency and the co-incidental co-existence of ITPase deficiency

the later presenters having much lower concentrations of dATP, with normal ATP levels, as did these two sisters. Both patients had no detectable ADA activity in their lysed erythrocytes but a low level of activity in their lymphocytes: 68 nmol/h/ mg protein for Case 1 and 55 nmol/h/mg protein for Case 2 (control range 1162 - 4500).

Fig1 demonstrates the autosomal recessive inheritance of ADA in this kindred, the parents and two siblings being carriers for the defect. A surprising finding was the co-existence of ITPase deficiency in the kindred, but the segregation suggested this co-existence was co-incidental..

The sisters were commenced on PEG-ADA therapy (4) and dATP levels have been monitored regularly, the objective being to reduce them to less than 10μmol/l. A serial reduction in dATP levels resulted, falling to concentrations below 10μmol/l at 5 weeks accompanied by some clinical and immunological improvement. Twelve months later both patients still have low IgG2 levels and fail to respond to vaccination with pneumococcal polysaccharide (Table 1). Case 1 has retained a high level of anti tetanus toxoid antibodies, while patient 2 has lost most of these antibodies.

DISCUSSION

We have studied a total of 33 patients with ADA deficient SCID. Until now they have all been children, 60% of them being severely affected early presenters. The children in our series who presented later, between 4 months and 2 years, had less severe immunological and biochemical parameters (1).These two latest cases of ADA deficiency, both in their thirties, are the oldest in our series and the first adults identified with ADA deficiency (4). The three pregnancies in these women are also the first to be described in ADA deficient patients. As with the other late presenters we have studied, these two adults had low but detectable ADA activity in their peripheral blood lymphocytes and the clinical picture was not characteristic of ADA deficient SCID. The fact that these two adult patients had the lowest erythrocyte dATP concentrations we have encountered confirms the hypothesis that erythrocyte dATP is a sensitive indicator of the severity of the defect. The fall in dATP post therapy is also used as a marker of the response to therapy with PEG-ADA (5), and indeed there was a rapid fall in dATP in these sisters, which has been sustained over twelve months.

This rapid improvement in dATP was not reflected in a similar rapid improvement in immunological and clinical parameters. Although, there is some improvement in the numbers of circulating CD4+ T cells, particularly in Case 2, the number of circulating B cells remains low (<0.05x10^9/l) in both sisters, which may explain the failure to respond to polysaccharides, as these antigens may have to be presented by B cells in order to generate a response within the lymphoid apparatus. These results question whether dATP accumulation is the principal mechanism of lymphotoxicity to dividing cells and suggests some additional, or alternative mechanism is involved. One explanation for the lymphotoxicity in ADA deficiency has related to inhibition of SAHH and hence vital cellular transmethylation reactions (2). Since erythrocyte SAHH fell immediately (within one week) to normal following commencement of PEG-ADA, well before the reduction in dATP, this too does not fully concur with accepted dogma.

In addition to the autosomal recessive inheritance of this disorder confirmed in the study of other kindred members (Fig 1), there were several remarkable additional features. First, as in a previously published kindred (6), co-existence of ITPase deficiency was found, the segregation likewise being apparently co-incidental. Second,the three children of these women, who would be obligate heterozygotes, had erythrocyte ADA activities within the normal range. One explanation suggested for this (7) is the possibility that up-regulation of the normal allele occurred *in utero* to compensate for the ADA deficiency in the mother. This kindred has been the subject of detailed analysis at the molecular level and the results are discussed in detail in a separate publication (7).Inheritance of a highly polymorphic marker showed that both Case 1 and Case 2 were compound heterozygotes at the ADA locus.

The important message in this report is that ADA deficiency should always be ruled out in adults as well as children with late onset immunodeficiency, and may explain some cases hitherto reported as 'non-HIV AIDS' (4). The studies in this kindred not only confirm that ADA deficiency displays variable expression but also suggest that the severity of the phenotype is influenced by factors additional to the structural gene mutations.

REFERENCES

1. G. Morgan, R.J. Levinsky, K. Hugh-Jones, L.D. Fairbanks, G.S. Morris and H.A. Simmonds. Heterogeneity of biochemical, clinical and immunological parameters in severe combined immunodeficiency due to adenosine deaminase deficiency . Clin exp Immunol 70: 491 (1987).
2. I. Santisteban, F.X. Arredondo-Vega, S Kelly, A. Mary, A. Fischer, D.S. Hummell, A. Lawton, R.U. Sorensen, E.R. Steihm, L.Uribe, K. Weinberg and M.S. Hershfield. Novel splicing, missense, and deletion mutations in seven adenosine deaminase-deficient patients with late/delayed onset of combined immunodeficiency disease. Contribution of genotype to phenotype. J Clin Invest 92:2291 (1993).
3. A. Goday, H.A. Simmonds, G.S. Morris and L.D. Fairbanks. Human B lymphocytes and thymocytes but not peripheral blood mononuclear cells accumulate high dATP levels in conditions simulating ADA deficiency. Biochem Pharmacol 34: 3561 (1985).
4. C.L. Shovlin, A.D.B. Webster, L.D. Fairbanks, H.A. Simmonds, S. Deacock, R. Lechler, I. Roberts and J.M.B. Hughes. Adult presentation of inherited adenosine deaminase deficiency. Lancet 341:1472 (1993).
5. M.S. Hershfield, S. Chaffee and R.U. Sorensen. Enzyme replacement therapy with polyethylene glycol-adenosine deaminase and adenosine deaminase deficiency. Ped Res 33:S42 (1993).
6. J.A. Duley, H.A. Simmonds, D.A. Hopkinson and R.J. Levinsky..Inosine triphosphate pyrophosphohydrolase deficiency in a kindred with adenosine deaminase deficiency. Clin. Chim. Acta.188:243 (1990).
7 C.L. Shovlin, H.A. Simmonds, L.D. Fairbanks, S. Deacock, J.M.B. Hughes, R. Lechler, A.D.B. Webster, Xi-Ming Sun, J.C. Webb and A.K. Soutar. Adult onset immunodeficiency due to inherited adenosine deaminase deficiency. J. Immunol. in press (1994).

SERUM GUANASE ACTIVITY IN HEPATITIS C VIRUS INFECTION

T. Shaw[1,2], J. Li[1,3], D. S. Bowden[1], G. Cooksley[4] and S. A. Locarnini[1]

Victorian Infectious Diseases Reference Laboratory[1] & Macfarlane Burnet Centre for Medical Research[2], Fairfield Hospital, Victoria 3078, Australia and Royal Brisbane Hospital Research Foundation Clinical Research Centre, Brisbane, Queensland 4029, Australia[4]

INTRODUCTION

Guanase (guanine aminohydrolase, EC 3.5.4.3), catalyses the hydrolytic deamination of guanine. The reaction is important because of its potential to infuence intracellular guanine nucleotide pools: guanine is directly salvagable to GMP, whereas its product xanthine is not (see Figure 1). Amongst human tissues, liver contains the highest guanase activity; activity is also detectable in brain and kidney [1,2]. In most assays, guanase activity in other tissues and in normal plasma and serum is close to or below the detection limit [3-5]. Gross increases in serum guanase are considered a specific indication of hepatocellular damage and have been shown to have prognostic and diagnostic value in a variety of clinical situations [2-9]. However, routine guanase measurement is not widely available due to lack of sufficiently facile and accurate assays. We have developed a sensitive assay for guanase using reversed-phase ion-pair HPLC. Using this assay, we found that guanase activity in normal sera was lognormally distributed about a geometric mean of 0.85 units/litre, with no significant gender difference. When sera submitted to us for testing by polymerase chain reaction (PCR) for suspected active hepatitis C virus (HCV) infection were screened in parallel for guanase activity, increases above the 95% percentile of the normal range were found to be predictive of HCV positivity. In two separate groups of patients being treated with interferon-alpha for chronic HCV infection, guanase activity was elevated before treatment and decreased significantly during treatment. We conclude that serum guanase activity is a useful surrogate marker for HCV infection and that serial measurements may have prognostic value.

---

[1] Tel: 61 3 280 2615, Fax : 61 3 481 3816. [3]Present address Jiangsu Province Hospital, Nanjing, People's Republic of China. Abbreviations: HCV - hepatitis C virus; PCR - polymerase chain reaction; DMSO - dimethyl sulphoxide.

MATERIALS AND METHODS

Reagents and chromatography

Chemicals and biochemicals were purchased from Sigma. Solvents and reagents for chromatography were obtained from local suppliers and were analytical or HPLC grade. Water was glass-distilled , deionised and filtered using a Millipore system. Chromatographic analyses were performed using a 50 x 4 mm pH-stable C18 reverse-phase column (Intersil ODS-2™, particle size 5 µM, from SGE International, Australia) protected by a guard column packed with the same material. The column was installed in an HPLC system from Bioanalytical Systems Inc. (West Lafayette, IN, USA), equipped with an inbuilt variable wavelength UV detector and driven by computer software which enabled automatic peak integration, calibration and analysis. Analytes were detected by UV absorption at 267 nm after isocratic elution with 0.5% (v:v) DMSO, 5% (v:v) ethanol and 10 mM heptane sulphonic acid in 20 mM potassium acetate, pH 4.5. The solvent flow rate was 1.0 ml/min.

Guanase assay

Assays were set up in 1.5 ml microfuge tubes. The assay mixture, which was preheated to 37°C, contained (final concentrations) 100 µM guanine, 2% (v:v) DMSO, 150 mM NaCl and 10 mM Tris-HCL, pH 7.4 in a total volume of 490 µl. The reaction was started by adding 5-50 µl serum and terminated by adding 500 µl of 5% (v:v) perchloric acid. After pelleting the protein precipitate, 5 µl aliquots of the supernatant were introduced into the HPLC using an autoinjector system (Carnegie Medicin, Sweden). One unit (U) of guanase activity was defined as the enzymatic activity required to deaminate 1 µM guanine to xanthine in 1 min under the conditions of assay. Controls with known guanase activity were included in each set of assays to facilitate between-run comparisons.

Sera and HCV detection by PCR

Control sera were selected randomly from samples sent to the VIDRL, Fairfield, for routine immunological testing unrelated to liver disease or dysfunction. The remaining sera (some of which were from HCV-infected patients undergoing treatment with recombinant human interferon 2a as part of a controlled clinical trial at the Royal Brisbane Hospital), were sent to our laboratory for routine HCV testing. Active HCV infection, indicated by the presence of HCV RNA in sera was detected using a procedure based on that of Okomoto and colleagues [10], which uses a reverse transcription (RT) step followed by nested PCR. Precautions were taken to reduce the possibility of contamination and positive, negative and reagent controls were included in every RT-PCR run.

Data Analysis

Statistical analyses were performed using the FASTAT® computer software package from SYSTAT Inc., Evanston, IL, USA.

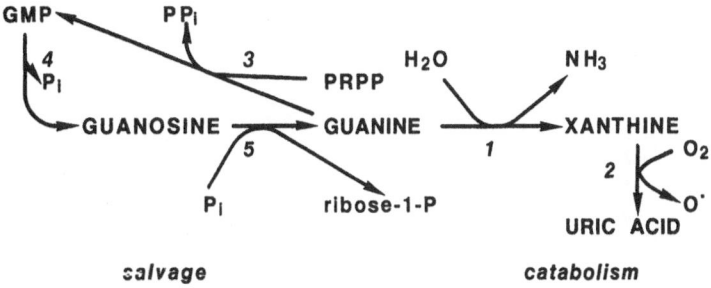

Figure 1. The role of guanase in guanine metabolism. Key enzymes: 1, guanase; 2, xanthine oxidase/dehydrogenase; 3, hypoxanthine-guanine phosphoribosyl-transferase; 4, IMP-GMP 5'-nucleotidase; 5, purine nucleoside phosphorylase.

RESULTS

Guanase assay and correlation with HCV infection

Xanthine and guanine were rapidly separated by RP-HPLC in less than 1 min, with retention times of 31 sec and 54 sec respectively. Addition of the ion-pairing agent heptane sulphonic acid reversed the normal elution order, resulting in better resolution of substrate and product and allowing more accurate quantitation of the earlier, usually smaller, xanthine peak (see Figure 2). The reaction rate remained linear until more than 75% of the substrate was depleted. The time taken for this to occur varied with enzyme activity, but for activities in the range of interest (c. 0.1-6 U/l), linearity was maintained for at least 24 h. For routine screening, 5 μl serum with overnight incubation (c.1000 min) was chosen for convenience. Pathological levels of guanase activity, which are often of the order of 10 - 100 fold higher than the normal upper limit, could be detected after correspondingly shorter incubation times; normal activity could quantitated after shorter incubations if larger serum volumes were used. Results of application of the assay are illustrated graphically in Figures 3-5.

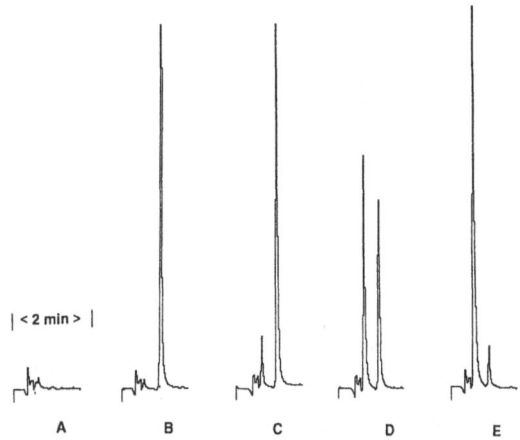

Figure 2. Guanase assay by ion-pair RP-HPLC. Representative chromatograms are shown: A - serum blank, B - substrate blank, C - end of assay, 1 U/l guanase, D - end of assay, 5 U/l guanase, E - end of assay, > 20 U/l guanase.

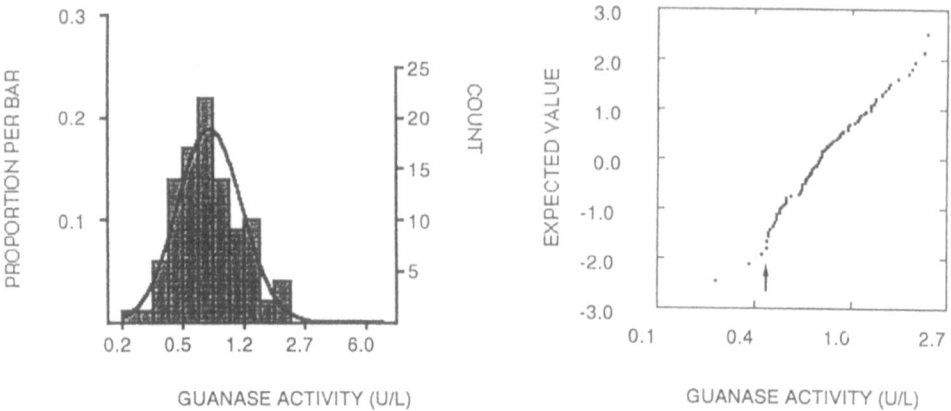

Figure 3. Distribution of guanase activity in normal sera. A, Frequency distribution histogram showing approximate lognormal distribution of guanase activity in 100 normal serum samples (52 females, 48 males). The population's geometric mean activity was c. 0.85 U/l with no significant gender difference in activity. The superimposed curve shows a predicted lognormal distribution. The normal upper limit (95 percentile) occurs at about 2.25 U/l. B, Logormal probabilty plot of the same data (guanase activity plotted on the x axis versus theoretical values predicted from normal distribution on the y axis). The arrow indicates an antimode in the frequency distribution distinguishing a possible subpopulation with low activity (see ref. 11].

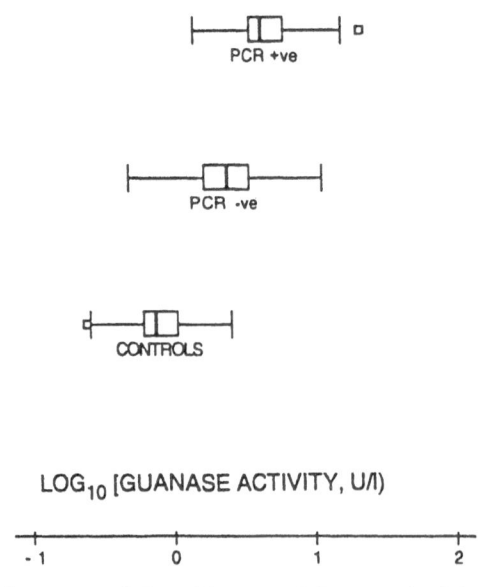

Figure 4. Correlation of guanase activity with serum HCV status by PCR. "Box and whiskers" plots of distribution of guanase activity in serum samples screened by RT-PCR for HCV RNA compared to control group. Note that the scale is logarithmic. Despite overlaps in activity, differences between means of each group were statistically significant at the $p < 0.05$ level.

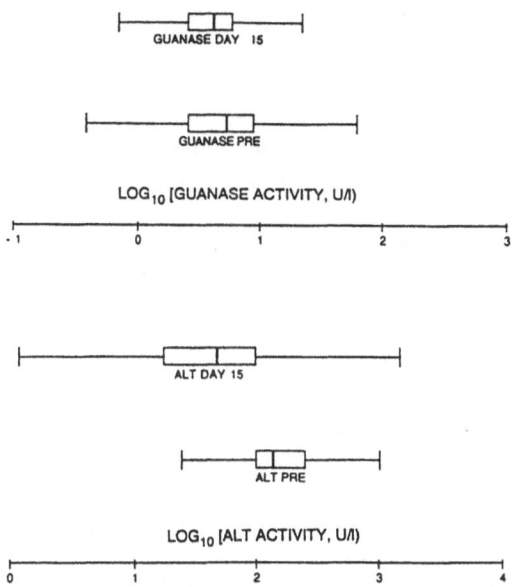

Figure 5. Effects of interferon treatment. "Box and whiskers" plots of results from a group of 12 patients treated with recombinant interferon. Similar results were obtained from a second, similar patient group. Paired serum samples taken before treatment and after 15-19 days of interferon treatment were assayed for guanase, alanine aminotransferase (ALT) and HCV RNA. The decrease in serum ALT creates the impression that liver damage (and presumably viraemia) had decreased significantly. The decrease in guanase activity is much less significant and may be a better indicator of viraemia, which at least half the patients was shown not to have resolved on the basis of PT-PCR. On a case by case basis, no clear correlations between any of the measured parameters emerged, although the non-quantitative nature of PCR results makes interpretation difficult.

DISCUSSION

Elevation of guanase activity in plasma of patients with liver damage was first reported more than thirty years ago by Passanati [1]. Numerous subsequent reports have established the clinical significance of serum guanase as an indicator of liver function and high serum guanase activity is now considered a specific indicator of hepatocellular damage [1-9]. A variety of methods have been used for its assay, including those based on changes in the UV absorbance, use of radiolabelled substrate, colorimetric estimation of ammonia liberated by guanine deamination as well as coupling to a variety of secondary reactions. Each method has characteristic advantages and disadvantages; few are suitable for use with automatic analyzers. Considering the substantial advances in HPLC technology made during the past decade, the apparent reluctance to exploit them for guanase assay is perhaps surprising, but may have been due to inherrent difficulties such as poor aqueous solubility of both substrate and product and their tendency to co-elute on normal RP-HPLC. The only direct HPLC-based method developed to date [9], although adequate for measuring the large increases in guanase activity associated with severe liver injury, is neither sufficiently sensitive to accurately quantify normal serum activity nor sufficiently rapid (run time = 12 min) to make its routine use for screening large numbers of samples feasible. The method outlined here (development and validation of which will be described in detail elsewhere [5]) overcomes these problems and appears to have a several potentially useful applications.

One application briefly explored here relates to non-A, non-B hepatitis. Most parenterally acquired non-A, non-B hepatitis is believed to be caused by HCV. The rate of carriage and the pool for infectivity of HCV are still unknown in most affected populations [12]. Epidemiological, biochemical and virological studies of these populations are important because infection by HCV is associated with high risk of subsequent development of serious liver damage, including cirrhosis and cancer [13]. At present, diagnosis of HCV infection is largely dependent on second-generation immunological tests, but anti-HCV antibodies may appear late or not at all [13,14]. Although recently developed assays can detect a variety of anti-HCV antibodies in chronic infections, direct tests for HCV antigens in serum are not yet available. Detection of HCV RNA by RT-PCR, which the only certain way to confirm viraemia, is slow, technically demanding, time consuming and expensive. Conventional biochemical indicators of liver damage, in particular serum aminotransferase activity have been used as surrogate markers of HCV activity but discrepancies between biological and virological tests are common, particularly in patients undergoing interferon therapy [13,15,16].

Two main lines of evidence seemed to indicate that assay of guanase activity might be a useful diagnostic and prognostic aid in HCV disease. First, guanase appears to be a more specific indicator of hepatocellular damage than alternative biochemical tests [4-9]. Second, the incidence of post-transfusional non-A, non-B hepatitis can be dramatically reduced by excluding potential blood donors with higher-than normal serum guanse activity [4]. Preliminary results presented here confirm expectations that assaying guanase is a useful indicator of active HCV infection. Normal serum guanase was approximately lognormally distributed (an important consideration in definining the upper limit). Subpopulations and gender differences were not detected. Although the probability plot (Figure 3B) suggests the presence of a possible a minor subpopulation with low guanase activity, analogous to the situation for liver xanthine oxidase [11], confirmation of its presence as well as the absence of significant gender differences must await analysis of larger population samples. Sera from individuals with active or recent HCV infection contained significantly higher-than-normal guanase activity, which decreased during treatment with interferon (see Figures 4 and 5). Because it is much simpler and economical than RT-PCR (which it cannot, however, be expected to replace), it is anticipated that the assay described here will be useful for screening individuals in high risk groups (e.g. injecting drug users, recipients of blood donations and organ transplants) besides being used to follow disease progression and response to therapy.

Specific guanase inhibitors, few of which are presently known, might also be identified using this assay. Such inhibitors would be extremely useful in investigation of all aspects of guanine nucleotide metabolism and may eventually help to define and perhaps counteract the presumed physiological consequences of leakage of hepatic guanase. These aspects of guanine metabolism have been generally neglected and deserve much more extensive investigation.

Acknowledgement

This work was partly supported by a grant from the National Health and Medical Research Council of Australia to S. Bowden, T. Shaw and S.A. Locarnini.

REFERENCES

1. G. Passanati, Enzyme tests identify specific diseases, *Med World Newslett.* 4:84-88 (1963).

2. T. Ando, T. Muraoka and H. Okuda, A sensitive spectrophotometric assay for guanase activity, *Analyt. Biochem.* 130:295-301 (1983).

3. G. Ellis, R.J. Spooner and D.M. Goldberg, Automated kinetic assays for routine determination of adenosine deaminase and guanase activities of human serum, *Clin. Chim. Acta* 47:75-87 (1973).

4. S. Ito, Y. Tsuji, N. Kitagawa, I. Akihiko, J. Syundo, Y. Tamura, S. Kishi and H. Mori, Clinical value of the guanase screening test in donor blood for prevention of posttransfusional non-A, non-B hepatitis, *Hepatology* 8:383-384 (1988).

5. T. Shaw, Facile assay for guanase in biological samples using ion-pair HPLC, (manuscript in preparation, May 1994).

6. G. Ellis, D.M. Goldberg, R.J. Spooner and A.M. Ward, Serum enzyme tests in diseases of the liver and biliary tree, *Amer. J. Clin. Pathol.* 70:248-258 (1978).

7. G.S. Crary, W.G. Yasmineh, D.C. Snover and W. Vine, Serum guanase: a biochemical indicator of rejection in liver transplant recipients, *Transplantation Proc.* 21:2315-2316 (1989).

8. K. Fukumoto and Y. Nishikawa, Epidemiological study of occupational exposure to hepatitis B virus and liver function tests, *Clin. Biochem.* 22:309-312 (1989).

9. S. Canepari, V. Carunchio, A.M. Girelli and A. Messina, New method for guanase activity measurement by high-performance liquid chromatography, *J. Chromatogr.* 616:25-30 (1993).

10. H. Okamoto, S. Okada, Y. Sugiyama, T. Tanaka *et al.*, Detection of hepatitis C virus RNA by a two-stage polymerase chain reaction with two pairs of primers deduced from the 5'-noncoding region, *Japan. J. Exp. Med.* 60:315-322 (1990).

11. R. Guerciolini, C. Szumlanski and R.M. Weinshilboum, Human liver xanthine oxidase: nature and extent of individual variation, *Clin. Pharmacol. Ther.* 50:663-672 (1991).

12. M.J. Alter, S.C. Hadler, F.N. Judson, A. Mares, *et al.*, Risk factors for acute non-A, non-B hepatitis in the United States and association with hepatitis C virus antibody, *JAMA* 264: 2231-2235 (1990).

13. L.M. Aledort, Consequences of chronic hepatitis C: a review article for the hematologist, *Amer. J. Hematol.* 44:29-37 (1993).

14. H. Schmilovitz-Weiss, M. Levy, N. Thompson and G. Dusheiko, Viral markers in the treatment of hepatitis B and C, *Gut* 34(Suppl 2):S26-S35 (1993).

15. J.Y. Lau, M. Mizokami, T. Ohno, D.A. Diamond, J. Kniffen and G.L. Davis, Discrepancy between biochemical and virological responses to interferon-alpha in chronic hepatitis C, *Lancet* 342:1208-1209 (1993).

16. S. Kakumu, K. Yoshika, K. Tanaka, Y. Higashi, S. Kurokawa, H. Hirofuji and A. Kusakabe, Long-term carriage of hepatitis C virus with normal aminotransferase after interferon treatment in patients with chronic hepatitis C, *J. Med. Virol.* 41:65-70.

GANCICLOVIR NUCLEOTIDE ANALYSIS IN HUMAN MYOCARDIAL TISSUE. USEFULNESS IN THE DIAGNOSIS OF CYTOMEGALOVIRUS MYOCARDITIS

Roselyne Boulieu [1,2], Nathalie Bleyzac [1,2], Olivier Bastien [3]

[1] Laboratoire de Pharmacie clinique
Institut des Sciences Pharmaceutiques et Biologiques
8 avenue Rockefeller - 69373 Lyon - France

[2] Service Pharmaceutique
Hôpital NeuroCardiologique
Lyon, France

[3] Département d'Anesthésie Réanimation
Hôpital NeuroCardiologique
Lyon, France

INTRODUCTION

Ganciclovir is a purine analog commonly used for treating cytomegalovirus (CMV) infection. Its antiviral activity is a result of intracellular phosphorylation of ganciclovir to the triphosphate form[1]. The initial phosphorylation step to ganciclovir monophosphate is catalysed by cellular enzymes induced by CMV.

As a consequence, in vitro studies have demonstrated that intracellular concentration of ganciclovir triphosphate in CMV-infected cells are higher than those in uninfected cells[2,3].

Data on cellular concentrations of ganciclovir nucleotides in vivo are not available. However, the lack of available standards of ganciclovir nucleotides represents a major drawback in the development of analytical methods. Therefore, we have developed a method for the determination of ganciclovir nucleotides in myocardial tissue to investigate ganciclovir nucleotide pool in heart biopsy from five patients with severe CMV infection.

METHODS

Patients

Five heart transplant patients who suffered from CMV infections and needed ganciclovir therapy were included in this study. Patients received intravenous infusion of ganciclovir at a dosage of 2.5 mg/kg/12h or 5.0 mg/kg/24h. A CMV myocarditis was diagnosed in two patients based on histological criteria and immunochemical stain for CMV in heart biopsy.

Purine and Pyrimidine Metabolism in Man VIII, Edited by
A. Sahota and M. Taylor, Plenum Press, New York, 1995

Sample collection and storage

Myocardial tissue samples were collected within the framework of graft rejection monitoring. In four patients, biopsy was collected four to five hours after the infusion and in one patient, tissue sample was carried out after death. Tissue samples were immediately frozen in liquid nitrogen to preserve nucleotide pool and stored at - 80 °C until analysis. Simulteanously, blood samples of 5 ml were collected in heparinized tubes and centrifuged without delay at low temperature. Plasma was stored at - 20 °C until analysis.

Chromatographic analysis

Ganciclovir was analysed using a high performance liquid chromatographic method described previously [4]. The chromatographic system consisted of a hypersil ODS 3 μm as a stationary phase and a 0.02 M potassium dihydrogenophosphate, pH 3.50 as a mobile phase. The flow rate was 1.5 ml/min and the detection was performed at 254 nm.

Sample treatment procedure

Myocardial tissue was added to 0.6 M perchloric acid (700 μl/10 mg, V/W) precooled in ice water and homogenized for 2 min 30 s at 6000 rpm in a stirpark homogenizer. After centrifugation at 2000 g for 15 min at 4 °C, the supernatants were adjusted to pH 7-8 with sodium hydroxide. 200 μl of the supernatant were removed for ganciclovir analysis. An other aliquot of 200 μl supernatant was incubated with 10 μl of alkaline phosphatase (type VII-NT, 10.000 glycine units/mg of protein) for 30 min at 37 °C. This procedure hydrolyses the mono, di and triphosphate nucleosides of ganciclovir to ganciclovir. Then, the supernatant for ganciclovir analysis and the supernatant incubated with alkaline phosphatase were evaporated at room temperature with a RC 1022 centrifugal evaporator. The residues were reconstituted with 30 μl of mobile phase and 20 μl was injected on to the column. Plasma samples were deproteinized with perchloric acid and analysed according to the previously published method[4].

RESULTS AND DISCUSSION

The sample treatment procedure for analysis of ganciclovir nucleotides is based on perchloric acid deproteinisation and enzymatic hydrolysis of nucleotides to ganciclovir.

Due to the lack of availability of ganciclovir nucleotides standards , the enzymatic conversion of nucleotide into nucleoside by alkaline phosphatase was evaluated using guanosine triphosphate (GTP) as nucleotide standard. In the conditions used, an average of 96 % of GTP was hydrolysed to guanosine.

Mean recovery of ganciclovir in myocardial tissue was 101 ± 2% (mean ± SD, n=5) and the minimum detectable amount was 2 pmol for ganciclovir that corresponds, using the sample treatment procedure described to 0.9 pmol/mg of tissue. The intraday and interday coefficients of variation (n = 5) were 1.0 % and 4.4 % respectively.

The concentrations of ganciclovir nucleotides in myocardial tissue are given in table 1. Ganciclovir nucleotides were recovered in three patients at concentrations ranging from 3.6 to 47.1 pmol/mg of tissue. The concentrations were higher in the two patients who demonstrated signs of CMV myocarditis, 42.6 and 47.1 pmol/mg of tissue respectively while ganciclovir nucleotides were not detectable or recovered at low levels (3.6 pmol/mg of tissue) in patients without signs of CMV myocarditis. Besides, no ganciclovir was found in tissue before enzymatic hydrolysis.

As shown in table 1, ganciclovir plasma concentrations were not related to myocardial ganciclovir nucleotide concentrations. So, ganciclovir distribution in myocardial tissue cannot be predicted using ganciclovir concentrations from plasma samples.

These results suggest that ganciclovir is preferentially phosphorylated in CMV infected cells than in uninfected as reported from in vitro studies. Moreover, this study also suggests that high concentrations of ganciclovir nucleotides in myocardial tissue could be considered as an indicator of the presence of CMV in the tissue.

Table 1. Concentration of ganciclovir in plasma and concentration of nucleotides of ganciclovir in myocardial tissue from five CMV-infected patients.

Patients	Dosage	Sampling time after infusion (hours)	Ganciclovir concentration in plasma (μmol/l)	Ganciclovir nucleotide concentration in myocardial tissue (pmol/mg of tissue)
1	2.5 mg/kg/12h	4	9.0	< 0.9
2	5.0 mg/kg/24h	4	21.6	3.6
3	5.0 mg/kg/24h	4	13.0	< 0.9
4[1]	5.0 mg/kg/24h	5	9.4	47.1
5[1]	5.0 mg/kg/24h	_[2]	-	42.6

[1] patients with CMV myocarditis

[2] post-mortem tissue samples

In conclusion, the sample treatment procedure described for the analysis of ganciclovir nucleotides in tissue samples, represents a suitable method for the investigation of intracellular ganciclovir nucleotide pool. Furthermore, the method described should be useful for the diagnosis of CMV myocarditis in patients with severe CMV disease.

REFERENCES

1. Y. Cheng, S.P. Grill, G.E. Dutschman, K. Nakayama, K.F. Bastow, Metabolism of 9 (1,3-dihydroxy-2-propoxymethyl) guanine, a new anti-herpes virus compound, in Herpes Simplex Virus infected cells, *J. Biol. Chem.* 258 : 12460 (1983).
2. D. F. Smee, R. Boehme, M. Chernow, B. P. Binko, T. R. Matthews, Intracellular metabolism and enzymatic phosphorylation of 9 (1,3-dihydroxy-2-propoxy methyl) guanine and acyclovir in Herpes Simplex Virus infected and non infected cells, *Biochem. Pharmacol.* 34 : 1049 (1985).
3. K. Biron, S.C. Stanat, J.B. Sorrell, J.A. Fyfe, P.M. Keller, C.U. Lambe, D.J. Nelson Metabolic activation of the nucleoside analog 9 (2-hydroxy-1(hydroxymethyl) ethoxy) methyl guanine in human diploïd fibroblasts infected with human cytomegalovirus, *Proc.Natl. Acad. Sci.* 82 : 2473 (1985).
4. R. Boulieu, N. Bleyzac, S. Ferry, Modified high performance liquid chromatographic method for the determination of ganciclovir in plasma from patients with severe renal impairment, *J. Chromatogr. Biomed. Appl.* 571 : 331 (1991).

TOXICITY OF ADENOSINE ANALOGUES AGAINST

HUMAN MALARIA (*Plasmodium falciparum*)

Annette M. Gero, Andrew M. Wood and David W. Coomber

School of Biochemistry and Molecular Genetics
University of New South Wales
Sydney, NSW 2052 Australia

INTRODUCTION

The transport and metabolism of purine nucleosides differ considerably between the human erythrocyte and the erythrocyte when infected with *Plasmodium falciparum*.[1-3]. In addition, malarial parasites appear to be unable to synthesize their purines *de novo* and thus rely on salvage pathways to obtain preformed purines required for growth and division . Therefore, deprivation of essential purines by the blockage of purine salvage or metabolic pathways by various purine nucleoside analogues can lead to inhibition of parasite growth and the potential development of new antimalarials.

In an attempt to exploit the differences between the host tissues and the intraerythrocytic parasite, analogues of purine nucleosides have been investigated for their potential antimalarial activity. In particular, deoxyadenosine analogues were considered to be of interest as potential antimalarial compounds because of their recognised antiviral and antitumour activities[4]. The structure-function relationships of these compounds against *P. falciparum* in culture were compared with the toxicities of many these compounds against human melanoma cells.

Further, analysis by HPLC of the metabolic pools from infected cells exposed to toxic nucleosides determined the site of action of the cytotoxic nucleosides on the purine salvage pathway of the parasite. Comparison of the toxicity of these compounds against *P. falciparum* with the toxicity towards human melanoma cell lines, indicated a differential toxicity, in that many of the compounds toxic towards *P. falciparum* were relatively non-toxic towards human melanoma cell lines and vice versa .Thus, the mechanism of toxicity of the deoxyadenosine and adenosine analogues, whose normal metabolism involves transport, metabolism and incorporation into nucleic acids appeared to vary significantly in *P. falciparum* compared to mammalian cells.

EXPERIMENTAL

Cell cultures and toxicity studies. *Plasmodium falciparum*. FCQ-27 isolate, was maintained in culture using the techniques described by Trager & Jensen[5.] The incorporation of [G-^{3}H]hypoxanthine into the nucleic acids of *P. falciparum* was used to assess the growth of the parasite *in vitro,* essentially as described by Desjardins ,*et al* (1979)[6]. Microculture plates (96 well)were prepared with each well containing 225 µl of a 2% haematocrit culture of asynchronous parasitized erythrocytes (1% parasitized cells), with varying concentrations of the drug to be studied (up to 100 μM). Each plate was incubated for 24 h in a gas mixture of 5% O_2, 5% CO_2 and 90% N_2 at 37°C, at which point 3.7 KBq of [G-^{3}H]hypoxanthine was added to each well and the incubation continued under the same conditions for a further 18-20 h. The infected cells in the control wells, without any drug, routinely reached a parasitaemia of 6-8% before harvesting. Cells were harvested and the nucleic acids isolated as described previously[7]. The IC_{37} value was calculated as the concentration at which [G-^{3}H]hypoxanthine incorporation had been reduced to 37 % of a parasitized control with no added drug. The results were confirmed by microscopic examination of a Giemsa stained blood smear obtained from each well.

Preparation of Trophozoites for HPLC: 50μl of packed *P.falciparum* trophozoites[7] were incubated at 37°C with 10 mM glucose and 100 μM of the toxic adenosine analogue for 3 h. After the completion of the incubation, the supernatants, prepared as described previously[7], were analysed by HPLC using an Altex Model 110 high performance liquid chromatograph incorporating a gradient programmer and a UV detector at 254 nm. The nucleotides were separated using a 30 min concave gradient from 10 mM to 500 mM $NH_4H_2PO_4$ (pH 4) at a flow rate of 2 ml min^{-1} on a Partisil 10-SAX column (Phenomenex, CA, USA).

RESULTS

The results of the toxicity screening *in vitro* of a large number of nucleoside derivatives tested in cultures of *P.falciparum* using [G-^{3}H]hypoxanthine incorporation as a measure of cell viability are summarised in Table 1. The chemical structures of the compounds tested fell into three main categories viz; (i) a modification of the purine ring to include substituents on positions 1 to 8 , (ii) a modification of the purine ring to include 7 deaza and 8 aza analogues and (iii) deoxy analogues of the above. Each compound was tested over a range of concentrations against *P.falciparum* to determine the IC_{37} value.

Eight compounds exhibited *in vitro* toxicity against *P.falciparum* with IC_{37} values over a range of 0.3 -50 μM. Any compound with an IC_{37} values over over 100μM was regarded as non toxic. Those compounds with the greatest toxicity were those with an aza group at the 8 position or a deaza alterations at the 7 position of the purine ring, compounds such as tubercidin (IC_{37} = 0.7μM), sangivamycin (IC_{37}= 0.3 μM) and 8-aza-2-amino-deoxyadenosine (IC_{37}= 11μM) . Removal of the aza group at the 8 position, producing the compound 2-amino-deoxyadenosine, reduced the toxicity by a factor of three (IC_{37}= 37μM). Other active compounds included 6-methylamino-deoxyadenosine and 6-methylamino-adenosine had IC_{37} values of 10 μM and 20 μM respectively.

Of particular interest was that the majority of deoxyadenosine analogues of those active compounds with substitututions at the 2 position were non-toxic at concentrations less than 100 μM. The toxicity of the adenosine analogues substituted with halogens at the 2 position (such as 2-chloro-adenosine and 2-bromo-adenosine) disappeared in the corresponding deoxyadenosine analogues. Those compounds with substitions at the 8 position and modifications in their linkage with the ribose moiety of adenosine or deoxyadenosine also proved non-toxic towards *P.falciparum*. (Table 1).

The majority of compounds were tested for toxicity against human melanoma lines and their IC_{37} values were compared with the toxicities in *P.falciparum* . The results indicated an obvious difference in the susceptibility of the malarial parasite and the melanoma cell lines to the toxicity of the various nucleoside analogues. This may reflect differences in either the metabolism and/or the transport of nucleosides between the two different species of cells.

In order to analyse the action of the toxic nucleoside analogs within the malaria infected cells, toxic nucleosides were incubated with the infected cells and HPLC analysis of the nucleotide pools of malaria infected erythrocytes was carried out.

The nucleotide profile of the erythrocyte infected with *P.falciparum* (Fig 1i) reflected the presence of the purine salvage and the pyrimidine *de novo* enzyme pathways of the intracellular parasite. There was a significant increase in several peaks appearing within the first 3 min of the elution profile which corresponded to the free bases and nucleosides found in the parasitized cell profile.

Table 1. IC37 values for adenosine analogues against *P.falciparum* and melanoma cell lines.

Compound	IC$_{37}$(μM) [1]	
	P. falciparum	Melanoma [2]
Adenosine	>100	450
2' deoxyadenosine	>100	250
Tubercidin (7-deazaadenosine)	0.7	0.02[3]
Sangivamycin (7-deaza amidoadenosine)	0.3	0.3 [4]
8-aza-2-amino-deoxyadenosine	11	0.30
8-aza-7-deaza-deoxyadenosine	> 100	>20
6-methylamino-adenosine	20	-
6-methylamino-deoxyadenosine	10	>20
6-dimethylamino-deoxyadenosine	> 100	>20
2-chloro-adenosine	11	-
2-chloro-deoxyadenosine	> 100	0.022
2-bromo-adenosine	50	-
2-bromo-deoxyadenosine	>100	0.18
2-fluoro-deoxyadenosine	> 100	0.11
2-CF$_3$-deoxyadenosine	> 100	>20
2-amino-deoxyadenosine	37	>20
2-hydroxy-deoxyadenosine	> 100	>20
8-bromo-adenosine	> 100	-
8-bromo-deoxyadenosine	> 100	>20
8-bromo-3-(2'-deoxyribosyl) adenine	> 100	>20
8-CF$_3$-3- (2'-deoxyribosyl) adenine	> 100	>20

[1] Average IC$_{37}$ values for each test compound against the FCQ-27 isolate of *P. falciparum*. Values are the mean of triplicate wells of each dose used over a range of five or more concentrations.

The IC$_{37}$ value is the concentration at which the compound reduces cell survival to 37%.

[2] Determined by Parsons *et al*. (4) and Parsons & Hayward (8) for the MM96 human melanoma.

[3] IC$_{50}$ value determined by Smith *et al*. (9) in KB strain of human carcinoma cells.

[4] IC$_{50}$ value determined by Glazer and Hartman (10) in Sarcoma 180 cells.

When the infected erythrocytes were incubated with 2-chloro-adenosine (Fig 1ii) there was only a small change in the elution profile when compared to the infected control suggesting that this compound may not exert its toxic effect by interfering with the parasite purine nucleotide metabolism but by some another mechanism. There was however considerable difference between the nucleotide profiles for the control infected cells and those incubated with tubercidin (Fig iv). The elution profile of cells incubated with tubercidin indicated there was a decrease in the levels of the nucleotide sugars and IMP whilst the levels of ATP increased significantly. There was also the appearance of two significant peaks, C and D, which eluted before the adenosine nucleotides, ADP and ATP which represented the di- and tri- phosphate derivatives of tubercidin. However, for sangivamycin (Fig 1 iii) only one new peak (E) was observed,whilst the concentration of ATP was significantly lowered.

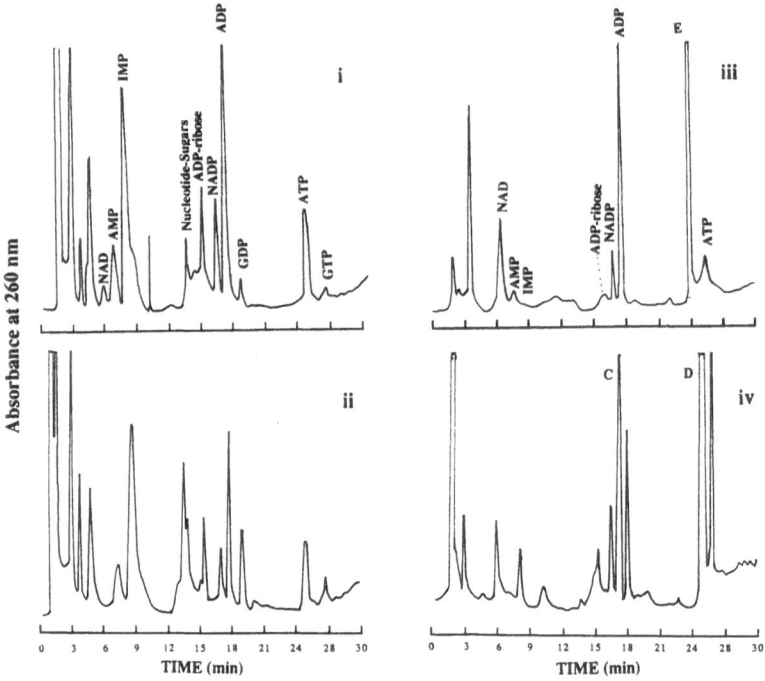

Fig. 1. Effect of various adenosine analogues on the nucleotide profile of trophozoite infected human erythrocytes. (i) control ; (ii) 2-chloro-adenosine; (iii) sangivamycin and (iv) tubercidin . 50 μl of packed trophozoite infected erythrocytes were incubated with each of the drugs (100 μM) for 3 h and the nucleotides isolated and separated by HPLC. Peaks are labelled once and may be identified in other traces by aligning the elution times.

DISCUSSION

The major objective of this work has been to determine whether the parasite purine salvage enzymes may serve as targets for potential chemotherapeutic agents. This has been achieved by assessing the loss of viability of *P. falciparum* grown in the presence of nucleoside analogues and the analysis of the metabolic pools of the parasites under these conditions. The present results demonstrated that eight of the nucleoside analogues tested were found to be highly toxic to *P. falciparum in vitro* with IC_{37}, values in the micromolar range (Table 1). Analysis by HPLC of the purine pools from *P. falciparum* infected cells incubated with tubercidin suggested that toxic activity may be due to the formation of the di- and tri- nucleotides of tubercidin via the parasite adenosine kinase. However, the highly toxic halogeno-deoxy adenosine analogs , which were active against melanoma cells were inactive against *P. falciparum* infected cells due to the possible lack of deoxycytidine kinase, which is known to phosphorylate these compounds in human cells.

The results demonstrate clearly the significant differences in toxicity that exist between different organisms. Of particular significance is the fact that those nucleosides selected for chemotherapy against other diseases such as cancer have been found to be ineffective against *P. falciparum* and vice versa. These observations present the opportunity to reevaluate many nucleoside analogues which have been unsuccessful as anticancer or antiviral agents for their potential activity against *P. falciparum*.

ACKNOWLEDGEMENTS

This work was supported by the National Health and Medical Research Council of Australia and the UNDP/World Bank/WHO Special Program for Research and Training in Tropical Diseases.

REFERENCES

1. A.M. Gero, E.M.A. Bugledich, A.R.P. Paterson, and G. Jamieson, Stage specific alterations of nucleoside permeability and nitrobenzylthioinosine insensitivity in *Plasmodium falciparum* -infected erythrocytes. *Mol.Biochem. Parasitol.* 27:159 (1988).
2. A.M. Gero and J. M. Upston, Altered membrane permeability: A new approach to malaria chemotherapy. *Parasitol. Today,* 8:283 (1992).
3. A.M. Gero and W.J. O'Sullivan. Purines and pyrimidines in malarial parasites, *Blood Cells*. 16: 467 (1990).
4. P.G.Parsons, E.P.W.Bowman, and R.L.Blakely, Selective toxicity of deoxyadenosine analogues in human melanoma cell lines. *Biochem Pharmacol*. 35:4025 (1986).
5. W. Trager and J.B. Jensen. Human malaria parasites in continuous culture.*Science* 193: 673 (1976).
6. R.E. Desjardins, C.J. Canfield, D. Haunes, and J.D. Chulay. Quantitative assessment of anti-malarial activity by a semiautomated microdilution technique *Antimicro. Ag. Chemo.*16:710 (1979).
7. A.M. Gero, H.V. Scott, W.J. O'Sullivan, and R.I. Christopherson, Antimalarial action of nitrobenzylthioinosine in combination with purine nucleoside antimetabolites. *Mol. Biochem Parasitol*. 34:87 (1989).
8. P.G. Parsons and I.P. Hayward. Human melanoma cells sensitive to deoxyadenosine and deoxyinosine for the MM96 human melanoma. *Biochem .Pharmacol*. 35:655 (1986).
9. C.G. Smith, W.L. Lummis and J.E. Grady. An improved tissue culture assay. II. Cytotoxic studies with antibiotics, chemicals, and solvents. *Cancer Research* 19: 847 (1959).
10. R.I. Glazer and K.D. Hartman. Cytokinetic and biochemical effects of sangivamycin in human colon carcinoma cells in culture. *Molec. Pharmacol.* 20:657 (1981).

ALTERED PURINE NUCLEOSIDE TRANSPORT

AS A TARGET FOR MALARIA CHEMOTHERAPY

Annette M. Gero and Joanne M. Upston

School of Biochemistry and Molecular Genetics
University of New South Wales
Sydney, N.S.W., 2052, Australia

INTRODUCTION

In order to meet its metabolic requirements during growth and reproduction, the intraerythrocytic human malarial parasite, *Plasmodium falciparum* has been shown to induce changes in the transport properties of the host erythrocyte membrane.

We have determined that in order to obtain purines from the extracellular milieu, the intraerythrocytic malaria parasite induces specific nucleoside transport sites into the membrane of the infected host erythrocyte of both normal and drug resistant strains of malaria, mainly at the trophozoite stage of the malarial parasite development[1-4]. Purine nucleosides cannot be synthesised *de novo* by either the erythrocyte or the malarial parasite and therefore the parasite relies solely on purine salage for its survival. The new parasite-induced nucleoside transport sites which comprises about 40% of the total nucleoside transport in the infected cells were distinguished from those of the normal host erythrocytic nucleoside transporter by their insensitivity to the classical nucleoside transport inhibitors, nitrobenzylthioinosine (NBMPR) or dilazep. The endogenous nucleoside transporter, which is potently inhibited by NBMPR, remained functional in the infected cells.

The induced nucleoside transporters in parasite infected erythrocytes presented the possibility of the utilization of cytotoxic nucleosides against *P. falciparum* infection in conjunction with a nucleoside transport inhibitor to protect the host tissue. The feasibility of this regime has been successfully demonstrated in mice infected with the rodent malaria *P.yoelii* or *P.berghei* by the co-administration of a lethal concentration of a cytotoxic nucleoside, such as tubercidin or sangivamycin in combination with NBMPR. This drug combination administered over four consecutive days was non toxic to non infected control mice and in the infected mice , the treatment significantly decreased the parasitic infection and doubled the survival time of the treated animals[5,6].

However, a more successful antimalarial compound would be one which affected the parasite but with minimal effect on the host. Thus we have investigated the possibility of the development of nucleosides which do not normally penetrate the host cell by the endogenous nucleoside transporter, but which meet the selectivity properties of the new nucleoside pathway in the infected erythrocyte and therefore can be targeted specifically only into the parasitised erythrocyte.

Purine and Pyrimidine Metabolism in Man VIII, Edited by
A. Sahota and M. Taylor, Plenum Press, New York, 1995

METHODS

P. falciparum, FCQ-27 isolate, was maintained in culture using the standard techniques described by Trager and Jensen[7]. Infected erythrocytes in culture were synchronised using sterile D-sorbitol[8]. Mature trophozoite infected cells were concentrated using Percoll (Pharmacia, Sweden) gradients as described previously[2]. The resulting cells (>80% parasitised with mature trophozoites) were washed 3 times to remove Percoll in phosphate buffered saline containing glucose (PBSg) (1.8 mM KH_2PO_4, 5 mM K_2HPO_4, 0.9% NaCl w/v, 5 mM D-glucose, pH 7.4) and resuspended to a concentration of 2 x 10^8 cells/ml in PBSg. *Crithidia luciliae* was grown aerobically in RPMI 1640 medium containing 10% (v/v) foetal calf serum and 1% (w/v) L-glutamine at pH 7 in a 26°C incubator[9]. *Trichomonas vaginalis*, (WAA 38 strain) was cultured in modified TYM medium[10]. *Giardia intestinalis* trophozoites (Portland 1 stock ATCC 30888;) were grown under microaerophilic conditions at 37°C using TYI-S-33 medium with the modification that pooled human serum replaced foetal calf serum[11]. Parasitic protozoa at mid logarithmic phase were harvested and resuspended in PBSg. Cell numbers were assessed microscopically using a haemocytometer with Improved Neubauer ruling.

Nucleoside transport measurements. Nucleoside transport over short time intervals at 22°C was measured by an established procedure as reported previously [14]. Transport assays were initiated when the cells (2 x10^7 in 100 μl) were added to Eppendorf centrifuge tubes containing 150 μl of an oil mixture with 100 μl of radiolabelled nucleoside permeant layered on top of the oil.

Transport assays at 22°C, were terminated between 2 and 10 s by fast centrifugation (16 000 x g for 15 s) which pelleted cells under the oil layer. Under the experimental conditions of the transport assay, using 1 μM L- or D-adenosine as the permeant, less than 5% of the nucleoside was metabolised within the cell, as determined by PEI chromatography.The cell pellet was processed as described previously [14] and counted in a Packard Model 1900TR liquid scintillation spectrometer.

Transport inhibitor studies. All compounds (obtained from NCI, Washington DC) tested as potential inhibitors were dissolved in PBSg and preincubated with the cells for 15 min at 37°C. The cells were returned to room temperature before assaying for transport. A saturated solution of NBMPR (Sigma Chemical Co. USA) in PBSg was obtained by stirring overnight at room temperature. The IC_{50} values for inhibition experiments were defined as the concentration of reagent required to inhibit transport by 50%.

RESULTS AND DISCUSSION

An initial comparison of biochemical characteristics of the transport of nucleosides in normal and infected cells revealed that there were some important structural components of a nucleoside which are necessary for their ability to be transported through the parasite induced nucleoside transporter.These consisted of (i) alteration of the 6 position of the purine ring of a nucleoside to a bulky component (compounds such as in (N6-nitrobenzyl -thioinosine(NBMPR); N6-(2-methylfuranyl)adenosine, (NCI 20); N6-(benzyl)adenosine, (NCI 108); or N6-(2-chloro-prop-1-ene)adenosine, (NCI 189)). These compounds , which caused inhibition of the endogenous transporter were not inhibitors of the infected cell transporter (Table 1) and in addition were transported into the infected cells.

(ii) Secondly it appears that the conformation of the sugar moiety of the nucleoside was important for selective transport of the compound into the parasite infected cell as demonstrated by the transport of the nonphysiological adenosine isomer, L-adenosine (9-β-L-ribofuranosyl), a compound which could not enter normal uninfected erythrocytes.

Table 1. Comparison of D- and L-adenosine influx in different cell types

Compound 1μM	% Inhibition of 1μM Adenosine Transport	
	Erythrocytes	Malaria infected cells
NBMPR	95	27 [1]
NCl 20	92	31
NCl 108	81	32 [1]
NCl 189	92	28
Dilazep	92	28

[1] Shown to be toxic and metabolised by the infected cells

The transport of L-adenosine into *P. falciparum* infected erythrocytes is represented in Fig. 1. *P. falciparum* infected cells containing 85% trophozoites demonstrated significant transport of L-adenosine at 1μM over 10 s. The initial rate of transport of L-adenosine was 0.08 ± 0.006 pmol L-adenosine /μl cell water per s. In addition NBMPR (1μM) at concentrations sufficient to fully inhibit 1μM D-adenosine transport in normal uninfected erythrocytes, had no effect on L-adenosine in *P. falciparum* infected erythrocytes. In comparison, L-adenosine was totally impermeable to uninfected human erythrocytes (Fig. 1). These results indicate that an unusual nucleoside transport function existed in *P. falciparum* infected erythrocytes which allowed L-adenosine permeation similar to the transport of this compound into *P.yoelii* mouse infected cells[12].

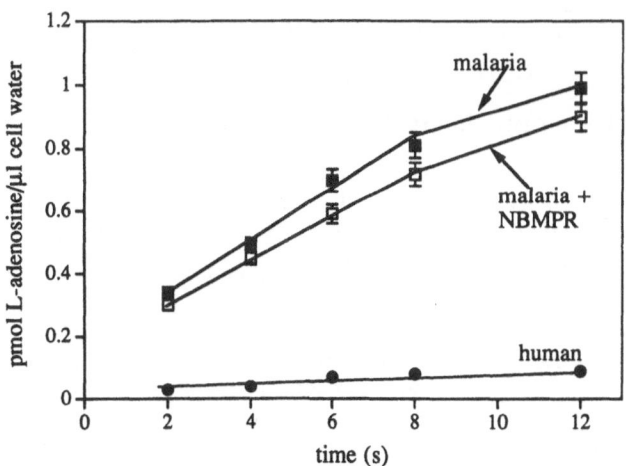

Fig. 1. Transport of L-adenosine in *P. falciparum* infected human erythrocytes. Transport of 1 μM [³H] L-adenosine measured over 12 s in *P. falciparum* trophozoite infected erythrocytes (85% parasitised) was compared to the transport of 1 μM [³H] L-adenosine in normal human erythrocytes (●) and 1 μM [³H] D-adenosine in the presence of 1 μM NBMPR in infected cells (□). Influx was measured by exposure of the cells to [³H] L-adenosine followed by centrifugation of the cells through an inert oil layer, which was calculated to take an additional 3 s. The results represent the mean of triplicate experiments ± S.E.

To determine whether L-adenosine transport was unique to malaria, influx was also examined in a variety of other cell types. In contrast to *P. falciparum* infected cells , other parasitic protozoa, such as *Crithidia luciliae, Trichomonas vaginalis* and *Giardia intestinalis*, and the mammalian HeLa cell line, failed to demonstrate any transport of 1 μM L-adenosine (Table 2). In all instances the transport rate of L-adenosine was negligible with a rate less than 0.003 pmol L-adenosine/μl cell water per s. However, all the above cells transported D-adenosine (Table 2). Thus, the L-adenosine transport function represents an unusual characteristic of Plasmodium infection ,which is characteristically different to the normal erythrocytes and other cell types .

Table 2. Comparison of D- and L-adenosine influx in different cell types

Cell Type	Initial rate of transport (pmol/μl cell water per s.) x 10^2	
	D-adenosine[1]	L-adenosine[1]
P. falciparum + 1 μM NBMPR	17 ± 0.9	8 ± 0.6
Human erythrocyte	20 ± 1	0.3 ± 0.03
C. luciliae	69 ± 3	0.3 ± 0.02
G. intestinalis	16 ± 0.8	<0.1
T. vaginalis	11 ± 0.6	<0.1
HeLa 229	3 ± 0.2	<0.1

[1] Influx rates of [^{3}H]L- and D-adenosine (1 μM) in different cell types were determined by calculation of initial transport over 3 - 5 s. The values represent the mean of at least 3 identical experiments ± S. E.

Properties of the parasite-induced transporter

The potential for the development of cytotoxic nucleosides which would selectively enter only malaria infected cells led us to a further characterisation of the L-adenosine transport system. In addition, compounds which block this pathway in *P. falciparum* infected cells may also deprive the parasite of essential nutrients needed for replication.

L-Adenosine transport showed nonsaturability in *P. falciparum* trophozoite infected cells up to concentrations of approximately 1 mM with a rate of 0.14 ± 0.01 pmol/μl cell water per s. In comparison, adenosine transport in normal mammalian cells is saturable. Thus the effects of various reagents on the rate of L-adenosine transport in *P. falciparum* infected erythrocytes was examined to determine whether influx occurred via a channel-like process, which would exhibit inhibition, or via a simple diffusion mechanism. As shown in Fig 2. furosemide, phloridzin and piperine were excellent inhibitors of L-adenosine transport in *P. falciparum* infected cells with IC_{50} values of 1, 3 and 13 μM respectively . These reagents were required in at least 10-1000 fold greater concentrations to block the transport of D-adenosine in normal erythrocytes, which had IC_{50} values of 1 mM, 500 μM and 100 μM respectively (data not shown). By contrast, quinine appeared to inhibit L-adenosine transport in *P. falciparum* -infected cells (IC_{50} = 1 mM,) and D-adenosine transport in normal human erythrocytes equally (IC_{50} = 1 mM). The results suggested that L adenosine transport influx occurred in *P. falciparum* - infected erythrocytes via a channel-like process.

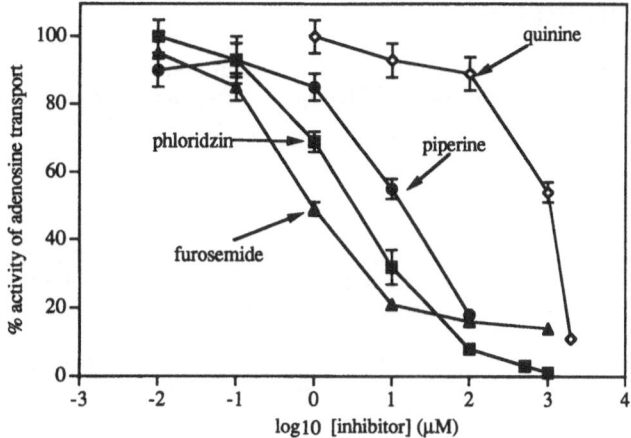

Fig. 2. Inhibition of L-adenosine transport. Furosemide (▲), phloridzin (■), piperine (●) and quinine (◇) were tested for inhibitory action against 1 μM [^{3}H] L-adenosine influx (5 s.) in *P. falciparum* (80% parasitised) over a wide concentration range (10 nM - 1 mM). The IC$_{50}$ concentrations, representing that concentration causing 50% inhibition of [^{3}H] L-adenosine influx were found to be; furosemide, 1 μM, phloridzin, 3 μM, piperine, 13 μM and quinine, 1000 μM. The experimental values represent the average of 3 determinations ± S.E.

The transport of L-adenosine into *P. falciparum* -infected cells was not inhibited by the sulphydryl reagents, p-chloromercuriphenylsulphonic acid (2mM), iodoacetamide (2mM) or N-ethylmaleimide (2mM) , suggesting the absence of thiol group/s involvement in L-adenosine influx. Incubation of infected cells with KCN (2mM) to deplete energy reserves also had no effect on L-adenosine transport.

CONCLUSION

We have demonstrated that the the substrate specificity of the parasite induced purine transporter in *P. falciparum*-infected erythrocytes varies significantly from that of normal mammalian cells. Studies with a variety of nucleoside analogs demonstrated that in *P. falciparum*-infected erythrocytes purine analogues were transported, some of which (those which contained a large N6 component on the purine ring or nucleosides with an altered ribose moiety) were not accepted by the mammalian nucleoside transporter and were thus impermeable to normal mammalian cells indicating that of the parasite transport mechanism was significantly different to the normal human nucleoside transporter. Further characterisation of the parasite purine transporter in infected cells indicated that it had the properties of a channel as it was blocked by a variety of channel inhibitors such as furosemide, piperine and phloridzin.

The altered substrate specificity of the nucleoside transport mechanisms found in malaria infected erythrocytes appears to be unique to malaria infected cells and hence allow the design of potential new antimalarials which are specific for infected cells and are nonpermeable to normal mammalian cells.

ACKNOWLEDGEMENTS

This work was supported by the National Health and Medical Research Council of Australia and the UNDP/World Bank/WHO Special Program for Research and Training in Tropical Diseases.

REFERENCES

1. A.M. Gero, E.M.A. Bugledich, A.R.P. Paterson, and G. Jamieson, Stage specific alterations of nucleoside permeability and nitrobenzylthioinosine insensitivity in *Plasmodium falciparum* -infected erythrocytes. *Mol.Biochem. Parasitol,* 27:159 (1988).
2. A.M. Gero, H.V. Scott, W.J. O'Sullivan, and R.I. Christopherson, Antimalarial action of nitrobenzylthioinosine in combination with purine nucleoside antimetabolites. *Mol. Biochem Parasitol .* 34:87 (1989).
3. A.M. Gero and J. M. Upston, Altered membrane permeability: A new approach to Malaria chemotherapy. *Parasitol. Today,* 8:283 (1992).
4. A.M. Gero and W.J. O'Sullivan. Purines and pyrimidines in malarial parasites, *Blood Cells .* 16: 467 (1990).
5. W.P. Gati, A.F. Stoyke, A.M. Gero, and A.R.P. Paterson, NBMPR insensitive nucleoside permeation in mouse erythrocytes infected with *Plasmodium yoelii, Biochim. Biophys. Res. Comm.* 145:1134 (1987).
6. A.M. Gero and A.M. Wood, New nucleoside transport pathways induced in the host erythrocyte membrane of malaria and babesia infected cells. In: Purine and Pyrimidine Metabolism in Man.VII, Part A. (Eds. R.A. Harkness G. B. Ellion & N. Zollner), Plenum Press, New York, 169 (1992).
7. W. Trager and J.B. Jensen. Human malaria parasites in continuous culture.*Science* 193: 673 (1976).
8. C. Lambros and J.P. Vanderberg, Synchronisation of *Plasmodium falciparum* erythrocyte stages in culture. *J. Parasitol.* 65:418 (1979).
9. S.T. Hall, G.A. Odgers, and A.M. Gero, Nucleoside transport in *Crithidia luciliae. Int. J. Parasitol.* 23:1039 (1993).
10. L.S. Diamond, The establishment of various trichomonads of animals and man in axenic culture. *J. Parasitol.* 43: 488 (1957).
11. G.F. Vitti, W.J. O'Sullivan, and A.M. Gero, The biosynthesis of uridine 5-monophosphate in *Giardia lamblia, Int. J. Parasitol.*17:805 (1987).
12. W.P. Gati, A.N. Lin, T.I. Wang, J.D. Young, and A.R.P. Paterson, Parasite-induced processes for adenosine permeation in mouse erythrocytes infected with the malarial parasite *Plasmodium yoelii. Biochem. J.* 272: 277 (1990).

QUANTITATIVE DETERMINATION OF *TRYPANOSOMA CRUZI* GROWTH INSIDE HOST CELLS IN VITRO AND EFFECT OF ALLOPURINOL

Takashi Aoki, Junko Nakajima-Shimada, and Yumiko Hirota

Department of Parasitology, Juntendo University School of Medicine, Hongo, Bunkyo-ku, Tokyo 113, Japan

INTRODUCTION

Trypanosoma cruzi, the parasitic protozoan flagellate that causes Chagas' disease against which a prominent chemotherapy is urgently needed, occurs as two different developmental stages in mammalian hosts (Brener, 1973; de Souza, 1984). The nondividing and infective trypomastigote form, which possesses a flagellum for locomotion, circulates in the bloodstream and the amastigote form, which has no free flagellum, proliferates in the cytoplasm of host cells, resulting in the destruction of infected cells. To identify the basis of a rational chemotherapy, it is of importance to analyze the biological and biochemical differences between these parasitic forms and host cells, since little is known about the molecular and cellular mechanisms involved in their morphological changes and the interactions that exist between the parasites and host cells. For this purpose, we have constructed an in vitro culture system of host cells infected with *T. cruzi* for the quantitative determination of time course of parasite growth. This system has enabled us to test allopurinol and other agents for their efficacies in inhibiting the parasite growth inside the host cells.

METHODS

Exponentially growing HeLa cells, 5×10^3 in 1-ml suspension, were inoculated into 24-well plates, followed by incubation for 2 days at 37ºC, and then infected with Tulahuene strain of *T. cruzi* trypomastigotes. On day 2, 3, 4, 6, and 8 after infection, the medium was withdrawn to measure the number of trypomastigotes. The host HeLa cells were fixed and stained with Diff-Quik (Kokusai Seiyaku Co., Ltd., Tokyo) for the determination of the percentage of infected cells and the mean number of amastigotes per infected cell. Details of the materials (Kaneda et al., 1986; Taliaferro and Pizzi, 1955) and methods will be described elsewhere.

RESULTS AND DISCUSSION

Figure 1A shows that the rate of infection by *T. cruzi* of HeLa cells increased constantly in an inoculum-size- and time-dependent manner and, on day 8 after infection, reached 84 and 53% in the cases of the inoculation with 1×10^5 and 1×10^4 trypomastigotes, respectively. Infection with 1×10^6 trypomastigotes caused much damage to the host cells by day 4, and this inoculum was thus thought to be inadequate for the time-course observation of *T. cruzi* infection to HeLa cells under the conditions used.

Figure 1B shows the time course of changes in the average number of amastigotes per infected HeLa cell. The number of amastigotes reached a peak of about 20 in an infected host cell on day 6, after which a decrease was observed. These results suggest that the amastigotes proliferate rapidly by binary fission in the host cells until day 6, and that the parasites then transform into trypomastigotes, resulting in the destruction of host cells and the decrease of amastigotes in number per infected HeLa cell. Although inoculum sizes of trypomastigotes differing by 10-fold were applied, only about 1.2-fold difference in the maximum number of amastigotes was obtained on day 6 (Fig. 1B).

When 1×10^5 trypomastigotes were inoculated, a large number of parasites still remained in the culture medium on days 2 and 3 and were transformed into extracellular amastigotes spontaneously (date not shown). In a separate experiment, we confirmed that these amastigotes did not infect HeLa cells and were not transformed into trypomastigotes. On the other hand, the

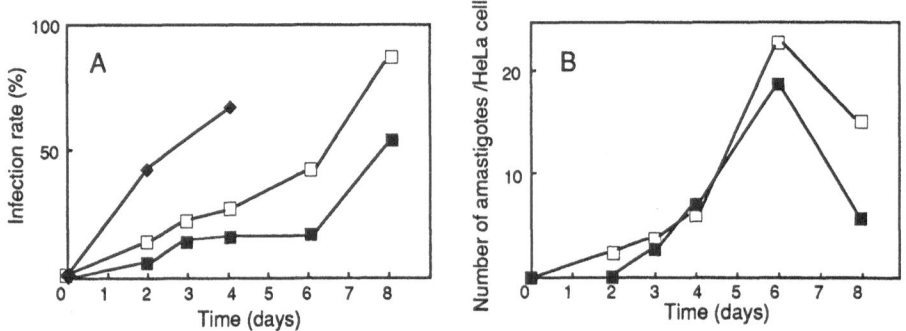

Figure 1. **(A)** Changes in the rate of infection of HeLa cells. HeLa cells (5×10^3) were precultured for 2 days, and on day 0, different numbers of *T. cruzi* trypomastigotes were inoculated.   , 1×10^4 parasites;   , 1×10^5 parasites;   , 1×10^6 parasites. A typical result of 3 separate experiments is presented.   **(B)** Changes in the number of *T. cruzi* amastigotes per infected HeLa cell. Trypomastigotes (, 1×10^4 and   , 1×10^5) were inoculated as in **(A)**. A typical result of 3 separate experiments is presented.

infection with 1×10^4 trypomastigotes did not yield the parasites remaining in the medium on days 2, 3, and 4. On day 6, trypomastigotes appeared again in the culture medium and , on day 8, freshly appearing 1.7×10^5 trypomastigotes (17 times more than the inoculum) were recovered.

These results are consistent with the following view. Within one day after the infection, trypomastigotes invaded HeLa cells and changed morphologically to amastigotes which proliferated up to day 6 after infection. Meanwhile, the alteration of morphology from amastigotes to trypomastigotes took place, the latter disrupting the host cells and again appearing in the culture medium. Consequently, we have succeeded in determining quantitatively the time course of growth of *T. cruzi* amastigotes and trypomastigotes inside mammalian cells.

Table 1. Effect of allopurinol and various compounds on the rate of infection of HeLa cells and on the average number of amastigotes per infected HeLa cell on day 4 after infection by *T. cruzi* trypomastigotes

Compound added	Concentration (µM)	Infection rate (% of control)	No. of amast. per HeLa cell (% of control)	/ED$_{50}$ (µM)
None (control)	0	100	100	
Allopurinol	1	93.9 ± 7.2[1]	87.7 ± 4.1[1]	
	10	62.3 ± 4.4	29.3 ± 4.5	/ 3
	100	60.2 ± 7.7	9.9 ± 0.6	
3'-Deoxyinosine	0.5	89.8 ± 4.5	94.2 ± 6.6	
	5	84.9 ± 7.0	74.1 ± 9.6	/ 10
	50	53.8 ± 8.4	23.4 ± 4.6	
3'-Deoxyadenosine	0.5	82.4 ± 4.9	88.4 ± 6.6	
	5	59.2 ± 6.7	46.8 ± 12.8	/ 3
	50	45.6 ± 5.1	11.5 ± 4.6	
3'-Deoxyguanosine	0.5	93.0 ± 7.6	98.8 ± 11.1	
	5	87.9 ± 4.2	87.9 ± 9.0	/ 50
	50	74.7 ± 5.7	39.6 ± 5.9	
Pentostam[2]	10	91.8	73.9	
Pentamidine[2,3]	0.5	59.0	39.8	
Acivicin[3]	0.5	69.3	54.4	
Pyronaridine[3]	0.5	100.5	89.4	

[1]Mean ± S.D. of 5 separate determinations.

[2]Existing drugs for trypanosomatid parasites.

[3]These compounds at 0.5 µM are harmful to the host HeLa cells.

As an application of this culture system, we have tested the efficacy of allopurinol, added at the same time of *T. cruzi* infection, on the parasite growth in HeLa cells (Table 1). Addition of 1, 10, and 100 μM allopurinol lowered the rate of infection of HeLa cells, and also markedly decreased the number of amastigotes per infected host cell to 88, 29, and 10% of the control, respectively, on day 4, yielding an ED_{50} value of 3 μM. High concentration of allopurinol (100 μM) completely inhibited the propagation of trypomastigotes transformed from amastigotes in HeLa cells; this may indicate that the treatment by 100 μM allopurinol gave a severe damage to the apparently surviving amastigotes (10% of the control). Since allopurinol did not affect the infectivity of trypomastigotes (data not shown), the primary target of this purine analog is likely to be a purine salvage capacity in the amastigote stage of the parasite. This compound afforded no inhibitory effect on the normal HeLa cells. 3'-Deoxyadenosine was also highly effective, with a ED_{50} value of about 3 μM (Table 1). These results are consistent with the fact that trypanosomes are incapable of synthesizing purines de novo and relay solely on salvaging preformed purines. The results also indicate that our in vitro culture system is, as a primary screening method, applicable to the search for potential drugs against Chagas' disease and can be utilized to collect substantial amounts of trypomastigotes and amastigotes for molecular and biochemical analyses.

ACKNOWLEDGMENTS

Pyronaridine was kindly provided from Dr. Chen Chang, Institute of Parasitic Diseases, Shanghai, China. This work received a support from the Ministry of Education, Science and Culture of Japan, and from the Japan Private School Promotion Foundation.

REFERENCES

Brener, Z., 1973, Biology of *Trypanosoma cruzi.*, *Annu. Rev. Microbiol.* 27: 347-382.

de Souza, W., 1984, Cell Biology of *Trypanosoma cruzi. Int. Rev. Cytol.* 86: 197-283.

Kaneda, Y., Nagakura, K., and Goutsu, T., 1986, Lipid composition of three morphological stages of *Trypanosoma cruzi. Comp. Biochem. Physiol.* 83B: 533-536.

Taliaferro, W.H., and Pizzi, T., 1955, Connective tissue reaction in normal and immunized mice to a reticulotropic strain of *Trypanosoma cruzi. J. Infect. Dis.* 96: 199-226.

POLYAMINE CONTENT OF *ACANTHAMOEBA POLYPHAGA* AT DIFFERENT STAGES OF DEVELOPMENT AND THE EFFECT OF PENTAMIDINE

Suad A. Asiri[1], Patrick O.J. Ogbunude[1]*, and David C. Warhurst[2]

[1]MBC-03, Department of Biological and Medical Research, King Faisal Specialist Hospital and Research Centre, P.O. Box 3354, Riyadh 11211, Kingdom of Saudi Arabia

[2]Department of Medical Parasitology, London School of Hygiene and Tropical Medicine, London, United Kingdom

INTRODUCTION

The low molecular weight biologically active polyamines which include putrescine, spermidine and spermine are essential for cell multiplication and differentiation and in addition act as co-factors for the biosynthesis of macromolecules (Pegg and McCann, 1988). Polyamines metabolism has been studied in various organisms including the free living pathogenic amoebae, *A. culbertsoni* (Gupta *et al.*, 1987) and *A. castellanii* (Poulin *et al.*, 1984; Kim *et al.*, 1987a). The polyamine contents of the free living amoebae measured at the stationary phase of the growth cycle revealed differences between strains. While Poulin *et al.* (1984) reported presence of only 1,3 diaminopropane and possibly norspermidine in Neff strain of *A. castellanii*, spermidine and putrescine were additionally detected in *A. culbertsoni* (Kim *et al.*, 1987b; Gupta *et al.*, 1984; Srivastava and Shukla, 1982). In another report where polyamine contents of unidentified pathogenic *Acanthamoeba* isolates were measured in growing and quiescent cells, Zhu *et al.* (1989) detected relatively high levels of 1,3 diaminopropane and spermidine in the growing cells. The polyamine levels decreased as the amoeba differentiated to cysts. N[8]-acetylspermidine and

*Author for correspondence

acetylspermidine were found in both developmental stages while acetylcadaverine was detected only in growing amoebae and N^1-acetylspermidine detected only in cysts.

Little if any thing is known of the polyamine content of the medically important free living amoeba, *A. polyphaga*, which is found associated with corneal ulceration (Jones *et al.*, 1975; Wright *et al.*, 1985). In addition, the polyamine content of this organism and indeed many *Acanthamoeba* organisms at different stages of growth have not been studied. The information from such studies would have potential benefit in developing metabolic inhibitors for the treatment and/or prophylaxis of the infections since some protozoans are known to be sensitive to interference in polyamine biosynthesis (Pegg and McCann, 1988; Bacchi and McCann, 1987).

MATERIALS AND METHODS

Chemicals: The polyamines 1,3 diaminopropane, putrescine (1,4 diaminobutane), spermidine (N-(3 aminopropyl)-1,4-diaminobutane), spermine (N,N'-bis(3 aminopropyl)-1,4-diaminobutane), N^8-acetylspermidine and N^1-acetylspermidine were obtained from the Sigma Chemical Co., St. Louis, MO. USA. The fine chemicals were of the highest quality available and were obtained from Fisher Chemicals, NJ. USA. Pentamidine isethionate was from Rhone-Poulenc-Rorer.

Culture of A. polyphaga: A. polyphaga (Ryd) isolated from human corneal ulcer, was obtained from the laboratory of one of the authors (DCW). The organism was subcloned and propagated at 28°C in the medium described by Visvesvara and Balamuth (1975). Growth curve of the trophozoites was determined by daily counting of the organisms attached to the culture flasks by the method previously described (Ogbunude *et al.*, 1991).

Measurement of polyamine content: The polyamine contents of *A. polyphaga* at different developmental stages (8, 16, 32, 48 hrs) were measured. Polyamines were extracted from the organisms by the method of Kim *et al.*, (1987a).

A Waters HPLC system Model 501 with Waters 990 photodiode array detector and Water's C_{18} 5 μ radial pak liquid chromatography cartridge was used for the analysis of the benzoylated polyamines. Samples were passed through 0.2 μm pore membrane filters (Gelman) and used immediately. Samples were run isocratically with methanol/water mixture (65/35 v/v) mobile phase at a flow rate of 1 ml/min and pressure of 1000 psi (Kim *et al.*, 1987a). Polyamine peaks were detected at 254 nm.

Drug studies: The effect of pentamidine on the intracellular distribution of polyamines in *A. polyphaga* at different developmental stages was studied by including in the culture medium, the IC_{50} concentration of pentamidine (12.4 μM). This is the concentration of the drug that inhibited by fifty percent the *in vitro* growth of *A. polyphaga*. The control flasks contained only the solution in which the drug was prepared. Polyamines were extracted at various times.

RESULTS

The growth curve of *A. polyphaga* trophozoites plated in a 96-well microtiter plate at a density of 1.5 x 10^5 organisms/well and maintained in Visvesvara and Balamuth medium at 28°C showed that the organism became confluent after 48 hr of incubation (data not shown).

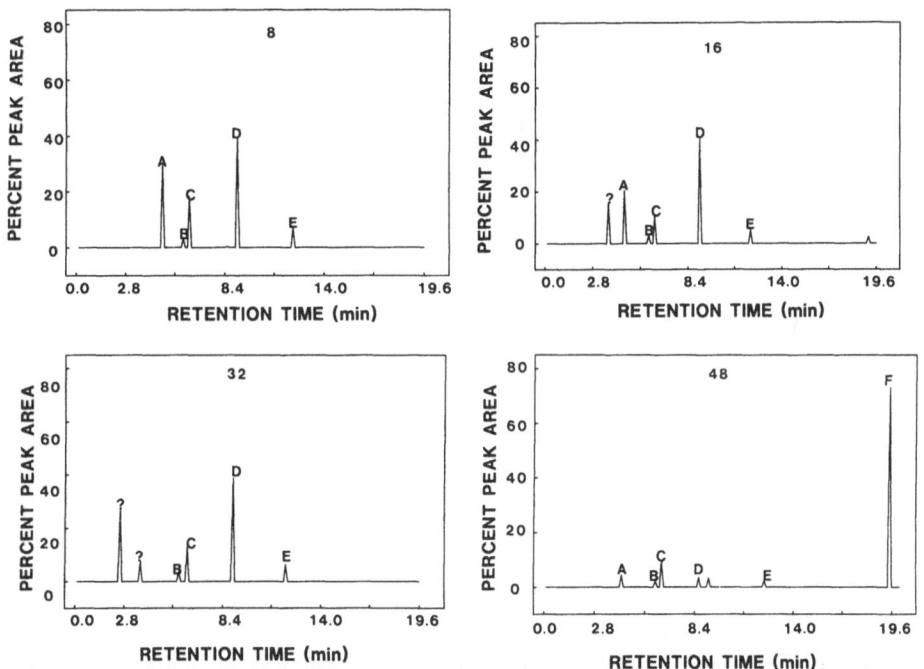

Figure 1: A HPLC run of benzoylated polyamine extracts from *A. polyphaga*.
The intracellular distributions of polyamines in *A. polyphaga* at different stages of growth were measured as described in Materials and Methods. A=N^1- or N^8-acetylspermidine, B=putrescine, C=1,3 diaminopropane, D=spermidine, E=spermine, F=unknown.

Figure 1 shows the polyamine content of *A. polyphaga* at different developmental stages (8, 16, 32, 48 hr). The major polyamines seen at 8 hr of culture were N^1-acetylspermidine or N^8-acetylspermidine (A), 1,3 diaminopropane (C) and spermidine (D). The HPLC system used did not separate properly the N^1-acetylspermidine from N^8-acetylspermidine. Low amounts of putrescine (B) and spermine (E), constituting less than 10% of the total extracts, were also detected. Essentially the 16 hr culture showed a similar polyamine profile to the 8 hr culture, however an additional peak which was not identified appeared before the acetylated polyamine(s) on HPLC runs. At 32 hr, the polyamines C and D constituted about 55% of the total polyamine extract. Two new peaks which constituted about 35% of the total extracts, however appeared. At 48 hr, a peak (F) which accounted for about 80% of total polyamine extract was detected. The other polyamines were now barely detectable except C which constituted about 10% of the extract.

The effect of IC_{50} concentration of pentamidine on the intracellular polyamine distribution is shown in Figure 2. Pentamidine decreased not only the levels of A,C and D of the 8 hr culture but also induced the organism to produce F which was seen only after 48 hr incubationin of organism in drug free medium. As the culture developed from early logarithmic (8 hr) to early stationary (32 hr) phase, F decreased from 60% to 20%. However, at 48 hrs, virtually all the polyamine extract (> 95%) was F. It is of interest to note that putrescine was not detected at any stage of development in the drug treated organism.

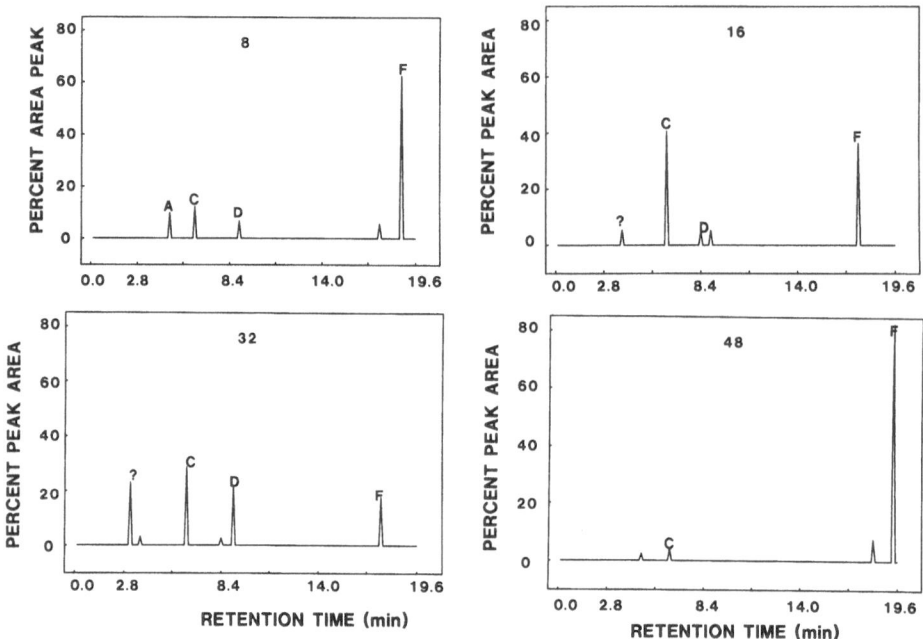

Figure 2: Effect of pentamidine on polyamines distribution in *A. polyphaga*.
The effect of pentamidine on polyamines distribution at different stages of growth were studied.

DISCUSSION

We have studied the intracellular distribution of polyamines in *A. polyphaga* and the effect of pentamidine on the polyamine distribution at various stages of growth in culture. Pentamidine is an analogue of dibromopropamidine and propamidine isethionate which were used successfully to treat *Acanthamoeba* infection due to *A. polyphaga* (Wright *et al.*, 1985). In general, our results on the polyamine content of *A. polyphaga* at stationary phase agree with that of Kim *et al.* (1987a) on *A. castellanii* in that 1,3 diaminopropane, spermidine, spermine and putrescine were detected in the organism and order of abundance was 1,3 diaminopropane> spermidine>N^1 or N^8-acetylspermidine>spermine >putrescine. However, in the 16 hr culture when the organism was still in logarithmic phase, the relative abundance of the polyamines were in the order spermidine> N^1- or N^8-acetylspermidine> 1,3 diaminopropane> spermine> putrescine suggesting that changes occur in the relative importance of different polyamines with stage of growth. We also observed that in the extract of 48 hr culture, a major peak which constituted approximately 75% of the total polyamine extract came out at 19.2 min in the HPLC run. Eukaryotes have beside the enzymes that catalyse the forward reaction of conversion of putrescine to spermine, other enzymes (spermidine/spermine acetyltransferase (SSAT) and polyamine oxidase (PAO)) that catalyse back synthesis of spermidine from spermine and of putrescine from spermidine (Seiler, 1987; Pegg, 1986). The latter enzyme PAO was demonstrated in *A. culbertsoni* (Shukla *et al.*, 1992). It preferentially utilizes N^8-acetylspermidine but could also accept other acetylated polyamines as substrates. In *Acanthamoeba*, PAO further splits spermidine to diaminopropane (Shukla *et al.*, 1992). Earlier work by Zhu *et al.*, (1989) had shown the presence of high levels of 1,3 diaminopropane and spermidine in the growing cells of *Acanthamoeba* organisms. As the amoebae differentiated to cysts, their levels decreased and more acetylated polyamines (N^8- and N^1-acetylspermidine) were detected. Thus it is possible that F is yet another unidentified acetylated polyamine perhaps resulting from combined activities of PAO, SSAT and/or other enzymes.

The most significant feature of the effect of pentamidine was its apparent induction of production of F by the early logarithmic phase organism. The induced polyamine (F) persisted the whole length of the culture period although the level decreased gradually until at 48 hr when it increased to above 85% of the total extract. The remaining 5% was 1,3 diaminopropane. Pentamidine has been shown to affect S-adenosylmethionine decarboxylase from the rat liver (Balana-Fouce *et al.*, 1986), *Trypanosoma brucei brucei* (Bitonti *et al.*, 1986) and *A. culbertsoni* (Gupta *et al.*, 1987). Pentamidine is also a competitive inhibitor of murine polyamine oxidase and human spermidine/spermine acetyltransferase (Libby and Porter, 1992), but the effect on *Acanthamoeba* SSAT or PAO (Shukla *et al.*, 1992) has not been reported. Thus, if F were an acetylated polyamine, it suggested that pentamidine might have instead of inhibiting the activities of either SSAT or PAO of *A. polyphaga*, stimulated the enzymes and hence resulted in increased acetylation of the polyamine intermediates or the products from polyamine oxidase reaction.

The significance of F is not clear at this time, however, the fact that only the stationary phase organism have above 75% of this polyamine and also that pentamidine in the medium caused the logarithmic stage trophozoites to produce F is suggestive of a role for the polyamine in the *A. polyphaga* survival. One possible role is that it may be involved in amoeba encystation. Under unfavourable condition, amoeba encyst. That polyamines are involved in amoeba encystation was demonstrated by Zhu *et al.*, (1989) who observed that exogenous polyamines, particularly spermidine, spermine and 1,3 diaminopropane were strong inhibitors of amoeba encystment. Hence, F may be a form of polyamine that will rather induce amoeba to undergo encystation under unfavourable condition rather than inhibit its normal survival activities. The actual function of F may be known when the identity is established and with further experimentation.

REFERENCES

Bacchi, C.J. and McCann, P.P. Parasitic Protozoa and Polyamines. In: Inhibition of polyamine metabolism, pp. 317, McCann, P.P., Pegg, A.E. and Sjoerdsma, A. (eds) Academic Press, Inc., Orlando (1987).

Balaña-Fouce, R; Pulido, T.G; Ordónez-Escudero, D. and Garride-Perterra, A. Inhibition of diamine oxidase and S-adenosylmethionine decarboxylase by diminasene aceturate (berenil). *Biochem. Pharmacol.* 35:1597 (1986) .

Bitonti, A.J; Dumont, J.A. and McCann, P.P. Characterization of *Trypanosoma brucei brucei* S-adenosyl-L-methionine decarboxylase and its inhibition by berenil, pentamidine and methylglyoxal bis(guanyl-hydrazone). *Biochem. J.* 237:685 (1986).

Gupta, S; Kishmore, P. and Shukla, O.P.   Polyamine metabolism as a target for growth inhibition of *Acanthamoeba culbertsoni*. *Am. J. Trop. Med. Hyg.* 37:550 (1987).

Jones, D.B; Visvesvara, G.S. and Robinson, N.M.   *Acanthamoeba polyphaga* keratitis and *Acanthamoeba* uveitis associated with fatal meningoencephalitis. *Trans. Ophthamol. Soc. Uk* 95:221 (1975).

Kim, B.G; Sobota, A; Bitonti, A.J; McCann, P.P. and Byers, T.J. Polyamine metabolism in *Acanthamoeba*: polyamine content and synthesis of ornithine, putrescine and diaminopropane. *J. Protozool.* 34:278 (1987a)

Kim, B.G; McCann, P.P. and Byers T.J.   Inhibition of multiplication in *Acanthamoeba castellani* by specific inhibitors of ornithine decarboxylase. *J. Protozool.* 34:264 (1987b)

Libby, P.R. and Porter, C.W. Inhibition of enzymes of polyamine back-conversion by pentamidine and berenil. *Biochem. Pharmacol.* 44:830 (1992)

Ogbunude, P.O.J; Asiri, S; Baer, H.P. and Warhurst, D.C.   New method for detachment and quantitation of *Acanthamoeba* trophozoites in culture. *Trop. Med. Parasitol.* 42:415 (1991)

Pegg, A.E. Recent advances in the biochemistry of polyamines in eukaryotes. *Biochem. J.* 234:249 (1986).

Pegg, A.E and McCann, P.P. Polyamine metabolism and function in mammalian cells and protozoans. *ISI Atlas of Science: Biochem.* 1:11 (1988).

Poulin, R; Laroclle, J. and Nadeau, P.   Polyamines in *Acanthamoeba castellani*: Presence of an usually high, osmotically sensitive pool of 1,3 diaminopropane. *Biochem. Biophys. Res. Commun.* 122:388 (1984).

Seiler, N. Functions of polyamine acetylation. *Can. J. Physiol. Pharmacol.* 65:2024 (1987).

Shukla, O.P; Müller, S. and Walter, R.D.   Polyamine oxidase from *Acanthamoeba culbertsoni* specific for N^8-acetylspermidine. *Mol. Biochem. Parasitol.* 51:91 (1992).

Srivastava, D.K. and Shukla, O.P.   Polyamines of *Acanthamoeba culbertsoni* and their effect on encystation. *Indian J. Parasitol.* 6:211 (1982).

Visvesvara, G. S. and Balamuth, W.   Comparative studies on related free-living and pathogenic amoebae with special reference to *Acanthamoeba*. *J. Protozool.* 22:245 (1975).

Wright, P; Warhurst, D.C. and Jones, B.R.   *Acanthamoeba* keratitis successfully treated medically. *Brit. J. Ophthalmol.* 69:778 (1985).

Zhu C.M; Cumaraswamy, A. and Henney, H.R. Jr.   Comparison of polyamine and S-adenosylmethionine contents of growing and encysted *Acanthamoeba* isolates. *Mol. Cellular Biochem.* 90:145 (1989).

SPECIFIC ENZYME SYNTHESIZING ADENOSINE FROM ADENINE AND RIBOSE-1-PHOSPHATE IN INVERTEBRATES

Halina Trembacz and Maria M. Jeżewska

Institute of Biochemistry and Biophysics
Polish Academy of Sciences, 02-532 Warsaw
36 Rakowiecka St., Poland

INTRODUCTION

An enzymic activity synthesizing and phosphorolysing adenosine is known to occur in *Bacilli* sp., Mycoplasmatales, and some parasitic protozoa (cf.Trembacz and Jeżewska, 1993), and in the parasitic trematode *Schistosoma mansoni* (Miech et. al., 1975). We have found the activity of adenosine phosphorylase (AdoPho) in several other species of trematodes as well as in their intermediate hosts – terrestrial and fresh--water gastropods of various species (Trembacz and Jeżewska, 1994); this evidences that the occurrence of AdoPho activity is not limited to parasitic organisms. A specific adenosine phosphorylase has been isolated from *B. subtilis* (Senesi et al., 1976; Jensen, 1978), *Acholeplasma laidlawii* (McElvain et al.,1988), *Helix pomatia* and *Fasciola hepatica* (Trembacz and Jeżewska, 1993, and the present report, respectively).

In contrast, a specific enzyme catalyzing adenosine synthesis from adenine and ribose-1-phosphate has not been found till now in vertebrates although adenine is produced in their tissues, probably by 5'-methylthioadenosine phosphorylase from 5'-methyl-thio-adenosine and deoxyadenosine, and/or by purine nucleoside phosphorylase exhibiting a low activity towards deoxy- and adenosine.

Vertebrates (from fish to man) are the final hosts in the trematode life cycle. The infections by these parasites cause the severe diseases (as fascioliasis, schistosomiasis, etc.), and are quite common (in the case of schistosomiasis 200 – 300 million people affected). Bearing in mind that the specific trematode AdoPho could serve as a target enzyme for the chemotheraphy and in diagnosis of parasitic infection, we investigated the properties of AdoPho isolated from the adult form of *F. hepatica*.

Purine and Pyrimidine Metabolism in Man VIII, Edited by
A. Sahota and M. Taylor, Plenum Press, New York, 1995

RESULTS AND DISCUSSION

Using the radiochemical methods (Trembacz and Jeżewska, 1993) we determined activities of several enzymes taking part in purine metabolism. The adenosine synthesizing activity was found in trematodes, gastropods and insects, at various stages of their life, therefore this activity seems to occur commonly in invertebrates. The AdoPho activity was compared (Table 1) to other enzymic activities investigated: inosine phosphorylase (InoPho), adenine and hypoxanthine phosphorib- osyltransferases (APRT and HGPRT, respectively), and adenos- ine deaminase (ADA). In insects the AdoPho activity was the lowest one exceeding only the HGPRT activity, whereas in the terrestrial gastropod, *H. pomatia*, and in the adult form of trematode parasite, *F. hepatica*, the AdoPho activity was several times higher than the activity of all other enzymes investigated. This seems to point to important role of AdoPho in these two invertebrate species. The high level of AdoPho in parasitic trematode is probably related to the indispens- able uptake of egzogenous adenine; the role of AdoPho in in the free-living gastropod remains obscure.

Table 1. Adenosine phosphorylase and other purine- -metabolizing enzymes in some invertebrate species.

Invertebrate Species Tissue[1]	Activity ratio: AdoPho[2] to			
	InoPho[2]	APRT	HGPRT	ADA
Insects				
Tenebrio molitor				
whole larvae	0.17	0.13	2.7	0.020
whole pupae	0.11	0.11	2.7	0.100
Drosophila melanogaster				
whole larvae	0.03	0.01	0.0	0.001
whole adult	0.04	0.07	3.8	0.002
Gromphadorfina coquereliana				
fat body of adult	0.46	0.39	11.7	0.010
Gastropods				
Helix pomatia				
hepatopancreas	2.1	4.2	16.7	4.360
Lymnaea stagnalis				
hepatopancreas	0.2	1.6	12.3	0.270
Coretus sp.				
hepatopancreas	0.14	0.72	75.0	0.370
Trematodes				
Echinoparyphium aconiatum				
rediae	0.04	0.28	0.19	0.480
Fasciola hepatica				
whole body of adult	7.9	7.6	87.5	35.000

[1]Enzymic activity was determined in crude extracts
[2]Nucleoside synthesis was measured.

The adult form of *F. hepatica* was isolated from the bile duct of cattle, rinsed with a special Krebs-Ringer solution pH 7.4, and then the whole parasite bodies were homogenized in 100 mM HEPPS – KOH buffer pH 8.0, containing 1 mM PMSF, 10 mM DTT and 20 % glycerol (v/v) with the use of a Potter-Elvehjem homogenizer, in an ice-bath. The homogenate was centrifuged at 14,000 g for 30 min in a cold room, and the resulting supernatant was centrifuged at 100,000 g for 1 hr at 4°C. Obtained supernatant was fractionated with ammonium sulphate (Table 2). All three protein fractions exhibited the nucleoside synthesizing activity of AdoPho. Desalted protein fractions (AS1) and (AS2) were applied onto the Cellulose DE-52 column, and gradient elution was performed with 50–450 mM NaCl in 50 mM HEPPS-KOH buffer pH 8.0. Fraction (AS1) gave one active peak, but fraction (AS2) – two peaks exhibiting AdoPho activity. Protein fractions of peak 1 and peak 2 were separately pooled, desalted on a Sephadex G-25 column, and concentrated in an Amicon Ultrafiltration Cell. The resulting solutions:(AS1, DE -52, peak 1) and (AS2, DE-52, peak 2) were chromatographed on Sephadex G-150 or Sephadex G-100 columns. In each of the two elution profiles obtained, the two peaks exhibiting AdoPho activity were found: peaks 1a, 1b, and peaks 2a, 2b, respectively. The molecular weights corresponding to these peaks are presented in Table 3.

The total separation of the AdoPho activity from that of InoPho by the above procedure evidences that the specific adenosine phosphorylase occurs in trematode *F. hepatica*. This enzyme is different from adenosine phosphorylase of the gastropod *H. pomatia* (Trembacz and Jeżewska, 1993), which precipitates mainly in 0.3 – 0.45 ammonium sulphate fraction, gives only one active peak in the elution profile from the Cellulose DE-52 column, and also single active peak during the chromatography on the Sephadex G-150 column.

Table 2. The separation of *F. hepatica* adenosine and inosine phosphorylases.

	AdoPho activity (munit/mg)	AdoPho to InoPho	Purific- ation fold	Yield (%)
Homogenate	14.8	6.5	1	100
Supernatant 14,000 g	30.2	13.7	2	95
Supernatant 100,000 g	31.1	13.1	2	77
Ammonium sulphate fractions				
0.30 – 0.45	6.6	20.0	–	4
0.45 – 0.60 (AS1)	32.1	36.8	2	20
(AS1) desalted	65.6	–	4	–
Cellulose DE-52 peak 1	243.6[1]	–	16	–
0.60 – 0.80 (AS2)	77.6	16.0	5	33
(AS2) desalted	160.2	13.0	11	–
Cellulose DE-52 peak 1	245.1[1]	–	16	–
peak 2	185.9[1]	–	12	–

[1]Most active fraction, InoPho not detected.

Table 3. Molecular weight of *F. hepatica* adenosine phosphorylase

Enzyme in peak	Molecular weight
AS1, DE-52,peak 1, Sephadex G-150 peak 1a	76 800
peak 1b	25 300
AS2, DE-52,peak 2, Sephadex G-100 peak 2a	109 100
peak 2b	29 140

It was interesting to compare the molecular weight of adenosine phosphorylase from various sources. *F. hepatica* adenosine phosphorylase seems to occur in several molecular forms: in the (AS1, DE-52) peak 1 we found a small form (25 300) which could be a monomer giving the second trimeric form (76 800); this last form resembles *H. pomatia* adenosine phosphorylase (66 000 - 77 600), however we have not found a monomer in this case (Trembacz and Jeżewska, 1993). The two molecules of the enzyme in peak 1a may give a molecule corresponding to a hexameric molecule of *B.subtilis* adenosine phosphorylase (153 000, Jensen, 1978). In turn, the (AS1, DE-52) peak 2 gave also two peaks: 2a and 2b corresponding to molecular weights of 109 100 and 29 140 (tetramer and dimer?) The AdoPho form in peak 2a resembles in respect of the molecular weight adenosine phosphorylase from *A. laidlawii* (McElvain *et al.*, 1988). however, this last enzyme exhibits also a hydrolytic activity.

The equilibrium of the reaction catalyzed by *F. hepatica* AdoPho is shifted very strongly towards adenosine synthesis; this must be taken into account when the presence of AdoPho is tested.

REFERENCE

Jensen.K.F., 1978, Two purine nucleoside phosphorylases in *B. subtilis*. Purification and some properties of adenosine -specific phosphorylase, Biochim. Biophys. Acta, 525:346.

McElvain,M.C., Wiliams.M.V., and Pollack.J.D.,1988, *Acholeplasma laidlawii* B-PG9 adenine-specific purine nucleoside phosphorylase that accepts ribose-1-phosphate, deoxyribose-1-phosphate, and xylose-1-phosphate, J.Bact.170:564.

Miech.R.P., Senft.A.W., and Senft.D.G.,1975, Pathway of nucleotide metabolism in *Schistosoma mansoni* - VI Adenosine phosphorylase, Biochem. Pharmacol. 24:407.

Senesi.S., Falcone.G., Mura.U., Sgarella.F., and Ipata.P.L., 1976, A specific adenosine phosphorylase, distinct from purine nucleoside phosphorylase, FEBS Lett.,64:359.

Trembacz.H., and Jeżewska.M.M., 1993, Specific adenosine phosphorylase from hepatopancreas of gastropod *Helix pomatia*, Comp. Biochem. Physiol. 104B:481.

Trembacz.H., and Jeżewska.M.M., 1994, Adenosine phosphorylase and other enzyme of purine salvage in Pulmonata snails and their Trematoda parasites, Comp. Biochem. Physiol. 107B:135.

MOLECULAR CHARACTERIZATION OF A CARBAMOYL-PHOSPHATE SYNTHETASE II (CPS II) GENE FROM *TRYPANOSOMA CRUZI*

Takashi Aoki, Rieko Shimogawara, Kaoru Ochiai, Hiroshi Yamasaki, and Junko Shimada

Department of Parasitology, Juntendo University School of Medicine, Hongo, Bunkyo-ku, Tokyo 113, Japan

INTRODUCTION

A number of protozoan and helminth parasites are incapable of synthesizing purines de novo and rely on salvaging preformed purines. In contrast, few organisms lack de novo pyrimidine synthesis, the pathway of which is subjected to an exquisite regulation through inhibition and stimulation of carbamoyl-phosphate synthetase II (CPS II) activity. Previous works in this laboratory (Aoki, 1994) showed that, in *Schistosoma mansoni* and *Ascaris suum*, as well as in mammals, CPS II occurs as the first enzyme in a multifunctional protein with the second and third enzymes of the pathway, aspartate carbamoyltransferase (ACT) and dihydroorotase. However, our attempt to purify these enzymes from trypanosomatids resulted in the separation of CPS II and ACT activities, suggesting the enzymes to be independent proteins (Aoki and Oya, 1987a). The partially purified CPS II showed kinetic and regulatory properties different from those of prokaryotic and eukaryotic CPSs II. To gain an insight into these biochemical differences, we have partially determined the sequence of CPS II cDNA from *Trypanosoma cruzi* , a protozoan flagellate that causes Chagas' disease in man in Latin America, and the deduced amino acid sequence has been characterized. We report here that *trans*-splicing may bring about the maturation of pre-mRNA and that a short polypeptide links the glutaminase and carbamoyl-P synthetase components, homologues of the light and heavy subunits of *Escherichia coli* CPS II.

MATERIALS AND METHODS

Culture forms of *T. cruzi* were propagated and harvested essentially as described (Aoki and Oya, 1987a). From the freshly collected protozoan cells, a

Purine and Pyrimidine Metabolism in Man VIII, Edited by
A. Sahota and M. Taylor, Plenum Press, New York, 1995

513

mRNA fraction was prepared using QuickPrep mRNA Purification Kit of Pharmacia. Approximately 30 μg of the purified mRNA was obtained from 1 ml of the packed cells. cDNA-PCR was carried out essentially as described (Kita et al., 1992). The amplified PCR-fragments were cloned into pT7Blue T-Vector (Novagen), and determined for the nucleotide sequences in a DNA sequencer, model 373A, using Taq Dye Primer Cycle Sequencing Kit (Applied Biosystems). The nucleotide and deduced amino acid sequences were analyzed by the computer program GENETYX.

RESULTS AND DISCUSSION

cDNA-PCR using the primers, which were designed for the *T. cruzi* spliced leader (SL) and a highly conserved N-terminal region of CPS II, provided a specific fragment. At its 5'-terminus, there was the SL of 39 bp, 5'-AAC TAA CGC TAT TAT TGA TAC TGT TTC TGT ACT ATA TTG-3'. This was followed by an intermediary sequence under the analysis at the present time, and then by an open reading frame (ORF) of the glutaminase component of the parasite CPS II. The result is consistent with the view that SL genes entail the production of small SL primary transcripts (SL-RNA) and that the SL is added by *trans*-splicing to an acceptor site upstream of the translational start site of the pre-mRNA (Agabian, 1990).

Figure 1 shows an alignment of the deduced amino acid sequences of a carboxyl part in glutaminase components of *T. cruzi* and various CPSs II. Four highly conserved regions were recognized. Region I may serve a structural role, whereas regions II and III form the glutaminase active site (Simmer et al., 1990); particularly, the trypanosomal Cys-254, His-339, and Glu-341 may directly participate in catalysis. The protozoan active-site Cys-254 region (box II), PIFGICMGNQ, well resembled the major consensus sequence PVFGICLGHQ, but Met-255 and Asn-257 in *T. cruzi* occurred only in *Dictyostelium discoideum* (Figure 1). We suspect a possible relationship between this characteristic sequence and the affinity of the glutaminase for the substrate glutamine and a glutamine analog, acivicin, since the parasite CPS II was more susceptible to acivicin than mammalian enzyme (Aoki, 1994). The CPS II-specific sequence, the region IV, identified by Simmer et al. (1990), also extends in *T. cruzi* from Ile-292 to Asp-302. The protozoan sequence of 123 amino acid residues in Figure 1 had the identical residues of 45, 49, 52, 53, and 53%, respectively, against *carA*, *URA2*, hamster *CAD*, *CPA1*, and *D. discoideum pyr 1-3*.

An N-terminal region of about 40 residues, which may start with Ala-30, in the parasite glutaminase component showed high homology, probably indicating the importance of this region for the interdomain interaction with the synthetase component of CPS II, as reported for *E. coli* CPS II (Guillou et al., 1989). Simmer et al. (1990) proposed an amino acid alignment of various linkers between the glutaminase and synthetase components. Our *T. cruzi* linker sequence, VKESKVKEASKYKPR, was the shortest, contained 6 basic and 2 acidic amino acids out of 15 residues, and is probably highly hydrophilic.

The cDNA sequence of the *T. cruzi* synthetase component was partially determined (data not shown). There may be two ATP-binding sites in the synthetase component, one of which be involved in the activation of CO_2 by

```
                              I
E. coli (carA)    234   DGIFLSNGPGDPAPCDYAITAIQKFLETD
Yeast (CPA1)      225   DGIFLSNGPGNPELCQATISNVRELLNNP
Yeast (URA2)      266   DGLFYSNGPGDPSVLDDLSQRLSNVLEAK
Hamster (CAD)     215   DGLFLSNGPGDPASYPGVVATLNRVLSEP
D. discoideum     199   DGVFISNGPGDPSLCGKAIENIRKVLALP
T. cruzi          219   DGLFISNGPGDPQMCTKTIEHVRWAITQD
```

```
            II
263   ----IPVFGICLGHQLLALASGAKTVKMKFGHHGGNHPVKDVEKNVVMIT
254   VYDCIPIFGICLGHQLLALASGASTHKLKYGNRAHNIPAMDLTTGQCHIT
295   K---TPVFGICLGHQLIARAAVQSTLKLKFGNRGHNIPCTSTISGRCYIT
244   N--PRPVFGICLGHQLLALAIGAKTYKMRYGNRGHNQPCLLVGTGRCFLT
228   V--AKAVFGVCMGNQLLGLAAGAQTHKMAFGNRGLNQPCVDQISGRCHIT
248   K----PIFGICMGNQILALAAGGSTYKMKYGHRGQNQPSTCRSDGHVFIT
```

```
      IV                                     III
309   AQNHGFAVDEATL-PAN-LRVTHKSLFDGTLQGIHRTDKPAFSFQGHPEA
304   SQNHGYAVDPETL-PKDQWKPYFVNLNDKSNEGMIHLQRPIFSTQFHPEA
342   SQNHGFAVDVDTL-TSG-WKPLFVNANDDSNERFYHSELPYFSVQFHPES
292   SQNHGFAVDADSL-PAG-WTPLFTNANDCSNEGIVHDSLPFFSVQFHPEH
276   SQNHGFVIDSNSLPAGSGWKTYFINANDASNEGIYHESKPWFSVQFHPEA
294   TQNHGFAVDFKSV-SQDEWEECFYNPNDDSNEGLRHRTKPFFSAQFHPEG
```

Figure 1. Amino acid alignment of a carboxyl part of glutaminase components of various CPSs II. The numbers correspond to the residues counted from Met-1 of the open reading frame of the glutaminase components. The numbers for *T. cruzi* are tentative. Highly conserved residues in the *T. cruzi* CPS II are double-underlined, when compared with the *E. coli* CPS II (*carA*), *Saccharomyces cerevisiae* arginine-specific CPS II (*CPA1*), *S. cerevisiae* pyrimidine-specific CPS II (*URA2*), hamster *CAD*, and *D. discoideum pyr1-3* gene products. The three regions involved in the glutaminase activity are labeled with I, II, and III. The number IV box represents the CPS II-specific high homology region.

ATP and the regulation of CPS II activity. The protozoan ATP-binding site, including its adjacent region (90 residues), was extremely highly conserved and much more homologous to the eukaryotic enzymes (~75%) than to the *E. coli* enzyme (~55%). The same sequencing strategy applied for the *T. cruzi* genomic DNA predicted the absence of intron in this protozoan CPS II gene, so far examined. The completion of the sequencing is awaited for further discussion.

ACKNOWLEDGMENTS

This work was supported in part by a Grant-in-Aid for Scientific Research (No. 04454192) from the Ministry of Education, Science, and Culture of Japan, and from the Japan-US Medical Cooperative Program on Parasitic Diseases.

REFERENCES

Agabian, N., 1990, *Trans* splicing of nuclear pre-mRNAs, *Cell* 61: 1157-1160.

Aoki, T., 1994, Initial steps of *de novo* pyrimidine nucleotide biosynthesis in parasites and mammalian tissues: Purification, regulation, adaptation, and evolution, *Jap. J. Parasitol.* 43: 1-10.

Aoki, T., and Oya, H., 1987a, Kinetic properties of carbamoyl-phosphate synthetase II (glutamine-hydrolyzing) in the parasitic protozoan *Crithidia fasciculata* and separation of the enzyme from aspartate carbamoyl-transferase, *Comp. Biochem. Physiol.* 87B: 143-150.

Aoki, T., and Oya, H., 1987b, Inactivation of *Crithidia fasciculata* carbamoyl-phosphate synthetase II by the antitumor drug acivicin, *Mol. Biochem. Parasitol.* 23: 173-181.

Guillou, F., Rubino, S.D., Markovitz, R.S., Kinney, D.M., and Lusty, C.J., 1989, *Escherichia coli* carbamoyl-phosphate synthetase: Domains of glutaminase and synthetase subunit interaction, *Proc. Natl. Acad. Sci. USA* 86: 8304-8308.

Kita, K., Mizuchi, D., Wang, H., Takamiya, S., Aoki, T., and Kojima, S., 1992, cDNA sequence of three cysteine-rich clusters in the iron-sulfur subunit of complex II (succinate-ubiquinone oxidoreductase) from *Caenorhabditis elegans* determined by automated sequencer, *Electrophoresis*, 13: 506-511.

Simmer, J.P., Kelly, R.E., Rinker, Jr., A.G., Scully, J.L., and Evans, D.R., 1990, Mammalian carbamyl phosphate synthetase (CPS). cDNA sequence and evolution of the CPS domain of the Syrian hamster multifunctional protein CAD, *J. Biol. Chem.* 265: 10395-10402.

DEVELOPMENT OF A MOUSE MODEL FOR THE STUDY OF HUMAN PURINE METABOLISM

Tristan S. Barnes,[1] Gary L. Brodsky,[1] George J. Barela,[1] John H. Bleskan[1,2] and David Patterson[1,2,3]

[1]Eleanor Roosevelt Institute
[2]University of Colorado Health Science Center
[3]University of Colorado Cancer Center
  Denver, CO

INTRODUCTION

Patients with Down syndrome are known to be hyperuricemic and this condition is thought to result from the over-production of purines (Pant *et al.*, 1968). Only one purine metabolism gene is located on human chromosome 21 in the critical region associated with Down syndrome. This gene encodes for the trifunctional protein composed of phosphoribosylglycinamide synthetase (GARS), phosphoribosylglycinamide transformylase (GART) and phosphoribosylaminoimidazole synthetase (AIRS). Increased levels of this protein, known collectively as GART, have also been observed in the cultured fibroblasts of Down Syndrome patients (Scoggin *et al.*, 1980) It is therefore highly likely that the increased levels of purines described in these patients is directly due to their over-expression of the GART trifunctional protein.

A definitive technique for studying the physiological importance of the human purine metabolism genes and in particular the human trifunctional GART protein, would be the transplantation of the relevant human genes into an animal model. Unfortunately, the end product of purine degradation in mice, the choice model for such transgenic experiments, is not the same as in humans. In mice, urate, the end product of purine degradation in humans, is oxidized to allantoin by the enzyme uricase (EC 1.7.3.3; urate oxidase; see Figure 1). However, a colony of mice in which the uricase gene has been "knocked out" by the insertion of a neomycin gene into the third exon has been established by Dr. Thomas Caskey

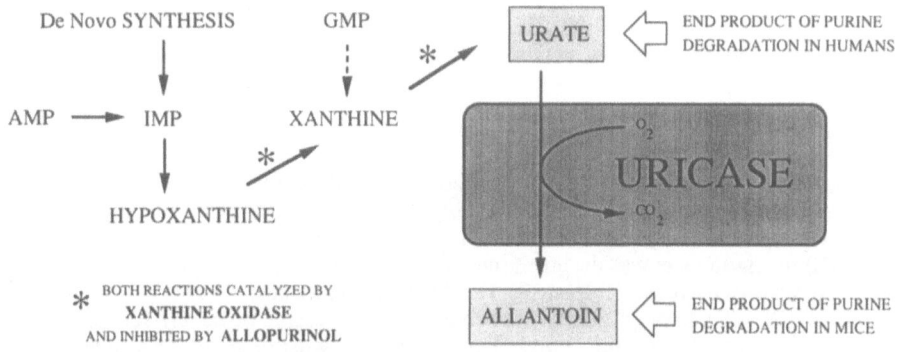

Figure 1. Pathway for purine degradation in humans and mice.

Purine and Pyrimidine Metabolism in Man VIII, Edited by
A. Sahota and M. Taylor, Plenum Press, New York, 1995

and co-workers (Wu *et al.*, 1994). A colony of these same mice has now been established in this laboratory. Since the level of urate in these mice can be controlled by the administration of allopurinol (Figure 1), the mice should prove to be a good model for the study of human purine metabolism as well as human hyperuricemia.

We report here a simple PCR test for the routine determination of uricase genotype in the "knock out" mice. Furthermore, we also report the generation of monoclonal and polyclonal antibodies to the various domains of the GART trifunctional protein, as well as two previously described monoclonal antibodies to phosphoribosyl formylglycinamide (FGAR) amidotransferase (Barnes *et al.*, 1994). These antibodies will be invaluable in determining the relevant protein levels in the transgenic mice.

METHODS AND RESULTS

Genotyping for Uricase

Before the uricase deficient mice can be used as hosts for the study of human purine metabolism transgenes, any defects in these mice associated purely with the loss of uricase must first be determined and characterized. In order to determine the uricase genotype of individual mice, either *in utero* or *postpartum*, we have developed a rapid PCR assay. Three PCR primers were designed and synthesized. Two of these primers flank the neomycin insertion site in the Uricase exon 3 and one is internal to the neomycin gene (Figure 2). Thus, blood or tissue from a normal mouse will give rise to a single PCR product of 240bp; tissue from a homozygous uricase minus mouse will give rise to two PCR products, one of 500bp and one of 1400bp; and tissue from a heterozygous uricase minus mouse should give rise to the 240, 500 and 1400kb PCR products. In practice, the 1400bp product is often absent from heterozygous samples. This, is probably the result of running all three primers in the same PCR reaction since small PCR products are often produced at the expense of larger products. Even with the loss of this one band, the uricase genotypes are clearly distinguishable using this assay (Figure 3).

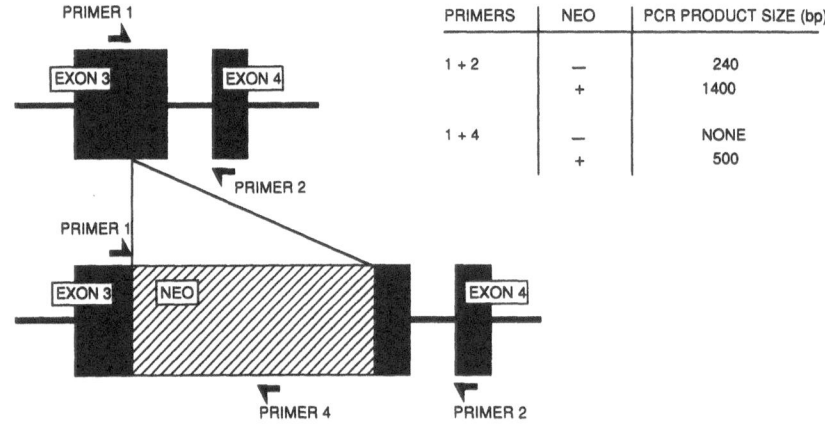

Figure 2. Position of primers for genotyping uricase-deficient mice.

Cloning and Expression of GART Domains

Oligonucleotide primers were synthesized so as to allow for the PCR amplification and subsequent subcloning of each of the three GART domains into a bacterial expression vector (Figure 4). A BamH1 (B) restriction site was incorporated into each of the 5' primers and an EcoR1 (E) restriction site was designed into each of the 3' primers. A plasmid containing the GART cDNA was used as a template for the three PCR reactions.

The pGEX-KT vector (Figure 4) was chosen for expression of the GART domains for a number of reasons. The vector's polylinker is located downstream of a sequence coding for glutathione S-transferase (GST). Thus, inserts are expressed as polypeptides fused to GST

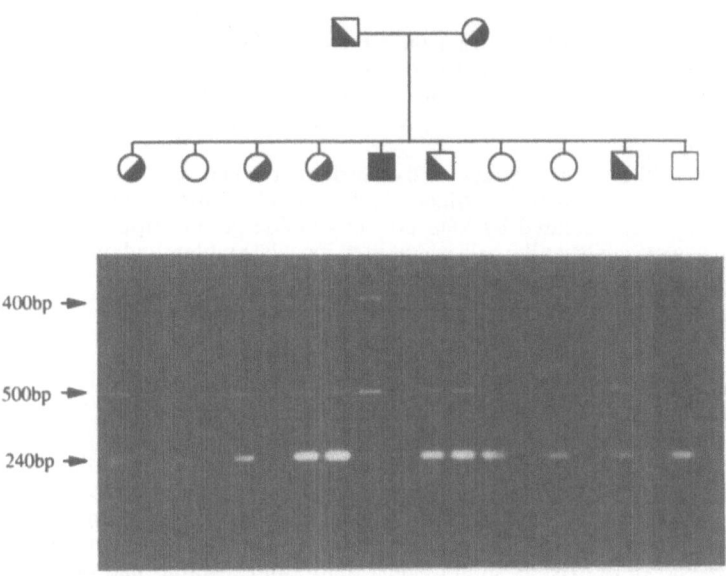

Figure 3. PCR assay for genotyping the uricase-deficient mice. PCR assays (25µl) were performed using a standard protocol using all three uricase primers simultaneously (80ng each). Approximately 100ng of template DNA was used for each reaction.

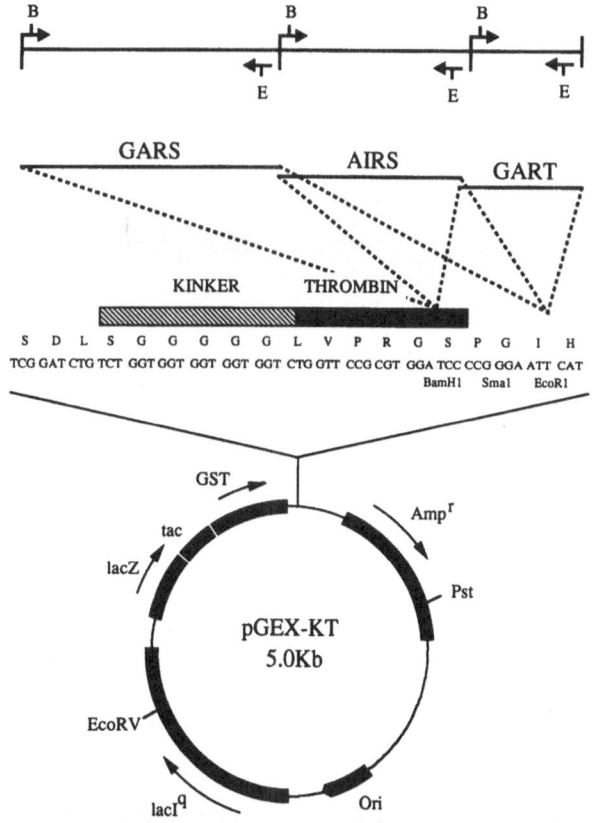

Figure 4. Subcloning of GART domains into the pGEX-KT bacterial vector. cDNAs corresponding to the individual GART domains were obtained by PCR using domain-specific primers. Subcloning of the cDNAs into the pGEX-KT plasmid was accomplished using standard procedures (see text).

and as such can be readily purified on glutathione Sepharose resin. Since the vector incorporates a tac promoter, transcription of the DNA insert can be induced by IPTG. A thrombin cleavage site allows for the rapid separation of the cloned protein sequence from the GST leader. The presence of a polyglycine "kinker" region increases the efficiency of thrombin proteolysis.

The three PCR DNA fragments were cloned into the pGEX-KT vector and their protein products expressed in bacteria. Since all three GST fusion proteins were found to be insoluble, they were isolated by SDS-polyacrylamide gel electrophoresis. The relevant protein bands were cut from the gels and used to inoculate rabbits and mice for the generation of polyclonal and monoclonal antibodies.

Generation of Antibodies

Using standard protocols (Barnes *et al.*, 1987) polyclonal antibodies were generated to all three GART domains and monoclonal antibodies were generated to the GARS domain and to FGAR amidotransferase. Figures 5 and 6 illustrate the recognition of human and CHO proteins by the various antibodies. All of the polyclonal antibodies and the GARS monoclonal antibody recognize a protein of approximately 120kDa molecular mass in both CHO and human cells. The GARS polyclonal and monoclonal recognize an additional protein of approximately 52 kDa molecular mass. Unlike GART and AIRS, the GARS sequence is known to exist as an individual transcript as well as part of the GARS-AIRS-GART transcript. The lower molecular mass protein recognized by the GARS antibodies may result from the translation of this other transcript.

The FGAR amidotransferase monoclonals recognize a single protein of 150kDa molecular mass in CHO cells. One of the antibodies, DD2, cross reacts with a human protein of apparently identical molecular mass.

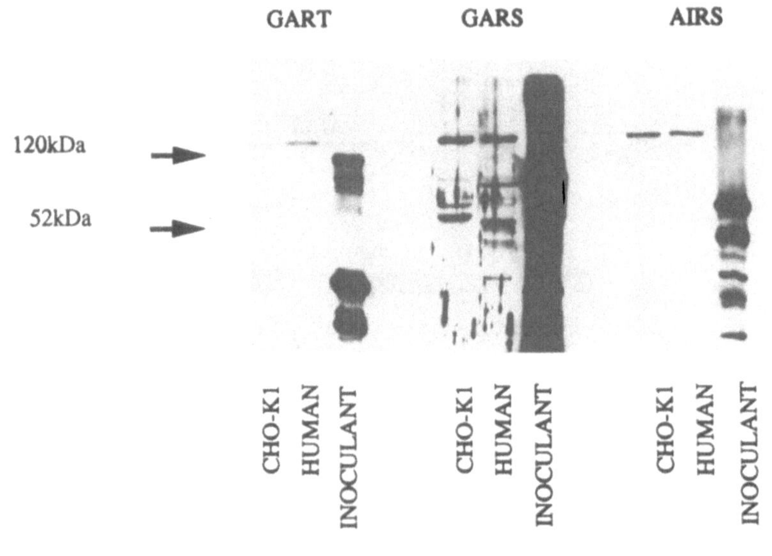

Figure 5. Polyclonal antibodies to the various GART domains. Proteins from CHO or human cell lines (10μg) or from the immunogen (1μg) were resolved by SDS-polyacrylamide gel electrophoresis and electroblotted onto nitrocellulose paper. The paper was then immunoblotted with the relevant polyclonal antibody.

CONCLUSIONS

Dr C. Thomas Caskey and co-workers have produced and bred a uricase-deficient mouse that may prove to be a good model for purine metabolism and hyperuricemia in humans. If so, these mice can be used for the study of human purine metabolism transgenes In order to characterize these mice further a rapid PCR assay has been developed that can be used *in*

GARS MAb   FGARAT MAbs

                        DD2   BD4

120kDa →   ← 150kDa

52kDa →

CHO-K1   HUMAN   FRAGMENT   CHO-K1   HUMAN   CHO-K1   HUMAN

Figure 6. Monoclonal antibodies to the GARS domain of GART and to FGAR amidotransferase. Proteins from CHO or human cell lines (10μg) or from the immunogen (1μg) were resolved by SDS-polyacrylamide gel electrophoresis and electroblotted onto nitrocellulose paper. The paper was then immunoblotted with the relevant monoclonal antibody.

utero or *postpartum* to determine uricase genotype. A number of monoclonal and/or polyclonal antibodies have been generated to the proteins catalyzing four of the first five steps of human *de novo* purine biosynthesis. These antibodies will be invaluable in the assessment of protein level in the uricase minus mice after the incorporation of purine metabolism transgenes.

ACKNOWLEDGMENTS

This is a publication (#1411) of the Eleanor Roosevelt Institute for Cancer Research. This work was supported by the NICHD (HD 17749), the Markey Charitable Trust and the University of Colorado Cancer Center.

REFERENCES

Barnes, T. S., Shaw, P. M., Burke, M. D. and Melvin, W. T., 1987, Monoclonal antibodies against human cytochrome P-450 recognizing different pregnenolone16a-carbonitrile-inducible rat cytochromes P-450, *Biochem. J.* 248:301.
Barnes, T. S., Bleskan, J. H., Hart, I. M., Walton, K. A., Barton, J. W. and Patterson, D., 1994, Purification of, generation of monoclonal antibodies to, and mapping of phosphoribosyl *N*-formylglycinamide amidotransferase, *Biochemistry*, 33:1850.
Pant, S. S., Moser, H. W. and Krane, S. M., 1968, Hyperuricemia in Down's syndrome, *J. Clin. Endocrinol.* 28:472.
Scoggin, C. H., Bleskan, J., Davidson, J. N. and Patterson, D., 1980, Gene expression of glycinamide ribonucleotide synthetase in Down syndrome, *Clin. Res.* 28:31.
Wu, X., Wakamiya, M., Vaishnav, S., Geske, R., Montgomery, C., Jr., Jones, P., Bradley, A. and Caskey, C. T., 1994, Hyperuricemia and urate nephropathy in urate oxidase-deficient mice, *Proc. Natl. Acad. Sci. USA* 91:742

EFFECT OF GALACTOSAMINE ON ADENINE AND URACIL NUCLEOTIDE LEVELS IN ISOLATED HEPATOCYTES OF YOUNG AND OLD RATS

Z. Kmiec[1], M. Marlewski[2], R.T. Smolenski[2], H.A. Simmonds[3]

[1]Department of Histology and Immunology, and
[2]Department of Biochemistry, Medical School Gdansk, Poland
[3]Purine Research Laboratory, Guy's Hospital, London, UK

INTRODUCTION

The aging process is associated with changes at cellular and tissue levels which may lead to deterioration of organ functions. Under basal conditions the majority of hepatic functions is well compensated in aging, however, under exogenous or endogenous stimulation age-related decrease in liver functional capacity might become evident.[1] We found previously[2] that upon incubation with galactosamine (GalN) protein synthesis in hepatocytes isolated from old rats was suppressed almost twice as much as in hepatocytes from young animals. In the present investigation we looked at nucleotide content of control and aminosugar-exposed hepatocytes of both age groups.

METHODS

Male Wistar rats 4-6 and 24-29 months old ("young", and "old", respectively) were used. This strain of rats has been used in our laboratory for several years and the characteristics of the animals have been described.[3]

Isolated hepatocytes were prepared by *in situ* collagenase perfusion of rat liver as described previously.[3] 3-6x10^6 hepatocytes (>90% excluding trypan blue) were preincubated for 15 min at 37°C in Krebs-Ringer-bicarbonate buffer, pH 7.4, containing 10 mM glucose, and then GalN was added for 60 min. Samples of cell suspensions were extracted at 4°C with 12 M perchloric acid. The precipitates were centrifuged, supernatant was neutralized with 2 M KOH to pH 6.6-8.0 and stored at -70°C.

Samples were analyzed by an anion-exchange HPLC method[4] using APS-Hypersil (5μm, 25/4.6 cm) column and Merck-Hitachi automated HPLC system.

Purine and Pyrimidine Metabolism in Man VIII, Edited by
A. Sahota and M. Taylor, Plenum Press, New York, 1995

Protein was determined by the method of Lowry. Because the protein content of isolated liver cells of young and old rats was similar (2.53 ± 0.12 and 2.89 ± 0.14 mg protein/10^6 cells), all results have been expressed *per* mg protein as the mean $\pm$ SEM. Statistical significance was determined by Student's t-test.

RESULTS

The data presented in Table 1 and Figure 1 show that the content of adenine nucleotides was slightly smaller in isolated hepatocytes of old rats, however, these differences were not statistically significant. To the contrary, the level of UTP was more than twice lower in control cells of old animals, while there was no age-related difference in the cellular content of UDP-glucose and UDP-glucuronic acid.

The incubation of cells in the presence of 10 mM GalN led in both age groups to a dramatic decrease in the cellular content of UTP, UDP-glucose and UDP-glucuronic acid, which was below the sensitivity of the assay. The amount of UDP-hexosamines, which were not detectable in control cells, was similar in hepatocytes of both age groups incubated for 60 min with GalN.

Tabel 1. UTP, UDP-glucose, UDP-glucuronic acid and UDP-hexosamines content in hepatocytes of young and old rats

	UTP	UDPG	UDP-GA	UDP-HexN
	nmoles/mg protein			
Liver cells of 4-6 mo old rats				
Control	0.45 ± 0.09	0.93 ± 0.18	1.41 ± 0.09	< 0.01
GalN 10 mM	< 0.01	< 0.01	< 0.01	1.67 ± 0.19
Liver cells of 24-29 mo old rats				
Control	$0.18\pm0.03^*$	0.82 ± 0.07	1.53 ± 0.21	< 0.01
GalN 10 mM	< 0.01	< 0.01	< 0.01	1.28 ± 0.21

$^*p < 0.01$ *vs* cells of young animals, n=5.

Galactosamine diminished ATP content of hepatocytes of young and old rats by 28% and by 24%, respectively. This was accompanied by a significant decrease of the adenylate energy charge, however, no age-related differences were observed (Fig. 1). The NAD content of hepatocytes was similar in both age groups and did not change significantly on the incubation with GalN.

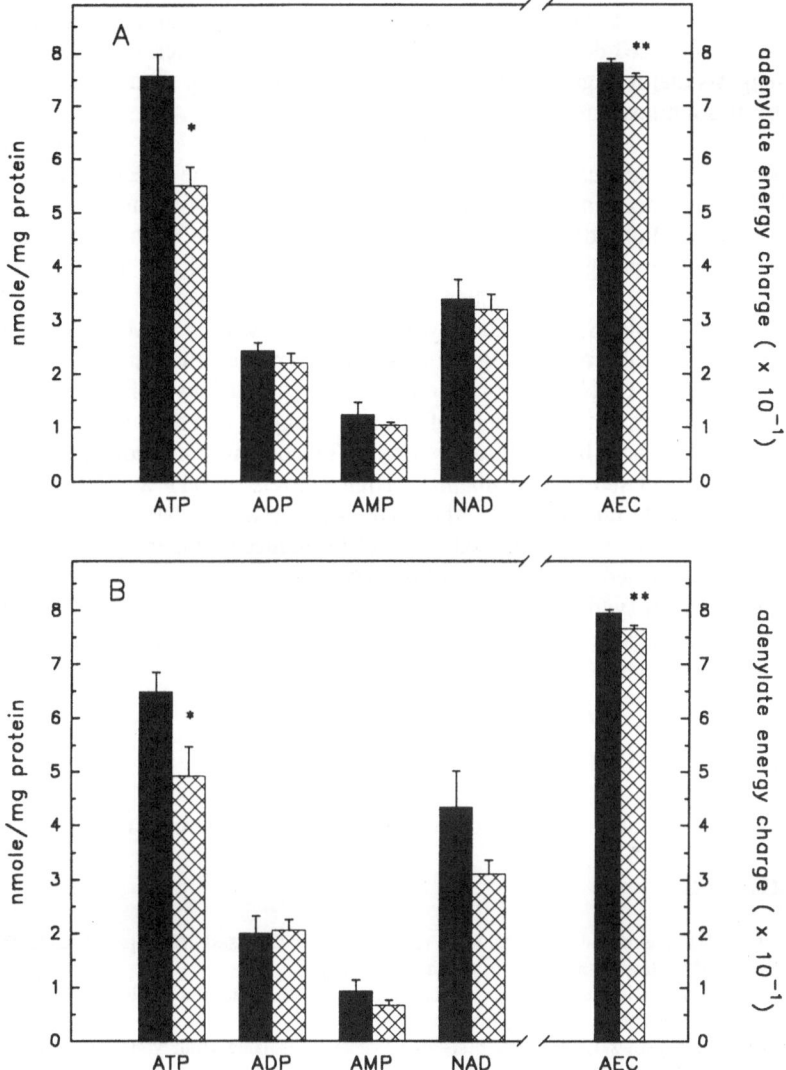

Figure 1. Hepatocytes isolated from young (A) and old (B) rats were incubated for 1h in the absence (black bars) or presence (crossed bars) of 10 mM GalN. Values represent means ± SEM, n=5. * p<0.05, ** p<0.01.

DISCUSSION

Galactosamine has been widely used as a model hepatotoxin because it produces reversible liver damage which morphologically and biochemically resembles human hepatitis. Biochemical lesions induced in liver by GalN involve depletion of uridine nucleotides and accumulation of UDP-sugars resulting in the inhibition of transcription and translation, and suppression of glycogen and UDP-glucuronic acid synthesis.[5] Galactosamine-induced decrease in adenylate energy charge which was observed in this study, suggests that it might be an additional factor contributing to the metabolic defects caused by this aminosugar.

Despite frequent use of GalN in toxicological studies, the effect of aging on GalN hepatotoxicity has been investigated by only few authors. *In vivo* studies showed either increased hepatocellular damage caused by GalN in old female Wistar rats as compared to young animals or no age-related changes in male Fischer 344 rats. In an *in vitro* study Abdul and Mehendale[6] observed better survival and higher levels of uridine nucleotides in hepatocytes of old Sprague-Dawley rats cultured with 5 mM galactosamine as compared to liver cells from young animals. However, we found that the inhibition of protein synthesis by GalN was more pronounced in suspensions of hepatocytes isolated from old male Wistar rats. The results of this study suggest that this age-dependent effect of GalN might be caused by significantly lower UTP content of hepatocytes from old rats leading to decreased substrate availability. Higher levels of UTP and UDP in cultured hepatocytes from old rats found by Abdul-Hussain and Mehendale might be caused by the dedifferantion process of cultured cells. GalN induced similar decrease of cellular ATP content in isolated hepatocytes of both young and old rats what might precipitate inhibition of protein synthesis. As the age-dependent decrease in basal UTP level might be caused by changes in UTP synthesis or degradation, the effect of aging on uridine nucleotide metabolism in rat liver has yet to be elucidated.

ACKNOWLEDGMENT

This study was partially supported by the Collaborative Research Grant from NATO.

REFERENCES

1. Popper H. (1986) Aging and the Liver, in: "Progress in Liver Diseases", H. Popper F. Schaffner, eds., 8, 659-683, Greene and Stratton, Orlando (1986).
2. Z. Kmiec, Prostaglandin cytoprotection of galactosamine-incubated hepatocytes isolated from young and old rat, Annals N.Y. Ac. Sci., in print (1994).
3. Z. Kmiec, A. Mysliwski, Urea synthesis in hepatocytes isolated from young and old rats. Exp. Gerontol. 20:271(1985).
4. H.A. Simmonds, J.A. Duley, P.M. Davies, Analysis of purines and pyrimidines in blood, urine and other physiological fluids, in: "Techniques in Diagnostic Human Biochemical Genetics: a Laboratory Manual", Wiley-Liss, New York (1991).
5. Decker K., Keppler D. (1974) Galactosamine hepatitis: key role of the nucleotide deficiency period in the pathogenesis of cell injury and cell death. Rev. Physiol. Biochem. Pharmacol. 71, 78-106.
6. Abdul-Hussain S.K., Mehendale H.M. (1991) Studies on the age-dependent effects of galactosamine in primary rat hepatocyte cultures. Toxicol. Appl. Pharmacol. 107, 504-513.

INFLUENCE OF PURINES ON ENDOTHELIAL HIGH ENERGY PHOSPHATES

Andrea Griesmacher[1], Ronney A. De Abreu[2] and Mathias M. Müller[3]

[1]Dept. of Cardiothoracic Surgery, University Hospital Vienna, A-1090
Vienna, Austria
[2]Dept. of Pediatrics University Hospital Nijmegen, NL-6500 HB Nijmegen,
The Netherlands
[3]Dept. of Laboratory Diagnostics, Kaiser Franz Josef Hospital Vienna,
A-1100 Vienna, Austria

INTRODUCTION

The endothelium shows high metabolic activity and thus takes influence on a great number of physiological processes. The role of the vascular endothelial cell in regulating the transport and the reutilisation of purine and pyrimidine metabolites has not, so far, been well defined. In regard to a possible and often discussed cellular compartmentation of adenine nucleotides and their breakdown products, the endothelial cells deserve particular attention, since they are responsible for an actively controlled metabolites' transport to the underlying strata of the blood vessels. Recently, it could be shown that endothelial cells predominately uptake and salvage radioactive adenosine [1,2,3]. Since under several pathophysiological conditions (e.g. ischemia or aggregation of platelets) many of purines are released to a greater extent into the blood stream, not only their uptake and salvage but also their influence on the absolute levels of high energy phosphates in endothelial cells seems to be of interest. In this study the effects of extracellularly added purines on the intracellular contents of the high energy phosphates were investigated. Moreover, the incorporation of [14]C- labelled adenine guanine, adenosine, hypoxanthine or inosine into the nucleotide pool should give information on the relative rates of purine metabolism.

MATERIALS AND METHODS

Cell Culture

Endothelial cells (HUVECs) were isolated from human umbilical veins and cultivated according to a standard procedure [4]. After reaching confluence the cells were seeded into precoated 24 well culture plates. The cells were identified as endothelial cells by the typical "cobblestone" contact-inhibited morphology [5] and by factor VIII (FVIII: vWF) staining [6].

Incubation Experiments

Confluent HUVECs of the first subculture were incubated for 60 min with 250 μl MEM-Medium containing 5.2 μmol/l unlabelled or ^{14}C-labelled adenine (2.04 GBq/mmol),guanine (2.07 GBq/mmol), adenosine(1.96 GBq/mmol), hypoxanthine (1.86 GBq/mmol) or inosine (2.11 GBq/mmol). After withdrawing the supernatants, cells were washed twice and lysed with 0.5 mol/l perchloric acid. The cell lysates were neutralised by adding 4 mol/l K_2HPO_4. Neutralised cell extracts were stored at -70°C.

Determination of High Energy Phosphates

ATP, ADP, GTP and GDP were measured by HPLC on a Partisil 10 SAX column using a K_2HPO_4-gradient (0.05 -0.5 mol/l).

Since in all experiments AMP and GMP were below the detection limit of the HPLC system, the total adenine nucleotide content (ΣAN) and the total guanine content (ΣGN) are only the sums of ATP and ADP or GTP and GDP, respectively.

Uptake and Incorporation of Labelled Purines

After addition of unlabelled tracers the purine metabolites were separated by thin layer chromatography on TLC aluminium silica gel 60 F254 precoated sheets (MERCK, FRG) using n-propanol, methanol, ammonia (33%), H_2O, (45:15:30:10) for nucleotides as solvent. The zones corresponding to standards were located under UV-light, cut out and their radioactivities measured.

Statistical Analysis

To compare control values with those obtained during the incubation experiments a matched pairs t test was used. $p < 0.05$ was taken to indicate a significant difference.

RESULTS AND DISCUSSION

The concentrations of ATP, ADP, GTP and GDP are summarized in Table 1 and Table 2.

Table 1: The influence of purines on the intracellular concentrations of ATP, ADP, and the total adenine nucleotide (ΣAN) content

PURINE	ATP	ADP	ΣAN
Control	11.90 ± 1.61	1.25 ± 0.90	13.19 ± 1.33
Adenine	12.85 ± 1.93*	1.28 ± 0.93	14.12 ± 1.34*
Guanine	11.99 ± 2.22	1.20 ± 0.96	13.16 ± 1.36
Adenosine	14.10 ± 2.36*	1.33 ± 0.60*	15.43 ± 1.56*
Hypoxanthine	12.30 ± 2.34	1.31 ± 0.51*	13.61 ± 1.58*
Inosine	12.07 ± 2.19	1.25 ± 0.99	13.32 ± 1.43

ATP, ADP, ΣAN: nmol/10^6 cells; *=statistically significant compared to controls (p<0.05); mean ± SD. (n=25)

Table 2: The influence of purines on the intracellular concentrations of GTP, GDP, and the total guanine nucleotide (ΣGN) content

PURINE	GTP	GDP	ΣGN
Control	1.76 ± 0.29	0.19 ± 0.05	2.10 ± 0.17
Adenine	1.69 ± 0.29	0.18 ± 0.03	2.07 ± 0.16
Guanine	1.88 ± 0.40	0.18 ± 0.04	2.34 ± 0.15*
Adenosine	1.94 ± 0.38 *	0.19 ± 0.03	2.39 ± 0.21*
Hypoxanthine	1.73 ± 0.35	0.18 ± 0.04	2.15 ± 0.18
Inosine	1.69 ± 0.39	0.18 ± 0.05	2.16 ± 0.19

GTP, GDP, ΣGN: nmol/10^6 cells; *=statistically significant compared to controls (p<0.05); mean ± SD (n=25)

As can be seen from our control experiments ATP is the main nucleotide in HUVECs. Intracellular GTP and UTP (data not shown) were found to be 6.8-fold lower than ATP, whereas CTP was under the detection limit. The ratio between ADP and ATP was 1:9.52, that between GDP and GTP 1:9.26. AMP and GMP as well as CDP, CMP UDP and UMP could not be detected in any of the experiments performed.

Although at the end of the incubation experiments considerable amounts of purines in the supernatants were detected (data not shown), indicating an excess of purine bases for the uptake and salvage processes, no toxicity was observed as described [7].

Adenine and guanine are directly salvaged to their corresponding mononucleotides by the adenine phosphoribosyltransferase (APRT) or the hypoxanthine guanine phosphoribosyl-transferase (HGPRT), respectively. Both purines caused only a slight increase by approximately 8% in their corresponding trinucleotides (ATP, GTP). The total adenine nucleotide content increased by 7% in presence of adenine, the total guanine nucleotide content by 11% in presence of guanine. The adenine nucleotides were found to contain 65% of ^{14}C- adenine, the guanine nucleotides 41% of ^{14}C-guanine (Table 3).

Table 3: Incorporation of ^{14}C-labelled purines into acid-soluble nucleotides

PURINE	AD-NT	IN-NT	GN-NT	ΣNT
Adenine	944 ± 63	353 ± 59	148 ± 38	1445 ± 235
Guanine	319 ± 58	402 ± 66	598 ± 199	1319 ± 145
Adenosine	4479 ± 210	1028 ± 145	546 ± 177	6053 ± 355
Hypoxanthine	614 ± 176	577 ± 192	359 ± 73	1550 ± 155
Inosine	500 ± 89	427 ± 75	358 ± 65	1285 ± 100

AD-NT=adenine nucleotides, IN-NT=inosine nucleotides, GN-NT=guanine nucleotides, ΣNT= sum of all ^{14}C-labelled nucleotides. Values are given as pmol/10^6 cells; mean ± SD (n=12)

Incubation with adenosine, which is converted to AMP by the enzyme adenosine kinase (AK), resulted in a pronounced increase in the formation of ATP (+18%) and a small rise in intracellular ADP. The total adenine nucleotide content was 17% higher compared to controls. 74% of ^{14}C-adenosine was found to be incorporated into the adenine nucleotides. These observations are in accordance with our recent study, in which a threefold higher activity of AK compared to the activities of APRT and HGPRT was found in intact HUVECs [1].

40% of ^{14}C-hypoxanthine (39% of ^{14}C-inosine) incorporated into nucleotides was metabolised to adenine nucleotides and 23% (28% of ^{14}C-inosine) to guanine nucleotides. Nevertheless, hypoxanthine and inosine failed to influence ATP or GTP levels. Only ADP was found to be slightly elevated in presence of hypoxanthine. Hypoxanthine and inosine (via its degradation to hypoxanthine) can both be converted to IMP by HGPRT. During the conversion of IMP to AMP or GMP, GTP or ATP are consumed. An enhanced ATP or GTP formation seems to be inhibited through this mechanism. This hypothesis is confirmed by the fact, that in case of ^{14}C-hypoxanthine and ^{14}C-inosine approx. 22% (other purines: lesser than 8%) of the labelled adenine nucleotides was AMP and more than 35% (other purines: lesser than 25%) of the labelled guanine nucleotides was GMP.

SUMMARY

It was demonstrated that extracellularly available adenosine, which is reported to protect the endothelium during ischaemia and reperfusion [8] , is predominately uptaken and salvaged by HUVECs, thereby enriching the adenine phosphate pool. In this way adenosine could contribute to a regeneration of metabolically disturbed cells via increasing the intracellular ATP. In contrast, all other purines offered can hardly be used by HUVECs to increase the pool of energy rich phosphates.

REFERENCES

1. A. Griesmacher, G. Weigel, A. Windischbauer, and M.M. Müller, Purine metabolism in human endothelial cells, Inter. J. Purine Pyrimidine Res. 2: 123 (1991).

2. A . Griesmacher, G. Weigel, I. Schimke A. Windischbauer, and M.M. Müller, The H$_2$O$_2$ induced effects on purine metabolism in human endothelial cells, Free Rad. Biol. Med. 15: 603 (1993).

3. A. Deussen, B. Bading, M. Kelm, and J. Schrader, Formation and salvage of adenosine by macrovascular endothelial cells, Am. J. Physiol. 264: H692 (1993).

4. E.A. Jaffe, R.L. Nachman, C.G. Becker, and C.R. Minck, Culture of human endothelial cells derived from umbilical veins, J. Clin. Invest. 52: 2745 (1973)

5. C.C. Haudenschild, R.S. Cotran, M.A Gimbrone, and J. Folkman, Fine structure of vascular endothelium in culture, J. Ultrastruct. Res. 50: 611 (1975).

6. E.A. Jaffe, L.W. Hoyer, and R.L. Nachman, Synthesis of antihaemophilic factor antigen by cultured human endothelial cells, J. Clin. Invest. 52: 2757 (193).

7. F.F. Snyder, J. Mendelsohn, and J.E. Segmiller, Adenosine metabolism in phytohemagglutinin - stimulated human lymphocytes, J. Clin. Invest. 58: 654 (1976).

8. S.W. Ely, and R.M. Berne, Protective effects of adenosine in myocardial ischemia, Circ. 85: 893 (1992).

RIBOSE 1-P-DEPENDENT ADP AND ATP FORMATION IN RAT LIVER

R.Leoncini, D.Vannoni, E.Marinello and R.Pagani

Institute of Biochemistry and Enzymology
University of Siena, Italy

INTRODUCTION

Different pathways lead to the formation of ATP or ADP. ATP is formed by: 1) oxidative phosphorylation; 2) substrate level phosphorylation; 3) myokinase reaction. In all these cases, ATP is derived from ADP. In reactions 2) and 3), the energy for the gamma-bond of ATP is derived from other high energy bonds, while, in reaction 1), it is obtained from electron flux through the respiratory chain.

The synthesis of ADP occurs through specific pathways, all of which involve high energy bonds; for example the degradation of ATP by ATPase, and the reverse myokinase reaction:

$$ATP + AMP \leftrightarrow 2ADP$$

In both cases, the formation of ADP involves the loss of a high energy bond by ATP. This is a puzzling aspect of energy metabolism, since the formation of ATP always starts with ADP, and the formation of ADP always involves loss of ATP, in a kind of endless circle with loss of high energy bonds:

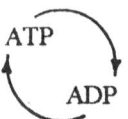

The de novo synthesis of ADP, starting from lower level energy compounds has never been reported. In recent experiments[1-3] we incubated AMP-succinate (AMP-S) at 37°C for 60 min in mixtures containing Tris, K-phosphate or pyrophosphate buffer (pH 7.5), 2.5 mM magnesium chloride, 2.5 mM potassium fluoride and liver supernatant (obtained by centrifuging homogenate at 45,000 rpm for 45 min), which had been treated with Norite A to eliminate endogenous nucleotides. The products of AMP-succinate metabolism were identified by HPLC as AMP and fumaric acid, along with other products of purine catabolism such as IMP, nucleosides, bases and uric acid. In the chromatograms we observed a peak that was not at any known mononucleotide (IMP, AMP, GMP) or fumaric acid. The absorption spectra and the release of adenine and adenylic acid (identified by

Address for correspondence: Prof. Enrico Marinello, Istituto di Biochimica e di Enzimologia, Università di Siena, Pian dei Mantellini, 44, 53100 SIENA Italy, Tel. +39-577-298026, FAX +39-577-298057

HPLC) on hydrolysis suggested that the compound was a derivative of adenylic acid. Its retention time was the same as that of ADP. The main objection to the identification of the compound as ADP was that no explanation of its formation could be advanced under these conditions. When AMP was added to the incubation mixtures, the product obtained was identical.

We have clearly identified the formed compound as ADP, through a series of all coincident tests, which are not reported here for the sake of brevity[4]. We further demonstrated that also ATP was formed under the same conditions.

A new aspect in the interpretation of the phenomena emerged when we observed that the formation of ADP from AMP under such conditions occurred when Ribose 1-P was added to the incubation mixtures[4,6-9]. Under the same conditions, we also observed that no PRPP was present. We therefore are came to the conclusion that we had discovered an hitherto unexpected phenomenon: the formation of ADP from AMP, in the presence of R-1-P. We called this reaction "ribose 1-P dependent ADP and ATP formation", a reaction which has no current explanation.

Here we report some further results demonstrating that the reaction not only occurs in the rat liver, but also in other tissues. Some properties related to the stability and kinetics of the enzyme are also reported.

METHODS

1 - Male albino Wistar rats (250 g b.w. - 9 weeks of age) were killed by decapitation. The livers and other organs were rapidly excised and homogenized (25%) at 60,000 rpm for 60 min. The supernatant was treated with Norite A for 15 min, centrifuged at 3,000 rpm for 10 min, and dialyzed against the medium of the homogenate. The supernatant was also fractionated, with solid ammonium-sulphate at 50-80% saturation, and the protein precipitate was dissolved in isotonic KCl (P_{80}).

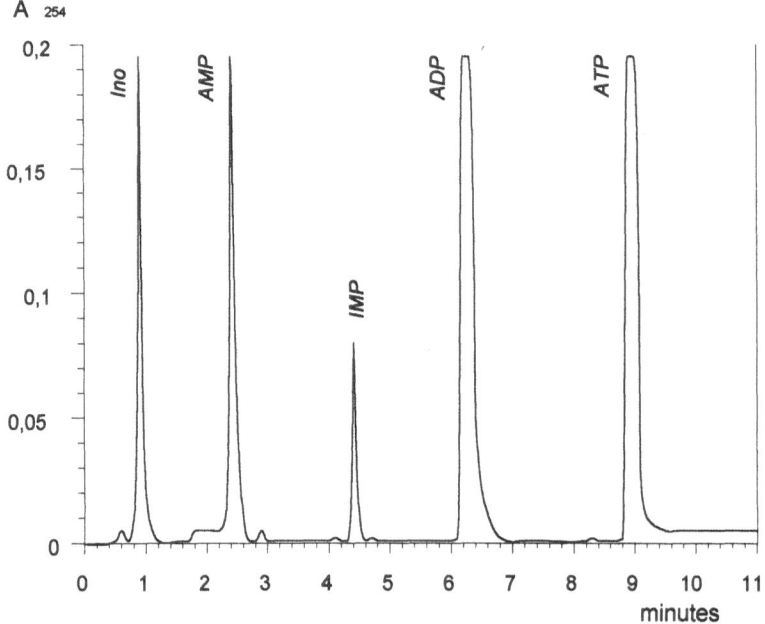

Figure 1. Typical HPLC chromatogram of standard nucleotide mixture.

2 - The assay mixtures contained 2.5 mM AMP-S or AMP, 60 mM Tris (pH 6), 2.5 mM KF, 2.5 mM MgCl$_2$ and protein extracts containing 4 mg protein. They were incubated for 60 min at 37°C, deproteinized with PCA and neutralized with KOH; 20 µl of clear supernatant was separated by HPLC.

3 - For HPLC separation we used a Varian Instruments Vista mod.5500 equipped with a variable-wavelength mod.2550 UV detector at 254 nm and a mod.4290 electronic integrator. A ready-to-use prepacked column (Partisil 5 SAX 100 x 4.6 mm), washed with 0.005 M ammonium phosphate (pH 2.8) (buffer A) completed the analytical system. Elution was carried out using a gradient with 0.5 M ammonium phosphate buffer (pH 4.8) (buffer B) as follows: from 0 to 80% in 3.5 min; 80% B for 0.5 min; from 80 to 100% B in 0.5 min; 100% B for 6.5 min. The flow rate was 1.5 ml/min. A typical chromatogram of standard compounds (which appear under tested conditions) is shown in Figure 1.

All reagents of the highest commercially available purity were from Merck or Fluka. Standard compounds and nucleotides (inosine, AMP, ADP and ATP) were obtained from Sigma.

RESULTS

Table 1 compares ribose 1-P-dependent ADP formation in the spleen, kidney, heart and liver. The phenomenon is evident in all tissues, especially the heart.

Table 1. Distribution of "Ribose 1-P-dependent ADP forming activity" in different rat tissues. Specific activity is expressed as nmol/h/mg protein.

Source	Protein mg/ml	nmoles/h	Specific activity	Total activity/g tissue
Liver	12.40	6.73	9.50	397
Spleen	11.66	3.25	2.00	150
Kidney	6.70	2.03	5.50	158
Heart	4.10	11.25	76.00	2177

We followed ADP and ATP formation at two different temperatures and at different times (Figure 2) demonstrating that the reaction has an initial lag time.

Table 2 reports the stability of the P_{80} fraction at 4°C, and shows a rapid decrease in the activity.

Table 2. Stability of P_{80} fraction at 4°C after dialysis and addition of different compounds. Values are reported as percentage of initial values.

P_{80}	100
P_{80} 24 h	79
P_{80} 48 h	58
P_{80} 72 h	24
P_{80} 1 h dialysis	86
P_{80} 4 h dialysis	128
P_{80} 24 h dialysis	154
P_{80} 24 h with 1mM AMP	80
P_{80} 24 h with 1mM R 1-P	28

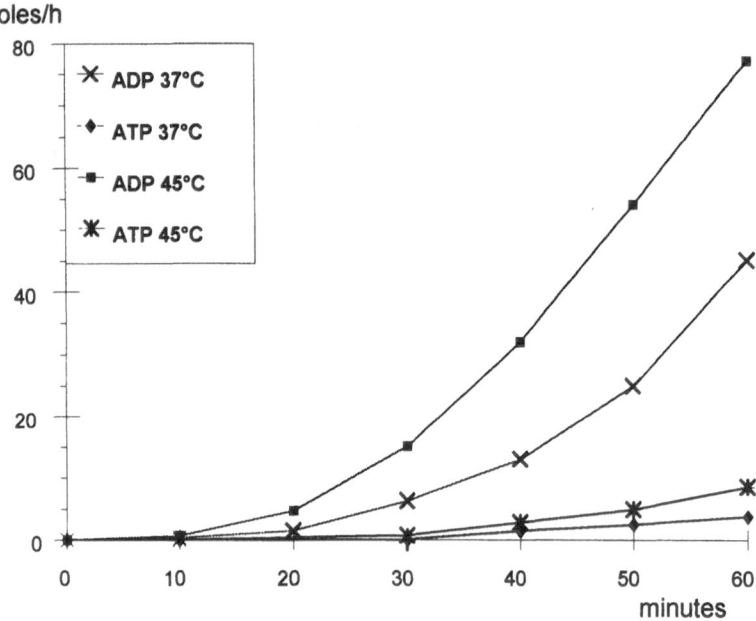

Figure 2. Behavior of ADP and ATP formation at different temperatures.

To prevent this inactivation we tested different compounds and treatments. First of all we checked the effect of substrate (1mM AMP), and cofactor (ribose 1-P 1mM) and various durations of dialysis (2, 4 and 24 hours). Only lengthy dialysis preserved enzyme activity.

Table 3 shows that the addition of glycerol at different percentages (5%, 10%) did not influence the stability of the enzyme preparation.

Table 3. Effect of different concentrations of glycerol on percentage enzyme activity in time.

	0 h	24 h	48 h	72 h
0% glycerol	100	95	81	73
5% glycerol	91	89	83	80
10% glycerol	83	80	77	74

DISCUSSION

The present experiments extend our knowledge on the phenomenon of "ribose 1-P-dependent ADP and ATP formation". Although ADP and ATP formation has yet to be clarified, this study shows that it occurs in many different tissues, besides its preferred organ, the liver. It is therefore a phenomenon of general importance.

REFERENCES

1. E.Marinello and D.Vannoni, Metabolism of AMP-S succinate in rat liver, *It.J.Biochem.* 38:386 (1989).
2. E.Marinello, D.Vannoni, B.Porcelli, A.Tabucchi, G.Cinci and R.Pagani, Metabolismo dei nucleotidi: studi sull'acido AMP-S. *Atti II Giornata di Facoltà*, Siena, 3 novembre 1990.
3. D.Vannoni, R.Leoncini, A.Tabucchi, M.Molinelli, P.Valerio, E.Marinello and R.Pagani, Formation of ADP in liver and in lymphocyte extracts. *It.J.Biochem.* 40:40 (1991).
4. R.Leoncini, D.Vannoni, C.Y.Lai, R.Guerranti, R.Pagani and E.Marinello, ADP and ATP formation in rat liver (Note I), *It.J.Biochem.* 41/5:331 (1992).
5. D.Vannoni, R.Leoncini, R.Guerranti, S.Righi, R.Pagani and E.Marinello, ADP and ATP formation in rat liver (Note II), *It.J.Biochem.* 41/5:332 (1992).
6. D.Vannoni, R.Leoncini, S.Righi, C.Y.Lai, R.Pagani and E.Marinello, ADP and ATP formation in rat liver (Note III), *It.J.Biochem.* 41/5:333 (1992).

MYOCARDIAL 5'DEOXY-5'METHYLTHIOADENOSINE PHOSPHORYLASE

Katarzyna Ruckemann, Piotr Jagodzinski
and Ryszard T. Smolenski

Department of Biochemistry
Academic Medical School of Gdansk
80-211 Gdansk, Debinki 1, Poland

INTRODUCTION

5'deoxy-5'methylthioadenosine phosphorylase (MTAP) is the enzyme involved in cleavage of 5'deoxy-5'methylthioadenosine (MTA) generated in the course of polyamine synthesis[1,2]. The enzyme is present in tissues with active polyamine pathway, but information on its myocardial activity and on the type of cells in the heart which express this activity is limited. In this study the activity of the enzyme in the homogenates of rat and human heart as well as in cardiomyocytes isolated from rat heart using collagenase perfusion technique was measured. MTAP activity was compared with that of purine nucleoside phosphorylase (PNP). Possible adenosine (ADO) breakdown via this route was evaluated by assay with ADO as the substrate.

MATERIALS AND METHODS

Specimens of human myocardium were obtained during cardiac surgery from papillary muscles obtained in the course of mitral valve replacement or from donor heart not used for transplantation. Samples were frozen in liquid nitrogen at the time of collection and were stored at -70°C. Rat heart samples were obtained from Wistar rats and were briefly flushed with 0.9 % saline to remove residual blood followed by freezing and storage as above. Rat cardiomyocytes were obtained using collagenase perfusion technique as described in detail previously[3]. Isolated cells were washed with albumin free buffer before freezing and storage. Tissue or cells were homogenized in buffer consisting of 150 mM KCl, 1 mM EDTA, 1 mM dithiothreitol and 20 mM TRIS (pH 7.0) at the tissue weight:buffer volume proportion of 1:9. During enzyme assay homogenates were further diluted eight times with the incubation buffer consisting of 50 mM $NA_2 HPO_4$, 1 mM EDTA and 1 mM dithiothreitol (pH 7.0). The substrate - MTA was present at 1 mM concentration. Incubation was carried out at 37°C for 30 min and was terminated by placing tubes in boiling waterbath for 5 min. The samples were then centrifuged at 4°C and supernatant was subjected to HPLC analysis[4]. Activity was evaluated by the measurement of adenine concentration increase. The reaction was found to be linear with the respect to incubation time and homogenate dilution within the values used in this study. Freezing and storage has no effect on MTAP activity as similar values were found in fresh tissue, both in human and rat heart. PNP was assayed using the method described[5]. Adenine formation from adenosine was evaluated under conditions of MTAP assay.

RESULTS AND DISCUSSION

Data presented here show substantal activity of MTAP in the heart (Table 1). This activity was especially high in human heart while the opposite was shown for PNP.

Table 1. Activities of MTAP, PNP and adenine formation from adenosine in heart and cardiomyocytes

	MTAP	PNP	Adenine formation from adenosine
(nmol/min/g wet wt)			
HUMAN HEART	34.4±5.2	230±28	8.1±0.9
RAT HEART	14.1±1.3	1117±273	3.7±0.3
RAT CARDIOMYOCYTES	8.0±1.5	44±41	2.8±0.4

Values represent mean ±S.E.M. (n=4-6)

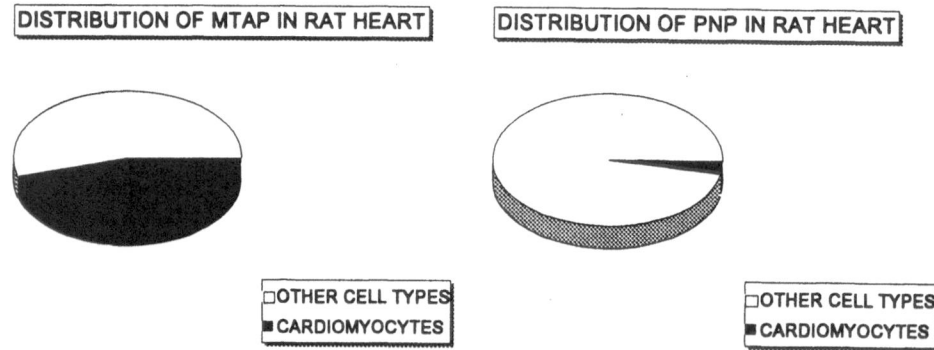

Figure 1. Contribution of cardiomyocytes to the total activity of MTAP and PNP in the rat heart

MTAP activity in rat cardiomyocytes was comparable to the total rat heart activity. This again contrasts with PNP. An estimate of the contribution of the cardiomyocyte compartment to the total activity in the heart is presented in Fig. 1. Almost half of the myocardial activity of MTAP is located in myocytes while the contribution of these cells to total heart PNP was only several percent. It is known that PNP is predominantly of endothelial location[6]. Relatively higher activity of MTAP in cardiomyocytes may be a consequence of active of polyamine synthesis, which may play an important regulatory role in the heart. ADO was shown also to release adenine under MTAP assay conditions. The rate of this release was highest in the human heart, in parallel to MTAP but not to PNP activity. Thus, MTAP seems to be responsible for small quantities of adenine which are released from the heart after ischemia, possibly from accumulated ADO.

Significance of the much higher MTAP activity in the human heart is of specific interest since most enzyme activities of purine metabolism are usually lower in man. This may imply even greater importance of polyamine metabolism in the human heart. Another possibility relates to the presence of MTAP activity which may be specialy important in the human heart is the supply of adenine nucleotide precursors.

ACKNOWLEDGMENTS

This study was supported by the Polish Committee for Scientific Research (Grant No: 4 S 402 016 04). The authors wish to thank to Dr. H.A. Simmonds and to Professor Mariusz M. Zydowo for many helpfull comments concerning this work.

REFERENCES

1. A. Sahota, D.R. Webster, C.F. Potter, H.A. Simmonds, V.A. Rodgers, and T. Gibson, Methylthoadenosine phosphorylase activity in human erythrocytes, *Clin. Chim. Acta* 128:283(1983).
2. F. Flamigni, C. Rossoni, C. Stefanelli, and C.M. Caldarera, Polyamine metabolism and function in the heart, *J. Mol. Cell. Cardiol.* 18:3(1986).
3. R.T. Smolenski, J. Schrader, H. de Groot, and A. Deussen, Oxygen partial pressure and free intracellular adenosine of isolated cardiomyocytes, *Am. J. Physiol.* 260:C708(1991).
4. R.T. Smolenski, D.R. Lachno, S.J.M. Ledingham, and M.H. Yacoub, Determination of sixteen nucleotides, nucleosides and bases using high-performance liquid chromatography and its application to the study of purine metabolism in hearts for transplantation, *J. Chromatogr.* 527:414(1990).
5. J.W. de Jong, R.T. Smolenski, M. Janssem, D.R. Lachno, M.M. Zydowo, and M.H. Yacoub, Uridine and purine nucleoside phosphorylase activity in human and rat heart, *Adv. Exp. Med. Biol.* 309B:185(1991).
6. R. Rubio and R.M. Berne, Lokalisation of purine and pyrimidine nucleoside phosphorylases in heart, kidney and liver, *Am. J. Physiol.* 239:H721(1980).

A NUCLEOSIDE MIXTURE AND ITS SPARING EFFECT ON *DE NOVO* PURINE NUCLEOTIDE SYNTHESIS

Hiroomi Yokoyama, Keiichi Okamoto, Hiroyuki Nogawa, Shinsaku Naitou and Mitsuo Itakura*

Otsuka Pharmaceutical Factory, Inc., Naruto City, Tokushima 772, Japan
*Otsuka Department of Clinical and Molecular Nutrition, School of Medicine
The University of Tokushima, Tokushima City, Tokushima 770, Japan

INTRODUCTION

Total parenteral nutrition (TPN) is a well-accepted method of managing patients to protect body composition and stimulate recover from lean body mass. While amino acids, carbohydrates and fats are essential to nutrition, nucleic acid precursors such as bases, nucleosides and nucleotides are considered less important because they are endogenously synthesized to meet the bodily requirements. However, these precursors may be necessary when the *de novo* synthesis of nucleotides is inadequate. This inadequate situation may occur during periods of severe surgical stress, infection, or long-term TPN treatment.[1] The use of nucleic acid components in TPN was investigated in rats by Ogoshi et al., using OG-VI comprising inosine, 5'-sodium guanylate, cytidine, uridine and thymidine.[2,3] Several effects of OG-VI, including a protein sparing effect in hepatectomized rats[4], have been reported as a result of the provision of nucleotide precursors.[4-7] Yet, it is not fully clear whether the OG-VI ends up sufficiently being utilized to spare the *de novo* synthesis of nucleotides in response to bodily need. The aim of this study was to investigate whether parenteral administration of OG-VI spares the *de novo* purine nucleotide synthesis which is closely associated with protein metabolism. Two experimental rat models were used which were stimulated for the *de novo* synthesis by continuous infusion of glucagon (glucagon model)[8] or by 70% hepatectomy (hepatectomized model).[9]

METHODS

Animals, OG-VI and TPN solution   Seven-week-old male Wistar rats weighing about 185 g (Charles River Japan Inc., Yokohama, Japan) were used. OG-VI (Otsuka Pharmaceutical Factory, Inc., Tokushima, Japan) contains inosine 0.80%w/v, 5'-sodium guanylate

1.22%w/v, cytidine 0.73%w/v, uridine 0.55%w/v and thymidine 0.18%w/v. TPN solution was prepared to contain 19.5% glucose, 3.3% amino acid, vitamins and electrolytes using commercial preparations.

Experiment in the glucagon model   After 24-hr infusion of 0.9% saline, animals in group G+O (n=4) received glucagon at 5 mg/kg/24 hr, TPN solution at a dose of 150 ml/kg/24 hr (137 kcal/kg/day and 0.79 g nitrogen/kg/day) and OG-VI at 0.3 mmol nucleoside (2.5 ml OG-VI), group G (n=4) received glucagon and TPN solution, and group T (n=4) received TPN solution alone. Glucagon and OG-VI were mixed in TPN solution and infused. The animals were sacrificed under anesthesia at 24 hr following TPN administration to sample liver tissue. Liver samples were stored frozen until analysis.

Experiment in the hepatectomized model   After an overnight fast, animals in group 1 (n=4) and group 2 (n=4) underwent 68% hepatectomy by the procedure of Higgins and Anderson[10] and group 3 (n=4) were sham-operated. TPN solution was infused at a speed of 150 ml/kg/24 hr following hepatectomy or sham operation. Group 1 received OG-VI mixed in TPN solution at 0.3 mmol nucleoside/24 hr. The animals were sacrificed under anesthesia at 0, 6, 12, 24 and 48 hr after hepatectomy to sample liver tissue.

Assay of de novo purine nucleotide synthesis   For the pulse-labeling of purine nucleotides and proteins, 185 KBq of 14-C-glycine (New England Nuclear, Boston, MA) was administered intravenously at 30 min before the end of TPN treatment in the experiment in a glucagon model, and 14-C-formate (New England Nuclear) was administered in the same manner in the experiment in hepatectomized model. The liver tissues sampled at the end of TPN treatment were measured for the radioactive incorporation to an acid-soluble purine fraction and acid-insoluble protein fraction,[8,9] and the ratio of purine to protein was determined.

Assay of purine nucleotide content   Purine nucleotides were extracted from liver samples with an appropriate volume of perchloric acid, and after centrifugation the supernatant was neutralized in potassium bicarbonate. The neutralized supernatant was measured for purine nucleotide by a high performance liquid chromatography system using TSK gel DEAE-2SW column (Tosoh Co., Ltd., Tokyo, Japan).

RESULTS AND DISCUSSION

In the glucagon model, incorporation of 14-C-glycine to an acid-soluble purine fraction (Fig. 1, A) in group G was significantly increased by 90% (P<0.01) at 24 hr after glucagon infusion compared with that of group T, whereas specific activity ratio of purine to protein was increased by 30% (not statistically significant). On the other hand, incorporation of 14-C-glycine to an acid-soluble purine and specific activity ratio of purine to protein in group G+O were both suppressed by 19.6% and 14.4%, respectively, compared with group G (not statistically significant). Overall, the *de novo* synthesis of purine nucleotides was suppressed by OG-VI administration, and this suggests that OG-VI is likely to spare the *de novo* purine nucleotide synthesis by suppressing the effect of glucagon. However, it is essential to examine the effect of OG-VI in relation to a beneficial metabolic effect of glucagon so that whether a decrease in the *de novo* synthesis reflects an appropriate sparing effect of the *de novo* synthesis will be understood.

The *de novo* purine nucleotide synthesis is thought to be regulated at the 5-phosphoribosyl-1-pyrophosphate (PRPP) synthetase reaction under inhibition by purine

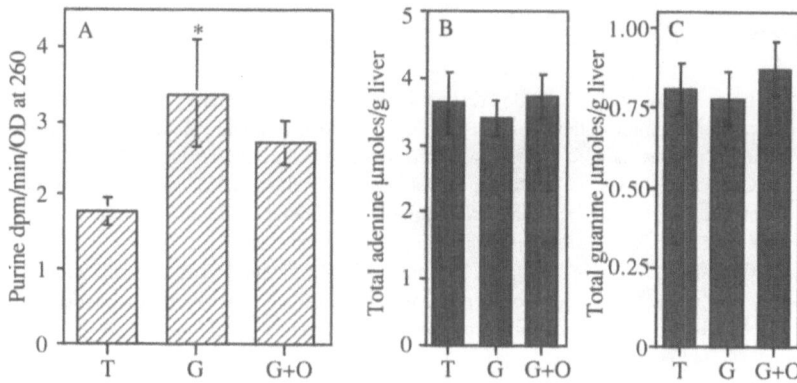

Figure 1. Effects of OG-VI on *de novo* purine nucleotide synthesis in glucagon-infused rats. A shows incorporation of 14-C-glycine to purine nucleotides. B and C show total adenine nucleotides and total guanine nucleotides, respectively. The rats receiving TPN alone, TPN+glucagon, and TPN+glucagon+OG-VI are shown by T, G and G+O, respectively. Data are means±SD. * Significant difference from T at P<0.01 (Tukey-Kramer test).

ribonucleotides,[11] and at the amidophosphoribosyltransferase (ATase) reaction under activation by substrate PRPP, feedback inhibition by purine ribonucleotides and amount of enzyme protein.[12-14] It has been reported that glucagon infusion increased PRPP concentration by 50% but did not change the purine ribonucleotide concentrations.[8] Therefore, it seems likely that if some of the contributing factors are modified, then the *de novo* synthesis by glucagon may be changed.

In the glucagon model, OG-VI slightly increased both of the contents of total hepatic adenine and guanine nucleotides (Fig 1, B, C) with no changes in adenylate energy charge. Thus, an increase in ribonucleotide content may act as feedback inhibitor on PRPP synthetase and ATase. In addition, PRPP which might be consumed in the salvage pathways for pyrimidine nucleotides, cytidine, and uridine is likely to downregulate an ATase reaction induced by glucagon infusion.

In agreement with reported results,[9] the rate of *de novo* purine synthesis in 68% hepatectomized rats (group 2), as measured by 14-C-formate incorporation to an acid-soluble purine fraction and specific activity ratio of purine to protein, was increased significantly compared with that of sham-operated rats (group 3). However, the effect of OG-VI on the rate in 68% hepatectomized rats (group 1) was not obvious at any time point of measurements as compared to that in group 2 (data not shown). OG-VI did not reduce *de novo* purine synthesis in hepatectomized rats, presumably because the effects caused by an increased metabolic flow mediated by hepatic ATase overwhelmed the effects caused by OG-VI in regulating the *de novo* purine synthesis.

CONCLUSION

OG-VI suppressed an increase in the *de novo* purine nucleotide synthesis in glucagon-infused rats, suggesting a sparing of the *de novo* purine synthesis. Thus, OG-VI may be beneficial when *de novo* synthesis of purine nucleotides is inadequate.

REFERENCES

1. S. Iijima, T. Tsujinaka, K. Kido, Y. Hayashida, H. Ishida, T. Homma, H. Yokoyama, and T. Mori, Intravenous administration of nucleosides and a nucleotide mixture diminishes intestinal mucosal atrophy induced by total parenteral nutrition, *J Paren Ent Nutr.* 17:265 (1993).

2. S. Ogoshi, M. Iwasa, T. Yonezawa, Y. Ohmori, and T. Tamiya, Effect of nucleotides and nucleoside mixture on total parenteral nutrition - Preliminary studies on ratio of nucleotides in normal rats, *Jpn J Paren Ent Nutr.* 6:53 (1984) (in Japanese).

3. S. Ogoshi, M. Iwasa, S. Kitagawa, Y. Ohmori, S. Mizobuchi, Y. Iwasa, and T. Tamiya, Effect of nucleotide and nucleoside mixture on total parenteral nutrition - Comparison of nucleotides with nucleosides mixture in hepatectomized rats, *Jpn J Paren Ent Nutr.* 6:937 (1985) (in Japanese).

4. S. Ogoshi, S. Mizobuchi, M. Iwasa, and T. Tamiya, Effect of a nucleoside nucleotide mixture on protein metabolism in rats after seventy percent hepatectomy, *Nutrition* 5:173 (1989).

5. A.A. Adjei, F. Takamine, H. Yokoyama, S-Y. Chung, L. Asato, S. Shinjo, T. Imamura, and S. Yamamoto, Effect of intraperitoneally administered nucleoside-nucleotide on the recovery from methicillin-resistant *Staphylococcus aureus* strain 8985N infection in mice, *J Nutr Sci Vitaminol.* 38:221 (1992).

6. K. Sato, T. Nakai, K. Hoshi, and K. Ichihara, Limitation of stunning in dog myocardium by nucleoside and nucleotide mixture, OG-VI, *Coronary Artery Disease* 4:1007 (1993).

7. H. Jyonouchi, L. Zhang-Shanbhag, Y. Tomita, and H. Yokoyama, Nucleotide-free diet impairs T-helper cell function in antibody production in response to T-dependent antigens in normal C57Bl/6 mice, *J Nutr.* 124:475 (1994).

8. M. Itakura, N. Maeda, M. Tsuchiya, and K. Yamashita, Glucagon infusion increases rate of purine synthesis *de novo* in rat liver, *Am J Physiol.* 253:E684 (1987).

9. M. Itakura, N. Maeda, M. Tsuchiya, and K. Yamashita, Increased rate of *de novo* purine synthesis and its mechanism in regenerating rat liver, *Am J Physiol.* 251:G585 (1986).

10. G.M. Higgins and R.M. Anderson, Experimental pathology of the liver I. Restoration of the liver of the white rat following partial surgical removal, *Arch Pathol.* 12:186 (1931).

11. M.A. Becker and M. Kim, Regulation of purine synthesis *de novo* in human fibroblasts by purine nucleotides and phosphoribosylpyrophosphate, *J Biol Chem.* 262:14531 (1987).

12. E.W. Holmes, J.B. Wyngaarden, and W.N. Kelley, Human glutamine phoshoribosylpyrophosphate amidotransferase: two molecular forms interconvertible by purine ribonucleotides and phosphoribosylpyrophosphate, *J Biol Chem.* 248:6035 (1973).

13. M. Itakura, R.L. Sabina, P.W. Heald, and E.W. Holmes, Basis for the control of purine biosynthesis by purine ribonucleotides, *J Clin Invest.* 67:994 (1981).

14. J.B. Wyngaarden and W.N. Kelley, Chapt.50, Gout, *in:*"The Metabolic Basis of Inherited Disease,"J.B. Stanbury, J.B. Wyngaarden, and D.S. Fredrickson, J.L. Goldstein, and M.S. Brown, ed., McGraw-Hill, New York (1983).

15. N. Prajda, N. Katunuma, H.P. Morris, and G. Weber, Imbalance of purine metabolism in hepatomas of different growth rates as expressed in behavior of glutamine PRPP amidotransferase, *Cancer Res.* 35:3061 (1975).

DE NOVO PYRIMIDINE NUCLEOTIDE BIOSYNTHESIS
IN SYNCHRONIZED CHINESE HAMSTER OVARY CELLS

Kiflai Bein and David R. Evans

Department of Biochemistry
Wayne State University
School of Medicine
Detroit, MI 48201

INTRODUCTION

Cellular requirements for pyrimidine nucleotides are fulfilled through the *de novo* or salvage biosynthetic pathways. In the *de novo* biosynthetic pathway, UMP, the first major end-product of the pathway, is synthesized from bicarbonate, glutamine, ATP, aspartate and phosphoribosyl pyrophosphate catalyzed by six major enzymes: carbamyl phosphate synthetase, aspartate transcarbamylase, dihydroorotase, didhydroorotate dehydrogenase, orotate phosphoribosyl transferase and orotidylate decarboxylase[1]. Mitchell and Hoogenraad[2] studied *de novo* pyrimidine biosynthesis in synchronized rat hepatoma and mouse embryo fibroblast (3T3) cells. They reported that the carbamyl phosphate synthetase and aspartate transcarbamylase enzyme activities and the incorporation of [14C] bicarbonate into acid soluble pyrimidine nucleotides and RNA were high during the S phase of the cell cycle and low during the G_2 and M phases.

In a more recent study on nucleic acid metabolism, in concanavalin A stimulated rat thymocytes, maximal incorporation rates of [14C]bicarbonate into nucleotides were detected during the S phase[3]. We wanted to investigate the mechanisms underlying the decrease in [14C]bicarbonate incorporated during the M phase. When the cell divides there is a concomitant decrease in cell size. In this study, we report that the rate of [14C]bicarbonate incorporation in Chinese hamster ovary cells arrested in the M phase by colcemid treatment is higher than in cells blocked in the S phase by thymidine blocking. The overall rate of *de novo* pyrimidine biosynthesis increases by a factor of approximately 2 over the cell cycle. As the cell increases in cell size, in preparation for cell division, there is a gradual increase in the rate

Purine and Pyrimidine Metabolism in Man VIII, Edited by
A. Sahota and M. Taylor, Plenum Press, New York, 1995

of [^{14}C]bicarbonate incorporation. Our results indicate that, unlike DNA synthesis which occurs periodically, *de novo* pyrimidine biosynthesis, like RNA and protein synthesis, occurs continuously.

METHODS AND MATERIALS

Cells and Culture Conditions

The Chinese hamster ovary cell line CHO-K1 was obtained from the American Type Culture Collection and maintained in DMEM (GIBCO) supplemented with 10% fetal calf serum (FCS), 100 µg/ml penicillin, 100 µg/ml streptomycin, and 0.1 mM nonessential amino acids. The cultures were grown at 37°C in a 5% CO_2 incubator.

Synchronization of CHO-K1 Cells

CHO-K1 cells blocked in S or G_2-M phase were obtained by double thymidine or colcemid blocking, respectively[4]. Alternatively, asynchronous CHO-K1 cells were fractionated by centrifugal elutriation using a Beckman elutriator[5]. The centrifuge was set at a rotor speed of 2000 rpm. Cells (8.7×10^7/50 ml) were introduced into the system at a pump setting of 1.6. Collection of ten fractions of 50 ml each was started at a pump setting of 2.0 and increased by 0.2 for subsequent fractions. Synchronization was verified using DNA flow cytometric analysis of CHO-K1 cells, stained with propidium iodide, at the immunology/microbiology facility, WSU.

Determination of [^{14}C]bicarbonate Incorporated into
Acid Soluble Nucleotides and Precursors

Incorporation of [^{14}C]bicarbonate into acid soluble pyrimidine biosynthesis products was measured as previously described[6] with slight modifications. The reaction was terminated by adding perchloric acid to 0.5 N. The acid lysate was processed as detailed before[6] and applied to a 1.0 ml Dowex 50X8, H$^+$ (200 to 400 mesh size) ion exchange column equilibrated with H_2O. The flow-through and wash which contained [^{14}C] incorporated in pyrimidine biosynthesis products was determined for radioactivity.

RESULTS AND DISCUSSION

The incorporation of [^{14}C]bicarbonate has been used for the simultaneous determination of the rates of pyrimidine and purine nucleotide synthesis *de novo*[6]. This method allows measurement of all the pyrimidine biosynthesis end-products and intermediates, but carbamyl phosphate, because of their acid-stable properties. Since the level of carbamyl phosphate in the cells is low[7], exclusion of the [^{14}C] incorporated in carbamyl phosphate from the acid stable counts per minute would not affect the determination.

A high degree of synchrony of CHO-K1 cells enriched for S (75%) or G2-M (70%) phase was obtained by using double thymidine and colcemid blocking, respectively. The

Table 1. *De novo* pyrimidine biosynthesis in double thymidine and colcemid blocked cells. CHO-K1 cells were incubated for 15 hrs in the presence of 2.5 mM thymidine followed by replacing with medium without inhibitor for 9 hrs, and were then incubated for another 14 hrs in the presence of 2.5 mM thymidine. After release from thymidine blocking the cells were assayed directly or further cultured in colcemid (50 μg/ml) for 9 hrs before *de novo* pyrimidine biosynthesis analysis in duplicate tubes.

Blocking agent	[^{14}C] incop.(cpm/10^6 cells/hr)
None	11115 ± 7
Thymidine	13540 ± 594
Colcemid	18671 ± 45

results show that the rate of *de novo* pyrimidine biosynthesis in CHO-K1 cells blocked in G2-M phase using colcemid is higher than in cells blocked in the S phase using thymidine (Table 1). To rule out the possibility of metabolic perturbation by chemical treatment, partial synchronization was obtained by centrifugal elutriation. An increase of about 2-fold in the rate of *de novo* pyrimidine biosynthesis was observed as the cell profile progressed from

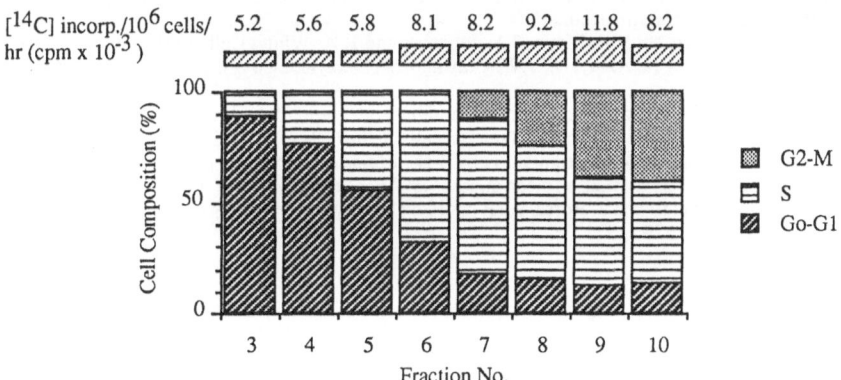

Figure 1. Flow cytometric analysis and *de novo* pyrimidine biosynthesis in different elutriator fractions. Following centrifugal elutriation, the cells were assayed for *de novo* pyrimidine biosynthesis as determined by [^{14}C]bicarbonate incorporation. The relative percents of G$_1$, S, and G2-M-phase cells obtained using FACScan (Becton & Dickinson) and ModFit (Verity) program are presented.

cells enriched for G_0-G_1 to G_2-M (Figure 1). The increase in the incorporated [^{14}C]bicarbonate correlates with the increase in cell size. In a study conducted to measure the ATP pool levels in synchronously growing Chinese hamster cells[8], Chapman *et al* showed that the ATP level increased approximately 2-fold over the cell cycle, and, the mean cell volume increased in a near-linear manner until cell division was detected. On the other hand, Mitchell & Hoogenraad[2] and Schobitz *et al*[3] concluded that maximal incorporation rates of [^{14}C]bicarbonate into nucleotides occur during the S phase. It is possible that this increase does not represent a cell cycle stage-specific activation of *de novo* pyrimidine biosynthesis. Although the rate of [^{14}C]bicarbonate incorporation increases during the S phase, the rate of [^{14}C]bicarbonate incorporation during the G2-M phase, before cell division occurs, is high as well. Just as replication of the DNA is required for cell propagation and genetic continuity, a continuous supply of nucleotides, RNA and protein is essential for normal growth and cellular functions. We conclude that, unlike DNA synthesis which occurs periodically, *de novo* pyrimidine biosynthesis occurs continuously during the cell cycle. The S phase-specific increase in the rate of [^{14}C]bicarbonate incorporation, if it occurs, is small. (Supported by ACS Grant BE-141)

REFERENCES

1. M.E. Jones, Pyrimidine nucleotide biosynthesis in animals: genes, enzymes, and regulation of UMP biosynthesis, *Annu Rev Biochem* 49: 253 (1980).
2. A.D. Mitchell and N.J. Hoogenraad, *De novo* pyrimidine nucleotide biosynthesis in synchronized rat hepatoma (HTC) cells and mouse embryo fibroblast (3T3) cells, *Exp Cell Res* 93: 105 (1975).
3. B. Schobitz, S. Wolf, R.I. Christopherson, and K. Brand, Nucleotide and nucleic acid metabolism in rat thymocytes during cell cycle progression, *Biochim Biophys Acta* 1095: 95 (1991).
4. T. Simmons, P. Heywood, S. Taube, and L. D. Hodge, Approaches to the study of the regulation of nuclear RNA synthesis in synchronized mammalian cells, *in:* "Cell Cycle Controls," G.M. Padilla, I.L. Cameron, and A. Zimmerman, ed. (1974).
5. D.J. Grdina, M.L. Meistrich, R.E. Meyn, T.S. Johnson, and R.A. White, Cell synchrony techniques, *in:* 'Techniques in Cell Cycle Analysis," J.W. Gray and Z. Darzynkiewicz, ed., Human Press, Clifton (1987).
6. W.H. Huisman, K.O. Raivio, and M.A. Becker, Simultaneous estimation of rates of pyrimidine and purine nucleotide synthesis *de novo* in cultured human cells, *J Biol Chem* 254: 12595 (1979).
7. M. Tatibana, K. Kita, and T. Asai, Stimulation by 6-azauridine of carbamoyl phosphate synthesis for pyrimidine biosynthesis in mouse spleen slices, *Eur J Biochem* 128: 625 (1982).
8. J.D. Chapman, R.G. Webb, and J. Borsa, ATP pool levels in synchronously growing Chinese hamster cells, *J Cell Biol* 49: 229 (1971).

THE CATALYTIC MECHANISM OF HAMSTER DIHYDROOROTASE

Neal K. Williams, Elizabeth L. Isaac, Yin Peide, and Richard I.
Christopherson

Department of Biochemistry
University of Sydney
Sydney NSW 2006
Australia

INTRODUCTION

Dihydroorotase catalyses the third reaction of the pathway for *de novo* biosynthesis of pyrimidine nucleotides. The reaction catalysed involves an intramolecular ring closure of N-carbamyl-L-aspartate (CA-asp) to form L-dihydroorotate (DHO), a dihydropyrimidine. The interconversion of CA-asp and DHO is freely reversible (Christopherson and Jones, 1979) and the pH-rate profiles for the biosynthetic and degradative reactions and inhibition of the enzyme by L-cysteine suggested that mammalian dihydroorotase has a catalytic mechanism similar to carboxypeptidase A (Christopherson and Jones, 1980) with a zinc atom at the active site. This zinc atom would polarise the $=C=O$ group of the scissile peptide bond in the dihydropyrimidine ring of DHO and stabilise the tetrahedral, dioxy anionic transition state of the reaction (Fig. 1). Taylor *et al.* (1976) had purified the dihydroorotase from *Clostridium oroticum* and found two gram atoms of zinc per subunit and similar pH-rate profiles. Kelly *et al.* (1986) subsequently demonstrated that hamster dihydroorotase also contained an atom of zinc which was required for catalysis. It therefore seems likely that mammalian dihydroorotase does have a catalytic mechanism which resembles that of zinc protease.

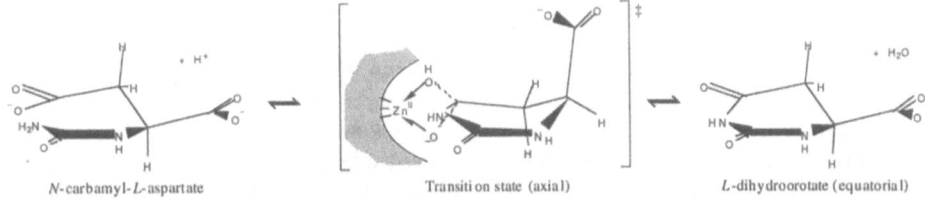

Figure 1. Interconversion of N-carbamyl-L-aspartate and L-dihydroorotate catalysed by dihydroorotase.

In mammals, dihydroorotase is the central domain of a trifunctional protein called CAD or DHO synthetase, which also contains the first two enzymic activities of the *de novo* pyrimidine pathway, carbamyl phosphate synthetase and aspartate transcarbamylase

(Simmer *et al.*, 1990; Williams *et al.*, 1990). The cDNA encoding the hamster dihydroorotase domain has been sub-cloned from the plasmid pCAD142, expressed in *Escherichia coli* SØ1263 *pyrC⁻* and the recombinant domain was purified (Williams *et al.*, 1993); similar results were reported by Zimmermann and Evans (1993). With the cDNA encoding the active dihydroorotase domain, and a suitable expression system, we have used chemical modification and site-directed mutagenesis to determine which amino acid residues are at the active site and whether they are involved in binding of the zinc atom or substrate; or participate in subsequent catalysis.

EXPERIMENTAL PROCEDURES

1-Cyclohexyl-3-(2-morpholinoethyl)-carbodiimide metho-p-toluenesulfonate (CMC), 1,2-cyclohexanedione and citraconic anhydride were purchased from the Sigma Chemical Company, USA. Tetranitromethane was obtained from the Aldrich Chemical Company Inc., USA. L-[^{14}C]dihydroorotate (57.0 Ci/mol) was synthesised from [^{14}C]bicarbonate (Christopherson *et al.*, 1989). [α–^{35}S]dATPαS (>1000 Ci/mmol) was from Amersham International, UK, Sequenase^{TM} sequencing kit was from United States Biochemical Corporation, USA and all restriction enzymes were from Boehringer Mannheim, FRG. The Oligonucleotide-directed *in vitro* Mutagenesis System was from Amersham International plc (Amersham, UK).

Assay of Dihydroorotase

For chemical modification experiments, assay mixtures contained in a total volume of 25 µl: 50 mM K. Hepes (pH 7.4), 10% (v/v) glycerol and 100 µM [^{14}C]DHO (5.0 Ci/mol; Christopherson *et al.*, 1978). For determination of the K_m and V_{max} values of mutant dihydroorotases; 10 concentrations of DHO were used over the range 2-200 µM and data were fitted by non-linear regression to the Michaelis-Menten equation using the program DNRP53 (Duggleby, 1984).

Purification of DHO Synthetase and Recombinant Dihydroorotase

DHO synthetase was purified from a mutant hamster cell line (165-23) which over-produces the trifunctional protein by more than 100-fold. Confluent cells were harvested, extracted and DHO synthetase was purified in a single step by adsorption to and elution from a column of Procion Blue MX-4GD-Sepharose 4B (Crofts *et al.*, 1990). The recombinant dihydroorotase domain was expressed in *E. coli* SØ1263 *pyrC⁻* transformed with the plasmid pCW27 and purified as described by Williams *et al.* (1993).

^{65}Zn-labelling of Recombinant Mutant Dihydroorotases

E. coli SØ1263 *pyrC⁻* transformed with the "wild-type" pCW27 expression plasmid or one of a variety of mutant plasmids, was grown in M9 medium with appropriate supplements (3 ml). Expression of the dihydroorotase domain was induced with IPTG (10 µM) and ^{65}ZnCl$_2$ (600 pmol, 54 kCi/mol) was also added, the culture was incubated with shaking at 30˚C for 4.5 h, cells were then harvested, extracted, run on non-denaturing polyacrylamide gel electrophoresis and autoradiographed.

RESULTS AND DISCUSSION

The zinc protease, carboxypeptidase A, contains Glu, Arg, Tyr and Lys residues at the active site which are involved in binding the peptide substrate and/or in subsequent catalysis (Christianson and Lipscomb, 1989). To determine whether dihydroorotase also used these residues to catalyse a similar reaction, purified DHO synthetase was modified with 1-cyclohexyl-3-(2-morpholinoethyl)-carbodiimide metho-p-toluenesulfonate (CMC), 1,2-cyclohexanedione (CHD), tetranitromethane (TNM) and citraconic anhydride (CAA, Table I) which are specific for Glu or Asp, Arg, Tyr and Lys residues, respectively, under the conditions used (Lundblad and Noyes, 1988). The presence of the substrate DHO (5 mM) had little effect upon the relative rate of inactivation of dihydroorotase by CMC, but enhanced inactivation by CHD 6.3-fold and by TNM 7.6-fold. DHO partially protected dihydroorotase against inactivation by CCA, there was a 4.9-fold reduction in the value of k/k_o, but the rate of inactivation was still 13-fold greater than that of the control (Table I).

Table I. Modification of Amino Acid Residues of Dihydroorotase.

Preincubation mixtures contained purified DHO synthetase (0.10 μg), modification reagents as indicated, 50 mM K.Hepes (pH 7.2) or K.Mes (pH 6.4), and [^{14}C]DHO (5 mM, 5.0 Ci/mol) as required. Reactions were at 25°C and samples (2 μl) were taken at 4 appropriate times for determination of dihydroorotase activity as described in Experimental Procedures. Decay of dihydroorotase activity due to modification of amino acid residues is expressed as a ratio (k/k_o) of the first-order rate constants in the presence (k) and absence (k_o) of the modification reagent.

Amino acid residue	pH	Preincubation conditions[a]	k (min^{-1}) x 10^3	k/k_o
Glu or Asp	6.4	Control	3.6	1
		CMC (50 mM)	9.5	2.7
		CMC (50 mM) + DHO (5 mM)	14	3.9
Arg	7.2	Control	20	1
		CHD (5 mM)	43	2.2
		CHD (5 mM) + DHO (5 mM)	270	14
Tyr	7.2	Control	21	1
		TNM (5 mM)	28	1.4
		TNM (5 mM) + DHO (5 mM)	210	10
Lys	7.2	Control	3.7	1
		CCA (5 mM)	240	65
		CCA (5 mM) + DHO (5 mM)	49	13

[a] The modification reagents used were: CMC, 1-cyclohexyl-3-(2-morpholinoethyl)-carbodiimide metho-p-toluenesulfonate; CHD, 1,2-cyclohexanedione; TNM, tetranitromethane; CCA, citraconic anhydride.

Chemical modification of DHO synthetase has shown that Glu (or Asp), Arg, Tyr and Lys residues are required for dihydroorotase activity (Table I). The data are consistent with Glu (or Asp), Arg and Tyr residues which are essential for dihydroorotase activity but are not at the active site because DHO does not protect dihydroorotase from inactivation. Indeed, for Arg and Tyr residues, the rate of inactivation is greater in the presence of DHO, consistent with a conformational change of the dihydrootoase domain during catalysis. Inactivation of dihydroorotase by CCA is partially protected against by DHO suggesting the presence of a Lys residue at the active site which interacts with the substrate (Table I). Modification of additional Lys residues remote from the active site would explain why the protection from inactivation by DHO is not complete.

While chemical modification experiments may give an indication of which amino acid residues are at the active site of an enzyme, such reagents are not completely specific and more than one residue may be modified. Site-directed mutagenesis of the cDNA encoding an enzyme is a more powerful technique for determining the catalytic roles of amino acids providing it is known which amino acids may be found at the active site. Alignment of the 8 known sequences for dihydroorotases from *E. coli, Salmonella typhimurium, Saccharomyces cerevisiae, Ustilago maydis, Bacillus subtilis, Dictostelium discoideum, Drosophila melanogaster* and hamster has revealed 22 amino acid residues which are totally conserved between species. Such residues may be used for binding of the substrate and/or catalysis, binding of the zinc atom at the active site or maintenance of the conformation of the active site. Modification of DHO synthetase with citraconic anhydride provided evidence for a Lys residue at the active site and one of the 22 conserved residues is Lys239 where amino acids are numbered from the N-terminal Thr1 of the dihydroorotase domain liberated after proteolysis of trifunctional DHO synthetase with elastase (Williams *et al.*, 1990). Using site-directed mutagenesis, the mutant recombinant dihydroorotases with the substitutions Lys239→Arg and Lys239→Gly were prepared, purified and their catalytic properties (K_m and V_{max}) were compared with the pure wild-type enzyme (Table II).

Table II. Catalytic Properties of Mutant Dihydroorotases.

Dihydroorotase	K_m (μM)	V_{max} (μmol min^{-1} mg^{-1})
Wild-type	4.02±0.71	1.15±0.06
Lys239→Arg	15.5±1.2	1.03±0.02
Lys239→Gly	454±45	2.41±0.11

The conservative substitution Lys239→Arg increased the K_m value for DHO 3.9-fold without significant effect upon V_{max} while Lys239→Gly increased the K_m value 110-fold and the V_{max} 2.1-fold. These data indicate that Lys239 is involved in binding DHO at the active site and is not directly involved in catalysis. The increase in the K_m of 110-fold suggests that the positively charged Lys239, or Arg239 of the first mutant, forms a hydrogen bond with the dihydropyrimidine ring of DHO. The loss of an electrostatic bond to the substrate would produce a far greater increase in K_m, and perhaps an essentially inactive enzyme. Such inactivation was observed for the mutant Arg19→Gly where an electrostatic bond has been proposed with the substrate (see N.K. Williams, S. O'Donoghue, and R.I. Christopherson, in this volume).

For carboxypeptidase A, the ligands of the zinc atom at the active site are two His residues and a Glu. The amino acid sequence of the hamster dihydroorotase domain contains 5 His residues which are totally conserved through the 8 known dihydroorotase sequences (His15,17,158,186 and 234), several other His residues are partially conserved (e.g. His 134 and 185). To determine which His residues coordinate the zinc atom at the active site of dihydroorotase, the following mutant enzymes were prepared: His15→Gly, His17→Gly, His134→Gly, His186→Ala and His234→Ala and transformed strains of *E. coli* SØ1263 *pyrC$^-$* were grown in the presence of ^{65}ZnCl$_2$ as described in Experimental Procedures. The mutant recombinant dihydroorotases His15→Gly and His17→Gly did not bind ^{65}Zn, His186→Ala showed only partial incorporation of ^{65}Zn while the remaining 2 mutant proteins bound ^{65}Zn like the wild-type enzyme. It is concluded that His15 and His17 directly bind the zinc atom at the active site but the indentity of the third zinc ligand is uncertain. Likely roles of other amino acid residues which have been conserved through evolution will be proposed in a second chapter (N.K. Williams, S. O'Donoghue, and R.I. Christopherson, in this volume) along with a proposed arrangement of residues at the active site.

Acknowledgements

We thank Kristen K. Seymour for technical assistance. This research was supported by a grant from the University of Sydney Cancer Research Fund.

REFERENCES

Christianson, D.W., and Lipscomb, W.N. 1989, Carboxypeptidase A, *Acc. Chem. Res.* 22:62.

Christopherson, R.I., and Jones, M.E. 1979, Interconversion of carbamyl-L-aspartate and L-dihydroorotate by dihydroorotase from mouse Ehrlich ascites carcinoma, *J. Biol. Chem.*, 254:12506.

Christopherson, R.I., and Jones, M.E., 1980, The effects of pH and inhibitors upon the catalytic activity of the dihydroorotase of multienzymatic protein *pyr1-3* from mouse Ehrlich ascites carcinoma, *J. Biol. Chem.*, 255:3358.

Christopherson, R.I., Matsuura, T., and Jones, M.E., 1978, Radioassay of dihydroorotase utilizing ion-exchange chromatography, *Anal. Biochem.*, 89:225.

Christopherson, R.I., Schmalzl, K.J., Szabados, E., Goodridge, R.J., Harsanyi, M.C. Sant, M.E., Algar, E.M., Anderson, J.E., Armstrong, A., Sharma, S.C. Bubb, W.A., and Lyons, S.D., 1989, Mercaptan and dicarboxylate inhibitors of hamster dihydroorotase, *Biochemistry*, 28:463.

Crofts, L., Yin, P. Woodhouse, A., Algar, E.M., and Christopherson, R.I., 1990, Purification of hamster dihydroorotate synthetase using Procion Blue-Sepharose, *Prot. Exp. Purif.*, 1:45.

Duggleby, R.G., 1984, Regression analysis of non-linear Arrhenius plots: an empirical model and a computer program, *Comput. Biomed. Res.*, 14:447.

Kelly, R.E., Mally, M.I., and Evans, D.R., 1986, The dihydroorotase domain of the multifunctional protein CAD. Subunit structure, zinc content, and kinetics, *J. Biol. Chem.*, 261:6073.

Lundblad, R.L., and Noyes, C.M. 1988, "Chemical Reagents for Protein Modification", CRC Press, Boca Raton, Florida.

Simmer, J.P., Kelly, R.E., Rinker Jr., A.G., Zimmermann, B.H., Scully, J.L., Kim, H., and Evans, D.R. 1990, Mammalian dihydroorotase: nucleotide sequence, peptide sequences, and evolution of the dihydroorotase domain of the multifunctional protein CAD, *Proc. Natl. Acad. Sci. USA*, 87:174.

Taylor, W.H., Taylor, M.L., Balch, W.E., and Gilchrist, P.S., 1976, Purification and properties of dihydroorotase, a zinc-containing metalloenzyme in *Clostridium oroticum*, *J. Bacteriol.*, 127:863.

Williams, N.K., Peide, Y., Seymour, K.K., Ralston, G.B., and Christopherson, R.I., 1993, Expression of catalytically active hamster dihydroorotase domain in *Escherichia coli*: purification and characterisation, *Protein Eng.*, 6:333.

Williams, N.K., Simpson, R.J., Moritz, R.L. Peide, Y., Crofts, L., Minasian, E., Leach, S.J. Wake, R.G., and Christopherson, R.I., 1990, Location of the dihydroorotase domain within trifunctional hamster dihydroorotate synthetase, *Gene*, 94:283.

Zimmermann, B.H., and Evans, D.R., 1993, Cloning, overexpression, and characterisation of the functional dihydroorotase domain of the mammalian multifunctional protein CAD, *Biochemistry*, 32:1519.

STIMULATION BY 6-AZAURIDINE OF *DE NOVO* PYRIMIDINE BIOSYNTHESIS IN BHK 165-23 CELLS

Kiflai Bein and David R. Evans

Department of Biochemistry
Wayne State University
School of Medicine
Detroit, MI 48201

INTRODUCTION

The uridine analog 6-azauridine is taken up by cells and phosphorylated to 6-azaUMP which acts on the OMP decarboxylase of the UMP synthetase, thus inhibiting *de novo* biosynthesis of pyrimidine nucleotides[1]. Upon 6-azauridine treatment, significant increases in the level of the *de novo* pyrimidine biosynthesis intermediates and derivatives, orotic acid and orotidine, have been noted in tissue culture cells, the tissues and urine of mice, and the urine of humans[2,3,4,5,6,7,8]. The accumulation of the intermediates has been explained by the inhibition of the decarboxylation of OMP by 6-azaUMP. However, Tatibana *et al* showed that 6-azauridine treatment of mouse spleen cells also stimulated carbamyl phosphate synthesis[3]. They suggested that an increase in the concentration of PRPP was the most likely regulatory factor stimulating *de novo* pyrimidine biosynthesis. Alternatively, with a decrease in the nucleotide pool caused by 6-azauridine treatment[8], the feedback inhibition on the carbamyl phosphate synthetase could be released resulting in enhanced *de novo* pyrimidine biosynthesis.

In the present study the cell line BHK 165-23 which expresses 115 times the CAD concentration in wild type cells was used for studying the effect of 6-azauridine on *de novo* pyrimidine biosynthesis. Unlike CHO-K1 cells which show approximately two-fold stimulation, treatment of BHK 165-23 cells with 0.1 mM 6-azauridine for 1 hr caused a 10-fold increase in the sum of acid stable [^{14}C]bicarbonate incorporated into pyrimidine biosynthetic pathway intermediates. It appears that neither a release of the feedback inhibition nor an increase in the concentration of PRPP is responsible for the stimulation observed.

Purine and Pyrimidine Metabolism in Man VIII, Edited by
A. Sahota and M. Taylor, Plenum Press, New York, 1995

METHODS AND MATERIALS

Cells and Culture Conditions

The CAD overproducing cell line BHK 165-23 and the Chinese hamster ovary cell line (CHO-K1) were used for studying the effect of 6-azauridine on *de novo* pyrimidine biosynthesis. The cell line BHK 165-23 was a gift from Dr. G.R. Stark. The cell line CHO-K1 was obtained from the American Type Culture Collection. The cells were maintained in DMEM (GIBCO) supplemented with 10% fetal calf serum (FCS), 100 μg/ml penicillin, and 100 μg/ml streptomycin. In addition, the CHO-K1 cells were supplemented with 0.1 mM nonessential amino acids. The cultures were grown at 37°C in a 5% CO_2 incubator.

Determination of [14C]bicarbonate Incorporated into Acid Soluble Nucleotides and Precursors

Incorporation of [14C]bicarbonate into acid soluble pyrimidine biosynthetic products was measured as previously described[9] with slight modifications. The reaction was terminated by adding perchloric acid to 0.5 N. The acid lysate was processed as detailed before[9] and applied to a 1.0 ml Dowex 50X8, H^+ (200 to 400 mesh size) ion exchange column equilibrated with H_2O. The flow-through and wash which contained [14C] incorporated into pyrimidine biosynthesis products was determined for radioactivity.

RESULTS AND DISCUSSION

The rate of [14C]bicarbonate incorporation into pyrimidine biosynthetic pathway products was determined as described before[9]. In asynchronously growing cells assayed at pH 7.4, in a suspension containing 7.5 mM NaHCO3 and 10 μCi [14C]NaHCO3 (58 mCi/mmol), the rate of *de novo* pyrimidine biosynthesis was 11921 ± 645 cpm/10^6 cells/hr in CHO-K1 and 11342 ± 923 cpm/10^6 cells/hr in BHK 165-23 cells.

The cell line BHK 165-23 was selected by exposing BHK cells to increasing concentrations of the bisubstrate analog phosphonacetyl-L-aspartate, an inhibitor of the aspartate transcarbamylase domain of CAD[10]. The cells contain 115-fold higher levels of CAD than the wild type cells. Despite the expression of such high levels of the rate-limiting enzyme CAD, the rate of [14C]bicarbonate incorporation into the pyrimidine biosynthetic pathway is not increased. This indicates that the rate of *de novo* pyrimidine biosynthetic pathway in BHK 165-23 cells operates at about 1% of its theoretical maximal capacity. The total acid soluble pyrimidine fraction prepared as described[9] represents the total amount of [14C]bicarbonate incorporated. Assay of cells in the presence and absence of 6-azauridine shows that 6-azauridine stimulates incorporation of [14C]bicarbonate into pyrimidine precursors. In BHK 165-23 cells the [14C]bicarbonate incorporated in the presence of 0.1 mM 6-azauridine was 10 times greater than in its absence (Table 1). In CHO-K1 cells only

Table 1. Effect of 6-azauridine on $[^{14}C]NaHCO_3$ incorporation in CHO-K1 and BHK 165-23 cells. Cells were resuspended in 0.5 ml suspension medium containing 10% FCS and 7.5 mM $NaHCO_3$ with or without 0.1 mM 6-azauridine. After preincubating at 37° for 60 min the cell suspension was chased with 10 μCi $[^{14}C]NaHCO_3$ for 1 hr. The reaction was terminated by adding $HClO_4$ to 0.5 N and analyzed for pyrimidine biosynthesis as described in methods and materials.

Cell Type	Number of cells (x 10^{-6})	$[^{14}C]$ incorporated* (cpm/10^6 cells/hr)		Ratio
		-AzU	+AzU	(+/-)
CHO-K1	0.5	6298	14178	2.3
	1.0	14870	35710	2.4
	2.0	28650	57468	2.0
BHK165-23	0.5	4263	43785	10.1
	1.0	10158	118895	11.7
	2.0	26548	208515	7.8

* The cells were incubated without (-) or with (+) 0.1 mM 6-azauridine.

about two-fold increase was obtained. The level of stimulation obtained with CHO-K1 cells (2.0 to 2.4) is comparable to the previous 2.1 to 2.3-fold increase in the rate of pyrimidine biosynthesis obtained upon treatment of mouse spleen slices with 0.5 mM 6-azauridine[3]. This observation suggests that although an increase in the concentration of the rate limiting enzyme CAD *per se* may not increase the rate of *de novo* pyrimidine biosynthesis significantly, addition of 6-azauridine leads to release of the regulatory mechanism(s). Galactosamine treatment has been shown to be an effective way of reducing the concentration of uridine nucleotides[11]. When BHK 165-23 cells were treated with 0.001 to 4 mM galactosamine the maximum incorporation of $[^{14}C]$bicarbonate obtained was about 1.5 times greater than in control cells suggesting that release of feedback inhibition is not responsible for the significant increase in pyrimidine biosynthesis detected. On the other hand, a 40-fold induction in PRPP concentration using 50 mM K_2HPO_4 did not affect the rate of $[^{14}C]$bicarbonate incorporation. More studies are needed to explain the mechanism of 6-azauridine-mediated activation of *de novo* pyrimidine biosynthesis. (Supported by ACS Grant BE-141)

REFERENCES

1. C.A. Pasternak and R.E. Handschumacher, The biochemical activity of 6-azauridine: interference with pyrimidine metabolism in transplantable mouse tumors, *J Biol Chem* 234: 2992 (1959).
2. J-J. Chen and M.E. Jones, Effect of 6-azauridine on *de novo* pyrimidine biosynthesis in cultured Ehrlich ascites cells: orotate inhibition of dihydroorotase and dihydroorotase dehydrogenase, *J Biol Chem* 254: 4908 (1979).

3. M. Tatibana, K. Kita, and T. Asai, Stimulation by 6-azauridine of carbamoyl phosphate synthesis for pyrimidine biosynthesis in mouse spleen slices, *Eur J Biochem* 128: 625 (1982).

4. R.E. Handschumacher, Orotidylic acid decarboxylase: inhibition studies with 6-azauridine 5'-phosphate, *J Biol Chem* 235: 2917 (1960).

5. V. Habermann, Elimination by the urine of orotic acid and orotidine in man after application of 6-azauracil, *Biochim Biophys Acta* 43:137 (1960).

6. H.J. Fallon, E. Frei, J. Block, and J.E. Seegmiller, The uricosuria and orotic aciduria induced by 6-azauridine, *J Clin Investig* 40: 1906 (1961).

7. H.J. Fallon, E. Frei, and E.J. Freireich, Correlations of the biochemical and clinical effects of 6-azauridine in patients with leukemia, *Am J Med* 33: 526 (1962).

8. C.M. Janezay and S. Cha, Effects of 6-azauridine on nucleotides, orotic acid, and orotidine in L5178Y mouse lymphoma cells *in vitro*, *Cancer Res* 37: 4382 (1977).

9. W.H. Huisman, K.O. Raivio, and M.A. Becker, Simultaneous estimation of rates of pyrimidine and purine nucleotide synthesis de novo in cultured human cells. *J Biol Chem* 254: 12595 (1979).

10. R.A. Padgett, G.M. Wahl, P.F. Coleman, and G.R. Stark, N-(phosphonacetyl)-L-aspartate-resistant hamster cells overaccumulate a single mRNA coding for the multifunctional protein that catalyzes the first steps of UMP synthesis. *J Biol Chem* 254: 12595 (1979).

11. J. Pausch, J. Wilkening, J. Nowack, and K. Decker, Control of pyrimidine biosynthesis in the perfused liver: feedback inhibition of glutamine-dependent carbamyl phosphate synthetase, *Eur J Biochem* 53: 349 (1975).

12. A. Petersen and B. Quistorff, Induction of millimolar amounts of 5-phosphoribosyl-1-pyrophosphate in human erythrocytes by incubation in inosine-pyruvate-phosphate medium. A ^{31}P-NMR study, *Biomed Biochim Acta* 49, S111 (1990).

OBSERVED RESISTANCE TO PYRIMIDINE ANALOGS AND SENSITIVITY TO URACIL IN *Drosophila* IS ATTRIBUTED TO DEREGULATION OF PYRIMIDINE METABOLISM

Jure Piškur[1], Leif Søndergaard[1], Zoran Gojkovic[1], Birgitte Stokbro[1], Charlotte Hjulsager[1], Jeffrey Davidson[2], Edward DeMoll[2], John Rawls[2] and Erik Bahn[1]

[1]Department of Genetics
Øster Farimagsgade 2A
University of Copenhagen
DK-1353 Copenhagen K, Denmark
[2]University of Kentucky
Lexington, KY 40536, USA

INTRODUCTION

Pyrimidine nucleotides play a central role in cellular metabolism and regulation. In most organisms two pathways provide pyrimidines: the *de novo* biosynthetic pathway and the salvage pathway. *Drosophila melanogaster,* the fruit fly, is an ideal organism for study of the genetic basis and regulatory mechanisms of various metabolic pathways. *De novo* pyrimidine biosynthesis in the fruit fly is a six-step pathway, which is catalyzed by enzymes encoded by three separate genes (Freund and Jarry, 1987; Rawls et al., 1993; Eisenberg et al., 1993). The gene *rudimentary (r)* is *Drosophila's* equivalent of the mammalian gene for CAD (Freund and Jarry, 1987). *De novo* pyrimidine biosynthesis is important for the proper development of flies. However, the salvage pathway can suffice when the external supply of pyrimidines is very high (Falk and Nash, 1974).

In higher organisms pyrimidines are degraded in a three-step catabolic pathway to beta-alanine. In *Drosophila* beta-alanine is crucial for proper maturation of the adult cuticle. The *black* (*b*) gene controls beta-alanine catabolism, such that *b* mutants have lower levels of beta-alanine, leading to black cuticle (Hodgetts and Choi, 1974). The *b* phenotype can be reversed by injection of either beta-alanine or uracil into flies (Jacobs, 1974; Weber et al., 1992). Presumably, when the pyrimidine pool is enlarged in the mutant flies, and/or the beta-alanine pool is increased, the *b* phenotype is normalized.

Purine and Pyrimidine Metabolism in Man VIII, Edited by
A. Sahota and M. Taylor, Plenum Press, New York, 1995

It appears that, in relatively advanced organisms, the size of the pyrimidine pool is regulated through a balance among *de novo* anabolic, salvage and catabolic pathways (Figure 1). Therefore, defects in any of these pathways may result in an imbalanced pyrimidine pool. It is the purpose of this manuscript to describe the genetic basis of various phenotypes occuring as a result of deregulation of the pyrimidine pool in the fruit fly.

MATERIALS AND METHODS

Drosophila strains and screening for mutants

The wild type strain of *Drosophila melanogaster* was Oregon[R]. The new mutations employed in this study are all located on the X-chromosome. They were kept in homozygous strains or in males which were in combination with *C(1)RM* females. This combination always gives male progeny genetically identical to the father and female progeny identical to the mother. Other details about strains and EMS mutagenesis can be found elsewhere (Piskur et al., 1993, Ashburner, 1989). Standard yeast sucrose medium was used, and the flies were maintained at 25°C. In feeding experiments 5-fluoro uracil (5-FU) or uracil was added to the standard medium (Tables 1, 2 and 3). Comparisons of survival of mutant and wild type larvae/flies arising on a test medium was taken as a measure of resistance or sensitivity to the added analog or uracil.

Several approaches were used for obtaining strains with a putative defect in pyrimidine metabolism: (1) *b* flies were screened for suppressor mutations that would revert the *b* phenotype. Such mutations, as *Su(b)¹* and *Su(b)^{DK⁻}* were mapped (for details see Piskur et al., 1993) and examined for resistance to 5-FU (Table 1). (2) Wild type and *b* flies were mutagenized and allowed to lay eggs on medium containing low concentrations of 5-FU. A few individuals, such as *fur¹*, appeared on the selective medium. (3) Various collections of mutant flies were examined for strains with apparent cuticle/beta-alanine deficiency. Such strains, among them *sable (s)* (Lindsley and Zimm, 1992), were examined for 5-FU resistance.

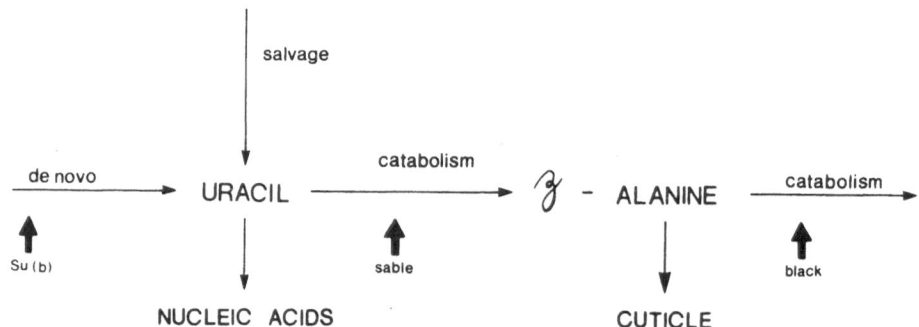

Figure 1. Metabolism of pyrimidines in *Drosophila*. Three mutations affecting the pyrimidine and beta-alanine pools are shown: *Su(b)* is characterized by diminished feedback inhibition by the end product, *sable* has presumably a defect in the pyrimidine catabolic pathway, and *black* has lower levels of beta-alanine.

HPLC determination of compounds in the pyrimidine pool

UTP levels in 3 day old male flies were determined. The extraction of nucleotides was performed employing a modification of the method reported by Ichiba et al. (1992). 100 flies were homogenized and deproteinized in 1 ml of 0.6M $HClO_4$. The sample was centrifuged and the supernatant (0.5ml) neutralized with 0.5 ml 0.2M K_2CO_3. Precipitate was removed by centrifugation and filtration. Samples (0.05 ml, roughly corresponding to two flies) were injected onto an anion exchange column (Partisil 5 SAX, 5μm particle size, Whatman Inc.), and a linear gradient (buffer A: 0.045M $NH_4OH/HCOOH$ adjusted to pH=4.6; buffer B: 0.5M NaH_2PO_4 adjusted to pH=2.7) was used to elute pyrimidine nucleotides. The linear gradient was from 0 to 100 % buffer B in 26 minutes, at a flow rate of 0.7ml/min at room temperature. The wavelength was set at 260 nm. Chromatographic peaks were identified by comparision with reference standards. Levels of pyrimidine nucleotides were quantitated by determining the area under the peaks, and comparison to a standard curve.

Fitness of the mutant alleles

Fly crosses were made between heterozygous females and wild type males. Therefore the frequencies of a mutant and the wild type allele and thereby phenotypes should be equal in the male progeny (Table 3). If the frequency of the wild type allele was significantly higher this was taken as evidence for reduced fitness of flies with the mutant phenotype.

A description of the Bennett population cage experiment can be found elswhere (Ashburner, 1989). In the present case the population experiment was started with females with the wild type allele on both X-chromosomes and males with a mutant allele on their X-chromosome. At time intervals, approximatelly 3 to 4 generations long, samples of adult flies were taken and the frequencies of mutant alleles were determined (Table 4).

RESULTS AND DISCUSSION

Resistance to 5-FU

The three screening procedures mentioned above provided several mutants, among them $Su(b)^{DK^-}$, fur^1 and s which were thought to have defects in pyrimidine metabolism. The $Su(b)^1$ was previously shown to have deregulated pyrimidine biosynthesis (Piskur et al., 1993). The $Su(b)^{DK^-}$ mutation exhibits a phenotype similar to $Su(b)^1$ and it is located within the r locus (Figure 1). The s mutants resemble b mutants. Their cuticles are darker than those of adult wild type flies, which implies that the beta-alanine pool is insufficient. The fur^1 flies have no special visual phenotype.

The above mutants and wild type and b flies were tested on media containing various concentrations of 5-FU (Table 1 and data not shown). The crosses were made so that in the progeny only males carried a mutant allele, and the females had the wild type one. In Table 1 the percentages of males, among individuals that developed from eggs laid on the selective media, is taken to represent the extent of resistance to the analog. Wild type larvae were sensitive to 0.05 mM 5-FU, and neither female or male progeny surviveed on this medium (Table 1). A similar result was obtained for the b mutation (Piskur et al., 1993). However, males of other strains with mutant alleles

Table 1. Sensitivity of various strains strains to 5-fluorouracil.[1]

Concentration of 5-FU (µM)	Oregon[R]	*Su(b)[1]*	*Su(b)[DK-]*	*sable*	*fur[1]*
5	45	37	59	49	48
10	48	38	53	56	71
25	39	47	52	55	71
50	no flies	100	95	89	100
100	no flies	100	100	80	no flies
250	no flies	no flies	no flies	no flies	no flies
500	no flies	no flies	no flies	no flies	no flies

[1]Percentages of males in the progenies appearing on selective media are given. Mutant and Oregon[R] males were crossed to *C(1)RM* females. In the absence of selection the progenies should have equal numbers of homozygous females with the wild type allele and males carrying a mutant allele.

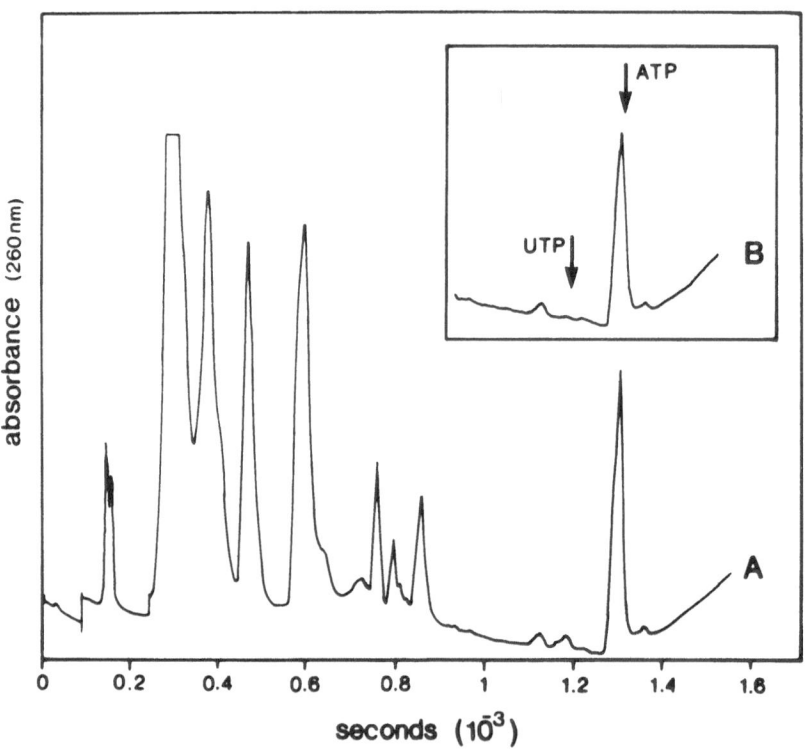

Figure 2. UTP and ATP profiles of the *Su(b)[1]* (A) and wild type (B) males as determined by high performance liquid chromatography. The UTP and ATP chromatographic peaks are indicated in the inset.

exhibited relatively higher resistance than their wild type allele females. Male flies carrying the *Su(b)* or *s* alleles can complete development on up to four times higher concentrations of 5-FU than wild type. This phenotype clearly distinguishes the *s* from the *b* mutation.

Pyrimidine pool

Resistance to toxic pyrimidine analogs can be due to larger pyrimidine pools. To test this hypothesis the nucleotide levels were measured in various mutants by high performance liquid chromatography. The UTP pool in *Su(b)* mutants was clearly higher than in the wild type or *b* flies (Figure 2). The average difference was approximately two fold (data not shown). Previously it was shown that in the *Su(b)¹* mutant the CPSase activity of the *r* enzyme was inhibited to a lesser extent by the end product, UTP (Piskur et al., 1993). Therefore, it seems that feedback inhibition plays a major role in the regulation of the pyrimidine pool in *Drosophila*. When negative feedback regulation is relaxed, higher levels of pyrimidines result.

The nucleotide levels seemed to be normal in adult males of the *fur¹* and *s* strains. However, preliminary measurements indicated that the pool is increased several-fold in the 3rd instar larvae of the *s* strain. An attractive hypothesis is that *s* impairs pyrimidine catabolism, thus leading to higher pyrimidine levels and lower beta-alanine concentrations, which subsequently results in a blackish cuticle phenotype (Figure 1).

Sensitivity to uracil

The salvage pathway represents an important source of pyrimidines in *Drosophila* (Falk and Nash, 1974). Therefore it was investigated whether or not there is any interaction between the elevated pyrimidine pool in some of our mutants and external pyrimidines present in the medium. Various mutant strains were allowed to lay eggs on media with added uracil, and fly development was followed (Tables 2 and 3). It is apparent that the *Su(b)* and *s* mutants, but not *fur¹*, exhibited greater sensitivity to higher concentrations of uracil than wild type. Wild type and *fur¹* individuals could tolerate and complete their development on concentrations of uracil four-fold higher than *Su(b)* mutants. It seems that high concentrations of external pyrimidines enter cells and saturate the catabolic pathway so that nucleotide imbalance occurs, which leads to lethality. Mutants that have elevated pyrimidine pools are therefore even more sensitive to external pyrimidines than those that do not.

Tight regulation of the pyrimidine pool is required

Various mutations that alter pyrimidine metabolism should not be of great disadvantage for development and survival if great fluctuations in the pyrimidine pool can occur. However, in genetic crosses, which should in theory give a progeny having similar frequencies of the wild type and mutant allele, discrepancies were observed (Table 3). A cross was constructed, which should give equal numbers of + and *s* males. However, the frequency of the *s* allele was substantially lower (Table 3). This result is interpreted to mean that the *s* mutation represents a disadvantage for an organism carrying it.

Similar results were obtained in experiments to determine whether there is a competitive advantage or disadvantage confered by the *Su(b)* alleles. A population box was started with known frequencies of the wild type and *Su(b)* alleles (Table 4). After

Table 2. Sensitivity of various strains to uracil.[1]

Concentration of uracil (mM)	Oregon[R]	Su(b)[1]	Su(b)[DK-]	fur[1]
0	53	34	49	58
10	52	35	46	51
25	50	22	41	58
100	33	0	0	50
250	no flies	no flies	no flies	no flies

[1]Percentage of males in the progenies appearing on uracil media are given. The progenies are from crosses that should give equal numbers of homozygous females carrying the wild type allele and males carrying a mutant allele in the absence of selection.

Table 3. Sensitivity of *sable* to uracil.[1]

Concentration of uracil (mM)	number of wild type males	number of *sable* males	percentage of *sable* males in the ♂ progeny
0	117	69	37
10	123	72	37
20	114	76	40
50	7	3	30
75	3	0	0
100	2	0	0

[1] The results show male progeny from a cross between +/s females and + males (approximately 250 eggs were laid). Such a cross should in theory give the same frequency of + and s alleles among the male progeny.

Table 4. Two population cage experiments showing a spontaneous decrease in the gene frequency of the *Su(b)* allele during the time period of 7 1/2 months.[1]

Mutant alleles initially present in the population	start	2 months	3 1/2 months	4 1/2 months	6 1/2 months	7 1/2 months
Su(b)[1]	33.3%	35.7%	14.0%	11.0%	5.1%	3.1%
Su(b)[DK7]	33.3%	19.2%	7.5%	3.9%	2.7%	1.8%

[1] Each population was started with females carrying the wild type alleles, +/+, and males carrying the *Su(b)* allele, so that the initial frequency of the *Su(b)* allele was 33.3%. A sample of flies was taken at different time intervals and examined for the presence of the *Su(b)* allele.

three and a half months the frequencies of alleles eliciting higher pyrimidine levels had already dropped significantly. After an equivalent of approximately fifteen generations the frequencies of the *Su(b)* alleles had decreased more than a factor of ten. It is apparent that *Su(b)* is outcompeted by wild type because the mutants have either lower viability or fertility.

We conclude that any deregulation of pyrimidine metabolism results in altered pyrimidine and associated pools. This causes various phenotypes, and such organisms are at a disadvantage in competition with wild type organisms.

ACKNOWLEDGMENTS

This work was supported by grants from the Danish Natural Science Research Council and the Plasmid Fond (Copenhagen) to J.P. and the National Science Foundation (grant DMB91-05826) to J.D. The authors acknowledge P.Eriksen and M. Mortensen for technical help and Dr. A.Kahn for comments on the manuscript.

REFERENCES

Ashburner, M., 1989, "Drosophila. A Laboratory Handbook," Cold Spring Harbor Laboratory Press, Cold Spring Harbor.

Eisenberg, M., Kirkpatrick, R. and Rawls, J., 1993, Structure of the rudimentary-like gene and UMP synthase in Drosophila melanogaster, *Gene* 124: 263-267.

Falk, D.R. and Nash, D., 1974, Sex-linked auxotrophic and putative auxotrophic mutants of Drosophila melanogaster, *Genetics* 76: 755-766.

Freund, J.N. and Jarry, B.P., 1987, The rudimentary gene of Drosophila melanogaster encodes four enzymic functions, *J.Mol.Biol.* 193:1-13.

Hodgetts, R. and Choi, A., 1974, Beta-alanine and cuticle maturation in Drosophila, *Nature* 252: 710-711.

Ichiba, M., Tomokuni, K. and Mori, K., 1992, Erythrocyte nuclotides in lead workers, *Int.Arch.Occup.Env iron.Health,* 63:419-421.

Jacobs, M., 1974, Beta-alanine and adaptation in Drosophila, *J.Insect Physiol.* 20: 859-866.

Lindsley, D.L. and Zimm, G.G., 1992, "The Genome of Drosophila melanogaster," Academic Press, San Diego.

Piskur, J., Kolbak, D., Søndergaard, L. and Pedersen, M.B., 1993, The dominant mutation Suppressor of black indicates that de novo pyrimidine biosynthesis is involved in the Drosophila tan pigmentation pathway, *Mol.Gen.Genet.* 241: 335-340.

Rawls, J., Kirkpatrick, R., Yang, J. and Lacy, L., 1993, The dhod gene and deduced structure of mitochondrial dihydroorotate dehydrogenase in Drosophila melanogaster, *Gene* 124: 191-197.

Weber, J.P., Bolin, R.J., Hixon, M.S. and Sherald, A.F., 1992, Beta-alanine transaminase activity in black and suppressor of black mutations of Drosophila melanogaster, *Biochim. Biophys. Acta* 1115: 181-186.

EXPRESSION OF *DE NOVO* PYRIMIDINE BIOSYNTHESIS GENES DURING SPERMATOGENESIS IN *DROSOPHILA MELANOGASTER*

Lawrence Porter, Jun Yang, and John Rawls

Molecular Cell Biology Group
T. H. Morgan School of Biological Sciences
University of Kentucky
Lexington, KY 40506 USA

SUMMARY

The *dhod* gene encodes dihydroorotate dehydrogenase, the fourth step of *de novo* pyrimidine biosynthesis. In addition to the common 1.5 kb *dhod* RNA expressed by embryos and females, adult males produce a group of slightly longer RNAs, arising from alternative sites of transcription initiation. Testis-limited expression of *dhod* conforms to an established pattern of expression of other genes that function during spermiogenesis: transcription in spermatocytes, storage of translationally inactive RNA through meiosis, translation of the RNA during spermiogenesis. Very similar expression of a testis promoter-*lacZ* fusion transgene indicates that sequences required for the spermatogenesis transcription and translation patterns are confined to the 5′ end of the *dhod* gene. We have extended these studies to the *D. melanogaster r* and *r-l* genes which encode, respectively, steps one through three (CAD protein) and steps five through six (UMP synthase) of *de novo* pyrimidine biosynthesis. The *in situ* hybridization pattern for *r* RNA and the pattern of expression of an *r* promoter-*lacZ* fusion transgene in testis is indistinguishable from the patterns observed for *dhod*. However, *r-l* RNA does not accumulate in spermatogenesis and *r-l* promoter-*lacZ* fusion transgene constructs give no distinct expression pattern in spermiogenesis. Thus, it appears that the first four enzymes of the *de novo* pyrimidine biosynthetic pathway participate in a novel function during spermatogenesis that does not involve synthesis of pyrimidine nucleotides. Rather, we hypothesize that these enzymes contribute electrons to unknown redox processes that are important for the final stages of spermiogenesis.

INTRODUCTION

In *Drosophila melanogaster*, three genes encode the proteins involved in pyrimidine biosynthesis: *r*, CAD protein (1); *dhod*, DHOdehase protein (2); and, *r-l*, UMP synthase (3). Mutations in these genes result in the expression of a common set of phenotypes including defective adult cuticle, defective embryogenesis, and female sterility. Similarities of transcript accumulation patterns observed during embryogenesis and oogenesis provides evidence for the coordinate regulation of these genes in proliferating tissues and in tissues involved in chitin biosynthesis (Porter and Rawls, in preparation). We have discovered a male-limited form of *dhod* RNA that arises solely in spermatogenic cells (Yang, Porter, and Rawls, in preparation) and the studies reported here were carried out to determine the patterns of expression of other genes of this pathway during spermatogenesis.

MATERIALS AND METHODS

Genetic strains and crosses. Flies were cultured on standard medium at 23° C (4). Most genetic markers are described elsewhere (5,6).

In situ **hybridizations.** Single-stranded, digoxygenin-labelled DNA probes were prepared by PCR-mediated, asymmetric amplification of linearized templates as described (7), with the following modifications: the primer annealing was 56°C for 30 sec; the extension temperature was 72°C for 1 min; the products were not size reduced; the products were twice precipitated with ethanol, then resuspended in hybridization buffer (8). To prepare antisense probe specific for testis RNA, plasmids containing the 5′ ends of *r, dhod,* or *r-l* genes were linearized with an appropriate restriction enzyme, annealed with a sense-oriented primer, and extended in the direction of the restriction site in order to generate 100-200 nucleotide single-stranded probes. Likewise, the sense stranded controls were prepared using an antisense-oriented primer.

The procedures of Tautz and Pfeifle (8) were followed for tissue preparation, hybridization, and detection with the modifications in Proteinase K digestion and post-hybridization washes as described previously (9; Urs Kloter, personal communication). After development of the alkaline phosphatase reaction, tissues were mounted in Euparal. All digoxygenin reagents used were from Boehringer Mannheim Biochemicals.

Germline Transformation. P element-mediated germline transformation was performed using methods described by Spradling (10). DNA used for injection included 300 μg/ml of transformant DNA (constructed by inserting a 5′ end fragment of the *r, dhod,* or *r-l* gene into the polylinker of pCaSpeR-AUG-βgal transformation vector (11)) and 100 μg/ml of pπ25.7wc helper plasmid. Multiple, independent transformant lines were generated for each gene fusion construct.

Histochemical staining of tissues. For β-galactosidase staining, tissues were dissected from adult males, fixed in 0.25% glutaraldehyde and stained for the enzyme according to Glaser et al. (12). Stained samples were washed in buffer, fixed in 3:1 ethanol/acetic acid for 10 min, washed in 2-propanol for 10 min, then mounted in Euparal for microscopic examination. Non-transposon-bearing siblings served as controls.

A histochemical staining method for DHOdehase was adapted from the succinic dehydrogenase method devised by Lawrence (13). Stain solution consists of 50 mM Tris (pH 8), 4 mM $MgCl_2$, 5 mM NaCN, 0.4 mg/ml nitro blue tetrazolium, 4 mM L-dihydroorotate and 4 mM DL-carbamylaspartate. Controls included tissues derived from mutants as well as staining mixtures from which dihydroorotate and carbamylaspartate were omitted.

RESULTS

A novel, male-limited form of *dhod*-specific RNA is expressed abundantly in testis and appears to be unique to the genes of the pyrimidine pathway

The common form of *dhod* is a 1.5 kb form found in all developmental stages, except adult males (3). The male-limited form is 1.6 kb and represents approximately 75% of all *dhod* RNA in adult males. S1 nuclease protection and primer extension assays have shown that this male-limited RNA contains a 100 nucleotide 5′ extension; reverse transcriptase-coupled PCR analysis of isolated tissues and mutants lacking germline cells shows that the male-limited RNA is confined to spermatogenic cells (Yang, Porter and Rawls, in preparation).

Both *r* and *dhod* RNA accumulate during the spermatocyte growth phase of spermatogenesis; *r-l* RNA accumulation is limited to the stem and gonial proliferative phase of spermatogenesis

In situ hybridization of single-stranded, antisense probes to wild-type testes revealed several interesting patterns of expression. In the first pattern (Fig 1A), representative of *r* and housekeeping *dhod* RNA distribution, the immunohistochemical staining pattern is

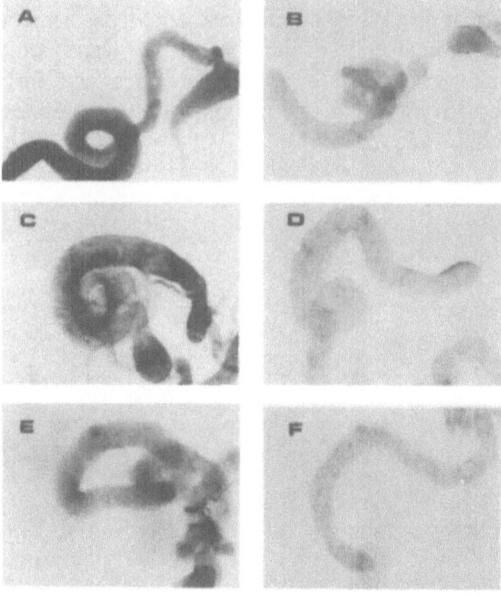

Figure 1. *In situ* hybridization to testes of wild-type flies using various digoxygenin-labelled probes. Panel A: an *r* antisense probe complementary to nucleotides +1262 through +1072 on the RNA (1). Panel B: an *r* sense probe corresponding to nucleotides +1014 through +1243 on the RNA. Panel C: a *dhod*, male-specific, antisense probe complementary to nucleotides +99 through +8 on the major spermatogenesis RNA (Yang, Porter and Rawls, in preparation). Panel D: a *dhod*, male-specific, sense probe corresponding to nucleotides +3 through +89 on the RNA. Panel E: an *r-l* antisense probe complementary to nucleotides +111 through +7 on the RNA (3). Panel F: an *r-l* sense probe corresponding to nucleotides +13 through +75 on the RNA.

confined to the cytoplasm within pre-meiotic cells and early post-meiotic cells. Specifically, distinct hybridization is observed in gonial cells in the distal tip of the testis, increasing in intensity throughout spermatocyte maturation, peaking in the cell mass along the inner face of the testes coil which contains cysts of late spermatocytes, meiotic stages, and early spermatids (14). The hybridization signal decreases considerably in elongated spermatids and subsequent stages. It is possible that signal diminution in these later stages is due to poor probe penetration. Sense strand controls (Fig 1B) show no distinct hybridization pattern.

The second pattern (Fig 1C) is displayed by the male-limited *dhod* specific RNA. Hybridization signal is confined to growing cysts in a pattern that is nearly identical to that of the common form of *dhod* RNA as described above, except that signal is much less prominent among early gonial cells within the distal tip of the testis. Again, sense strand controls (Fig 1D) show no distinct hybridization pattern. The third pattern (Fig 1E) is unique for *r-l* RNA. Hybridization signal is confined to gonial cells and very early spermatocyte cysts at the distal tip of the testis. No hybridization is detected in subsequent stages. Sense strand controls (Fig 1F) yield no distinct hybridization in testes.

DHOdehase enzymatic activity is expressed only in late spermiogenesis

Direct histochemical staining of testes reveals strong DHOdehase activity in late spermatids, especially in those exhibiting cystic bulge processes (Fig 2A). The latter are systolic constrictions that progress the length of each spermatid bundle, resulting in a cytoplasmic bolus termed the "waste bag" (14). This process also individualizes the formerly syncytial spermatids. Weak DHOdehase activity is sometimes observed among the mitotically dividing gonial cells at the apex of the testes; otherwise, no activity is found in other spermatogenic cells (*e.g.*, spermatocytes and early spermatids). This appearance of enzyme activity in late spermatids follows the appearance of the RNA in spermatocytes by at least three days (14).

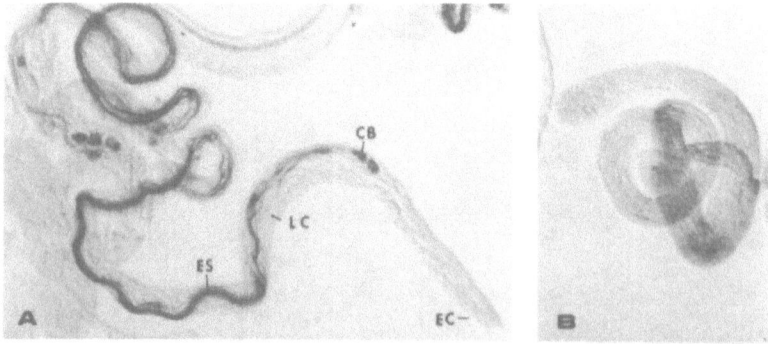

Figure 2. Histochemical staining of DHOdehase activity in testes of adult males. Structures indicated are early spermatocyte cysts (EC), late spermatocyte cysts (LC), elongated spermatids (ES), cystic bulges (CB), and waste bag structures (WB). Panel A: DHOdehase in testis of a wild-type fly. Panel B: DHOdehase in testis of a *dhod*-null *(Df(3R)p²⁵/dhod^CA2)*fly.

Spermatogenesis-specific and delayed expression of a heterologous reporter transgene is directed by 5′ end sequences of both the *r* and *dhod* genes, but not the *r-l* gene

We have created a variety of gene fusions in which the *dhod* spermatogenesis promoter is fused to an *E.coli lacZ* ORF within the *Drosophila* transformation vector pCaSpeR-AUG-βgal (11). Following germline transformation, strains were isolated which contain single copy insertions of the various fusions. β-galactosidase histochemical staining of larvae and adults shows that staining is limited to the testis (Fig 3A) and, specifically, to elongated spermatids and associated "waste bags" that are formed from cystic bulge processes (14). Analysis of various deletion constructs shows that this pattern of expression is specified by an 89 bp DNA fragment that extends from -54 to +35, relative to the major transcription initiation site of *dhod* in testes (Yang, Porter and Rawls, in preparation).

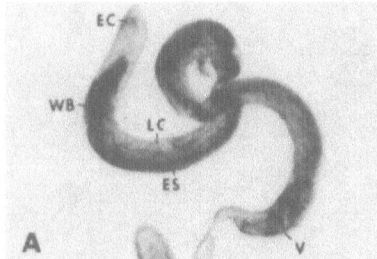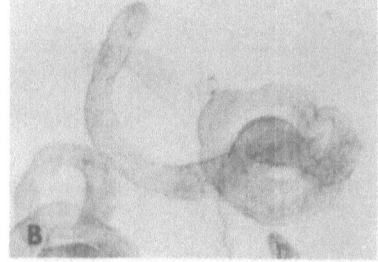

Figure 3. β-galactosidase staining of testes from adult males. Labels are the same as Fig. 2. Panel A: Male bearing the *[ds1]* transposon in which a *dhod* 5′ fragment extending from -1028 through +35, relative to the major spermatogenesis transcription start site, is fused to *lacZ*. Panel B: Wild-type male sibling of the male in Panel A.

We have also created *r* promoter-*lacZ* transgenic lines in which *r* 5′ flanking and 5′ untranslated sequences drive β-galactosidase expression. Each of these lines display a staining pattern similar to that of *dhod*, although a few of the lines show some staining of earlier spermatogenesis stages. Overall, these results indicate that sequences directing delayed expression of the CAD protein might reside within the 5′ end of the *r* gene and we are carrying out additional experiments to test whether CAD is expressed in spermatids.

We have created independently-derived, *r-l* promoter-*lacZ* transgenic lines in the same manner as above but have failed to detect β-galactosidase staining in any of those lines. Within transgenic animals, β-galactosidase is expressed in other tissues in a pattern essentially similar to that seen in *r* promoter-*lacZ* transgenics; so, we are confident that the constructs are correct. Testis staining is the sole obvious discordance in staining of transgenic animals bearing *r* and *r-l* promoter fusions.

DISCUSSION

Our results show that both the *r* and *dhod* genes are abundantly transcribed during spermatocyte growth, but synthesis of the protein products of these genes appears to be delayed until late spermatid differentiation, several days later. A variety of observations indicate that this pattern of gene expression is common in *Drosophila* spermatogenesis (14,15). Figure 4 diagrams the stages of spermatogenesis and summarizes our observations on the timing of transcription and presumptive translation of the genes of *de novo* pyrimidine biosynthesis.

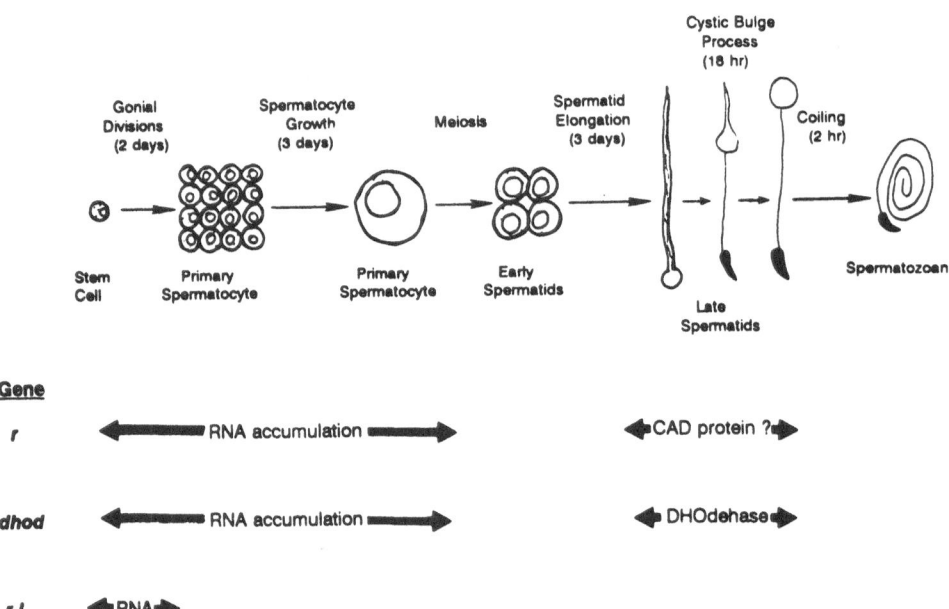

Figure 4. Expression of *de novo* pyrimidine biosynthesis pathway genes during spermatogenesis in *Drosophila melanogaster*. The top of the figure represents the stages of spermatogenesis. Below is an interpretation of our results depicting the timing of transcription and presumed translation of the *r*, *dhod* and *r-l* genes during spermatogenesis.

During gonial cell divisions, RNAs for all three genes accumulate, presumably reflecting coordinate expression to meet needs of cell proliferation that are shared with developing somatic cells. During this period, these genes are probably transcribed exclusively via their common, housekeeping promoter.

During primary spermatocyte growth, *r* and *dhod* RNAs accumulate to high levels, whereas *r-l* RNA declines to undetectable levels (Fig. 4). Transcription of *dhod* is limited to the spermatocyte promoter and the resulting RNA is sequestered for expression in late spermiogenesis. We do not know whether *r* transcription is directed by a novel promoter nor do we know whether this RNA is translated during this period. However, delayed expression of *r* promoter-*lacZ* transgenes suggests that the *r* RNA might be sequestered in a manner very similar to that seen for *dhod* RNA.

At meiosis and during early spermatid differentiation, *r* and *dhod* RNA levels peak and, probably, transcription ceases. A variety of previous work indicates that general transcription ceases with meiosis in *Drosophila* spermatogenesis and that subsequent spermiogenesis is directed by stored RNAs that were synthesized during spermatocyte growth (15). Our observations on the *r* and *dhod* genes are fully consistent with this model.

During late spermatid differentiation, DHOdehase expression accompanies the final formation of the sperm head and tail. In these cells, formation of the axoneme and associated mitochondrial derivatives is virtually complete and final extrusion of excess cytoplasm is occurring, resulting in individualized spermatozoa. Although we have not directly measured expression of the CAD protein during this period, *r* RNA appears to have sequences that direct delayed expression of the reporter protein of *r* promoter-*lacZ* transgenes. Spermatid expression of β-galactosidase in transgenics bearing only 5′ flanking and untranslated segments of *r* or *dhod* suggests that delayed expression is not a protein-level phenomenon; rather, this is probably due to delayed translation of *r* and *dhod* RNAs sequestered since the period of spermatocyte growth.

Abundant expression of *r* and *dhod* during spermatogenesis in the absence of *r-l* expression implies that the CAD and DHOdehase proteins participate in an unusual metabolic process during late spermatid differentiation. The nature of this metabolism is unclear at this time. Presumably, this truncated pathway should lead to accumulation of orotate and free electrons in late spermatids. Several factors argue against orotate and its derivatives as the objectives of this truncated pathway. The late stage in spermatogenesis at which DHOdehase activity is expressed would appear to limit subsequent metabolism. Also, there are relatively few known metabolic roles of orotate and we suspect that this relatively insoluble compound might be efficiently eliminated from spermatids by the cystic bulge process. Lastly, the fertility of *r* and *dhod* mutant males argues that CAD protein and DHOdehase functions are not essential for spermatogenesis. Because DHOdehase is the only known means of orotate biosynthesis, orotate must also be dispensable for functional spermatogenesis. We believe that electron metabolism in final spermatozoan formation is a more likely role for the truncated pathway. Fertility of *r* and *dhod* males could be explained by redundancy in the electron-generating systems of spermatids; that is, the CAD protein-DHOdehase combination might be one of a number of systems providing electrons for late spermiogenesis processes. The ultimate acceptor(s) for electrons generated by DHOdehase could be any of a number of compounds: quinones, fumarate, molecular oxygen. Experiments are in progress to explore these provocative possibilities.

REFERENCES

1. W. Zerges, A. Udvardy, and P. Schedl, Molecular characterization of the 5′ end of the *rudimentary* gene in *Drosophila* and analysis of three P element insertions. *Nucl. Acids Res.* 20:4639-4647 (1991).
2. J. Rawls, R. Kirkpatrick, J. Yang, and L. Lacy, The *dhod* gene and deduced structure of mitochondrial dihydroorotate dehydrogenase in *Drosophila melanogaster*. *Gene* 124:191-197 (1993).
3. M. Eisenberg, R. Kirkpatrick, and J. Rawls, Structure of the *rudimentary-like* gene and the UMP synthase in *Drosophila melanogaster*. *Gene* 124:263-267 (1993).
4. W.K. Jones, R. Kirkpatrick, and J. Rawls, Molecular cloning and transcript mapping of the dihydroorotate dehydrogenase *dhod* locus in *Drosophila melanogaster*. *Molec. Gen. Genet.* 219:397-403 (1989).
5. D. Lindsley and K. Tokuyasu. "The Genome of *Drosophila melanogaster*," Academic Press, NY (1992).
6. M. Ashburner. "*Drosophila*. A Laboratory Handbook," Cold Spring Harbor Laboratory Press, Cold Spring Harbor, NY (1989).

7. N. Patel and C. Goodman, Detection of *even-skipped* transcripts in *Drosophila* embryos with PCR/digoxygenin-labeled DNA probe, *in:* "Nonradioactive *In Situ* Manual," Boehringer Mannheim Biochemicals, Mannheim (1992).

8. D. Tautz and C. Pfeifle, A nonradioactive *in situ* hybridization method for the localization of specific RNAs in *Drosophila* embryos reveals translational control of the segmentation gene *hunchback. Chromosoma* 98:81-85 (1989).

9. A. Michelson, S. Abmayr, M. Bate, A. Martinez-Arias, and T. Maniatis, Expression of *MyoD* family members prefigures muscle pattern in *Drosophila* embryos. *Genes and Development* 7:2086-2097 (1991).

10. A.C. Spradling, P element-mediated transformation, *in:* "*Drosophila*: A Practical Approach," D. Roberts, ed., IRL Press, Oxford, pp. 175-198 (1986).

11. C.S. Thummel, A.M. Boulet, and H.D. Lipshitz, Vectors for *Drosophila* P element-mediated transformation and tissue culture transfection. *Gene* 74:445-456 (1988).

12. R.L. Glaser, M.F. Wolfner, and J.T. Lis, Spatial and temporal pattern of *hsp26* expression during normal development. *EMBO J.* 5:447-454 (1986).

13. P.A. Lawrence, A general cell marker for clonal analysis of *Drosophila* development. *J. Embryol. Exp. Morphol.* 64:321-332 (1981).

14. D. Lindsley and K. Tokuyasu, Spermatogenesis, in: "The Genetics and Biology of *Drosophila*," Vol. 2d, M. Ashburner and T. Wright, eds., Academic Press, NY, pp. 225-294 (1980).

15. M. Fuller, Spermatogenesis, *in:* The "Development of *Drosophila melanogaster*," M. Bate and A. Martinez-Arias, eds., Cold Spring Harbor Laboratory Press, Cold Spring Harbor, NY, pp. 71-148 (1993).

REGULATION OF CALF THYMUS CYTOSOLIC 5'-NUCLEOTIDASE/NUCLEOSIDE PHOSPHOTRANSFERASE

M. G. Tozzi[1], M. Camici[2], R. Pesi[2], S. Allegrini[2], C. Baiocchi[2], M. Turriani[2], C. Scolozzi[2], P. L. Ipata[2]

[1]Istituto di Chimica Biologica, Facoltà di Farmacia, Università di Sassari, Italy
[2]Dipartimento di Fisiologia e Biochimica Università di Pisa Italy

INTRODUCTION

Many phosphatases with different specificity and cellular location have been described to be able to catalyze a phosphotransferase reaction[1,2] as well as several enzymes named phosphotransferases have been demonstrated to hydrolyze phosphoesters[3,4]. Even though a number of papers and very accurate reviews have been published on phosphohydrolases, particularly on nucleotidases classified following their substrate specificity or cellular location, still additional information is needed on the molecular characteristics and on the regulation of the phosphohydrolase/phosphotransferases in order to understand the role, if any, of their potential bifunctionality.

Cytosolic 5'-nucleotidase specific for IMP, GMP and their deoxyderivatives, is an ubiquitous enzyme which is regulated by 2,3 bisphosphoglycerate, ATP, ADP, and phosphate[5]. This nucleotidase purified from several sources, behaves as a bifunctional enzyme since it catalyzes the transfer of phosphate from a nucleoside monophosphate to a nucleoside acceptor, probably forming an enzyme-phosphate intermediate, thus operating a mononucleotide interconversion[1]. In the absence of a suitable nucleoside the enzyme-phosphate complex is hydrolytically cleaved thus functioning as a phosphatase.

The phosphotransferase activity of the IMP preferring cytosolic 5'-nucleotidase represents the only cellular enzyme activity able to phosphorylate inosine and guanosine analogs, which are not substrates of known cellular kinases[6,7]. The study of the regulation of the two activities borne by this enzyme, is of great importance for the understanding of its role both in the purine metabolism and in the mechanisms of "activation" and "inactivation" of purine pro drugs.

The cytosolic 5'-nucleotidase has been purified from calf thymus. The substrate specificity and molecular characteristics of the enzyme were similar to those described for human colon carcinoma cytosolic 5'-nucleotidase[8]. In this poster, the enzyme kinetic and regulatory characteristics as a function of pH, energy charge and concentration of nucleoside acceptor are presented.

METHODS

Enzyme Purification

The enzyme was purified from calf thymus in 5 steps including ammonium sulphate fractionation, ionic exchange chromatography, pentyl-agarose chromatography, Matrex Green A chromatography and finally an affinity chromatography on ADP-agarose. The final enzyme preparation displayed a specific phosphotransferase activity of 7.45 U/mg, with a purification of 1000 fold and a yield of 6%. The purified 5'-nucleotidase/nucleoside phosphotransferase was supplemented with 1 mg/ml albumin and kept at -80 °C in the presence of 1 mM dithiothreitol (DTT).

Enzyme assays

In the presence of a nucleoside monophosphate and a nucleoside as substrates, three product may be formed: nucleoside, phosphate and nucleoside monophosphate. Two different methods have been utilized to determine enzyme activity: a) determination of the rate of labelled nucleoside production from labelled nucleoside monophosphate (referred to as nucleotidase activity), b) determination of the rate of labelled nucleoside monophosphate production from labelled nucleoside (referred to as phosphotransferase activity)[8]. To measure the enzyme activity as a function of pH a new method was applied, based on the use of deoxyGMP and inosine as substrates. The concentration of the reaction products (deoxyguanosine and IMP) was determined after analysis of the reaction mixture on Capillary Electrophoresis with a modification of the method described by Lecoq et al[9].

Phosphate was determined according to Chifflet et al[10].

RESULTS AND DISCUSSION

Calf thymus 5'-nucleotidase is activated by ADP, ATP and bisphosphoglycerate and is inhibited by millimolar concentration of phosphate (results not shown). This observation prompted us to study the effect exerted by variations of adenylate energy charge on the bifunctional enzyme. Our results indicate that the increasing of energy charge caused an enhancement of both enzyme activities but the overall effect was depending on the presence of phosphate and inosine. In fact, in our assay conditions, the activatory effect of inosine on nucleotidase activity was markedly enhanced by adenylate energy charge (Fig. 1A). The presence of 5 mM phosphate in the assay mixture shifted the curve of the enzyme activities dependence on adenylate energy charge toward higher and more physiological values.

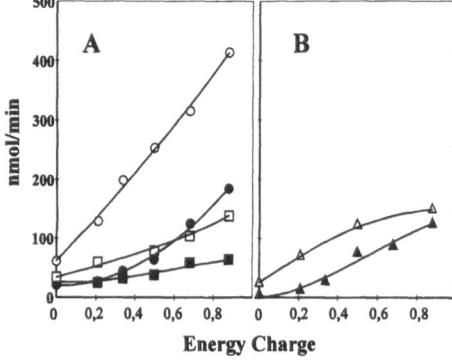

Fig. 1. Dependence of enzyme activities on adenylate energy charge, in the absence (open symbols) or in the presence (closed symbols) of 5 mM phosphate. A: 5'-nucleotidase activity in the absence (□) or in the presence of 1.4 mM inosine (○). B: (△) nucleoside phosphotransferase activity.

Furthermore our results demonstrated that at physiological adenylate energy charge values and in the presence of millimolar concentration of phosphate the possibility of an in vivo nucleoside phosphorylation should depend only on the availability of a suitable nucleoside. In fact in the presence of phosphate and inosine the rate of inosine production from IMP and of IMP production from inosine are superimposable (Fig. 1A and B).

The dependence of enzyme activity on the pH was also investigated measuring the rate of formation of the three reaction products in the presence of deoxyGMP and inosine as substrates. Fig. 2B shows that the optimum pH for nucleoside monophosphate hydrolysis (6.5) is slightly more acidic than the optimum pH for the transfer of the phosphate, (7.2). Furthermore, Fig. 2A shows that between pH 6.2 and 7.8, in our assay conditions, inosine exerts a moderate activatory effect and actually causes a modification of the pH dependence profile. In fact, inosine, the best phosphate acceptor [8], caused the maximal activation in the same conditions where the phosphotransferase activity was favored (see also Fig. 1). This observation is in line with the hypothesis that the stimulatory effect exerted by the phosphate acceptor implies that the hydrolysis of the phosphoenzyme intermediate is the rate limiting step of the whole process.

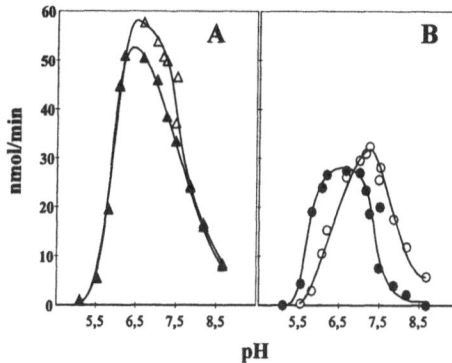

Fig 2. Dependence of enzyme activity on the pH. A: rate of deoxyguanosine production from deoxyGMP in the absence (▲) or in the presence (△) of 1.4 mM inosine. B: rate of phosphate (●) or IMP (○) production in the presence of 2 mM deoxyGMP and 1.4 mM inosine.

To better investigate on the molecular basis of the different optimum pH displayed by the two enzyme activities, Vmax and Km for deoxyGMP were determined at different pH values from double reciprocal plots. From the analysis of the plot of log Vmax vs. pH (Fig. 3A) a dissociation process with a pK_a of 8 was observed. Two additional dissociation processes (pK_a of 6.4 and 7.2) were evaluated from the pKm vs. pH curve (Fig. 3B) related with ionizable groups either of the free enzyme or of the free substrate. The analysis of log V_0 of IMP formation in the presence of subsaturating concentration of inosine and saturating concentrations of deoxyGMP vs. pH curve (Fig 3C), revealed that the formation of the enzyme-phosphate-inosine complex was depending on two dissociation processes with pK_a of 6.4 and 7.8. It is difficult to predict the groups involved in the catalysis only from the effects of pH on the kinetic parameters, however, a critical presence of a cysteine residue at the active site is supported by the thiol reagent sensitivity of 5'-nucleotidase which was completely inactivated after 1 h incubation in the presence of 0.3 mM 5,5'-dithio-bis(2-nitrobenzoic acid) (DTNB), while the activity was recovered in the presence of 5 mM DTT (results not shown). Even though the pK_a 6.4 evaluated from pKm curve might well be related with the pK_a of the substrate (GMP) dissociation, it might indicate the involvement of an histidine in the formation both of enzyme-GMP complex and enzyme-phosphate-inosine complex (Fig. 3B and C). Furthermore the difference of optimum pH between hydrolysis and transfer of phosphate might be explained on the basis of the hypothesis that a cysteine residue with a peculiarly low pK_a (7.2) is present in the free enzyme and that the environment of this residue, being significantly affected by the formation of the enzyme-phosphate intermediate, causes the change of its pK_a value to about 8 in the complex, shifting the dependence of the rate of the phosphotransferase reaction toward higher pHs.

Finally, on the light of our results we might speculate that at physiological pH, high adenylate energy charge and in the presence of phosphate, the phosphotransferase reaction is favored.

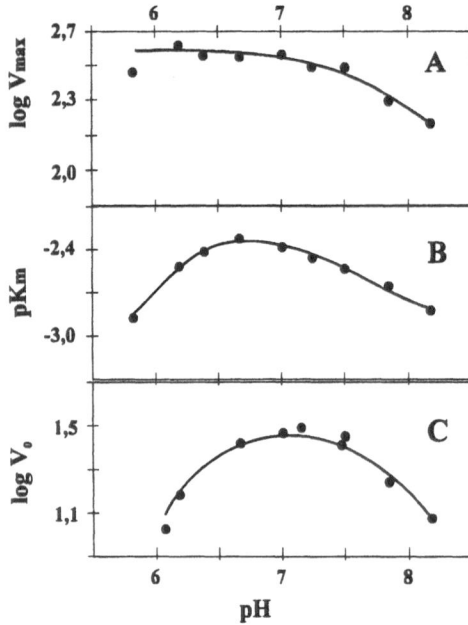

Fig.3. Effect of pH on the kinetic parameters of cytosolic 5'-nucleotidase. In A and B log Vmax and pKm are plotted as a function of pH with deoxyGMP as variable substrate. in C log of the rate v_0 of the IMP formation from saturating deoxyGMP concentration and subsaturating inosine concentration as a function of pH is also reported. Solid curves are computer assisted plots obtained by fixing a pK_a for the enzyme substrate complex at 8, two pK_a values for the free enzyme or substrate at 6.4 and 7.2 and two pK_a values for the enzyme-phosphate complex at 6.4 and 7.8.

On the contrary, in the case of ATP depletion and a slight acidification, the enzyme should catalyze mainly an hydrolytic reaction.

ACKNOWLEDGEMENTS

This work was supported by a grant from Italian CNR Target Project "Biotechnology and Bioinstrumentation".

REFERENCES

1 -Worku Y., Newby A.C., Nucleoside exchange catalized by the cytoplasmic 5'-nucleotidase, *Biochem. J.*, 205:503 (1982)

2 -Cirri P., Chiarugi P., Camici G., Manao G., Raugei G., Cappugi G., and Ramponi G., The role of cys12, cys17 and arg18 in the catalytic mechanism of low-M_r cytosolic phosphotyrosine protein phosphatase, *Eur. J. Biochem.*, 832:1, (1993).

3 - Brunngraber E.F., and Chargaff E., Nucleoside phosphotransferase from carrot: kinetic studies and exploration of active sites *I. Biol. Chem*, 245:4825 (1970).

4 -Tesoriere G., Vento R., Tesoriere L., and Giuliano M., The purification and properties of nucleoside phosphotransferase from mucosa of chicken intestine, *Biochim. Biophys. Acta*, 786:231, (1984).

5 -Itoh R., IMP-GMP 5'-nucleotidase, *Comp. Biochem. Physiol.*, 105B:13, (1993).

6 -Keller P.M., McKee S., and Fyfe J.A., Cytoplasmic 5'-nucleotidase catalyzes acyclovir phosphotylation, *J. Biol. Chem.* 260: 8664 (1985).

7 -Johnson M.A., and Frdland A., Phosphorylation of 2',3'-dideoxyinosine by cytosolic 5'-nucleotidase of human lymphoid cells, *Molec. Pharmac.* 36:291, (1989).

8 -Tozzi M.G., Camici M., Pesi R., Allegrini S., Sgarrella F., and Ipata P.L., Nucleoside phosphotransferase activity of human colon carcinoma cytosolic 5'-nucleotidase, *Arch. Biochem. Biophys.*, 291:212, (1991).

9 -Lecoq A.F., Leuratti C., Marafante E., Di Biase S., Analysis of nucleic acid derivates by micellar electrokinetic capillary chromatography, *J. Hi. Res. Chromatography*, 14:667, (1991).

10-Chifflet S., Torriglia A., Chiesa R., Tolosa S., A method for the determination of inorganic phosphate in the presence of labile organic phosphate and high concentrations of protein: application to lens ATPases, *Anal. Biochem.* 168:1, (1988).

DIVERSE GENETIC REGULATORY ELEMENTS ARE REQUIRED TO DIRECT THE PROPER TISSUE-SPECIFIC AND DEVELOPMENTAL EXPRESSION OF THE MURINE ADENOSINE DEAMINASE GENE

John H. Winston[1], Lyhna Hong[1], Simon Akroyd [1], Gerri Hanten[2], Katrina Waymire[2], Paul Overbeek[2] and Rodney E. Kellems[1]

[1]Verna and Mars McLean Department of Biochemistry
[2]Department of Cell Biology
Baylor College of Medicine
Houston, TX 77030

INTRODUCTION

Development of an organism from a single cell to a multitude of diverse cell types requires the differential expression and repression of 75% of the genome and the ubiquitous expression of the remaining 25%. As an initial step in defining the biochemical mechanisms governing differential gene expression, we wish to define the cis-acting regulatory sequences involved. The enzymes of purine and pyrimidine metabolism offer the opportunity to investigate both the mechanisms of ubiquitious and tissue-specific gene expression. Most of these genes are expressed at low levels in most cell types and the expression of some are significantly upregulated in a tissue-specific and developmentally regulated manner. An excellent example of this pattern of expression is the murine adenosine deaminase (ADA) gene. Expressed in every tissue of the mouse, ADA expression is more than 100-fold higher in tissues at the fetal-maternal interface (decidua and placenta), the keratinizing epithelium of the upper gastrointestinal tract and the absorptive epithelium of the duodenum (1). Two phases of enhanced ADA expression occur at the fetal/maternal interface (2). From day 6 to 9 of gestation, high levels of ADA are found in maternal decidual cells. From day 9 through birth high levels of ADA are found in fetally derived cells primarily in the basal zone of the placenta. Following birth, ADA levels increase in the gastrointestinal epithelium (1).

This pattern could be produced by a single regulatory element recognized in all tissues, separate regulatory elements for each tissue and/or a combination of positive and negative elements. In order to distinguish among these possibilities, our immediate goal is to define the genetic regulatory elements at the murine ADA locus.

RESULTS

To identify genetic regulatory elements at the ADA locus, we first searched for candidate regions of the gene by locating DNase I hypersensitive sites (4), then tested these regions for the presence of genetic regulatory elements in transgenic mice. Over 50 kb of genomic DNA surrounding the ADA locus was surveyed in four tissues expressing variable ADA levels: decidua, placenta, thymus and liver (5). This analysis revealed a major site at the promoter in all tissues examined regardless of the level of ADA expression (Fig. 1). Two thymus specific sites were observed in intron 1 (Fig 1).

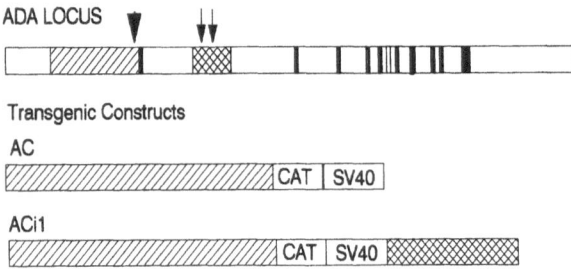

Fig. 1. Murine ADA locus and Transgenic Constructs. The murine ADA gene consists of 12 exons (black bars) spread over 23 kb of genomic DNA. The location of the DNase I hypersensitive site at the promoter (large arrow) and the two thymus specific sites (small arrows) are shown. The indicated regions of the gene were used to prepare transgenic constructs AC and ACi1.

To determine whether DNA sequences within 6.5 kb 5' of the transcription start site including the promoter HS site contained functional genetic regulatory elements, this region was fused to a chloramphenicol acyl transferase (CAT) reporter gene generating construct AC (Fig. 1). This construct contained the SV40 small t intron and early polyA signal. Transgenic mice were prepared by microinjection of this construct into the male pronucleus of the fertilized oocyte. Nine lines carrying variable numbers of the AC transgene integrated in at a single location were produced. CAT activity was measured first in tissues at the fetal/maternal interface. In seven of nine lines, this construct directed high levels of CAT expression to the placenta relative to the embryo on day 13 of gestation (Table I). CAT expression was position-dependent and was roughly corelated with copy number. No CAT expression was detected in the maternal decidua.

Table I. CAT activity in day 13 embryos and placentas.

Line	Copy number	ADA activity (nmol/min/mg) Embryo	Placenta	CAT activity (pmol/min/mg) Embryo	Placenta
52	50	30	1400	<1	170
37	37	30	1150	<1	5400
66	17	20	1600	<1	1300
53	10	10	470	<1	90
77	7	20	780	<1	100
19	5	50	860	<1	150
58	5	20	660	<1	50
25	1	10	510	2	3
68	1	10	560	5	3

CAT and ADA values for each sample represent the average of three determinations. (Reprinted with permission from reference 3.)

In tissues of the mature mouse, CAT activity was detected in the forestomach in all lines (Fig. 2). Much lower levels were observed in tongue and esophagus. CAT expression was not consistently detected in any other tissues examined (Fig 2). This section of 5' flanking DNA contained placenta and forestomach genetic regulatory elements, but lacked information for decidua, intestine, and ubiquitous expression respectively.

The 3.6 EcoRI fragment containing the two thymus-specific sites was added to the AC construct to produce ACi1 (Fig 1). Three lines of ACi1 transgenic mice showed CAT expression in all tissues examined (Table II). Addition of a control DNA fragment from intron 2 had no effect on the pattern of activity observed for the AC construct. CAT activity in normally low ADA expressing tissues such as liver was 100 fold lower than that measured in forestomach. Since this was the pattern of expression observed for the endogenous ADA gene, the 3.6EE fragment appeared to contain sequences required for ubiquitous expression. CAT activity in the thymus was comparable to that of forestomach indicating that this fragment also contained a thymus-specific enhancer.

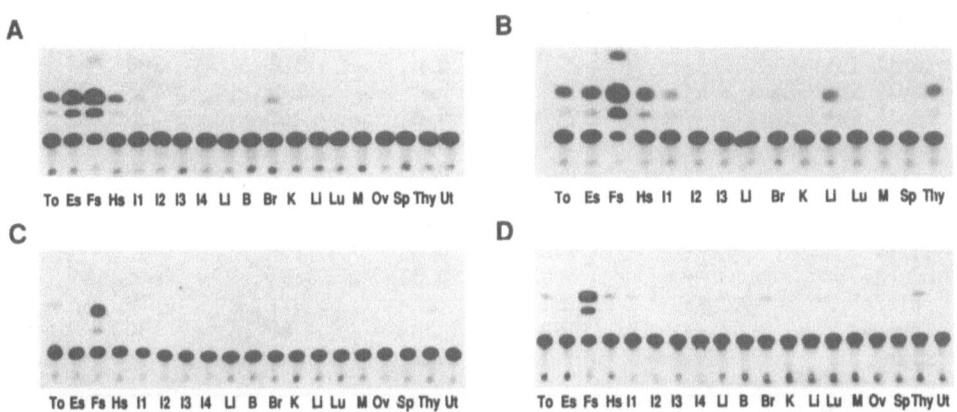

Fig. 2. CAT activity in adult tissues of AC transgenic mice. Homogenates were prepared from tissues isolated from a 2-month-old F1 female mouse from four transgenic lines: A, 52; B, 19; C, 68; and D, 77, respectively. CAT activity was measured in a 2-h incubation utilizing 0.5 mg of protein from each tissue extract. The tissues are: To, tongue; Es, esophagus; Fs, forestomach; Hs, hindstomach; B, bladder; Br, brain; K, kidney; Li, liver; Lu, lung; M, cardiac muscle; Ov, ovaries; Sp, spleen; Thy, thymus; Ut, uterus. The small intestine was divided into four parts from duodenum to ileum: I1, I2, I3 and I4. (Reprinted with permission from reference 3.)

Results based upon CAT activity measurements did not address whether absolute levels of expression were comparable. To directly compare transgene and ADA expression, steady-state CAT and ADA mRNA levels were measured. On a per gene basis, forestomach CAT mRNA level was close to that of ADA for two of the three ACi1 lines (Table III). CAT message was at least three-fold lower than that of ADA in the forestomach from AC mice. The CAT mRNA levels in 3 low ADA expressing tissues were higher than that produced from the endogenous gene (Table IV). These results suggest that the transgene contains the information to reproduce the correct quantitative pattern of ADA expression. Thus the level of CAT expression measured in low ADA expressing tissues such as liver may truly reflect basal expression.

Table II. Comparison of CAT activity to ADA activity in tissues of transgenic mice.

	CAT Activity (pmoles/min/mg)				ADA activity (nmoles/min/mg)
construct	AC	ACi1	ACi1	ACi1	ACi1
Line	52	M1	522	523	M1
copy #	50	70	50	40	70
Animal	475	546	654	1312	546
Tongue	*	3.4	9.0	1.8	3400
esophagus	--	1.0	1.9	1.0	1900
forestomach	50	150	154	70	5200
hindstomach	--	0.1	1.5	0.2	20
small int. 1	--	3.1	5.6	2.8	2000
small int. 2	--	3.3	5.1	2.8	530
small int. 3	--	2.7	4.8	2.2	200
small int. 4	--	2.4	4.4	3.3	160
Large int.	--	0.1	0.9	1.1	90
Brain	--	0.1	0.3	0.07	10
kidney	--	0.1	2.6	1.7	20
liver	--	1.2	4.0	1.9	10
muscle	--	0.05	0.5	0.03	2
spleen	--	21	11	9.1	20
Thymus	--	180	73	60	300
skin	--	1.7	9.0	0.9	3

--= < 0.001

Table III. Ratios of CAT to ADA mRNA per gene in the forestomach of transgenic mice.

AC			ACi1			ACi2		
Line	Copy#	CAT/ADA	Line	Copy#	CAT/ADA	Line	Copy#	CAT/ADA
58	5	0.4	523	40	0.9	57	80	0.3
19	7	0.19	522	50	1.6			
66	17	0.33	M1	70	0.1			
52	50	0.36						

Table IV. Ratio of CAT/ADA mRNA per gene in tissues from ACi1 transgenic mouse.

Tissue:	Duodenum	Liver	Spleen	Skin
CAT/ADA:	0.02	4	9	3

DISCUSSION

To produce the complex pattern of murine ADA expression, our results show that more than one type of genetic regulatory element residing in different portions of the gene are required. Sequences within a 3.6kb EcoRI fragment from intron 1 are required in combination with the promoter to establish ubiquitous expression. Enhanced expression is achieved by tissue-specific enhancers located in other regions of the gene. Placenta and forestomach enhancers are located within 6.5 kb 5' flanking region. At least some of the genetic regulatory elements required for tongue, esophagus, intestine and decidua expression appear to reside in other portions of the gene. The unusually high thymus expression from the ACi1 transgene may be due to the absence of negative regulatory elements. A region from intron 1 of the human ADA gene also contains both thymus-specific and ubiquitious genetic regulatory elements (6,7).

The lack of detectable reporter gene activity in most tissues from the AC construct may be due to lack of transcription of the transgene promoter in these tissues. Since the presence of a DNase I HS site at the promoter of the endogenous gene corelates with a transcriptionally active promoter in all tissues, we looked for this site in the AC and ACi1 transgenes respectively. As reported for the endogenous gene, the promoter site is observed in both liver and thymus from ACi1 mice in which reporter gene activity is detected but not in liver and thymus carrying the AC transgene which fails to show CAT activity (5). This difference in the chromatin structure at the two transgene promoters may be due to repression of the AC promoter.

Since the presence of the basal activator sequences leads to altered chromatin structure at the ADA promoter, its function may be to maintain a chromatin conformation at the promoter favorable to allow the interaction with the transcription machinery. In the abscence of the basal activator, the transgenes may be sequestered in heterochromatin. Although neither the placenta nor the forestomach enhancers require the presence of the basal activation activity to produce expression, the inclusion of the activator appears to increase the efficiency of these enhancers.

Deletion analysis of the 5' flanking region and of the 3.6EE fragment designed to precisely define sequences required for basal activation may lead to a better understanding of how these elements interact in an in vivo developemental context to generate the complex pattern of ADA expression.

Acknowledgements. Some of the research reported here was supported by Public Health Service grants GM30204, HD21452, and AI25255 from the National Institutes of Health and grant Q-893 from the Robert A. Welch Foundation.

REFERENCES

1. J.M. Chinsky, V. Ramamurthy, W.C. Fanslow, D.E. Ingolia, M.R. Blackburn, K.T. Shaffer, H.R. Higley, J.J. Trentin, F.B. Rudolph, T.B. Knudsen, and R.E. Kellems, Developmental expression of adenosine deaminase in the upper alimentary tract of mice, Differentiation 42:172-183(1990).

2. T.B. Knudsen, M.R. Blackburn, J.M. Chinsky, M.J. Airhart, and R.E. Kellems, Ontogeny of adenosine deaminase in the mouse decidua and placenta: immunolocalization and embryo transfer studies, Biol. Reprod. 44:171-184(1991).

3. J.H. Winston, G.R. Hanten, P.A. Overbeek, and R.E. Kellems, 5' Flanking sequences of the murine adenosine deaminase gene direct expression of a reporter gene to specific prenatal and postnatal tissues in transgenic mice, J. Biol. Chem. 267:13472-13479(1992).

4. D.S. Gross and W.T. Garrard, Nuclease hypersensitive sites in chromatin, Ann. Rev. Biochem. 57:159-197(1988).

5. L. Hong, The relationship between chromatin structure and gene expression at the murine adenosine deaminase locus, Ph.D. Dissertation, The Graduate School, Baylor College of Medicine, Houston(1992).

6. B. Aronow, D. Lattier, R. Silbiger, M. Dusing, J. Hutton, G. Jones, J. Stock, J. McNeish, S. Potter, D. Witte and D. Wigington, Evidence for a complex regulatory array in the first intron of the human adenosine deaminase gene, Genes Dev. 3:1384-1400(1989).

7. B. Aronow, R. Silbiger, M. Dusing, J. Stock, K. Yager, S. Potter, J. Hutton and D. Wiginton, Functional Analysis of the human adenosine deaminase gene thymic regulatory region and its ability to generate position-independent transgene expression, Mol. Cell. Biol. 12:4170-4185(1992).

THE MUSCLE AND NONMUSCLE ISOZYMES OF ADENYLOSUCCINATE SYNTHETASE ARE ENCODED BY SEPARATE GENES WITH DIFFERENTIAL PATTERNS OF EXPRESSION

Oivin M. Guicherit[1], Bruce F. Cooper[2],
Frederick B. Rudolph[2], and Rodney E. Kellems[1]

[1]The Department of Biochemistry
Baylor College of Medicine, Houston, TX 77030
[2]The Department of Cell Biology and Biochemistry
Rice University, Houston, TX 77005

INTRODUCTION

Adenylosuccinate synthetase (AdSS) catalyzes the first of two steps leading to the synthesis of AMP from IMP. Two isoforms of AdSS have been observed in all mammalian species examined (Stayton et al., 1983). One isoform, termed the nonmuscle isozyme, is widely distributed among mammalian tissues and functions at the branchpoint of purine nucleotide metabolism in *de novo* synthesis of AMP (Figure 1). The other isoform, termed the muscle isozyme, is highly abundant in striated muscle tissues and is part of the purine nucleotide cycle (Figure 2; Stayton et al., 1983, Van Waarde, 1988; Lowenstein, 1990; Van den Berghe et al., 1992). This cycle, which involves AdSS as well as AMP deaminase and adenylosuccinate lyase, is active in cardiac and skeletal muscle where it is believed to play a role in muscle energy metabolism (Van Waarde, 1988; Lowenstein, 1990). Lower levels of the muscle isozyme of AdSS are found in kidney, brain, and testes, but the role of this isozyme in these tissues is unclear at this time.

As expected from their distribution and different metabolic roles, the mammalian AdSS isozymes have been observed to differ in their functional and physical properties (see Stayton et al. (1983) for review). The physiological significance, however, of these different biochemical properties of the isozymes is still not completely understood. Because of its abundance in skeletal muscle, most previous biochemical studies (Stayton et al., 1983) have focused primarily on the muscle isozyme. Comparisons between these studies, however, have been difficult because of variable assay conditions utilized. Furthermore, the analysis of the ubiquitous nonmuscle isozyme has been hampered by low abundance and instability, making protein purification very difficult. Future comparisons of muscle and nonmuscle isozymes of AdSS will be aided by a convenient source of the nonmuscle isozyme.

The genetic basis for two AdSS isozymes had so far not yet been determined. The distinct protein distribution of the isozymes indicates that their synthesis is most likely controlled by different genetic signals. On the other hand, as mentioned before, the isozymes have been detected in the same tissues (kidney, brain, testes) suggesting that at least some of these signals controlling their expression are responsive to similar environmental cues. However, it cannot be ruled out that the expression of each isozyme is restricted to different cell-types within the shared tissues. To fully appreciate the

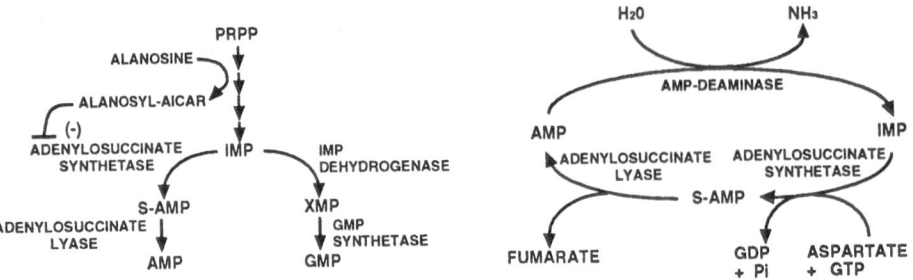

A. DE NOVO PURINE BIOSYNTHESIS

B. THE PURINE NUCLEOTIDE CYCLE

Figure 1. Pathways of purine nucleotide metabolism which involve the adenylosuccinate synthetase reaction. **A**, *de novo* biosynthesis of purine nucleotides; The aspartate-analog alanosine is converted by SAICAR synthetase to alanosyl-AICAR, a compound that competes for IMP in the AdSS reaction. **B**, The purine nucleotide cycle, a cyclic pathway that interconverts AMP and IMP. This pathway is postulated to play a role in muscle energy metabolism.

physiological significance of the AdSS isozymes in mammalian tissue, it will therefore be crucial to also investigate their pattern of expression as well as their precise cellular localization.

To investigate the functional roles of the mammalian AdSS isozymes and allow characterization of the molecular mechanisms that govern their synthesis, we have begun to isolate isozyme-specific cDNAs and antibodies. Here we describe the generation of a genetically enriched cell line that facilitated the purification of the nonmuscle enzyme. In addition, we present the isolation of mouse cDNAs specific for each isozyme. The cDNAs enabled us to show that the muscle and nonmuscle AdSS isozymes are encoded by separate genes with differential patterns of expression.

EXPERIMENTAL PROCEDURES

Details concerning materials and methods utilized in the experiments have been described elsewhere (Guicherit et al., 1991 and 1994).

RESULTS

Selective Overproduction of the Nonmuscle AdSS Isozyme in Murine T-Lymphoma Cells

The purification of the nonmuscle isozyme of AdSS has been hampered by its low abundance in tissues and cell lines. To circumvent this problem we isolated a cultured cell line that produces large quantities of this enzyme (Guicherit et al., 1994). This was accomplished by selecting a murine T-lymphoma cell line, YAC-1, for resistance to increasing concentrations of alanosine, an aspartic acid analog. Alanosine is metabolized to alanosyl-AICAR by SAICAR synthetase (Figure 1), an enzyme of *de novo* purine nucleotide biosynthesis (Tyagi and Cooney, 1984). Alanosyl-AICAR functions as an IMP analog and competitively inhibits AdSS (Tyagi and Cooney, 1984; Stayton et al., 1983). We used a stepwise selection procedure to obtain a population of cells, YAC-A16, that grew in 16 mM alanosine. As shown by the killing curves in Figure 2A the drug-resistant cells were approximately 1,000-fold more resistant to alanosine than the parental cells. Furthermore, the YAC-A16 cells were characterized by the presence of a very high level of AdSS activity (~500 nmol/min/mg) as compared to the original YAC-1 population (not measurable, with lower limit of detection < 5 nmol/min/mg). SDS-PAGE analysis of lysates from YAC-1 and YAC-A16 cells showed that the alanosine resistant cells were

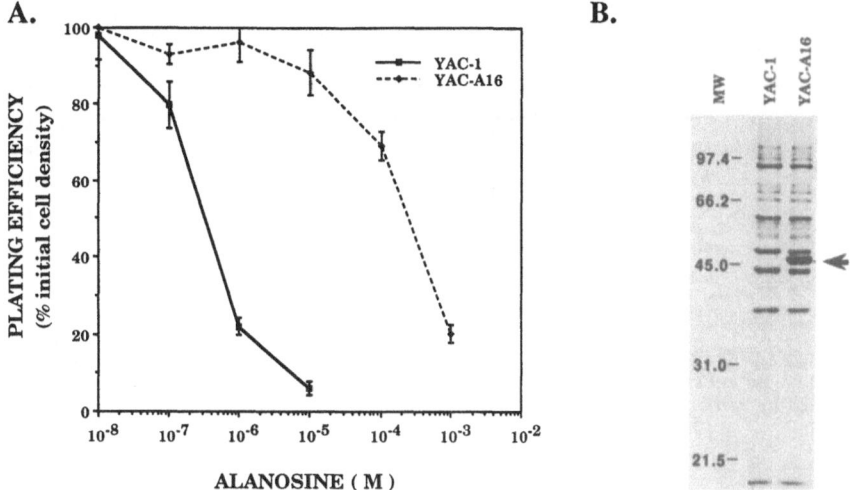

Figure 2. Selective overproduction of the nonmuscle AdSS isozyme in alanosine-resistant cells. **A,** Survival curves for the parental cell line (YAC-1) and its alanosine-selected derivative (YAC-A16) in increasing alanosine concentrations. Cell survival was plotted as a percentage of the initial cell density. **B,** SDS-polyacrylamide gel (10%) analysis of high speed supernatants from parental (YAC-1) cells and alanosine-selected (YAC-A16) cells. *Arrow* indicates position of nonmuscle AdSS. (Figure adapted from Guicherit et al., 1994, with permission)

specifically enriched in a protein of approximately 50-kDa (Figure 2B, *arrow*), the size expected for AdSS (Stayton et al., 1983). As shown in Figure 2B, the 50-kDa protein appears to be the only protein enriched in the alanosine resistant cell line (YAC-A16). The chromatographic profile and kinetic properties of the synthetase activity in the drug-resistant cells were characteristic of those expected for the nonmuscle isozyme. These results indicate that YAC-A16 cells are considerably enriched for the nonmuscle isozyme of AdSS and that these cells could be a convenient source from which to purify this enzyme.

Purification of the Muscle and Nonmuscle Isozymes of AdSS

Because of the high abundance of synthetase activity in rat skeletal muscle, and our previous success in using this tissue, rat muscle was utilized for isolating the muscle AdSS isozyme. This procedure involved high speed centrifugation (20,000 x g), ammonium sulfate precipitation (45-60%), and two column chromatographic steps (phosphocellulose and GTP-agarose column matrices; Guicherit et al., 1991). Similarly, the nonmuscle isozyme was purified from the alanosine-resistant cells by a combination of high speed centrifugation (100,000 x g), ammonium sulfate precipitation (40-70%), and column chromatography (DEAE and Procion Red column matrices; Guicherit et al., 1994).

Cloning and Analysis of Mouse cDNAs Encoding the Muscle and Nonmuscle Isozymes of AdSS

For the cloning of isozyme-specific cDNAs, first several peptides were isolated and sequenced from both the purified muscle (rat) and nonmuscle (mouse) isozymes. This allowed the design of degenerate oligonucleotide probes based on peptide sequence specific for each isozyme. The isozyme-specific oligonucleotide mixtures were end-labeled and used to screen cDNA libraries from mouse skeletal muscle and mouse kidney. Sequence analysis of AdSS cDNAs isolated from the mouse muscle library revealed an open reading frame encoding a polypeptide of 457 amino acids, with a calculated mass of ~50 kDa (Guicherit et al., 1991). Similarly, AdSS cDNAs isolated from the mouse kidney library encoded a polypeptide of 456 amino acids, also with a calculated mass of ~50 kDa (Guicherit et al., 1994). Figure 3 shows an alignment of the amino acid sequences of the

```
adss1 MSGTRASNDR PPGTGGVKRG RLQQEAAATG SRVTVVLGAQ WGDEGKGKVV DLLATDADIV SRCQGGNNAG HTVVVDGKEY DFHLLPSGII NTKAVSFIGN 100
         ||  |       |   |   | |  |||||||||| |||| ||||  |||| ||||  |||||||||||  |||||  || |||||||||| ||  ||||||     ||||
adss2 MSISESSPAA TSLPNGDC-G RPR--ARSGG MRVTVVLGAQ WGDEGKGKVV DLLAQDADIV CRCQGGNNAG HTVVVDSVEY DFHLLPSGII NFNVTAFIGN  97

adss1 GVVIHLPGLF EEAEKN--EK KGLKDWEKRL IISDRAHLVF DFHQAVDGLQ EVQRQAQEGK NIGTTKKGIG PTYSSKAART GLRICDLLSD FDEFSARFKN 198
      |||||||||| ||||||  ||  |||  ||||| |||||||||| || |||| ||  | ||| | |  ||||||| | |||||| ||| |||  ||| || || || |  ||
adss2 GVVIHLPGLF EEAEKNVQKG KGLDGWEKRL IISDRAHIVF DFHQAADGIQ EQQRQEQAGK NLGTTKKGIR PVYSSKAARS GLRMCDLVSD FDGFSERFKV 197

adss1 LAHQHQSMFP TLEIDVEGQL KRLKGFAERI RPMVRDGVYF MYEALHGPPK KVLVEGANAA LLDIDFGTYP FVTSSNCTVG GVCTGLGIPP QNIGDVYGVV 298
      | | | | |  |||||  | |   ||  |||  || ||| |||||  ||||||||| | ||||||||| |||||||||| |||||||||| |||||||| ||  | |||||
adss2 LTNQYKSIYP TLEIDIEGEL QQLKGYMERI KPMVKDGVYF LYEALHGPPK KILVEGANAA LLDIDFGTYP FVTSSNCTVG GVCTGLGMPP QNVGEVYGVV 297

adss1 KAYTTRVGIG AFPTEQINEI GDLLQNRGHE WGVTTGRKRR CGWLDLMILR YAHMVNGFTA LALTKLDILD VLSEIKVGIS YKLNGKRIPY FPANQEILQK 398
      |||||||||| |||||| ||| | ||  |  | | ||||||||| ||||| ||  || ||||  |||||||||||  ||| |  ||  || |||||| ||||| ||
adss2 KAYTTRVGIG AFPTEQDNEI GELLQTRGRE FGVTTGRKRR CGWLDLVSLK YAHMINGFTA LALTKLDILD MFTEIKVGVA YKLDGETIPH FPANQEVLNK 397

adss1 VEVEYETLPG WKADTTGARK WEDLPPQAQS YVRFVENHMG VAVKWVGVGK SRESMIQLF 457
      |||  | ||||| ||   |   ||  ||  ||||  | |      |||  |||| ||||||||||
adss2 VEVQYKTLPG WNTDISNART FKELPVNAQN YVRFIELELQ IRVKWIGVGK SRESMIQLF 456
```

Figure 3. Comparison of the amino acid sequences of the muscle and nonmuscle isozymes of AdSS. The sequences of the mouse muscle (adss1) and nonmuscle (adss2) isozymes are aligned and identical amino acids are indicated by *vertical dashes*. The most significant region (NH2-terminus) of non-homology is indicated in *bold*.

mouse muscle and nonmuscle isozymes as deduced from the muscle and kidney cDNAs, respectively. This alignment shows an extensive sequence conservation throughout the polypeptide sequence (>70% identity). On closer examination several regions of low homology were observed of which the most striking is the NH2-terminal stretch of ~30 residues (Figure 3, *bold* region) that showed virtually no homology between the mammalian isozymes. Interestingly, this region is completely absent from the *E. coli* (Wolfe and Smith, 1988) and *D. discoideum* synthetases (Wiesmuller et al., 1990), suggesting that it might fulfill a functional role specific for mammalian species. Further studies are required to determine the functional relevance of these NH2-terminal sequences.

The Muscle and Nonmuscle Isozymes of AdSS are Encoded by Separate Genes that are Differentially Regulated in Mouse Tissues

Although the AdSS isozymes are clearly related by sequence similarity (Figure 3), their comparison revealed a significant number of mismatches throughout the proteins (as well as the cognate cDNAs), suggesting that the isozymes are not likely to be the products of the same gene. This was confirmed by duplicate genomic Southern blots probed with full-length isozyme-specific cDNAs. These showed different hybridization patterns for each cDNA, indicating that the two isozymes are encoded by separate genes (Guicherit et al., 1994). We proposed to designate the gene encoding the muscle enzyme, AdSS1, and the gene encoding the nonmuscle enzyme, AdSS2.

Northern analysis with a muscle isozyme-specific cDNA probe revealed a single transcript of about 1.8 kb which is very abundant in all striated muscle tissues tested (Figure 4A; skeletal muscle, heart, tongue, and esophagus). This muscle (AdSS1) transcript was barely detectable in liver and kidney and virtually nonexistent in the other non-muscle tissues. In contrast, Northern analysis with a cDNA probe specific for the nonmuscle isozyme showed three transcripts (1.7, 2.8, and 3.4 kb) which were present in all non-muscle tissues tested, but were barely detectable in both cardiac and skeletal muscle (Figure 4B). The structural relationship among the three nonmuscle (AdSS2) transcripts remains to be determined.

CONCLUSION

Here we describe the purification of both mammalian isozymes of adenylosuccinate synthetase, the subsequent isolation of their cognate cDNAs, and the genetic analysis of their expression pattern. For this the highly abundant muscle isozyme was purified from rat skeletal muscle. Because of its low abundance in mammalian tissues and available cell lines, we developed a genetic selection scheme to isolate a cell line that is enriched for the nonmuscle synthetase. The availability of this cell line significantly facilitated the purification of this isozyme. Degenerate oligonucleotide probes based on peptide sequences

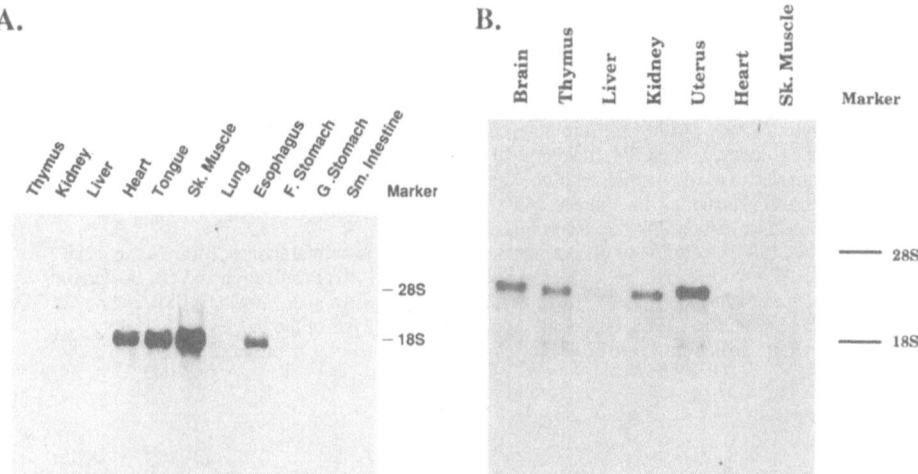

Figure 4. Tissue distribution of the muscle and nonmuscle AdSS mRNAs. Total RNA from several mouse tissues was fractionated on a denaturing 1.4% agarose gel and transferred to a nylon membrane. The Northern blot was probed with a cDNA encoding the muscle isozyme (A) or a cDNA encoding the nonmuscle isozyme (B). The relevant mouse tissues are listed above each lane. Ribosomal markers are included (18S, 28S). (Figure adapted from Guicherit et al., 1991 and 1994, with permission)

derived from the purified AdSS isozymes were used to isolate mouse cDNAs specific for each isozyme. The deduced amino acid sequences of the mouse muscle and nonmuscle isozymes show extensive similarity (>70% identity) throughout their polypeptide sequence. The present study shows for the first time that the muscle and non-muscle isozymes of AdSS are encoded by different genes, which have distinct and almost mutually exclusive patterns of expression. Northern analysis showed that the muscle AdSS1 gene encodes a single transcript (1.8 kb) which is predominantly present in cardiac and skeletal muscle, while the nonmuscle AdSS2 gene encodes three transcripts (1.7, 2.8, 3.4 kb) which are expressed at lower levels in a wide range of tissues. The AdSS2 gene expression is virtually undetectable in striated muscle. The apparently contrasting expression patterns of the AdSS genes may indicate the importance of keeping the isozymes separate because they could potentially interfere which each others role in metabolism, as a consequence of their regulatory differences.

ACKNOWLEDGMENTS

This work was supported by Grant CA14030 from the National Cancer Institute, Grant GM42436 from the National Institute of Health, Grants C-1041 and Q-893 from the Robert A. Welch Foundation, and a Grant from the Muscular Dystrophy Association. Alanosine was kindly supplied to us by the Drug Synthesis and Chemistry Branch, Developmental Therapeutics Program, Division of Cancer Treatment, National Cancer Institute.

REFERENCES

Guicherit, O.M., Rudolph, F.B., Kellems, R.E., and Cooper, B.F., 1991, Molecular cloning and expression of a mouse muscle cDNA encoding adenylosuccinate synthetase, J. Biol. Chem. 266:22582

Guicherit, O.M., Cooper, B.F., Rudolph, F.B., and Kellems, R.E., 1994, Amplification of an adenylosuccinate synthetase gene in alanosine-resistant murine T-lymphoma cells, J. Biol. Chem. 269:22582

Lowenstein, J.M., 1990, The purine nucleotide cycle revisited, Int. J. Sports Med. 11:S36

Stayton, M.M., Rudolph, F.B., and Fromm, H.J., 1983, Regulation, genetics, and properties of adenylosuccinate synthetase: A review, Curr. Top. Cell. Regul. 22:103

Tyagi, A.K. and Cooney, D.A., 1984, Biochemical pharmacology, metabolism, and mechanism of action of L-alanosine, a novel, natural, antitumor agent, Adv. Pharm. Chemother. 20:69

Van den Berghe, G., Bontemps, F., Vincent, M.F., and Bergh, F.V.d., 1992, The purine nucleotide cycle and its molecular defects, Prog. in Neurobiol. 39:547

Van Waarde, A., 1988, Operation of the purine nucleotide cycle in animal tissues, Biol. Rev. 63:259

Wiesmuller, L., Wittbrodt, J., Noegel, A.A. and Schleicher, M., 1991, Purification and cDNA-derived sequence of adenylosuccinate synthetase from *Dictyostelium discoideum*, J. Biol. chem. 266:2480

Wolfe, S.A. and Smith, J.M., 1988, Nucleotide sequence and analysis of the *pur*A gene encoding adenylosuccinate synthetase of *Escherichia coli* K12, J. Biol. Chem. 263:19147

EXPRESSING ENZYMATIC DOMAINS OF HAMSTER CAD IN

CAD-DEFICIENT CHINESE HAMSTER OVARY CELLS

Jeffrey N. Davidson and Robert S. Jamison

Department of Microbiology and Immunology
University of Kentucky
Lexington, KY 40502

INTRODUCTION

The first three enzymes of *de novo* pyrimidine biosynthesis, carbamyl phosphate synthetase (CPSase), aspartate transcarbamylase (ATCase) and dihydroorotase (DHOase), are domains of a single polypeptide (called CAD) in species from *Dictyostelium* to humans. One approach to demonstrate the independence of the enzymatic domains of CAD has been to express portions of the hamster *CAD* cDNA in *E. coli* defective in synthesizing pyrimidines *de novo*. For example, the *CAD* sequence encoding the ATCase domain cloned into a bacterial expression vector was transformed into ATCase-deficient *E. coli* (Maley and Davidson, 1988). The transformants were able to grow in the absence of uracil and contained hamster protein with ATCase activity. This system is proving to be a powerful strategy for studying the structure and evolution of the enzymatic domains of CAD (e.g. Musmanno et al., 1991; Zimmermann and Evans, 1993; Williams et al., 1993). However, in order to study the significance of CAD structure to pyrimidine metabolism in mammals, it may be helpful to express the enzymatic domains of CAD separately in a hamster cell line deficient in CAD. This was first achieved with the successful expression of the CPSase domain in a CAD-deficient hamster cell line (Musmanno et al., 1992). In this report the ATCase and DHOase domains are expressed as separate proteins in a CAD-deficient cell line, and the application of this approach in generating a carbamyl phosphate synthetase activity unresponsive to feedback inhibition by UTP is described.

MATERIALS AND METHODS

Cells

E. coli strains HB101 and JM109 were used in all cloning manipulations. Chinese hamster ovary K1 cells (CHO-K1) are the wild-type line and synthesize a full-length CAD protein. G9c is a derivative of CHO-K1, has little CPSase, ATCase, or DHOase activity and requires 30 μM uridine for growth. Mammalian cells were grown in Ham's F12 medium supplemented with 10% fetal calf serum or dialyzed fetal calf serum.

Plasmid Constructs

Portions of the Syrian hamster *CAD* cDNA were cloned into the mammalian expression vector pHβAPr-1-neo (Gunning et al., 1987), which contains 3 kb of β-actin 5' flanking sequence including promoter, 5' untranslated region in exon 1, intervening sequence, short polylinker downstream of 3' splice junction to exon 2 and an SV40 cleavage-polyadenylation sequence. Polymerase chain reaction was used to amplify the appropriate portion of the *CAD* cDNA, to generate unique restriction sites that would allow for cloning into the multiple cloning site of the vector and to create a sequence around an ATG codon that matches well the consensus established by Kozak (1987) for efficient translation. An example of the strategy used is presented in Musmanno et al (1992). The construct encoding a carboxyl truncation of the CPSase was generated by substituting the 3' end of the cDNA encoding the CPSase with a smaller polymerase chain reaction product that contained a stop codon. This modified construct was missing 387 bases as compared to the construct described by Musmanno et al (1992).

Transfections

Plasmid DNA (10-50 μg) was transfected into tissue culture cells by the calcium phosphate precipitation method (Kingston et al., 1984). Clones were selected either by growth on the neomycin analog G418 (600-800 μg/ml) or in the case of isolating G9c(C,A,D) on medium lacking uridine. The portion of the CAD protein present in a transfectant is indicated within a parenthesis; for example, G9c(DA) carries stably integrated DNA encoding a bifunctional protein with DHOase and ATCase domains.

Western Analysis and Enzyme Assays

Cells in culture were lysed by sonication as described previously (Davidson et al., 1979). Cleared lysates were separated by SDS polyacrylamide gel electrophoresis and transferred to nylon membranes. Western blotting was performed by the methods of Burnette (1981) and Harlow and Lane (1988), and polyclonal anti-CAD antibody was provided by George Stark. Enzyme were assayed as described previously (Davidson et al., 1979; Rumsby et al., 1984).

RESULTS AND DISCUSSION

Plasmids encoding the ATCase, DHOase, and the linked DHOase-ATCase domains of Syrian hamster *CAD* were transfected into G9c. Several G418-resistant colonies were selected and assayed for CPSase, ATCase, and DHOase activities. G9c(A) clones had as much ATCase activity as 70% of wild-type CHO-K1 levels, G9c(D) clones had as much DHOase activity as 210% of CHO-K1 levels and G9c(DA) clones had as much DHOase and ATCase activities as 200% of CHO-K1 levels. Protein from the cell extract from the best clone of each type of transfectant was separated by SDS polyacrylamide gel electrophoresis and tested by western analysis using polyclonal antibody against Syrian hamster CAD (Figure 1A). Only a single band was observed in each lane and the size the protein was identical to what was predicted for the polypeptide encoded by the cDNA segments in each transfectant type: 38 kDa for ATCase, 43 kDa for DHOase and 85 kDa

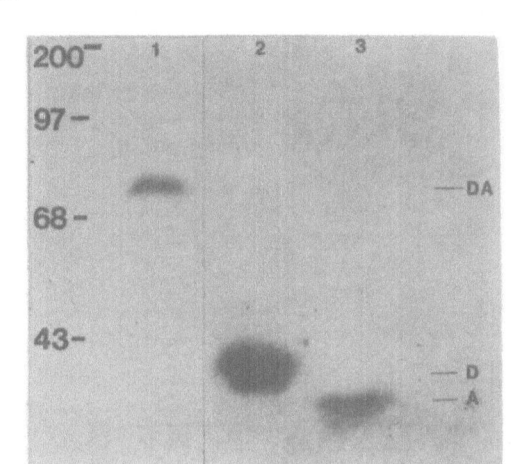

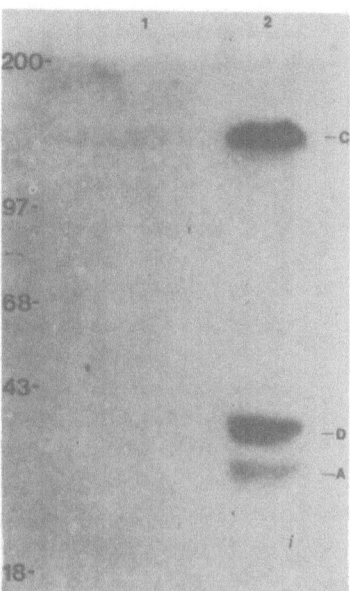

Figure 1. Western blots of transfected and control cell lines probed with anti-CAD holoenzyme polyclonal antiserum. Equal amounts of total cell protein were loaded into each lane. A. lane 1: G9c(DA); lane 2: G9c(D); and lane 3: G9c(A). B. lane 1: G9c; and lane 2: G9c(C,A,D)

for linked DHOase-ATCase. The G9c transfectants like the parent G9c cell line were incapable of growth in the absence of uridine and had <1% of wild-type levels of CPSase. These experiments along with previously published data for the CPSase (Musmanno et al., 1992) demonstrated that each of the enzymatic domains of CAD can be expressed as a stable protein with enzymatic activity in CAD-deficient CHO cells. Such findings set-up the possibility of examining the synthesis of pyrimidines *de novo* where the enzymes are on a single polypeptide or on separate polypeptides.

To study the cellular consequences of having the three enzymes of CAD on separate proteins, the G9c(A) clone with the highest activity of ATCase was co-transfected with plasmids separately encoding the CPSase and DHOase domains. Cells were plated onto medium without uridine. A single colony grew in the absence of uridine and had CPSase, ATCase, and DHOase activities of 103%, 66% and 88%, respectively, of wild-type levels. To confirm that the clone, G9c(C,A,D), was expressing the three enzymes as separate proteins, cell extract from this clone was loaded onto an SDS-polyacrylamide gel. Figure 1B shows that three proteins with the expected sizes including the ~150 kDa band for CPSase are visible by western analysis. When approximately 100 cells of G9c(C,A,D) were plated onto medium with or without uridine, the average number of clones observed were 67 and 79, respectively. These results indicated that linkage of the three enzymes on a single polypeptide was required neither for growth in the absence of uridine (i.e. synthesis of pyrimidines by the *de novo* pathway) nor for cell viability. However, G9c(C,A,D) cells do appear to grow more slowly in the absence of uridine, a result that may imply some channeling or proximity effect in the normal CAD protein or some build-up of potentially harmful intermediates such as carbamyl phosphate or carbamyl aspartate when the enzymes are on separate proteins. Further investigation is necessary to clear-up this last point.

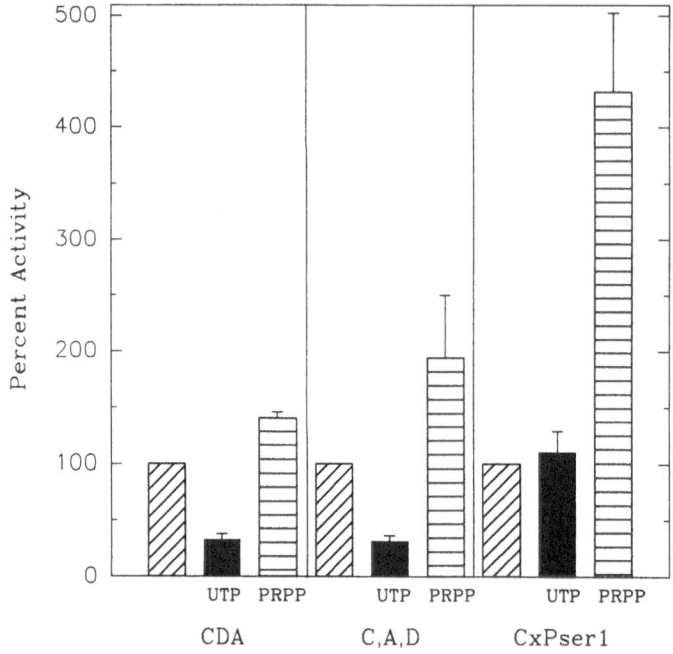

Figure 2. *In vitro* assayed allosteric effects of 2 mM UTP and 2 mM PRPP on glycerol gradient-purified CPSase activity. Activities for CHO-K1 and for G9c(CDA), carrying a construct encoding the full-length CAD protein, are shown in panel labelled "CDA"; activities for G9c(C,A,D) are shown in panel labelled "C,A,D"; and activities for G9c(CxPser1), carrying a construct predicted to lack the site for UTP binding, are shown in panel labelled "CxPser1". The first bar in each panel is activity in the absence of allosteric effectors. Enzyme activities shown in bars two and three of each panel are calculated in comparison to activity seen in the absence of UTP and PRPP, where that activity is taken as 100%.

A last set of experiments was aimed at examining the allosteric site of the CPSase. Previous work indicated that the subdomain responsible for feedback inhibition by UMP in the *E. coli* enzyme is at the carboxyl end of the CPSase (Rubio et al., 1991). The plasmid encoding the CAD CPSase was modified at the 3' end of the cDNA by deleting the codons for the last 129 amino acids. When this plasmid, CxPser1, was transfected into G9c cells, cell extract was prepared a few days after transfection, and CPSase was partially purified by glycerol gradient centrifugation. Figure 2 shows that while normal CAD and a complete CPSase are inhibited by UTP and stimulated by PRPP, the truncated CPSase is unresponsive to UTP and shows significantly greater stimulation by PRPP. This result confirmed that the UTP binding site is located at the carboxyl end of the CPSase domain of CAD and that, while the PRPP and UTP binding sites may overlap, they are not synonymous. No stable G9c transfectants expressing CPSase activity were obtained with the modified CPSase construct. Similar transfections with G9c(DA) and CHO-K1 also failed to yield a single stable transfectant expressing the truncated CPSase. The preliminary conclusion from these attempts is that a CPSase unresponsive to UTP inhibition is lethal. If correct, then the regulation of *de novo* synthesis of pyrimidines is crucial, and the reason for the lethality of unregulated pyrimidine biosynthesis is unknown. Supporting data for this conclusion comes from recent work on *Drosophila* mutants with reduced response of their CPSase to UTP inhibition (Piskür et al., 1993; Piskür et al., this book).

ACKNOWLEDGEMENTS

G9c cells were kindly provided by David Patterson. This work was supported by NSF grant DMB91-05826.

REFERENCES

Burnette, W.H., 1981, "Western blotting" electrophoretic transfer of proteins from sodium dodecyl sulfate-polyacrylamide gels to unmodified nitrocellulose and radiographic detection with antibody and radioiodinated protein A, *Anal. Biochem.* 112:195.

Davidson, J.N., Carnright, D.V., and Patterson, D., 1979, Biochemical genetic analysis of pyrimidine biosynthesis in mammalian cells: III. Association of carbamyl phosphate synthetase, aspartate transcarbamylase, and dihydroorotase in mutants of cultured Chinese hamster cells, *Somat. Cell Genet.* 5:175.

Gunning, P., Leavitt, J., Muscat, G., Ng, S.Y., and Kedes, L., 1987, A human β-actin expression vector system directs high-level accumulation of antisense transcripts, *Proc. Natl. Acad. Sci. USA.* 84:4831.

Harlow, E., and Lane, D., 1988, Antibodies: A Laboratory Manual, Cold Spring Harbor Laboratories, Cold Spring Harbor.

Kingston, R.E., Kaufman, R.J., and Sharp, P.A., 1984, T1 regulation of transcription of the adenovirus EII promoter by EIa gene products: absence of sequence specificity, *Mol. Cell. Biol.* 4:1970.

Kozak, M., 1987, An analysis of 5'-noncoding sequences from 699 vertebrate messenger RNAs, *Nucl. Acids Res.* 15:8125.

Maley, J.A., and Davidson, J.N., 1988, The aspartate transcarbamylase domain of a mammalian multifunctional protein expressed as an independent enzyme in *Escherichia coli*, *Mol. Gen. Genet.* 213:278.

Musmanno, L., Maley, J.A., and Davidson, J.N., 1991, Expression of an active hamster dihydroorotase domain in *E. coli*, *Gene* 99:211.

Musmanno, L.A., Jamison, R.S., Barnett, R.S., Buford, E., and Davidson, J.N., 1992, Complete hamster CAD protein and the carbamyl phosphate synthetase domain of CAD complement mammalian cell mutants defective in *de novo* pyrimidine biosynthesis, *Somat. Cell Mol. Genet.* 18:309.

Piskür, J., Kolbak, D., Sondergaard, L., and Pedersen, M.B., 1993, The dominant mutation suppressor of black indicates that *de novo* pyrimidine biosynthesis is involved in the *Drosophila* tan pigmentation pathway, *Mol. Gen. Genet.* 241:335.

Rubio, V., Cervera, J., Lusty, C.J., Bendala, E., and Britton, H.G., 1991, Domain structure of the large subunit of *Escherichia coli* carbamoyl phosphate synthetase: location of the binding site for the allosteric inhibitor UMP in the COOH-terminal domain, *Biochemistry* 30:1068.

Rumsby, P.C., Campbell, P.C., Niswander, L.A., and Davidson, J.N., 1984, Organization of a multifunctional protein in pyrimidine biosynthesis: a domain hypersensitive to proteolysis, *Biochem. J.* 217:435.

Williams, N.K., Peide, Y., Seymour, K.K., Ralston, G.B., and Christopherson, R.I., 1993, Expression of catalytically active hamster dihydroorotase domain in *Escherichia coli*: purification and characterization, *Prot. Engineer.* 6:333.

Zimmermann, B.H., and Evans, D.R., 1993, Cloning, overexpression, and characterization of the functional dihydroorotase domain of the mammalian multifunctional protein CAD. *Biochemistry* 32:1519.

HOMOLOGY AND MUTAGENESIS STUDIES OF HAMSTER DIHYDROOROTASE

Neal K. Williams, Séan O'Donoghue, and Richard I. Christopherson

Department of Biochemistry
University of Sydney
Sydney NSW 2006
Australia

INTRODUCTION

The third reaction in *de novo* pyrimidine biosynthesis is catalyzed by dihydroorotase (for details see Williams *et al.*, in this volume). By screening a variety of structural analogues of N-carbamyl-L-aspartate (CA-asp) and L-dihydroorotate (DHO), Christopherson and Jones (1980) found dihydroorotase to be highly specific for its natural substrates. Orotate and 5-substituted derivatives, such as 5-fluoroorotate, are effective inhibitors, but dihydrouracil and the CA-asp analogues, N-carbamyl-β-alanine, N-carbamyl-α-alanine, and N-acetyl-L-aspartate are not inhibitory. These observations suggest the identity of essential attachment points in the enzyme-substrate complex. Dihydrouracil lacks the carboxylate group at position 4 of dihydroorotate, and N-carbamyl-β-alanine lacks the corresponding α-carboxylate of CA-asp demonstrating that this group is required for substrate binding, possibly by interacting with a positively charged enzymic group (Christopherson and Jones, 1980). N-carbamyl-α-alanine is lacking the β-carboxylate of CA-asp and N-acetyl-L-aspartate does not possess the terminal ureido nitrogen of the substrate, indicating attachments at these locations. The β-carboxylate of CA-asp may form a coordination complex with the active site zinc atom (see Williams *et al.*, in this volume). We have used knowledge of the cDNA sequence of hamster dihydroorotase in combination with site-directed mutagenesis in an attempt to identify the amino acids involved in these substrate attachments.

The amino acid sequence of dihydroorotase has not been conserved strongly during evolution. The percentage identity between the amino acid sequences of the *pyrC* genes of *Escherichia coli* and *Salmonella typhimurium* and the dihydroorotase domain of the *Drosophila melanogaster rudimentary* and hamster *CAD* genes is in the order of 20-25% (Freund and Jarry, 1987; Simmer *et al.*, 1990). It has been noted however that there appear to be two families of dihydroorotase sequences (Faure *et al.*, 1989). The products of the *pyrC* genes share approximately 40% identity with the *Saccharomyces cerevisiae URA4* gene-product (Guyonvarch *et al.*, 1988), and there is also strong similarity (~55%) between the dihydroorotase domains of the multifunctional proteins of the higher eukaryotes. Thus, the monofunctional enzymes and dihydroorotase domains of multimeric proteins are

Purine and Pyrimidine Metabolism in Man VIII, Edited by
A. Sahota and M. Taylor, Plenum Press, New York, 1995

evolutionarily distinct (see Simmer *et al.*, 1990). The alignment of the known dihydroorotase sequences has revealed 4 conserved regions (Guyonvarch *et al.*, 1988; Williams *et al.*, 1990) containing a small number of invariant amino acid residues. It is likely that these residues play important roles in enzyme function. Using site-directed mutagenesis to replace such residues, and observing the effects of the changes, it is possible to identify key amino acids and their functions in the enzyme.

EXPERIMENTAL PROCEDURES

The recombinant dihydroorotase domain and mutant enzymes were expressed in the *pyrC⁻ E. coli* strain SØ1263 transformed with the plasmid pCW27 and derivatives as described by Williams *et al.* (1993). Site-directed mutagenesis was performed using the Amersham International plc Oligonucleotide-directed *in vitro* Mutagenesis System. The dihydroorotase assay has been described (Williams *et al.*, in this volume).

Computer Molecular Modelling

Modelling of the dihydroorotase substrate-binding site was based on residues Ile91 to Ser99, His119, and the zinc atom of the crystal structure of human carbonic anhydrase II (1ca2 in the Brookhaven Protein Databank; Eriksson *et al.*, 1988a). The crystal structure of DHO methyl ester (Hambley *et al.*, 1993) was used to model the dihydroorotase reaction intermediate or transition state. To convert the carbonic anhydrase sequence to that of hamster dihydroorotase, residues Gln92 and Gly98 were changed to Asp and Arg, respectively, using Insight II (Biosym Inc., San Diego, USA). Insight II was also used to adjust the orientation of the substitute Asp residue, and to add a second oxygen atom to C6 of the model for the dihydroorotase transition state. All subsequent modelling was performed in X-PLOR (Brunger, 1993) using the X-PLOR param19x.pro and toph19x.pro files. Structures were viewed using Midas Plus (Ferrin *et al.*, 1988).

RESULTS AND DISCUSSION

In order to identify potential zinc ligands and functional sequence motifs in the absence of an X-ray structure for dihydroorotase, sequence homology was sought with zinc enzymes of known crystal structure. Figure 1 shows a comparison of the hamster sequence from the first and most conserved region of the dihydroorotases with a region of human carbonic anhydrases I and II. Striking sequence homology is evident between these sequences. The X-ray structure for carbonic anhydrase shows that the homologous region is part of a central β-sheet and contains 2 of the 3 His residues that coordinate the zinc atom at the active site (Kannan *et al.*, 1984; Eriksson *et al.*, 1988a). The opposite face of the β-sheet to the active site forms part of an extended hydrophobic core of carbonic anhydrase (Eriksson *et al.*, 1988a). The histidyl zinc ligands are adjacent to three large hydrophobic residues, 2 Phe and a Trp, which participate in the hydrophobic core. This arrangement probably helps stabilize the molecular architecture of the active site of the enzyme (Eriksson *et al.*, 1988a). The dihydroorotase residues corresponding to the carbonic anhydrase zinc ligands are His15 and His17. Zinc-binding studies of replacement mutants at these His residues, which are invariant among the dihydroorotases, support their assignment as zinc ligands (see Williams *et al.*, in this volume). His15 and His17 are also surrounded by hydrophobic residues, in this case 2 Val and a Leu .

His94 of carbonic anhydrase is hydrogen-bonded to the amide carbonyl of the Gln92 side-chain, forming a "carbonyl-His-zinc triad" (Christianson and Alexander, 1989).

hamster dihydroorotase	12	I D V H V H L R E
human carbonic anhydrase I	91	F Q F H F H W G S
human carbonic anhydrase II	91	I Q F H F H W G S

Figure 1. Comparison of a 9 amino acid sequence from hamster dihydroorotase with regions of human carbonic anhydrases I and II.

Christianson and Alexander have identified this triad, more commonly involving the carboxylate of an Asp residue, in all zinc enzymes with known 3-dimensional structures, and by homology, in a total of at least 36 zinc enzymes. This zinc triad ($-C=O \cdots His \rightarrow Zn^{II}$) from the active site of carbonic anhydrase is likely to occur in dihydroorotase. The residue of hamster dihydroorotase corresponding to Gln92 of carbonic anhydrase is Asp13 which could form a $-C=O \cdots His \rightarrow Zn^{II}$ triad with His15. The conservative replacement of Asp13 with Asn yielded a dihydroorotase mutant with no activity, suggesting that this residue is of critical importance to enzyme function. It is significant that the reactions catalyzed by these two enzymes are also chemically similar. In carbonic anhydrase, an hydroxide ion formed by ionization of a zinc-bound water molecule undergoes nucleophilic attack on a polarized carbonyl carbon of the substrate, carbon dioxide (Eriksson *et al.*, 1988b; Håkansson *et al.*, 1992). In the reverse reaction catalyzed by DHOase as proposed by Christopherson and Jones (1980), an hydroxide ion acts as a nucleophile in the hydrolysis of DHO to form CA-asp. In carbonic anhydrase, the pK_a of the zinc-bound water is approximately 7. The pH profiles for the forward and reverse reactions of DHOase suggest the involvement of a catalytic group with a pK_a of 7.1 (Christopherson and Jones, 1979). It seems likely, therefore, that these two enzymes have similar structures for their zinc-binding site, and exhibit similarities in their catalytic mechanisms.

The most significant difference between the sequences in Figure 1 is the substitution of Gly98 of carbonic anhydrase, with Arg19 of dihydroorotase. It would appear that these corresponding residues have different functions in the 2 enzymes. It is possible that in dihydroorotase the Arg is involved in substrate binding, and as a result has no direct counterpart in carbonic anhydrase which has a very different substrate. To investigate this possibility, part of the active site from hamster dihydroorotase was modelled from the X-ray crystallographic coordinates of human carbonic anhydrase II. The atomic coordinates for residues 91-99 of human carbonic anhydrase II, plus the third zinc ligand, His119, and the zinc atom were obtained from the X-ray structure (Eriksson *et al.*, 1988a). Residues Gln92 and Gly98 of the carbonic anhydrase structure were replaced with Asp13 and Arg19 of the DHOase sequence, and the side chains of the alternating amino acids from the hydrophobic face of the β-sheet were deleted for clarity. The model, thus, represents residues 12-20 of hamster DHOase (Figure 1) and includes the bound zinc atom, a third, remote hystidyl ligand, but ignores the side-chains of the intervening amino acid residues not proposed to be in the active site (Ile12, Val14, Val16, Leu18, Glu20). The proposed tetrahedral transition state was modelled on the crystal structure of DHO-methyl ester (Hambley *et al.*, 1993).

Figure 2 shows a model of the hamster dihydroorotase active site. It appeared likely from the early modelling that, if the nonapeptide of DHOase was part of a β-sheet as suggested by the homology with carbonic anhydrase, the side-chain of Arg19 was too remote from the proposed catalytic site, centred on the Zn^{2+} ion, to be directly involved in catalysis. For this reason, the distance was constrained between the positively-charged side-chain of Arg19 and the negative carboxylate of C4 on DHO, which was the most likely group on the substrate to be in a position to interact. The peptide back-bone from the human carbonic anhydrase II crystal structure deviates considerably from regular antiparallel β-sheet, particularly at the second His ligand (His17, Figure 2). The effect of the deviation is to rotate the direction of the C^α-C^β bond of Arg19 away from the substrate. The back-bone of the model was, therefore, allowed some flexibility in the His17 to Arg19 region to bring the Arg19 side-chain and the substrate carboxylate to within an interacting distance of approximately 2 Å.

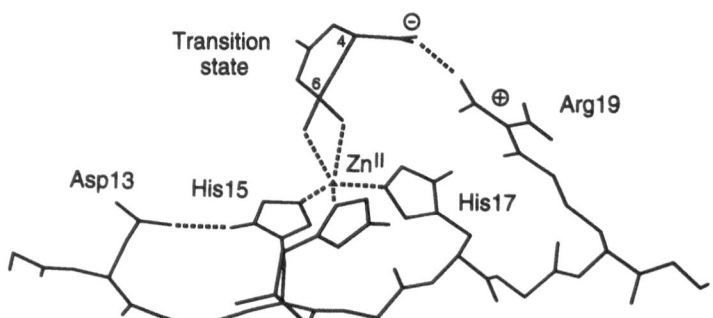

Figure 2. A proposed structure for part of the active site of hamster dihydroorotase.

Two dihydroorotase replacement mutants at Arg19 were constructed, Arg19→Gly and Arg19→Lys. Neither mutant displayed dihydroorotase activity, even at high (millimolar) substrate concentrations. This result is, however, consistent with the loss of an electrostatic interaction between an enzyme and substrate. In addition, inhibitor studies have shown that the C4 carboxylate of DHO is necessary for substrate binding (see Introduction). Thus, these results suggest that the modelled nonapeptide from hamster dihydroorotase contains active site residues involved in both zinc and substrate binding. Further experiments are in progress to identify the third zinc ligand, and residues involved in catalysis.

REFERENCES

Christianson, D.W., and Alexander, R.S., 1989, Carboxylate-histidine-zinc interactions in protein structure and function, *J. Am. Chem. Soc.* 111:6412.

Christopherson, R.I., and Jones, M.E., 1979, Interconversion of carbamyl-L-aspartate and L-dihydroorotate by dihydroorotase from mouse Ehrlich ascites carcinoma, *J. Biol. Chem.* 254:12506.

Christopherson, R.I., and Jones, M.E., 1980, The effect of pH and inhibitors upon the catalytic activity of the dihydroorotase of multienzymatic protein *pyr1-3* from mouse Ehrlich ascites carcinoma, *J. Biol. Chem.* 255:3358.

Eriksson, A.E., Jones, T.A., and Liljas, A., 1988a, Refined structure of human carbonic anhydrase II at 2.0 Å resolution, *Proteins* 4:274.

Eriksson, A.E., Kylsten, P.M., Jones, T.A., and Liljas, A., 1988b, Crystallographic studies of inhibitor binding sites in human carbonic anhydrase II: a pentacoordinated binding of the SCN⁻ ion to the zinc at high pH, *Proteins* 4:283.

Faure, M., Camonis, J.H., and Jacquet, M., 1989, Molecular characterization of a *Dictyostelium discoideum* gene encoding a multifunctional enzyme of the pyrimidine pathway, *Eur. J. Biochem.* 179:345.

Ferrin, T.E., Huang, C.C., Jarvis, L.E., and Langridge, R., 1988, The {MIDAS} display system, *J. Mol. Graphics* 6:13.

Freund, J.N., and Jarry, B.P., 1987, The *rudimentary* gene of *Drosophila melanogaster* encodes four enzymic functions, *J. Mol. Biol.* 193:1.

Guyonvarch, A., Nguyen-Juilleret, M., Hubert, J.C., and Lacroute, F., 1988, Structure of the *Saccharomyces cerevisiae URA4* gene encoding dihydroorotase, *Mol. Gen. Genet.* 212:134.

Håkansson, K., Carlsson, M., Svensson, L.A., and Liljas, A., 1992, Structure of native and apo carbonic anhydrase II and structure of some of its anion-ligand complexes, *J. Mol. Biol.* 227:1192.

Hambley, T.W., Phillips, L., Poiner, A.C., and Christopherson, R.I., 1993, A crystallographic and molecular mechanics study of inhibitors of dihydroorotase, *Acta Cryst.* B49:130.

Kannan, K.K., Ramanadham, M., and Jones, T.A., 1984, Structure, refinement, and function of carbonic anhydrase isoenzymes: refinement of human carbonic anhydrase I, *Ann. N.Y. Acad. Sci.* 429:49.

Simmer, J.P., Kelly, R.E., Rinker Jr., A.G., Zimmermann, B.H., Scully, J.L., Kim, H., and Evans, D.R., 1990, Mammalian dihydroorotase: Nucleotide sequence, peptide sequences, and evolution of the dihydroorotase domain of the multifunctional protein CAD, *Proc. Natl. Acad. Sci. USA* 87:174.

Williams, N.K., Peide, Y., Seymour, K.K., Ralston, G.B., and Christopherson, R.I., 1993, Expression of catalytically active hamster dihydroorotase domain in *Escherichia coli*: purification and characterization, *Protein Eng.* 6:333.

Williams, N.K., Simpson, R.J., Moritz, R.L., Peide, Y., Crofts, L., Minasian, E., Leach, S.J., Wake, R.G., and Christopherson, R.I., 1990, Location of the dihydroorotase domain within trifunctional hamster dihydroorotate synthetase, *Gene* 94:283.

EVOLUTION OF THE GATase, CPSase, DHOase-LIKE, ATCase MULTIFUNCTIONAL PROTEIN IN EUKARYOTES: GENETIC AND MOLECULAR APPROACHES WITH YEASTS *S. cerevisiae* AND *S. pombe*

M. Lollier, L. Jaquet, T. Nedeva, F. Lacroute[*], S. Potier and J.L. Souciet

Laboratoire de microbiologie et génétique
URA nº1481 Université Louis-Pasteur /CNRS
28, rue Goethe
F-67083 Strasbourg France
* Centre de génétique moléculaire du CNRS
Avenue de la Terrasse - Bâtiment 26
F-91198 GIF-sur-YVETTE France

INTRODUCTION

Multiple enzymatic activities encoded by a single polypeptide chain are identified in numerous eukaryotic pathways. The organization of these multifunctional proteins is compatible with the theory proposed by Yourno et al. (1970), suggesting that the generation of these kinds of enzymes result from the fusion of genes encoding monofunctional proteins. At the opposite end of the spectrum, the corresponding activities in prokaryotes are mostly due to monofunctional proteins.

The enzymes of the pyrimidine pathway have been biochemically characterized and the corresponding coding sequences are depicted in numerous prokaryotes and eukaryotes. This offer a good opportunity for sequence comparison studies permitting the investigation of the relationship between structure and function. In yeast, the first two reactions of the pyrimidine pathway (encoded by the *URA2* gene) are catalyzed by a multifunctional protein carrying two catalytic activities (see Figure 1): carbamoylphosphate synthetase (CPSase ; EC   6.3.5.5) and aspartate transcarbamylase (ATCase; EC 2.1.3.2). The third step dihydroorotase (DHOase ; EC 3.5.2.3) is encoded by the *URA4* gene which is both physically and genetically independent of the *URA2* gene. The CPSase-function is accomplished in two steps, by a glutamine-dependant amidotransferase activity followed by a synthetase (see Figure 1).

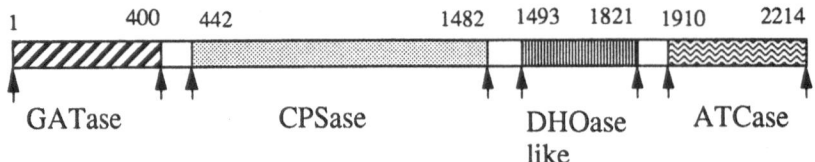

Figure 1. Organization of the protein CPSase-ATCase of *S. cerevisiae*.

Both CPSase and ATCase are subject to feedback inhibition by UTP, the final product of the pathway (Jaquet et al., 1993).

Comparison of the aminoacid sequence of yeast CPSase-ATCase protein and equivalent enzymes of higher eukaryotes led us to propose that the CPSase and the ATCsae are linked by a large apparently non-functional domain (approximately 345 aa long) named DHOase-like. The reason was that by size, position and amino acids identity, it is related to the CAD gene (Souciet et al., 1989), except for the aminoacids corresponding to the putative active site. The purpose of this work is to understand why this domain is conserved even when it seems to be inactive. This question must be addressed in three ways.

1- Is the DHOase-like domain a vestige of a putative active DHOase ?

2- Is this domain involved in another function for instance in feedback inhibition?

3- Is the presence of this domain necessary for protein architecture?

Moreover, we are checking by cloning and sequencing the equivalent gene of *S. pombe* whether the organization of the *S. cerevisiae* protein is a unique case in eukaryotic cells or if it can be found in another eukaryote.

RESULTS AND DISCUSSION

Recovery of DHOase function

By a genetic approach, we have tried to positively select clones which have recovered a functional DHOase from the DHOase-like sequence.

First, we have verified that the *URA2* gene inserted in a low copy number plasmid is unable to complement a *ura4* mutant.

Secondly, we have cloned by PCR amplification the DHOase-like domain in a yeast expression vector and transformed a *ura4* deleted strain in order to test a putative activity of the separate domain. As a control, we have cloned by this process DHOase coding sequence of the hamster protein CAD and we have established that it is capable to complement a DHOase deficient yeast strain. On the contrary, in the same conditions, the isolated cryptic yeast domain is unable to complement.

Moreover, we are unable to select prototrophic clones by: random mutagenesis, site-directed mutagenesis and ectopic recombination with the corresponding CAD domain sequence.

These experiments suggest that the differences in amino acids sequence between an active DHOase and a inactive one as DHOase-like domain are too important to be overcome by few mutational events. In other words, this sequence does not probably act as a storage sequence for a putative DHOase activity.

Role of the DHOase-like domain in feedback process.

We have selected 16 feedback mutants and none mapped in this domain (Jaquet et al., 1994). This observation suggest that this part of the protein is not involved in the process or that a mutation in this domain leads to an inactive ATCase.

Role of the DHOase-like domain in the protein architecture

In another poject in our group, we investigated yeast genome stability (Roelants et al., 1994 submitted for publication). For this purpose, we have used a mutant allele of the *URA2* gene carrying three nonsense mutations in the CPSsae coding region. Even when the ATCase coding sequence is not mutated, this domain is non functional because of the polarity of nonsense mutation which leads to mRNA instability (Souciet et al., 1982). Starting from this allele, we have selected 37 ATCase[+] revertants. Three kinds of events lead to the recovery of a functional ATCase:

1- Insertion of a yeast transposon (Ty) upstream the ATCase coding sequence. The Ty acts as a promotor.

2- Deletion of the part carrying the three mutations

3- Duplication of the ATCase domain and fusion to a resident promotor on the same chromosome or on another chromosome.

TheDHOase-like domain is involved in the process whatever event in all three cases occurs :

1- Ty insertion occurs upstream the DHOase-like domain.

2- Deletions do not affect cryptic domain

3- DHOase-like and ATCase domains are duplicated.

This result strongly suggests that the DHOase-like domain is linked to the function of the ATCase. In other words, the association between these two domains should be essential for the ATCase activity.

Organization of CPSase-ATCase of *S. pombe*

In order to check if in eukaryotic cells, the presence of DHOase-like domain is particular to *S. cerevisiae*, we cloned and sequenced the corresponding *URA1* gene of *S. pombe*, which is an organism considered closer to higher eukaryotes on an evolutionary level (Egel et al., 1980).

The general organization of the protein is the same as *S. cerevisiae* including the presence of a DHOase-like domain. Moreover, amino acids of the sequence of putative active site is degenerate.

Using computer analyses (programs PILEUP and DISTANCE; Devereux et al., 1984), we performed comparison studies between active and inactive DHOase. In summary, 4 classes may be defined and CPSase-ATCase yeast proteins fall in another class than equivalent higher eukaryote ones. But It must be noticed that inactive DHOases are always associated with ATCase whereas active DHOase may be associated to ATCase or not.

CONCLUSION

Yeast DHOase-like domains are probably derivatives of a functional DHOase domain, that are a result of mutational events at the active sites. During this functional divergence, the yeast DHOase-like domains retain structural properties, that are required for the ATCase function. Moreover, this domain is not involved in the feedback process.

ACKNOWLEDGEMENTS

We thank S. Sen Gupta for correcting this manuscript. Marc Lollier and Laurence Jaquet were supported by french "Ministère de la recherche et de l'enseignement supérieur". Tania Nedeva was supported by an Erasmus fellowship.

REFERENCES

Devereux J., Haeberli, P. and Smithies O. (1984) A comprehensive set of sequence analysis programs for the VAX. *Nucl. Acids. Res.* **12**: 387-395.

Egel, R., Kohli, J., Thuriaux, P. and Wolf K. (1980). Genetics of the fission yeast *Schizzo saccharomyces pombe*. *Ann. Rev. Genet.* **14**: 77-108.

Jaquet, L., Lollier, M., Souciet, J.L. and Potier, S. (1993). Genetic analysis of yeast strains lacking negative feedback control: a one-step method for positive selection and cloning of carbamoylphosphate synthetase-aspartate transcarbamylase mutants unable to respond to UTP *Mol. Gen. genet.* **241**: 81-88.

Jaquet, L., Lollier, M., Navratil, O., Schoendorf, A., Brondani, V., Souciet, J.L. and Potier, S. (1984) Feedback of yeast CPSase-ATCase: selection, cloning and sequencing of mutants alleles. In: *Advances in Experimental Medicine and Biology*. This volume.

Roelants, F., Potier, S. Souciet, J.L. and de Montigny Jacky. (1994) Reactivation of the ATCase domain of the *URA2* gene complex: a positive selection of Ty insertions and chromosomal rearrangements in *Saccharomyces cerevisiae*. *Mol. Gen. Genet.* sub mitted

Souciet, J.L., Hubert, J.C. and Lacroute, F. (1982). Cloning and restriction mapping of the yeast *URA2* gene coding for the carbamoylphosphate synthetase - aspartate transcarbamylase complex. *Mol. Gen. Genet.* **186**: 385-390.

Souciet, J.L., Nagy, M., Le Gouar, M., Lacroute, F. and Potier, S. (1989). Organization of the yeast *URA2* gene : identification of a defective dihydroorotase-like domain in the multifunctional carbamoylphosphate synthetase - aspartate transcarbamylase complex. *Gene* **79**, 59-70.

Yourno, J.T., Kohno, T. and Roth, J.R. (1970) Enzyme evolution: generation of a bifunctional enzyme by fusion of adjacent genes. *Nature*. **228**: 820-824.

NUCLEOSIDE DIPHOSPHATE KINASE:
AN OLD ENZYME WITH NEW FUNCTIONS?

Michel Véron[1], Annemiek Tepper[1], Martin Hildebrandt[1], Ioan Lascu[1],
Marie-Lise Lacombe[2], Joël Janin[3], Solange Moréra[3], Jaqueline Cherfils[3],
Christian Dumas[3], and Max Chiadmi[3]

[1]Unité de Biochimie Cellulaire, Institut Pasteur, 75724, Paris, France
[2]INSERM, U402, Faculté Médecine St. Antoine, 75011, Paris, France
[3]Biologie Structurale, CNRS, 91198, Gif-sur-Yvette, France

This short review summarizes the recent work that we have performed on the biochemical and structural properties of nucleoside diphosphate (NDP) kinase.

INTRODUCTION

NDP kinase exchanges a phosphate between a (deoxy-) nucleoside triphosphate and a nucleoside diphosphate. The mechanism of the reaction is ping-pong, with the formation of a phospho-histidine intermediate during the catalytic cycle according to the following reaction scheme:

$$N_1TP + E <--> N_1DP + E{\sim}P$$
$$E{\sim}P + N_2DP <--> E + N_2TP$$

NDP kinase is not specific. It accepts purine and pyrimidine as well as oxy- or deoxy-nucleotides as substrates. It will even use thiophosphates albeit at a much slower rate. NDP kinase is ubiquitous. While enzylomogical studies on NDP kinase flourished in the 1970's, little work was performed in the 1980's, probably due to a decreased interest for this enzyme, because of its lack of specificity. The presence of two isozymes of NDP kinase in erythroytes (NDPK-A and NDPK-B) was demonstrated (Presecan et al., 1989), but sequence data became available only in 1990 when NDP kinase was cloned from the procaryote *Myxococcus Xanthus* by Munoz-Dorado (Munoz-Dorado et al., 1990), and by us in *Dictyostelium discoideum* (Lacombe et al., 1990). Genes for NDP kinases have now been identified from several species including spinach, budding yeast, flruit fly and mammals. The sequence conservation among NDP kinases is high, with *Dictyostelium* and *Drosophila* enzymes showing 62% and 73% similarity respectively with human NDP kinase. Thus, the sudy of the biochemical and structural properties of the NDP kinase from *Dictyostelium* constitutes a framework for the study of similar enzymes from higher eucaryotes.

Purine and Pyrimidine Metabolism in Man VIII, Edited by
A. Sahota and M. Taylor, Plenum Press, New York, 1995

THE STRUCTURE OF NDP KINASE FROM *DICTYOSTELIUM*

We solved the X-ray structure of *Dictyostelium* NDP kinase to 2.2 Å resolution using recombinant protein. The first structure (Dumas et al., 1992) was that of an inactive enzyme in which the active site histidine had been substituted for a cysteine, allowing to obtain heavy atoms derivatives. The structure of the wild type enzyme has now been solved at 1.8 Å using molecular replacement (Moréra et al., 1994). NDP kinase is a symetrical hexamer made of identical 17 kDa subunits arranged as three vertical dimers or two horizontal trimers. The fold of the protein is original and shows no similarity with other nucleotide binding proteins such as adenylate kinase, Ras, or the catalytic subunit of the cAMP dependent protein kinase. The monomer has an α/β structure with an antiparallel β sheet surounded by α helixes. A loop, that we called the K-pn loop, plays a crucial role in forming the contacts that allow for the formation of trimers. The histidine residue phosphorylated during the catalytic reaction (His 122 in *Dictyostelium* NDP kinase) is buried. Apart from an unexpected similarity with the regulatory subunit of aspartate transcarbamylase, NDP kinase has no strong structural homology with other known protein structures.

We also obtained crystals of the *Drosophila* NDP kinase and its structure was resolved using the molecular replacement method. It is also an hexamer whose structure is almost superimposable with that of *Dictyostelium* (Chiadmi et al., 1993). No structure is available yet for the human NDP kinases. However, they are likely to be similar to the *Dictyostelium* and the *Drosophila* enzymes, given their high sequence homology and the fact that they also form hexamers (Gilles et al., 1991). The structure of *Myxococcus xanthus* NDP kinase has also been determined (Williams et al., 1993). Although the fold of its monomer is very similar to that of *Dictyostelium* NDP kinase, and this similarity extends to the surface areas generating the dimer contacts. However, the enzyme from *Myxococcus* is a tetramer and the dimers show a totally different way to associate than both in the *Dictyostelium* and in the *Drosophila* enzymes.

Recently, we have determined the X-ray structures of two binary complexes in which ADP, a purine nucleotide and dTDP, a pyrimidine deoxy-nucleotide were bound to *Dictyostelium* NDP kinase. The binding of dTDP (Cherfils et al., 1994) is very similar to that of ADP (Moréra et al., 1994). Each subunit carries one active site. There is no kinetic evidence of cooperativity between these sites. The mode of binding of the nucleotide is different from that observed in other nucleotide binding proteins. The base moiety is not hydrogen bonded to the main chain and stacks between hydrophobic residues (a phenylalanine and a valine), almost at the surface of the protein, explaining the lack of discrimination of NDP kinase between purine and pyrimidine nucleotides. The nucleotides make several contacts with the side chains of Lys16, Arg 92, Thr 98, Arg 109, and Asn 119 both through the ribose moeity and with the α and β phosphates. An unusal H-bond is found within the substrate between 3' OH of the ribose and the β phosphate. Lys 16, and Tyr 56 are involved in catalysis while Glu 133 contributes to fix a proton at the position Nϵ of His 122 on which phosphotansfer occurs. We have yet not been able to determine the structure of the enzyme with a bound triphospho-nucleotide. However, in the nucleotide diphosphates structure, a water molecule is present between the β-phosphate and the Nδ of His 122, at a position likely to correspond to the γ-phosphate.

The active site residues mentionned above are identical in all NDP kinases known to date, including those from prokaryotes. In addition, the active site model of the NDP kinase from *Myxococcus* determined from the crystal structure of the complex with ADP (Williams et al., 1993) is almost identical to that of *Dictyostelium* described above. Thus, this conserved struture is likely also to e found in humna NDP kinases A and B.

Using *in vitro* mutagenesis, we have substituted all of the active site amino acids strictly conserved among known NDP kinases. The catalytic properties of the mutants were established with purified recombinant proteins.They provide a strong experimental support for the active site model proposed on the basis of the X-ray structure. A provisional scheeme for the transition state (Morera et al., 1994) is also supported by the greatly reduced activity of mutants K16A and Y56F. The importance of Glu 133 is shown by the very low activity in mutants E133A, E133D and E133Q, while E133K substitution leads to a totally inactive enzyme (Tepper et al., in preparation).

Another important motif found in all NDP kinase srtructures is the Kpn loop : In *Drosophila,* the P97S substitution in this loop is responsible for the conditional dominant lethal phenotype of the *killer of prune* mutant. The NDP kinase from the *Kpn* mutant was purified and shown to have altered stability (Lascu et al., 1992). However, the reason fo the lethal phenotype is not yet understood. However, the Kpn loop is directly involved in the inter-subunits contacts that allow for trimer formation. In addition, it carries several residues that contribute to substrate binding, in particular Arg109 which is important in maintaining the β-phosphate in place. Thus the correct positioning of the *Kpn* loop seems a condition for a fully active enzyme. The increased flexibility of the K-pn loop when the hexamer dissociates into dimers or monomers may explain why monomers of NDP kinase have very low activity although they are able to auto-phosphorylate on His 122 (Lascu et al., 1993).

NDPK-A IS THE PRODUCT OF THE HUMAN GENE *nm23 -H1*

In recent years, NDP kinase has been involved in unexpected regulatory processes, since it has been identified to the product of several important genes in mammals.

In human, the gene *nm23-H1* was cloned by differential screening of a cDNA library from a melanoma cell line and proposed to be a metastasis suppressor gene on the basis of its decreased expression in highly metastic cells (Steeg et al., 1988). Because of the high similarity of sequence of the protein encoded by *nm23* with the NDP kinase cloned from *Dictyostelium,* we first proposed that *nm23* (now called *nm23-H1*) was encoding a human NDP kinase (Wallet et al., 1990). A second gene from human, *nm23-H2,* was later identified (Stahl et al., 1991). The identity of sequence between isoforms A and B of NDP kinase purified from human erythrocytes and *nm23-H1* and *nm23-H2* cDNA sequences, definitively established that the genes *nm23-H1* and *nm23-H2* encode NDPK-A and NDPK-B respectively (Gilles et al., 1991).

The actual status of *nm23* gene(s) as metastasis suppressor(s) is controversial : We, and others observed the consistent over-expression of *nm23-H1* in several tumors types including breast tumors (Lacombe et al., 1991; Sawan et al., 1994), neuroblastoma (Hailat et al., 1991) and proliferating lymphocytes (Keim et al., 1992) . However, in these studies, an inverse correlation with the metasatic status of the tumor was not observed. In contrast, lower expression of *nm23-H1* has been reported in melanoma cell lines (Rosengard et al., 1989) and transfection of the gene in these cells rescued, at least partially, the invasive phenotype (Leone et al., 1991). In conclusion, while an increase in NM23/NDP kinase is consistently observed in neoplasic proliferating cells, a decreased expression of *nm23-H1* may only occur in certain types of metastasing cells.

NDPK-B HAS DNA BINDING ACTIVITY

A new unexpected line of research on NDP kinase has recently appeared with the identification of a DNA binding protein (PuF) with the product of the human gene *nm23-H2,* encoding the NDPK-B. Indeed, PuF , a factor previously characterized as binding to purine rich sequences, plays a role in the regulation of the transcription of *c-myc*. PuF has recently been cloned and identified to NDPK-B. Moreover, pure recombinant NDPK-B was able to bind to an oligonucleotide representing the *c-myc* promoter, and to activate the transcription of the gene in *vitro* (Postel et al., 1993).

This unexpected property for a metabolic enzyme led us to investigate the binding of NDP kinase to DNA in more details, by the techique of gel shift. Using recombinant human NDPK-B, we demonstrated that the DNA-binding to the *c-myc* oligonucleotide is specific for this isozyme, since practically no signal was observed when human recombinant NDPK-A,

Drosophila or *Dictyostelium* NDP kinases were used. Moreover, our experiments indicate that NDP kinase binds with greater affinity to single stranded DNA than to double stranded DNA (Hildebrandt et al., submitted). The biological significance of the interaction between DNA and NDP kinase remains to be established.

ACKNOWLEDGEMENTS

This work was supported in part by grants from "Association de la Recherche contre le Cancer" (ARC), The "Comité de Paris de la Ligue contre le Cancer" and from "Association Nationale de la Recherche contre le SIDA" (ANRS).

REFERENCES

Biggs, J., E. Hersperger, P. S. Steeg, L. A. Liotta and A. Shearn, 1990, A Drosophila gene that is homologous to a mammalian gene associated with tumor metastasis codes for a nucleoside diphosphate kinase. *Cell.* 63: 933-940.

Cherfils, J., S. Moréra, I. Lascu, M. Véron and J. Janin, 1994, X-Ray structure of Nucleoside Diphosphate Kinase complexed with dTDP and Mg2+ at 2Å resolution. *Biochemistry.* in press:

Chiadmi, M., S. Moréra, I. Lascu, C. Dumas, G. LeBras, M. Véron and J. Janin, 1993, The Awd Nucleoside Diphosphate Kinase ,from Drosophila. *Structure.* 1: 283-293.

Dumas, C., I. Lascu, S. Morena, P. Glaser, R. Fourme, V. Wallet, M. Lacombe and M. J. Véron J., 1992, X-ray structure of Nucléoside diphiophate kinase. *EMBO J.* 11: 3203-3208.

Gilles, A. M., E. Presecan, A. Vonica and I. Lascu, 1991, Nucleoside diphosphate kinase from human erythrocytes. Structural characterization of the two polipeptide chains responsible for heterogeneity of the hexameric enzyme. *J. Biol. Chem.* 266: 8784-8789.

Hailat, N., D. R. Keim, R. F. Melhem, X. Zhu, C. Eckersckorn, G. M. Brodeur, C. P. Reynols, R. C. Seeger, F. Lottspeich, J. R. Strahler and S. M. Hanash, 1991, High levels of p19/nm23 protein in neuroblastoma are associated with advanced stage disease and with N-myc gene amplification. *J. Clin. Invest.* 88: 341-345.

Hildebrandt, M., M. Lacombe, S. Passeron and M. Veron, 1994, Human NDP-Kinase B binds specifically to single-stranded poly-pyrimidine DNA sequences. *submitted*

Keim, D., N. Hailat, R. Melhem, X. X. Zhu, I. Lascu, M. Veron, J. Strahler and S. M. Hanash, 1992, Proliferation related expression of P19/NM23 Nucleoside Diphosphate Kinase. *J. Clinic. Invest.* 89: 919-924.

Lacombe, M. L., V. Wallet, H. Troll and M. Veron, 1990, Functional cloning of a nucleoside diphosphate kinase from Dictyostelium discoideum. *J Biol Chem.* 265: 10012-8.

Lacombe, M., X. Sastre-Gareau, I. Lascu, V. Wallet, J. Thiery and M. Veron, 1991, Overexpression of Nucleoside Diophosphate Kinase (Nm23) in Human Solid Tumors. *Eur. J. Canc.* 27: 1302-1307.

Lascu, I., A. Chaffotte, B. Limbourg-Bouchon and M. and Véron, 1992, A Pro/Ser substitution in nucleoside diphosphate kinase of Drosophila melanogaster (mutation Killer of prune) affects stability but not catalytic efficiency of the enzyme. *J. Biol. Chem.* 267: 12775-12781.

Lascu, I., D. Deville-Bonne, P. Glazer and M. Véron, 1993, Equilibrium dissociation and unfolding of nucleoside diphosphate kinase from Dictyostelium discoideum. Role of proline 100 in the stability of the hexameric enzyme. *J. Biol. Chem.* 268: 20268-20275.

Leone, A., U. Flatow, C. Richterking, M. Sandeen, I. M. K. Margoulies, L. A. Liotta and P. S. Steeg, 1991, Reduced Tumor Incidence, Metastatic Potential, and Cytokine Responsiveness off Nm23 Transfected Melanoma Cells. *Cell.* 65: 25-35.

Munoz-Dorado, J., M. Inouye and S. Inouye, 1990, Nucleoside diphosphatase kinase from *Myxococcus xanthus*. I. Cloning and sequencing of the gene. *J. Biol. Chem.* 265: 2702-2706.

Postel, E. H., S. J. Berberich, S. J. Flint and C. A. Ferrone, 1993, Human c-myc transcription factor PuF identified as nm23-H2 nucleoside diphosphate kinase, a candidate suppressor of tumor metastasis. *Science.* 261: 478-480.

Presecan, E., A. Vonica and I. Lascu, 1989, Nucleoside diphosphate kinase from human erythrocytes. Purification, molecular mass and subunit structure. *FEBS lett.* 250: 629-632.

Rosengard, A. M., H. C. Krutzsch, A. Shearn, J. R. Biggs, E. Barker, I. M. K. Margulies, C. R. King, L. A. Liotta and P. S. Steeg, 1989, Reduced Nm23/Awd protein in tumor metastasis and aberrant *Drosophila* development. *Nature.* 342: 177-180.

Sawan, A., I. Lascu, M. Veron, J. J. Anderson, C. Wright, C. H. W. Horne and B. Angus, 1994, NDP-K/nm23 Expression in Human Brest Cancer in Relation to relapse, Survival, and other Prognostic Factors: an Imuunohistochemical Study. *J. Pathology.* 172: 27-34.

Stahl, J. A., A. Lerone, A. M. Rosengard, L. Porter, C. R. King and P. S. Steeg, 1991, Identification of a second human nm23 gene, nm23-H2. Cancer Res. 51:

Steeg, P. S., G. Bevilacqua, L. Kopper, U. P. Thorgeirsson, J. E. Talmadge, L. A. Liotta and M. E. Sobel, 1988, Evidence for a novel gene associated with low tumor metastasic potential., *J. Natn. Cancer Inst.* 80: 200-204.

Wallet, V., R. Mutzel, H. Troll, O. Barzu, B. Wurster, M. Veron and M. L. Lacombe, 1990, Dictyostelium nucleoside diphosphate kinase highly homologous to Nm23 and Awd proteins involved in mammalian tumor metastasis and Drosophila development. *J Natl Cancer Inst.* 82: 1199-202.

Williams, R. L., D. A. Oren, J. Munoz-Dorado, S. Inouye, M. Inouye and E. Arnold, 1993, Crystal structure of Myxococcus xanthus nucleoside diphosphate kinase and its interaction with nucleotide substrate at 2.0Å resolution. *J. Mol. Biol.* 234: 1230-1247.

PRODUCTION OF ADENOSINE AND NUCLEOSIDE ANALOGUES BY AN EXCHANGE REACTION CATALYZED BY ADENOSINE KINASE

Mohsine Mimouni, Françoise Bontemps and Georges Van den Berghe

Laboratory of Physiological Chemistry, International Institute of Cellular and Molecular Pathology, and University of Louvain Medical School, B-1200 Brussels

INTRODUCTION

We have shown previously that rat liver adenosine kinase can catalyze an exchange reaction between adenosine (Ado) and AMP in the absence of ATP (1). This exchange reaction was potently stimulated by ADP. When measured in the absence of added ADP, the exchange could be due to a slight (0.001 %) contamination by ADP of analytical grade AMP (2). The ADP requirement of the Ado-AMP exchange intervenes in an ordered Bi Bi mechanism in which ATP is the first substrate to bind to the enzyme, and ADP the last product to dissociate. In the present work we have investigated (*i*) if Ado or AMP could be replaced by, respectively, nucleoside or nucleoside monophosphate analogs and (*ii*) if the Ado-AMP exchange is restricted to liver.

METHODS

Experiments with analogs were performed with adenosine kinase purified to homogeneity from livers of fed male Wistar rats as described in (2). For the study of the exchange reaction in various tissues, liver, kidney cortex, heart and cerebral hemispheres were excised rapidly, rinsed with physiologic saline and homogeneized in 4 vol. of 25 mM Hepes (pH 7.0) containing 1 mM dithiothreitol, 1 mM EDTA (buffer A), 100 mM KCl, 0.25 M sucrose, 10 µg/ml leupeptin and antipain. The homogenate was centrifuged for 10 min at 8 000 x g, and the resulting supernatant was further centrifuged for 60 min at 100 000 x g. This high-speed supernatant was dialysed against 300 vol. of buffer A and filtered on Sephadex G-25 fine to remove nucleotides and other small molecules. The absence of nucleotides in the filtrate was confirmed by HPLC (3). Blood was collected on heparin. Rat and human haemolysates were prepared as described in (4), except that haemolysis was followed by high-speed centrifugation as given above.

Adenosine kinase activity was measured in the presence of 3.4 mM MgATP and 1 μM [2-^{3}H]Ado as described in (1). With purified liver adenosine kinase, the [^{14}C]Ado-AMP and [^{14}C]AMP-Ado exchange reactions were measured in the presence of 0.1 mM ADP as described in (2). The [^{14}C]Ado-AMP exchange activity in the filtered high-speed supernatant of various tissues was measured in the absence of added ADP to avoid the formation of ATP by adenylate kinase.

RESULTS

Exchange reaction between [^{14}C]Ado and nucleoside monophosphates

Adenosine kinase catalyzes the phosphorylation of adenosine and of several related nucleosides and analogues into the corresponding monophosphates, using ATP or GTP as phosphoryl donor (5). As shown in table 1, [^{14}C]AMP could also be synthetized from [^{14}C]Ado in the presence of the unlabelled monophosphates of tubercidin, 6-chloropurineriboside and N6-methyl-Ado at rates of, respectively, 50, 28 and 20 % of that with AMP. Other nucleoside monophosphates were ineffective as phosphate donors.

Table 1. Formation of [^{14}C]AMP from [^{14}C]Ado in the presence of various nucleoside monophosphates

Nucleoside monophosphate (1 mM)	[^{14}C]AMP formed (nmol/min/mg protein)
AMP	1550
Tubercidin-monophosphate	750
6-Chloropurineriboside-monophosphate	425
N6-Methyl-AMP	300
ZMP	2
d-AMP	2
GMP	< 1
IMP	< 1
CMP	< 1
UMP	< 1
6-Mercaptopurineriboside-monophosphate	< 1
N6-Etheno-AMP	< 1

Exchange reaction between [^{14}C]AMP and nucleosides

In the absence of Ado, adenosine kinase did not catalyze the hydrolysis of AMP. However, in the presence of various unlabelled analog nucleosides, [^{14}C]Ado was released from [^{14}C]AMP and the corresponding analog nucleoside monophosphate was formed. As shown in table 2, tubercidin and

N6-methyl-Ado were the most efficient analog substrates, [14C]Ado being formed at 60 % of the rate with 0.2 mM unlabelled Ado. The rate with 6-methylmercaptopurine riboside was 45 % of that with Ado. With the other nucleosides, little or no detectable exchange activity was measured.

Table 2. Production of [14C]Ado from [14C]AMP in the presence various unlabelled analog nucleosides

Nucleoside (0.2 mM)	[14C]Ado formed (nmol/min/mg protein)
Adenosine	730
Tubercidin	442
N6-Methyladenosine	440
6-Methylmercaptopurine riboside	325
AICAriboside	3
2'-Deoxyadenosine	1
6-Mercaptopurine riboside	1
Inosine	< 1
Adenine arabinoside	< 1
Ribavirin	< 1
N6-Ethenoadenosine	< 1
Guanosine	< 1

Table 3. Exchange reaction and adenosine kinase activities in various rat and human tissues

	[14C]Ado-AMP exchange activity (nmol/min/mg protein)	Adenosine kinase activity (nmol/min/mg protein)
Rat		
Liver	0.075	13.7
Heart	0.008	1.4
Kidney	0.015	5.0
Brain	0.003	1.8
Erythrocytes	0.007	0.4
Human		
Erythrocytes	0.003	0.3

[14C]Ado-AMP exchange in other tissues

The cytosol fractions of all tissues and cell types investigated were able to catalyze an Ado-AMP exchange (table 3). The exchange activity was most pronounced in rat liver, followed by in this order, kidney, heart, erythrocytes and brain, and human erythrocytes. The same activity profile was observed with respect to adenosine kinase activity.

CONCLUDING REMARKS

Our studies indicate that adenosine kinase cannot only catalyze an exchange reaction between Ado and AMP but also between Ado and nucleotide analogues, and between AMP and nucleoside analogues. The Ado-AMP exchange discovered in rat liver, is not restricted to this cell type, but may be a common property of the enzyme in all tissues.

Interestingly, a recently described placental adenosine phosphotranferase (6) which is different from adenosine kinase, has several common features with the exchange activity we report for adenosine kinase from rat liver. Thus both activities have strong preference for the adenine moiety in both acceptor and donor, and are strongly stimulated by ADP. They differ, however, in the hydrolysis of AMP that can be realized in the absence of an acceptor nucleoside with the placental adenosine phosphotransferase but not with adenosine kinase. IMP-GMP cytosolic 5'-nucleotidase has also been shown to catalyse a nucleoside exchange but it occurs preferentially between inosine and IMP (7,8).

REFERENCES

1. Bontemps F, Mimouni M and Van den Berghe G. Phosphorylation of adenosine in anoxic hepatocytes by an exchange reaction catalysed by adenosine kinase. *Biochem J* 290: 679-684, 1993.
2. Mimouni M, Bontemps F and Van den Berghe G. Kinetic studies of rat liver adenosine kinase. Explanation of the exchange reaction between adenosine and AMP. *J Biol Chem* (in press).
3. Hartwick RA and Brown PR. The performance of microparticle chemically-bonded anion-exchange resins in analysis of nucleotides. *J Chromat* 112: 651-662, 1975.
4. Bontemps F, Van den Berghe G and Hers HG. 5'-Nucleotidase activities in human erythrocytes. Identification of a purine 5'-nucleotidase stimulated by ATP and glycerate 2,3-bisphosphate. *Biochem J* 250: 687-696, 1988.
5. Anderson EP. Nucleoside and nucleotide kinases. The Enzymes (Boyer PD, ed) Vol 9: 49-96, 1973. Academic Press, New York.
6. Garvey EP and Krenitsky TA. A novel human phosphotransferase highly specific for adenosine. *Arch Biochem Biophys* 296: 161-169, 1992.
7. Worku Y and Newby AC. Nucleoside exchange catalysed by the cytoplasmic 5'-nucleotidase. *Biochem J* 205: 503-510, 1982.
8. Tozzi MG, Camici M, Pesi R, Allegrini S, Sgarrella F and Ipata PL. Nucleoside phosphotransferase activity of human colon carcinoma cytosolic 5'-nucleotidase. *Arch Biochem Biophys* 291: 212-217, 1991.

DIFFERENT SUBSTRATE SPECIFICITY OF TWO ISOZYMES OF CYTOSOLIC 5'-NUCLEOTIDASE FROM RABBIT HEART

A.C. Skladanowski,[1] C.S. Hoffmann,[2] J.D. Krass,[2] W. Makarewicz,[1] and B. Jastorff[2]

[1]Department of Biochemistry, Gdansk School of Medicine, 80-211 Gdansk, Poland and
[2]Department of Biology/Chemistry, University of Bremen, 28359 Bremen, Germany

INTRODUCTION

Two isozymes of soluble cytoplasmic 5'-nucleotidase (5'-NT) have been found in the heart of several species[1,2], including man (Skladanowski, submitted). AMP- and IMP-specific forms (also coded N-I and N-II), having K_M's at milimolar range, are thought to be involved in the control of intracellular levels of AMP and IMP[3]. AMP-specific isozyme has been thoroughly characterized after purification from pigeon[4], rabbit[5] and dog[6] hearts. Physiological roles plausibly played in the heart by two distinct isoforms could be: production of 'retaliatory' adenosine[7] by AMP-specific 5'-NT, and controlling of the overall purine degradation rate by IMP-specific 5'-NT (for the review on 5'-NT see ref.8).

Studying the structure-activity relationship for homolog substrates and AMP-specific 5'-NT, could be useful in designing potential modulators for adenosine production. A special "test-kit" of analogs was designed where the natural substrate for enzyme has been modified in almost every position[9].

In the present study, we determined the activity of two forms of cytoplasmic soluble 5'-NT isolated from rabbit heart towards 20 various AMP analogs. The results confirm that AMP-specific 5'-NT is distinct from its IMP-specific isozyme and provide some structural informations about the requirements in substrate molecule for effective binding to the active site.

MATERIALS AND METHODS

Materials. [2-^{3}H]AMP (47 MBq/mmol) was obtained from The Radiochemical Centre (Amersham, Bucks., U.K.). AMP (sodium salt), ADP (potassium salt) and ATP (sodium salt) were purchased from Boehringer-Mannheim (Germany). TDP, D,L-dithiotreitol, magnesium phosphate, HEPES, 2'-deoxy-adenosine-5'-monophosphate (2'-dAMP), guanosine-5'-monophosphate (GMP), xanthosine-5'-monophosphate (XMP), 6-mercapto-purineribofuranoside-5'-monophosphate (6-SH-PuMP), 7-deaza-adenosine-5'-monophosphate (7-CH-AMP), 1,N^6-etheno-adenosine-5'-monophosphate (1,N^6-etheno-AMP), adenosine-2'-monophosphate (2'-AMP), adenosine-3'-monophosphate (3'-AMP), adenosine-5'-monophosphorothioate (AMPS), adenosine-5'-monophosphoric acid - morpholidate (AMP-morpholine) were purchased from Sigma (St.Louis, MO, U.S.A.).
8-Bromoadenosine-5'-monophosphate (8-Br-AMP) was supplied by ICN-Chemicals and 8-amino(aminoethyl)-adenosine-5'-monophosphate (8-NH-Et-NH$_2$-AMP) by BIOLOG Life Science Institute (Germany). The following derivatives were generous gift from Dr. John P.Miller at the S.R.I. (Stanford, U.S.A.): 8-hydroxy-adenosine-5'-monophosphate (8-OH-AMP), 8-methylamino-adenosine-

-5'-monophosphate (8-NH-Me-AMP), 8-aza-adenosine-5'-monophosphate (8-N-AMP). The following derivatives were synthetized in our laboratory: adenosine-N^1-oxide-5'-monophosphate (1-NO-AMP), 2-amino-purineribofuranoside-5'-monophosphate (2-amino-PuMP), purineribofuranoside-5'-monophosphate (PuMP) and (6-S-methylo)mercapto-purineribofuranoside (6-S-Me-PuMP). Imidazole and high performance liquid chromatography columns were delivered by Merck (Darmstadt, Germany). Reagents used for purification of cytosolic 5'-nucleotidase isozymes were as reported earlier[4].

5'-NT assays. Activities of both isozymic forms of 5'-NT were determined from the amount of riboside produced after incubation with substrate analogs. The reaction was followed by 30 min or 16 hrs in a total volume of 0.1 ml in Eppendorf tubes. The incubation mixture contained: 0.1 M HEPES, pH 7.0, 0.03 M NaCl, 0.01 M $MgCl_2$ or 0.05 M imidazole, pH 6.5, 500 mM NaCl, 50 mM $MgCl_2$, for AMP-specific or IMP-specific isozyme respectively. Activators (1 mM ADP in the hydrolysis rate or 0.1 mM TDP in the kinetic experiments), substrates (AMP, IMP or analogs in the concn. range 0.1 - 10 mM) were added as indicated in Tables 1-2 and Fig. 1. Reaction was started by the addition of 40 µl of appropriately diluted enzyme (12.8 or 11 munits and/or 28.4 or 10.4 g of protein for AMP- or IMP-specific isozyme respectively). It was stopped by addition of 50 µl of 1.6 M $HClO_4$ and placement the tubes on ice for 30 min. After deproteinization and neutralization, samples were subjected to product estimation by reverse-phase HPLC. The HPLC procedure described by Smolenski et al.[10] was used. In the inhibition experiments, the activity of pigeon heart AMP-specific 5'-nucleotidase was measured radiochemically with 10 mM AMP and 2-3 kBq of [2-^{3}H]AMP according to the method described by Skladanowski & Newby[4].

Enzyme preparations. Cytoplasmic 5'-NT isozymes were isolated from the hearts of rabbits purchased from the local breeder. Purification of 5'-NT isozymes was performed according to the method described by Yamazaki et al.[2] The isozymes of cytoplasmic 5'-NT obtained by this procedure had the specific activities 450 and 1060 munits per mg of protein for AMP-specific and IMP-specific form respectively. One unit is defined as an amount of activity dephosphorylating 1 µmol of the preferred substrate during 1 min.

Kinetics. Kinetic constants were evaluated using the linear regression method with a Hanes-Woolf transformation of the Michaelis-Menten equation. Checks were made beforehand to ensure a hyperbolic-type of the substrate saturation curve

RESULTS

Hydrolysis of AMP analogs. Ranges of hydrolysis rates catalysed by two isozymes of cytosolic 5'-NT from rabbit heart with various AMP analogs as substrates are presented in Table 1. The investigated compounds were categorized arbitrarily into four groups A-D. Group A gathers the best substrates for both isozymes and others with a relative reaction rate higher than 80%. Neither of the compounds in this category was common for both isozymes of rabbit heart 5'-NT. In the category B (compounds dephosphorylated within the range 25-55% of the maximum rate), we found two common analogs, 6-(S-methyl)-mercaptopurineribofuranoside-5'-monophosphate and xanthosine-5'-monophosphate, which were dephosphorylated similarly by both isozymes. Poor substrates for two forms, however still hydrolysable, in the range 2-18% of the maximum, are placed in category C. This group comprises less numerous compounds for IMP-specific than for AMP-specific form of 5'-NT. The lack of 6-oxo group in an analog makes it handicapped for the IMP-specific isozyme. The same compounds for both forms are: 6-mercaptopurine-5'-monophosphate and 8-aza-adenosine-5'-monophosphate. The last category D, forms a set of analogs unhydrolysable by both 5'-NTs. Five of such compounds were found common for two forms. They are AMP analogs modified within phosphoric acid moiety: adenosine-2'- and -3'-monophosphates, adenosine-5'-monophosphorothioate, adenosine-5'-phosphoric acid - morpholidate and one analog with a frozen *syn*-conformation, 8-amino(aminoethyl)-adenosine-5'-monophosphate.

The differences in specificity of both isozymes of rabbit heart 5'-NT are presented in Fig. 1. Relative preferences are very much dependent on the type of substituent in 6-position of purine ring but the absence of such substituent (purine riboside-5'-monophosphate and 2-amino-purine riboside 5'-monophosphate) gives priority rather to AMP-specific (N-I) 5'-NT. The positive effect of methylation of 6-thio-purine riboside-5'-monophosphate as substrate for both forms is unclear. One could explain this as a protection of the compound against forming -SS- bounds with the enzyme protein.

Table 1. SPECIFIC ACTIVITIES OF TWO FORMS OF SOLUBLE CYTOPLASMIC 5'-NUCLEOTIDASE FROM PIGEON HEART WITH VARIOUS ANALOGS OF AMP

Category (specific activity) μmol/30 min per mg protein,	AMP-specific 5'-NT (N-I)	IMP-specific 5'-NT (N-II)
A (8-6)	5'-AMP 8-OH-AMP 2'-dAMP 7-CH-AMP	GMP IMP
B (4-2)	6-S-Me-PuMP 8-Br-AMP 2-amino-PuMP GMP IMP 1,N⁶-etheno-AMP XMP 1-NO-AMP PuMP	6-S-Me-PuMP XMP 7-CH-AMP
C (<2)	8-NH-Me-AMP 6-SH-AMP 8-N-AMP	AMP 1,N⁶-etheno-AMP 8-OH-AMP 2'-dAMP 1-NO-AMP 8-Br-AMP 6-SH-AMP 8-N-AMP PuMP
D (<<1)	2'-AMP 3'-AMP 5'-AMPS 8-NH-Et-NH$_2$-AMP AMP-morpholine	2'-AMP 3'-AMP 5'-AMPS 8-NH-Et-NH$_2$-AMP AMP-morpholine 8-NH-Me-AMP 2-amino-PuMP

Reactions were followed in duplicates by 30 min with compounds selected in categories A and B, and 16 hrs with compounds in categories C and D. All the compounds were used at 5 mM concentration.

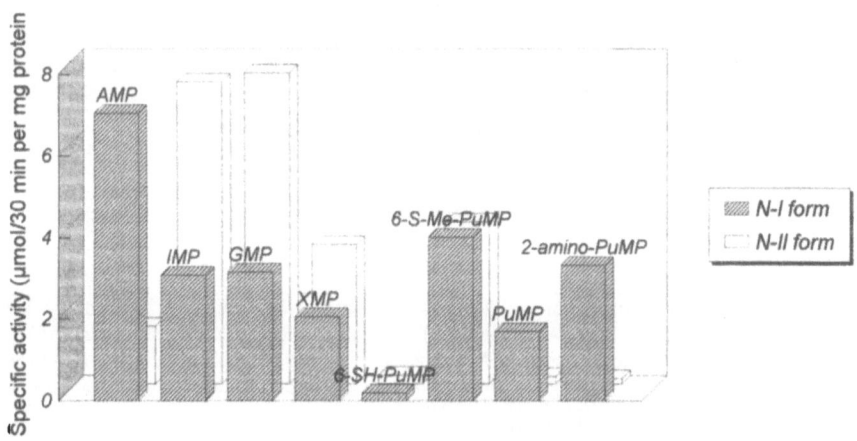

Figure 1. Comparison of N-I with N-II form.

Kinetics of the reaction catalysed by isozymes of cytosolic 5'-NT with various AMP analogs. The compounds selected into categories A and B (Table 1) with some exceptions (2-amino-PuMP and 1,N⁶-etheno-AMP for AMP-specific and 7-CH-AMP for IMP-specific isozyme of 5'-NT) were used for kinetic experiments and estimation of K_M and V_{max} constants (Table 2).

Table 2. KINETIC CONSTANTS OF TWO ISOZYMES OF RABBIT HEART 5'-NT WITH AMP ANALOGS

	AMP-specific 5'-NT (N-I)			IMP-specific (N-II)		
Compound	V_{max} (units per mg protein)	K_M (mM)	correlation coefficient	V_{max} (units per mg protein)	K_M (mM)	correlation coefficient
AMP	91.2	8.3	0.99	22.57	18.41	0.9
8-OH-AMP	28.2	6.5	0.95			
2'-dAMP	6.45	1.2	0.99			
7-CH-AMP	67.3	14.4	0.9			
6-S-Me-AMP	40.8	13.6	0.99	51.1	4.2	0.99
8-Br-AMP	6.6	2.2	0.97			
GMP	15.4	6.1	0.93	82.43	7.22	0.89
IMP	20.9	8.3	0.82	50	2.7	0.97
XMP	34.5	13.6	0.83	37.1	5.3	0.97
1-NO-AMP	21.5	6.2	0.87			
PuMP	35.6	6.2	0.77			

The reactions were followed in duplicates by 30 min.

The maximum velocity V_{max} reveals the effect of chemical modification of AMP on the rate of the catalytic reaction, once the analog is bound to the enzyme. The K_M gives the information on the binding to the enzyme. Since both parameters were affected by most of modifications of AMP, the interpretation of these values should be careful. Among 10 analogs investigated with AMP-specific 5'-nucleotidase, two: 2'-deoxy-adenosine-5'-monophosphate and 8-bromoadenosine-5'-monophosphate appeared to have much lower values of K_M compared to others including natural substrate, adenosine-5'-monophosphate. It may indicate on the higher affinity of these compounds to the enzyme however the maximum velocities were also lower and the effect of modification seems to be more complex. Besides, these two analogs were found the most potent competitive inhibitors of AMP dephosphorylation by AMP-specific 5'-nucleotidase from pigeon heart (not shown) what further suggests their high affinity to the substrate-binding site.

The Michaelis constant K_M estimated for IMP-specific isozyme with compounds classified in categories A and B were all lower (2.5-7 times) than for AMP itself. Interestingly, K_Ms for AMP and IMP were identical for AMP-specific isozyme but 5 fold higher V_{max} with AMP was observed. The other isozyme showed both lower K_M and higher V_{max} with preferred substrate - IMP. One may suppose that the differences in binding of either AMP or IMP to AMP-specific isozyme is secondary and the conversion of enzyme-substrate complex to product should play a decisive role in overall dephosphorylation rate.

DISCUSSION

Interpretation of the influence of modifications for specific activity with both forms of 5'-NT is limited as most of the compounds were AMP analogs. Two single changes should be then considered when discussing affinity to the IMP-specific isozyme. Nevertheless, two types of modifications within 5'-AMP make the reaction impossible for both 5'-NTs from rabbit heart. They are: 1) displacing of the phosphoric acid on either 2'- or 3'-position of ribose (2'-AMP, 3'-AMP), and 2) changing of the electron density distribution by substitution of oxygen for sulphur atom (AMPS) or by formation of phosphoramide bond with morpholine (AMP-morpholine). The first fact indicates 5'-regiospecificity of binding. The last changes can disturb either the correct steric orientation of phosphate residue required for hydrolysis and/or formation of salt-like bridges with the enzyme. The changes that can drasticly decrease hydrolyzability of AMP analogs are: 1) introducing a positively charged, large group in 8-position of purine ring (8-NH-Et-NH$_2$-AMP) which can either fix the base moiety in *syn*-conformation because of steric hinderance or interact electrostatically with enzyme's amino acids, and 2) exchange of 6-NH$_2$-group for sulphydryl function (6-SH-PuMP). A number of modifications in the base moiety had a limited impact on the rate of dephosphorylation by the AMP-specific 5'-NT. However, the negative influence had these modifications which potentially could delete a hydrogen bond formation at 6-NH$_2$-position (i.e. PuMP, 1-NO-AMP, 1,N^6-etheno-AMP and 6-oxo-analogs).

Scarce information is available on the species and organ distribution of the AMP-specific isozyme of cytosolic 5'-NT. Till now, heart is the only organ where it was detected. Comparing to AMP, other naturally occuring nucleotides like: dAMP, GMP, dGMP, IMP, dIMP, UMP and CMP pronounce 1.5-3 times lower rates of dephosphorylation[4]. TMP and dCMP are 7-8 times slower hydrolysed by the analogous form of 5'-NT in pigeon[4] or dog heart[6]. Well pronounced activity of AMP-specific 5'-NT contrasts with a relatively high K$_M$ towards its substrate. Because a free AMP in myocardium does not exceed 10^{-6} mole/l[11], we suggest that AMP-specific 5'-NT is significantly activated only in the situations when the concentration of AMP is peaking far above its physiological level. The enzyme sets the upper limit of AMP concentration and might then take a part in an AMP-overflow safety system during ATP breakdown.

ACKNOWLEDGMENTS

Supported by the Polish Committee for Scientific Research with grant 4 4034 91 02 and grants from the Gdansk School of Medicine.

REFERENCES

1.   V.L. Truong, A.R. Collinson, and J.M. Lowenstein, 5'-Nucleotidase in rat heart. Evidence for the occurence of two soluble enzymes with different substrate specificities, *Biochem J.* 253:117 (1988)
2.   Y. Yamazaki, A.R. Collinson, V.L. Truong, and J.M. Lowenstein, Regulation of soluble 5'-nucleotidase from rabbit heart, in "Purine and Pyrimidine Metabolism in Man. VI. Part B: Basic Research and Experimental Biology," K. Mikanagi, K. Nishioka and W.N. Kelley, eds., Plenum Press, New York and London (1989), pp 107-111.
3.   A.C. Skladanowski and A.C. Newby, 5'-Nucleotidases involved in adenosine formation, in: "Role of Adenosine and Adenine Nucleotides in the Biological Systems," S. Imai and M. Nakazawa, eds., Elsevier, Amsterdam, New York, Oxford (1991), pp.289-299.
4.   A.C. Skladanowski and A.C. Newby, Partial purification and properties of an AMP-specific soluble 5'-nucleotidase from pigeon heart, *Biochem. J.* 268:117 (1990)
5.   Y. Yamazaki, V.L. Truong and J.M. Lowenstein, 5'-Nucleotidase I from rabbit heart, *Biochemistry* 30:1503 (1991)
6.   A. Darvish and P.J. Metting, Purification and regulation of an AMP-specific cytosolic 5'-nucleotidase from dog heart, *Am J Physiol.* 264:H1528 (1993)
7.   A.C. Newby, Adenosine and the concept of 'retaliatory metabolites', *Trends in Biochem Sci.* 9:42 (1984)
8.   H. Zimmermann, 5'-Nucleotidase: molecular structure and functional aspects, *Biochem J.* 285:345 (1993)
9.   B. Jastorff, J. Hoppe, J.M. Mato and T.M. Konijn, Ligand analogs - a tool for the elucidation of protein-ligand interactions. Model system: cAMP, in: "Molecular Mechanisms of Biological Recognition," M. Balaban, ed., Elsevier/North Holland Biomedical Press (1979), pp. 107-117
10. R.T. Smolenski, D.R. Lachno, S.J.M. Ledingham and M.H. Yacoub, Determination of sixteen nucleotides, nucleosides and bases using high-performance liquid chromatography and its application to the study of purine metabolism in hearts for transplantation, *J Chromatogr.* 527:414 (1990)
11. R. Bunger and S. Soboll, Cytosolic adenylates and adenosine release in perfused working heart: comparison of whole tissue with cytosolic nonaqueous fractionation analyses, *Eur J Biochem* 159:203 (1986)

HUMAN PLACENTAL CYTOSOLIC PURINE 5'-NUCLEOTIDASE IS EFFECTIVELY

REACTIVE WITH AN ANTI-CHICKEN LIVER ENZYME ANTIBODY

Jun Oka

The National Institute of Health and Nutrition
Shinjuku-ku, Tokyo 162
Japan

INTRODUCTION

Dephosphorylation of nucleoside 5'-monophosphates by 5'-nucleotidase (5'-ribonucleotide phosphohydrolase, EC 3.1.3.5.) is the first irreversible step in the nucleotide degradation pathway. Its activity has been found in different subcellular fractions such as the plasma membrane (1), the cytoplasm (2-7), the mitochondria (8), and the lysosome (9,10) (Fig. 1). While ecto-5'-nucleotidase localized on the plasma membrane degrades extracellular purine/pyrimidine 5'-nucleotides with micromolar $\underline{Km}$ values, intracellular degradation of 5'-nucleotides is catalyzed possibly by several distinct cytosolic 5'-nucleotidases co-existent in the cytoplasm (11,12). Namely, purine 5'-nucleotidase (2,3), pyrimidine 5'-nucleotidase (4), deoxyribonucleotidase (5'(3')-nucleotidase (EC 3.1.3.31.)) (5), and AMP-selective 5'-nucleotidase (6,7) are cytosolic enzymes with millimolar $\underline{Km}$ values for various nucleotides, and may have a critical role in the maintenance of a constant composition of intracellular purine/pyrimidine 5'-nucleotides (4,13).

Among these cytosolic 5'-nucleotidases, purine 5'-nucleotidase hydrolyzes IMP and other purine nucleotides, and was immunocytochemically located in the cytoplasmic matrix (14,15). This enzyme is allosterically regulated by various activators including ATP, and by an inhibitor, Pi (2,3). The primary role of cytosolic purine 5'-nucleotidase is suggested to generate inosine from IMP, which arises in turn from AMP by AMP deaminase. Both AMP deaminase and purine 5'-nucleotidase are activated by ATP and inhibited by Pi (2,3,16), suggesting that the two enzymes function in a coordinated way to set the upper limit for the intra-cellular adenine nucleotide pool (13). In uricotelic animals, a physiological role of purine 5'-nucleotidase in uric acid production through IMP is suggested by the fact that it is particularly abundant in chicken liver, and in response to a high protein diet its activity increased simultaneously with those of purine nucleoside phosphorylase and xanthine dehydrogenase (2,3).

Cytosolic purine 5'-nucleotidase is widely distributed across animal species, and has been purified from various sources; birds, mammals and invertebrates (3), but a structural relationship between the enzymes from these various sources mentioned above remains obscure. Itoh and Yamada discussed (17), and Itoh et al. showed (18) that the anti-chicken liver enzyme antibody raised in a rabbit did not effectively precipitate the enzyme activity from mammals. In this study, I re-examined the reactivity of the anti-chicken liver enzyme antibody with purified mammalian purine 5'-nucleotidase, and demonstrated that the enzyme purified from human placenta effectively cross-reacted with the antibody. This result indicated that the mammalian and avian enzymes were closely similar in their structures.

EXPERIMENTAL PROCEDURES

From chicken liver and heart were purified cytosolic purine 5'-nucleotidases (19), both of which were assumed to be products of proteolysis (20,21). Human placental cytosolic purine 5'-nucleotidase was partially purified by essentially the same method as above (19). An antibody against purified chicken liver cytosolic purine 5'-nucleotidase was raised in a rabbit (14,15). Ouchterlony double diffusion analysis with the antibody and enzymes was done as described previously (14).

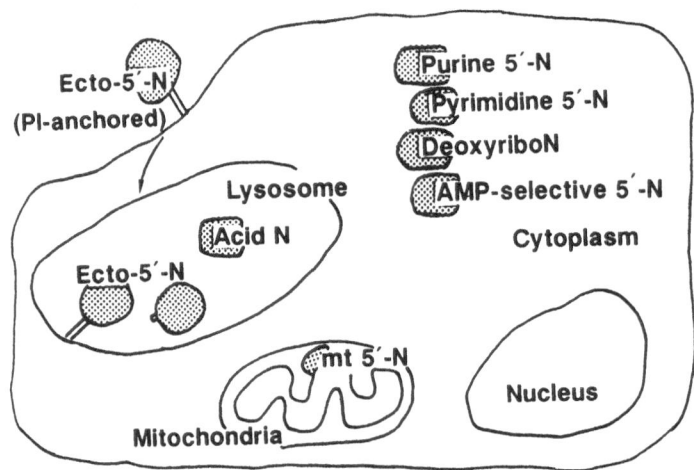

Figure 1. Subcellualr distributions of various nucleotidases (Ns). In the lysosome, 73% of the total ecto-5'-nucleotidase activity was found in the membrane fraction and 24% in the soluble contents fraction (9); it is still uncertain whether the enzyme is transported from the plasma membranes via endosomes as indicated by an arrow or from the trans-Golgi regions via primary lysosomes. In the mitochondria, the 5'-nucleotidase is located on the outer surface of the inner mitochondrial membrane (8). **Nomenclature.** Some authors (3) proposed the name of IMP-GMP 5'-nucleotidase for purine 5'-nucleotidase. However, that name is of tautology and if IMP-GMP is to be used, it should be changed to IMP-GMP hydrolase.

624

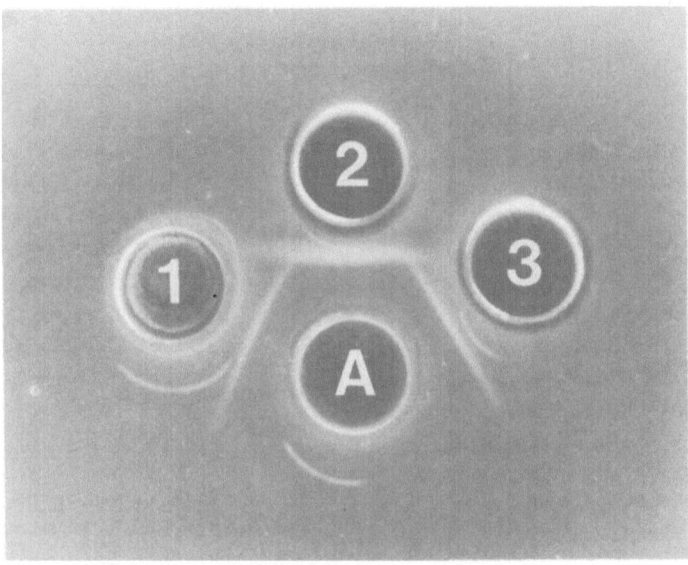

Figure 2. Ouchterlony double diffusion. Well A, the antibody (200 μg); Well 1, 2, and 3, the enzymes purified from human placenta (1 unit), chicken liver (0.5 unit), and chicken heart (0.2 unit), respectively.

RESULTS AND DISCUSSION

Cytosolic purine 5'-nucleotidase was partially purified from human placenta by essentially the same method as from chicken liver and heart (19). On SDS-gel electrophoresis, purified human placental enzyme preparation gave several bands, but by immunoblot analysis using the anti-chicken liver enzyme antibody (20) it developed a band of 53 kDa (data not shown). Spychala et al. also reported that the subunit molecular mass of the purified human placental enzyme is 53 kDa (22).

In Ouchterlony double diffusion, it was demonstrated that a single precipitin line was produced between the antibody, and human placental cytosolic purine 5'-nucleotidase and the chicken liver enzyme, respectively (Fig. 2). These data suggested that the anti-chicken liver cytosolic purine 5'-nucleotidase antibody effectively cross-reacted with the purified human enzyme. As previously reported (23) two precipitin lines between the antibody, and the chicken liver and heart enzymes were fused each other with a small spur (Fig. 2), indicating that they are structurally related isozymes.

The above result indicated that the mammalian and avian cytosolic purine 5'-nucleotidases are closely similar in their structures, and the anti-chicken liver enzyme antibody will be a good and sufficient probe to clone the human cytosolic purine 5'-nucleotidase.

ACKNOWLEDGMENTS

This work was supported in part by a Research Grant from Gout Research Foundation of Japan.

REFERENCES

1. DePierre, J. W., and Karnovsky, M. L. (1974) <u>J. Biol. Chem.</u> <u>249</u>, 7111–7120
2. Tsushima, K. (1986) <u>Adv. Enzyme Regul.</u> <u>25</u>, 181–200
3. Itoh, R. (1993) <u>Comp. Biochem. Physiol.</u> <u>105B</u>, 13–19
4. Paglia, D. E., and Valentine, W. N. (1975) <u>J. Biol. Chem.</u> <u>250</u>, 7973–7979
5. Hoglund, L., and Reichard, P. (1990) <u>J. Biol. Chem.</u> <u>265</u>, 6589–6595
6. Yamazaki, Y., Truong, V. L., and Lowenstein, J. M. (1991) <u>Biochemistry</u> <u>30</u>, 1503–1509
7. Darvish, A., and Metting, P. J. (1993) <u>Am. J. Physiol.</u> <u>264</u>, H1528–H1534
8. Henke, W., Lang, M., Dubiel W., Holzhutter, H.-G., and Gerber, G. (1989) <u>Biochem. Int.</u> <u>18</u>, 833–844
9. Wada, I., Eto, S., Himeno, M., and Kato, K. (1987) <u>J. Biochem. (Tokyo)</u> <u>101</u>, 1077–1085
10. Arsenis, C., and Touster, O. (1968) <u>J. Biol. Chem.</u> <u>243</u>, 5702–5708
11. Iizasa, T., Takeuchi, F., Honda, Z., Nishida, Y., Kamatani, N., and Miyamoto, T. (1986) <u>Biochim. Biophys. Acta</u> <u>882</u>, 228–233
12. Bomtemps, F., Van den Berghe, G., and Hers, H. G. (1988) <u>Biochem. J.</u> <u>250</u>, 687–696
13. Carson, D. A., Carrera, C. J., Wasson, D. B., and Iizasa, T. (1991) <u>Biochim. Biophys. Acta</u> <u>1091</u>, 22–28
14. Yokota, S., Oka, J., Ozasa, H., and Itoh, R. (1988) <u>J. Histochem. Cytochem.</u> <u>36</u>, 983–989
15. Oka, J., Ozasa, H., Itoh, R., and Yokota, S. (1989) <u>Adv. Exp. Med. Biol.</u> <u>253B</u>, 113–118
16. Chapman, A. G., and Atkinson, D. E. (1973) <u>J. Biol. Chem.</u> <u>248</u>, 8309–8312
17. Itoh, R., and Yamada, K. (1991) <u>Int. J. Biochem.</u> <u>23</u>, 461–465
18. Itoh, R., Echizen, H., Higuchi, M., Oka, J., and Yamada, K. (1992) <u>Comp. Biochem. Physiol.</u> <u>103B</u>, 153–159
19. Itoh, R., and Oka, J. (1985) <u>Comp. Biochem. Physiol.</u> <u>81B</u>, 159–163
20. Oka, J., Ozasa, H., and Itoh, R. (1988) <u>Biochim. Biophys. Acta,</u> <u>953</u>, 114–118
21. Oka, J., Itoh, R., and Ozasa, H. (1991) <u>Int. J. Purine Pyrimidine Res.</u> <u>2</u> (Suppl.1), p.73 (Abstr.)
22. Spychala, J., Madrid-Marina, V., and Fox, I. H. (1988) <u>J. Biol. Chem.</u> <u>263</u>, 18759–18765
23. Oka, J., Itoh, R., and Ozasa, H. (1991) <u>Adv. Exp. Med. Biol.</u> <u>309B</u>, 151–154

CHARACTERIZATION OF THE ADENINE PHOSPHORIBOSYLTRANSFERASE FROM *SACCHAROMYCES CEREVISIAE*

J. Alfonzo-Garcia[1], Amrik Sahota[2]
and Milton W. Taylor[1]

[1] Department of Biology, Indiana University,
  Bloomington , IN 47405

[2] Department of Medical Genetics, Indiana
  University , Indianapolis, IN 46223

INTRODUCTION

Adenine phosphoribosyltransferase(APRT) catalyzes the conversion of adenine and 5'-phosphoribosylpyrophosphate(prpp) into the nucleotide 5'-adenosine monophosphate. This enzyme enables the utilization of free adenine derived from nucleotide catabolism or from exogenous sources to help mantain nucleotide pool balance.

The APRTs from various organisms have been extensively studied at both the biochemical and molecular levels(1)(2)(3). These enzymes share higher than 70% similarity at the amino acid level. All APRTs are encoded by a single gene and the enzymes natively exist as dimers of identical subunits.

We have been studying the APRT from *S. cerevisiae*. Preliminary analysis shows this enzyme to be the product of two similar, but not identical, genes. The purified enzyme has a molecular weight of 50kD.

In the present report, we show that *S. cerevisiae* APRT is a heterodimeric enzyme, consisting of two sub-units with molecular weights of 34kD and 20kD, respectively. We also present evidence for a magnesium-dependent behavior of this enzyme when separated by gel-filtration chromatography. Similar behavior for the APRT from *S. pombe* has been previously described by Nagy and Ribet(4). In *S. cerevisiae* APRT, this behavior correlates with different affinities of this enzyme for PRPP.

METHODS AND RESULTS

APRT was purified from cakes of Fleischmann's yeast (ATCC 24903). Five pounds of yeast cakes were resuspended in water to make a thick paste. The paste was frozen in liquid nitrogen and homogenized in a Waring blender. The homogenized powder was resuspended in extraction buffer (50 mM Tris-Cl pH 7.4, 5mM $MgCl_2$, 20 mM KCl) and stirred for 1 hour at 4^oC. The mixture was centrifuged at 20,000 rpm for 30 min. and the supernatant was fractionated with ammonium sulfate. The 40-60% salt fraction was collected as a pellet and assayed for activity as described by Hershey and Taylor(5) . This fraction contained more than 85% of the activity found in the crude extract. The resulting pellet was treated with acetic acid until the pH dropped to 5.0 . The pH 5.0 lysate was centrifuged and the pellet discarded. The clarified lysate was treated with

Table 1. Purification of APRT from *S. cerevisiae* (as described in text).

Purification Step	Protein	Total Activity	Yield	Specific Activity	Purification
	mg	*umol/min.*	*%*	*umol/mg/min.*	*fold*
1. Crude	13504.80	124.20	100.00	0.01	1.00
2. MnCl$_2$	13347.10	222.90	179.50	0.02	1.80
3. NH$_2$SO$_4$ (40-60%)	6026.23	119.32	96.10	0.02	2.15
4. Alumina Cγ gel	87.73	28.34	22.80	0.32	35.11
5. Calcium Phosphate Gel	39.74	23.40	18.80	0.59	64.10
6. Mono Q	5.59	23.10	18.60	4.14	450.00
7. Mono P	0.58	8.72	7.02	15.14	1645.10

A.

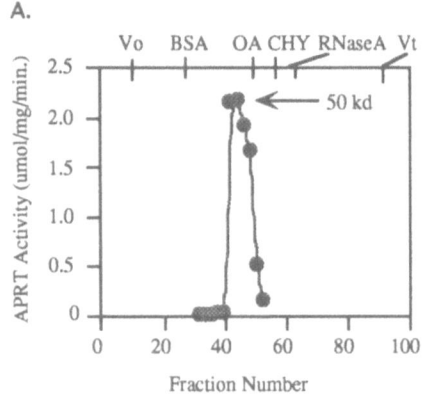

B.

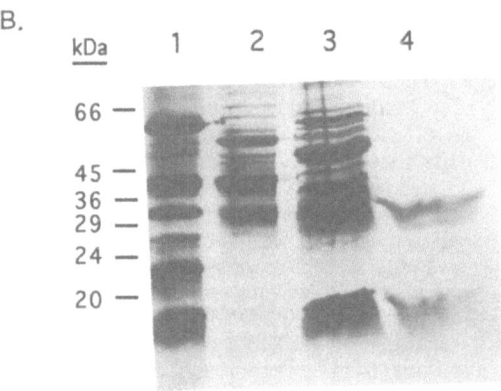

Figure 1. Superose 12 chromatography of the purified APRT.
A) The purified enzyme separated by gel filtration on a Superose 12 column in buffer containing 10 mM Mg^{2+}. The column was calibrated with bovine serum albumin (BSA, Mr = 66,000), ovalbumin (OA, Mr = 43,000), chymotrypsinogen A (CHY, Mr = 25,000), ribonuclease A (RNase A, Mr = 13,700).
B) Fractions 43-48 were concentrated to 0.5 ml and a sample was analyzed by SDS - polyacrylamide gel electrophoresis. Lane 1, molecular weight markers; Lane 2, Calcium Phosphate gel fraction; Lane 3, Mono Q pooled fractions; Lane 4, Mono P purified enzyme.

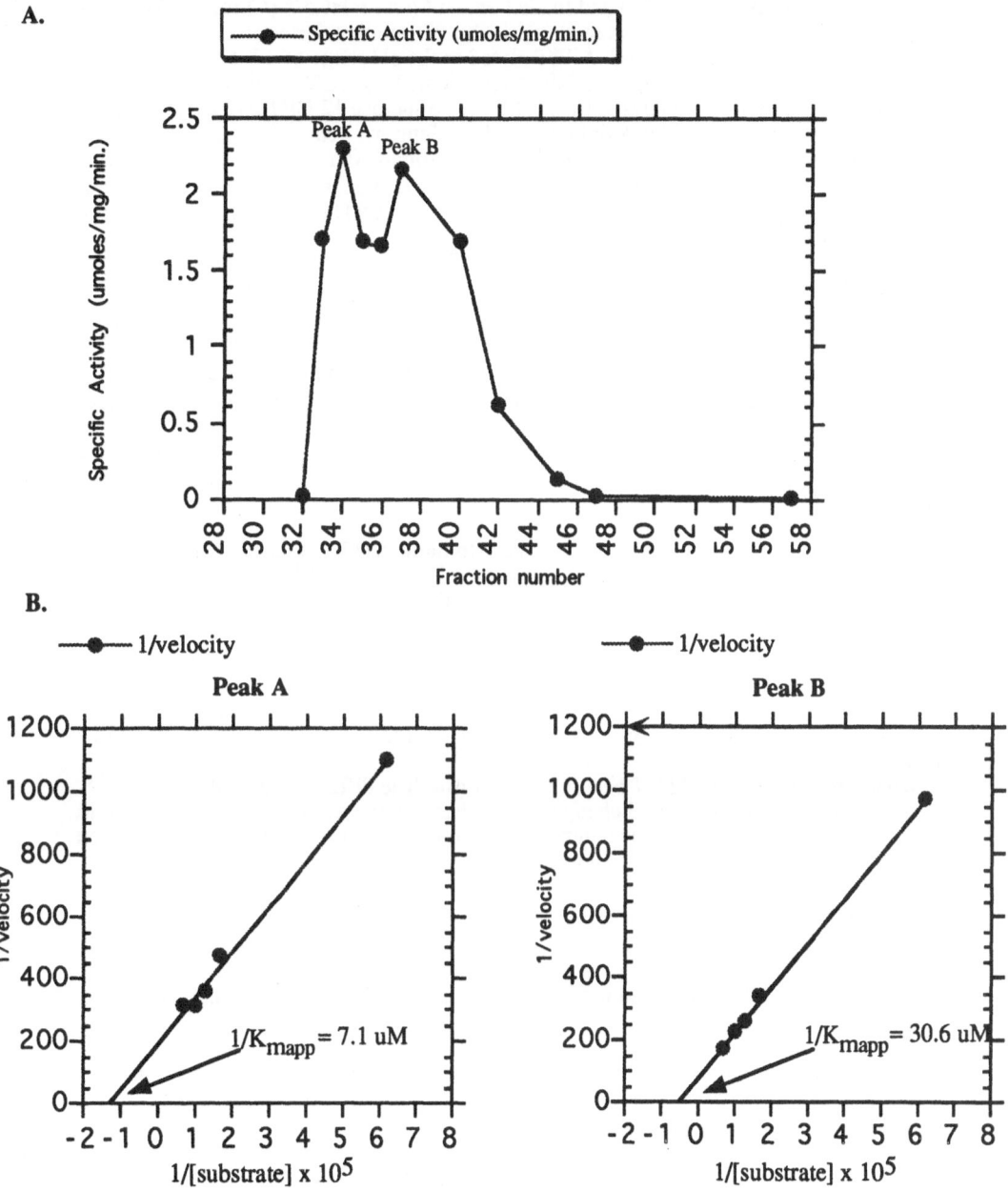

Figure 2. Superose 12 chromatography in the absence of Mg 2+. The purified enzyme was fractionated, as described in the text, in buffer lacking divalent cations. A. Elution pattern of the enzyme. B. Double reciprocal plots of peaks A and B, reactions were carried out in 5.0 uM adenine and varying concentrations of PRPP from 0.8 -15.0 uM.

alumina C gel (25mg of gel/ ml), followed by ccalcium phosphate gel as described by Woods (personal comm.) . The resulting supernatant was further purified by FPLC. The partially-purified supernatant was loaded onto a Mono Q HR5/5 (anion exchange) column (Pharmacia) and eluted with a linear gradient of 50mM-1.0M KCl. Fractions with highest activity were pooled and loaded onto a Mono P HR5/5 (Chromatofocussing) column (Pharmacia). Fractions were eluted with a pH gradient between 6.0-4.0, peak fractions were pooled and assayed. APRT eluted in the pH range of 4.8-5.0. This fraction contained approximately 4% of the original activity with over 1200-fold purity (table 1). The previous fraction was analyzed by a calibrated Superose 12 HR10/10 (Pharmacia) column for molecular weight determination. The column was equillibrated with buffer containng 10mM tris-Cl pH7.4, 10mM $MgCl_2$ and 20mM KCl. The activity eluted with a native molecular weight of 50kDa (fig.1-a). The purified enzyme was also resolved on a 12% denaturing gel, Two bands of molecular weights of 20kDa and 34 kDa, respectively, were observed(fig.1-b).

The purified enzyme was also analyzed on the Superose 12 column in the presence and absence of magnesium (10mM). The affinity for adenine of peak fractions from both conditions was determined . In buffer containing Mg^{2+}, the enzyme elutes as a single peak of activity. The enzyme on this fraction has a Km of 30 uM for PRPP . Fractionation in the absence of Mg^{2+} separated the activity into two distinct peaks (fig.2) with Km_{app} of 7uM for the fastest eluting peak (peak A) and 30uM for the slowest eluting peak (peak B).

DISCUSSION

We used FPLC to purify *S. cerevisiae* APRT to homogeneity. This enzyme is a dimer of non-identical subunits, structurally different to all the other APRTs described so far. Sequence comparison of the two genes coding for the subunits show similarities to other aprts (data not shown).

In 1977 Nagy and Ribet, described two active forms for the APRT from *S. pombe* . These forms were easily separated when chromatographed in buffer lacking Mg^{2+}. We have observed a similar behavior for the *S. cerevisiae* enzyme. In the absence of Mg^{2+} we have shown that these two distinct species have different affinities for the substrate adenine.

We suspect that native APRT occurs in two forms. The different Km values for adenine could indicate structural changes of this enzyme *in vivo* depending on substrate concentration. The heterodimeric nature of this enzyme could suggest a novel regulation of this activity at the level of protein folding .

REFERENCES

1. M.W. Taylor, A.E. Simon and R.M. Kothari. " The APRT System," Wiley,New York (1985).

2. J. G. De Boer and B.W. Glickman, Mutational analysis of the structure and function of the adenine phosphoribosyltransferase enzyme of chinise hamster, J. Biol. Chem. 221:163 (1991).

3. H.V. Hershey and M.W. Taylor, Nucleotide sequence and deduced aminoacid sequence of Escherichia coli adenine phosphoribosyltransferase and comparisson with other analogous enzymes, Gene. 43:287 (1986).

4. M. Nagy and A.M. Ribet, Purification and comparative study of adenine and guanine phosphoribosyltransferases from *Schizosaccharomyces pombe* , Eur. J. Biochem. 77:77(1977).

5. H.V. Hershey and M. W. Taylor, Purification of adenine phosphoribosyltransferase by affinity chromatography, Prep. Biochem. 8: 453 (1978).

DIFFERENTIAL SUBSTRATE PROPERTIES OF MAMMALIAN RIBONUCLEOTIDE REDUCTASE

Joseph G. Cory, Ann H. Cory and Deborah L. Downes

Department of Biochemistry
East Carolina University School of Medicine
Greenville, NC 27858

INTRODUCTION

Ribonucleotide reductase catalyzes the reaction in which ribonucleoside 5'-diphosphates are reduced to 2'-deoxyribonucleoside 5'-diphosphates. This is the rate-limiting step in the generation of 2'-deoxyribonucleoside 5'-triphosphates required for DNA replication. The active enzyme consists of two non-identical protein subunits that are encoded by different genes. One subunit contains non-heme iron and a tyrosyl free-radical (NHI subunit); the other subunit has the effector-binding sites for the nucleoside 5'-triphosphate allosteric effectors.[1] While it is reported that one enzyme catalyzes the reduction of all four substrates (CDP, UDP, GDP and ADP)[2], there are data that show that the activity of ribonucleotide reductase toward the various substrates can be differentially affected by a variety of inhibitors or agents[3-5] or show different ratios of activities in hydroxyurea-resistant cell lines derived from various sources. [6-8]

In this study the effects of two specific reagents that interfere with the interactions of the NHI and EB subunits to form the holoenzyme are studied with respect to the substrate specificity of ribonucleotide reductase.

It has been previously shown that the monclonal antibody to yeast tubulin, YL 1/2, recognizes a specific epitope on the NHI subunit of ribonucleotide reductase and as a result, inhibits CDP reductase activity.[9,10]

Other studies showed that a nonapeptide (NSFTLDADF), corresponding to the C-terminus of the NHI subunit, inhibited CDP reductase activity by binding to the effector-binding (EB) subunit and aborting the interaction between the NHI and EB subunits that is required to generate the active enzyme species.[11]

The results of our current studies show that perturbations of the interactions of the NHI and EB subunits using these reagents result in differential effects on the substrate specificity of mammalian ribonucleotide reductase.

METHODS AND MATERIALS

Methods

Preparation and assay of ribonucleotide reductase: Ribonucleotide reductase was prepared from Ehrlich tumor cells as previously described[5] and the non-heme iron and effector-binding subunits separated on blue dextran-Sepharose.[12] CDP reductase activity was assayed by the method of Steeper and Steuart.[13] ADP reductase was assayed by the method of Cory et al.[14] GDP and UDP reductase activities were assayed by the method of Sato, et al.[15]

Materials

[^{14}C]CDP (476 mCi/mmole) and [^{3}H]ADP (29.7 Ci/mmole) were purchased from New England Nuclear. [^{14}C]GDP and [^{14}C]UDP were prepared enzymatically from [^{14}C]GMP and [^{14}C]UMP by the method of Cory and Bacon.[16] [^{14}C]GMP (450 mCi/mmole) and [^{14}C]UMP (272 mCi/mmole) were purchased from Research Products International and ICN, respectively. The other biochemicals were purchased from Sigma Chemical Company, St. Louis, MO. The YL 1/2 antibody was purchased from Accurate Chemical and Scientific Corporation, Westbury, NY. The nonapeptide was synthesized by Dr. David Klapper in the Core Facility at the University of North Carolina, Chapel Hill, NC.

RESULTS AND DISCUSSION

The monoclonal antibody raised against yeast tubulin, YL 1/2, inhibited CDP reductase activity in a manner which was dependent on the amount of YL 1/2 antibody and on the amount of ribonucleotide reductase. Increasing the amount enzyme at a fixed concentration of YL 1/2 antibody reduced the degree of inhibition of CDP reductase activity. Conversely, increasing the concentration of YL 1/2 antibody at a fixed concentration of enzyme caused increased inhibition of CDP reductase activity. These data are consistent with the results reported by Thelander, et al.[10] The effects of the YL 1/2 antibody on the reductase activities utilizing CDP, ADP, UDP and GDP as substrates were determined. The titration of the reductase activities with YL 1/2 antibody caused a marked inhibition of CDP reductase activity with much less effect on ADP reductase activity (Figure 1). At the highest concentration of YL 1/2 antibody used in this study, there was a shift in the ratio of CDP to ADP reductase activities from 1.22 in the control untreated enzyme, to 0.61 in the YL 1/2-treated enzyme. The effects of YL 1/2 antibody on UDP and GDP reductase activities were intermediate to those seen for CDP reductase and ADP reductase activities. These data show that YL 1/2 which specifically binds to the NHI subunit results in differential alterations in the enzyme activity toward the various substrates.

There is a great deal of specificity in the YL 1/2 antibody. Thelander, et al.[10] showed that another monoclonal antibody to yeast tubulin, YOL 1/34, had no effect on CDP reductase activity. Further, Spector, et al.[17] reported that ,whereas, the mammalian enzyme was inhibited by YL 1/2 antibody, the ribonucleotide reductase from varicella zoster virus was much less inhibited, even at five times the concentration of the YL 1/2 antibody required to completely inhibit the CDP reductase activity from mammalian cells. The inhibition of the activity of CDP reductase from herpes simplex virus types 1 and 2 was intermediate between the inhibition of mammalian reductase and the varicella zoster virus ribonucleotide reductase.

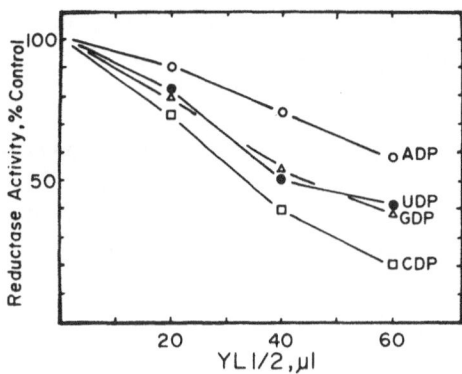

Figure 1. Titration of CDP, ADP, UDP and GDP reductase activities with yeast tubulin antibody (YL 1/2). The ribonucleotide reductase fraction was incubated on ice with varying concentrations of YL 1/2 antibody (4.87 µg of protein/10 µl) for 1hr on ice. The substrate mixes were added and assayed as described in the "Methods" section. The control activities for CDP, ADP, GDP and UDP reductions were 1.91, 1.56, 1.50 and 0.53 nmoles/30 min/mg protein, respectively. All assays were carried out in triplicate.

Our studies suggest that the epitope recognized by the YL 1/2 antibody, as defined by Wehland, et al.[18] is more accessible in CDP reductase than in ADP, UDP or GDP reductases from Ehrlich tumor cells since the CDP reductase activity is inhibited to a greater extent. Whether this represents a difference in the availability of the C-terminus due to conformational changes is not known. However, since the YL 1/2 antibody was incubated with the enzyme in the absence of the positive effectors, there would be little reason to suspect tertiary or quaternary structural changes in the subunits which could account for the differences in susceptibilities of the reductase activities to the YL antibody.

Following up on the studies of Cosentino, et al.[11] who showed that peptides corresponding to the C-terminal region of the NHI subunit inhibited reductase activity, we carried out studies with the enzyme from Ehrlich tumor cells. The nonapeptide (NSFTLDADF) corresponding to the C-terminus of the mammalian NHI subunit inhibited CDP reductase activity. The inhibition was dependent on the concentration of the nonapeptide and the concentrations of the enzyme subunits. Increasing the concentrations of the protein subunits resulted in the decreased inhibition of CDP reductase by the nonapeptide. The effect of the nonapeptide on CDP and ADP reductase activities was then compared. ADP reductase activity was more sensitive to the effects of the nonapeptide then was CDP reductase activity. We had previously shown that the presence of ATP or AMP PNP as a positive effector of CDP reductase caused a large increase in the S_{20} and molecular weight of the active ribonucleoside reductase species,[19,20] but that these corresponding changes were not caused by dGTP as a positive effector of ADP reductase. We therefore considered whether the differences in sensitivities of CDP reductase and ADP reductase to the nonapeptide were due to the conformational state of the enzyme. The presence of AMP-PNP which can effectively replace ATP as the positive effector of CDP reductase did not alter the inhibition of ADP reductase by the nonapeptide. Further, increasing the concentration of dGTP in the ADP reductase assay did not decrease the inhibition of ADP reductase by the nonapeptide.

The two reagents used in these studies are very specific, recognizing defined peptide sequences. The antibody YL 1/2 recognizes a specific sequence in the C-terminal region on the NHI subunit which in turn alters the binding of the EB subunit to form the active holoenzyme. The nonapeptide which corresponds to the C-terminal region of the NHI subunit competes with the NHI for binding to the EB subunit. This is a very specific interaction since the nonapeptides corresponding to the C-terminus of the NHI subunits from herpes simplex virus or E. coli have no effect on the mammalian ribonucleotide reductase.[11]

These studies taken together show that the interaction between the NHI, and EB subunits takes place through the C-terminal region of the NHI subunit and an undefined region on the EB subunit and that alterations in these interactions result in differential substrate activities that are independent of the allosteric effectors.

ACKNOWLEDGEMENTS

This research was supported by grant funds from the Public Heath Service, National Cancer Institute, CA 55540 and the Phi Beta Psi Sorority.

REFERENCES

1.   J.G. Cory. "Role of ribonucleotide reductase in cell division," in: Inhibitors of Ribonucleoside Diphosphate Reductase Activity, J.G. Cory and A.H. Cory, ed., Pergamon Press, New York (1989).

2.   S. Eriksson, L. Thelander and M. Akerman, Allosteric regulation of calf thymus ribonucleoside diphosphate reductase, *Biochemistry* 18:2948-2952 (1979).

3.   J.G. Cory, Properties of ribonucleotide reductase from Ehrlich tumor cells; multiple nucleoside diphosphate activities and reconstitution of activities from components, *Adv. Enzyme Regul.* 17:115-131 (1979).

4.   J.G. Cory and A.E. Fleischer, Mode of inhibiton of tumor cell ribonucleotide reductase by 2,3-dihydro-1H-pyrazolo (2,3-A) imidazole (NSC51143), *Cancer Res.* 40:3891-3894 (1980).

5.   J.G. Cory and M.M. Mansell, Studies on mammalian ribonucleotide reductase inhibition by pyridoxal phosphate and the dialdehyde derivatives of adenosine, adenosine 5'-monophosphate and adenosine 5'-triphosphate, *Cancer Res.* 35:390-396 (1975).

6.   J.A. Wright, Altered forms of mammalian nucleoside diphosphate reductase from mutant cell lines, *Pharmac. Ther.* 22:81-102 (1983).

7.   S.E. Koropatnick and J.A. Wright, Ribonucleotide reductase activity in drug-resistant mammalian cells detected by an assay procedure for intact cell, *Enzyme* 25:220-227 (1980).

8.   W.H. Lewis and J.A. Wright,. Isolation of hydroxyurea-resistant CHO cells with altered levels of ribonucleotide reductase, *Somat. Cell Gent.* 5:83-96 (1979).

9.   N.M. Standart, S.J. Bray, E.L. George, T. Hunt, J.V. Ruderman, The small subunit of ribonucleotide reductase is encoded by one of the most abundant translationally regulated maternal RNAs in clam and sea urchin eggs, *J.Cell. Biol.* 100:1968 (1968).

10.   M. Thelander, A. Grasland and L. Thelander. Subunit M2 of mammalian ribonucleotide reductase. Characterization of a homogeneous protein isolated from M2-overproducing mouse cell, *J. Biol. Chem.* 260:2737 (1985).

11. G. Cosentino, P. Lavallee, S. Rakkit, R. Plante, J. Baudette, C. Lawetz, P.W. Whithead, J.-S. Duceppe, C. Lepine-Frenelty, M. Dansereau, C. Grilbault, Y. Langelir, P. Gandreau, L. Thelander and Y. Grindon, Specific inhibition of ribonucleotide reductases by peptides corresponding to the C-terminal of their second subunit, *Biochem. Cell. Biol.* 69:79 (1991).

12. J.G. Cory, A.E. Fleischer and J.B. Munro, III,. Reconstitution of the ribonucleotide reductase enzyme from Ehrlich tumor cells, *J. Biol. Chem.* 253:2898 (1978).

13. J.R. Steeper and C.E. Steuart, A rapid assay method for CDP reductase activity in mammalian cell extracts, *Anal. Biochem.* 34:123 (1970).

14. J.G. Cory, F.A. Russell and M.M. Mansell. A convenient assay for ADP reductase activity using Dowex-1-borate columns, *Anal. Biochem.* 55:449 (1973).

15. A. Sato, A.E. Fleischer and J.G. Cory, Assay fo UDP, GDP, CDP and ADP reductase activities of column chromatography on polyethyleneimine cellulose, *Anal. Biochem.* 135:431 (1983).

16. J.G. Cory and P.E. Bacon,. Preparation of [^{14}C]uridine 5'-diphosphate and [^{14}C]guanosine 5'-diphosphate, *Prep. Biochem.* 14:231 (1984).

17. T. Spector, J.G. Stonehuerner, K.K. Brion, D.R. Averett, Ribonucleotide reductase induced by Varicella Zoster Virus. Characterization, and potentiation of acyclovir by its inhibition, *Biochem. Pharm.* 36:4341 (1987).

18. J. Wehland, H.C. Schroder and K. Weber. Amino acid sequence reqirements in the epitope recognized by the alpha-tubulin-specific rat monoclonal antibody YL 1/2, *EMBO J.* 3:1295 (1984).

19. G.L. Klippenstein and J.G. Cory. Ribonucleotide reductase: Association of the regulatory subunit in the presence of allosteric effectors, *Biochem. Biophys. Res. Commun.* 83:252 (1978).

20. J.G. Cory and A.E. Fleischer, The molecular weight of Ehrlich tumor cell ribonucletoide reductase and its subunits: Effector-induced changes, *Arch. Biochem. Biophys,* 217:546 (1982).

ALTERED KINETIC PROPERTIES OF RECOMBINANT HUMAN CYTOSOLIC THYMIDINE KINASE (TK1) AS COMPARED WITH THE NATIVE FORM

Helle Kock Jensen and Birgitte Munch-Petersen

Department of Life Sciences and Chemistry
DK 4000 Roskilde, Denmark

INTRODUCTION

Human cytosolic thymidine kinase (TK1) is a cell cycle regulated key enzyme in the salvage pathway of the nucleoside metabolism catalyzing the first phosphorylation step in dTTP synthesis. Thymidine kinase (TK) is an enzyme of high interest, primarily because of the specific cell cycle regulated expression. In addition, TK phosphorylates a number of nucleoside analogs which in the phosphorylated form interfere with DNA replication and repair and hence are useful in the chemotherapeutic treatment of cancer and viral infections.

A more detailed knowledge of TK will be valuable for constructing more TK specific and efficient nucleoside analogs. A main problem in the structural an physical investigation of TK is the extremely low level of TK protein in mammalian cells. In this work we describe an expression vector for human TK1. Surprisingly, the enzymatic properties of the recombinant TK1 differ from those of the native TK1 with respect to the recently described specific regulatory effect of ATP[1]. This may indicate that native TK1 is subject to posttranslational modifications, such as phosphorylation, modifications that are not operable in *E.coli*.

METHODS

Constuction of TK1-pet3a vector

The TK1-pet3a vector was constructed by standard methods[2] by inserting the amino-acid coding region of the human TK1 into the polylinker of the pet3a vector[3] which is under control of a T7 RNA polymerase promotor. The aminoacid coding region was PCR amplified from the pTKII[4] containing the complete human TK1 cDNA. The two primers in the PCR flank the aminoacid coding sequence and each has a restriction site in the 5' end (NdeI and BamHI, respectively). The PCR product was blunt-end ligated into the EcoRV site of pBR322. This cloning was nessesary because it was not possible to cut the amplified product directly by NdeI and BamHI. The pBR322 with the insert was cleaved by BamHI and NdeI and the insert was purified by agarose gel electrophoresis and band purification by standard methods[2]. The insert and the pet3a vector were ligated and used to transform a host E.coli (HB101). The TK1-pet3a vector was isolated and sequenced by T7 DNA polymerase to control that no mutations had occurred during the PCR.

Expression of TK1

E.coli strain BL21(DE3)lysS containing T7 RNA polymerase under control of the IPTG inducible UV5 promotor was transformed by the TK1-pet3a vector. The expression was carried out in LB medium (10 g tryptone, 5 g yeast, 10 g NaCl/liter) at 25°C. When OD_{600} was 0,2, IPTG was added to a final concentration of 0,4 mM to induce the UV5 promotor. The cells were harvested by centrifugation (14,000 g, 3 min), resuspended in 1/4 vol of lysis buffer (50 mM Tris-HCl (pH 8), 1 mM EDTA, 100 mM NaCl, 0.5% Nonidet P-40, 5 mM DTT, 0.5 mM PMSF, 5 mM Benzamidine, 10% glycerol, 50 mM NaF) and sonicated 3 times 15 sec (40 Watt) on ice. Cell debris was removed by centrifugation 20,000 g at 4°C for 30 min.

Purification of TK1

All steps in the purification were performed essentially as decsribed[5]. Enzyme extract from 1 liter culture was filtered on a G-25 column (5.5 x 20 cm) and the enzyme chromatographed on a 3'-dTMP-Sepharose column (1 x 4 cm).

For enzyme kinetics studies, thymidine was removed by CM-sepharose chromatography as described[1]. *E.coli* TK is removed together with thymidine in this step, because in contrast to the human TK, it does not bind to the negative carboxy groups[6].

Enzyme kinetics

The thymidine kinase activity was assayed by the DEAE-cellulose 81 paper method[7]. Standard assay conditions were: 50 mM Tris-HCl, pH 8, 2.5 mM $MgCl_2$, 10 mM dithiothreitol, 3 mg/ml BSA, 2.5 mM ATP, 10 μM radiolabelled thymidine. One unit of enzyme activity is defined as the amount of enzyme that can phosphorylate 1 μmol of nucleoside per min at 37°C under standard assay conditions. K_m values and Hill coeficients were determined from Hill plots (log v/(V-v) versus log s).

Sephadex G-200 chromatography

Sephadex G-200 superfine was swelled and packed according to standard procedures (600 x 20 mm). V_e values were determined for five marker proteins (beta-amylase (200 KD), alcohol dehydrogenase (150 KD), bovine serum albumin (66 KD), carbonic anhydrase (29 KD), cytochrome C (12,4 KD)) and for TK1 in absence and presence of ATP. Before determinating the V_e value in the presence of ATP the enzymes were preincubated with 2.5 mM ATP and the column was pre-equilibrated with buffer containing 2.5 mM ATP.

RESULTS AND DISCUSSION

An expression-vector was constructed to ensure unmodified ends of the TK1. When BL21(DE3)lysS is transformed by this vector the yield is around 1 mg TK1 per liter culture (25°C, 24 hours).

The recombinant TK was purified by affinity chromatography on a 3'-dTMP-Sepharose column to more than 95% homogeneity as estimated from a SDS-PAGE (figure 1). The subunit molecular weight was as expected, similar to the native TK1 - around 24 KD.

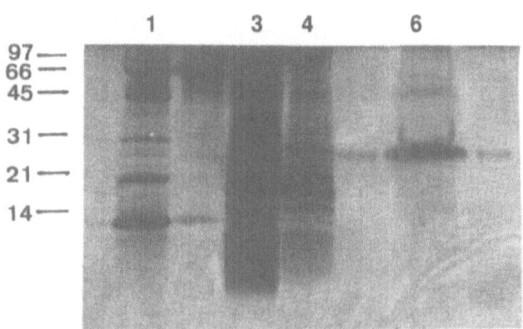

Figure 1.SDS-PAGE of purified recombinant TK1. Lane 1: Proteinmarker (97 KD, 66 KD, 45 KD, 31 KD, 21 KD, 14 KD), lane 3: Bacteria lysate, lane 4: Wash fractions of affinity column, lane 6: Eluate from affinity column.

Table 1. Comparison of the properties of lymphocyte TK1 and recombinant TK1.
 * Hill coefficient = n

		Lymphocyte TK1	Recombinant TK1
Spec. act. nmol/min/mg		9.500	10.000
K_m (μM)	+ATP	0.5	0.4
	-ATP	12	0.4
Hill coeffi cient	+ATP	1.25	1.5
	-ATP	0.7	1.4
Subunit molecular mass (KD)		24	24
Native molecular mass	+ATP	150	150
	-ATP	70	150

The kinetic data for the native TK1 (Table 1) are in agreement with our previously reported data, and show that incubation or storage with ATP induces a kinetically slow transition from a low affinity form (Km=12μM, n=0.7) to a high affinity form of TK1 (Km=0.5 μM, n=1.25)[1]. For the recombinant enzyme, the Km and n values were 0.4 μM and 1.4 - 1.5, respectively, for both the -ATP and +ATP form. This indicates that the recombinant enzyme only occurs as the high affinity form with the low Km value. In addition the Hill coefficients of 1.4 - 1.5 indicate a slightly higher degree of positive homotropic cooperativity than found for the +ATP form of the native TK1.

639

The different kinetics of the +ATP and -ATP forms were related to the ability of the native TK1 to appear as a dimer in the absence of ATP and a tetramer in the presence of ATP[1]. When the apparant molecular weights of the native and recombinant TK1 were determined in absence and presence of ATP (figure 2), we found that recombinant TK1 was a tetramer also in absence of ATP.

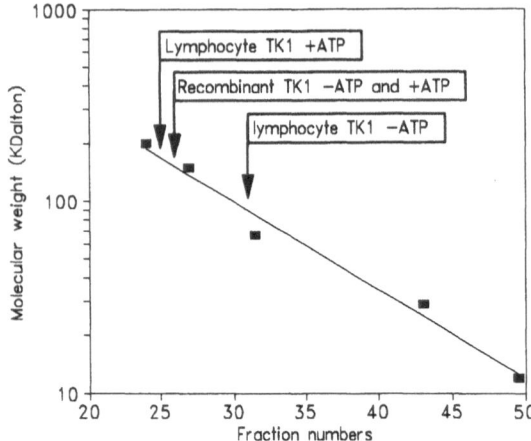

Figure 2. The molecular weight determined by G-200 superfine Sephadex. The column is calibrated by protein standards (12 - 200 KD) marked as black squares. The recombinant and lymphocyte TK1 preincubated with and without ATP are marked with black arrows.

Two conclusions are implicated in these results: Firstly the ATP induced kinetic transition of the native TK1 is associated with the dimer-tetramer transition. Secondly, the missing ability of recombinant TK1 to undergo the ATP induced transition between two distinct molecular forms, may be due to the lack of post-translational modifications when expressed in E.*coli*. Such modifications may enable native TK1 to undergo the reversible transition, and we find it likely that this mechanism is involved in the cell cycle regulated expression of TK1 as previously suggested[1,8].

REFERENCES

1. B.Munch-Petersen, G.Tyrsted and L.Cloos, Reversible ATP-dependent transition between two forms of human cytosolic thymidine kinase with different enzymatic properties,The Journal of Biological Chemistry.268:15621(1993).
2. J.Sambrook, E.F.Fritsch, and T.Maniatis, Molecular cloning: A laboratory manual. 2nd edn. Cold Spring Habor Laboratory Press, New York(1989).
3. F.W.Studier, A.H.Rosenberg, J.J.Dunn and J.W.Dubendorff, Use of T7 RNA polymerase to direct expression of cloned genes, Methods of enzymology.185:60(1990).
4. D.H.Bradshaw and P.L.Deininger, Human thymidine kinase gene: Molecular cloning and nucleotide sequence of a cDNA expressible in mammalien cells, Molecular and Cellular Biology.4:2316(1984).
5. B.Munch-Petersen, L.Cloos, G.Tyrsted and S.Eriksson, Diverging substrate specificity of pure human thymidine kinases 1 and 2 against antiviral dideoxynucleosides. Journal of Biological Chemistry.266:9032(1991).
6. R.Okazaki and A.Kornberg, Deoxythymidine kinase of Escherrichia coli: 1. Purification and some properties of the enzyme. Journal of Biological Chemistry.239:269(1964).
7. B.Munch-Petersen, Differences in the kinetic properties of thymidine kinase isoenzymes in unstimulated and phytohemagglutinin stimulated human lymphocytes,Mol. Cell. Biochem.64:173(1984).
8. M.G.Kauffman and T.J.Kelly, Cell cycle regulation of thymidine kinase: residues near the carboxyl terminus are essential for the specific degradation of the enzyme at mitosis. Molecular and Cellular Biology.11:2538(1991).

THE PROMOTER OF THE CHO APRT GENE CONTAINS THREE REGULATORY REGIONS

Bin-Ru She and Milton W. Taylor

Department of Biology
Indiana University
Bloomington, IN 47405

INTRODUCTION

APRT (Adenine phosphoribosyltransferase) catalyzes the formation of AMP from adenine and PRPP. The CHO APRT gene is expressed from a 2.7 Kb Bam HI/ XbaI fragment (Tang and Taylor, 1992). Multiple transcription initiation sites were detected by S1 nuclease mapping and primer extension (Park and Taylor,1988). A strong initiation site, which is 64 nucleotides upstream of the translation start codon, is denoted as +1. Deletion of 5' region of the gene down to -89 has no effect on gene expression, while deletion extended to -60 decreases gene expression to 40% (Park and Taylor, 1988). CHO APRT promoter is therefore defined as the -89 downstream region. The APRT promoter is GC-rich and does not contain obvious TATA or CCAAT boxes. Sp1 binding sites are located at -96, -86 and +43 (She and Taylor, 1991). The promoter also contains two GCF binding sequences. One overlaps with the -86 Sp1 binding sequence and the other one is located at -40.

We have further characterized the APRT promoter sequences by mutational analysis and defined three regulatory regions (see Fig 4). Deletion of region I decreases gene expression to 40%, while linker-scanning mutations in this region do not significantly decrease expression. Mutations in region II reduce gene expression to 14% to 60%. Although deletions of region II greatly reduce gene expression, extensive deletions which eliminates region II and its flanking sequence (hence bringing region I and III into close proximity) restores gene expression to 100%.

In addition, the transcription initiation sites of the APRT gene were re-examined with a more sensitive assay, RNase protection, and were found to be very heterogeneous.

RESULTS AND DISCUSSION

The Transcription Initiation Sites of the APRT Gene

RNase protection is a very sensitive method for mapping the 5' ends of RNA. The probes used in our RNase protection assay corresponds to the -329 to +85 region of the APRT template strand. The negative control was the poly A-RNA isolated from S225 cells (a CHO mutant with both APRT alleles deleted). The positive control was an in vitro synthesized RNA corresponding to the -39 to +159 region of the APRT non-template strand. No band was detected in the lane of negative control and a band was detected at -39 in the lane of the positive control, as expected. Multiple bands were detected in the lane of CHO RNA, with the strongest loci in the +42 to +46 region. Less strong loci were at +33 and +1 (fig 1). The pattern

Purine and Pyrimidine Metabolism in Man VIII, Edited by
A. Sahota and M. Taylor, Plenum Press, New York, 1995

of bands is reproducible through different experiments. The strong bands could also be detected by primer-extension (data not shown).

The extreme heterogeneity of the APRT transcription initiation sites is rare, but not unique. Several genes, such as thymidylate synthase (Jolliff et al. 1991) also have a large number of initiation sites.

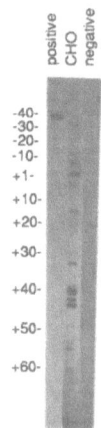

Figure 1 Mapping the transcription initiation sites of the CHO APRT gene by RNase protection. See text for "positive con. " and "Negative con.".

Analysis of the APRT Promoter Sequence by Mutagenesis

A series of linker-scanning and deletion mutations were introduced into the APRT promoter. To study the effect of the mutations on APRT activity, each mutant was cotransfected with pSV2cat into S225 cells. Two days later APRT and CAT enzyme activities were measured, with the latter as an internal control. For selected mutants, the APRT mRNA was analyzed by Northern blot and RNase protection. The APRT enzyme activity correlates well with the APRT RNA quantitatively (data not shown). Base on these results, three regions of the promoter are defined as discussed below.

Region I

Region I is located between -101 to -53. Deletion of this region (-97 to -44) decreases gene expression (APRT enzyme activity and the level of APRT mRNA) to 44%. RNase protection assay shows that the 5' ends of the APRT mRNA remains normal, indicating the decreased activity is due to lowered transcription. Although cumulatively this region is important for the level of transcription, linker-scanning mutations in this region have little effect on gene expression with the exception of mutations around -86.

Linker-scanning mutation around the -86 region (-90 to -78) disrupts Sp1 binding on both the -96 and -86 sites (data not shown), yet increases gene expression two fold. Since this mutation also disrupts a putative binding site for GCF, an ubiquitous transcription repressor (Kageyama and Pastan 1989), it is probable that the increased expression results from the loss of a GCF binding site, instead of Sp1 binding sites. Mutation (-48 to -40) on the other putative GCF binding site also increase gene expression two fold. Disruption of the -96 or +43 Sp1 binding sites has little effect on gene expression (She and Taylor 1991).

Region II

Region II is located between -33 to +19, within which all mutations (with one exception) greatly decrease gene expression. Replacing this region with similar length of linkers reduces the expression to 14%. Linker-scanning mutations in -23 to -14 and -33 to -14 regions reduce gene expression to 21% and 14% respectively. RNase protection assay shows that the former mutation results in a complete loss of transcription initiated in the -23 downstream region (data not shown). Other linker-scanning mutations examined only change the pattern of initiation sites within, but not outside, the mutated region. A specific protein / DNA complex was detected in the -23 to -14 region by gel-shift assays (data not shown).

Region III

Region III is located between +56 and +85. This region is defined by the observation that regions I and III alone (when in proximity) drive transcription as efficiently as the entire 5' region of APRT gene (discussed below).

As mentioned above, region II is important for the APRT transcription. Deletion of region II reduces gene expression to 33%. However, with deletion extended further downstream or upstream the expression increases (Fig 2). Deletion of the -39 to +55 region completely restores gene expression. Gene expression is reduced again when the deletion is extended beyond -52. When fused to a promoter-less CAT gene, regions I and III (but not region I or region III alone) drive transcription as efficiently as the entire 5' region of APRT gene (-329 to +85). These results suggest an interaction between regions I and III, which can be disrupted by the -39 to +55 sequence.

	Element I	Element II	Element III	APRT activity	bp between I and III
Wt				100%	105
Δ -33/+19				33%	74
Δ -33/+49				52%	44
Δ -33/+55				56%	38
Δ -39/+55				104%	23
Δ -48/+55				100%	14
Δ -52/+55				100%	10
Δ -66/+55				38%	*
Δ -75/+55				23%	*
Δ -97/+55				0%	*
Δ ..-102; -52/+55				88%	10

Figure 2 Gene expression of the APRT mutants with region II and the flanking region deleted. Hatched boxes indicate regions I (-101 to -53), II (-33 to +19), and III (+56 to +85). Solid boxes indicate the linkers used for construction, which are shown in order to accurately present the distance between regions I and III. * In these mutants deletion was extended into region III, hence the distance between regions I and III is not shown.

The effect of the proximity of regions I and III on gene expression

The -39 to +55 sequence can disrupt the interaction between regions I and III by a sequence-specific mechanism (binding of a specific protein, e.g.) or by simple spatial separation of the two regions. To differentiate between these two mechanisms, we first removed the sequence between regions I and III (Δ-52 / +55), then inserted different lengths of linkers (spacer) and examined the APRT expression. As shown in Fig 3, spacers of 10 and 20 bp have little effect on gene expression, however when the lengths of spacers are longer than 34 bp the expression decreases. A 74 bp spacer decreases gene expression to 4%. Since the possible palindromic structure of linkers may reduce RNA stability or translation efficiency, we tested another spacer (Δ-52/+55 (random)), which is a randomly picked sequence that does not contain obvious secondary structure or specific protein binding site. This spacer reduces expression to similar level as a linker spacer of the same length (Δ-52/+55(64)). Therefore the distance between regions I and III is important for the APRT expression when region II is deleted.

The initiation sites of transcription driven by regions I and III in the absence of region II are also heterogeneous. The strongest loci are located in the -75 to -88 region, which is a GC box.

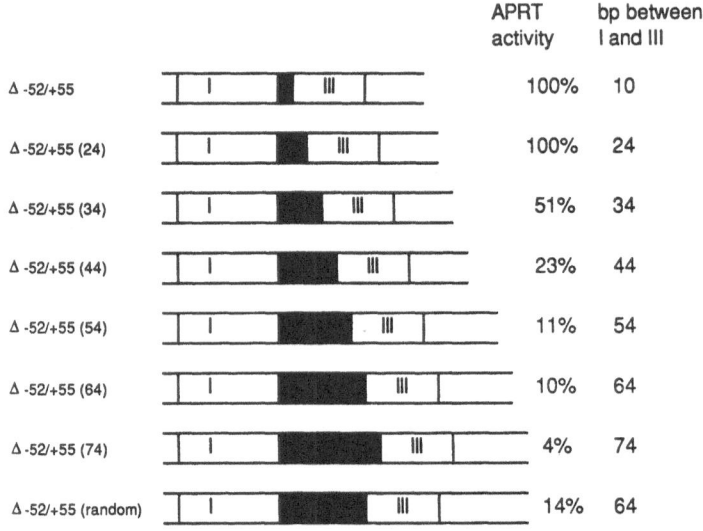

Figure 3 The effect of the proximity between regions I and III on gene expression. Solid boxes indicate the foreign sequences inserted between regions I and III.

The importance of Sp1 binding sequence in gene expression driven by regions I and III

Region I contains two Sp1 binding sites. Although Sp1 binding does not appear to play an essential role in the wild type APRT promoter, it could be important for the expression directed by regions I and III. To test this possibility, the cytosine nucleotides at -95 and -85 were mutated to adenines, which destroys both -96 and -86 Sp1 binding sites. This mutation (pImu-F) indeed greatly reduces the expression driven by regions I and III (Fig 4). It is possible that Sp1 bound on region I interacts with other transcription factors bound on region III to activate transcription and in a wild type APRT promoter it is difficult to achieve such interaction due to the long distance between the two.

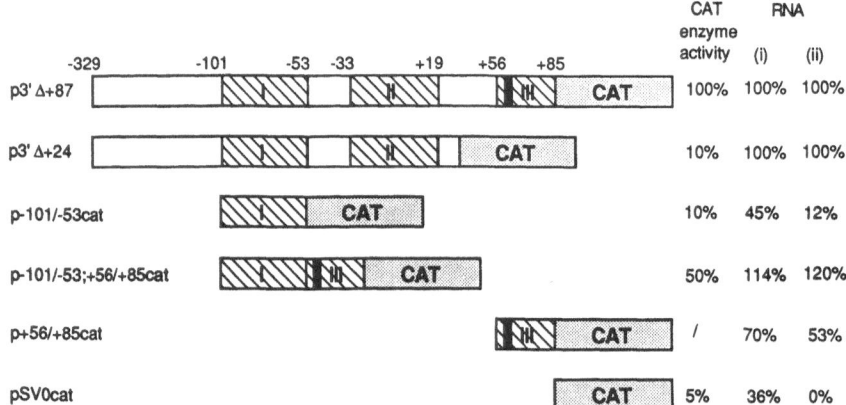

Figure 4 The importance of the Sp1 binding sequences on gene expression driven by regions I and III. The hatched boxes indicate regions I, II and III. The translation start codon (+64) is shown as a solid bar. The dots in region I represent the mutated Sp1 binding sites. In pImu-R, the orientation of region I relative to region III is inverted.

REFERENCES

Jolliff, K., Li, Y., and Johnson L.F., 1991, Multiple protein-DNA interactions in the TATAAA-less mouse thymidylate synthase promoter, *Nucleic Acid Res.* 19:2267.

Kageyama, R., and Pastan, I., 1989, Molecular cloning and characterization of a human DNA binding facttor that represses transcription, *Cell* 59:815.

McKnight, S.L., 1982, Functional relationships between transcriptional control signal of the thymidine kinase gene of Herpes simplex virus, *Cell* 31:355.

Park, J.H., and Taylor, M.W., 1988, Analysis of signals controlling expression of the Chinese hamster ovary APRT gene, *Mol.Cell.Biol.* 8:2536.

Segal, R., and Berk, A.J., 1991, Promoter activity and distance constraints of one versus two Sp1 binding sites, *J. Biol. Chem.* 266, 20406.

She, B.R., and Taylor, M.W., 1991, Analysis of the promoter region of the cho aprt gene. *Advances in Experimental Medicine and Biology*, 309b. Ed. Harkness, Elion and Zollnerr, Plenum Press.

Tang, D.C., and Taylor, M.W., 1990, Transcriptional activation of the adenine phosphoribosyltransferase promoter by an upstream butyrate-induced Moloney murine sarcoma virus enhancer-promoter element, *J.Virol.* 64:2907.

THE MOUSE *APRT* GENE AS A MODEL FOR STUDYING EPIGENETIC GENE INACTIVATION

Mitchell S. Turker[1,2], Padmaja Mummaneni[1], and Gregory E. Cooper[2]

[1]Department of Pathology
[2]Department of Microbiology and Immunology
University of Kentucky
Lexington KY 40536

INTRODUCTION

Somatic cell variation is a broad term that covers heritable changes in phenotype due to mutational or epigenetic changes (Siminovitch, 1976). Mutational changes represent alterations in the primary structure of the DNA sequence and range from events as small as base-pair substitutions to those as large as loss of entire chromosomes. In contrast, epigenetic changes do not affect the primary structure of the DNA sequence, which makes them somewhat more difficult to analyze.

One of the best studied epigenetic mechanisms in mammalian cells is DNA methylation, which is a modification that affects a subset of cytosine residues that are followed by guanine residues. This dinucleotide pair, commonly termed CpG, is relatively rare in the mammalian genome except in the promoter regions of housekeeping and some specialized genes (Bird, 1986). These CpG clusters are termed CpG islands. Methylation of CpG islands has been correlated with loss of transcription (Cedar, 1988). This event is considered normal for genes on the inactive X-chromosome such as *hprt* (hypoxanthine phosphoribosyltransferase) (Grant and Chapman, 1988). However, abnormal methylation of CpG islands has also been reported, such as the observations that the promoter regions of the Wilms' tumor (Royer-Pokora and Schneider, 1992) and retinoblastoma (Sakai et al, 1991) genes are methylated in some human cancers. A number of issues regarding abnormal DNA methylation are still unanswered. These include the origin of the methylation signal, how methylation induces gene inactivation, and the mechanisms, if any, used by promoter regions to block methylation-associated gene inactivation.

The mammalian *aprt* gene has proven to be a versatile locus for the study of somatic cell variation. It is a relatively small gene (2.3 Kbp for the mouse *aprt* gene, Fig. 1A) that is both dispensable and selectable in cultured cells. These features have proven quite useful for mutational studies and a large body of literature exists describing mutations in the hamster, human, and mouse genes (see Khattar et al, 1992 and references therein). Far less common, however, are reports documenting epigenetic events that can affect aprt expression.

Purine and Pyrimidine Metabolism in Man VIII, Edited by
A. Sahota and M. Taylor, Plenum Press, New York, 1995

This manuscript is intended to summarize recent work from our laboratory which has led to the development of the mouse *aprt* gene as a model system for the study of methylation-associated gene inactivation.

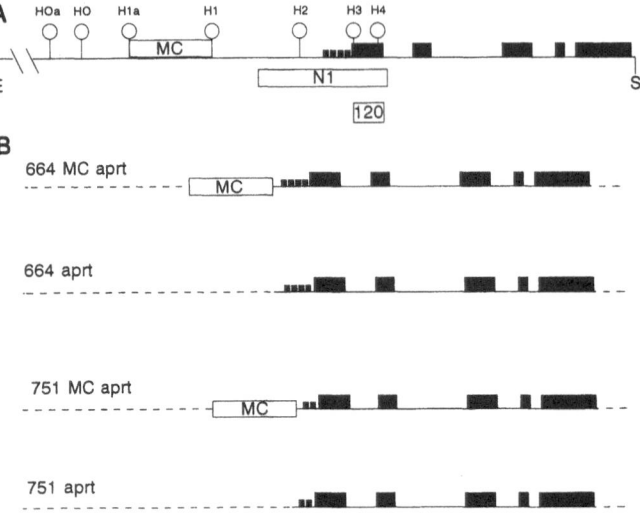

Figure 1. A. The pSam6.3 construct. This construct contains the mouse *aprt* gene, consisting of 5 exons (large closed boxes) and an additional 3.8 Kbp of upstream sequence. The *aprt* promoter, consisting of 4 Sp1 binding sites (small closed boxes, not drawn to scale) is shown in front of the first exon. The bubble figures represent HpaII/MspI restriction sites that are discussed in the text or other figure legends. The 838 bp methylation center is located between the H1a and H1 sites. EcoR1 (E) and Sph1 (S) restriction sites bracket the pSam6.3 insert. The relative locations of the N1 and 120 probes are shown.
B. Deletion constructs used in this study. See text for detail.

RESULTS AND DISCUSSION

We took advantage of allelic variation for *aprt* in the mouse P19 embryonal carcinoma cell line (Turker et al, 1989) to isolate cells with hemizygous deficiencies (i.e with a deletion of one *aprt* allele). During a study in which cells with complete aprt deficiencies were isolated from the *aprt* hemizygotes, we identified a subset that reacquired aprt expression at very high frequencies (10^{-2} to 10^{-3}) (Cooper et al, 1991). A methylation analysis with the restriction enzymes HpaII and MspI revealed that the promoter region of the *aprt* gene was hypermethylated in the non-expressing cells and that this event was reversed in the cells that had reacquired aprt expression (Fig. 2). Further analysis of the methylated *aprt* allele revealed the following: 1) the methylated promoter region was in a closed chromatin conformation (Cooper et al, 1992), 2) Sp1 protein binding did not occur when the promoter region was methylated (Fig. 2), and 3) treatment of cells with 5-azacytidine, a potent demethylating agent, could restore aprt expression to 75-90% of the cells containing the methylated *aprt* allele (Cooper et al, 1993). In light of data from other laboratories demonstrating that methylation of Sp1 binding sites is not sufficient to block Sp1 protein binding in vitro (Harrington et al, 1988), our results suggest that methylation of the *aprt* promoter region alters the chromatin conformation which, in turn, prevents Sp1 protein binding.

Given that methylation-associated inactivation of the mouse *aprt* gene can occur in the P19 cells, it will be important to ultimately determine how this event occurs. One clue comes from our observation that a cis-acting element located upstream of the *aprt* gene provides a bidirectional de novo methylation signal (Mummaneni et al, 1993). This element is termed a methylation center (Fig. 1A). As seen in Fig. 3, the methylation pattern for a transfected pSam6.3 construct (see Fig. 1A) is markedly similar to that observed normally in tissues such as kidney and liver and in the P19H22 cells. This pattern consists of fully methylated H1a and H1 HpaII/MspI sites and partially methylated H0a, H0, and H2 sites. The 838 bp methylation center, located between the H1a and H1 sites, is responsible for this conserved methylation pattern. The exception to the conserved methylation pattern is seen for sperm DNA, which contains fully methylated H0a, H0, H1a, and H1 sites and a non-methylated H2 site as revealed by a 3.4 Kbp hybridization band (Fig. 3).

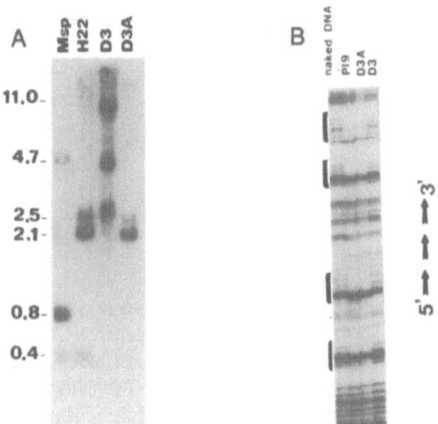

Figure 2. A. Epigenetic inactivation of the mouse aprt gene is associated with hypermethylation. DNA preparations from the *aprt* hemizygote P19H22 (H22), the non-expressing cell line D3 (derived from H22), and the revertant cell line D3A were digested with HpaII, Southern blotted, and hybridized with the N1 probe (Fig. 1A). MspI digested P19H22 DNA is also shown (Msp). B. The methylated promoter region does not bind the Sp1 protein. The method of in vivo footprinting (Garrity and Wold, 1992) was used to demonstrate that the four the Sp1 binding sites that comprise the *aprt* promoter (in brackets) are protected in cells that express aprt (P19 and D3A), but they are not protected in the non-expressing D3 cell line. The dark band in each protected region represents a guanine residue that becomes hypersensitive when bound to Sp1.

Based on the above data, we hypothesized that the methylation center signal could contribute to epigenetic inactivation of the *aprt* gene. To test this hypothesis several constructs were created in which the methylation center was moved next to the *aprt* promoter. These constructs are shown in Fig. 1B. The 664MC construct contains the methylation center placed in front of all 4 Sp1 binding sites. Two of these binding sites were removed to create 751MC. We note that only 2 Sp1 binding sites are necessary for maximal aprt expression (Dush et al, 1988). As controls, the methylation center was removed from the 664MC and 751MC constructs, to create 664 and 751, respectively. For a first experiment, all 4 of the above constructs were transfected via electroporation into DelTG3 cells (P19 cells in which both *aprt* alleles are deleted) and the transfection efficiency for aprt expression measured by selecting transfectants with medium containing azaserine and adenine (AzA). The transfection efficiencies for 664, 664MC, and 751 were similar, being approximately 12-15 X 10^{-5}. However, there was a 3-4 fold reduction in the transfection efficiency for the 751MC construct (data not shown). Two tentative conclusions were drawn from this result. The first is that inactivation of the 751MC construct occurs approximately 75% of the time when it is integrated into the DelTG3 genome and that this event is induced by the presence of the cis-acting methylation center. The second is that the presence of 4 Sp1 binding sites is sufficient to block the inactivation event.

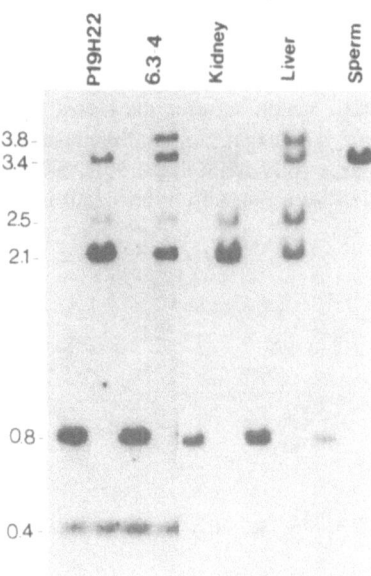

Figure 3. The methylation pattern for the region upstream of the mouse *aprt* gene. DNA preparations from the P19H22 cell line, the 6.3-4 cell line, mouse kidney, liver, and sperm DNA were digested with EcoR1 and MspI (left hand lane) or EcoR1 and HpaII (right hand lane), Southern blotted, and hybridized with the N1 probe (Figure 1A). The 6.3-4 cell line represents the DelTG3 cell line (P19 cells that lack both *aprt* alleles) transfected with the pSam6.3 plasmid. This result demonstrates that the endogenous methylation pattern can be reproduced via de novo methylation in cultured cells. The interpretation of the HpaII hybridization bands is as follows: 1) the 2.1 Kbp hybridization band contains methylated H1a and H1 sites, 2) the 2.5 Kbp hybridization band contains methylated H1a, H1, and H2 sites, 3) the 3.4 Kbp hybridization band contains methylated H0a, H0, H1a, and H1 sites, and 4) the 3.8 Kbp hybridization band contains methylated H0a, H0, H1a, H1, and H2 sites. The relative locations of these sites is shown in Fig. 1A.

To demonstrate directly epigenetic inactivation of the 751MC construct, the 751 and 751MC constructs were again introduced into the DelTG3 cells, except that this time they were cotransfected with the bacterial *neo* gene and G418 resistant cells selected. This approach, therefore, did not bias for or against aprt expression. The G418 resistant cells were then screened with a Southern blot analysis for the presence of the 751 or 751MC constructs. In this way, 6 transfectants containing the 751 construct and 7 transfectants containing the 751MC construct were identified. A cell-free assay for aprt enzymatic activity revealed that all 6 of the transfectants containing the 751 construct expressed high levels of aprt activity, but that only 2 of 7 transfectants containing the 751MC construct expressed detectable levels of the aprt enzyme. This result confirmed that inactivation of the 751MC construct could occur and that the methylation center was responsible. One transfectant (751MC-302) was then chosen for further analysis. The 751MC-302 cells were treated with 5-azacytidine and expressing cells were selected with AzA medium. Clones capable of growth in the AzA medium were found in the azacytidine treated cells, but not in untreated cells, demonstrating that inactivation of the 751MC construct in the 751MC-302 cells was reversible and that DNA methylation was involved in this event. We next examined DNA preparations from the 751MC-302 cells and two azacytidine induced revertants for DNA methylation using a promoter region specific probe. As controls, DNA preparations from the D3 cell line, which contains a methylated *aprt* allele, and the D3A cell line, which contains a non-methylated *aprt* allele (see Fig. 2), were also examined. Fig. 4 shows significant methylation for the 751MC construct in the 751MC-302 cell line, but not for the revertant cell lines R2 and R4. This result confirms that epigenetic inactivation of the 751MC construct was associated with high levels of DNA methylation.

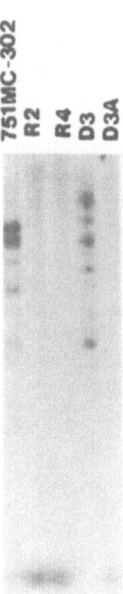

Figure 4. Methylation of a transfected 751MC construct correlates with expression. DNA preparations from the 751MC-302 cell line, the azacytidine induced revertants R2 and R4, the D3, and the D3A cell lines were digested with HpaII and Southern blotted with the 120 probe (Fig. 1A).

In summary, we have demonstrated that epigenetic inactivation of the mouse *aprt* gene can be induced by a cis-acting methylation center that normally is found 1.2 Kbp upstream of the *aprt* promoter. Our results have also suggested that the presence of 4 Sp1 binding sites in the *aprt* promoter may have evolved to prevent this inactivation event. Our work further expands the versatility of the mammalian *aprt* gene as a model for the analysis of somatic cell variation.

ACKNOWLEDGEMENTS

This work was supported by NIH grants AG08199 and CA56383.

REFERENCES

Bird, A.P., 1986, CpG-rich islands and the function of DNA methylation, *Nature* 321:209.

Cedar, H., 1988, DNA methylation and gene activity, *Cell* 53:3.

Cooper, G.E., DiMartino, D.L., and Turker, M.S., 1991, Molecular analysis of APRT deficiency in mouse P19 teratocarcinoma stem cell line, *Somat. Cell Mol. Genet.*, 17:105.

Cooper, G.E., Khattar, N.H., Bishop, P.L., and Turker, M.S. 1992 At least two distinct epigenetic mechanisms are correlated with high frequency "switching" for APRT phenotypic expression in mouse embryonal carcinoma stem cells. *Somat. Cell. Mol. Genet.* **18**:215-225.

Cooper, G.E., Bishop, P.L., and Turker, M.S., 1993, Hemidemethylation is sufficient for chromatin relaxation and transcriptional activation of a methylated *APRT* gene in mouse P19 embryonal carcinoma cell line, *Somat. Cell Mol. Genet.*, 19:221.

Dush, M.K., Briggs, M.R., Royce, M.E., Schaff, D.A., Khan, S.A., Schaff, D.A., Tischfield, J.A., and Stambroodk, P.J., 1988, Identification of DNA sequences required for mouse APRT gene expression, *Nucl. Acids. Res.* 16:8509.

Garrity, P.A., and Wold, B.J., 1992, Effect of different polymerases in ligation mediated PCR: enhanced genomic sequencing and in vivo footprinting, *Proc. Natl. Acad. Sci.* 89:1021.

Grant, S.G. and Chapman, V.M., 1988, Mechanisms of X-chromosome regulation, *Ann. Rev. Genetics* 22:199.

Harrington, M., Jones, P.A., Imagawa, M., and Karin, M., 1988, Cytosine methylation does not affect binding of transcription factor Sp1, *Proc. Natl. Acad. Sci.* 85:2066.

Khattar, N.H., Cooper, G.E., DiMartino, D.L., Bishop, P.L., and Turker, M.S., 1992, Molecular and biochemical elucidation of a cellular phenotype characterized by adenine analog resistance in the presence of high levels of adenine phosphoribosyltransferase activity, *Biochem. Genet.* **30**:635.

Mummaneni, P., Bishop, P.L., and Turker, M.S., 1993, A cis acting element accounts for a conserved methylation pattern upstream of the mouse adenine phosphoribosyltransferase gene, *J. Biol. Chem.* **268**:552.

Royer-Pokora, B. and Schneider, S., 1992, Wilms' tumor-specific methylation pattern in 11p13 detected by PFGE, *Genes Chromosomes Cancer* 5:132.

Sakai, T., Toguchida, J., Ohtani, N., Yandell, D.W., Rapaport, F.M., and Dryja, T.P., 1991, Allele-specific hypermethylation of the retinoblastoma tumor-suppressor gene, *Am. J. Hum. Genet.* 48:880.

Siminovitch, L., 1976, On the nature of hereditable variation in cultured somatic cells, *Cell* 7:1.

Turker, M.S. Stambrook, P.J., Tischfield. J.A., Smith, A.C., and Martin, G.M., 1989, Allelic variation linked to adenine phosphoribosyltransferase locus in mouse teratocarcinoma cell line and feral-derived mouse strains, *Somat. Cell Mol. Genet.* 15:159.

ANALYSIS OF IN VIVO SOMATIC MUTATIONS AT THE APRT LOCUS

P.K. Gupta[1], A. Sahota[1], S.A. Boyadjiev[1], S. Bye[1], J.P. O'Neill[2], T.C. Hunter[2], R.J. Albertini[2], and J.A. Tischfield[1]

Department of Medical & Molecular Genetics, Indiana University School of Medicine, Indianapolis, Indiana[1]; Genetics Laboratory, University of Vermont, Burlington, Vermont[2]

INTRODUCTION

Adenine phosphoribosyltransferase (*APRT*; E.C.2.4.2.7) catalyzes the synthesis of AMP from adenine and 5-phosphoribosyl-1-pyrophosphate. A deficiency of this enzyme leads to 2,8-dihydroxyadenine stone formation in the kidney. The APRT gene is located at 16q24.3 and its genomic sequence size is 2.6 kb and coding sequence 0.54 kb. It has been used extensively to examine mammalian mutagenesis (1-4). Earlier studies have focused on spontaneous and induced mutations in cultured cells. We have recently analyzed germline mutations from more than 33 patients (5,6). Germline mutations seem to cluster at the intron 4 splice site and few other sites different than the *in vitro* mutagenesis results in cultured cells. These studies suggested differences in mutagenesis pathways. Therefore, it was of interest to examine another group of mutations which also arise *in vivo* in human cells. We initiated studies to examine the nature of *in vivo* somatic mutations in T lymphocytes from heterozygote (*APRT* [+/-]) individuals.

Germline mutations exhibited predominantly intragenic point mutations. In contrast, *in vivo* somatic mutations which may arise later in life showed a predominance of the APRT allele loss (5,6). This observation parallels the loss of heterozygosity (LOH) pathway commonly found during carcinogenesis (7,8). Therefore, our *in vivo* somatic mutation studies in the APRT locus could also be expanded to examine the molecular mechanisms of loss of heterozygosity.

Purine and Pyrimidine Metabolism in Man VIII, Edited by
A. Sahota and M. Taylor, Plenum Press, New York, 1995

MATERIALS AND METHODS

T Lymphocyte Assay

Lymphocyte mutation assays are well characterized and have been used to enumerate mutations in human populations (9,10). T lymphocytes with *in vivo* somatic mutation at the *APRT* $^{-/-}$ in the heterozygote individual (*APRT* $^{+/-}$) were selected as follows. Lymphocytes were isolated from peripheral blood by density sedimentation and pre-existing *APRT* $^{-/-}$ T lymphocytes were selected by immediately plating the isolated lymphocytes in a media containing 100 μM 2,6-diaminopurine (DAP). Individual DAP resistant clones were identified, scored and further propagated to isolate genomic DNA for analysis. Isolated lymphocytes were also plated in parallel cultures in media containing 10 μM 6-thioguanine (TG) to identify TG resistant clones and to analyze mutations in the HPRT gene.

Molecular Analysis of in vivo Mutations

In vivo somatic mutations may comprise of intragenic point mutations or large chromosomal changes that encompass the entire APRT gene. Therefore, DAP resistant clones were first analyzed to distinguish between localized and large chromosomal changes. We utilized CA microsatellite repeat polymorphism to initially classify these changes. Numerous CA mirosatellite repeat markers have been identified on chromosome 16 and were used as follows (11-13).

Primers were either obtained from Research Genetics or synthesized in our laboratory. The PCR reaction conditions were as follows: 1 μg of genomic DNA, 100 ng of each primer, 50mM KCl, 10mM Tris-HCl (pH 8.3), 1.5mM MgCl$_2$, 0.2mM each dATP, dGTP, dCTP and dTTP, and 0.5 unit Taq polymerase (Perkin/Elmer Cetus). Samples were overlaid with mineral oil and were processed through 30 temperature cycles consisting of 60 sec at 95°C, 60 sec at 65°C and 30 sec at 72°C. PCR products were first visualized on 1% agarose gel. Aliquots of the amplified DNA were mixed with 2 volumes of loading gel buffer containing 95% formamide, heat denatured at 98°C for 5 min, and electrophoresed on a 5% denaturing polyacrylamide DNA sequencing gel. Following electrophoresis, DNA fragments were visualized using a chemiluminescent probe (CA)$_{11}$ bound to horseradish peroxidase enzyme. All the reagents for chemiluminescent probe detection were obtained from Amersham Co. and reaction conditions as described by the supplier were followed.

Retention of heterozygosity of the markers, D16S303 and D16S305 which closely flank the APRT locus suggested intragenic point mutations, the exact nature was determined by amplification, cloning and sequencing of the APRT gene. On the other hand, a loss of heterozygosity of these markers suggested large chromosomal changes. These clones were analyzed with additional CA microsatellite repeats markers proximal and distal to the APRT gene.

RESULTS

The *APRT* and *HPRT* mutant frequencies in two families, father (*APRT* $^{+/-}$) / mother (*APRT* $^{+/-}$) / child (*APRT* $^{-/-}$) were examined. Obligate heterozygote parents showed DAP-resistant (*APRT* $^{-/-}$) frequencies which were different and varied from 1- to 7- folds among the four individuals. However, as expected the TG-resistant frequency exhibited a range of 2- to 4- fold in same individuals and is consistent with previous population data (10). Though this data is preliminary, results indicate a greater variation of DAP-resistant frequency as compared to TG-resistant frequency and these differences may be, in part, due to the nature of mutation at these loci.

All of the CA repeat markers were first tested for their heterozygosity using genomic DNA from the obligate heterozygote as shown in Figure 1. D16S261 is heterozygous in MD and FD cells but homozygous in HTD114 cells while D16S283 is homozygous in MD cells and heterozygous in FD and HTD114 cells. Preliminary analysis of 84 DAP resistant T lymphocyte clones indicated that 30% may elicit intragenic point mutations while large chromosomal changes were found in the remaining clones. Studies are in progress to further characterize the nature of mutation in these clones.

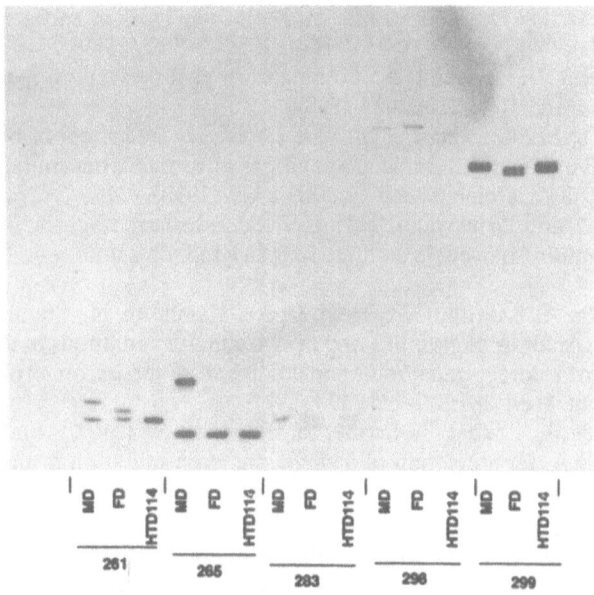

Figure 1. Analysis of CA microsatellite repeat polymorphism. Two bands indicate two alleles (heterozygous) and one band (homozygous).

DISCUSSION

We have chosen to utilize human T lymphocytes from normal individuals who are heterozygous at the APRT locus. The germline mutation in these individuals is known and the origin of *in vivo* somatic mutation in the remaining APRT allele has been examined. The choice of T lymphocytes is ideal because they can be easily obtained in a sample of peripheral blood and can be propagated *in vitro* as clones from a single cell. T lymphocytes grown *in vitro* maintain a normal karyotype and have both DNA and mRNA for genetic analysis. Furthermore, we have also measured mutation at X-linked *HPRT* in the same assay to complement the *APRT* data.

The APRT gene provides a suitable mutagenesis assay. Both germline and *in vivo* somatic mutations in human cells can be examined. *In vitro* mutagenesis studies in mammalian cells differ significantly than mutations found in human *APRT* arising *in vivo* in both somatic and germline cells (5,6). Therefore, it seems reasonable to assume that mutational mechanisms in human will be elucidated by studying human populations and analyzing results in the context of other species. Our *in vivo* somatic mutation studies in human as well as additional animal model mutagenesis assays may provide insights into the mechanism of mutagenesis.

Acknowledgements

Supported by NIH grants DK38125, CA30688, and USDOE FG028760502.

REFERENCES

1. Phear, G., Armstrong, W., and Meuth, M. Molecular basis of spontaneous mutation at the APRT locus of CHO cells. J. Mol. Biol. 209:577-582 (1989).
2. Hakoda,M., Hirai, Y., Akiyama, K. Cloning of in vivo-derived thioguanine-resistant human B cells. Mut. Res. 210:29-34 (1989).
3. Ward, M.A., Yu, M., Glickman, B.M., and Grosovsky, A.J. Loss of heterozygosity in mammalian cell mutagenesis: molecular analysis of spontaneous mutations at the APRT locus in CHO cells. Carcinogenesis 11:1485-1490 (1990).
4. Klinedinst, D.K., and Drinkwater, N.R. Reduction to homozygosity is the predominant spontaneous mutational event in cultured human lymphoblastoid cells. Mut. Res. 250:365-374 (1991).
5. Chen, J., Sahota, A., Maritn, G.F., Hakoda, M., Kamatani, N., Stambrook,P.J. and Tischfield, J.A. Analysis of germline and in vivo somatic mutations in the human adenine phosphoribosyltransferase gene: mutational hotspots at the intron 4 splice donor site and at codon 87. Mut. Res. 287:217-225 (1993).
6. Chen, J., Sahota, A., Laxdal, T., Scrine, M., Bowman, S., Cui, C., Stambrook,   P.J. and Tischfield, J.A. Identification of a single missense mutation in the adenine phosphoribosyltransferase (APRT) gene from five Icelandic patients and a British patient. Am. J. Hum. Gen. 49:1306-1311 (1991).
7. Knudson, A.G. Heriditary, cancer, oncogenes and antioncogenes Can. Res. 45:1437-1443 (1985).
8. Stanbridge, E.J. Human suppressor genes. Ann. Rev. Gen. 24:615-650 (1990).
9. Albertini, R.J., Nicklas, J.A., O'Neill, P.J. and Robison, S.H. In vivo somatic mutations in humans: measurement and analysis. Ann. Rev. Gen. 24:305-326 (1990)
10. Branda, R.F., Sullivan, L.M., O'Neill, J.P., Falta, M. T., Nicklas, J.A., Hirsch, B., Vacek, P.M., and Albertini, R.J. Measurement of HPRT mutant frequencies in T lymphocytes from healthy human populations. Mut. Res. 285:267-281 (1993).
11. Weber, J.L. Informativeness of human (dC-dA)n.(dG-dC)n polymorphisms. Genomics 7:524-530 (1990).
12. Thompson, A.D., Holman, K., Sutherland, G.R., Callen, D.F., and Richards, R.I., Isolation and charaterization of (ACOn microsatellite genetic markers from human chromosome 16. Genomics 13:402-408 (1992).
13. Weissenbach, J., Gyapay, G., Dib, C., Vignal, A., Morissette, J., Milasseau, P., Vaysseix, G., and Lathrop, M. A second generation of linkage map of the human genome. Nature 359:794-801 (1992).

IDENTIFICATION OF POLYMORPHIC MARKERS FLANKING THE HUMAN APRT GENE

Simeon A Boyadjiev, Amrik Sahota, and Jay A Tischfield

Department of Medical and Molecular Genetics, Indiana University
School of Medicine, Indianapolis, IN 46202, USA

INTRODUCTION

Because of its small size and selectable nature, the human adenine phosphoribosyltransferase (APRT) gene has been widely used as a model system for studying mutational events in somatic cells *in vitro* (1). We are currently using this gene as a model to study mutational events in somatic cells *in vivo*. Because of the autosomal nature of *APRT*, these events are best studied in obligate heterozygotes (eg, parents of APRT-deficient patients). This can be done by culturing T lymphocytes from the heterozygotes in the presence of diaminopurine (DAP) and then analyzing drug-resistant clones for mutations.

The above studies require the availability of informative markers to distinguish the *APRT* allele with the germline mutation from the allele that has undergone mutation in somatic cells *in vivo*. The markers currently available are: an *Sph*I RFLP 5' of *APRT* (2); a *Taq*I RFLP in intron 2 (3); and *Bgl*II (4) and *Taq*I (5) RFLPs downstream of the gene, but only the first two markers have high enough allele frequencies to be useful.

We have initiated a search for additional markers closely flanking *APRT*. A 3.0 kb region (1.2 upstream of *APRT* and 1.8 kb downstream) from *APRT* heterozygotes was amplified by PCR and subjected to heteroduplex and SSCP analyses. Fourteen polymorphisms were detected in these regions and, together with microsatellite markers further away from *APRT*, they are being used to study *in vivo* somatic mutations in *APRT*.

SUBJECTS

Our group consists of 16 *APRT* obligate heterozygotes and a wild-type control.

METHODS

A 1.2 kb region upstream of the initiation codon and a 1.8 kb region downstream of the termination codon was analyzed by PCR (6). The upstream region was amplified in three reactions and the downstream region in six reactions. P32-a-dCTP was included in the reaction mixture to label the PCR products.

The PCR products (after dilution) were analyzed by SSCP on nondenaturing gels (5% acrylamide, 5% glycerol) at room temperature (7). For HD analysis (8), an additional reannealing step was performed and the samples analyzed as above. All fragments showing

Purine and Pyrimidine Metabolism in Man VIII, Edited by
A. Sahota and M. Taylor, Plenum Press, New York, 1995

abnormal migration were sequenced directly from asymmetric PCR products or after subcloning into M13mp18.

RESULTS

DNAs from 17 individuals (16 *APRT* heterozygotes, and one control) were examined for polymorphisms by SSCP and HD. Eleven individuals, including the control, were of Caucasian origin and six were Japanese. The result are summarized in Table 1.

Polymorphisms in the 5'-flanking region of *APRT*

The 5'-flanking region of *APRT* has been sequenced previously (Chen J., unpublished data) and submitted into GENBANK under accession number U09817. The 5' primer and polymorphism locations are numbered according to this sequence. Six polymorphisms were identified in this region. Three of the polymorphisms in the 5'-1 region were located in an *Alu*-like sequence, and they were found to segregate together. A G2104C transversion, which abolishes an *Hph*I site, was the only polymorphism in the 5' region amenable to restriction enzyme analysis of the PCR product. The T2526G and C2736T substitutions were present at high frequency (80%) in the analyzed samples, suggesting that *APRT* from the control individual is a sequence variant.

Table 1. Summary of the polymorphisms in the 5-' and 3-' regions of *APRT*.

Region	SSCP	HD	Polymorphism	Caucasian	Japanese	Freq.
5'-1	-	+	A2100G	+	+	9/26
	-	+	G2104C	+	+	9/26
	-	+	C2172A	+	+	9/26
	-	-	T2526G	+	+	8/10
5'-2	+	-	C2736T	+	+	16/20
5'-3	+	+	C3110A	+	+	6/24
3'-1	-	+	A9G	+	-	2/30
3'-2	-	+	A266G	+	-	2/26
3'-3	+	+	Tins680	+	-	2/30
	+	+	G687T	+	-	2/30
3'-5	-	-	C1048T	+	-	1/30
	+	+	G1157A	+	+	7/30
	+	+	G1206C	+	-	1/30
	+	+	G1303C	+	+	9/30

The C2736T and C3110A polymorphisms were also observed in two Icelandic heterozygotes. This would allow more detailed genetic analyses of the *APRT* locus in this population which contains the Asp-65-Val mutation (possibly due to a founder effect).

The last 5' polymorphism (C3110A) was found in the GC box closest to the first codon of *APRT*, suggesting that this GC box may not be an important regulatory element.

Polymorphisms in the 3'-flanking region of *APRT*.

The sequence of the 3' region was obtained from plasmid p8.6 (9), which contained an 8.6 kb *Hinc*II genomic fragment cloned into plasmid pIBI20. By sequencing the double stranded plasmid DNA, a total of 1656 bp of new 3'-flanking genomic sequence was obtained (GENBANK accession number U04709).

We analyzed 1847 bp of 3'-sequence: 324 bp 3' of the end of *APRT* exon 5 (containing the polyadenylation site) and 1523 bp of new sequence. This region was amplified in 6 separate PCR reactions yielding overlapping fragments between 217 and 450 bp .
The 3' polymorphism locations are numbered according to the GENBANK sequence.

Eight polymorphisms were identified in this region (Table 1). Two of the polymorphisms abolished restriction sites, including the previously described *Bgl*II site (4). Two polymorphisms were located within an inverted *Alu*-like repeat sequence, one was a single base insertion, and the remainder were single base substitutions.

Four of the polymorphisms (C1048T, G1157A, G1206C, and G1303C) were located in the same PCR fragment (region 3'-5) and they gave rise to characteristic heteroduplex migration patterns , depending on the number of substitutions present in each sample.

The allele frequencies in the 5' and 3' regions varied from 3.3% to 80%. One of the alleles from a control individual contained all eight polymorphisms. Seven individuals (41%) were heterozygous for at least one of the polymorphisms.

DISCUSSION

HD and SSCP are useful and rapid screening procedures for determining sequence variations in DNA. In this study 12/14 variations (85.7%) were detectable by HD but only seven (50.0%) by SSCP. No false positive migration shifts were observed. Two sequence changes were found by DNA sequencing only.

Our *APRT* allele frequency data is limited, but it suggests that the observed variations are more likely to be due to polymorphisms rather than to rare variants or mutations. The polymorphisms occurred with an average frequency of 1/200 bp, suggesting that these regions are likely to contain non-coding rather than coding sequences.

No association between a given polymorphism and a germline mutation was observed. Most of the *APRT* polymorphisms were present in both Caucasian and Japanese heterozygotes as well in the control individual. This suggests that the polymorphisms arose before the mutations and probably before racial divergence.

The combined use of these polymorphisms enabled us to distinguish the two *APRT* alleles in seven of the 16 heterozygotes (41%). No polymorphisms were found in seven subjects, and two Icelandic *APRT* heterozygotes were found to be homozygous for some of the 5' region polymorphisms. In addition to their application in somatic mutation studies, these polymorphisms may also be useful for studying population migrations and racial divergence.

REFERENCES

1. De Jong,P. J., A. J. Grosovsky and B.W. Glickman (1988) Spectrum of spontaneous mutation at the APRT locus of Chinese hamster ovary cells: An analysis at the DNA sequencing level. Proc. Natl. Acad. Sci. USA,85, 3499-3503.

2. Arrand J. E. Murray A. M. Spurr N. (1987) *Sph* I restriction fragment length polymorphism on human chromosome 16 detected with an APRT gene probe. Nucl. Acids Res. 15(22):9615.

3. Stambrook P. J., Dush M. K., Trill J. J. Tischfield J. A. (1984) Cloning of a functional human adenine phosphoribosyltransferase (APRT) gene: identification of a restriction fragment length polymorphism and preliminary analysis of DNAs from APRT-deficient families and cell mutants. Somatic Cell Mol. Genet. 4: 359-367.

4. Ogasawara N., Goto H. (1989): Restriction fragment length polymorphisms of HPRT and APRT genes in Japanese population. Adv.Exp. Med.Biol. 253A: 461.

5. Kamatani N., Kuroshima S., Hakoda M., Papella T. D., Hidaka Y. (1990) Crossovers within a short DNA sequence indicate a long evolutionary history of *APRT**J mutation. Hum. Genet.85:600.

6.Erlich H. A. (Ed.) (1989) PCR technology: Principles and Applications of DNA amplification, Stockton Press, New York.

7. Orita M., Suzuki Y., Sekiya T., and Hayashi K. (1989) Rapid and sensitive detection of point mutations and DNA polymorphisms using the polymerase chain reaction. Genomics 5,874-879.

8. Nagamine C. M., Chan K., and Lau Y. C. (1989) A PCR artefact: generation of heteroduplexes. Am. J. Hum. Genet. 45,337-339.

9. Chen J., Sahota A., Stambrook P.J., and Tischfield J. (1991) Polymerase chain reaction amplification and sequence analysis of human mutant adenine phosphoribosyltransferase genes: The nature and frequency of errors caused by *Taq* DNA polymerase. Mut. Res., 249,169-176.

GERMLINE AND SOMATIC MUTATION AT THE APRT LOCUS OF MICE AND MAN

J.A. Tischfield,[1] S.J. Engle,[1] P.K. Gupta,[1] S. Bye,[1] S. Boyadjiev,[1] C. Shao,[1] P. O'Neill,[2] R.J. Albertini,[2] P.J. Stambrook,[3] and A.S. Sahota[1]

[1]Medical and Molecular Genetics, Indiana University School of Medicine, Indianapolis IN 46202
[2]Genetics Laboratory, University of Vermont, Burlington VT 05401
[3]University of Cincinnati, College of Medicine, Cincinnati OH 45267

INTRODUCTION

Somatic cell genetics has its origins in the notion that it is possible to understand gene expression in mammalian cells by *in vitro* manipulations similar to those used with bacterial cultures. For example, cultures can be propagated in logarithmic fashion and single cell phenotypic variants can be subsequently isolated as clones. It was further assumed that phenotypic variation in cultured mammalian cells would arise by mechanisms similar to those proposed for such events *in vivo*. Thus it was anticipated that epigenetic variation similar to that observed in organismal ontogeny would be common. However, most early reports described phenomena that have now been demonstrated to be the result of mutations rather than epigenetic changes.

Foremost in determining this outcome was the choice of phenotypes studied. Most of the mammalian cell drug resistance phenotypes (1) that have been the traditional grist for somatic cell geneticists are the consequence of the expression of genes that exhibit minimal regulation during development. Thus, it is unlikely that these genes would be regulated in the same manner as, for example, muscle myosin genes. In retrospect, the work of Chu and Malling (2), showing that the frequency of 8-azaguanine "variants" was increased by mutagen treatment, and that of Lieberman and Ove (3) and Atkins and Gartler (4), demonstrating by independent methods that 2,6-diaminopurine (DAP) resistant variants arise independently of DAP selection, suggested that mutation would likely be a primary mechanism for these kinds of drug resistance. While many recent studies of DAP resistance *in vitro* demonstrate changes in the APRT gene DNA sequence (point mutations) there are other studies which implicate larger chromosomal events (*e.g.*, 5,6,7,8) or epigenetic mechanisms (*e.g.*, 9). We shall review our own work which demonstrates both point mutations and chromosomal events in humans and cultured human cells and describe our efforts to produce a mouse model for further, more controlled studies.

Germline and Somatic Cells *In Vivo*

Hakoda *et al.*(7) studied T cells of *APRT* heterozygotes and reported that loss of heterozygosity (LOH) at *APRT in vivo* was due to "loss" of the normal *APRT* in about 80 percent of clones. Chen *et al.* (10) subsequently sequenced genomic *APRT* in 10 clones of the remaining 20 percent and described a series of point mutations that were similar in nature and position to those observed in 18 different mutant alleles from APRT-deficient patients. Mutant *APRT*s from 7 additional T cell clones derived from two non-Japanese heterozygotes show a similar spectrum of point mutations (Boyadjiev *et al.*, unpublished). APRT deficient T cells have an incidence of about 10^{-4} in heterozygotes, although it may be up to five fold lower in certain individuals, whereas in homozygotes their incidence is less than 10^{-7}.

T cells from the above referenced non-Japanese were observed to undergo "loss" of the normal *APRT* in about 75 percent of clones as determined by analysis of the *Taq*I RFLP in *APRT* intron 2 or an upstream RFLP (Boyadjiev *et al.*, this volume). To date, 85 such clones have been analyzed for their extent of LOH through the use of polymorphic CA microsatellite repeats spaced at various intervals along the entire length of chromosome 16. The data indicate that LOH is restricted to 16q and that it can be as small as several Mb of DNA encompassing *APRT* and nearby markers, or it can be as large as about 90 percent of 16q. Furthermore, LOH may begin within a large number of regions along the length of 16q (Gupta *et al.*, this volume). Chromosome studies of 10 of these T cell clones using Giemsa banding or FISH with a cocktail of "painting" probes indicate the presence of two normal chromosomes 16. In at least 7 of the 10 instances, analysis with the microsatellite repeats indicates that deletion or translocation should be microscopically detectable were it the cause of the relatively extensive LOH. Using FISH with cosmid probes specific for the *APRT* region, however, we have demonstrated a submicroscopic deletion in one clone (Shao *et al*, unpublished). Thus, these data suggest that mitotic recombination is the primary cause of LOH in T cells *in vivo*.

The above data suggest that *APRT* is a good surrogate for studying the types of events that result in LOH at tumor suppressor loci. Its position on distal 16q, its small size (540 bp coding region and 2.2 kb gene), and the ease with which APRT-deficient clones can be selected support this notion. It may also be significant that regions close to *APRT* frequently exhibit LOH in advanced hepatocellular carcinoma (11) and breast cancers of various stages (12).

A Human Transformed Cell Line *In Vitro*

Analyses similar to those described above for T cell clones have been done with a HT-1080-derived, fibrosarcoma cell line heterozygous at *APRT* (Gupta *et al.*, unpublished). Twenty spontaneously arising APRT-deficient clones were analyzed. In 9 clones, heterozygosity of *APRT* and two flanking markers about 9 cM apart was retained. In 3 clones point mutations within *APRT* were demonstrated, and we presume that this is likely to be the case for the remainder (13). In 11 clones, major chromosomal changes such as chromosome loss, mitotic recombination, interstitial deletion, and deletion of 16q extending up through the centromere with translocation of the distal part of 16p were observed. As a group, these clones can be distinguished from the T cell clones by the occurrence of translocations, large deletions, or chromosome loss. We propose that these differences arise because these clones, which are derived from a permanent cell line, are significantly less karyotypically stable than the primary T cells in short-term culture. Most of the fibrosarcoma cell clones show a large subtetraploid cell population in addition to pseudodiploid cells. It is unclear as to whether or not these cells are a good model for LOH events in tumors.

A Mouse Model for *APRT* LOH

A mouse heterozygous at *aprt* would permit the study of spontaneous as well as *induced* LOH in T cells and *other tissues*. In addition, such a mouse could, presumably, be bred to homozygosity for other kinds of mutation analyses involving *aprt* (*e.g.*, reversion) and for studies of purine metabolism. For example, it was reported that the chemical inhibition of APRT in *hprt* mice produces behavioral abnormalities homologous to those seen with Lesch Nyhan syndrome (14). An *hprt* , *aprt* , mouse could test this observation and be a valuable model for studying the bizarre behavior associated with this inherited human disease.

To the above end, we have produced mouse D3 embryonic stem (ES) cells in which one allele of *aprt* is inactivated by virtue of a bacterial *neo* gene interrupting exon 3. This was accomplished by homologous recombination with an introduced vector containing a *neo*-interupted *aprt* exon 3 flanked by approximately 2.5 kb of genomic DNA on each side. A terminal HSV thymidine kinase gene was included in the targeting vector to permit gancyclovir selection which enriches for clones with proper *aprt* homologous recombination (15). These gene targeted ES cells were injected into host blastocysts which were then implanted into pseudopregnant mice to produce chimeric pups (16). Several chimeras had ES cell-derived sperm that produced pups that were *aprt* heterozygotes. These pups appear normal and fertile and are being bred with each other or with *hprt* mice in an attempt to produce completely APRT-deficient homozygotes or animals deficient in both salvage pathways, respectively.

CONCLUSIONS

APRT will be of continuing interest to those studying purine metabolism and related inherited diseases. It is also likely to utilized in studies of the molecular nature of point mutations and larger chromosomal events that may also be involved in LOH in some tumors. Research in all of these areas will be facilitated by the availability of mice with mutations in *aprt*.

Acknowledgements

Supported by NIH grants DK 38185, ES 05652, ES 05204, NCI CA 30688, and USDO E FG 028760502.

REFERENCES

1. L. Siminovitch, On the nature of hereditable variation in cultured somatic cells. *Cell* 7:1-11 (1976).
2. E. Chu, and H. Malling, Mammalian cell genetics, II. Chemical induction of specific locus mutations in Chinese hamster cells *in vitro*. *Proc. Natl. Acad. Sci. USA* 61:1306-1312 (1968).
3. I. Lieberman, and P. Ove, Enzyme studies with mutant mammalian cells, *J. Biol. Chem.* 235:1765 (1960).

4. J.H. Atkins, and S.M. Gartler, Development of a non-selective technique for studying 2,6-diaminopurine resistance in an established murine cell line, *Genetics*, 60:781 (1968).

5. E.M. Eves, and R.A. Farber, Expression of recessive Aprt-mutations in mouse CAK cells resulting from chromosome loss and duplication. *Somatic Cell Genet.* 9:771-778 (1983).

6. P. Dewyse, and W.E. Bradley, A very large spontaneous deletion at aprt locus in CHO cells: sequence similarities with small aprt deletions. *Somat. Cell Mol. Genet.* 17:57-68 (1991).

7. M. Hakoda, H. Yamanaka, and N. Kamatani, Diagnosis of heterozygous states for adenine phosphoribosyltransferase deficiency based on detection of *in vivo* somatic mutants in blood T cells: application to screening of heterozygotes. *Am. J. Hum. Genet.* 48:552-562 (1991).

8. L.E. Smith, and A.J. Grosovsky, Genetic instability on chromosome 16 in a human B lymphoblastoid cell line. *Somat. Cell Mol. Genet.* 19:515-527 (1993).

9. G.E. Cooper, N.H. Khattar, P.L. Bishop, and M.S. Turker, At least two distinct epigenetic mechanisms are correlated with high-frequency "switching" for APRT phenotypic expression in mouse embryonal carcinoma stem cells. *Somat. Cell Mol., Genet.* 18:215-225 (1992).

10. J. Chen, A. Sahota, G.F. Martin, M. Hakoda, N. Kamatani, P.J. Stambrook, and J.A. Tischfield, Analysis of germline and *in vivo* somatic mutations in the human adenine phosphoribosyltransferase gene: mutational hot spots at the intron 4 splice donor site and at codon 87. *Mutat. Res.* 287:217-225 (1993).

11. H. Tsuda, W.D. Zhang, Y. Shimosato, J. Yokota, M. Terada, T. Sugimura, Allele loss on chromosome 16 associated with progression of human hepatocellular carcinoma. *Proc. Natl. Acad. Sci. USA.* 87:6791-6794 (1990).

12. H. Tsuda, D.F. Callen, T. Fukutomi, Y. Nakamura, S. Hirohashi, Allele loss on chromosome 16q24.2-qter occurs frequently in breast cancers irrespectively of differences in phenotype and extent of spread. *Cancer Res.* 54:513-517 (1994).

13. Y. Zhu, P.J. Stambrook, J.A. Tischfield, Loss of heterozygosity: the most frequent cause of recessive phenotype expression at the heterozygous human adenine phosphoribosyltransferase locus. *Mol. Carcinogenesis* 8:138-144 (1993).

14. D.W. Melton, Production of a model for Lesch-Nyhan syndrome in hypoxanthine phosphoribosyltransferase-deficient mice. *Nature Genet.* 3:235-240 (1993).

15. S.L. Mansour, K.R. Thomas, and M.R. Capecchi, Disruption of the proto-oncogene int-2 in mouse embryo-derived stem cells: a general strategy for targeting mutations to non-selectable genes. *Nature* 336:348-352 (1988).

16. B. Hogan, F. Constantini, and E. Lacy, Manipulating the mouse embryo: A Laboratory Manual, Cold Spring Harbor Laboratory, Cold Spring Harbor, NY (1986).

664

ISOLATION AND CHARACTERIZATION OF MUTATIONS IN THE MOUSE APRT GENE THAT ENCODE FUNCTIONAL ENZYMES WITH RESISTANCE TO TOXIC ADENINE ANALOGS

Nada H. Khattar[1], C. Darrell Jennings[2], Kimberly A. Walker[2], and Mitchell S. Turker[1,2]

[1]Department of Microbiology and Immunology
[2]Department of Pathology
University of Kentucky
Lexington, KY 40536

INTRODUCTION

Adenine phosphoribosyltransferase (APRT, EC 2.4.2.7) is a purine salvage enzyme that catalyses the reaction between adenine and 5-phosphoribosylpyrophosphate (PRPP) to form AMP. Expression of the *aprt* gene is not required for cell survival in culture, therefore, it may be selected against by growing cells in the presence of toxic adenine analogs such as 2,6-diaminopurine (DAP). In general, DAP-resistant cell lines do not exhibit significant levels of APRT activity (Taylor and Sahota, 1989).

The mammalian *aprt* gene is relatively small (ranges from 2.2 to 2.6 kilobases) and consists of five exons. The sequences of the hamster (De Boer *et al.*, 1989), human (Wilson *et al.*, 1986) and mouse (Dush *et al.*, 1985) *aprt* genes have been reported. These features along with its selectability make the *aprt* gene a prime target for mutational studies (Phear *et al.*, 1987, 1989; De Boer and Glickman, 1989; De Jong *et al.*, 1988). As a result of these studies and sequence comparisons with other nucleotide- and PRPP-binding proteins it has been suggested that sequences in exon three contribute to the adenine binding site (Hershey and Taylor, 1986; De Boer and Glickman, 1991) while the PRPP binding site is encoded by sequences in exon four (Dush *et al.*, 1985; Hershey and Taylor, 1986; Broderick *et al.*, 1987; De Boer and Glickman, 1991). Most mutational studies have targeted the Chinese hamster (De Boer and Glickman, 1991) or the human (Hidaka *et al.*, 1987; Sahota *et al.*, 1991) *aprt* genes. Only two base-pair substitutions have been reported in the mouse gene (Khattar *et al.*, 1992).

Starting with an embryonal carcinoma cell line hemizygous for *aprt*, we have isolated a total of 45 DAP-resistant cell lines that appear to have base-pair substitutions, small insertions or small deletions in *aprt* causing the DAP resistance. Here we present a molecular and biochemical characterization of a subset of these cell lines that are distinguished by significant levels of APRT enzymatic activity in cell-free extracts.

MATERIALS AND METHODS

Cell Culture

The mouse embryonal carcinoma cell lines used in this study were derived from the parental cell line P19 (McBurney and Rogers, 1982; Cooper *et al.,* 1991). H22 and H29 were obtained as P19 clones capable of growth in 6 µg/ml DAP. H29D1 was obtained from H29 for its ability to grow in 80 µg/ml DAP. The cell lines 17UV, 18UV, 60UV and 109UV were derived from H22 after treatment with ultraviolet light (UV; 8.5 Joules/m^2) and the cell lines 114E and 206E were derived from H22 after treatment with methanesulfonic acid ethyl ester (EMS; 600µg/ml for 20 hours). P19 contains two active *aprt* alleles, whereas all the other cell lines carry a single *aprt* allele.

Cells were cultured at 37°C and 5% CO_2 in Dulbecco's minimal essential medium (DMEM; GIBCO, Grand Island, NY) supplemented with 5% heat-inactivated fetal bovine serum (FBS; GIBCO), 5% serum-plus (SP; JRH Biosciences, Kansas City, MO) and when necessary 80 µg/ml DAP.

APRT Enzyme Assay

This assay was performed as described by Turker *et al.* (1984). When used to determine apparent K_m for adenine, apparent K_m for PRPP and AMP inhibition all components of the reaction mix were kept constant while the concentrations of adenine, PRPP and AMP were varied, respectively.

Southern Blot Analysis

These methods were as described by Davis *et al.* (1986). Blots were probed with a 3.1kb fragment encompassing the mouse *aprt* gene (Dush *et al.*, 1985).

DNA Sequencing

The nucleotide sequences of exons three and four were determined as described in Khattar *et al.* (1992).

RESULTS

Southern Blot, APRT Enzyme Assay and DNA Sequencing

A total of 23 UV-induced and 14 EMS-induced DAP-resistant cell lines maintaining intact *aprt* genes were identified by Southern blots. Cell-free extracts were prepared from these cell lines and assayed for APRT enzyme activity. Six of these cell lines, namely 17UV, 18UV, 60UV, 109UV, 114E and 206E, showed significant levels of APRT activity. The specific activities are shown in table 1. Sequencing results for exons three and four from cell lines with APRT enzyme activity are also shown in table 1.

Table 1. Summary of the biochemical and molecular data for wildtype and DAP resistant cell lines.

cell line	APRT specific activity[1]	AMP inhibition at 300 µM	Km for adenine in µM	Km for PRPP in µM	mutation	amino acid change
P19	1.10	14%	4.52 ± 0.84	13.12 ± 2.46	none	
H22	0.32	13%	ND[2]	13.99 ± 2.33	none	
17UV	0.38	19%	39 ± 4.5	43.21 ± 4.64	unknown, wildtype in exons 3 & 4	
18UV	0.19	24%	sigmoidal	3.13 ± 0.44	T to A at 2277 in exon 3	Tyr to Asn at amino acid 101
60UV	0.56	64%	sigmoidal	238.5 ± 34.86	C to A at 2546 in exon 4	Asp to Glu at amino acid 128
109UV	0.07	47%	sigmoidal	ND	unknown, wildtype in exons 3 & 4	
114E	0.19	10%	sigmoidal	6.72 ± 1.69	C to A at 2485 in exon 4	Ala to Val at amino acid 108
206E	0.20	10%	sigmoidal	7.47 ± 0.78	G to A at 2293 in exon 3	Gly to Glu at amino acid 106
H29D1	0.30	71%	57 ± 4.3	14.87 ± 3.68	G to A at 2236 in exon 3	Arg to Gln at amino acid 87

[1] APRT specific activity is expressed in nmole of adenine per minute per mg protein

[2] ND = not determined

K_m and AMP Inhibition Assays

The affinity of the APRT enzymes from the wildtype P19, hemizygous H22 and the DAP-resistant cell lines was determined for both adenine and PRPP. In the assays done to determine the apparent K_m for adenine the PRPP concentration was kept constant at 400 µM while the adenine concentrations were varied from 0 to 600 µM. In determining the apparent K_m for PRPP the concentration of adenine was kept at 94 µM and PRPP concentrations were varied from 0 to 1000 µM. The results are shown in table 1. All APRT enzymes from DAP-resistant cell lines were found to have decreased affinity for adenine. H29D1 and 17UV maintained Michaelis-Menten kinetics however the enzymes from the rest of the DAP-resistant cell lines showed sigmoidal kinetics. As for the affinity for PRPP it was found to be significantly altered in only two cell lines, 17UV and 60UV. Both retained Michaelis-Menten kinetics.

The inhibition of APRT activity by AMP was also assayed for all cell lines. The results are shown in table 1 and figure 1. APRT enzymes from H29D1, 109UV and 60UV were found to be more sensitive to inhibition by AMP than the wildtype enzyme.

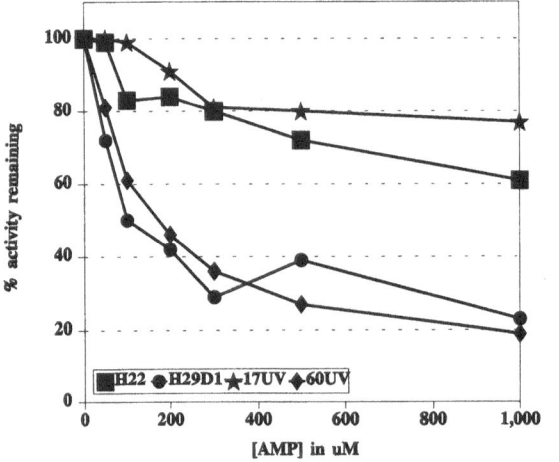

Figure 1. Demonstration of altered sensitivity to AMP inhibition of APRT enzyme activity. Cell-free extracts from the different cell lines were used in an APRT enzyme assay in the presence of varying amounts of AMP. Percentage activity remaining is plotted against AMP concentration (µM).

DISCUSSION

We have isolated 7 DAP-resistant cell lines that maintain significant levels of APRT activity. This study determined reduced affinity for adenine in all of the DAP-resistant mutants examined. If this is accompanied by a concomitant reduction in affinity for toxic adenine analogs, which presumably bind at the same site as adenine, then this would explain the ability of cells carrying these mutations to survive in the presence of these analogs. De Boer and Glickman (1991) have speculated that altered affinity for PRPP alone may account for resistance to toxic adenine analogs, however, from our data it appears that reduced affinity for adenine is essential for cells to exhibit the described phenotype. In the cell lines, 17UV and 60UV, where the APRT enzyme shows altered affinity for PRPP it also presents a reduced affinity for adenine.

It is worth noting that the mutation in 60UV is in codon 128 which encodes the most highly conserved amino acid within the PRPP binding site. Another DAP-resistant mutant, 22UV, carrying a mutation in the same codon which results in a change of amino acid from Asp to Tyr has no detectible APRT activity. The reason the enzyme from 60UV retains any ability to bind PRPP is probably because the amino acid change that results, Asp to Glu, is a very conservative one.

We have described four new base-pair substitutions in the mouse *aprt* gene that result in resistance to the toxic adenine analog DAP in the presence of an active APRT enzyme. Based on previous work on H29D1 (Khattar *et al.*, 1992) we believe that these base-pair substitutions are responsible for the phenotype observed.

Acknowledgement

This work was supported by NIH grant CA56383.

REFERENCES

Broderick, T.P., Schaff, D.A., Bertino, A.M., Dush,M.K., Tischfield, J.A., and Stambrook, P.J., 1987, Comparative anatomy of the human APRT gene and enzyme: Nucleotide sequence divergence and conservation of a non-random CpG dinucleotide arrangement, *Proc. Natl. Acad. Sci. USA* 84:3349.

Cooper, G.E., DiMartino, D.L., and Turker, M.S., 1991, Molecular analysis of APRT deficiency in mouse P19 teratocarcinoma stem cell line, *Somat. Cell Mol. Genet.* 17:105.

Davis, L.G., Dibner, M.D., and Battey, J.F., 1986, "Basic Methods in Molecular Biology", Elsevier, New York.

De Boer, J.G., and Glickman, B.W., 1989, Sequence specificity of mutation induced by the anti-tumor drug cisplatin in CHO aprt gene, *Carcinogenesis* 10:1363.

De Boer, J.G., Drobetsky, E.A., Grosovsky, A.J., Mazur, M., and Glickman, B.W., 1989, The Chinese hamster aprt gene as a mutational target. Its sequence and an analysis of direct and inverted repeats, *Mutat. Res. Letters* 226:239.

De Boer, J.G., and Glickman, B.W., 1991, Mutational analysis of the structure and function of the adenine phosphoribosyltransferase enzyme of Chinese hamster, *J. Mol. Biol.* 221:163.

De Jong, P.J., Grosousky, A.J., and Glickman, B.W., 1988, Spectrum of spontaneous mutation at the APRT locus of Chinese hamster ovary cells: An analysis at the DNA sequence level, *Proc. Natl. Acad. Sci. USA* 85:3499.

Dush, M.K., Sikela, J.M., Khan, S.A., Tischfield, J.A., and Stambrook, P.J., 1985, Nucleotide sequence and organization of the mouse adenine phosphoribosyltransferase gene: Presence of a coding region common to animal and bacterial phosphoribosyltransferases that has a variable intron/exon arrangement, *Proc. Natl. Acad. Sci. USA* 82:2731.

Hershey, H.V., and Taylor, M.W., 1986, Nucleotide sequence and deduced amino acid sequence of Escherichia coli adenine phosphoribosyltransferase and comparison with other analogous enzymes, *Gene* 43:287.

Hidaka, Y., Palella, T.D., O'Toole, T.E., Tarle, S.A., and Kelly, W.N., 1987, Human adenine phosphoribosyltransferase, *J. Clin. Invest.* 80:1409.

Khattar, N.H., Cooper, G.E., DiMartino, D.L., Bishop, P.L., and Turker, M.S., 1992, Molecular and biochemical elucidation of a cellular phenotype characterized by adenine analogue resistance in the presence of high levels of adenine phosphoribosyltransferase activity, *Biochem. Genet.* 30:635.

McBurney, M.W., and Rogers, B.J., 1982, Isolation of male embryonal carcinoma cells and their chromosome replication patterns, *Dev. Biol.* 89:503.

Phear,G., Nalbantoglu, J., and Meuth, M. 1987, Next-nucleotide effects in mutations driven by DNA precursor pool imbalance at the aprt locus of Chinese hamster ovary cells, *Proc. Natl. Acad. Sci. USA* 84:4450.

Phear, G., Armstrong, W., and Meuth, M., 1989, Molecular basis of the spontaneous mutation at the aprt locus of hamster cells, *J. Mol. Biol.* 209:577.

Sahota, A., Chen, J., Stambrook, P.J., and Tischfield, J.A., 1991, Mutational basis of adenine phosphoribosyltransferase deficiency, *Adv. Exp. Med. Biol.* 309B:73.

Taylor, M.W., and Sahota, A., 1989, Adenine analogs, in: "Drug Resistance in Mammalian Cells, Vol.I," R.S. Gupta (ed.), CRC Press, Boca Raton, FL, pp. 111-123.

Turker, M.S., Smith, A.C., and Martin, G.M., 1984, High frequency "switching" at the adenine phosphoribosyltransferase locus in multipotent mouse teratocarcinoma stem cells. *Somat. Cell Mol. Genet.* 10:55.

Wilson, J.M., O'Toole, T.E., Argos, P., Shewach, D.S., Daddona, P.E., and Kelly, W.N., 1986, Human adenine phosphoribosyltransferase: Complete amino acid sequence of the erythrocyte enzyme, *J. Biol. Chem.* 261:13677.

ANALYSIS OF *APRT* MUTATIONS BY REVERSE-TRANSCRIPTION PCR

Steve Bye, Amrik Sahota, Ju Chen, and Jay A Tischfield

Department of Medical and Molecular Genetics, Indiana
University School of Medicine, Indianapolis, IN 46202

INTRODUCTION

Adenine phosphoribosyltransferase (APRT) is an enzyme of purine metabolism and
its deficiency can lead to 2,8-dihydroxyadenine urolithiasis (1). The APRT gene is small
(genomic region 2.6 kb, coding region 0.54 kb), contains five exons, and is located at
16q24. By sequence analysis of PCR-amplified genomic DNA, we have identified 18
different mutations in patients with APRT deficiency (2). Nine of these mutations,
including two nonsense mutations, are located in exon 3. Several of the *APRT* mutations,
including the TGG-to-TGA nonsense mutation at trp_{98} in both alleles from patient ASA1
(3), have now been analyzed by reverse-transcription (RT-PCR). The results are presented
here. RNA from the second patient with a nonsense mutation in both alleles in exon 3
(CGA-to-TGA, arg_{87}) (4) was not available for analysis.

SUBJECTS

The subjects included three patients with complete APRT deficiency, three APRT
heterozygotes, and ten controls. Two of the patients were homozygous for mutations in
exon 3; one (ASA1) had a nonsense mutation at codon 98 and the other had a missense
mutation at codon 65. The third patient was a compound heterozygote, with a 7 bp deletion
in exon 3 (codon 93) in one allele and a missense mutation in the translation initiation
codon in the other allele. Two of the heterozygotes were parents of ASA1 and the third had
a missense mutation in exon 3 (codon 89) in one allele.

Purine and Pyrimidine Metabolism in Man VIII, Edited by
A. Sahota and M. Taylor, Plenum Press, New York, 1995

METHODS

RNA was extracted from frozen cell pellets of EBV-tranformed lymphoblast cell lines using the Promega RNAgents total RNA isolation system. cDNA synthesis and PCR amplification were carried out using the Perkin Elmer RNA PCR kit and specific *APRT* primers P1 (ATG GCC GAC TCC GAG CTG CA) and P2 (GTC ACT CAT ACT GCA GGA GAG AG) (Fig. 1). The PCR amplification conditions were 95°C for 20 sec, 60°C for 15 sec, and 72°C for 75 sec for 30 cycles. The initial denaturation was at 95°C for 5 min. The PCR products were electrophoresed on a 1% agarose gel and stained with ethidium bromide.

The gel was then transferred to a nylon membrane and sequentially hybridized with P32 end-labeled oligonucleotide probes P3 (GAG GCT TCC TCT TTG GC) and P5 (GAA CCA TGA ACG CTG CC) specific for *APRT* exons 3 and 5, respectively (Fig. 1). The hybridization conditions were 16 h at 45°C in 6X SSPE, 5X Denhardt's solution, 0.5% SDS, 100 ug/ml denatured salmon sperm DNA. The final wash was in 1x SSC + 0.1% SDS for 30 min at 45°C. Both short (5 h) and long (16 h) film exposures were made. The short PCR product from ASA1 was purified from agarose using a gel extraction kit (Qiagen), subcloned into M13mp18, and sequenced using the Sequenase V2.0 kit (USB) with the M13 universal primer.

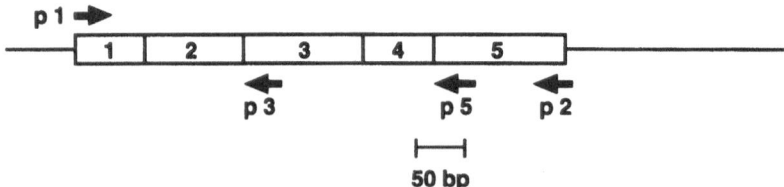

Fig. 1. Diagram showing RT-PCR of *APRT*. Primers P1 and P2 were used to amplify the coding region. P3 and P5 are the hybridization probes used.

RESULTS

The *APRT* coding region (540 bp) from APRT-deficient patients, heterozygotes, and controls was amplified by RT-PCR. A 0.54 kb fragment was detected in all subjects by ethidium bromide staining, but a 0.4 kb fragment (estimated at about 10% of the intensity of the 0.54 kb fragment) was also detected in the sample from ASA1. Transfer of the gel to a nylon membrane and hybridization with probe P5 confirmed that the 0.54 kb fragment from all subjects, and the 0.40 kb fragment from ASA1, contained *APRT* (exposure time 5 h). Low levels of the 0.4 kb fragment were also observed in the other samples when the exposure time was increased to 16 h (Fig. 2). The 0.4 kb fragment did not give a hybridization signal with probe P3 even after a long exposure, suggesting that exon 3 was missing from this fragment. The 0.54 kb fragment showed a strong signal with this probe at both short and long exposures.

The 0.40 kb fragment from ASA1 was then subcloned and sequenced. Exon 3 (134 bp) was completely missing from this fragment, but all the other exons were present. There were no mutations in the splice junctions or in the introns, as documented by our previous *APRT* sequence data from this patient (3). The skipping of exon 3 is expected to lead to a change in the *APRT* reading frame which results in premature termination after the addition of one incorrect amino acid at the end of exon 2.

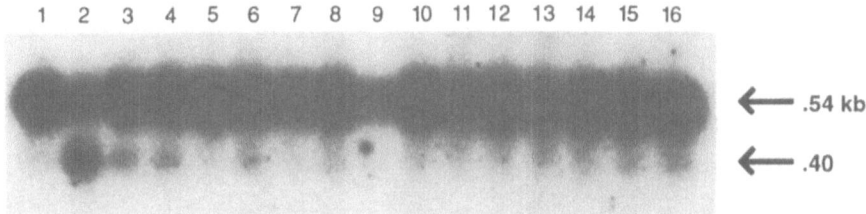

Fig. 2. Southern blot showing hybridization with probe P5 after a 16 h exposure. Lane 1, APRT heterozygote with missense mutation in exon 3 (codon 89) in one allele. Lane 2, ASA1. Lane 3. APRT-deficient patient homozygous for missense mutation in exon 3 (codon 65). Lanes 4 and 6. Father and mother of ASA1. Lane 5. Wild type father of the subject in lane 1. Lane 9. APRT-deficient patient with 7 bp deletion in exon 3 (codon 93) in one allele and missense mutation in initiation codon in the other allele. All other lanes are control subjects. All 16 lanes show the 0.54 kb fragment. The 0.40 kb fragment is clearly visible in lane 2 (ASA1), but is also present at low levels in other lanes.

DISCUSSION

A transcript of the expected size (0.54 kb) was observed in the APRT-deficient patients and heterozygotes we examined by RT-PCR. In addition, patient ASA1 (who is homozygous for a nonsense mutation in exon 3), also contained a 0.4 kb transcript that lacked exon 3. Low levels of this transcript were also observed in mRNA from other patients and controls after a long exposure.

It is well documented that mutations at splice sites can lead to skipping of exons during mRNA splicing. Recently, nonsense mutations in a number of genes including fibrillin in Marfan syndrome, ornithine-delta-aminotransferase in gyrate atrophy, p53 in the rat hepatoma cell line FAA-HTC1, factor VIII in hemophilia A, and *FACC* in Fanconi anemia group C have also been shown to lead to exon skipping (5-8). In most genes with nonsense mutations, loss of an exon results in an in-frame transcript that might be translated into a protein with reduced function. In the case reported here, loss of exon 3 results in a transcript with a frameshift. This is expected to lead to a protein with a stop codon immediately after exon 2 with the addition of one incorrect amino acid. Skipping of exons in wild type transcripts is not unique to *APRT*, as it has been observed in exon 13 of the FACC gene and exon 9 of the Duchenne/Becker (DMD/BMD) gene (8,9).

The skipping of exon 9 in the DMD/BMD gene is reported to be due to alternative splicing, because of the presence of a weak 3' splice site at the 3' end of this exon (9). Alternative splicing may also account for skipping of exon 3 in *APRT*, but this remains to be verified. It is also possible that the creation of a premature termination codon *per se* effects the efficiency of splicing, as has been suggested for the p53 gene (6).The observed increase in levels of the shorter transcript in ASA1 may be explained by preferential amplification of this transcript, especially if the transcript with the nonsense mutation (0.54) kb) is unstable.

ACKNOWLEDGMENT

This work was supported by NIH grant DK38135.

REFERENCES

1. Simmonds HA, Sahota AS, Van Acker KJ (1989) In Scriver CR, Beaudet AL, Sly WS, Valle D (Eds), The metabolic basis of inherited disease, vol I, 6th edn, McGraw-Hill, New York, pp 1029-1044
2. Sahota A, Chen J, Stambrook PJ, Tischfield JA (1991) Adv Expt Med Biol 309B:73-76
3. Sahota A, Chen J, Asako K, Takeuchi H, Stambrook PJ, Tischfield JA (1990) Nuc Acids Res 18:5915-5916
4. Chen J, Sahota A, Martin GF, Hakoda M, Kamatani N, Stambrook PJ, Tischfield JA (1993) Mutat Res 287:217-225
5. Dietz HC, Valle D, Francomano CA, Kendzior RJ, Pyetz RE, Cutting GR (1993) Science 259:680-682
6. Fukuda I, Ogawa K (1992) Cell Structure Function 17:427-432
7. Naylor JA, Green PM, Rizza CR, Giannlli F (1993) Hum Molec Genet 2:11-17
8. Gibson RA, Hajianpour A, Murer-Orlando M, Buchwald M, Mathew CG (1993) Hum Molec Genet 2:797-799
9. Riess J, Rininsland F (1994) Hum Molec Genet 3:295-298

MOLECULAR CHARACTERIZATION OF A NOVEL MUTATION IN *APRT* HETEROZYGOTES

Amrik Sahota[1], Steve Bye[1], Ju Chen[1], Nada H. Khattar[2], Mitchell S. Turker[2], Fernando Moro[3], H. Anne Simmonds[3], Brian T. Emmerson[4], Ross B.Gordon[4], and J. A. Tischfield[1]

[1]Department of Medical and Molecular Genetics, Indiana University School of Medicine, Indianapolis, IN 46202; [2]Department of Pathology, University of Kentucky, Lexington, KY 40536; [3]Purine Research Laboratories, Guy's Hospital, London SE1 9RT, UK; [4]Department of Medicine, Princess Alexandra Hospital, Brisbane, Queensland 4102, Australia

INTRODUCTION

Partial deficiency of adenine phosphoribosyltransferase (APRT) (enzyme activity in the heterozygote range in erythrocyte lysates) was discovered prior to recognition of the complete enzyme defect, during screeing for HPRT deficiency in patients with gout (1-4). This was an accidental finding and the association between partial APRT deficiency and gout has not been substantiated. The same clinical features were found with equal frequency in gouty patients with or without partial APRT deficiency. More recent studies, however, have identified a number of families with familial juvenile hyperuricemic nephropathy (FJHN) that also have reduced APRT activity in erythrocyte lysates (5). The nature of the association is unknown. FJHN is a dominant disorder charcterized by early onset of gout and/or hyperuricemia that affects males and females equally (McKusick 162000).

Complete deficiency of APRT is an autosomal recessive disorder in which the major clinical manifestation is 2,8-dihydroxyadenine (DHA) urolithiasis (McKusick

102600). Although this is a relatively rare metabolic disorder, measurement of erythrocyte enzyme activity in large population groups has suggested that 0.4 to 1.2% of the general population may be heterozygous for *APRT* (1). This is a relatively high frequency and would suggest that 1 in 250,000 to 1 in 28,000 individuals may be affected with this disorder, but the number of observed cases is much less than expected. APRT deficiency has been identified in over 100 Japanese families and over 50 families from other countries (1; also unpublished data). In earlier investigations of this disease, DHA stones were often misidentified as uric acid stones, but this has not been a problem in more recent cases (1). Thus problems with diagnosis are unlikely to explain the discrepancy between the number of APRT-deficient cases observed and that expected from erythrocyte enzyme activity measurements.

In an attempt to determine the molecular basis of the observed association between some cases of FJHN and partial APRT deficiency, we have analyzed genomic DNAs from some of these families, as well as from families with other purine enzyme defects, for mutations in the APRT gene. A missense mutation in codon 3 (arg 89) was identified in three families with familial nephropathy, in one family with severe combined immunodeficiency disease (SCID), and in an unaffected family. We have also screened a large number of random DNA samples to determine if this mutation is responsible for the reduced APRT activity seen in hemolysates in the general population.

METHODS

APRT activity was measured in erythrocyte lysates using standard procedures (data not shown). Bloods were collected from eight individuals (from five families in Britain) with reduced APRT activity in hemolysates. Four of the individuals were from three families with FJHN (two affected brothers from Family 1 and one affected female from each of Families 2 and 3). Two individuals (affected son and unaffected mother) were from a family with SCID (Family 4). Two individuals (mother and son) were from an unaffected family (Family 5). The son in this family was found to have reduced APRT activity during routine screening, but his father had normal enzyme activity. DNAs were isolated directly from blood or from EBV-transformed lymphoblastoid cell lines. DNAs from siblings with reduced APRT activity from another family with familial gout were received from Australia (Family 6).

DNAs from the Australian siblings were amplified by PCR to give a 2.4 kb fragment containing *APRT*, and this was subcloned into M13mp18 and sequenced completely (6). A single base substitution was identified in exon 3 in one allele from both siblings. DNAs from families 1-5, 5 additional FJHN families (45 individuals), as well as 500 DNA samples randomly selected from our DNA Repository, were screened for the above substitution initially by allele specific oligonucleotide hybridization and later by allele specific PCR. For allele specific hybridization, the 2.4 kb fragment was amplified as above and then blotted onto duplicate nylon membranes. One membrane was probed with a sequence specific for the wild type *APRT* allele and the other with a sequence specific for the mutant allele. For allele specific PCR, three primers located in exon 3 were used, and the fragment sizes for the amplified wild type and mutant alleles were 328 and 195 bp, respectively.

The same substitution was also found in one allele from the unaffected son with reduced APRT activity (Family 5). The variant and wild type alleles from this individual were subcloned into a plasmid vector, electroporated into an APRT-deficient mouse cell line and the transfectants selected on alanosine-adenine medium.

RESULTS

A single base substitution was identified by DNA sequencing in exon 3 in one *APRT* allele from two Australian siblings (Family 6). The gene sequence was otherwise identical to our wild type sequence. This substitution leads to the replacement of arg by gln at codon 89. The same substitution was found in the two FJHN brothers from Family 1, the FJHN female in family 2, the mother and son in Family 4 (SCID), and the mother and son in Family 5 (unaffected). As expected from enzyme assay data, the father in this family had the wild type *APRT* sequence. The base change was not found in the FJHN female in Family 3 or in 5 additional FJHN families, and it was also not found in the 500 randomly selected DNA samples. The growth of the transfectants with the variant allele from the son in Family 5 was slower than their wild type counterparts, suggesting that they had reduced APRT activity.

DISCUSSION

A single base substitution in codon 89 was found in individuals with partial APRT deficiency from three of nine families with familial nephropathy (including the Australian family). In FJHN Family 2, however, the patient inherited the disease from her mother and the variant *APRT* allele from her unaffected father. From these studies it is not possible to say whether the association between FJHN and partial APRT deficiency is significant or whether it is incidental, as has been suggested previously for gout (1-4). Further studies are in progress to determine if the female from FJHN Family 3 has sequence alterations elsewhere in the APRT gene. The finding of the same base change in a family with SCID may be incidental.

The same substitution was found in one allele from an unaffected family, but not in 500 randomly selected DNA samples. This suggests that the base change is a mutation rather than a simple polymorphism. This mutation has also not been found in the 100 or so APRT-deficient families that have been analyzed at the DNA level. The codon 89 mutation is thus unlikely to explain the observed incidence of *APRT* heterozygosity in the general population. It is likely that other mutations will be identified in *APRT* heterozygotes.

Nine of the 18 *APRT* mutations (including one poymorphism) we have identified are located in exon 3 (134 bp), and 7 of the nine mutations are near the middle of this exon (codons 86-99) (7). This region is thus more prone to mutation than the rest of the gene. A mutation identical to the one described here has been found at *aprt* codon 87 (arg-to-gln) in a mouse embryonal carcinoma cell line (8). The mutant cell line was resistant to DAP, accumulated near wild type levels of adenine, had high levels of APRT activity, and was able to grow in azaserine-adenine medium. The mutant protein had reduced affinity for adenine and increased sensitivity to inhibition by AMP compared with the parental cell line. Transfectants containing the human *APRT* mutant allele were also able to grow in alanosine-adenine medium, indicating that they contained functional APRT activity. Further characterization of the mutant protein is in progress.

ACKNOWLEDGMENT

We thank the patients and their families for donating blood or DNA samples for this study. This work was supported by NIH grant DK38185.

REFERENCES

1. Simmonds HA, Sahota AS, Van Acker KJ (1989) In Scriver CR, Beaudet AL, Sly WS, Valle D (Eds) The metabolic basis of inherited disease, 6th ed, Vol 1, McGraw-Hill, New York, pp 1029-1044
2. Kelly WN, Levy RI, Rosenbloom FM, Henderson JF, Seegmiller JE (1968) J Clin Invest 47:2281-2289
3. Emmerson BT, Gordon RB, Thompson L (1975) Aust N Z J Med 5:440-446
4. Fox IH, Lacroix S, Planet G, Moore M (1977) Medicine 56:515-526
5. Moro F, Ogg CS, Simmonds HA, Cameron JS, Chantler C, McBride MB, Duley JA, Davies PM (1991) Clin Nephrol 35:263-269
6. Chen J, Sahota A, Stambrook PJ, Tischfield JA (1991) Mutat Res 249:169-176
7. Sahota A, Chen J, Stambrook PJ, Tischfield JA (1991) Adv Expt Med Biol 309B:73-76
8. Khattar NH, Cooper GE, DiMartino DL, Bishop PL, Turker MS (1992) Biochem Genet 30:635-648

DIRECT EVIDENCE FOR A HOT SPOT OF GERMLINE MUTATION AT HPRT LOCUS

Shin Fujimori,[1] Tetsuo Tagaya,[1] Noriko Yamaoka,[1] Hirobumi Saito,[1] Naoyuki Kamatani,[2] and Ieo Akaoka[1]

[1]Second Department of Internal Medicine, Teikyo University School of Medicine, Tokyo 173, Japan
[2] Institute of Rheumatology, Tokyo Women's Medical College, Tokyo, Japan

INTRODUCTION

Hypoxanthine-guanine phosphoribosyltransferase (HPRT) is a well-studied purine metabolic enzyme which converts hypoxanthine or guanine into IMP or GMP in the presence of 5-phosphoribosyl-1-pyrophosphate. A genetic deficiency of this enzyme causes either Lesch-Nyhan syndrome characterized by neurological symptoms and self-mutilation behavior or severe overproduction-type gout, both of which show X-linked inheritance. The germline mutations causing this enzyme deficiency have been studied extensively at the molecular level (Sculley et al. 1992). Typically, germline mutations associated with HPRT deficiencies differ from family to family. These data imply that the defective mutant genes at this locus do not expand in human populations and that mutations in patients from different families are a reflection of different germline mutations. Although the mutations in a limited number of families with this disease have shown the same molecular alteration, conclusive evidence that these mutations reflected independent events has been lacking (Davidson et al.1991, Sculley et al.1991, Marcus et al. 1992, Peterson et al. 1993).

The HPRT locus has also been a target of extensive studies in the investigation of in vivo somatic mutations. Somatic mutations causing HPRT deficiency are present in adult peripheral blood T cells at an average frequency of $5\text{-}10 \times 10^{-6}$ (Albertini et al. 1982, Morley et al. 1983). In the present investigations, we obtained evidence that an identical mutation in two separate families with Lesch-Nyhan syndrome was caused by two independent new germline mutations. Analysis of the the HPRT mutational data base showed that this base substitution is rather common in in vivo somatic mutations, indicating that this site is a hot spot for both germline and in vivo somatic HPRT mutations in humans.

PATIENTS AND METHODS

Patients: Patient T.H. is a previously reported classical case of Lesch-Nyhan syndrome with a nonsense mutation in exon 3 of the HPRT locus (Fujimori et al. 1990). K.F. is a Japanese male who is unrelated to T.H. His behavior is compatible with Lesch-Nyhan syndrome and he has a complete deficiency of HPRT activity in red cells. Patients Y.Y., T.S., and N.T. are from separate families and all exhibit typical Lesch-Nyhan symptoms. Patients H.K. and D.S. are gouty cases with partial HPRT deficiency.

Detection of mutation: Cytoplasmic RNA was isolated from lymphoblastoid cells derived from each patient and cDNA was generated using oligo d(T)15 as a primer. The coding region of HPRT cDNA was enzymatically amplified, cloned and sequenced, as described elsewhere (Fujimori et al. 1990). Total DNA was extracted from white blood cells derived from patients T.H. and K.F., as well as from their family members, using standard procedures. These DNA samples were used as templates for the polymerase chain reaction (PCR) amplifying an 1059-bp fragment, including exon 3 of the HPRT gene. The primers and other conditions used for the PCR reaction were as described by Gibbs et al.(1990). The amplified genomic DNA was digested with restriction enzyme *Taq*I and electrophoresed on 2% agarose gel.

Analysis of data base: A data base of mutations at the human HPRT locus constructed by Cariello et al. (1992) was generously provided by Dr. N.F. Cariello.

RESULTS AND DISCUSSION

Table I shows a list of mutations in the HPRT deficient patients identified in our laboratory. Six different mutations at the HPRT locus have been observed in seven separate families. A single base substitution of T for C at base position 151 generating a stop codon at 51 (CGA to TGA; Arg to End) was seen in patient T.H. and K.F. from two separate families (designated HPRT Kanagawa and HPRT Fujimi).

Table 1. Mutations identified in our Japanese HPRT deficient patients.

mutation (patient)	mutation in cDNA	amino acid change	mutation site
Kanagawa (K.F.)	C_{151} to T	Arg_{51} to End	exon 3
Shinagawa[a] (T.S.)	G_{538} to A and 77 bp deletion (skipping of exon 8)	Gly_{180} to Arg and loss of 36 aa	exon 8
Adachi (N.T.)	291 bp deletion	loss of 97 aa	deletion of exon 2 and 3
Fujimi[b] (T.H.)	C_{151} to T	Arg_{51} to End	exon 3
Tokyo (Y.Y.)	G_{419} to A	Gly_{140} to Asp	exon 6
Tachikawa (H.K.)	A_{215} to G	Tyr72 to Cys	exon 3
Niigata (D.S.)	G_{472} to T	Val158 to Phe	exon 6

[a]Two types of HPRT cDNA were amplified from RNA derived from T.S. One was a full-length HPRT cDNA with a substitution of A for G at base position 538, and the other was cDNA shorter than normal by 77 bp probably generated by skipping exon 8.
[b]The molecular defects of mutant HPRT from T.H. have already been reported (Fujimori et al. 1990,1992).

The independence of the germline mutation observed in HPRT Kanagawa and HPRT Fujimi was investigated. This C to T transition should cause a loss of a *Taq*I recognition site in exon 3 of the normal HPRT gene. The PCR procedure described in the Methods amplified a 1059-bp DNA fragment when genomic DNA was the template. As expected, this fragment was cleaved into two pieces with sizes of 691 and 368 bp when the template DNA was from normal subjects. When the genomic DNA from either of the two Lesch-Nyhan patients was used as template, the 1059-bp fragment was resistant to cleavage with *Taq*I (Figure 1). The amplified fragments from each of the mothers were completely cleaved with this enzyme, thereby indicating that the peripheral blood cells of the mothers did not have the mutant allele (Figure 1). Thus, the mothers of the two probands were not heterozygotes, and the mutations in HPRT Kanagawa and HPRT Fujimi were probably de novo events.

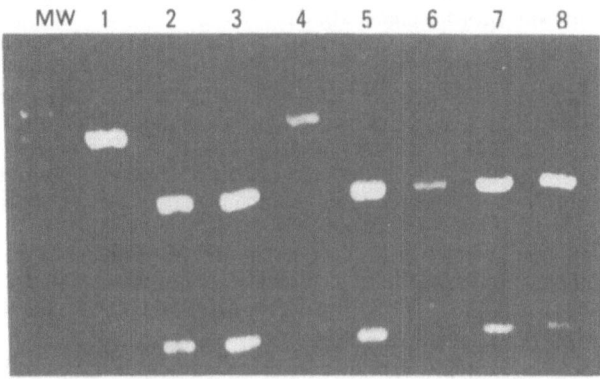

Figure 1. Cleavage of PCR-amplified exon 3 of the genomic HPRT gene. The PCR-amplified DNA was digested with *Taq*I, applied to agarose gel electrophoresis, and stained with ethidium bromide. The 1059-bp amplified fragment was cleaved in two fragments of 691 bp and 368 bp in case of the normal sequence. MW: φX-174 digested with Hae III. Lane 1: T.H. Lane 2: mother of T.H. Lane 3: sister of T.H. Lane 4: K.F. Lane 5: mother of K.F. Lane 6: father of K.F. Lane 7: normal male. Lane 8: normal female.

The finding that two independent new germline mutations show an identical base change strongly suggests that this position is a hot spot for germline mutations. The same base substitution has been previously detected in four Caucasian patients with Lesch-Nyhan syndrome (Davidson et al.1991, Tarle et al.1991, Peterson et al. 1993), although the origins of these mutations were not given. Since it is unlikely that the germline mutations seen in the Caucasian patients share their origin with either of our two Japanese families, this is additional evidence that this site is a hot spot for mutations.

Recently, a data base of mutations at the human HPRT locus has been developed (Cariello et al., 1992). By analyzing this data base, we found that the C to T transition mutation at position 151 observed in both HPRT Fujimi and HPRT Kanagawa is not rare in in vivo somatic mutations in humans. Thus, 4 (3.5%) out of 114 in vivo somatic base substitution mutations in the HPRT coding regions were C to T change at position 151, causing a codon change from CGA to TGA. Therefore, this position is likely to be a hot spot in both germline and in vivo somatic mutations in humans.

Data from studies on the molecular evolution of various vertebrate DNA sequences suggest that the CpG doublet is a favored target for germline mutations. Many mutant sequences have shown CpG to TpG or CpG to CpA substitutions when compared with their normal counterparts (Cooper and Youssoufian 1988). Since the 5 position of cytosine in some CpG dinucleotides in genomic DNA is methylated, this tends to undergo deamination to thymine (Coulondre et al. 1978).

In the present work, we have identified that a CpG doublet in exon 3 of the human HPRT gene is a hot spot for both germline and in vivo somatic cell mutations. Considering the mechanism of mutations at the CpG dinucleotide, base C of the CpG doublet in exon 3 of the human HPRT gene may have a methyl residue at position 5 in both germline and in vivo somatic cells. Despite the great importance of germline and somatic cell gene mutations in human diseases and in the evolution of species, mechanisms remain to be elucidated. Our present observations of the similarity between the two kinds of mutations should aid in a better understanding of the nature of these biologically important events in humans.

REFERENCES

Albertini,R.J., Castle, K.L., Borcherding, W.R., 1982, T-cell cloning to detect the mutant 6-thioguanine-resistant lymphocytes present in human peripheral blood. Proc. Natl. Acad. Sci. USA 79: 6617.

Cariello, N.F., Craft ,T.R., Vrieling, H., van Zeeland, A.A,. Adams, T., and Skopek,T.R., 1992, Human HPRT mutant database: Software for data entry and retrieval. Environ. Mol.Mutagen. 20: 81.

Cooper, D., and Youssoufian, H., 1988, The CpG dinucleotide and human genetic disease. Hum. Genet. 78:151.

Coulondre, C., Miller, J.F., Farabaugh, P.J., and Gilbert, W., 1978, Molecular basis of base substitution hotspots in Escherichia coli. Nature 274:775.

Davidson, B.L., Tarle, S.A., Antwerp, M.V., Gibbs, D.A., Watts, R.W.E., Kelley, W.N., and Palella, T.D., 1991, Identification of 17 independent mutations responsible for human hypoxanthine-guanine phosphoribosyltransferase (HPRT) deficiency.Am. J. Hum. Genet. 48:951.

Fujimori, S., Kamatani, N., Nishida, Y., Ogasawara, N., and Akaoka, I., 1990, Hypoxanthine guanine phosphoribosyltransferase deficiency :nucleotide substitution causing Lesch-Nyhan syndrome identified for the first time among Japanese. Hum.Genet. 84:483.

Fujimori, S., Tagaya, T., Kamatani, N., and Akaoka, I., 1992, A germline mutation within the coding sequence for the putative 5-phosphoribosyl-1-pyrophosphate binding site of hypoxanthine-guanine phosphoribosyltransferase (HPRT) in a Lesch-Nyhan patient: missense mutations within a functionally important region probably cause disease. Hum. Genet. 90: 385.

Gibbs, R.A., Nguyen, P.N., Edwards, A.L., Civitello, A.B., and Caskey, C.T., 1990, Multiplex DNA deletion detection and exon sequencing of the hypoxanthine phosphoribosyltransferase gene in Lesch-Nyhan families. Genomics 7:235.

Marcus, S., Steen, A.M., Andersson, B., Lambert, B., Kristoffersson, U., and Francke, U., 1992, Mutation analysis and prenatal diagnosis in a Lesch-Nyhan family showing non-random X-inactivation interfering with carrier detection tests. Hum. Genet. 89:395.

Morley, A.A., Trainor, K.J., Seshadri, R., and Ryall, R.J., 1983, Measurement of in vivo mutations in human lymphocytes. Nature 302: 155.

Peterson, K.S., Chambers, J., Jones, O.W., and Nyhan, W.L., 1993, Characterization of mutations in phenotypic variants of hypoxanthine phosphoribosyltransferase. Hum. Mol. Genet. 1: 427.

Sculley, D.G., Dawson, P.A., Beacham, I.R., Emmerson, B.T., and Gordon, R.B., 1991, Hypoxanthine-guanine phosphoribosyltransferase deficiency: analysis of HPRT mutations by direct sequencing and allele-specific amplification. Hum. Genet. 87:688.

Sculley, D.G., Dawson, P.A., Emmerson, B.T., and Gordon, R.B., 1992, A review of the molecular basis of hypoxanthine-guanine phosphoribosyltransferase (HPRT) deficiency. Hum. Genet. 90: 195.

Tarle, S.A., Davidson, B.L., Wu, V.C., Zidar, F.J., Seegmiller, J.E., Kelley, W.N., and Palella, T.D., 1991, Determination of the mutations responsible for the Lesch-Nyhan syndrome in 17 subjects. Genomics 10:499.

REGULATION OF LOW Km (Ecto-) 5'-NUCLEOTIDASE GENE EXPRESSION IN LEUKEMIC CELLS

Jozef Spychala and Beverly S. Mitchell

Department of Pharmacology and Internal Medicine
University of North Carolina at Chapel Hill
Chapel Hill, NC 27599-7365

INTRODUCTION

Ecto-5'-NT has been considered a biochemical marker in leukemias and lymphomas (1). The enzyme activity is low in lymphocytes (2) and decreased in B-cells in patients with X-linked agammaglobulinemia (3), decreased or not detectable in lymphocytes from chronic lymphocytic leukemia and infectious mononucleosis (4) and decreased in CD8+ lymphocytes in AIDS (5). In general, low activity has been observed during development and in undifferentiated cells which increases markedly in some cell types at the final stages of differentiation and during senescence and aging. Although there are some data showing correlation of low ecto-5'-NT activity with efficacy of nucleoside analogues of anti-cancer activity, no specific role which would explain the pattern of distribution of this enzyme has been proposed.

The cDNA's for rat liver, human placenta and mouse liver ecto-5'-NT has been cloned recently (6-8) and the gene for ecto-5'-nucleotidase has been asigned to chromosome 6 (9)

The highly variable distribution of ecto-5'-NT in human and animal tissues suggests a tissue specific regulation of expression of this enzyme. No data are available to date on the mechanisms of this expression or on the promoter structure of ecto-5'-NT gene. Therefore, as a first step in evaluating the functional significance of variable distribution of ecto-5'-NT, the present paper investigates the molecular basis of 5'-NT gene expression at the transcriptional level.

MATERIALS AND METHODS

Materials. Fetal calf serum (FCS) was from HiClone. TRI Reagent, High Efficiency Hybridization System and Northern/Southern Transfer Solution were obtained from Molecular Research Center Inc., Cincinnati, OH. Restriction enzymes, Klenow fragment of DNA Polymerase, Taq Polymerase and Polynucleotide Kinase were from Promega, Madison, WI. Sequenase 2.0 for dideoxy-sequencing was from U.S. Biochem. Corp. D-threo[dichloroacetyl-1-^{14}C] Chloramphenicol (56 mCi/mmol) was from Amersham.

Cells. K562 erythroleukemic cells (ATCC CCL-243), Jurkat T-leukemia and U-937 promyelocytic (ATCC CRL-1593) cells were maintained in RPMI-1640 medium supplemented with 10% FCS.

Genomic library screening. A human genomic clone designated 5NT-21 was obtained by screening 5 x 10^5 phage clones from a human genomic DNA library constructed in lambda ZAP (gift of Dr John Lowe, University of Michigan). An ecto-5'-NT probe was produced by PCR using primers encompassing the 5'-end of the ecto-5'-nucleotidase cDNA extending from bp -15 to bp (7); sense primer (primer 1): 5'CCAGTTCACGCGCCACAGCTATGTGTCC3' and antisense primer (primer 2): 5'CCATTATCAAATTCATGATTTCCCAGTGCCATGG 3'. Plaques that were positive on primary screening were hybridized to P^{32}-labeled oligonucleotide probe (primer 3, encompassing first 7 bases of coding region and 14 bases of 5'untranslated region). Five positive plaques from this screen were purified and shown to be identical to each other by restriction enzyme digestion and Southern blot analysis. The 18 kb insert of the genomic clone 5NT-21 was digested with Sac I, liberating restriction fragments ranging from 1.2 to 5.6 kb in size that were subcloned into pGEM 7Zf+ vector (Promega).

Plasmid construction. The pCAT-Basic vector (Promega) was used to test for ecto-5'-NT promoter activity. The subcloning strategy is depicted in Fig.1. Since there was no convenient restriction site downstream of the transcription start site (designated +1), the promoter fragment encompassing bp -260 to +111 (designated 5NT-371) was produced by PCR using primers 1 and 2. Primers 1 and 2 were designed to include an Xba I and Hind III sites at the respective 5' and 3' ends of the PCR product, which was subsequently subcloned into the pCR II vector (Invitrogen) and sequenced to confirm the fidelity of amplification. The promoter fragment encompassing bp -87 to +111 (designated 5NT-198) was produced by cutting out the Pst I / Nsi I fragment from the clone 5NT-371 (in PCR II vector) and subcloning it into Pst I digested pCAT-Basic vector. To construct a promoter fragment encompassing bp -851 to +111 (designated 5NT-962), the original 2.9 kb Sac I fragment (in pGEM vector) was digested with Pst I and Xba I to remove the bulk of first intron and part of first exon containing the ATG codon.

This fragment was replaced with a 198 bp fragment from pCR II 5NT-371. Subsequently, the 5.6 kb Sac I subclone, adjacent to 2.9 kb fragment in the original phage 5NT-21 and containing further 5' elements, was digested with BamHI (in pGEM multicloning site) and Nde I to remove the bulk of 5' upstream sequences and the ends were filled-in with Klenow DNA Polymerase and religated. A resulting 460 bp Hind III and Sac I fragment was cut out from this clone and linked to a derivative of the 2.9 clone described above (in pGEM) using a Sac I common site. The 5NT-371 and 5NT-962 promoter fragments were subcloned into pCAT-Basic vector in 5' to 3' direction using the Hind III and Xba I sites.

Cells transfections. Cells were transfected by electroporation using a Bio-Rad Gene-Pulser. Approximately 1.0 x 10^7 cells were transfected at room temperature in RPMI-1640 medium with equimolar amounts of plasmid DNA containing the promoter constructs and 30 to 40 µg of herring sperm DNA to bring the total amount of DNA to 70 µg. To control for the electroporation efficiency, separate aliquots of cells were transfected with 30 µg of CMV β-Galactosidase plasmid and 40 µg of herring sperm DNA. Electroporation was performed at 250 V and 960 µF and cells were subsequently suspended in 22 ml of RPMI-1640 medium containing 20% horse serum. After 36 hours, cells were harvested, washed twice with cold PBS and extracted with 150 µl of 0.25 M Tris pH 8.0 by triple freeze-thawing. For the chloramphenicol acetyltransferase (CAT) activity assay, extracts were subsequently treated at 60° C for 10 min.

CAT and β-Galactosidase assays. Chloramphenicol acetyltransferase activity was measured at 37° C in a reaction medium containing 1 mM dibutyryl CoA (25 µg per assay), 10 µM D-threo[dichloroacetyl-1-^{14}C] Chloramphenicol (56 mCi/mmol, 0.25 µCi per assay) in 0.25 M Tris, pH 8.0 buffer. The assay was initiated by the addition of 25 to 50 µl of cell extract (125 µl of total volume) and terminated after 2 to 4 hr with addition of 300 µl Xylenes. After repeated extractions, reaction products in the organic phase were counted in a scintillation counter. Reaction velocity was expressed as cpm converted per hr and per mg protein. Reaction time and extract amount was adjusted so that no more than 20% of substrate was utilized. The CAT background activities of cells

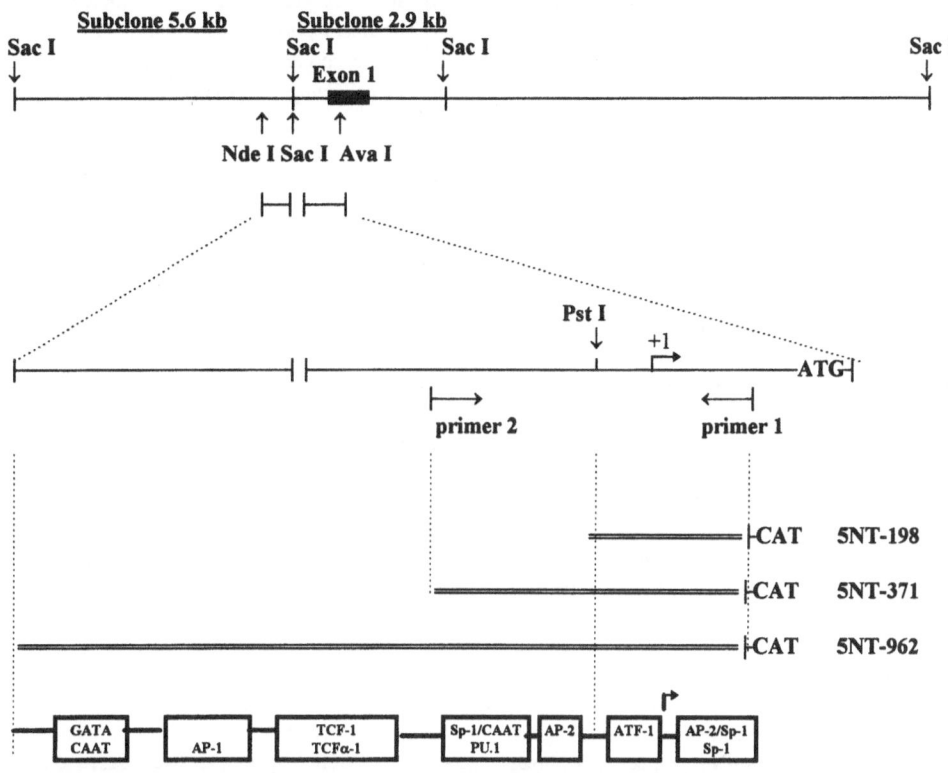

Figure 1. Scheme of the 5' region of the ecto-5'-NT gene and subcloning strategy to generate promoter constructs.

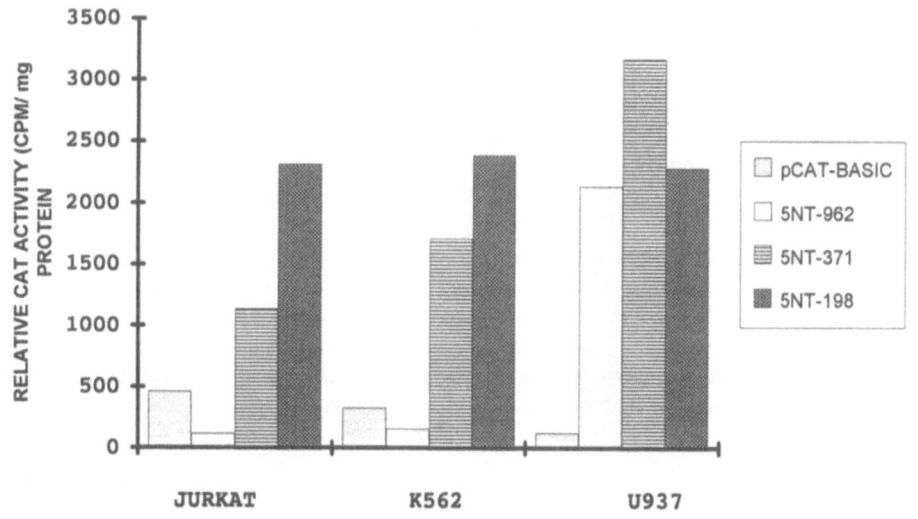

Figure 2. Relative CAT activity resulting from transfection of ecto-5'-NT promoter constructs into leukemic cell lines.

transfected with CMV β-gal was subtracted for all samples and results expressed as cpm of product produced per hr and per mg protein. To normalize for the variable transfection efficiency, the CAT activity was divided by β-galactosidase activity obtained for each cell batch. β-Galactosidase activity was assayed spectrophotometrically at 564 nm in 50 mM phosphate buffer, 10 mM $MgCl_2$ and 3 mg/ml chlorophenol red-β-D-galactopyranoside (CPRG) at 37° C. The reaction was initiated by the addition of 5 to 25 μl of cell extract and run for 1 min. The linear portion of the curve was used to compute zero order rate and expressed as μmoles of substrate used per min and per mg protein. The β-galactosidase activity varied by less than 10% in duplicate cell samples. Protein concentration was assayed by the Lowry method.

RESULTS AND DISCUSSION

Isolation of the 5'-end of the 5'-nucleotidase gene. A single genomic clone 5NT-121, 18 kb in length, was isolated during genomic library screening. Sac I digest of this clone produced 6 fragments of 5.60; 4.80; 4.20; 2.90; 1.40 and 0.8 kb sizes which were subcloned into the pGEM 7Z vector. Further analysis which included Southern blot hybridization with end-labeled oligonucleotide (primer 3), restriction mapping and sequencing identified a subclone of 2.9 kb (Fig.1) which contained the whole 5'-untranslated region, part of first intron and 413 bp of 5' upstream sequence. Similarly, subclone 5.6 kb was identified to contain further upstream sequences.The transcription initiation site was identified approxamately 162 bp upstream of the ATG initiation codon by a combination of primer extension analysis using end-labeled primer 3 and RNase protection assay (Spychala and Mitchell, in preparation).

The presumed promoter region does not contain a TATA box. However, a number of other potential binding elements, including two "CAAT" boxes (at -206 and at -628), were identified. Eleven bases upstream from the transcription start site, there is an c-fos/ATF-1 site. At position -148 there is an AP-2 site and at -198 a cluster of three adjacent sites: PU.1 box (-198), CAAT box (-206) and Sp-1 (-218). Further upstream (bp -317 to -388) there are three subsequent TCF-1 binding elements followed by a single AP-1 (-550). An additional cluster of potential binding elements is located between bp -619 and -645 and includes CAAT and GATA sites (Fig. 1). To analyze the 5'-flanking region of the ecto-5'-nucleotidase gene for promoter activity, three constructs were generated (Fig. 1). The selection of respective fragment sizes within putative promoter was based on the following assumptions: promoter fragment 5NT-371 is devoid of the upstream cluster of three TCF-1 binding elements which potentially may be responsible for inhibiting promoter activity in Jurkat T-cells, but includes a cluster of Sp-1/CAAT box/PU.1 elements which may be important in myeloid cells (U937). Construct 5NT-198 constitutes the shortest fragment potentially exhibiting basal 5'-NT promoter driven activity (Fig.1). Data presented in Fig. 2 show that indeed, in Jurkat and K562 cell lines, fragment 5NT-962 exhibited very low activity, significantly below promoterless pCAT-Basic levels, and there was a significant increase in promoter activity when fragments 5NT-371 and 5NT-198 were used (Fig.2). On the other hand, all promoter fragments were relatively active in U937, with fragments 5NT-371 and 5NT-198 exhibiting similar activity. The endogenous activity of ecto-5'-NT in leukemic cell lines has been measured and shown to be 0.97; 0.12 and 0.40 nmoles/min mg protein in Jurkat, K562 and U937 respectively.

These data suggest that in Jurkat T-cell and erythroleukemic K562 cell lines, there is a significant negative regulation mediated by regulatory elements located upstream of the core promoter fragment (5NT-198) and especially upstream of the Sp-1/CAAT/PU.1 cluster. This low level of promoter activity agrees well with our and published data on low endogenous ecto-5'-NT activity in these cell lines (10, 11). Which particular elements mediate this suppression, whether it is TCF-1/TCF-1α in Jurkat cells or GATA in K562 cells, remains to be established. Surprisingly, significantly higher CAT activity was detected when myeloid U937 cells were transfected with the same constructs. This result is in sharp contrast with measured endogenous ecto-5'-NT activity which appeared to be in the same range as in Jurkat and K562. This discrepancy could be explained by the existence of other negative elements active in this cell line and located further upstream of the promoter fragment we studied or further 3' in the gene. Alternatively, ecto-5'-NT mRNA processing could differ in these cells.

ACKNOWLEDGMENTS

This study was supported by an Investigator Award from Arthritis Foundation to Jozef Spychala and by an NIH grant RO1-CA34085.

REFERENCES

1. Gutensohn, W. and E. Thiel, Prognostic Implication of Ecto-5'-Nucleotidase Activity in Acute Lymphoblastic Leukemia, Cancer. 66;1755 (1990)
2. Edwards, N. L., J. T. Magilavy, J. T. Cassidy and I. H. Fox, Lymphocyte ecto-5'-nucleotidase deficiency in agammaglobulinemia, Science. 201;628 (1978)
3. Thompson, L. F., G. R. Boss, H. L. Spiegelberg, I. V. Jansen, R. D. O'Connor, T. A. Waldmann, R. N. Hamburger and J. E. Seegmiller, Ecto-5'-nucleotidase activity in T and B lymphocytes from normal subjects and patients with congenital X-linked agammaglobulinemia, J. Immunol. 123;2475 (1979)
4. Quagliata, F., D. Faig, M. Conklyn and R. Silber, Studies on the lymphocyte 5'-nucleotidase in chronic lymphocytic leukemia, infectious mononucleosis, normal subpopulations, and phytohemagglutinin-stimulated cells, Canc. Res. 34;3197 (1974)
5. Salazar-Gonzales, J. F., D. J. Moody, J. V. Giorgi, O. Martinez-Maza, R. T. Mitsuyasu and J. L. Fahey, Reduced ecto-5'-nucleotidase activity and enhanced OKT10 and HLA-DR expression on CD8 (T suppressor/cytotoxic) lymphocytes in the acquired immune defficiency syndrom: Evidence of CD8 immaturity, J. Immunol. 135;1778 (1985)
6. Misumi, Y., S. Ogata, S. Hirose and Y. Ikehara, Primary Structure of Rat Liver 5'-Nucleotidase Deduced from the cDNA, J. Biol. Chem. 265;2178(1990)
7. Misumi, Y., Ogata, S., Ohkubo, K., Hirose, S. Ikehara, Y. Primary structure of human placental 5'-nucleotidase and identification of the glycolipid anchor in the mature form, Eur. J. Biochem. 191:563 (1990)
8. Resta, R., S. W. Hooker, K. R. Hansen, A. B. Laurent, J. L. Park, M. R. Blackburn, T. B. Knudsen and L. F. Thompson, Murine ecto-5'-nucleotidase (CD73) - cDNA cloning and tissue distribution, Gene. 133;171 (1993)
9. Boyle J. M., Y. Hey, A. Guerts van Kessel and M. Fox, Assignment of ecto-5'-nucleotidase to human chromosome 6, Hum. Genet. 81;88 (1988)
10. Dornand, J., J.-C. Bonnafous, J. Favero and J.-C. Mani, Inverse relationship of 5'-nucleotidase and adenosine deaminase activities among human lymphoid cells, Biochem. Med. 28;144 (1982)
11. Spychala, J., V. Madrid-Marina, P. J. Nowak and I. H. Fox, AMP and IMP dephosphorylation by soluble high- and low-Km 5'-nucleotidases., Am. J. Physiol. 256;E386 (1989)

ECTO-5'-NUCLEOTIDASE (CD73): GENOMIC CLONING AND CHARACTERIZATION OF REGIONS UPSTREAM OF THE TRANSLATION START SITE

K.R. Hansen, R. Resta, C.F. Webb and L.F. Thompson

Immunobiology and Cancer Program
Oklahoma Medical Research Foundation
825 NE 13th St.
Oklahoma City, OK 73104, USA

INTRODUCTION

Ecto-5'-nucleotidase (5'-NT, CD73) catalyzes the dephosphorylation of ribo- and deoxyribonucleoside monophosphates to the corresponding nucleosides, which are then able to enter cells via facilitated diffusion and participate in purine salvage or are catabolized to uric acid. 5'-NT is expressed on a variety of cell types including lymphocytes, transitional cell mucosa, the basal layer of nonkeratinizing squamous epithelium, and the stromal cells of human breast cancers.[1] There are at least three examples where 5'-NT and adenosine deaminase (ADA), a second purine salvage pathway enzyme, are expressed in a reciprocal fashion in adjacent layers of cells: the thymus[2,3], the developing mouse embryo[4], and the retina[5]. Additionally, an examination of the specific cell types which express 5'-NT and ADA in murine gastrointestinal tract tissues revealed that ADA was localized to the superficial layers of the squamous epithelium of the tongue and esophagus[6], while 5'-NT was found in the basal layer of human oral mucosa, a similar tissue[1]. Because 5'-NT and ADA are expressed in a reciprocal manner in some tissues, we hypothesize that these molecules may share genetic elements which regulate their expression in a coordinated manner. In order to evaluate this hypothesis extensive characterization of the 5'-NT promoter region is necessary. This initial characterization will create a framework upon which additional experiments can directly assess the interactions between factors which regulate the expression of 5'-NT and ADA.

MATERIALS AND METHODS

Genomic Cloning. 5'-NT genomic clones were isolated from a human placenta genomic library in the lambda FIX II vector (Stratagene) with various cDNA[7] probes. Approximately 10^6 plaques were plated for each screening. Nylon filters (Amersham) with lifted plaques were incubated in prehybridization solution (6X SSC, 2X Denhardts, 4mM EDTA, 0.1mg/ml salmon sperm DNA) for three to six hours at 68° C. Hybridization was conducted in the same solution with radiolabeled probe (1-2 x 10^6 cpm/ml) for 16 h at 68° C. The filters were washed three times in 2X SSC, 0.5% SDS, 4mM EDTA. Autoradiography was performed at -80° C overnight with an intensifying screen.

DNA sequencing. Sequencing of DNA was performed by the dideoxy method using [α-^{35}S]dATP and Sequenase (US Biochemical) with primers synthesized by the Molecular Biology Core Facility at the University of Oklahoma.

RNase Protection Assay. Body-labeled RNA probes corresponding to genomic DNA segments (illustrated in Figure 1) were prepared by subcloning genomic DNA fragments into pSKII+ (Stratagene) and subsequent *in vitro* transcription with T3 or T7 RNA polymerase (Promega). Probes were purified on a 4% polyacrylamide-8.3 M urea gel. Human placental RNA or tRNA (15μg) was hybridized overnight with 1-5 x 10^5 cpm in 30μl hybridization buffer (80% formamide, 40 mM Pipes pH 6.4, 400 mM NaCl, 1 mM EDTA) between 55° C and 75° C. Probe-RNA mixtures were then digested with RNase T1 (Boehringer-Mannheim) at 570 u/ml, treated with 0.5% SDS and 0.13mg/ml Proteinase K (Boehringer-Mannheim), extracted with phenol/chloroform, ethanol precipitated with tRNA, and run on a 6% polyacrylamide-8.3M urea gel. Autoradiographs were exposed for 1-14 days at -80° C with an intensifying screen.

RACE (Rapid Amplification of cDNA Ends) cloning of the 5'-end of the 5'-NT message. RACE-Ready cDNA from human placenta (Clontech) was amplified in a primary PCR reaction with an antisense, gene-specific 5'-NT primer and a primer complementary to the synthetic oligonucleotide anchor on the 5'-end of the RACE-Ready cDNA. A diluted aliquot of this primary PCR product was re-amplified with the anchor primer and a second, nested antisense gene-specific 5'-NT primer. The resulting amplified product was ligated into the pGEM-T vector (Promega) and transformed into competent *E. coli*.

RESULTS AND DISCUSSION

Genomic Cloning and Sequencing

A total of twelve 5'-NT genomic clones were isolated from three separate screens of a human placenta genomic library utilizing 5'-NT cDNA probes. Three of these clones hybridized to a probe from the most upstream region of the cDNA (Figure 1). These three clones overlapped significantly, spanning ~ 26kb of genomic DNA, with ~ 13.7kb upstream of the translation start site. All three clones were mapped with several restriction enzymes. An EcoRI - BamHI restriction fragment between -3.8kb and +0.6kb (with 0 = the translation start site) is illustrated in Figure 1. Sequencing of 2kb upstream of the translation start site indicates that this region is remarkably GC rich, with the region from 0 to -400bp having a GC content of 74%. The first 500bp upstream of the translation start site contains five Sp1 and one CREB consensus binding sites. A CAAT sequence is present at -376, but no corresponding TATA box is found downstream of this site. Significant secondary structure may exist in this region of genomic DNA, as PCR reactions are incapable of amplifying this region in the absence of 10% DMSO, an organic solvent thought to facilitate PCR ampifications by destablizing strand annealing.

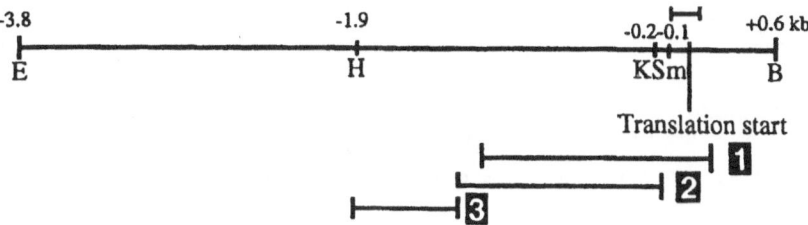

Figure 1. Schematic representation of a region of 5'-NT genomic DNA between -3.8kb and +0.6 kb, with 0= the translation start site. E= EcoRI, H= HindIII, K= KpnI, Sm= SmaI and B= BamHI. The bars labeled 1,2 and 3 represent regions of genomic DNA utilized as probes in the RNase protection assay, while the bar above the translation start site represents the 101bp cDNA probe utilized for isolating genomic clones.

RNase Protection Assay

To identify the transcription start site, we performed the RNase protection assay with probes corresponding to genomic DNA. Three separate probes (Figure 1) were synthesized by *in vitro* transcription and individually hybridized to total human placenta RNA or a tRNA control as described in Methods. Probe 1 protected fragments detectable following overnight autoradiography were between 178 and 199bp in size. The ends of these protected fragments fell between -58 and -80bp in the genomic DNA (0 = the translation start site). A fourteen day exposure of the same gel revealed the presence of a slightly larger protected band between 207 and 218bp in size, the end of which fell between -88 and -99 in the genomic DNA. The protected fragments were not seen in a control experiment in which tRNA was substituted for human placental RNA. These protected fragments map either transcription start sites, or intron/exon junctions (splice sites). RNase protection assays utilizing probes 2 and 3 (which do not overlap the region identified with probe 1) gave no protected fragments, indicating that the corresponding genomic regions either fall within an intron or are upstream of the transcription start site.

RACE (Rapid Amplification of cDNA Ends) Cloning

As the RNase protection assay is incapable of distinguishing between transcription start sites and splice sites, RACE cloning was utilized to identify a transcription start site for 5'-NT. This method allows for the confirmation of transcription start sites due to the occasional reverse transcription of the methyl-G cap found at the 5'-end of most eukaryotic mRNAs. The reverse transcription of the methyl-G cap results in a " non-template encoded" or extra G/C pair in the amplified 5'-end sequence[8]. RACE cloning of the 5'-NT cDNA was performed as described in Methods. RACE clone inserts of varying size were chosen for DNA sequencing. Of a total of twenty-two clones that were sequenced, four contained an "unencoded", or extra C between positions -63 and -64bp in the antisense strand, suggesting a transcription start site at -63bp (Figure 2). Of the remaining eighteen RACE clones sequenced, ten longer than -63bp and eight shorter, none contained an unencoded C between the known genomic DNA sequence and the anchor oligonucleotide sequence. The shorter clones (ending between -18 and -50bp) may represent incomplete reverse transcription of the 5'-NT mRNA or additional transcription start sites, whereas the presence of RACE clones with sequences upstream of -63bp (ending between -86 and -89bp) would suggest the presence of a transcription start site upstream of -63bp.

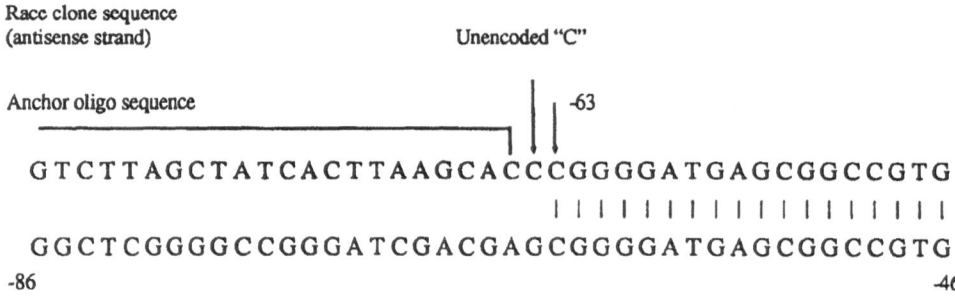

Figure 2. Sequence of the antisense strands of 5'-NT genomic DNA and a 5'-NT RACE clone containing an extra C not found in the genomic DNA.

As both the RNase protection assay and RACE cloning have identified the same region of genomic DNA, it is highly probable that a transcription start site exists at -63. A second transcription start site may also exist slighty upstream of the first, between -86 and -89bp. Future studies will evalutate the promoter activity of this region of genomic DNA by creating reporter constructs in which genomic DNA segments are placed upstream of a luciferase reporter gene. Preliminary experiments indicate that 5'-NT genomic DNA upstream of the translation start site and including the transcription start identified here, are capable of directing expression of the reporter gene.

Acknowledgments

This work was supported by NIH grant AI18220.

References

1. Thompson, L.F., Ruedi, J.M., Glass, A., Moldenhauer, G., Moller, P., Low, M.G., Klemens, M.R., Massaia, M. and Lucas, A.H. Production and characterization of monoclonal antibodies to the glycosyl phosphatidylinositol-anchored lymphocyte differentiation antigen ecto-5'-nucleotidase (CD73). Tissue Antigens 35:9-19, (1990).
2. Ma, D.D.F., Sylwestrowicz, T.A., Granger, S., Massaia, M., Franks, R., Janossy, G., and Hoffbrand, A.V. Distribution of terminal deoxyribonucleotidyl transferase and purine degradative and synthetic enzymes in subpopulations of human thymocytes. J. Immunol. 129:1430-1435, (1982).
3. Chechik, B.E., Schrader, W.P., and Minowada, J. An immunomorphologic study of adenosine deaminase distribution in human thymus tissue, normal lymphocytes and hematopoietic cell lines. J.Immunol. 126:1003-1007, (1981).
4. Blackburn, M.R., Gao, X., Airhart, M.J., Skalko, R.G., Thompson, L.F. and Knudsen, T.B. Adenosine levels in the postimplantation mouse uterus: quantitation by HPLC-fluorometric detection and spatiotemporal regulation by 5'-nucleotidase and adenosine deaminase. Dev.Dyn. 194:155-168, (1992).
5. Blazynski, C. and Perez, M-T.R. Adenosine in vertebrate retina: Localization, receptor characterization and function. Cell. Mol. Neurolbiol. 11:463-484, (1991).
6. Chinsky, J.M., Ramamurthy, V., Fanslow, W.C., Ingolia, D.E., Blackburn, M.R., Shaffer, K.T., Higley, H.R., Trentin, J.J., Rudolph, F.B., Knudsen, T.B. and Kellems, R.E. Developmental expression of adenosine deaminase in the upper alimentary tract of mice. Differentiation 42:172-185, (1990).
7. Misumi, Y., Ogata, S., Ohkubo, K., Hirose, S., and Ikehara, Y. Primary structure of human placental 5'-nucleotidase and identificiation of the glycolipid anchor in the mature form. Eur. J. Biochem. 191:563-569, (1990).
8. Hirzmann, J., Luo, D., Hahnen, J. and Hobom, G. Determination of messenger RNA 5'-ends by reverse transcription of the cap structure. Nucleic Acids Res. 21:3597-3598, (1993).

692

MOLECULAR CLONING AND SEQUENCE ANALYSES OF RAT LIVER DIHYDROOROTATE DEHYDROGENASE

Andrea Rotgeri and Monika Löffler

Department of Physiological Chemistry
School of Medicine, Philipps-University
D-35033 Marburg, Germany

INTRODUCTION

Dihydroorotate dehydrogenase (DHOdehase) catalyzes the fourth step of *de novo* pyrimidine synthesis, the conversion of dihydroorotate to orotate. DHOdehase diverges in many aspects from the other enzymes of UMP biosynthesis: In mammals the enzyme is located in the inner mitochondrial membrane whereas the CAD protein and the UMP synthase are cytosolic. It is the only redox reaction of this pathway and is linked to the respiratory chain via ubiquinone. The enzymes of *de novo* pyrimidine synthesis are organized as multifunctional proteins that arose through ancestral gene fusions and DHOdehase is the only monoenzymic protein of the pathway. Biochemical and structural characteristics of mammalian DHOdehase are not fully understood, nor are details of its gene regulation known so far. The genes of different mammalian CAD proteins and UMP synthases have been cloned and mapped to different chromosomes. The sole sequence information concerning mammalian DHOdehases available so far is the truncated cDNA encoding for human DHOdehase[1]. Recently the rat and porcine enzymes were purified and characterized in our lab[2-4]. In order to further investigate structural characteristics of the protein, the cDNA of rat liver DHOdehase was cloned and analysed.

METHODS

Hybridization screening with the cDNA of human DHOdehase

The human cDNA of DHOdehase (a gift from Dr. Michèle Minet, Centre de Génétique Moléculaire C.N.R.S., Gif sur Yvette, France) was multiprime-labeled with digoxigenin (DIG). 10^6 pfu of a rat liver λZapII cDNA library (Stratagene) prepared from adult male Sprague Dawley rat liver was screened with the DIG labeled probe. Positive clones were purified and isolated. *In vivo* excision of the pBluesript SK plasmid from the λZapII vector was performed and sequencing was done using the sequential primer method.

Purine and Pyrimidine Metabolism in Man VIII, Edited by
A. Sahota and M. Taylor, Plenum Press, New York, 1995

Rapid Amplification of cDNA Ends (RACE)

Investigation of the 5′ end of the mRNA from rat liver DHOdehase was performed by a PCR based method using the 5′-RACE-Kit (Gibco-BRL). RNA from rat liver was prepared conventionally. First strand cDNA synthesis was performed using an antisense GSP1 primer binding to a region at the 5′end of the isolated cDNA clone. The RNA was digested by RNAseH, and a homopolymeric deoxycytidine tail was added by a terminal deoxynucleotide transferase reaction. PCR amplification was performed using GSP2 primer (binding upstream of GSP1) and an anchor primer complementary to the poly-dC tail. The restriction sites of the anchor primer allow cloning of the PCR product, if another restriction site is present at the cDNA upstream of the binding site of GSP2. Specific RACE products were confirmed by Southern blot analyses, which were performed with a DIG labeled *EcoRI* -*NcoI* fragment representing the 5′end of the isolated rat DHOdehase cDNA. The DNA corresponding to Southern blot positive bands was subcloned to pBluescript vector and sequenced. Computational sequence analyses was done by means of HUSAR (EMBL, Heidelberg) and PC-Gene (IntelliGenetics) programs.

RESULTS

One single positive clone was isolated from 10^6 phages of a rat liver λZapII cDNA library. First, sequence analysis revealed extensive sequence homology with the cDNA of the human enzyme. This evaluation allowed the conclusion that this cDNA encoded for rat liver DHOdehase but apparently did not represent full length cDNA. In order to complete the sequence information of the mRNA 5′end of DHOdehase, the RACE method was used with rat liver RNA as template for the PCR. The Southern blot analyses of the RACE products can be seen in Fig. 1.

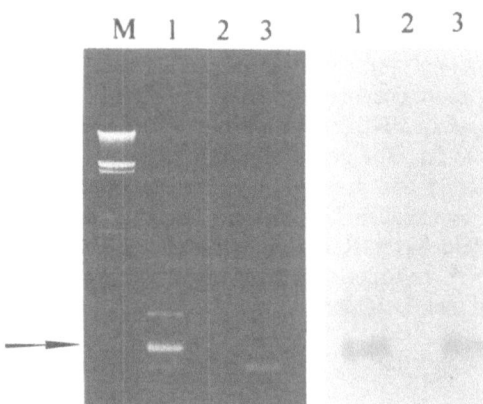

Figure 1. Agarose gel electrophoresis and Southern blot analyses of the RACE products.
Line 1: RACE product
Line 2: negative control without tailing reaction of the terminal deoyxnucleotide transferase
Line 3: reamplification of the RACE products of line 1

The broad band (arrow), representing a specific signal in the Southern blot analysis, was subcloned into pBluescript and sequenced. Sequence analyses revealed 5 clones of different length corresponding to the 5′end of rat liver DHOdehase. The additional sequence information obtained by this method were 71 bp of the 5′end comprising a coding region of 1242 bp (Fig. 2). Since this additional sequence does not contain any start codon the cDNA is still not complete.

```
   1 AGCCAGGTTTTCCAATGGAATGCAGGCAGAGGCGAGCTTAGTGGTAGAAGAAGCATGGCG    60
   1  S  Q  V  F  Q  W  N  A  G  R  G  E  L  S  G  R  R  S  M  A     20
                                          +           +  +
  61 TGGAGACAGCTGAGAAAGCGGGCCCTGGATGCAGTCATCATCCTTGGAGGTGGAGGACTT   120
  21  W  R  Q  L  R  K  R  A  L  D  A  V  I  I  L  G  G  G  G  L     40
         +           +  +  +              ------------------------------
 121 CTCTTCACCTCTTACCTGACGGCCACAGGGGATGACCATTTCTATGCTGAGTACCTGATG   180
  41  L  F  T  S  Y  L  T  A  T  G  D  D  H  F  Y  A  E  Y  L  M     60
     --------------------
 181 CCAGGTCTGCAGAGGCTGCTTGATCCAGAGTCAGCCCACCGGCTAGCTGTTCGAGTCACC   240
  61  P  G  L  Q  R  L  L  D  P  E  S  A  H  R  L  A  V  R  V  T     80
 241 TGGGTGGGGCTCCTTCCTCGAGCTACATTTCAGGACTCCGACATGCTGGAAGTGAAAGTC   300
  81  W  V  G  L  L  P  R  A  T  F  Q  D  S  D  M  L  E  V  K  V    100
 301 CTGGGCCATAAATTCCGAAATCCAGTAGGAATTGCTGCAGGATTTGACAAGAACGGCGAA   360
 101  L  G  H  K  F  R  N  P  V  G  I  A  A  G  F  D  K  N  G  E    120
 361 GCTGTGGACGGACTCTACAAGCTGGGCTTTGGTTTCGTTGAGGTAGGAAGTGTCACTCCC   420
 121  A  V  D  G  L  Y  K  L  G  F  G  F  V  E  V  G  S  V  T  P    140
 421 CAGCCTCAGGAAGGAAACCCCAGGCCTAGAGTATTCCGTCTCCCCGAGGACCAAGCTGTC   480
 141  Q  P  Q  E  G  N  P  R  P  R  V  F  R  L  P  E  D  Q  A  V    160
 481 ATTAACAGGTATGGATTCAACAGCCATGGGCTCTCGGTGGTGGAACACAGGCTACGGGCC   540
 161  I  N  R  Y  G  F  N  S  H  G  L  S  V  V  E  H  R  L  R  A    180
 541 AGACAGCAGAAGCAGGCCCAGCTCACTGCAGATGGGCTGCCTCTTGGAATAAACCTGGGG   600
 181  R  Q  Q  K  Q  A  Q  L  T  A  D  G  L  P  L  G  I  N  L  G    200
 601 AAGAATAAGACTTCGGAGGATGCTGCTGCAGACTATGCAGAGGGTGTTCGAACCCTGGGC   660
 201  K  N  K  T  S  E  D  A  A  A  D  Y  A  E  G  V  R  T  L  G    220
 661 CCCTTGGCTGACTACCTGGTGGTAAACGTGTCCAGCCCCAACACTGCTGGTCTGAGGAGC   720
 221  P  L  A  D  Y  L  V  V  N  V  S  S  P  N  T  A  G  L  R  S    240
 721 CTACAGGGGAAGACCGAACTGCGCCACCTGCTGTCCAAGGTGCTGCAGGAGAGGGACGCC   780
 241  L  Q  G  K  T  E  L  R  H  L  L  S  K  V  L  Q  E  R  D  A    260
 781 TTGAAGGGCACACGGAAGCCAGCAGTGCTGGTGAAGATTGCCCCTGACCTCACGGCCCAG   840
 261  L  K  G  T  R  K  P  A  V  L  V  K  I  A  P  D  L  T  A  Q    280
 841 GACAAGGAGGACATTGCCAGCGTGGCGAGAGAGCTGGGCATCGATGGACTGATTGTCACA   900
 281  D  K  E  D  I  A  S  V  A  R  E  L  G  I  D  G  L  I  V  T    300
 901 AACACCACAGTGAGTCGCCCTGTTGGTCTCCAGGGTGCTCTGCGCTCTGAGACAGGAGGA   960
 301  N  T  T  V  S  R  P  V  G  L  Q  G  A  L  R  S  E  T  G  G    320
 961 CTGAGTGGGAAGCCACTCCGAGATCTGTCGACTCAGACCATCCGGGAGATGTACGCCCTT  1020
 321  L  S  G  K  P  L  R  D  L  S  T  Q  T  I  R  E  M  Y  A  L    340
1021 ACCCAAGGCAGGATTCCCATTATCGGGGTTGGTGGTGTGAGCAGCGGACAGGACGCCTTG  1080
 341  T  Q  G  R  I  P  I  I  G  V  G  G  V  S  S  G  Q  D  A  L    360
1081 GAGAAGATCCAGGCAGGGGCCTCCCTGGTGCAGCTCTACACAGCCCTCATCTTCCTGGGA  1140
 361  E  K  I  Q  A  G  A  S  L  V  Q  L  Y  T  A  L  I  F  L  G    380
                                          ------------------------------
1141 CCACCAGTCGTGGTCAGGGTCAAGCGTGAACTAGAGGCCCTTCTAAAAGAGCGGGGTTTT  1200
 381  P  P  V  V  V  R  V  K  R  E  L  E  A  L  L  K  E  R  G  F    400
     ---------------
1201 ACCACAGTTACAGATGCCATTGGAGCAGATCATCGGAGGTGACCATACGTGCCAGAAGTC  1260
 401  T  T  V  T  D  A  I  G  A  D  H  R  R  *                      420
1261 CCATCCAGATTGTGCCTTCCAACTCAGGTGAGCTGTGTGACTGCACCGTGAGGAAGAGCC  1320
1321 GTCTCAATCCATGTCCCTTGAACTGATGGCCTGGACGGACTGCACAGGCCAGTCGTGGGC  1380
1381 ATCACATGCCAGTGAAGAACTTCTCTAACCACTTGAGGAGACCACAAATCCCACTGTCAC  1440
1441 TCCCTAGATCTAAATCCTGGGATTGATCAGTATCAGAAGGACATTGGCTTCTTGGGAGGA  1500
1501 AAAATCGTGGAGAAAATAAAGCCATGTAAACCTGTAAAAAAAAAAAAAAA   1550
```

Figure 2. Nucleotide and deduced aminoacid sequence of truncated rat liver DHOdehase. Amino acids of the putative transmembrane domains are underlined. Basic amino acids belonging to the putative mitochondrial targeting signal are marked by a cross.

The deduced amino acid sequence reveals a protein of about 45 kDa and shows significant homology to most of the currently known DHOdehases (see Table 1).

Table 1. Comparison of DHOdehase amino acid sequences from different organisms with rat DHOdehase.

DHOdehase from	Identity* to rat DHOdehase %	DHOdehase from	Identity* to rat DHOdehase %
Homo sapiens	87	*Samonella typhimurium*	41
Arabidopsis thaliana	53	*Plasmodium falciparum*	38
Drosophila melanogaster	50	*Bacillus subtilis*	35
Agrocybe aegerita	47	*Bacillus caldolyticus*	31
Schizosaccharomyces pombe	47	*Saccharomyces cerevisiae*	27
Escherichia coli	41	*Dictyostelium discoideum*	19

DISCUSSION

We have isolated and sequenced a truncated cDNA of DHOdehase from rat liver by means of hybridizationscreening and 5′-RACE experiments. Although the 5′ end of DHOdehase mRNA does not appear to be complete, the sequence data as obtained here, provides the most complete coding sequence (1242 bp) of a mammalian DHOdehase so far.

A comparison of the deduced amino acid sequences of rat liver DHOdehase revealed significant homologies to most of the currently known DHOdehases. The minor homology to DHOdehase from *Dictyostelium discoideum* and *Saccharomyces cerevisiae* are presumably due to divergent evolution leading to a different localization of DHOdehase in the cytosolic compartment of these species. In the other organisms cited in table 1, DHOdehase is bound either to the cytoplasmic or the mitochondrial membrane.

Hydrophobicity analyses of the aminoacid sequence according the algorithm of Kyte and Doolittle[6] revealed two hydrophobic domains: one of them near the N-terminus and the other one at the C-terminus of the protein (underlined in Fig. 2). These hydrophobic stretches could be two putative transmembrane domains responsible for membrane attachment of DHOdehase.

At the N-terminus a stretch of 27-29 residues is found, which resembles a typical targeting signal for the import into the mitochondria. Mitochondrial targeting signals normally consist of 15-70 residues at the N-terminus, predominantly comprising basic, hydrophobic, and hydroxylated residues, but lacking acid residues[7]. This signal is flanked by the N-terminal hydrophobic domain. This characteristic structure is found in most of the mitochondrial proteins of the intermembrane space or at the surface of the inner mitochondrial membrane[8].

Whether our data of DHOdehase primary structure support the stop-transfer hypothesis of mitochondrial protein import - as discussed for the *Drosophila* DHOdehase[9] - or the hypothesis of complete translocation of the new polypeptide chain to the matrix and succeeding re-export to the intermembrane space, has to be elucidated by future work.

---

*Alignments were used according the algorithm of Needlman and Wunsch[5] with a gap opening penalty of 3 and a gap extension penalty of 0.1. Identities of aminoacids are given in %.

ACKNOLEDGEMENTS

We thank Dr. M. Minet for kindly providing the cDNA of human DHOdehase.
This work was supported by W. Sander-Stiftung and Deutsche Forschungsgemeinschaft.

REFERENCES

1. M. Minet, M.-E. Dufour, F. Lacroute, Cloning and sequencing of a human cDNA coding for dihydroorotate dehydrogenase by complementation of the corresponding yeast mutant, *Gene* 121: 393 (1992).

2. G. Lakaschus and M. Löffler, Differential susceptibility of dihydroorotate dehydrogenase/oxidase to brequinar sodium (NSC 368390) *in vitro, Biochem. Pharmacol.* 43: 1025 (1992)

3. G. Lakaschus, D. Altekruse, M. Löffler, Preparation and characterization of two mammalian dihydroorotate dehydrogenases (EC 1.3.3.1), *Biol. Chem. Hoppe-Seyler* 372: 702 (1991)

4. M. Löffler, A. Rotgeri, C. Becker, G. Lakaschus, G. Schuster, Dihydroorotate dehydrogenase from rat liver: localization, characterization and sequencing of cDNA-clones, *Z. Gastroenterol.* 32: 53 (1994)

5. S.B. Needleman and C.D. Wunsch, A general method applicable to the search for similarities in the amino acid sequence of two proteins, *J. Mol. Biol.* 48: 443 (1970)

6. J.Kyte and R.F. Doolittle, A simple method for displaying the hydropathic character of a protein, *J. Mol. Biol.* 157: 105 (1982)

7. G. von Heijne, Mitochondrial targeting sequences may form amphiphilic helices, *EMBO J.* 5: 1335 (1986)

8. B.S. Glick, E.M. Beasley, G. Schatz, Protein sorting in mitochondria, *Tr. Biochem. Sci.* 17: 453 (1992)

9. J. Rawls, R. Kirkpatrick, J. Yang, L. Lacy
The *dhod* gene and deduced structure of mitochondrial dihydroorotate dehydrogenase in *Drosophila melanogaster, Gene* 124: 191 (1993)

MUTAGENESIS STUDIES OF CONSERVED RESIDUES IN MAMMALIAN DIHYDROOROTASE

Barbara H. Zimmermann[1], Nancy M. Kemling[2], and David R. Evans[2]

[1]Department of Biochemistry
University of Puerto Rico
Medical Sciences Campus
San Juan, PR 00936-5067
[2]Department of Biochemistry
Wayne State University
Detroit, MI 48201

INTRODUCTION

Dihydroorotase (DHOase[3]) (L-5,6-dihydroorotate amidohydrolase, EC 3.4.2.3) catalyzes the reversible cyclization of carbamyl aspartate to form dihydroorotate, the third step in *de novo* pyrimidine biosynthesis. In mammals, the activity is carried by the multifunctional protein CAD, which also carries the activities of the first two steps in the pathway, carbamyl phosphate synthetase (CPSase) and aspartate transcarbamylase (ATCase). The hamster CAD molecule is a hexamer consisting of six identical polypeptides, each having a molecular weight of 242 kDa. Limited proteolysis and sequence analysis have shown that the DHOase activity is associated with a discrete domain of the protein (Kelly et al., 1986; Simmer et al., 1990).

A 44 kDa fragment containing the DHOase domain has been purified from controlled elastase digests of hamster CAD (Kelly et al., 1986). The CAD DHOase domain exhibits kinetic behavior similar to the intact protein. The isolated domain exists in solution as a dimer of ~ 88 kDa. Each monomer contains a tightly bound zinc which may be involved in catalysis (Christopherson & Jones, 1979, 1980; Kelly et al., 1986). Monofunctional zinc-containing dihydroorotases have been isolated from *E. coli* (Washabaugh & Collins, 1984) and *C. oroticum* (Pettigrew et al., 1985).

---

[1]Abbreviations: CAD, multifunctional protein that catalyzes the first three steps in pyrimidine biosynthesis in higher eukaryotes; DEPC, diethylpyrocarbonate; DHOase, dihydrorotase (EC 3.5.2.3); HEPES, N-(2-hydroxyethyl)piperazine-N'-ethanesulfonic acid); kDa, kilodaltons; MES, 2-(N-Morpholino)ethanesulfonic acid.

Purine and Pyrimidine Metabolism in Man VIII, Edited by
A. Sahota and M. Taylor, Plenum Press, New York, 1995

The DHOase coding region of CAD has been sequenced (Simmer et al., 1990), and a comparison of this sequence with other DHOase sequences shows three highly conserved regions which are likely to contain active site residues and zinc ligands. Recently we reported the cloning and overexpression of the CAD DHOase domain in *E. coli*. The recombinant protein has been purified, and its kinetic parameters are similar to those of the DHOase domain produced by proteolysis (Zimmermann & Evans, 1993). Below we report our initial attempts to determine the roles of different conserved residues by site-directed mutagenesis of the recombinant DHOase.

EXPERIMENTAL PROCEDURE

E. coli strain EK1104 and pEK81, the pEK2 derived expression vector, were gifts of Dr. E. R. Kantrowitz (Nowlan & Kantrowitz, 1985). The dihydroorotase-deficient *E. coli* cell strain , X7014, cgsc 5358 (Semple & Silbert, 1975), was obtained from the Yale *E. coli* Genetic Stock Center. The minimal media was the same used in Zimmermann & Evans (1993) with minor changes. Complementation experiments were performed as described (Zimmermann & Evans, 1993).

Site-directed mutations were constructed by the method of Kunkel (1985), using a Muta-Gene Phagemid In Vitro Mutagenesis Kit, Version 2, from BioRad, and pBZ22, a plasmid encoding the recombinant CAD dihydroorotase domain (Zimmermann & Evans, 1993), and the following oligonucleotides: 5'AAGGTGCACAGCGACGTCGATC3', H1471A; 5'CTCCCGAAGGGCCACATGGACG3', H1473A; 5'GCAGGGGCATTAATGATGGGG3', D1512N; 5'CGCTCTGCATTGGCCACAATG3', H1590N; 5'AGGAAGAGGTTGTGGGGTGCG3', H1642N; 3'TCCAGGTATTGGGAGCGTGG5', H1690N; where nucleotides differing from the wild type sequences are underlined. Oligonucleotides for two of the mutants were designed to add (D1512N, Ase I) or destroy (H1473A, Apa L1) restriction sites, and the presence of the mutation was confirmed by restriction. Mutants H1590N, H1642N, and H1690N were confirmed by nucleotide sequencing. Oligonucleotides used for sequencing and for site-directed mutagenesis were synthesized by the Wayne State Macromolecular Core facility.

Protein was measured by the method of Bradford (1976) with reagents from BioRad, using bovine serum albumin as a standard. Enzyme activity was measured using the colorimetric assay (Prescott & Jones, 1969). Kinetic parameters were obtained by nonlinear least square fit to the Michaelis-Menton equation using the program MINSQ (Micromath). For the pH studies, the assay mixtures were buffered with 100 mM HEPES (pH 7.0, 7.5, 8.0) or 100 mM MES (pH 6.5, 7.0).

RESULTS AND DISCUSSION

Mutant H1471A, H1473A, H1590N, H1642N, H1690N, and D1512N were transformed into *E. coli* X7014, a DHOase deficient strain. The growth curves of cells transformed with the mutant D1512N were similar to those of cells transformed with pBZ22, the plasmid encoding the wild type recombinant DHOase (Figure 1). Cells transformed with H1642N consistently grew slightly less well than cells transformed with pBZ22. Cells transformed with mutants H1471A,

H1473A, H1590N, and H1690N did not grow, demonstrating that these mutants are unable to complement the host's deficiency.

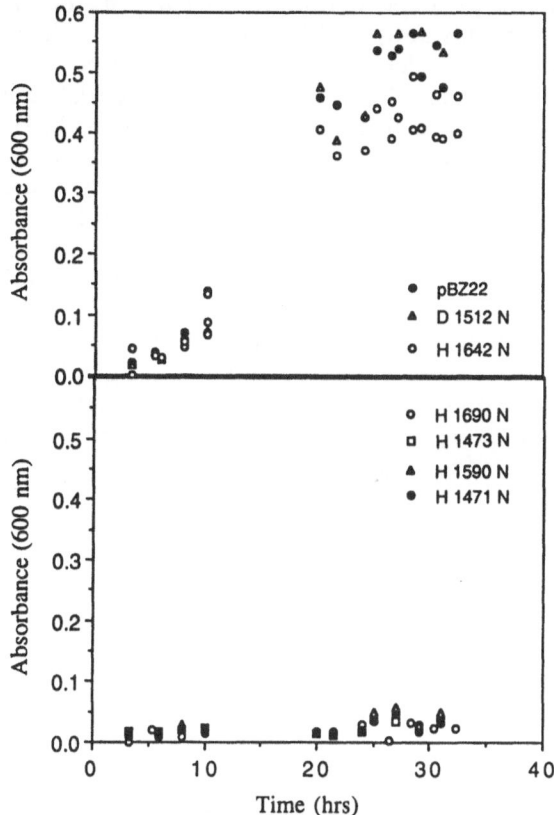

Figure 1. Growth curves showing complementation of DHOase deficient *E. coli* by wild type recombinant and mutant CAD DHOases. Cell were grown in a minimal media lacking uracil. Data points represent the combined results from three experiments.

Mutant proteins were purified following the procedure developed for the wild type protein (Zimmermann & Evans, 1993). The kinetic parameters for three of the purified mutant proteins were measured in the degradative direction at pH 8.3. Mutant H1642N exhibited 12 %, and mutant H1690N exhibited 3 % of the activity of the wild type recombinant DHOase. Both of these mutants had K_m values higher than the wild type recombinant; H1642N exhibited a 3-fold increase, and H1690N exhibited a 9-fold increase. Mutant H1590N had no measurable activity. Mutants H1471A and H1473A eluted at higher salt concentrations than the wild type recombinant, and their kinetic parameters were not determined because of contamination by *E. coli* DHOase (detected by antibodies raised against *E. coli* DHOase). Since neither of these two mutants were able to complement the DHOase deficient host, we expect that their specific activities are in the same range as H1690N and H1590N.

The work of Christopherson & Jones (1979) showed that the DHOase activity of mouse CAD is pH dependent in both the degradative and the biosynthetic directions. In the degradative direction, they found the enzyme to be 17-fold more active at pH 8.0 than at pH 6.0. To see whether either histidines 1642 or 1690 could play a role in the pH dependence, we measured the degradative activity of the corresponding mutants between pH 6.0 and pH 8.0. The activity of mutant H1642N exhibits a pH dependence similar to the wild type recombinant. The pH dependence of the activity of H1690N is greatly reduced (data not shown). Only a 2-fold difference was found in the activities at measured at pH 6.0 and pH 8.0.

Our observations can be summarized as follows. The spacing of glutamate 1512 relative to histidines 1471 and 1473 made us initially consider it as a potential zinc ligand (Vallee & Auld, 1990). However, since mutant D1512N is able to complement the DHOase deficient strain as efficiently as the wild type recombinant, this glutamate residue is clearly not involved in zinc coordination. Changing any of the five conserved histidine residues in recombinant CAD DHOase results in mutant proteins with impaired activities. Histidines 1471, 1473, and 1590, should still be considered potential zinc ligands because of the low activities of the corresponding mutants. While more detailed studies are in order, our preliminary data strongly suggest that histidine 1690 is partially responsible for the pH dependence of the DHOase.

ACKNOWLEDGMENTS

We thank Dr. E. R. Kantrowitz (Boston College, Boston, MA) for the gifts of pEK81 and E. coli strain EK1104. We thank J. Cruz for excellent technical assistance.

REFERENCES

Bradford, M. (1976) *Analyt. Biochem. 72*, 248-254.
Christopherson, R. I. & Jones, M. E. (1979) *J. Biol. Chem. 254*, 12506-12512.
Christopherson, R. I. & Jones, M. E. (1980) *J. Biol. Chem. 255*, 3358-3370.
Kelly, R. E., Mally, M. I., & Evans, D. R. (1986) *J. Biol. Chem. 261*, 6073-6083.
Kunkel, T. A. (1985) *Proc. Nat. Acad. Sci. U.S.A. 82*, 488-492.
Nowlan, S. F. & Kantrowitz, E. R. (1985) *J. Biol. Chem. 260*, 14712-14716.
Prescott, L. M., & Jones, M. E. (1969) *Anal. Biochem. 32*, 408-419.
Semple, K. S., & Silbert, D. F. (1975) *J. Bacteriol. 121*, 1036-1046.
Simmer, J. P., Kelly, R. E., Scully, J. L., Grayson, D. R., Rinker, A. G., Jr., Zimmermann, B. H., Kim, H., & Evans, D. R. (1990) *Proc. Nat. Acad. Sci. U.S.A. 87*, 174-178.
Vallee, B. L. & Auld, D. S. (1990) *Biochemistry 29*, 5647-5659.
Zimmermann, B. H. & Evans, D. R. (1993) *Biochemistry 32*, 21519-21527.

GENE MUTATIONS FOR HUMAN ERYTHROCYTE AMP DEAMINASE DEFICIENCY

Yasukazu Yamada,[1] Haruko Goto,[1] Takaji Murase,[2] and
Nobuaki Ogasawara[1]

[1]Department of Genetics, Institute for Developmental Research,
  Aichi Prefectural Colony, Kasugai, Aichi 480-03, Japan
[2]Aichi Red-Cross Blood Center, Seto, Aichi 489, Japan

INTRODUCTION

AMP deaminase (EC 3.5.4.6) is widely distributed in various mammalian cells and tissue-specific isozymes were found. In human, there are three fundamental isoforms, isozyme M found in skeletal muscle, isozyme L in liver, and isozyme E1 existing as the major isozyme in erythrocyte.[1,2] Three different genes for AMP deaminase have been identified recently: *AMPD-1* encoding isozyme M,[3,4] *AMPD-2* encoding L,[5,6] and *AMPD-3* encoding erythrocyte AMP deaminase.[7,8] The human erythrocyte AMP deaminase deficiency identified firstly by us, is clinically completely asymptomatic. The inheritance of the deficiency was autosomal recessive and the heterozygote frequency was estimated at about 1/30 in Japan, Korea and Taiwan.[9,10] The deficiency was also reported in Europe[11] and the frequency in northern Poland was almost the same as that in east Asia.

In recent study,[12] a homogeneous point mutation of C to T at nucleotide position (nt.) 1717, resulting in an amino acid change of Arg to Cys had been identified on the *AMPD-3*; two individuals with complete deficiency were homozygous for the detected mutation (C1717T), and two individuals with partial deficiency were diagnosed as heterozygous. The C1717T seems to be very frequent at least in Japanese. In this study, we screened the mutant gene to clarify the frequency and existence of other mutants.

MATERIALS AND METHODS

Screening of AMP deaminase activity. Blood samples were prepared according to the methods described previously.[9,10] AMP deaminase activity was assayed by estimating the production of ammonia.[1] One unit of enzyme activity was defined as the amount of enzyme which catalyzed the formation of 1 μmole of ammonia/min.

Purine and Pyrimidine Metabolism in Man VIII, Edited by
A. Sahota and M. Taylor, Plenum Press, New York, 1995

cDNA synthesis. cDNA synthesis was carried out according to the method described previously,[12] from B-lymphoblast (1 x 10^8 cells).

PCR amplification. The PCR mixture was consisted of 50 mM Tris-HCl (pH 8.8), 50 mM KCl, 2.5 mM MgCl$_2$, 0.2 mg/ml gelatin, 0.2 mM of each deoxyribonucleotide triphosphate, 1 μM of each primer, 2 units of *Taq* DNA polymerase (Ampli-Taq, Perkin-Elmer Cetus), and 250 ng of template DNA, cDNA or genomic DNA, in the final volume of 0.1 ml. Standard PCR amplification was carried out for 35 cycles (94°C, 1 min; 55°C, 1 min; 70°C, 2 min).

DNA sequencing. The protocol for direct sequencing was slightly modified from that of Sequenase Version 2.0 (United State Biochemical Co.), as described previously.[13] The pUC (pUC18, Boehringer Mannheim) recombinants were sequenced according to the standard protocols. DNA sequences were recorded into personal computer and analyzed with a software of gene analysis, GENETYX version 9.0 (SDC, Japan).

RESULTS AND DISCUSSION

In screening about 2600 Japanese blood samples, 61 samples showed about a half of the enzyme activity, 3.5 - 7.5 units/mg hemogrobin (Hb), compared with the control value (12.5 ± 1.5 units/mg Hb[9]) and 2 samples no enzyme activity (< 0.3 units/mg Hb). In one allele of the 45 individuals, the C1717T was detected by the analysis of genomic PCR amplification and digestion at the *Pst* I site created in the mutant gene (Table 1). But the remained 16 samples diagnosed as heterozygous from the enzyme activity, were not digested by *Pst* I, demonstrated the existence of other mutations. Two individuals with complete deficiency found in this study were homozygotes of C1717T, similar to those in previous study.[12] The frequency of the major mutation, C1717T, was calculated as about 75 % of all the mutations on *AMPD-3*.

The 16 samples, which have not the C1717T, should have other mutations on *AMPD-3* in one of alleles. We obtained four B-lymphoblast cell lines, B01, B02, B17, and B22, from the samples diagnosed as heterozygous but not having the C1717T. And, only genomic DNA samples were prepared from the other 12 samples, A01, A08, A11, A12, A20, A33, A35, A36, A37, A38, A41, and A42. Using these samples, other mutations were analyzed.

Table 1. Screening of the mutant gene (C1717T) in enzyme deficient individuals. Genomic DNA fragments including the major mutation site were amplified and then digested by *Pst* I at the site created in the mutant gene.

Deficiency (enzyme activity*)	Total	*Pst* I site		
		+ +	+ –	– –
				No.
Partial (3.5 - 7.5 units/mg Hb)	61	0	45	16
Complete (< 0.3 units/mg Hb)	2	2	0	0

* AMP deaminase activities in erythrocytes from about 2600 blood samples were assayed and the normal ranges were estimated as 9 - 15 units/mg Hb.

(A) Major mutation

```
Normal
   1690 ATG.GTG.CTC.AAC.AAC.CTC.CGC.AGG.GAG.CGC.GGC.CTG.AGC.ACG 1740
    564 Met-Val-Leu-Asn-Asn-Leu-Arg-Arg-Glu-Arg-Gly-Leu-Ser-Thr  577
C1717T                              #
   1690 ATG.GTG.CTC.AAC.AAC.CTC.TGC.AGG.GAG.CGC.GGC.CTG.AGC.ACG 1731
    564 Met-Val-Leu-Asn-Asn-Leu-Cys-Arg-Glu-Arg-Gly-Leu-Ser-Thr  577
```

(B) Other mutations

a) Deletion

```
Normal
    592 TTC.CAC.CCT.CCT.CCA.CTG.........GTG.GAC.ATG.AGC.CAC.ATC. 783
    198 Phe-His-Pro-Pro-Pro-Leu---------Val-Asp-Met-Ser-His-Ile- 261
T600D           #
    592 TTC.CAC.CC C.CTC.CAC.TGC.........TGG.ACA.TGA.GCC.ACA.TCC.783
    198 Phe-His-Pro- Leu-His-Cys---------Trp-Thr-stop               258
```

b) Point mutations

```
Normal
    955 CAT.GCG.GCC.GCC.TGC.ATG.AAC.CAA.AAG.CAT.CTG.CTG.CGC.TTC 996
    319 His-Ala-Ala-Ala-Cys-Met-Asn-Gln-Lys-His-Leu-Leu-Arg-Phe 332
C959T        #
    955 CAT.GTG.GCC.GCC.TGC.ATG.AAC.CAA.AAG.CAT.CTG.CTG.CGC.TTC 996
    319 His-Val-Ala-Ala-Cys-Met-Asn-Gln-Lys-His-Leu-Leu-Arg-Phe 332
T971C                    #
    955 CAT.GCG.GCC.GCC.TGC.ACG.AAC.CAA.AAG.CAT.CTG.CTG.CGC.TTC 996
    319 His-Ala-Ala-Ala-Cys-Thr-Asn-Gln-Lys-His-Leu-Leu-Arg-Phe 332
C991T                                               #
    955 CAT.GCG.GCC.GCC.TGC.ATG.AAC.CAA.AAG.CAT.CTG.CTG.TGC.TTC 996
    319 His-Ala-Ala-Ala-Cys-Met-Asn-Gln-Lys-His-Leu-Leu-Cys-Phe 332
```

Figure 1. Gene mutations for human AMP deaminase deficiency. The mutant with 1 bp deletion was found in B01 and B17. The point mutations were detected; T971C in B02, C991T in B22 and A12, and C959T in A36, respectively.

Analyzing cDNA by the RT-PCR method described previously[12] using B-lymphoblast cell lines, three other mutations have been identified (Figure 1). From the analysis of B01 and B17, one base deletion of T at nt. 600 (T600D) was found in one of alleles. By this mutation, a stop codon (TGA) appears at codon 258 because of frame-shift of codon from mutation site (codon 200). Single nucleotide substitution of T to C at nt. 971 (T971C) resulting in an amino acid change of Met to Thr, and that of C to T at nt. 991 (C991T) generating the change of Arg to Cys, were detected in B02 and B22, respectively.

The mutations identified by the cDNA analysis in this study were screened in genomic DNA from the other 12 samples (A series). We amplified DNA fragment including the mutation sites of T971C and C991T by genomic PCR methods,[13] and the mutations were detected by the digestion of restriction enzymes; *Bsg* I site was created by the mutation T971C and *Bbv* I site was lost by the mutation C991T. The C991T was detected in only A12 but the T971C was not found in the other samples. From sequencing analysis of the amplified DNA fragments, a mutation named as C959T has been firstly identified in A36: a single nucleotide substitution of C to T at nt. 959 results in an amino acid change of Ala to Val (Figure 1). The C959T lost *Not* I restriction site on *AMPD-3*, but it was not detected in the other samples. The T600D was also not detected in any other samples by PCR amplification and following direct sequencing. The mutations in the remained 10 samples are unclear. We are trying now to identify these mutations by genomic PCR.

The molecular basis of human myoadenylate deaminase (muscle type AMP deaminase) deficiency has been determined recently by Morisaki et al.[14] The mutant allele was found in 12 % of Caucasians and 19 % of African-Americans, whereas none of the 106 Japanese subjects surveyed has the mutant allele. In contrast, the gene frequency of the deficiency of human erythrocyte AMP deaminase shows no significant differences among Asians, such as Japanese, Korean and Taiwanese,[9,10] and Europeans represented by northern Poland.[11] The erythrocyte AMP deaminase deficiency in Japanese seems to be associated with 75 % major homogeneous mutation and 25 % other heterogeneous mutations. It is interesting to know whether the same mutation occurs in Caucasian mutants, which will give further information on the origin of the mutation. We are trying to screen mutant genes worldwide. The identification of the complete genomic structure of *AMPD-3* is also now in progress, which is necessary for the analysis of the other mutants.

ACKNOWLEDGMENTS

This work was supported by a grant from the Ministry of Education, Culture and Science of Japan, a Gout Research Foundation grant, and an Intractable Disease grant from the Ministry of Health and Welfare of Japan.

REFERENCES

1. N. Ogasawara, H. Goto, Y. Yamada, T. Watanabe and T. Asano, AMP deaminase isozymes in human tissues., *Biochim. Biophys. Acta* 714:298 (1982).
2. N. Ogasawara, H. Goto, Y. Yamada, and T. Watanabe, Distribution of AMP deaminase isozymes in various human blood cells., *Int. J. Biochem.* 16:269 (1984).
3. R.L. Sabina, R. Marquetant, N.M. Desai, K. Kaletha, and E.W. Holmes, Cloning and sequence of rat myoadenylate deaminase cDNA: evidence for tissue-specific and developmental regulation., *J. Biol. Chem.* 262:12397 (1987).
4. R.L. Sabina, T. Morisaki, P. Clarke, R. Eddy, T.B. Shows, C.C. Morton, and E.W. Holmes, Characterization of the human and rat myoadenylate deaminase genes., *J. Biol. Chem.* 265:9423 (1990).
5. T. Morisaki, R.L. Sabina, and E.W. Holmes, Adenylate deaminase: a multigene family in humans and rats., *J. Biol. Chem.* 265:11482 (1990).
6. M.T. Bausch-Junken, D.K. Mahnke-Zizelman, T. Morisaki, and R.L. Sabina, Molecular cloning of AMP deaminase isoform L. Sequence and bacterial expression of human *AMPD2* cDNA., *J. Biol. Chem.* 267:22407 (1992).
7. Y. Yamada, H. Goto, and N. Ogasawara, Cloning and nucleotide sequence of the cDNA encoding human erythrocyte-specific AMP deaminase., *Biochim. Biophys. Acta* 1171:125 (1992).
8. D.K. Mahnke-Zizelman, and R.L. Sabina, Cloning of human AMP deaminase isoform E cDNAs. Evidence for a third AMPD gene exhibiting alternatively spliced 5'-exons., *J. Biol. Chem.* 267: 20866 (1992).
9. N. Ogasawara, H. Goto, Y. Yamada, I. Nishigaki, T. Itoh, and I. Hasegawa, Complete deficiency of AMP deaminase in human erythrocytes., *Biochem. Biophys. Res. Commun.* 122:1344 (1984).
10. N. Ogasawara, H. Goto, Y. Yamada, I. Nishigaki, T. Itoh, I. Hasegawa, and K.S. Park, Deficiency of AMP deaminase in human erythrocytes., *Hum. Genet.* 75:15 (1987).
11. M.M. Zydowo, J. Purzycka-Preis, and N. Ogasawara, Deficiency of AMP deaminase in human erythrocytes., *Adv. Exp. Med. Biol.* 253A:31 (1989).
12. Y. Yamada, H. Goto, and N. Ogasawara, A point mutation responsible for human erythrocyte AMP deaminase deficiency., *Hum. Mol. Genet.* 33:331 (1994).
13. Y. Yamada, H. Goto, K. Suzumori, R. Adachi, and N. Ogasawara, Molecular analysis of five independent Japanese mutant genes responsible for hypoxanthine guanine phosphoribosyltransferase (HPRT) deficiency., *Hum. Genet.* 90:379 (1992).
14. T. Morisaki, M. Gross, H. Morisaki, D. Pongratz, N. Zollner, and E.W. Holmes, Molecular basis of AMP deaminase deficiency in skeletal muscle., *Proc. Natl. Acad. Sci. USA* 89:6457 (1992).

POINT MUTATIONS IN <u>PRPS1</u>, THE GENE ENCODING THE PRPP SYNTHETASE (PRS) 1 ISOFORM, UNDERLIE X-LINKED PRS SUPERACTIVITY ASSOCIATED WITH PURINE NUCLEOTIDE INHIBITOR-RESISTANCE

M.A.Becker, J.M.Nosal, R.L.Switzer, P.R.Smith, T.D.Palella, B.J.Roessler

University of Chicago, Chicago, IL; University of Illinois, Urbana, IL; University of Michigan, Ann Arbor, MI, USA

INTRODUCTION

The aim of these studies was to establish the genetic basis of PRPP synthetase (PRS) superactivity, an X chromosome-linked disorder of purine metabolism characterized by gout, uric acid overproduction[1], and, in some families, neurologic and/or developmental impairment[2]. PRPP, an allosteric activator and substrate in the pathway of purine nucleotide synthesis de novo, is synthesized from ATP and Rib-5-P in a reaction dependent on Pi and Mg^{2+} both as cofactors and activators. The reaction is inhibited by purine, pyrimidine, and pyridine nucleotide products of the pathways of PRPP utilization as well as AMP and PRPP, the products of the PRS reaction, and certain additional phosphorylated compounds[3].

Kinetic defects underlying PRS superactivity (reviewed in reference 2) include: 1) regulatory defects characterized by purine nucleotide inhibitor-resistance and increased apparent affinity of PRS for Pi; 2) catalytic defects in which maximal reaction velocity is increased but substrate and activator affinity and inhibitor responsiveness are normal; 3) combined regulatory and catalytic defects; and 4) increased affinity for the substrate Rib-5-P.

Catalytic defects have been the most common class of kinetic alteration encountered in the nearly 30 families with PRS superactivity identified to date[4]. Phenotypic expression of catalytic superactivity is most often limited to gout and/or uric acid urolithiasis in young adult males. In contrast, neurodevelopmental abnormalities, infantile or childhood-onset in males, and milder but definite clinical expression of the metabolic and/or neurologic features in heterozygous women characterize 5 of the 6 families in which regulatory defects constitute all or part of the kinetic aberration underlying PRS superactivity[3,5,6]. The exception is the family described by Drs. Sperling and deVries[1] in which purine nucleotide resistance is associated with the metabolic features only. In fibroblasts cultured from affected males, the key metabolic derangements, accelerated PRPP and purine nucleotide synthesis, are more severe in cells with regulatory defects than in those with catalytic superactivity[7]. The precise genetic defects accounting for PRS superactivity and the clinical consequences have not to date been defined.

In 1987, Professor Tatibana and his group reported the cloning and characterization of highly homologous but distinct rat PRS cDNAs and provided evidence that each was encoded by a separable gene[8]. They also reported that of the 3 corresponding human genes, 2 mapped to the X chromosome and the third to chromosome 7p[9]. Subsequent cloning and characterization of human PRS cDNAs have provided the basis for defining the genetic defects in human PRS superactivity. Human PRS cDNAs (referred to as PRS1 through PRS3 cDNAs) identify mRNA transcripts of 2.3, 2.7, and 1.4 kb, respectively[10,11]. PRS1 and PRS2 transcripts are encoded by genes (<u>PRPS1</u> and <u>PRPS2</u>) that map, respectively, to the long arm (Xq22-q24) and the tip of the short arm (Xp22.2-p22.3) of X. In the 954-base pair translated regions of PRS1 and PRS2 cDNAs, sequence identity is 81% and the predicted amino acid sequence identity is 96%. Expression of human PRS1 and PRS2 cDNAs has been achieved in E. coli, and the purified recombinant isoforms have been compared with regard to enzymatic, physical, and protein chemical properties[12]. Both recombinant isoforms are active in aggregates of multiple identical subunits and are distinguishable in some kinetic and protein chemical properties.

Tissue-differential expression of human PRS transcripts has been demonstrated[10], with findings compatible with the view that the PRPS1 gene is constitutively expressed and predominates in cells and tissues with modest growth rates while PRPS2 gene expression may be enhanced in rapidly growing tissues and in response to growth-promoting factors and viral transformation. Autosomal PRPS3 gene expression appears to be limited to testicular tissue.

EXPERIMENTAL PROCEDURES

Toward our aim of establishing the molecular pathology underlying PRS superactivity, we examined first the X chromosome-encoded PRS1 and PRS2 transcripts in cells cultured from affected males in families in which defects in PRS regulation comprise at least a portion of the respective kinetic abnormalities[1,3,5,6]. Cultured fibroblasts and, where available, B lymphoblasts from 6 unrelated families were studied. The experimental approach involved: isolation of total cellular RNA from cultures in late log phase growth; reverse transcription of RNA into cDNA, primed with a PRS1-or PRS2-specific antisense primer; and PCR amplification of PRS cDNAS, catalyzed by Taq DNA polymerase and primed with appropriate oligodeoxynucleotide primer pairs corresponding to the respective 5' and 3' untranslated sequences of normal human PRS1 and PRS2 cDNAs[13,14]. With this approach, the complete translated regions of the PRS1 and PRS2 cDNAs were amplified. In initial studies, the cDNAs were sequenced directly from PCR mixtures by a modification of the method of Sanger[14], utilizing both the specific amplification primers and a series of PRS consensus primers, which hybridize to translated sequences of both PRS1 and PRS2 cDNAs[13].

For the preparation of PRS1 cDNAs for insertion into the plasmid expression vector and confirmation of the sequence of these constructs, a second round of PCR was performed, utilizing the nested primer approach[14]. Amplified PRS1 cDNA sequences with appropriate restriction sites were cloned into the corresponding restriction sites in the plasmid vector pSPRBS[13]. E. coli strain DH5 ∝ was transformed with these constructs, and plasmids with the appropriate PRS1 cDNA sequences were in turn used to transform E. coli BL21 (DE3)/pLysS, a bacterial strain lysogenized 2with DE3 phage bearing the T7 RNA polymerase gene under the control of the IPTG-inducible lac UV5 promoter[15]. Transformed host cells were grown at 37°C to an optical density of 0.4 prior to addition of 0.4 mM IPTG. Two hours later, bacterial cells were harvested, resuspended, and disrupted in a lysis buffer. Centrifuged bacterial supernatants were serially diluted in lysis buffer[12] containing bovine serum albumin and were assayed for PRS activity. Appropriate dilutions of the extracts were employed to study the kinetics of the PRS1 reaction, including affinities for substrates, activation by Pi, and purine nucleotide inhibitor responsiveness. Portions of the extracts were also subjected to SDS PAGE electrophoresis followed by Coomassie blue staining and by immunoblotting[12,13].

RESULTS AND DISCUSSION

The complete translated regions of normal and patient-derived PRS1 and PRS2 cDNAs were sequenced. All PRS2 cDNA sequences were identical throughout the entire translated region. In contrast, PRS1 cDNA sequences derived from the transcripts of each of the 6 unrelated affected patients differed from one another and from normal PRS1 cDNA, in each instance by a single base substitution predicting replacement of the corresponding amino acid residue in the primary sequence of normal PRS1 with a different amino acid residue (Table 1). Each alteration in PRS1 cDNA sequence was confirmed on both strands of the respective cDNA and on multiple independent isolates of total cellular RNA from cultured cells. We next sought to establish, by means of expression of PRS1 cDNAs in E. coli, the relationship between the single base substitutions in the patient-derived PRS1 cDNAs and the abnormal kinetic properties of the respective PRSs, previously identified in the cells of the affected individuals. For this purpose, an IPTG-inducible T7 RNA polymerase-T7 promoter system utilizing a plasmid expression vector was employed[15]. Expression of normal and patient-derived PRS1 cDNAs was achieved, with recombinant human PRS1 representing from 3 to 5% of total bacterial cell extract protein, as assessed by direct assay of PRS specific activity[12] and by results of SDS-PAGE analysis of bacterial cell extracts under conditions of IPTG induction. Induction of cells transformed with the pSPRBS vector devoid of PRS1 cDNA, yielded a protein pattern with a zone clear of identifiable bands in an area corresponding to about 34-35 kDa after Coomassie blue staining of SDS-PAGE gels. When, however, normal or patient-derived PRS1 cDNA was included in the vector construct, a band corresponding to a protein of molecular weight 34.5 kDa occupied this zone. The molecular weight of this recombinant protein corresponds closely to previous estimates of human PRS1, including that predicted from human PRS1 cDNA sequence. Immunoblotting with the IgG fraction of rabbit polyclonal antiserum raised in response to immunization with purified normal human erythrocyte PRS confirmed the identity of the recombinant protein as PRS1.

TABLE I. Mutations in PRS1s

Patient	Base substitution	Amino acid replacement
NB	A341→G	Asn113→Ser
SM	G547→C	Asp182→His
AL	C569→T	Ala189→Val
RD	C385→A	Leu128→Ile
VRG	C579→G	Phe192→leu
OG	G154→C	Asp 51→His

Activities of recombinant normal and patient-derived PRS1s were studied at 1.0 mM Pi in bacterial cell extracts prepared after IPTG induction. E. coli PRS activity generally constituted less than 1% of total PRS activity under conditions of these studies. No major differences between recombinant normal and patient-derived PRS1s were found with respect to dissociation constants for the substrates, MgATP and Ribe-5-P. In contrast, significant resistance to inhibition of the PRS reaction by ADP and GDP was demonstrable for each of the recombinant PRS1s from the 6 affected patients in whom single base substitutions were identified in the respective PRS1 cDNAs (Table 2). In all instances, the concentrations of the inhibitors at which 50 percent inactivation of the PRS reaction was achieved ($I_{0.5}$) were at least several-fold higher than the corresponding concentrations necessary to inhibit recombinant normal PRS1, and inhibitor resistance was particularly striking for GDP.

Patterns of purine nucleotide inhibitor resistance varied among the 6 mutant PRS1s but generally paralleled those found in measuring PRS inhibitor responsiveness in fibroblast extracts from the respective patients[1,3,5,6]. The degree of resistance of recombinant mutant PRS1s to ADP and GDP inhibition was, however, 2-to 3-fold greater than was the case for PRS from extracts of cell cultured from the respective patients. Human fibroblasts and lymphoblasts express both PRS1 and PRS2 transcripts[10], and it is likely that PRS activities in these cells represent a composite of the independent activities of these isoforms. Of note in this regard is the finding that purified recombinant PRS2 is more resistant to both ADP and GDP inhibition than is recombinant PRS1[12]. The greater resistance to nucleotide inhibition seen in recombinant PRS1s compared with PRSs in cultured cells may thus provide a more accurate assessment of the effects of the respective point mutations on the allosteric control of PRS1 activity. Overall, the association of point mutations in PRS1 cDNAs with alterations in PRS1 regulatory properties strongly supports the concept that PRS superactivity with purine nucleotide inhibitor resistance is a consequence of point mutation in the PRPS1 gene.

The mechanisms mediating the allosteric control of PRS1 activity by nucleotide inhibitors await delineation, which is likely to require a better understanding of the relationship between the structure of PRS1 and the binding of substrates, activators, and inhibitors to this complex enzyme. Nevertheless, a rather widespread array of mutations in PRS1 in this cohort of patients results in nucleotide inhibitor resistance, suggesting that a major portion of the PRS1 polypeptide is involved in the transmission of allosteric effects to the active site. Whether the location of mutations along the length of the polypeptide correlates with either the severity of regulatory derangements in enzyme function or the clinical phenotype encountered in the involved families are points currently under investigation, as is the genetic basis of catalytic superactivity of PRS.

TABLE 2. Purine Nucleotide Inhibition of rPRS1s

rPRS1	$I_{0.5}$ for	
	ADP	GDP
	uM	
Normal	20	50
NB	76	700
SM	140	810
AL	70	730
RD	106	>1000
VRG	107	---
OG	75	220

ACKNOWLEDGEMENTS

This research was supported by United States Public Health Service Grants DK-28554, DK-13488, and DK-38932 and by a Grant from the Arthritis Foundation, Illinois Chapter. The authors acknowledge the excellent manuscript preparation by Ms. Sandy Crane.

REFERENCES

1. O.Sperling, P.Boer, S.Brosh, E.Zoref, and A.deVries, Superactivity of phosphoribosylpyrophosphate synthetase, due to feedback resistance, causing purine overproduction and gout, in: "Purine and Pyrimidine Metabolism" Ciba Foundation Symposium 48, Elsevier, Amsterdam (1977).
2. M.A.Becker, J.G.Puig, F.A.Mateos, M.L.Jimenez, M.Kim, and H.A.Simmonds, Inherited superactivity of phosphoribosylpyrophosphate synthetase: association of uric acid overproduction and sensorineural deafness, Am. J. Med. 85:383 (1988).
3. I.H.Fox and W.N.Kelley, Human phosphoribosylpyrophosphate synthetase: kinetic mechanism and end product inhibition, J. Biol. Chem. 47:2126 (1972).
4. M.A.Becker, M.J.Losman, A.L.Rosenberg, I.Mehlman, D.J.Levinson, and E.W.Holmes, Phosphoribosylpyrophosphate synthetase superactivity. A study of five patients with catalytic defects in the enzyme, Arthritis Rheum. 29:880 (1986).
5. M.A.Becker, M.J.Losman, J.Wilson, and H.A.Simmonds, Superactivity of phosphoribosylpyrophos-phate synthetase due to altered regulation by nucleotide inhibitors and inorganic phosphate, Biochem. Biophys. Acta 882:168 (1986).
6. M.A.Becker, K.O.Raivio, B.Bakay, W.B.Adams, and W.L.Nyham, Variant phosphoribosylpyrophos-phate synthetase altered in regulatory and catalytic functions, J. Clin. Invest. 65:109 (1980).
7. M.A.Becker, M.J.Losman, and M.Kim, Mechanisms of accelerated purine nucleotide synthesis in human fibroblasts with superactive phosphoribosylpyrophosphate synthetases, J. Biol. Chem. 262:5596 (1987).
8. M.Taira, S.Ishijima, K.Kita, K.Yamada, T.Iizasa, and M.Tatibana, Nucleotide and deduced amino acid sequences of two distinct cDNAs for rat phosphoribosylpyrophosphate synthetase, J. Biol. Chem. 262:14867 (1987).
9. M.Taira, J.Kudoh, S.Minoshima, T.Iizasa, H.Shimada, Y.Shimizu, M.Tatibana, and N.Shimizu, Localization of human phosphoribosylpyrophosphate synthetase subunit I and II genes (PRPS1 and PRPS2) to different regions of the X chromosome and assignment of two PRPS1-related genes to autosomes, Somatic Cell Mol. Genet. 15:29 (1989).
10. M.Taira, T.Iizasa, K.Yamada, H.Shimada, and M.Tatibana, Tissue-differential expression of two distinct genes for phosphoribosylpyrophosphate synthetase and existence of the testis-specific transcript, Biochem. Biophys. Acta 1007:203 (1989).
11. M.A.Becker, S.A.Heidler, G.I.Bell, S.Seino, M.M.LeBeau, C.A.Westbrook, W.Neuman, L.J.Shapiro, T.K.Mohandas, B.J.Roessler, and T.D.Palella, Cloning of cDNAs for human phosphoribosylpy-rophosphate synthetases 1 and 2 and X chromosome localization of PRPS1 and PRPS2 genes, Genomics 8:555 (1990).
12. J.M.Nosal, R.L.Switzer, and M.A.Becker, Overexpression purification and characterization of recombinant human 5-phosphoribosyl-1-pyrophosphate synthetase isozymes I and II, J. Biol. Chem. 268:10168 (1993).
13. B.J.Roessler, J.M.Nosal, P.R.Smith, S.A.Heidler, T.D.Palella, R.L.Switzer, and M.A.Becker, Human X-linked phosphoribosylpyrophosphate synthetase superactivity is associated with distinct point mutations in the PRPS1 gene, J. Biol. Chem. 268:26476 (1993).
14. F.M.Ausubel, R.Brent, R.E.Kingston, D.D.Moore, J.G.Seidman, J.A.Smith, and K.Struhl. "Current Protocols in Molecular Biology," Wiley Interscience, New York (1987).
15. F.W.Studier, A.H.Rosenberg, J.J.Dunn, and J.W.Dubendorff, Use of T_7 polymerase to direct expression of cloned genes, Methods Enzymol. 185:60 (1990).

2', 3'-DIDEOXYCYTIDINE PHOSPHORYLATION BY RECOMBINANT MOUSE AND HUMAN DEOXYCYTIDINE KINASE

Magnus Johansson and Anna Karlsson

Medical Nobel Institute for Biochemistry
Department of Medical Biochemistry and Biophysics
Karolinska Institute
S-171 77 Stockholm, Sweden

INTRODUCTION

Deoxycytidine kinase (dCK, EC 2.7.1.74) is an important enzyme for activation of several anti-viral and anti-cancer nucleoside analogs [1,2]. One of the dCK activated compounds is 2',3'-dideoxycytidine (ddC), presently in clinical use for treatment of infection caused by the human immunodeficiency virus (HIV). The phosphorylation of ddC by dCK is a prerequisite for the anti-HIV activity of this compound [3]. Mouse cells and mouse models are extensively used to study the metabolism and toxicity of nucleoside analogs of interest for clinical use. The relevance of these data for the metabolism in human cells is dependent on detailed knowlege on the metabolism of the analogs by the different species.

Earlier studies have indicated differences between human and mouse cells in the metabolism of several nucleosides and nucleoside analogs phosphorylated by dCK [3,4]. In order to determine if there are differences at the molecular level between mouse and human dCK we have cloned and expressed both enzymes in E.coli. Here we report on similarities and differences in the phosphorylation of ddC by recombinant mouse and human dCK using ATP or UTP as phosphate donor.

MATERIALS AND METHODS

The cDNA encoding mouse and human dCK were cloned and expressed in the QIAexpress System (Qiagen) as described earlier [5].

Deoxycytidine kinase activity was assayed with 100 µM [^{3}H]-dCyd or [^{3}H]-ddC (Moravek Inc.) as substrates in a reaction mixture containing 50 mM Tris-HCl, pH 7.6, 5 mM MgCl$_2$, 5 mM dithiothreitol, 10 mM NaF and 5 mM ATP or UTP. 2 mg human or 0.8 mg mouse dCK was added to a total volume of 50 µl. After incubation for 40 min at 37° C, 10 µl was transferred to a PEI-cellulose thin layer chromatography (TLC) sheet. The monophosphate products were separated from the nucleosides by a buffer of butyric acid:H$_2$O:ammonia (66:33:1). The TLC sheets were subsequently cut in pieces, eluted in 0.1 M HCl and 0.2 M KCl and finally counted in a scintillation counter.

RESULTS

The kinetic constants for mouse and human dCK were determined (data not shown). Their ability to phosphorylate deoxycytidine (dCyd) was very similar with a K$_m$ of 0.1 µM for both enzymes. The K$_m$ for ddC differed, however, between the two species studied. While human dCK had a K$_m$ of approximately 700 µM the K$_m$ for the human enzyme was instead 200 µM.

Since both ATP and UTP can act as phosphate donor, a comparison of mouse and human dCK was performed using 100µM ddC as substrate and 5 mM ATP or 5 mM UTP as phosphate donor (figure 1). The phosphorylated product of ddC was separated by TLC and was easily distinguishable from phosphorylated dCyd. For both mouse and human dCK the activity was higher using UTP as phosphate donor. The activity was overall higher in the mouse enzyme preparation both for dCyd and ddC phosphorylation.

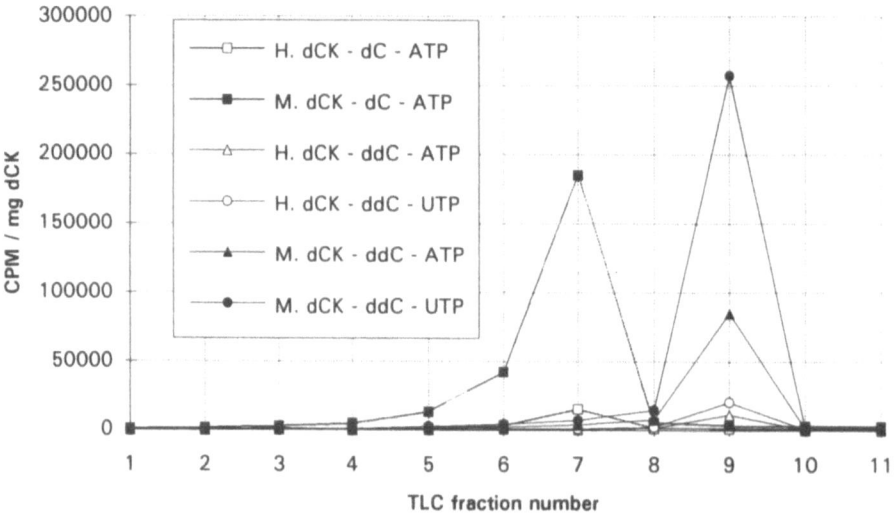

Figure 1. TLC chromatogram of dCyd and ddC phosphorylated products by mouse (M) and human (H) deoxycytidine kinase (dCK) using ATP or UTP as phosphate donor.

Since the kinetic parameters previously have been determined showing similar K_m and V_{max} values for dCyd phosphorylation by mouse and human dCK, the ability for the enzymes to phosphorylate ddC was calculated in relation to their ability of dCyd phosphorylation (table 1). These results show that at a substrate concentration of 100μM ddC the most efficient phosphorylation for both mouse and human dCK was obtained by using 5 mM UTP as phosphate donor. Using ATP as phosphate donor, however, the activity decreased and the two enzymes showed different phosphorylation rates as compared to dCyd phosphorylation. Of the two species studied, the human enzyme showed a higher relative phosphorylation of ddC as compared to the mouse enzyme.

Table 1. The relative phosphorylation of ddC and dCyd by mouse and human deoxycytidine kinase.

Enzyme substrate 100μM ddC	Phosphate donor 5.0 mM	CPM_{ddC} / mg dCK	CPM_{ddC} / CPM_{dC}
H. dCK	ATP	11100	0.73
	UTP	20500	1.3
M. dCK	ATP	85300	0.46
	UTP	257000	1.4

DISCUSSION

We recently cloned the cDNA encoding mouse dCK [5]. The human dCK cDNA sequence was published earlier [6]. Great homology between the two enzymes was revealed with only 16 of the 260 amino acids being different in the mouse enzyme as compared to the human enzyme. Both mouse and human dCK were expressed in the same vector system to enable a direct comparison of the substrate specificity of the two enzymes.

Earlier reports on differences in the phosphorylation of the anti-HIV compound ddC between mouse and human dCK was based on studies with purified or partially purified enzyme preparations from human or mouse tissue [4.] We presently study the substrate specificity and kinetic constants for several nucleoside analogs for recombinant mouse and human dCK in order to elucidate if any differences are determined by the primary sequence of the proteins. Preliminary data suggest differences between the two species in their affinity for the different phosphate donors ATP and UTP (Johansson and Karlsson, unpublished data).

The importance of UTP as the most critical phosphate donor for dCK has recently been revealed [7,8]. We now conclude that this is valid also for the phosphorylation of ddC. We further demonstrate differences in kinetic parameters for ddC phosphorylation between mouse and human dCK that may be important for different *in vitro* effects of this compound determined by the cell species.

ACKNOWLEDGMENTS

This work was supported by grants from the Medical Faculty of the Karolinska Intitute and the Swedish Society for Medical Research.

REFERENCES

1. Plunkett W and Saunders P.P, Metabolism and Action of Purine Nucleoside Analogs, *Pharmacol. Ther.* 49 (3), 239-68 (1991)
2. Eriksson S, Kierdaszuk B, Munch-Petersen B, Öberg B and Johansson N-G, Comparison of the Substrate Specificities of Human Thymidine Kinase 1 and 2 and Deoxycytidine Kinase Toward Antiviral and Cytostatic Nucleoside Analogs, *Biochem. Biophys. Res. Commun.* 176(2); 586-592 (1991)
3. Balzarini J., Pauwels R., Baba M., Herdewijn P., De Clercq E., Broder S. and Johns D.G. The *In Vitro* and *In Vivo* Anti-Retrovirus Activity and Intracellular Metabolism of AZT and ddC are Highly Dependent on the Cell Species, *Biochem. Pharmacol.* 37:897-903 (1988)
4. Habteyesus A., Nordenskjöld A., Bohman C. and Eriksson S. Deoxynucleoside Phosphorylating Enzymes in Monkey and Human Tissues Show Great Similarities, While Mouse Deoxycytidine Kinase has a Different Substrate Specificity, *Biochem. Pharmacol.* 42: 1829-36 (1991)
5. Karlsson A., Johansson M. and Eriksson S. Cloning and Expression of Mouse Deoxycytidine Kinase, *Submitted* (1994)
6. Chottiner E.G., Shewach D.S., Datta N.S., Ashcraft E., Gribbin D., Gingsburg D., Fox I.H. and Mitchell, B.S. Cloning and Expression of Human Deoxycytidine Kinase cDNA, *Proc. Natl. Acad. Sci. USA,* 88:1531-5 (1991)
7. White J.C. and Capizzi R.L., A Critical Role for Uridine Nucleotides in the Regulation of Deoxycytidine Kinase and the Concentration Dependence of 1-D-Arabinofuranosylcytosine Phosphorylation in Human Leukemia Cells, *Cancer Res,* 51: 2559-2565 (1991)
8. Shewach D.S., Reynolds K.K. and Hertel L., Nucleotide Specificity of Human Deoxycytidine Kinase, *Mol. Pharmacol,* 42: 518-524 (1992)

FEEDBACK OF *S. CEREVISIAE* CPSase-ATCase : SELECTION, CLONING AND SEQUENCING OF MUTANT ALLELES

L. Jaquet, M. Lollier, O. Navratil, A. Schoendorf, V. Brondani, J.L. Souciet and S. Potier

Laboratoire de microbiologie et génétique
URA n°1481 Université Louis-Pasteur /CNRS
28, rue Goethe
F-67083 Strasbourg France

INTRODUCTION

Allosteric enzymes are implicated in regulatory mechanisms, which include negative feedback control (Wiley and Lipscomb, 1968; O'Donovan and Neuhard, 1970; Messenger and Zalkin, 1979; Zalkin *et al.*, 1984). In this case, enzymes are generally regulated by effectors such as the final product of a pathway that differ from the substrates of the enzyme. By this mechanism, an increase in the concentration of the final product leads to a subsequent inhibition of its production.

A key step in the understanding of this kind of enzyme regulation process has been accomplished with the studies of the *E. coli* allosteric aspartate transcarbamylase (ATCase; EC 2.1.3.2). This enzyme catalyses the second step of the *de novo* pyrimidine pathway, the carbamylation of the amino group of aspartate by carbamyl phosphate to produce *N*-carbamyl-L-aspartate and inorganic phosphate. It consists of two different polypeptides, a catalytic one of 34 kDa which is an aggregate that is composed of two catalytic trimers (2c3) and a regulatory one of 17 kDa which is an aggregate composed of three dimers (3r2) (Wiley and Lipscomb, 1968). Catalytic subunits are enzymatically active, however, they exhibit no allosteric transitions and regulatory subunits bind allosteric effectors but are inactive. ATCase is subject to negative feedback by CTP which is synergistic with UTP (Wild et al, 1989) and is activated by ATP (Gerhart and Pardee, 1962). Recently, crystallographic studies provided evidence for the molecular basis of these allosteric transitions (Kantrowitz and Lipscomb; 1990) and the identification of amino acids involved in the binding of the allosteric effector and in the transmission of the signal to the ATCase catalytic site (Kosman et al., 1993).

In higher eukaryotes, ATCase is a multifunctional protein and is associated with the first and the third enzymatic activities of pyrimidine pathway, namely the carbamoylphosphate synthetase (CPSase ; EC 6.3.5.5) and dihydroorotase (DHOase ; EC 3.5.2.3). This is arranged as CPSase - DHOase - ATCase (Padgett et al., 1982 ; Freund and Jarry, 1987).

In *Saccharomyces cerevisiae*, the same organization was found except for the fact that the DHOase domain is non functional (Souciet et al., 1989). The protein encoded by the *URA2* gene called CPSase - ATCase (Mr = 240 000) is oligomeric, comprising a unique polypeptide chain carrying both catalytic and regulatory functions (Lee et al, 1985). The ATCase domain exhibits 42% amino acid identity with the *E. coli* catalytic subunit (Nagy et al, 1989). Negative feedback control is the most effective regulatory mechanism of the *de novo* pyrimidine pathway (Potier et al., 1990); both CPSase and ATCase are inhibited by UTP (Aitken et al., 1973 ; Liljelund and Lacroute, 1986) and there is no activation.

Our goal is to study the yeast feedback inhibition mechanism on a molecular level as a eukaryotic model of an allosteric process, with respect to the identification of amino acids important in negative feedback control. For this purpose, we have chosen to select missense mutations that are affected in feedback control by UTP without altering enzymatic activities. In yeast, it is possible to positively select these mutant strains *in vivo* as those resistant to 0.01 M 5-fluorouracil (5-FU, an analog of uracil), because the lack of feedback inhibition leads to an enrichment of the UTP pool and a dilution of toxic 5-FUTP (Jund and Lacroute, 1970; Denis-Duphil and Lacroute, 1971). Moreover, CPSase-ATCase protein has not been purified yet and it was irrelevant to perform site directed mutagenesis on the cloned *URA2* gene.

RESULTS AND DISCUSSION

Selection procedure

Resistance to 5-FU could potentially be due to mutations in three other loci, *FUR1*, *FUR3* and *FUR4*, but all these mutations are recessive or semi-dominant. By contrast, previous *URA2* feedback mutants were described as dominant (Jund and Lacroute, 1970). As in *S. cerevisiae*, it is possible to select mutants in haploid as well as in diploid strains, we have used a diploid one in order to counterselect *fur1*, *fur3* or *fur4* mutants.

Moreover, the ability to clone in a single step the corresponding alleles is also added to the positive selection procedure. A diploid strain was deleted for a large part of both the *URA2* chromosomic alleles, and was transformed by a low copy number plasmid carrying a unique functional *URA2* allele. Mutated alleles are then easily recovered by preparing the plasmid. Furthermore, even when CPSase activity may be complemented by the CPSase of arginine pathway encoded by genes *CPA1* and *CPA2* (Lacroute et al, 1965), the mutants might retain sufficient ATCase activity in order to grow on minimal medium. However, since the ATCase domain is located on the C-terminus of CPSase-ATCase protein, the only kind of mutations to be identified are missense ones.

In the first screening of mutants, we have identified 2 spontaneous and 1 UV-induced mutants among 14 5-FU resistant strains into which resistance was linked to the plasmid (Jaquet et al, 1993). Assays on the ATCase of crude extracts have revealed that enzymatic activity was conserved but feedback control by UTP was abolished. CPSase was functionally active and feedback on this activity was affected *in vivo*. In the 11 other mutants, resistance was due to undetermined chromosomic mutations.

Sequence of mutant alleles

Complete sequencing of the three *URA2* plasmid alleles called *URA2-119*, *URA2-200* and *URA2-202* confirmed that mutations are missense. Moreover, only one amino acid change occurred in each allele in CPSase as well as in ATCase domains. These single mutations are specifically responsible for the lack of feedback without drastic alteration of catalytic activities since mutants conserve both a CPSase[+] and ATCase[+] phenotype.

Genetic mapping

At this step, it was demonstrated that our selective scheme was efficient. Nevertheless, it can be improved so that the sequencing of the entire *URA2* allele (7kb) of each mutant can be avoided. Since there is no easy way to genetically map a putative dominant mutation, we have successfully developed a functional mapping test. Haploid strains containing different chromosomal deleted alleles previously constructed (Potier et al., 1987) were transformed by 2 kb linear restriction fragments of plasmid mutant allele of *URA2-119*, *URA2-200* and *URA2-202*. Integrative events by homologous recombination at the *URA2* locus following a one step replacement method (Rothstein, 1991) were selected. The specific fragment containing the feedback mutation confered both prototrophy for uracil and 5-FU resistance. Therefore, sequencing of 2kb is sufficient to identify functional feedback mutation.

Moreover, in this construction, the mutant allele is reintroduced within the chromosome in its original locus. Expression of the gene may be then studied in a physiological situation by contrast with a gene carried by a plasmid. For instance, crossing of the subsequent strains with a haploid wild type one allowed us to confirm that the 3 alleles are dominant.

Selection of 13 new mutants

In a second round of spontaneous mutant selection, we collected 13 other independent strains affected in the feedback process in a population of about 100 5-FU resistant strains. In all of these, as in the previous three, a single amino acid was changed and catalytic activities were retained.

By genetic mapping, 5 of the 16 mutations selected in this round and in the previous one were located within the CPSase domain and 11 within the ATCase domain. This result is not surprising considering that both activities are subject to feedback inhibition. Furthermore, since no mutations were located within the DHOase-like domain, it is obvious that this domain does not exert a role in the process.

All of the 16 mutants are dominant or semi-dominant. This observation suggests an association between two or more oligomers.

Role of the modified amino acids in oligomeric organization

The role of the modified amino acids in feedback inhibition must be discussed in relation to a possible role in the oligomeric organization of the CPSase-ATCase protein.

Oligomeric association is necessary for the ATCase activity in *E. coli* (Kantrowitz and Lipscomb, 1988) and in hamster cells (Lee et al., 1985). In *E. coli*, ATCase catalytic sites are formed by the contact of two catalytic chains of the same trimer and involve amino acids of these two chains. Active site residues are perfectly conserved in *S. cerevisiae* ATCase domain (Nagy et al., 1989), hamster protein (CAD) ATCase domain (Simmer et al., 1989) and *E. coli* catalytic subunit (Kantrowitz and Lipscomb, 1988).

In other words, oligomeric association is probably necessary for ATCase activity in yeast. Since mutants are functionally ATCase$^+$, we suggest that amino acid changes do not disturb oligomeric association of CPSase-ATCase. These observations are strengthened by mutants that have been selected which are dominant or semi-dominant.

Amino acid changes seem to directly affect the feedback UTP site or to interrupt signal transmission between regulatory and catalytic sites.

PERSPECTIVES

The selection scheme permits us to positively select *in vivo* single amino acid changes specifically affecting feedback. It is a suitable scheme in order to study structure-function relationships. An efficient mapping tool combined with the screening system will allow us not only to study protein regions important in the feedback process with an enlarged collection of mutants, but also to study the significance of dominance from a structural point of view or to study mutant alleles expression in different genetic backgrounds.

At present, we use a second screening system in order to focus specifically on feedback control of CPSase and to clone corresponding alleles in the same step. The goal is to select mutants affected in the allosteric regulation of only one activity. Mapping and identification of mutated amino acids are currently in progress.

If we replace in the plasmid *URA2* sequences by corresponding cDNA of higher eukaryotes, it could be possible to select in *S. cerevisiae* 5-FU resistant feedback mutants of eukarytic proteins equivalent to CPSase-ATCase. Molecular studies on the process in other eukaryotic cells could then be undertaken.

ACKNOWLEDGEMENTS

We thank S. Sen Gupta for correcting this manuscript. Laurence Jaquet, Marc Lollier and Older Navratil were supported by french "Ministère de la recherche et de l'enseignement supérieur".

REFERENCES

Aitken, D.M., Bhatti, A. R. and Kaplan, J. G. (1973). Characterization of the aspartate carbamoyltransferase subunit obtained from a multienzyme aggregate in the pyrimidine pathway of yeast : activity and physical properties. *Biochim. Biophys. Acta* **309**, 50-57.

Denis-Duphil, M. and Lacroute, F. (1971). Fine structure of the *URA2* locus in *Saccharomyces cerevisiae*. I. *In vivo* complementation studies. *Mol. Gen. Genet.* **112**, 354-364.

Freund, J.N. and Jarry, B.P. (1987). The *rudimentary* gene of *Drosophila melanogaster* encodes four enzymatic functions. *J. Mol. Biol.* **193**, 1-13.

Gerhart, J. and Pardee, A.B. (1962). The enzymology of control by feedback inhibition. *J. Biol. Chem.* **237**: 891-896.

Jaquet, L., Lollier, M., Souciet, J.L. and Potier, S. (1993). Genetic analysis of yeast strains lacking negative feedback control: a one-step method for positive selection and cloning of carbamoylphosphate synthetase-aspartate transcarbamylase mutants unable to respond to UTP *Mol. Gen. genet.* **241**: 81-88.

Jund, R. and Lacroute, F. (1970). Genetic and physiological aspects of resistance to 5-fluoropyrimidines in *Saccharomyces cerevisiae. J. Bacteriol.* **102**, 607-615.

Kantrowitz, E.R. and Lipscomb, W.N. (1988). *Escherichia coli* aspartate transcarbamylase : the relation between structure and function. *Science* **241**, 669-674.

Kantrowitz, E.R. and Lipscomb, W.N. (1990). *Escherichia coli* aspartate transcarbamylase: the molecular basis of concerted allosteric transition. Trends Biochem. Science. **15** : 53-59

Kosman, R.P., Gouaux,J.E. and Lipscomb, W.N. (1993). Crystal structure of CTP-ligated T state aspartate transcarbamoylase at 2.5A resolution: implications for ATCase mutants and the mechanism of negative cooperativity. Proteins. **15**: 147-176.

Lee, L., Kelly, R.E., Pastra-Landis, S.C. and Evans, D.R. (1985). Oligomeric structure of the multifunctional protein CAD that initiates pyrimidine biosynthesis in mammalian cells. *Proc. Natl. Acad. Sci. USA* **82**, 6802-6806.

Lacroute, F., Piérard, A., Grenson, M. and Wiame, J.M. (1965). The biosynthesis of carbamoylphosphate in *Saccharomyces cerevisiae.* J. Gen. Microbiol. **40**: 127-142.

Liljelund, P. and Lacroute, F. (1986). Genetic characterization and isolation of the *Saccharomyces cerevisiae* gene coding for the uridine monophosphokinase. *Mol. Gen. Genet.* **205**, 74-81.

Messenger, L.J. and Zalkin, H. (1979). Glutamine phosphoribosylpyrophosphate amidotransferase from *Escherichia coli* : purification and properties. *J. Biol. Chem.* **254**, 3382-3392.

Nagy, M., Le Gouar, M., Potier, S., Souciet, J.L. and Hervé, G. (1989). The primary structure of the aspartate transcarbamylase region of the *URA2* gene product in *Saccharomyces cerevisiae* ; features involved in activities and in nuclear localization. *J. Biol. Chem.* **14**, 8366-8374.

O'Donovan, G.A. and Neuhard, J. (1970). Pyrimidine metabolism in microorganisms. *Bacterial Rev.* **34**, 278-343.

Padgett, R.A., Wahl, G.M. and Stark, G.R. (1982). Structure of the gene for CAD, the multifunctional protein that initiates UMP synthesis in syrian hamster cells. *Mol. Cell. Biol.* **2**, 293-301.

Potier, S., Souciet, J.L. and Lacroute, F. (1987). Correlation between restriction map, genetic map and catalytic functions in the gene complex *URA2. Mol. Gen. Genet.* **209**, 283-289.

Simmer, J.P., Kelly, R.E., Scully, J.L., Grayson, D.R., Rinker, A.G., Bergh, S.T. and Evans, D.R. (1989). Mammalian aspartate transcarbamylase (ATCase) : Sequence of the ATCase domain and interdomain linker in the CAD multifunctional polypeptide and properties of the isolated domain. *Proc. Natl. Acad. Sci. USA* **86**, 4382-4386.

Souciet, J.L., Nagy, M., Le Gouar, M., Lacroute, F. and Potier, S. (1989). Organization of the yeast *URA2* gene : identification of a defective dihydroorotase-like domain in the multifunctional carbamoylphosphate synthetase - aspartate transcarbamylase complex. *Gene* **79**, 59-70.

Wiley, C.D. and Lipscomb,W.N. (1968). Crystallographic determination of symmetry of aspartate transcarbamylase. *Nature* **218**, 1119-1121.

Zalkin, H., Paluh, J.L., Van Cleemput, M., Moye, W.S. and Yanovsky, C. (1984). Nucleo tide sequence of *Saccharomyces cerevisiae* genes *TRP2* and *TRP3* encoding bifunct tional anthranilate synthetase : indole-3-glycerol phosphate synthetase. *J. Biol. Chem.* **259**, 3985-3992.

GENE STRUCTURE AND REGULATION OF THE EXPRESSION OF THE R1 AND R2 SUBUNITS OF MOUSE RIBONUCLEOTIDE REDUCTASE

Erik Johansson, Stefan Björklund and Lars Thelander

Department of Medical Biochemistry and Biophysics
Umeå University
S-901 87 Umeå, Sweden

INTRODUCTION

Ribonucleotide reductase catalyzes the first unique reaction in the cell leading to DNA synthesis[1,2]. Therefore, enzyme activity is regulated at many different levels. One level is the allosteric control of the enzyme. Both substrate specificity and overall activity are carefully regulated by a complicated pattern of binding of nucleoside triphosphates to two different classes of effector binding sites. This ensures a balanced supply of deoxyribonucleotides for error-free DNA synthesis.

Furthermore, there is a close correlation between ribonucleotide reductase activity and rate of DNA synthesis. No activity can be detected in resting cells but just before cells enter S phase, ribonucleotide reductase activity appears and shows maximal values during the S phase. The proliferation/cell cycle specific regulation of enzyme activity is not controlled by enzyme activation but instead at the gene level. There is also specific induction of ribonucleotide reductase by DNA damage[3]. All these properties makes ribonucleotide reductase an obvious target for antiproliferative treatment[4].

ENZYME STRUCTURE

Mammalian ribonucleotide reductase is a heterotetramer formed by a 1:1 complex between the two non-identical homodimers proteins R1 and R2[5] (Fig. 1). Protein R1 contains the binding sites for ribonucleoside diphosphate substrates and nucleoside triphosphate effectors. In addition, it harbors a number of redox active disulfides participating in the catalytic reaction.

Recently the three-dimensional structure of the E. coli R1 protein was solved showing that the catalytic site is buried deep inside the R1 structure[6].

Protein R2 contains per polypeptide chain one binuclear ferric iron center which, during its formation, generates a tyrosyl free radical required for catalysis. X-ray crystallography studies on the E. coli R2 protein has shown this radical to be buried about 10Å from the surface of the protein. Therefore, long-range electron transfer between the substrate binding site of the R1 protein to the tyrosyl radical of the R2 protein has been suggested to be essential for forming a catalytically active enzyme[7].

The C-terminal end of the R2 polypeptide could not be detected in the X-ray structure but was later shown by NMR to be very flexible in solution for the mouse R2 protein[8]. On binding to the R1 protein this flexibility is lost. The C-terminal tail is very important for subunit interaction. Peptides corresponding to this region have been shown to inhibit ribonucleotide reductase activity in a species specific way by competing with the R2 protein for its cognate site on the R1 protein[4].

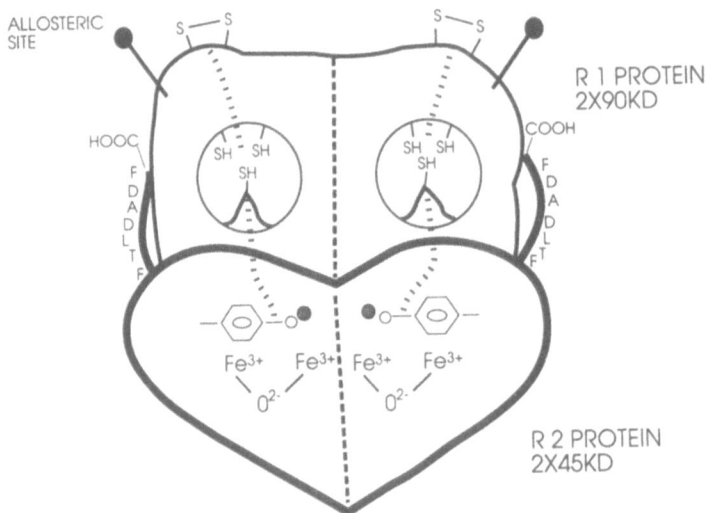

Fig 1. Model of mouse ribonucleotide reductase.

S PHASE SPECIFIC EXPRESSION OF RIBONUCLEOTIDE REDUCTASE

The increase in holoenzyme activity during S phase was shown to correlate with de novo synthesis of the R2 protein. In contrast, the R1 protein in proliferating cells is present in excess and in constant amounts all through the cell cycle[5]. The cloning of full-length cDNAs encoding the mouse R1 and R2 proteins enabled determination of the levels of the corresponding mRNAs in synchronized cells during the cell cycle. Using a very sensitive solution hybridization, RNase protection assay, we could show that both transcripts were undetectable in G_0 cells but started to appear just before S phase. They increased in parallel during the S phase reaching similar levels (about 70 molecules per cell) and finally declined when cells progressed into G_2+M. Similar results were obtained with serum starved and elutriated cells indicating a true cell cycle control of ribonucleotide reductase expression[9].

R1 AND R2 GENE STRUCTURE

To study the cell cycle control of ribonucleotide reductase at the gene level, mouse R1 and R2 genes were cloned and characterized. The R1 gene, localized to mouse chromosome 7, was compiled from overlapping λ clones and contains 19 exons spanning around 26 kb of DNA[10]. All exon/intron boundaries were sequenced as well as the 5' and the 3' regions. Multiple transcription starts within a limited area were mapped by S_1 nuclease protection experiments. The sequence upstream from the transcription start showed no TATA homology.

DNase I footprinting experiments with nuclear extracts and end-labelled probes from the promoter region identified two protected, almost identical 23-mers at position -189 to -167 and -98 to -76. Gel shift assays demonstrated binding of three different protein complexes to the 23-mer oligonucleotide. One of these DNA-protein complexes showed S phase specific binding suggesting a direct role in the cell cycle dependent R1 gene expression.

The active R2 gene was isolated as one 13 kb Hind III fragment from mouse chromosome 12[11]. The transcribed region spans about 6 kb and contains 10 exons. Transfecting mouse cells with this DNA fragment made the cells less sensitive to hydroxyurea showing that indeed this represented the active gene and not any of the 3 pseudogenes. The transcription start site was mapped by primer extension experiments and the region upstream from the start was sequenced showing a TATA homology at the expected position (-29 to -24).

S PHASE SPECIFIC R2 GENE EXPRESSION IS CONTROLLED BY RELEASE FROM A TRANSCRIPTIONAL BLOCK

R2 promoter-luciferase reporter gene constructs showed a very low promoter activity in serum starved G_0 cells but an early activation when cells started to proliferate after the readdition of serum. This R2 promoter activation occurred long before any R2 transcripts started to appear. The early activation of the R2 promoter was confirmed by in vitro nuclear run-on experiments. However, due to a transcriptional block, the early activation only results in very short transcripts until cells enter the S phase. Then synthesis of mature full-length R2 mRNA occurs by an S phase specific release from the transcriptional block.

The position of the block was localized to a DNA sequence about 87 bp downstream from the first exon/intron boundary. These results identify S phase specific release from a transcriptional block as a mechanism to control cell cycle regulated gene expression[12].

ACKNOWLEDGEMENTS

This work was supported by a grant from the Swedish Natural Science Research Council.

REFERENCES

1. L. Thelander and P. Reichard, Reduction of ribonucleotides, Ann. Rev. Biochem. 48:133 (1979).
2. P. Reichard, Interactions between deoxyribonucleotide and DNA synthesis, Ann. Rev. Biochem. 57:349 (1988).
3 S.J. Elledge, Z. Zhou, J.B. Allen and T.A. Navas, DNA damage and cell cycle regulation of ribonucleotide reductase, BioEssays 15:333 (1993).
4. G. Cosentino, P. Lavallée, S. Rakhit, R. Plante, Y. Gaudette, C. Lawetz, P.W. Whitehead, J.-S. Duceppe, C. Lépine-Frenette, N. Dansereau, C. Guilbault, Y. Langelier, P. Gaudreau, L. Thelander and Y. Guindon, Specific inhibition of ribonucleotide reductases by peptides corresponding to the C-terminal of their second subunit, Biochem. Cell Biol. 69:79 (1991).
5. L. Thelander and A. Gräslund, Ribonucleotide reductase in mammalian systems, in: "Metal ions in biological systems, Vol. 30," H. Sigel, ed., Marcel Dekker, Inc., New York (1994).
6 U. Uhlin and H. Eklund, Ribonucleotide reductase contains a new type of barrel domain and has a unique way of using cysteine residues in its catalysis, submitted (1994).
7. P. Nordlund and H. Eklund, Structure and function of the Escherichia coli ribonucleotide reductase protein R2, J. Mol. Biol. 232:123 (1993).
8. P.-O. Lycksell, R. Ingemarson, R. Davis, A. Gräslund and L. Thelander, 'H NMR studies of mouse ribonucleotide reductase: The R2 protein carboxyl-terminal tail, essential for subunit interaction, is highly flexible but becomes rigid in the presence of protein R1, Biochemistry 33:2838 (1994).
9. S. Björklund, S. Skog, B. Tribukait and L. Thelander, S-phase specific expression of mammalian ribonucleotide reductase R1 and R2 subunit mRNAs, Biochemistry 29:5452 (1990).
10. S. Björklund, K. Hjortsberg, E. Johansson and L. Thelander, Structure and promoter characterization of the gene encoding the large subunit (R1 protein) of mouse ribonucleotide reductase, Proc. Natl. Acad. Sci. USA 90:11322 (1993).
11. M. Thelander and L. Thelander, Molecular cloning and expression of the functional gene encoding the M2 subunit of mouse ribonucleotide reductase: a new dominant marker gene, EMBO J. 8:2475 (1989).
12. S. Björklund, E. Skogman and L. Thelander, An S-phase specific release from a transcriptional block regulates the expression of mouse ribonucleotide reductase R2 subunit, EMBO J. 11:4953 (1992).

MOLECULAR CHARACTERIZATION OF IMP DEHYDROGENASE IN

ACQUIRED RESISTANCE TO MYCOPHENOLIC ACID

Floyd F. Snyder, Therese Lightfoot and Stephen D. Hodges

Departments of Paediatrics and Medical Biochemistry
Faculty of Medicine, University of Calgary
Calgary, Alberta T2N 4N1, Canada

IMP dehydrogenase is recognized as a rate limiting step in the de novo synthesis of guanine nucleotides and is a target for chemotherapeutic intervention. We selected a mouse neuroblastoma (NB) cell line capable of growth in 1 mM mycophenolic acid (NB-Myco cells) by incremental increases in drug concentration[1]. This represents a 10,000-fold increase in resistance to the inhibitor of IMP dehydrogenase. Approximately 20% of cytoplasmic protein from the NB-Myco cell line was comprised of a 57 kDa protein demonstrated to be IMP dehydrogenase, which corresponded to a 500-fold increase in abundance. Measurements of enzyme activity revealed a lessor 25-fold increase in IMP dehydrogenase activity for NB-Myco cells as compared to the parental NB cell.

A number of features of this cell line are consistent with the mechanism of resistance being due to gene amplification. These include: the increased activity and marked increase in protein abundance, a 3-fold decrease in IMP dehydrogenase activity when NB-Myco cells were cultured for 80 days in the absence of mycophenolic acid, and karyotypic analysis which revealed the presence of double minute chromosomes.

Other parameters suggested the gene encoding IMP dehydrogenase had undergone mutation as well as amplification. Principally the 25-fold increase in activity does not correlate well with the 500-fold increase in protein abundance. Kinetic analysis revealed that IMP dehydrogenase from NB-Myco cells had a remarkable 2400-fold increase in K_i for mycophenolic acid as compared to the enzyme from NB cells[1]. Further, there were 4-fold increases in the K_m for NAD and the K_i for XMP. Together these findings suggested that IMP dehydrogenase had undergone both amplification and mutation in the resistant cells and that resistance was therefore due to both the increase in enzyme amount and the decreased sensitivity towards mycophenolic acid.

Molecular evidence for both amplification and mutation of IMP dehydrogenase from NB-Myco cells has now been obtained[2]. IMP dehydrogenase mRNA levels of the NB-Myco cells are 500-fold increased over NB cells, which is in agreement with the change in protein abundance. A lessor but marked 25-33 fold increase in IMP dehydrogenase gene copy number was also apparent for the resistant cell line. The cDNA for IMP dehydrogenase from NB-Myco cells was cloned and sequenced, revealing two single base substitutions when compared to the parental cell line (Table 1)[2]. These changes were verified independently by reverse transcriptase-PCR and restriction enzyme analysis.

Table 1. **Nucleotide and deduced amino acid changes in IMP dehydrogenase from mycophenolic acid resistant NB-Myco cells.**

Nucleotide Position	Codon Change	Amino Acid Position	Residue Change
998	ACC → ATC	333	Thr → Ile
1052	TCT → TAT	351	Ser → Tyr

The two nucleotide substitutions resulting in codon changes predict the substitution of threonine 333 by isoleucine and serine 351 by tyrosine. Of considerable interest is the conservation of threonine 333 across all species (Table 2). This residue is also near to cysteine 331 which has been recently located at the IMP binding site. 6-Chloropurine riboside 5'-monophosphate inactivates human IMP dehydrogenase by covalent modification of cysteine 331[3]. Although threonine 333 is immediate to cysteine 331, the change to isoleucine 333, had no effect on the K_m for IMP. Thus the K_m for IMP was 13 μM for IMP dehydrogenase from both NB and NB-Myco cells[1]. By contrast serine 351 is conserved only among mouse, hamster, human type II, and Drosophila IMP dehydrogenases (Table 2).

Table 2. **Comparative amino acid sequence for IMP dehydrogenases in the vicinity of the mutant residues 333 and 351.**

Organism	Amino Acid Sequence	Reference
Escherichia coli	....T.RI.TGV.V..I...ADAV.ALEG	4
Bacillus subtilis	....T.RV.AGV.V..I..I.DCATE..K	5
Tritrichronomas foetus	......R.QKGI..G.....ID.VAERNK	6
Leishmani donovani	.................G......AQ.CAS	7
Drosophila melanogaster	.........M...C.......Q..T...Q	8
Human type I	.........M......G......A.....	9
Human type II	....	9,10
Hamster	....	10
Mouse	GSICITQEVLACGRPQATAVYKVSEYARR	11
NB-Myco	.....I.................Y.....	2
Position	333 351	

The transition from bacterial to protozoan to mammalian IMP dehydrogenases reveals a general trend characterized by progressive increases in enzyme affinity for NAD and increased sensitivity to inhibition by mycophenolic acid (Table 3). The mutant IMP dehydrogenase from NB-Myco cells has both a decreased affinity for NAD and a remarkably increased K_i for mycophenolic acid. Human IMP dehydrogenase type I, which has alanine in place of serine 351, has a significant 4.5-fold increase in K_i for mycophenolic acid as compared to human IMP dehydrogenase type II[12], but this change in K_i is not of the three orders of magnitude difference observed between the NB-Myco mutation as compared to the wild type enzyme. The conservation of cysteine 331 and threonine 333 among IMP dehydrogenases is striking whereas serine 351 is highly variable (Table 2). Further analysis is required to determine if the two amino acid substitutions act in concert with regard to the kinetic changes, or if substitution at either residue 333 or 351 alone can produce the observed alterations. Full consideration of these alternatives awaits the expression and kinetic analysis of the single mutants at residues 333 and 351 respectively.

Table 3. Comparative differences in the kinetic constants for NAD and mycophenolic acid among IMP dehydrogenases.

Organism	NAD K_m (μM)	Mycophenolic Acid K_i (μM)	Reference
Escherichia coli	330	"insensitive"	13
Bacillus subtilis	300	0.5	14,15
Tritrichomonas foetus	340	9; 1-18	16,17
Leishmania donovani	200	-	7
Eimeria tenella	150	0.2-0.7	18
Hamster (V79)	29	0.016	19
Human Type I	46	0.033	12
Human Type II	32	0.007	12
Mouse (NB cells)	25	0.0014	1
Mouse (NB-Myco cells)	94	3.4	1

Acknowledgements

This work was supported by the Medical Research Council of Canada.

REFERENCES

1. S.D. Hodges, E. Fung, D.J. McKay, B.S. Renaux, and F.F. Snyder, Increased activity, amount, and altered kinetic properties of IMP dehydrogenase from mycophenolic acid-resistant neuroblastoma cells. J. Biol. Chem. 264: 18137 (1989)
2. T. Lightfoot and F.F. Snyder, Gene amplification and dual point mutations of mouse IMP dehydrogenase associated with cellular resistance to mycophenolic acid. Biochim. Biophys. Acta 1217: 156 (1994)
3. L.C. Antonino, K. Straub, and J.C. Wu, Probing the active site of human IMP dehydrogenase using halogenated purine riboside 5'-monophosphates and covalent modification reagents. Biochemistry 33: 1760 (1994).

4. A.A. Tiedman and J.M. Smith, Nucleotide sequence of the guaB locus encoding IMP dehydrogenase of Escherichia coli K12. Nucleic Acids Res 13:1303 (1985)

5. N. Kanazaki and K. Miyagawa, Nucleotide sequence of the Bacillus subtilis IMP dehydrogenase gene. Nucleic Acids Res 18:6710 (1990)

6. J.T. Beck, S. Zhao, and C.C. Wang, Cloning, sequencing, and structural analysis of the DNA encoding inosine monophosphate dehydrogenase (EC 1.1.1.205) from tritrichomonas foetus. Exp. Parasitol. 78: 101 (1994)

7. K. Wilson, F.R. Collart, E. Huberman, J.R. Stringer and B. Ullman, Amplification and molecular cloning of the IMP dehydrogenase gene of Leishmania donovani. J. Biol. Chem. 266: 1665 (1991).

8. C.D. Sifri, K. Wilson, S. Smolik, M. Forte, and B. Ullman, Cloning and sequence analysis of a Drosophila melanogaster cDNA encoding IMP dehydrogenase. Biochim. Biophys Acta 1217: 103 (1994).

9. Y. Natsumeda, S. Ohno, H. Kawasaki, Y. Konno, G. Weber and K. Suzuki, Two distinct cDNAs for human IMP dehydrogenase. J. Biol. Chem. 265:5292 (1990)

10. F.R. Collart and E. Huberman, Cloning and sequence analysis of the human and Chinese hamster inosine-5'-monophosphate dehydrogenase cDNAs. J. Biol. Chem. 263:15769 (1988)

11. A.A. Tiedman and J.M. Smith, Isolation and sequence of a cDNA encoding mouse IMP dehydrogenase. Gene 97:289 (1991)

12. S.F. Carr, E. Papp, J.C. Wu, and Y. Natsumeda, Characterization of human type I and type II IMP dehydrogenases. J. Biol. Chem. 268: 27286 (1993)

13. H.J. Gilbert, C.R. Lowe, and W.T. Drabble, Inosine-5'-monophosphate dehydrogenase of Escherichia coli. Purification by affinity chromatography, subunit structure and inhibition by guanosine-5'-monophosphate. Biochem. J. 183:481 (1979)

14. H. Yokosawa, T. Tobita, and T. Yamada, Studies on inosine-5'phosphate dehydrogenase of bacillus subtilis. Purification and general properties. Biochim. Biophys Acta 227: 538 (1971)

15. T.-W. Wu and K.G. Scrimgeour, Properties of inosinic acid dehydrogenase from bacillus subtilis. II. Kinetic properties. Can. J. Biochem. 51:1391 (1973).

16. R. Verham, T.D. Meek, L. Hedstrom and C.C. Wang, Purification, characterization and kinetic analysis of inosine-5'-monophosphate dehydrogenase of Tritrichomonas foetus. Mol. Biochem. Parasitol. 24:1 (1987)

17. L. Hedstrom and C.C. Wang, Mycophenolic acid and thiazole adenine dinucleotide inhibition of Tritrichomonas foetus inosine-5'-monophosphate dehydrogenase: Implications on enzyme mechanism. Biochemistry 29:849 (1990)

18. D.J. Hupe, B.A. Azzolina and N.D. Behrens, IMP dehydrogenase from the intracellular parasitic protozoan Eimeria tenella and its inhibition by mycophenolic acid. J. Biol. Chem. 261:8386 (1986)

19. E. Huberman, C.K. McKeown and J. Friedman, Mutagen induced resistance to mycophenolic acid in hamster cells can be associated with increased inosine-5'-monophosphate dehydrogenase activity. Proc. Natl. Acad. Sci. USA 78:3151 (1981)

FUNCTION OF THE POLYPEPTIDE CHAIN SEGMENT CONNECTING THE DIHYDROOROTASE AND ASPARTATE TRANSCARBAMYLASE DOMAINS IN THE MAMMALIAN MULTIFUNCTIONAL CAD

Hedeel I. Guy and David R. Evans

Department of Biochemistry
Wayne State University School of Medicine
Detroit, Michigan 48201, U.S.A.

INTRODUCTION

The multifunctional protein CAD catalyzes the first three steps in *de novo* pyrimidine biosynthesis in mammalian cells[1-3]. Glutamine dependent carbamyl phosphate synthetase (Gln CPSase), aspartate transcarbamylase (ATCase) and dihydroorotase (DHOase) activities are carried by a 243 kDa polypeptide chain that is organized into discrete functional domains connected by interdomain linkers[4-6]. One of these linkers, the DA linker[7] which bridges the ATCase and DHOase domains, is very large, 12 kDa, very hydrophilic and has an unusual amino acid composition, rich in proline, hydroxy amino acids, and positively charged residues. The length but not the sequence of the DA linker is conserved in other multifunctional proteins in this family. We suggested that this linker may function to bring the CPSase and ATCase domains into proximity to allow channeling of carbamyl phosphate[7]. A deletion mutant of the plasmid pCK-CAD10[8] which expresses the fully functional CAD polypeptide in *E. coli*, has been constructed which lacks the entire DA linker. While the catalytic activities of the CAD domains were not altered by the removal of the DA linker, the deletion reduced the stability of the complex and effectively eliminated carbamyl phosphate channeling[8]. It is interesting that one of the two cAMP dependent protein kinase phosphorylation sites, identified by Carrey et al.[9] is located within the DA linker although its function is not known.

EXPERIMENTAL PROCEDURES

The cloning, expression and partial purification of CAD and the CAD deletion mutant have been described previously[8], as have the methods for measuring carbamyl phosphate channeling[8,10]. CAD was also isolated from an overproducing strain of Syrian hamster cells[3,11]. Glutamine dependent carbamyl phosphate synthetase activity was measured using a radiometric method[3,4]. The assay mixtures contained 2 mM ATP, 4 mM $MgCl_2$, 5 mM [^{14}C] sodium bicarbonate (0.80 mCi/mmol), 3.5 mM glutamine, 20.2 mM aspartate in 0.10 M Tris Cl, pH 8.0, 0.10 mM KCl, 1 mM dithiothreitol, 7.5% dimethyl sulfoxide and 2.5% glycerol in the presence or absence of 2 mM UTP or 10 μM PRPP. The cAMP dependent kinase mediated phosphorylation of CAD was carried out using a slight modification[12] of the method described by Carrey[9]. The reaction CAD (100 μg) with 10 units of alkaline phosphatase (Boehringer Mannheim) was carried out as described by the supplier for 30 minutes at 37°.

Purine and Pyrimidine Metabolism in Man VIII, Edited by
A. Sahota and M. Taylor, Plenum Press, New York, 1995

RESULTS

The structural organization of the CAD polypeptide chain and the location of the DA linker and phosphorylation sites are shown in Figure 1.

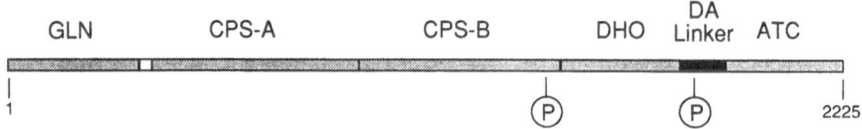

Figure 1. Schematic Representation of the CAD Polypeptide.
CAD consists of the glutaminase (GLN), carbamyl phosphate synthetase comprised of two subdomains (CPS.A and CPS.B), dihyroorotase (DHO) and aspartate transcarbamylase (ATC) domains. The diagram also shows the DA linker (dark shaded area) and the location of phosphorylation site I at the end of the CPS.B subdomain, and site II in the DA linker.

Effect of the Phosphorylation State on Carbamyl Phosphate Channeling

The partial channeling of carbamyl phosphate between the active sites of the CPSase and the ATCase domain, was abolished by the removal of the DA linker[8]. Since phosphorylation site II is located within this linker, we considered the possibility that the phosphorylation state of the protein might also alter the channeling of this intermediate. Channeling was assayed by measuring the extent of dilution of the endogenously synthesized $[^{14}C]$ carbamyl phosphate by unlabeled exogenous carbamyl phosphate[8,10]. With CAD isolated from Syrian hamster cells (Figure 2), the addition of 210 μM unlabeled carbamyl phosphate to the assay mixture reduced the conversion of carbamyl phosphate synthesized by the CPSase domain into carbamyl aspartate by 50%. We previously have shown[8] that under these conditions, 50% dilution would be expected at 60-80 μM exogenous carbamyl phosphate if there is no channeling.

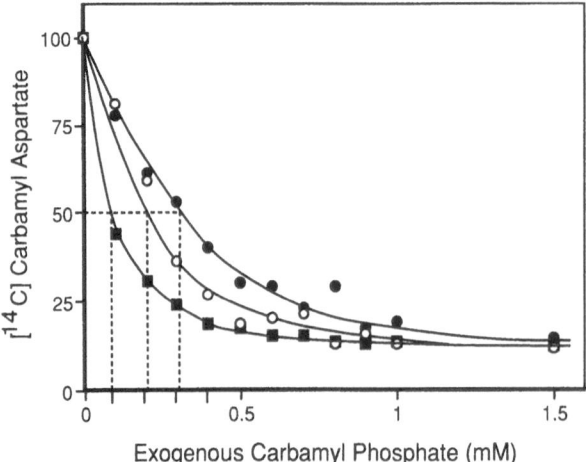

Figure 2. Effect of Phosphorylation and Dephosphorylation on Carbamyl Phosphate Channeling.
This plot shows the amount of the $[^{14}C]$ radiolabeled carbamyl aspartate produced during a 5 minute assay period by purified CAD (O), CAD phosphorylated with cAMP dependent protein kinase (■) and CAD treated with alkaline phosphatase (●) as a function of unlabeled exogenous carbamyl phosphate added to the assay mixture. The concentration of radiolabeled carbamyl aspartate is expressed relative to the value obtained when no exogenous carbamyl phosphate was added.

When CAD was phosphorylated with ATP and cAMP dependent protein kinase, the endogenously synthesized intermediate was diluted out at a lower concentration, 85 μM, indicating that channeling was reduced or abolished. CAD isolated from exponentially growing Syrian hamster cells is partially phosphorylated. Approximately 0.84 mole of phosphate per mole of CAD, although it is not known how the phosphate is distributed between the two sites[9]. Although the effect is not large, pre-treatment of purified CAD with alkaline phosphatase resulted in an enhancement of channeling. The concentration of exogenous carbamyl phosphate required for 50% dilution was increased to 310 μM.

Phosphorylation of the CAD Deletion Mutant

Previous studies[9] showed that cAMP dependent protein kinase mediated phosphorylation activates CAD and reduces feedback inhibition by UTP. The effect correlates best with the modification of site I located at the carboxyl end of the CPSase domain[9]. To determine whether site II, located in the DA linker, is involved in regulation, the effect of phosphorylation on the regulatory properties of the CAD deletion mutant were measured (Table I). Phosphorylation of CAD purified from Syrian hamster cells results in a decrease of the percent UTP inhibition from 46.0% to 10.3% and a decrease in PRPP activation from 145.2% to 24.9%. Similar results were obtained for recombinant CAD isolated from *E. coli* transformed with pCK-CAD10. Removal of the DA linker and phosphorylation site II, did not alter the effect of phosphorylation on feedback inhibition or activation suggesting that the attenuation of the allosteric transitions can be attributed solely to phosphorylation of site I.

Table I. The Effect of Phosphorylation on the Regulatory Properties of CAD and the Deletion Mutant

	UTP Inhibition %	PRPP Activation %
Purified CAD		
unphosphorylated	46.0	145.2
phosphorylated	10.3	24.9
Recombinant CAD		
unphosphorylated	23.7	181.1
phosphorylated	6.8	1.8
Recombinant CAD Deletion Mutant		
unphosphorylated	24.2	163.0
phosphorylated	2.6	-5.1

DISCUSSION

The elimination of partial carbamyl phosphate channeling by the removal of the 111 amino acid residue chain segment which connects the dihydroorotase and aspartate transcarbamylase domains was attributed[8] to the close physical proximity of the active sites catalyzing consecutive reactions (Figure 3) and the high intrinsic activity of the ATCase domain. The combination of these two factors prevents the intermediate from completely equilibrating with the bulk solvent. Directly fusing the ATCase and DHOase domains together by removal of the DA linker might be expected to alter the juxtaposition of the domains and prevent the active sites from closely approaching one another. The proximity of active sites might also be expected to be sensitive to any factor which alters the conformation of the DA linker. Phosphorylation introduces two negative charges into the otherwise positively charged linker which might alter the folding of this region of the molecule (Figure 3). The decrease in channeling exhibited by phosphorylated CAD and its enhancement following alkaline phosphatase treatment supports this interpretation and provides further evidence for the involvement of the DA linker in the partial sequestering of the intermediate.

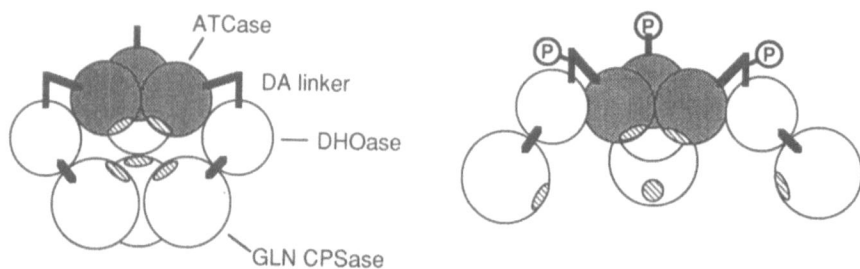

Figure 3. Schematic Representation of Carbamyl Phosphate Channeling by CAD.
Channeling is thought to occur between the active sites (hatched region) of the CPSase domain (Gln CPSase) and aspartate transcarbamylase domain (ATCase) because of their physical proximity. Removal of the DA linker (not shown) or phosphorylation (P) is thought to change the juxtaposition of the domains and disrupt channeling.

Phosphorylation of the CAD deletion mutant, which lacks site I, affects the allosteric transitions to the same extent as the native molecule which has both phosphorylation sites. This constitutes the first direct evidence that the effect of phosphorylation on regulation can be accounted for entirely by the modification of the protein kinase A consensus sequence located within the CPSase domain. The phosphorylation site located within the DA linker does not appear to be important in regulation.

ACKNOWLEDGEMENTS

This research was supported by a grant from the U.S. Public Health Service GM47399.

REFERENCES

1.   Shoaf, W.T., and Jones, M.E., Uridylic Acid Synthesis in Ehrlich Ascites Carcinoma. Properties, Subcellular Distribution, and Nature of Enzyme Complexes of Six Biosynthetic Enzymes, Biochemistry 12, 4039-4051 (1973)

2.   Mori, M., Ishida, H., and Tatibana, M., Aggregation States and Catalytic Properties of the Multienzyme Complex Catalyzing the Initial Steps of Pyrimidine Biosynthesis in Rat Liver, Biochemistry 14, 2622-2630 (1975)

3.   Coleman, P.F., Suttle, D.P., and Stark, G.R., Purification from Hamster Cells of the Multifunctional Protein that Initiates *de novo* Synthesis of Nucleotides, J. Biol. Chem. 252, 6379-6385 (1977)

4.   Mally, M.I., Grayson, D.R., and Evans, D.R., Controlled Proteolysis of the Multifunctional Protein that Initiates Pyrimidine Biosynthesis in Mammalian Cells, Proc. Nat. Acad. Sci. U.S.A. 78, 6647-6651 (1981)

5.   Davidson, J.N., Rumsby, P.C., and Tamaren, J., Organization of a Multifunctional Protein in Pyrimidine Biosynthesis. Analyses of Active Tryptic Fragments, J. Biol. Chem. 256, 5220-5225 (1981)

6.   Kim, H., Kelly, R.E., and Evans, D.R., The Structural Organization of the Multifunctional Protein CAD: Domains, Linkers and Controlled Proteolysis, J. Biol. Chem. 267, 7177-7184 (1992)

7.   Simmer, J.P., Kelly, R.E., Scully, J.L., Grayson, D.R., Rinker, A.G., Jr., Bergh, S.T., and Evans, D.R., Mammalian Aspartate Transcarbamylase (ATCase): Sequence of the ATCase Domain and the Interdomain Linker in the CAD Multifunctional Polypeptide and Properties of the Isolated Domain, Proc. Natl. Acad. Sci. U.S.A. 86, 4382-4386 (1989)

8.   Guy, H.I., and Evans, D.R., Cloning and Expression of the Mammalian Multifunctional Protein CAD in *E. coli*: Characterization of the Recombinant Protein and a Deletion Mutant Lacking the Major Interdomain Linker, J. Biol. Chem. 269, in press (1994)

9.   Carrey, E.A., Campbell, D.G., and Hardie, D.G., Phosphorylation and Activation of Hamster Carbamyl Phosphate Synthetase II by cAMP dependent Protein Kinase. A Novel Mechanism for the Regulation of Pyrimidine Biosynthesis, EMBO Journal 4, 3735-3742 (1985)

10.   Christopherson, R.I., and Jones, M.E., The Overall Synthesis of L-5,6 Dihydroorotate by the Multienzymatic Protein Pyr 1-3 from Hamster Cells, J. Biol. Chem. 255, 11381-11395 (1980)

11.   Mally, M.I., Grayson, D.R. and Evans, D.R., Catalytic Synergy in the Multifunctional Protein the Initiates Pyrimidine Biosynthesis in Syrian Hamster Cells, J. Biol. Chem. 255, 11372-11380 (1980)

12.   Liu, X., Guy, H.I., and Evans, D.R., The Identification of the Regulatory Domain of the Multifunctional Protein CAD by Construction of an *E. coli* Mammalian Carbamyl Phosphate Synthetase Hybrid, J. Biol. Chem. 269, in press (1994)

MAPPING A GENE THAT DETERMINES ERYTHROCYTIC GUANOSINE-5'-TRIPHOSPHATE CONCENTRATION (Gtpc) ON MOUSE CHROMOSOME 9

Floyd F. Snyder, Jack P. Jenuth, Janet L. Noy and Ernest Fung

Departments of Paediatrics and Medical Biochemistry,
Faculty of Medicine, University of Calgary,
Calgary, Alberta, Canada T2N 4N1

INTRODUCTION

Guanosine-5'-triphosphate (GTP) is a nucleotide serving as an energy source and substrate in multiple reactions including protein synthesis, cellular signal transduction, as a precursor for pterin cofactor synthesis, and a substrate in RNA synthesis. Erythrocytes from inbred strains of mice exhibit a 10-14 fold variation in GTP concentration[1], but a comparison of enzymes involved in guanine metabolism did not reveal the basis for the variation in cellular GTP concentration . We describe here a genetic analysis of the erythrocytic guanine nucleotide concentration determining trait which we have designated Gtpc.

RESULTS AND DISCUSSION

Nine inbred mouse strains were surveyed for erythrocyte nucleotide profiles by high performance liquid chromatography[2], and found to fall into two groups (Fig. 1). Strains having low GTP levels, between 1.4-3.4 nmole/10^9 cells are represented by C3H/HeJ and WB/ReJ. Strains having high GTP levels between 11.0-14.8 nmole/10^9 cells are represented by C57BL/6J and CBA/J. Erythrocytic ATP levels did not vary significantly among these strains, 63-87 nmole/10^9 cells. Student-Neuman-Keuls[3] and Rankit[4] analysis confirmed that the strain survey of GTP concentration fell into two distinct groups. For ease of discrimination between groups, the GTP/ATP ratio was compared.

Crosses between low and high GTP strains gave F_1 progeny having intermediate levels of GTP (Table 1). In addition, the mating of the F_1's with the parental strains gave backcross progeny which segregated in a 1:1 ratio for GTP concentration characteristic of the F_1 and parental strains (Table 1). These findings

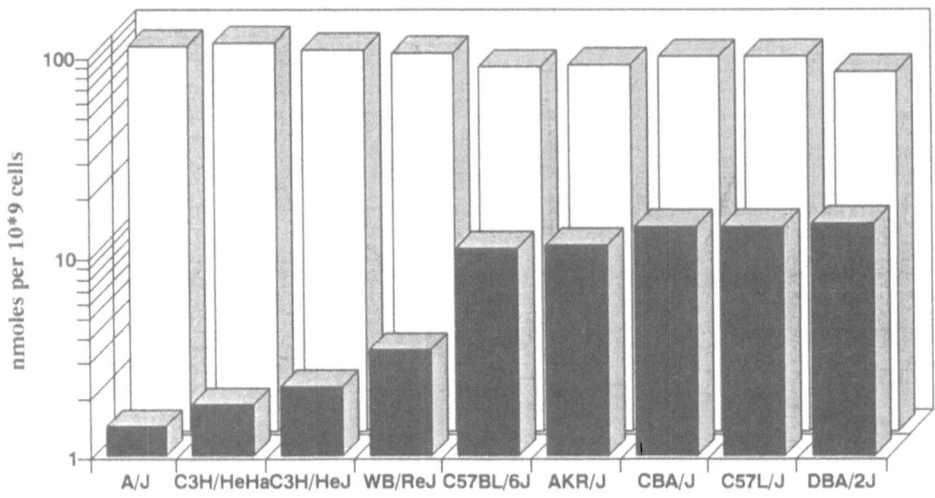

Figure 1.   Erythrocyte GTP (filled bar) and ATP (open bar) concentrations for various inbred mouse strains.

are consistent with there being a single locus that determines erythrocytic GTP concentration which is inherited in an autosomal co-dominant manner.

Series of recombinant inbred strains derived from two inbred mouse parental strains are powerful tools in genetic analysis and mapping. We screened the available 12 recombinant inbred BXH strains derived from C57BL/6J (B) and C3H/HeJ (H) (Jackson Laboratories) which differ in GTP concentration. Linkage for the GTP concentration determining trait (Gtpc), was assessed by comparison of the GTP results for each of the BXH strains to the strain distribution patterns for previously typed loci on all chromosomes using the data base and programs of Manley[5]. The GTP concentration determining trait showed possible linkage with loci on both chromosomes 5 and 9, having 0/12 strain distribution pattern differences with Gtpc (probability of linkage = 0.93[6]).

Linkage of Gtpc to loci on chromosomes 5 and 9 was examined by backcross analysis. The visual marker, dominant white spotting, W, on chromosome 5 was tested for linkage to Gtpc, by crossing WB/ReJ W/+ having dominant white spotting and low GTP, with C57BL/6J, which is black and has high GTP. The backcross gave mice having F_1 like and C57BL/6J like GTP levels in a 1:1 ratio (Table 1). Typing of backcross progeny gave 16/45 recombinants between Gtpc and W which does not differ significantly from the unlinked 1:1 ratio of parental to recombinant classes (rejected by the G test at alpha=0.01[3]). These results do not support linkage between Gtpc and W on chromosome 5.

Linkage between Gtpc and transferrin (Trf) on chromosome 9 was also assessed by backcross analysis. Transferrin was typed by serum electrophoresis[7] and strains CBA/J and WB/ReJ are polymorphic for both transferrin and Gtpc. Backcross progeny segregated into the expected two groups with respect to GTP concentration, F_1 like and WB/ReJ like (Table 1). Backcross analysis gave 6/41 recombinants between Gtpc and transferrin. The recombination frequency of 14.6 ± 5.5 is consistent with linkage of Gtpc to Trf on chromosome 9 (consistent with the G-test at alpha = 0.001[3]).

Table 1. Segregation of erythrocytic GTP levels in F_1 and Backcross Progeny

Strain/Cross	GTP/ATP (x10^2)	N
A. Chromosome 5 Testmating Cross		
C57BL/6J	16.6 ± 0.7 [a]	5
WB/ReJ	4.28 ± 0.92[b]	5
(C57BL/6JxWB/ReJ)F_1	9.39 ± 0.75[c]	10
(F_1 x C57BL/6J)Backcross		
C57BL/6J like	17.7 ± 2.1 [a]	21
F_1 like	8.76 ± 0.84[c]	25
B. Chromosome 9 Testmating Cross		
CBA/J	19.1 ± 2.4 [1]	6
WB/ReJ	4.28 ± 0.92[2]	5
(WB/ReJxCBA/J)F_1	11.7 ± 1.5 [3]	5
(F_1 x WB/ReJ)Backcross		
F_1 like	11.1 ± 2.3[3]	23
WB/ReJ like	4.52 ± 1.02[2]	19

Results are the mean± SD for N number of animals. Nucleotide ratio comparisons were made by the Student-Newman-Keuls test at P = 0.01[3]. Superscript letters and numbers indicate significant differences between groups.

Simple sequence repeat length polymorphisms have recently been described in the mouse[8] that can be conveniently typed by polymerase chain reaction. We utilised the primer pair describing the length polymorphism, D9MIT14, which is located 1 cM distal to Trf on mouse chromosome 9[9]. CBA/J and WB/ReJ are polymorphic for this DNA marker. DNA was prepared from mouse tails and the length polymorphism was typed by agarose electrophoresis of the polymerase chain reaction products. Typing of the backcross progeny of [(WB/ReJ x CBA/J)F_1 x WB/ReJ] for D9MIT14 and Gtpc gave 8/75 recombinants for a map distance of 10.7±3.6, thereby confirming the assignment of Gtpc to mouse chromosome 9. The loci having 0/12 strain distribution pattern differences with Gtpc in the recombinant inbred study were all were distal to Trf on chromosome 9, thereby placing Gtpc on the telomeric side of both Trf and D9MIT14.

Although the gene product responsible for the GTP concentration determining trait has not been elucidated, activities which produce or utilize guanine nucleotides are potential candidates. Possible candidate genes in the region of mouse chromosome 9 to which Gtpc has been mapped include a cluster of G protein genes. These are the guanine nucleotide binding proteins: alpha transducing subunit, Gnat-1; and alpha inhibitory subunit, Gnai-2. Gnai-2 has been localized at 3.1 cM distal of Trf[10]. The region of mouse chromosome 9 where Gtpc has been mapped is syntenic to human chromosome 3p[9,11]. The gene encoding human IMP dehydrogenase type II has recently been localized to human chromosome 3p21.2 - p24[12], thereby providing indirect evidence for the existence of a homolog on mouse chromosome 9. This observation is of relevance as IMP dehydrogenase is the rate limiting step in the synthesis of GTP from IMP[13]. Additional studies will focus on examining these potential candidate genes.

We have shown that a cellular GTP concentration determining trait, Gtpc, is linked to transferrin and D9MIT14 on mouse chromosome 9.

Acknowledgements

This work was supported by the Medical Research Council of Canada.

REFERENCES

1. J.F. Henderson, G. Zombor, M.M. Johnson, and C.M. Smith, Variation in erythrocyte purine metabolism among mouse strains. Comp. Biochem. Physiol. 76B:419 (1983).
2. S.D. Hodges, E. Fung, D.J. McKay, B.S. Renaux, and F.F. Snyder, Increased activity, amount and altered kinetic properties of IMP dehydrogenase from mycophenolic acid resistant neuroblastoma cells. J. Biol. Chem. 264:18137 (1989).
3. R.R. Sokal and F.J. Rohlf. "Biometry", W.H. Freeman and Co., San Francisco, First Edition (1969).
4. R.R. Sokal and F.J. Rohlf. "Biometry", W.H. Freeman and Co., San Francisco, Second Edition (1981).
5. K.F. Manley, A Macintosh program for storage and analysis of experimental genetic mapping data. Mamm. Genome 4:303 (1993).
6. J. Silver, and C.E. Buckler, Statistical considerations for genetic linkage analysis using recombinant inbred strains and backcrosses. Proc. Natl. Acad. Sci. USA 83:1423 (1986).
7. B.L. Cohen, Genetics of plasma transferrins in the mouse. Genet. Res. 1:431 (1960).
8. W. Dietrich, H. Katz, S.E. Lincoln, H.-S. Shin, J. Friedman, N.C. Dracopoli, and E.S. Lander, A genetic map of the mouse suitable for typing intraspecific crosses. Genetics 131:423 (1992).
9. D.M. Kingsley, Mouse chromosome 9. Mamm. Genome 4:S136 (1993).
10. T.M. Wilkie, D.G. Gilbert, A.S. Olsen, X.N. Chen, T.T. Amatruda, J.R. Korenberg, B.J. Trask, P. deJong, R.R. Reed, M.I. Simon, N.A. Jenkins, N.G. Copeland, Evolution of the mammalian G protein alpha subunit multigene family. Nature Genet. 1:85 (1992).
11. J.H. Nadeau, M.T. Davisson, D.P. Doolittle, P. Grant, A.L. Hillyard, M.R. Kosowsky, T.H. Roderick, Comparative map for mice and humans. Mamm. Genome 3:480 (1992).
12. D. Glesne, F. Collart, T. Varkony, H. Drabkin, and E. Huberman, Chromosomal localization and structure of the human type II IMP dehydrogenase gene (IMPDH2). Genomics 16:274 (1993).
13. F.F. Snyder and J.F. Henderson, A kinetic analysis of purine nucleotide synthesis and interconversion in Ehrlich ascites tumor cells in vitro. J. Cell. Physiol. 82:349 (1973).

REGULATION OF MAMMALIAN SERINE HYDROXYMETHYLTRANSFERASE

Suzy K. Whitehouse,[1] Paula C. Byrne,[1] Peter Sanders[2] and Keith Snell[1]

[1]Institute of Cancer Research, Sutton,
[2]University of Surrey, Guildford,
Surrey, U.K.

Serine hydroxymethyltransferase (SHMT) is a pyridoxal 5' phosphate dependent enzyme that directs serine towards nucleotide synthesis by catalysing the interconversion of serine and glycine. The products of this reaction, glycine and 5, 10-methylene tetrahydrofolate, are involved directly and indirectly in pyrimidine and purine biosynthesis. SHMT is therefore an important target for enzyme directed anticancer agents, as inhibition of the enzyme would block DNA synthesis by action on the two parallel biosynthetic pathways resulting in a combination chemotherapeutic effect from a single inhibitor. Both cytosolic and mitochondrial isozymes of SHMT exist in the cell. However the individual contributions of each isozyme to nucleotide biosynthesis remains to be elucidated.

Examination of the 5' untranslated region (UTR) sequence of rabbit cytosolic SHMT reveals a potential mechanism for regulating SHMT translation. The presence of an upstream ATG codon would provide a block to a scanning ribosomal subunit. The effect of this ATG and overlapping open reading frame (ORF) on SHMT expression was examined. Removal of the upstream ATG codon, and hence the upstream ORF, resulted in a 50-fold increase of SHMT activity over the unmutated recombinant cDNA when expressed in COS-1 cells. This suggests that this ATG codon is involved in regulation of gene expression and supports the scanning model of translation.

cDNA coding for rabbit mitochondrial SHMT has recently been isolated. Degenerate oligonucleotides specific to the mitochondrial isoform of rabbit SHMT were designed by comparing the amino acid sequences of rabbit mitochondrial and cytosolic SHMT and used in PCR to amplify part of the rabbit mitochondrial cDNA. This fragment was then used to screen a rabbit cDNA library and positive clones were purified. These clones were sequenced and were found to contain cDNA coding for the entire mature protein, at least part of the mitochondrial signal sequence and a 3' UTR. The cDNA has 72% identity to human mitochondrial SHMT over the coding sequence and 45% identity over the 3' UTR which included a region of 42 nucleotides with 98% identity. The amino acid sequence preceeding the mature protein sequence has the properties of a mitochondrial signal sequence. Highly conserved amino acids in the active site and substrate binding regions of the protein are present in rabbit mitochondrial SHMT. The isolation of both mitochondrial and cytosolic isoforms of SHMT will allow comparative studies to be carried out to further define the roles of each isoform.

Purine and Pyrimidine Metabolism in Man VIII, Edited by
A. Sahota and M. Taylor, Plenum Press, New York, 1995

REGULATION AND ROLE OF INOSINE-5'-MONOPHOSPHATE DEHYDROGENASE IN CELL REPLICATION, MALIGNANT TRANSFORMATION, AND DIFFERENTIATION

E. Huberman, D. Glesne, and F. Collart

Center for Mechanistic Biology and Biotechnology, Argonne National Laboratory, Argonne, IL 60439 and Department of Molecular Genetics and Cell Biology, University of Chicago, Chicago, IL 60637

INTRODUCTION

Inosine-5'-monophosphate dehydrogenase (EC 1.1.1.205, IMPDH) is a branch point enzyme in the synthesis of adenine and guanine nucleotides and is the rate-limiting enzyme in the *de novo* synthesis of guanine nucleotides.[1] This enzyme has an essential role in providing necessary precursors for DNA and RNA biosynthesis, a role that can be verified by the abrupt cessation of DNA synthesis when cells are treated with IMPDH inhibitors and by the circumvention of this effect with the addition of exogenous guanosine.[2]

INCREASED IMPDH GENE EXPRESSION IN HUMAN TUMOR CELLS

In general, normal tissues that exhibit increased cell proliferation have increased IMPDH activity.[3] A similar relationship between increased proliferation and increased enzyme activity has also been found in a panel of rat hepatomas having varied growth rates. Moreover, these rat hepatomas had IMPDH activities that were disproportionately higher than those of normal tissue.[1] These studies suggest that increased IMPDH activity is associated with cell proliferation and may be linked with malignant transformation or tumor progression. Although there are two human IMPDH isozymes,[4,5] only type II expression is suggested to be involved in malignant cell growth.[6] The gene coding for type II IMPDH is localized to the p21.2 → p24.2 portion of human chromosome 3,[7] where a number of tumor suppressor genes have been localized.

To assess the involvement of this enzyme in malignant transformation, we examined IMPDH expression in a series of human tumor tissues and cell lines in comparison to their normal counterparts.[8] A reduced steady-state level of IMPDH mRNA was consistently observed in RNA isolated from lymphocytes relative to the leukemic cell lines, which had markedly increased expression of a 1.9-kilobase transcript corresponding to the type II IMPDH message. A similar pattern of increased IMPDH expression was observed for the amounts of cellular IMPDH detected with the specific IMPDH antibody using the Western blotting technique and for measurements of the IMPDH activity in these cells.

Purine and Pyrimidine Metabolism in Man VIII, Edited by
A. Sahota and M. Taylor, Plenum Press, New York, 1995

We extended these analyses by examining the IMPDH gene expression in tissue samples from neuronal tumors and normal brain cells.[8] The tumors used in this study included neuroblastomas, astrocytomas, and an Ewing's sarcoma. Northern blot analysis of total RNA indicated increased levels of IMPDH II mRNA in all the tumor samples (Fig. 1). The level of IMPDH mRNA was similar in all the tumors except the CM-6 (Ewing's sarcoma) sample, in which the IMPDH mRNA level was approximately eightfold higher. Southern blot analysis of genomic DNA isolated from the CM-6 tumor tissue indicated that there were the same number of copies of the IMPDH gene in DNA prepared from the CM-6 tumor and from several normal tissue samples, eliminating the possibility that the increased IMPDH mRNA is a consequence of IMPDH gene amplification.

These marked differences in expression, amount, and activity of IMPDH between the normal and tumor cells may be associated with the low level of cell proliferation in the normal cells vs. the active proliferation of the tumor cells. To examine this possible relationship, we extended our analysis to an examination of IMPDH expression in sarcoma cells and proliferating cultured normal human fibroblasts. The sarcoma cells also had higher levels of mRNA, larger amounts of enzyme, and greater amounts of IMPDH activity than the normal fibroblasts. These differences may be attributable in part to a difference in the growth rate of the various cell types, because the doubling time of the normal fibroblasts is greater than that of the sarcoma cells. However, other factors also appear to influence the IMPDH expression, because an absolute correlation between IMPDH activity and cellular growth rate was not always observed for the tumor cell lines. Although we do not exclude such additional regulatory controls, our data suggest that regulation of the steady-state level of IMPDH mRNA, type II in particular, is a primary means of modulating cellular IMPDH activity and is the basis for the increased IMPDH activity observed in tumors.

INDUCTION OF DIFFERENTIATION IN HUMAN TUMOR CELLS BY IMPDH INHIBITORS

Another aspect of the function of IMPDH is indicated by the ability of IMPDH inhibitors such as mycophenolic acid (MPA) and tiazofurin (TZ) to induce terminal differentiation in a variety of human tumor cell tissues. These included leukemia,[9-14] breast tumor cells,[15,16] and melanoma,[17] (Table 1). The MPA- and TZ-induced inhibition of cell replication and expression of the various maturation markers was dose and time- dependent. We have also found that this induction is a consequence of MPA- or TZ-evoked reduction in guanine nucleotides, because exogenously added guanosine was capable of abrogating the effects of MPA or TZ.[11,14,17]

These studies indicate that IMPDH activity and guanine nucleotide levels affect cell replication and differentiation, and they suggest that treatment with effective IMPDH inhibitors may provide a useful means of overcoming the differentiation block that exists in tumor cells.

REGULATION OF IMPDH GENE EXPRESSION BY ITS END PRODUCTS, GUANINE NUCLEOTIDES

In our differentiation studies,[11,17] we observed that shortly after treatment of human leukemia and melanoma cells with MPA, there was an increase in the steady-state level of IMPDH mRNA. This increase was preceded by a decrease in the level of cellular guanine nucleotides, a consequence of MPA-mediated inhibition of IMPDH enzyme activity. These observations suggest that the level of purine end products may influence the expression of the IMPDH gene.[18] To examine this possibility, we treated human and Chinese hamster cell lines with nucleosides, nucleotides, nucleotide analogues, or MPA and analyzed their effects on IMPDH gene expression. Our results

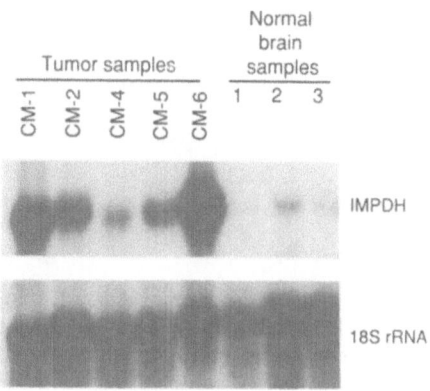

Fig. 1. Northern blot analysis of IMPDH steady-state mRNA levels in human tumor and normal brain tissue. The same filter was used for hybridization to both the IMPDH and 18S ribosomal RNA probes. (From Ref. 8)

Table 1. Differentiation of Human Tumor Cells Caused by IMPDH

Tumor Cell Type	Markers	Ref.
Myeloid leukemia	Nitroblue tetrazolium reduction; maturation antigens	9–11
Erythroid leukemia	Hemoglobin synthesis	12, 13
T-lymphoid leukemia	Maturation antigens	14
Breast carcinoma	Lipid and casein formation	15, 16
Melanoma	Tyrosinase; melanin synthesis	17

indicated that alterations in the level of guanine ribonucleotides indeed influence the expression of IMPDH mRNA (Fig. 2). A decrease in guanine nucleotide pools causes an increase in the expression of the gene, while an increase in these pools causes a decrease in IMPDH gene expression. This regulation also influences the enzyme level, since immunoblot analyses indicated that IMPDH protein levels were repressed in response to guanosine treatment and increased in response to MPA treatment. The elevation of IMPDH protein levels by MPA was counteracted by cotreatment with guanosine.

To determine which mechanism is responsible for the guanine-nucleotide-mediated regulation of IMPDH gene expression, we first used nuclear run-on assays to measure transcriptional activity of the IMPDH gene. We found that a reduction in guanine nucleotides as a consequence of MPA treatment or an increase in these nucleotides as a consequence of guanosine treatment had little or no detectable effect on transcriptional activity of the IMPDH gene. Another point at which regulation of eukaroytic mRNA levels can be exerted is at the level of mRNA stability. The kinetics of decay of IMPDH mRNA were measured in response to MPA or guanosine by treating cells with actinomycin D and measuring RNA levels at subsequent time points by Northern analysis. The half-life of IMPDH mRNA was determined to be approximately 1 h in untreated cells and did not change as a result of MPA or guanosine treatment. Furthermore, IMPDH regulation was not due to translational mechanisms, since cycloheximide or puromycin did not influence the regulation of IMPDH mRNA levels.

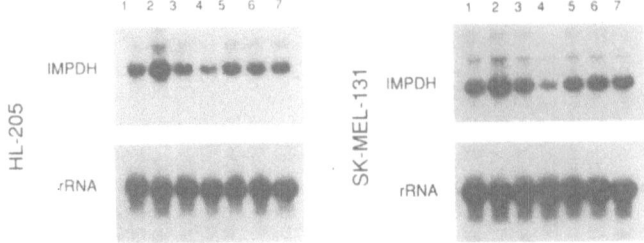

Fig. 2. Northern blot analysis of IMPDH gene expression in HL-205 (human myeloid leukemia) cells (left) and SK-MEL-131 (human melanoma) cells (right) after treatment with MPA or nucleosides. Cells were not treated (lane 1) or treated with 1.6 μM MPA, 300 μM adenosine, 300 μM guanosine, 300 μM hypoxanthine, 300 μM inosine, or 300 μM xanthine (lanes 2 to 7, respectively) for 24 h. (From Ref. 22)

Because IMPDH regulation by guanosine and MPA was shown to be nontranscriptional and noncytoplasmic, we sought to determine whether or not the regulation occurred within the nucleus. To that end, we purified nuclear RNA from human leukemia cells treated with either guanosine or MPA for time periods that elicit a maximal effect. Our results indicated that the regulation of IMPDH mRNA levels does indeed occur within the nucleus. Thus, IMPDH gene expression is regulated by a post-transcriptional nuclear event in response to fluctuations in the intracellular level of guanine ribonucleotides, the end product of the enzyme.

IMPDH SEQUENCE CONSERVATION

The crucial role that IMPDH plays in nucleic acid biosynthesis, cell replication, and differentiation may perhaps explain the high degree of amino acid sequence conservation of IMPDH. Of the 514 amino acids that make up eukaryotic IMPDH II, only eight differences are noted between the human and Chinese hamster proteins; five of these changes are conservative in terms of the chemical nature of the amino acid.[5] The mouse and Chinese hamster cDNAs differ in only five amino acid residues.[19] The IMPDH gene isolated from *Leishmania donovani*[20] specifies a protein having a 53% identity with the mammalian (human, hamster, or mouse) enzymes. Considering conservative substitutions between similar amino acids, the leishmanial and mammalian IMPDH proteins match at more than 70% of the residues. A similar relationship exist between IMPDH amino acid sequences of yeast[21] and plant (*Arabidopsis thaliana*)[22] genes and the mammalian IMPDH amino acid sequence. The bacterial IMPDH proteins exhibit a 70% similarity among themselves and about 60% with those of mammalian cells.[20] The amino acid sequence deduced from an IMPDH gene from *Pyrococcus furiosus*,[22] a hyperthermophilic microorganism belonging to the kingdom Archea, further establishes the conserved nature of this enzyme. The degree of conservation of all these sequences permits a comparison of the relationship between IMPDH from various organisms which is consistent with accepted evolutionary trees (Fig. 3).

This conservation in amino acid sequence is also reflected in the characteristics of IMPDH from a variety of sources. The molecular mass, kinetic parameters, and hydrophobic nature are similar in all of the eukaryotic and prokaryotic IMPDH enzymes thus far characterized. There are, however, some differences in the susceptibility to specific IMPDH inhibitors among enzymes belonging to bacteria, yeast, protozoa, and mammalian cells that may be exploited for therapy of infectious diseases.

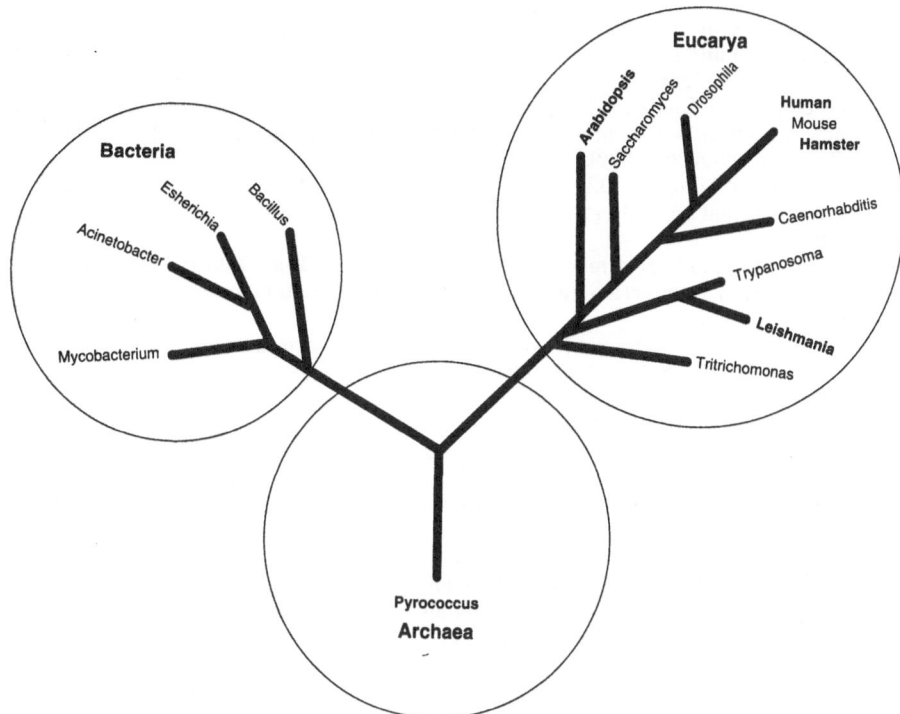

Fig. 3. Relationship of IMPDH protein sequences from different organisms.

The information derived from sequence comparisons of the various IMPDH enzymes will provide a rational basis for the design of more selective and effective chemotheraputic agents.

ACKNOWLEDGMENTS

This work was supported by the U.S. Department of Energy, Office of Health and Environmental Research, under Contract No. W-31-109-Eng-38.

REFERENCES

1. G. Weber, Biochemical strategy of cancer cells and the design of chemotherapy: G.H.A. Clowes memorial lecture, *Cancer Res.* 43:34666–3492 (1983).
2. M.B. Cohen and W. Sadee, Contributions of the depletions of guanine and adenine nucleotides to the toxicity of purine starvation in the mouse T lymphoma cell line, *Cancer Res.* 43:1587–1591 (1983).
3. D.A. Cooney, Y. Wilson, and E. McGee, A straightforward radiometric technique for measuring IMP dehydrogenase, *Analyt. Biochem.* 130:339–345 (1983).
4. F.R. Collart and E. Huberman, Cloning and sequence analysis of the human and Chinese hamster inosine 5'-monophosphate dehydrogenase, *J. Biol. Chem.* 263:15769–15772 (1988).
5. Y. Natsumeda, S. Ohno, H. Kawasaki, Y. Konno, G. Weber, and K. Suzuki, Two distinct cDNAs for human IMP dehydrogenase, *J. Biol. Chem.* 265(9):5292–5295 (1990).
6. M. Nagai, Y. Natsumeda, Y. Konno, R. Hoffman, S. Irino, and G. Weber, Selective up-regulation of type II inosine 5'-monophosphate dehydrogenase messenger RNA expression in human leukemias, *Cancer Res.* 51:3886–3890 (1991).

7. D. Glesne, F. Collart, T. Varkony, H. Drabkin, and E. Huberman, Chromosomal localization and structure of the human type II IMP dehydrogenase gene), *Genomics* 16:274–277 (1993).

8. F.R. Collart, C.B. Chubb, B.L. Mirkin, and E. Huberman, Increased inosine-5'-phosphate dehydrogenase gene expression in solid tumor tissues and tumor cell lines, *Cancer Res.* 52:5826–5828 (1992).

9. R.D. Knight, J. Mangum, D.L. Lucas, D.A. Cooney, E.C. Khan, and D.G. Wright, Inosine monophosphate dehydrogenase and myeloid cell maturation, *Blood* 69:634–639 (1987).

10. S.M. Khabanda, M.L. Sherman, D.R. Spriggs, and D.W. Kufe, Effects of tiazofurin on protooncogene expression during HL-60 cell differentiation, *Cancer Res.* 48:5965 (1988).

11. F.R. Collart and E. Huberman, Expression of IMP dehydrogenase in differentiating HL-60 cells, *Blood* 75:570–576 (1990).

12. E. Olah, Y. Natsumeda, T. Ikegami, Z. Kote, M. Horanyi, J. Szelenyim, E. Paulik, T. Kremmer, S.R. Hollan, J. Sugar, and G. Weber, Induction of erythroid differentiation and modulation of gene expression by tiazofurin in K-562 leukemia cells, *Proc. Natl. Acad. Sci. USA* 85:6533–6537 (1988).

13. J. Yu, V. Lemas, T. Page, J.D. Connor, and A.L. Yu, Induction of erythroid differentiation in K562 cells by inhibitors of inosine monophosphate dehydrogenase, *Cancer Res.* 49:5555–5560 (1989).

14. Kiguchi, F.R. Collart, C. Henning-Chubb, and E. Huberman, Cell differentiation and altered IMP dehydrogenase expression induced in human T-lymphoblastoid leukemia cells by mycophenolic acid and tiazofurin, *Exp. Cell Res.* 187:47–53 (1990).

15. Y. Sidi, C. Panet, L. Wasserman, A. Cyjon, A. Novogrodsky, and J. Nordenberg, Growth inhibition and induction of phenotypic alterations in MCF-7 breast cancer cells by an IMP dehydrogenase inhibitor, *Br. J. Cancer* 58:61–63 (1988).

16. S. Bacus, K. Kiguchi, D. Chin, C.R. King, and E. Huberman, Differentiation of cultured human breast cancer cells (AU-565 and MCF-7) associated with loss of cell surface HER-2/neu antigen, *Mol. Carcinog.* 3:350–362 (1990).

17. K. Kiguchi, F.R. Collart, C. Henning-Chubb, and E. Huberman, Induction of cell differentiation in melanoma cells by inhibitors of IMP dehydrogenase: altered patterns of IMP dehydrogenase expression and activity, *Cell Growth Differen.* 1:259–270 (1990).

18. D.A. Glesne, F.R. Collart, and E. Huberman, Regulation of IMP dehydrogenase gene expression by its end products nucleotides, *Mol. Cell Biol.* 11:5417–5425 (1991)

19. A.A. Tiedeman and J.M. Smith, Isolation and sequence of a cDNA encoding mouse IMP dehydrogenase, *Gene* 97:289–293 (1991).

20. K. Wilson, F.R. Collart, E. Huberman, J.R. Stringer, and B. Ullman, Amplification and molecular cloning of the IMP dehydrogenase gene of *Leishmania donovani*, *J. Biol. Chem.* 266:1665–1671 (1991).

21. Genbank accession number L28920.

22. F.R. Collart, J. Osipiuk, J.D. Trent, and E. Huberman (unpublished information).

RECIPROCAL ALTERATIONS OF ENZYMIC PHENOTYPE OF PURINE AND PYRIMIDINE METABOLISM IN INDUCED DIFFERENTIATION OF LEUKEMIA CELLS

Yasufumi Yamaji, Taiichi Shiotani, Hiroyuki Nakamura,Yuuki Hata,Yasuko Hashimoto, Masami Nagai, Jiro Fujita, and Jiro Takahara

First Department of Internal Medicine
Kagawa Medical School
Ikenobe, Miki, Kagawa 761-07, Japan

INTRODUCTION

12-O-tetradecanoylphorbol-13-acetate(TPA),dimethyl sulfoxide (DMSO) and retinoic acid have been shown to induce morphologic and functional differentiation in various cultured leukemic cell lines.[1,15] The human promyelocytic leukemia cell line, HL-60, can be differentiated to macrophage/monocyte-like cells by TPA and DMSO, and also differentiated to granulocyte-like cells by retinoic acid.[1] Induced differentiation in leukemia cells was accompanied by an arrest of the proliferation of the cells and changes in the cell cycle kinetics. Although these cellular response to the common differentiation inducers are thought to be triggered by specific receptors on the cells and/or by directly activating signal transduction pathways,[1] induced differentiation may be also controlled to a critical degree by the metabolic involvement with the cell. Several nucleotides of purine and pyrimidine and antimetabolites have been shown to induce differentiation of leukemic cells,[2,3] suggesting a possible role of DNA metabolism in induced differentiation. This brief review focus on the changes of purine and pyrimidine metabolism in induced differentiation of leukemic cells.

PYRIMIDINE METABOLISM

Thymidine(dThd) is salvaged by dThd kinase(TK), providing an alternate route for dTMP production for DNA biosynthesis. dThd also may be degraded by dThd phosphorylase(TP) and through the rate-limiting enzyme of dThd catabolism, dihydrothymine dehydrogenase(DHT DH), for eventual catabolism to CO_2 and ammonia.[4,5] Since TP is an equilibrium enzyme, the balance of TK and DHT DH may determine the metabolic routing of dThd.[4,5] Thymidylate, an important precursor of DNA

Purine and Pyrimidine Metabolism in Man VIII, Edited by
A. Sahota and M. Taylor, Plenum Press, New York, 1995

synthesis, also may be produced by the *de novo* pathway through dTMP synthase. In human leukemia-lymphoma cell lines, a reciprocal regulation in behavior of the activities of these opposing enzymes of dThd metabolism has been demonstrated. The activities of the anabolic enzymes, TK and dTMP synthase, increased, and those of catabolic enzymes, TP and DHT DH, decreased compared with those of normal lymphocytes.[4] These studies indicate that the enzymic balance of anabolism *versus* catabolism, especially the activity of DHT DH, is a primary factor determining dThd utilization. Therefore, the behavior of the enzymic capacities in the opposing pathways of dThd metabolism during induced differentiation has been interesting.

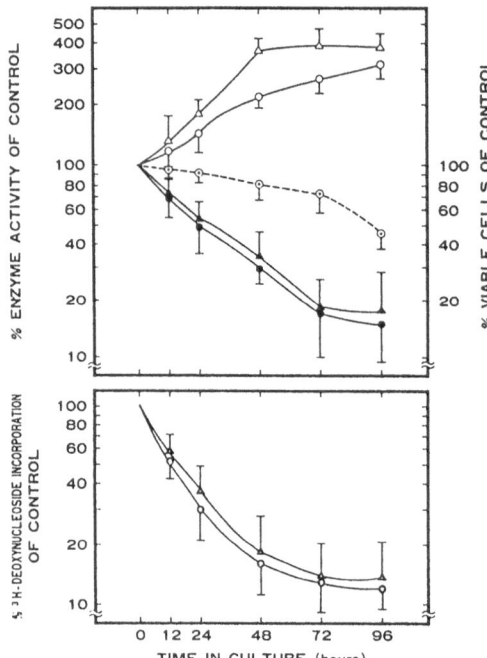

Fig.1. Behavior of the dThd metabolic capacities during differentiation by TPA.
Enzymic activities, rate of 3H-deoxynucleoside incorporation and cell growth inhibition are expressed as percentage of those of cells cultured in absence of TPA for the same periods.Points are means of four determinations.
(Upper Fig.)Enzyme activities and growth inhibition curve: TK (•),TS(▲),TP(Δ),DHTDH(o), % vaible cells of control(⊙), (Lower Fig.) Rate of 3H-deoxynucleoside incorporation : 3H-dThd(⊙),3H-deoxyuridine(▲).

Behavior of the dThd metabolic capacities and cell cycle distribution during induced differentiation

Exposure of HL-60 cells to TPA resulted in the inhibition of cell growth in a time-dependent manner(Fig.1)[14]. After 96 h exposure to TPA(1.62nM), cell growth was inhibited by 45% compared to that of untreated cells cultured for the same period. Activities of TK and TS declined within 12 h and reduced linearly by 72 h to 17% and 19% of values of untreated cells(Fig.1). The specific activities of TK, TS, TP and DHT DH in the cells treated with TPA for 96 h were 3.78±0.95,25.05±3.51, 498.0±139.4 and 3.02±0.32 nmol/h/mg protein, respectively. In contrast, the catabolic enzyme activities elevated within 12 h after TPA treatment. The activities of TP increased for the first 48 h to 390% of control value. DHT DH activities progressively elevated over a 96 h incubation period to 318% of the control value(Fig.1). This elevation of DHT DH activities is dependent on the enzyme amount in the differentiated cells.[14]

As a result, reciprocal alterations of dThd anabolic and catabolic enzyme activities were observed during differentiation in HL-60 cells induced by TPA.

Regulatory roles of dThd anabolic enzymes and catabolic enzymes in the balance of intracellular pyrimidine nucleotides

We have demonstrated previously a positive or negative relationship between levels of the catabolic or anabolic enzymes and the degree of differentiation of the cell in normal adult and fetal liver tissues.[5] It is likely that an elevated level of dThd catabolic capacity and a decline in that of anabolism might be required for the maintenance of the differentiated state.

PURINE METABOLISM

Guanine nucleotides are involved in many important anabolic processes including biosynthesis of RNA, DNA and protein and transmembrane signaling.[6] Guanylate synthesis is increased during cell proliferation,[7] whereas the inhibition of guanylate synthesis followed by depletion of GTP pool in leukemia cells triggers cell maturation.[1,8] *De novo* GMP synthesis from IMP is catalyzed by IMP DH and GMP synthase, whereas GMP reductase(R) converts GMP to IMP, constituting a cycle comparable with the purine nucleotide cycle catalyzed by adenylosuccinate synthase, adenylosuccinate lyase and AMP deaminase.[9] It is generally noticed that the increase in the activity of IMP DH, which is the rate-limiting and a key enzyme for *de novo* GTP biosynthesis, has been shown to linked with proliferation, transformation and progression.[5]

Table 1. Effect of differentiation on IMP DH and GMPR activities

Agents	Enzyme activities: nmol/h/mg protein			Cell differentiation % of NBT-positive cells
	IMPDH	GMPR	IMPDH/GMPR	
None,control	107±1 (100)	13.3±0.5 (100)	8.0 (100)	3.3±0.4
Ethanol (0.1%)	109±1 (102)	14.9±0.5 (112)	7.3 (91)	4.1±0.8
TPA (33nM)	65.0±0.8* (61)	30.2±0.48* (227)	2.2 (28)	70.3±6.5*
DMSO (1.5%)	76.6±3.8* (72)	25.3±0.6* (190)	3.0 (38)	64.0±3.2*
Retinoic acid (1µM)	48.1±1.2* (45)	31.2±0.3* (235)	1.5 (19)	69.6±3.5*

* Statistically different from values of control(p<0.05).
The cells were treated with the respective agents for 72 h.
Means+S.E. of 3 or more samples are given. Values in parentheses are % of control

Reciprocal behavior of IMPDH and GMPR activities during cell differentiation

As shown in Table 1,[10] in HL-60 cells induced differentiation by TPA was accompanied by a decrease in IMPDH activity to 35% of the initial value at 72 h culture. By contrast, during differentiation GMPR activity increased to 235% over the control level. These alterations occurred in a time-dependent manner. A similar patter was observed during maturation induced by other differentiation inducers. The differentiation induced by the drugs was associated with retardation of cell proliferation.

Intracellular concentration of guanine nucleotides has been shown to decline during induced differentiation in HL-60 cells. This depression has been considered to be linked with the decrease in IMP DH activity. However, our study[10] reveals an increase in GMPR activity at the same time during differentiation. The joint action of the reciprocal alterations in GMPR and IMP DH activities should result in a decrease in the intracellular guanylate pool.

Regulatory roles of GMPR and IMP DH in the balance of intracellular purine nucleotides

IMP is located at the central position for adenine and guanine nucleotide interconversion. The enzymes involved in interconversion between IMP and the respective nucleotides are regulated by similar allosteric mechanisms. The initial step enzymes for IMP utilization, adenylosuccinate synthase and IMP DH, are inhibited by the corresponding pathway products, AMP and GMP, respectively.[11,12] Furthermore, AMP deaminase and GMPR activities which directly provide IMP from AMP and GMP are accelerated by ATP and GTP,[9,13] respectively. These mechanisms contribute, at least in part, to controlling the balance between intracellular adenine and guanine nucleotide pools. The pool balance, however, changes during cell proliferation and these changes depend on the growth rate of the tumor cells.[5] Subsequently, it was also shown that IMP was preferentially channelled into guanylate synthesis during transition from resting to log phases in rat hepatoma 3924A cells and that the GTP pool expansion was more marked than that of the ATP pool.[7] In proliferating cells the purine metabolic imbalance is due chiefly to the increase in IMP DH activity.[5] On the other hand, the guanylate pool was reduced in HL-60 cells during differentiation[1] and our study[10] showed that the reduction is accompanied not only with decreased IMP DH activity but also with an increase in GMPR activity. The combined actions of reciprocal behavior of IMP DH and GMPR activities might play an important role in the control mechanisms for the balance of intracellular adenine and guanine nucleotide pools during differentiation.

REFERENCES

1. S.J.Collins, The HL-60 promyelocytic leukemia cell line:proliferation,differentiation, and cellular oncogene expression, Blood. 70:1233(1987).
2. D.L.Lucus,H.K.Webster,and D.G.Wright, Purine metabolism

in myeloid precursor cells during maturation.Studies with the HL-60 cell line, J. Clin. Invest. 72: 1889 (1983).

3. A.J.Bodner,R.C.Ting,and R.C.Gallo, Induction of differentiation of human promyelocytic leukemia cells(HL-60) by nucleosides and methotrexate, J.Natl. Inst. 67: 1025 (1981).

4. T.Shiotani,Y.Hashimoto,T.Tanaka, and S.Irino, Behavior of activities of thymidine metabolizing enzymes in human leukemia-lymphoma cells, Cancer Res.49:1090 (1989).

5. G.Weber, Biochemical strategy of cancer cells and design of chemotherapy: G. H. A. Clowes Memorial Lecture, Cancer Res.43:3466(1983).

6. M.L.Pall, GTP: A central regulator of cellular anabolism, Curr.Top.Cell.Regl. 25:1(1985).

7. Y.Natsumeda,T.Ikegami,K.Murayama, and G.Weber, *De novo* guanylate synthesis in commitment to replication in hepatoma 3924A cells, Cancer Res.48:507(1988).

8. D.G.Wright, A role of guanine ribonucleotides in the regulation of myeloid cell maturation, Blood. 69: 334 (1987).

9. J.M.Lowenstein, Ammonia production in muscle and other tissue:the purine nucleotide cycle, Physiol. Rev.52: 383 (1972).

10.H.Nakamura, Y.Natsumeda, M.Nagai, J.Takahara, S.Irino, and G.Weber, Reciprocal alterations of GMP reductase and IMP dehydrogenase activities during differentiation in HL-60 leukemia cells, Leuk.Res.16:561(1992).

11.R.C.Jackson, H.P.Morris, and G.Weber, Partial purification, properties and regulation of inosine 5'-phosphate dehydrogenase in normal and malignant rat tissues, Biochem.J.166:1(1977).

12.M.M.Stayton, Regulation, genetics, and properties of adenylosuccinate synthetase:a review, Curr.Top.Cell Regul. 22:103(1983).

13.T.Spector,T.E.Jones, and R.L.Miller, Reaction mechanism and specificity of human GMP reductase, J.Biol.Chem.254:2308(1979).

14.Y.Hashimoto, T. Shiotani, J. Fujita, Y.Yamaji, H.Futami, N.Yamanouchi,M.Bungo,H.Nakamura,T.Tanaka, and S.Irino, Reversal of enzymic phenotype of thymidine metabolism in induced differentiation of HL-60 cells, Leuk.Res. 13: 1123 (1989).

15.H.P.Koeffler, Induction of differentiation of human acute myelogenous leukemia cells:therapeutic implications, Blood.62:709(1983).

EXTRACELLULAR CONCENTRATIONS OF OXYPU-
RINES IN XANTHINE OXIDASE-DEFICIENT
HEPATOMA-DERIVED CELL LINE HuH-7

Tetsuya Yamamoto, Yuji Moriwaki, Oluyemi E. Agbedana, Sumio
Takahashi, Yumiko Nasako, Yuji Yokoyama and Kazuya Higashino

Third Department of Internal Medicine, Hyogo College of Medicine
Mukogawa-cho 1-1, Nishinomiya, Hyogo, 663 Japan

INTRODUCTION

Patients with xanthinuria have biochemical abnormalities which include high concentrations of hypoxanthine and xanthine in plasma, higher concentration of xanthine than hypoxanthine and the low concentration of uric acid in plasma and urine [1-3]. However, it is unresolved why the concentration of xanthine is higher than that of hypoxanthine in plasma and urine.   In the present study, using xanthine oxidase-deficient hepatoma-derived cell line HuH-7 cells, we tried to elucidate the mechanism that regulated oxypurines in patients with xanthinuria.

METHODS AND MATERIALS

Materials

The human hepatoma-derived cell line HuH-7 was established by Nakabayashi et al[4]. Cells of this line were propagated in RPMI 1640 supplemented with final concentrations of 3×10^{-8} M Na_2SeO_3 and 30 µg/ml of kanamycin in a humidified atmosphere of 5 % CO_2 in air and subcultured at 1:2 dilution every 10 days.   Inocula of 2.5×10^6 cells were seeded into 25 cm^2 tissue culture flasks (Corning Glass Works, U.K.) with 5 ml of the medium described above.   Twelve hours later, the conditioned medium was replaced by 5 ml of the medium described in the Figures except Figure 4.   In Figure 4, in the 75 cm^2 flasks, 7.5 x 10^6 cells were seeded and then 12 hours later, the medium in the flasks was replaced with 7.5 ml of respective fresh medium containing 0.21 µCi/ml ^{14}C-hypoxanthine.   Four hours after replacement, cells were harvested.

Analysis of Samples

Purine bases in the conditioned medium of HuH-7 cells were measured by the method of Yamamoto et al [5] using high-performance liquid chromatography (HPLC).   The activities of hypoxanthine-guanine phosphoribosyl transferase (HGPRT) and adenine phosphoribosyl transferase (APRT) in HuH-7 cells were measured by the method of Sakuma et al.[6]
The concentrations of purine nucleotides in cells were determined as described previously[7].

Purine and Pyrimidine Metabolism in Man VIII, Edited by
A. Sahota and M. Taylor, Plenum Press, New York, 1995

The activities of purine nucleoside phosphorylase (PNP), adenosine deaminase (ADA) and xanthine oxidase (XO) were determined using HPLC, because the turbidity of the superna tant of the lysate of HuH-7 cells affected the spectrophotometric detection of the activities of ADA, PNP and XO.   The spectrophotometric methods for the assays of PNP, ADA and XO activities[8,9] were modified as described below.   In brief, as for PNP, at 2 minutes and 5 minutes after the enzyme reaction was initiated, an aliquot was taken and deproteinized by 20 % perchloric acid.   After centrifugation, the supernatants were neutralized by 1 M K_2CO_3.   As for ADA, at 5 minutes and 15 minutes after the enzyme reaction was initiated, reaction mixtures were also treated as described above and as for XO, at 5 minutes and 15 or 35 minutes, reaction mixtures were treated as described above.   The activity of guanase was determined by HPLC as follows.   One hundred μl of the dialyzed supernatant was added to 2.9 ml of the mixture containing 40 μM guanine in 50 mM phosphate buffer (pH 7.4) and to 2.9 ml of 50 mM phosphate buffer (pH 7.4).   The enzyme reaction and the treatment of mixtures were the same as in the measurement of the activity of PNP.   In the measurement of the activities of these enzymes, elution conditions of HPLC were as de-scribed previously[5].   The activity of cytosolic 5'-nucleotidase was determined by the modi-fied method of Abd-Elfattah et al.[10] as described previously[7], using α, β-methyleneadenosine 5'-diphosphate.

RESULTS AND DISCUSSION

We found by chance that HuH-7 cells did not produce uric acid at detectable levels, but produced considerable amounts of oxypurines, the results suggesting that HuH-7 cells did not possess the activity of xanthine oxidase.   Therefore, we measured the activity of xanthine oxidase and those of the other enzymes related to purine metabolism in HuH-7 cells and demonstrated that HuH-7 cells possessed no activity of xanthine oxidase but pos-sessed the activities of other enzymes related to purine metabolism (Table I).   The activities of guanase in HuH-7 cells were lower, while that of HGPRT was higher than the corre-sponding level in human liver (Table 1).   Other enzymes (PNP, ADA, cytosolic 5'-nu-cleotidase and APRT) were not markedly different from those in human liver.

Table 1. Enzyme activity in HuH-7 cells, PLC/PRF/5 cells and human liver

	Enzyme activity (nmol/min/mg protein) (mean±s.d., N=3)		
	HuH-7	PLC/PRF/5	human liver
PNP	15.5±1.0	96.4±6.2	18.1±2.0
ADA	4.7±0.4	13.5±1.8	2.5±0.4
guanase	0.9±0.2	3.2±0.4	6.6±0.4
HGPRT	0.37±0.05	0.21±0.05	0.09±0.01
APRT	0.07±0.01	0.07±0.01	0.04±0.01
cytosolic 5'-nucleotidase	1.81±0.31	1.99±0.48	1.70±0.30
xanthine oxidase	N.D.	0.011±0.002	1.35±0.17

The activities of these enzymes in PLC/PRF/5 cells and human liver were provided for comparison. N.D., not detected.

The concentration of hypoxanthine in the conditioned medium ranged from 45 μg/L during 24 hour-incubation of HuH-7 cells. On the other hand, the concentration of xanthine in the conditioned medium increased gradually during 24-hour incubation of HuH-7 cells and rose to 60 µg/L after the 24-hour incubation. Uric acid was below the limit of detection in the conditioned medium during 24-hour incubation. The concentration of hypoxanthine in the conditioned hypoxanthine medium(Hx-medium) was lower than the concentration of hypoxanthine added to the medium plus the concentration of hypoxanthine in the control conditioned medium of 12-hour incubation (Fig 1), while the concentration of xanthine in the Hx-medium increased during incubation of HuH-7 cells (Fig 1).

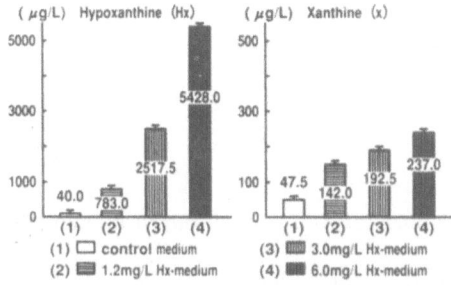

Fig. 1. Concentrations of hypoxanthine and xanthine in hypoxanthine medium (Hx-medium) after 12-hour incubation

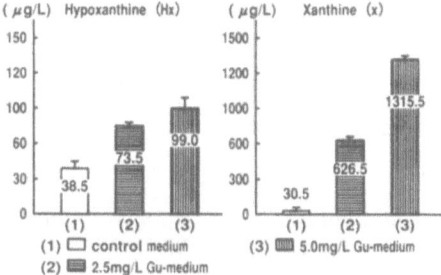

Fig. 2. Concentrations of hypoxanthine and xanthine in guanine medium (Gu-medium) after 12-hour incubation

The concentration of hypoxanthine in the conditioned medium did not change with the addition of 5 mg/L xanthine (Fig. 2). The concentration of xanthine in the conditioned xanthine medium (X medium) was approximately the same as that of xanthine added to the medium plus that of xanthine in the control medium (data not shown). The concentrations of hypoxanthine and xanthine increased in the conditioned guanine medium after 12-hour incubation (Fig 2).

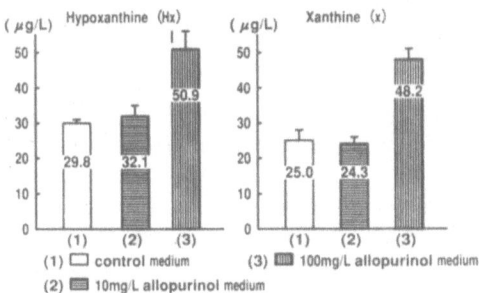

Fig. 3. Concentrations of hypoxanthine and xanthine in allopurinol medium after 12-hour incubation

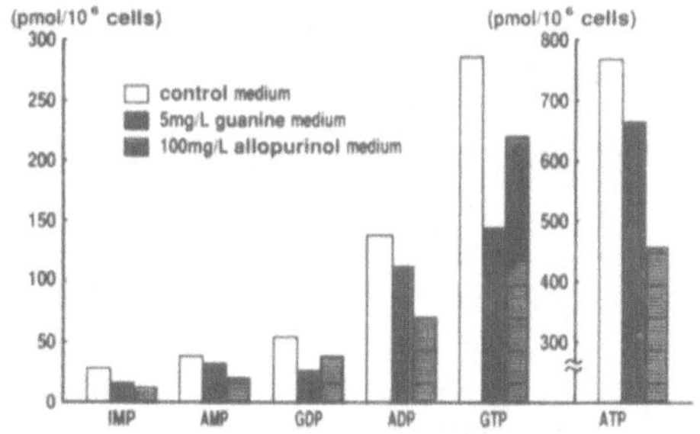

Fig. 4. Effect of guanine or allopurinol on uptake of 0.21 µCi/ml ^{14}C-hypoxanthine by respective nucleotides

In the presence of 100 mg/L allopurinol, the concentrations of hypoxanthine and xanthine increased, when compared with the corresponding values in the allopurinol-free control medium (Fig. 3).   These results suggested that HuH-7 cells possessed capacity to salvage hypoxanthine and hypoxanthine competitively inhibited the action of HGPRT which converts guanine, the precursor of xanthine to GMP and/or consumed PRPP, a substrate of HGPRT. Therefore, we investigated ^{14}C-hypoxanthine to purine nucleotides in HuH-7 cells and demonstrated that both allopurinol and guanine prevented the uptake of ^{14}C-hypoxanthine by purine nucleotides in HuH-7 cells (Fig. 4).   These findings also suggested that an increase in the concentration of xanthine in the conditioned medium was due to the competitive inhibition of HGPRT and/or the consumption of PRPP. Although the activity of HGPRT in HuH-7 cells is higher than human liver, our findings seem to indicate a likely explanation for the usually reported higher concentration of xanthine than hypoxanthine in both plasma and urine in xanthinuric patients[1,3].

REFERENCES

1. CE. Dent and GR. Philpot Xanthinuria, an inborn error (or deviation) of metabolism. *Lancet*, 1: 182 (1954)
2. RWE. Watts, K. Engelman, JR. Klinenberg, JE. Seegmiller and A. Sjoerdsma The enzyme defect in xanthinuria. *Biochem. J.*, 90,4P, (1964)
3. T. Yamamoto, K. Higashino, N. Kono, M. Kawachi, M. Nanahoshi, S. Takahashi, M. Suda, and T. Hada Metabolism of pyrazinamide and allopurinol in hereditary xanthine oxidase deficiency. *Clin. Chim. Acta*, 180: 169 (1989)
4. H. Nakabayashi, K. Taketa, K. Miyano, K. Yamane and J. Sato Growth of human hepatoma cell lines with differentiated functions in chemically defined medium. *Cancer Res.* 42: 3858 (1982)
5. T. Yamamoto, Y. Moriwaki, S. Takahashi, T. Hada, and K. Higashino Separation of hypoxanthine and xanthine from pyrazinamide and its metabolites in plasma and urine by high-performance liquid chromatography. *J. Chromatgr.* 382: 270 (1986)
6. R. Sakuma, T. Nishina, M. Kitamura, H. Yamanka, N. Kamatani and K. Nishioka Screening for adenine and hypoxanthine phosphoribosyl transferase deficiencies in human erythrocytes by high-performance liquid chromatography. *Clin. Chim. Acta* 170:, 281 (1987)
7. T. Yamamoto, Y. Moriwaki, M. Suda, S. Takahashi, K. Hiroishi and K. Higashino Theophylline-induced increase in plasma uric acid - purine catabolism increased by theophylline -. *Int. J. Clin. Pharmacol. Ther. Toxicol.* 29: 257 (1991)
8. DA. Hopkinson, PJC Cook and H. Harris Further data on the adenosine deaminase (ADA) polymorphism and a report of a new phenotype. *Ann. Hum. Genet.* 32: 361 (1969)
9. ED. Corte and F. Stirpe The regulation of rat liver xanthine oxidase Involvement of thiol groups in the conversion of the enzyme activity from dehydrogenase (type D) into oxidase (type O) and purification of the enzyme. *Biochem. J.* 126: 739 (1972)
10. AS. Abd-Elfattah and AS. Wechsler Supriority of HPLC to assay for enzymes regulatung adenine nucleotide pool intermediates metabolism, 5'-nucleotidase, adenylate deaminase and adenylosuccinate lyase - a simple and rapid determination of adenosine. *J. Liq. Chromatogr.* 10: 2653 (1987)

INDUCTION OF CELL DIFFERENTIATION BY IMPDH ANTISENSE OLIGOMER IN HL-60 AND K562 HUMAN LEUKEMIA CELL LINES

Hiroshi Tsutani, Kunihiro Inai, Shin Imamura, Takanori Ueda
Toru Nakamura

First Department of Internal Medicine
Fukui Medical School
Matsuoka, Fukui 910-11, Japan

INTRODUCTION

Inosine-5'-monophosphate dehydrogenase (IMPDH, EC 1.1.1.205) is the enzyme that catalyzes the formation of xanthosine-5'-monophosphate from inosine-5'-monophosphate (IMP). In the purine *de novo* synthetic pathway, IMPDH is positioned at the branch point in the synthesis of adenine and guanine nucleotides and is thus the rate-limiting enzyme in the *de novo* synthesis of guanine nucleotides. Alterations in the activity of IMPDH and the levels of guanine nucleotides have been implicated in the regulation of cell growth and differentiation[1]. The addition of IMPDH inhibitors, such as mycophenolic acid and tiazofurin, to cultures induces terminal differentiation in a variety of human leukemia cells, including K562 erythroid leukemia cells, and HL-60 myeloid leukemia cells as well as human breast cancer cells and melanoma cells[2-4]. To verify the role of this enzyme in controlling differentiation, we determined the ability of specific IMPDH antisense oligomers to induce maturation in the K562 and HL-60 human leukemia cell lines. From these experiments, we concluded that a specific reduction in IMPDH results in inhibition of cell replication and induction of terminal differentiation in the two human leukemia cell lines.

MATERIALS AND METHODS

Synthesis of Antisense Oligomers

Unmodified 18-base oligodeoxynucleotides were synthesized by means of β-cyanoethyl phosphoamidite chemistry and were purified by a nucleic acid purification column,

Purine and Pyrimidine Metabolism in Man VIII, Edited by
A. Sahota and M. Taylor, Plenum Press, New York, 1995

NENSORB PREP (Du Pont Co., Boston, MA). An antisense oligomer sequence was based on the human IMPDH type II cDNA sequence reported by Collart *et al.* [5] and is described as follows: 5'-ACTAATCAGGTAGTCCGC-3' (codon 2-7). This sequence is identical with that of IMPDH type I. A nonsense oligomer having "scrambled" sequence corresponding to the antisense oligomer was synthesized as a control oligomer and is described as follows: 5'-CTTGAGCATTCGCAGAAC-3'.

Cell Culture and Differentiation Marker

K562 erythroleukemia cell lines and HL-60 myeloid leukemia cell lines were grown in RPMI 1640 media supplemented with 10% fetal bovine serum and 150 mg/ml kanamycin sulfate at 37°C in an atmosphere of 5% CO_2 humidified air. Some cultures received 200 μM of guanosine or inosine in addition to the antisense oligomer. Periodically, viability of the cells was determined using trypan blue exclusion. Morphology was assessed by cytocentrifugation followed by May-Gruenwald-Giemsa staining. Monoclonal antibodies for CD15 (LeuM1) and glycophorin A were used to detect granulocytic and erythroid differentiation, respectively. The acetone/methanol-fixed cytocentrifuged preparation was used for immunocytochemistry by alkaline phosphatase anti-alkaline phosphatase (APAAP). The percentage of positive cells was assessed by counting 200 cells.

RESULTS

Effect of Antisense Oligomer on Growth of Leukemia Cells

The inhibition of the cell growth by the oligomers was dose dependent in K562 cells. The inhibitory effect was observed over a range of 0.1 μM to 100 μM of the antisense oligomer and a range more than 10 μM using the control oligomer. The inhibitory effect of the antisense oligomer became more distinct with a second addition of antisense oligomer to the medium at 50% of the initial dose after 24 hr (Fig. 1). Exposure of K562 and HL-60 cells to the antisense oligomer resulted in decrease in cell density relative to untreated cells, while the control oligomer had a negligible effect on the growth of the cell lines. In comparison to untreated cells, the mean growth rates of the antisense-treated cells were 23% in K562 cells and 32% in HL-60 cells after 72 hours of culture, while those of the nonsense-treated cells were 97% in K562 cells and 90% in HL-60 cells. Addition of guanosine to media containing the antisense oligomer abrogated, to a large degree, the reduction in cell number in both cell lines.

Effect of Antisense Oligomer on Differentiation of Leukemia Cells

Incubation of K562 cells and HL-60 cells with the antisense oligomer for 3 days produced significant changes in the morphology of 60% to 80% of the cells. In general, these cells developed a decreased nuclear cytoplasmic ratio, looser chromatin, and less prominent nucleoli than did K562 cells and HL-60 cells treated with the control oligomer. K562 cells treated with the antisense oligomer had more basophilic cytoplasm than the untreated controls and those treated with the control oligomer. Fourteen percent of untreated K562 cells were positive for reactivity with a glycophorin A monoclonal antibody, while 84% of K562 cells

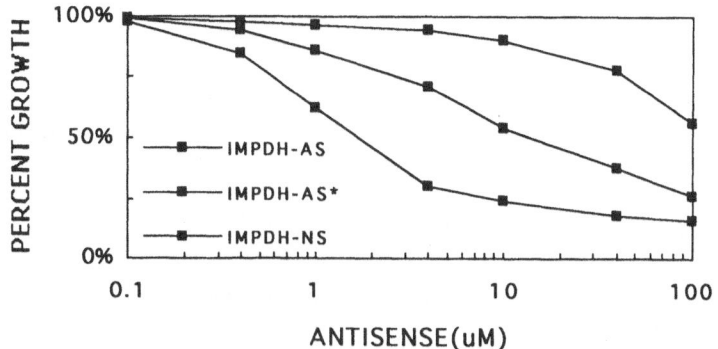

Fig. 1. Dose-effect of IMPDH antisense oligomer on growth of K562 cells. K562 cells at an initial density of 50 cells/μl for three days to an antisense oligomer complementary to human IMPDH cDNA (IMPDH-AS) and a nonsense oligomer having "scrambled" sequence corresponding to IMPDH-AS (IMPDH-NS). Each point represents the mean percentage growth compared to that of untreated cells in three replicate cultures. *Additional exposure of the oligomer at 24 hr.

treated with the antisense oligomer showed reactivity with the antibody. Only 10% of K562 cells treated with the control oligomer showed positive reactivity. Similarly, 58% of HL-60 cells treated with the antisense oligomer showed positive reactivity with CD15 antibody compared to 11% in cells treated with the control oligomer and 8% in untreated cells.

DISCUSSION

Sequence-specific antisense oligomers have been reported[6] to inhibit the synthesis of specific proteins. We used a specific antisense oligomer complementary to human IMPDH cDNA to directly demonstrate the function of encoded protein. The findings described in this report show that addition of the antisense oligomer to cultures resulted in inhibition of cell growth in HL-60 and K562 human leukemia cell lines. The inhibition of cell growth could be blocked by the addition of guanosine to the culture medium, which is consistent with circumvention of the *de novo* block of guanine nucleotide biosynthesis imposed by the antisense oligomer inhibition of IMPDH protein synthesis. A specific reduction in IMPDH caused inhibition of cell replication and induction of maturation markers, such as glycophorin A in K562 cells and CD15 in HL-60 cells, as well as morphological maturation.

Human IMP dehydrogenase consist of two molecular species (types I and II)[7]. Although the type II IMPDH is much more closely correlated with cell growth and differentiation than type I[8], the exact role of both types of IMPDH is still not known. Since the antisense oligomer used in this study is complementary to the sequence of both type I and II IMPDH, we cannot distinguish the relative contributions of the two enzymes. More specific antisense to each type of enzyme could clarify their roles.

ACKNOWLEDGEMENTS

We thank Miss Yuki Takahashi for her technical assistance. This work was supported partly by grants from the Ministry of Education, Japan and Uric Acid Research Foundation, Japan.

REFERENCES

1. Collart, F.R. and E. Huberman, Expression of IMP dehydrogenase in differentiating HL-60 cells, *Blood*. 75: 570-6 (1990).
2. Yu, J., *et al.*, Induction of erythroid differentiation in K562 cells by inhibitors of inosine monophosphate dehydrogenase, *Cancer Res*. 49: 5555-60 (1989).
3. Kiguchi, K., *et al.*, Induction of cell differentiation in melanoma cells by inhibitors of IMP dehydrogenase: altered patterns of IMP dehydrogenase expression and activity, *Cell Growth Differ*. 1: 259-70 (1990).
4. Sidi, Y., *et al.*, Growth inhibition and induction of phenotypic alterations in MCF-7 breast cancer cells by an IMP dehydrogenase inhibitor, *Br J Cancer*. 58: 61-3 (1988).
5. Collart, F.R. and E. Huberman, Cloning and sequence analysis of the human and Chinese hamster inosine-5'-monophosphate dehydrogenase cDNAs, *J Biol Chem*. 263: 15769-72 (1988).
6. Weintraub, H., J.G. Izant, and R.M. Harland, Anti-sense RNA as a molecular tool for genetic analysis, *Trends Genet*. 1: 22-25 (1985).
7. Natsumeda, Y., *et al.*, Two distinct cDNAs for human IMP dehydrogenase, *J Biol Chem*. 265: 5292-5 (1990).
8. Collart, F.R., *et al.*, Increased inosine-5'-phosphate dehydrogenase gene expression in solid tumor tissues and tumor cell lines, *Cancer Res*. 52: 5826-8 (1992).

INHIBITION OF CTP SYNTHETASE INDUCES DIFFERENTIATION OF HL-60 CELLS AND DOWN-REGULATION OF THE C-MYC ONCOGENE

André B.P. Van Kuilenburg, A. André Van den Berg, J. Rutger Meinsma, Robert J. Slingerland, and Albert H. Van Gennip

Academic Medical Center
University of Amsterdam
Departments of Pediatrics and Clinical Chemistry
PO Box 22700
1100 DE Amsterdam
The Netherlands

INTRODUCTION

CTP synthetase and IMP dehydrogenase are the two 'key-enzymes' for the *de novo* biosynthesis of CTP and GTP, respectively. Inhibition of IMP dehydrogenase depletes the intracellular guanine nucleotide pools and induces granulocytic differentiation of HL-60 cells. Furthermore, the induction of HL-60 differentiation to granulocytes is associated with specific changes in the expression of protooncogenes[1,2]. These findings have suggested that guanine ribonucleotides play an important role in the regulation of the differentiation of leukemic-myeloblastic cells. So far, the role of CTP synthetase and cytosine nucleotides in the process of differentiation and regulation of oncogene expression in HL-60 cells has hardly been studied. Therefore, we studied the pattern of *c-myc* expression in HL-60 cells after differentiation induction with either DMSO or with 3'-deazauridine (DAU) an inhibitor of CTP synthetase.

MATERIALS AND METHODS

Cell Culture

HL-60 cells were grown in RPMI 1640 media supplemented with 10% (v/v) foetal calf serum. HL-60 cells at a density of 2 x 10^6 cells/ml were diluted to a concentration of 1 x 10^6 cells/ml prior to the addition of DMSO to a final concentration of 1.25% (v/v). HL-60 cells at an initial density of 1.2 x 10^6 cells/ml were diluted to a density of 0.6 x 10^6 cells/ml prior to the addition of 3'-deazauridine (DAU) to a final concentration of 25 µM. After 2, 6, 10, 24, 48 and 72 h of incubation with either DMSO or DAU, appropriate cell samples were taken for the isolation of RNA and the determination of nucleotides, cell numbers and the viability of the cells. Cell numbers were determined using a Coulter Counter Z 1000 and the viability of the cells was determined by the trypan blue exclusion method. The number of cells capable to generate superoxide after activation with phorbol myristate acetate was determined with a nitro blue tetrazolium (NBT) test, as described before[3]. The amount of nucleotides was determined after extraction of the cells with 0.4 M HClO$_4$ at 4 °C and subsequent analysis with an anion-exchange HPLC procedure[4].

Preparation of RNA and hybridisation conditions

Total cellular RNA was isolated with the urea-LiCl method, as described before[5]. RNA (10 µg) was subjected to electrophoresis through a 1% agarose, 2.2 M formaldehyde gel and transferred to Nytran filters. The Northern blots were hybridised to ^{32}P-labelled cDNA probes. The *c-myc* probe consisted of the 1.2 kb Sac I fragment containing the major part of exon II plus part of the 5' flanking sequence of the human *c-myc* gene, cloned in the recombinant pKH 47 plasmid[6]. The γ-actin probe consisted of the 1 kb *Bam*H1-*Hind*III fragment from the pHF1 plasmide[7] which was further subcloned into the pSP 64 plasmid. The γ-actin probe contained part of the coding region plus part of the 3' untranslated sequence of the cDNA clone. The Northern blots were hybridised at 42 °C for 16-24 h in the presence of 50% (v/v) formamide, 6 x SSC (0.15 M NaCl, 0.015 M sodium citrate, pH 7.0), 1 x Denhardt's solution, 1% (w/v) SDS, 10% (w/v) dextran sulphate and 200 µg/ml salmon sperm DNA. The filters were washed and exposed to a Fuji RX film using an intensifying screen at -80 °C. The densitometric scans of the films were made with a laser densitometer.

RESULTS AND DISCUSSION

When HL-60 cells were incubated with 25 µM 3'-deazauridine a rapid depletion of the CTP ribonucleotides was observed within 2 h which persisted at least until 24 h (Fig. 1). The decline in the CTP levels reflects the inhibition of CTP synthetase by deazaUTP which is a competitive inhibitor of the enzyme. Furthermore, only a small increase in the UTP pool was observed which can be explained by the fact that uridine and deazauridine are both metabolised by the same uridine-cytidine kinase. The progressive increase in the levels of ATP and GTP until 24 h, is probably due to stimulation of the purine *de novo* pathway by PRPP which accumulates after inhibition of the pyrimidine *de novo* pathway[8].

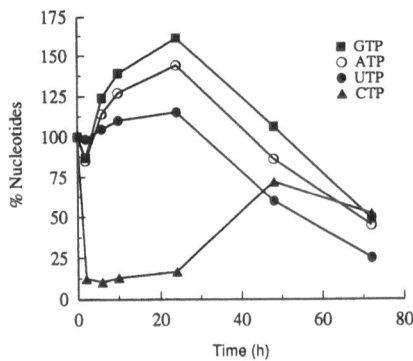

Figure 1. The nucleotide levels in HL-60 cells treated with 25 µM DAU relative to those observed in exponentially growing HL-60 cells.

The mRNA levels of the *c-myc* oncogene was followed up to 72 h of exposure of HL-60 cells to either DMSO or DAU. Figure 2 shows that the treatment of HL-60 cells with DMSO resulted in a very rapid decline of the *c-myc* oncogene transcript and that no *c-myc* mRNA could be detected after 2 h. Determination of the cell numbers revealed that the population of HL-60 cells was doubled after 5 days. Furthermore, the proportion of HL-60 cells capable of generating an oxidative burst, a phenomenon characteristic of mature granulocytic cells, increased from 2% prior to the start of the experiment to approximately

80% after 5 days. Therefore, the down-regulation of the *c-myc gene* expression is probably a necessary, but not a sole prerequisite, for the onset of proliferation arrest when HL-60 cells were induced to differentiate with DMSO.

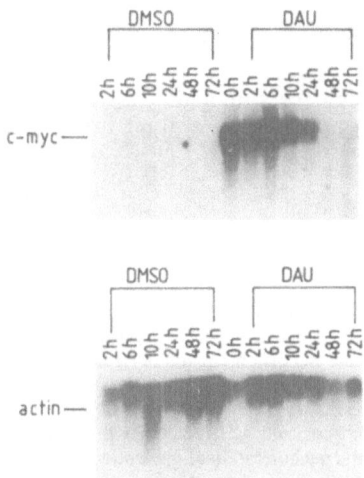

Figure 2. Northern blot analysis of RNAs isolated from HL-60 cells during the course of granulocytic differentiation induced by DMSO or DAU.

The inhibition of CTP synthetase by deazaUTP resulted in a progressive decline of the *c-myc* transcript to undetectable levels within 48 h (fig. 2). The densitometric analysis of the amount of *c-myc* mRNA and γ-actin mRNA showed that the inhibition of CTP synthetase induced a rapid decline of the *c-myc* transcript to approximately 40% of that observed in uninduced exponentially growing HL-60 cells (fig. 3). Furthermore, the total cell number of the HL-60 cell population did not increase. Since CTP synthetase has an important role in providing the necessary precursors for RNA and DNA biosynthesis, the observed decrease in proliferation rate might be due to the depletion of cytosine (deoxy) ribonucleotides. The proportion of HL-60 cells capable of generating an oxidative burst increased from 2% prior to the start of the experiment to approximately 50% after 5 days.

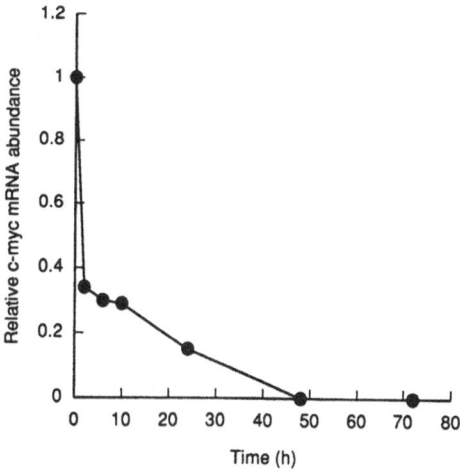

Figure 3. The relative *c-myc* mRNA abundance compared to that observed in uninduced exponentially growing HL-60 cells.

Our results demonstrate that the inhibition of CTP synthetase is associated with the down-regulation of the *c-myc* oncogene which is a prerequisite in the process of terminal differentiation of HL-60 cells. Therefore, drugs that inhibit the activity of CTP synthetase might provide a powerfull tool to overcome the differentiation block of leukemic cells.

Acknowledgements

We thank Dr. Rogier Versteeg for stimulating discussions and critical reading of the manuscript. This study was supported by a grant from the Maurits and Anna de Kock foundation.

REFERENCES

1. S.M. Kharbanda, M.L. Sherman, D.R. Spriggs, and D.W. Kufe, Effects of Tiazofurin on Protooncogene Expression during HL-60 Cell Differentiation, *Cancer Res.* 48: 5965 (1988).
2. L.S. Mitchell, R.A. Neill, and G.D. Birnie, Temporal relationships between induced changes in *c-myc* mRNA abundance, proliferation, and differentiation in HL60 cells, *Differentiation*, 49: 119 (1992).
3. L.J. Meerhof, and D. Roos, Heterogeneity in chronic granulomatous disease detected with an improved nitro blue tetrazolium slide test, *J. Leuk. Biol.* 39: 699 (1986).
4. D. De Korte, W.A. Haverkort, D. Roos, and A.H. Van Gennip, Anion-exchange high performance liquid chromatography method for the quantification of nucleotides in human blood cells, *Clin. Chim. Acta* 148: 185 (1985).
5. C. Auffray, and F. Rougeon, Purification of Mouse Immunoglobulin Heavy-Chain Messenger RNAs from total Myeloma Tumor RNA, *Eur. J. Biochem.* 107: 303 (1980)
6. W.W. Colby, E.Y. Chen, D.H. Smith, and A.D. Levinson, Identification and nucleotide sequence of a human locus homologous to the v-*myc* oncogene of avian myelocytomatosis virus MC29, *Nature* 301: 722 (1983).
7. P. Gunning, P. Ponte, H. Okayama, J. Engel, H. Blau, and L. Kedes, Isolation and Characterization of Full-Length cDNA Clones for Human α-, β-, and γ-Actin mRNAs: Skeletal but Not Cytoplasmic Actins Have an Amino-Terminal Cysteine that Is Subsequently Removed, *Mol. Cell. Biol.* 3: 787 (1983).
8. A.A. Van den Berg, P.A.W. Mooyer, H. Van Lenthe, E.H. Stet, A.B.P. Van Kuilenburg, and A.H. Van Gennip, Antagonising effects of the IMP dehydrogenase inhibitor mycophenolic acid and the CTP synthetase inhibitor 3'-deazauridine on Molt-3 human leukemia cells: a central role for PRPP, *submitted for publication.*

ADENOSINE TRANSPORT AND METABOLISM
IN CARDIOMYOCYTES OF HYPERTHYROID RAT

Ryszard T. Smolenski, Anne-Marie L. Seymour and Magdi H. Yacoub

M.R.S. Group
National Heart and Lung Institute at Harefield Hospital
Harefield, Middx UB9 6JH
U.K.

INTRODUCTION

Hyperthyroidism produces a number of metabolic and morphological changes in the heart including impaired energetics, increased protein synthesis, a decrease in cell surface/cell volume ratio and a reduction in endothelial density due to hypertrophy[1-4]. These changes may also affect cardiomyocyte adenosine (ADO) transport and metabolism which, in turn may alter the capacity to release physiologically active adenosine and nucleotide synthesis. This study evaluated changes in ADO transport capacity in isolated cardiomyocytes and enzyme activities involved in ADO metabolism in normal and hyperthyroid myocardium.

MATERIALS AND METHODS

Hyperthyroidism was induced in male Wistar rats weighing 250-350 g by daily intraperitoneal injections of thyroxine (T_4) (0.35 μg/100g body weight) over seven days[1]. As a consequence of this treatment, a 30% increase in heart weight/body weight ratio was observed. To study ADO transport and metabolism, cardiomyocytes were isolated by a collagenase perfusion technique as described previously[5,6] from animals treated with T_4 and from normal rat heart. Cellular preparations contained 60-70 % of cardiomyocytes with rod-shaped morphology. Cells were incubated with [2,8-^{3}H] ADO for 1 min at 37°C. [U-^{14}C] sucrose was included as marker of extracellular space[7]. Final ADO concentrations were 1, 10, 100 or 1000 μM. After the incubation cell suspensions were quickly layered on the top of Eppendorf tube containing 0.1 ml of 2 M perchloric acid covered by layer of bromododecane, followed by centrifugation to separate cells and medium[8]. As a zero time control, myocyte suspensions were subjected to a similar procedure without 1 min incubation. Cell extracts and medium were used directly for scintillation counting or were subjected to HPLC separation[9], followed by radioactivity analysis of separated fractions corresponding to the various nucleotide metabolites. To study enzymes of adenosine metabolism, heart homogenates were prepared from T_4 treated and normal rats. Enzyme assay procedures were based on HPLC separation of substrates and products[9,10]. The activities of ADO deaminase (ADA), ADO kinase (AK), AMP-deaminase (AMP-DA), SAH-hydrolase (SAHH), purine nucleoside phosphorylase (PNP), membrane-5'-nucleotidase (M5'N) and cytosolic 5'ucleotidases: AMP specific (A5'N) and IMP-specific (I5'N) were evaluated.

RESULTS

The ADO transport rate was significantly higher in T_4 treated group than in controls especially when expressed per number of cells. At 10 µM concentration, ADO transport was also significantly higher in relation to total cellular protein content (Table 1). Further analysis of cell and medium extract by HPLC revealed that at concentrations 1 and 10 µM ADO was almost completely incorporated into ATP whilst at 100 and 1000 µM concentration only a small fraction of ADO was phosphorylated in both cell types. However, at 10 µM concentration ADO incorporation rate into ATP was significantly higher in the T_4 treated group. Among the enzyme activities measured M5'N and ADA activities were lower in T_4 treated rat while AK activity was higher. PNP tended to be more active in T_4 treated group but this difference did not reach statistical significance

Table 1. ADO uptake, phosphorylation and intracelluar ADO accumulation in cardiomyocytes of T_4 treated and control rat.

	ADO concentration							
	1 µM		10 µM		100 µM		1,000 µM	
	T_4 treated	control	T_4 treated	control	T_4 treated	control	T_4 treated	control
ADO uptake (pmol/min/mg prot)	38.9 ±2.6	32.8 ±2.1	*275.3 ±20.5	197.4 ±26.2	285.5 ±16.7	247.2 ±17.5	630.6 ±111.9	752.9 ±207.7
ADO uptake (pmol/min/10^6 cells)	*126.4 ±13.5	73.1 ±13.1	*934.7 ±125.7	462.5 ±112.7	998.1 ±72.0	561.9 ±90.4	2,066.6 ±516.3	1,609.3 ±394.7
ADO incorporation into ATP (pmol/min/mg prot)	33.1 ±2.3	31.1 ±2.1	*250.2 ±16.9	180.1 ±23.8	68.7 ±12.6	57.8 ±9.3	65.6 ±16.1	61.1 ±22.3
intracellular ADO accumulation (pmol/min/mg prot)	0.9 ±0.3	1.6 ±0.1	5.8 ±3.2	-0.9 ±4.4	156.9 ±25.1	213.8 ±42.9	539.1 ±123.7	665.8 ±185.4

Values represent the mean ±S.E.M (n=6). *p<0.05 in comparison with control

Table 2. Myocardial activities of enzymes of purine metabolism in control and hyperthyroid rat heart (μmol/min/g wet wt)

	CONTROL	T_4TREATED
AMP-deaminase	4.58 ±0.48	4.87 ±0.27
adenosine deaminase	1.56 ±0.05	1.39* ±0.04
purine nucleoside phosphorylase	1.12 ±0.12	1.40 ±0.09
ecto-5'-nucleotidase	9.11 ±0.28	6.93** ±0.15
IMP cytosolic-5'-nucleotidase	0.179 ±0.038	0.179 ±0.018
AMP cytosolic-5'-nucleotidase	0.181 ±0.036	0.141 ±0.016
SAH hydrolase	0.0080 ±0.0002	0.0073 ±0.0002
adenosine kinase	0.238 ±0.007	0.283* ±0.015

Values represent the mean ±S.E.M (n=4).
*p<0.05, **p<0.001 in comparison to control.

DISCUSSION

ADO transport capacity increases in cardiomyocytes of hyperthyroid rat which is evident if expressed per cell number or per total cellular protein. Increased transport capacity was associated with a higher activity of AK measured in homogenate and a faster rate of adenosine phosphorylation. Thus, all ADO reutilization steps were induced in hyperthyroid cardiomyocytes. Lower M5'N and ADA activities in the T_4 treated group may reflect an increased membrane/cell volume ratio and/or a lower density of endothelial cells which are typical consequences of hypertrophy since these enzymes are membrane bound and of much higher activity in the endothelium[11]. However, PNP, located in the endothelium[12] tended to be more active after T_4 treatment which is difficult to explain according to available data. The substantial decrease in E5'N activity may be of physiological significance because this enzyme is involved in the production of ADO from extracellular nucleotides which may provide an important mechanism for endogenous vascular protection[13,14]. The reduction in E5'N may shift the balance between proaggregatory nucleotides and antiaggregatory adenosine and thus facilitate intravascular thrombus formation and may promote the damages caused by catecholamines and leukocytes which are also counteracted by ADO. The combination of lower ADO production capacity with accelerated reutilization rate may result in a substantial decrease in interstitial ADO concentration in the heart, further affecting myocardial homeostatic mechanisms.

ACKNOWLEDGMENT

This study was supported by the British Heart Foundation (Grant No: 91/167)

REFERENCES

1. A-M.L. Seymour, H. Eldar, and G.K. Radda, Hyperthyroidism results in increased glycolytic capacity in the rat heart.A ^{31}P-NMR study, *Biochim. Biophys. Acta* 1055:107(1990).
2. H. Yamamoto and M. Avkiran, Left ventricular pressure overload during postnatal development. Effects on coronary vasodilator reserve and tolerance to hypothermic global ischemia, *J. Thorac. Cardiovasc. Surg.* 105:120(1993).
3. J. Aussedat, S. Lortet, A. Ray, A. Rossi, M. Heckman, H.G. Zimmer, M. Vincent, and J. Sassart, Energy metabolism of the hypertrophied heart studied by 31P nuclear magnetic resonance, *Cardioscience.* 3:233(1992).
4. G. Olivetti, F. Quaini, C. Lagrasta, R. Ricci, G. Tiberti, J.M. Capasso, and Anversa P, Myocyte cellular hypertrophy and hyperplasia contribute to ventricular wall remodeling in anemia-induced cardiac hypertrophy in rats, *Am. J. Pathol.* 141:227(1992).
5. R.T. Smolenski, J. Schrader, H. de Groot, and A. Deussen, Oxygen partial pressure and free intracellular adenosine of isolated cardiomyocytes, *Am. J. Physiol.* 260:C708(1991).
6. R.T. Smolenski, A. Suitters, and M.H. Yacoub, Adenine nucleotide catabolism and adenosine formation in isolated human cardiomyocytes, *J. Mol. Cell Cardiol.* 24:91(1992).
7. D.A. Ford and M.J. Rovetto, Rat cardiac adenosine transport and metabolism, *Am. J. Physiol.* 252:H54(1987).
8. T. Geisbuhler, R.A. Altschuld, R.W. Trewyn, A.Z. Ansel, K. Lamka, and G.P. Brierley, Adenine nucleotide metabolism and compartmentalization in isolated adult rat heart cells, *Circ. Res.* 54:536(1984).
9. R.T. Smolenski, D.R. Lachno, S.J.M. Ledingham, and M.H. Yacoub, Determination of sixteen nucleotides, nucleosides and bases using high-performance liquid chromatography and its application to the study of purine metabolism in hearts for transplantation, *J. Chromatogr.* 527:414(1990).
10. R.T. Smolenski, J.W. de Jong, M. Janssen, D.R. Lachno, M.M. Zydowo, M. Tavenier, T. Huizer, and M.H. Yacoub, Formation and breakdown of uridine in ischemic hearts of rats and humans, *J. Mol. Cell. Cardiol.* 25:67(1993).
11. J.W. de Jong, E. Keijzer, T. Huizer, and B. Schoutsen, Ischemic nucleotide breakdown increases during cardiac development due to drop in adenosine anabolism/catabolism ratio, *J. Mol. Cell. Cardiol.* 22:1065(1990).
12. M. Borgers and F. Thone, Species differences in adenosine metabolic sites in the heart, *Histochem. J.* 24:445(1992).
13. M.B. Grisham, L.A. Hernandez, and D.N. Granger, Adenosine inhibits ischemia-reperfusion-induced leukocyte adherence and extravasation, *Am. J. Physiol.* 257:H1334(1989).
14. B.N. Cronstein, R.I. Levin, M. Philips, R. Hirschhorn, S.B. Abramson, and Weissmann G, Neutrophil adherence to endothelium is enhanced via adenosine A1 receptors and inhibited via adenosine A2 receptors, *J. Immunol.* 148:2201(1992).

NUCLEOSIDE TRANSPORT AND METABOLISM IN LYMPHOCYTES, POLYMORPHONUCLEAR CELLS AND CEREBRAL SYNAPTOSOMES

Maria Staub[1], Maria Sasvari-Szekely[1], Magda Solymossy[1] and K. Szikla[2]

[1]Institute of Biochemistry I, Semmelweis Medical University, Budapest
[2]Institute of Biochemistry II, Semmelweis Medical University, Budapest
H-1444 Budapest 8, POB 260, Hungary

SUMMARY

The salvage of nucleosides dominates over *de novo* biosynthesis in lymphocytes, polymorphonuclear cells (PMN) and in central neutral nervous system (CNS) in higher organisms. Earlier works in our laboratory have shown that the salvage of deoxycytidine (dCyd) did not correlate with DNA synthesis. The uptake and metabolism of dCyd was higher in undifferentiated germinal center lymphocytes and in follicles comparing to more differentiated cells.

Recently we have compared the transport of thymidine (dThd), dCyd, uridine (Urd) and adenosine (Ado) in the three cell systems in which the salvage of nucleosides is dominating. It was found that dCyd was transported 30 times more effectively into lymphocytes than into PMN and synaptosomes, while Urd was transported about the same rate into the two cells and into synaptosomes. All transport processes could be inhibited by dipiridamole, NBRPR, papaverine and dilazep.

The dCyd and dThd was phosphorylated even at 0°C up to TTP and dCTP without incorporation into DNA and into liponucleotides. Our results show that the processes of transport-phosphorylation, as well as the processes of DNA-CDP-phospholipid synthesis are tightly coupled to each other in intact cells and organelles.

INTRODUCTION

Cells and tissues might be extremely different in the capability of reutilization of purine and pyrimidine bases and nucleosides. The highest capacity of nucleoside salvage can be found in lymphoid cells, including monocytes and macrophages, in polymorphonuclear cells (PMN), as well as in cells of central nerves system (CNS). In these cell types, the *de novo* biosynthesis of nucleotides are negligible, so the rate of nucleic acid synthesis is highly

Manipulation of adenosine transport is a potential therapeutic approach in many cardiovascular and CNS disorders, like in ischemic injury[1,2], and in ischemia induced neuronal degradation[3]. Similar effects of transport inhibiting drugs (e.g. nitrobenzylthioinosine, dilacep) support the idea about a common nucleoside carrier system in different cell types[4,5], including leukemic cells[6,7]. The only exception seems to be thymidine (dThd), which has a separate carrier[8]. As the normal human plasma contains thymidine and thymine in micromolar range, its transport by lymphoid cells might have a physiological significance. Moreover, much higher concentrations of dThd might be generated locally by excretion at sites of rapid cell turnover[9,10,11].

Following uptake of nucleosides, they are metabolized by phosphorylation through the action of deoxycytidine kinase, an enzyme with broad substrate specifity[12,13] and through thymidine kinase[14]. On the other hand, *de novo* synthesized nucleotides are continuously degraded and excreted as nucleosides[15,16]. These mechanisms provides a fine tuning of NTP and dNTP pools[17,18]. Any disturbance in nucleotide pool balance should have serious consequences, like in the case of adenosine deaminase defficiency[19,20].

In this paper further evidences are presented for the significance of pyrimidine transport and metabolism in lymphocytes, in PMN and in synaptosomes. Moreover, a tight connection between transport and further metabolism of these pyrimidine nucleosidesis is suggested by experiments performed at 0^0C.

METHODS

Tonsillar lymphocytes were prepared from tonsils of children[20]. Cerebral-cortical synaptosomes were made as described[21]. Transport of nucleosides was measured by inhibitor-stop filtration method[21]. Metabolism of nucleosides was characterized by its uptake into the acid soluble pool of the cells as well as its incorporation into the acid insoluble fraction, as described earlier[22,23]. Liponucleotides and nucleotides were separated by thin layer chromatography on Kieselgel F60 in case of ^{3}H-dCyd labeling, ^{3}H-Thd labeled nucleotides were separated on DEAE-cellulose[23].

RESULTS AND DISCUSSION

Transport of different pyrimidine nucleosides was measured in tonsillar lymphocytes, in polymorphonuclear cells (PMN) and in synaptosomes (Fig.1,2), as a model for cell types having a characteristically high level of nucleoside salvage. On the other hand, PMN was chosen as a model of non-dividing cells, tonsillar lymphocytes as stimulated cells, while the synaptosomes represent the function of nucleoside transport in the CNS. Lymphocytes were shown to have a much higher rate of dCyd than any other cell type (e.g. PMN, see Fig.1.) in accordance with the high expression of deoxycytidine kinase in lymphoid cells[12]. PMN had no significant dThd transport, but with a similar rate of Urd transport as measured in lymphocytes (data not shown). Adenosine (Ado) transport is difficult to measure because of a high level of adenosine deaminase in these cells. Synaptosomes was shown to have a similar rate of Ado and Urd transport (see Fig.2.). All of the pyrimidine transport processes could be inhibited by dipiridamol and its analogue compounds (E1 and E2), in accordance with the suggested common transporter for nucleosides[4-7].

Separation of the transport process from the subsequent metabolism is often made by ccooling the samples, supposing that there is no phosphorylation at 0^oC. However, experiments are presented on Fig.3-4 suggesting an intensive metabolism of nucleosides at

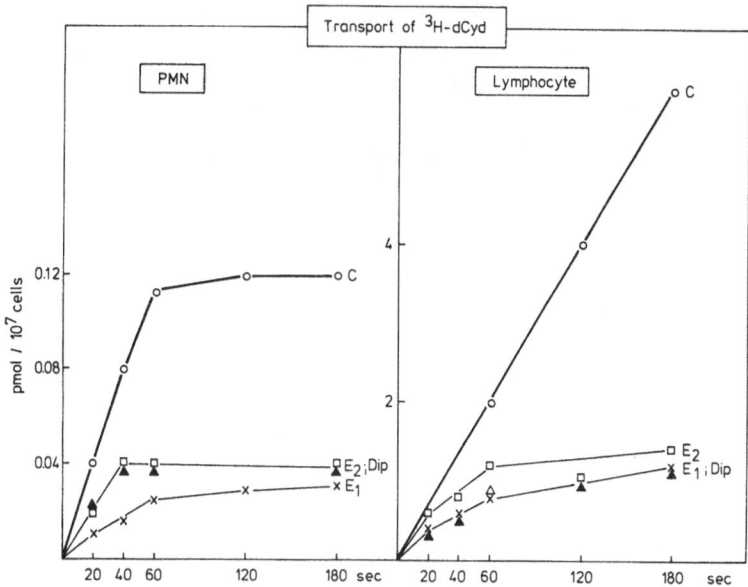

Fig.1. Transport of ^{3}H-dCyd into polymorphonuclear cells (PMN) and into lymphocytes. Transport rate was measured according to[21] by inhibitor-stop filtration method in the absence(C) or presence of different inhibitors like dipiridamol(Dip) and its analogue compounds (E1 and E2) in 10 uM final concentration.

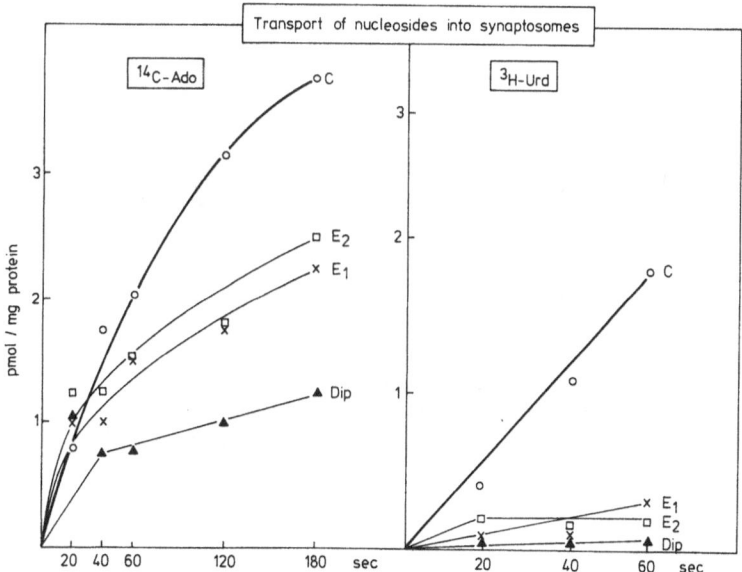

Fig.2. Transport of ^{14}C-Ado and ^{3}H-Urd into synaptosomes. Conditions are the same as under Fig.1.

0°C, including their phosphorylation. Fig.3. shows the distribution of intracellular metabolites in case of ^{3}H-dCyd labeling at 0°C as well as at 37°C. Phosphorylation of nucleosides (Nucleotides) has about the same speed at 2 hours at 0°C and at 37°C, while there is no significant liponucleotide and DNA labeling at low temperature. Thin layer chromatography of labeled nucleotides (Fig. 4) proved the presence of significant amount of dCTP at low temperature, but no labeling into liponucleotides (LN) and into DNA.

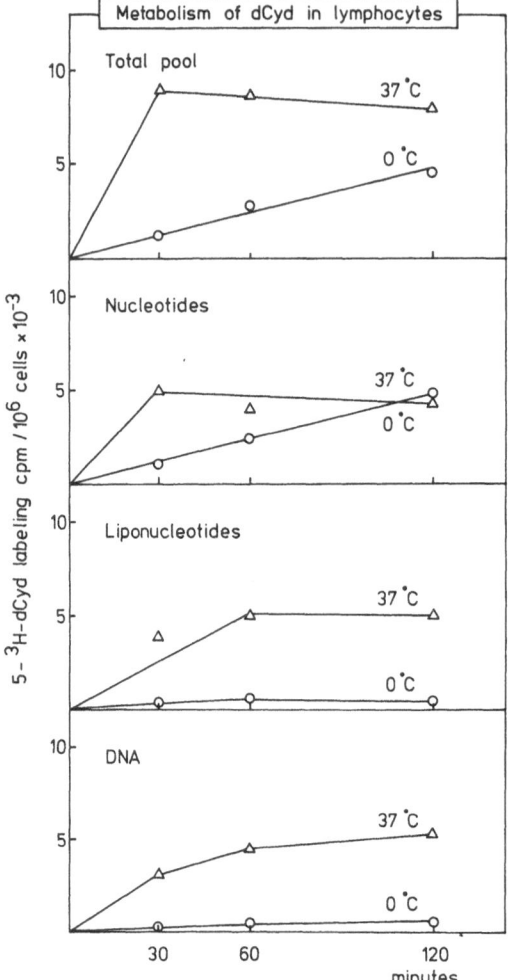

Fig. 3. Metabolism of dCyd at 0°C and at 37°C in lymphocytes. Lymphocytes were labeled with ³H-dCyd at different temperatures, than different intracellular fractions were analyzed for radioactivity as indicated.

These results rise the question whether it is advisable to believe that a low temperature, transport or binding to different receptors could be measured without the disturbing effect of the metabolism.

AKNOWLEDGEMENT

The authors wish to thank to S. Virga for the excellent technical assistance. this work was supported by OTKA T 006229, PHARE 0430 and COST-PECO CIPDCT925055.

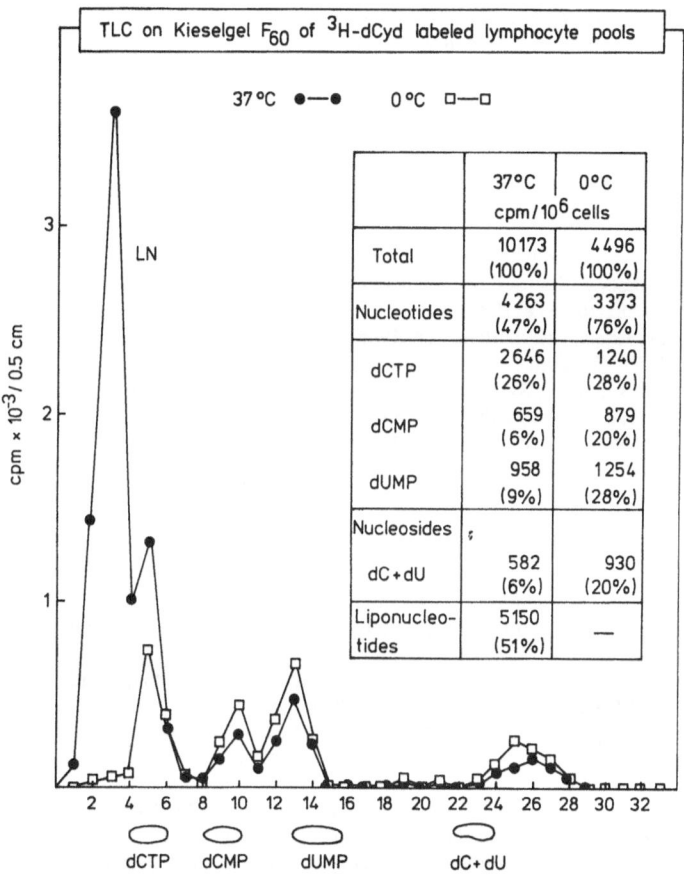

Fig. 4. **Separation of nucleosides and nucleotides synthesized at 0°C and at 37°C in lymphocytes.** After 1 hour of labeling of lymphocytes with ³H-dCyd at different temperatures, the ethanol soluble pool was separated using thin layer chromatography on Kiselgel.

REFERENCE

1. De Jong, J.W., Van der Meer, P. Owen and Opie, L. H., Prevention and treatment of ischemic injury with nucleosides, *Bratisl. Lek. Listy* 92:165-173 (1991)
2. G.E. Zhang, Paul H. Franklin and Thomas F. Murray, Manipulation of endogeneous adenosine in the rat prepiriform cortex modulates seizure susceptibility, *J. Pharmacology and Experimental Therapeutics* 264:1415-1424 (1992).
3. Philips, J.W. and O'Reagan, M.H., Deoxycoformycin antagonizes ischemia induced neuronal degeneration, *Brain. Res. Bull* 22:537-540 (1969).
4. Belt, J.A., Marina, N.M., Phelps, D.A. and Crawford, C.R., Nucleoside transport in normal and neoplastic cells, *Adv. Enzyme Regul.* 33, 235-252 (1993)
5. Plagemann P.G., Wohlhueter, R.M. and Woffendin, C., Nucleoside and nucleobase transport in animal cells, *Biochem. Biophys. Acta* 947:405-443 (1988).
6. Takimoto, T., Kioth, T., Tanizawa, A., Akiyama, Y., Kiriyama, Y., Kubota, M. and Mikawa, H., Characterization of nucleoside transport during leukemic cell differentiation, *in: "Purine and Pyrimidine Metabolism in Man,"* Mikanagi, K., Mishioka, K. and Kelley W.N. ed., Plenum Publ. Corp., (1989).
7. Tanaka, M. and Yoshida, S., Formation of cytosine arabinoside-5'-triphosphate in cultured human leukemic cell lines correlates with nucleoside transport capacity, *Jpn. J. Cancer Res. (Gann)* 78:851-857 (1987).
8. Taljanidisz, J., Sasvari-Szekely, M., Spsokukotskaja, T., Antoni, F. and Staub, M., Reversible permeabilization of lymphocytes destroys the incorporation of deoxythymidine but not of deoxycytidine. *Biochim. Biophys. Acta* 885:266-271 (1986).
9. Hamatani, K. and Amano, M., Different labelling patterns in mouse lymphoid tissues with [^{3}H]-deoxycytidine and [^{3}H]-thymidine, *Cell Tiss. Kinet.* 13:435-443 (1980).
10. Iizasa, T. and Carson, D.A., Synthesis and release of deoxycytidine by human B and T lymphoblasts, *Biochim. Biophys. Acta* 888:249-251 (1986).
11. Penit, C. and Papiernik, M., Regulation of thymocyte proliferation and survival of deoxynucleotides. Deoxycitidine produced by thymic accesory cells protects thymocytes from deoxyguanosine toxicity and stimulates their spontaneous proliferation, *Eur. J. Immunol.* 16:257-263 (1986).
12. Eriksson, S., Kierdaszuk, B., Munch-Petersen, B., Oberg, P. and Johansson, N.G., Comparison of the substrate specificities of human thymidine kinase 1 and 2 and deoxycytidine kinase toward antiviral and cytostatic nucleoside analogs. *Biochem. Byophys. Res. Commun* 176:586-92 (1991).
13. Weber, G., Signal, R.L., Abonyi, M., Parja, N., Hata, Y., Szekeres, T., Yeh, A.. and Look, K.Y., Regulation of deoxycytidine kinase activity and inhibition by DFDC, *Adv. Enzyme Regul.*, 33:39-59 (1993)
14. Xu, Y.Z. and Plunkett, W., Regulation of thymidine kinase and thymidilate synthase in intact human lymphoblast CCRF-CEM cells, *J. Biol. Chem.* (1993)
15. Nicander, B. and Reichard, P., Dynamics of pyrimidine deoxynucleosidё triphosphate pools in relationship to DNA synthesis in 3T6 mouse fibroblasts. *Proc. Natl. Acad. Sci. USA* 80:1347-1351 (1983).
16. Reichard, P., Regulation of deoxyribotide synthesis. *Biochemistry* 26:3245-3248 (1987).
17. Staub, M., Salvage of nucleosides and nucleotide balance in higher organisms, *Path to Pyrimidines* 2: 7
18. Carson D.A., Kaye, J. and Seegmiller, J.E., Lymphospecific toxicity in adenosine deaminase deficiency and purine nucleoside phosporylase deficiency: Possible role of nucleoside kinase(s). *Proc. Natl. Acad. Sci. USA.* 74:5677-5681 (1977).
19. Carson D.A., Kaye, J. and Wasson, D.B., The potential importance of soluble deoxynucleotidase activity in mediating deoxyadenosine toxicity in human lymphoblasts. *J. Immunol.* 126:348-352 (1981).
20. Staub, M., Antoni, F. and Sellyei, M., DNA synthesis in tonsil lymphocytes. I. Changes in cell population during culture, *Biochemical Med.* 15:246-253 (1976).
21. Lee, C.W. and Jarvis, S.M., Nucleoside transport in rat cerebral-cortical synaptosomes. Evidence for two types of nucleoside transporters, *Biochem. J.* 249:557-564 (1988).
22. Staub, M., Sasvari-Szekely, M., Spsokukotskaja, T., Antoni, F. and Meretey, K., Differences between human lymphocyte subpopulations with respect to the uptake and incorporation of [^{3}H]-uridine and ribonucleoside triphosphate pools, *Biochemical Medicine* 19:218-130 (1978).
23. Spasokukotskaja, T., Spyrou, G. and Staub, M., Deoxycitidine is salvaged not only into DNA but also into phospholipid precursors. *Biochem. Biophys. Res. Commun.* 155:923-927 (1988).

INHIBITION OF NUCLEOSIDE TRANSPORT BY REACTIVE OXYGEN SPECIES IN BOVINE HEART MICROVASCULAR ENDOTHELIAL CELLS

Jerzy Barankiewicz[1], Jon Uyesaka[1], Wilhelm Kossenjans[2], Zbigniew Rymaszewski[3]

[1]Gensia Inc., 9360 Towne Centre Drive, San Diego, CA 92121.,[2]Dept. Obstetrics and Gynecology,[3] Dept. Internal Medicine, University of Cincinnati, Cincinnati, Ohio 45267, USA

Cellular homeostasis depends on the structural and functional integrity of the membrane bilayer. Damage to the membrane can interfere with vital cellular processes, including signal transduction, molecular recognition, maintenance of the membrane potential, cellular metabolism and transport of molecules. Modification of the membrane bilayer by reactive oxidative species (ROS) is a major contributor to membrane damage and has been implicated in many pathological processes. In a number of diseases, injury to endothelial cells is mediated by oxidative species generated during ischemia/reperfusion or by activated neutrophils. During ischemia and oxidant injury, several metabolic events occur, including depletion of intracellular ATP and formation of a number of purine catabolites including inosine (Ino), hypoxanthine (Hyp) and adenosine (Ado) (Halliwell and Gutteridge, 1990). Accumulated Hyp can be oxidized by xanthine oxidase when the tissue is oxygenated, causing rapid generation of superoxide and hydrogen peroxide. On the other hand, formation of Ado is especially important because it has vasodilatory and anti-inflammatory activity. Alterations in nucleoside transport (NT) by ROS might have an effect on extracellular Ado concentration *in vivo*. Reduced Ado transport could elevate Ado concentrations, especially when Ado is formed extracellularly. Inhibition of NT could also diminish extracellular nucleoside uptake required for the reconstitution of intracellular nucleotide pools after cellular stress. Decreased nucleoside efflux might protect cells by allowing more efficient intracellular nucleoside salvage.

Endothelial cells are not only a source of ROS during anoxia/reoxygenation, but are also targets for ROS-mediated injury. To evaluate the effect of endogenous and exogenous ROS on NT, a bovine heart microvascular endothelial cell (BHMEC) line established at Gensia was studied. BHMEC were maintained in alpha MEM with 10% heat-inactivated fetal bovine serum and 2 mM L-glutamine. Cells were grown at 37°C under 5% CO_2 and split every 4-5 days. The BHMEC maintained endothelial cell morphology and expressed factor VIII and acetylated LDL after more than two hundred passages.

NUCLEOSIDE TRANSPORT IN BHMEC

To determine sensitivity of the Ado transport in BHMEC to NT inhibitors, cells were incubated with different concentrations of nitrobenzylthioinosine (NBMPR) and dipyridamole (DIP). Influx of Ado in BHMEC was inhibited by nanomolar concentrations of NBMPR as well as by DIP, indicating that Ado is transported into these cells by NBMPR-sensitive type of NT system (Table 1.)

Although Na^+ dependent NT systems have been found in various cells (Dagnino, et al., 1991), Na^+ dependent nucleoside (Ado) permeability was not found in BHMEC (data not shown).

Purine and Pyrimidine Metabolism in Man VIII, Edited by
A. Sahota and M. Taylor, Plenum Press, New York, 1995

Table 1. Inhibition of Ado transport (influx) by NBMPR and DIP in BHMEC

NT inhibitor	concentrations (nm)	inhibition %
NBMPR	1	20.7±2
	10	92.4±7
dipyridamole	10	60.0±12
	500	100.0±0

Cells (2×10^6/plate) were incubated in 0.5 ml alpha MEM at 37°C for 60 sec with NBMPR or DIP followed by incubation with 1 uM [2,8-^{3}H] adenosine (36 Ci/mmole) for a 10 sec. Cells were then rapidly washed with 10 ml medium (5 sec) and extracted on ice with 100 ul of 0.4 M perchloric acid. Adenosine transport into the cells was evaluated by determining radioactivity in the cell extracts. Control cells incorporated 63,841±2267 cpm of adenosine/2×10^6 cells.

EFFECT OF REACTIVE OXYGEN SPECIES ON NUCLEOSIDE TRANSPORT

To evaluate the effect of exogenous ROS on NT, cells were exposed to exogenous hydrogen peroxide (H_2O_2) or hypochlorous acid (HClO). To evaluate the effects of intracellularly generated ROS on nucleoside transport, cells were preincubated with menadione (2-methyl-1,4 naphthoquinone).

Addition of hydrogen peroxide and hypochlorous acid to BHMEC resulted in a concentration-dependent inhibition of NT with IC_{50} values of 300 and 500 μM, respectively (Table 2). The inhibitory effect of H_2O_2 but not hypochlorous acid on NT was completely reversed by first adding catalase (0.2 units/0.5 ml) to the cells for 1 min.

Table 2. Effect of hydrogen peroxide or hypochlorous acid on adenosine transport (influx) in BHMEC.

Hydrogen peroxide (uM)	Ado transport (% inhibition)	Hypochlorous acid (μM)	Ado transport (% inhibition)
3	5±1	5	12±8
30	5±4	50	28±13
300	48±8	100	36±8
3000	77±2	500	54±4

Cells (2×10^6/plate) were incubated with different concentrations of H_2O_2 or hypochlorous acid for 10 min. Adenosine transport was measured using a 10 sec assay. For additional details see Table 1.

Menadione is a quinone that generates intracellular ROS. After entering the cell, the flavoprotein catalyzes redox cycles with dioxygen and forms large amounts of intracellular superoxide, H_2O_2, and hydroxyl radicals (Thor, et al., 1982). To confirm ROS formation in BHMEC in the presence of menadione, we measured superoxide, hydrogen peroxide and hydroxyl radical formation. Hydrogen peroxide was determined by the peroxidase method (Miki, et al., 1988) and fluorescent technique (Salata, et al., 1983). Superoxide formation was determined using superoxide dismutase-inhibitable cytochrome c reduction (Matsubara and Ziff, 1986). Hydroxyl radicals were measured by HPLC using hydroxyl adduction to salicylate to form 2,5 dihydroxybenzoic acid (Onodera and Ashraf, 1991). Detection of ROS was also performed by electron spin resonance spectroscopy using 5,5-dimethyl-1-pyrroline-N-oxide as a spin trap.

Table 3. Effect of menadione on ROS formation in bovine heart microvessel endothelial cells.

ROS	(nm/mg protein/hr)
Hydrogen peroxide	<4.3
Superoxide	68.0±2.4
Hydroxyl radical	0.03±0.003

BHMEC (2×10^6 cells/plate) were incubated with 100 uM menadione for 30 min. Hydrogen peroxide, superoxide and hydroxyl radicals were measured in the incubation medium.

Incubation with menadione (30 min, 100 μM) resulted in the formation of cellular lipoperoxides as measured by MDA-TBA adduct (139 pmoles/mg protein). Scavenging of superoxide with superoxide dismutase (SOD; 100ug/ml) diminished the production of lipid peroxides by 65% indicating that the menadione treatment led to superoxide production. H_2O_2 production was confirmed using leucodiacetyl 2′,7′-dichlorofluorecein, a probe which fluoresces after oxidation to dichlorofluorescin by nanogram amounts of H_2O_2.

Incubation (30 min) of BHMEC with different concentrations of menadione resulted in a strong dose-dependent inhibition of Ado transport (influx; $IC_{50}=20uM$) (Figure 1).

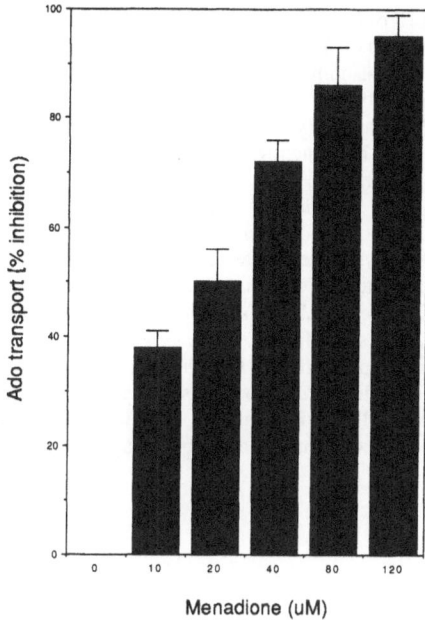

Menadione (uM)

Figure 1. Effect of menadione on Ado transport in BHMEC. Cells ($2x10^6$ cells/plate) were incubated with different concentrations of menadione for 30 min and then Ado transport was assayed during 10 sec period.

The menadione-related decrease in nucleoside influx was partially reversed (by 30%) after addition of 10 mM dimethylthiourea (DMTU), which is known as a highly permeable, nontoxic scavenger of hydrogen peroxide. This suggested that hydroxyl radicals participated in nucleoside transport inhibition. When superoxide dismutase (100 ug/ml) was also added to the cells, nucleoside transport was further decreased compared to cells treated with menadione only. Hence, superoxide formed inside cells during O_2 reduction by semiquinone radicals was released from cells. After conversion into hydrogen peroxide by exogenous superoxide dismutase, additional damage to cell membranes occurred.

Transport of other nucleosides, inosine and uridine, was also inhibited by menadione (Table 4).

Table 4. Effect of menadione on inosine and uridine transport in BHMEC

Menadione (uM)	Inosine	Uridine
	($cpm/2x10^6$ cells)	
0	31,274	71,520
100	11,596	36,544

Cells ($2x10^6$/plate) were incubated for 30 min with 100 uM menadione, then medium was changed and transport of 0.2 uM [^{3}H]inosine (36 Ci/mmole) or 0.2 uM uridine (35 Ci/mmole) was assayed during 10 sec period.

Our data indicate that ROS, generated extracellularly or intracellularly, markedly reduced transport of Ado into endothelial cells, which might result in elevation of extracellular Ado concentrations. However, inhibition of NT by ROS could also delay reconstitution of the intracellular pool of nucleotides from extracellular nucleosides. Inhibition of nucleoside transport by ROS appears to be a relatively early modification of cellular membrane function, because cell membrane integrity (measured by LDH release) remained intact during a 5 hour incubation of BHMEC with 200 uM menadione (Kossenjans, et al., 1994).

REFERENCES

1. Dagnino, L, Bennett, Jr. L.L., and Paterson A.R.P., Sodium dependent nucleoside transport in mouse leukemia L1210 cells, *J.Biol.Chem.* 266:6308 (1991).
2. Halliwell, B., and Gutteridge, M.C., Role of free radicals and catalytic metal ion in human disease: an overview in oxygen radicals in biological systems *in*: "Methods in Enzymology," Packer, L. and Glazer, A.N. eds. 186:1 (1990).
3. Kossenjans, W., Ashraf, M., Barankiewicz, J., Bobst, A., and Rymaszewski, Z. Menadione-induced oxidative stress in bovine heart microvascular endothelial cells, in preparation (1994).
4. Matsubara,T., and Ziff, M., Superoxide anion release by human endothelial cells: synergism between a phorbol ester and calcium ionophore, *J. Cell. Physiol.* 127:207 (1986).
5. Miki,S., Ashraf,M., Salka, S.. and Sperelakis, N., Myocardial dysfunction and ultrastructural alteration mediated by oxygen metabolites, *J. Mol. Cell. Cardiol.* 20:1009 (1988).
6. Onodera,T., and Ashraf, M., Detection of hydroxyl radicals in the post-ischemic reperfused heart using salicylate as a trapping agent, *J. Mol. Cell. Cardiol.* 23:365 (1991).
5. Salata, R.A., Sullivan, J.A., and Mandell, G.L., Visualization of hydrogen peroxide in living polymorphonuclear neutrophils utilizing leucodiacetyl 2′,7′-dichlorofluorescin: photomicrographic and microphotometric studies, *Trans. Assoc. Am. Phys.* 96:383 (1983).
6. Thor, H., Smith, M.T., Hartzell, P., Bellomo, G., Jewell, S.A., and Orrenius S., The metabolism of menadione (2-methyl-1,4-naphthoquinone) by isolated hepatocytes, *J. Biol. Chem.* 257:12419 (1982).

NOVEL NUCLEOSIDE TRANSPORT INHIBITORS OF NATURAL ORIGIN

Yong-Su Zhen, Jian Su, Yu-Chuan Xue,
Chang-Qing Qi and Ji-Lan Hu

Institute of Medicinal Biotechnology,
Chinese Academy of Medical Sciences,
Beijing 100050, China

INTRODUCTION

The salvage pathway of nucleotide biosynthesis is one of the attractive targets for cancer chemotherapy. Dipyridamole, a nucleoside transport inhibitor, may be used for blocking the salvage pathways. Studies have demonstrated that dipyridamole can reduce the rescue effect of exogenous nucleosides and potentiated the cytotoxicity of acivicin, an antimetabolite of *de novo* pathways of nucleotide biosynthesis, in hepatoma cells[1,2]. The inhibitory effect of dipyridamole was cell growth phase -dependent. Cultured hepatoma cells in lag and log phase were highly sensitive to dipyridamole; by contrast, cells in stationary phase were insensitive[3,4]. The combination of acivicn and dipyridamole administered in rats bearing s.c. transplanted hepatoma yielded a summation effect, decreasing NTP and dNTP pools[5]. The synergistic effect of dipyridamole and acivicin was also demonstrated in colon cancer cells[6]. There have been reports that dipyridamole potentiated the cytotoxicity of several antimetabolites including methotrexate, 5-fluorouracil and N-phosphonacetyl-L-aspartate (PALA)[7,8,9]. In addition to potentiating antimetabolites, dipyridamole was found to enhance the cytotoxicity of doxorubicin, etoposide , vincristine and mitoxantrone[10,11,12]. These synergistic effects of dipyridamole and drugs which are not antimetabolites were due to, at least in part, increased intracellular concentrations of anticancer drugs. Drug sensitivity of multidrug resistant cells may be modulated by dipyridamole[13]. The use of dipyridamole in cancer chemotherapy has drawn much attention and clinical trials have been underway[14,15.] The study of dipyridamole indicates that blocking nucleoside salvage pathways by inhibition of nucleoside transport may be one of the effective ways in chemotherapy. By examination of a series of compounds of natural origin, we have found that green tea polyphenols (GP), antibiotics C3368-A (CA) and C3368-B (CB) are highly active in blocking nucleoside transport in cancer cells. Investigations have demonstrated that GP, CA or CB show synergistic effects with anticancer drugs.

Purine and Pyrimidine Metabolism in Man VIII, Edited by
A. Sahota and M. Taylor, Plenum Press, New York, 1995

MATERIALS AND METHODS

GP was extracted from green tea. The sample of polyphenols was composed of 4 components: (-)-epicatechin, (-)-epicatechin gallate, (-)-epigallocatechin and (-)-epigallocatechin gallate. CA and CB were produced by a fungus strain from a soil sample which was collected in Antarctica. CA and CB were isolated from the fermentation liquors, and the chemical properties of CA and CB are to be reported separately.

Nucleoside transport was assayed with radiolabeled thymidine and uridine in L1210 leukemia cells or Ehrlich ascites carcinoma cells. The sample and cells were preincubated for 5 min. After adding the radiolabeled precursor for 30 sec, the radioactivity in the whole cell was determined by a method as described[16].

Clonogenic assay. Cultured cancer cells in exponentially growing phase were harvested and seeded in 24-well plates. Drugs were added 24 h after seeding and colonies were counted under an inverted microscope after 6 days of incubation.

Growth inhibition assay. Cultured tumor cells in exponentially growing phase were seeded in 24-well plates. Drugs were added 24 h later. After 72 h of culture, viable cells were counted by trypan blue exclusion.

Inhibition of tumor growth in mice. For the leukemia model, L1210 or P388 cells were inoculated i.p. in DBA/2 mice and the drugs were administered i.p. daily for 10 days. Therapeutic effects were evaluated by survival time. For the solid tumor model, colon carcinoma 26 cells were inoculated s.c. in BALB/c mice and drugs were given i.p. or p.o. for 10 days. Therapeutic effects were evaluated by tumor weight.

RESULTS AND DISCUSSION

Inhibition of nucleoside transport. GP, CA and CB displayed inhibitory effect on nucleoside transport. The IC50 values of GP for thymidine and uridine transport were 3.2 uM and 8.0 uM, respectively. Those of CA and CB for thymidine transport were 4.6 uM and 7.5 uM; for uridine transport, 7.7 uM and 9.6 uM.

Potentiation of antimetabolite cytotoxicity. In hepatoma BEL-7402 cells GP potentiated the cytotoxicity of methotrexate. At 0.02 uM of methotrexate, colony inhibition reached 100%. It reduced to 11% by addition of hypoxanthine plus thymidine. Notably, combination of methotrexate and GP (15 uM) raised the inhibition up to 86%. GP also enhanced cytarabine cytotoxicity in hepatoma cells. At 0.5 uM of cytarabine, inhibition reached 100%, which was decreased to 14% by addition of deoxycytidine; however, GP at 10 uM almost completely eliminated the nucleoside rescue effect, raising the inhibition to 97%. CA displayed synergism with methotrexate in KB cells. When used separately, methotrexate (0.007 uM) and CA (10 uM) exerted 30% and 67% of colony inhibition; the combination of 2 drugs yielded 97% inhibition.

Modulation of drug activity in multidrug resistant cells. CB was found to potentiate vincristine cytotoxicity. By growth inhibition assay, CB (20 uM) showed no effect on L1210/MDR cells. Vincristine at 0.5 uM and 1.0 uM inhibited the growth by 8% and 23%; in combination with CB, the growth inhibition was raised to 68% and 87%, respectively.

Augmentation of antitumor effect of cytarabine *in vivo*. Cytarabine (10 mg/kg) and GP (50 mg/kg) given separately showed no antitumor effect on leukemia L1210 in mice, yielding T/C (%) values of 90% and 76%; the combination of 2 drugs raised the T/C (%) value to 224% (*P* < 0.05). In leukemia P388 bearing mice, the T/C (%) values for cytarabine, GP and cytarabine + GP were 122%, 103% and 233%, respectively.

Augmentation of antitumor effect of mitomycin *in vivo*. GP potentiated mitomycin antitumor effect against colon carcinoma 26 in mice. Inhibition rates for mitomycin (1.0 mg/kg), GP (300 mg/kg), and mitomycin + GP were 82%, 23% and 92%, respectively; the combination yielded synergism. CA was also found to potentiate mitomycin antitumor effect. Inhibition rates against colon carcinoma 26 for mitomycin (0.5 mg/kg), CA (50 mg/kg) and mitomycin + CA were 29%, 3% and 57%, respectively.

To control nucleoside salvage pathways is one of the strategic goals in cancer chemotherapy. Nucleoside transport as possible target to block salvage has drawn much attention and the studies of dipyridamole have provided evidence that nucleoside transport inhibitor may effectively potentiate the antitumor effects of various drugs. It is of interest to develop novel nucleoside transport inhibitors and to determine their possible use in modulating the effects of anticancer drugs. The present studies of GP, CA and CB demonstrate that nucleoside transport inhibitors of natural origin may be useful in cancer therapy.

ACKNOWLEDGEMENTS

Supported by the National Natural Sciences Foundation of China.

REFERENCES

1. G. Weber, Biochemical strategy of cancer cells and the design of chemotherapy: G. H. A. Clowes Memorial Lecture, *Cancer Res* 43:3466 (1983).
2. Y.-S. Zhen, M.S. Lui and G. Weber, Effects of acivicin d dipyridamole on hepatoma 3924A cells, *Cancer Res.* 43:1616 (1983).
3. Y.-S. Zhen, M.A. Reardon and G. Weber, Amphotericin B renders stationary phase hepatoma cells sensitive to dipyridamole, *Biochem. Biophys. Commun.* 140:434 (1986).
4. Y.-S. Zhen, M.A. Reardon and G. Weber, Amphotericin B: a biological response modifier in targeting against the salvage pathways, *Biochem. Pharmacol.* 36:3641 (1987).
5. G. Weber, M.S. Lui, Y. Natsumeda and M.A. Faderan, Salvage capacity of hepatoma 3924A and action of dipyridamole, *Advan. Enzyme Regul.* 21:53069 (1983).
6. P.H. Fischer, R. Pamukcu, G. Bittner and J.K.V. Willson, Enhancement of the sensitivity of human colon cancer cells to growth inhibition by acivicin achieved through inhibition of nucleic acid precursor salvage by

dipyridamole, *Cancer Res.* 44:3355 (1984).

7.   J.A. Nelson and S. Drake, Potentiation of methotrexate toxicity by dipyridamole, *Cancer Res.* 44:2493 (1984).

8.   J.L. Grem and P.H. Fischer, Augmentation of 5-fluorouracil cytotoxicity in human colon cancer cells by dipyridamole, *Cancer Res.* 45:2967 (1985).

9.   T.C.K. Chan and S.B. Howell, Mechanism of synergy between N- phosphonacetyl-L-aspartate and dipyridamole in a human ovarian cell line, *Cancer Res.* 45:3598 (1985).

10.   H. Kusumoto, Y. Maehara, H. Anai, T. Kusumoto and K. Sugimachi, Potentiation of adriamycin cytotoxicity by dipyridamole against HeLa cells *in vitro* and sarcoma 180 cells *in vivo*, *Cancer Res.* 48:1208 (1988).

11.   S.B. Howell, D. Hom, R. Sanga, J.S. Vick and I.S. Abramson, Comparison of the synergistic potentiation of etoposide, doxorubicin, and vincristine cytotoxicity by dipyridamole, *Cancer Res,* 49:3178 (1989).

12.   M. Sotomatsu, S. Yugami, T. Shitara and T. Kuroume, Dipyridamole enhancement of drug sensitivity in acute lymphoblastic leukemia cells, *Am. J. Hematol.* 43:251 (1993).

13.   D.R. Shalinsky, M. Andreef and S.B. Howell, Modulation of drug sensitivity by dipyridamole in multidrug resistant tumor cells *in vitro, Cancer Res.* 50:7537 (1990).

14.   J.K.V. Willson, P.H. Fischer, S.C. Remick, K.D. Tutsch, J.L. Grem, L. Nieting, J.Bruggink,J.M. Koeller and D.L. Trump, Methotrexate and dipyridamole combination therapy based upon inhibition of nucleoside salvage in humans, *Cancer Res,* 49:1866 (1989).

15.   S.C. Remick, J.L. Grem, P.H. Fischer, K.D. Tutsch, D.B. Alberti, L.M. Nieting, M.B.Tombes, J. Bruggink and J.K.V. Willson, Phase I trial of 5-fluorouracil and dipyridamole administered by 72 h concurrent continuous infusion, *Cancer Res,* 50:2667 (1990).

16.   Y.S. Zhen, S.S. Cao, T.C. Xue and S.Y. Wu, Green tea extract inhibits nucleoside transport and potentiates the antitumor effect of antimetabolites, *Chin. Med. Sci. J.* 6:1 (1991).

MODIFICATION OF TUMOR NECROSIS FACTOR (TNF) PRODUCTION AND SURVIVAL RATE BY A NUCLEOSIDE MIXTURE IN LIPOPOLYSACCHARIDE-INJECTED RATS

Hiroomi Yokoyama, Seiichiro Kano, Keiichi Okamoto and Yoshiyuki Shinagawa

Otsuka Pharmaceutical Factory, Inc.
Naruto City, Tokushima 772, Japan

INTRODUCTION

Nutritional support is beneficial in the treatment of malnutrition and in facilitating postoperative recovery partly by potentiating bio-defence activity of patients.[1] Although nucleosides are not regarded as essential nutrients, it has been suggested that they may potentiate the bio-defence activity of patients with compromised immune systems.[2]

The use of nucleic acid components in total parenteral nutrition (TPN) was first attempted in rats in 1984 by Ogoshi et al.[3] They reported that intravenous administration of preparations containing inosine, 5'-sodium guanylate, cytidine, uridine and thymidine improved the nutritional condition and promoted recovery of partially hepatectomized rats suffering from postoperative malnutrition.[4] OG-VI is one of the preparations they formulated, and was designed to provide nucleic acid components to surgically stressed patients to speed their recovery from postoperative malnutrition.

Further studies indicate that OG-VI may exhibit other pharmacological actions, other than providing nucleic acid metabolites incorporated into the salvage pathway. Namely, OG-VI was shown to restore the decrease in hepatic ATP levels due to liver ischemia in rats,[5] to improve myocardial contraction which is deteriorated following reperfusion after ischemia (stunning) in dogs,[6] and to maintain resistance against methicillin-resistant *Staphylococcus aureus* (MRSA) infection in mice fed a nucleotide-free diet.[7]

It is well known that tumor necrosis factor (TNF) increases substantially in all of the above-referred conditions.[8-11] Thus it may be hypothesized that OG-VI may exerts its beneficial actions partly through modulating TNF production. This study was formulated to examine this hypothesis: the effects of OG-VI on the TNF-α production and survival was examined in rats administered with lipopolysaccharide (LPS).

Purine and Pyrimidine Metabolism in Man VIII, Edited by
A. Sahota and M. Taylor, Plenum Press, New York, 1995

MATERIALS AND METHODS

Experiment 1

Male Wistar rats (approximately 360 g; Nissin Tokushima Institute for Animal Reproduction, Tokushima, Japan) were randomly divided into 3 groups (15 rats/group), each group received intraperitoneally (*ip*) either normal saline (20 ml/kg) or OG-VI (2 or 20 ml/kg; inosine 0.80%w/v, 5'-sodium guanylate 1.22%w/v, cytidine 0.73%w/v, uridine 0.55%w/v and thymidine 0.18%w/v, Otsuka Pharmaceutical Factory, Inc.). Forty min after administration, the rats were injected *ip* with *Escherichia coli* LPS (O111:B4; 11 mg/kg; Difco Laboratories, Detroit, MI) which was dissolved (4 mg/ml) in pyrogen-free distilled water. The survival of the rats was examined until 66 hr after LPS administration. Rat chow and drinking water were given *ad libitum* during the experiment.

Experiment 2

Male Wistar rats (approximately 340 g; Charles River Japan, Inc.,Yokohama, Japan) were cannulated into the jugular vein under anesthesia by *ip* administration of pentobarbital (35 mg/kg; Abbott Laboratories, North Chicago, IL). On the following day, these animals were divided into 2 groups, each group received either normal saline (15 ml/kg) or OG-VI (15 ml/kg) *ip*. Sixty min later, each group was divided into 2 subgroups (5 or 4 rats/ subgroup), one subgroup receiving 2.5 mg/kg of LPS, and the other 10 mg/kg. Blood (0.6 ml) was sampled through the cannulated catheter immediately before LPS administration, and at 30, 60, 90, 120 and 180 min after administration. Serum samples were stored at -80 °C until TNF-α was determined by enzyme-linked immunosorbent assay (ELISA) using a mouse tumor necrosis factor ELISA kit (Factor-Test mTNF-α; Genzyme, Boston, MA).

RESULTS

As shown in Fig. 1, A, the survival rates of OG-VI groups (2 and 20 ml/kg) were higher than that of the control group (by the Cox-Mantel test, P=0.280 and 0.059, respectively).

OG-VI significantly suppressed the increase in the serum TNF-α level which peaked at 90 min after LPS administration (Fig. 1, B). The survival rate of the rats in experiment 2, which received OG-VI followed by 10 mg/kg of LPS was 50%(2/4), whereas, animals in the control group which received normal saline followed by 10 mg/kg of LPS all died. All rats which received 2.5 mg/kg of LPS survived during the study period.

DISCUSSION

Many biological activities of LPS has been shown to be caused through the action of TNF-α which was produced by monocyte-macrophage lineage cells.[12,13,14] The results shown in this study indicate that OG-VI prevented the decline of survival rate of LPS-injected rats. This correlated well to the decreased rise of TNF-α in the serum after LPS challenge, which takes place shortly after the LPS administration (max at 90 min), indicating that the improved survival of rats given OG-VI may be caused by the inhibition of TNF-α production. The mechanism for this inhibition is to be clarified. One possible mechanism is that nucleosides formulated in OG-VI may modulate the interaction between LPS complex (a complex of LPS and LPS-binding protein)[15,16] and LPS receptor (CD14) expressed on

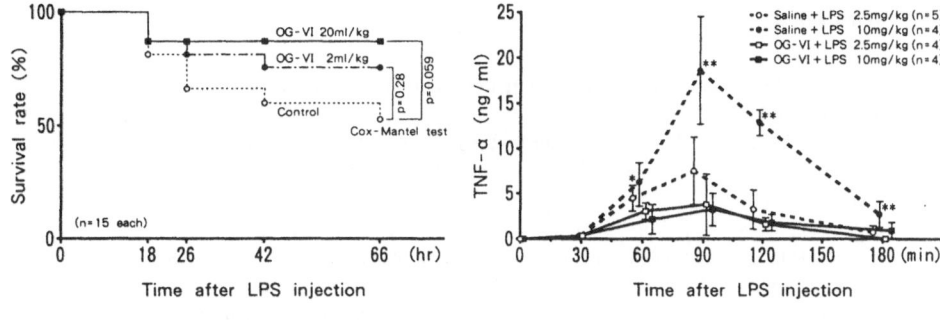

(A) Survival rate of LPS – injected rats.   (B) Serum TNF – α level after LPS injection.

Figure 1. A; Effect of OG-VI administration on the survival rate of LPS-injected rats. The rats were injected ip LPS (O111:B4;11 mg/kg) 40 min after receiving ip either normal saline (20 ml/kg; Control) or OG-VI (2 or 20 ml/kg). B; Effect of OG-VI administration on the serum TNF- α level stimulated by LPS injection. The rats were injected ip LPS (O111:B4;2.5 mg/kg or 10 mg/kg) 1 hr after receiving ip either normal saline (15 ml/kg) or OG-VI (15 ml/kg). Blood was sampled from the same rats serially at the indicated points. Values are means+SD and statistically analyzed by the Tukey-Kramer test.   *P<0.05, **P<0.01 Statistically different from OG-VI+LPS 10 mg/kg.

TNF-producing cells. Another possibility is that OG-VI directly neutralizes the pharmacological activity of LPS, stimulator of TNF production. To determine the exact mechanism, it will be necessary to determine whether OG-VI modifies TNF- α production potentiated by stimulants other than LPS.

In a rat liver ischemic reperfusion model, the involvement of LPS was not demonstrated: no LPS was detected in the serum. However, TNF- α appeared to be involved in the pathogenesis of reperfusion liver damage, since in vivo administration of anti-TNF- α inhibited the rise of serum glutamic pyruvic transaminase, which occurs in parallel to the systemic increase of TNF production.[8] In the same model, OG-VI restored the deteriorated liver function as determined by cellular ATP levels as a parameter for liver damage in the ischemic reperfusion liver, although the blood TNF level was not determined.[5]

The possible involvement of TNF- α in the increased resistance of OG-VI administered to immune compromised mice against MRSA[7] seems complicated. OG-VI may also suppress the synthesis of TNF- α in this model; however, enhanced synthesis of this cytokine after MRSA infection may prevent further spread of the microorganism like the case of *Staphylococcus aureus* infection.[11] On the other hand it is also reported that OG-VI may suppress LPS-induced non-specific polyclonal B-cell activation, perhaps through restoring regulatory T-helper cell functions or other unknown mechanisms.[17] In understanding the exact mechanisms of increased resistance against MRSA by OG-VI and the role of OG-VI in modulation of TNF production, further studies are very much needed.

CONCLUSION

OG-VI, a mixture of four nucleosides and a nucleotide, improved the survival rate of rats injected LPS probably by inhibiting TNF- α production.

REFERENCES

1. F.B. Cerra, Nutrient modulation of inflammatory and immune function, *Am J Surg* . 161:230 (1991).

2. F.B. Rudolph, A.D. Kulkarni, W.C. Fanslow, R.P. Pizzini, S. Kumar, and C.T. Van Buren, Role of RNA as a dietary source of pyrimidines and purines in immune function, *Nutrition* 6:45 (1990).

3. S. Ogoshi, M. Iwasa, T. Yonezawal, Y. Ohmori, and T. Tamiya, Effect of nucleotides and nucleoside mixture on total parenteral nutrition - Preliminary studies on ratio of nucleotides in normal rats, *Jpn J Paren Ent Nutr.* 6:53 (1984) (in Japanese).

4. S. Ogoshi, S. Mizobuchi, M. Iwasa, and T. Tamiya, Effect of a nucleoside-nucleotide mixture on protein metabolism in rats after seventy percent hepatectomy, *Nutrition* 5:173 (1989).

5. M. Ogino, M. Usami, H. Ohyanagi, H. Kasahara, S. Nishimatsu, K. Furuchi, I. Yasuda, K. Sun, and Y. Saitoh, Effect of nucleosides and a nucleotide mixture on ischemic and reperfusion injury of the liver evaluated by in vivo P-31 NMR spectroscopy, *J Jpn Soc Paren Ent Nutr.* 6:242 (1991) (in Japanese).

6. K. Sato, T. Nakai, K. Hoshi, and K. Ichihara, Limitation of stunning in dog myocardium by nucleoside and nucleotide mixture, OG-VI, *Coronary Artery Disease* 4:1007 (1993).

7. A.A. Adjei, F. Takamine, H. Yokoyama, S-Y. Chung, L. Asato, S. Shinjo, T. Imamura, and S. Yamamoto, Effect of intraperitoneally administered nucleoside-nucleotide on the recovery from methicillin-resistant *Staphylococcus aureus* strain 8985N infection in mice, *J Nutr Sci Vitaminol.* 38:221 (1992).

8. L.M. Colletti, D.G. Remick, G.D. Burtch, S.L. Kunkei, R.M. Strieter, and D.A. Campbell, Jr, Role of tumor necrosis factor-α in the pathophysiologic alterations after hepatic ischemia/reperfusion injury in the rat, *J Clin Invest.* 85:1936 (1990).

9. M.S. Finkel, C.V. Oddis, T.D. Jacob, S.C. Watkins, B.G. Hattler, and R.L. Simmons, Negative inotropic effects of cytokines on the heart mediated by nitric oxide, *Science* 257:387 (1992).

10. A. Nakane, T. Minagawa, and K. Kato, Endogenous tumor necrosis factor (cachectin) is essential to host resistance against *Listeria monocytogenes* infection, *Infect Immun,* 56:2563 (1988).

11. A. Nakane, M. Okamoto, M. Asano, M. Kohanawa, and T. Minagawa, Role of endogenous cytokines in host defence against *Staphylococcus aureus* infection, *Bio Defence (Tokyo)* 4:41 (1993) (in Japanese).

12. B. Beutler, I.W. Milsark, and A.C. Cerami, Passive immunization against cachectin/tumor necrosis factor protects mice from lethal effect of endotoxin, *Science* 229:869 (1985).

13. K.J. Tracey, Y. Fong, D.G. Hesse, K.R. Manogue, A.T. Lee, G.C. Kuo, S.F. Lowry, and A. Cerami, Anti-cachectin/TNF monoclonal antibodied prevent septic shock during lethal bacteraemia, *Nature* 330:662 (1987).

14. H.R. Michie, K.R. Manogue, D.R. Spriggs, A. Revhaug, S. O' Dwyer, C.A. Dinarello, A. Cerami, S.M. Wolff, and D.W. Wilmore, Detection of circulating tumor necrosis factor after endotoxin administration, *N Eng J Med.* 318:1481 (1988).

15. R.R. Schumann, S.R. Leong, G.W. Flaggs, P.W. Gray, S.D. Wright, J.C. Mathison, P.S. Tobias, and R.J. Ulevitch, Structure and function of lipopolysaccharide binding protein, *Science* 249:1429 (1990).

16. S.D. Wright, R.A. Ramos, P.S. Tobias, R.J. Ulevitch, and J.C. Mathison, CD14, a receptor complexes of lipopolysaccharide (LPS) and LPS binding protein, *Science* 249:1431 (1990).

17. H. Jyonouchi, L. Zhang-Shanbhag, Y. Tomita, and H. Yokoyama, Nucleotide-free diet impairs T-helper cell functions in antibody production in response to T-dependent antigens in normal C57Bl/6 mice, *J Nutr.* 124:475 (1994).

THE EFFECT OF DIETARY NUCLEIC ACID DEFICIENCY AND THE ADMINISTRATION OF A NUCLEOTIDE AND NUCLEOSIDES MIXTURE SOLUTION ON ENDOTOXIN SHOCK IN RATS

Seiji Haji, Makoto Usami, George Kotani, Atsunori Iso, Kyosuke Ohta, Kazuya Sakata, Enmei Sou, Kai Sun, Taichi Kanamaru, Hiroshi Kasahara, and Yoichi Saitoh

First Department of Surgery, Kobe University School of Medicine, 7-5-2 Kusunoki-cho, Chuo-ku, Kobe, 650 Japan

INTRODUCTION

The importance of dietary nucleic acid in modulation of immune function without disease has been reported in nutritional disorder except under surgical stress. Dietary nucleic acid deficiency decreases T lymphocytes function[1] and dietary ribonucleic acid (RNA) enhances immune response[2]. However, it is not known whether dietary nucleic acid influences on macrophage function. Recently, monocytes/macrophages are considered as key factors in the pathogenesis of systemic inflammatory response leading to septic shock by producing humoral mediators under endotoxin stimulation[3,4]. Therefore, the aim of this study is to evaluate the influence of dietary nucleic acid deficiency on endotoxin shock and macrophage functions stimulated with lipopolysaccharide (LPS), which is the major component of endotoxin. Then, the effect of intraperitoneal administration of a nucleotide and nucleosides mixture solution (OG-VI)[5] was evaluated.

MATERIALS AND METHODS

Six-weeks-old male Wistar rats were maintained on the nucleic acid free diet (NFD : AIN-B. Oriental Corp.) or standard chow (CE-2, CREA) for 7 days. A nucleotide and nucleosides mixture solution (OG-VI) (2.8 ml/kg, OG group) or saline (2.8 ml/kg, NF group) were injected two times intraperitoneally on 6th and 7th days. The chow group rats were maintained on standard chow with saline injection. OG-VI composed of 5'-GMP-2Na 30mM, inosine 30mM, cytidine 30mM, uridine 22mM, and thymidine 7.4mM was a kind gift from Otsuka Pharmaceutical Co., Tokushima, Japan (Table 1).

In an in vitro experiment, rats in the NF group and the chow group were killed by cardiac puncture on 8th day and peritoneal macrophages were prepared. The peritoneal cavity was lavaged with Ca^{2+} and Mg^{2+} free Hanks balanced salt solution (HBSS) to obtain resident peritoneal cells. The cells were washed and were plated onto plastic tissue culture dishes in RPMI 1640 containing 10% (w/v) heat inactivated fetal calf serum, penicillin G (10000 unit/ml), and streptomycin (100 μg/ml). The cells were incubated for 1 hour at 37 °C in 5% CO_2 and 100% humidified atomosphere to allow peritoneal macrophages attaching to culture dishes. The dishes were washed and then peritoneal macrophages, over 95% purity and viability assessed by esterase staining and trypan blue staining, were obtained. After 24 hours pre-culture, peritoneal macrophages (1×10^5 cells/ml) were cultured with 10 μg/ml of LPS (E.coli. 011: B4, Difco.) for 18 hours. Then, tumor necrosis factor - alpha (TNF) level, by ELISA (ELISA kit, Genzyme), and

6-keto-prostaglandin (PG) F$_{1\alpha}$, the stable metabolite of PGI2, thromboxane (TX) B$_2$, the stable metabolite of TXA2, and PGE2 levels in the medium were measured by RIA (^{125}I RIA kit, NEN) to investigate humoral mediators production from macrophages. To investigate LPS uptake of macrophages, peritoneal macrophages (5 x 10^5 cells/ml) were cultured with culuture medium containing fluorescein isothiocyanate LPS (FITC-LPS. E.coli. 011: B4) at a concentration of 50 μg/ml for 1 hour. Then, the cells were analyzed using flow cytometry (FACScan, Becton Dickinson) and FITC - positive cell numbers and mean fluorescence intensity were measured.

In an in vivo experiment, 10 mg/kg LPS was injected intraperitoneally in three groups, the NF, the OG and the chow group at 1 hour after second injection of OG-VI or saline. The survival was evaluated for 72 hours. Blood sample was obtained from abdominal aorta at 30, 60, 120, 240 minutes after LPS injection and endotoxin levels (Endospec-SP test kit, SEIKAGAKU corp.) at each time and TNF levels at 60 minutes were measured.

All data were expressed as mean ± S.D. For stastical analysis, data were analyzed by the two tailed Student's t test, one-way analysis of variance, and generalized Wilcoxon rank sum determinations.

RESULTS

IN VITRO EXPERIMENT : All humoral mediators in the medium were elevated under LPS stimulation in both the NF group and the chow group. 6-keto PGF$_{1\alpha}$, PGE2, TXB2 levels were significantly lower in NF group compared with the chow group (p<0.05, Tale 1). TNF level was lower in the NF group compared with the chow group, but both groups had no significant difference. As shown in Figure 1, the percentage of FITC-LPS positive cells was significantly lower in the NF group (53.2 ± 2.3%) than the chow group (58.0 ± 2.1%) (p<0.01). But, mean fluorescence intensity was not different in both groups.

IN VIVO EXPERIMENT: Blood endotoxin levels in the OG and the NF group were remarkably elevated by 120 minutes after LPS administration (OG: 33.6 ± 9.8 μg/ml, NF: 39.0 ± 8.6 μg/ml) and were recovered at 240 minutes. In contrast, blood endotoxin level in the chow group were gradually elevated by 240 minutes. Blood endotoxin levels at 120 minutes were higher in the OG group and the NF group than in the chow group (8.9 ± 4.2 μg/ml) with significant difference (p<0.05) (Figure 2). Blood TNF levels at 1 hour after LPS administration were lower in the OG group (2547 ± 1463 pg/ml) than those of the NF group (4191 ± 826 pg/ml) and the chow group (4120 ± 1205 pg/ml). The administration of OG-VI reduced blood TNF- α level after LPS injection significantly (p<0.02). But, dietary nucleic acid did not affect TNF- α level. The survival rates in 72 hours were 100% in the OG group and 0% in the chow group respectively with statistically significance (p<0.01), and was 60% in the NF group (Table 2).

DISCUSSION

Clinically, the nutritional support for the patients after severe illness and gastrointestinal surgery has been performed mainly by total parenteral nutrition (TPN). But, most parenteral formulas do not contain nucleic acids and the effect of their deficit on macrophage function has not been reported. To determine the significance of the administration of nucleic acids, the present study focused on the effect of dietary nucleic acid deficiency on peritoneal macrophage functions and shock state after LPS injection.

Our data revealed that dietary nucleic acid deficiency during 7 days reduced in vitro production of humoral mediators from peritoneal macrophages stimulated by LPS and also reduced in vitro uptake of LPS and also in vivo clearance of circulating endotoxin. In humoral mediators produced from macrophages, TNF is considered as the earliest and the most important endogenous mediator in endotoxemia[4]. Arachidonic acid metabolite PGI2, vasodilator and platelet antiaggregator, and TXA2, a potent vasoconstrictor and aggregator, are induced by endotoxin[3,6,7,8]. Therefore, in vitro results in peritoneal macrophages are considered to correspond with the difference of the survival rate after LPS injection in between the NF group and the chow group. The study in the mechanism of inhibition in LPS uptake by dietary nucleic acid deficiency requires further investigation, since the steps of LPS stimulation by macrophages itself is still controvertial and many different

Table 1. Humoral mediators production by peritoneal macrophages under LPS (10 μg/ml) stimulation (n=10)

	TNF-α	6-keto PGF1 α	PGE2	TXB2
NF group	459.2±707.4	1059.6±1460.2 ※	760.1±1029.3 ※	3432.5±1967.6 ※
Chow group	555.2±690.0	2353.4±1197.8	1523.1±1127.1	5920.2±4485.2

※ ; p<0.05 vs Chow group,   mean ± S.D. (pg/ml)

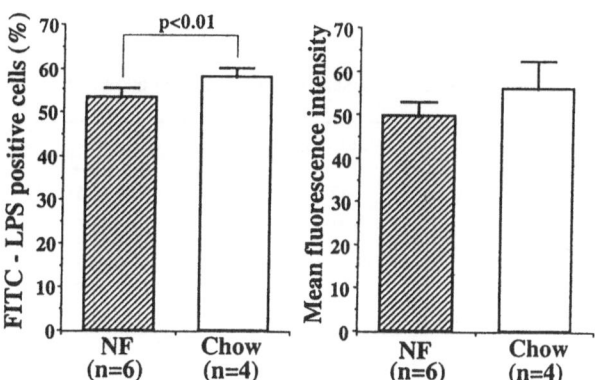

Figure 1. FITC-LPS uptake of peritoneal macrophages

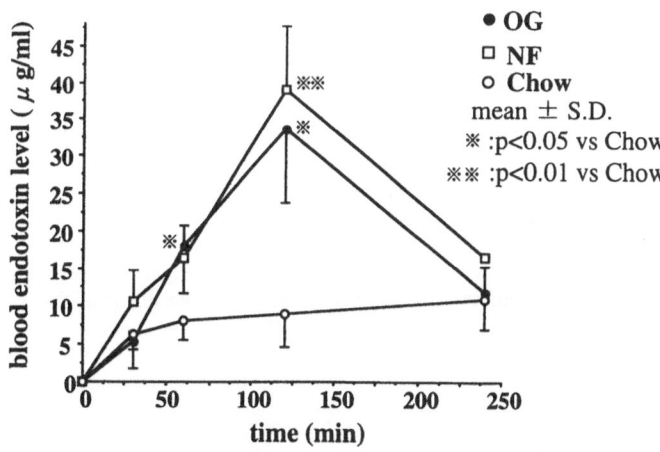

Figure 2. Changes of blood endotoxin levels after LPS administration in rats

mechanisms are postulated[8]. However, the in vitro results are in accordance with the in vivo results of improved survival (without statistically difference) and increased blood endotoxin levels. It is suggested the importance of this new area to research and also suggested that dietary nucleic acid deficiency during TPN management may influence on disease condition under endotoxin stimulation.

As a method to supply nucleotide and nucleoside during parenteral nutrition, the effect of administration of a nucleotide and nucleosides mixture solution (OG-VI) was investigated in the same endotoxin shock model. Our result demonstrated that the intraperitoneal administration of OG-VI reduced blood TNF level and improved the survival rate followed LPS injection even under nucleic acid free diet. OG-VI solution is considered as the different source of purine and pyrimidine from dietary supplementation by means of direct supply of nucleosides for purine and pyrimidine salvage synthesis. Our data are in accordance with the reports showing that intravenous administration shows advantage in tissue incorporation of purine from orally administration[9] and that intraperitoneal administration of OG-VI improved the survival of methicillin - resistant Staphylococcus aureus (MRSA) infected mice[10].

In conclusion, dietary nucleic acid deficiency showed the in vitro and in vivo effect on inhibition of LPS uptake and humoral mediators secretion by peritoneal macrophages.

Table 2. blood TNF level and survival rate after LPS administration in rats

	TNF-α (pg/ml)	survival rate (%)
OG group	2547 ± 1463 ※	100 (5/5) #
NF group	4191 ± 826	60 (3/5)
Chow group	4120 ± 1205	0 (0/5)

TNF-α level (n=8) values as mean $\pm$ S.D.

※ : $p<0.02$ OG group vs NF group and Chow group

: $p<0.01$ vs Chow group

REFERENCES

1. C.T. Van Buren, A.D. Kulkarni, et al, Dietary nucleotides, a requirement for helper/inducer T lymphocytes, *Transplantation*. 40:694 (1985).
2. C.T. Van Buren, F.B.Rudolph, et al, Reversal of immunosuppression induced by a protein-free diet, *Crit Care Med*. 18(2):S114 (1990).
3. E.A. Deitch, Multiple organ failure, *Ann Surg*. 216:117 (1993).
4. S.F. Lowry, Cytokine mediators of immunity and inflammation, *Arch Surg*. 128:1235 (1993).
5. S. Ogoshi, M. Iwasa, S. Kitagawa, et al, Effects of total parenteral nutrition with nucleoside and nucleotide mixture on D-Galactosamine-induced liver injury in rats, *J Paren Ent Nutr*. 12:53 (1988).
6. J.A.Cook, P.V.Halushka, W.C.Wise, Modulation of macrophage arachidonic acid metabolism, *Circ Shock*. 9:605 (1982).
7. S. Ghosh, R.D.Latimer, B.M.Gray, et al, Endotoxin-induced organ injury, *Cric Care Med*. 21:S19 (1993).
8. B.P. Giroir, Mediators of septic shock, *Crit Care Med*. 21:780 (1993).
9. D.A. Saviano, C.Y. Ho, V. Chu, et al, Metabolism of orally and intravenously administered purines in rats, *J Nutr*. 110:1793 (1980).
10. A.A. Adjei, F. Takamine, et al, The effects of oral RNA and intraperitoneal nucleoside-nucleotide administration on Methicillin-resistant Staphylococcus aureus infection in mice, *J Paren Ent Nutr*. 17:148 (1993).

RADIOIMMUNOASSAY OF FOLIC ACID AND ITS CORRELATION WITH AGE

L.Lorenzini,[1] A.De Martino,[1] W.Testi,[1] F.Sorbellini,[1] L.Bisozzi,[2]
L.Terzuoli,[2] R.Leoncini,[2] M.Pizzichini,[2] E.Marinello,[2] R.Pagani,[2] and
B.Porcelli[2]

[1]Institute of General Surgery
[2]Institute of Biochemistry and Enzymology
University of Siena, Italy

INTRODUCTION

Folic acid is a very well known coenzyme that acts as carrier of monocarbon units and has a great importance in the synthesis of purine nucleotides. It occurs in vegetables as tri- and hepta-glutamate derivatives. The human body is absolutely unable to synthesize it.

It is absorbed in the first section of the proximal intestine after depolymerization, which is carried out by a carboxypeptidase that eliminates most of the glutamate residues. Folate then penetrates the enterocyte, where it is first reduced and then methylated. Only in this final form, it is able to exert its function, namely the transport of monocarbon units. Folic acid finally enters the blood, where it binds to an α-globulin and an albumin, which transport it to the tissues.

In the different cells of the body, it is conjugated to penta- and hexaglutamates, which are stored in the cells as protein-conjugates; their function is not yet well known[1]. Polymerization prevents folates from crossing cell membranes.

Another regulatory mechanism of folate absorption and distribution is the entero-hepatic cycle of folic acid: folates are absorbed by the intestinal mucosa and carried to the liver, which reduces and methylates it, and returns it to the intestine with the bile.

Intestinal processing and storage, the enterohepatic cycle, binding to plasma proteins, distribution and storage in the tissues are different ways in which the human body regulates the homeostasis of folates.

Daily folate requirements are estimated at 50-400 µg. A daily intake of 150-200 µg is estimated avoid any symptoms of anemia and cover all requirements even in old people. Doses calculated to evoke a favourable hematological response in anemic subjects are around 5-20 mg/day taken orally. The amount of folate stored by the body is calculated to be sufficient for one month's requirements.

Folate deficiency can occur under the following conditions: 1) deficient intake, 2) poor

Address for correspondence: Prof. Enrico Marinello, Istituto di Biochimica e di Enzimologia, Università di Siena. Pian dei Mantellini, 44, 53100 SIENA Italy, Tel. +39-577-298026, FAX +39-577-298057

Purine and Pyrimidine Metabolism in Man VIII, Edited by
A. Sahota and M. Taylor, Plenum Press, New York, 1995

absorption, 3) defective utilization, 4) increased requirements. The clinical syndrome of folate deficiency, commonly known as "macrocytic anemia", is characterized by a variety clinical symptoms and peculiar erythrocyte morphology.

Despite this body of knowledge of folate biochemistry, many aspects are still unclear, for instance, the pattern of folates with age. Many observations suggest that folate intake decreases with age: this may be due to the inability of the elderly to cook, prolonged periods in hospital, or at home, with inadequate care.

Three main factors impede folate absorption in elderly people: 1) atrophic gastritis, which has an incidence of 50% in the elderly: it stimulates the proliferation of abnormal bacterial flora, producing less active folates which hinder the absorption of "true" folates[2,3]; 2) the age-related decrease in folate depolymerizing enzymes; 3) drugs[4,5] such as inhibitors of H_2 receptors for histamine and alcohol[6]. Ethanol affects the liver, inhibiting the release of methyl-folate derivatives by hepatocytes. All these factors affect plasma folate levels in old people.

Here we report a series of determinations of plasma folate levels in elderly subjects of different ages.

Folic acid may be assayed by microbiological, enzymatic and chromatographic (HPLC = high performance liquid chromatography) procedures. All these methods suggest that 6-21 ng/ml in serum and 160-640 ng/ml in the erythrocytes are normal values. Functional tests include the parenteral administration of folate and monitoring of its urinary excretion. Folic acid deficiency may also be verified by the histidine load test: in the absence of folate, the degradation of histidine is blocked at the stage of formiminoglutamic acid. The urinary excretion of this metabolite is monitored.

The most sensitive and precise procedure has proved to be the radioimmunological method, which we have adopted in the present study. The purpose of the present research is to standardize the normal levels according to age, in order to carry out an extensive comparison with tumor bearing patients.

METHODS

Clinical data

The subjects were 65 males and 35 females, varying in age from 50 to more than 80 years (Table 1). We excluded all subjects on folate supplements.

Table 1. Subjects divided according to sex and age range.

Age range	Male	Female	Total
50-55	7	4	11
56-60	11	9	20
61-65	8	3	11
66-70	12	8	20
71-75	9	4	13
76-80	9	6	15
>80	9	1	10

For folic acid determination, heparinized blood samples were taken at 8 a.m. As folic acid is light-sensitive, the samples were not exposed to direct light.

The percent at low risk decreases with age, and those at medium and high risk increases. Elderly subjects are therefore at risk for hypofolemia, for reasons that range from, physiological to psychological and sociological.

ACKNOWLEDGEMENTS

Thanks are due to Progetto Finalizzato A.C.R.O. for financial supports.

REFERENCES

1. I.Chanarin, "Folic acid. Nutritional aspects. The megaloblastic anemias". Chanarin Ed. Blackwell Sci. Pube, Oxford (1979).
2. C.H.Halsted, Folate deficiency and the small intestine, *in*: "Folic acid. Neurology, Psychiatry and Internal Medicine", Ed. Batez M.I., Reynolds E.H., Raven Press, New York (1979).
3. J.Runcie, Folate deficiency, in the elderly, *in*: "Folic acid. Neurology, Psychiatry and Internal Medicine", Ed. Batez M.I., Reynolds E.H., Raven Press, New York (1979).
4. I.H.Rosemberg, B.B.Bowman, B.A.Cooper, C.H.Haldsted and J.Lindenbaum, Folate nutrition in the elderly, *Am.J. Clin.Nutr.* 36:1060 (1982).
5. H.Baker and O.Frank, Sub clinical vitamin deficits in various age groups, *Int.J.Vit.Nutr.Res.* (Suppl) 27:47 (1985).
6. C.G.Alsted, Folate deficiency in alcoholism, *Am.J.Clin. Nutr.*, 33:2736 (1980).

URIC ACID AND ALLANTOIN IN RAT LIVER AFTER OXONIC ACID AND ^{14}C-FORMATE

M.Pizzichini, L.Arezzini, G.Cinci, M.L.Pandolfi, L.Terzuoli, A.Tabucchi, E.Marinello and R.Pagani

Institute of Biochemistry and Enzymology
University of Siena, Italy

INTRODUCTION

Uric acid and allantoin are the key catabolites of purine degradation: their quantity and patterns of radioactive labelling from precursors (such as ^{14}C-glycine, ^{14}C-formate, ^{14}C-adenine) are two parameters used in the evaluation of purine catabolism[1-5]. The correct evaluation of both compounds has always presented several difficulties, due to the fact that many interfering metabolites are present in tissues.

There is very little information on this subject as far as the rat liver is concerned. The incorporation of ^{14}C-formate into uric acid and allantoin has been widely used as an index of purine nucleotide catabolism.

Uric acid can be determined with uricase in trichloroacetic acid (TCA) extracts. The Rimini-Schryver reaction is the only available method for the determination of allantoin[6], but some metabolites can interfere with the assay. Purification of uric acid and allantoin from TCA extracts is necessary before quantitative estimation of both compounds and their labelling can be performed.

In our simple procedure for the assay of liver uric acid and allantoin and their specific radioactivity after incorporation of a labelled precursor[7], uric acid is first quantified by the uricase reaction in liver trichloroacetic acid (TCA) extracts. The 'true' allantoin content of the liver can only be estimated after precipitation with Hg-acetate, a step by which the standard allantoin is also quantitatively recovered. For the determination of specific radioactivity, the Hg-acetate precipitate is further purified by ion-exchange chromatography. The purity of the two metabolites is confirmed by ultraviolet absorbance spectra, HPLC, constancy of specific radioactivity and the absence of amino acids.

We studied the incorporation of ^{14}C-formate into uric acid and allantoin in the liver by this procedure and found that the radioactivity in allantoin was several-fold higher than that in uric acid, up to 60 min after administration of the ^{14}C-formate precursor[7]. This quite unexpected result is not easily explained, and cannot be ascribed to a different pool of the

Key words: Uric acid; allantoin; separation procedure; purine catabolism; formate; (rat liver)

Address for correspondence: Prof. Enrico Marinello, Istituto di Biochimica e di Enzimologia, Università di Siena, Pian dei Mantellini, 44, 53100 SIENA Italy, Tel. +39-577-298026, FAX +39-577-298057

two catabolic compounds, since their content in the liver is similar. No contaminant was present in hepatic allantoin prepared by our original purification procedure.

We also demonstrated that the phenomena was exclusively hepatic because it also occurs in perfused rat liver[8]. For a complete interpretation of our results, we have formulated many hypotheses: the formation of a modified allantoin (formyl-allantoin); an exchange reaction of C_2 or C_8 of uric acid and allantoin with ^{14}C-formate; an intermediate metabolite of de novo synthesis producing allantoin, independently from uric acid; and finally, that uricase might be involved in the phenomena observed, for instance through different compartmentalization and presence of isoenzymes.

We eventually intend to verify all the above hypotheses.

In order to verify whether the higher specific radioactivity of allantoin is linked to the uricase reaction, we studied the effect of oxonic acid, a competitive inhibitor of uricase, on hepatic levels of uric acid and allantoin and on the specific radioactivity of these catabolic compounds after administration of ^{14}C-formate.

MATERIALS AND METHODS

Animals

We used male Albino rats, weighing 250 g, previously kept on a standard diet. All animals were fasted for 12 h before decapitation; the livers were immediately removed, washed in saline solution and homogenized at 5°C with 5% trichloroacetic acid (1 g tissue + 5 ml 5% TCA).

Materials

Allantoin, uric acid and oxonic acid were purchased from Sigma. Uricase was from of Boehringer Mannheim (Germany). All other products were of the highest commercially available purity.

General procedures

Incorporation of ^{14}C-formate into uric acid and allantoin. Sodium ^{14}C-formate (54.8 mCi/mmol) was injected (10 µCi/100 g body wt.) and the animals were killed 15 min later.

Determination of uric acid by uricase reaction. Uric acid was determined according to Praetorius and Poulsen[9]. In the various purification steps, 50-200 µl of the different extracts were employed, adding 1 ml of 0.1 M sodium borate (pH 9.5) and 10 µl of uricase. The decrease in absorbance was followed at 293 nm.

Determination of allantoin by the Rimini-Schryver reaction. Allantoin content was measured using the Rimini-Schryver reaction as modified by Dietrich and Borries[10]. This involves treatment with 0.1 M alkali (7 min at 100°C), formation of allantoic acid, subsequent addition of final 0.025 M HCl to produce urea and glyoxylic acid and determination of the latter as a 2,4-dinitrophenylhydrazine derivative[10].

Uric acid and allantoin were determined directly in TCA extracts and after purification, which was carried out as reported below.

Separation and purification of uric acid and allantoin. Separation and purification of uric acid and allantoin were accomplished according to the following diagram.

5% TCA extract (6 g tissue + 30 ml 5% TCA)
↓
Hg-acetate precipitation of uric acid and allantoin
↓
Removal of Hg as HgS

The separation of uric acid and allantoin was achieved by two chromatographic runs, the first on Dowex AG 50W-X8, where catabolic compounds were eluted with distilled water. This eluate was concentrated, brought to pH 2 and applied on Dowex 1-X8, as previously[7].

Administration of oxonic acid. Oxonic acid was administered i.p. (25 mg/100 g b.w.), 2 hours before killing the rats.

Determination of radioactivity. This was carried out on a Nuclear Chicago Delta Scintillation Counter, using Instagel as the liquid scintillation cocktail.

RESULTS

Results are reported in Figure 1 and Tables I and II. Figure 1 shows the chromatographic separation on a Dowex 1-X8 column, of uric acid and allantoin, from livers of rats treated or otherwise with oxonic acid.

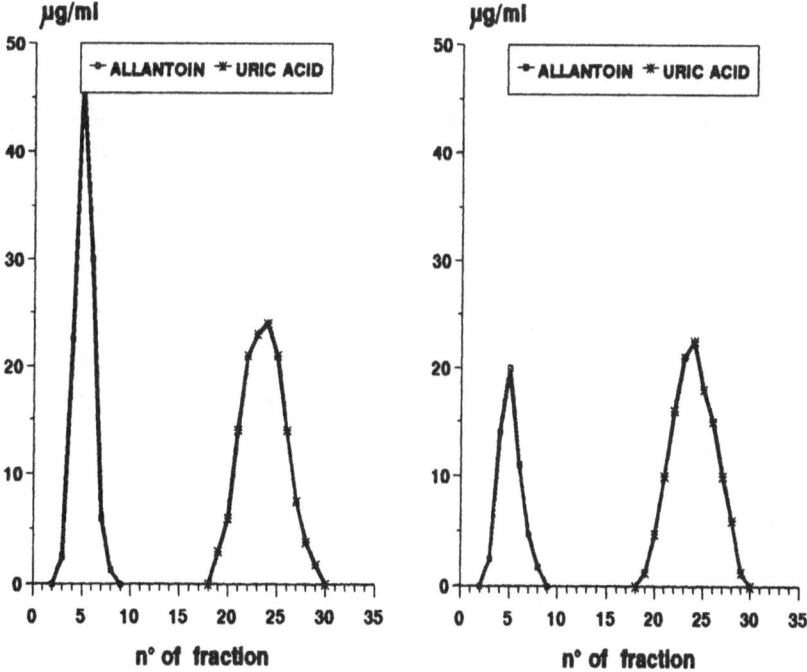

Figure 1.
Dowex 1-X8 chromatography of allantoin and uric acid from normal (left) and oxonic acid treated rat liver (right).

Table 1 and 2 show the quantity and specific radioactivity of the fractions containing uric acid and allantoin (after chromatography on a Dowex 1-X8 column), without and with oxonic acid treatment.

Table 1. Effect of i.p. injection of oxonic acid (25 mg/100 g.b.w.) on uric acid and allantoin levels in rat liver. Data of a typical experiment.

Treatment	Uric acid		Allantoin	
	μmol/g	mg/g	μmol/g	mg/g
None	0.113	0.019	0.158	0.025
N + oxonic acid	0.107	0.018	0.095	0.015

Table 2. Effect of i.p. injection of oxonic acid (25 mg/100 g.b.w.) on hepatic uric acid and allantoin after ^{14}C-formate administration. Data of a typical experiment.

Treatment	Uric acid		Allantoin	
	dpm/μmol	dpm/g	dpm/μmol	dpm/g
Normal (N)	7,332	829	17,437	2,759
N + oxonic acid	12,095	1,295	5,818	552

The i.p. injection of oxonic acid (25 mg/100 g.b.w.) in normal rats, 2 hours before killing, decreased the allantoin content of the liver, but no significant variation in uric acid was found. When ^{14}C-formate (10 μCi g.b.w.) was administered 15' before killing rats treated with oxonic acid, the specific radioactivity of uric acid was higher than that of allantoin; that of allantoin was lower than when obtained from rats not treated with oxonic acid.

These results confirm that no formation of allantoin occurs independently of uric acid, as demonstrated by our previous experiments[11] with ^{14}C-glycine labeling (the S.A. of allantoin and uric acid were similar). Moreover, the high specific radioactivity of allantoin after ^{14}C-formate labelling is at least partially linked to the uricase reaction. In fact, when oxonic acid is administered, the quantity of allantoin decreased, and the specific radioactivity of uric acid was higher than that of allantoin. This shows that under normal conditions (no administration of oxonic acid), the higher specific radioactivity of allantoin could somehow be, at least partially, linked to the uricase reaction. Xanthine oxidase known to occur in soluble and particulate form and uricase is shown to be localyzed in peroxisomes in rat liver[12,13]. This might be due to one single enzyme or different isoenzymes and would involve different pools of uric acid: one very hot, rapidly transformed into allantoin, the other colder and subject to the slow action of uricase. When TCA extracts are prepared, the resulting allantoin pool is a mixture of both and might contain an overall higher specific radioactivity of allantoin than of uric acid. Only one of the uricase compartements would be sensitive to oxonic acid, and once this is blocked, the specific radioactivity of allantoin would fall to lower levels.

REFERENCES

1. E.Zoref-Shani, A.Shainberg and O.Sperling, Characterization of purine nucleotide metabolism in primary rat muscle cultures, *Biochim.Biophys.Acta* 716:324 (1982).
2. G.P.Wheeler, J.A.Alexander and H.P.Morris, in "Advances in Enzyme Regulation", Wheber G. ed., Vol.2, pp.347-369, Pergamon, Oxford (1964).
3. G.Van den Berge, F.Bontemps and H.Hers, Purine catabolism in isolated rat hepatocytes, *Biochem.J.* 188:913 (1980).
4. A.Di Stefano, M.Pizzichini, G.Resconi and E.Marinello, Determination of allantoin in mammalian tissues, *Pediatr.Res.* 24:111 (1988).
5. M.Pizzichini, M.L.Pandolfi, B.Porcelli, L.Arezzini, R.Pagani and R.Fulceri, Labelling of urinary uric acid and allantoin, *Biochem.Soc.Trans.* 20:379S (1992).
6. E.Young and C.F.Conway, On the estimation of allantoin by the Rimini-Schryver reaction, *J.Biol.Chem.* 142:839 (1942).
7. A.Di Stefano, M.Pizzichini, R.Leoncini, D.Vannoni, R.Pagani and E.Marinello, Quantitative separation of uric acid and allantoin from rat liver tissue, *Biochim.Biophys.Acta* 117:1 (1992).
8. R.Pagani, R.Leoncini, M.Pizzichini, G.Cinci, E.Marinello and R.Fulceri, Production of uric acid and allantoin, *Adv.Ex.Med.Biol.* 309A:255 (1991).
9. E.Praetorius and H.Poulsen, Enzymatic determination of uric acid with detailed directions, *Scand.J.Clin. Lab.Invest.* 5:273 (1953).
10. L.S.Dietrich and E.Borries, On the determination of xanthine oxidase activity in animal tissues, *J.Biol. Chem.* 208:287 (1954).
11. M.Pizzichini, M.L.Pandolfi, G.Cinci, L.Terzuoli, B.Porcelli, R.Pagani and R.Fulceri, Study of purine catabolism using ^{14}C-glycine as tracer, *Biochem.Soc. Trans.* 20:376S (1992).
12. S.Angermuller, G.Bruder, A.Volkl, H.Wesch and H.D.Fahimi, Localization of xanthine oxidase in crystalline cores of peroxisomes. A cytochemical and biochemical study, *Eur.J.Cell Biol.* 45:137 (1987).
13. A.Volkl, E.Baumgart and H.D.Fahimi. Localization of urate oxidase in the crystalline cores of rat liver peroxisomes by immunocytochemistry and immunoblotting, *J.Histochem.Cytochem.* 36(4):329 (1988).

THE MEASUREMENT OF DEOXYNUCLEOTIDE
(dNTP) POOLS BY RADIOIMMUNOASSAY (RIA)

Wynne Aherne, Anthea Hardcastle, Lloyd Kelland and
Ann Jackman

CRC Centre for Cancer Therapeutics at the Institute
of Cancer Research, Sutton
Surrey, SM2 5NG, UK

INTRODUCTION

Inhibition of the pathways of DNA synthesis by purine and pyrimidine antimetabolites will result in perturbations of intracellular deoxynucleotide (dNTP) pools. The extent and duration of dNTP pool perturbations can provide information on the mechanisms of action of antimetabolites, are important determinants of cell death and may also prove to be useful pharmacodynamic indicators of effective chemotherapy.

The measurement of dNTP pools presents a challenging analytical problem because of their low intracellular concentrations and the simultaneous presence of 100-1000 fold higher levels of the corresponding ribonucleotides. HPLC is frequently used for the measurement of dNTPs but generally requires relatively high cell numbers and often lacks resolution. Improved sample clean-up protocols and detectors have more recently improved resolution and sensitivity. The DNA polymerase assay is also widely used and is an extremely sensitive assay for dNTPs although the polymerase cannot distinguish between TTP and dUTP. This is important when the effects of inhibitors of thymidylate synthase (TS), which converts dUMP to TMP prior to the formation of TTP, are being investigated as significant amounts of dUTP can be formed behind the enzyme block[1].

RADIOIMMUNOASSAYS FOR dNTPs

As an alternative method of analysis we have developed novel sensitive high throughput radioimmunoassay (RIA) procedures for the measurement of dNTPs. Antisera to TTP[2], dUTP[3], dCTP[4] and recently dGTP have been produced in sheep or rabbits in response to immunisation with conjugates prepared by water soluble carbodiimide condensation to ovalbumin. Specificity studies (Table 1) showed that each antiserum has a similar pattern of cross-reactivity with closely related compounds. There is minimal cross-reaction with the corresponding ribonucleotide and to ensure

Purine and Pyrimidine Metabolism in Man VIII, Edited by
A. Sahota and M. Taylor, Plenum Press, New York, 1995

that these much larger pools do not cause interference a periodate oxidation step is included in sample preparation. Deoxynucleosides and the non-relevant purine or pyrimidine do not cross react significantly. However, the antibodies always show a degree of recognition of the relevant diphosphate and, to a lesser extent, the monophosphate. Direct RIA of the cell extract therefore, represents "immunoreactive" dNTP levels although specific measurement of the triphosphate can be obtained using a QAE Sephadex chromatography step to separate it from the di- and monophosphate[2].

Table 1. Percent cross-reactivity of dNTP antisera.

Compounds	Antiserum			
	TTP (S1524)	dUTP (R9)	dCTP (R6)	dGTP (R8764)
dNTP	100	100	100	100
dNDP	89	42.2	7.1	33.3
dNMP	0.2	3.1	2.1	0.6
non-relevant dNTPs	<2.0	<0.4	<0.03	<1.0
NTP	0.3	3.0	2.7	0.9
deoxynucleoside	0.2	<0.5	<0.01	<0.6

Each antiserum was incubated for 2h with the appropriate H³-dNTP in the presence of the standard dNTP or cross-reactant. Antibody-bound radiolabel was separated using dextran-coated charcoal (2.5%). Percent cross-reactivity was calculated at 50% inhibition of binding at Bo.

The RIAs are extremely sensitive eg the TTP standard curve is sensitive to the addition of 25fmole and can be carried out on a PCA extract of 1×10^6 cells or less with a limit of detection of $\sim$1pmole/10^6 cells. The results obtained by RIA compare well with values previously reported for HPLC and DNA polymerase methods.

THE EFFECT OF TS INHIBITORS ON dNTP POOLS

TS has received a great deal of attention in recent years as a target for chemotherapy. We required a rapid pharmacodynamic measure of TS inhibition during the development of the specific TS inhibitors ICI(Z)D1694 (Tomudex[R]) and ZD9331. D1694, presently undergoing Phase III clinical studies, is a potent quinazoline antifolate inhibitor of TS which exerts its effect primarily through the formation and retention of intracellular polyglutamates[5]. ZD9331, the glutamyl-γ-tetrazole analogue of 2,7-dimethyl, 2'F-N[10]-propargyl-5,8-dideazafolic acid is also a potent TS inhibitor and antitumour agent but is not dependent on polyglutamation (Jackman *et al* accompanying paper). Compounds such as ZD9331 should be able to overcome resistance to D1694 which is mediated through reduced or defective polyglutamation. The lack of drug retention through polyglutamation also suggests that as soon as drug administration has ceased, TS activity should recover rapidly reducing the possibility of prolonged toxicity.

The extent, duration and recovery of TS inhibition following exposure of cells to D1694 and ZD9331 were therefore compared by the measurement of dNTP pools either side of the enzyme block. TTP was measured directly on cell extracts as results obtained on control cells compared well with the specific measurement of TTP following chromatography. The dUTP RIA was adapted to measure "dUMP" pools in cell extracts by exploiting the cross-reactivity of the antiserum with dUMP and re-optimising the assay to include dUMP standards and radiolabel. Results represent deoxyuridine nucleotide immunoreactivity and will be subject to interference from dUTP and dUDP formed as a result of dUMP accumulation. A specific measure of dUTP was also obtained on appropriate samples by including the chromatography step.

dNTP pools in untreated cells

TTP and "dUMP" pools were measured in untreated L1210, human W1L2 lymphoblastoid and ovarian tumour cell lines and in variants of the same lines with acquired resistance to D1694 (Table 2). TTP pools were comparable in the L1210 and W1L2 cell lines but were up to 4 times greater in the 2 ovarian lines. There was no difference between the parent cells and the variants with acquired resistance to D1694.

Table 2. TTP and "dUMP" pools (pmole/10^6 cells) in tumour cell lines with acquired resistance to D1694.

Cell line	Resistance mechanism	TTP pools	"dUMP" pools
L1210 mouse leukaemia		22.6 ± 6.8	3.0 ± 0.8
L1210:R^{D1694}	Polyglutamation	20.4/19.8	3.1/3.0
W1L2 human lymphoblastoid		28.2 ± 6.9	8.3 ± 5.9
W1L2:R^{D1694}	TS overexpression	19.2 ± 6.2	5.1 ± 2.7
CH1 human ovarian		87.0 ± 11	12.1 ± 5.0
CH1:R^{D1694}	TS overexpression	99.9 ± 7.8	18.7 ± 9.1
41M human ovarian		67.6 ± 22	8.1 ± 5.6
41M:R^{D1694}	Transport defect	74.9 ± 6.8	8.2 ± 5.6

TTP and "dUMP" pools were measured in control cells (n = 2-3) in duplicate. Cells were extracted with 0.4M PCA, neutralised with 0.73M KOH in 0.16M KHCO$_3$ and treated with 0.5M sodium periodate before RIA.

dNTP pools in TS inhibited cells

Exposure of L1210 cells to D1694 caused a dose and time-dependent depletion of TTP pools with a concomitant increase in "dUMP" immunoreactivity. For example the results obtained following a 4h exposure to multiples of the IC$_{50}$ (10nM) and resuspension in DFM are shown in Table 3. At the higher doses of D1694 pools were not normalised even after 16h in DFM consistent with the retention of polyglutamated forms of D1694. dCTP pools were also perturbed following exposure to D1694. A 4h exposure to 100 x IC$_{50}$ caused dCTP pools to fall to 55% of controls (28.0 ± 2.3pmoles/10^6 cells) with recovery to 70% following resuspension in DFM for 16h. Significant formation of dUTP (1-2pmole/10^6 cells) was observed only at late time-points at 0.5 and 1.0μM D1694. The "dUMP"/TTP ratio obtained from these studies provides a useful comparative index of TS inhibition. In untreated cells from different origins the pool ratio was 0.1-0.3 and a ratio of >1 was indicative of TS inhibition.

The perturbations in TTP and "dUMP" pools caused by D1694 in L1210 cells were also observed in the human W1L2 cell line. Table 4 compares the "dUMP"/TTP

Table 3. Effect of D1694 on TTP and "dUMP" pools (pmole/10^6 cells) in L1210 cells.

Dose multiple (4h)	4h exposure		+ 4h DFM		+ 16h DFM	
	TTP	"dUMP"	TTP	"dUMP"	TTP	dUMP
10x	9.2	65	6.1	35	33	30*
50x	1.8	36	1.9	43	6.7	51
100x	1.2	54	<1.0	45	4.1	48

L1210 cells in logarithmic growth were exposed to D1694 for 4h at multiples of the IC$_{50}$ (0.01μM), resuspended in DFM for 4 or 16h. Results are the mean of 2-4 experiments in duplicate (* n=1). Control levels were 22 ± 3.6 (n=11) and 6.8 ± 7.2 (n=8) for TTP and "dUMP" respectively.

ratios obtained in W1L2 cells following exposure to D1694 and ZD9331. Both compounds cause depletion of TTP and elevation of "dUMP" as indicated by the high pool ratio but in the case of ZD9331 the ratio is completely normalised when the cells are resuspended in DFM for 4h. This result was to be expected in view of the inability of this compound to undergo polyglutamation.

Table 4. The "dUMP"/TTP ratios in W1L2 cells treated with D1694 or ZD9331.

Dose	D1694			ZD9331	
	4h	4h + 4h DFM	4h + 16h DFM	4h	4h + 4h DFM
IC$_{50}$	6.0	3	0.5	23	0.2
10xIC$_{50}$	58	19	6.2	116	0.1
100xIC$_{50}$	155	76	128	251	0.1
1000xIC$_{50}$	114	118	280	223	0.2

W1L2 cells were exposed to D1694 or ZD9331 at multiples of their IC$_{50}$ (5nM and 8nM respectively). TTP and "dUMP" were measured by RIA in cell extracts as before. Results are the mean of duplicate flasks in 1 experiment. Controls were 15 and 1.5 pmole/10^6 cells for TTP and "dUMP" respectively (ratio = 0.1).

SUMMARY

Radioimmunoassay provides an alternative, sensitive and reproducible high throughput method for the measurement of dNTP pools. The extent and duration of inhibition of TS can be investigated by determination of the TTP and "dUMP" pools and the effects of D1694 and ZD9331 have confirmed their biochemical profiles. The RIAs will be useful in providing information for the design of treatment protocols for TS inhibitors and with the specific assay of dUTP, on mechanisms of cell death in different cell lines.

REFERENCES

1. N.J. Curtin, A.L. Harris and G.W. Aherne, Cancer Res. 51:2346 (1991).
2. G.W. Aherne, E. Piall, S. Aitkenhead and N.J. Curtin, Biochem. Soc. Trans. 17:1052 (1989).
3. E.M. Piall, N.J. Curtin, G.W. Aherne, et al, Anal. Biochem. 177:347 (1989).
4. E. Piall, G.W. Aherne and V. Marks, Anal. Biochem. 154:276 (1986).
5. A.L. Jackman, G.A. Taylor, W. Gibson et al, Cancer Res. 51:5579 (1991)

RAPID SIMULTANEOUS MEASUREMENT OF NUCLEOTIDES, NUCLEOSIDES AND BASES IN TISSUES BY CAPILLARY ELECTROPHORESIS

Tilman Grune[1] and David Perrett[2]

[1]Clinics of Physical Therapy & Rehabilitation, Medical Faculty (Charite) Humboldt University, Berlin, Germany
[2]Dept of Medicine, St Bartholomew's Hospital Medical College, London, U.K.

INTRODUCTION

Capillary electrophoresis (CE) is now one of the most efficient separation techniques. It has been rapidly developed since Jorgenson et al.[1] used small diameter capillaries for the separation of ions in an electrical field. One of the greatest advantages of CE is the use of small sample amounts - from some few nl of sample to the cytosolic fluid of single cells[2]. These small sample volumes make it possible to solve a number of biological and clinical problems, where the amount of biological material is limited.

The separation of nucleotides has been performed by a number of authors[3-12]. In general micellar electrokinetic chromatography (MECC) mode has been used. Other groups including ourselves[13] have separated the purine bases by free solution CE whilst others have used MECC for these compounds. However only a few of the interesting purine metabolites for investigation of energy metabolism or purine degradation studies were analyzed in these reports.

Under the appropriate conditions CE is also able to analyze cations and anions in the same run, which is relevant to the separation of nucleotides and bases. The aim of the present study was to develop a free solution capillary electrophoresis, for the separation of common naturally occurring purines in both their nucleoside and nucleotide forms and to apply the developed technique for the analysis of purine energy state in biological tissues.

Purine and Pyrimidine Metabolism in Man VIII, Edited by
A. Sahota and M. Taylor, Plenum Press, New York, 1995

MATERIAL AND METHODS

Materials

Purine reference standards were purchased from Sigma (Poole, U.K). Sodium borate was obtained from Merck (Poole, U.K.). Fused silica capillary was obtained from Polymicro Technologies (Az, USA).

Instrumentation and separation conditions

For all electrophoretic separations a SpectraPhoresis 1000 (Spectra Physics, San Jose, CA, USA) instrument was used. Analyses were performed in uncoated silica capillary of 44 cm length and 75 μm internal diameter. The window was located at a distance of 37 cm from the anode. UV scanning detection over the range of 200 to 300nm was performed. Hydrodynamic injection mode was used. For separation a 20mM sodium borate buffer pH 9.2 was employed. The separation was optimized with respect to pH, temperature, applied potential and injection time as described below.

Sample treatment

Mouse liver was rapidly frozen in liquid nitrogen. 100 mg portions of frozen liver were extracted with the twofold amount (w/v) of trichloroacetic acid (12% w/v). The extracts were centrifuged for 10 min at 2000 g. The supernatant was neutralized by a 3-fold extraction with diethyl ether. The final extract was stored at -20°C until analysis.

RESULTS

For biochemical investigations of the cellular energy status and for estimation of metabolic changes quantitation of the following purines is necessary: adenine (Ade), guanine (Gua), hypoxanthine (Hyp), guanosine (Guo), inosine (Ino), uric acid (UA), inosine monophosphate (IMP) and the mono-, di- and triphosphorylated compounds of adenine and guanine. These compounds employed in the following optimization of separation procedure.

Influence of pH

The influence of pH on the separation characteristics was investigated in the range of pH values from 8.5 to 10.5 since at these pH values the nucleotides, nucleosides and bases should possess similar -ve charges. Changes of the pH were accompanied by great differences in the selectivity. The optimal pH for separation of all peaks was in the range between 9 and 9.5, with the best separation at pH 9.2. for the standard mixtures investigated.

Temperature effects

The effect of capillary temperature on the separation quality at a pH of 9.2 were investigated in the range between 15 and 35°C. High temperatures caused faster migration times but the migration order of the selected compounds was not influenced by temperature. However, the optimal temperature for the analysis of the complex purine mixture was 15°C, giving a good separation quality and a separation time below 18 minutes.

Applied Potential

The dependence of voltage on migration time was investigated. There was a drastic reduction of migration times with increasing voltage. Although high voltages led an improved separation. The increase of voltage was limited: firstly, by the increasing current levels, and their associated Joule heating effects and secondly, by the peak resolution. The optimal separation was achieved at a voltage of 20 kV.

Under optimized conditions a separation of 19 purine and related compounds was achieved Figure 1.. The final overall analysis time was less than 18 minutes. The separation efficiency expressed as the achieved theoretical plates for the optimized separation conditions ranged from 226,000 for uracil to 80, 000 for IMP.

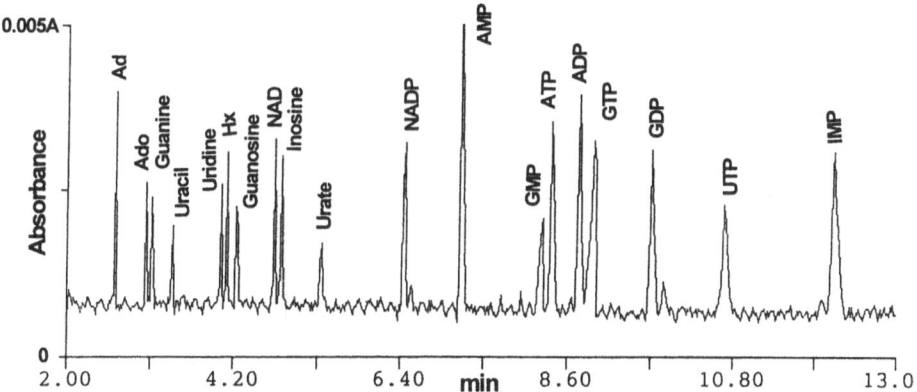

Fig.1. Separation of a standard mixture of 19 purines, pyridines and pyrimidines. Separations were performed in an uncoated silica capillary (75 μm i.d. x 44cm, 37 cm to window) in a 20mM borate buffer pH 9.2, 25kV, 15°C and 5 sec hydrodynamic load. The detection wavelength was 260 nm.

Calibration and reproducibility

Detection was performed with wavelength scanning from 200 to 300nm, which aided peak tracking and peak identification in actual samples. Detection at 210nm was significantly more sensitive than at 260nm. Also for uric acid the detection at 200nm shows a higher detection limit than at 292 nm. Monitoring at these low wavelength is usually not possible with HPLC.

The correlation coefficients for the peak areas of these compounds for injection times between 1 and 20 sec exceeded 0.96 for all peaks except IMP which was 0.93. There were no differences in the correlation coefficients for detection at 260 and 200 nm. So it seems better to detect the separated compounds at 200 nm because of the higher detector signal. For an injection times of 1 sec there were correlation coefficients higher than 0.93

for a concentration range from 2 µmol/l to 120 µmol/l. The relative standard derivation (in %) of ten injections of a standard mixture of 60 µmol/l for retention times and peak areas ranges from 2 to 8% and 2 to 5% respectively. The optimised system shows high reproducibility of retention times and peak areas and the CE data is comparable to that found with HPLC..

Analysis of biological samples

Fig.2 shows the separation of liver TCA extracts before and after ischemia. The chromatograms demonstrated the possibility of analyzing biological samples by the described method: less than 5nl of sample were actually analyzed.. In Fig.4b the drastic increase of nucleosides and -bases during organ ischemia is clearly demonstrated whilst in the same run the fall in nucleotide concentrations is quantifiable.

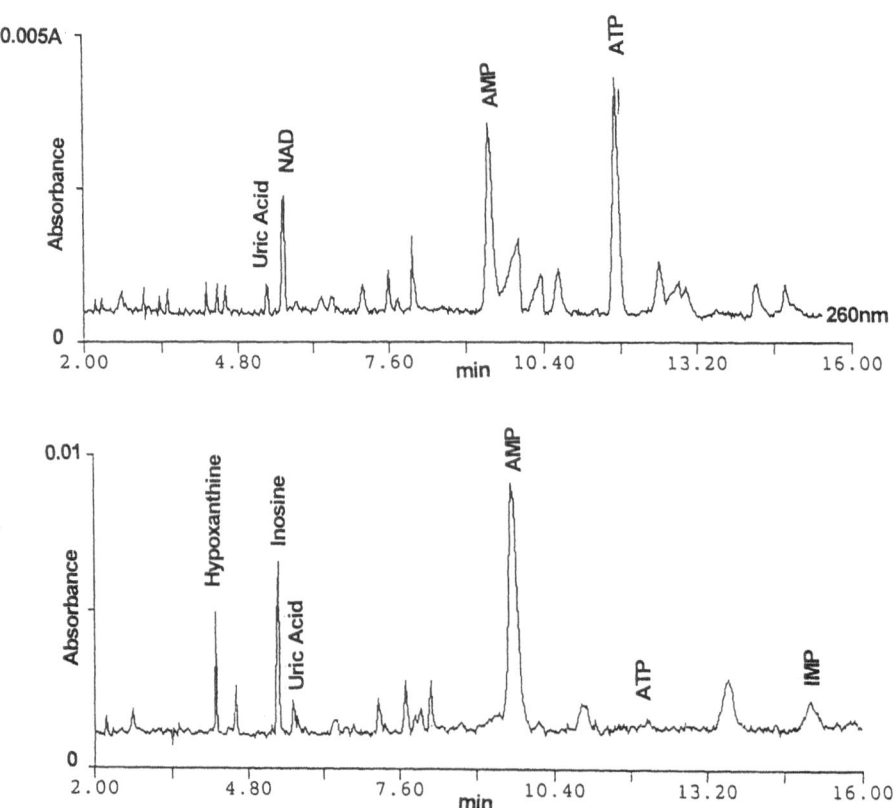

Fig.2 Separation of a TCA-liver extract (upper trace) fresh sample and (lower trace) 10 minutes of ischemia. Separation conditions and peak numbers as in Fig.1.

CONCLUSION

Capillary electrophoresis techniques has been used to separate various as well naturally occurring purine compounds, as ribonucleotides[9,11], deoxyribonucleotides[3,7,12], nucleosides and bases[13,14], cyclic nucleotides[15], as also modified compounds[16] and pharmacological interesting~ derivatives [17].

Various electrophoretic separation techniques, as micellar electrokinetic capillary chromatography [7-9,12], isotachophoresis and capillary zone electrophoresis[3,11], has been used. Few of the mentioned papers document the detection of even simple mixture of nucleotides, nucleosides and bases.

So Huang et al.[11] demonstrated a separation of at least 14 compounds mainly mono-, di- and triphosphorylated ribonucleotides of adenine, guanine, uracil and cytosine and additional XMP and IMP. The separation was performed within 50 minutes. So it takes more time than HPLC separations of an equal number of substances. Ng. et al.[4] documented an analysis of 8 purine and pyrimidine nucleotides in 25 minutes of separation.

The method described in this paper allows the separation of the major naturally occurring purines within 18 minutes. That means that the separation time is comparable with that of powerful HPLC separations applicable to either nucleosides and bases or nucleotides[20] but not usually both. A major advantage, in comparison with other capillary electrophoresis methods[4,11] and almost all HPLC methods is the ability of this optimised separation to determine nucleotides, nucleosides and bases in a single separation in biological samples.

The present method therefore allows the separation of the major purine compounds in TCA cell extracts and can be used as analytical tool for purine determination in small biological samples.

ACKNOWLEDGEMENTS

This study was supported by a grant from the Medical Faculty of the Humboldt University Berlin for T.G. D.P. was supported by grants from the Joint Research Board of St. Bartholomews Hospital and North East Thames Regional Health Authority (LORS Scheme).

REFERENCES

1. J.W. Jorgenson and K.D. Lukacs, Zone electrophoresis in open-tubular glass capillaries. *Anal Chem*. 53: 1298(1981).
2. T.M. Olefirowicz and A.G. Ewing. Capillary electrophoresis in 2 and 5μm diameter capillaries: Application to cytoplasmic analysis. *Anal Chem*, 62: 1872 (1990)
3. R. Takigiku and R.E. Schneider. Reproducibility and quantitation of separation for ribonucleoside triphosphates and deoxyribonucleoside triphosphates by capillary zone electrophoresis. *J Chromatogr*, 559: 247 (1991)
4. M. Ng, T.F. Blaschke, A.A. Arias and R.N. Zare. Analysis of free intracellular nucleotides using high-performance capillary electrophoresis. *Anal Chem*, 64: 1682 (1992)
5. T. Tsuda, G. Nakagawa, M. Sato and K. Yag. Separation of nucleotides by

high-voltage capillary electrophoresis. *J Appl Biochem*, 5: 330 (1983)

6.   T. Tsuda, K. Takagi, T. Watanabe and T. Satake. Separation of nucleotides in organs of guinea pig by capillary zone electrophoresis. *HRC & CC J High Resolut Chromatogr Chromatogr Comm*, 11: 721 (1988)

7.   W.H. Griest, M.P. Maskarinec and K.H. Row. Evaluation of micellar electrokinetic capillary chromatography for the separation and detection of normal and modified deoxyribonucleosides and deoxyribonucleotides. *Sep Sci Technol*, 23: 1905 (1988)

8.   J. Liu, F. Banks and M. Novotny. High-speed micellar electrokinetic capillary chromatography of common phosphorylated nucleosides. *J Microcol Sep*, 1: 136 (1989)

9.   D. Perrett and G. Ross. Capillary electrophoresis for the analysis of cellular nucleotides. In Purine and Pyrimidine Metabolism in Man VII Part B, ed. G.B. Elion, R.A. Harkness and N. Zollner,, 1-5. New York. Plenum Press.(1991)

10.   A.L. Nguyen, J.H.T. Luong and C. Masson. Determination of Nucleotides in Fish Tissues using Capillary Electrophoresis. *Anal Chem*, 62: 2490 (1990)

11.   M.X. Huang, S.F. Liu, B.K. Murray and M.L. Lee. High resolution separation and quantitation of ribonucleotides using capillary electrophoresis. *Anal Biochem*, 207: 231 (1992)

12.   T. Lee, E.S. Yeung and M. Sharma.   Micellar electrokinetic capillary chromatographic separation and laser-induced fluorescence detection of 2'-deoxynucleoside 5'-monophosphate of normal and modified bases. *J Chromatogr*, 565: 197 (1991)

13.   T. Grune, G.A. Ross, H. Schmidt, W. Siems and D. Perrett. Optimized separation of purine bases and nucleosides in human cord plasma by capillary zone electrophoresis. *J Chromatogr*, 636: 105 (1993)

14.   A.C. Schoots, T.P.E.M. Verheggen, P.M.J.M. de Vries and F.M. Everaerts. Ultraviolet-absorbing organic anions in uremic serum separated by capillary zone electrophoresis. *Clin Chem*, 36: 435 (1990)

15.   L. Hernandez, B.G. Hoebel and N.A. Guzman. Analysis of cyclic nucleotides by capillary electrophoresis using ultraviolet detection. In Analytical Biotechnology. Capillary Electrophoresis and Chromatography (ACS Sym Series No 434), ed. C. Horvath and J. Nikelly,, 50-59. Washington DC. American Chemical Society.(1990)

16.   I.Z. Atamna, G.M. Janini, G.M. Muschik and H.J. Issaq. Separation of xanthines and uric acid by capillary zone electrophoresis and micellar electrokinetic capillary electrophoresis. *J Liquid Chromatogr*, 14: 427 (1991)

17.   D.K. Lloyd, A.M. Cypess and I.W. Wainer.   Determination of cytosine-ß-D-Arabinoside in plasma using capillary electrophoresis. *J Chromatogr*, 568: 117 (1991)

18.   A.F. Lecoq, L. Montanarella and S. Di-Biase. Separation of nucleosides and nucleotide-3'-monophosphates by micellar electrokinetic capillary chromatography. *J High Res Chromatogr*, 14: 667 (1991)

19.   R.P. Singhal, D. Hughbanks and J. Xian. Separation of dideoxyribonucleosides in trace amounts by automated liquid chromatography and capillary electrophoresis. *J Chromatogr*, 609: 147 (1992)

20.   D. Perrett, L. Bhusate, J. Patel and K. Herbert, Comparative performance of ion-exchange and reversed-phase ion-pair HPLC for the determination of nucleotides in biological samples. *Biomed. Chromatogr.* 5: 207 (1991)

APPLICATION OF CAPILLARY ZONE ELECTROPHORESIS FOR RAPID SCREENING OF ENZYMATIC DEFECTS IN PURINE METABOLISM

Costantino Salerno[1] and Carlo Crifò[2]

[1]Department of Human Biopathology,
[2]Department of Biochemical Sciences
University of Roma La Sapienza, Italy

INTRODUCTION

Confirmation of diagnostic suspicion of inherited disorders in purine metabolism generally requires direct assay of the defective enzyme activity in crude extracts of blood cells or tissues obtained by biopsy. The available methods involve, in most of the cases, tedious multi-step procedures that need standardization protocols even in expert laboratory centers, as the analysis is required very seldom. The majority of the methods employ radioactive compounds and differ from one another only in the way substrate and product are separated.[1-4] Procedures based on high-pressure liquid chromatography with and without radiolabelled substrates are also used.[5] Protein removal is necessary to prevent column degradation. This is achieved by addition of acid precipitating agents followed by extraction and/or neutralization of the protein-free phase.[6]

Since, with the recent development of commercial instruments, capillary zone electrophoresis became more widely available, we developed a general procedure that is applicable, with minor modifications, to assess most of the purine-related enzymatic defects. This microanalytical technique, routinely employed in our laboratory, can be proposed as an useful tool for determining the enzyme activity in biological specimens in relatively short times without pretreatment steps.

EXPERIMENTAL PROCEDURES

A commercially available instrument (P/ACE System 2000, Beckman, Palo Alto, CA) was used for standardizing the method. The data were collected at 5 Hz frequency and stored on the hard disk of an IBM PS/2 mod. 55-SX personal computer. The electropherograms were UV monitored at 254 nm. The detector rise time was 1 s. The capillary (uncoated fused silica) had a 75 μm i.d., an effective length of 50 cm, and a total length of 57 cm. The temperature of the capillary cartridge was maintained at 25°C. The capillary was cleaned by flushing sequentially with 100 mM NaOH, deionized water, and separation buffer for 10 min each. Between the analyses, the capillary was flushed at

Purine and Pyrimidine Metabolism in Man VIII, Edited by
A. Sahota and M. Taylor, Plenum Press, New York, 1995

0.5 psi for 4 min with the separation buffer prior to injection of the next sample. Samples were pressure-injected at 0.5 psi for 5 s. The applied potential was increased in 12 s to the final running voltage (30 kV), where the potential was held constant for the duration of the analysis. All separations were performed in 50 mM sodium borate, pH 8.5, with the current stable around 30-35 μA. Electroosmotic flow was stronger than electrophoretic migration, so all the solutes migrated toward the negative electrode.

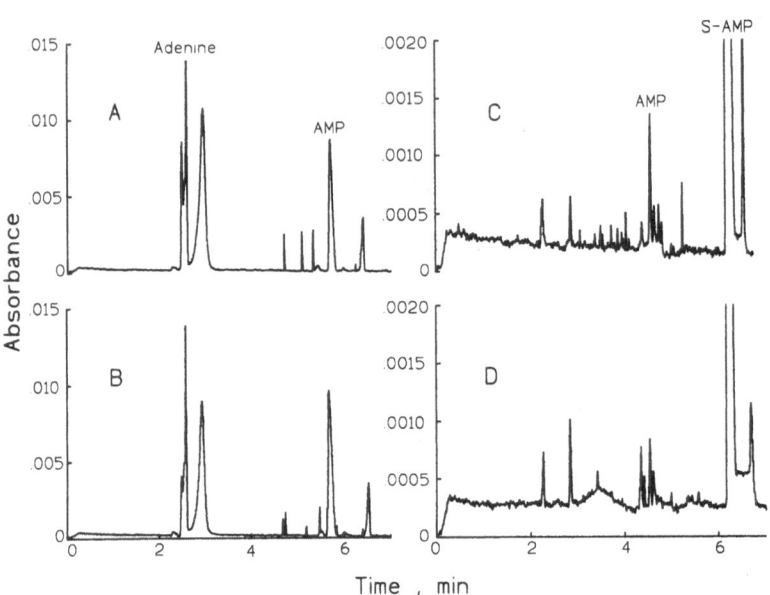

Figure 1. *Left:* adenine phosphoribosyltransferase-catalyzed reaction. The reaction mixtures contained erythrocyte lysates (5 mg hemoglobin / ml). The electropherograms were obtained after 1.5 hr incubation at 37°C. Other conditions were as described in reference 3. **A:** control healthy subject. **B:** patient (3 years-old boy with massive bladder stones and hyperuricosuria) who had been erroneously suspected for the enzyme defect. *Right:* adenylosuccinate lyase-catalyzed reaction. The reaction mixtures contained lysates of mixed peripheral blood lymphocytes (5000 cells / μl). The electropherograms were obtained after 4 hr incubation at 37°C. Other experimental conditions were as described in reference 4. **C:** control healthy subject. **D:** patient (10 years-old girl with severe psychomotor retardation) who had been found to excrete anomalous amounts of succinylnucleosides in the urine.

Purified human erythrocytes and mixed peripheral blood lymphocytes, as well as whole venous blood, were used as source of enzymes. The cell-free extracts, obtained by osmotic lysis and/or freezing and thawing, were added last to substrate solutions that had been prepared according to published procedures.[1-5] The reaction mixtures (usually 100-500 μl) were incubated at 37°C. At different periods of time, 20-μl aliquots were withdrawn from the incubating mixture, added into the P/ACE microvials with a syringe, and employed for the electrophoretic analysis.

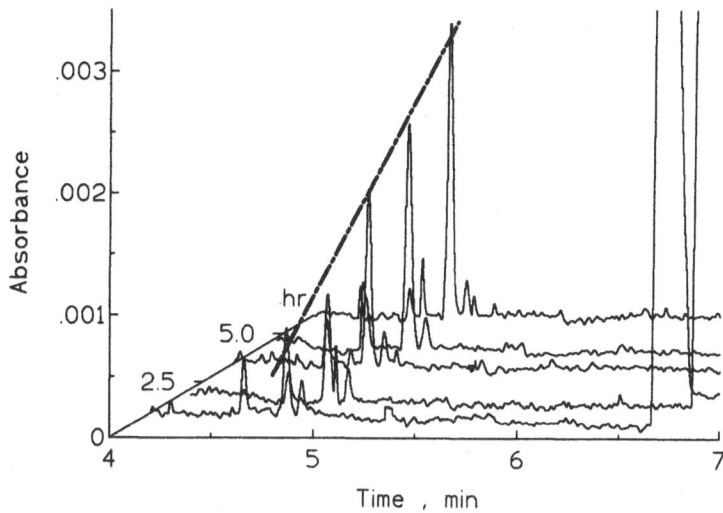

Figure 2. Time course of adenylosuccinate lyase-catalyzed reaction (*dot-dash-dot*). The electropherograms were obtained by analyzing, at different periods of time, samples from a reaction mixture containing erythrocyte lysate (0.12 mg hemoglobin / ml) from the patient with the congenital enzyme defect. Enzyme activity was 40% of that of a control healthy subject. Other experimental details are reported in Figure 1D.

RESULTS AND DISCUSSION

Capillary zone electrophoresis offers a new approach to quantitative determination of purine enzyme activity. Technically it is simpler than gradient elution high-pressure liquid chromatography, it is equally fast if not faster, it needs much less sample and it does not require protein removal.

We have used this system to determine quantitatively the reaction products in mixtures containing up to 5 mg protein/ml. Figure 1 shows typical electropherograms obtained for two different enzyme assays (adenine phosphoribosyltransferase and adenylosuccinate lyase) using undialized extracts of erythrocytes and of mixed peripheral blood lymphocytes. By increasing protein concentration, we observed a shift in electrophoresis times. Thus, we used always controls with about the same protein content and appropriate calibration curves for each assay. Compared to the anion-exchange high-pressure liquid chromatography,[5] the capillary zone electrophoresis allowed higher resolution of substrates and products in the phosphoribosyltransferase-catalyzed reaction. As far as the lyase-catalyzed reaction is concerned, our method is certainly easier than that described so far,[4] involving separation of radiolabelled product by thin-layer chromatography.

Using the assay conditions already reported in the literature, we obtained linear time course for the enzymatic reactions (Figure 2). In each electropherogram, Rt reproducibility ranged from 2-5% while the variation coefficient for AMP area was less than 6% (n = 5). Capillary zone electrophoresis can be successfully employed for determination of defective enzyme stability in crude cellular extracts by using a minimal amount of biological samples (the curve reported in Figure 3 has been obtained with less

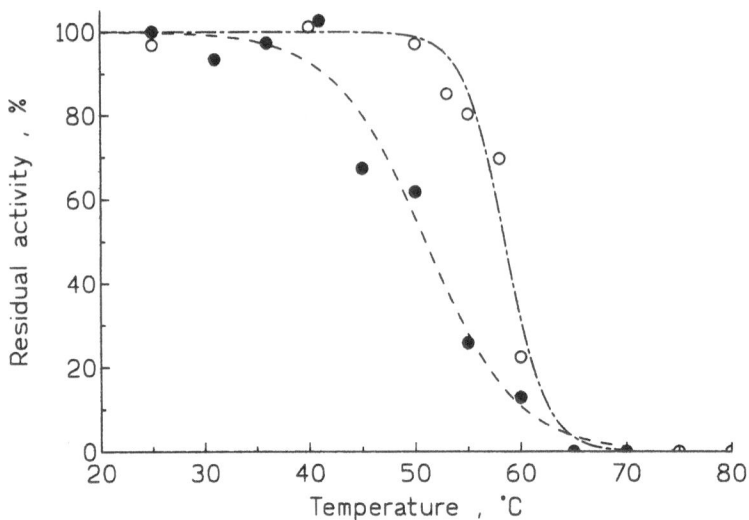

Figure 3. Heat denaturation curve of adenylosuccinate lyase from erythrocyte lysates. O : control healthy subject. ● : patient with the congenital enzyme defect described in Figure 1D. The packed erythrocytes were diluted in 300 volumes of distilled water, incubated for 15 min at different temperatures, and then tested for enzyme activity. Other experimental conditions were as reported in Figure 2.

than 2 μl of packed erythrocytes from the patient with the inherited defect), thus this technique could be the method of choice for the biochemical characterization of inherited metabolic diseases when few amounts of biological samples are available.

REFERENCES

1. W.N. Kelley, F.M. Rosenbloom, J.F. Henderson, and J.E. Seegmiller, A specific enzyme defect in gout associated with overproduction of uric acid, *Proc. Natl. Acad. Sci. USA* 57:1735 (1967).
2. R. Schmidt, M. Forêt, and U. Reichert, Improved microscale assay for purine phosphoribosyltransferase activities, *Clin. Chem.* 22:67 (1976).
3. M.A. Becker, P.J. Kostel, L.J. Meyer, and J.E. Seegmiller, Human phosphoribosylpyrophosphate synthetase: increased enzyme specific activity in a family with gout and excessive purine synthesis, *Proc. Nat. Acad. Sci. USA* 70:2749 (1973)
4. F. Van den Bergh, M.F. Vincent, J. Jaeken, and G. Van den Berghe, Radiochemical assay of adenylosuccinase: demonstration of parallel loss of activity toward both adenylosuccinate and succinylaminoimidazole carboxamide ribotide in liver of patients with the enzyme defect, *Anal. Biochem.* 193:287 (1991).
5. L.D. Fairbanks, A. Goday, G.S. Morris, M.F.J. Brolsma, H.A. Simmonds, and T. Gibson, Rapid determination of purine enzyme activity in intact and lysed cells using high performance liquid chromatography with and without radiolabelled substrates, *J. Chromatogr.* 276:427 (1983).
6. M. Zakaria and P.R. Brown, High-performance liquid column chromatography of nucleotides, nucleosides and bases, *J. Chromatogr.* 226:267 (1981).

A REVERSE-PHASE HPLC METHOD FOR CYCLIC NUCLEOTIDE PHOSPHODIESTERASES ACTIVITY AND CLASSIFICATION

Giuseppe Spoto, Sandro Berardi, Giuseppe Ajerba, and Vittorio De Laurentiis

Institute of Biochemical Sciences, University of Chieti "G.D'Annunzio,"
66013 Chieti

INTRODUCTION

Cyclic AMP and cyclic GMP phosphodiesterases (PDEs) exist as a family of isoenzymes which possess distinct primary amino acid sequences and are encoded by distinct genes (Beavo & Refsnyder, 1990). These multiple enzymes have been classified into five major groups according to kinetic, physical, immunological and regulatory properties (Beavo and Refsnyder, 1990). The phosphodiesterase plays a pivotal role in the cellular signalling process, limiting the cellular response to agents which utilize cyclic nucleotides as second messengers. However, these isoenzymes do not solely act as a brake on cyclic nucleotide synthesis, since their activity states are also regulated by a variety of hormones such as insulin, adrenaline, glucagon, prostaglandin I_2 and prostaglandin E_1 (Zinman & Hollenberg,1974; Loten et al.,1978; Makino& Kono, 1980; Heyworth et al.,1983; Macphee et al.,1988). Cyclic AMP plays a dominant role in the neuro-hormonal regulation of cardiac contractility (Hartzell,1988; Lindemann and Watanabe,1989).

The physiological role of cGMP in the heart muscle is less well characterised. Early works suggested that cGMP was the key messenger involved in the parasympathetic negative modulation of cardiac contractility (Hartzell,1988; Lindemann and Watanabe, 1989).

The existence of multiple forms of PDE in cardiac tissues has been reviewed previously (Well & Hardman, 1977; Beavo et al.,1982). In general, cardiac- ventricle sources have been found to contain three isoforms of PDE: a Ca^{2+}/calmodulin-stimulated activity, a cyclic GMP-stimulated activity, and an activity demonstrating substrate selectivity and high affinity for cyclic AMP ('low-Km' form).

The best known and most sensitive method for assay of phosphodiesterase is based on the isotopic conversion of cyclic to non-cyclic nucleotides, which includes hydrolysis of the product by 5' nucleotidase and chromatographic separation of the labelled adenosine, and/or guanosine. We have published a different and less sensitive reverse-phase HPLC method for cAMP fosphodiesterase activity (Spoto et al.,1991).

This method allows the simultaneous assay of cAMP PDE and cGMP PDE since after incubation with cAMP and cGMP, substrates and products of both reactions can be separated in the same chromatographic run.

In one brief study on human cardiac ventricle PDEs, Hidaka et al.,(1977) have reported three typical phosphodiesterase activities. In addition, a new PDE activity was found in guinea-pig cardiac ventricle. It is probable that this new PDE activity is a contaminant of 'low-Km' PDE preparations previously reported, thus contributing to the observed cooperation with cyclic AMP hydrolysis by this enzyme.

METHODS

Materials. The materials employed were obtained from commercial sources as indicated: adenosine-5'-monophosphate (AMP), adenosine-3':5'-monophosphate (cAMP),

Purine and Pyrimidine Metabolism in Man VIII, Edited by
A. Sahota and M. Taylor, Plenum Press, New York, 1995

phosphodiesterase (PDE) from beef heart were purchased from (Boehringer- Mannheim). All other reagents were of analytical grade from Merck.

Protein assay; Protein content was determined using a bicinchoninic acid protein determination kit from Sigma with bovine serum albumin as standard.

PDE assay. The enzymatic reaction was carried out in 0.1 M Tris-HCl buffer, pH 8.3, 10 mM $MgCl_2$, 0.1 M KCl at 37°C. Alfa casein (2 mg) may be used as carrier for protein precipitation when protein concentration is low. The final volume ranged between 500 and 100 ml. The reaction was initiated by the addition of cAMP or/and cGMP in a final concentration of $0.5*10^{-4}$ M. Enzyme and time of incubation may vary and if are not indicated in the specific experimental procedure is 60 minutes. The reaction was terminated by transferring the tubes with the reaction mixture in boiling water bath for 3 min. The sample was then centrifuged and filtered through a nylon-66 filter, 0,2 mm (Rainin corporation). The clear filtrate obtained may be used directly for HPLC assay or stored at -20°C. A blank with protein, denatured by boiling water bath for 3 minutes, with and without substrate was performed. Both incubation time and enzyme concentration were adjusted so that no more than 25% of the substrate was hydrolyzed under the assay conditions. All assays were done in duplicate.

In some experiments sodium nitroprusside was added to the enzymatic reaction in three different concentration: 50, 100, 200 uM.

Chromatographic apparatus. The HPLC system was from Beckman and consisted of a two 110A pumps; a variable wavelength spectrophotometer Spectroflow 783 (Kratos Analytical) measuring at 254 nm; a autosampler Promis (Spark Holland).

Chromatographic conditions. The column used was a 5-mm Li-Chrospher 100 CH 18/2 Merck (250x4 mm). The mobile phase employed for the separation of nucleotides consisted in 200 mM ammonium acetate pH 6.0 with 2% acetonitril (v/v) . The flow rate was 1 ml/min ; the detection was performed at 254 nm. Peak identities were confirmed by coelution with standards. Quantitative measurements were carried out by comparison using standard solutions of known concentrations.

Separation of cardiac PDE activities. Samples of PDE from beef heart (Boehringer-Mannheim) were applied to a column Resource ^{TM} Q pre-equilibrated with 20 ml of 10 mM Tris-HCl buffer pH 7.8 (buffer A). The flow rate was 4 ml/min. The PDE activities were eluted with a linear gradient of 0,0-0,5 M NaCl in 10 mM Tris-HCl pH 7.8; proteic fractions were collected and assayed for PDE activity. For long-term storage, one mg of alpha casein was added and fractions concentrated on Amicon were stored at -20°C.

RESULTS AND DISCUSSION

Nucleotides separation, as described in methods, is shown in Figure 1. Four nucleotides were separated in about 30 minutes consenting a quantitative value of nucleotides in biological solutions. Using this chromatographic separation we assayed phosphodiesterase activity. As described in the methods, PDE was incubated with cGMP; cAMP; both cGMP and cAMP alone and in the presence of Ca^{2+}/Calmodulin. There are five PDE families but only four of them, type I,II,III, V, are detectable using cGMP as substrate.

Kinetic studies of PDE activity produced a decreased concentration of cGMP with simultaneous increase of GMP concentration. This comparison makes sure that quantitative values are not imputable to casual variations.

It is possible to obtain an analogous separation from PDE families like type I, II, III, IV using cAMP as a substrate , as shown by Spoto et al. (1991).

It is possible by this method of separation to exam a simultaneous decrease of cyclic nucleotide substrates of PDE and contemporaneously increasing nucleotide monophosphates versus time, see figure 2. Experiments with beef heart PDE with single substrate and with both cyclic nucleotides show linearity with similar slope versus time. The results obtained show no significant presence either of PDE III , inhibited by cGMP, or PDE II, cGMP-stimulated, in beef heart.

Experiments in the cAMP or cGMP range 5 to 40 μM give rise to linear double reciprocal plots , with resulting Km of 15.4 and 28.6 uM . These values should be modified in the presence of PDE II or PDE III (Murray and England,1992).

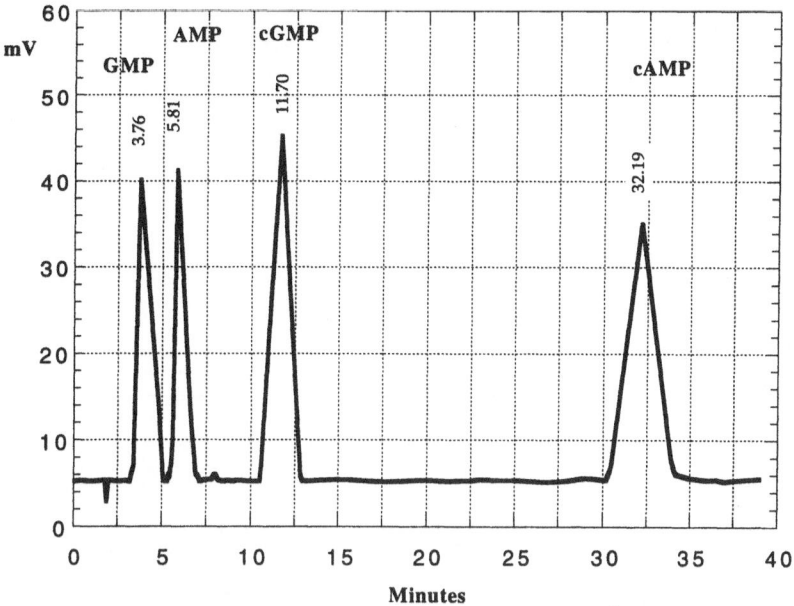

Figure 1. Elution profiles of nucleotides (GMP, AMP, cGMP, cAMP) after Li-Chrospher 100 CH 18/2 chromatography. Absorbance at 254 nm.

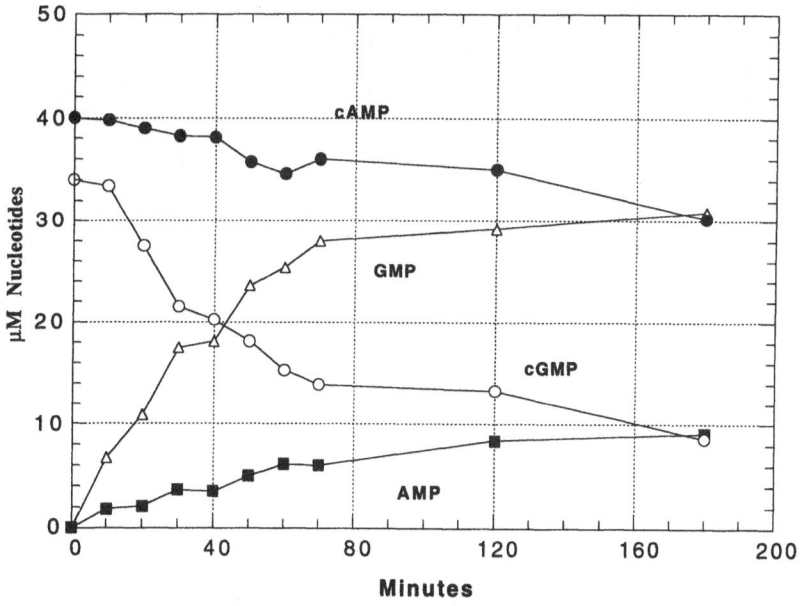

Figure 2. Concentration values of cGMP and cAMP, GMP and AMP versus time in presence of 5 µg beef heart PDE by HPLC.

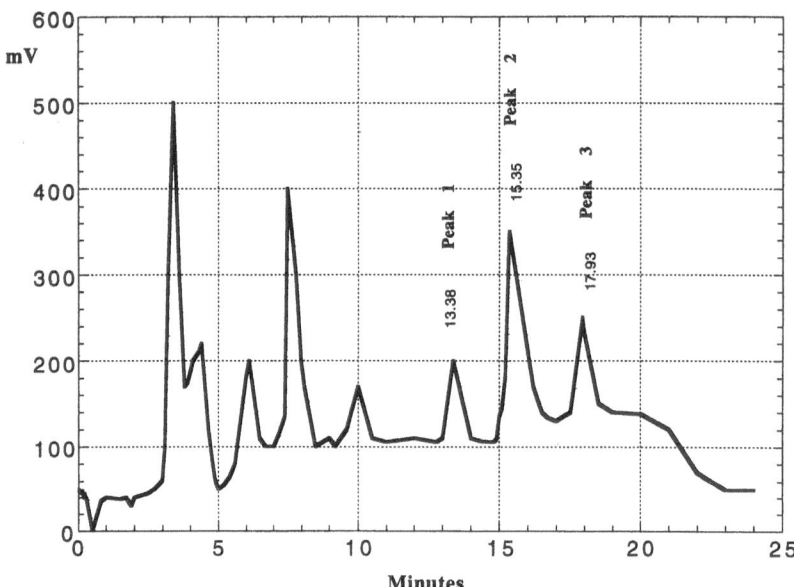

Figure 3. Absorbance profile of elution at 280 nm from beef heart after Resource Q chromatography.

In order to classify beef heart PDEs we separeted the isoenzymes, as described in methods, and showed it in figure 3. We obtained 3 active proteic fractions. Each of these gives different affinity for cyclic nucleotides as described above. Peak 1 with RT 13,38 was classified as PDE I; peak 2 with RT 15.36 was classified as PDE V; peak 3 with RT 17,93 was classified as PDE IV. Therefore without a Km value we can classify the PDE. The same procedure was utilized to study PDE activities in the human heart and only four isoenzymes were identified. Much attention has been focused on cardiac-ventricle PDE isoenzymes since the recognition of a range of compounds with positive inotropic properties, potently and specifically inhibit the 'low-Km' PDE activity. In several reports the 'low-Km' PDE activity isolated from cardiac ventricle showed non-Michaelis-Menten kinetics for cAMP, which were attributed to negative cooperation of substrate binding (Russell et al.,1972; Weishaar et al.,1986). However, similar kinetic activity would also have been obtained if other PDE isoenzymes were present in the 'low-Km' PDE fraction, however incomplete resolution during the isolation procedure could prevent this. Harrison et al.,(1986) reported the purification of a PDE activity from bovine heart with a low Km for cyclic AMP, which has Michaelis-Menten kinetics. In this study we confirm this last result.

Some compounds such as sodium nitroprusside and glycerol trinitrate generate nitric oxide (NO) which enhances the catalytic activity of the soluble guanylyl cyclase, presumably after binding to the heme moiety of the enzyme (Ignarro,1990; Stamleret al.1992). However, studies using NO donors have led to conflicting results as to the involvement of NO in the spontaneous cardiac contractility or in its regulation by muscarinic agonists (Thelen et al.,1992). In order to confirm the NO effect we used sodium nitroprusside, as a generator of NO, at 3 different concentrations with PDE beef heart, as described in Methods. The results obtained show that NO does not significantly alter phosphodiesterase activity (results not shown). In addition, inhibition of NO synthase activity by arginine analogs produces no effect on spontaneous cardiac contractility (Garcia et al.,1992) , although a reduction in cardiac cGMP content has been demonstrated (Klabunde et al.,1992). Moreover, arginine analogs were also found to reverse the chronotropic effects of muscarinic agonists in isolated rat cardiac myocytes (Balligandet al.,1993a). It should be noted, however, that arginine analogs may not only act as NO synthase inhibitors since they have been demonstrated recently to act also as muscarinic receptor antagonists (Buxton et al.,1993) . Activation by NO donors of a membrane form of guanylyl cyclase modulates I_{Ca} via the regulation of two cAMP phosphodiesterases, the cGI- and the cGS-PDEs.

This method is useful to study PDE from human cardiac ventricle.

REFERENCES

Balligand,J.L., Kelly,R.A.,Marsden,P.A.,Smith,T.W.,and Michel,T.1993 *Proc. Natl. Acad. Sci.* 90:347-351

Beavo,J.A., Hansen,R.S., Harrison, S.A.,Hurwitz, R.L., Martins,T.J. and Mumby, M.C., 1982 *Mol. Cell. Endocrinol.* 28:387-410

Beavo,J.A., and Refsnyder, D.H., 1990 *Trends Pharmacol.Sci.* 11:150-155

Buxton,I.L.O., Cheek,D.J., Eckman,D., Wesfall, D.P., Sanders, K.M.,and Keef, K.D., 1993 *Circ.Res.* 72:387-395

Garcia, J.L., Fernandez, N., Garcia-Villalon, A.L., Monge, L., Gomez, B., and Dieguez, G., 1992 *Br.J. Pharmacol.* 106:563-567

Harrison, S.A., Chang, M.L., and Beavo, J.A., 1986 *Circulation* 73:109-116.

Hartzell, H.C., 1988 *Prog.Biophys.Mol.Biol.* 52:165-247.

Heyworth, C.M., Wallace, A.V. and Houslay, M.D., 1983 *Biochem.J.* 214:99-110.

Hidaka, H., Yamaki, T., Ochiai, Y., Asano, T.,and Yamabe, H.1977 *Biochim.Biophys.Acta 484: 398-407.*

Ignarro, L.J., 1990 *Annu.Rev.Pharmacol. Toxicol.* 30:535-560.

Klabunde, R.E., Kimber, N.D., Kuk, J.E., Helgren, M.C., and Forstermann, U.1992 *Eur.J.Pharmacol.* 223:1

Lindemann, J.P., and Watanabe, A.M. 1989 Physiology and Pathophysiology of the Heart, 2nd ed.,pp.423-452, M.Nijhoff Publishing, Boston.

Loten, E.G., Assimacopoulos-Jeannet, F.D., Exton, J.H., and Park, C.R. 1978 *J.Biol.Chem.* 253:746-757.

Macphee, C.H., Reifsnyder, D.H., Moore, T.A., Lerea, K.M., and Beavo, J.A. 1988 *J.Biol.Chem.* 265:10353-10358.

Makino, H., and Kono, T., 1980 *J.Biol.Chem.* 255:7850-7854.

Murray,K.J.,England,P.J.,1992 *Biochem.Soc. Transactions* 20:460-463.

Russell, R., Thompson, W.J., Schneider, F.W., and Appleman, M.M.1972 *Proc.Natl.Acad.Sci. U.S.A.* 69:1791-1795.

Spoto, G., Whitehead, E., Ferraro, A., Di Terlizzi, P.M., Turano, C., and Riva, F., 1991 *Anal. Biochem.* 196:207-210.

Stamler, J.S., Singel, D., and Loscalzo, J. 1992 *Science* 258:1898-1902.

Thelen, K.I., Dembinskakiec, A., Pallapies, D., Simmet, T., and Peskar, B.A. 1992 *Naunyn-Schmied. Arch. Pharmacol.*. 345:93-99.

Weishaar, R.E., Burrows, S.D., Kobylarz, D.C., Quade, M.M., and Evans, D.B. 1986 *Biochem. Pharmacol.* 35:787-800.

Wells, J.N. and Hardman, J.G., 1977 *Adv.Cyclic Nucleotide Res.* 8:119-143.

Zinman, U., and Hollenberg, C.H. 1974 *J.Biol.Chem.* 249:2182-2187

DETERMINATION OF THE ACTIVITY OF RECOMBINANT HUMAN PHOSPHORIBOSYLPYROPHOSPHATE SYNTHETASE ISOFORM 1 BY A NON-ISOTOPIC, ONE-STEP METHOD

Rosa Torres, Felícitas Mateos, Juan G. Puig, and *Micheal A. Becker

Divisions of Clinical Biochemistry and Internal Medicine "La Paz" University
Hospital , Madrid, Spain
*Dept. of Medicine, Rheumatology Section, University of Chicago,
Illinois, USA

INTRODUCTION

Phosphoribosylpyrophosphate synthetase (Ribophosphate pyrophosphokinase, EC 2.7.6.1., PRS) catalyses the synthesis of phosphoribosylpyrophosphate (PRPP) from ribose 5-phosphate and MgATP in the reaction:

$$\text{Ribose 5-phosphate} + \text{MgATP} \longrightarrow \text{PRPP} + \text{AMP}$$

Superactivity of PRS is an X-chromosome-linked inborn error of metabolism in which excessive enzyme activity, due to a variety of kinetic alterations, is associated with hyperuricemia, gout and, in some instances, neurodevelopmental impairment (1).

To date, two distinct X-chromosome-linked loci for PRS, PRS 1 and PRS 2, have been identified (2). Indeed, two different point mutations have been found in PRS 1 cDNA of two patients with PRS superactivity (3). Recently, recombinant human PRS isoform 1 has been expressed in a bacterial host (4).

Methods, currently employed to measure PRS activity and allow complete kinetic study, require an auxiliary enzyme reaction, a radioactive nucleobase, and multiple incubations. Recently, we have developed a new simplified nonisotopic method to determine PRS activity in hemolysates (5). The aim of the present study is to modify this non-isotopic method to measure the activity of recombinant human PRS isoform 1, expressed in extracts of transformed *E. Coli*.

MATERIALS AND METHODS

Sample Preparation:

Bacterial extract with human recombinant PRS isoform 1 was obtained in NaPi 50 mmol/l, MgCl$_2$ 6 mmol/l, Dithiothreitol (DTT) 1 mmol/l, ATP 0.3 mmol/l, EDTA 1 mmol/l, Phenylmethylsulfonyl fluoride (PMSF) 0.1 mmol/l, pH = 8.0 (4), with a protein concentration of 2.04 mg/ml.

The extract was appropriately diluted for assay at pH 8.0 in Tris-HCl 50 mmol/l, $MgCl_2$ 6 mmol/l, DTT 1 mmol/l, EDTA 1 mmol/l, BSA 1 mg/ml, and appropriate Pi (to assay the effects of different Pi concentrations on PRS activity). Then, the diluted extract was chromatographed on Sephadex G-25, eluted with the above mentioned dilution buffer, and the material eluting in the void volume was collected and assayed for PRS activity.

PRS Activity Assay:

One hundred microliters of the chromatographed extract (with a protein concentration of 10-40 μg/ml) were incubated for 30 min at 37°C, with 100 μl of a pH 7.4 reaction mixture containing: Tris-HCl 50 mmol/l, $MgCl_2$ 5 mmol/l, EDTA 1 mmol/l, DTT 1 mmol/l, NaPi 32 mmol/l (or dilutions), saturating concentrations of both substrates ATP (0.5 mmol/l) and Ribose 5-phosphate (0.15 mmol/l), and $P^1 P^5$ diadenosine pentaphosphate (Ap5A, an inhibitor of adenylate kinase) 0.25 mmol/l. In some experiments the inhibitor GDP (0.02 mmol/l) was also added. A control reaction mixture without ribose 5-phosphate was employed for each sample, due to the presence of contamination (about 2%) of ATP with AMP (no AMP was formed in the control mixture). Upon complexion of the incubation, an excess of EDTA was added, samples were centrifuged in Amicon cones (30,000 mol. wt. membrane cutt-off, Centricon™[30], Amicon Inc, Beverly, MA, USA), and generated AMP was separated by HPLC as previously described (5), and measured by spectrophotometry.

Linear regression analysis and Student's t test were applied to experimental data, as appropriated.

RESULTS

In the absence of Ap5A addition, ADP was generated rather than AMP, due to the presence of adenylate kinase (which catalyses the reaction: ATP + AMP = 2ADP) in the bacterial extract. However, under assay conditions no ribose 5-phosphate-dependent ADP was formed.

Under assay conditions, there were linear relationships between the incubation time and both the ribose 5-phosphate-dependent generation of AMP (r= 0.984, p < 0.005) and the ribose 5-phosphate-dependent consumption of ATP (r=0.986, p < 0.005) (Figure 1). Also, linear relationships were found between the protein concentration and both the ribose 5-phosphate-dependent generation of AMP (r= 0.986, p < 0.005) and the ribose 5-phosphate-dependent consumption of ATP (r=0.973, p < 0.01) (Figure 1). When either incubation time (r= 0.990, p < 0.005) or protein concentration (r= 0.976, p < 0.005) was varied, there was a linear correlation between the ribose 5-phosphate-dependent AMP generated and ATP consumed.

No detectable ribose 5-phosphate-dependent IMP or adenosine generation was observed. Finally, the hyperbolic curve relating Pi concentration to initial reaction velocity, obtained with chromatographed (free from inhibitors) extracts, changed to a sigmoidal function when crude extract was assayed and the inhibitor GDP, 0.02 mmol/l, was added to the reaction mixture (Figure 2).

CONCLUSIONS

This convenient method can be employed to measure PRS 1 isoform activity and regulation with several advantages over the previously employed methods: a) it does not require coupling to an auxiliary phosphoribosyltransferase, b) it requires only a single incubation, c) it measures PRS 1 activity directly, through determination of the rate of product (AMP) formation, and, finally, d) it avoids the use of a radioactive nucleobase.

Preliminary results suggest that this method should permit determination of PRS activity in a variety of human cells allowing simplified screening of PRS activity disorders.

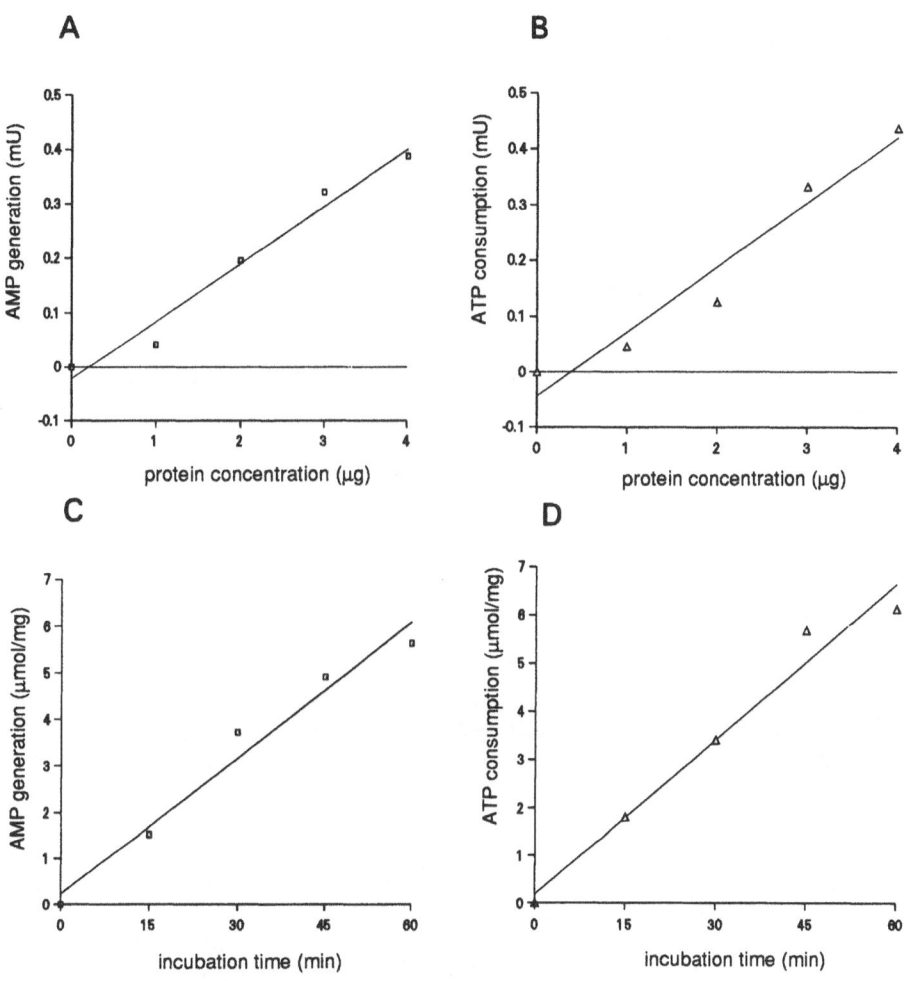

Figure 1. Linear relationship between protein concentration and both the ribose 5-phopshate-dependent AMP generation (**A**; $y= 0.104x - 0.02$, $r=0.986$, $p<0.005$) and ATP consumption (**B**; $y= 0.114x -0.04$, $r=0.973$, $p<0.01$), and between incubation time and both ribose 5-phosphate-dependent AMP generation (**C**; $y=0.098x + 0.23$, $r=0.984$, $p<0.005$) and ATP consumption (**D**; $y=0.108x + 0.18$, $r=0.986$, $p<0.005$).

ACKNOWLEDGEMENTS

This work has been supported by a grant from the Spanish Institute of Health (FIS 92/0622). Dra Torres is a fellow of The Spanish Institute of Health FISss.

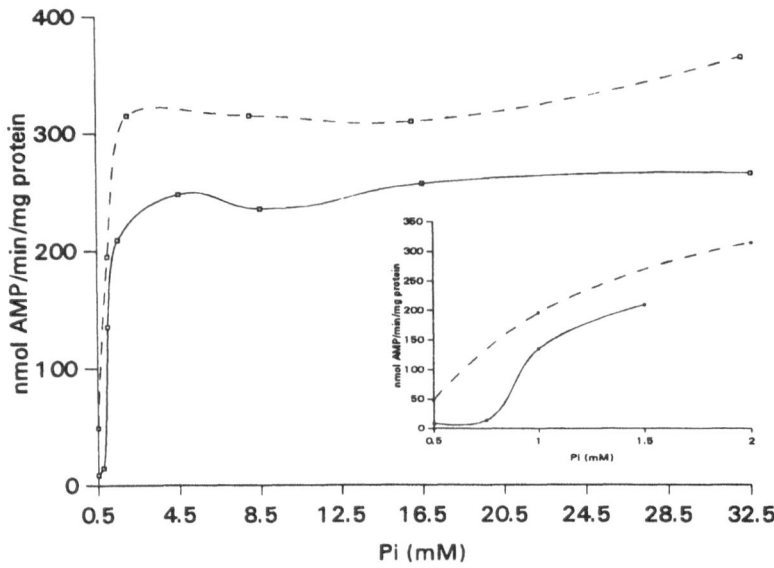

Figure 2. Curves relating Pi concentration to initial reaction velocity of PRS1, in chromathographed extract (dotted line) or in crude extract with 0.02 mmol/l GDP in the reaction mixture (solid line). Inset presents expanded X-axis to emphasize differing contours of respective activation curves.

REFERENCES

1. Becker MA, Puig JG, Mateos FA, Jimenez ML, Kim M, Simmonds A. Inherited superactivity of phosphoribosylpyrophosphate synthetase: association of uric acid overproduction and sensorial deafness. Am J Med 1988; 85: 383

2. Becker MA, Heidler SA, Bell GI, et al. Cloning of cDNAs for human phosphoribosylpyrophosphate synthetase 1 and 2 and X chromosome localization of PRS1 and PRS2 genes. Genomics 1990; 8: 555.

3. Roessler BJ, Golovoy N, Palella TD, Heidler S, Becker MA. Identification of distincts PRS1 mutations in two patients with X-linked phophoribosylpyrophosphate synthetase superactivity. Adv Exp Med Biol 1991; 309B: 125.

4. Nosal JM, Switzer RL, Becker MA. Overexpression, Purification, and Characterization of Recombinant Human 5-Phosphoribosyl-1-pyrophosphate Synthetase Isoenzymes I and II. J Biol Chem 1993; 268: 10168.

5. Torres RJ, Mateos FA, Puig JG, Becker MA. A simplified method for the determination of phopshoribosylpyrophosphate synthetase activity in hemolysates. Clin Chem Acta 1994; 224: 55.

AUTHOR INDEX

SUBJECT INDEX

Gene expression, 249, 683
Gene therapy, 391
Germline mutation, 653, 661, 679
Gliomas, 219
Glomerular filtration rate, 31, 70
Glutaminase, 513, 730
Glutathione, 296, 299
Glycinamide ribonucleotide formyltransformylase, 100, 179
Glycinamide ribotide, 152
Glycosuria, 42
Gout, 1, 15, 27, 31, 35, 42, 47, 53, 61, 65, 69, 73, 79, 83, 331, 342, 345, 357, 379, 675, 679, 707, 821
Granulocyte colony stimulating factor, 120
Green tea polyphenols, 779
Guanine aminohydrolase, 475
Guanine phosphoribosyltransferase, 148, 484, 759, 806, 815
Guanosine triphosphate, 484, 735

Hemolysate, 439, 614, 676, 821
Heart, 431, 617
Heart muscle, 815
Heart transplant, 483
Hemangioma, 11
Hematopoiesis, 395
Hemoglobin, 207
Hepatectomized rats, 541, 783
Hepatic regeneration, 11
Hepatic triglyceride lipase, 83
Hepatitis C virus, 475
Hepatocellular carcinoma, 662
Hepatocellular disease, 57
Hepatocytes, 313, 323, 523
Hepatoma cell, 779
Hepatotoxicity, 526
Hereditary oroticaciduria, 1
Herpes virus, 459
Heteroduplex, 657
High-density lipoprotein cholesterol, 79
High performance liquid chroatography, 15, 40, 70, 383, 446, 453, 484, 542, 735, 753
Histidine load rest, 792
Histoimmunochemical staining, 11
Homeostasis, 791
Homologous recombination, 663
HL-60 cells, 761
Human breast cancer cell, 757
Human colon carcinoma cell lines, 275
Human erythrocyte AMP deaminase deficiency, 703
Human gastric cancer cells, 131
Human immunodeficiency virus, 443, 453, 459, 465, 711
Human lymphoblastic cell lines, 115
Human melanoma cells, 487
Human placental cytosolic purine 5'-nucleotidase, 623
Human promyelocytic leukemia cells, 252
Human T-lymphocytic cells, 261
Human umbilical venous endothelial cells, 299

9-(3-Hydroxy-2-phosphonylmethyl-propyl) adenine, 459
5-Hydroxypyrazinamide, 43
Hydroxyurea-resistant cell lines, 631
Hyperglycemia, 42
Hyperlipidemia, 79, 87
Hypermethylation, 649
Hypertension, 73
Hyperthyroidism, 765
Hypertriglyceridemia, 81
Hyperuricemia, 1, 27, 31, 53, 61, 65, 73, 77, 81, 90, 331, 337, 341, 345, 517, 675, 821
Hypothermia, 293
Hypouricemia, 27, 57
Hypoxanthine, 17, 40, 47, 53, 69, 98, 115, 132, 174, 295, 345, 350, 353, 358, 364, 375, 387, 437, 441, 487, 679, 753, 775, 806
Hypoxanthine-guanine phosphoribosyltransferase, 1, 74, 98, 102, 115, 209, 219, 331, 337, 338, 341, 345, 349, 350, 353, 357, 358, 365, 388, 510, 529, 679, 753,
Hypoxia, 299, 313

Immunodeficiency, 1, 249, 387, 391
Immunosuppression, 157, 169, 275, 439
Inhibitory concentrations, 445
Inosine, 133, 167, 173, 324, 345, 350, 357, 363, 367, 371, 437, 441, 462, 576, 616, 623, 775, 806
Inosine monophosphate dehydrogenase, 100, 115, 146, 155, 219, 283, 725, 737, 741, 749, 757, 761
Insosine phosphorylase, 510
Inosinic branch point, 213
Interferon, 479
Interleukin-3, 122
Intestinal mucosa, 215
Intestine, 581
Intracellular nucleotide pool, 749, 761, 775
Intragenic point mutations, 654, 661
Intravenous fructose tolerance test, 37, 372
5-Iodotubercidin, 324, 417, 439
Ischemia, 81, 291, 319, 346, 423, 428, 770, 775, 807
Isoenzyme, 227, 231, 796, 815

Kelley-Seegmiller syndrome, 331
Kidney size, 2, 65, 73

Leiomyoma, 243
Leishmania donovani, 744
Lesch-Nyhan syndrome, 1, 331, 337, 341, 353, 663, 679
Leukemia, 97, 119, 125, 195, 227, 231, 237, 249, 253, 275, 279, 283, 683, 742, 747, 758, 770, 780
Leukocytes, 237, 412, 767
Linker-scanning mutation, 641
Lipid peroxidation, 299, 303, 313
Lipopolysaccharides, 783, 787
Lipoprotein lipase, 83